AA002459

2014 15th International Conference on Thermal, Mechanical and Multi-Physics Simulation and Experiments in Microelectronics and Microsystems

(EuroSimE 2014)

Ghent, Belgium
7-9 April 2014

IEEE Catalog Number: CFP14566-POD
ISBN: 978-1-4799-4789-8

Copyright © 2014 by the Institute of Electrical and Electronic Engineers, Inc
All Rights Reserved

Copyright and Reprint Permissions: Abstracting is permitted with credit to the source. Libraries are permitted to photocopy beyond the limit of U.S. copyright law for private use of patrons those articles in this volume that carry a code at the bottom of the first page, provided the per-copy fee indicated in the code is paid through Copyright Clearance Center, 222 Rosewood Drive, Danvers, MA 01923.

For other copying, reprint or republication permission, write to IEEE Copyrights Manager, IEEE Service Center, 445 Hoes Lane, Piscataway, NJ 08854. All rights reserved.

***This publication is a representation of what appears in the IEEE Digital Libraries. Some format issues inherent in the e-media version may also appear in this print version.

IEEE Catalog Number: CFP14566-POD
ISBN 13:   978-1-4799-4789-8

Additional Copies of This Publication Are Available From:

Curran Associates, Inc
57 Morehouse Lane
Red Hook, NY 12571 USA
Phone: (845) 758-0400
Fax: (845) 758-2633
E-mail: curran@proceedings.com
Web: www.proceedings.com

TABLE OF CONTENTS

Applications Of Laser Welding To Electric Connectors 1
    C. Liao, S. Liu, K. Liao

Thermo-Electrical And Structural Coupled Simulations Of Buckling Beam Microprobes In High Temperature/High Current Conditions 7
    D. Eckhaut, E. Bertarelli, D. Acconcia, R. Vallauri, G. Cocchetti, A. Corigliano

Limitations And Accuracy Of Steady State Technique For Thermal Characterization Of Solid And Composite Materials 11
    M. AboRas, B. Wunderle, D. May, R. Schacht, T. Winkler, S. Rzepka, B. Michel

A Product Based Lap Shear Fatigue Testing Of Electrically Conductive Adhesives 18
    B. Ozturk, A. Youssef, P. Gromala, C. Silber, K. Jansen, L. Ernst

Process And Reliability Of SF_6/O_2 Plasma Etched Copper TSVs 23
    L. Filipovic, R. Orio, S. Selberherr

Characterization And Post Simulation Of Thin-Film PZT Actuated Plates For Haptic Applications 27
    F. Casset, J. Danel, P. Renaux, C. Chappaz, G. Le Rhun, C. Dieppedale, M. Gorisse, S. Basrour, S. Fanget, P. Ancey, A. Devos, E. Defay

Using Molecular Modeling To Uncover The Origins Of Subtle Solvation-Based Film Defects. 31
    N. Iwamoto, T. Baldwin

Description Of The Thermo-Mechanical Properties Of A Sn-Based Solder Alloy By A Unified Viscoplastic Material Model For Finite Element Calculations 32
    A. Kabakchiev, B. Metais, R. Ratchev, M. Guyenot, P. Buhl, M. Hossfeld, X. Schuler, R. Metasch, M. Roellig

Failure Mechanisms In Chip-Metallization In Power Applications 38
    C. Durand, M. Klingler, D. Coutellier, H. Naceur

Thermo-Mechanical Stress Investigation Of Integrated SAW Strain Sensors 43
    J. Hempel, S. Anees, J. Wilde, L. Reindl

Software Reliability And Its Interaction With Hardware Reliability 47
    W. Driel, M. Schuld, R. Wijgers, W. Kooten

Electromigration Reliability Of Cylindrical Cu Pillar $SnAg_{3.0}Cu_{0.5}$ Bumps 55
    L. Meinshausen, K. Weide-Zaage, B. Goldbeck, A. Moujbani, J. Kludt, H. Fremont

Reliability Assessment Of Discrete Passive Components Embedded Into PCB Core 61
    R. Schwerz, M. Roellig, S. Osmolovskyi, K. Wolter

Development And Validation Of Dye-Sensitized Solar Cell Finite Element Model For Sealing Failure Investigation 68
    C. Han, S. Park

Measuring Young's Modulus Of Polysilicon Via Cantilever Microbeam Arrays 73
    S. Kehrberg, M. Dorwarth, S. Gunther, S. Markisch, C. Geckeler, J. Mehner

Degradation Of Silicone In White LEDs During Device Operation: A Finite Element Approach To Product Reliability Prediction 77
    S. Watzke, P. Altieri-Weimar

Creep And Fatigue As Main Degradation Phenomena In Reliability Of Solder Joints 82
    K. Jankowski, A. Wymyslowski

Investigation Of Temperature And Moisture Effect On Interface Toughness Of EMC And Copper Using Cohesive Zone Modeling Method 87
    X. Ma, G. Zhang

Comparation Of Thermal And Hygro Effects On The Degradation Of LED Package 92
    X. Ma, G. Zhang

A Lifetime Prediction Method For Solid State Lighting Power Converters Based On SPICE Models And Finite Element Thermal Simulations 96
    B. Sun, X. Fan, L. Zhao, C. Yuan, S. Koh, G. Zhang

Understanding Delamination For Fast Development Of Reliable Packages For Automotive Applications. A Consideration Of Robustness For New Packages Based On Simulation 100
    R. Pufall, M. Goroll, G. Reuther

Evaluation Of The Residual Stress Distribution In Thin Films By Means Of The Ion Beam Layer Removal Method 106
    D. Kozic, R. Treml, R. Schongrundner, R. Brunner, D. Kiener, T. Antretter, H. Ganser

Thermo-Mechanical Analyses Of Printed Board Assembly During Reflow Process For Warpage Prediction 111
    S. Chung, S. Oh, T. Lee, M. Park

Design Independent Lifetime Assessment Method For PCBs Under Low Cycle Fatigue Loading Conditions 116
P. Fuchs, G. Pinter, T. Krivec

Determination Of Cyclic Mechanical Properties Of Thin Copper Layers For PCB Applications 121
K. Fellner, P. Fuchs, T. Antretter, G. Pinter, R. Schongrundner

Drop Impact Simulations For Lifetime Assessment Of PCB/BGA Assemblies Regarding Pad Cratering 128
G. Simo, H. Shirangi, M. Nowottnick, R. Dudek, E. Kaulfersch, S. Rzepka, B. Michel

Predictive Reliability Using FEA Simulations Of Power Stacked Ceramic Capacitors For Aeronautical Applications 138
W. Benhadjala, B. Levrier, I. Bord-Majek, L. Bechou, E. Suhir, Y. Ousten

An Investigation Of The Tensile Deformation And Failure Of An Epoxy/Cu Interface Using Coarse-Grained Molecular Dynamics Simulations 144
S. Yang, J. Qu

Methodology For Supporting Electronic System Prototyping Through Semiautomatic Component Selection 152
R. Swierczynski, K. Urbanski, A. Wymyslowski

An Accelerated Method For Characterization Of Bi-Material Interfaces In Microelectronic Packages Under Cyclic Loading Conditions 156
E. Poshtan, S. Rzepka, B. Michel, C. Silber, B. Wunderle

Simulation Of Stress Distribution In Assembled Silicon Dies And Deflection Of Printed Circuit Boards 163
K. Macurova, P. Angerer, R. Schongrundner, T. Krivec, M. Morianz, T. Antretter, R. Bermejo, M. Pletz, M. Brizoux, W. Maia

Acquisition Unit For In-Situ Stress Measurements In Smart Electronic Systems 170
A. Palczynska, F. Pesth, P. Gromala, T. Melz, D. Mayer

Modelling Of Non-Stationary Processes In Optomechanical Thermal Microsensors 174
A. Kozlov

Multiphysics Modeling For Current Carrying Capability Of A Power Package 181
Q. Qian, Y. Liu, Y. Liu

Comparison Of Bondwire Life With Effective Strain Method And Cohesive Zone Method For A Power Package 187
J. Zhang, Y. Xu, Y. Liu

Adhesion Work Analysis By Molecular Modelling And Wetting Angle Measurement 194
K. Allaf, D. Krol, A. Wymyslowski, I. Zubel, K. Rola

Modeling And Simulation Of A MEMS Thermal Actuator With Polysilicon Heater 199
D. Dellaert, J. Doutreloigne

Thermo-Mechanical Stress Of Underfilled 3D IC Packaging 205
M. Wang, M. Wu

Investigation Of Color Shift Of LEDs-Based Lighting Products 210
S. Koh, H. Ye, M. Mehr, J. Wei, W. Driel, L. Zhao, G. Zhang

Crosstalk Phenomena Analysis Using Electromagnetic Wave Propagation By Experimental An Numerical Simulation Methods 215
A. Palczynska, A. Wymyslowski, T. Bieniek, G. Janczyk, D. Pasquet, T. Dinh

Size Effect On The Microbridges Quality Factor Tested In Free Air Space 225
M. Pustan, C. Birleanu, F. Rusu, C. Dudescu, O. Belcin

FEM Simulations For Built-In Reliability Of Innovative Liquid Crystal Polymer-Based QFN Packaging And Sn96.5Ah3Cu0.5 Solder Joint 231
W. Chenniki, I. Bord-Majek, B. Levrier, K. Wongtimnoi, J. Diot, Y. Ousten

Framework To Extract Cohesive Zone Parameters Using Double Cantilever Beam And Four-Point Bend Fracture Tests 235
S. Raghavan, I. Schmadlak, G. Leal, S. Sitaraman

Measuring The Mechanical Relevant Shrinkage During In-Mold And Post-Mold Cure With The Stress Chip 240
F. Schindler-Saefkow, F. Rost, A. Rezaie-Adli, K. Jansen, B. Wunderle, J. Keller, S. Rzepka, B. Michel

CMOS Stress Sensor For 3D Integrated Circuits: Thermo-Mechanical Effects Of Through Silicon Via (TSV) On Surrounding Silicon 245
K. Ewuame, V. Fiori, K. Inal, P. Bouchard, S. Gallois-Garreignot, S. Lionti, C. Tavernier, H. Jaouen

Structure Design And Reliability Assessment Of Double-Sided With Double-Chip Stacking Packaging 253
Y. Su, C. Lin, T. Kuo, K. Chiang

2D Micro-Chamber For DC Plasma Working At Low Power 257
V. Rochus, V. Samara, B. Vereecke, P. Soussan, B. Onsia, X. Rottenberg

The Shear Strength Of Nano-Ag Solders And The Use Of Ag Interconnects In The Design And Manufacture Of SiGe-Based Thermo-Electric Modules 263
M. Edwards, K. Brinkfeldt, M. Silva, D. Andersson

Analysis Of RF-MEMS Switches In Failure Mode: Towards A More Robust Design 272
T. Kuenzig, T. Muschol, J. Iannacci, G. Schrag, G. Wachutka

Thermo-Mechanical Simulation Of Plastic Deformation During Temperature Cycling Of Bond Wires For Power Electronic Modules 278
A. Wright, A. Hutzler, A. Schletz, P. Pichler

Electrical Characteristics Evolution Of The Deep Trench Termination Diode Based On A Finite Elements Simulation Approach 283
F. Baccar, F. Henaff, L. Theolier, S. Azzopardi, E. Woirgard

Molecular Dynamics Simulation Of Adhesion Performance And Conformation Transition Of SAM-Modified Cu/Epoxy Interface Under Electric Field 290
S. Kwok, M. Yuen

Multiphysics Study Of RF/Microwave Planar Devices: Effect Of The Input Signal Power 295
M. Sanchez-Soriano, M. Edwards, Y. Quere, D. Andersson, S. Cadiou, C. Quendo

Investigation Of A Finned Baseplate Material And Thickness Variation For Thermal Performance Of A SiC Power Module 302
Y. Zhang, I. Belov, M. Bakowski, J. Lim, P. Leisner, H. Nee

Thermal Characteristics Of SiC Diode Assembly To Ceramic Substrate 310
R. Kisiel, M. Guziewicz, M. Mysliwiec, J. Krasniewski, W. Janke

Theoretical And Experimental Study Of Thermal Management In High-Power AlInGaN LEDs 314
A. Chernyakov, A. Zakgeim, K. Bulashevich, S. Karpov, V. Smirnov, V. Sergeev

Characterizing And Modelling The Behaviour Of An Isotropic Conductive Adhesive In View Of Electronic Assembly Fatigue Life Studies Under Thermal Cycling 319
S. Pin, M. Sartor, L. Michel, J. Parain, S. Dareys

Thermo-Mechanical Stress Induced By CPI On 3D Interposer Package 326
M. Lofrano, M. Gonzalez

Theoretical And Experimental Investigations On Failure Mechanisms Occuring During Long-Term Cycling Of Electrostatic Actuators 331
R. Behlert, T. Kunzig, G. Schrag, G. Wachutka

A Model In Predicting Color Of LED Packages With Different Phosphor Layer Dimensions 338
C. Wong, S. Leung, Y. Xiong, C. Yuan, G. Zhang

Design Aspects For CPI Robust BEOL 343
M. Gonzalez, L. Kljucar, B. Vandevelde, I. Wolf, Z. Tokei

Numerical Simulations Of Piezoelectric MEMS Energy Harvesters 348
G. Gafforelli, R. Ardito, A. Corigliano, C. Valzasina, F. Procopio

Analytical Tool For Electro-Thermal Modelling Of Microbolometers 357
P. Zajac, C. Maj, M. Szermer, M. Lobur, A. Napieralski

Correlation Of Activation Energy Between LEDs And Luminaires In The Lumen Depreciation Test 363
G. Lu, C. Yuan, X. Fan, G. Zhang

Modelling The Lifetime Of Aluminum Heavy Wire Bond Joints With A Crack Propagation Law 366
A. Grams, T. Prewitz, O. Wittler, S. Schmitz, A. Middendorf, K. Lang

Thermo-Mechanical Characterization Of Passive Stress Sensors In Si Interposer 372
B. Vianne, P. Bar, V. Fiori, S. Gallois-Garreignot, K. Ewuame, P. Chausse, S. Escoubas, N. Hotellier, O. Thomas

Molecular Dynamic Simulations Of Maximum Pull-Out Forces Of Embedded CNTs For Sensor Applications And Validating Nano Scale Experiments 380
S. Hartmann, O. Holck, T. Blaudeck, S. Hermann, S. Schulz, T. Gessner, B. Wunderle

The Underfill-Microbump Interaction Mechanism In 3D ICs: Impact And Mitigation Of Induced Stresses 386
A. Ivankovic, V. Cherman, M. Gonzalez, B. Vandevelde, D. Vandepitte, G. Beyer, E. Beyne, I. Wolf

Characterization And Modeling Of The AuCuSn Thin Solder Joint Under Thermal Cycling 394
T. Pelisset, B. Karunamurthy, R. Otremba, T. Antretter

Challenges Of Viscoelastic Characterization Of Low TG Epoxy Based Adhesives For Automotive Applications In DMA And Relaxation Experiments 400
I. Maus, H. Preu, M. Niessner, M. Fink, K. Jansen, R. Pantou, B. Michel, B. Wunderle

Improvement Of Freestanding CMOS-MEMS Through Detailed Stress Analysis In Metallic Layers 407
S. Orellana, B. Arrazat, P. Fornara, C. Rivero, A. Giacomo, S. Blayac, K. Inal, P. Montmitonnet

Electronic Control Package Model Calibration Using Moiré Interferometry 413
D. Kim, B. Han, A. Yadur, P. Gromala

A New Life-Test Equipment Designed For Medium-Duty Electromagnetic Contactors 418
S. Biyik, M. Aydin

Experimental Investigation And Interpretation Of The Real Time, In Situ Stress Measurement During Transfer Molding Using The Piezoresistive Stress Chips 425
A. Adli, K. Jansen, F. Schindler-Saefkow, F. Rost

B-Field Characterization And Equivalent Circuit Modeling Of A Poly-SiGe-MEMS Based Xylophone Bar Magnetometer 429
M. Farghaly, V. Rochus, X. Rottenberg, U. Mohammed, H. Tilmans

Computationally Efficient And Stable Order Reduction Method For A Large-Scale Model Of MEMS Piezoelectric Energy Harvester 435
M. Kudryavtsev, E. Rudnyi, T. Bechtold, J. Korvink

Multiphysical Modeling Of Nanosecond Laser Dicing On Ultra-Thin Silicon Wafers 440
G. Galasso, M. Kaltenbacher, B. Karunamurthy, H. Eder, T. Polster

An Analytical Model For Thermal Failure Analysis Of 3D IC Packaging 446
J. Lan, M. Wu

Mechanical Analysis Of Encapsulated Metal Interconnects Under Transversal Load 451
B. Keymeulen, M. Gonzalez, F. Bossuyt, J. Baets, J. Vanfleteren

Design, Technology, Numerical Simulation And Optimization Of Building Blocks Of A Micro And Nano Scale Tensile Testing Platform With Focus On A Piezoresistive Force Sensor 459
P. Meszmer, K. Hiller, D. May, S. Hartmann, A. Shaporin, J. Mehner, B. Wunderle

Study On Configuration Design Of Interconnection In High Power Module 469
L. Liao, T. Hung, C. Liu, Y. Su, K. Chiang

2-Gb/s/pin DDR3 Memory Channel Design And Simulation For Carbon Reduction 473
N. Chen

System-Level-Model Development Of An SWCNT Based Piezoresistive Sensor In VHDL-AMS 481
V. Kolchuzhin, J. Mehner, E. Markert, U. Heinkel, C. Wagner, J. Schuster, T. Gessner

FEM Simulation And Measurement Validation Of A cMUT Cell 487
S. Mao, X. Rottenberg, V. Rochus, B. Nauwelaers, H. Tilmans

Microstructure Simulation Of Grain Growth In Cu Through Silicon Via Using Phase-Field Modeling 494
N. Nabiollahi, N. Moelans, M. Gonzalez, J. Messemaeker, C. Wilson, K. Croes, E. Beyne, I. Wolf

Compact Thermal Modeling Of Microbolometers 498
M. Janicki, P. Zajac, M. Szermer, A. Napieralski

MedeA®: Atomistic Simulations For Designing And Testing Materials For Micro/Nano Electronics Systems 502
A. France-Lanord, D. Rigby, A. Mavromaras, V. Eyert, P. Saxe, C. Freeman, E. Wimmer

FEM Stress Analysis Of Various Solar Module Concepts Under Temperature Cycling Load 510
F. Kraemer, S. Wiese

FEM Wire Bonding Simulation For Sensor Chip Applications 518
F. Kraemer, S. Wiese

Mechanical Stress Analysis In Photovoltaic Cells During The String-Ribbon Interconnection Process 524
F. Kraemer, J. Seib, E. Peter, S. Wiese

Analytical Stress Model For Tin Based Solder Material 531
M. Guyenot, A. Fix

A Crack Analysis Model For Silicon Based Solar Cells 538
J. Ahmar, S. Wiese

Effect Of Laminar Air Flow On Probe Burn For Spring Probe 543
B. Zafer, B. Tunaboylu

Thermal Management Of Electrical Overload Cases Using Thermo-Electric Modules And Phase Change Buffer Techniques: Simulation, Technology and Testing 547
M. Springborn, B. Wunderle, D. May, R. Mrossko, C. Manier, H. Oppermann, M. Ras, R. Mitova

Simulation And Measurement Of Pressure Dependent Q-Factors In NEMS Resonators 558
J. Manz, G. Schrag, G. Wachutka

Electro-Thermal Characterization Of Through-Silicon Vias: 563
A. Todri-Sanial

Experimental Investigation Of The Visco-Plastic Mechanical Properties Of A Sn-Based Solder Alloy For Material Modelling In Finite Element Calculations Of Automotive Electronics 569
R. Metaseh, M. Roellig, A. Kabakchiev, B. Metais, R. Ratchev, K. Meier, K. Wolter

Analysis Of Mechanical Properties Of Thermal Cycled Cu Plated-Through Holes (PTH) 577
H. Walter, A. Kaltwasser, M. Broll, S. Huber, O. Wittler, K. Lang

Failure Mode Analysis And Optimization Of Assembled High Temperature Pressure Sensors 584
R. Zeiser, S. Ayub, M. Berndt, J. Muller, J. Wilde

Reliability Study On Chip Capacitor Solder Joints Under Thermo-Mechanical And Vibration Loading 590
K. Meier, M. Roellig, A. Schiessl, K. Wolter

Modeling Of SiC Power Modules With Double Sided Cooling 597
K. Brinkfeldt, K. Neumaier, A. Mann, O. Zschieschang, A. Otto, E. Kaulfersch, M. Edwards, D. Andersson

Modeling And Simulation Of Monolithic Integration Of Rectifiers For Solid State Lighting Applications 603
M. Veknatesh, P. Liu, H. Zeijl, G. Zhang

Microactuator Modeling To Develop A New Template For The Braille 609
S. Soulimane, M. Nigassa, B. Bouazza, H. Camon

Multiphysics Modelling Of The Fabrication And Operation Of A Micro-Pellistor Device 612
F. Biro, Z. Hajnal, A. Pap, I. Barsony

Determination Of Residual Stress With High Spatial Resolution At TSVs For 3D Integration: Comparison Between HR-XRD, Raman Spectroscopy And fibDAC 618
D. Vogel, U. Zschenderlein, E. Auerswald, O. Holck, P. Ramm, B. Wunderle, R. Pufall

Modelling, Simulation and Optimization For A SThm Nanoprobe 626
B. Yang, M. Lenczner, S. Cogan, F. Menges, H. Riel, B. Gotsmann, P. Janus, G. Boetch

Combined Experimental - And FE - Studies On Sinter-Ag Behaviour And Effects On IGBT-Module Reliability 632
R. Dudek, R. Doring, P. Sommer, B. Seiler, K. Kreyssig, H. Walter, M. Becker, M. Gunther

Fluid Damping In Compliant, Comb-Actuated Torsional Micromirrors 641
R. Mirzazadeh, S. Mariani, A. Ghisi, M. Fazio

New Equivalent Stress Describes The Dicing Caused Anisotropic Breaking Strength Of Silicon Dies 648
M. Steiert, J. Wilde

Accurate Prediction Of SnAgCu Solder Joint Fatigue Of QFP Packages For Thermal Cycling 656
M. Niessner, G. Schuetz, C. Birzer, H. Preu, L. Weiss

Thermo-Mechanical Properties Of Underfills At Partial And Full Filler Percolation - Sub-Layering The Underfill. 662
G. Schlottig, M. Haupt, S. Zimmermann, J. Zurcher, T. Brunschwiler

Hidden Head-In-Pillow Soldering Failures 669
B. Vandevelde, G. Willems, B. Allaert

Reliability And Accelerated Test Methods For Plastic Materials In LED-Based Products 675
M. Mehr, W. Driel, G. Zhang

Finite Element Multi-Physics Modeling For Ohmic Contact Of Microswitches 680
H. Liu, D. Leray, P. Pons, S. Colin

GaN-Based LEDs: State Of The Art And Reliability-Limiting Mechanisms 688
E. Zanoni, M. Meneghini, N. Trivellin, M. Lago, G. Meneghesso

System Reliability For LED-Based Products 693
J. Davis, K. Mills, M. Lamvik, R. Yaga, S. Shepherd, J. Bittle, N. Baldasaro, E. Solano, G. Bobashev, C. Johnson, A. Evans

Assessment Of Microelectronics Interconnect Reliability - Current Practice And Trends 700
P. Borgesen

Design For Thermo-Mechanical Reliability Of A 3D Microelectronic Component Using 3D FEM 705
S. Belhenini, A. Tougui, F. Dosseul

Numerical Modeling Of Flexible Actuator For Dynamic Lighting 711
T. Ma, X. Li, J. Wei, G. Zhang, P. Sarro

Thermal Performance Of Embedded Heat Pipe In High Power Density LED Streetlight Module 715
H. Tang, J. Zhao, B. Li, S. Leung, C. Yuan, G. Zhang

Author Index

2014 15th International Conference on Thermal, Mechanical and Multi-Physics Simulation and Experiments in Microelectronics and Microsystems

Edited by

G.Q. Zhang
Philips Lighting, Eindhoven, The Netherlands
Delft University of Technology, The Netherlands

W.D. Van Driel
Philips Lighting, Nijmegen, The Netherlands
Delft University of Technology, The Netherlands

P. Rodgers
The Petroleum Institute, United Arab Emirates

C. Bailey
The University of Greenwich, United Kingdom

O. de Saint Leger
Astefo, France

EuroSimE 2014

IEEE

15ʰ International conference on

Thermal, Mechanical and Multi-Physics Simulation and Experiments in Microelectronics and Microsystems

EuroSimE 2014

Gent, Belgium – April 7-8-9, 2014

Edited by

> **G.Q. Zhang**
> **W.D. Van Driel**
> **P. Rodgers**
> **C. Bailey**
> **O. de Saint Leger**

Technically sponsored by **IEEE - CPMT**

Conference financially co-sponsored by
> AMIC GmbH, China
> DS Simulia, The Netherlands
> Infinite Simulation Systems, The Netherlands
> Philips Lighting, The Netherlands
> Robert Bosch GmbH, Germany
> State Key Laboratory SSL, China

> and with logistics support of IMEC vzw, Belgium

Organization Committees of EuroSimE 2014:

General chair:	**G.Q. Zhang**	*Philips Lighting & Delft University of Technology, The Netherlands*
Co-chairs:		
(Local organizer)	**B. Vandevelde**	*IMEC, Belgium*
(Financial affairs)	**O. de Saint Leger**	*Astefo, France*
Technical Committee:		
Chairs:	**W.D. Van Driel**	*Philips Lighting, Nijmegen, The Netherlands; Delft University of Technology, The Netherlands*
Co-chairs	**P. Rodgers**	*The Petroleum Institute, United Arab Emirates*
	C. Bailey	*The University of Greenwich, United Kingdom*
Exhibition Committee:		
Chair:	**W.D. Van Driel**	*Philips Lighting, The Netherlands*
	A. Wymyslowski	*Wroclaw University of Technology, Poland*
Short Course Committee:		
Chair:	**R. Dudek**	*Fraunhofer Institute IZM, Germany*
	B. Wunderle	*Fraunhofer Institute IZM, Germany*
Publicity Committee:		
	B. Vandevelde	*IMEC, Belgium*
	D. Andersson	*Swerea IVF, Sweden*
Industry Liaison Committee:		
Chair:	**M. Guyenot**	*Robert Bosch, Germany*
	Y. Liu	*Fairchild Semiconductor, USA*
	B. Schwarz	*Siemens, Germany*
International advisory committee:		
	T. Luk	*Fairchild Semiconductor, USA*
	R. Mertens	*IMEC, Belgium*
	B. Michel	*Fraunhofer Institute IZM, Germany*
	S. Pienimaa	*Nokia, Finland*
	A.J. van Roosmalen	*Philips, The Netherlands*
	B. Vigna	*STMicroelectronics, Italy*
	P. Wesling	*IEEE/CPMT, USA*

Preface

On behalf of the Organizing and Technical Committees, it is our pleasure to welcome you to the thirteenth international conference on "Thermal, Mechanical and Multiphysics Simulation and Experiments in Microelectronics and Microsystems" (EuroSimE 2014), held at the Marriott Hotel of Gent, Belgium, on April 7-8-9, 2014.

The annual EuroSimE conference was created as the only international conference with a focus on Thermal, Mechanical and Multi-Physics Simulation and Experiments in Micro-Electronics and Micro-Systems. EuroSimE was initiated in 2000 by the COMPETE network, with sponsorship from the European Commission, to meet research and development needs in the fields of Microelectronics and Microsystems. Since then, EuroSimE has gained worldwide appeal with participants from more than thirty countries, spanning all continents, and has become a fully sponsored IEEE – CPMT event. The conference proceedings are part of the IEEE conference publication program and can be found in both the IEL and XPLORE systems.

The EuroSimE 2014 Technical Committee has made substantial efforts to provide a high quality technical program covering the latest advances in its fields of expertise, with approximately one hundred and thirty technical papers presented in twenty one sessions. As per our tradition, a special Technology Keynote Session will be given by Invited Speakers holding leading industrial positions, to highlight major technological and industrial development trends, challenges and roadmaps. In addition, three Technical Keynote Sessions, comprising of Invited Talks, will address pressing issues in modeling and experimentation for micro-electronics and micro-systems. EuroSimE now has a well-established, dedicated multi-physics track in addition to the thermo-mechanical and thermal tracks. Extensive opportunities will be offered for technical exchange between participants and exhibitors of computer simulation software.

It is our pleasure to acknowledge the contributions of the organizing and technical committees, authors and co-authors, speakers and exhibitors. We would also like to thank all of our sponsors, including IEEE-CPMT, and financial co-sponsors. We are especially grateful for the unique contribution made by IMEC, who served as this year's local organizer.

Finally we wish to thank you all for attending this year's conference and making it such a successful event, and look forward to hopefully seeing you at next year's EuroSimE conference!

G.Q. Zhang, W.D. Van Driel, P. Rodgers, C. Bailey and O. de Saint Leger

EuroSimE 2014 Technical Committee

Technical Chair, co-Chair for thermomechanical track	Van Driel, W.D.	Philips Lighting Delft University of Technology	The Netherlands
Technical co-Chair for multi-physics track	Bailey, C.	The University of Greenwich	United Kingdom
Technical co-Chair for thermal track	Rodgers, P.	The Petroleum Institute	United Arab Emirates

Members

Andersson, D. R. Swerea IVF - Sweden

Bailey, C. University of Greenwich - UK

Baelmans, M. K.U. Leuven - Belgium

Chiang, K. National Tsing Hua University/ITRI - Taiwan

Corigliano, A. Politecnico di Milano - Italy

Dasgupta, Abhijit, University of Maryland -USA

Van Der Sluis, Olaf, Philips Applied Technologies, The Netherlands

Van Driel, W.D., Philips Lighting, Delft University of Technology -The Netherlands

Dudek, R. Fraunhofer IZM - Germany

Elata, David, Technion – Israel

Ernst, L.J., The Netherlands

Eveloy, V. The Petroleum Institute -UAE

Fan, X.J. Lamar University, USA

Gao, Feng, Osaka University -Japan

Gonzalez, M. IMEC - Belgium

Geers, M. G.D. Eindhoven University of Technology - Netherlands

Hanreich, G. Vienna University of Technology - Austria

Iwamoto, N. Honeywell – USA

Jansen, K., technical University of Delft, The Netherlands

Keulen, F. van, Delft University of Technology - The Netherlands

Korvink, J., University of Freiburg - Germany

Lee, R. Hong Kong Univ. of Science & Technology - China

Liu, Y. Fairchild Semiconductor -USA

Malik, Tahir ICI - UK

Noritake, Chikage, Denso Corporation – Japan

Pandolfi, A., Politecnico di Milano, Italy

Pape, H., Germany

Perpinya, X. Centro Nacional de Microelectrónica, Spain

Plaza, J.A., Centro Nacional de Microelectrónica, Spain

Rantala, J. Nokia - Finland

Rodgers, P. The Petroleum Institute – UAE

Shirangi, Hossein – Robert Bosch GmbH, Germany

Schwarz, B. Siemens - Germany

Rejaei, B., Delft University of Technology - The Netherlands

Rochus, V., Université de Liège, Belgium

Sitaraman, S. K. Georgia Tech - USA

Tay, A.O. NUS - Singapore

Tee, T.Y. Amkor Technology -Singapore

Vandevelde, B. IMEC - Belgium

Vellvehi, M.H., Centre Nacional de Microelectrònica (CNM) - Spain

Wang, Bo Ping, Univ. of Texas at Arlington - USA

Wang, Z.P. Philips - China

Wiese, S. Technical University of Dresden - Germany

Wymyslowski, A. Wroclaw University of Technology - Poland

Yang, Daoguo, Guilin University, China

Zhang, Guang-Ping, Shenyang National Laboratory for Materials Science/ Institute of Metal Research (IMR), Chinese Academy of Sciences - China

Zhang, G.Q. Delft University of Technology -The Netherlands

Zhong Zhaowei, Nanyang Technology University - Singapore

Zhou, J. Lamar University - USA

Applications of Laser Welding to Electric Connectors

Chung-Fu Liao, Shih-Po Liu, Kuo-Chi Liao*
Department of Bio-Industrial Mechatronics Engineering
National Taiwan University, Taipei, Taiwan
*kokki@ntu.edu.tw

Abstract

Characteristics of a pulsed Nd:YAG laser spot-welding procedure applied to electronic connectors are investigated in the current study. A finite element analysis is carried out to evaluate the weld pool profile of SUS304 stainless steel sheets under various laser intensities. Bead shapes including the pool diameter and the penetration depth as well are predicted and compared with the associated experiments. The Gurson-type damage model is implemented into the numerical analysis to describe the failure of spot-welded structures. Different spot-welded specimens subjected to the tensile-dominated and shear-dominated loading conditions were conducted to verify the suitability of parameters implanted in the damage model. Structure strength of a stainless steel sheet assembled with a metallic shell of an electronic connector utilizing laser spot-welding is then examined, and results based on the simulation are in fair agreement with those based on the measurements.

Keywords: laser welding, electronic connector, finite element analysis

1. Introduction

Laser welding is a reliable, efficient, and precise technique commonly used in industrial manufacturing processes. Low heat input to relatively limited weld region based on laser welding generally induces less distortion in components compared to other welding approaches. Dimensions of electronic connectors are usually required to be reduced due to the size constraint of electronic devices. As shown in Figure 1, a SUS304 stainless steel sheet is therefore integrated with a metallic shell of a micro USB connector using pulsed Nd:YAG laser spot-welding to enhance the structure strength.

Chang and Na [1] developed a combined model of a finite element analysis and a neural network algorithm to predict laser spot-welding bead shapes of metallic sheets with various gaps between workpiece pairs. Differences between numerical results and the corresponding experimental measurements were remarked to be rather limited. Chang and Na [2] further proposed a heat source equation to more precisely estimate the laser spot-welding bead shape of stainless steel sheets. He et al. [3] adopted the transient heat transfer and fluid flow model to investigate the fusion zone geometry of stainless steels under the laser spot welding procedure. Weld pool depth and width for various laser power densities based on the calculations were reported to be in good agreement with those based on the corresponding experiments. Kazemi

and Goldak [4] estimated the transient temperature profile and the dimensions of the fusion zone during the welding processes. They proposed a modified heat source model combining a circular disk source with a Gaussian distribution of thermal flux on the top surface and a line source through the thickness of workpiece to accurately predict the weld cross section in deep penetration laser welding. Kong et al. [5] used an inverse modeling methodology to characterize mechanical properties of three fusion zones of spot-welded joints. Constitutive laws of different fusion zones were implemented into a three-dimensional finite element model with a Gurson damage criterion to capture the deformation and failure of spot-welded joints under the tensile-shear loading condition. Ventrella et al. [6] applied pulsed Nd:YAG laser to bond stainless steel thin foils, and stated that the pulse energy plays an important role on tensile-shear strength of welded joints and hardness of multiple fusion zones as well. As indicated in Lee et al. [7], higher effective stress-strain curves employed for the weld and heat affected zones and the geometry of the weld protrusion result in the necking/shear failure mode of laser welded lap-shear specimens of high strength low alloy steel sheets. They also reported that the introduction of void nucleation and growth mechanisms to the material model is able to identify the ductile failure initiation site which agrees well with the experimental observations.

A heat source equation implemented into a user subroutine DFLUX of a finite element commercial package ABAQUS [8] is adopted to estimate the weld pool profile of stainless steel sheets under the laser spot-welding procedure. Non-homogeneous mechanical properties of the base metal and the fusion zone as well are then explored. Lately developed the extension to the shear modified Gurson model is used to describe the damage progress of welded components. Finally, load carrying capacity of the reinforced connector structure is investigated in the current study.

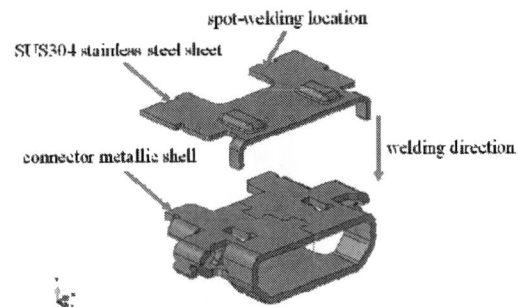

Figure 1 Assembly of a stainless steel sheet with a metallic shell of a micro USB connector

978-1-4799-4789-8/14 $31.00 © 2014 IEEE

2. Experiments

All welds were performed using a pulsed Nd:YAG laser welding machine with shielding gas of argon at the frequency of 1.1 Hz. Laser intensity was set to rise from zero to the designated work power within 0.3 ms, then decrease linearly down to 90% of the selected power in the subsequent 2.3 ms. A specimen for the tensile-dominated test and a lap-shear specimen for the shear-dominated test were comprised of a vertical workpiece spot-welded with a horizontal one and two partly-overlap spot-welded workpieces, respectively. Schematic drawing of the experimental set-up for the strength investigation of the micro USB connector is shown in Figure 2. A solid plate was inserted into the metallic shell and compressed against the structure until its load carry capacity dropped significantly. Five specimens were prepared for each test mentioned above, and associated reaction forces with respect to the increasing controlled displacement were then recorded.

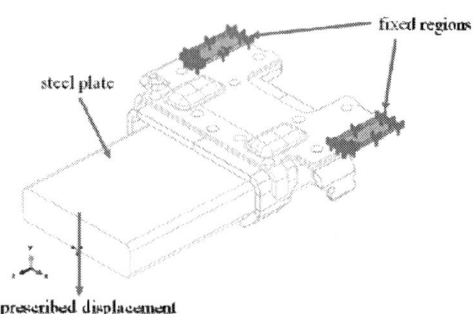

Figure 2 Schematic drawing of the experimental set-up for the strength investigation of the micro USB connector

3. Numerical Simulations

Chang and Na [2] stated that the intensity of the laser beam can be treated as a volumetric heat source determined by the beam penetration depth. Heat flux Q with a Gaussian-like distribution can then be expressed as a function of the radial distance from center of the beam r, the depth from the surface z, and time t.

$$Q(r,z,t) = \frac{2P(t)}{\pi r_{f_0}^2 d}\left(\frac{r_f}{r_{f_0}}\right)\exp\left(-\frac{2r^2}{r_f^2}\right)u(z), \quad (1)$$

where d is the vaporization depth determined by the drilling velocity, $P(t)$ is the time-dependent laser power at the center point, r_{f_0} and r_f are respectively the beam's focal radius at the surface and at the depth, $u(z) = 1$ for $0 \le z \le d$, and $u(z) = 0$ otherwise. Equation (1) is implemented into a user subroutine DFLUX as a heat source here.

Basic mechanical properties including Young's modulus, Poisson's ratio, and the yield stress of SUS304 are listed in Table 1. Relationships of the uniaxial tensile stress and plastic strain of the base metal shown in Figure 3 (solid line) are extracted from Zhou and Ling [9]. Since it is not an easy task to obtain the tensile stress-plastic

strain curve of the fusion zone, it however can be estimated by scaling the tensile stress of the base metal with a ratio of the measured hardness based on the fusion zone and that based on the base metal for a given plastic strain. The indentation tests were conducted to evaluate the Vickers hardness of the fusion zones processed using different laser energy levels. Measurements are listed in Table 2 while the corresponding stress-plastic strain curves of the fusion zones are also demonstrated in Figure 3. As in Antunes and Menezes [10], values of the Vickers hardness of the fusion zones based on the predicted stress-plastic strain behaviors are further calculated to validate the appropriateness of the dotted and dashed curves displayed in Figure 3. Table 2 shows that the simulated magnitudes of the hardness agree fairly with the associated experiments.

Gurson [11] proposed a closed-form yield criterion using an upper bound approach for porous materials with isotropic matrices. Tvergaard [12] later introduced three fitting parameters into the Gurson model to obtain reasonable results of shear band instability based on a finite element model for porous materials compared with those based on the continuum model using the Gurson model. The fitting parameters of the proposed Gurson model can be obtained via either experiments or numerically simulations. This modified Gurson model for the isotropic material with spherical voids can be described as

$$\Psi = \left(\frac{\sigma_e}{\sigma_0}\right)^2 = 2q_1 f \cosh\left(q_2\frac{3\sigma_m}{2\sigma_0}\right) - 1 - q_3 f^2 = 0, \quad (2)$$

where σ_0 is the matrix yield stress under uniaxial loading conditions, f is the void volume fraction, q_i ($i = 1, 3$) are fitting parameters, σ_m is the macroscopic mean stress, and σ_e is macroscopic von Mises equivalent stress.

Recently, Nahshon and Hutchinson [13] proposed the modified Gurson model to account for damage growth under the shear-dominated loading condition with low triaxiality straining. Nielsen and Tvergaard [14] further extended the shear modification model to capture the damage development of spot-welded lap-shear specimens under low triaxiality shearing without increasing the damage rate in regions under moderate to high stress triaxiality. Evolution of the void volume fraction, which arises from the growth of existing voids, the nucleation of new voids, and the mechanism of void softening in shear, can then be expressed as

$$\dot{f} = (1-f)(\dot{\varepsilon}_{11}^P + \dot{\varepsilon}_{22}^P + \dot{\varepsilon}_{33}^P) + A\dot{\bar{\varepsilon}}^P + k_\omega f\omega_0\frac{s_{ij}\varepsilon_{ij}^P}{\sigma_e}, \quad (3)$$

The first term of right hand side of Equation (3) describes the void growth rate originated with the incompressibility assumption of the matrix material. Here $\dot{\varepsilon}_{ij}^P$ ($i, j = 1$ to 3) is the macroscopic plastic strain rate. Furthermore, a so-called plastic strain controlled void nucleation model is chosen here as illustrated in the second term. A parameter

978-1-4799-4789-8/14 $31.00 © 2014 IEEE

A, chosen to have a nucleation following a normal distribution, is related to the volume fraction of void nucleating particles f_N, the mean strain for nucleating ε_N, and the standard deviation of the normal distribution s_N as

$$A = \frac{f_N}{s_N\sqrt{2\pi}}\exp\left[-\frac{1}{2}\left(\frac{\bar{\varepsilon}^p - \varepsilon_N}{s_N}\right)^2\right] \text{ for } \dot{\bar{\varepsilon}}^p > 0, \quad (4)$$

The last contribution of $\dot{f}$ primarily accounts for the phenomenon of shear-dominated failures. A constant k_ω represents the magnitude of the damage growth rate under pure shear states. A parameter ω_0 is specified as

$$\omega_0 = \omega(\sigma)\Omega(T), \quad (5)$$

where

$$\omega(\sigma) = 1 - \left(\frac{27J_3}{2\sigma_e^3}\right)^2, \quad (6)$$

with J_3 a third invariant of stress and a linear interpolation function

$$\Omega(T) = \begin{cases} 1 & T < T_1 \\ (T - T_2)/(T_1 - T_2) & \text{for } T_1 \leq T \leq T_2 , \\ 0 & T > T_2 \end{cases} \quad (7)$$

Here T denotes the stress triaxiality while s_{ij} ($i, j = 1$ to 3) indicates the deviatoric stress state.

Constitutive responses for porous materials are formulated in a user subroutine VUMAT. To be concise of the presentation, comprehensive derivations can be found in Nielsen and Tvergaard [14].

Table 1 Basic mechanical properties of SUS304 stainless steel

Young's modulus (MPa)	Poisson's ratio	yield strength (MPa)
199000	0.285	286

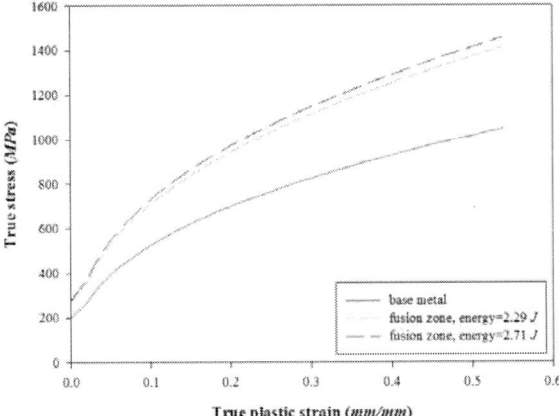

Figure 3 Relationships of the uniaxial tensile stress and plastic strain of the base metal and fusion zone processed using different laser energy levels

Table 2 Vickers hardness of the base metal and fusion zones processed using different energy levels based on the experiments and the simulation results

	base metal	fusion zone	fusion zone
energy (J)	-	2.29	2.71
experimental hardness (HV)	289.4	390.4	403.2
ratio	-	1.35	1.39
simulated hardness (HV)	295.6	377.3	388.4

4. Results and Discussion

Figure 4 shows temperature contours near weld regions of two different specimen thicknesses processed using several energy levels at the end of the processing duration. Simply the temperatures above the melting point of $1500^\circ C$ of the stainless steel treated as the weld pool are displayed in the figure. Corresponding pool diameters and penetration depths of the specimen thickness of 0.15 mm and 0.25 mm based on the experimental measurements and the numerical simulations are illustrated in Figure 5. Calculation results are generally in good agreement with experiments.

Bead shapes are then extracted from the results of the thermal analysis mentioned above, and they are subsequently imported into new models respectively subjected to the tensile-dominated and shear-dominated loading conditions. Parameters required for the extended shear-modified Gurson model are tabulated in Table 3. Figures 6(a) and 6(b) reveal prescribed displacement–reaction force curves, based on five measurements and the simulations, of the tensile-dominated specimens processed using energy levels of 2.29 J and 2.71 J, respectively. These comparisons demonstrate that the simulation results approximately exhibit similar trends to the measured ones. Snapshots of predicted rupture joints further match well with the corresponding micrographs as displayed in Figure 7. Simulated relationships of the displacement and force of the lap-shear specimen processed using energy level of 2.29 J also roughly catch the tendency of the experimental ones as depicted in Figure 8. Suitability of parameters listed in Table 3 can then be concluded along with the results shown in Figures 6 and 8.

Figure 9 shows an analysis model for the strength investigation of the micro USB connector with appropriate boundary conditions. The solid plate having much larger stiffness than other components is considered as a rigid body in the simulations. Reaction forces of the solid plate as a function of the prescribed displacement based on the experiments and the simulations are displayed in Figure 10. Measured and calculated relationships exhibit relatively similar oscillation patterns, and fluctuation in the figure are due to the sequential rupture of weld joints.

978-1-4799-4789-8/14 $31.00 © 2014 IEEE

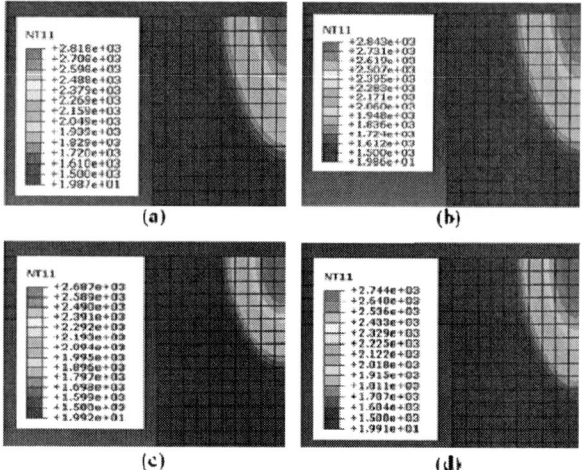

Figure 4 Temperature contours near weld regions of specimen thicknesses of (a) 0.15 *mm* processed using energy level of 2.29 *J*, (b) 0.15 *mm* processed using energy level of 2.71 *J*, (c) 0.25 *mm* processed using energy level of 2.29 *J*, and (d) 0.25 *mm* processed using energy level of 2.71 *J* at the end of the processing duration

Table 3 Parameters required for the extended shear-modified Gurson model

φ	ε_N	S_N	σ_N	f_c	f_f
0.01	0.2	0.1	2.2 σ_0	0.022	0.152

q_1	q_2	q_3	k_ω	T_1	T_2
1.5	1.0	2.25	3.5	0.2	0.7

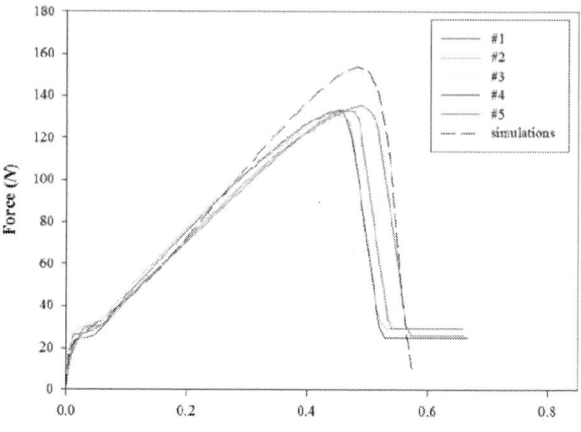

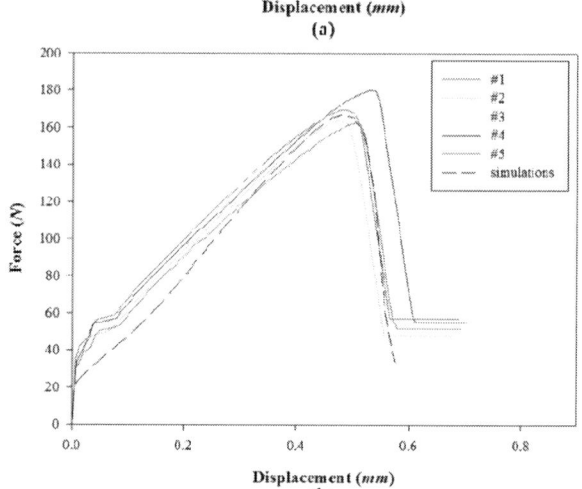

Figure 6 Prescribed displacement–reaction force curves of the tensile-dominated specimens processed using energy levels of (a) 2.29 *J* and (b) 2.71 *J* based on the measurements and the simulations

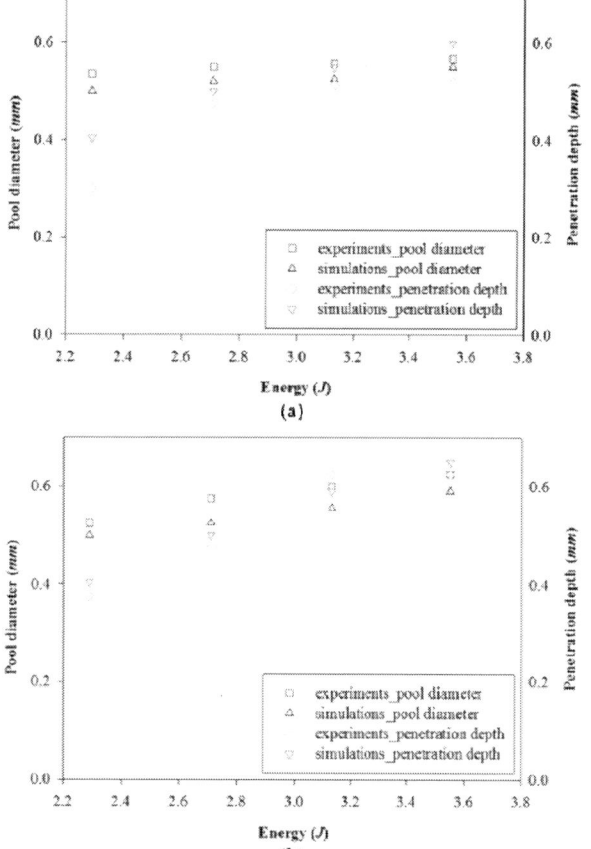

Figure 5 Pool diameters and penetration depths of the specimen thickness of (a) 0.15 *mm* and (b) 0.25 *mm* based on the measurements and the simulations

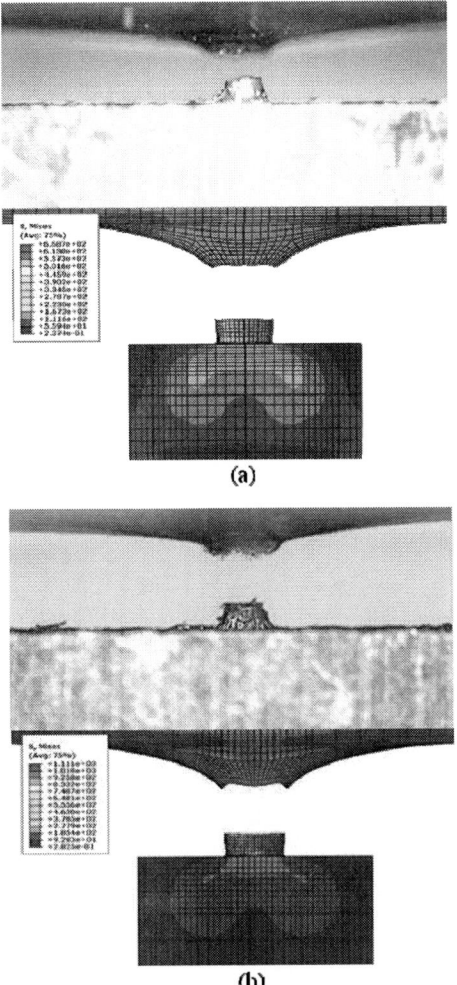

(a)

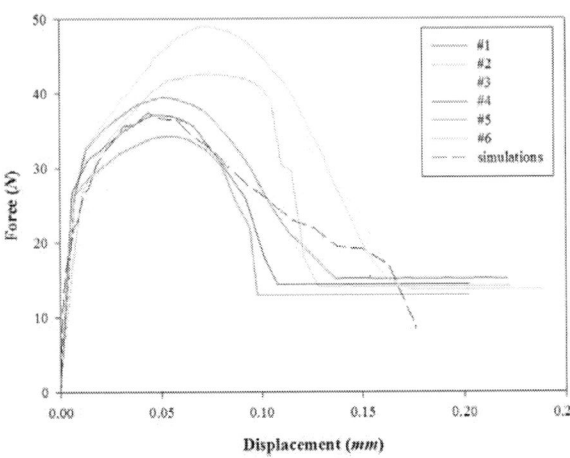

(b)

Figure 7 Rupture joints of the tensile-dominated specimens processed using energy levels of (a) 2.29 J and (b) 2.71 J based on the micrographs and the simulations

Figure 8 Prescribed displacement–reaction force curves of the shear-dominated specimens processed using energy levels of (a) 2.29 J and (b) 2.71 J based on the measurements and the simulations

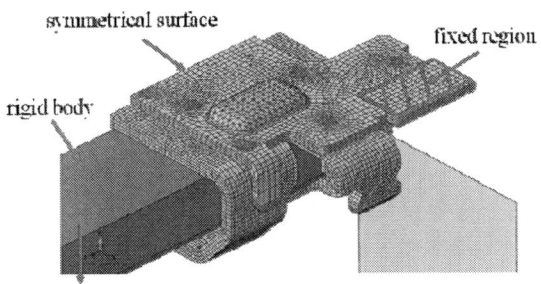

Figure 9 An analysis model for the strength investigation of the micro USB connector

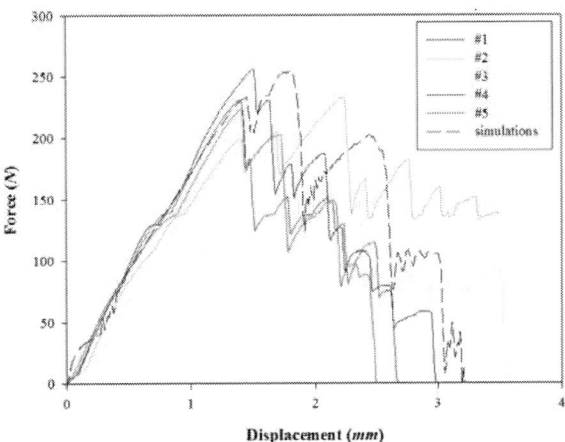

Figure 10 Prescribed displacement–reaction force curves of the micro USB connector based on the experiments and the simulations

5. Conclusions

1. The volumetric heat source coded in the subroutine DFLUX is used to successfully evaluate the bead shape in the present study.

2. The extended shear modified Gurson model implemented into the subroutine VUMAT is adopted to describe the damage evolution of the porous material.

3. Stiffness of the stainless steel sheet laser-welded with the metallic shell of the micro USB connector can be reasonably predicted utilizing the current systematic procedures. Numerical simulations can then be applied to assess the efficient layout of the spot-weld location in the future.

Acknowledgments

The authors are grateful for the financial support from National Science Council, Taiwan under Contract Number NSC-102-2221-E-002-044.

References

[1] Chang, W. S. and Na, S. J., "Prediction of Laser-Spot-Weld Shape by Numerical Analysis and Neural

Network," *Metallurgical and Materials Transactions*, Vol. 32 (2001), pp 723-731.

[2] Chang, W. S. and Na, S. J., "A Study on the Prediction of the Laser Weld Shape with Varying Heat Source Equations and the Thermal Distortion of a Small Structure in Micro-Joining," *Journal of Materials Processing Technology*, Vol. 120 (2001), pp 208-214.

[3] He, X. *et al.*, "Heat Transfer and Fluid Flow During Laser Spot Welding of 304 Stainless Steel," *Journal of Applied Physics*, Vol. 36 (2003), pp 1388-1398.

[4] Kazemi, K. and Goldak, A. J., "Numerical Simulation of Laser Full Penetration Welding," *Computational Materials Science*, Vol. 44 (2007), pp 841-849.

[5] Kong, X. Q. *et al.*, "Numerical Study of Strengths of Spot-Welded Joints of Steel," *Materials and Design*, Vol. 29 (2008), pp 1554-1561.

[6] Ventrella, V. A. et al., "Pulsed Nd:YAG Laser Seam Welding of SUS316L Stainless Steel Thin Foil, " Journal of M*aterials Processing Technology*, Vol. 14 (2010), pp 1838-1843.

[7] Lee, J. *et al.*, "Modeling of Failure Mode of Laser Welds in Lap-Shear Specimens," *Engineering Fracture Mechanics*, Vol. 78 (2010), pp 374-396.

[8] Hibbit, H. D. *et al.*, "ABAQUS User Manual," Version 6.8, (RI, 2008).

[9] Zhou, Z. and Ling, X., "Ductile Damage Analysis for Small Punch Specimens of Type 304 Stainless Steel Based on GTN Model," *Journal of Testing and Evaluation*, Vol. 37(2013), pp 1-7.

[10] Antunes, J. M. *et al.*, "Three-Dimensional Numerical Simulation of Vickers Indentation Tests," *International Journal of Solids and Structures*, Vol. 43 (2005), pp 784-806.

[11] Gurson, A. L., "Continuum Theory of Ductile Rupture by Void Growth: Part I – Yield Criteria and Flow Rules for Porous Ductile Media," *J. Eng. Mater. Tech.*, Vol. 99 (1977), pp. 2-15.

[12] Tvergaard, V., "Influence of Voids on Shear Band Instabilities under Plane Strain Conditions," *Int. J. Fract.*, Vol. 17 (1981), pp. 389-407.

[13] Nahshon, K. and Hutchinson, J. W., "Modification of the Gurson Model for Shear Failure," *European J. Mechanics A/Solids*, Vol. 27 (2008), pp. 1-17.

[14] Nielsen, K. L. and Tvergaard, V., "Ductile Shear Failure or Plug Failure of Spot Welds Modelled by Modified Gurson Model," *Eng. Fracture Mechanics*, Vol. 77 (2010), pp. 1031-1047.

Thermo-electrical and structural coupled simulations of buckling beam microprobes in high temperature/high current conditions

D. Eckhaut[1], E. Bertarelli[2], D. Acconcia[2], R. Vallauri[2], G. Cocchetti[1], A. Corigliano[1]

[1] Department of Civil and Environmental Engineering, Politecnico di Milano,
Piazza L. da Vinci 32 - 20133 Milano, Italy
[2] R&D Department, Technoprobe SpA,
Via Cavalieri di Vittorio Veneto, 2 - 23870 Cernusco Lombardone (LC), Italy
emanuele.bertarelli@technoprobe.com, raffaele.vallauri@techncoprobe.com, alberto.corigliano@polimi.it

Abstract

To design effective and reliable probe heads, it is crucial to have a deep understanding of the thermo-electro-mechanical coupled behavior of these complex systems. This work aims to investigate the behavior of vertical type microprobes for high-end wafer probing applications in high temperature / high current regimes, by adopting coupled thermo-electrical and structural numerical simulations. Probe electro-thermal heating, probe force degradation and current carrying capability are studied. The results obtained through the modelling framework introduced in this work are successfully compared to experimental data.

1. Introduction

Probe heads are the fundamental constituent of the devices – commonly referred as probe cards – designed and manufactured with the aim to test IC and MEMS on wafers before bonding, cutting and back-end processes [1]-[4]. A typical vertical type probe head is made of ceramic, drilled guiding plates and of several microprobes (Fig. 1).

Fig 1: Example of vertical type probe card head, with a dense full-array probe layout.

A single probe head can contain more than 20.000 microprobes, in the shape of extremely slender micro-needles; each of them have to establish an effective electrical contact with target structures on the Device Under Test (DUT), such as metallic pads or bumps. In service, microprobes are loaded to several ($> 10^6$) mechanical cycles and at the same time submitted to high current density (> 0.5 mA/μm^2) and possibly high temperature probing environment. It is worth to underline that, while testing temperatures can be up to 200 °C, the further probe heating due to current flowing through compact arrays of probes can generate a considerable localized increase of the temperature.

To design effective and reliable probe heads, it is crucial to have a deep understanding of the thermo-electro-mechanical coupled behavior. In this work, a multi-physics modeling approach is proposed and validated through comparison with experimental measurements on prototypes and real systems.

In the first part of the work the microprobe structural response under different temperature and current boundary conditions is modeled. The main goal at this stage is to evaluate the micro-needle stiffness degradation and the consequent contact force decrease that leads to a higher contact electric resistance (commonly referred as C_{res}) when the current intensity increases.

The second part of the work deals with micro-needle cooling mechanisms inside the microprobe head. Natural and forced convection conditions are studied. While in a closed probe head natural convection is expected, the possibility to achieve needle cooling driven by forced convection is also considered. Various Finite Element models are formulated, considering both a single micro-needle and micro-needle arrays. The results obtained are in line with analytical solutions (where available) and are coherent with the experimental measurements.

2. Thermo-electro-mechanical modelling

The temperature distribution along the microprobe is computed through a fully-coupled electro-thermal model. Due to the very high aspect ratio of the micro-needle and low Biot number $Bi \ll 1$, thermal conduction is assumed to be mono-dimensional with constant temperature over the needle cross section. Material resistivity $\rho(T)$ is assumed to be a linear function of temperature T, namely:

$$\rho(T) = \rho_0 \left[1 + \alpha_{res} \left(T - T_{ref} \right) \right], \qquad (1)$$

where ρ_0 is the electrical resistivity at reference temperature, α_{res} is the temperature coefficient of resistivity, used to account for a linear variation of the resistivity with the temperature, while T_{ref} is the reference temperature.

The power lost by radiation is neglected, while concerning convection the coefficient h_{conv} is assumed to be a constant. The thermal balance for an unit length needle element is then:

$$\lambda \frac{d^2 T}{dx^2} A + 2 h_{conv} \left(T_{ext} - T \right)\left(a + b \right)$$

$$+ \rho_0 \left[1 + \alpha_{res} \left(T - T_{ref} \right) \right] \frac{I^2}{A} = 0 \qquad (2)$$

where λ is the thermal conductivity, A is the needle section surface, T_{ext} is the ambient temperature, a and b are the dimensions of the needle cross section, I is the current intensity.

The buckling beam vertical microprobe mechanical response (here considered as a compressed micro-column) is described according to von Karman hypotheses for the strain:

$$\varepsilon_x = \frac{du}{dx} + \frac{1}{2}\left[\left(\frac{dw}{dx} \right)^2 + \left(\frac{dv}{dx} \right)^2 \right] - y \frac{d^2 w}{dx^2} - z \frac{d^2 v}{dx^2}, \quad (3)$$

where ε_x is the strain in longitudinal direction, u is the longitudinal displacement, v and w are the transversal displacements.

Exact shape functions are implemented in a Finite Element formulation. To account for the needle stiffness degradation with temperature, a Young's modulus and a yield stress temperature-dependent behaviour is implemented adopting a linear variation. The resulting mechanical model is sequentially, weakly coupled with the electro-thermal model.

The equilibrium equations are [5]:

$$\begin{cases} \dfrac{d^2}{dx^2}\left(E(T) I_y \dfrac{d^2 w}{dx^2} \right) + \dfrac{d}{dx}\left(P \dfrac{dw}{dx} \right) = 0 \\[2mm] G(T) J_t \cdot \theta' = 0 \\[2mm] \dfrac{d^2}{dx^2}\left(E(T) I_z \dfrac{d^2 v}{dx^2} \right) + \dfrac{d}{dx}\left(P \dfrac{dv}{dx} \right) = 0 \end{cases}, \qquad (4)$$

where E and G are the Young's modulus and the shear modulus. I_y, I_z and J_t are the needle bending moments of inertia in the two bending principal direction and torsional moment of inertia, respectively. It can be noticed that the mechanical problem is completely uncoupled, since it is the perfect superposition of two simple flexural-buckling in the principal directions y and z. The torsion is constant along the needle.

The head displacement in the x direction is the linear superposition of three parts, namely first order contribution, second order contribution and heat dilatation:

$$u(L) = \int_0^L \alpha\, \Delta T\, dx - P \int_0^L \frac{1}{EA} dx +$$

$$- \frac{1}{2} \int_0^L \left(\frac{dw}{dx} \right)^2 + \left(\frac{dv}{dx} \right)^2 dx \qquad (5)$$

The plastic behavior of the material is modeled introducing plastic hinges; these are activated when the local stress is higher than yield stress. The contacts between the micro-needle and the intermediate guiding ceramic plates are considered in the model as simple supports.

3. Thermo-electro-mechanical modelling: results

In Figures 2 and 3 the variation of temperature along the needle axis computed from the electro-thermal model is shown. The temperature trend is parabolic for the region which has a constant section. At both ends, the needle has section reductions which are responsible for the linear trend near the two ends. The peak is located approximately at the midpoint. The maximum value of the needle temperature strongly depends on the current value, on the temperature at the end points of the needle, and on the needle ends geometry.

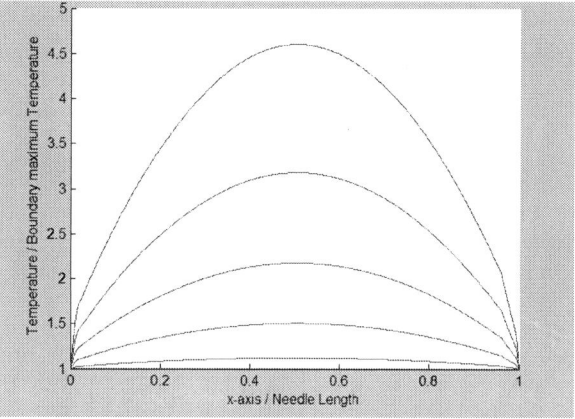

Fig. 2: Temperature along the needle axis for different current values

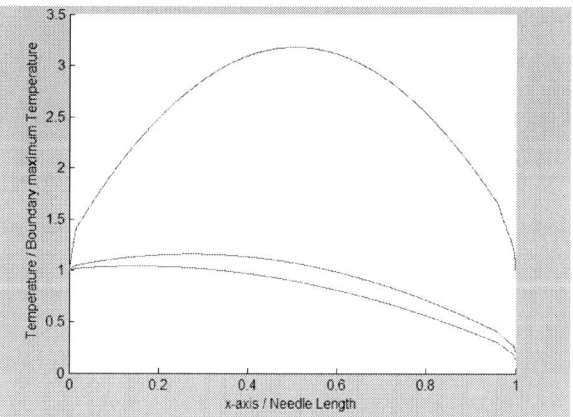

Fig. 3: Temperature along the needle axis for different current values, with different temperatures at needle ends.

The temperature distribution along the needle axis is then used in the mechanical simulation of the probe needle. A needle subjected to an imposed displacement in the axial direction (working overdrive) and a force degradation due to increasing current levels is simulated.

978-1-4799-4789-8/14 $31.00 © 2014 IEEE

This corresponds to a simulated Current Carrying Capability (CCC) test, as shown in Fig. 4.

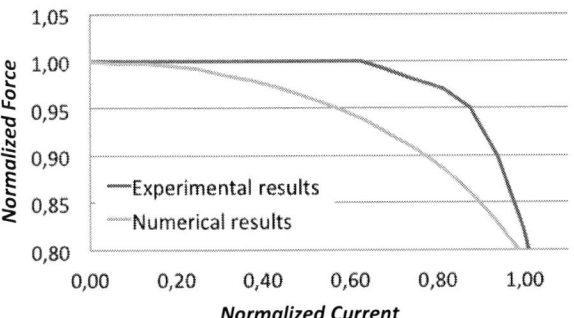

Fig. 4 : Simulation of CCC Test.

4. Microprobe forced cooling

As discussed in the Introduction, to decrease the needle temperature and to consequently reduce the force degradation, a viable solution can be needle cooling. To this aim, the potential effectiveness of forced convection with respect to natural convection should be studied by modeling needle arrays undergoing thermal and current loads.

4.1 Natural Convection

To define the needle geometry, the needle's shape has been simplified to a rectangular prism and the intermediate plates, which do not provide any thermal contribution, have been neglected. The needle grid is then defined as an array of 16 by 16 items. The superior plate is Gold whereas the inferior one is Aluminum. Simulations have been carried out by means of the commercial code COMSOL Multiphysics®.

Fig. 5 shows an example of the adopted model for the study of natural convection. The surface to surface radiation between the needles is neglected. The temperature of the two plates is fixed and a potential difference is applied between them.

Rayleigh's Number study led to the conclusion that the dominant mechanism of thermal exchange is conduction inside the needle array, whereas the natural convection is dominant outside the array [6], [7].

It is expected that central needles in the array have a higher temperature than external ones, resulting in a higher electrical resistance. This has been checked by analyzing the temperature and the current distribution among the needles.

4.2 Forced convection

In the forced convection case Reynolds' Number remains at low values, due to the reduced dimensions of the needles section (characteristic dimension of the system). Then, the air flow can be assumed as laminar and an air speed of 0.1 m/s to 0.5 m/s can be considered. This leads to a convective coefficient h of up to 400 W/(m²K), a typical value for forced convection case [6], [7].

By means of the numerical model here introduced it is possible to evaluate the effects of the forced convection on the system under study in different situations. This allows one to compare the current capacity of the needles in the different cases as a function of air flow temperature, air flow velocity, housing temperature (i.e., temperature of the enclosure) and other system physical parameters.

Figures 6 and 7 show typical results that can be obtained. An air flow, even with no temperature control, can lead to a notable improvement in the current that can be carried by micro-needles.

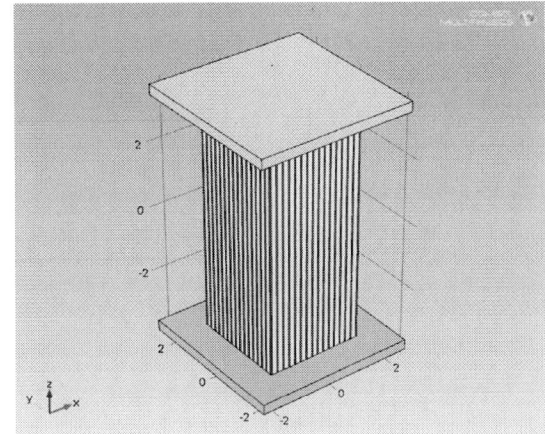

Fig. 5: COMSOL model for natural convection.

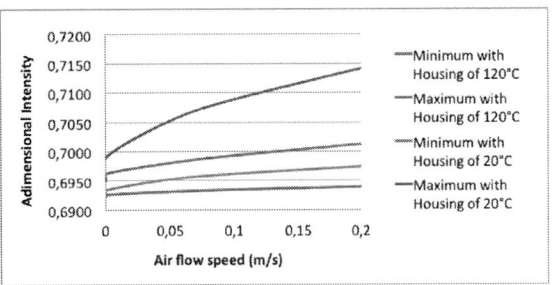

Fig. 6: Carried current intensity for several air flow speed in a needle grid.

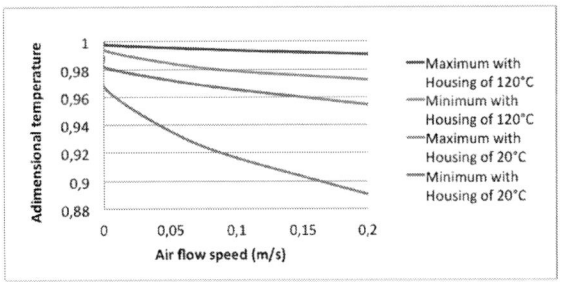

Fig. 7: Extreme temperature for several air flow speed in a needle grid.

5. Conclusions

A tool for the numerical simulation of CCC testing for micro-needles is discussed in this paper; the model reproduces the thermal-electrical-mechanical behaviour of a needle in its work environment. The proposed model has been validated by a comparison with experimental results and a good agreement is found.

In a second part of this study, forced convection is taken into consideration as a possible solution to improve probing performances. The modeling approach has been validated by comparison with theoretical results from the literature. It represents a useful tool to evaluate needle array performances in terms of current capability.

References

1. Gonzales D. E., Kister J., "Advancements in Performance of Buckling Beam Probes", *Proc. South West Test Workshop* 1999.

2. Kirby R, Yan H. F., "New Methodology for Probe CCC Characterization", *Proc. South West Test Workshop* 2004.

3. Liu Y., Desbiens D., Luk T., Irving S., "Parameter Optimization for Wafer Probe Using Simulation", *Proc. EuroSimE* 2005.

4. Corigliano A., Courard A., Cocchetti G., Magagnin L., Vallauri R., Acconcia D., "Multi-Physics Simulations for the Design of Probe-Heads Micro-Needles", *Proc. COMSOL Conference, Milan 10-12 October,* 2012.

5. Corradi dell'Acqua, L., Meccanica delle strutture, Vol. 3, McGraw-Hill, Milan (1994).

6. Staton D. A., Cavagnino A., "Convection Heat Transfer and Flow Calculations Suitable for Electric Machines Thermal Models", *IEEE T. Ind. Electron.,* Vol. 55, No. 10 (2008), pp. 3509-3516.

7. J. H. Lienhard, A Heat transfert textbook, Plogiston Press (2008).

Limitations and Accuracy of Steady State Technique for Thermal Characterization of solid and composite materials

Mohamad AboRas[1,2], Bernhard Wunderle[2,4], Daniel May[4], Ralph Schacht[2,3], Thomas Winkler[1], Sven Rzepka[2], Bernd Michel[2]

[1] Berliner Nanotest and Design GmbH, Berlin, Germany (E-mail: aboras@nanotest.org | Phone: +49 30 6392 3880)
[2] Fraunhofer Institute for Electronic Nano Systems ENAS, Chemnitz, Germany
[3] Brandenburg University of Technology, Cottbus-Senftenberg, Germany
[4] Chemnitz University of Technology, Chemnitz, Germany

Abstract

The steady state method is a commonly used and in principle simple way to measure thermal resistance and conductivity of thermal interface materials (TIMs). A heat flow through the TIM has to be generated and the temperature gradient across the TIM has to be measured. This is also defined by the ASTM standard ASTM D5470 [4]. To generate the heat flow the TIM must be positioned between a hot and a cold plate. However, for the new generation of highly conductive and thin TIMs the resolution of the common steady state technique often reaches its limit. To increase the resolution of the steady state equipment beyond the state-of-the-art the test systems must be analyzed and parasitic effects be studied. Accuracy and resolution depend not only on the precision of the setup, but decisively on the selection and execution of the measuring method conformed to the specific measurement task. In this paper we will present a test stand for thermal characterization of TIMs, die attachs and substrates based on the mentioned steady state method. It has been developed as a platform which allows the integration of various modules for characterization of different materials under different conditions, e.g. mated surface, finish, operation temperature, pressure, aging etc.

1. Introduction

Interfaces between electronic packaging materials or components have a significant impact on the thermal impedance of electronic systems and can be the dominant factor in achieving an effective thermal transfer. TIMs are used to connect an electronic device to the thermal transfer medium such as substrate, heat pipe and heat sink, or the thermal management components to each other. In some cases they are important to perform the tasks of attachment, stress/strain relief and thermal transfer simultaneously [1].

Fig. 1 shows a schematic of a typical application of thermal interface materials (TIMs) in an electronic package. Two types of TIM application can be seen. The so called TIM1 has been applied between silicon chip and copper heat spreader. The other TIM is the TIM2 which fills the gap between copper heat spreader and aluminum heat sink. As roughness and waviness of Si, Cu and Al surfaces are different, the type of TIM used and its bond line thickness (BLT) have to fit the application.

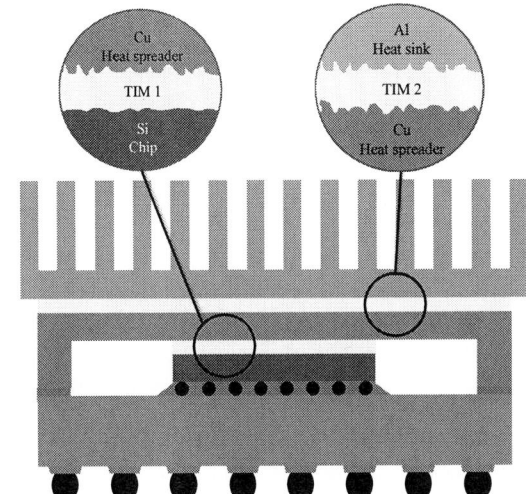

Fig. 1: Schematic application of thermal interface materials in electronic packages

Characterization of thermal interface materials becomes an even tougher challenge at low BLTs and higher thermal conductivities of the TIMs as more accurate measurement techniques are required. Thermal characterization methods for TIMs narrate a long story of confusion, as results from different characterization methods often disagree formidably. Even worse, actual thermal conductivity values will be different when applied to the real device, likely to cause over-, or even worse fatally, under-designed thermal heat paths. The reason for this misjudgment is often that TIM characterization is done under laboratory conditions (e.g. polished surfaces, excessive pressure conditions) or disregarding technological influences (e.g. cure regime for adhesives, dissimilar surfaces) [2, 3].

Thermal interface materials are traditionally tested under standard test conditions such as the ASTM D5470 tester [4]. However, such testers quite often disregard the actual use conditions.

For reliable statements for life time simulation of electronic packages the properties of all materials and components used (TIM, substrate, chip, heat spreader, etc.) are needed with respect to the real application and processing conditions.

In this paper we will present a test stand based on the steady state technique for thermal characterization of materials and components. We will show the limitation and calculate the accuracy of the steady state method. Some results of selected materials will be given to demonstrate the test stand and its accuracy.

2. Measurement Principle of the Steady State Method

The standard method ASTM D5470 [4] represents the measuring principle of the steady state techniques. Materials are tested between a hot and a cold plate. To determine the thermal resistance of the sample, the temperature gradient along the sample and the heat flow through the sample have to be measured (Fig. 2).

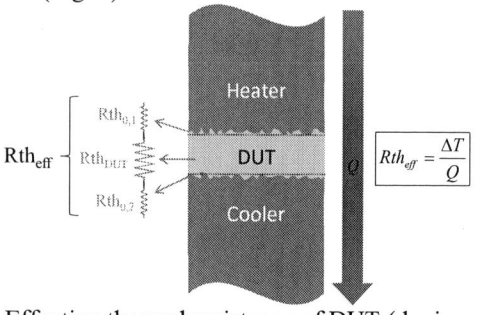

Fig. 2: Effective thermal resistance of DUT (device under test), with ΔT: temperature gradient along the sample and Q: heat flux through sample

The measured thermal resistance Rth_{eff} is the sum of the thermal resistance of the bulk sample Rth_{bulk} and the thermal interface resistances of the contacting mediums $Rth_{0,1}$ and $Rth_{0,2}$

$$\text{with}: Rth_{0,1} + Rth_{0,2} = Rth_0 \qquad \ldots\ldots\ldots (1)$$

$$\Rightarrow \quad Rth_{eff} = Rth_{bulk} + Rth_0 \qquad \ldots\ldots\ldots (2)$$

3. The Test Stand *TIMA Tester*

TIMA Tester is a test stand for thermal characterization of TIMs, die attachs and substrates doing analysis in steady state. It has been developed as a variable platform allowing the integration of various modules for characterization of different materials under different conditions (e.g. mated surface, finish, operation temperature, pressure, aging etc.) The very first generation of *TIMA Tester* was reported in [5].

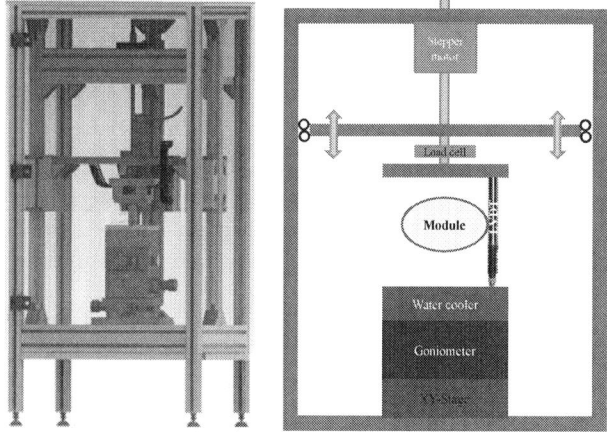

Fig. 3: Photography and schematic of TIMA platform

For standard characterizations two measuring variants are available for TIMs and substrates:

- Variant 1: Silicon-metal mating surfaces

For characterization of greases, pads, PCMs, gap fillers and other non-adhesive TIMs as well as substrates and coatings the following test method is used:

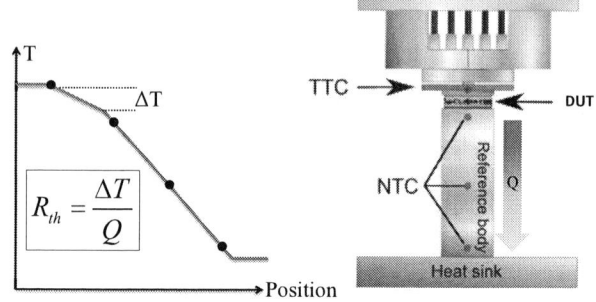

Fig. 4: Measuring principle of *TIMA Tester* using the silicon-metal mating surfaces

TIMs are tested between a silicon chip and a reference body (usually aluminium or copper) as they would be used in real assemblies. The chip is a thermal test chip assembled in flip-chip technology and used as heat source as well as temperature sensor to measure the top temperature of the sample [6]. The actual heat flow through the sample is estimated by using three NTC (Negative Temperature Coefficient Thermistors) sensors at defined positions in the calibrated reference body. The reference body is fixed on the heat sink; the thermal test chip is fixed on a stepper motor which can be moved vertically providing the possibility of inducing defined pressure (e.g. constant or periodically). Additionally after initial zeroing of the built-in position sensor the DUT thickness is also measurable.

- Variant 2: Metal-metal mating surfaces

This measuring variant is developed as low cost variant for characterization of adhesive TIMs, sinter material and solder.

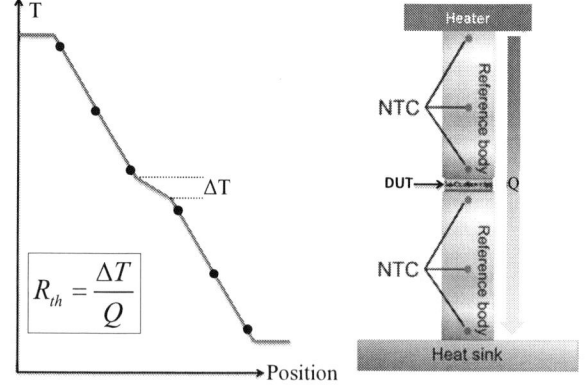

Fig. 5: Measuring principle of *TIMA Tester* using the metal-metal mating surfaces

A stack of two metal reference bodies of aluminium, copper or any desired metal and the DUT (most likely one of the examples given above) can be assembled, cured and measured representing a case of application.

4. Accuracy and Limitation of Steady State Method

To give an exact statement about the accuracy of the test system the method was precisely analyzed. In following the systematic error of the thermal resistance measurement will be analyzed.

The effective thermal resistance of the sample results from the temperature gradient (ΔT) on the sample and the heat flow (Q) through the sample.

$$Rth_{eff} = \frac{\Delta T_{DUT}}{Q} \qquad \ldots\ldots\ldots (3)$$

The maximal relative error of the effective thermal resistance can be calculated by the following equation:

$$\delta Rth_{eff} = \delta(\Delta T_{DUT}) + \delta Q \qquad \ldots\ldots\ldots (4)$$

The heat flow is measured by integrated temperature sensors (NTC) in the reference bodies (RB). The geometries and thermal conductivity of reference body are known.

$$Q = \frac{\Delta T_{RB}}{l_{RB}} A_{RB} \cdot \lambda_{RB} \qquad \ldots\ldots\ldots (5)$$

The maximal relative error of the heat flow can be calculated by the following equation:

$$\delta Q = \delta(\Delta T_{RB}) + \delta l_{RB} + \delta A_{RB} + \delta \lambda_{RB} \ldots\ldots\ldots (6)$$

As the manufacturing tolerance of the reference bodies (length and cross section area) can be disregarded due to high fabrication accuracy and their bulk thermal conductivity can be determined precisely most of the total error of the thermal resistance results from the error in temperature measurement. The following calculation shows the percentage error source within the measurement of a TIM with 10 mm²K/W of thermal resistance using following parameters:

	Symbol	Value	Abs. error	Rel. error
Thermal cond. of RB	λ_{RB}	380 W/mK	5 W/mK	1,3 %
Length RB	L_{RB}	25 mm	0,1 mm	0,3 %
Area RB	A_{RB}	100 mm²	0,1 mm²	0,1 %
Temperature gradient on RB	ΔT_{RB}	37,2 K	0,4 K	1,1 %
Heat flow	Q	56,5 W	1,6 W	2,9 %
Temperature gradient on TIM	ΔT_{TIM}	5,6 K	0,4 K	7,1 %
Thermal resistivity of TIM	**Rth**	**10 mm²K/W**	**1 mm²K/W**	**10 %**

It can be seen in the calculation table that 38% of the total error of the heat flow cause through the temperature error and about 70% of the total error of thermal resistance is coming from the error of the temperature measurement.

For the *TIMA Tester* high precise temperature sensors (NTC) with maximum error below 0.2 K in the range between room temperature and 80°C were chosen. Reference bodies were fabricated precisely (manufacturing tolerance below

10µm) and their thermal conductivity was measured using different methods (e.g. laser flash).

The total relative error of the measurement is dependent on the value of the thermal resistance. So the relative error of highly conductive materials is high because of the small temperature gradient along the sample. Also it is high for materials with very low thermal conductivity due to less heat flow.

The diagram in Fig. 6 shows the maximum error for different thermal resistance values. It can be seen that thermal resistances between 10 mm²K/W and 900 mm²K/W can be measured within measuring error below 10% by using Cu reference bodies. By using reference bodies with lower thermal conductivity e.g. Al the resolution increases for samples with higher thermal resistance due to the higher heat flow, but it decreases for highly conductive materials.

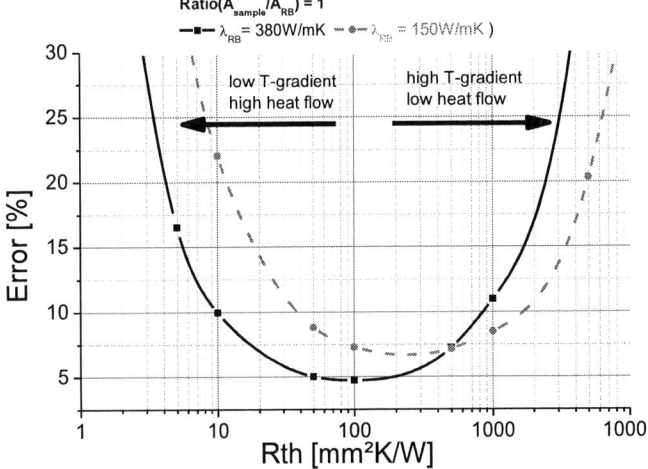

Fig. 6: Maximal measuring error as function of the thermal resistance values

To extend the resolution of the steady state method and to enable characterization of new generations of thermal interface materials and substrates, the cross section area of the sample should be kept as small as possible to generate a higher temperature gradient on the samples and the cross section area of the reference body should be kept as large as possible to generate higher heat flow as can be seen in equation 3.

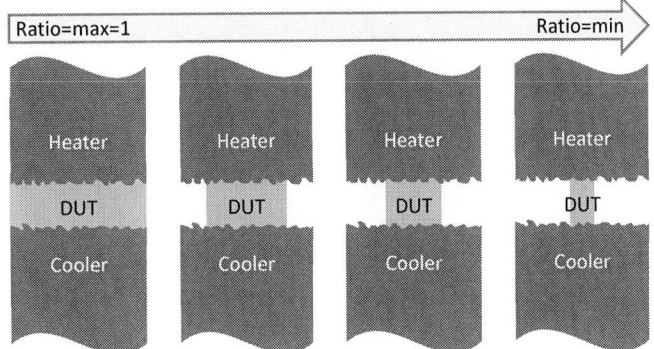

Fig. 7: Schematic of the variation of the ratio (A_{sample}/A_{RB})

978-1-4799-4789-8/14 $31.00 © 2014 IEEE

For the study of limitations of the steady state methods the maximal error of the thermal resistance was calculated for different ratios between the cross section areas of the reference bodies and the sample and for different reference body materials (Fig. 7).

Fig. 8 shows the maximum relative error as function of the thermal resistance values for different ratio between reference body area and sample area using Cu reference bodies with 380 W/mK thermal conductivity. It can be seen inter alia that through reducing the ratio of the contact area to 0.1 thermal resistance of 1 mm²K/W can be measured with a maximal measuring error below 10%.

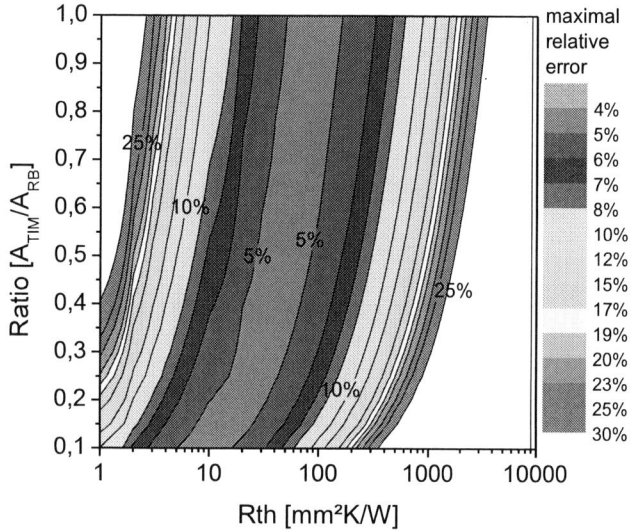

Fig. 8: Maximal measuring error as function of thermal resistance values for different ratio using Cu reference bodies

The same calculation has been done using Al reference bodies with 150 W/mK thermal conductivity. It can be seen in Fig. 9 that the resolution decreases for small thermal resistance values and increases for large values.

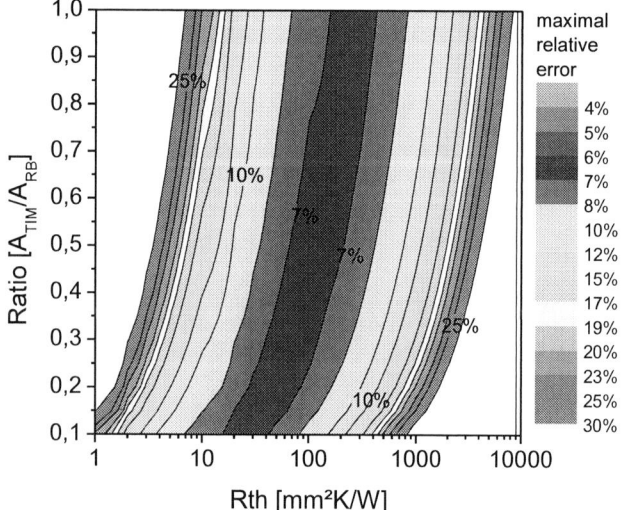

Fig. 9: Maximal measuring error as function of thermal resistance values for different ratio using Al reference bodies

To make the results of the study clearer the maximal measuring error for ratio = 1 and ratio = 0.1 is plotted in Fig. 10 using Cu and Al reference bodies.

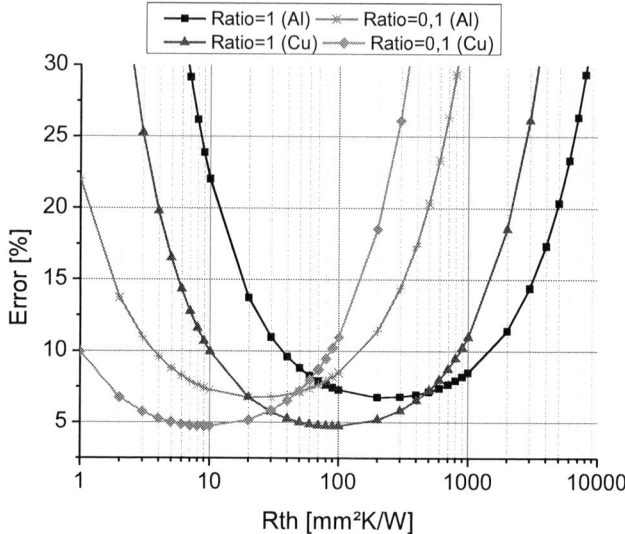

Fig. 10: Comparison between ratio 1 and 0.1 for Cu and Al reference bodies

It can be seen that the resolution of the thermal resistance can be increased from 10 mm²K/W to 1 mm²K/W with a maximal measuring error below 10 % by reducing the cross section area of the sample to 10 % of the cross section area of the reference bodies.

In other words: samples with a BLT of 100 µm and a thermal conductivity of 100 W/mK (or samples with a BLT of 10µm and a thermal conductivity of 10 W/mK respectively) can be measured by steady state method.

5. Measurements

Owing to the modularity of the test system, almost any thermal interface material, substrate, bulk metal, sintered metal etc. can be characterized with it. Following properties, Interdependencies and influences can be performed:
- Effective thermal conductivity
- Bulk thermal conductivity
- Thermal interface resistance
- Thermal transition coefficient
- Thermal resistance as function of pressure and temperature
- Influence of the temperature on thermal conductivity
- Thermal interface resistance as function of surface roughness
- Influences of curing conditions on thermal conductivity of thermal adhesives
- Influence of mechanical and thermal loading on thermal properties of TIMs

In this section, selected studies of different materials are presented in order to demonstrate the functionality and the accuracy of the test stand.

5.1. Measurement of bulk thermal conductivity and thermal interface resistance of standard thermal interface materials

The thermal resistance of a thermal interface material between two contact media is an effective value and depends on thermal resistance of the bulk TIM and the interface resistance between TIM and the mating surfaces.

$$Rth_{eff} = Rth_{TIM} + Rth_0 \qquad \dots\dots\dots (7)$$

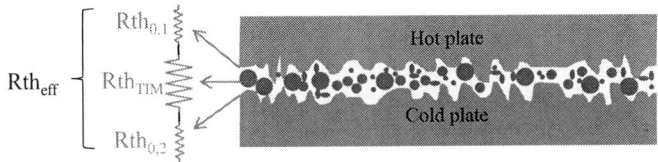

The thermal resistance of a TIM is dependent on its thickness, its bulk thermal conductivity and its cross section area. It can be calculated by following equation:

$$Rth_{TIM} = \frac{1}{\lambda_{TIM} \cdot A_{TIM}} \cdot BLT \qquad \dots\dots\dots (8)$$

$$\Rightarrow Rth_{eff} = \frac{1}{\lambda_{TIM} \cdot A_{TIM}} \cdot BLT + Rth_0 \qquad \dots\dots\dots (9)$$

Obviously equation 9 describes a linear function where $(\lambda_{TIM} \cdot A_{TIM})^{-1}$ is the slope and Rth_0 is the y-axis intercept of the function. This means, by measuring the TIM under the same conditions but with different thicknesses the bulk thermal conductivity and the thermal interface resistance can be determined. The diagram in Fig. 11 shows the effective thermal resistance of standard thermal grease in different bond line thickness measured between silicon thermal test chip and aluminum reference body. Each measured value is accompanied by an error bar, which has been calculated in situ from the thermal state of the measuring setup.

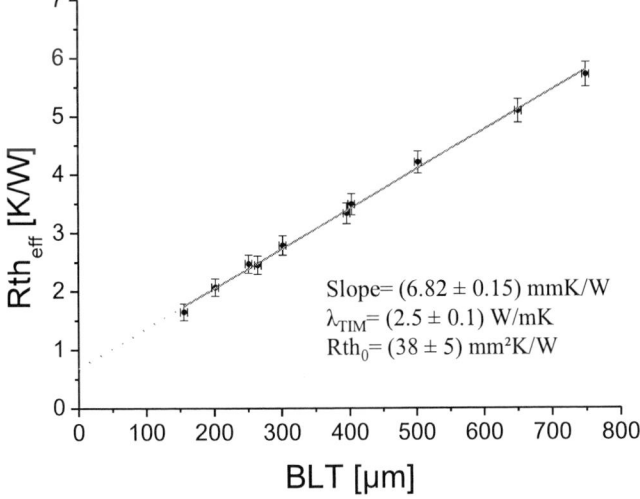

Fig. 11: Effective thermal resistance of standard grease as function of the bond line thickness

5.2. Characterization of substrates

Usually the thermal conductivity of substrates is measured by transient methods such as the laser-flash method [7]. But using this method the thermal diffusivity is measured. So for calculating the thermal conductivity the specific heat capacity and the density are needed in addition.

$$\alpha = \frac{\lambda}{\rho \cdot c} \qquad \dots\dots\dots (10)$$

Further this method is limited to bulk samples. For multilayer substrates such as IMS or for substrates with electrical or thermal vias such as LTCC, HTCC or FR4 the steady state method could be the better choice, as this method will rather make a statement about the whole sample's thermal conductivity. There are two challenges within the substrate characterization by steady state techniques: On one hand it is the relatively high thermal interface resistance between the sample and the matted surfaces and on the other hand it is the high thermal conductivity.

In contrast to characterization of TIMs, the thermal interface resistance heavily deteriorates the characterization of highly thermal conductive substrates using the steady state techniques, due to the high surface roughness and the lack of a gap filler. The following figure shows schematically the contact between matted surfaces and substrate.

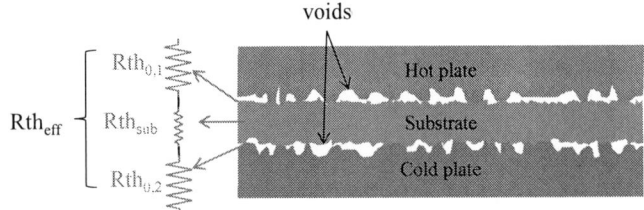

To eliminate the thermal interface resistance highly thermally conductive liquid metal is used, which can perfectly fill the gap between the sample and the matted surfaces whilst only having a negligible influence on the measurement result's credibility.

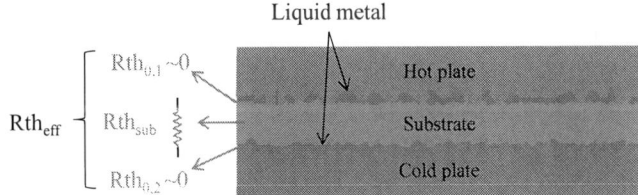

As the roughness of the surfaces usually is in micro meter range the thermal resistance of the liquid metal layer is insignificant. It can be shown within a simple experiment done by the test setup shown in Fig. 5, that the thermal resistance of liquid metal is nearly zero. So the effective thermal resistance corresponds to the thermal resistance in the measured sample. In the following characterization results of a ceramic substrate with thermal vias will be presented:

To increase the resolution of the measurement the samples were cut in 2 mm x 2 mm. A Cu reference body with 3.5 mm diameter was chosen. As interface medium the mentioned liquid metal was used.

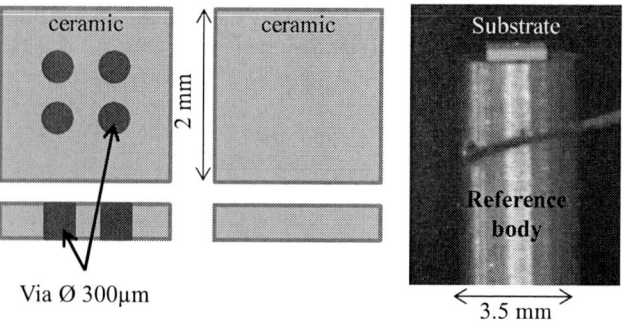

Fig. 12: Test setup and samples. Right: setup; middle: bulk ceramic substrate; left: ceramic substrate with thermal vias

Substrates with vias and other without vias were measured. The substrate thickness was 400 μm.

1st measurement:
Ceramic substrate w/o vias: $Rth_{ceramic}= (172 \pm 5)$ mm²K/W → $\lambda_{bulk}= (2.5 \pm 0.1)$ W/mK

2nd measurement:
Ceramic substrate with 4 thermal vias á 300μm diameter:
$Rth_{total}= (17.3 \pm 0.6)$ mm²K/W → $\lambda_{eff}= (23 \pm 1)$ W/mK
To determine the thermal conductivity of the metal vias, following equivalent circuit and equations have been used:

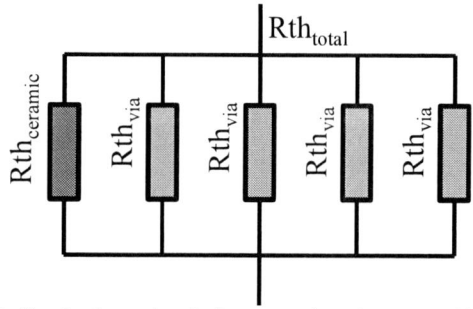

Fig. 13: Equivalent circuit for ceramic substrate with four thermal vias (parallel circuit)

$$Rth_{via} = \frac{4}{\left(\dfrac{1}{Rth_{total}}\right)-\left(\dfrac{1}{Rth_{ceramic}}\right)} \quad \dots\dots\ (11)$$

$$\Rightarrow \lambda_{via} = \frac{BLT_{via}}{A_{via} \cdot Rth_{via}} \quad \dots\dots\ (12)$$

$Rth_{via}= (1.36 \pm 0.06)*10^{-6}$ mm²K/W → $\lambda_{via}= (295 \pm 10)$ W/mK
The thermal conductivity value of the vias is in the expected range for sintered silver.

	Thermal resistance [mm²K/W]	Thermal conductivity [W/mK]
Substrate w/o vias	172 ± 5	2.5 ± 0.1
Substrate with four vias	17.3 ± 0.6	23 ± 1
Thermal via with 300μm diameter	$(1.36 \pm 0.06)*10^{-6}$	295 ± 10

5.3. Characterization of a nano adhesive TIM assembled between two Cu substrates

To characterize the adhesive with respect to the process conditions the adhesive was preassembled between two Cu substrates in three different BLTs.

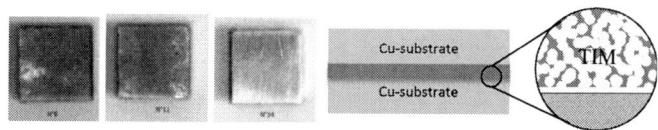

The samples were measured by the test stand *TIMA Tester* using liquid metal to eliminate the interface resistances between the Cu substrate and the reference body and between the Cu substrate and the thermal test chip.

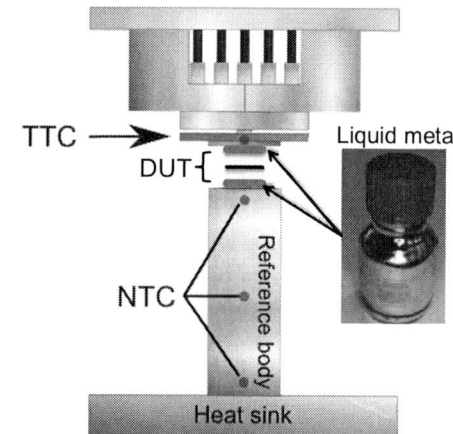

Fig. 14: Test setup used for characterization of the nano adhesive samples

Fig. 15 shows the results of the thermal resistances for the three samples. As the thickness of the Cu substrates is the same for all samples, the bulk thermal conductivity of the adhesive can be calculated from the slope while a y-axis intercept is the thermal interface resistance Rth_0 between the adhesive and the substrate including the thermal resistance of the Cu substrates.

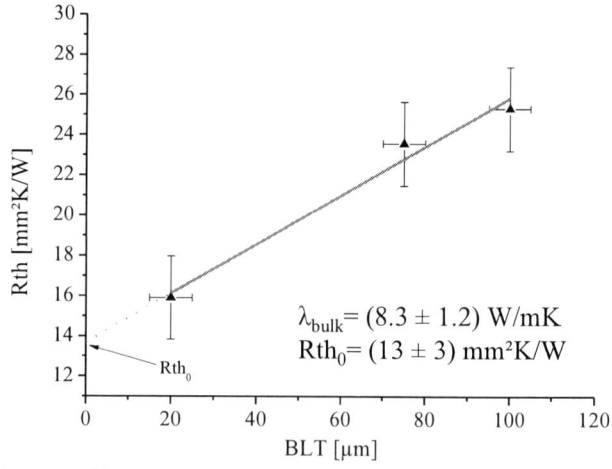

Fig. 15: Effective thermal resistance of nano adhesive and two Cu substrates as function of the bond line thickness

978-1-4799-4789-8/14 $31.00 © 2014 IEEE

6. Summary & Conclusions

In this paper we have presented a test stand for the thermal characterization of the most common thermal interface materials, substrates, bulk metal and sintered metal as well as components. The test stand is based on the steady state technique. The accuracy of the method was analyzed and all parasitic effects were studied by systematic error calculation.

We have shown the limitation of the method and presented methods to improve the resolution of the test stand to be able to characterize materials with very low thermal resistance. We have showed that even samples with thermal resistances below 1 mm²K/W can be characterized using the descript steady state methods.

Some measurement examples were presented and discussed to demonstrate functionality accuracy of the test stand. It has been shown that the method is able to measure standard thermal interface materials but also the thermal conductivity of sintered silver could be measured as well.

In summary shall be said, that by understanding the steady state method and by using the right parameters within the right setup the resolution can be increased significantly.

Acknowledgments
The authors appreciate the support of the EU FP 7 Integrated Project "Nanotherm". We would like to thank Enrico Merten, Orlando Gamarra and Tobias von Essen for their encouragement.

References

[1] Young R, et al (2006) Developments and trends in thermal management technologies- a mission to the USA.
www.lboro.ac.uk/research/iemrc/documents/CB2007.pdf

[2] B. Wunderle, J. Kleff, D. May, M. Abo Ras, R. Schacht, H. Oppermann, J. Keller and B. Michel. „In-situ measurement of various thin Bond-Line-Thickness Thermal Interface Materials with Correlation to Structural Features". Proc 14th Therminic 2008, Sept 24-26, Rome, Italy.

[3] M. Abo Ras, B. Wunderle, D. May, H. Oppermann, R. Schacht, B. Michel. „Influences of technological processing and surface finishes on thermal behaviour of thermal interface materials", 16th Therminic, 2010 Oct 6-8, Barcelona, Spain.

[4] ASTM: Standard Test Method for Thermal Transmission Properties of Thin Thermally Conductive Solid Electrical Insulation Materials, Designation D 5470-01

[5] R. Schacht, D. May, B. Wunderle, O. Wittler, A. Gollhardt, B. Michel and H. Reichl. „Characterization of Thermal Interface Materials to Support Thermal Simulation". Proc. 12th Therminic 2006, Sep 27-29, Nice, Côte d'Azur, France.

[6] M. Abo Ras, G. Engelmann, D. May, M. Rothermund, R. Schacht, B. Wunderle, H. Oppermann, T. Winkler, S. Rzepka, B. Michel. "Development and Fabrication of a Thin Film Thermo Test Chip and its Integration into a Test System for Thermal Interface Characterization", 19th Therminic, 2013 Sep. 25-27, Germany, Berlin.

[7] ASTM: Standard Test Method for Thermal Diffusivity by the Flash Method, Designation E 1461 – 07

[8] Dean, N.F. and A.L. Gettings, "Experimental Testing of Thermal Interface Materials with Non-Planar Surfaces", Proceedings of the Institute of Electrical and Electronics Engineering SemiTherm Conference, San Diego, CA, 1998, IEEE.

[9] Tzeeng, J. J. W., Weber, T. W. and Krassowski, D. W., "Technical Review on Thermal Conductivity Measurement Techniques for Thin Thermal Interfaces". Sixteenth IEEE Semi-Therm Symposium, 2000.

[10] Teerstra et al, "Thermal Conductivity and Contact Resistance Measurements for Adhesives", Proceedings of InterPACK 2007, Vancouver, Canada July 8-12 2007.

[11] Kempers, R., et al. "Development of a high-accuracy thermal interface material tester."11th Intersociety Conference on Thermal and Thermomechanical Phenomena in Electronic Systems", ITHERM 2008, IEEE, Orlando, USA 2008.

A Product Based Lap Shear Fatigue Testing of Electrically Conductive Adhesives

B. Öztürk[1,2], A. Youssef[1], P. Gromala[1], C. Silber[1], K.M.B. Jansen[2], L.J. Ernst[3,4]

[1] Robert Bosch GmbH, Automotive Electronics, Tübingerstr. 123, 72762, Reutlingen, Germany
[2] Delft University of Technology, IO, Landbergstraat15, 2628 CE Delft, The Netherlands
[3] Emeritus Professor of Delft University of Technology, The Netherlands
[4] Ernst Consultant, Schoonhoven, The Netherlands Amperelaan 8, 2871 ZC, Schoonhoven, The Netherlands
Phone: +49-(0)7121-35-1472, e-mail: berkan.oeztuerk@de.bosch.com

Abstract

Thermoset-based adhesives are used as thermal and electrical interfaces. These adhesives are filled with different particles in order to meet heat transfer and electrical properties. In automotive applications, they are required to have excellent adhesion since delamination may precipitate other electrical, thermal or mechanical failure mechanisms. A vast amount of literature is available on the investigation of solder die-attach reliability where lap shear experiments are frequently used in microelectronics industry. Both environmental and performance requirements resulted in replacing solder die-attach with lead-free alternatives like electronically conductive adhesives. However, only very few studies so far focus on fundamental understanding of fatigue degradation of these materials. **The present paper addresses the above issue. To authors' best knowledge; it is the first time that in lap shear testing of adhesives, the specimens are obtained directly from production line and tested for their fatigue behavior.** Authors present a novel, 28 mm long lap shear sample which is made by identical fabrication processes as in the microelectronic component. Thin quad flat packages (TQFP) are inspected with scanning acoustic microscope for possible initial defects right after the production. A cutting process, using a dicing saw, is developed to produce lap shear samples from the packages. The cut samples are investigated by optical and scanning acoustic microscope to improve cutting and test results. Mechanical response of the electrically conductive adhesive is investigated under cyclic loading conditions at 25°C and 100°C for different stress ratios and frequencies. Finally, the stiffness degradation during testing is analyzed and the cross-sections of the samples are examined on bulk cracking. The presented method can be used to test different adhesives under production conditions in a fast manner which will decrease product development time and the dependence on time-consuming temperature cycle tests.

Introduction

As electronic devices become increasingly smaller and more complex, reducing the size of the components becomes more important. Nowadays, epoxy and bismaleimide (BMI) formulations filled with different particles are available on the market with promising features [1]. Examples are electrically (ECA) and thermally conductive adhesives (HCA). During temperature cycles, thermo-mechanical property mismatch (e.g. Elastic moduli, coefficient of thermal expansion etc.) between various materials used in an assembly results in thermo-mechanical stresses [2]. The residual and thermo-mechanical stresses may result in adhesive rupture and delamination of adhesive interfaces which may induce other electrical or mechanical failure mechanisms.

An important issue to address is the determination of long-term mechanical behavior (e.g. fatigue behavior) of the thin layer adhesives under different conditions (e.g. temperature, humidity, loading etc.) as well as developing a methodology for fast pre-qualification of adhesives. *Sauer et al.* [3] discussed the factors affecting fatigue behavior of polymers and concluded that stress amplitude (R), frequency (f), temperature, molecular weight and viscoelastic characteristics are important factors. *Kriese et al.* [4] investigated the adhesion strength by using nano-indentation. *Florando et al.* [5] developed a micro beam bending technique for studying the stress-strain relations for metal thin films on silicon wafer substrates. Both approaches require specially prepared samples and are not suitable for fatigue loading.

Another testing method, which is widely used in testing of thin films, is lap shear testing [6]. *Broughton et al.* [7] made a study to investigate the effect of lap joint configurations on the fatigue behavior of the joints and concluded that fatigue resistance is strongly affected by the joint geometry. In the design of lap shear specimens, peel and cleavage stresses should be minimized while shear stresses should be maximized [6]. *Bonk et al.* [8] stated that asymmetric constructions lead to bending moments during tensile loading and this leads to the deformation of the samples, which in turn leads to concentration of shear and peel stresses (six times more than the average applied stress) at the bond edges. *Lee et al.* [9] developed an analytical and a finite element model to calculate the shear and peel stresses at the bond ends of a lap shear joint for symmetrical and asymmetrical cases. They stated that both peel and shear stresses are increased when the geometry is asymmetrical. *Sancaktar et al.* [10, 11, 12] tried to understand the effect of loading and environmental conditions on the life time of a silver filled epoxy adhesive which is to be used for electronic applications. When possible, results from *Sancaktar et al.*'s studies are compared to current findings in this study.

During qualification of IC-packages or die-attach adhesives, long-term temperature cycle tests are performed in order to assess package/die-attach reliability. The present study aims to implement a pre-qualification method for die-attach materials which will help determining the best adhesive candidates by comparing the high cycle fatigue behavior under different loading and environmental conditions in a fast, reliable manner. It is known that there are three factors affecting the static and dynamic response of an adhesively bonded joint, namely, adherent, adhesive and environmental or loading conditions. Therefore, if fatigue behavior of the adhesive is to be evaluated, the results will be strongly affected by the adherent type. Using different adherents compared to the actual components in integrated circuits might lead to wrong results. That is why, for the first time in adhesive lap shear testing, specimens are directly taken from the assembly line to include effects of processing.

2. Experimental issues

2.1. Package structure

TQFP is available in different sizes. The package used in this study has the dimensions of 20 x 20 x 1 mm. Figure 1 shows the cross-section of a TQFP. The package consists of four different materials, namely, epoxy molding compound, silicon die, die-attach and copper lead-frame.

Figure 1: Cross-section of TQFP, #1: molding compound, #2: silicon die, #3: die-attach, #4: copper lead-frame.

2.2. Sample preparation

Sample preparation consists of different steps which can be summarized as in Figure 2.

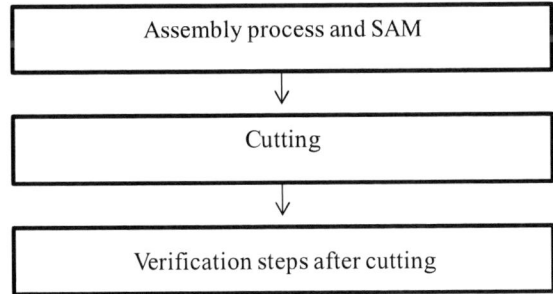

Figure 2: Sample preparation scheme.

The package assembly process involves a wide range of cutting edge technologies and processes. It includes wafer processes, introduction of die-attach on the lead frame, placement of the silicon die on the die-attach, bonding processes and encapsulation with molding compound. In order to ensure the reproducibility of the results, once the assembly is completed, the packages are checked with a scanning acoustic microscope (SAM) for possible initial defect (e.g. voids, delamination etc.). No initial defects were observed in any of the packages. Afterwards, samples are cut with a dicing saw. Cutting is divided into 4 steps as shown in Figure 3. The first cut is done through the molding compound, silicon and die-attach. Then the package is flipped and the second and third cuts are done through the whole package thickness. Before removing the sample from the sample holder, the fourth cut is done through the copper exposed pad (ePAD), until half of the thickness of adhesive. After the fourth cut, the samples look like as in Figure 4.

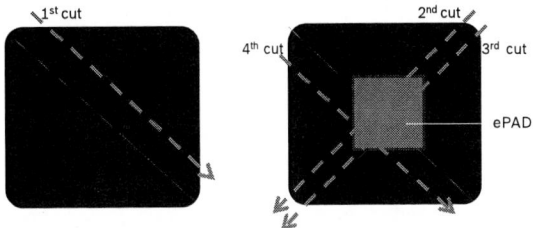

Figure 3: Illustration of cutting sequence.

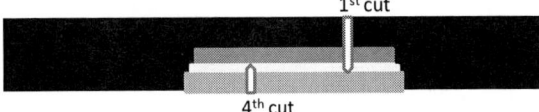

Figure 4: Cross-section illustration of the samples.

At the end of the cutting process, samples are dried. Obtained specimens have the dimensions of 28 mm length, 3 mm width and 1 mm thickness, with the following thickness values of individual components: 150μm of ePAD thickness, 50μm of adhesive thickness and 380μm of silicon die thickness.

After the cutting process, a sample selection criterion is introduced and only the samples which do not have initial damage due to the cutting are selected. Figure 5 illustrates the observed issues after cutting. In cases where better resolution is needed, a scanning electron microscope is used for investigation in order to determine if the sample can be tested or not.

978-1-4799-4789-8/14 $31.00 © 2014 IEEE

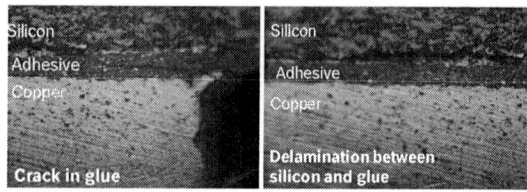

Figure 5: Examples of different issues appearing after cutting: Crack in glue (left) and delamination between silicon and glue (right).

The final step of the sample preparation is to store the samples under vacuum to avoid moisture absorption in the die-attach.

3. Testing

3.1. Static tests

Lap shear samples are subjected to static loading at 25°C and 100°C with 0.5 N/s crosshead speed to find the load at failure. The load at failure, P_{max}, will be used to calculate minimum load during a fatigue cycle, P_{min}, with the help of load ratio, R. That is $R=P_{min}/P_{max}$. Figure 6 illustrates that the maximum load the samples can carry dropped 21 % when the temperature is increased from 25°C to 100°C, whereas the displacement at failure is increased by 38% at 100°C, compared to the maximum displacement value at 25°C. The observed increase in the displacement is attributed to the increased motion of the polymer chain segments at higher temperatures.

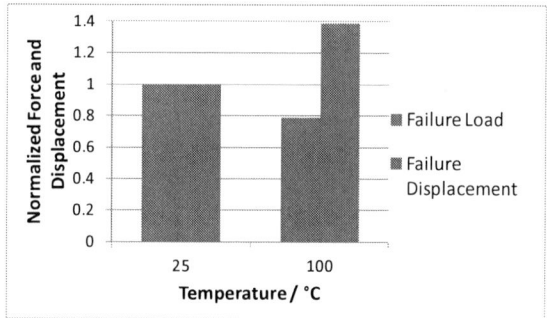

Figure 6: Normalized load and displacement values at the time of failure at 25°C and 100°C.

3.2. Fatigue Tests

Static test results are used to plan subsequent fatigue tests. Maximum load values are used to calculate minimum force values by the help of previously defined load ratios. At 25°C, the load ratios 0.1 and 0.3 are defined for 25°C at 10 Hz and 0.1, 0.3 and 0.5 are defined for 100°C at 10 Hz.

Figure 7 shows the room temperature fatigue behavior of the adhesive depending on P_{max} and R.

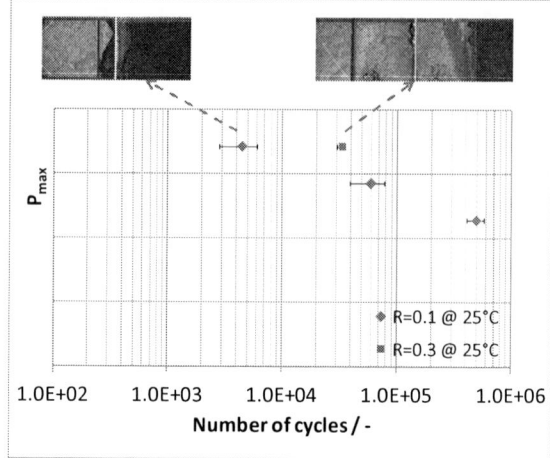

Figure 7: P_{max} vs. number of cycles at 25°C for R=0.1 and 0.3 at f=10 Hz. The fracture surfaces for R=0.1 and R=0.3 are given. When R=0.1, adherent failure is observed, whereas, R=0.3 resulted in cohesive failure.

It is seen that R has a strong effect on the lifetime of the lap shear samples. For the same P_{max} value for different R, adhesive lifetime tends to be higher with R values (smaller amplitude). On the other hand, as P_{max} is decreased, a longer life time is observed. Fracture surfaces revealed that at low R and high P_{max}, adherent fracture happens. In this study, adherent fracture of silicon will be interpreted as interfacial fracture since, if a conventional lap shear specimen would be used, we would have a ductile material instead of silicon (e.g. steel, aluminum) which would not be broken and the crack would propagate at the interface. However, in this study, conventional lap shear specimens are replaced with the specimens coming from the assembly line which contains brittle materials. It is also observed that as R is increased, fracture becomes more cohesive dominant whereas as P_{max} is decreased, the fracture type becomes interfacial dominated. Similar results are also observed in 13, 14, 7, 15, 16. However, authors of the above references did not observe any adherent fracture (fracture of silicon).

At 100°C (Figure 8), the adherent fracture type is not observed, in other words, only cohesive and interfacial failures are observed. When P_{max} is high, the fracture surfaces exhibit a mixture of adhesive and cohesive fracture characteristics. As P_{max} is decreased within the same load ratio, the fracture type tends to be more interfacial dominant unlike the case at 25°C. This is explained by the fact that as the temperature is increased, the interfacial strength is weakened at a higher rate compared to the situation for the bulk strength [12, 17].

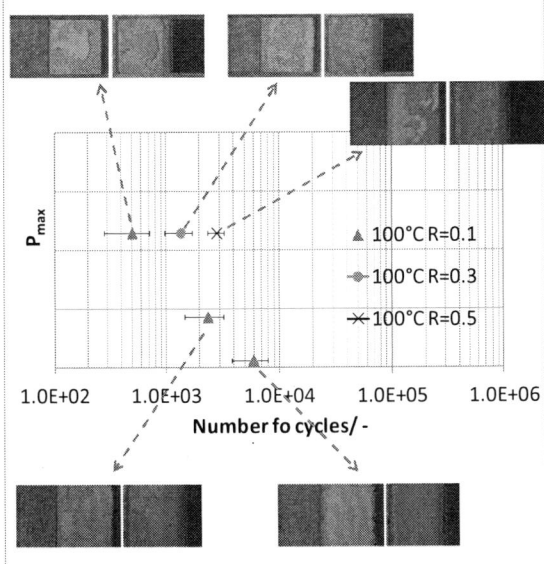

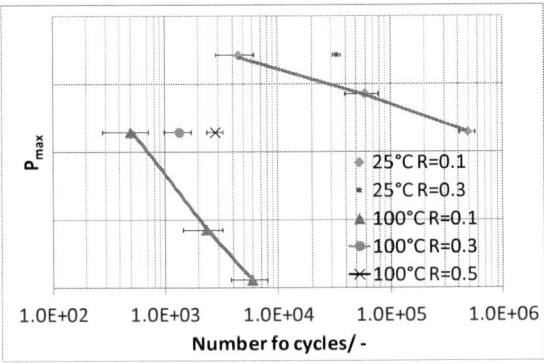

Figure 8: P_{max} vs. number of cycles at 100°C for R=0.1, 0.3 and 0.5 at f=10 Hz. The fracture surface are given for R=0.1 at different P_{max} values and R=0.3 and 0.5.

It is observed that the lifetime decreases as temperature is increased. This is due to the reduction in the bulk adhesive strength and in the silicon/die-attach interface strength. Both bulk strength and silicon/die-attach interface strength are functions of temperature and both strengths decrease with increased temperature. S-N curves are combined and summarized in Figure 9.

Figure 9: Fatigue behavior of the adhesive at 25°C and 100°C under different R and P_{max} values.

After each experiment, force vs. displacement curves are drawn for different cycle numbers to investigate the stiffness change during the experiments. At 25°C, there is only 3% stiffness change calculated, whereas at 100°C, the stiffness change becomes 6%. With respect to the initial stiffness, the authors conclude that during lap shear fatigue investigation, the samples do not show considerable amount of stiffness change. Very small or negligible stiffness reductions are also reported in 10 and 12.

In order to check the effect of frequency, a similar experimental procedure is applied to the lap shear samples at 50Hz at 25°C and 100°C. It is observed that as frequency is increased (for constant R and P_{max}), the number of cycles to failure is increased independent on the temperature.

4. Conclusions and future work

Aim of this work is to develop a fast experimental pre-qualification method for adhesives used in electronic packages. Up to now, S-N curves are generated by using different R, P_{max}, frequency and temperature. It is shown that at room temperature, R has a strong effect on the fatigue life and failure mode. At high temperatures, R is shown to affect the failure mode less. For both temperatures, as P_{max} is decreased, the adhesive shows a longer fatigue life. Upon increasing the frequency from 10 Hz to 50 Hz, the number of cycles to failure is observed to be increasing. The stiffness evolution evaluation with the number of cycles shows that there is almost no stiffness change for the samples investigated at 25°C and 100°C.

Future work will focus on completion of the S-N curves at 25°C and 100°C for different combinations of R and P_{max}. At the same time, the effect of creep on fatigue damage will be investigated. Particularly, attention will be given to damage evolution by checking the cross-section of the samples before they fail in order to observe micro structural changes.

References

1. Tong, X.C., Advanced Materials for Thermal Management of Electronic Packaging, Springer (Heidelberg, 2011), pp. 201-221.
2. Calder, J.C., Harrigan, D.R., Riley, W.C., "Graphite fiber composites for CTE control in thermal management," *Proc 1998 International Conference on Multichip Modules and High Density Packaging*, Denver, Colorado, USA; April, 1998.p.160-4.
3. Sauer, J.A.; Richardson, G.C. Fatigue of polymers. *Int. J. Fract.* **1980**, *16*, 499-532.
4. Kriese, M.D.; Boismier, D.A.; Moody, N.R.; Gerberich, W. W. Nanomechanical fracture-testing of thin films. *Eng. Fract. Mech.* **1998**, *61*, 1–20.
5. Florando, J.N.; Nix, W.D. A microbeam bending method for studying stress–strain relations for metal thin films on silicon substrates. *J. Mech. Phys. Solids.* **2005**, *53*, 619–638 .
6. Banea, M.D.; da Silva, L. F. M. Adhesively bonded joints in composite materials: an overview *Proc. Inst. Mech. Eng. Part L: J. Mater. Des. Appl.* **2009**, *223*, 1–18.

7. Broughton, W.R.; Mera, R.D.; Hinopoulos, G. "Project PAJ3 - Combined Cyclic Loading and Hostile Environments 1996-1999 Cyclic Fatigue Testing of Adhesive Joints Test Method Assessment," Teddington, Middlesex, UK, 1999.

8. Bonk, R. B.; Osterndorf, J. F.; Pettenger, B. L.; Ambrosio, A.M. Methods for evaluating adhesive systems and adhesion. U.S. Army Armament Research, Development and Engineering Center, 1996.

9. Lee, J.; Kim, H. Stress Analysis of Generally Asymmetric Single Lap Adhesively Bonded Joints. *J. Adhes.* **2005**, *81*, 443–472.

10. Gomatam, R.; Sancaktar, E. Dynamic fatigue and failure behavior of silver-filled electronically conductive adhesive joints at ambient environmental conditions. *Adhes. Sci. Technol.* **2004**, *18*, 731-750.

11. Gomatam, R.; Sancaktar, E. The effects of stress state, loading frequency and cyclic waveforms on the fatigue behavior of silver-filled electronically-conductive adhesive joints. *J. Adhes. Sci. Technol.* **2006**, *20*, 53-68.

12. Gomatam, R.; Sancaktar, E. Fatigue and Failure behavior of silver-filled electronically conductive adhesive joints subjected to elevated temperatures. *J. Adhes. Sci. Technol.* **2004**, *18*, 849-881 .

13. Tian, X.; Prince, L. J. Electronic packaging adhesive fatigue life prediction using thermal cycling step-stress testing. *Annu. Reliab. Maintainab. Symp.* **2002**, 628–635.

14. Gladkov, A.; Bar-Cohen, A. Parametric Dependence of Fatigue of Electronic Adhesives. Adhesive Joining and Coating Technology in Electronics Manufacturing, Binghamton, NY, 1998.

15. Tomblin, J.; Seneviratne, W.; Escobar, P.; Yap, Y. Fatigue and Stress Relaxation of Adhesives in Bonded Joints. U.S. Department of Transportation, Federal Aviation Administration, 2003.

16. Kinloch, A.J.; Osiyemi, S. O. Predicting the Fatigue Life of Adhesively-Bonded Joints. *J. Adhes.* **1993**, *43*, 79-90.

17. Jurf, R.A.; *Adhesively Bonded Joints: Testing, Analysis, and Design*, 7th ed. Philadelphia, 1988.

Process and Reliability of SF_6/O_2 Plasma Etched Copper TSVs

Lado Filipovic, Roberto Lacerda de Orio, and Siegfried Selberherr

Institute for Microelectronics, Technische Universität Wien, Gußhausstraße 27-29/E360, A-1040 Wien, Austria

Email: {filipovic | orio | selberherr}@iue.tuwien.ac.at

Abstract

The formation of a TSV for three-dimensional interconnects using SF_6/O_2 plasma is explored. Adjusting the O_2 gas concentration to 45 sccm, while the SF_6 concentration is set to 35 sccm, produced the best combination of chemical and physical etching to provide sidewall angles of 88°. Three TSV aspect ratios are etched (5/58, 10/100, and 20/100 μm) and subsequently analyzed using the finite element method. The TSVs' series resistance, current density, thermo-mechanical stress, and electromigration induced stress after 300 hours of operation at a $2MA/cm^2$ current density are analyzed. An additional comparison to ideal TSVs with sidewall angles at 90° is performed.

1. Introduction

The major focus of the semiconductor industry over the last decades has been to continue integrated circuit (IC) miniaturization along with Moore's law. It is expected that a physical scaling limit will be reached around the 6nm node; however, even before that limit is reached, the increased process equipment and factory costs for scaling will require other means of "more Moore" and "more than Moore" integration [1]. A major development in this direction has been the through silicon via (TSV), which allows for vertical integration through three-dimensional (3D) stacking of ICs. The main advantages of 3D integration is the reduction of RC delay for submicron circuits, reduced power consumption, and the ability for heterogeneous integration of chips [2].

The two main methods to etch the silicon layer for TSV implementation are plasma etching and the Bosch process [1]. Each process has its own flaws and reliability concerns. Problems specific to the Bosch process are a rough, scalloped TSV sidewall, notch formation at the TSV bottom, and potential step coverage issues relating to depositing layers on a scalloped wall [3]. The plasma etching of silicon results in angled sidewalls and an added wall curvature due to the via taper edge [3]; however, the rough scallops are avoided. This work examines by means of simulation the potential of using an SF_6/O_2 plasma in order to etch TSV structures and compares their thermal and electrical performance to an ideal cylindrical TSV.

2. Silicon Etching

The etching of silicon wafers with a SF_6/O_2 plasma has been described in [4]. The etch rate is mainly governed by the applied bias voltage, pressure, and the ratio of O_2

to SF_6 in the ambient. To test the RF bias, the pressure, and the SF_6/O_2 ratio effects on the etch, each parameter is varied while the others remained constant. The constant values are set to -20V RF bias, 25mTorr pressure, and an SF_6/O_2 ratio of 1. [4]. The observations, shown in Fig. 1, lead to the conclusion that the fastest etch rate can be reached with an RF bias of -120V, pressure of 25mTorr, and an SF_6/O_2 ratio of 1. However, when etching through silicon as a step in the manufacture of TSVs, the sidewall angle is an important aspect which must not be overlooked, even at the cost of a reduced etch rate.

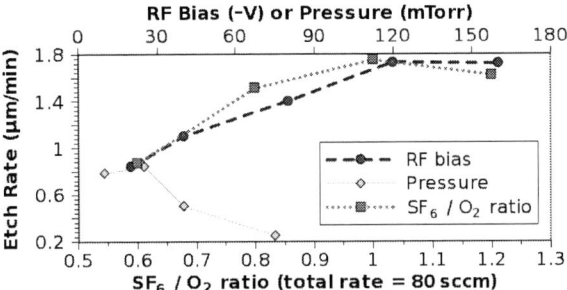

Figure 1. Effects of process parameters on the Si etch rate. When testing the effects of one parameter on the rate, the other two are kept constant.

It has previously been determined that the effect of O_2 on the SF_6 plasma is a dramatic increase in the F atom concentration and a subsequent decrease in lateral etching [5]. Controlling the F atom concentration is essential to generating desired sidewall angles. Therefore, an etching simulation is performed for several desired TSV diameters while varying the O_2 concentrations, resulting in the profiles shown in Fig. 2. Table 1 describes the etch conditions for profiles A–D. From Fig. 2, it is evident that varying the O_2 concentration results in the variation of the lateral extents of the etched hole, but also in a variation of the overall etch rate. Increasing the presence of O_2 results in an overall reduction of the etch rate, both in the lateral and vertical directions.

A. Silicon Etching Model

The model for Si etching is described in [6] and it has been implemented in a level set environment, as described in [7]. From the observed profiles from Fig. 2, it is evident that the O_2 concentration plays a major role in determining the lateral etching and subsequent TSV sidewall angle. By increasing the O_2 concentration, lateral etching is

Table 1. Experimental parameters for the simulation results shown in Fig. 2.

Parameter	A	B	C	D
SF$_6$ concentration	45 sccm	40 sccm	35 sccm	30 sccm
O$_2$ concentration	35 sccm	40 sccm	45 sccm	50 sccm
Pressure	25mTorr			
Total gas flow rate	80 sccm			
RF bias voltage	-120V			
Wafer temperature	5°C			

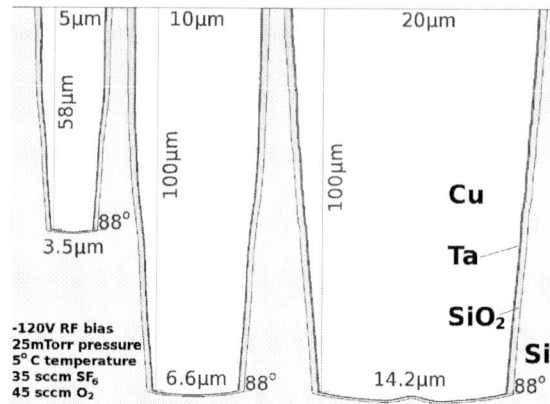

Figure 3. TSV structures after Si etching and SiO$_2$/Ta deposition. 5μm/59μm, 10μm/100μm, and 20μm/100μm are the TSV profiles.

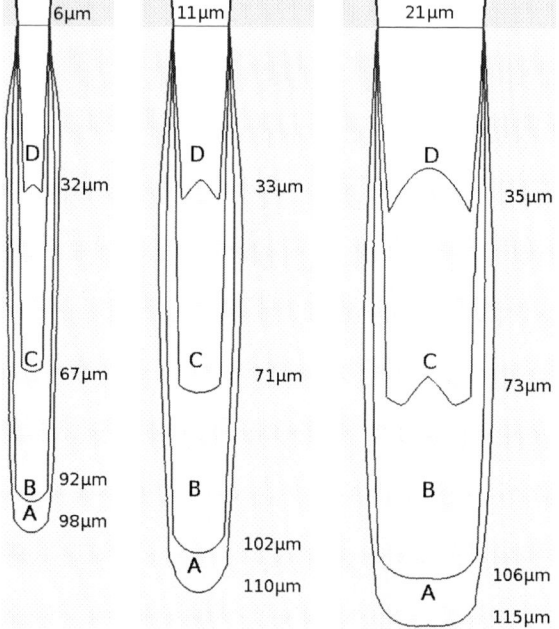

Figure 2. Etch profiles after a 30-minute etch in SF$_6$/O$_2$ plasma. The etch profiles are prepared for (left) a 5μm TSV diameter, (middle) a 10μm TSV diameter, and (right) a 20μm TSV diameter. The additional 1μm in diameter length is required in order to deposit an insulating layer of SiO$_2$ and a tantalum liner, prior to copper deposition. The processes (A–D) are described in Table 1.

reduced; however, at the cost of a reduced etch rate. For the purposes of TSV etching, option C with an O$_2$/SF$_6$ ratio of 1.3 is selected. The three TSV geometries tested have a diameter/depth ratio of 5μm/58μm (27 minute etch), 10μm/100μm (43 minute etch), and 20μm/100μm (43 minute etch). The resulting profiles, including the copper (Cu), tantalum (Ta), silicon dioxide (SiO$_2$), and silicon (Si) layers are shown in Fig. 3. The SiO$_2$ and Ta layers have been deposited using constant rates to obtain the desired thicknesses (400nm for SiO$_2$ and 100nm for Ta). The resulting hole can then be filled with copper.

The TSVs with the chosen aspect ratios can all be filled without appearing seam voids. For the 5μm/58μm TSV filling can be performed using electrochemical deposition of Cu with chemical vapor deposition of tungsten and a sputtered TiW/Cu seed layer [8]. The TSVs with the other aspect ratios can be filled using electroplating with an added 50ppm of chloride ions, as suggested in [9]. The presence of seam voids or poor sticking of the individual

materials can lead to delamination under increased stress conditions. The stress can be due to a large temperature drop which occurs during structure cooling after a thermal processing step. An additional source of stress for the structure which is analyzed is the electromigration-induced stress.

3. TSV Characterization

The TSVs are characterized according to several electrical parameters and stress distributions. It is expected that a TSV with a smaller radius will have an increased resistance. Additionally, etched structures experience some tapering (about 88°), which results in copper thinning along the depth of the TSV. This thinning is expected to increase the overall TSV resistance and the current density on the TSV bottom, when compared to an ideal cylindrical geometry. The simulated parameters are compared to the ideal cylindrical TSVs in order to estimate the effects of SF$_6$/O$_2$ etching for Cu TSVs. In addition, the thermo-mechanical and electromigration-induced stress distribution in the structure is analyzed.

Thermo-mechanical simulations of the TSVs are carried out in order to calculate the von Mises stress. This stress serves as a yield criterion for mechanical reliability and is caused by the subsequent chip cooling after an annealing step from a high temperature (∼300°C) down to room temperature. The variation in the coefficient of thermal expansion (CTE) between the Cu, Ta, SiO$_2$, and Si layers causes the Si layer to experience tensile stress. The main concern is the CTE mismatch between Cu and Si, while the Ta liner is ignored in many studies due to its thickness being negligible compared to the SiO$_2$ liner. In this study, all materials which comprise the TSV structure are included in the analysis.

In addition to thermo-mechanical stress, the electromigration (EM) induced stress can be a cause for concern. EM can trigger a chip failure because of the formation and growth of voids in a metal line, such as a TSV structure. Whether or not a void will nucleate depends on how much stress develops on the Cu line in the presence of a high current density. This current causes a movement of vacancies which are concentrated at material interfaces and grain boundaries. The stress is calculated using the model described in [10], dealing with vacancy accumulation at material interfaces.

4. Results

The TSV structures from Fig. 3 were placed between layers of tantalum (100nm) and copper and the finite element method (FEM) is used to calculate the TSVs' electrical parameters, thermo-mechanical stress, and EM induced stress.

A. Electrical Performance

The electrical performance of the TSVs is analyzed by observing the current density and resistance of the final structures. Fig. 4 shows the difference in the current density distribution through an ideal cylindrical TSV and an etched TSV. It is evident that the etched TSV, due to the tapered sidewalls, experiences an increased electric current density through the thinned Cu region.

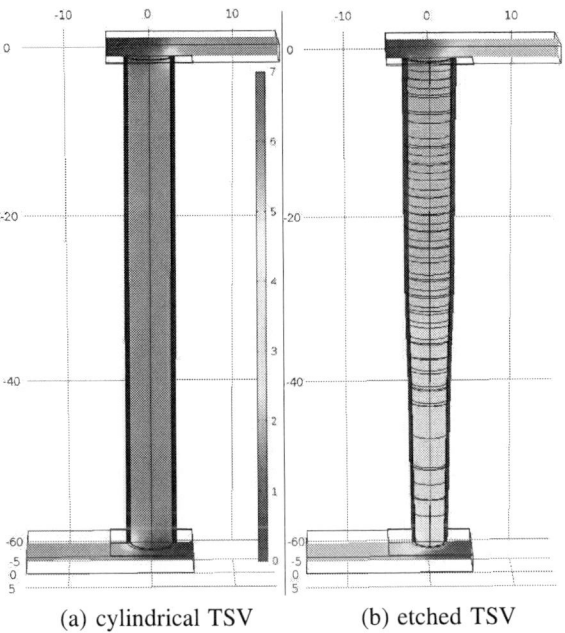

(a) cylindrical TSV   (b) etched TSV

Figure 4. Current density distribution in the $5\mu m/58\mu m$ TSV. The increased current density in the etched TSV is expected to result in an increased electromigration-induced stress.

Table 2 lists several characteristics of the tested TSVs. The TSV resistance is directly related to the Cu volume, while the maximum Cu and SiO_2 stresses are increased with increasing TSV diameter. The resistances of etched TSVs are also approximately 1.3–1.4 times higher than the resistances of the ideal TSVs. This difference is also noted in the approximate reduction in volume that the etched TSVs experience due to sidewall tapering.

Table 2. Simulation results for the TSVs.
5/58, 10/100, and 20/100 refer to the TSV aspect ratios.

	Volume (μm^3)			Resistance ($m\Omega$)		
	5/58	10/100	20/100	5/58	10/100	20/100
Ideal	1128	7788	31150	109	47.5	11.93
Etched	614	6080	24450	140	65.1	15.61

	Max Cu stress (GPa)			Max SiO_2 stress (GPa)		
	5/58	10/100	20/100	5/58	10/100	20/100
Ideal	1.50	1.70	1.80	1.32	1.53	1.69
Etched	1.80	1.89	2.07	1.73	1.81	2.11

B. Thermo-Mechanical Stress

The thermo-mechanical stress is an indication of the effects of device cooling after a thermal processing steps. Table 2 lists the maximum stress experienced by the copper and silicon dioxide layers for the ideal and etched TSV structures. It can be concluded that larger structures experience an increased thermal stress. This is due to a larger contact area between materials with varying CTE values. Therefore, it is important to note that, although larger TSVs allow for higher current densities and have a smaller overall resistance, they experience a higher thermal stress. In Fig. 5 the stress distribution along the top of the TSV is shown. The von Mises stress is concentrated near the Ta/SiO_2 and SiO_2/Si interfaces, while the stress is significantly reduced at a distance of approximately $5\mu m$ from the SiO_2/Si interface, through the silicon. Fig. 6 shows the thermo-mechanical stress through the silicon layer, moving away from the interface. The etched and ideal TSVs show only a slight variation, while an increased TSV size results in a higher von Mises stress through the silicon. The maximum stress in the Cu and SiO_2 for the etched TSVs is also slightly higher than that of the ideal cylindrical TSV.

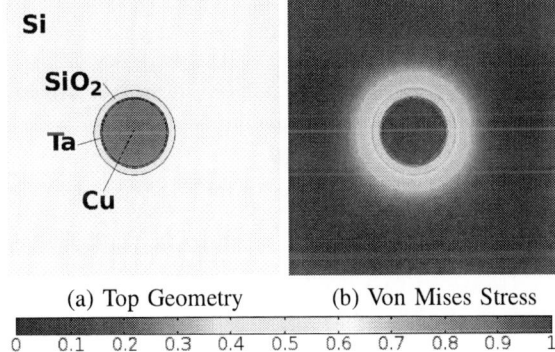

(a) Top Geometry   (b) Von Mises Stress

Figure 5. Normalized stress distribution at the top of the TSV structure. (a) Top view of the TSV with materials labeled, (b) The von Mises stress distribution (Maximum stress is about 1GPa)

C. Electromigration-Induced Stress

Electromigration is a major reliability issue for modern integrated circuits. It normally triggers a chip failure due

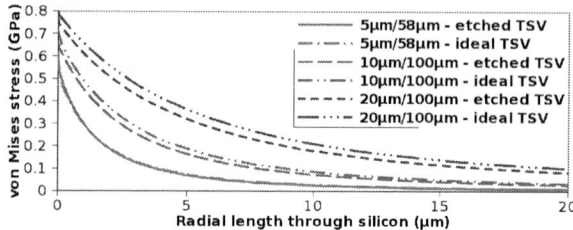

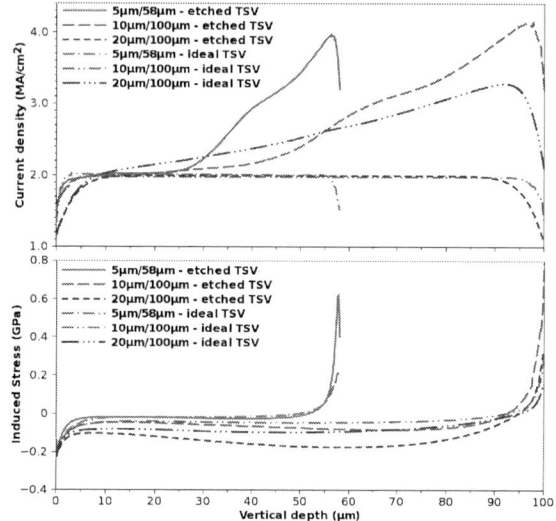

Figure 6. Maximum thermo-mechanical stress through silicon. Radial length of $0\mu m$ refers to the SiO_2/Si interface.

Figure 7. Current density and EM induced stress through the center of the TSV.

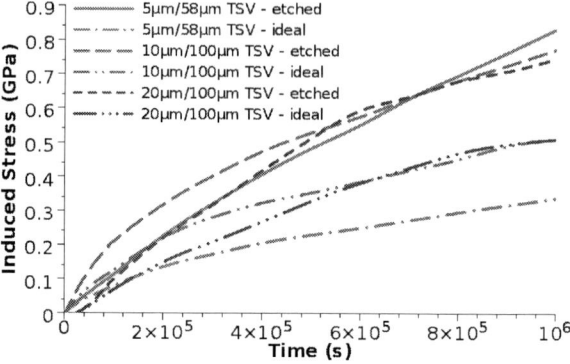

Figure 8. Current density and EM induced stress through the center of the TSV.

structure experiences the highest stress increase. Although all etched TSVs experience approximately the same stress level, the maximum electromigration-induced stress in the smallest ideal TSV ($5\mu m/58\mu m$) is significantly lower.

to formation and growth of voids in a metal line of an interconnect structure [10]. The model which is used for this study measures the stress generated in a structure prior to void formation. When the stress reaches a critical level, a small flaw could potentially be nucleated to form a void. The void formation is the first step towards void growth, resistance increase, and eventually device failure. The model used to analyze the induced stress in presented in [10].

In Fig. 7 the distribution of the current density and the EM induced stress through the center of each TSV is shown. The simulations were performed to show the level of current density and stress induced after introducing a $2MA/cm^2$ current through the top of the structure for approximately 300 hours. The link between copper thinning, and hence the current density increase, and the induced stress is evident. Fig. 8 shows the maximum stress build-up in the structure over time. The increased stress in the etched structures is noticeable; however, the smallest

5. Conclusion

In this work several TSV geometries have been generated using an SF_6/O_2 plasma model, implemented in a level set framework. The resulting structures were then extracted and electrical parameters, thermo-mechanical stresses, and electromigration induced stresses were simulated. A comparison between the etched TSV and ideal cylindrical TSVs has been performed. With this work, a direct link between simulations related to TSV processing and TSV thermo-mechanical stress and reliability is realized.

Acknowledgment

The research leading to these results has received funding from the European Union Seventh Framework Programme (FP7/2007–2013) under grant agreement no. 318458 SUPERTHEME.

References

[1] "The International Technology Roadmap for Semiconductors (ITRS)," 2011
[2] R. Li et al. 2008 *J. Micromech. Microeng.* **18** (12) 125023–125030
[3] N. Ranganathan et al. 2011 *IEEE Trans. Comp., Packag., Manufact. Technol.* **1** (10) 1497–1507
[4] S. Gomez et al. 2004 *J. Vac. Sci. Technol., A* **22** (3) 606–615
[5] R. Figueroa et al. 2005 *J. Vac. Sci. Technol., B,* **23** (5) 2226–2231.
[6] R. J. Belen et al. 2005 *J. Vac. Sci. Technol., A* **23** (5) 1430–1439
[7] O. Ertl and S. Selberherr 2009 *Comput. Phys. Commun.* **180** (8) pp. 1242–1250
[8] M. J. Wolf et al. 2008 *Proc. 58th ECTC 2008* pp. 563–570
[9] S. Wang and S. R. Lee 2011 *Proc. IMECE 2011* pp. 1–9
[10] R. Orio et al. 2011 *Microelectron. Reliab.* **51** (9) 1573–1577

Characterization and post simulation of thin-film PZT actuated plates for haptic applications

F. Casset[1], JS. Danel[1], P. Renaux[1], C. Chappaz[2], G. Le Rhun[1], C. Dieppedale[1], M. Gorisse[3], S. Basrour[3], S. Fanget[1], P. Ancey[2], A. Devos[4], E. Defay[1]

[1]CEA-LETI, MINATEC Campus, 38054 Grenoble, France
[2]STMicroelectronics, 850 rue Jean Monnet, 38926 Crolles, France
[3]TIMA (CNRS-Grenoble INP-UJF), 38000 Grenoble, France
[4]IEMN/ISEN-CNRS 8520, Av. Poincaré, 59652, Lille, France
+33 438 922 756, fabrice.casset@cea.fr

Abstract

The tremendous development of tactile interface in many customers' applications such as Smartphone, tablet PC or touch pad leads industrials to study "haptic interfaces" or "touch screen" solutions. This technology is already used but with limitations such as high power consumption and limited feedback effect (simple vibration). PZT is a good candidate for many actuator applications due to its high piezoelectric coefficient. In particular, it can be used for haptic interfaces to create squeeze-film effect. It consists in changing the friction between the finger and a plate resonator. It provides high granularity level of haptic sensation (texture rendering), using low power consumption compared to existing solutions. We manufactured demonstrators using a generic technology. We proved the concept through electro-mechanical characterizations and haptic feedback effect was noticeable with one's finger on our thin-film PZT demonstrators. In this paper, we presented the characterization and post-simulation of PZT-actuated plates with various actuator configurations. Measurement results, in good agreement with simulation, indicate that 2 actuator columns separated by a wavelength allow obtaining the highest substrate displacement amplitudes.

1. Introduction

Many applications can be concerned by high performances, low voltage haptic interfaces which allow the user to interact with its environment by the sense of touch. We developed sol-gel thin-film $Pb(Zr_{0.52}, Ti_{0.48})O_3$ (PZT) actuated plates allowing the feeling of textured surfaces. It consists in changing the friction between the finger and a plate resonator due to the squeeze-film effect, which occurs when the plate displacement amplitude reaches about 1μm in a flexural anti symmetric Lamb mode [1, 2]. It provides high granularity level of haptic sensation (texture rendering), using low power consumption compared to existing solutions.

We manufactured demonstrators using a generic technology compatible with RF MEMS [3], loudspeakers [4] or micro mirror [5]. We proved the concept through electro-mechanical characterizations. Laser vibrometer measurements indicate that our PZT actuator design promotes the desired mode with the targeted amplitude. Haptic feedback effect was noticeable with one's finger on our thin-film PZT demonstrators [6].

This paper reports on the design, the characterization and the post-simulation of thin-film PZT actuated haptic plates. We developed predictive model, and in particular we studied the best actuator configuration in order to promote the highest substrate displacement amplitude.

2. Design

We designed high performances thin-film PZT actuated haptic plates using Finite Element Method (FEM) approach (*CoventorWare*[TM]). All simulations are performed using an unclamped 725μm thick Silicon substrate as resonant plate. Our model neglects top and bottom electrodes which sandwiched the 2μm thick PZT layer due to their weak impact on simulation results. We used *Manhattan* bricks to mesh the haptic plate. As input data, we used the thin-film PZT d_{31} piezoelectric coefficient, namely 165 pm/V [7], the PZT Young's modulus E_{PZT} = 82 GPa and Poisson's ratio v = 0.39 [8].

We already proved that the maximum substrate displacement amplitude of our haptic plate can be obtain by taking the deformed shape of the selected mode into account, and matching the actuators position with the substrate maximum displacement amplitude as shown in Figure 1. PZT actuators promote the desired mode by bimorph effect [6].

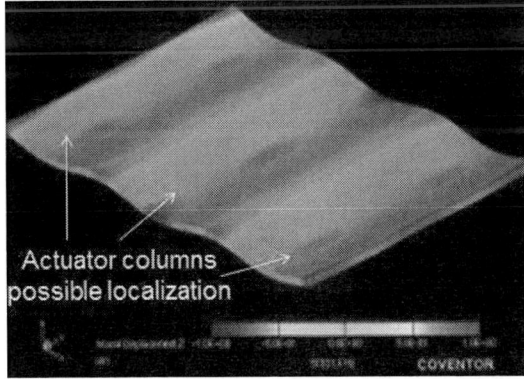

Figure 1: Modal simulation (51.89 kHz) on a 40×30 mm² plate and PZT actuator columns possible localization.

978-1-4799-4789-8/14 $31.00 © 2014 IEEE

As shown in Figure 1, various PZT actuator configurations are possible. In particular, we studied:

- 1 actuator column,
- 2 actuator columns separated by a wavelength: in this case the actuator columns are located on successive in-phase maximal substrate displacement amplitude areas,
- 2 actuator columns located at the extremities of the plate: the actuator columns are located on in-phase maximal substrate displacement areas near the ends of the plate.

For each studied configuration, we first performed a modal simulation in order to define the frequency of the desired mode. Then relative harmonic simulations were performed using arbitrary damping parameters and a given actuation voltage. As shown in Figure 2, two actuators columns separated by a wavelength induce the highest substrate displacement amplitude for both 60×40 and 40×30 mm² plates.

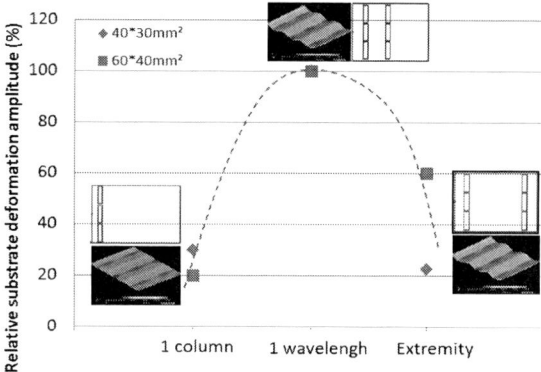

Figure 2: Harmonic simulations with arbitrary damping parameter study – Actuator columns configuration study for both, 60×40 and 40×30 mm² plate.

3. Technology

Devices were manufactured out of 200 mm standard silicon wafers. First step is a 500 nm thick thermal oxidation (SiO₂). Then, the piezoelectric stack, which consists of 2 µm thick sol-gel PZT in between 100 nm thick Pt bottom electrode and 100 nm thick Ru top electrode, was deposited and etched. We deposited and patterned a passivation silicon oxide layer followed by gold lines and pads. Figure 3 gives a cross section view of the technological stack as well as a SEM cross section of the PZT actuator. Only 5 photolithography masks were necessary to build the demonstrators.

Finally, we sawed the substrate to transfer the individual plate onto a mechanical support allowing to hold the vibrating plate and to connect it electrically. The plate was fixed by an adhesive tape and the actuators were wire bonded to electrical connectors. Figure 4 gives an example of a 60×40 mm² plate ready for electrical characterization.

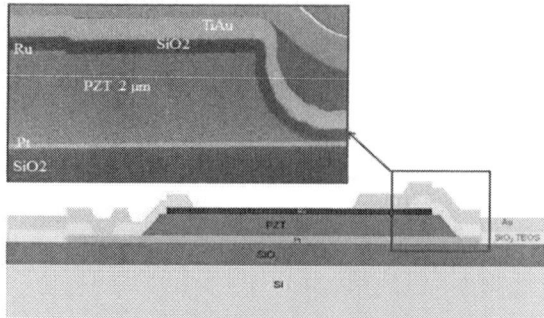

Figure 3: Schematic cross section of the technological stack and SEM cross section of the PZT actuator.

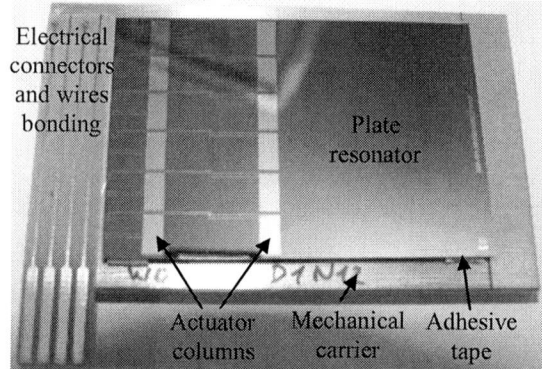

Figure 4: 60×40 mm² plate fixed on a carrier using adhesive tape and with PZT actuators bonded to electrical connectors.

4. Electromechanical characterization and post-simulation

Optical measurements were performed using a laser vibrometer (*POLYTEC*) by applying a dc-ac mixed voltage to the top electrode of plates (60×40 or 40×30 mm²). Laser vibrometer measurements give the substrate displacement amplitude for a given actuation voltage. Several measurements have been necessary to build the plate's whole picture given in Figure 5. This picture allows us to validate that the studied mode is the flexural antisymmetric Lamb mode required for haptic applications. We measured the desired resonant mode at 32.2 kHz for the 60×40 mm² plate. The agreement between measurements and simulations is satisfactory. Indeed a discrepancy lower than 1% is noticed. Moreover, we note that the substrate displacement amplitude reached 500 nm under 4V ac/peak to peak. This value is very satisfactory for such a low actuation voltage.

Using FEM approach, we adjusted a damping parameter to fit the substrate displacement amplitude previously measured. Figure 6 gives the comparison between the simulated substrate displacement amplitude and the measured one under 4V. It allows us to determine that a damping parameter of $2.5.10^{-8}$ is necessary to fit measurement for 60×40 mm² plates.

Only one damping parameter per plate sizes was used as reported in Table 1.

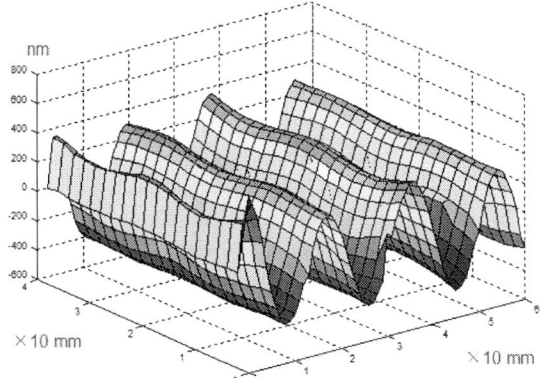

Figure 5: Laser vibrometer measurement on the 60×40 mm² plate (12V$_{dc}$+4V$_{ac/pp}$). Resonant frequency measured at 32.2 kHz.

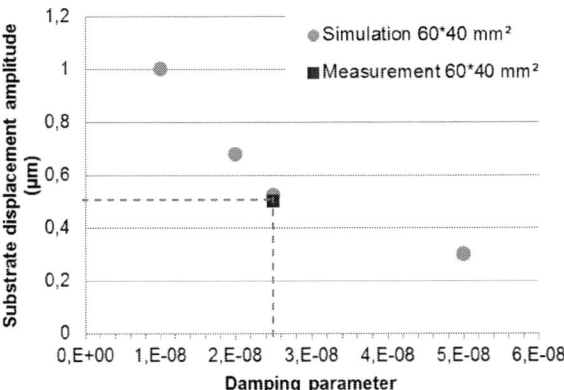

Figure 6: Comparison between the simulated and the measured substrate displacement amplitude of a 60×40 mm² plate under 4V for damping parameter identification.

Plate size (mm²)	Damping parameter
60×40	$2.5.10^{-8}$
40×30	10^{-8}

Table 1: Damping parameter function of the plate size.

For example, with a damping parameter of $2.5.10^{-8}$, we can post-simulate a 60×40 mm² plate as shown in Figure 7 (discrepancy < 5%). One can note the micrometric displacement amplitude, namely 1.1 µm, obtained using only 8V ac/peak to peak. A haptic feedback effect was felt with the finger on this plate by using an actuation signal modulated at 10 Hz in amplitude [6].

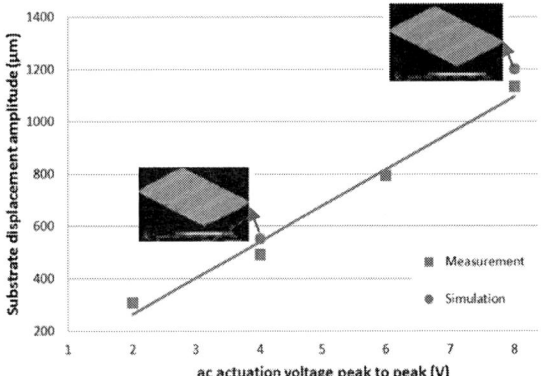

Figure 7: Laser vibrometer measurements and simulated substrate vertical displacement (pictures) amplitude comparison on a 60×40 mm² plate (12V$_{dc}$+variable V$_{ac/pp}$).

Both on 60×40 and 40×30 mm² plates, the best actuator columns configuration were studied. As shown in Figure 8, measurements results obtained using 12V$_{dc}$ + 4V$_{ac/pp}$, in good agreement with simulation, indicates that 2 actuator columns separated from each other by a wavelength allow obtaining the highest substrate displacement amplitude. Unfortunately, most of the targeted applications require transparency, at least of the center of the plate, and this actuator configuration induces a large opaque zone (dead area). Consequently it will only be retained if transparency is not required by the focused application.

We note on Figure 8 that the measured substrate displacement amplitude with 2 actuator columns located at the plate ends is only 5% smaller for 40×30 mm² plates and 11% smaller for 60×40 mm² plates, compared to the best configuration discussed previously. This configuration presents a larger area dedicated to touch experience. It will be favored, for example for Smartphone applications, for which we have to preserve the largest transparent surface as possible, dedicated for touch experience.

Finally the last configuration with only one actuator column presents the smaller dead area. It induces measured substrate displacement amplitude respectively 46 and 88% smaller for 40×30 and 60×40 mm² plates, compared to the best configuration discussed previously. In the case of the characterized 40×30 mm² plate, the measured substrate displacement amplitude is all the same satisfactory for such a low actuation voltage. FEM simulation indicates that we can reach the micrometric substrate displacement amplitude required for haptic application under 14 V$_{ac}$. In the case of the characterized 60×40 mm² plate, the low measured substrate displacement amplitude can result from a malfunction of the device. Further measurements must be performed to confirm this experimental result. This configuration can be retained for applications that need the smallest non-transparent dead area, to the detriment of low actuation voltage.

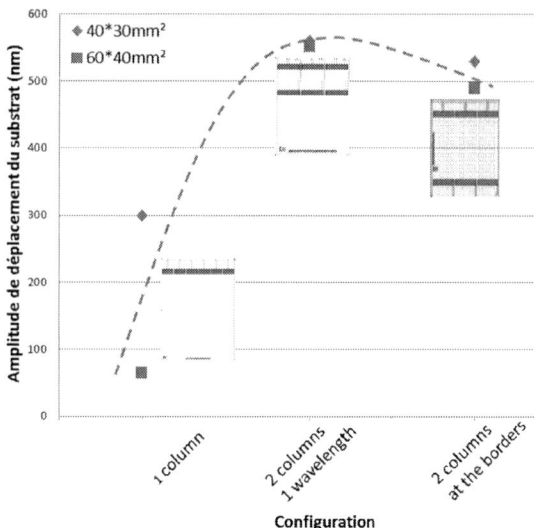

Figure 8: Laser vibrometer measurements -Substrate vertical displacement amplitudes on 60×40 and 40×30 mm² plate presenting various actuator configurations ($12V_{dc}+4V_{ac/pp}$).

5. Conclusion

We have developed a unique solution based on PZT thin-film actuated plates allowing haptic feedback effect using low bias voltage due to squeeze-film effect. In this paper, we presented the characterization and post-simulation of PZT-actuated plates with various actuator configurations. Measurement results, in good agreement with simulations, indicate that 2 actuator columns separated by a wavelength allow the highest substrate displacement amplitudes. But this configuration, inducing a large non-transparent zone, occupied by actuators, will be retained if transparency in not required by the focused application. The second configuration, with 2 actuator columns located at the plate ends presents substrate displacement amplitude slightly smaller compared to the best configuration, but exhibits a larger area dedicated to touch experience. This configuration will be favored for example for Smartphone applications for which we have to preserve the largest transparent surface as possible. Finally, the configuration with only one actuator column presents the smaller dead area. It induces smaller substrate displacement amplitude compared to the best configuration. This configuration will be favored for the applications requiring the smallest non-transparent dead area, to the detriment of low actuation voltage.

ACKNOWLEDGEMENTS

The authors wish to acknowledge all participants at the MINALOGIC project *Touch It*. Moreover we wish to acknowledge the OSEO funding to give us the opportunity to work on this subject.

References

1. P. Sergeant, F. Giraud, B. Lemaire-Semail, "Geometrical optimization of an ultrasonic tactile plate", *Sensors and Actuators A 161,* pp. 91-100, 2010.
2. M. Biet, F. Giraud, B. Lemaire-Semail, « Squeeze film effect for the design of an ultrasonic tactile plate », *IEEE transactions on ultrasonics, Ferroelectrics and Frequency control, vol. 54, n° 12,* December 2007, pp. 2678-2688.
3. M. Cueff, E. Defaÿ, P. Rey, G. Le Rhun, F. Perruchot, C. Ferrandon, D. Mercier, F. Domingue, A. Suhm, M. Aïd, L. Liu, S. Pacheco, M. Miller, "A fully package piezoelectric switch with low voltage actuation and electrostatic hold", *IEEE Int. Conf. on Micro Electro Mechanical Systems (MEMS) Conference,* pp. 212-215, 2010.
4. R. Dejaeger, F. Casset, B. Desloges, G. Le Rhun, P. Robert, S. Fanget, Q. Leclere, K. Ege, JL. Guyader, "Development and characterization of a piezoelectrically actuated MEMS digital loudspeaker", *European Conference on Solid-State Transducers (Eurosensors),* September 9-12, 2012.
5. T. Bakke, A. Vogl, O. Zero, F. Tyholdt, IR. Johansen, D. Wang, "A novel ultra-planear, long-stroke and low-voltage piezoelectric micromirror", *Journal of Micromechanics and Microengineering 20, 064010,* 1st June 2010.
6. F. Casset, JS. Danel, C. Chappaz, Y. Civet, M. Amberg, M. Gorisse, C. Dieppedale, G. Le Rhun, S. Basrour, P. Renaux, E. Defaÿ, A. Devos, B. Semail, P. Ancey, S. Fanget, "Low voltage actuated plate for haptic applications with PZT thin-film", *The 17th International Conference on Solid-State Sensors, Actuators and Microsystems,* June 16-20, 2013.
7. F. Casset, M. Cueff, A. Suhm, G. Le Rhun, J. Abergel, M. Allain, C. Dieppedale, T. Ricart, S. Fanget, P. Renaux, D. Faralli, P. Ancey, A. Devos, E. Defaÿ, "PZT piezoelectric coefficient extraction by PZT-actuated micro-beam characterization and modeling", *IEEE Thermal, Mechanical and Multiphysics Simulation and Experimentsin Micro-Electronics and Micro-Systems conference (EuroSime),* April 16-18, 2012.
8. Casset, Devos, Sadtler, Le Louarn, Emery, Le Rhun, Ancey, Fanget, Defay, « Thin film PZT Young's modulus and Poisson's ratio characterization using Picosecond Ultrasonics », IEEE International Ultrasonics Symposium (Ultrasonics), October 7-10, 2012.

Using Molecular Modeling to Uncover the Origins of Subtle Solvation-Based Film Defects.
Nancy Iwamoto, Teri Baldwin

A series of lithographic metallization steps are used to build up the interconnect network on the chip during wafer fabrication and one often overlooked film technology that is an integral part of the process is the bottom anti-reflective coating (BARC), used to enable the production of uniform fine features in the resist. In order to create these features, near-perfect thin films must be produced, and any surface defect found during the spin-coating process is unacceptable, especially as interconnect densities become finer and finer.

Recently, a defect was found that seemed to be sporadic in manifestation and proved to be difficult to explain from normal inspection of the BARC manufacturing process, which included inspection of every step of the process from incoming raw materials through final filtration and packaging. At that point, molecular modeling was called in to simulate a mechanism that could cause the defect. This molecular modeling encompassed a variety of methods, including thermodynamic calculations of the reaction steps using quantum mechanics and several molecular dynamics techniques to calculate solvation compatibilities (a cohesive energy density calculation), cohesive modulus on suspect oligomers, and simulation of temperature history-dependent pathways of the cohesive energy density on a high suspect oligomer. This final simulation uncovered a temperature sensitivity which explained the origin of the sporadic defects. This discovery initiated corrective actions that led to a final resolution. This paper will describe the modeling that was done and show the often forensic nature of uncovering the origin of this defect using molecular modeling. This example demonstrates how molecular modeling's role can be expanded by being used in a purely investigative manner rather than a predictive one.

978-1-4799-4789-8/14 $31.00 © 2014 IEEE

Description of the thermo-mechanical properties of a Sn-based solder alloy by a unified viscoplastic material model for Finite Element calculations

A. Kabakchiev[1*], B. Métais[1], R. Ratchev[1], M. Guyenot[1],
P. Buhl[2], M. Hossfeld[2], X. Schuler[2], R. Metasch[3], M. Roellig[3]

[1] Robert Bosch GmbH, Corporate Research Division, Postfach 300240, 70442 Stuttgart, Germany
[2] Materialprüfungsanstalt (MPA) Universtät Stuttgart, Pfaffenwaldering 32, 70569 Stuttgart, Germany
[3] Fraunhofer Institute for Ceramic Technology and Systems, Material Diagnostics, Dresden, Germany
* Corresponding author, email: alexander.kabakchiev@de.bosch.com, phone: +49711-811-27090

Abstract

Automotive electronic devices are exposed to substantially harsher thermo-mechanical loads compared to commercial consumer electronic products. As a consequence, solder joints carrying out the electrical interconnection between the components undergo deformation and degradation under thermal cycling, which can determine the lifetime of the electronic assembly in long term operation. In the past decade, lifetime prediction methods for solder joints based on finite element (FE) simulations are increasingly employed in the process of product design. However, constitutive FE models for solder alloys capable of describing their mechanical behavior at the relevant conditions of automotive applications are still not widely established. Here, we employ a unified viscoplastic material model initially proposed by Chaboche et al. in order to address the mechanical properties of an as-casted Sn-based solder alloy under a cyclic mechanical load. Extensive experimental investigations at temperatures from -40°C up to +150°C reveal a complex nonlinear interplay between hardening, recovery and thermally activated inelastic deformation processes in the material studied. We identified the necessary constitutive model terms and performed parameter calibration according to their specific functionality. A very good agreement between the numerical calculations and experimental data is achieved, which renders the constitutive model used a very promising approach for a wide use in FE simulations of lead-free solder alloys.

1. Introduction

During their operation, automotive control units are exposed to substantial temperature changes occurring at their location, especially if placed close to the car engine. As a result, solder joints undergo a cyclic mechanical deformation due to the mismatch of the thermal expansion between their neighboring components. A typical lead-free eutectic solder comprises more than 95 wt% Sn, 1-4 wt% Ag as well as low amounts of copper < 0.6 wt%. Such compositions have a melting temperature T_m of about 217-221°C. Under operation or during lab qualification tests, automotive electronic devices can be subjected to a temperature cycling from -40°C up to 150°C, which

corresponds to a range of $0.47 < T_h < 0.86$ in terms of the homologous temperature $T_h = T/T_m$ (K). A common occurrence in metallic materials under an applied force at temperatures above $\frac{1}{2}T_h$ is plastic deformation accompanied by creep, which is a thermal activated inelastic deformation with a pronounced time-dependence [1]. Numerous works on lead-free Sn-based solders report on creep properties and especially the evolvement of its primary and secondary creep stages [2]–[4]. Based on experimental data on the secondary creep regime, phenomenological FE-material models were proposed assuming stationary creep as the main inelastic deformation mechanism within the full temperature range of applications [2], [5]–[7]. FE-models taking into account only the elastic and secondary creep domains neglect strain hardening behavior as well as recovery processes being out of equilibrium during the primary stage [8]–[10]. Furthermore, creep experiments are usually performed under a unidirectional load of a constant force magnitude, which strongly differs from typical automotive operating conditions. These imply cyclic thermal load leading to mechanical deformation of varying strain rates, on the one hand, and dwell time periods at a constant temperature, where stress relaxation is expected, on the other hand. Similar operating conditions are taking place in the area of gas turbine components where high temperature steel- and Ni-based alloys are exposed to temperatures above $\frac{1}{2}T_h$. Such applications were the driving force for the development of unified viscoplastic constitutive models capable of mapping the complex interplay of plastic deformation, hardening and creep. These models are being actively extended during the past 30 years and form the state of the art in FE-material modeling [11]–[15]. However, a wide application of unified viscoplastic models on solder materials is still not achieved in both research and industry. One of the main obstacles for that is often the lack of an extensive experimental data base, which maps the strain rate and temperature dependence of deformation as well as stress relaxation behavior. A further difficulty is the tedious identification of model terms necessary for an accurate phenomenological description of the observed stress-strain profiles as a function of time [16]. Here, a unified viscoplastic material model by Chaboche et al. [17] is employed in order to address the mechanical properties of a SnAg3,5 solder alloy under isothermal cyclic conditions prior to degradation. We developed an experimental program covering cyclic loads at different strain rates as well as stress relaxation stages. Strain amplitudes applied were located in the regime of primary creep and the onset of secondary creep. According to the material response obtained we identified a set of model terms, capable of mapping the experimental results. The determination of the material model parameters renders a challenging task for achieving a consistent simulation of the rate-dependent cyclic material response as well as the evolution of stress-relaxation. A systematic procedure for model calibration with a special emphasis on the parameter temperature dependence was worked out. Finally, we discuss the unified constitutive model framework with respect to its application to the specific solder alloy properties in the temperature range from -40°C up to 150°C.

2. Experimental techniques

Test specimens were prepared by melting and casting of a SnAg3,5 alloy supplied by Stannol GmbH into a pre-heated metal mould. Reproducible solder solidification was achieved by a controlled cooling of the mould by a rate of 20K/min. The sample quality was X-ray monitored in order to insure that defect free specimens are used for the mechanical testing. The as-cast samples were annealed for 24 h at 125°C and were not subjected to any machining operations. Uniaxial cyclic mechanical tests at room temperature and above were carried out on a custom made set-up. For measurements at -40°C, a commercial material testing machine from Zwick GmbH equipped with a liquid nitrogen cool chamber was adapted with appropriate measurement sensors. In all cyclic tension-compression tests, sample elongation was recorded by a linear variable differential transformer (LVDT) with a sub-micrometer resolution. The analog LVDT-sensor output was fed into a closed-loop drive system of the material testers, which implement strain rate controlled mechanical measurements. The specimens have a cylindrical test region with a diameter of 4 mm and a length of 6 mm. We carried out isothermal experiments at temperatures of -40°C, 25°C, 75°C, 125°C and 150°C. Each specimen was subjected to a three-level mechanical cycling within total strain amplitudes of ±0.25%, ±0.50% and ±1.00%. At each amplitude-level, a series of tension-compression cycles at four different strain rates of 1×10^{-3}, 1×10^{-4}, 1×10^{-5} and 1×10^{-6} (s^{-1}) were performed including two to three repetitions at each strain rate. Stress relaxation in tension and compression was measured at the end of each strain amplitude level by keeping the sample elongation constant over a period of up to 4 hours. A detailed description of the measurement setup is published in [18].

3. Material model formulation

The constitutive model used in the present study belongs to the domain of the unified viscoplastic phenomenological theory. It incorporates strain hardening and time recovery effects formulated in a kinematic back stress tensor X_{ij}. The viscous properties are depicted in the over-stress σ_v which represents an equipotential stress state scaled from the yield surface and dependent on the deformation rate. Finally, the dissipation potential Ω (s. eq. (1)) links the whole formulation into a consistent thermodynamic theory [12]

$$\Omega = \frac{Km}{m+1} \langle \frac{\sigma_v}{K} \rangle^{\frac{m+1}{m}} \qquad (1)$$

Assuming plastic strain incompressibility, the model follows the von Mises yield criterion in J2 plasticity. The von Mises stress is a scalar equivalent quantity expressed by the deviatoric part of stress tensor S_{ij} and the back stress tensor X_{ij} (s. eq (2)). The sign of the right hand side of eq. (2) determines an elastic (< 0) or inelastic deformation (> 0).

$$\sigma_v = J_2(\sigma - X) - \sigma_0 \qquad (2)$$

$$= \sqrt{\frac{3}{2}(S_{ij} - X_{ij}):(S_{ij} - X_{ij})} - \sigma_0 \qquad (3)$$

The inelastic strain rate tensor eq. (4) can be derived from the dissipation potential Ω following the normality rule for associative plasticity. The plastic strain rate and the plastic flow direction n_{ij} are depicted in eq. (5).

$$\dot{\varepsilon}_{ij}^p = \frac{3}{2} \langle \frac{\sigma_v}{K} \rangle^{\frac{1}{m}} \frac{S_{ij} - X_{ij}}{J_2(\sigma - X)} \qquad (4)$$

$$\dot{\bar{\varepsilon}}^p = \langle \frac{\sigma_v}{K} \rangle^{\frac{1}{m}} \quad \text{and} \quad n_{ij} = \frac{3}{2} \frac{S_{ij} - X_{ij}}{J_2(\sigma - X)} \qquad (5)$$

The drag stress K and the viscous exponent m in eq. (1), (4), (5) are referred to as viscous parameters. The Macaulay brackets $\langle \rangle$ ensure the positivity of the inelastic flow rule (eq. (4), (5)) and states that the plastic strain increment becomes zero in case of elastic deformation $\sigma_v < 0$ in eq. (2). The kinematic strain hardening X_{ij} comprises a superposition of multiple kinematic models as proposed by [17].

$$X_{ij} = X_{ij}^{(1)} + X_{ij}^{(2)} + X_{ij}^{(3)} \qquad (6)$$

We implemented three kinematic models (eq. (7), (8), (9)) comprising the Armstrong-Fredericks strain hardening formulation by a linear Prager hardening term and a dynamic recovery term [19]. We extended two of the kinematic models (eq. (7), (8)) by static recovery terms $d_{1,2}J_2(X^{(1,2)})X_{ij}^{(1,2)}$, which enable a time-dependent relaxation of the accumulated kinematic hardening [15].

978-1-4799-4789-8/14 $31.00 © 2014 IEEE

$$\dot{X}_{ij}^{(1)} = \frac{2}{3}c_1\dot{\varepsilon}_{ij}^p - \gamma_1\dot{\bar{\varepsilon}}^p X_{ij}^{(1)} - d_1 J_2(X^{(1)})X_{ij}^{(1)} \quad (7)$$

$$\dot{X}_{ij}^{(2)} = \frac{2}{3}c_2\dot{\varepsilon}_{ij}^p - \gamma_2\dot{\bar{\varepsilon}}^p X_{ij}^{(2)} - d_2 J_2(X^{(1)})X_{ij}^{(2)} \quad (8)$$

$$\dot{X}_{ij}^{(3)} = \frac{2}{3}c_3\dot{\varepsilon}_{ij}^p - \gamma_3\dot{\bar{\varepsilon}}^p X_{ij}^{(3)} \quad (9)$$

The kinematic hardening models (eq. (7), (8), (9)) are expressed in differential formulations with respect to time. $\dot{\varepsilon}_{ij}^p$ and $\dot{\bar{\varepsilon}}^p$ denote the time derivative of the equivalent strain tensor and the norm of the plastic strain rate, respectively. We implemented the constitutive model by an implicit Backward Euler integration scheme as a function routine using the commercial software Matlab® [20]. The differential non-linear equations are successively solved by a Newton gradient procedure where all state variables are implicitly derived for an infinitesimal strain and time increment.

4. Material model calibration

The unified constitutive model describes non-linear interplay between viscous deformation and plastic hardening by a set of parameters, which has to be successively determined at each temperature. In the following, the calibration procedure is described, which comprises both direct graphical determination methods, as well as an iterative optimization technique. The solder alloys Young's Modulus was obtained based on the measurement at 25°C (298K). For this, we evaluated the slope of the stress-strain curve at the onset of unloading, i.e. after the maximal strain within each cycle has been reached. Young's moduli E (eq. (10)) for the other temperatures (Table 1) were calculated by taking the graphically determined value at 25°C as a reference, and applying the temperature dependence reported by [21]. A graphical determination method was further applied for describing the stress dependency on strain rate [22]. This effect is mapped by the evolution of the viscous stress σ_v according to eq. (5). It can be evaluated from a stress measurement during a load built up to a predefined strain and a subsequent relaxation step, during which the sample elongation is kept constant (s. fig. 1 b). For sufficiently large strain amplitudes applied prior to the relaxation step, the stress rate decreases, and the total strain increment $\dot{\varepsilon}$ can be assumed to be nearly plastic $\dot{\varepsilon}^p$ (s. eq. (10)).

$$\dot{\varepsilon} = \dot{\varepsilon}^e + \dot{\varepsilon}^p = \frac{\dot{\sigma}}{E} + \dot{\varepsilon}^p \approx \dot{\varepsilon}^p \quad (10)$$

In uniaxial loading configuration, the viscous stress σ_v (eq. (2)) is equivalent to

$$\sigma_v = \sigma - (X + \sigma_0) \quad (11)$$

The quantity $X + \sigma_0$ is a measure of the hardening accumulated during the loading phase. Assuming, that the

kinematic back-stress X does not change significantly during stress relaxation [22], the quantity $X + \sigma_0$ manifests in a rest-stress σ_R measured after sufficiently large relaxation time (s. Fig. 1 b). Therefore, the viscous stress σ_v can be evaluated from the total stress σ_{max} after the initial strain elongation. According to eq. (10), (11), σ_v results in

$$\sigma_v = \sigma_{max} - \sigma_R = K(\dot{\varepsilon})^m \quad (12)$$

Thus, the material parameters K and m can be directly determined from a linear regression according to eq. (12).

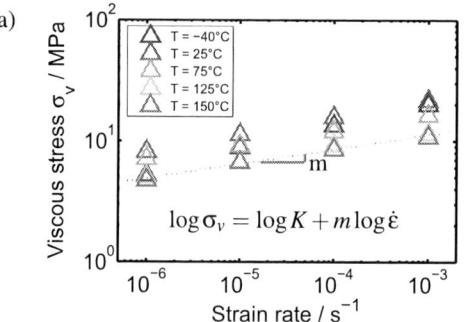

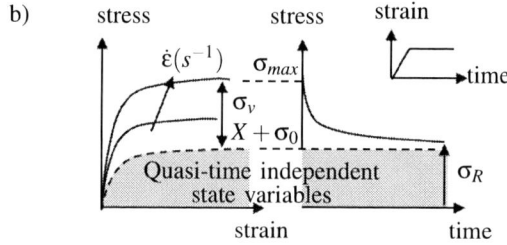

Figure 1. Viscous stress as a function of strain rate a) and a schematic representation of the superposition of time independent hardening and viscous stress evolution b)

Figure 1 a shows the results of the viscous stress evaluation on SnAg3,5. We used the total uniaxial stress σ measured after a strain elongation of 1% at four different strain rates from 1×10^{-3} to 1×10^{-6} (s^{-1}). The corresponding rest-stress σ_R was read out at the end of the relaxation step at a time of 4 h. In our experiments we did not observe an elastic region at the initial tensile load limited by a yield strength. We assumed a small yield stress of 0.5 MPa of constant magnitude for all temperatures, as well as a Poisson ratio of 0.36. Due to the graphical determination of the viscous function and elastic modulus (Table 1), the number of parameters to be further calibrated reduces down to the constitutive model part of kinematic terms.

Table 1. Material model parameters obtained by a graphical determination method

Temperature	-40°C	25°C	75°C	125°C	150°C
E (MPa)	26459	22353	19039	15592	13818
K (MPas^m)	71,43	56,71	37,66	26,09	23,95
m (-)	0,184	0,140	0,122	0,117	0,114

The identification of the parameters corresponding to the linear hardening c_1, c_2, c_3 dynamic recovery γ_1, γ_2, γ_3 and static recovery d_1, d_2 was achieved by an iterative optimization procedure [23]. A simulation of the uniaxial stress state evolution upon the strain-time profiles from the experiments as boundary conditions was realized on a single Gauß point level in the software Matlab. The results are shown on Fig. 2 for the measurement at 25°C.

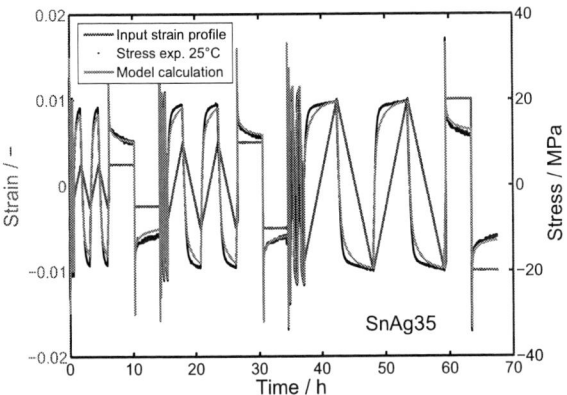

Figure 2. Measured and simulated stress-time profiles including the complete three-level cycling program

A comparison of the measured and calculated cyclic stress-strain profiles as well as stress relaxation curves for all temperatures studied is represented in Fig. 3.

The kinematic hardening calibration plays a significant role for an accurate description of the non-linear stress evolution within a tension-compression cycle. Close after each reversal of load direction (s. stress-strain hysteresis in Fig. 3 a, b), stress builds up rapidly due to the strain controlled sample elongation. After a short transition period of continuous decrease, the stress rate tends to saturation towards a steady state region. These regions of stress-evolution can be identified in the shapes of all cyclic hysteresis shown in Fig. 3 a and b. The superposition of the kinematic models (eq. (7), (8), (9)) employed in the present study has the goal of describing the regions of non-linear stress rate evolution. However, for a consistent calibration, parameter values have to be initially identified for use as a starting parameter set in the optimization routine. For this, a graphical evaluation of the stress-strain slopes before, and after the time-points of load reversal was performed. Since the Armstrong-Fredericks formulation comprises a set of six parameters, the possibility exists, that a minimization of the objective function is reached but the resulting parameter values do not provide a perfect match to the data in all parts of the experimental program. This may be the reason for the observed deviation between the simulated and measured curve of the cyclic hysteresis obtained at -40°C in Fig. 3 a.

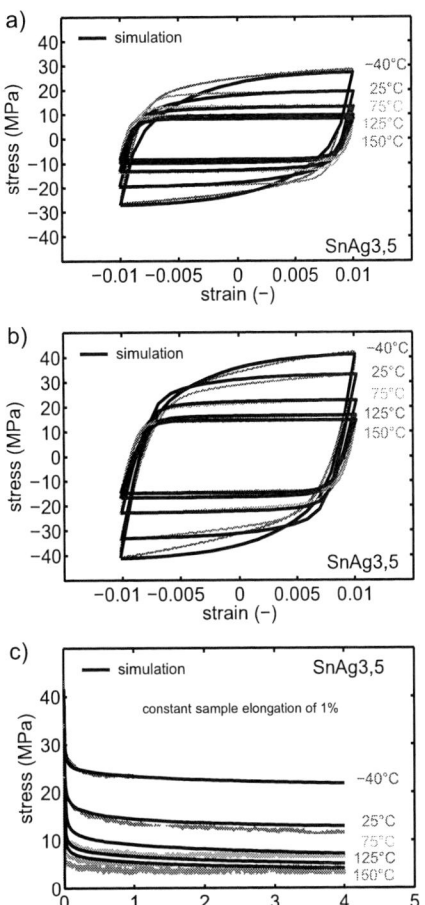

Figure 3. Experimental and simulated stress-strain hysteresis at different temperatures for a strain rate of a) 1×10^{-6} (s^{-1}); b) 1×10^{-3} (s^{-1}) and c) stress relaxation measurements. Relaxation times in the experiments: 1 h at -40°C, 4 h for all other temperatures.

A better parameter determination might be achieved by using advanced optimization procedures based on genetic algorithms [21]. As can be further seen from Fig. 3 a,b, there is a substantial increase of absolute stress within the stress strain hysteresis with decreasing temperatures. This naturally leads to significant variation of model parameter values. In order to describe the temperature dependence of each model parameter P_i we employed a parabolic expression (s. eq. (13)) [17].

$$P_i(T) = a_i T^2 + b_i T + c_i, \qquad (13)$$

where the temperature is expressed in Kelvin. Figure 4 gives an overview on the parameter evolution with temperature.

It should be noted, that the temperature dependence of the model parameters listed in Table 2 was determined on the basis of the data obtained on one specimen at each

Table 2. Temperature dependence of the material model parameters P_i by function (13).

P_i	a_i	b_i	c_i
c_1 (MPa)	$3{,}10E^{-1}$	$-3{,}04E^2$	$8{,}15E^4$
γ_1 (-)	$9{,}51E^{-3}$	$-9{,}32$	$9{,}79E^3$
c_2 (MPa)	$2{,}57E^{-1}$	$-2{,}52E^2$	$6{,}62E^4$
γ_2 (-)	$3{,}99E^{-2}$	$-3{,}91E^1$	$1{,}17E^4$
c_3 (MPa)	$8{,}73E^{-2}$	$-7{,}56E^1$	$1{,}64E^4$
γ_3 (-)	$6{,}70E^{-4}$	$-6{,}57E^{-1}$	$3{,}77E^2$
d_1 (s^{-1})	$1{,}26E^{-7}$	$-5{,}80E^{-5}$	$6{,}67E^{-3}$
d_2 (s^{-1})	$1{,}72E^{-9}$	$-7{,}22E^{-7}$	$7{,}59E^{-5}$
K (MPas^m)	$8{,}03E^{-4}$	$-7{,}87E^{-1}$	$2{,}14E^2$
m (-)	$1{,}12E^{-6}$	$-1{,}10E^{-3}$	$3{,}73E^{-1}$

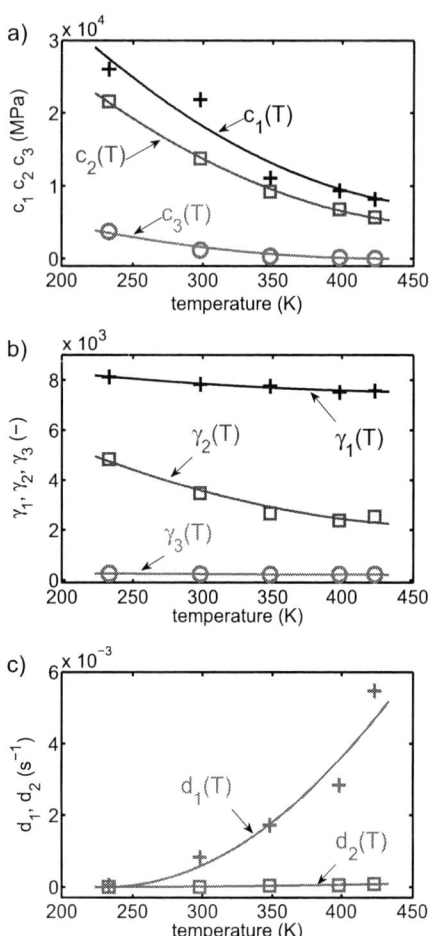

Figure 4. Values of the kinematic hardening parameters and modeling according to eq. (13), of the linear hardening a), dynamic recovery b), and static recovery c).

temperature. Since the crystal structure of Sn shows a strong anisotropy of elastic and thermal properties [24], [25], a variation of strength can be expected between different samples. However, the study of the statistical variation of the mechanical as well as degradation properties of the SnAg3,5 is beyond the scope of the present study.

An analytical description of the temperature dependence (s. eq. (13), Table 2) of each phenomenological term provides means for a continuous parameter value extraction during non-isothermal FE-calculations. This can substantially improve the performance and numerical stability in simulations of solder joints subjected to varying thermal load profiles.

5. Discussion

The simulation results by the viscoplastic constitutive model correspond very well to the measured stress profiles for the strain rates and temperatures studied (s. Fig. 2). Besides the viscous time dependent deformation, hardening properties are of utmost importance for a correct description of the material behavior. The stress-strain hysteresis (Fig. 3) gives strong evidence for nonlinear stress evolution consistent with the Bauschinger effect. We did not observe a cyclic hardening, i.e., there was no evidence of a cyclic increase of the materials elastic domain, as reported also for SnAgCu [21]. Thus, the use of kinematic terms (eq. (7), (8), (9)) for the model's hardening low is a worthwhile approach in the phenomenological description of the SnAg3,5 properties. However, a more complex behavior cannot be excluded for other solder alloys, which may show a cyclic hardening especially in the low temperature regime. In such a case, the constitutive model should be extended by isotropic hardening. The experiments on SnAg3,5 reveal pronounced dependence on temperature. For an increasing temperature, we identify a continuous decrease of strain hardening values (Fig. 4 a,b) whereas static recovery parameters increase (Fig. 4 c). The viscous function and static recovery formulations describe creep processes, which determine both the strain rate dependence and relaxation behavior of stress. Thus, the temperature dependence of the constitutive kinematic and viscous model parameters can be interpreted as a tool for quantifying the interplay between hardening and creep processes. The ratio between both phenomena varies continuously in a large range of homologous temperature. Apart from absolute strength, the solder alloy shows a behavior consistent with high temperature steel- and Ni-alloys at $T_h > 0.5$ for which viscoplastic behavior has been numerously reported [11], [13], [14], [26]. The present results render the unified Chaboche viscoplastic model as a very promising approach for an accurate simulation and thus for a wide application on Sn-based solder alloys. Last but not least, an accurate calculation of the intrinsic mechanical properties of solder alloys is highly important in the domain of lifetime prediction models, which take into account stress, accumulated strain or dissipated plastic work. Furthermore, the viscoplastic phenomenological framework, which is highly adaptive and extendable, opens up new opportunities for the development of advanced lifetime-prediction techniques based on damage mechanics [9], [21]. These may be capable of achieving a reliability assessment independent of specific thermal load profiles, local stress-states and different geometries of the various solder joint types.

6. Acknowledgements

The authors would like to thank to Dr. M. Spraul, N. Schafet, Dr. U. Becker, S. Bollwerk, T. Heinrich, Dr. D. Krätschmer and Dr. M. Rauch from the Automotive Electronics Division of Robert Bosch GmbH for valuable discussions during the project.

References

[1] J.H. Frost and M.F. Ashby. *Deformation-Mechanism Maps, The Plasticity and Creep of Metals and Ceramics*. Pergamon Press, 1982.

[2] S. Wiese and K.-J. Wolter. Microstructure and creep behaviour of eutectic SnAg and SnAgCu solders. *Microelectronics Reliability*, 44:1923–1931, 2004.

[3] S. Deplanque, W. Nuechter, M. Spraul, B. Wunderle, R. Dudek, and B. Michel. Relevance of primary creep in thermo-mechanical cycling for life-time prediction in sn-based solders. In *6th. Int. Conf on Thermal, Mechanical ond Multiphysics Simulation and Experiments in Micro-Electronics and Micm-System, EuroSimE*, pages 71–78, Berlin, Germany, April 2005.

[4] R. Metasch, J.C. Boareto, M. Roellig, S. Wiese, and K.-J. Wolter. Primary and tertiary creep properties of eutectic SnAg3.8Cu0.7 in bulk specimens. In *10th. Int. Conf on Thermal, Mechanical and Multiphysics Simulation and Experiments in Micro-Electronics and Micro-Systems, EuroSimE*, Delft, The Netherlands, April 2009.

[5] A. Schubert, R. Dudek, E. Auerswald, A. Gollbardt, B. Michel, and H. Reichl. Fatigue life models of snagcu and snpb solder joints evaluated by experiements and simulations. In *53th Electronic Components and Technology Conference ECTC*, pages 603–610, New Orleans, USA, May 2003.

[6] Q. Zhang. Viscoplastic constitutive properties and energy-partioning model of lead-free Sn3.9Ag0.6Cu solder alloy. In *53th Electronic Components and Technology Conference ECTC*, pages 1862–1868, New Orleans, USA, May 2003.

[7] A. Syed. Accumulated creep strain and energy density based thermal fatigue life prediction models for SnAgCu solder joints. In *54th Electronic Components and Technology Conference ECTC*, pages 737–746, Las Vegas, June 2004.

[8] J.P. Clech. An extension of the omega method to primary and tertiary creep of lead-free solders. In *55th Electronic Components and Technology Conference, ECTC*, pages 1261–1271, Orlando, FL, May 2005.

[9] M. Grieu, G. Massiot, O. Maire, C. Munier, Y. Bienvenu, and J. Renard. Modelling of Sn3.0Ago.5Cu thermo-mechanical behaviour by a continuum damage approach. In *10th. Int. Conf on Thermal, Mechanical and Multiphysics Simulation and Experiments in Micro-Electronics and Micro-Systems, EuroSimE*, pages 1–6, Delft, The Netherlands, April 2009.

[10] D. Shierley, H.R. Ghorbani, and J.K. Spelt. Effect of primary creep and plasticity in the modeling of thermal fatigue of SnPb and SnAgCu solder joints. *Microelectronics Reliability*, 48:455–470, 1960.

[11] D. Nouailhas. Unified modelling of cyclic viscoplasticity: Application to austenitic stainless steels. *International Journal of Plasticity*, 5:501–520, 1989.

[12] J.L. Chaboche. Constitutive equations for cyclic plasticity and cyclic viscoplasticity. *International Journal of Plasticity*, 5:247–302, 1989.

[13] E. Roos, H. Yu, A. Klenk, and K. Maile. Description of deformation and failure behavior of a nut-bolt-assembly by means of a viscoplastic constitutive equation. In *ASME PVP Conference*, pages 95–102, Boston, USA, August 1999.

[14] M. Ringel, E. Roos, K. Maile, and A. Klenk. Constitutive equations of adapted complexity for high temperature loading. In *ECCC Creep Conference*, pages 638–648, London, UK, September 2005.

[15] E. Kullig and S. Wippler. Numerical integration and FEM-implementation of a viscoplastic Chaboche-model with static recovery. *Computational Mechanics*, 38:491–503, 2006.

[16] S. Wippler and M. Kuna. Experimental and numerical investigation on the reliability of leadfree solders. *Engineering Fracture Mechanics*, 75:3534–3544, 2008.

[17] J.L. Chaboche. A review of some plasticity and viscoplasticity constitutive theories. *International Journal of Plasticity*, 24:1642–1693, 1966.

[18] R. Metasch, M. Roellig, A. Kabakchiev, B. Métais, R. Ratchev, and K.-J. Wolter. Experimental investigation of the visco-plastic mechanical properties of a Sn-based solder alloy for material modeling in finite element calculations for automotive electronics. In *15th. Int. Conf on Thermal, Mechanical and Multiphysics Simulation and Experiments in Micro-Electronics and Micro-Systems, EuroSimE*, Ghent, Belgium, April 2014.

[19] P.J. Armstrong and C.O. Frederick. A mathematical representation of the multiaxial bauschinger effect. *Report RD/B/N731, CEGB, Central Electricity Generating Board, Berkeley, UK*, 1966.

[20] The MathWorks Inc. Matlab, Version R2012b.

[21] M. Kuna and S. Wippler. A cyclic viscoplastic and creep damage model for lead free solder alloys. *Engineering Fracture Mechanics*, 77:3635–3647, 2010.

[22] J. Lemaitre and J.L. Chaboche. *Mechanics of Solid Materials*. Pergamon Press, 1990.

[23] J.A. Nelder and R. Mead. A simplex method for function minimization. *Computer Journal*, 7:308–313, 1965.

[24] J.A. Rayne and B.S. Chandrasekhar. Elastic constants of beta-tin from 4.2k to 300k. *Physical Review*, 120:1658–1663, 1960.

[25] T.R. Bieler, B. Zhou, L. Blair, A.Z.P. Darbandi, F. Pourboghrat, T.K. Lee, and K.C. Liu. The role of elastic and plastic anisotropy of sn in recrystallization and damage evolution during thermal cycling in SAC305 solder joints. *Journal of Electonic Materials*, 41:283–301, 2012.

[26] M. Rauch. *Entwicklung eines Lebensdauerkonzeptes für Schaufel-Welle-Verbindungen stationärer Turbinen aus Nickelbasis- und 10%-Chromlegierungen*. Ph.D thesis, Materialprüfungsanstalt (MPA), Universität Stuttgart, 2006.

978-1-4799-4789-8/14 $31.00 © 2014 IEEE

Failure Mechanisms in Chip-Metallization in Power Applications

C. Durand[1,2], M. Klingler[1], D. Coutellier[2], H. Naceur[2]

[1] Robert Bosch GmbH, Automotive Electronics, Tübingerstr. 123, 72762
Reutlingen, Germany
[2] LAMIH UMR CNRS 8201, PRES Lille Nord de France, University of
Valenciennes, Le Mont Houy, 59313 Valenciennes Cedex 9, France
e-mail: camille.durand@de.bosch.com

Abstract

Degradation of the chip-metallization layer in power electronic packages under Active Power Cycling is still a major reliability concern. During Active Power Cycling tests, the chip acts as a heat source and temperature gradients develop within the package inducing stress and plastic deformation in aluminum metallization. This study is conducted on a power module using a copper clip soldered on the top side of the chip, instead of aluminum wire bonds. Both experiments and simulations are performed, to better understand mechanisms of chip-metallization degradation.

In this paper, experimental Power Cycling tests are performed on power packages and a 2D Finite Elements model of MOSFET is used for thermo-mechanical simulation. Modules are monitored during tests and metallographic specimens are made at the end of tests in order to examine changes in the metallization layer. Thermo-mechanical analysis allows us to monitor the evolution of stress and strain in aluminum during power pulses. A study of the sensitivity of various test parameters is also simulated and the influence of those parameters on the mechanical behavior of power metallization is quantified. Knowledge of degradation phenomena gained with simulation helps to improve product design.

1. Introduction

Degradation of the metallization layer is known as one of the most frequent failure mechanisms observed to affect mainly power modules, with bond wire lift off and solder fatigue… [1].

Under Active Power Cycling, the chip is powered and dissipates heat within the entire module. This power dissipation leads to a spatially inhomogeneous time-dependent temperature rise with peak temperatures on the chip. The thermo-mechanical stress induced by these temperature cycles causes plastic deformation and disruption of the metallization. Indeed the Coefficient of Thermal Expansion (CTE) mismatch between Silicon (2ppm/K) and Aluminum (25ppm/K) is responsible for the generation of high compressive stress in the aluminum layer. This stress is likely to exceed by large the elastic limit. Under these conditions, the stress relaxation occurs mainly by plastic deformation at the grain boundaries, leading to layer degradation with a subsequent open circuit.

This metallization degradation phenomenon is well known and some work has been dedicated to understand the electro-thermal and thermo-mechanical behavior of aluminum [2,3,4], or to develop new designed layers to reduce failure generation [5].

This work aimed at a better understanding of the thermo-mechanical behavior of the metallization layer in the new designed MOSFET under Active Power Cycling with varying load parameters by means of experiments and simulations.

2. Design of power module

The power module used in this study is a newly designed MOSFET package. It has an electric connection achieved by a copper clip soldered on top of the chip instead of using wire bond (see figure 1).

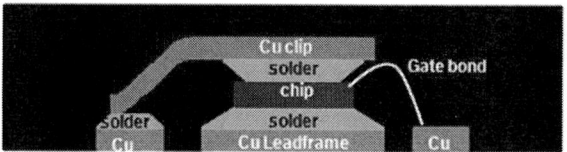

Figure 1: Inner structure of a MOSFET with a copper clip

This structure is more reliable because it avoids wire bond fatigue failures, often the root cause for the device failure. The inner structure of this MOSFET is a bit more complicated than the one of standard power modules. It has some additional material layers on top of the others between the chip and the copper clip (see figure 2).

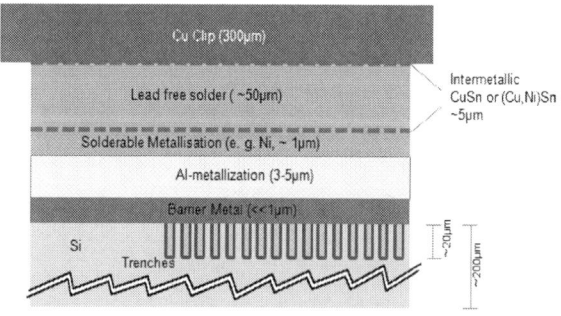

Figure 2: Scheme of layers composition between the chip and the copper clip of a MOSFET

978-1-4799-4789-8/14 $31.00 © 2014 IEEE

3. Experimental and numerical methodologies

Experiments

Active Power Cycling tests were performed on MOSFETs with copper clips. The device under test is mounted onto a heat sink as in a real application. A load current is conducted by the power chip and the power losses heat up the package. When the maximum target temperature within the chip is reached, the load current is switched off and the system cools down to a minimum temperature. The end of the cycle is achieved when the minimal temperature is reached. The next cycle begins by starting the load current again. The characteristic parameters for Active Power Cycling tests are: the start Temperature T_{start}, the temperature swing ΔT (given by the temperature difference between maximal junction temperature at the end of the heating phase and the minimal junction temperature at the end of the cooling interval) and the pulse width t_{pulse}. Active current pulses range from 200A up to 600A, temperature range can vary from 60K to 120K, start temperature ranges from -40°C to 130°C, and pulse width varies from 0.2s to 60s.

Before starting the test and at the end of it, power modules are electrically and thermally characterized by measuring leak currents, the resistance R_{DSon} and the thermal impedance Z_{th}. Metallographic specimens of modules which have reached end of life criteria were carried out and defects investigated into.

Simulations

A 2D Finite Element Model has been created for a MOSFET with an axisymmetry condition (see figure 3). Such a simplified model of the power package is possible because MOSFETs included in the package are thermally decoupled, meaning that every MOSFET acts as a single one. Moreover, this 2D axisymmetric model with chip-midpoint as symmetry axis provides results reliable and precise enough as the center of the chip is the neutral point NP in terms of thermal expansion.

Thermo-mechanical analysis is performed, which means that thermal results of the transient thermal simulation are imported as loads in the thermo-mechanical simulation. Three cycles are simulated after an initial cooling down from the stress free temperature of the module to the ambient temperature (see figure 4).

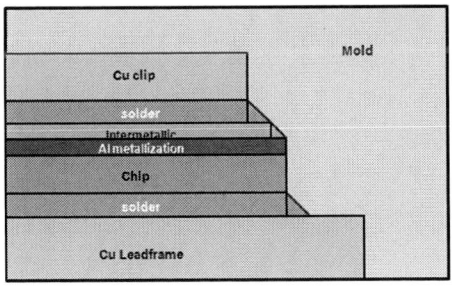

Figure 3: Schematic 2D section of the power device used for simulation

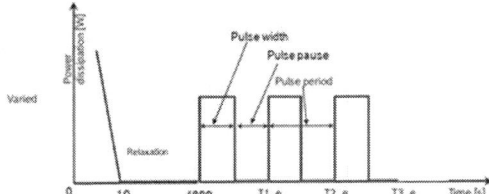

Figure 4: Schematic of the three cycles simulated

The material properties used are temperature dependent. Aluminum properties are modeled with a temperature dependent bilinear kinematic hardening, in order to properly simulate the phenomenon of metallization degradation (see figure 5). σ_{y0} on the graph corresponds to the yield stress at 20°C.

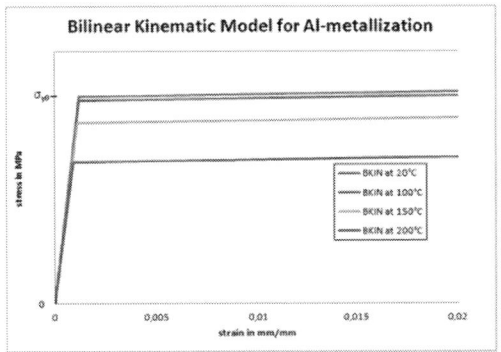

Figure 5: Bilinear kinematic hardening of aluminum

With the thermo-mechanical simulation, evolutions of stress and strain in aluminum are monitored during power pulses. A study of the sensitivity of various test parameters (start Temperature T_{start}, temperature swing ΔT, and pulse width t_{pulse}) is also simulated and the influence of those parameters on the mechanical behavior of power metallization is quantified.

4. Failure mechanisms observed in chip metallization after APC tests

Some Active Power Cycling tests were performed on MOSFETs with different experimental parameters: T_{start} ranges from -30°C to 60°C, temperature swing ΔT varies from 60K to 120K and the pulse width t_{pulse} ranges from 0.2s to 10s. Modules fail after 200.000 to 5 million cycles. This proves the high reliability of such power modules with double sided soldered chips. Despite this very good endurance, modules are submitted to degradation phenomena. The chip-metallization is often affected by this thermally induced fatigue. Thanks to metallographic specimens of failed devices, different types of degradation in aluminum were highlighted: deformation of the layer (figure 6), migration of aluminum in the solder, delamination between aluminum and solder (figure 7), and cracks within the layer (figure 8).

978-1-4799-4789-8/14 $31.00 © 2014 IEEE

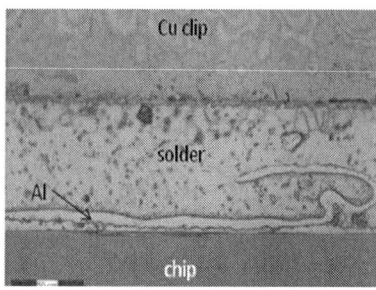

Figure 6: Optical microscope's image of a MOSFET with a deformed aluminum metallization

Those degradations often appear at the area of aluminum located under the top solder meniscus. Cracks and delamination between top solder and aluminum often start at this solder meniscus area, and then propagate to the center of the module. The degradation of the aluminum metallization seems to be more severe with an increase in temperature swing ΔT. It is more difficult to detect the influence of the other test parameters. But this is also due to the fact that the number of experiments performed is still insufficient to make strong assumptions.

The presence of damage in aluminum after Active Power Cycling is also correlated to the increase in the resistance R_{DSon} and the slight increase in thermal impedance Z_{th} at the beginning of the measurement. Indeed, the metallization layer is on top of the chip, and has to homogeneously distribute the current to the chip. When the metallization is degraded, the current is no longer perfectly distributed, thus generating an increase in the resistance R_{DSon} and in the thermal impedance.

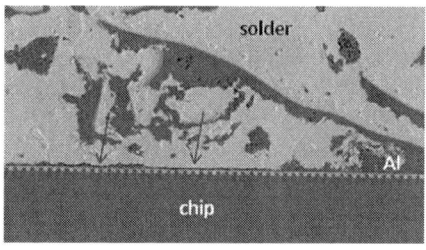

Figure 7: REM image of a MOSFET with a delamination between aluminum and solder

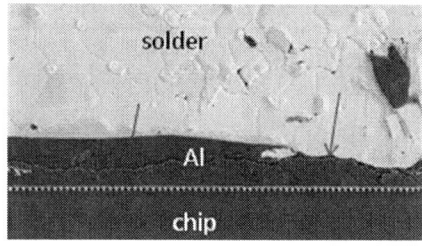

Figure 8: REM image of a MOSFET with a crack in aluminum

5. Simulation results

The general behavior of the module was already described [6]: after cycling, a bending of the module and shrinkage of mold are observed. The chip is entirely submitted to "in plane" compression, while chip metallization, copper clip and leadframe are submitted to tensile stresses.

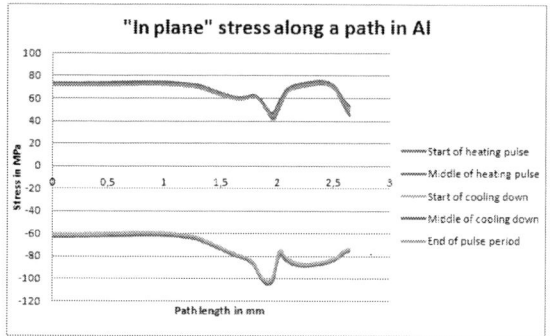

Figure 9: "In plane" stress along a path in aluminum

A path was defined in the middle of aluminum metallization, from its center to one of its extremity, to be able to observe the evolution of stresses and plastic strain in function of the position in aluminum. One simulation case (T_{start}= -40°C, ΔT=120K, t_{pulse}= 10s) is taken as an example to describe mechanisms that are taking place in the metallization layer. We first look at the "in-plane" stress, and plot it along the path at different times of the pulse period (see figure 9). At high temperatures (end of heating pulse/ start of cooling down), aluminum is under compressive stress, whereas at low temperatures tensile stress is dominant. The stress remains below +/-100MPa and is therefore not critical. An irregularity appears in these curves at about 2mm. It is interesting to note that this point corresponds to the top solder meniscus, which means that, up to this point the chip is only soldered on side. This irregularity appears also by plotting the "out of plane" stress (see figure 10). The stress values reached here are quite low, but a stress peak is observable at the top solder meniscus area.

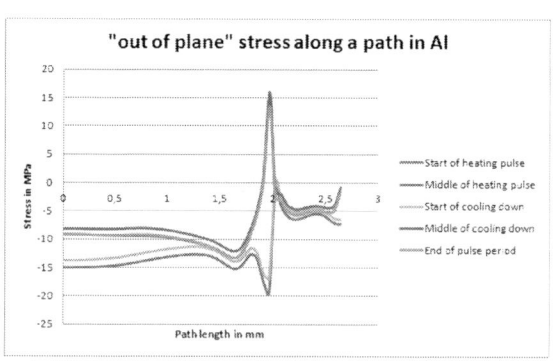

Figure 10: "Out of plane" stress along a path in aluminum

As well as for the shear stress, values reached are low but a peak appears at the same region. Finally the von Mises stress in aluminum is analyzed (see figure 11). At the beginning of the path, curves for high and low temperature follow two different trends: stress values reached at high temperatures are lower than at low temperatures. But at the location where the top solder meniscus takes place, curves join and then overlap. Based on this analysis one can see that von Mises stress is not critical and is driven by the "in plane" stress component, and that the area beneath the top solder meniscus is an important region in aluminum.

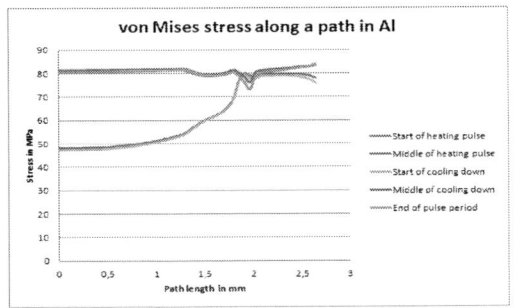

Figure 11: von Mises stress along a path in aluminum

Plastic strain curves along the aluminum path confirm the importance of the zone under the top solder meniscus, as a peak of plastic strain is observed exactly at this position (see figure 12). In order to better understand the mechanisms aluminum is submitted to, the accumulated plastic strain was calculated for one node of the metallization, and its evolution together with the von Mises stress evolution was plotted versus time (see figure 13). The von Mises stress at the beginning of the temperature pulse originates from the initial conditions: a 175°C stress free temperature is defined and the module is cooled down to room temperature. This gives the frozen internal stresses within the package. Now, when the system heats up, these inner stresses decrease and reach a minimum value. By continuing the heating up process, the stress changes its direction and the von Mises stress and the plastic strain increase simultaneously. Then the temperature remains constant, and a stress relaxation without an increase in the plastic strain can be observed. In the following cooling phase the same behavior is repeated, with the opposite deformation direction.

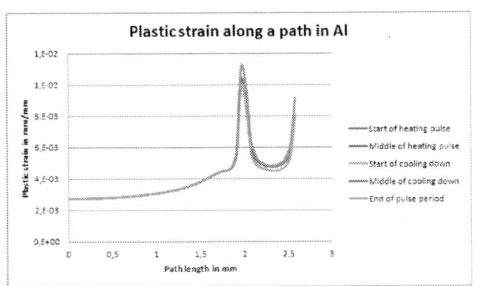

Figure 12: Plastic strain along a path in aluminum

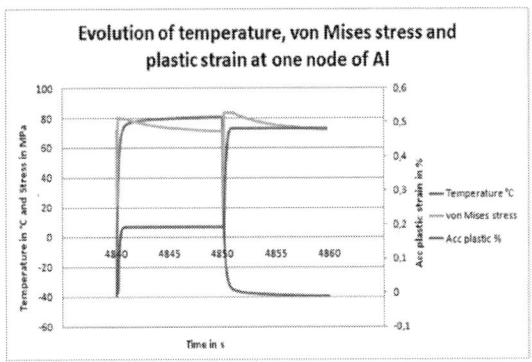

Figure 13: Evolution of temperature, von Mises stress in MPa and accumulated plastic strain at one node of aluminum

6. Interpretation and discussion

A sensitivity study of tests parameters (T_{start}, ΔT, and t_{pulse}) was simulated and the influence of those parameters on the "in plane" stress and accumulated plastic strain is quantified. For the "in plane" stress, 3 values are to be taken into account: the minimum and maximum stress state reached in aluminum in one cycle, and the stress amplitude which is the difference between the maximal and minimal states of stress. At high temperatures, the minimum stress state is reached and the aluminum is under compression. At low temperatures, the maximal stress state is attained and the aluminum is under tension. On the graph figure 14, one can see that the maximum stress slightly increases with the increase in start temperature and temperature swing. The compressive stress is higher with short pulses and increases with the increase in temperature swing. This is explainable by the fact that increasing these parameters amounts to an increase in the highest temperature reached in one cycle. Hence short pulses and high temperature swings are the most critical parameters in terms of stress amplitude in aluminum.

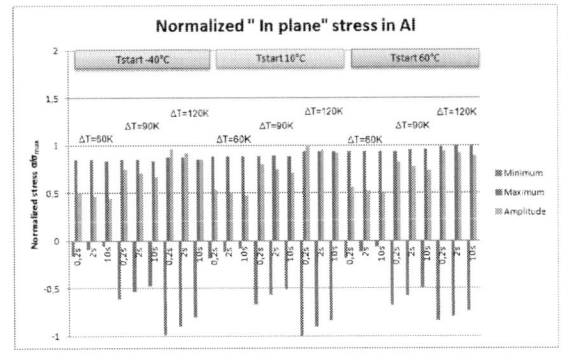

Figure 14: Histogram of normalized "in plane" stress in aluminum for different sets of test parameters

With both experiments and simulations, it was pointed out that there is a critical region in the chip metallization: the area located just beneath the top solder meniscus. The accumulated plastic strain in this zone for one cycle was calculated and plotted for different sets of parameters (see figure 15). The influence of test parameters on plastic strain is quite clear. First of all, the lower the start temperature is, the higher the percentage of plastic strain will be. High temperature swings are needed for the accumulation of plastic strain in one cycle. For small temperature swings like $\Delta T=60K$, there is no accumulated plastic strain at all. This confirms the trend observed with experiments. Finally, short pulses also generate more accumulation of plastic strains in aluminum. Consequently, low start temperatures with large temperature swings and short pulses are the most critical tests parameters in terms of plastic metallization deformation.

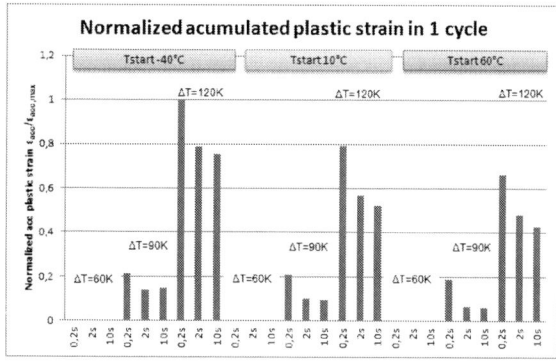

Figure 15: Histogram of normalized accumulated plastic strain in aluminum for different sets of test parameters

7. Conclusion

This paper reports failure mechanisms observed in the chip metallization of power MOSFETs during Active Power Cycling tests and manages to improve the understanding of these mechanisms with a thermo-mechanical 2D FEM simulation.

Active power cycling tests were performed on MOSFETs, with different load parameters. Metallographic specimens of modules which have reached end of life criteria were carried out and defects investigated into. Failure mechanisms were highlighted and a critical area was pointed out. Then a 2D model of the module was created and a thermo-mechanical analysis has been performed.

Results indicate that the aluminum metallization is mainly subjected to "in plane" stress and plastic strain. Once again the area beneath the top solder meniscus was found out to be critical. A sensitivity study on tests parameters was conducted and the influence of these parameters on "in plane" stress and accumulated plastic strain in aluminum was determined.

In order to better understand crack propagation, a thermo-mechanical simulation including a crack in the middle of the aluminum metallization layer will be run. It will allow us to determine critical fracture mechanical properties using standard methods of fracture mechanics in FEM.

References

1. Ciappa, M, "Selected Failure Mechanisms of Modern Power Modules", *Microelectronics reliability* 42 (2002), pp. 653-667

2. Kanert, W, *et al* "Reliability challenges for power devices under active cycling", *47th Annual International Reliability Physics Symposium*, Montréal, 2009

3. Kanert, W, *et al* "Modelling of Metal Degradation in Power Devices under Active Cycling Conditions", *12th Int Conf on Thermal, Mechanical and Multiphysics Simulation and Experiments in Microelectronics and Microsystems, EurosimE*, Linz, 2011

4. Ciappa, M. and Malberti, P., "Plastic strain of Aluminum Interconnections during Pulsed Operation of IGBT Multichip Modules", *Qual. Rel. Eng. Int.*, Vol. 12, (1996), pp.297-303

5. Alpern, P. *et al,* "On the Way to Zero Defect of Plastic-Encapsulated Electronic Power Devices-Part I: Metallization", *IEEE Trans-Device and materials reliability*, Vol. 9, No.2 (2009), pp. 269-278

6. Durand, C. *et al,* "Confrontation of Failure Mechanisms Observed during Active Power Cycling Tests with Finite Element Analyze Performed on a MOSFET Power Module", *14th Int Conf on Thermal, Mechanical and Multiphysics Simulation and Experiments in Microelectronics and Microsystems, EurosimE*, Wroclaw, 2013

Thermo-Mechanical Stress Investigation of Integrated SAW Strain Sensors

Jochen Hempel[1], Sohaib Anees[1],
Jürgen Wilde[2], Leonhard Reindl[1]
[1]Laboratory for Electrical Instrumentation
[2]Laboratory for Assembly and Packaging Technology
Department of Microsystems Engineering IMTEK
University of Freiburg, Germany
Email: hempel@imtek.de

Abstract

This paper presents investigations of thermo-mechanical stress generated due to the integration process of Surface Acoustic Wave (SAW) strain sensors. A 3D finite element (FE) model, based on visco-elastic material measurements, is developed for thermo-mechanical stress computation. The simulation results are compared with experiments. Therefore, SAW strain sensors were mounted, the sensor response and the sensor deformation measured. The deviation between the simulated and measured sensor chip deflection is $\leq 14.4\%$ for the full measurement range. Simulated thermo-mechanical stresses were used for the frequency shift computation of the SAW sensor device. The calculated frequency shift and the performed deformation measurement verified the correctness of the FE model.

1. Introduction

Thermo-mechanical stress occurs during the assembly and packaging process of MEMS devices due to the mismatch of the material properties [1]–[3]. For SAW sensors, this effect has been addressed in [4], [5]. Residual stress effects on integrated SAW strain sensors have been investigated in detail in [6]. Depending on utilized materials and the configuration of the sensor chip integration, there is a significant frequency shift of the SAW device due to the residual stress [7].

If the effect of thermo-mechanical stress due to sensor integration can be predicted, the performance of the SAW sensor can be matched in the design phase of the sensor for operating in the ISM band.

In this work, a 3D finite element (FE) simulation model was developed for thermo-mechanical stress computation. Therefore, the thermal load during the sensor assembly process was modeled. The material model of the bond material is based on *CTE* measurements with a dilatometer. The results of the simulation were compared with measurements. For this, three SAW strain sensors were bonded with utilized one-component particle filled epoxy adhesive on a steel carrier. The effect of thermo-mechanical stress was measured with white light interferometry. Additionally, the SAW strain sensor response was interrogated with a network analyzer before and after the bonding process. For the verification of the simulation model, the computed thermo-mechanical stresses were used for the frequency shift calculation.

2. FE Simulation

For the thermo-mechanical stress simulation, the simulation software ANSYS® was used. The cooling curve of the sensor chip assembly process from process temperature T_p

to room temperature was recreated and the resulting residual stress was computed. For this, the material parameters as the coefficient of thermal expansion *CTE*, the Young's Modulus E, the glass transition temperature T_g, and the process temperatures T_p for the bonding process were included. The anisotropy and the temperature dependent material properties of the SAW strain sensors chip were modeled as presented in [8]. The steel carrier and the bond material were modeled with isotropic material properties. At room temperature the Young's Modulus E of the steel carrier and the bond material is 210 GPa and 5 GPa, respectively. The coefficient of thermal expansion *CTE* and the Young's Modulus E of the steel carrier are linearly approximated up to the process temperature T_p of 150°C with -46 MPa/°C and 6.6 ppb/°C², respectively. The *CTE* of the bond adhesive was measured below T_g with a dilatometer, see Figure 1. The measurement has been extrapolated up to the process temperature T_p as presented in [9]. Since the glass transition temperature T_g of the adhesive is at 134°C, the dilatometer measurements were carried out up to 105°C. The Young's Modulus E of the bond material in Figure 1 was modeled as reported for particle filled epoxy composites in [10].

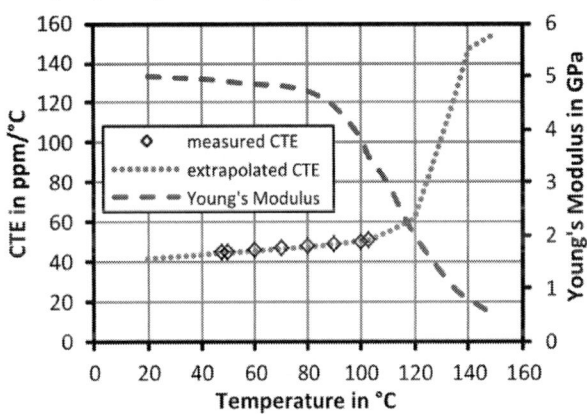

Figure 1: Implemented material data of the bond adhesive in dependence of the temperature for the thermo-mechanical stress simulation.

Figure 2 depicts the 3D FE model for the thermo-mechanical stress simulation. The bond layer thickness of the polymer adhesive was simulated with 100 μm, based on thickness measurements. The SAW sensor chip was modeled with a thickness of 350 μm, a width of 5.4 mm, and a length of 9 mm. The steel carrier was defined with a thickness of 1.2 mm, a width of 8 mm, and a length of 40 mm. The position of the sensitive sensor element, the SAW resonator area, is shown in Figure 2.

978-1-4799-4789-8/14 $31.00 © 2014 IEEE   43

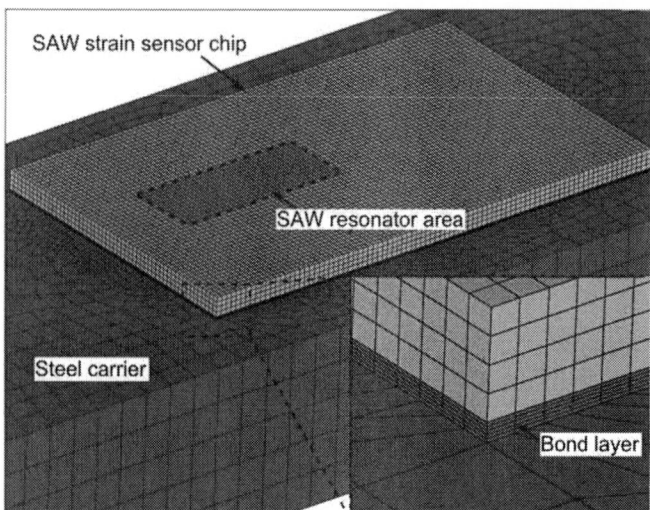

Figure 2: 3D FE model for thermo-mechanical stress simulation.

2.1 Thermo-mechanical Stress Computation

The thermo-mechanical stress in the area of the SAW resonator has been computed by averaging the stress components. The longitudinal and transversal stress T_l and T_t in the plane of the SAW resonator were considered. The average stress, longitudinal (T_l) and transversal (T_t) orientated to the SAW resonator is 6.06 ± 1.35 MPa and -6.39 ± 0.11 MPa, respectively. The significant standard deviation of the longitudinal stress T_l originates from the strong deformation of the sensor chip in longitudinal direction, see Figure 3. The maximum longitudinal stress T_l occurs close to the vertex of the parabolic sensor chip deformation. The transversal stress T_t in the area of the SAW resonator is approximately constant.

2.2 Simulated SAW Sensor Chip Deformation

The deformation normal to the sensor chip surface was extracted along a path, crossing the SAW resonator area, as depicted in Figure 4. Figure 3 depicts the simulated deformation of the sensor chip, normal to the sensor surface. The maximum deflection is 1.12 µm after cooling down from the process temperature T_P to room temperature.

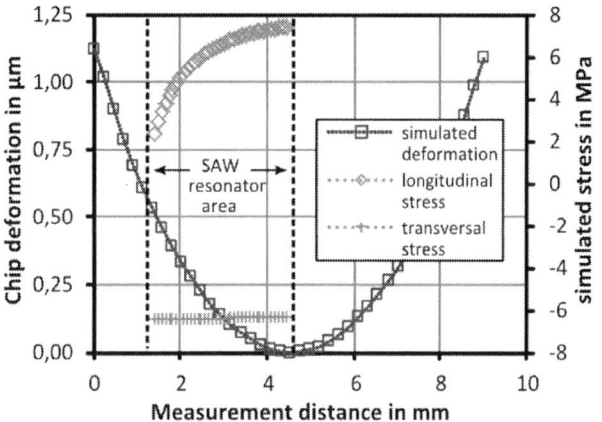

Figure 3: Simulated sensor chip deformation after assembly process and corresponding stresses in the SAW resonator area.

3. Experiments and Results

For the experimental evaluation of the thermo-mechanical stress, the deformation of the SAW sensor chip (normal to sensor chip surface) was measured with white light interferometry. Therefore, three SAW strain sensor chips were assembled with the one-component particle filled epoxy adhesive. The deformation of the sensor chip was extracted along a measurement path, as shown in Figure 4.

Before and after the mounting process of the SAW sensors, the sensor response was measured, using a probe station and a network analyzer.

3.1 SAW Sensor Chip Deformation Measurements

A white light interferometry measurement of an assembled SAW sensor chip is displayed in Figure 4. The measurement path utilized for sensor chip deformation investigation and the SAW resonator area are highlighted.

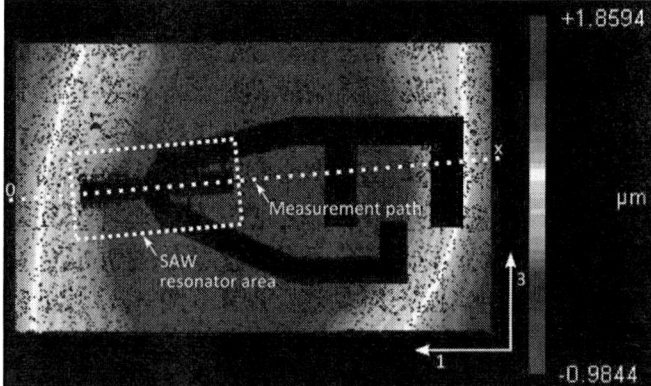

Figure 4: White light interferometry measurement of a SAW sensor chip after assembly, with marked SAW resonator area and the deformation measurement path.

The measured sensor chip deformation along the measurement path is depicted in Figure 5. The measurements were repeated three times. For better visualization of the measurement graphs, Figure 5 and 6 are plotted for every 20th measurement value. The maximum sensor chip deformation was extracted from the data fit of the measurements. A deflection of 1.68 ± 0.17 µm was obtained, close to the SAW resonator area.

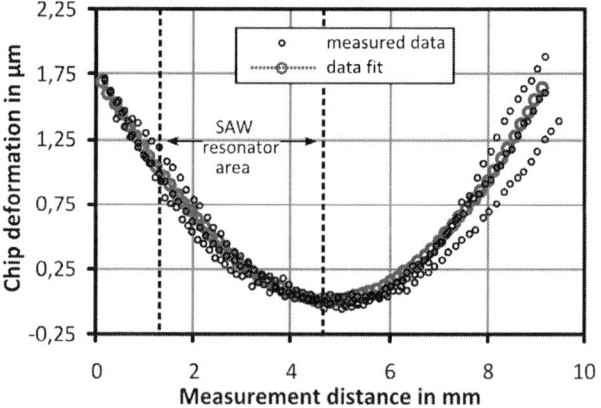

Figure 5: Sensor chip deformation after assembly process, measured with white light interferometry.

978-1-4799-4789-8/14 $31.00 © 2014 IEEE

As presented in [6], the sensor chip deformation consists of a superpositioning effect of the initial sensor chip deformation and the chip deformation due to the assembly process. The total deformation comprises of the initial sensor deformation after the chip dicing process and the deformation due to the mounting process of the SAW sensor chip. Figure 6 shows the initial sensor chip deformation measurement before the assembly process. The measurement was carried out along the measurement path depicted in Figure 4. An average initial chip deformation of 0.42 ± 0.07 µm was measured close to the SAW resonator area. The sensor chip deformation due to the integration process of the sensor chip is 1.26 ± 0.18 µm. It was obtained from the difference of the total chip deformation after the mounting process and the initial sensor chip deformation. The resulting sensor chip deformation of the SAW sensor due to integration process is shown in Figure 7.

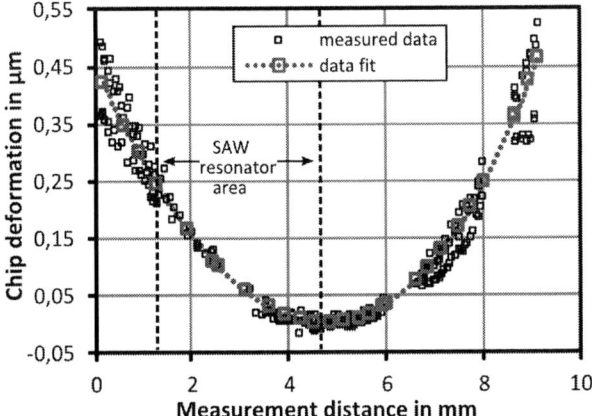

Figure 6: Initial sensor chip deformation before the assembly process, measured with white light interferometry.

Figure 7 depicts the simulated and mean measured sensor chip deformation after the bonding process. The maximum deviation between the measurement and the simulation is ≤ 14.4% over the entire measurement section.

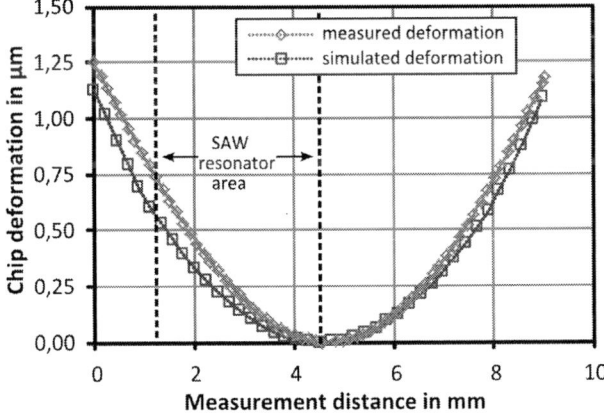

Figure 7: Comparison of the simulated and measured SAW sensor chip deformation due to the integration process.

3.2 SAW Strain Sensor Response Measurement

Utilized SAW strain sensors were interrogated before and after the integration process on the steel carrier at room temperature. For this, a probe station and a network analyzer were used for reproducible and reliable sensor read out.

The thermo-mechanical stress due to the bonding process causes a frequency shift Δf of the SAW strain sensor. The relative frequency shift $\Delta f / f_0$ was considered, to be independent of the load free resonance frequency f_0 of the SAW devices. The thermo-mechanical stress due to the integration process with the utilized one-component particle filled epoxy adhesive created a relative frequency shift $\Delta f / f_0$ of -85.9 ± 6.2 ppm.

3.3 SAW Strain Sensor Response Calculation

The relative frequency shift $\Delta f / f_0$ of a loaded SAW strain sensor can be in general computed, using equation (1) [11]. Provided that the occurring stresses and strains are homogeneously distributed and the perturbations are small. Here, α_{ij} represents the stress-sensitivity coefficients and $\overline{T_{ij}}$ the quasi-static stress tensor [11], [12].

$$\frac{\Delta f}{f_0} = \sum_{i,j=1}^{3} \alpha_{ij} \, \overline{T_{ij}} \tag{1}$$

If the planar stresses on the sensor surface are considered and the shear stresses $T_{13} = T_{31}$ are neglected, equation (1) becomes reduced and results in equation (2).

$$\frac{\Delta f}{f_0} = \alpha_{11} T_{11} + \alpha_{33} \, T_{33} \tag{2}$$

Using the measured stress sensitivities α_{11}, α_{33} published in [13] and the simulated stress components T_l and T_t the relative frequency shift $\Delta f / f_0$ due to the integration process of the SAW sensor chip can be computed with equation (2). Whereby, the longitudinal stress component T_l complies T_{11} and the transversal stress component T_t corresponds T_{33}. The same correlation applies to the stress sensitivities $\alpha_l = \alpha_{11}$ and $\alpha_t = \alpha_{33}$. Since, the measured stress sensitivities and the simulated stress components are subjected to variations; the relative frequency shift $\Delta f / f_0$ is calculated with the mean values and the error propagation approach for the maximum total error. A relative frequency shift $\Delta f / f_0$ of -85.7 ± 17.4 ppm has been computed.

4. Discussion and Conclusion

The implemented FE simulations and the measurements of the SAW sensor chip deformation due to the mounting process are in good agreement. The simulation model has been verified with the experimental results. Using the superpositioning approach, a maximum SAW sensor chip deformation close to the SAW resonator of 1.26 ± 0.18 µm was measured. At the same position a deflection of 1.12 µm was simulated, which correlates to a deviation of approximately 11%. The maximum deviation over the full measurement section between the simulation and the measurement is ≤ 14.4%, see Figure 7. From the simulation the resulting stress components due to the integration process were extracted. Longitudinal stress T_l of 6.06 ± 1.35 MPa and a transversal stress T_t of -6.39 ± 0.11 MPa were found. The significant variation of the longitudinal stress T_l originates from the strong deformation of the sensor chip in longitudinal direction, where the SAW resonator is located. The sensor chip deformation transversely orientated to the SAW resonator is approximately constant (Figure 4). Therefore, the transversal stress component T_t is approximately constant (Figure 3).

The relative frequency shift $\Delta f/f_0$ of the utilized SAW strain sensors is -85.9 ± 6.2 ppm after the integration process on a steel carrier. Due to the negative frequency shift, the SAW sensor device is apparently loaded with tensile stress, although the sensor surface with the SAW resonator is compressed.

The computed relative frequency shift is -85.7 ± 17.4 ppm, using the extracted stress components (T_l, T_t) of the simulation and the chip-level stress sensitivities presented in [13]. The maximum total error is significant. It is mainly affected by the high variation (~22%) of the longitudinal stress component T_l.

However, the presented simulation model in this work can be used for the first evaluation of the thermo-mechanical stresses due to the integration process of a SAW sensor device. This has been demonstrated with the performed deformation and sensor response measurements of the utilized SAW strain sensor.

Acknowledgments

The authors would like to thank the company SENSeOR GmbH for cooperation and Dr. A. Yousaf for fruitful discussions.

References

[1] P.-H. Tsao and A. S. Voloshin, „Manufacturing stresses in the die due to the die-attach process", *IEEE Transactions on Components, Packaging, and Manufacturing Technology, Part A*, vol. 18, no. 1, pp. 201–205, 1995.

[2] S. S. Walwadkar, P. W. Farrell, L. E. Felton, and J. Cho, „Effect of die-attach adhesives on the stress evolution in MEMS packaging", in *2003 International Symposium on Microelectronics*, vol. 5288, Washington: International Microelectronics& Packaging Society, pp. 847–852, 2003.

[3] X. Zhang, S. Park, and M. W. Judy, „Accurate Assessment of Packaging Stress Effects on MEMS Sensors by Measurement and Sensor-Package Interaction Simulations", *Journal of Microelectromechanical Systems*, vol. 16, no. 3, pp. 639–649, 2007.

[4] R. L. V. Kalinin, „Development of a calibration procedure for contactless torque and temperature sensors based on SAW resonators", pp. 1865 – 1868, 2008.

[5] B. Donohoe, *The Development of a Surface Acoustic Wave Strain Sensor*. Trinity College, 2011.

[6] J. Hempel, J. Wilde, and L. M. Reindl, „Effects of Residual Stress on Assembled SAW Strain Sensors", in *Proceedings of PIERS 2013*, Taipei, Taiwan, pp. 286–290, 2013.

[7] B. Donohoe, D. Geraghty, G. E. O'Donnell, and R. Stoney, „Packaging Considerations for a Surface Acoustic Wave Strain Sensor", *IEEE Sensors Journal*, vol. 12, no. 5, pp. 922 –925, 2012.

[8] E. Zukowski, T. Fellner, J. Wilde, and M. Berndt, „Parameter optimization of torque wireless sensors based on surface acoustic waves (SAW)", in *2012 13th International Conference on Thermal, Mechanical and Multi-Physics Simulation and Experiments in Microelectronics and Microsystems (EuroSimE)*, pp. 1-6, 2012.

[9] R. Dudek, D. Vogel, and B. Michel, „Mechanical failure in COB-technology using glob-top encapsulation", *IEEE Transactions on Components, Packaging, and Manufacturing Technology, Part C*, vol. 19, no. 4, pp. 232–240, 1996.

[10] T. B. Lewis and L. E. Nielsen, „Dynamic mechanical properties of particulate-filled composites", *Journal of Applied Polymer Science*, vol. 14, no. 6, pp. 1449–1471, 1970.

[11] E. Bigler, D. Hauden, and G. Theobald, „Stress-sensitivity mapping for surface acoustic waves on quartz", *IEEE Transactions on Ultrasonics, Ferroelectrics and Frequency Control*, vol. 36, no. 1, pp. 57 –62, 1989.

[12] S. Ballandras and E. Bigler, „Surface-acoustic-wave devices with low sensitivity to mechanical and thermoelastic stresses.", *Journal of Applied Physics*, vol. 72, no. 8, pp. 3272–3281, 1992.

[13] J. Hempel, D. Finke, M. Steiert, R. Zeiser, M. Berndt, J. Wilde, and L. Reindl, „SAW strain sensors - high precision strain sensitivity investigation on chip-level", in *Ultrasonics Symposium (IUS), 2013 IEEE International*, pp. 1942–1945, 2013.

Software Reliability and Its Interaction with Hardware Reliability

W.D. van Driel[1,3], M. Schuld[2], R. Wijgers[2], W.E.J. van Kooten[1]
[1]Philips Lighting, High Tech Campus, Eindhoven, The Netherlands
[2]CQM, Eindhoven, The Netherlands
[3]Delft University of Technology, DIMES-ECTM, The Netherlands
Email: willem.van.driel@philips.com

Abstract

Software reliability models can provide quantitative measures of the reliability of software systems during development processes. Research activities in software reliability engineering are conducted over the past four decades, and many software reliability models are proposed. In this paper we will present our results in predicting the reliability of software and how that relates to the reliability of hardware.

1. Introduction

Today, almost everyone in the world is directly or indirectly affected by electronic systems [1]. They are used in diverse areas for various applications including air traffic control, nuclear reactors, aircraft, industrial process control, automotive mechanical and safety control, and hospital health care, affecting many millions of people. As the functionality of computer operations becomes more essential and yet more complicated and critical applications increase in size and complexity, there is a great need for looking at ways to quantify and predict the reliability of such systems in various complex operating environments [1]. A (complex) system is a set of interacting or interdependent components forming an integrated whole. This implicates that two components together already form a system. When the number of components and their interactions hugely increase, so-called large or complex systems are formed. The types of components, their quantities, their qualities and the manner in which they are arranged within the system have a direct effect on the system's reliability. The commonly used description for system reliability [2] is given as:

The probability that a system, including all hardware, firm-ware, software, and their interactions will satisfactorily perform the task for which it was designed or intended, for a specified time and in a specified environment.

From a system reliability point of the view, the challenge is to master the reliability of all these components. Figure 1 shows two possible lighting applications where we see a large penetration of software enabled controls.

Figure 1: Lighting applications with software enabled controls.

Software failures are a primary cause of product reliability problems. Unlike hardware failures, software failures are not caused by faulty components, wear-out or physical environment stresses such as temperature and vibration. Software failures are caused by latent software defects that were introduced into the software as it was being developed, but were not detected and removed before the software was released to customers. The best approach to achieving higher software reliability is to reduce the likelihood that latent defects are in released software. Unfortunately, even with the most highly-skilled software engineers following industry best practices, the introduction of software defects is inevitable due to the inherent complexities of the software functionality and its execution environment. As a system is built out of sub-systems, which each consist out of modules, the system testing approach will focus on verification of the system and its interfaces to requirements. Often the V-model is used for product development. This model demonstrates the relationships between each phase of the development life cycle and its associated phase of testing. The horizontal and vertical axes represent time or project completeness (left-to-right) and level of abstraction, as depicted in Figure 2. In this approach, the software can be considered as a component as well and, thus, reliability of this component needs to be considered. In this paper we will present our results in predicting the reliability of software and how that relates to the reliability of hardware. We will present a use case in which we will demonstrate the use of these so-called software reliability models.

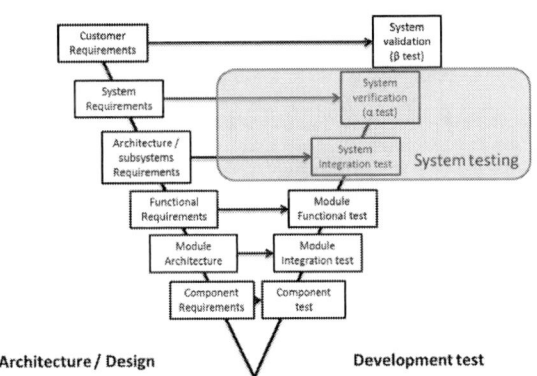

Figure 2: V-model approach.

2. Software versus Hardware Reliability

Software reliability or robustness is the probability of failure-free software operation for a specified period of

978-1-4799-4789-8/14 $31.00 © 2014 IEEE

time in a specified environment. Software failures may be due to errors, ambiguities, oversights or misinterpretation of the specification that the software is supposed to satisfy, carelessness or incompetence in writing code, inadequate testing, incorrect or unexpected usage of the software or other unforeseen problems. Software reliability is not a function of time. There is not something as software 'wear-out'; software will not change in time. Software reliability relates to errors that are induced by circumstances or contexts that are unforeseen/not addressed in the design phase. Typical questions that need to be addressed are:

- How many errors are left?
- What is the probability of no failures in a given time period?
- What is the expected time until the next failure?
- What is the expected number of errors in a given time period?

The development of hardware reliability theory has a long history and got established to improve hardware reliability greatly [4, 5]. The history of reliability as we know it now goes back to the 1950s, when electronics played a major role for the first time. It may seem strange today but at that time there was considerable resistance to recognizing the stochastic nature of the time to failure, and hence reliability. Software systems do not degrade over time unless modified. There are many differences between the reliability and testing concepts and techniques of hardware and software. Therefore, a comparison of software and hardware reliability would be useful in developing software reliability modeling. Table 1 shows the differences and similarities between the two.

Table 1: Commonalities and differences between soft- and hardware reliability [1].

Hardware reliability	Software reliability
Failure rate has a bathtub curve	Without considering program evolution, failure rate is statistically non-increasing
Material deterioration can cause failures even though the system is not used	Failures never occur if the software is not used
Failure data are fitted to some distributions. The selection of the underlying distribution is based on the analysis of failure data and experiences. Emphasis is placed on analyzing failure data	Most models are analytically derived from assumptions. Emphasis is on developing the model, the interpretation of the model assumptions, and the physical meaning of the parameters
Failures are caused by material deterioration, random failures, design errors, misuse, and environment	Failures are caused by incorrect logic, incorrect statements, or incorrect input data. This is similar to design errors of a complex hardware system
Hardware reliability can be improved by better design, better material, applying redundancy and accelerated life testing	Software reliability can be improved by increasing the testing effort and by correcting detected faults.
Hardware repairs restore the original condition	Software repairs establish a new piece of software
Hardware failures are usually preceded by warnings	Software failures are rarely preceded by warnings

3. Software Reliability Testing and Modeling

Software reliability engineering [6, 7] is a field of software development that relates to testing and modelling the software ability to function (or not), given environmental conditions, for a particular amount of time. No method of development can guarantee totally reliable software. A set of statistical modelling techniques are required that:

- Enable the achieved reliability to be assessed or predicted
- Is based on observation of system failures during system testing and operational use

It uses general reliability theory, but is much more than that:

o How reliable is the program/component now?
o Based on the current reliability of the software: can we accept it or should we reject it?
o Based on the current reliability of the hard-software system: can we stop testing and start shipping?
o How reliable will the system be, if we continue testing for some time?
o When will the reliability objective be achieved?
o How many failures will occur in the field (and when)?

Seven distinct steps can be marked in the software reliability engineering process (see also Figure 3):

1. Define reliability objective: express failure intensity as failures per natural unit (such as failures / Kpages printed, failures / Ktransactions, failures / Kcalls, etc). Per severity level, the objective may be established by contractual, warranty, or regulatory requirements.

2. Expected system usage: a complete set of operations with their probabilities of occurrences that represent 'field conditions':

- Gives information on how users will employ the product we are building so that we can focus on our development and test resources.
- Model how users will employ the software: environment, type of installation, distribution of inputs over input space.
- Indicate the relative usage of program modules. You can construct this table from the operational profile and the operation-usage matrix. The operation-module matrix indicates which operations use which modules.
- According to the usage model, test cases are selected randomly.

3. Prepare test cases
- Define test scheme, define what to test, for example:
  o Functionality test of each function
  o User interface tests
  o File input / output tests
  o Determine how many tests, for example
  o Create tests

4. Execute test - collect failure data and severity levels:

- This form of testing has the advantage of testing more intensively the system functions that will be used the most.
- Hence we differentiate testing that aims at finding defects (verification, α-tests) and testing whose purpose is reliability assessment (validation, β-test).

5. Reliability Growth Modeling, model-specific assumptions:
   o The system test reflects the intended usage of the system.
   o The failures are mutually independent.
   o The number of failures detected at any time is proportional to the current number of faults in the program.
   o Each time a failure occurs, the fault which caused it is immediately removed, and no new faults are introduced.
   o The (cumulative) number of failures by time follows a Poisson Process (NHPP) or General Order Statistics (GOS) model.
   - Data Requirements
     o Actual times when failures occurred or failure time intervals (test execution time), and severity levels (critical / major / average / minor).

6. Projection to field life:
   - Once the model parameters are estimated, it can be used to predict failure intensity in the future, not only to estimate its current value. From this, we can plan how much additional testing is likely to be needed.
   - The model allows for the realistic situation where fault correction is not perfect (infinite number of failures at infinite time).
   - When faults stop being corrected, the model reduces to a homogeneous Poisson process.

7. Monitor field performance
   - Logging and monitoring schemes are required to track the field performance of the system / software.

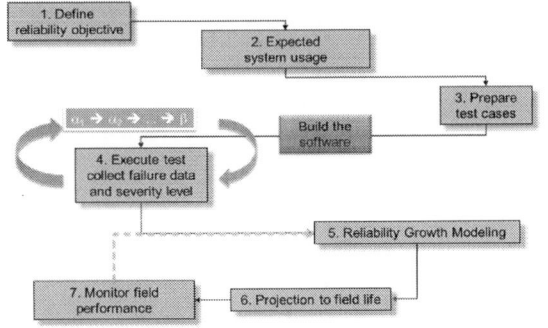

Figure 3: Software Reliability Engineering Process.

Software reliability growth models (SRGMs) enable the project leader / test manager to estimate the software reliability and the number of errors remaining in the software. This can help him to decide whether or not the code is suitable for customer use and how much more testing is required if it is not ready yet to release the software. It also provides an estimate of the number of faults that users will encounter when operating the software. These estimates also help to define the appropriate levels of support that will be required for fault correction after the software has been released.

During the testing phase of software development, faults are removed after they are detected. This reduces the number of total faults in the software and the fault-detection rate should decrease as more code is covered. In other words, the length of intervals between fault discoveries should increase. When the fault-detection rate reaches an acceptably low level, the software is deemed suitable to release to customers.

Software reliability growth models (SRGM) are mathematical functions that describe fault-detection and removal phenomenon. Some realistic issues such as imperfect debugging and learning phenomenon of software developers have been studied and incorporated in software reliability assessment. Among all SRGMs, a large class of stochastic reliability models is based on a *Non-Homogeneous Poisson Process*. These models are known as NHPP reliability models and have been widely used to track reliability improvement during software testing. Another popular class is the class of *General Order Statistics*, or GOS models. These two classes will be described in next sections.

Background NHPP [8, 9]

Software testing process has been widely modelled as a failure counting process. A counting process $\{N(t), t \geq 0\}$ is said to be a non-homogeneous Poisson process with intensity function $\lambda(t)$, if $N(t)$ follows a Poisson distribution with mean value function m(t), i.e.,

$$P\{N(t) = k\} = \frac{m(t)^k}{k!} e^{-m(t)}, k = 0,1,2,...$$

The mean value function m(t), which is the expected number of failures experienced up to a certain time t, can be expressed in terms of failure rate of the program, i.e.,

$$m(t) = \int_0^t \lambda(s)ds,$$

where $\lambda(s)$ is the failure intensity function. Software reliability R(x|t) is defined as the probability that no software failure is detected in the time interval $\{t, t+x\}$, given that the last failure occurred at testing time t (t ≥ 0, x >0). That is, $R(x \mid t) = e^{-(m(t+x)-m(t))}$

For special cases, when t=0 then R(x|0)=e$^{-m(x)}$, and t=∞ then R(x|∞)=1.

Background GOS [10]

In this section we describe an important class of software reliability growth models known as General Order Statistics (GOS). The main assumption for this class of models is that the times between failures of a software system can be defined as the differences between two consecutive order statistics. It is assumed that the initial number of failures, denoted by N, is unknown but fixed and finite. Thus, for any $n \leq N$, we can interpret the first n failure times $T_1 < T_2 < ... < T_n$ as the first n order statistics. The times between failures are defined as the difference of two order statistics, i.e., $X_i = T_i - T_{i-1}$ for all $i \geq 1$. In general, the times between failures of GOS models are not independent nor identically distributed. It can be proven that the random variable N(t) follows a binomial distribution. For that reason the class of GOS models is often called the class of binomial distributions. Different GOS models arise when one considers different distributions for the failure times. The most well-known GOS model is based on the exponential distribution. Many existing software reliability models are variants or extensions of this basic model. Other popular GOS models consider the Weibull and Pareto distribution for the order statistics.

Data analysis [11, 12]

It is highly recommendable to carefully look at the data before starting any kind of statistical analysis. For example, it is possible to gain some understanding about the nature of the process being studied simply by plotting the data as a function of time. Figure 4 shows a plot of the failure times against the cumulative number of observed failures.

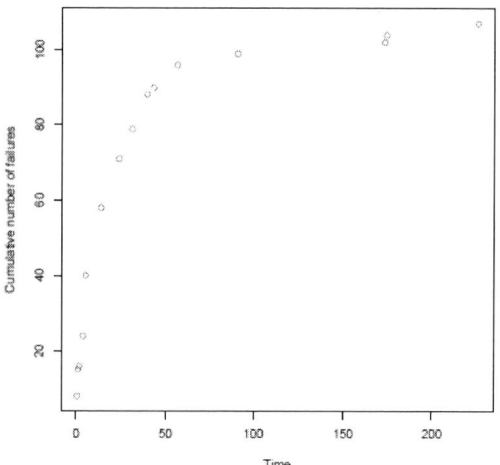

Figure 4: Failure times vs. cumulative number of observed failures

Note that a concave plot indicates that software becomes more reliable during testing, due to the fact that failures are repaired, so that more effort is required to find future failures. With this simple step we may detect that failure times follow certain patterns that may reveal some trend associated to a growth or a decrease in reliability. The well known Laplace and the Military Handbook (MIL-HDBK-189) tests can be used to test reliability growth.

The Military Handbook Test

This test performs well at finding significance when the choice is between no trend and a NHPP Power Law (Duane) model. In other words, if the failure process follows the Power Law, this test will generally do better than any other test in terms of finding significance.
Suppose we have r failure times $T_1, T_2, T_3, ...,T_r$ with the observation period ending at time $T_{end} > T_r$. Calculate

$$\chi_{2r}^2 = 2\sum_{i=1}^{r} \ln\left(\frac{T_{end}}{T_i}\right)$$ and compare this to percentiles of

the chi-square distribution with 2r degrees of freedom. For a one-sided improvement test, reject no trend (or HPP) in favour of an improvement trend if the chi square value is beyond the 90 (or 95, or 99) percentile.

The Laplace Test

This test performs well at finding significance when the choice is between no trend and a NHPP exponential model. In other words, if the data come from a failure process following the Exponential Law, this test will generally do better than any test in terms of finding significance. As before, we have r failure times $T_1, T_2, T_3, ...,T_r$ with the observation period ending at time $T_{end} > T_r$.

Calculate $z = \dfrac{\sqrt{12r}\sum_{i=1}^{r}\left(T_i - \dfrac{T_{end}}{2}\right)}{r T_{end}}$ and compare this

to percentiles of the standard normal distribution. The interpretation of the test statistic is the following: for small values of the test statistic the null hypothesis of HPP is rejected in favour of reliability growth.

Model type selection [12, 13]

The main problem that we find when trying to select a suitable model for a specific problem is that there are no general rules to select a model. Although it is possible to find a large variety of lists of assumptions and data requirements for software reliability models in the literature, this is often far from facilitating model selection. For example, there is no universal agreement in the literature on the list of assumptions for certain well-known models. Therefore we propose the following procedure:

1. Select the assumptions which are relevant to the testing data at hand, and make a subset of applicable models.
2. Fit the data to several growth models, and take the one that best fits the data. There are two criteria that are often used for comparison of goodness-of-fit:

- Mean Square Error (MSE) = $\dfrac{\sum_{i=1}^{n}\left(m(t_i) - y_i\right)^2}{n-k}$,

where y_i is the total number of failures at time t_i, $m(t_i)$ is the estimated cumulative number of failures at time

t_i for i=1,2,...,n, and k is the number parameters in the model.

- Akaike's Information Criterion (AIC) = -2log(max. likelihood value) + 2k. The AIC measures the ability of a model to maximize the likelihood function that is directly related to the degrees of freedom during fitting, increasing the number of parameters will usually result in a better fit. AIC criterion takes the degree of freedom into consideration by assigning a model with more parameters a larger penalty.

Estimation of parameters [14 – 17]

Once the analytical solution for m(t) is known for a given model, the parameters need to be determined. Estimation is achieved by applying the technique of Maximum Likelihood Estimation (MLE). This is the most widely used estimation technique. In many cases, the maximum likelihood estimators are consistent and asymptotically normally distributed as the sample size increases. We refer to [1, 14 - 17] for further details.

Goel-Okumoto

This is the most well-known NHPP model. Due to the important role that this model has played on the software reliability modelling history, it is often called "the" NHPP model. It is based on the following assumptions:

- All failures in a program are mutually independent from the failure detection point of view.
- The number of failures detected at any time is proportional to the current number of failures in a program.
- The isolated failures are removed prior to future test occasions.
- Each time a software failure occurs, the software error which caused it is immediately removed, and no new errors are introduced.

The mean-value function is given by:

$$m(t) = a(1 - e^{-bt}),$$ for all t ≥ 0, where a > 0 and b > 0. The parameter a is the expected number of failures to be eventually detected while b is the rate at which each individual failure will be detected during testing.

Musa-Okumoto

Musa-Okumoto observed that the reduction in failure rate resulting from repair action following early failures are often greater because they tend to the most frequently occurring once. This property has been taken into account in the model. The mean value function is given by:

$$m(t) = a \cdot \ln(1 + bt), a > 0, b > 0$$

where a is the expected total number of failures to be detected, and b is the detection rate.

Yamada S-shaped

The so-called S-shaped model was presented in Yamada and Osaki (1984). It receives the name S-shaped because the curve of the mean-value function is often S-shaped (in comparison with the exponential-shaped mean-value function of the Goel-Okumoto model). The mean-value function is given by:

$$m(t) = a(1 - (1 + bt)e^{-bt}),$$ for all t ≥ 0, where a > 0 and b > 0. The parameters of the model have the same interpretation as in the Goel-Okumoto model.

NHPP Imperfect Debugging models

Many existing models describe perfect debugging in previous section, that is, a(t) = a and where the error detection rate b(t) function is time-dependent. In this section, we discuss several software reliability models with imperfect debugging processes. The NHPP imperfect debugging model is based on the following assumptions:

- When detected errors are removed, it is possible to introduce new errors.
- The probability of finding an error in a program is proportional to the number of remaining errors in the program.

There are 2 types of the Yamada imperfect debugging model:

1. $$m(t) = \frac{ab}{b + \alpha}(e^{\alpha t} - e^{-bt}),$$ for all t ≥ 0, where a > 0, b > 0, and $\alpha \geq 0$.

2. $$m(t) = a(1 - e^{-bt})\left(1 - \frac{\alpha}{b}\right) + a\alpha t,$$ for all t ≥ 0,

   where a > 0, b > 0, and $\alpha \geq 0$.

In the same analogy the Pham-Nordmann-Zhang (PNZ) model is defined:

$$m(t) = \frac{a}{1 + \beta e^{-bt}}\left((1 - e^{-bt})(1 - \alpha/b) + \alpha at\right),$$

for all t ≥ 0, where a > 0, b > 0, α > 0, and β > 0. PNZ is a so-called imperfect debugging S-shaped model, where a is the expected total number of failures, α is the constant failure introduction rate, b and β are constants.

GOS model: Jelinski-Moranda

The assumptions in this model include the following:

- The program contains N initial failures which is an unknown but fixed.
- Each failure in the program is independent and equally likely to cause a failure during a test.
- Time intervals between occurrences of failure are independent of each other.
- Whenever a failure occurs, a corresponding failure is removed with certainty.
- The failure that causes a failure is assumed to be instantaneously removed, and no new failures are inserted during the removal of the detected failure.
- The software failure rate during a failure interval is constant and is proportional to the number of failures remaining in the program.

978-1-4799-4789-8/14 $31.00 © 2014 IEEE

The program failure rate at the ith failure interval is given by:

$$\lambda(t_i) = \phi(N - (i - 1)), i = 1, 2, \ldots, \quad N, \quad \phi \text{ is a}$$

proportional constant, i.e. the contribution of any one failure makes to the overall software. Hence, the software reliability function equals:

$$R(t_i) = \exp(-\phi(N - i + 1)t_i)$$

4. Use Case: Visual Basic software

A visual basic application developed to evaluate product designs is taken as our use case. The features of the software are:

- About 13500 codes lines
- Divided over 29 modules
- Dedicate GUI that interacts with the users

As a rule of thumb, research study has shown that professional programmers implement on average six software defects for every 1000 lines of code (LOC) written [1]. Following this number, then the case software would encompass 6*13500 / 1000 = 81 defects or failures.

In our case, the software itself is developed according to the flowchart given in Figure 3. In total, 3 α-test phases are executed and one β-test phase. Alpha testing is performed by experienced users at the developers' site. Alpha testing is a form of internal acceptance testing, before the software goes to beta testing. Beta testing comes after alpha testing and can be considered as a form of external user acceptance testing. Versions of the software, known as beta versions, are released to a limited audience outside of the programming team. After the β-test the software was released to groups of people so that further testing can ensure the product has few failures. The reliability objective of the software was set at 1 critical or major failure per 100 user sessions. The expected usage of the software was derived by determining both the users and their typical behavior with the software. Based on this operational profile, we developed more than 100 functionality tests for the first α phase. This test plan comprises the following consecutive parts:

1. Functionality & load tests
2. Security tests
3. User tests

The way of working for both the α- and β-tests is:

- Run the software
- Record the starting time
- Report any failure or event that occurs, including its level of severity
- Record the test time that was needed to detect the failure or event
- Close the software
- Repeat the above

For the testing itself, a scoring form is used. It is very important that this form is filled in a good manner and all details are very precisely logged. Screen shots of the

failures need to be included. The ranking of the failures is done a-priori by the tester and evaluated afterwards. In this evaluation, both the α & β testers and the software developer need to agree on the ranking. As given in the test release plan, four different ranks are used:

1. Critical
2. Major
3. Average
4. Minor

The explanation of these codes is given in Table 2. A typical response from a tester is shown in Figure 5. An overview of all the results is given in Table 3.

Table 2: Severity coding for testing the software.

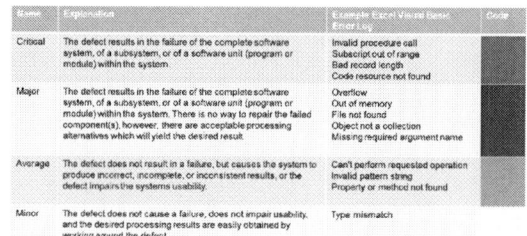

Figure 5: Screenshot example of the test results.

Table 3: Found failures and corrections during the different test phases of the software development process.

Test Phase	Number of test cases	# Critical	# Major	# Average	# Minor
α-2-phase	100	8	4	4	5
Solved		-8	-3		
α-3-phase	100	5	2	2	4
Solved		-2		-2	-4
β-phase	200	6	8	2	3
Solved		-8	-10	-2	-8
Release	300	1	2		1

New features were added after the α-2 and α-3 phases.

Based on the time series of identified failures during the α-2, α-3 and β tests, see Table 3, we fitted software models as described in the previous sections. At the α-1 phase, however, the software was not considered to be

978-1-4799-4789-8/14 $31.00 © 2014 IEEE   52

sufficiently mature for testing. Also, all found failures as shown in Table 3 are newly discovered. Software reliability can be estimated once the mean value function is determined. Based on Akaike's Information Criterion (AIC), the Mean Square Error (MSE) values and the significance of parameters, it appeared that the simple Goel-Okumoto model fits the failure rates best for three test series, see Table 4. Table 5 shows the estimated number remaining failures and their upper bounds of the 95% confidence intervals. After the β release we found another 3 failures in approximately 300 cases. This number is smaller than the 95% upper bound as determined at the end of test phase β. Figure 6 shows the failure intensity function $\partial m(t) / \partial t = b(a - m(t))$ of each phase. Note the (large) increases of the failure intensities by adding new features (with additional failures).

In the discussed α- and β- phases we discovered 57 failures (= 36 critical/majors + 21 average/minors). This number does not deviate hugely from the number of failures as predicted by the rule of thumb (81). We expect that users will discover some additional failures in the (near) future.

Table 4: Fitting results for the several software reliability models durign the consequtive testing phases.

Results α–2:

Model	MSE	AIC	remark
Goel-Okumoto	0.187	30.240	
Musa-Okumoto	0.108	30.049	parameter estimates are not significant
Yamada S-shaped	1.157	33.691	
Yamada Imperfect 1	0.059	31.128	parameter estimates are not significant
Yamada Imperfect 2	17.848	31.911	parameter estimates are not significant
Jelinski-Moranda	0.690	82.549	parameter estimates are not significant

The PNZ model could not be estimated due to lack of convergence.

Results α–3:

Model	MSE	AIC	remark
Goel-Okumoto	0.150	19.956	
Musa-Okumoto	0.133	20.031	parameter estimates are not significant
Yamada S-shaped	0.591	21.297	
Yamada Imperfect 1	0.546	20.492	parameter estimates are not significant
Yamada Imperfect 2	No convergence MLE		
Jelinski-Moranda	0.449	59.895	parameter estimates are not significant

The PNZ model could not be estimated due to lack of convergence.

Results β:

Model	MSE	AIC	remark
Goel-Okumoto	0.058	32.181	
Musa-Okumoto	0.277	32.835	parameter estimates are not significant
Yamada S-shaped	0.990	34.921	
Yamada Imperfect 1	0.074	34.091	parameter estimates are not significant
Yamada Imperfect 2	14.511	34.089	parameter estimates are not significant
Jelinski-Moranda	0.063	97.904	

The PNZ model could not be estimated due to lack of convergence.

Table 5: Remaining failures and 95% upper bound

End of test phase	Expected remaining Critical/Major failures	95% upper bound
α-2-phase	6	20
α-3-phase	8	14
β-phase	0	5

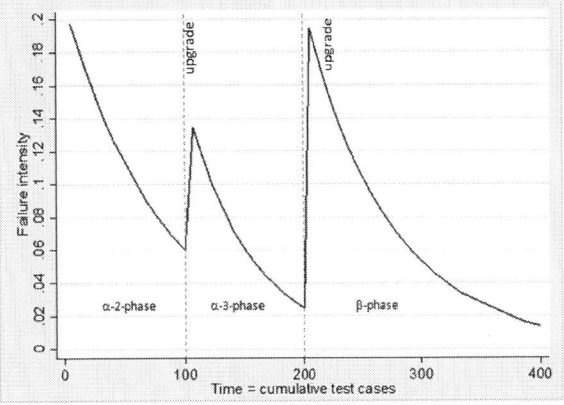

Figure 6: Failure intensity function

5. Interaction with Hardware: System Reliability

Using the V-model, one is able to consider the reliability of both the hardware and software components. By doing so, and we demonstrated in this paper that this is possible, one still needs to combine these two models in order to determine the reliability of the system. In section 2, we already described the huge differences between these two components. To make it even more complex, consider that on a system level, there is a distinct difference between the reliability and the availability of the system. As such, the definition of availability is:

The degree to which a system is operational and accessible, when required for use. Availability requirements for consumers are lower as for healthcare applications.

A system can have a low reliability but still a very high availability. But in general, a high reliability will always lead to a high availability. Both items belong to what is called the system dependability [20]. Dependability (see also Figure 6) is the ability of a system to avoid failures that are more frequent and more severe than acceptable. A dependable system is:

- having all its required properties,
- and does not show failures.

Much literature is available on the prediction of software reliability [1] and even more is available on the prediction of system reliability based on its hardware [2]. Little literature is available on the interaction of the above. In a consecutive study we will devote our investigation to this subject.

978-1-4799-4789-8/14 $31.00 © 2014 IEEE

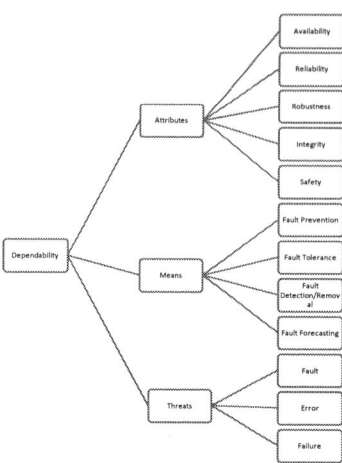

Figure 6: The dependability tree [19].

6. Conclusions

In this paper we will present our results in predicting the reliability of software and present a use case to demonstrate how software reliability models can be obtained. These software reliability models enable the developer to determine the optimal time to stop software testing and decide to release the software. Several other criteria, such as the number of remaining errors, failure rate, reliability requirements, or total system cost, may be used to determine optimal testing time.

Acknowledgments

This research work was funded by the European Union (EU) Artemis project Design, Monitoring, and Operation of Adaptive Networked Embedded Systems (DEMANES).

References

1. H. Pham, "System Software Reliability," in Springer series in Reliability Engineering, vol. 79, London, Springer, March, 2006, pp. 45-52.
2. W.D. van Driel, X.J. Fan, Solid State Lighting Reliability: Components to Systems, ISBN 978-1-4614-3066-7, 31 August 2012, Springer, 617 pages.
3. A. Avizienis, J.-C. Laprie, B. Randell, and C. Landwehr, "Basic Concepts and Taxonomy of Dependable and Secure Computing," IEEE Transactions on Dependable and Secure Computing, vol. 1, pp. 11-33, 2004.
4. James McLinn, A short history of reliability, The Journal of Reliability Information, January 2011, pp 8 – 15.
5. J.H. Saleh, K. Marais, Highlights from the early (and pre-) history of reliability engineering, Reliability Engineering and System Safety 91 (2006) 249–256.
6. M. Garzia, J. Hudepohl, W. Snipes, M. Lyu, J. Musa, C. Smidts and L. Williams, "How should software reliability engineering (SRE) be taught?," SIGSOFT Software Engineering Notes, vol. 31, no. 4, pp. 1-5, July, 2006.
7. M. Lyu, Handbook of Software Reliability Engineering, New York: McGraw-Hill and IEEE Computer Society, 1996.
8. D.R.Jeske and H.Pham, "On the Maximum Likelihood Estimates for the Goel-Okumoto Software Reliability Model," The American Statistician, vol. 55, no. 3, pp. 219-222, 2001.
9. L.Yin and K. Trivedi, "Confidence interval estimation of NHPP-based software reliability models," Proceedings of 10th International Symposium on Software Reliability Engineering (ISSRE), pp. 6-11, 1999.
10. H. Joe, "Statistical inference for General-Order-Statistics and Non-homogeneous Poisson-Process software reliability models," IEEE Transactions for Software Engineering, vol. 15, no. 11, p. 1485–1490, 1989.
11. M. Ohba, "Software reliability analysis models," IBM J. Res. Develop., vol. 28, no. 4, pp. 428-443, 1984.
12. N. Langberg and N. Singpurwalla, "A unification of some software reliability models," SIAM Journal of Science and Statistical Computing, vol. 6, no. 3, pp. 781-790, 1985.
13. A.L. Goel, "Software reliability models: Assumptions, limitations, and applicability," IEEE Trans. Soft. Eng., vol. 11, no. 12, pp. 1411-1423, 1985.
14. S. Hossain and R. Dahiya, "Estimating the parameters of a non-homogeneous Poisson-process model for software reliability," IEEE transactions on Reliability, vol. 42, no. 4, pp. 604-612, 1993.
15. G. Knafl and J. Morgan, "Solving ML equations for 2-parameter Poisson-process models for ungrouped software-failure data," IEEE transactions on Reliability, vol. 45, no. 1, pp. 42-53, 1996.
16. G. Knafl, "Solving Maximum Likelihood equations for two-parameter software reliability models for using grouped data," IEEE Proceedings of the International Symposium on Software Reliability Engineering, pp. 205-213, 1992.
17. L. Rongguan and F. Heliang, "Estimation of the parameters for the Musa–Okumoto and inverse linear models in software reliability," Chinese Journal of Applied probability and Statistics, vol. 18, no. 4, pp. 425-430, 2002.
18. M. Xie, Y. Dai and K. Poh, Computing Systems Reliability. Models and Analysis, New York: Kluwer, 2004.
19. A. Avizienis, J.-C. Laprie, B. Randell, and C. Landwehr, "Basic Concepts and Taxonomy of Dependable and Secure Computing," IEEE Transactions on Dependable and Secure Computing, vol. 1, pp. 11-33, 2004.

Electromigration Reliability of Cylindrical Cu Pillar SnAg$_{3.0}$Cu$_{0.5}$ Bumps

L. Meinshausen[a,b], K. Weide-Zaage[a], B. Goldbeck[a], A. Moujbani[a], J. Kludt[a], H. Frémont[b],

[a]*Information Technology Laboratory, Leibniz Universität Hannover, Schneiderberg 32, 30167 Hannover, Germany*
[b]*Laboratoire IMS, CNRS UMR 5218, Université Bordeaux, France*

Abstract

The main trends in consumer electronics are increasing performances of their products and a reduction of the costs. These trends lead to an ongoing integration on package level which leads to a decreasing size of the solder contacts. This goes along with a higher sensibility to thermal-mechanical stress and void formation due to electromigration (EM). Against this background copper pillar bumps were introduced, because they combine the robustness of metal wire bonds with the low bonding pressure of reflow soldering.

Experimental results have shown a longer lifetime of Cu pillar bumps during EM tests, but a continuative analysis is still needed for design optimization. Against this background a finite element analysis (FEA) was performed to compare the EM induced mass flux in conventional solder bumps and in two different designs for Cu pillar bumps. The thermal electrical simulations were performed with ANSYS®. Afterwards a user routine was used to calculate the EM induced mass fluxes and mass flux divergences. The simulation results are used to identify possible reasons for the increased EM performance of Cu pillar bumps and they enable the identification of preferable designs.

1. Introduction

To keep in touch with their customer needs, the micro electronics industry has to increase the performance of their (mobile) electronic devices. The common down scaling on transistor level and the related increase of the IC performance is limited by high costs due to complex physical issues [1-3]. In case of mobile electronic devices alternative solutions on package level have to take into account the clear limitations in space. Against this background electronic components being based on stacking of packages (e.g. Package on Package) and the stacking of ICs (e.g. Chip on Chip) were developed. As a consequence the miniaturization of electrical contacts on package level becomes relevant for the development of future electronic devices.

A decreasing size of the solder joints leads to increase in the current density. Under this conditions EM induced reliability issues can appear if conventional solder joints are used (Fig. 1, left) [4, 5]. A possible alternative to solder bumps is the realization of electrical contacts with Cu pillars, being connected with a relative thin SnAgCu layers (Fig. 1, right). The resulting contacts are called Cu pillar bumps. Compared to solder bumps they have shown an increased performance during EM tests [6-9]. Furthermore Cu pillar bumps do not need a high bonding

pressure like wire bonding or sintering with nano particles, and they are based on the same materials as the previous solder bumps [6]. Hence the introduction Cu pillar bumps goes along with relative low development risks.

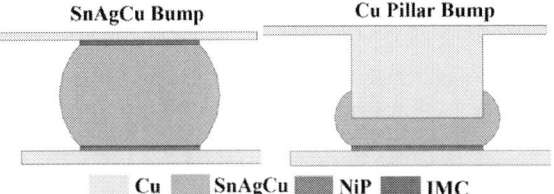

Figure 1: Scheme of conventional solder joints and Cu pillar bumps (Design: "A").

Current crowding (CC) at the corners of the contact surfaces is the main reason for EM induced void formation in solder joints [4, 5]. In conventional solder joints CC often appears at the interface between the SnAgCu solder and the intermetallic compound (IMC) layers [5]. The strength of Cu pillar bumps during EM tests is due the appearance of CC in the Cu pillar, because Cu (E_A=0.6-1.3eV [10]) has a low mobility compared to SnAgCu (E_A=0.56eV [11]). Furthermore the thin SnAgCu layers between the Cu pillars and the contact pad are easily transformed into Cu-Sn IMC joints [9]. In general Cu-Sn IMC joints are not affected by EM induced void formation, because the mobility of the single components (Cu and Sn) is too low (E_A=0.6-0.8eV [12]).

To verify the assumptions a finite element analysis (FEA) was made. Based on the results of a thermal electrical ANSYS® simulation the EM induced mass flux and mass flux divergences were calculated with a user routine, being described in [13]. The Cu pillar bumps and the conventional solder joints are integrated into a three dimension model of two ball grid arrays (BGA) for a Package on Package (PoP) structure. The lower BGA, with the higher contact density, was designed as cupper pillar bump array.

As a first step the weak points of the Cu pillar structure being indicated by the maxima of the EM induced mass flux divergences will be compared with the failure pictures of real EM tests. After the verification of the simulations results two different forms of Cu pillar bumps were compared with each other: the design "A" is based on free standing Cu pillars (Fig. 1, right), while the design "B" includes SnAgCu covered Cu pillars (Fig. 2). Both bump designs will also be compared with conventional SnAg$_{3.0}$Cu$_{0.5}$ solder bumps.

978-1-4799-4789-8/14 $31.00 © 2014 IEEE

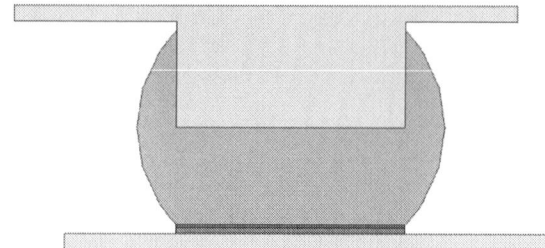

Figure 2: Alternative design for a Cu pillar bump Design: "B"

2. Model description and migration theory

The lower BGA of a PoP structure was chosen as possible application for the Cu pillar bumps (Fig 3). The finite element model is based on the geometry of an Amkor PoP [14]. The Cu contact pads of the BGAs and the metal lines are placed on FR-4 boards (violet). The packages are covered with an epoxy resin (green).

The environmental temperature was set to 100°C and a current of 1A (j_{avg}=18 µA/µm²) was applied to induce the EM. The single bumps are connected to a daisy chain. Hence the investigation of the up- and the downstream case was possible.

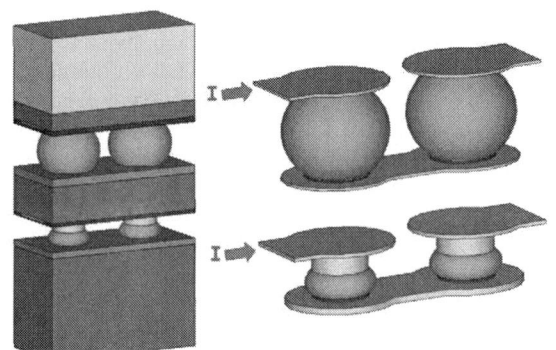

Figure 3: Finite element model of the PoP structure with Cu pillar bumps for the lower BGA (left). Principal direction of the current flow (right.)

During previous EM tests on Amkor® PoP structures an applied current of 1A led to a 30°C joule heating. Based on these results a heat transfer coefficient (HTC) of 60W/m²K was chosen for the upper and lower surfaces of the FE model. The high HTC is a virtual value that represents the free convection on a complete PoP structure [14]. For the surface of the solder joints a HTC of 11W/m²K was used [15].

The specific resistance and the thermal conductivity of the different materials, being needed for the thermal-electrical (TE) simulations, are given in table 1.

Material	ρ [Ωµm]	κ [W/(µmK)]
SnAg$_{3.0}$Cu$_{0.5}$ [8]	0.132	55.52x10^{-9}
Cu	17.4x10^{-3} [8]	0.42x10^{-3} [16]
Ni [16]	66.9x10^{-3}	0.11x10^{-3}
S.M. [15]	0.2x10^{18}	0.21x10^{-6}
FR-4 [17]	0.5x10^{15}	KXX = 1.05x10^{-6} KYY = 1.05x10^{-6} KZZ = 0.34x10^{-6}
M.C. [18]	10x10^{15}	0.167x10^{-6}
Cu$_6$Sn$_5$ [29]	0.175	34.1x10^{-6}
Resin [20]	1x10^{18}	0.2x10^{-6}

Table 1: Material properties for the thermal electrical simulations. S.M.: Solder mask, M.C.: Molding compound

The migration processes in the solder joints are thermally activated. Hence the diffusion coefficient (D) is described by an Arrhenius relationship (3). The Arrhenius includes the diffusion constant (D$_0$), the activation energy (E$_A$) and the product of the Boltzmann constant (k$_B$) with the temperature in Kelvin.

The mass flux due to EM is given by (4). In addition to D eq. 4 also includes the current density (j), the atomic density (N), the specific resistance ρ and the effective charge of the moving ion Z*.

$$D = D_0 \cdot exp\left(-\frac{E_A}{k_B T}\right) \quad (3)$$

$$\overrightarrow{J_{el}} = \frac{D \cdot N}{k_B \cdot T} \cdot Z^* \cdot e \cdot \rho \cdot \vec{j} \quad (4)$$

The material constants for the migration of SnAg$_{3.0}$Cu$_{0.5}$ are given in table 2.

Material Properties	SnAg$_{3.0}$Cu$_{0.5}$
α [1/K]	2.8x10^{-3} [8]
D$_0$ [m²/s]	27x10^{-3} [21]
Z*	-23 [22]
E$_A$ [eV]	0.56 [11]

Table 2: EM related material parameters.

The local difference in the mass flux is described by the mass flux divergence (5). A positive mass flux divergence shows a place where N is reduced over time. A negative mass flux divergence shows a place where N is increased. Finally a positive mass divergence can lead to a void formation, while a negative mass flux divergence indicates a possible hillock formation. The divergence of the EM induced mass flux is the scalar product of current density and temperature gradients (5). The value of the mass flux divergence depends on the magnitude of these two vectors and the angle between them.

$$div\left(\overrightarrow{J_{el}}\right) = \left(\frac{E_A}{k_B T^2} + \frac{\alpha_T \rho_0}{\rho} - \frac{1}{T}\right) \cdot \overrightarrow{J_{el}} \cdot grad(T) \quad (5)$$

3. Verification of the Simulation Results

The thermal and the electrical conductivity of Cu are better than the properties of SnAgCu. In case of Cu pillar bump this leads to a reduced temperature of the package compared to conventional solder bumps (Fig. 4). Due to the Arrhenius law (3) the EM induced mass flux is exponentially temperature dependant. As a consequence the reduced temperature leads to better EM performance and an increased current carrying capacity of Cu pillar joints. The same effect was observed during real EM tests [7].

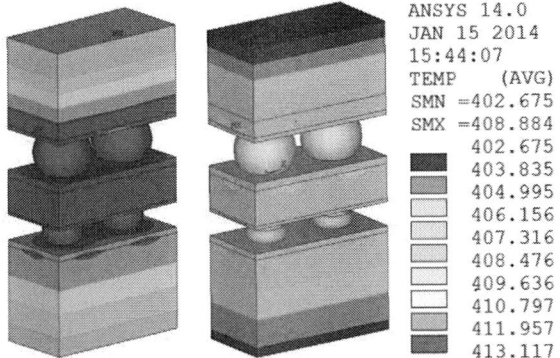

Sn-Ag-Cu; I=1A, Tsub: 100, EM-DC

Figure 4: Temperature distribution in a Pop with conventional bumps (left) and with Cu pillar bumps (right). T=100°C; I=1A.

The weak point of a conventional solder bump is its die sided contacts surface (Fig. 5, left) [5, 9]. In case of Cu pillar bumps this area is covered with Cu. Hence the CC is still present but the low mobility of Cu suppresses the EM induced void formation in the Cu pillar (Fig. 5, right).

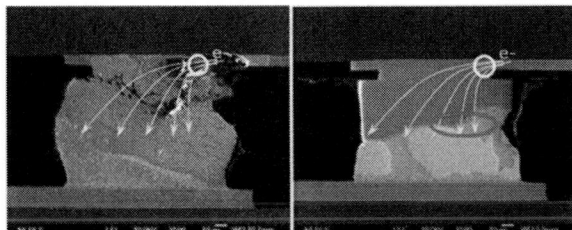

Figure 5: Failure pictures of downstream stressed solder bumps and Cu pillar bumps [9].

Figures 6 & 7 show the direction of the current flow and the temperature gradients in conventional solder bumps and in Cu pillar bumps. In case of conventional solder bumps (Fig. 6) the increased temperature gradients are combined with CC and the high mobility of SnAgCu. In case of Cu pillar bumps (Fig. 7) the CC is still present but the temperature gradients are smaller due to the better thermal and electrical conductivity of Cu. Further a CC accelerated void formation would have to appear in the high melting Cu, while in the low melting SnAgCu no relevant CC could be found. In both models high temperature gradients are found at the board side contacts surfaces. This is a consequence of the relative low thermal

and electrical conductivities SnAgCu compared to Cu. It leads to an increased temperature of the solder joints compared to their contact pads and as a consequence high temperature gradients at the interfaces appear.

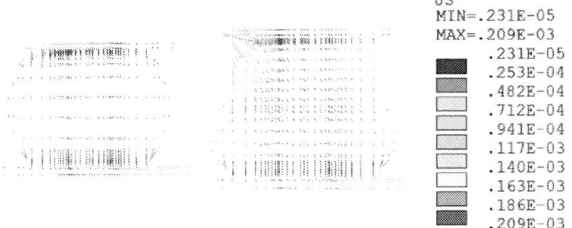

Figure 6: Current flow in a solder bump (left) and in a Cu pillar bump (right). [A/μm²]

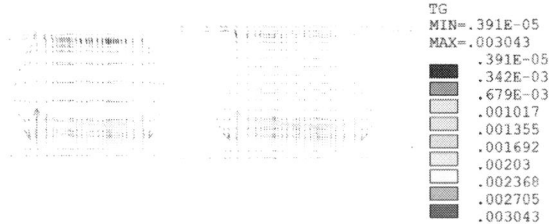

Figure 7: Temperature gradients in a solder bump (left) and in a Cu pillar bump (right). [K/μm]

In case of the conventional solder bumps the temperature gradients at the board side contact surfaces are not responsible for EM induced reliability issues, because at the die side contacts higher mass flux divergences appear as the consequence of CC going along with increased temperature gradients (Fig. 8, left). Nevertheless even in conventional solder joints increased mass flux divergences can be found at the board side contact surfaces (Fig. 8, right).

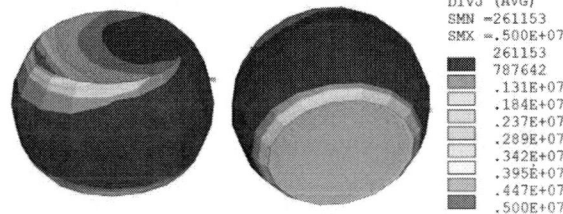

Figure 8: Mass flux divergences in a solder bump. [1/μm³s]

In case of the Cu pillar bumps the high temperature gradients at the board side contact surfaces become a relevant reliability issue. This has two consequences:

- The EM performance of Cu pillar bumps is better than the solder bump results, because the relative fast CC induced void formation at die side contact surface is suppressed. In the simulations the maximum mass flux divergence in Cu Pillar bumps (Fig. 9) was only 33% of the maximum value in the conventional solder bumps (Fig. 8).

- During EM tests Cu pillar bumps fail because of void formation at the board side contact surfaces (Fig. 10).

978-1-4799-4789-8/14 $31.00 © 2014 IEEE   57

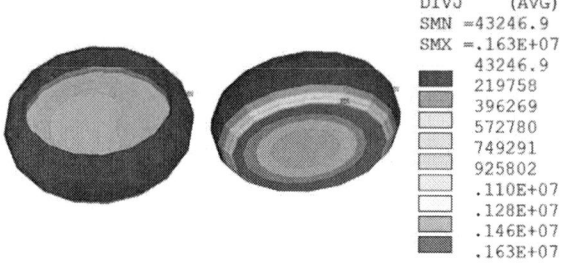

```
DIVJ    (AVG)
SMN =43246.9
SMX =.163E+07
         43246.9
         219758
         396269
         572780
         749291
         925802
         .110E+07
         .128E+07
         .146E+07
         .163E+07
```

Figure 9: Mass flux divergences in Cu pillar bump. [1/µm³s]

The failure pictures of several EM stress tests on Cu pillar bumps prove that the board side contact surfaces are the critical areas regarding to EM induced void formation [4, 7-9]. In difference to conventional solder bumps, in Cu pillar bumps high temperature gradients appear along the whole lower contact surface and they become more important than the local effects of CC. Hence the void formation begins along outer edges of the contact surfaces and moves to the center of the contact pad (Fig. 10). In case of CC induced failures the void formation begins at one corner and than goes along the contact surface (pancake void).

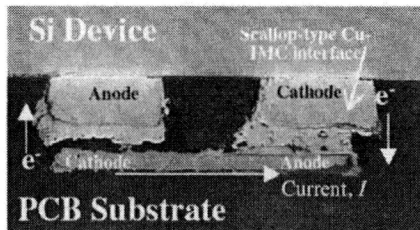

Figure 10: Failure picture of EM stressed Cu pillar bumps [7].T=150°C, 1.27x10⁴A/cm², t=300h

4. Comparison of Different Cu Pillar Designs

After the verification of the simulation results a comparison of different Cu pillar designs is possible. For the comparison two principle designs were modeled. The first design "A" is based on free-standing Cu pillars (Fig.1, right). For the second design "B" the Cu pillars were covered with $SnAg_{3.0}Cu_{0.5}$ (Fig. 2). Both Cu pillar designs will also be compared with a same sized conventional solder bumps (Fig 1, left).

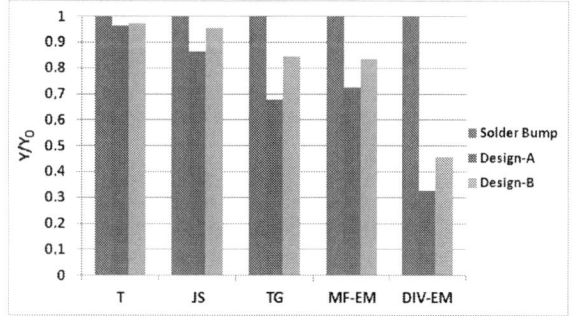

Figure 11: Comparison of the simulation results for conventional solder bumps and the two Cu pillar bump designs "A" and "B".

A comparison between the maximum model temperatures, the current densities, the temperature gradients (TG), the EM induced mass flux values ("MF-EM") and EM induced mass flux divergences ("DIV-EM") is shown in Fig. 11. The results of the conventional solder bumps are used as comparative values (Y_0). The related simulations results are given in table 3.

T_{max} [°C]	J_{max} [µA/µm²]	TG_{max} [mK/µm]	MF_{max} [1/µm²s]	DIV_{max} [1/µm³s]
139	33.7	0.41	4.07×10^{11}	5.0×10^{6}

Table 3: Simulation results for conventional solder bumps (Y_0 in Fig. 11).

Compared to the solder bumps, the two Cu pillar bump designs lead to clearly reduced maximum current densities and temperature gradients in the SnAgCu solder. the main reason is absence of CC at the sided contact surfaces. A consequence of this is a reduction of the mass flux divergences by more than 50%. The reduced mass flux divergences are one possible reason for the longer lifetimes of Cu pillar joints during EM test in [4, 7-9].

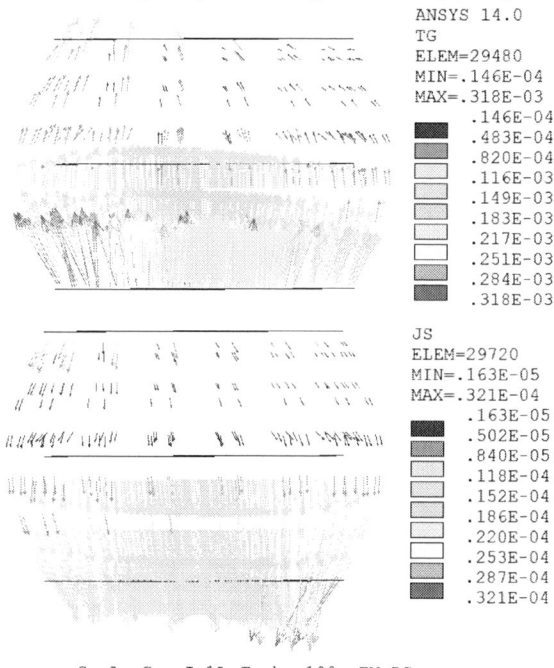

```
ANSYS 14.0
TG
ELEM=29480
MIN=.146E-04
MAX=.318E-03
         .146E-04
         .483E-04
         .820E-04
         .116E-03
         .149E-03
         .183E-03
         .217E-03
         .251E-03
         .284E-03
         .318E-03

JS
ELEM=29720
MIN=.163E-05
MAX=.321E-04
         .163E-05
         .502E-05
         .840E-05
         .118E-04
         .152E-04
         .186E-04
         .220E-04
         .253E-04
         .287E-04
         .321E-04
```

Sn-Ag-Cu; I=1A, Tsub: 100, EM-DC:

Figure 12: Temperature gradients [K/µm] and current density distribution [A/µm²] in the Cu pillar design B.

The comparison of the two cu pillar bump designs shows that free standing Cu pillar bumps do not lead to a significant reduction of the temperature compared to SnAgCu covered Cu pillars bumps. Nevertheless at the board side contact surface of design B, the EM induced mass flux vectors and the temperature gradients are increased due to a more strong CC effect, while their orientations are in opposite directions (Fig. 12). As a consequence relative high mass flux divergences can be observed. In design A the mass flux vectors and the temperature gradients are smaller and the angle between both vectors is smaller than 180°. Hence the mass flux divergences in design A are lower than in design B.

Overall the simulations show why Cu pillar bumps have a longer lifetime during EM stress tests than solder bumps. A comparison between two Cu pillar bump designs has shown that free standing Cu pillars with a relative flat SnAgCu connections lead to smaller mass flux divergences.

5. Conclusions

EM stress tests on conventional solder bumps and Cu pillar bumps have shown a longer lifetime of Cu pillar bumps. A finite element analysis (FEA) was performed to proof possible reasons for the better EM performance of Cu pillar bumps. Afterwards different designs of Cu pillar were compared.

The simulation results indicate that the better thermal and electrical conductivity of the Cu pillars lead to a reduced temperature of the packages and the solder joints. Independent from the layout a lower joint temperature reduces the risk of migration induced failures. Furthermore it was proven that the absence of SnAgCu at the CC affected die side contacts leads to smaller maximum mass flux divergences. The presence of Cu at the die sided contact surfaces suppresses the formation of pancake voids. Hence the presence of CC in the Cu pillar instead of the solder joint is one possible reason for the longer lifetime of CU pillar bumps during EM tests.

Like real EM stress tests, the simulation results have shown that the critical areas of the Cu pillar bumps are the board side contact surfaces. In difference to the void formation process in conventional solder joints the high mass flux divergences in Cu pillar joint were found around the whole lower contact surface. Hence the void growth can appear at several points.

After the verification of the simulation results by comparing them with failure pictures of real EM tests, two different Cu pillar bump designs were investigated. The comparison of free standing Cu pillar bumps (design "A") with SnAgCu covered Cu pillars (design "B") has shown lower mass flux divergences in design A. The main reason for the better EM results of the free standing Cu pillar bumps was a less strong CC effect. As a consequence the values of the EM induced mass flux and the temperature gradients were reduced and they have a more favorable orientation regarding to the resulting mass flux divergences.

Overall it was possible to proof the expected reasons for the better EM performance of Cu pillar bumps with FEA. Based on this, a simulation based investigation of different Cu pillar bump designs was possible. Via FEA free standing Cu pillars were identified as a promising design for more EM resistive Cu pillar bumps.

Acknowledgments

This collaboration was supported by PROCOPE PKZ 55888031and PHC 28245YB. This program is founded by the DAAD, financed by the BMBF and the French Ministry for Foreign Affairs.

References

1. R. H. Havemann, J. A. Hutchby, "High-Performance Interconnects: An Integration Overview", IEEE Proceedings Vol. 86 No. 5 (2001) pp. 586-601.

2. S. Borkar, "Design challenges of technology scaling", IEEE, Micro, Vol. 14 No. 4 (1999), pp. 23-29.

3. K. Rupp, S. Selberherr, "The Economic Limit to Moore's Law", IEEE, Transactions on Semiconductor Manufacturing, Vol. 24, No. 1 (2011), pp. 1-4.

4. C. Chen, H.M. Tong, K.N. Tu, "Electromigration and Thermomigration in Pb-Free Flip Chip Solder Joints", Annual Review of Materials Research, Vol. 40 (2010), pp. 531-555.

5. L. Meinshausen, K. Weide-Zaage, H. Frémont, "Electro- and Thermomigration induced Failure Mechanisms in Package on Package", IEEE, Microelectronics Reliability, Vol. 52 No 12 (2012), pp. 2889-2906.

6. A. L. X. Jiang, L. C. Ming, J. C. Y. Gao et al., „Pillar Bump Technology and Integrated Embedded Passive Devices", IEEE, 7th International Conference on Electronics Packaging Technology (ICEPT), Shanghai, August 2006, pp. 1-5.

7. S. Lee, Y. X. Guo and C.K. Ong, „Electromigration Effect on Cu-pillar(Sn) Bumps", IEEE, 7th Electronics Packaging Technology Conference (EPTC), Singapore, December 2005, pp. 135-139.

8. Y. S. Lai, Y. T. Chiu, C. W. Lee, et al. „Electromigration Reliability and Morphologies of Cu Pillar Flip-chip Solder Joints", 58th Electronic Components and Technology Conference (ECTC), Lake Buena Vista, Florida, May 2008, pp. 330-335.

9. J. H. Yoo1, I. S. Kang, G. J. Jung, „Analysis of Electromigration for Cu Pillar Bump in Flip Chip Package", 12th Electronics Packaging Technology Conference (EPTC), Singapore, December 2010, pp. 135-139.

10. A. Moujbani, J. Kludt, K. Weide-Zaage et al., "Dynamic Simulation of Migration Induced Failure Mechanism in Integrated Circuit Interconnects", IEEE, Microelectronics Reliability, Vol. 53 No 9-11, 2013, pp. 1365-1369.

11. M.-S. Yoon, M.-K. Ko, O.-H. Kim et al., „Electromigration Behaviors of SnAgCu Solder Lines", Electronic Materials Letters, Vol. 2, 2006, pp. 127-130.

12. L. Meinshausen, K. Weide-Zaage, H. Frémont, „Migration induced material transport in Cu-Sn IMC and SnAgCu micro bumps", IEEE, Microelectronics Reliability, Vol. 51 No 9-11, 2011, pp. 1860-1864.

13. K. Weide-Zaage: "Exemplified calculation of stress migration in a 90nm node via structure", IEEE, Mechanical & Multi-Physics Simulation, and Experiments in Microelectronics and Microsystems (EuroSimE), 2010.

14. L. Meinshausen, H. Frémont, K. Weide-Zaage et al., Influence of contact geometry variations on the lifetime distribution of IC packages during electromigration testing, IEEE, 14th Mechanical & Multi-Physics Simulation, and Experiments in Microelectronics and Microsystems (EuroSimE), Wroclaw, Poland, 15-17 April, 2013, pp. 1-7.

15. K. Weide-Zaage: "Simulation of Migration Effects in Solder Bumps", IEEE, Device and Materials Reliability (2008).

16. R. C. Weast: "CRC Handbook of Chemistry and Physics", CRC Press, Edition 62 (1981), Boca Raton, FL.

17. H. Ye, C. Basaran, C. Douglas, et. al: "Mechanical degradation of microelectronics solder joints under current stressing", International Journal of Solids and Structures, Vol. 40 (2003), pp. 7269-7284.

18. ESA Publications Division ESTEC: "Data for Selection of Space Materials", Paris (1994), European Space Agency.

19. R. J. Fields, S. R. Low, "Physical and mechanical properties of Intermetallic compounds commonly found in solder joints", NIST, 1991.

20. W. Feng, K. Weide-Zaage, F. Verdier: "Electrically driven matter transport effects in PoP interconnections", IEEE, Mechanical & Multi-Physics Simulation, and Experiments in Microelectronics and Microsystems (EuroSimE), 2009.

21. Y. Liu, L. Liang, S. Irving et. al: "3D modeling of electromigration combined with thermal-mechanical effect for IC device and package", Microelectronics Reliability, Vol. 48, 2008, pp. 811-824.

22. K. Weide-Zaage, H. Frémont, L. Wang: "Simulation of Migration effects in PoP", IEEE, Mechanical & Multi-Physics Simulation and Experiments in Microelectronics and Microsystems (EuroSimE), EuroSimE, 2008.

Reliability Assessment of Discrete Passive Components embedded into PCB Core

R. Schwerz[1], M. Roellig[1], S. Osmolovskyi[2], K.-J. Wolter[2]

1) Fraunhofer IKTS-MD, Dresden, Germany

2) Technische Universität Dresden, Electronics Packaging Laboratory, Dresden, Germany

robert.schwerz@ikts-md.fraunhofer.de; +49 351 / 88815-584

Abstract

This paper will present the research results for reliability of two embedding technologies in comparison to the current standard - surface mount technology.

The chosen embedding approaches utilize a cavity to place the necessary components into the PCB core. The difference is found in the way the component is connected to the PCB routing. For the first approach the circuit is first assembled on a carrier substrate using conventional surface mount technology (SMT). The solder paste is printed, the components are placed and the substrate board is soldered afterwards. The base substrate is then put together with prepared additional layers holding preformed cavities at the component locations. After another top layer has been added, the stack is finally laminated and the components are placed in the PCB core. The second approach is based on placing the components, then putting together the stack-up as described earlier and followed also with the laminating process. However the components have not been soldered. Instead an opening to the component's terminals is created through laservias. Then galvanic deposition is utilized to establish the connection to the PCB routing. For a comparison of the technologies samples with embedded resistors and ceramic capacitors in various sizes for the technologies as well as standard SMT have been prepared. To assess the reliability potential the samples have undergone temperature cycling tests. The testing is supported with FEM simulations which aided in the detection of critical design parameters and assess the residual manufacturing stress/strain states. The results of the investigations have shown that the damage mechanisms and predominant failure sites of the SMT & Cavity embedding variant is significantly different from conventional SMT. Here the resin material should be adopted to increase lifetimes even more. In the case of Microvia & Cavity the geometry of the microvia is essential towards the achievable reliability. Overall the results indicate the increased reliability potential of the novel approaches.

1 Introduction - component embedding

Aside to the well-known miniaturization efforts like System on Chip, System in Package or Package on Package the embedding technology has continuously been developed in the last two decades. Today passive and even IC components can be integrated into Printed Circuit Boards (PCB). The active and passive components embedding technology enables microsystems integration in various new ways and has several applications in electronics and sensor packaging. The main benefit is the reduction of the used surface area for the single components. Here two approaches are possible. It is possible to apply the embedding approach on the package level to reduce footprint sizes. Alternatively it can be used at the board level to create functional System-in-Board (SiB) solutions. Both variants lead to the miniaturization of the electronic package or the board size needed to incorporate the system functionality. In addition to the miniaturization the embedding technologies have a large potential to provide protection for the package towards environmental loading and to increase the overall system reliability and satisfy high demands in lifetime. As of today multiple embedding approaches are manufactured, however publications on reliability investigation in comparison to the current default surface mount technology are still rare.

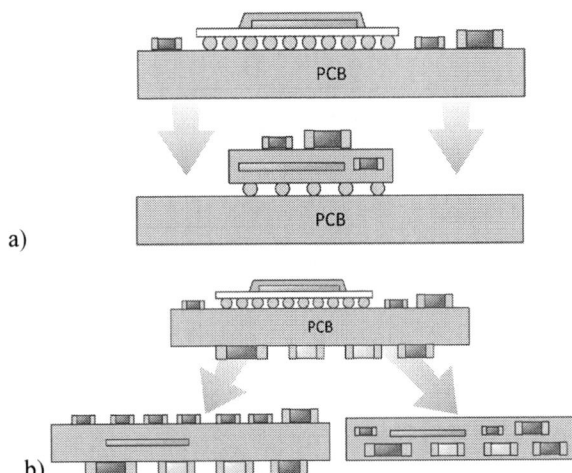

Figure 1: currently available general embedding approaches: (a) as enhancement to packaging or (b) to create functional System-in-Board solutions

2 Used embedding concepts

Integration approach using SMT & Cavity in PCB

The first concept utilizes PCB technology processing steps to embed the necessary components directly into a PCB core. In this case the circuit is first assembled on a thin carrier substrate using the conventional surface mount technology processing steps. This includes structuring of the PCBs routing, printing solder paste, placing the components and finally reflow soldering. Now layers, which have been prepared with cavities at each of the component locations, are stacked onto the substrate. Than an additional top layer is added as cover and the stack up is ready to be laminated (see Figure 2). Generally the combined thickness of the middle layers should exceed the maximum component height to prevent pad indentation or component damage during the lamination process. This method also offers very easy testability and accessibility of the circuit function on the carrier before lamination.

Afterwards the components are surrounded by isotropic matrix material and FR4 material without warped fibers.

978-1-4799-4789-8/14 $31.00 © 2014 IEEE

The planar assembly result adds the possibility of stacking other components or modules.

Electrical resistance measurement, x-ray analysis, ultrasound analysis and cross sectioning have been utilized to inspect and verify the quality of the created samples before testing the reliability in the following step. No delamination or voiding in the matrix materials were found (see Figure 3). The cross-sectioning also revealed very good overall mechanical assembly result. The standoff volume below the component has been completely filled by resin material, due to the applied vacuum and pressure during the lamination process. The embedding concept has been successfully applied.

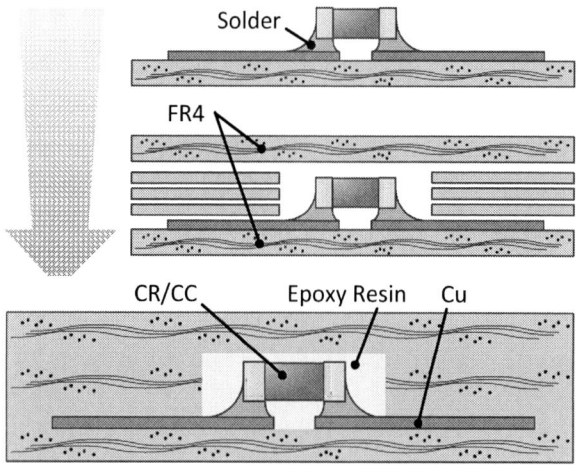

Figure 2: Schematics of the integration approach using SMT & cavity in PCB

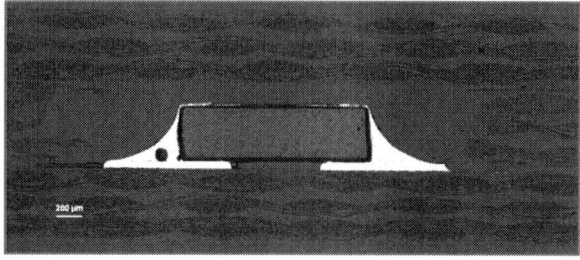

Figure 3: Example cross sections of a chip resistor embedded with SMT & Cavity in PCB

Integration approach using Microvia & Cavity

An alternative approach widely utilized in the industry already uses purely PCB processes to connect the embedded components to the routing. Here the processing steps are as follows: First the components are placed on a thin substrate. This can be a very thin PCB laminate or just a copper foil. The fixation is done through non-conductive adhesive, which in the case of a pure copper foil is solely responsible for the dielectric in between the routing and the component. Once all components are placed and fixated at their positions, the middle layers and cover are prepared as described earlier. The stack-up is ready for lamination. To achieve a connection between the now embedded component terminals to the routing of the PCB, vias need to be drilled through the adhesive and/or PCB material. Here

laser drilling is commonly applied. While UV-lasersystems can easily drill through copper traces, a CO2-laser can be used to ablate the organic material down to the terminal metallization since it has very little absorption on metallic surfaces such as copper. Alternatively the substrate can be structured beforehand to create the needed openings in the copper pads to access each via location. Once all microvias are prepared, interconnects are created with galvanic metallization. To successfully apply this step the terminal needs to be process compatible with the metallization procedure. Hence simple tin based terminals and finishes cannot be used. Instead pure copper terminals or PdAg are required. Currently component manufacturers are increasingly offering such products.

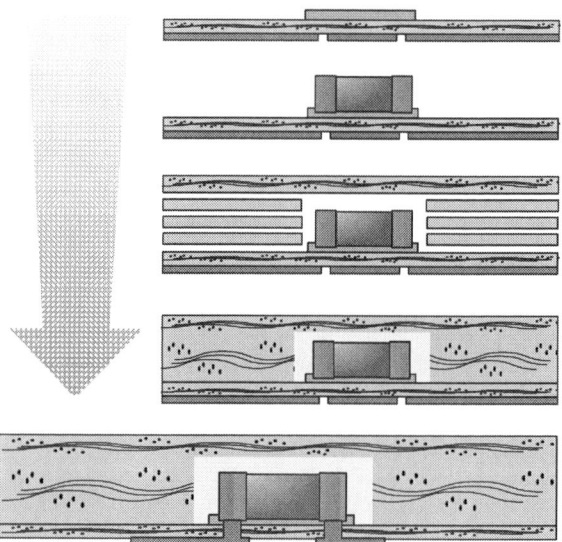

Figure 4: Schematics of the integration approach using microvia-drilling and metallization processing

3 Finite-Element-Modeling

For both embedding assembly approaches and the conventional SMD technology a parameterized 3D finite-element model has been created. The model uses the same routines to create the component and solder geometries. This ensures equivalent meshing for the three options. Because of symmetry conditions only a part of the assembly needs to be modeled. For convenience a half model has been used to help reduce calculation time.

To describe the material's behavior a number of published models have been implemented. The complete overview of the all utilized material properties is given in Table 1 while all corresponding coefficients have already been presented elsewhere and can be found in [3]. The critical solder material (SnAgCu type alloy) includes time dependent viscoplastic properties using the hyperbolic sine creep law for secondary creep. The plated copper in the microvia variant is modelled using an elastic-plastic approach with a kinematic hardening law. The stress-strain relationship for multiple temperatures is shown in Figure 5. Special care has also been taken to appropriately model the orthotropic behavior of the composite material FR4.

978-1-4799-4789-8/14 $31.00 © 2014 IEEE

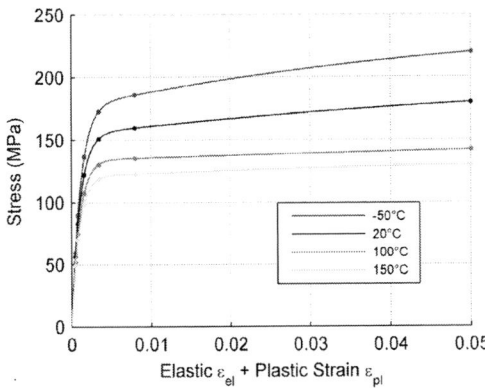

Figure 5: Elastic-plastic modeling of the copper material implementing a kinematic hardening law for different temperature, data sets from [5]

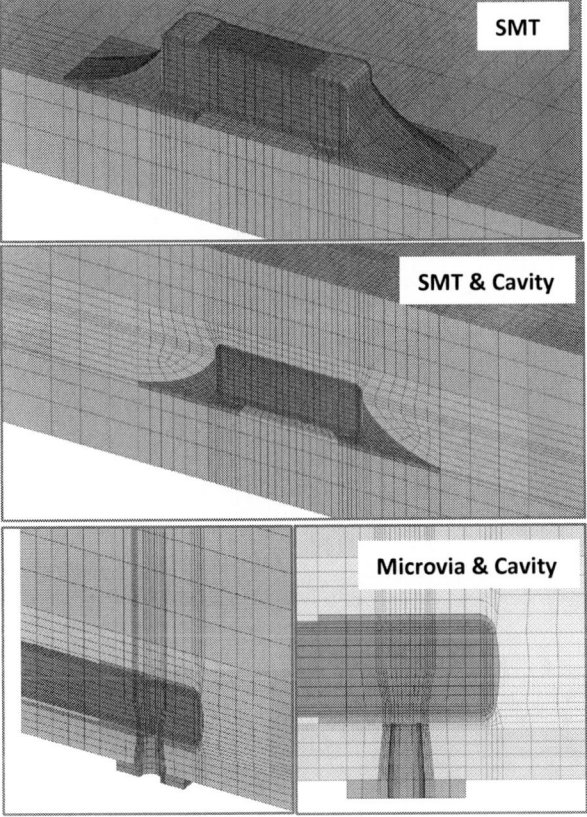

Figure 6: FEM geometries: Test sample stripe with integrated component, SMT chip component platform model (top) adapted to the SMT & Cavity PCB approach (middle) and microvia variant (bottom) using symmetry conditions and mapped meshing for uniform mesh

Table 1: Properties of the materials used in the simulations

Feature	Material	Properties
Component	Ceramic	elastic
Terminals	Nickel/Copper	elastic, plastic, temp. dependent
Pad	Copper	elastic, plastic, temp. dependent
Substrat	FR4	elastic, orthotropic, temp. dependent
Cavity	Epoxy	elastic
Solder	SAC	elastic, viscoplastic, temp. dependent

4   Results and Discussion

Manufacturing process

Both presented embedding methods involve the lamination processing step, which applies high temperatures and significant pressure to the layer construction. Consequently the processing steps itself require examination and understanding to control and reduce the manufacturing risk. To reduce the mechanical pressure on the component the cavity has to be carefully designed to avoid the lamination pressure on the component. Here testing of different spacing has shown that about 100µm between component and PCB cover is sufficient to effectively lift the loading.

Table 2: Manufacturing loads for the lamination process

Maximum Temperature	180°C - 200°C
Duration of Process	2h
Pressure	15-30 bar
Vacuum	x

During the lamination process heat and pressure cause the preformed composite materials (prepregs) with fibers and matrix start to join. At high temperature the polymer monomers and oligomers build crosslinks and the laminate and resins develop a fully cured polymer matrix. At this point the material attains its effective mechanical properties. Since the process is occurring at elevated temperatures, the cooling down from the high processing temperature to ambient temperature leaves residual stresses in the specimen-structures. It has been shown in previous publications that the residual stresses for varying embedding methods are significant and should be included during subsequent simulation steps [2][3][4].

In the case of microvias, the interconnect structures are not yet existent during the lamination process. Considerable influences for this method are the laser-processing and metallization steps. The laser process causes local heating and might exert deformation and change the involved geometries and can also causes post-curing effects in the polymers. This might alter the mechanical material properties.

978-1-4799-4789-8/14 $31.00 © 2014 IEEE   63

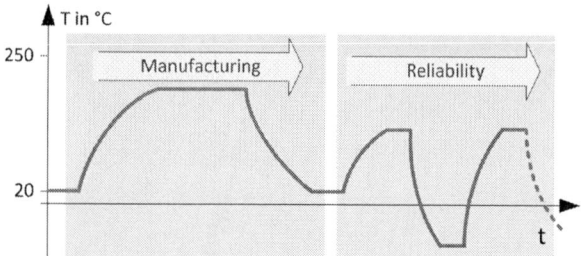

Figure 7: Schematic of prestress incooperation as effect of the manufacturing processes into the reliability consideration for the embedding concept SMT & Cavity

Temperature cycling test

Using the previously described FEM-Framework the temperature cycling load onto the component has been calculated. The complete simulation-run is depicted in Figure 7 and incoorperates the residual stress state due to manufacturing loads. It could be observed, that the low temperature induces more stress and strain into the specimen, which makes it the considerably more critical state. The Mises stress distribution of the SMT and SMT with cavity embedding variant at this critical low temperature point is shown in Figure 8. The most stressed regions for the SMT&Cavity embedding approach are at the top component edges in the epoxy volume. Additionally stress peaks at the inner component pad regions can be observed which also correspond to the crack findings in the cross sections (Figure 11).

During the temperature change the cte mismatch causes a cyclic displacement in the structure. The high cte of the resin material generates increased deformation in the embedded assembly around the component. Consequently more energy needs to be absorbed in the solder material which then increases the joint's accumulated creep strain (Figure 9). For conventional SMT assemblies this would translate to fewer cycles to failure and reduced lifetimes. However even though more creep energy is essentially accumulated in the interconnects of the embedding variant, neither the anticipated macro cracks nor micro cracks could be found for the same number of temperature cycles (Figure 10). This indicates a change in the damage mechanism for embedded components because the solder volume is completely fixated by the resin and composite material.

Instead cross sectioning revealed that the resin material at the predicted position (top component edges) starts to crack after a few hundred cycles. Until 2500 cycles the cracks starts to propagate through the resin towards the cover laminate and is deflected at this point by the glass fibers. Now the resin crack does not progress further into the cover and instead starts to widen. After sufficient opening the constant inelastic deformation slowly works the solder material into the geometric opening (see Figure 11). This probably takes place at elevated temperatures when the solder is very ductile and creep mechanisms are the most active. Thus for embedded discrete passives delamination and cracking in the matrix material in the cavity region and cover laminate become the major damage occurrences.

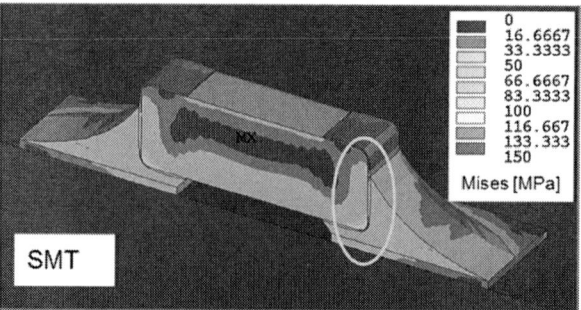

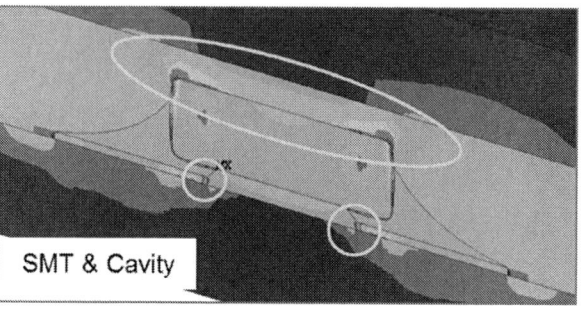

Figure 8: Mises stress and of the embedding approaches at the critical low temperature -55°C during reliability testing

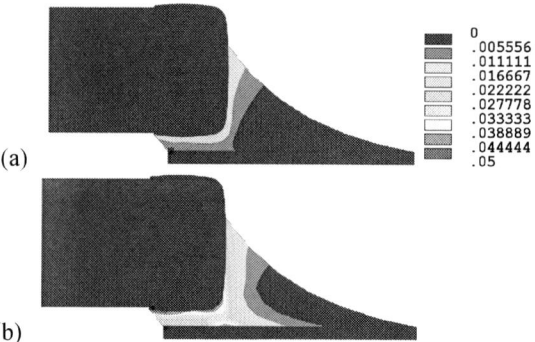

Figure 9: Accumulated equivalent creep strain for the SMT and SMT & Cavity embedding methods: clearly the encapsulation causes higher inelastic deformation

For the Microvias&Cavity embedding method the magnitude of the plastic strain in the copper signifies the tendency of forming plasticity-induced crystal defects like dislocations and vacancies, which can potentially promote void growth, cracking or debonding damage. According to Pucha et al. the Engelmaier low-cycle fatigue model for copper foils can be used to predict the fatigue life of a microvia [6]. This fatigue model correlates the accumulated plastic strain towards the expected mean cycles to failure. This is a very similar approach as the Coffin-Manson correlation for solder material which is based on creep strain rather than plastic strain accumulation. Qualitatively higher inelastic strain accumulation translates into lower lifetime expectations for both.

The prepared finite element model has been used to inspect the temperature dependent loading of the Microvia&Cavity embedding method during cycling scenarios. The simulation profile along with an example of the accumulation of the maximum plastic strain is depicted in Figure 12. Exemplary stress and strain field destributions

978-1-4799-4789-8/14 $31.00 © 2014 IEEE   64

are also given in Figure 13 and Figure 14. Both indicate a concentration at the via-bottom to component metallization. The maximum is located inward to the component. Hence fatigue failure in the form of barrel type cracking is anticipated before corner cracks.

Furthermore the results for geometric variations of the via's have been calculated for comparison. The applied aspect ratios reached from 0.5 to 1.5 while a wall thickness between 10µm to 25µm and completely filled vias have been included. In all cases the maximum stress and strain occurred at the same location already depicted. Also the qualitative distributions were independent of the geometric alterations calculated. Therefore the anticipated failure type is also assumed to stay the same. The Figure 15 and Figure 16 display the response surface plot of the maximum stress and accumulated strain for the calculated microvia geometries. It can clearly be seen that the shape is significantly influential on the expected reliability. Low aspect ratios in combination with a thick wall metallization reduced the maximum stress which reduced the plastic strain accumulation. The least loading is observed for fully filled microvia structures. However beyond a wall-thickness of 25µm the accumulated strain is already very close the minimum. The investigations have shown that the expected cycles to failure depend be significantly on applying geometric optimizations through the manufacturing processes.

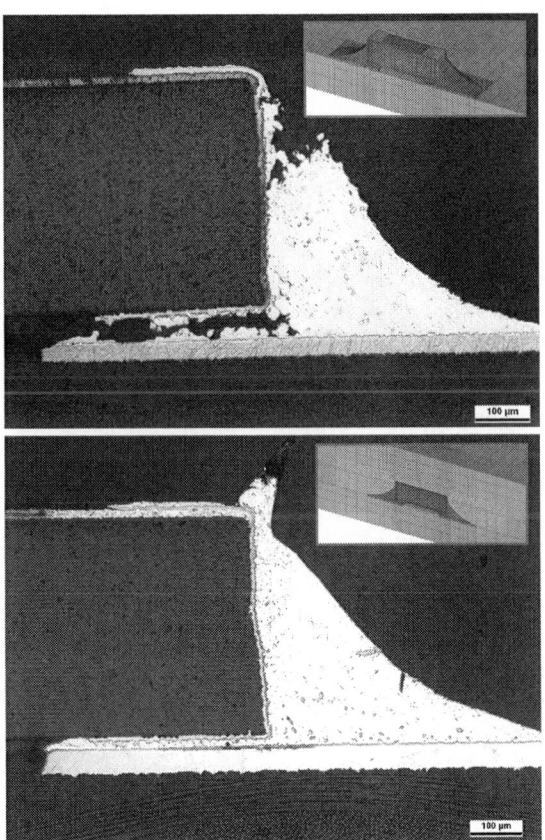

Figure 10: Representative sample cross sections of the conventional SMT (top) and SMT with cavity embedding approach (bottom) after 4500 temperature cycles

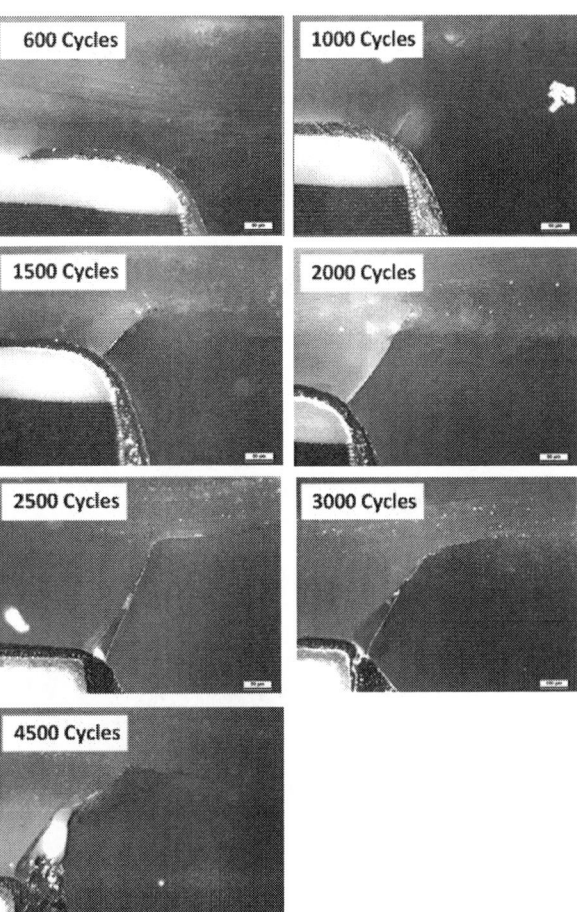

Figure 11: Propagation of the characteristic resin cracking the SMT & Cavity embedding variant throughout the temperature cycling test (-55°C/125°C)

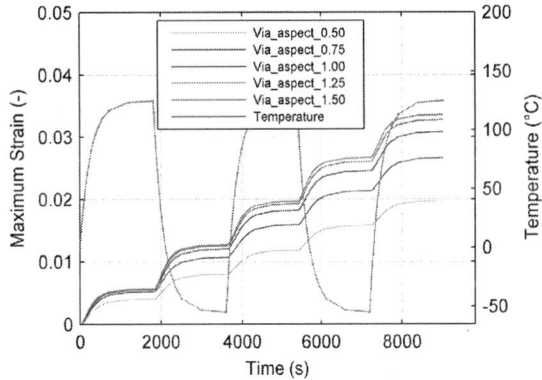

Figure 12: Simulation profile and exemplary plastic strain accumulation for 15µm wall-thickness and varying aspect ratios

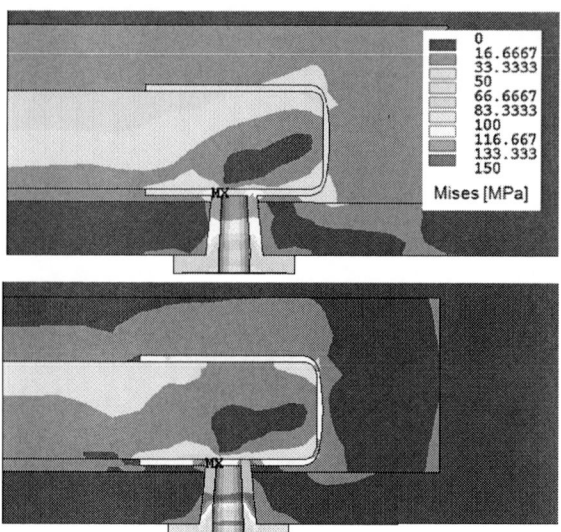

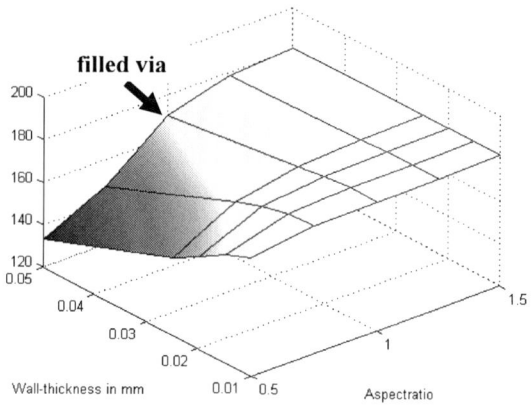

Figure 15: Maximum Mises stress in the microvia at the critical low temperature point -55°C

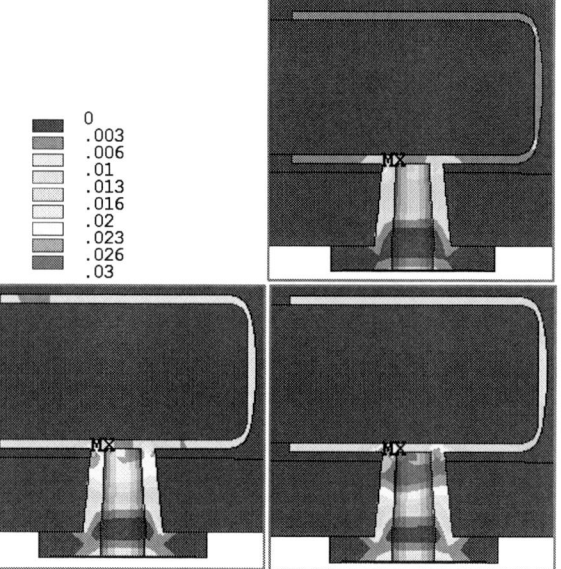

Figure 13: Mises stress distribution during the TCT at the high temperature 125°C (top) and the more critical low temperature -55°C (bottom)

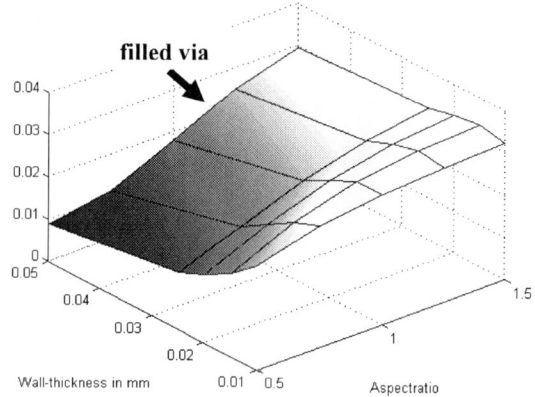

Figure 16: Maximum accumulated plastic strain in the microvia structures after 2.5 cycles. The highest wall thickness represents the completely filled results.

the predictions of the simulation results. However it has to be emphasized that the current understanding of fatigue in solder is not valid for embedded variants. Instead of the creep paths, cracking in the resin material and composite become major failure modes. Here further work has to be done to understand the crack initiation and propagation effects. For Microvia&Cavity embedding the failure modes are similar to the ones observed in regular microvia strucutres. In both cases the geometric dimensions defined through aspect ratio and metallization thickness are defining factors for lifetime.

Figure 14: Representative destribution of the accumulated plastic strain in the copper material of the microvia at different simulation stages: 2.5 cycles (top-right), 3.5 cycles (bottom-left) and 4.5 cycles (bottom-right)

5 Conclusions

The paper summarizes the research concerning the failure investigations of embedded electrical components and electronics. Two general approaches have been presented. Both utilize a cavity to bury the components into the PCB core. The first approach uses solder as interconnect structure while the second is based on microvia drilling with subsequent metallization. For both methods finite element analysis has been done. While manufacturing risks have been addressed, the temperature cycling load has been chosen for reliability comparisons. The observed crack initiation and crack propagation of the SMT&Cavity embedding variant has good correlation with

References

[1] B. Boehme, M. Roellig, G. Lautenschlaeger, M. Franke, J. Schulz and K.-J. Wolter, "Einbettung von Ultraschallwandler- und Elektroniksystemen in CFK-Strukturen für die sensorische Strukturüberwachung", in *Elektronische Baugruppen und Leiterplatten 2012 (EBL)*, Fellbach, 2012

[2] R. Schwerz, K. Meier, M. Röllig, A. Schießl, A. Schingale, K.-J. Wolter and N. Meyendorf, " Evaluation of embedded IC approach for automotive application", in *14th International Conference on Thermal, Mechanical and Multi-Physics Simulation*

and Experiments in Microelectronics and Microsystems (EuroSimE), Wroclaw, 2013

[3] R. Schwerz, B. Boehme, M. Roellig, K.-J. Wolter and N. Meyendorf, "Reliability of Embedding Concepts for Discrete Passive Components in Organic Circuit Boards", *64th Electronic Components and Technology Conference (ECTC)*, Las Vegas, 2013

[4] R. Schwerz, M. Roellig, M. Franke, G. Lautenschläger and K.-J. Wolter, "Zuverlässigkeitspotential von eingebetteten passiven und aktiven Bauelementen für die sensorische Strukturüberwachung", in *Elektronische Baugruppen und Leiterplatten 2014 (EBL)*, Fellbach, 2014

[5] F. Feustel, "FEM-Simulation der thermo-mechanischen Beanspruchung in Flip-Chip-Baugruppen zur Bewertung ihrer Zuverlässigkeit", *Fortschritt-Berichte VDI, vol. 9*, Issue 355, 2002

[6] R. V. Pucha, G. Ramakrishna, S. Mahalingam, S. K. Sitaraman, "Modeling spatial strain gradient effects in thermo-mechanical fatigue of copper microstructures", in *International Journal of Fatigue*, Volume 26, Issue 9, September 2004, Pages 947-957

Development and Validation of Dye-sensitized Solar Cell Finite Element Model for Sealing Failure Investigation

Changwoon Han and Seung Il Park
Korea Electronics Technology Institute
Seongnam, 463-816, Republic of Korea
cw_han@keti.re.kr

Abstract

Large-scaled Dye-sensitized solar cell (DSC) modules are recently developed for BIPV applications. In the modules, two glasses with electrodes and dye are joined for electrolyte and sealed together to prevent the leakage of electrolyte. Test of the module in a high temperature environment discloses several symptoms of electrolyte leakage through sealing failure. A finite element model is developed for the DSC module to investigate the sealing failure from the different thermal expansions. The accuracy of the model is validated by experimenting the deformation patterns of the module in high temperature and comparing with the model results. Finally, the sensitivities of the cell width and glass thickness are studied and optimal configuration is suggested for preventing the sealing failure of DSC.

1. Introduction

Dye-sensitized solar cells (DSC) have attracted many interests as one of the next generation solar cells. They have many relative advantages such as simple production process, cost-effective efficiency, using environmentally sustainable materials, wide range of color possible, semi-transparency, and working in cloudy weather and low-light conditions. DSCs are also considered for an important player in building-integrated photovoltaic (BIPV) applications.

However, the reliability, especially long-term durability, of DSC module is still in question. Many studies reported that the performance of DSC showed a rapid degradation after exposed in a high temperature condition. Kontos et al. [1] tested a small DSC cell in 80°C for 2,000 hour and reported 70% efficiency drop after the test. Bari et al. [2] also tested DSC cells in several high temperatures and obtained rapid efficiency drops such as 75% drop after 60 hour in 85°C. Sastrawan et al. [3] showed a 12% drop of efficiency after 1,000h in 85°C. They also analyzed the degraded cell and concluded the cause of the efficiency drop is mainly the loss of liquid electrolyte of the DSC cell during the high temperature tests. Note that all the reported tests were conducted for cell level smaller than 100 mm.

Recently, a large DSC module, sized 300x300 mm², is developed for BIPV application. The BIPV application requires more solid durability and the larger the module size, the more difficult to meet the durability requirements. In the study, a high temperature test is conducted to the DSC module and rapid efficiency degradation is reported. To investigate the degradation mechanism, a finite element model for the DSC module is developed. The model suggests the delamination of sealant in the module. The validation tests for the FE model are also conducted. Finally, using the model, parametric studies are conducted to provide the design rule to optimize the module design in durability view.

2. Durability Test of DSC Module

The tested large-scaled DSC module is presented in Fig. 1(a), which shows the back side of the module. Two glass plates coated transparent conductive oxide (TCO) layers are used for the top and bottom parts. Dye on nano particle and platinum electrode layers are printed on each glass plate. To produce electrical power effectively, the module is divided into several cells, which can be detected in Fig. 1(a) as purple areas are separated by white lines in the module. Cross section of each cell structure is illustrated in Fig. 1(b); note that the figure is not scaled. Each cell is divided by Ag electrode and sealing material and filled with iodide-based liquid redox electrolyte as shown in Fig. 1(b). A transparent thermoplastic adhesive material is used for the sealant.

(a)

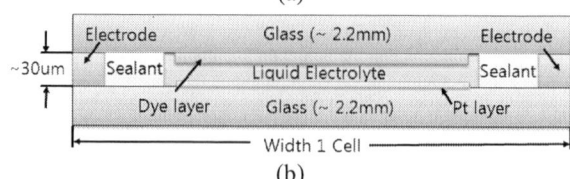

(b)

Fig. 1 Structure of DSC module; (a) back side view and (b) illustration of cell cross section (not scaled).

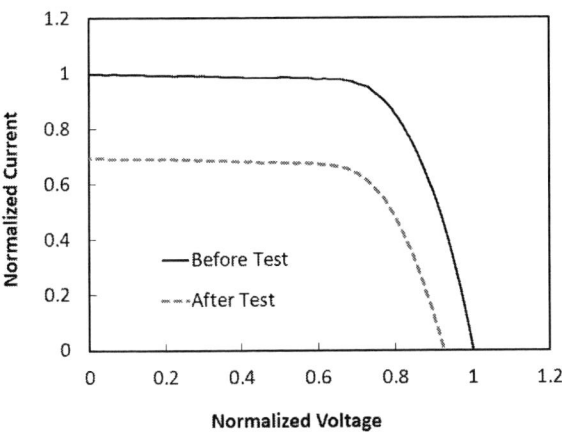

Fig. 2 Change of module electrical characteristics during the high temperature test.

Detected island-type defects

Fig. 3 Back sides of DSC modules after the high temperature test.

To test the durability of the developed module, the DSC module was stored in high temperature of 85°C for 200 hour. The current-voltage characteristics (IV curve) of the module were measured before and after the test and the results are shown in Fig. 2. After test, the power dropped into 64%, or 36% drop. The power drop is mainly from the decrease of short-circuit current.

Several island-type defects, where color changed into grey and shapes are irregular, are found after the rapid drop of efficiency as shown in Fig. 3. To investigate the origin of the defect, the tested module were placed on a hot plate and heated. As the temperature of the module increased, it was detected that the sizes of island-type defects become smaller. Some examples of the test are suggested in Fig. 4. The results mean that the defects are made from the empty liquid electrolyte; as the

temperature increases, the electrolyte expands and empty area get smaller. The induction is also agreed with the previous study results which mentioned the loss of electrolyte after high temperature test [1-3].

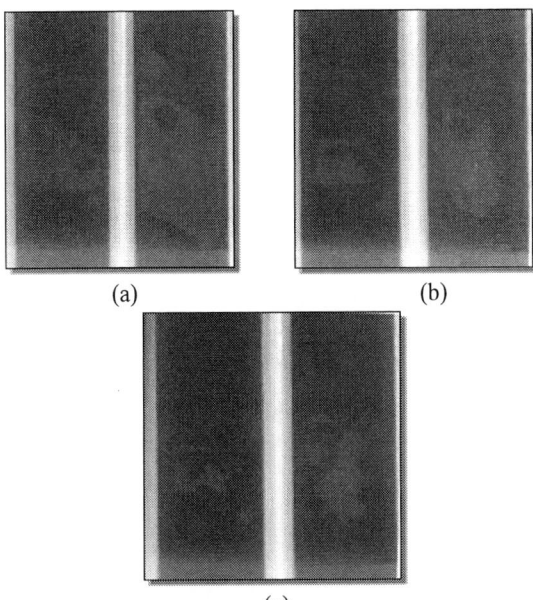

Fig. 4 Test results to analyze the defect origin; change of island-type defects size along the cell temperature of (a) 25°C, (b) 65°C, and (c) 80°C.

3. FE Analysis of DSC Cell

In the section, finite element models are developed to analyze how the electrolyte is lost from the DSC module at high temperature condition. A commercial FE package, ANSYS, is used to build a cross-sectional model for the half of one cell. Two glasses, dye layer, sealant, and Ag electrode are modeled by a 2D structural solid elements (plane182) with plain strain option and shown in Fig. 5. Electrolyte area is modeled by 2D hydrostatic fluid elements (hsfld241), which are not shown in Fig 5. Note that Pt counter electrode layer was not included in the model because the layer is thin enough to neglect.

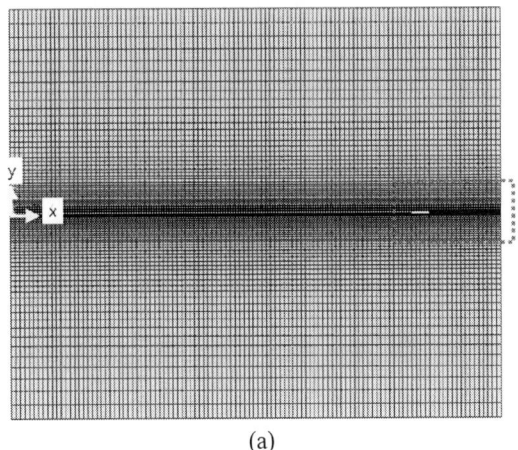

(a)

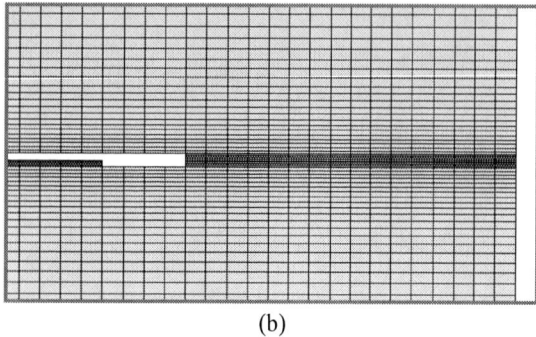

(b)

Fig. 5 2D FE model for half of a cell; (a) full model and (b) detailed view on sealant area.

All the material properties used in the model are listed in Table 1. In case of sealant, properties of Dupont's Surlyn™ 9020 are adapted, not because the material is used in the module (9020 is not designed for solar cell application), but because the Surlyn data are only practically available to authors. For electrolyte, the data of bulk modulus are averaged with several general solvent materials and the data of coefficient of thermal expansion (CTE) are taken from the solvent, acentonitrile (ACN). As there are many assumptions in the model properties, model validation process will be conducted in the next section.

Table 1 Material properties for FE Model

Material	Tensile Modulus (GPa)	CTE** $(10^{-6}/°C)$	Ref.
Soda-lime glass	74	9.5	[4]
Dye layer (TiO2)	65	7.14	[5,6]
Ag electrode	20	19.6	[7]
Sealant	0.1	170	[8]
Electrolyte (solvent)	1.0*	1397	[9,10]

* Bulk modulus
** Coefficient of Thermal Expansion

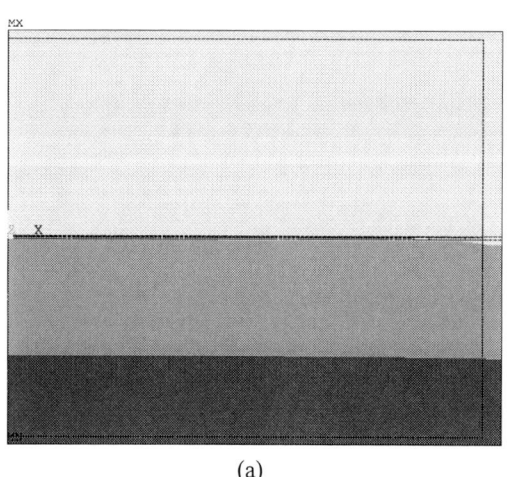

(a)

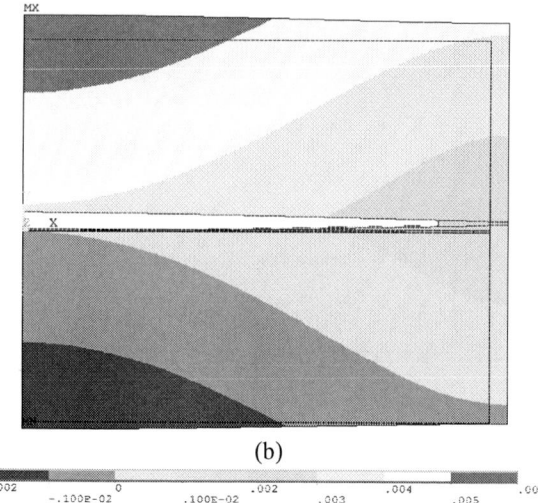

(b)

Fig. 6 Deformation contour in y-direction for the FE models (a) without and (b) with electrolyte elements.

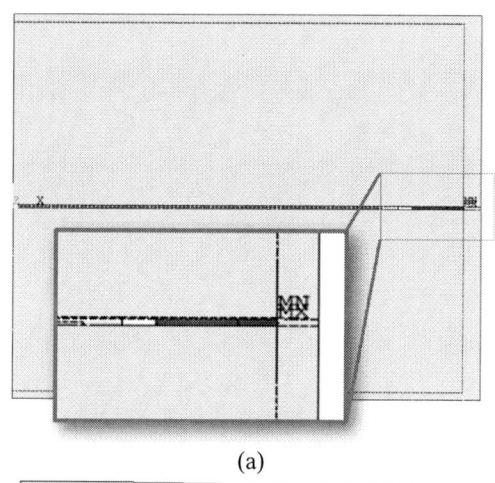

(a)

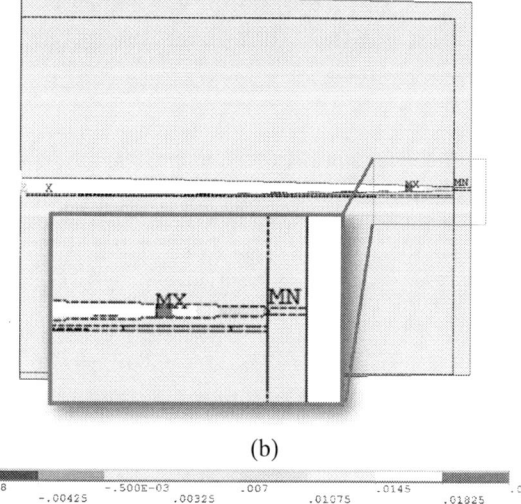

(b)

Fig. 7 Normal strain contours in y-direction for the FE models (a) without and (b) with electrolyte elements.

978-1-4799-4789-8/14 $31.00 © 2014 IEEE

Symmetry and coupling boundary conditions are respectively applied at left and right sides. 85°C temperature is applied to the model. Some analysis results are shown in Fig. 6 and 7. To emphasize the effect of electrolyte, results with and without electrolyte elements are compared. Figure 6 presents the y-direction deformation pattern for the two cases. The model without electrolyte elements has only uniform glass expansion, while the result with electrolyte elements has the deformation gradient; the cell swelled and about 2 μm of gap occurred between the center and edge of a cell at top and bottom glass. The consequences of the deformation are displayed at Fig. 7. Y-direction normal strains, which are principal strain, are maximized at the edge of sealant materials in the model with electrolyte elements, while the results without electrolyte model shows negative strain at the sealant area.

From the analysis results, the breakage or delamination of sealant material is expected at exposure to high temperature. The electrolyte would penetrate into the delaminated sealant and electrode also. The loss of electrolyte would drop the short circuit current and the contaminated electrode would drag fill factor of the DSC module.

4. FE Model Validation Tests

Two different tests were conducted to validate the results derived at the previous section. First, the failed module was disassembled in parts and the area nearby island-type defects was investigated using energy-dispersive X-ray spectroscopy (EDS). The results are shown in Fig. 8. At silver electrode and sealant area, element iodine was detected, which confirms that the iodide-based liquid electrolyte penetrated up to the electrode after the sealant delaminated.

Next, to confirm the deformation pattern of the FE analysis, the real-time deformation of DSC module was measured in high temperature by shadow moiré technology. Shadow moiré is comprised of white light, camera, grating, and temperature-controlled hot plate as illustrated in Fig. 9(a). Using the 0.1 mm pitch of grating and 63.4° of configuration, 50 μm per fringe sensitivity is acquired [11]. The basic sensitivity can be increased up to micro meter level with sub-fringe counting.

In the test, the deformations of two DSC modules were measured in 85°C; normal DSC module and electrolyte-empty module (DSC module without electrolyte was intentionally made for the test.). Two different deformation patterns for the modules are shown in Fig. 9(b) and (c). In the normal DSC module, sub-fringe pattern occurred along the cell line. The pitch of periodical sub-fringe pattern is equivalent to the cell width as illustrated at Fig. 9(b). Less than 5 μm of deformation is expected from the sub-fringe calculation of the picture. The deformation pattern of the electrolyte-empty module does not have the periodical patterns but uniform pattern as seen in Fig. 9(c). All the result corresponds to the analysis results of the previous section.

Both of the validation tests confirmed the effectiveness of the developed FE model.

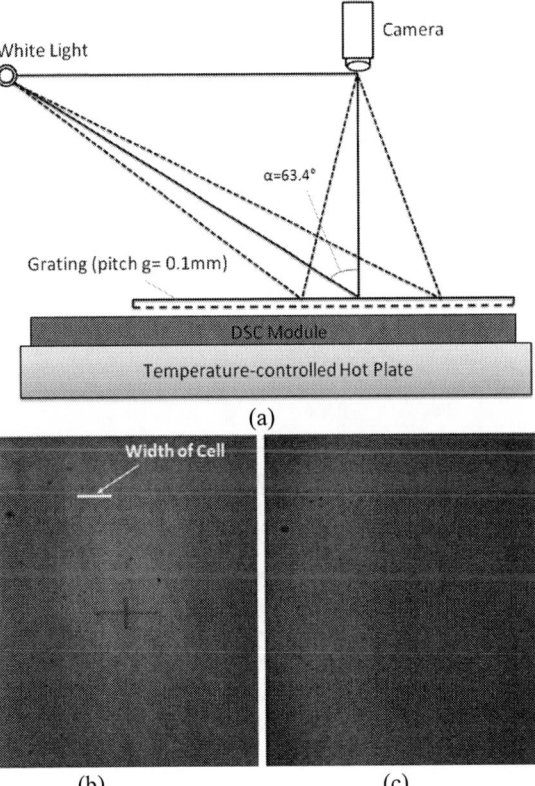

(a)

(b)   (c)

Fig. 9 (a) Shadow moiré setup for model validation and captured fringe results with (b) normal and (c) electrolyte-empty DSC module.

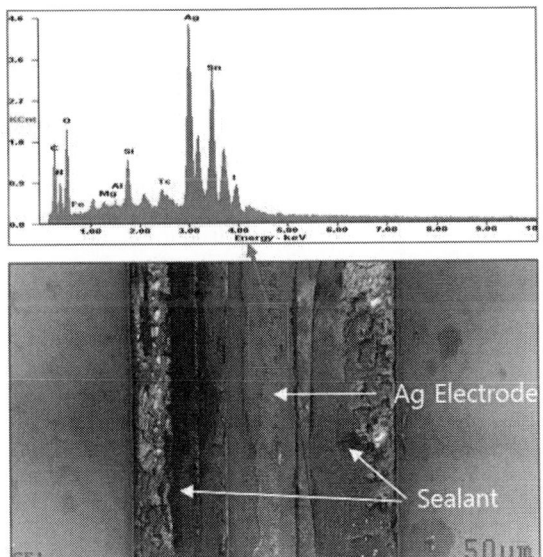

Fig. 8 Failure analysis for the failed DSC module; element iodine is detected in EDX spectra at Ag electrode area.

5. Parametric Study for Durability Improvement

In the section, parametric studies were carried out with the developed FE model to provide design information for durability improvement of DSC module. Two design parameters were selected; cell width and glass thickness. Maximum strains at sealant materials were calculated with the FE model and summarized at Fig. 10. It was observed in the figure that maximum strain decreases rapidly as cell width increases or glass thickness decreases. Therefore, it is suggested to reduce the glass thickness or enlarge the cell width to improve the durability of the module. But the decision must include the consideration of other factors such as efficiency and cost.

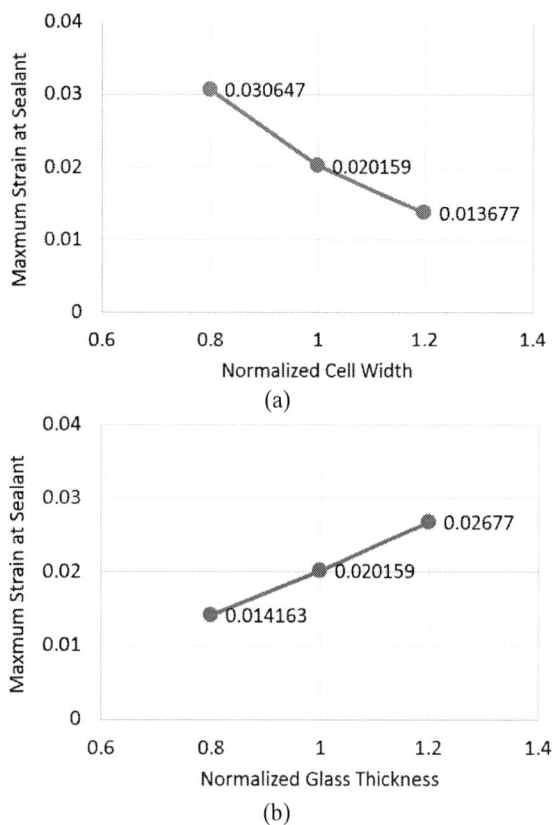

Fig. 10 Parametric study results to reduce the maximum sealant strain; the effect of (a) cell width and (b) glass thickness.

6. Conclusions

The durability of recently developed large-scaled DSC module was tested. High temperature test in 85°C was conducted for 200 hour and power drop of 36% was reported. To investigate the failure mechanism, a finite element model for the DSC module was developed. The model suggested the delamination of sealant. Validation tests for the FE model were conducted. The failed module was disassembled in parts and was investigated. The real-time deformation of DSC module was also measured in high temperature. Both of the validation tests confirmed the effectiveness of the developed FE model. Finally, with the model, parametric studies were conducted. It was suggested to reduce the glass thickness or enlarge the cell width to improve the durability of the module.

Acknowledgments

This study was supported by Korea Institute of Energy Technology Evaluation and Planning (KETEP) and Ministry of Trade, Industry and Energy (MOTIE), Korean government.

References

1. Kontos, A. G. *et al.*, "Long-Term Thermal Stability of Liquid Dye Solar Cells," *J. Phys. Chem. C*, Vol. 117, (2013), pp. 8636-8646.
2. Bari, D., "Thermal Stress Effects on Dye-Sensitized Solar Cells (DSSCs)," *Microelectronics Reliability*, Vol. 51, (2011), pp. 1762-1766.
3. Sastrawan, R. *et al.*, "New Interdigital Design for Large Area Dye Solar Modules Using a Lead-free Glass Frit Sealing," *Pro. Photovolt. Res. Appl.*, Vol. 14, (2006), pp. 697-709.
4. Seward III, T. P. *et al.*, High Temperature Glass Melt Property Database for Process Modeling, Wiley-American Ceramic Society (2005).
5. Anderson, O. *et al.*, "Density and Young's Modulus of Thin TiO2 Films," *Fresnius J. Anal. Chem.*, Vol. 358, (1997), pp. 315-318.
6. Lee, J. H. et al., "Investigation and evaluation of structural color of TiO2 coating on stainless steel," *Ceramics Internat.*, Vol. 38, (2012), pp. S661-S664.
7. Matweb, accessed February 10, 2014, from http://www.matweb.com/search/DataSheet.aspx?MatGUID=69b3f2f666e94095b9d45396a992d970.
8. Matweb, accessed February 10, 2014, from http://www.matweb.com/search/DataSheet.aspx?MatGUID=393b76bd15764059881dbf59dbc30b26.
9. Riddick, J. A. *et al.*, Organic Solvents: Physical Properties and Methods of Purification, Wiley-Interscience (New York, 1986).
10. Wikipedia, accessed Febrauary 10, 2014, from http://en.wikipedia.org/wiki/Bulk_Modulus.
11. Han, C.-W. *et al.*, "High Sensitivity Shadow Moiré Using Nonzero-Order Talbot Distance," *Exp. Mech.*, Vol. 46, (2006), pp. 543-554.

Measuring Young's Modulus of Polysilicon via Cantilever Microbeam Arrays

Steven Kehrberg[*], Markus Dorwarth[*], Sebastian Günther[*], Steffen Markisch[*], Carsten Geckeler[*] and
Jan Mehner[†]

[*]Robert Bosch GmbH, Automotive Electronics, P.O. Box 13 42, 72703 Reutlingen, Germany,
[†]Technical University Chemnitz, Faculty for Electrical Engineering and Information Technologies,
Dept. Microsystems and Precision Engineering, 09126 Chemnitz, Germany

E-Mail: steven.kehrberg@de.bosch.com

Abstract

Polysilicon is the most commonly used material for MEMS. Consequently, the Young's modulus of polysilicon is very important for MEMS design. Measurements of this material parameter can easily be erroneous due to fabrication tolerances, like etch loss variations. The presented method separates etch loss variations from the desired measurement by using an array of micro cantilevers. The measured Young's modulus of 164.1 GPa has a small standard deviation of 0.5 GPa, in comparison to other published data.

1. Introduction

Microelectromechanical systems (MEMS) like accelerometers or gyroscopes are normally made of polysilicon. The Young's modulus of polysilicon depends on the orientation of the grains [1]. For monocrystalline silicon, the Young's modulus can vary from 130 GPa to 188 GPa, depending on the crystal orientations [1]. Consequently, the Young's modulus of polysilicon can theoretically alter in the same range [1]. An accurate value of this material parameter is mandatory for the MEMS design process. For example, it defines the sensitivity of an accelerometer and the drive frequency of a vibratory gyroscope.

The easiest way to handle this problem is by assuming that all grain orientations are equally distributed. This leads to a Young's modulus of 161 GPa [2]. Different growth rates during the vapor deposition can lead to an anisotropic grain orientation and thereby to a different Young's modulus. Previous research showed an influence of deposition temperature, pressure and doping on the crystallographic structure [3].

Therefore, Test samples for investigating the Young's modulus should be fabricated identically in order to have the same distribution of grain orientations. This leads to strict boundary conditions in the geometry of test samples which have to be in the micrometer scale. There already exist different measurement methods to handle this problem like X-ray diffraction [4], tensile testing [5], bending [6] and resonance analyses [7].

In the following, a resonance measurement is used because frequencies can be measured more accurate than absolute displacements in the micro- or nanometer range. A major problem with resonance measurements is that the resonance frequency is not only depending on the Young's modulus. The deep reactive ion etching, which is used for fabrication, can have etch loss variations [8]. Consequently, the unpredictably changed geometry would lead to wrong results. Therefore, a new test sample is presented which allows separate measurements of Young's modulus and etch loss .

2. Test sample

The idea of the test sample is to have a structure as simple as possible to measure the Young's modulus E. Therefore, single clamped beams are used. The first in-plane eigenfrequency f_{eig} of the cantilever beam is analytically described by [9]:

$$f_{eig} \approx \frac{3.52}{2\pi} \sqrt{\frac{EI}{m(l-e_1)^3}} \; . \tag{1}$$

A rectangular cross section is assumed for the beam which leads to the following second moment of area I.

$$I = \frac{(w-e_w)^3 t}{12} \tag{2}$$

$$f_{eig} \approx \frac{3.52}{2\pi} \sqrt{\frac{E}{12\rho(l-e_1)^4}} (w-e_w) \tag{3}$$

The geometry parameters w and l are the width and length of the cantilever without etch loss. The thickness t of the cantilever does not influence the first eigenfrequency. The density ρ of polysilicon is assumed to be equal with the density of monocrystalline silicon [10]. A problem arises for the exact measurement of the etch losses e_1 and e_w. Both parameters can only be measured with an accuracy of >10 nm due to the sidewall roughness. This is not sufficient for an exact calculation of Young's modulus. Relatively long beams of 300 μm length are utilized to reduce the influence of e_1. Extending the beam width is not advisable as it would result in cantilever structures with too high eigenfrequencies out of the measurement range.

In order to handle this problem of etch loss variations, an array of cantilever microbeams with different widths (Fig. 1) instead of a single beam width is used. By placing beams with widths of 2, 3, 4, 5, 6, 7 and 8 μm and the

978-1-4799-4789-8/14 $31.00 © 2014 IEEE

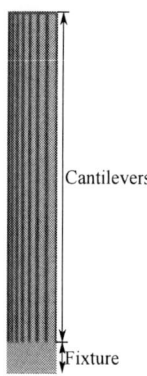

Figure 1. Microscope picture of the cantilever test sample including the fixture. The seven cantilevers are 300 μm long and 2, 3, 4, 5, 6, 7 and 8 μm wide.

same gap next to each other, it can be assumed that all beams have approximately the same e_w.

The expected first eigenfrequency of all cantilevers is in the range between ~25 kHz (for 2 μm wide beam) and ~110 kHz (for 8 μm wide beam). The first out-of-plane eigenmode of all beams (calculated with (3) and inserting the thickness of the beams instead of the width w) is at ~165 kHz. It can be distinguished easily from the in-plane modes due to the large frequency gap. The second in-plane eigenmode can be calculated by [9]:

$$f_{eig2} \approx \frac{22}{2\pi} \sqrt{\frac{E}{12\rho(l - e_1)^4}} (w - e_w) . \qquad (4)$$

The thinnest beam has an eigenfrequency of ~160 kHz. It is also higher than all modes of interest and does not disturb the measurements.

The fixture of the array was designed as massive as possible to reduce undesired influences. In (3) a perfectly stiff fixture is assumed.

3. Measurements

After fabrication all test chips were decapped. The chips were mounted in a vacuum chamber with an ambient pressure of 0.1 mbar to avoid frequency shifts induced by damping. Consequently, it is assumed that the measured resonance frequency is identical with the eigenfrequency.

A piezoelectric shaker, as in [11], was used for stimulation of the test chip. Electrostatic spring softening effects, which are produced by electrostatic stimuli, are prevented by using this mechanical stimulus [12]. Furthermore, this test bench does not have any resonances in the frequency range of interest. Although an out-of-plane shaker was used, a sufficient stimulation of the in-plane modes was achieved. This was obtained by a minor misalignment between shaker and test chip. A periodic chirp from 20 kHz to 120 kHz was utilized as stimulus signal.

The transfer function of the cantilever beams was measured by the laser Doppler vibrometer of the Polytec MSA 500. This 1D-vibrometer is made for measuring out-of-plane velocities. Nevertheless, it is possible to find in-plane resonances with the vibrometer by placing the laser spot at the edge of the beam. The intensity of the reflected laser beam is amplitude modulated with the in-plane frequency in the case of an in-plane movement.

The signal-to-noise ratio was further optimized by using complex averaging over ten measurements of every cantilever. In order to verify that the measured displacement is the desired in-plane eigenmode, the stroboscope of the MSA 500 was used. The in-plane eigenmode could be clearly identified using this stroboscopic video microscopy. The stroboscope was not used for all measurements because of time consuming manual measurements. Furthermore, the test sample is designed to have no other eigenmodes in the frequency range of interest which eliminates the risk of wrong eigenmode correlations.

A sum of 399 beams (57 blocks consisting of seven beams) from different chips has been semi-automatically measured using the vibrometer.

4. Analysis

All measured transfer functions were analyzed to find the first eigenfrequency. This has been accomplished by a least square fit of a Lorentzian function [13]. The measured resonance frequencies were verified by checking the bandwidth of the fitted Lorentzian function whether they are in a plausible range.

A. Analytical Calculation

Afterward, the measured first eigenfrequencies of the cantilever beams have been plotted versus the beam widths w without etch loss (Fig. 2). A linear correlation between the two plotted values is expected from (3). All linear fits of the eigenfrequencies showed small quadratic residuals (Fig. 3). The reason for this undesired behavior was found in the imperfect fixture of the cantilevers which was not stiff enough. The analytical equation (3) is only valid for perfect stiff boundary conditions at the face of the beam which can hardly be reached in reality. Using the analytical equation in combination with a too soft fixture would lead to an erroneously reduced Young's modulus by a few percent.

B. Finite Element Calculation

In order to handle this problem, a 3D finite element model (Fig. 4) was build which includes the fixture. Thereby, the influence of the too flexible fixture can be taken into account. The fully parametrized model was used for a parameter study. The first eigenfrequency $f_{eig,i}$ of all seven beams ($i = 1 \ldots 7$) was calculated as a function of

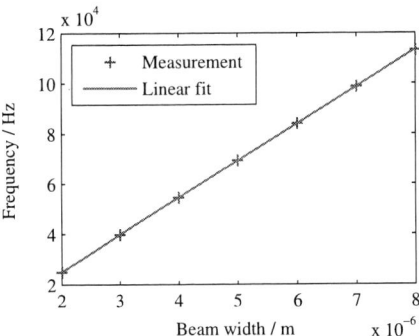

Figure 2. The seven measured first resonance frequencies of a test sample are plotted versus the beam width without etch loss. The expected linear correlation in (3) can be seen.

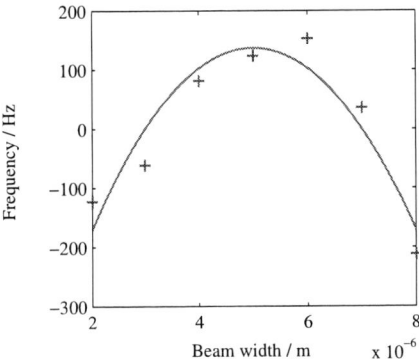

Figure 3. The residuals of the measured data and the linear fit (Fig. 2) are plotted versus beam widths without etch loss. The residuals show a quadratic behavior (green). The expected reason for the quadratic residuals is the too soft fixture.

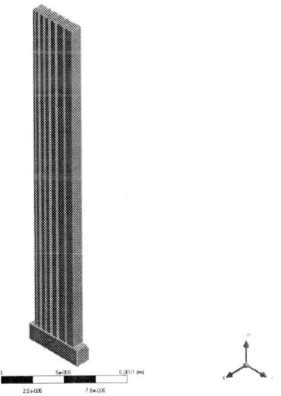

Figure 4. Finite element model of the cantilever test sample including the fixture to calculate the effect of the fixture on the first eigenfrequencies of all beams.

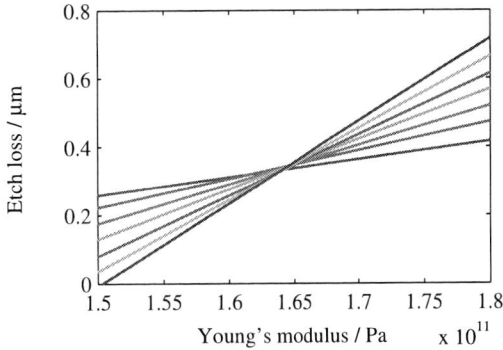

Figure 5. Etch loss is plotted as a linear function of Young's modulus as described in (7). The steepest line (black) represents the cantilever with a width of $2\,\mu m$ and the flattest line (blue) represents the cantilever with a width of $8\,\mu m$.

Young's modulus E and etch loss e_w using finite element modal analysis:

$$f_{\mathrm{eig},i} = f(E, e_\mathrm{w}) . \tag{5}$$

In order to get a mean value for Young's modulus and etch loss for every array consisting of seven beams, the inverse function has to be found. A plane fit (6) with the parameters a_i, b_i and c_i was utilized to find the desired function for every beam width.

$$f_{\mathrm{eig},i} = a_i + b_i E + c_i e_\mathrm{w} \tag{6}$$

$$e_\mathrm{w} = -\frac{b_i}{c_i} E + \frac{f_{\mathrm{eig},i} - a_i}{c_i} \tag{7}$$

The converted plane equation (7) was reduced to seven linear equations (for one block of seven cantilevers) by inserting the measured eigenfrequencies $f_{\mathrm{eig},i}$. Consequently, the etch loss is now a linear function of the Young's modulus. In Fig. 5, the seven plotted equations for all beams of one test sample can be seen. All straight lines cross each other as they have different slopes. If all beams would have the same Young's modulus and etch loss, they would cross in a single point. There is a minor variation between all crossing points due to fabrication tolerances and statistical errors.

It is not advisable to calculate all crossing points and average them in order to find a mean value for Young's modulus and etch loss. When two lines have only slightly different slopes they can cross at an implausible value of etch loss and Young's modulus (compare Fig. 5). This can be prevented by calculating the variance of etch loss between all lines. It is assumed that all beams have almost identical etch losses as they are situated next to each other with equal gaps.

When plotting this variance versus the Young's modulus (Fig. 6), a single minimum can be found. This minimum

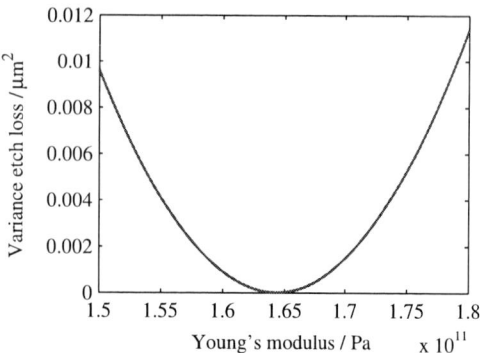

Figure 6. Variance of etch loss versus Young's modulus of the seven lines plotted in Fig. 5. It is assumed that all beams have almost identical etch losses. Thereby, an average Young's modulus can be found by searching the minimum of variance in etch losses.

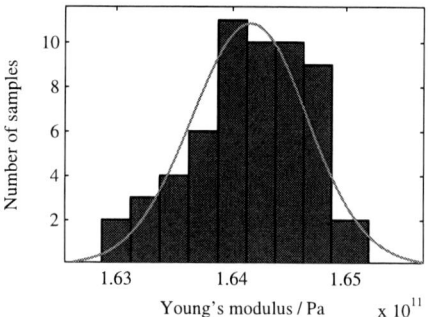

Figure 7. Measured distribution of the Young's modulus with Gaussian fit. The distribution has a mean value of 164.1 GPa and a standard deviation of 0.5 GPa (0.3%). 57 test samples (each consisting of seven beams with different widths) were successfully measured in order to obtain the provided data.

represents the mean Young's modulus of all seven beams of one block. The mean etch loss can be calculated by averaging all seven etch losses at this Young's modulus.

5. Results

The described method was successfully applied for 57 test samples (each consisting of seven beams). Thereby, a total number of 399 beams had been measured. The calculated data is presented in Fig 7. A mean value of 164.1 GPa is calculated for the Young's modulus. The standard deviation is 0.5 GPa (0.3%).

6. Conclusions

The novel method was used for the measurement of almost 400 beams which were made by a Bosch foundry process [14]. The calculated Young's modulus of 164.1±0.5 GPa is in the expected range for recent precise measurements of polysilicon [15]. The low standard deviation of 0.3% compared to other published data

underlines the reliability of the presented method. These very good results are achieved by reducing the influence of etch loss variations through a new evaluation method in combination with a newly developed test chip and an experimental measurement setup.

Acknowledgements

The authors thank all colleagues at Robert Bosch Automotive Electronics, Engineering Sensor Technology, for their valuable contributions to this work.

References

[1] M. A. Hopcroft, W. D. Nix, and T. W. Kenny, "What is the young's modulus of silicon?," *Microelectromechanical Systems, Journal of*, vol. 19, no. 2, pp. 229–238, 2010.

[2] H. Guckel, D. W. Burns, H. Tilmans, D. DeRoo, and C. Rutigliano, "Mechanical properties of fine grained polysilicon-the repeatability issue," in *Solid-State Sensor and Actuator Workshop, 1988. Technical Digest., IEEE*, pp. 96–99, IEEE, 1988.

[3] P. Joubert, M. Sarret, L. Haji, L. Hamedi, and B. Loisel, "Pressure dependence of in situ boron-doped silicon films prepared by low-pressure chemical vapor deposition. i. microstructure," *Journal of applied physics*, vol. 66, no. 10, pp. 4806–4811, 1989.

[4] P. R. Cantwell, H. Kim, M. M. Schneider, H.-H. Hsu, D. Peroulis, E. A. Stach, and A. Strachan, "Estimating the in-plane Young's modulus of polycrystalline films in MEMS," *Microelectromechanical Systems, Journal of*, vol. 21, no. 4, pp. 840–849, 2012.

[5] W. N. Sharpe Jr, B. Yuan, R. Vaidyanathan, and R. L. Edwards, "Measurements of young's modulus, poisson's ratio, and tensile strength of polysilicon," in *Micro Electro Mechanical Systems, 1997. MEMS'97, Proceedings, IEEE., Tenth Annual International Workshop on*, pp. 424–429, IEEE, 1997.

[6] G. McShane, M. Boutchich, A. S. Phani, D. Moore, and T. Lu, "Young's modulus measurement of thin-film materials using microcantilevers," *Journal of Micromechanics and Microengineering*, vol. 16, no. 10, p. 1926, 2006.

[7] L. Kiesewetter, J.-M. Zhang, D. Houdeau, and A. Steckenborn, "Determination of young's moduli of micromechanical thin films using the resonance method," *Sensors and Actuators A: Physical*, vol. 35, no. 2, pp. 153–159, 1992.

[8] Y.-H. Jang, J.-W. Kim, J.-M. Kim, and Y.-K. Kim, "Engineering design guide for etch holes to compensate spring width loss for reliable resonant frequencies," in *Micro Electro Mechanical Systems (MEMS), 2012 IEEE 25th International Conference on*, pp. 424–427, IEEE, 2012.

[9] H. Lang, M. Hegner, and C. Gerber, "Nanomechanical cantilever array sensors," *Springer Handbook of Nanotechnology, ISBN 978-3-540-29855-7. Springer-Verlag Berlin Heidelberg, 2007, p. 443*, vol. 1, p. 443, 2007.

[10] M. Fürtsch, *Mechanical properties of thick polycrystalline silicon films suitable for surface micromachining*. Shaker, 1999.

[11] S. Kehrberg, P. Wellner, C. Geckeler, and J. E. Mehner, "Modal analysis of MEMS using ultrasonic base excitation," in *Systems, Signals and Devices (SSD), 2012 9th International Multi-Conference on*, pp. 1–4, IEEE, 2012.

[12] S. Chowdhury, M. Ahmadi, and W. C. Miller, "Pull-in voltage study of electrostatically actuated fixed-fixed beams using a vlsi on-chip interconnect capacitance model," *Microelectromechanical Systems, Journal of*, vol. 15, no. 3, pp. 639–651, 2006.

[13] P. R. Bevington and D. K. Robinson, *Data reduction and error analysis for the physical sciences*, vol. 2. McGraw-Hill New York, 1969.

[14] F. Laermer and A. Urban, "Challenges, developments and applications of silicon deep reactive ion etching," *Microelectronic Engineering*, vol. 67, pp. 349–355, 2003.

[15] K. Hemker and W. Sharpe Jr, "Microscale characterization of mechanical properties," *Annu. Rev. Mater. Res.*, vol. 37, pp. 93–126, 2007.

Degradation of silicone in white LEDs during device operation: a finite element approach to product reliability prediction

S. Watzke, P. Altieri-Weimar
Osram Opto Semiconductors GmbH
Regensburg, Germany
stefan.watzke@osram-os.com, paola.altieri-weimar@osram-os.com

Abstract

Silicone, which is a very common material for Light Emitting Diode (LED) packaging components like lens, casting and housing, undergoes degradation during high temperature and current operation. Indeed, electrical and optical losses cause material shrinkage and hardening, inducing mechanical stress within the LED assembly, which can end up into crack formation in silicone. In order to evaluate the reliability of LED package regarding the silicone crack, a degradation material model is developed, which is based on the experimental investigation of the mechanical properties of silicone during degradation. A coupled thermo-optical model is used for the calculation of the temperature distribution in the device during steady-state operation. The crack reliability model, which is build combining the stress simulation results based on finite element approach and the visual inspection of the corresponding LED package during steady-state operation, is used to estimate the package lifetime depending on the operation conditions.

1. Introduction

Due to its high transparency in the visible wavelength range and long term optical stability against light radiation down to the deep blue [1,2] as well as the excellent workability for jetting and molding processes, methyl based silicone is widely used in packaging of LEDs [3]. In particular, this silicone is often employed in white LEDs as covering layer of the blue diode and is filled up with phosphor particles, which are responsible for blue light conversion into yellow and red wavelength, resulting in white light emission (Fig. 1).

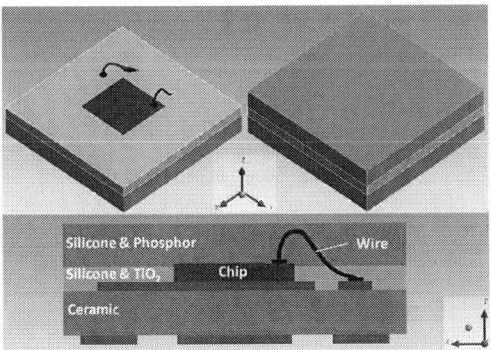

Fig. 1: CAD model of a white LED.

Despite the high efficiency of white LEDs, which is nowadays close to 110 lmW^{-1} for a color temperature of 2700K, some power is still lost, producing self-heating of the device. Heat is generated on one hand by electrical losses in the diode, due to limited diode quantum efficiency and to series resistance [4], on the other hand by optical losses in the phosphor particles, associated to the Stoke's shift from short to long wavelength as well as to limited conversion efficiency of the phosphor [5]. Whereas the first are merely responsible for heating up of the diode, which is usually good thermally connected to the driving board, resulting into moderate temperature increase, the heat generated at the phosphor particles can only be hardly dissipated, because of the pretty low thermal conductivity of silicone, and can induce severe warming. In some operation and environmental conditions the temperature of the silicone can increase far beyond 150°C, inducing degradation of the material. A typical appearance of the white LED investigated in this paper after long operation at high temperatures and driving current is shown in Fig. 2. Two features can be observed: a ring-shaped surface alteration (a) and a surface depression in the middle (b). The first is associated to micro-cracks and blistering in the silicone layer and is responsible for a shift of the chromaticity coordinate of the LED, thus affecting the reliability of the device.

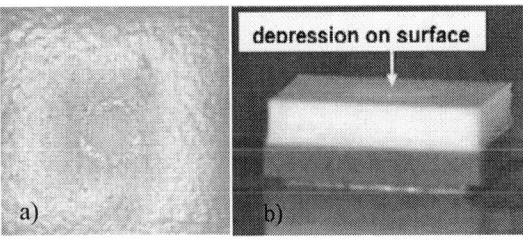

Fig. 2: Microscope pictures of a white LED after long operation at high temperature and current.

In order to investigate the root cause of this failure and to enable prediction of device lifetime for different operation conditions, an extensive set of experiments and simulations is carried out. In section 2 the heating up of the LED is illustrated, which is estimated by means of a coupled thermo-optical simulation of the device under steady-state operation. In section 3, the change of the mechanical properties of silicone test samples during high temperature storage is presented. Furthermore, a degradation material model is proposed, to be employed

978-1-4799-4789-8/14 $31.00 © 2014 IEEE

within the thermo-mechanical FE model for simulation of stress in silicone during device operation. In section 4, the results of the stress simulations are discussed and the stress based reliability model is elucidated, which is applied to predict the lifetime of the white LED under different operation conditions.

2. Temperature distribution under operation

During current operation, the blue light emitted by the diode is coupled into the silicone layer and is partially scattered and partially converted into yellow/red light by the phosphor particles. As already mentioned in the introduction, light conversion is associated with heat generation, which is indeed depending on the amount of absorbed blue light and on the light conversion efficiency of the phosphor particles embedded in the silicone, thus inducing locally warming in the silicone. Additional internal losses induce self-heating of the diode, which again affect the temperature distribution in the silicone. In order to take into account these different effects and determine correctly the temperature distribution in the silicone, which is responsible for silicone degradation on long term device operation, a coupled opto-thermal simulation is performed. A commercial ray-tracing tool is used to simulate the light propagation in the silicone filled with phosphor particles. Light scattering, absorption and conversion are calculated and the resulting distribution of the optical losses in the silicone is used as local heat generation rate in the FE thermal model, build in the commercial FE tool Ansys (Fig. 3). Additionally, the heat generated in the diode is considered in the thermal model, where the heat is dissipated in the cold plated located under the driving board.

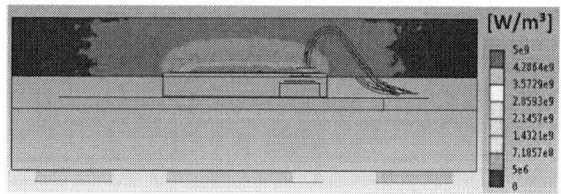

Fig. 3: Heat generation rate in converter silicone layer in a white LED at operation current of 1A.

The resulting temperature distribution at 1A steady-state current operation with a cold plate temperature of 85°C is shown in Fig. 4. It is worthy to note that the highest temperature in the silicone, which is about 230°C, does not occur there where the optical loss density is the most. This is due to the relative low thermal conductivity of silicone, which is in the range of 0.17-0.4 Wm^{-1}K^{-1} [6], depending on the degree of filling, and therefore much lower than that of the other LED component materials, which varies in the range of 100-400 Wm^{-1}K^{-1}. Moreover, the simulation shows that the temperature on the surface of the diode is about 110°C, which is close to the maximal allowed temperature of 125°C mentioned in the diode application specifications. Thus, the temperature range around 230°C is the one of interest, where the degradation of silicone inducing LED surface alteration must be investigated.

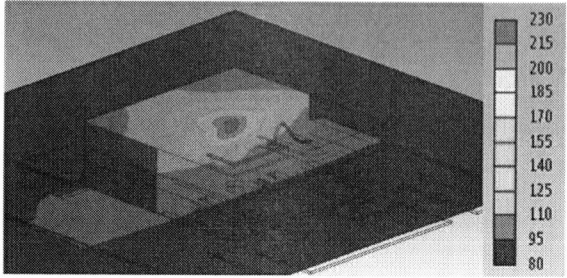

Fig. 4: Temperature distribution in white LED on metal core driving board 1A steady-state current operation mounted on a cold plate at 85°C.

3. Degradation of the mechanical properties

Storage of silicone in air at high temperatures causes depolymerization, which is a sum of several reactions [7,8]. The polymer chains are split by the thermal energy, ending with chain ends determinate by radicals. This thermo-mechanically instable configuration forces secondary reactions. The first one is a siloxane bond rearrangement leading to the elimination of volatile cyclic oligomers and the shortening of the residual chain length. The second reaction is the oxidation of the condensed phase leading to further tight crosslinking of the polymer. These reactions are associated to a weight and volume shrinkage and to an increase of material stiffness. In literature, degradation is often investigated by differential scanning calorimetry[7,8], where the reaction heat and degradation rate can be determined. However, since the degradation of silicone in the relevant temperature range for this study is pretty slow, this method is not practicable. Here, another approach is proposed to determine the degree of degradation, which is based on the amount of material volume shrinkage. Indeed, this quantity is more reliable than the reduction of break strain proposed by *Moghadam et al.* [9], which is more influenced by variation in sample preparation. As previously discussed, degradation comes along with evaporation of material and further cross-linking and therefore it is meaningful to use volume change to measure the degree of degradation implicitly. Since the investigated silicones are homogeneous and isotropic, only the sample length is monitored during degradation. The length measurements are carried out with an optical measurement system and the sample length is averaged out of 6 positions. All the measurements are carried out at the constant ambient temperature of 22°C. The samples are obtained by curing silicone at 150°C for 2 hours in PTFE profiles of dimensions 45 x 10 x 2 mm. The degradation is performed at 4 different storage temperatures between 150°C and 270°C. For this purpose, the samples are stored in a furnace without any mechanical constrains except gravity. In Fig.5 the shrinkage results are shown, where shrinkage is calculated as following:

$$s(t) = \frac{l_0 - l(t)}{l_0} \qquad (1)$$

l_0 being the initial sample length and $l(t)$ the sample length after a degradation time t at constant storage temperature. AS reported in literature [7], degradation rate increases with increasing aging temperature, which is also true for the measured sample length shrinkage. In order to determine the amount of shrinkage between the time readouts and eventually extrapolate it at temperatures just above 150°C for very long aging time, the experimental data are fitted by the following empirical function:

$$s(t) = 1 - \left(1 + m(T - T_0)\right)^{-t^{\left(B_0 + s(T - T_0)\right)}} \qquad (2)$$

where T is the absolute temperature in °C, while m, s, T_0 are B_0 are constant equal to 6.81e-5, 4.83e-3, 159.33 °C and 1.04e-1 respectively. Below the reference temperature T_0 no further thermal aging will occur on significant time scales.

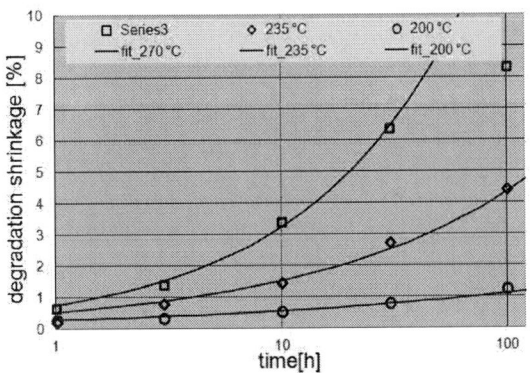

Fig. 5: Silicone length shrinkage at over time after storage at different aging temperatures (symbols).

Along with volume shrinkage, degradation induces hardening of the silicone, which is investigated in tensile tests. In Table 1 the evolution of the Young's modulus during time is shown for samples aged at 270 °C. The hardening of the material can be explained with an increasing number of cross-links, which reduces the mobility of the polymer chains, thus making the silicone more and more a glass-like material. Therefore it is assumed that the change in Young's modulus is coupled to the degradation shrinkage, which is representative for the amount of material degradation.

Table 1: Young's modulus and degradation shrinkage of silicone versus time for a degradation temperature of 270 °C.

Time [h]	0	1	3	10
Modulus [MPa]	0.97	1.93	19.21	57.03
Shrinkage [%]	0	0.66	1.38	3.40

In Fig. 6 the experimental Young's modulus data are associated with the degradation shrinkage measured for the appropriate aging times. Subsequently the data are fitted with a sigmoidal growth function, in order to determine the material stiffness between the time readouts.

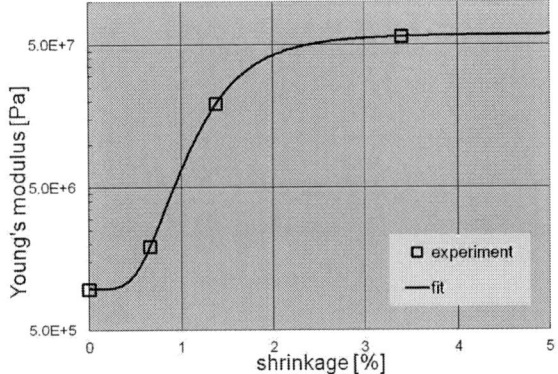

Fig. 6: Superposition of Young's modulus and shrinkage of silicone for different degradation temperatures and time.

The calibrated functions describing the time evolution of silicone shrinkage and hardening during degradation are implemented in a user subroutine within the Ansys FE code, which is used for calculation of deformation and stress in silicone during LED operation.

3. Stress simulation and reliability model

In Fig.7, the thermo-mechanical stress in silicone within a white LED at 1A steady-state current operation for a cold plate temperature of 85°C is shown after few seconds of operation (a) and after 24 hours operation (b). For the stress calculation, the temperature distribution depicted in Fig. 4 is used. It is worthy to note that the stress increases remarkably during one day steady-state operation and that the ring-shaped distribution of maximal principal stress on the surface of the silicone is comparable to the surface appearance of the white LED after high temperature and current degradation, shown in Fig. 2. Therefore, the micro-cracks in silicone, which cause high scattering of light and therefore are responsible for the "white" ring on the top of the LED, are induced by the stress accumulated in silicone, which is generated by the degradation volume shrinkage.

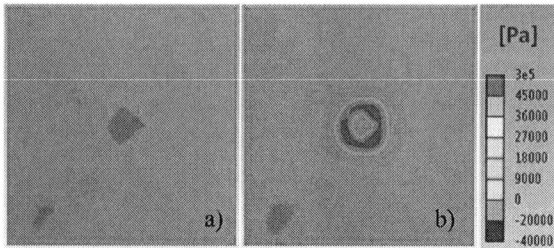

Fig.7: Maximum principle stress on the surface of the white LED after few seconds (a) and 24 hours (b) at 1A steady-state current operation with a cold plate temperature of 85°C.

The simulated maximal stress in the ring for different current operation is plotted over time for two different operation currents. In Fig. 8, the time points at which the "white" ring is observed during the corresponding reliability tests of the white LED are reported. The experimental time to "failure" is 24 hours for steady-state operation at 1A and 96 hours for 0.85A, both with cold plate temperature at 85°C. As a matter of fact, the micro-cracks associated to the "white" ring, occur at the same level of stress on the surface of the silicone, which is about 51000 Pa. This result provides a failure criterion for the micro-crack formation, which is based on the amount of stress on the surface of the silicone. Thus, it is possible to determine the lifetime to micro-crack formation for given operation conditions, by evaluating the time at which the stress on the surface of the silicone reaches the critical value of 51000 Pa. Based on this failure criterion, the FE thermo-mechanical model can be employed to generate the so called de-rating curves, which give indication to the customers about the lifetime to failure of the LED device. These are depending on the operating current and are usually referred to the solder point temperature, which is the temperature at the interconnect between the LED device and the driving board. In Fig. 9, the derating curves for the white LED regarding the micro-cracks failure in silicone are shown. In order to satisfy the typical required lifetime for LED application in solid state lighting of about 10000 hours, the white LEDs investigated in this study can be driven only at low current and moderate ambient temperatures.

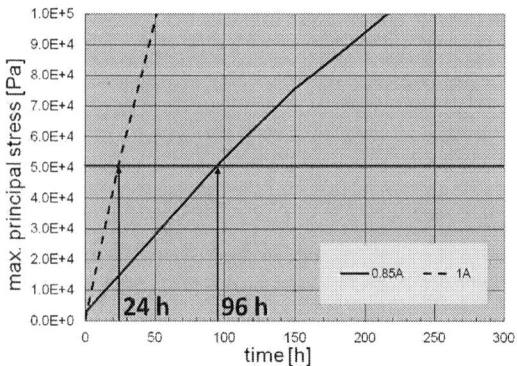

Fig. 8: Maximal stress at the ring on the top of silicone versus time, for two different operating currents.

Of course, optimization of converter silicone layer regarding its filling degree, influencing the thermal conductivity, its overall thickness and the choice of phosphor particles with higher conversion efficiency, can all contribute to a reduction of the maximal temperature in the silicone, thus enhancing the lifetime to micro-crack formation. Moreover, the implementation of more stable materials against high temperature degradation might be a decisive development step for long term reliable white LEDs in the high power application field.

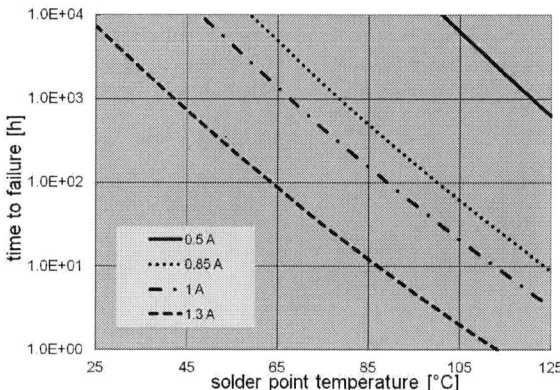

Fig. 9: De-rating curves of white LEDs regarding the micro-crack formation in silicone filled with phosphor particles.

4. Conclusions

It is shown that the thermal degradation of silicones comes along with the change of several material properties, like volume shrinkage and hardening. A material degradation model based on the time evolution of the material shrinkage, which is used to quantify the state of degradation, is developed and implemented in the FE thermo-mechanical model to calculate the stress in silicone during LED device operation. The temperature distribution in the device is calculated taking into account the electrical losses in the blue diode as well as the optical losses in the silicone filled with phosphor particles, which is determined by means of a ray-tracing optical simulation. A crack criterion based on the amount of stress on the surface of the silicone is validated by optical inspection of the white LED during reliability test experiments. The FE thermo-mechanical model is employed to generate the de-rating curves of the white LED for the determination of the lifetime to formation of micro-cracks in the converter silicone depending on the solder point and operation current.

Acknowledgments

We would like to thank Dr. Hans Walter from Fraunhofer IZM for their support with measurement equipment. Dr.

Michael Stoll and Dr. Alexander Linkov from OSRAM Opto Semiconductors for the support with reliability data and optical simulation results, respectively. Moreover, we would like to thank Prof. H. Müller from TU Berlin for the helpful suggestions and profitable discussions.

References

[1] Y.H. Lin, J.P. You, Y.C. Lin, N.T. Tran, and F.G. Shi, "Development of High-Performance Optical Silicone for the Packaging of High-Power LEDs," *IEEE Transactions on Components and Packaging Technologies* , vol. 33, no. 4, pp. 761-766, 2010.

[2] H.R. Fischer et al., "Degradation mechanism of silicone glues under UV irradiateion and options for designing materials with increased stability," *Paloymer Degradation and Stability*, vol. 98, pp. 720-726, 2013.

[3] M. Bahadur, A.W. Norris, A. Zarisfi, J.S. Alger, and C. Windiate, "Silicone materials for LED packaging," in *Sixth International Conference on Solid State Lighting*, San Diego, California, USA, 2006.

[4] Y. Narukawa et al., "White light emitting diodes with super-high luminous efficacy," *Journal of Physics D: Applied Physics*, vol. 43, 2010.

[5] S.J. Oh et al., "Enhanched phospor conversion efficiency of GaN-based white light-emitting diods having dichroic-filtering contacts," *Journal of Materials Chemistry C*, vol. 1, no. 36, pp. 5733-5740, 2013.

[6] C.H. Lui, H. Huang, Y. Wu, and S.S. Fan, "Thermal conductivity improvement of silicone elastomer with carbon nanotube loading," *Applied Physics Letters*, vol. 84, pp. 4248-4251, 2004.

[7] G. Camino, S.M. Lomakin, and M. Lazzari, "Polydimethylsilixan thermal degradation Part 1: Kinetic aspects," *Polymer*, vol. 42, pp. 2395-2402, 2001.

[8] T.H. Thomas and T.C. Kendrick, "Thermal Analysis of Polydimethylsilixanes. I. Thermal Degradation in Controlled Atmospheres," *Journal of Polymer Science: Part A-2*, vol. 7, pp. 537-549, 1969.

[9] M.K. Moghadam, J. Morshedian, M. Ehsani, and M. Bahrami, "Lifetime Prediction of HV Silicone Rubber Insulators based on Mechanical Tests after Thermal Ageing," *IEEE Transactions on Dielectrics and Electrical Insulation*, vol. 20, no. 3, pp. 711-716, 2013.

Creep and Fatigue as Main Degradation Phenomena in Reliability of Solder Joints

K. Jankowski, A. Wymysłowski

Wroclaw University of Technology, Faculty of Microsystem Electronics and Photonics,
11/17 Janiszewskiego Street, 50-372 Wrocław, Poland

Abstract

Observed from many years a rapid growth of electronic devices production, induced by introducing new functionalities for customers and moving towards environmentally friendly packaging caused that life prediction of solder joints is more prominent issue in electronic assembly.

Thanks to their properties solder alloys are a part of fundamental components of each printed circuit boards. They are responsible for mechanical (structural support, heat transportation) and electrical (signals transmission) connections. Therefore it is very important to estimate behavior of solder joint in various work conditions. One of the problems is failure due to the combined loading of creep and fatigue phenomena which initiate degradation process.

In daily engineer practice creep and fatigue phenomena are taken into account only in an independent way. As a result reliability prototyping is not very precise, long lasting and very costly. The usual tests are based only on one dominating failure mode, which can last for a number of months. Moreover all failure analysis are done under the simplified conditions, which means that obtained results are not so reliable. To accelerate that kind of tests simultaneously acting of creep and fatigue is needed. That combination can include analytical, experimental and numerical approach for understanding the combined multi-loading and multi-failure problem. Therefore, an appropriate understanding and development of analytical methods and experimental tools for multi-failure criteria analysis could be very helpful. In this work new method, which allows for combined failure of solder joints, in order to examine simultaneously creep and fatigue phenomena is presented.

Using this or similar method with numerical modeling and simulation technique can help establish the failure time, at which specimen is damaged due to accelerated tests and behaviour of the sample based on obtained material model describing shape and parameters of tested solder joints [1].

1. Problem description

The fundamental problem is related to low deformation and stress distribution which are consequences of difference between thermal expansion coefficients (**CTE**) of different individual component materials. Solder joint does not have adequate ductility to ensure the repeated relative displacements between the chip carrier and the circuit board [2]. The second problem is time needed to make reliability tests. It is too long in regard to progress, which is observed in nowadays electronic industry. As a result there is a strong request to develop a new method that can help in significant time decreasing of long-term reliability tests and accuracy of failure prediction.

The standard tests of solder joint life prediction are based on thermo-mechanical cycling due to elevated temperature and shearing loading conditions. Unfortunately these conditions are not enough to proper and faster determination of the reliability. Thereby, there are many researches which are concerned with replacing the standard thermal cycling towards the multi-failure numerical analysis. In fact such a method should include the multi-failure criteria problem as e.g. combination of creep and fatigue. Thus it would be possible to improve the reliability prediction based on real work conditions and the same to shorten the overall testing time and cost.

1.1 Creep

Creep is a physical phenomenon associated with long-term exposures occurred at high levels of stress and temperature, e.g. compressing. This causes irreversible deformation of the material due to constant loading applied during a sufficient period of time. Creep is dominating failure mode in case of elevated temperature conditions and corresponding dwell time.

1.2 Fatigue

Fatigue in opposition to creep is related to cycling loading conditions. In another words it is a weakening of the material caused by repeatedly applied loads. Repeated loading and unloading above a certain threshold creates microscopic cracks. When critical amount of cycles is reached the fracture occurs. There are two types of fatigue:

- High cyclic (about 10^4 to 10^8 cycles)
- Low cyclic (typically failure is occurred less than 10^4 cycles)

In case of electronic packaging and features of used measuring device only low cyclic method could be applied with the maximum test speed up to 5mm/s.

1.3 Combination of creep and fatigue

Independent tests, which are taken into account in today failure analysis bring in only some knowledge about the solder joints reliability. Thus, a simplified assumption that only one failure mode is dominating under selected loading conditions requires some corrections. In order to extract essential properties of solder alloys measurements were executed mainly for the analysis of conventional individual failure modes i.e. creep, fatigue. Then an integrated mode, typical for the multi-failure criteria, i.e. creep plus fatigue [3], allowed to study an appropriate model of damage accumulation and thus more accurate analysis and reliability forecasting due to an existence of multi-failure modes. In case of standard cycling tests the dominating failure criteria is assumed to be fatigue.

Unfortunately in case of accelerated thermal cycling (**ATC**), due to increased temperature (Fig. 1), the basic failure mode would be rather a combination of creep and fatigue, which is not taken into account in the standard analysis.

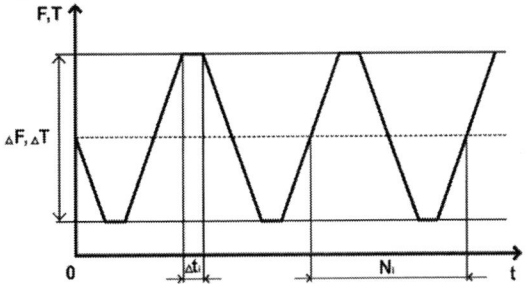

Figure 1. Typical ATC loading parameters.

Thus, during low cycling thermal loading conditions two failure progression modes exist at the same time and the same they should be investigated together, which is shown in the figure 2.

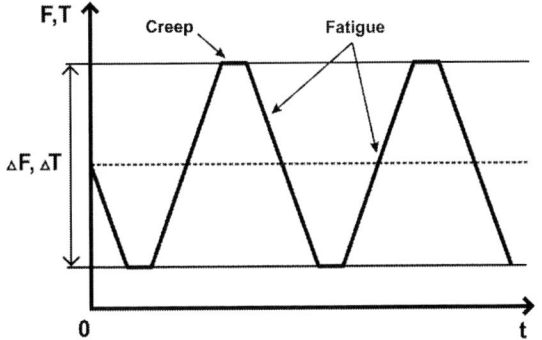

Figure 2. Failure progression modes due to simultaneously acting creep and fatigue.

The above problem can be considered analytically as the multi-loading and multi-failure state. Analytical analysis of this state could be done according to one of damage accumulation models [4], which in certain cases can be considered as linear and based on the following formula:

$$\sum_{i=1}^{n_f} \frac{N_i}{N_f} + \sum_{i=1}^{n_t} \frac{\Delta t_i}{t_r} = D \qquad (1)$$

where: n_f – number of fatigue cycles, N_i – number of cycles, N_f – fatigue limit, Δt_i – time at given stress and temperature, t_r – creep limit, n_t – number of dwell time periods, D – failure modes damage

According to the above equation (1) failure is assumed to happen when the sum of the portions of failure modes damage D reaches 1, which is schematically depicted in the figure 3.

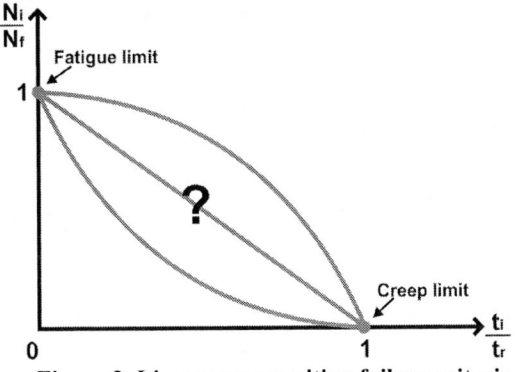

Figure 3. Linear superposition failure criteria.

Thus, failure will depend on separate portions of damage accumulated by one of the failure modes, which is either creep or fatigue.

Basing on the above relationships it is possible to determine a boundary points, between which reliability plot is placed, despite of the used failure mode. In order to recognize the reliability characteristic due to both failure modes one need a couple of points obtained from tests using simultaneously creep and fatigue modes. Thus, an estimation can be made on the correction and interaction of both failure phenomena and more reliable prediction of failure could be possible.

The total damage accumulation and criteria can be finally modeled by the following formula:

$$\sum_{i=1}^{n} \left(\frac{N_i}{N_f} + \frac{\Delta t_i}{t_r} \right) = 1 \qquad (2)$$

where: n – number of cycles, N_i – number of fatigue cycles, N_f – fatigue limit, Δt_i – dwell time at the elevated temperature or force, t_r – creep limit. In fact, depending on the load conditions, either force F or temperature T, the other parameter should be estimated accordingly.

2. Experimental analysis

Taken into account the above, special plan of experiments was carried out with Bondtester 4000+. It was equipped with specially designed cartridges, which allow make pull, push and shear tests on the basis of creep and fatigue phenomena.

While performing creep and/or fatigue tests Hot Bump Pull (HBP) method was used. Nordson Dage company HBP system is a totally industry unique solution, conforming to IPC-9708. The procedure consists of two parts:

- setting up the test parameters (type of test, value of applied load, temperature profile, etc.),
- positioning the pull pin over a solder bump.

In the first step a temperature profile is created, this allows user to input temperature and time criteria for reflow and test conditions [5]. An important issue in setting the temperature profile is to set soak level, which allows to heat the sample to temperature of the pin. In that way the solder joint is properly created.

The temperature profile consists of six dependent stages: Preheat (T1), Soak (T2), Rate of rise (T3), Reflow (4), Cool down (T5) and Test execution (T6) which is shown in the figure 4.

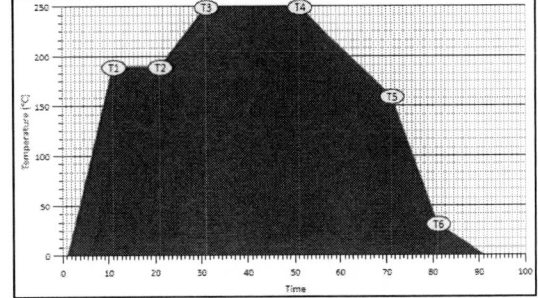

Figure 4. Temperature profile for melting the pin in a solder bump.

The next step is an attachment of a straight, copper pin into the cartridge and then bonding with the solder bump (Fig. 5).

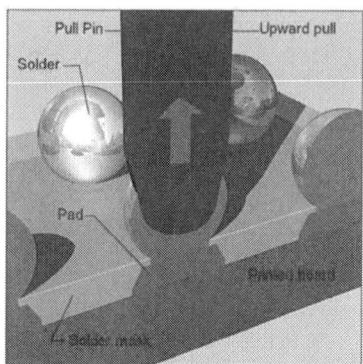

Figure 5. Cross section model of a solder bump and copper pin configuration [5].

In the initial stage the pin is set on the surface of the solder bump. Then a test is executed, the pin is heated up according to earlier defined temperature profile. At reflow point the pin drops down to a desired level performing properly solder joint. In this stage a clamping mechanism is engaged, which clamps the pin so it is ready to be pulled. After that a cooling operation is handled with pulsing compressed air. Once the test temperature (30 degrees of Celsius) is reached, the test is automatically executed, recording the force-time, force-distance results and energy values.

Unfortunately if values of temperature profile are not set properly delamination of the substrate could happen, which creates additional stresses causing weakness of connection between solder joint and substrate, as presented in the figure 6. In order to avoid this, a specimen should be warmed up (190 degrees of Celsius) through couple seconds at the beginning for compensating temperature difference between substrate and the pin and thus lead to a proper solder joint at soak stage. Afterwards, while cooling down phase is initiated, a solder joint can't be cooled too fast because that operation can cause stresses arising around the joint, which can lead to the delamination between copper substrate, which is bonded with solder alloy and FR4 (Fig. 6).

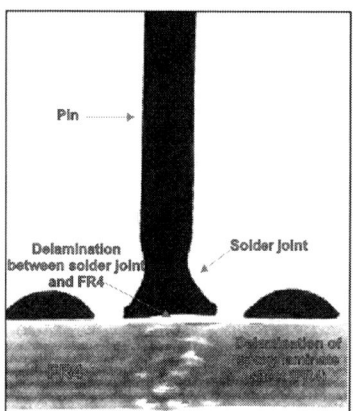

Figure 6. X-ray image of delamination of the proper substrate (FR4) and solder joint.

For all performed tests epoxy laminate substrate (FR4 – low value of Young's modulus) was used to create an array of copper plates (Fig. 7a). The samples were generated by covering copper plates with balls of 0.76 mm diameter made of tin-lead eutectic solder alloy Pb37Sn63 [6], such as used in BGA technique. After

applying a flux on the substrates solder balls were put on them - two balls on one substrate (Fig. 7b) to keep constant amount of solder which determine the final size of the joint in the final tests. Such structure were put into the oven in 240 degrees of Celsius for 30 seconds, what caused dissolving of the solder and a creation of spherical samples (Fig. 7c).

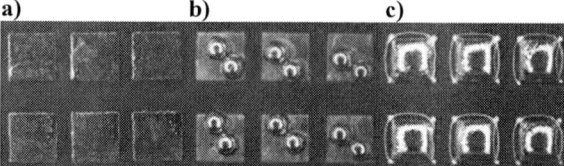

Figure 7. All stages of specimen creation: clear cooper substrates a), BGA balls as solder alloys located on substrates b), specimens ready to tests c).

The size of the plates was pointed by height of spherical sample which should be not flat (the joint will be not created) and not too high (the joint will be too strong and range of the cartridge will not allow to make a test).

Such shape of the plates and specimens was determined by the need of using single BGA ball located on BGA array, which in certain cases caused problem of testing not solder joint but bond between the substrate and BGA ball. In fact, that kind of operation was the main objective for of the investigation.

3. Destruct test

The first test which were performed before the final fatigue and creep tests was life destruct examination (Fig. 8a). It helped to estimate an average force that cause destruction of the samples. After obtaining that, a value of force for the following fatigue test was estimated. To receive average value of applied load there were 30 measurements done, which is depicted on the figure 8b. Additionally the Paragon Software, supplied with Bondtester, enabled creating the histogram of performed measurements. So, finally the average value of the applied force for destruct test was recognized as 59,5 N.

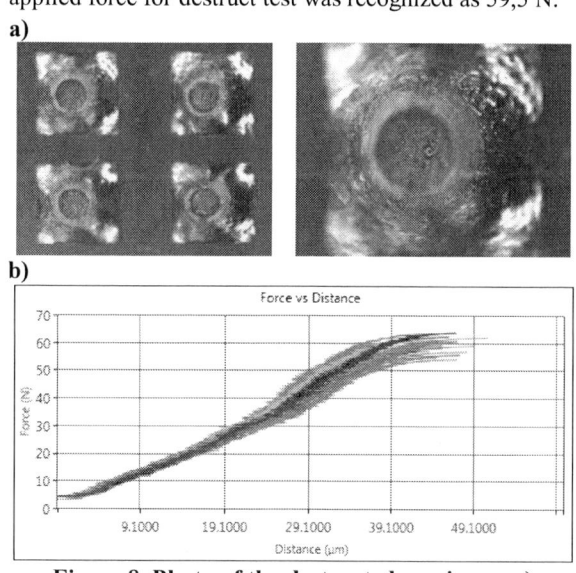

Figure 8. Photo of the destructed specimens a) and the plot of distance versus force in case of 30 destruct tests b).

4. Fatigue test

Initial fatigue tests were performed in order to estimate value of the force that can be used to make a final tests based on the combined load. The value was selected on the basis of the number of cycles to failure. It was assumed at the level of approximately 1000 cycles in order to initiate microscopic cracks propagation in solder joint. The force of 25 N was selected. The achieved results are shown in the figure 9.

a)

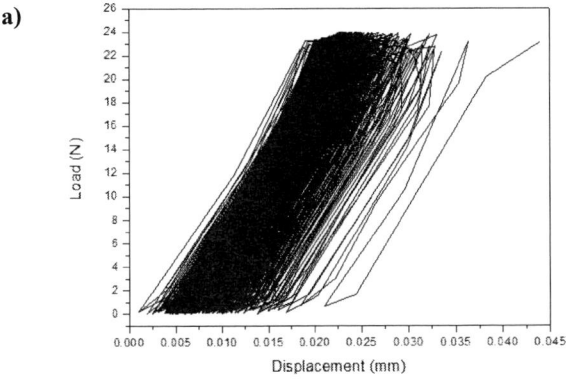

b)

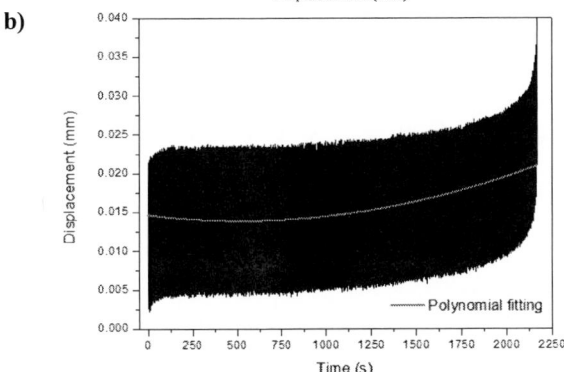

Figure 9. Fatigue test for 25N force results. Applied loading versus displacement a) and displacement versus time b).

5. Combined fatigue and creep test

The Paragon software allowed to make at the same time creep and fatigue loading, given in figure 10, where dwell time Δt_i can be changed in a range 1s up to 1000s.

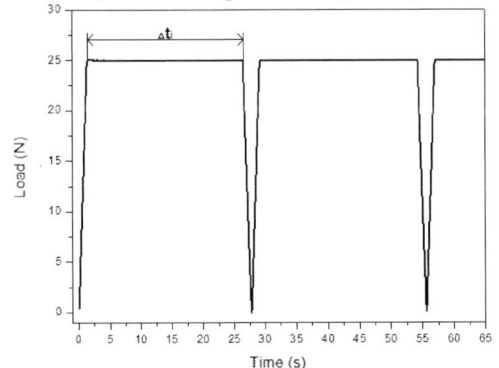

Figure 10. Simultaneously acting creep and fatigue phenomena in multi-failure test.

By setting a couple of parameters as: value of maximum and minimum applied force, dwell time at maximum and minimum using load, speed of force rising and decreasing in each cycles, a number of testes were performed and the results are collected in figure 11a,b.

a)

b)

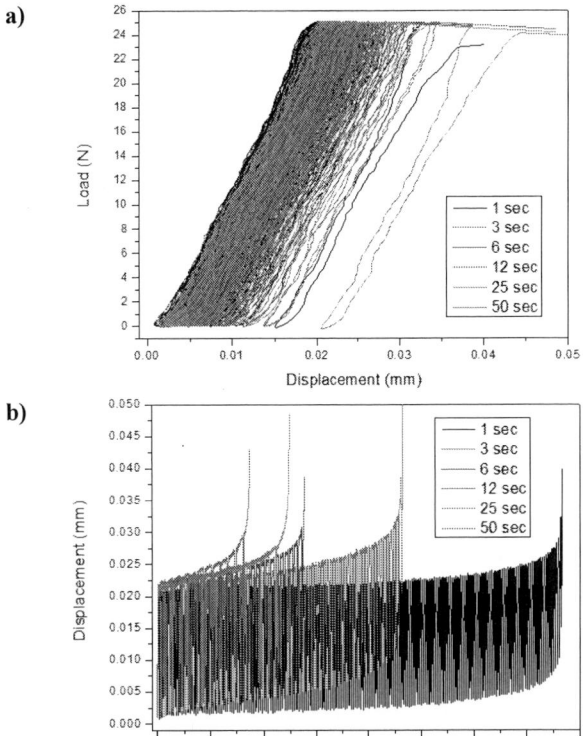

Figure 11. Fatigue with creep simultaneously acting in one test for different value of dwell time and constant value 25N of applied force. Applied loading versus displacement a) and displacement versus time b).

Results were gathered in table 1 and depicted in the figure 12. Taking into account the above results and the achieved relation between dwell time, corresponding to the creep phenomenon and amount of the cycles relative to the fatigue, a proper damage accumulation model can be estimated. The same reliability prediction of solder joints, depending on the accumulated amount of damage due to creep and fatigue could be possible, e.g. for solder joint based on PbSn alloy.

Table 1. Dwell time versus amount of obtained cycles.

Δt_i [s]	1	3	6	12	25	100	200	300	400
N_i [cycle]	500	140	66	33	11	4	3	2	1.5

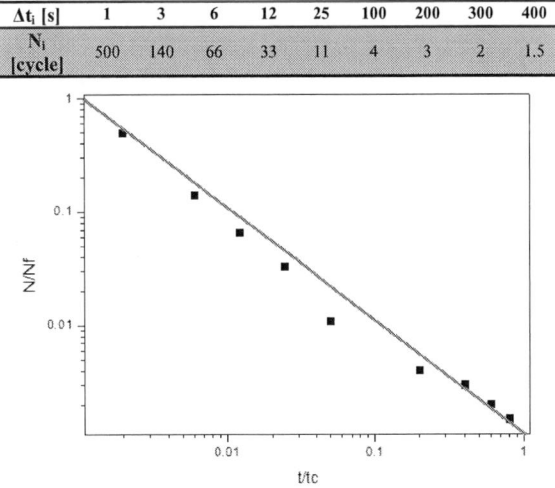

Figure 12. Linear superposition failure criteria of the obtained results.

According to the preliminary results, given on the figure 12, it can be concluded that the combined failure of solder alloys due creep and fatigue can be analyzed with one of damage accumulation models, which will be a next step of the presented research.

6. Conclusions

The above paper summarizes the work on the combined failure mode due to creep and fatigue of the solder joints. It was concluded that creep and fatigue must be considered together, what is observed on the figure 12. It is shown strongly dependence between these two phenomena. The combination helps to estimate reliability of solder joints in real working conditions in comparison to the so far engineering practice where both failure modes are considered independently. The above work will be continued in the future and it is planned the to concentrate on the both: precise failure model and corresponding material model, which would reflect the overall solder joints reliability.

Acknowledgments

This work was performed as part of the "Partitioning and modeling of system-in-package (PARSIMO)" project proposed by ENIAC Joint Undertaking.

References

1. D. Chicot, K. Tilkin, K. Jankowski, A. Wymysłowski „Reliability analysis of solder joints due to creep and fatigue in microelectronic packaging using microindentation technique", Microelectronics Reliability, Vol. 53, Iss. 3, pp. 761-766, May 2013

2. Liang Zhang, Ji-guang Han, Cheng-wen He, Yong-huan Guo, "Reliability behavior of lead-free solder joints in electronic components", Journal of Materials Science: Materials in Electronics, vol. 24, issue 1, p. 172-190, January 2013

3. K. Jankowski, R. Świerczyński, K. Urbański, A. Wymysłowski, D. Chicot, R. Dudek, „Maintenance Of Solder Joints On The Strength Of Simultaneously Acting Creep And Fatigue Phenomena By Using Microindentation Technique", Thermal, Mechanical and Multi-Pysics Simulation and Experiments in Microeletronics and Microsystems (EuroSime), 2013

4. K.Jankowski, A.Wymysłowski, D.Chicot, „Combined loading and failure analysis of lead-free solder joints due to creep and fatigue phenomena", Soldering & Surface Mount Technology, Vol. 26 Iss: 1, pp. 22-26, January 2014

5. Nordson Dage Team, "Hot Bump Pull/ Hot Pin Pull Technique", Application Note, Nordson Dage, 25 Faraday Road, Rabans Lane Industrial Area, Aylesbury, Bukinghamshire HP19 8RY, UK, November, 2011

6. Chin YT, Lam PK, Yow HK, Tou TY. Microelectron Reliab 2008;48:1079–86.

Investigation of Temperature and Moisture Effect on Interface Toughness of EMC and Copper Using Cohesive Zone Modeling Method

Xiaosong Ma, G.Q. Zhang
Guilin University of Electronic Technology, Guilin, China
No. 1 Jinji Road, Guilin, Guangxi, China, 541004
Tel: ++86-773-2290108; Fax: ++86-773-2290108
E-mail: glmaxiaosong@163.com

Abstract—Interface delamination is one of the most important issues in the microelectronic packaging industry. Epoxy molding compound (EMC) and copper interfaces are the most important interface mostly concerned by the industry and researchers. Delamination between EMC and copper will severely result in product failure. In order to predict this delamination, interface properties should be characterized. Bi-material, copper-EMC, samples are made according to the industrial package processes. A four point bending test system is established in order to perform delamination tests at different temperatures using a universal tester Zwick/Roell Z005. In addition, a Keyence optical system is mounted to capture a series of pictures during the delamination processes. Four point bending tests have been performed at room temperature, 85°C respectively. In addition pre conditioning sample are also tested at room temperature and 85°C respectively after 48 hours pre conditioned at 85°C/85%RH. . Experiments show that the "critical delamination load" decreases steadily with temperature increasing. Experiments also show moisture has effect on the "critical delamination load" compared with the dry samples tested at the same temperatures. This means that moisture has effects on the interface toughness between copper and EMC. To quantify the interface properties, numerical simulations of the four point bending test have been performed by using a finite element model comprising cohesive zone elements which will describe the transient delamination process during the four point bending tests. Correspondently, the interface toughness decreases from 21.2J/m^2 at room temperature to 3.7J/m^2 at 85°C as calculated from the cohesive zone element model. These results show that temperature has a large effect on the interface toughness. Saturated moisture, at 85°C/85%RH, decrease about 20% interface toughness between EMC-copper.

Keywords: interface toughness; cohesive zone method; four point bending

I. Introduction

Fracture mechanics has been studied for a long time. There are three modes of crack extension in fracture mechanics: opening mode, sliding mode and tearing mode (see Fig. 1). The opening mode, Mode I, is characterized by the symmetric separation of the crack surfaces with respect to the plane. The sliding mode, Mode II, is characterized by displacement in which the crack surfaces slide over one another perpendicular to the leading edge of the crack. The tearing mode, Mode III, finds the surfaces sliding with respect to another parallel to the leading edge. Mode I and II are common failure modes. It can be seen that different failures are caused by different loadings.

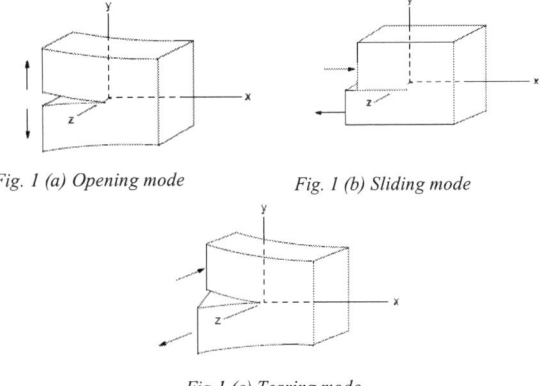

Fig. 1 (a) Opening mode   Fig. 1 (b) Sliding mode

Fig.1 (c) Tearing mode

Therefore, different failure modes are investigated by different test methods. Fig. 2 shows some different forms of modes [1] resulted by different loadings.

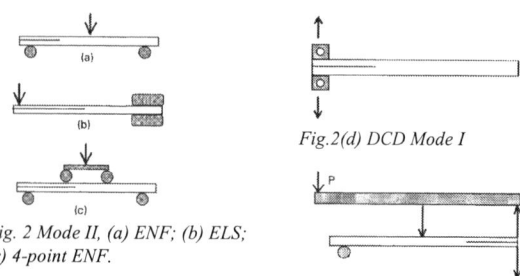

Fig. 2 Mode II, (a) ENF; (b) ELS;
(c) 4-point ENF.

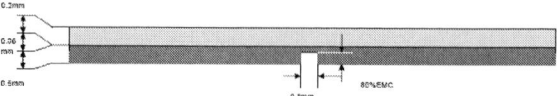

Fig.2(d) DCD Mode I

Fig. 2(e) Mixed mode I/II

Figure 2(d) shows the specimen of Double Cantilever Beam (DCB) for Mode I. The data analysis is based on the Irwin-Kies equation [1]. This equation is then modified by Williams and JIS (Japanese Industrial Standards) respectively [2, 3]. They are referred to as CBT (corrected beam theory) and modified compliance calibration (MCC) by Hojo [4]. The mode II test, see Fig. 2(a), (b), (c) remains controversial for practical reasons, such as unstable propagation in the ENF specimen, friction effects and difficulty in defining a starter defect. The specimen for mixed mode bending (MMB), Fig. 2(e) as introduced by Reeder, Crews and Reeder [5, 6] has become the most widely used specimen for the determination of mixed mode envelopes. These types of specimen are suitable for bi-material samples. For the tri-materials samples, some modifications are needed to ensure that the delamination occurs at the required interface. Based on Charalambides [7], a four point bending Mode I tool was designed and constructed. This test method shows stable delamination and reproducible results.

II. Experimental Samples and Equipment

A. Double layers material sample

In order to obtain the similar interface toughness as in the actual package, samples are made according to the packaging processes. So the interface properties results are almost same as in the electronic packages. The test samples are double layers strips consisting of copper layer (0.2mm in thickness), epoxy molding compound layer [EMC] (0.6mm in thickness). The sample dimension is 60x10x0.85 mm^3.

First, the copper leadframe is placed in the mold. EMC molding is finished in a pre-heated mold at 175°C within 60 seconds and post cured in the mold for 90 seconds. The final map mold is shown in Fig. 3(a). Before cutting off the sample, the map mold is post cured at 175°C for 4 hours to ensure that the epoxy molding compound is fully cured. Then the map molds are cut into 60x9mm² strips, see Fig. 3(b).

Fig.3(a) Map mold

Fig.3(b) Cut samples

To trigger the interface delamination, a pre-defined notch (0.5mm wide and 80% depth of EMC) is created in the epoxy molding compound materials by sawing. The geometry and dimensions of the sample are shown in Fig.4.

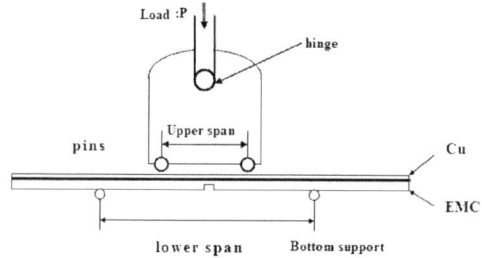

Fig. 4 Geometry and dimensions of the sample

B. Setup of Four Point Bending

A special four point bending tool is designed and manufactured to investigate the interface toughness. Fig.5 schematically shows the test setup while the actual four point bending tool, which consists three parts, is shown in Fig.6. The first part is the four point bending frame which is used to support the two rollers. The second part consists of two rollers which are used to support the test sample. For decreasing friction between the rollers and the moulding compound layer of the test sample, two rollers are allowed to rotate freely. Silicon lubricant grease is used in bearings for withstanding the high temperature effects. The third part is the loading head which applies displacement or loading to the sample.

Fig.5 Schematic overview of four point bending test setup

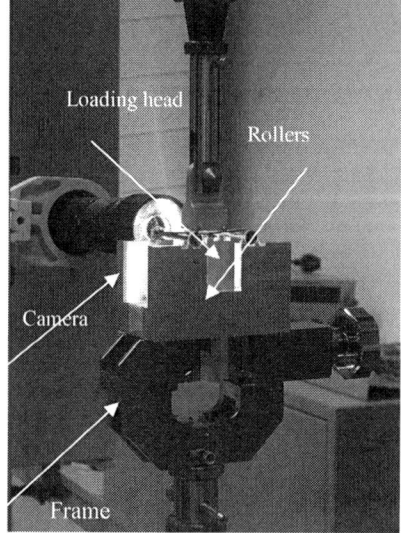

Fig.6 Four bending tool setup

C. Loading system and optical system

Fig. 7 shows the universal tester Zwick/Roell Z005. Loading head moves downwards at speed 100 to 200um/min and no differences between the test results. The reaction force of the loading head is measured by the force sensor when loading head touches the sample. A Keyence optical camera system is mounted at the back of the bending tool, see Fig. 6. The optical camera focuses on the notch and monitors the deformation and delamination between EMC and copper.

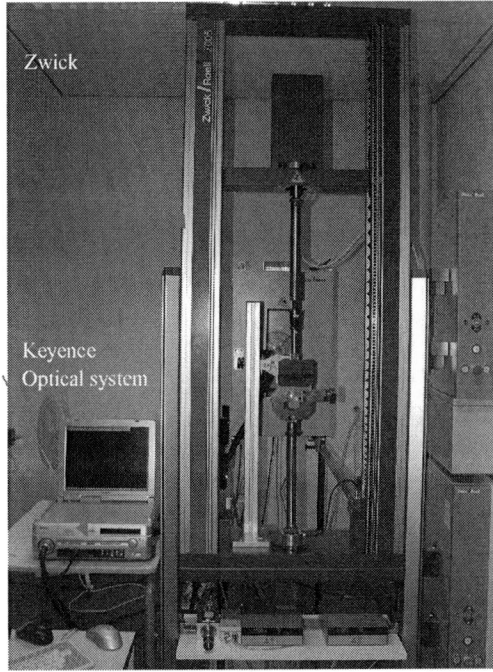

Fig.7 Displacement/loading and optical system

III. Four Point Bending Test Results

In order to obtain more reliable interface toughness properties, four point bending tests are performed at room temperature and 85°C. Loading speed is 0.1 to 0.2mm/min and it appears that the loading speed has no effects on the "critical delamination load" according to our tests at room temperature.

When the load head touches the test sample, the test sample deforms and the response is elastic. The applied load increases with the displacement and the load gradually reaches the highest point. At some critical displacement, the load suddenly drops. The EMC notch cracks and the load suddenly begins to decrease. After the delaminations at both sides have started, the crack propagation load "stabilizes" around a constant value, see Fig. 8. Fig.8 is the test results of four point bending at 85C. From this constant allowable load, the interface fracture

toughness value is derived with the aid of the results of numerical simulations .

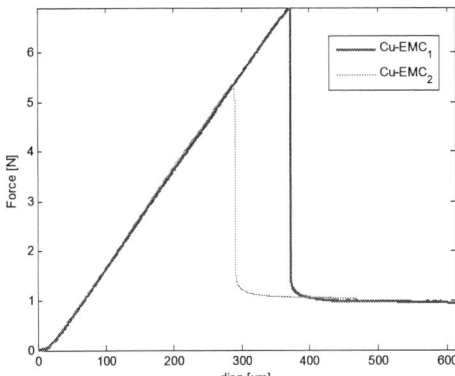

Fig. 8 Sample crack processes

A. Test result at different temperatures of dry samples

Table 1 Critical Loadsand G_c value of Equation/Simulation

Temp (°C)	Load (N)		G_c (J/m^2)	
	Dry	PRECON 85°C/85%RH	Eq. (2)	Dry Simulation/ Wet Simulation
25	1.9	1.7	5.1	5.2/4.8
85	1.1	X	3.5	1.9/X

Table 1 shows the critical crack propagation forces. The tests were performed at room temperature and 85°C respectively. These results show that temperature has a great effects on the critical crack propagation load. This means that the interface toughness decreases with increasing temperature.

B. Four Point Bending Test results of PRECON Samples

In order to investigate moisture effects on interface toughness, test samples are put in humidity oven at 85°C/85%RH for at least 48 hours in order to reach moisture saturation. Four point bending tests are performed at room temperature after this pre conditioning.

Table 1 shows the average crack propagation load for both dry and pre-conditioned samples. Test results show that the average crack propagate load of the pre conditioned sample at room temperature is about 11% lower than that of the dry sample tested at room temperature. This means that moisture has certain effect on the interface toughness between EMC and copper leadframe.

C. Calculation of interface toughness

1). Analytical model

The critical interface fracture toughness, G_c can be deduced analytically by recognizing that it is simply the difference in the strain energy in the uncracked and cracked beam. Since there is negligible strain energy in the beam above the crack, G_c can be deduced from the consideration of energies in the uncracked section, and the lower section below the crack. Using the Euler-Bernoulli theory and plane strain conditions, these energies can be expressed in terms of the applied moment M as [7]

$$U = (1-\upsilon^2)M^2/2EI \qquad (1)$$

where U is the strain energy per unit cross-section and I is the second moment of area per unit width, G_c can then derived into (2)

$$G_C = \frac{M^2(1-\upsilon_2^2)}{2E_2}\left(\frac{1}{I_2} - \frac{\lambda}{I_c}\right) \qquad (2)$$

where I_2 and I_c are second moment of inertia per unit cross-sectional area for the bottom layer and the composite beam, respectively, and

$$\lambda = E_2(1-\upsilon_1^2)/E_1(1-\upsilon_2^2) \qquad (3)$$

$$I_2 = \frac{1}{12}h_2^3 \qquad (4)$$

$$I_C = \frac{1}{12}h_1^3 + \frac{\lambda}{12}h_2^3 + \frac{\lambda h_1 h_2(h_1+h_2)^2}{4(h_1+\lambda h_2)} \qquad (5)$$

The subscript 1 indicates quantities relevant to the top layer, whereas the subscript 2 denotes the corresponding quantities for the bottom layer. Subscript c refers to the composite beam. Note that the moment per unit width $M = Pl/2B$, with P being the constant load and l the spacing between inner and outer span. According to Eq. (2) analytical G_c can be obtained. This G_c value can be used initially to estimate crack propagation load in the simulation model, (see Table 1), temperature effects on young's modules are roughly considered in analytical equation (2).

2). Numerical model

In addition to the analytical model, the interface toughness G_c can be obtained by changing the G_c value until crack propagation load equals the experimental propagation load.

A four point bending model is constructed in the finite element package Marc and its graphical user interface Mentat (see Fig.8). Due to the symmetry of the model, only half a specimen is modeled. Top layer is copper and bottom layer is EMC. Vertical displacements fixed at the bottom support (see arrow bottom supports) is the boundary condition. Top load applies a loading in downwards direction. Symmetry_dx fixes displacement of all symmetric nodes in horizontal direction. Thermal deformation is considered here. Symmetry_4PB fixes displacement of only symmetric nodes of copper layer in horizontal direction see Fig. 9.

Four stages are prescribed in the simulation model including initial thermal stresses. The first stage is cooling down from the initial condition 175°C. Temperature is cooled from 175°C to room temperature using boundary conditions at bottom support and symmetry_dx. The second stage is the heating up. Temperature is increased from room temperature to the test temperature using boundary conditions bottom support and symmery_dx. The third stage is the relaxation. Temperature is

kept at the test temperature using the boundary conditions of bottom support and symmetry_dx. The last one is four point bending stage. Temperature is kept at test temperature and other boundary conditions are bottom support, top load and symmetry_4PB. Top load is the downwards displacement applied to the sample. Symmery_dx fixes all nodes in x direction at the symmetry line while symmetry_4PB fixes only nodes on copper in x direction as shown in Fig. 9.

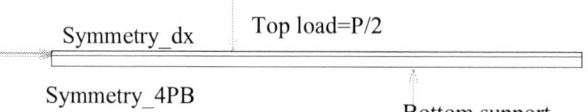

Fig. 9 2D four point bending FE model

3). Cohesive zone element

Marc has a library of interface elements, which can be used to simulate the onset and propagation of delamination. The constitutive behavior of these elements is expressed in the terms of tractions versus relative displacements between the top and bottom edge/surface of the elements, as shown in the Fig.10.

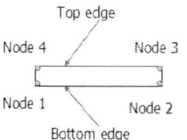

Fig.10 2D Interface element

The traction t is introduced as function of effective opening displacement and is characterized by an initial reversible response, followed by an irreversible response once the critical effective opening displacement v_c has been reached. An exponential function is used in the simulation.

$$t = G_c \frac{v}{v_c^2}e^{\frac{-v}{v_c}} \qquad (6)$$

in which G_c is the interface toughness and $e = 2.718$ [8].

It can easily be verified that the maximum effective traction t_c, corresponding to the critical effective opening displacement v_c is given by [8, 9]:

$$t_c = \frac{G_c}{ev_c} \qquad (7)$$

If the maximum effective traction is known, the critical or effective opening displacement can be determined by:

$$v_c = \frac{G_c}{et_c} \qquad (8)$$

Damage is defined as:

$$D = \frac{\int_{v_c}^{v_s} t\,dv}{\int_{v_c}^{\infty} t\,dv} \qquad (9)$$

$v_c < v_s$

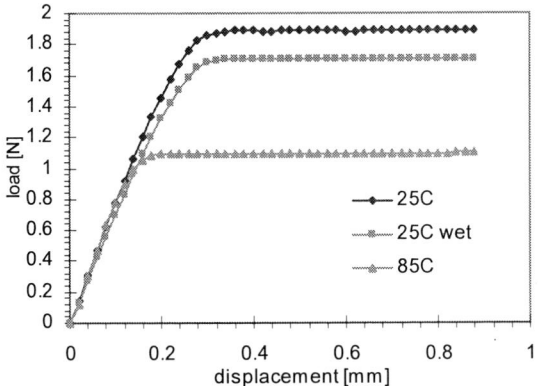

Fig. 11 Fitted load by changing G_c

If $D = 1$, the element is fully damaged.

G_c is obtained from 4 point bending tests combined with simulation of the fitted crack propagation load. G_c values from analytical equation is input as initial value. By adjusting the G_c value until simulation load equals to measured load, the actual G_c can be found (see simulation results in Fig. 11).

Fig. 11 shows some typical four point bending simulation results at different temperatures. In this simulation, the relationship between applied displacement and load are shown in this figure. The mesh size is chosen such that convergence of the results is obtained.

Table 1 shows the crack propagation load, G_c values of equation (2) and simulation. It can be seen that G_c value decreases rapidly with increasing temperature. As already discussed, tests results show that in our case, moisture has certain effects on G_c value.

IV.   Conclusion

From a series of four points bending tests, it is found that temperature has a large effect on the interface toughness. G_c value greatly decreases with increasing temperature. In addition, moisture has a relatively larger effect on interface toughness of copper and EMC in our samples.

This important interface property can be used in the future to predict the possibility of delamination combined with critical opening value (not measured and calculated) in this paper..

Acknowledgement:

This project is supported by Key Lab of Guangxi Manufacturing System and Advanced Manufacturing Technololy, Project No. PF12127X and Experimental Center of Guangxi Information Science, Project No. LD13113X..

References:

[1] Davies, P., Blackman, B.R.K., and Brunner, A.J., "Standard Test Methods for Delamination Resistance of Composite Materials: Current Status," Applied Composite Materials, Vol. 5, No. 6 (1998), pp. 345-364.

[2] Hashemi, S., Kinloch, A. J., and Williams, J. G., "Corrections needed in double-cantilever beam tests for assessing the interlaminar failure of fibre-composites," J. Mat. Sci. Letter 8 (1989), pp. 125-129.

[3] Hashemi, S., Kinloch, A. J., and Williams, J. G., "The Analysis of Interlaminar Fracture in Uniaxial Fibre-Polymer," Proc. Royal Soc. A427 (1990), pp. 173-199.

[4] Hojo, M., Kageyama, K. and Tanaka, K., "Pre-Standardization Study on Mode I Interlaminar Fracture Toughness Test for CFRP," Japan Composites, Vol. 26, No. 4 (1995), pp. 243-255.

[5] Crews, J. H. and Reeder, J. R., "A Mixed-Mode Bending Apparatus for DelaminationTesting," NASA TM 100662, 1988

[6] Reeder, J. R. and Crew, J. H., "Nonlinear Analysis and Redesign of the Mixed-Mode Bending Delamination Test," NASA TM 102777, 1991

[7] Charalambides, P.G., "A Test Specimen for Determining the Fracture Resistance of Bimaterial Interface," J. Appl. Mech., Vol. 56 (1989), pp. 77-82.

[8] MSC.Marc/Mentat, volue A, 2008 r1, 2008

[9] M.Ortiz and A. Pandolfi, "Finite-Deformation irreversible cohesive elements for three dimensional crack-propagation analysis, International Journal for numerical methods in engineering," Vol. 44 (1999), pp.1267-1282.

[10] O. van der Sluis, P.H.M. Timmermans, E.J.L. van der Zanden, J.P.M. Hoefnagels. Analysis of the three-dimensional delamination behavior of stretchable electronics applications. In M.H. Aliabadi,S. Abela, S. Baragetti, M. Guagliano, and H-S. Lee, editors, Key Engineering Materials, volume 417–418, pages 9–12. Trans Tech Publications, 2010. Special volume: Advances in Fracture and Damage Mechanics VIII, ISSN 1013–9826

Comparation of Thermal and Hygro Effects on the Degradation of LED Package

Xiaosong Ma, G.Q.Zhang
Guilin University of Electronic Technology,
NO. 1 Jinji Road, Guilin,China
Tel: ++86-773-2290108; Fax: ++86-773-2290108
glmaxiaosong@163.com

Abstract—*Humidity test is common to all the packaged integrated circuits, few works have been reported on the study of the humidity effects combined with high temperature on packaged LEDs. Moreover, the packaged LEDs are also subjected to moisture environment in many of their applications. In this work, the high temperature–humidity (95℃–85% RH) aging test is used to evaluate the reliability of the packaged LEDs with respect to their optical output properties.*

The degradation of two kinds of commercial packaged light-emitting diodes (LEDs) are investigated using high temperature aging combined with fast moisture diffusion method, which is quatitively calculated by simulation. First, the tested LED devises is put in the chamber for 84 hours at 95 ℃/85%RH , then the temperature is set to 85 ℃ in order to know the effects of moistute loss, and then the LED devises are under stress conditions 65℃/85%RH for 50 hours. At last sample are put in the oven for 25 hours at 105 ℃. An optical measurement system consisting of a one-meter diameter integrating sphere and a spectroradiometer is set up to measure the luminous flux of the LEDs. The spectral flux responsivity of the measurement system is calibrated at the start and the end of the measurement of LEDs to achieve high measurement accuracy. During the optical measurement, the LED under test is powered up with a nominal constant current as recommended by the manufacturers of the packaged LEDs. Temperature and moisture has effects on both type of epoxy packaged LED the luminous flux, temperature color but almost no effectsof the color purity, peak wave length and half wave width. Further study are still onging to investigated the relation between aging time and optical degradation and also relation between moisture content and optical degradation.

Keywords—*Thermal and Hygro, degradation, LED Package, Expoxy Packaging Material*

I. Introduction

Today, one the most promising applications of high-power LEDs is to produce white light with a broad spectrum [1]. The commonly used method of producing white light using LED is to use a phosphor material to convert monochromatic blue light from a GaN based chip to white light As the fourth-generation light source, LED has many advantages such as high luminous efficiency, low power consumption and long life. In the same time due to the long life of LED, the average life could up to tens of thousands of hours, the reliability of the test requires long time and high cost, so reliability tests and accelerated life test life prediction of LED has become the current focus of the study.

In particular, despite the fact that humidity test is common to all the packaged integrated circuits, few works have been reported on the study of the humidity effects on packaged white LEDs[2-7]. Moreover, the packaged white LEDs are also subjected to moisture environment in many of their applications.

In this work, the high temperature–humidity aging test is used to evaluatethe reliability of the packaged white LEDs with respect to their optical output properties.

II. The Thermal and Hygro Degradation test

A) Test Samples

In this test, two kinds of led of 1W LED sample with aluminum substrate are used, see Fig. 1. They are all packaged by epoxy resin, emitting angle is 120-140 degrees.

Fig. 1. LED test samples

(A),   (B)

B) Test Conditions

Tests are conducted at following different conditions: 95℃/85%RH for 84 hours, 85℃ for 72 hours, 65℃/85%RH for 50 hours and 105℃ for 25 hours . At 85℃ and 105℃, moisture desorption is used to see the optical parameters recovery situations.

C) Test Equipments

Currently, the lab has a high temperature and humidity degradation chamber which is used to conduct a LED high temperature degradation test, see Fig. 2.

Fig. 2. High temperature and humidity chamber
(A)   (B)

The chamber Model is H/GW-010L shown in Fig.2(A), which temperature range is RT +10 ℃ to 200 ℃, heating rate is 3 ~ 5℃ / min, , uniformity ≤ 2℃. Fig.2 (B) shows an high temperature and relative himidity aging chamber, which can monitor the optical parameters during the aging tests, so the light can be seen through the window. The Leds are not on during the tests due to the Led devise power limitations. Adjust the temperature of the experiment manually, set the test temperature and relative humidity as required and put the samples into the chamber. EVERFINE HAAS-2000, see Fig.3, high accuracy spectrtoradiometer and integrating sphere are used to measure the degradation parameters of LED samples. EVERFINE HAAS-2000 high accuracy spectrtoradiometer and integrating sphere can measure the LED color parameters (spectral distribution, color temperature, chromaticity coordinates, the peak wavelength, dominant wavelength and Ra etc.), photometric parameters (luminous flux, luminous power, etc.), electrical performance parameters (voltage, current, power, power factor). Using temperature and relative humidity as a stress of the accelerated life test, set the driving current as 300mA, aging time as required, all samples have different levels and other parameters flux attenuation.

Fig. 3. HAAS-2000 high accuracy spectrtoradiometer and integrating sphere

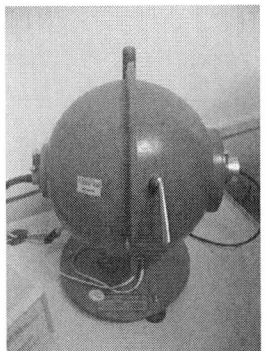

Fig. 5. The integrating sphere

Integrating sphere calibration process: first open the meter, and then turn spectral radiation dose, and then turn the integrating sphere, placed the calibration lamps on integrating sphere measured shelves, and then set the parameter as the label, then open the software, select the spectral calibration, zero calibration first, and then turn on the power switch, select Start calibration, calibration ends, and save the calibration.

III. Test plan design

All the optical parameters are measured and recorded before the tests.

Period A: after 84 hours of 95℃/85%RH aging, all aged samples are measured again.

Period B: in order to investigate the moisture effects on the, 85℃ dryness for 72 hours is set to everate the moistute dissolved in the Leds.

Period C: pre conditioning at 65℃/85%RH for 50 hours is set and all optical parameters are measured and recorded.

Period D: pre conditioning is conducted at 105℃ for 25 hours in order to remove all the moisture in the Leds.

IV. Experiment data and discussion

As shown in the Fig. 6(A) and Fig. 6(B), luminous flux change with pre condtioning time. There are 10 samples are used and tested in each group. The four deflected lines are tests represented the four test periods, A, B, C and D.

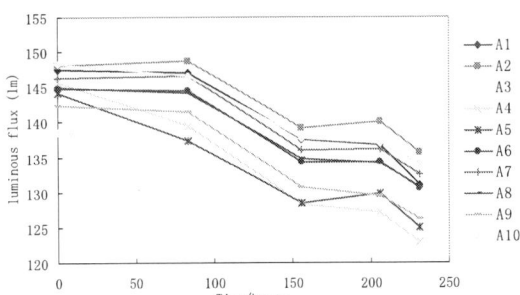

Fig. 6(A) The change of luminous flux

978-1-4799-4789-8/14 $31.00 © 2014 IEEE

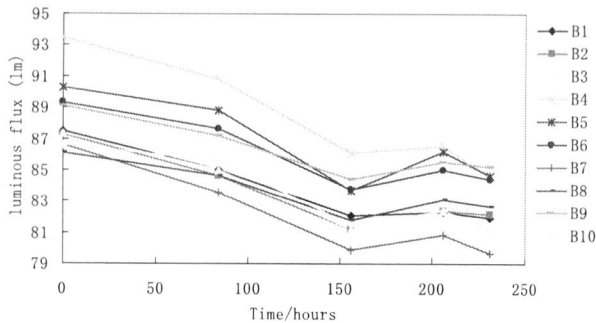

Fig. 6(B) The change of luminous flux

From the first stage of the two group tests results, it can be seen that thermal and hygro effects on luminous flux is not very large.

In period A, for samples A, the everage luminous flux decreases from 145 lm to 143.2 lm with -1.25% decrease, very small change and for sample B, the everage luminous flux decreases from 88.3 lm to 86.3 lm with -2.37% decrease, a little bigger than that of samples A.

In period B, for the following 72 hours of 85℃ dryness, the luminous flux of sample A deceases more than that of the first 84 hours of 95℃/85%RH. This is different from the expectation. The aging effects of first 84 hours at 95℃/85%RH condition should be biger than that of second 72 hours dryness at 85℃. The possible explanation can be that first is aging needs time and the effects on the luminous flux can be shown after certain time and the second reason maybe moisture can slow down the degradation. If the package forms are compared, the packaging form A is better than that of packaging form B.

In period C, pre conditioning at 65℃/85%RH for 50 hours. Both luminous flux of sample A and sample B increase a little. If period B and C are compared, it can be concluded that moisture can slow down the aging effect of packaging material on the luminous flux.

Period D is similar to period B.

As shown in the Fig. 7(A), at every pre condtioning , the Peak wave length variation trend of sample A. All kinds of ageing at different periods have no effects on the peak wave length of sample A.

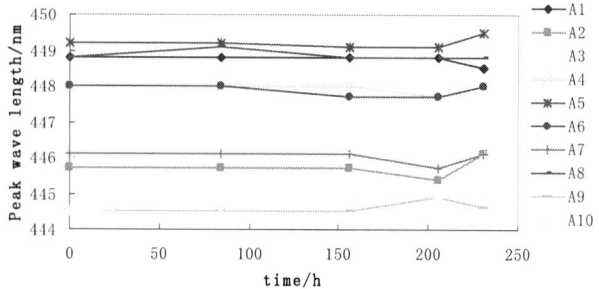

Fig. 7(A) Peak wave length vs time(hours)

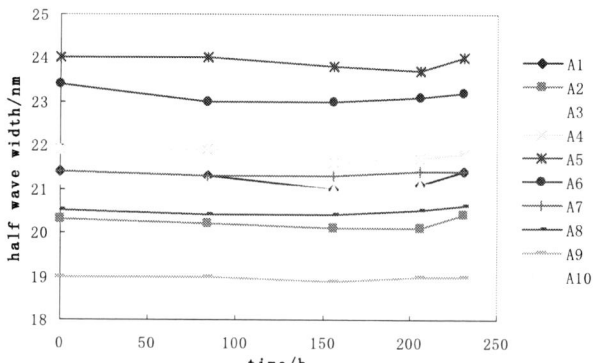

Fig. 7(B) Peak wave length vs time(hours)

As shown in the Fig. 7(B), at every pre condtioning , the Peak wave length variation trend of sample B. All kinds of ageing at different periods have no effects on the peak wave length of sample B, eventhough it seems a little different from samples A

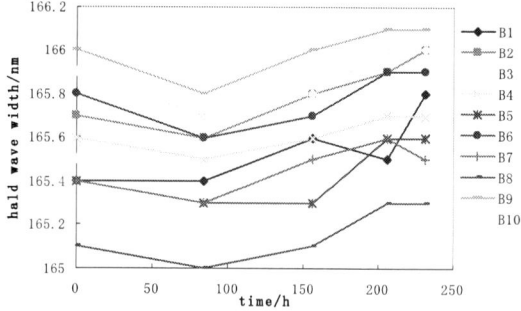

Fig. 8(A) Half wave width vs time of sample A

Fig. 8(B) Half wave width vs time of sample B

Fig. 8 (A) and Fig. 8 (B) show the half wave width changes vs the time period of different pre conditioning. Different pre conditioning no effects on half wave width of sample A but the color index of sample B is increased a little.

As shown in the Fig. 9(A) and Fig. 9(B), In the all condition for sample A, the color purity is always decreasing with all kinds of pre conditioning. But for sample B, the color purity is not obviously change in the all pre conditioning. So it can be concluded that the temperature and moisture has little effect on the color purity.

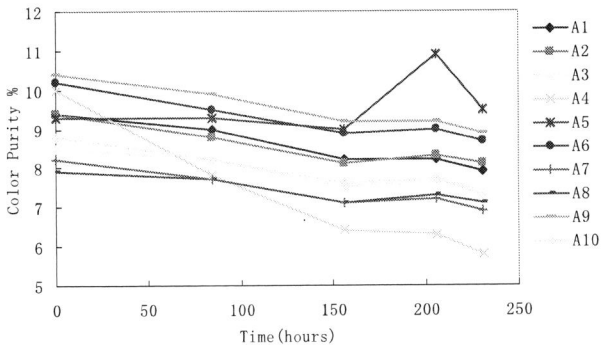

Fig. 9(A) The change of color purity of sample A

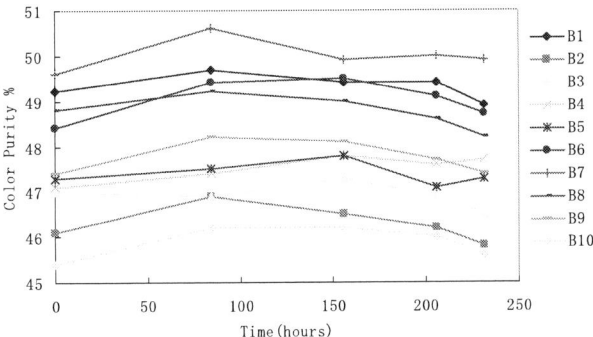

Fig. 9(B) The change of color purity of sample B

V. Conclusion

From all above experiments, following conclusions can be obtained.

For sample A and sample B, temperature aging has great effects on the luminous flux. The higher the temperature, the greater the effects on the luminous flux of the Leds.

Moisture also has effects on the luminous flux of Leds, but its effects on the luminous flux of Led is smaller. From those tests, it shows that moisture can slow down packaging materials effects on luminous flux of Led.

For sample A, color purity is always decreasing with variuous pre condition. But the same effects on sample B is quite small or even not obvious.

Temperature and hygro effects on the peak wave length and half wave width of sample A and sample B are not obvious.

Acknowledgement:

This project is supported by Key Lab of Guangxi Manufacturing System and Advanced Manufacturing Technololy, Project No. PF12127X and Experimental Center of Guangxi Information Science Project No.LD13113X.

References

[1] Tan CM, Chen B, Foo YY, Chan RY, Xu G, Liu YJ. Humidity effect on the degradation of packaged ultra-bright white LEDs. In: 10th IEEE electronicspackaging technology conference; 2008. p. 923–8.

[2] Miran Biirmen, Frango Pernus, "Accelerated estimation of spectral degration of white GaN based LEDs," Measurement science and technology. pp: 230-238.

[3] Marek Osinski, Joachim Zeller, "ALGaN/InGaN/GaNblue light emitting diode degradation under pulsed current stress." American Institute of Physics. pp: 898-900.

[4] CherMing Tan, Boon-Khai-Eric-Chen, "Analysis of bumidity effects on the degradation of high power white LEDs," Microelectronics Reliability. pp: 1227-1230.

[5] Ming Gong, Xiaosong Ma, "Reliability Assessment for LED luminaries Based on Step Stress Accelerated life Test." pp: 1546-1548.

[6] Jiajie Fan,Kam Chuen Yung, "Lifetme Estimation of High Power White LED Using Degerdation Data Driver Method." pp: 470-477.

[7] Zhanqiang Jia,Jinyan Cai, "Raliability assessment technology for electronic equipment based on step-up-stress accelerated degradation testing." pp: 1280-1285.

A Lifetime Prediction Method for Solid State Lighting Power Converters Based on SPICE Models and Finite Element Thermal Simulations

*Bo Sun[1,5], **Xuejun Fan[2,5], Lubing Zhao[5], C.A. Yuan[3,5], Sau Wee Koh[1,5], Guoqi Zhang[3,4]

[1]Beijing Research Center, Delft University of Technology, Haidian, Beijing, China
[2]Lamar University, Beaumont, Texas, USA
[3]Institute of Semiconductors, Chinese Academy of Sciences, Haidian, Beijing, China
[4]DIMES Center for SSL Technologies, Delft University of Technology, Delft, Netherlands
[5]State Key Laboratory of Solid-State Lighting, Haidian, Beijing, China
*bsun@sklssl.org, **xuejun.fan@lamar.edu

Abstract

Solid State Lighting (SSL) power converters are considered as the reliability bottleneck of a light emitting diode lighting system. This paper proposes a lifetime prediction method for solid state lighting power converter based on SPICE models and finite element thermal simulations. This method consists of four major parts: lifetime meter, components decay models, SPICE simulator and finite element simulator. The estimated lifetime can be obtained by a number of iterations of electronic and thermal performances of the interested power converter.

Key Words: SSL, LED, Power converter, Lifetime, SPICE, FEA

1. Introduction

As the next generation lighting approach, Solid State Lighting (SSL) has great advantages of energy-saving and smart lighting, provides a larger degree of freedom for lighting design than conventional lighting. With the help of new semiconductor materials and technologies, SSL light source, the light emitting diode, has a much longer lifetime than incandescent compact fluorescent lamps. However, due to the limits of cost and operation conditions, SSL power converter is an unreliable power electronic device in a SSL system. According to reports from both academia and industries, SSL power converters are considered as the reliability bottleneck of the whole system.

Firstly, a high performance SSL power converter has a complicated structure and thus a large number of failure modes [4], such as electrolytic capacitor failure, solder cracking, semiconductor device breaking down, etc. Although there are many researches on components' reliability, a few publications reported simulation methods considered reliability models of these components. Secondly, failure of SSL power converter is a multi-physical problem, stresses like voltage/ current, temperature, humidity can damage the SSL power converter, leading a system-level failure. Besides, design for reliability of the SSL system attracts increasing research interesting in recent years, the most significant part of design for reliability is lifetime prediction. Therefore, it is necessary to develop a lifetime prediction method for SSL power converter that could integrate component's reliability models in multi-physical conditions.

This paper proposes a lifetime prediction method for solid state lighting power converter based on SPICE models and finite element thermal simulations. Over three decades' development, the simulation tools for Finite Element Analysis (FEA) and Simulation Program with Integrated Circuit Emphasis (SPICE) are well developed and can provide accurate electronic and thermal simulation results with appropriate models for an SSL power converter. In this method, iterations combine electronic and thermal simulations will be carried out until the performance of the SSL power converter satisfies the failure criteria. With appropriate SPICE models, the heat generation and electronic characteristics of each component can be obtained by SPICE simulations; meantime, the core or junction temperatures of these components can be obtained by finite element thermal simulations. By substitution of core or junction temperatures and operation hours into components' reliability models, the SPICE parameters after degradation can be calculated for next iteration. Finally, the lifetime of the interested SSL power converter is the production of the operation time interval and number of iterations when its performance satisfies the failure criteria. In the following section, this lifetime prediction method for SSL power converters will be discussed in details.

2. The Lifetime Prediction Method

As mentioned in Section 1, this lifetime prediction method for SSL power converters integrates SPICE simulation, finite element thermal analysis, and components' reliability model. Figure 1 illustrates the flowchart of this lifetime prediction method, which consists with four major parts: lifetime meter, components decay models, SPICE simulator and finite element simulator, and two iterations: the major iteration and the secondary iteration. The major iteration consists with lifetime meter, component reliability models and SPICE simulator, which stops when performances of the SSL power converter exceed its failure criteria. The secondary iteration consists with component reliability models, SPICE simulator and finite element simulator, it stops when the SPICE parameters for performance simulations converge at a steady value. Obviously, if the SPICE models contain temperature dependent parameters

978-1-4799-4789-8/14 $31.00 © 2014 IEEE

(TDPs), more than one loop of secondary iteration should be carried out to obtain accurate SPICE parameters.

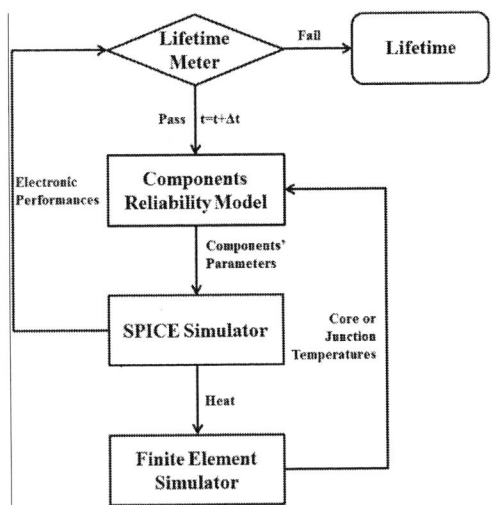

Figure 1: Flowchart of the Lifetime Prediction Method for SSL Power Converters

Although there are various failure modes, failure of SSL power converters are determined by its electronic performances. Essence of lifetime prediction is find the time that electronic performances of the interested SSL power converter have enough deviations which can be considered as failures. The lifetime meter contains a comparator which can compare electronic performances of an SSL power converter after certain hours' operation calculated by the SPICE simulator with their initial value. The operating hours of the target SSL power converter will be increased by a certain step Δt for each loop of the major iteration until its performances exceed their failure criteria. Total operation hours of the target SSL power converter is the production of the operation time interval and number of iterations. All iterations will be stopped when the performances exceed their failure criteria. The lifetime prediction result is between total operation hours of last iteration and the one before last iteration. Obviously, the lifetime prediction accuracy is controlled be the value of Δt.

Similar to other electronic devices, SSL power converter consists with many types of components, such as semiconductor devices, surface mount devices, solder joints, electrolytic capacitors, inductors and transformers, which have disparate operation conditions and failure modes. The decay model for a component which is a set of SPICE parameter distribution model over time and operation conditions, can describe failures of a component. Theoretically, a component decay model has several input variable, includes total operation duration Δt and operation conditions, such as temperature, humidity and etc. And the output of this model is the set of SPICE parameter degradation for the SPICE simulator to calculate the performance of the entire system. This method assumes degradation rate distribution which is determined by operating conditions is stable during each iteration step length Δt, thus the degradation amount of these SPICE parameters follow their own distribution. As a result, one possible estimated lifetime can be obtained at the end of the iteration, and estimated lifetime distribution can be obtained if the number of iterations is large enough.

Theoretically, degradations of all components in an SSL power can be considered in this method, but only major failures and critical components are considered in practice. On one hand, some types of components have a complicated SPICE code, but less impact on the system's performance. On the other hand, some types of failures have tiny possibility and hardened to be described by current reliability models. Hence, this method carries out the FMMEA for an SSL power converter, determining the major failures and critical components, before lifetime prediction.

The failures of an SSL power converter are defined by its electronic performances. Thus, in this method, a SPICE simulator is used for the calculation of its performances. SPICE (Simulation Program with Integrated Circuit Emphasis) is a mature general-purpose analog electronic circuit simulation tool based on electronics and physics, which used to predict circuit behavior in IC or board level. This method firstly establishes the net-list of the SSL power converter circuit by its circuit schematic, and then imports this net-list and SPICE parameters of each component from component decay models into a SPICE simulator which could calculate the performances at last. It is common sense that the junction temperatures or core temperatures of components which are determined by self-heating is significant to system's lifetime. The SPICE simulator can obtain active power which is usually considered as self-generated heat of each component for the junction temperatures or core temperature calculation by the finite element simulator.

Owing to the significant impact on reliability, many methods were developed for junction temperature or core temperature estimation in solid state lighting. The finite element stable-state thermal analysis is one of accurate approaches. In this method, the finite element model of an SSL power converter, consists with component models, is built by the finite element simulator, and then the active powers of each component and boundary conditions were applied to the finite element model to calculate the junction temperature or core temperature of each critical component for a next iteration loop. For SPICE parameters of each critical component were assumed to be stable in one iteration loop, its self-generated heat and thus junction temperatures or core temperatures keep constant during this iteration loop.

3. Validations and Discussion

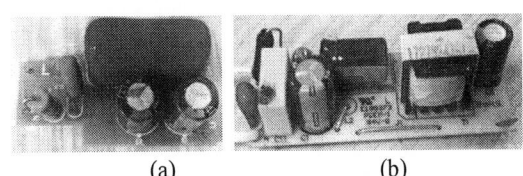

(a)   (b)

Figure 2 Power Converters for Validation

To validate this lifetime prediction method for SSL power converter, this work concerned with two basic types of SSL power converters: (a) an RC voltage-reducing converter and (b) a fly-back converter, as shown in Figure 2 and 3.

The RC voltage-reducing converter consists with surface mounted resistors, a bridge rectifier and electrolytic capacitors which are the most critical components; the fly-back converter consists with surface mounted resistors, a bridge rectifier, two power diodes, a transformer and electrolytic capacitors. Besides electrolytic capacitors, the transformer is also a critical component of the fly-back converter.

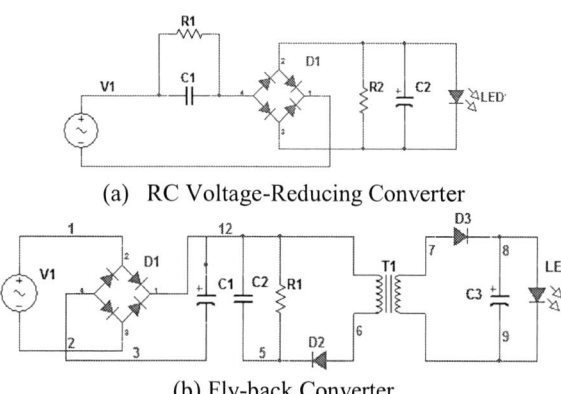

(a) RC Voltage-Reducing Converter

(b) Fly-back Converter

Figure 3 Schematics of Validation Power Converters

(1) Reliability models for critical components

According to industry experiences and academic studies [4, 5], major failure modes in such kind of SSL power converters are degradation of electrolytic capacitors and transformer. Degradation of the electrolytic capacitor at output end may lead current ripple increasing; degradation of the transformer has great impact on output character of a SSL power converter. The previous researches [2, 7] suggest that, as shown in Figure 4, an electrolytic capacitor has two major parameters: capacitance and equivalent series resistance (ESR). The equivalent series inductance is too small to be considered in this work.

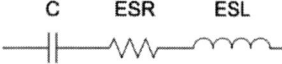

Figure 4 Reliability Models of Electrolytic Capacitor

As shown in Figure 5, for constant core temperature, the degradation of capacitance and ESR follow the model of:

$$C(t) = C_0 - At;$$
$$ESR(t) = ESR_0 \cdot \exp(B \cdot t);$$

Where, t is operation duration, C(t) and ESR(t) are capacitance and ESR at time t, C_0 and ESR_0 are initial capacitance and ESR value, A and B are degradation rates which follow the Arrhenius Equation:

$$R = \alpha \cdot exp(-E_a/\kappa T)$$

Where, E_a is the activation energy, k is the Boltzmann Constant and T is the junction or core temperature in Kelvin.

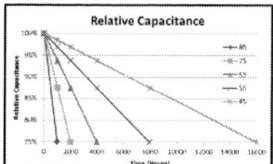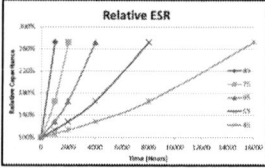

Figure 5 Reliability Model for Electrolytic Capacitors

Similar to an electrolytic capacitor, the major parameter of a transformer is the coupling value. As shown in Figure 6, this work supposes that the coupling value degradation follows the linear model:

$$K(t) = K_0 - At;$$

Where, it is operation duration, L (t) is inductance at time t, L_0 is initial inductance, A is a degradation rate which follow the Arrhenius Equation [8].

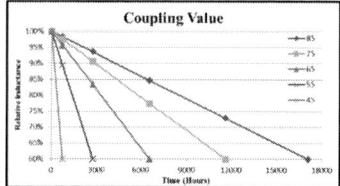

Figure 6 Reliability Model for Transformers

(2) SPICE and FE models for critical components

The SPICE net-lists can be established by their circuit schematics shown in Figure 2 with initial component parameters as circuit design. As shown in Figure 7, the finite element models were established by their system layouts with the material characteristics listed in Table-1.

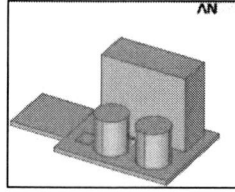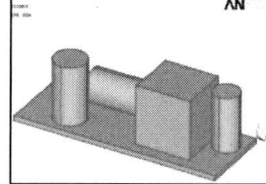

(a) RC Voltage-Reducing, (b) Fly-back

Figure 7 Finite Element Models of Power Converters Validation

Table-1 Material Characteristics of FE Models

Material	Thermal conductivity
FR4	0.25
Electrolytic Material	3~5e-5
Cu	377
Package Material	1.38~1.73

Considering the real operation condition, we assume the ambient temperature is 45℃. According to current test standards, three performance indicators are considered as the failure criteria for lifetime prediction: output voltage, output current and current ripple, which are significant for these SSL power converters. A power converter with output voltage/ current deviations larger than 10% of the

978-1-4799-4789-8/14 $31.00 © 2014 IEEE   98

initial value, or ripple content larger than 20% will be considered as failed [6].

(3) Lifetime Predictions

Figure 8 displays the core temperature simulation result of electrolytic capacitors by the finite element simulator. Operation in the system, the core temperature of these electrolytic capacitors increases with its degradation. More importantly, critical components' degradations have significant impact on the system's performance depreciation. Thus, system's lifetime is shorter than the expected lifetime of critical components.

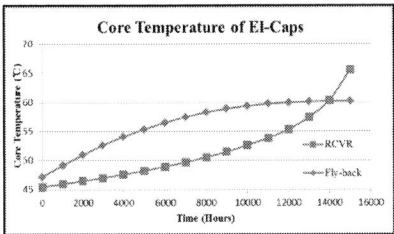

Figure 8 Core Temperatures of Electrolytic Capacitors

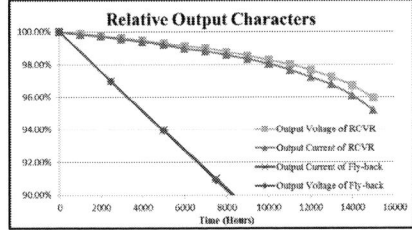

Figure 9 Relative Output Characters Depreciations

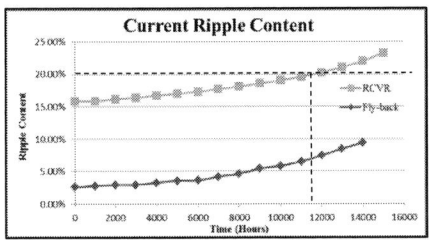

Figure 10 Current Ripple Content Increasing

As shown in Figure 9 and 10, this method well simulated failure modes of these two types SSL power converters. The failure mode of the RC voltage-reducing converter is current ripple increasing lead by degradation of the electrolytic capacitor at output end; the failure mode on the fly-back converter is output depreciation lead by degradation of the transformer. The RC voltage-reducing converter fails after 12'000 hour operation, which is shorter than an electrolytic capacitors' expected lifetime. Meanwhile, the lifetime of fly-back converter is about 8'000 hours. Compare to conservative weakest-link lifetime estimation method, this result is closer to industry experience. Obviously, prediction results by this method depend on the component reliability models. Accurate reliability models can give an estimated lifetimes close to realities. Thus, understanding of failure modes and establishing accurate reliability models are crucial for lifetime prediction of SSL power converter by this method.

4. Conclusions and Future work

This paper proposes a lifetime prediction method for solid state lighting power converter based on SPICE models and finite element thermal simulations. This method focuses on the interaction between electronic and thermal characters of an SSL power converter. According to validation results, this method well simulated failure modes of validation SSL power converters. Compare to conservative weakest link lifetime estimation method, this method gives prediction results closer to industry experience. However, the prediction results depends on the component reliability models. Understanding failure modes and establishing accurate reliability models are crucial to lifetime prediction. There are few data about the degradation of SPICE models of board-level components.

More validations are still undergoing on simulations and tests of more complicated SSL power converters with component reliability distribution models which consider more types of stresses in realities.

Acknowledges

Authors want to thank State Key Laboratory of Solid State Lighting (China) for financial support of this project, and the companies who provide the samples for the validation simulations and tests. The authors also thank to SKL colleagues, Ms. Hongyu Tang and Mr. Zenghui Fan, for the FEA simulations and regular electronic testing.

References

1. Tarashioon S., Baiano A., Henk van Zeijl, Cheng Guo, Sau Wee Koh, W.D. van Driel, Guoqi Zhang, "An approach to 'Design for Reliability' in solid state lighting systems at high temperatures," *Microelectronics Reliability*, 52 (2012), pp.783-793.
2. Lei Han, Nadarajah Narendran, "Developing an accelerated life test method for LED drivers," *IEEE Transactions on Power Electronics*, Vol. 26, No. 8 (2011), pp. 2249-2257.
3. W.D. van Driel, Xuejun Fan, Solid State Lighting Reliability: Components to Systems, Springer, 2013
4. RTI International, "Hammer Testing Findings for Solid-State Lighting Luminaires," Dec. 2013
5. IBM, "Why do Power Supplies Fail and What can be done about it," 2005
6. International Electro-technical Commission, IEC-34A/1622/NP: Principal Component Reliability Testing for LED-based Products, IEC (2012)
7. Xiang Li, Dailin Li, Xuerong Ye, Guofu Zhai, "Research on the Degradation Model of Aluminum Electrolytic Capacitors in LED SMPS," *Journal of Power Supply*, No.6, Nov. (2012)
8. US DoD, MIL-HDBK-217F: Reliability Prediction of Electronic Components, (1985)
9. Michael B. Steer, SPICE: User's Guide and Reference Edition 1.3, North Carolina State University, 2007
10. Bo Sun, Sau Wee Koh, C.A. Yuan, Xuejun Fan, Guoqi Zhang, "Accelerated Lifetime Test for Isolated Components in Linear Drivers of High-Voltage LED System" *EuroSime*, Wroclaw, Poland, April. 2013

978-1-4799-4789-8/14 $31.00 © 2014 IEEE

Understanding delamination for fast development of reliable packages for automotive applications. A consideration of robustness for new packages based on simulation

R. Pufall, M. Goroll, G.M. Reuther
Infineon Technologies AG
Am Campeon 1-12, 85579 Neubiberg, Germany
Email: Reinhard.pufall@infineon.com

Abstract

Thermo-mechanical stress caused by the mismatch of coefficients of thermal expansion (CTE) and temperature variations remain a major concern for the reliability of semiconductor components. This issue is usually addressed by exposing the component to temperature cycling stress tests for a certain number of cycles, followed by e.g. scanning acoustic microscopy (SAM) to investigate delamination. Discussions about specific cycling conditions, e.g. using -65 °C/+175 °C instead of -55 °C/+150 °C for the minimum and maximum temperatures of the cycles or even using liquid-liquid cycling instead of air to air to speed up investigations [1], are often moot, because no real understanding of the effect of the cycling conditions on the component is available.

Furthermore, it is almost a truism that testing alone does not suffice to ensure the reliability of a component. Reliability has to be built into the components from the beginning. As a consequence, the question should be turned around: It is not enough to look at delamination after a certain number of cycles in a stress test. The question is rather how the component should be designed and how the materials should be chosen to prevent delamination. Thus, the focus is changed from measuring delamination to measuring adhesion.

This paper presents an approach for a better understanding of adhesion in terms of possible material combinations, temperature influence (ageing, delamination due to critical induced stress) and topology of interfaces.

1. Introduction

First approaches to measure adhesion of moulding compounds on different interfaces have already been published [2, 3]. These data constitute the basis for predicting delamination, i.e. loss of adhesion. This step has to take into consideration the stress state generated by the temperature variation. Finite element simulation is crucial to understand this stress state and the effect of different material and design parameters. This is a tremendously challenging task, but the typical trial and error procedure being both time consuming and cost intensive, is by far no expedient alternative.

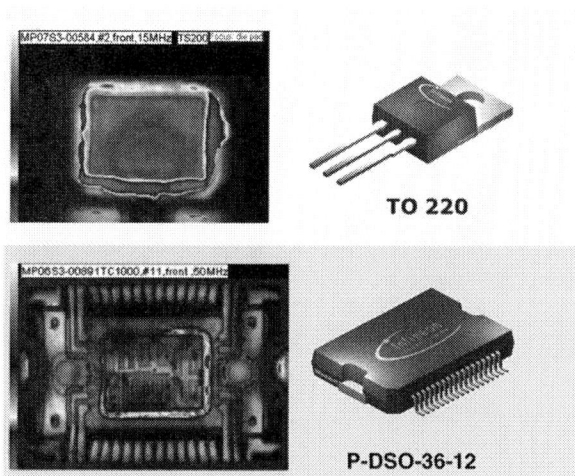

Fig. 1   SAM (scanning acoustic microscopy) image of the die-paddle moulding compound interface showing delamination (gap between the two interfaces)
Packages with big heat slugs show delamination after a certain number of temperature cycles but no crack reaches the outside for the designed lifetime

In Fig. 1 a typical SAM (scanning acoustic microscopy) image shows delamination after a certain number of temperature cycles between die pad and moulding compound. Performing this kind of stress tests it is not possible to isolate the root cause for the observed delamination. A limitation of the SAM method is its execution at room temperature. The stress distribution is definitely different, sometimes even changing from tensile to compressive stress thus reducing the interface gap and providing erroneous conclusions.

Packages with big heat slugs show delamination after a certain number of temperature cycles. However, no crack reaches the outside during the designed lifetime and exposes the silicon directly to harmful environments damaging the protecting moulding compound.

Test standards like AEC Q100 offer alternatives for different temperature cycling conditions, specifically for grade 0:

978-1-4799-4789-8/14 $31.00 © 2014 IEEE

1. **-65 ... +175°C, 500 cycles**
2. **-50 ... +175°C, 1000 cycles**
3. **-50 ... +150°C, 2000 cycles**

The first one offers the shortest test time but it must be verified that no different failure mechanism will be addressed at low temperatures (-50 °C → -65 °C).

In *Fig. 2* the development objectives for a new package type (exposed die pad) are listed. Here, the moulding compound does not anymore enclose chip and die pad completely for the exposed pad designs.

Development objectives

- •reduction of height
- •substitution of Au wire with Cu
- •exposed pad for better cooling
- •better adhesion at all interfaces
- •adaptation to printed circuit board (CTE 15 ppm/K)

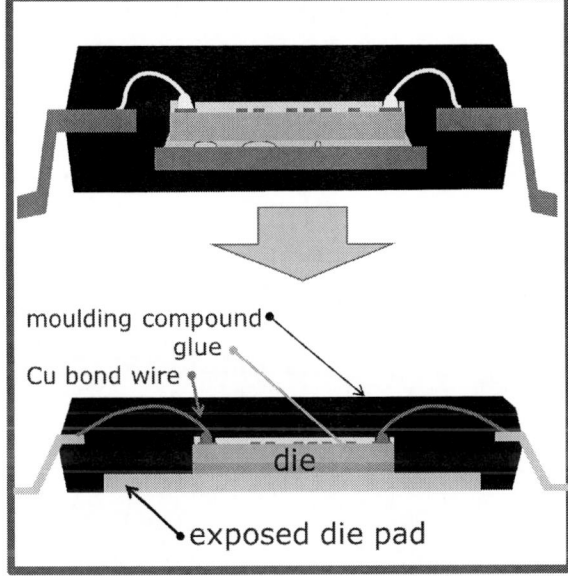

Fig. 2   *Cross section of a moulded device showing the development objectives in direction of exposed pad packages*

This paper presents an approach to increase the robustness of reliable packages during development, based on adhesion measurement, simulation and highly accelerated stress tests. This approach allows fast investigations of different material combinations and pre-selection of potential candidates. It can also provide insights into potential problems during temperature cycling. The method of use of just one half cycle (cooling down) to judge the susceptibility of delamination of known critical interfaces provides a very high speed-up of development cycles and influences strongly the selection of suitable material combinations for the required reliability. Also cohesive zone element simulation can be speeded up significantly by just calculating one half-cycle.

Ageing effects on materials, e.g. moulding compounds, can be incorporated into the investigations, too. By subjecting the moulded buttons (for shear tests) to environmental stresses, such as high temperature storage or humidity; these influences have been studied [2]. This approach is supplemented by finite element simulations, which are then used to investigate delamination behaviour of packages. The benefits demonstrated do not only comprise a better understanding of test accelerations, allowing use of liquid-liquid temperature shock tests instead of air to air temperature cycling tests. They also provides a more rapid and profound possibility for selecting or sorting out specific material combinations. This applies to the assessment of adhesion promoters, too. Influences of processing will be addressed in addition, in order to access their drawback on adhesion. This will be done with samples prepared under the influencing process or afterwards. Some basic ideas have been published [4], but the link to how much adhesion is needed can only be answered with a verified simulation model.

The principle idea to predict the delamination of real packages by simulation with cohesive zone modelling was already presented [5]. The experimental proof of the delamination hypothesis (failure mechanism) was lacking because of an appropriate method. It will be shown that the dye penetration test can be used to verify the existence of an interface crack reaching the outside of the package

2. Simulation and Experiments

The failure mode found in temperature cycle experiments is shown simplified in *Fig. 3*. A gap between the two interfaces, i.e. die pad and moulding compound, is marked by a red ring.

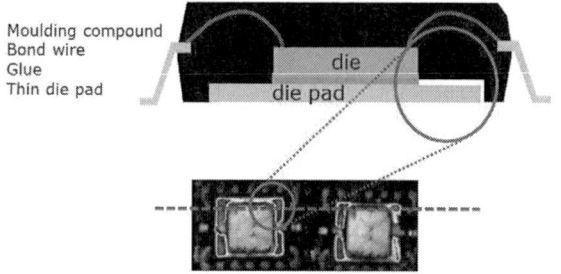

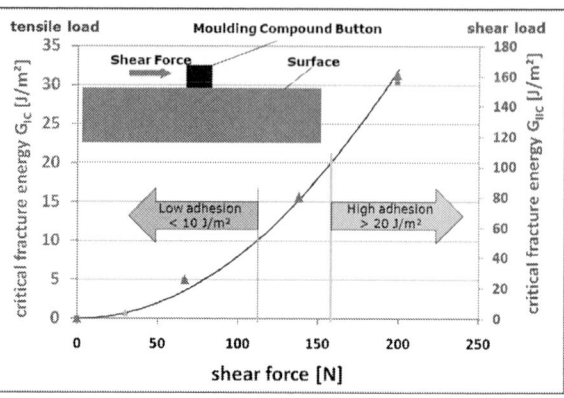

SAM (scanning acoustic microscopy) image (top view)

Fig. 3   SAM (scanning acoustic microscopy) image of the die-paddle moulding compound interface showing delamination (gap between the two interfaces)

The explanation for the delamination is not evident and the hypothesis of bad adhesion between moulding compound and die pad cannot be proven because of a missing failure mechanism. One possible explanation is adhesion or even lack of adhesion.

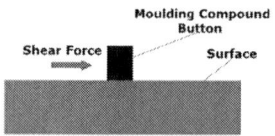

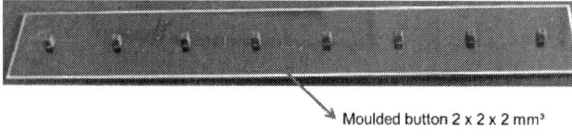

Fig. 4   Advanced button shear test to determine adhesion of moulding compounds on different surfaces

Fig. 4 depicts the setup of a button shear test, our method used to determine adhesion. This test has been improved towards the advanced button shear test: today it is not only possible to measure the maximum force but also the fracture energy and the initiation of the interface crack.

The energy concept allows a comparison of different methods to determine adhesion. There are several approaches available to determine fracture energies:

- Advanced button shear test (force, energy, crack initiation)
- Mixed-mode bending test
- Bi-material 3 - and 4 - point bending test

Fig. 5   Calibration of measured shear force (button shear test) with critical fracture energy. Assumption: $G_{IC} = G_{IIC}/5$ (relation between tensile and shear modus)

All these tests aim to determine adhesion and can also be used to show that adhesion promoters help to avoid delamination.

The advanced button shear test is calibrated to derive the critical fracture energies, as illustrated in *Fig. 5*. The relation between tensile critical fracture energy and shear mode critical facture energy is assumed as 1 : 5.

Gaps are difficult to detect when they are in the range of less than 1 µm. Dye penetration tests can be performed successfully when, as in the shown case, the gap is connected to the external of the package. If the gap is very small, which is expected for our package at room temperature, we doubt that the dye liquid can penetrate the gap in reasonable time frame.

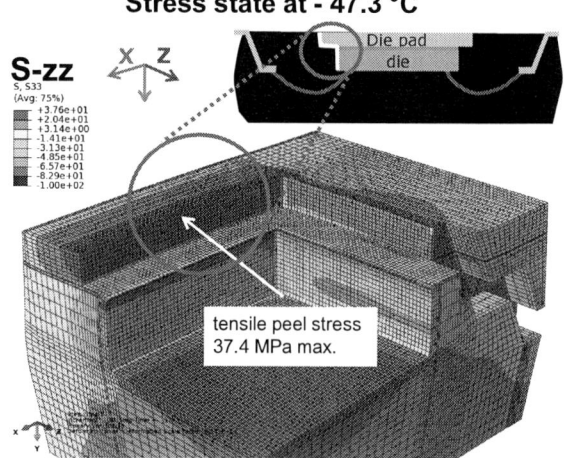

Fig. 6   Cross section of a moulded device showing critical locations of possible crack initiation

Often also different considerations are made to improve the robustness of a package. In

978-1-4799-4789-8/14 $31.00 © 2014 IEEE   102

Fig. 6 the two red rings show the location of possible initiation of interface cracks. At the chip edge the stress is high because of the CTE (coefficient of thermal expansion) mismatch between silicon (3 ppm/K) and moulding compound (6-12 ppm/K). Reducing the temperature during the cool-down cycle, increasing intrinsic stress can only be avoided when the CTE mismatch is small. If the stress causes delamination the robustness can be increased extending clearance x to a value such that crack propagation will not reach complete delamination for the desired lifetime of the component. If the chip is large and the die pad size cannot be increased adequately, robustness is bad.

In this paper we follow a different approach and describe adhesion by fracture toughness (critical adhesion energy) in combination with stress (tensile and shear). Assumption:

$$G_{Ic} = 10 \text{ J/m}^2 \text{ and } G_{IIc} = 50 \text{ J/m}^2$$

The delamination risk $D = G/G_c$ is used in the simulations ($0 <= D <= 1$) where $D = 1$ denotes complete delamination ($G = G_c$).

Delamination occurs by unstable crack growth at one certain temperature

T = - 47.4 °C

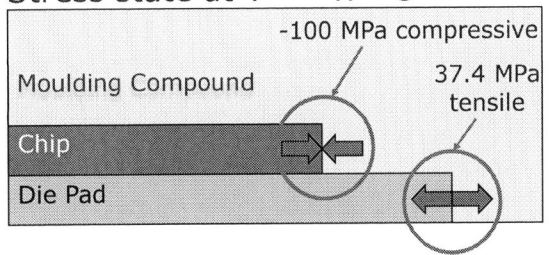

read area: 0.99 ≤ D ≤ 1 = delaminated

Fig. 7   Simulation of unstable crack growth (red area) initiated at the sidewall of the exposed pad on cooling down.

FEM simulation (here, quarter model shown) with cohesive zone elements can predict delamination by cooling down the sample (moulding is performed at 175°C, but cycle tests start at i.e -65 °C). The approach to define a delamination temperature is shown in Fig. 7 (at T = -47.4 °C the moulding compound on the die pad is delaminated). If the assumed fracture toughness values G_c (10/50 J/m²) are valid an unstable crack growth occurs between die pad and moulding compound but not near the chip corner where the stress is highest. In order to verify this delamination hypothesis the stress at different locations has been calculated. In *Fig. 6* the stress is shown for -47.3 °C at the critical interfaces just before

delamination occurs. In this view, lead-frame and chip have been removed, to allow the view onto the interfaces.

Just before the crack initiates, the tensile stress (trying to separate moulding compound and die pad) reaches 37.4 MPa. The shear stresses are low in the range of 15 MPa and not critical to exceed the shear toughness of materials and interfaces

In Fig. 8 the stress situation before delamination is shown for T = -47.3 °C: stress = -100 MPa compressive at the chip and 37.4 MPa tensile at the die pad. The critical stress is tensile and if not enough adhesion is available delamination will take place.

Stress state at T = -47 °C

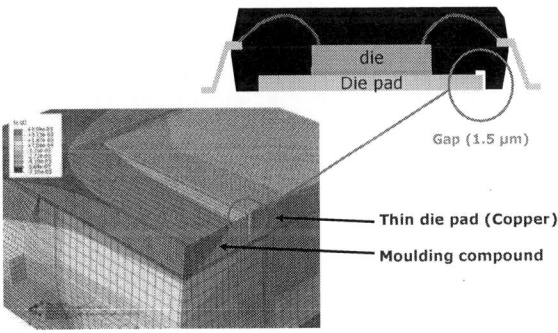

Fig. 8   Stress at -47.3°C at the interfaces chip - moulding compound and die pad - moulding compound

Cooling down to -65°C a gap between moulding compound and die pad is observable (*Fig. 9*). If the delamination occurs the stress is released and the package undergoes a warpage towards the outer edges opening the gap.

out of plane displacement u-y (mm)

Fig. 9   At -65°C a gap between die pad and moulding compound is clearly visible

For the selected package a root cause for delamination has been found when cooling down. The mechanism is shown in *Fig. 9*; tensile stress and insufficient adhesion will create a gap between moulding compounds and die pad just by cooling down to the delamination temperature.

Unfortunately no direct access to adhesion data is available for real packages. Performing button shear tests (*Fig. 4*) and recording force-displacement diagrams for different material combinations, the adhesion can be measured, performing shear tests for various material combinations [6]. The determined critical fracture energy is plotted in *Fig. 5*. We have found values between 10 N and 200 N (1 J/m² -- 30 J/m²). Selecting a material combination with critical fracture energy of 20 J/m² simulations shows no delamination even when cooling down to -65 °C.

In the following, a proposal to improve the robustness of packages is given:

1. Observe the failure mode,
2. Formulate a hypothesis for the possible failure mechanisms,
3. Verify the failure cause by simulation,
4. Answer the question whether better adhesion will avoid the delamination,
5. Change your design if this enables stress reduction,
6. Verify if one half-cycle (cooling down) can break the interface (delamination),
7. Decide which TC test is appropriate.

If this procedure is followed, new packages can fulfil the reliability requirements.

In order to complete the picture for robust packages an important aspect is still missing: the experimental proof of the proposed damage hypothesis.

By finite element simulation using cohesive zone modelling for the adhesion at a characteristic temperature (during the cool down process) the fracture toughness of the lead frame moulding compound interface is exceeded (see Fig. 8 and *Fig. 9*) and unstable crack growth occurs leaving a small gap between die paddle and moulding compound (~1.5 μm).

A method is needed to proof that an open gap, even not visible by optical means, exists and gives access of unwanted substances to the interior of the package.

A dye penetration test has been adapted (operated at -20 °C and filled with fluorescence dye and vacuum to avoid air trapping) to penetrate the small gap and fill it by capillary effect. In *Fig. 10* the proof is depicted for two sample showing delamination after the half cycle test (-65 °C). The delaminated samples were treated with the dye penetration test at -20 °C in order to have a bigger

gap than at room temperature. This results - after performing acoustic microscopy - in a disappearing delamination. This is only possible if the dye can penetrate into the gap from outside. After hardening the resin in the dye penetration recipe it is possible make cross sectioning to determine the penetration path.

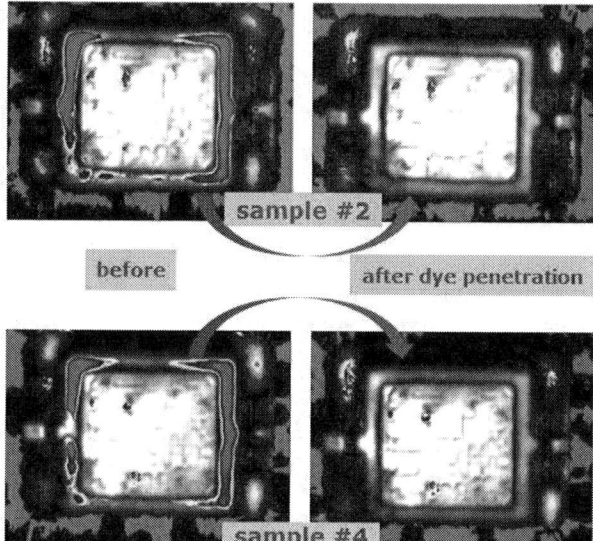

Fig. 10 Acoustic microscopy image of delaminated samples before and after dye penetration test with disappeared delamination.

Fig. 11 shows the fluorescence image of a moulding compound detecting sub-micron cracks which are not detectable with normal optical means

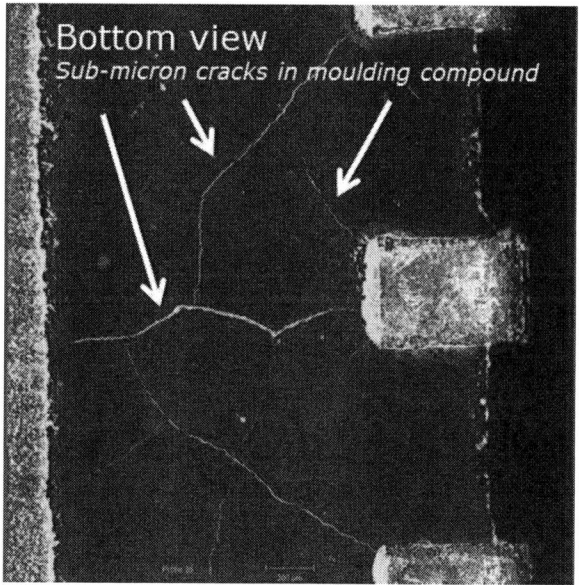

Fig. 11 Sub-micron cracks in moulding compounds visualised by fluorescence dye penetration.

The best way to avoid delamination is the selection of suitable material combinations considering CTE mismatch, adhesion and moisture sensitivity [1, 2, 4, 5]. Better adhesion can be achieved by using adhesion promoters,"nano lawn" or by surface structuring.

In *Fig. 12* a micro-structured lead frame material is tested with the advanced button shear test (*Fig. 4*). The wettability of the invested moulding compounds is very good and the adhesion of moulding compounds benefits from the interlocking of lead frame and moulding compound depicted in the cross section of *Fig. 12*. High adhesion values were measured (critical fracture energy up to 30 J/m², *Fig. 5*) and compared to the used adhesion energies, 10 J/m², no delamination would occur even at very low temperatures (< 65 °C).

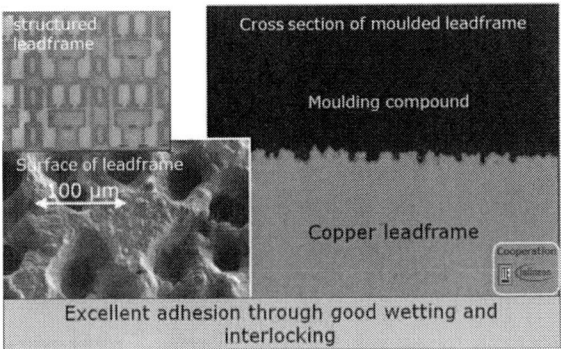

Fig. 12 Measurements at 25°C with a modified surface topography (micro structuring) showed increased adhesion (up to 30 J/m²)

Conclusions

Shear tests provide quantitative data on adhesion and the robustness of packages (cycle performance) can be assessed by performing one half cycle test (cooling down), however a method must be established to observe delamination a low temperatures. In future the selection of materials should be done considering also the parameter "adhesion".

Simulations with cohesive zone elements allow for understanding of adhesion, stress states and locations of design weaknesses of new packages.

The pre-selection of material combinations for high adhesion could improve the number of temperature cycles into ranges of zero delamination for the product lifetime.

Furthermore the dye penetration test with fluorescence materials even at low temperatures can be used to detect sub-micron gaps not detectable by acoustic microscopy or optical means.

In this paper low temperatures (high stress level at interfaces) have been considered. To have a complete picture of delamination risk of exposed pad packages the degradation of moulding compounds must be taken into account (change of material properties, loss of adhesion).

If this behaviour is known robust packages (no delamination during the lifetime) can be designed by finite element simulation considering adhesion (cohesive zone modelling) as the key factor and a pre-selection of material combinations can be performed according to the defined criteria.

References

1. Kanert, W. and Pufall, R., "Investigation of Chip-Package Interaction -Looking for More Acceleration in Product Qualification Tests," *Proc. 47th International Reliability Physics Symposium (IRPS)*, Montreal, Canada, April 2009.

2. Pufall, R. *et al*, "Degradation of moulding compounds during highly accelerated stress tests. A simple approach to study adhesion by performing button shear tests," *Proc. 12th Conference on Thermal, Mechanical and Multi-Physics Simulation and Experiments in Microelectronics and Microsystems (EuroSimE)*, Linz, Austria, April 2011.

3. Dudek, R. *et al*, "Studies on the reliability of power packages based on strength and fracture criteria," *Proc. 12th Conference on Thermal, Mechanical and Multi-Physics Simulation and Experiments in Microelectronics and Microsystems (EuroSimE)*, Linz, Austria, April 2011.

4. R. Pufall et al, "Adhesion of moulding compounds on various surfaces. A study on moisture influence and degradation after high temperature storage". (ESTC 2010, Berlin)

5. R. Pufall, M. Goroll, W. Kanert, R. Dudek. Increasing the robustness for reliable packages by prediction of delamination by cohesive zone element simulation, *2012 13th International Conference on Thermal, Mechanical and Multi-Physics Simulation and Experiments in Microelectronics and Microsystems, EuroSimE 2012*

Acknowledgments

This work has been partly supported by the European Commission Framework Programme 7 within the project e-BRAINS (Project-no FP7- ICT- 257488)

Evaluation of the Residual Stress Distribution in Thin films by Means of the Ion Beam Layer Removal Method

Darjan Kozic[1]*, Ruth Treml[2], Ronald Schöngrundner[1], Roland Brunner[1],
Daniel Kiener[2], Thomas Antretter[3], Hans-Peter Gänser[1]

[1] Materials Center Leoben, Roseggerstraße 12, 8700 Leoben, Austria
[2] Department Materials Physics, Montanuniversität Leoben, Jahnstraße 12, 8700 Leoben, Austria
[3] Institute of Mechanics, Montanuniversität Leoben, Franz-Josef-Straße 18, 8700 Leoben, Austria
*darjan.kozic@mcl.at, Tel.: +43 3842 45922 - 506, Fax.: +43 3842 45922 - 500, www.mcl.at

Abstract

A microelectronic device, designed from multiple structured thin films of different materials deposited on each other, can have a very complex shape. Such a structure can show relatively high residual stresses, which lead to malfunctions and a decrease in lifetime of the device. In this paper a numerical method relying on an inverse optimization algorithm and a finite element (FE) simulation for calculating these stresses is introduced. The evaluation of the residual stress distribution makes use of the so-called ion beam layer removal (ILR) method, where layers of material are removed from a specific region of a micro-cantilever. As a result it is shown that a thin film of material, deposited on a substrate, is occupied by evolving residual stresses through the layer thickness. The calculations and analysis are done automatically using an in-house developed graphical user interface (GUI).

1. Introduction

Manufacturing processes in the microelectronics industry have on the one hand the goal to produce electronic devices at ever smaller scales, and on the other hand to increase their functionality at the same time. A multi-layered microelectronic component is subject to a residual stress distribution which may vary highly throughout the different material layers. These stresses are mainly caused by the differences in thermal expansion and the conditions during the fabrication process. For ensuring higher functionality and reliability of microelectronic devices it is very important to describe the residual stress distribution in a spatially resolved manner throughout the production chain.

Research in recent years has shown an increase in interest for methods capable of evaluating the residual stresses in composite structures. Most approaches for calculating residual stress, for example from wafer curvature or X-ray diffraction (XRD) results [1-3], assume a homogeneous stress distribution through the material's depth. Techniques trying to resolve the naturally expected case of an inhomogeneous stress distribution approach the problem locally. One of these approaches is based on focused ion beam (FIB) milling together with digital image correlation (DIC) [4-6]. Here a specific region with characteristic features is imaged with a scanning electron microscope (SEM) before and after FIB milling. In the vicinity of the introduced FIB cut the internal stresses relax and therefore deformation takes place. Subsequently, a new SEM picture is taken and compared with the previous one to deduce the corresponding deformation field. With this information at hand, one can evaluate the stresses analytically or with a FE simulation. The calculations can be performed for a variety of geometrical shapes [7,8,10]. Like every other method also the FIB/DIC combination has its limitations. From the FIB perspective the shaping of complex geometries can be a very challenging task. Also, the created FIB cuts can have a negative impact on the behavior of the displacement and strain gradient, as the strain relief can lie within the FIB damaged region [9,10]. Investigations about FIB damage have shown that the ion damage commonly depends on the ions and materials involved, the ion current, their incident angle and acceleration voltage [15]. Regarding image analysis, the reliability of the DIC measurements depends on the precision of the imaging system and on the used correlation algorithm [11].

The ion beam layer removal (ILR) method, providing an analytical solution for the residual stress distribution, is a promising alternative for such calculations [12-14,16,17]. Although the approach utilizes the FIB, it does not share all the issues of the method mentioned above. Comparing the extent of the FIB milled region of the beam to the sample thickness, the FIB damage will not have a significant impact on the outcome. Here, the stress distribution through a material film is calculated from micro-cantilever bending experiments. It is assumed that the deflection of a beam depends on the stress distribution in the film. One suggestion how to measure the deflection is by using an SEM. In this work, a numerical solution is pursued which accounts for environmental conditions close to the experiment, such as realistic geometries and boundary conditions. However, there are more details which must be considered in order to reconstruct the experiment as accurately as possible [16]. Here we want to emphasize the importance of these model details. Therefore, our results will also present a simple and explicit, but less accurate analytical solution where the above-mentioned circumstances are not considered. The following chapter will present a detailed description of the developed solution, containing a three dimensional (3D) FE model coupled with an optimization program, and describe how to handle the GUI designed for this purpose.

2. Evaluation details

To the reader's convenience, the ILR method mentioned before shall be summarized here. For a full description and details about sample preparation we refer to Massl *et al* [12]. To calculate the residual stress distribution through a thin film deposited on a substrate material, a micro-cantilever is fabricated with a FIB workstation. This cantilever is initially fixed on both sides, as shown in Fig. 1a. If one side is cut free from the rest of the environment (Fig. 1b), it deflects due to the presence of residual stresses. The measurement procedure is performed in the following way:

i) The initial deflection of the cantilever is measured at its free end (Fig. 1b) by SEM imaging.

ii) As a second step a certain amount of the film thickness is milled by using the FIB at the ILR-Area (Fig. 1b). Again SEM imaging is used to measure the dimensions, especially layer thickness and length of the ILR-Area, as well as the deflection of the cantilever, required for the post-procedure evaluation of residual stresses for each layer of the film.

iii) In the following the second step is repeated until the whole film is removed from the substrate.

The sample considered here consists of a 1.58 μm thick Cu film electrodeposited on a 725 μm thick Si substrate in (100) crystallographic orientation. In a somewhat simplified setting one can assume isotropic material properties. For Si we have used a Young's modulus of 170 GPa and a Poisson's ratio of 0.28, evaluated as a mean value from different sources [18-21]. For the film material Cu the values are 130 GPa and 0.3, respectively, see the values reported in [22-24]. The Young's modulus and Poisson's ratio for Si vary for different crystal orientations, which is an important part of wafer specification. As the Si lattice has cubic symmetry, calculations with orthotropic material behavior can be conducted.

Therefore, we have additionally performed simulations with orthotropic behavior for the wafer material Si with the properties given in Table 1, [18].

E_{Si} [GPa]		G_{Si} [GPa]		v_{Si} [-]	
E_x	169	G_{xy}	50.9	v_{xy}	0.064
E_y	169	G_{xz}	79.6	v_{xz}	0.360
E_z	130	G_{yz}	79.6	v_{yz}	0.360

Table 1: Orthotropic elastic properties for (100) silicon wafers.

As far as the analytical solution is concerned, one can determine the residual stress distribution from a force and momentum balance. The calculation depends on the curvature in the ILR-Area, which changes for each of the removed layers. The curvatures are determined iteratively from the actual deflections, measured in the experiment. A detailed description of the analytical method can be found in [16,17].

In order to obtain suitable results with the numerical calculations, the real cantilever dimensions from the experiment are used to construct a proper 3D FE model with the software package ABAQUS [27] (Fig. 2). The FE model is connected to an inverse optimization algorithm, called Levenberg-Marquardt algorithm [25,26]. This is realized within an in-house developed GUI (see Fig. 3), utilizing the above-mentioned experimental data. Three tabs are providing automatic creation of the FE model and evaluation of the residual stress distribution for a desired structure. The "Settings" tab acts as a part of the pre-processor in the complete ABAQUS environment (CAE), where one defines the geometries, mesh requirements and material properties for the model.

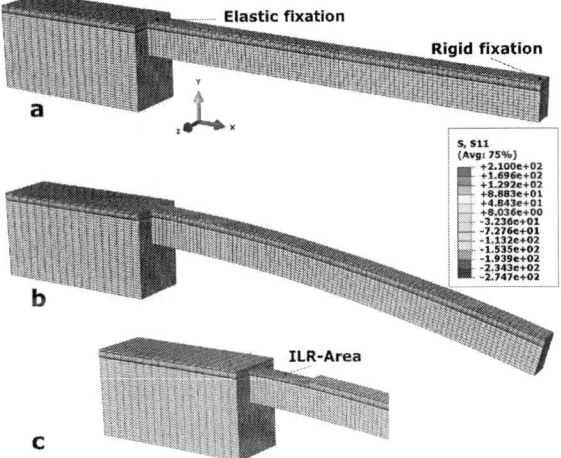

Figure 2: Model of the cantilever: a) Initial setting, where the cantilever is fixed from both sides. b) Released cantilever on the right side, leading to a deflection. c) A closer look at the ILR-Area, where the thin layers are removed. One can also see the "box", attached to the cantilever on the left side.

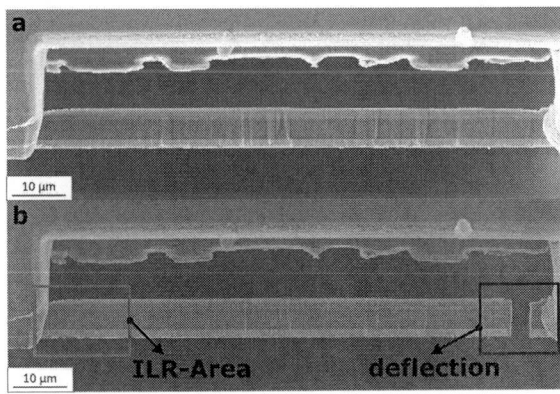

Figure 1: SEM images of the Cu/Si sample used for calculations: a) Initial cantilever, fixed from both sides. b) Released cantilever, cut free with the FIB. Also the approximate size of the ILR-Area and the deflection zone are highlighted.

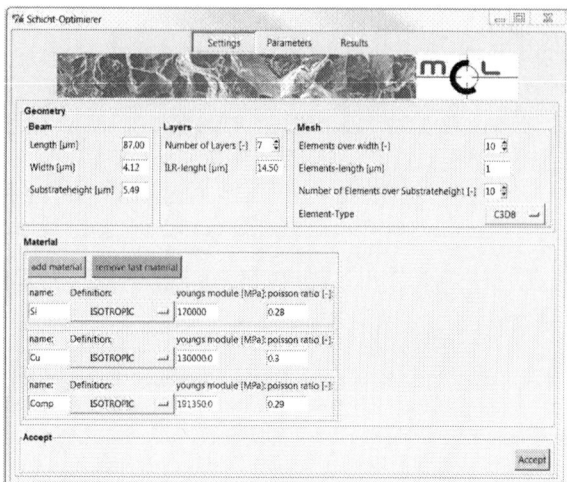

Figure 3: An example of how the "Settings" tab of the GUI looks like.

After accepting the entered data, one proceeds to the "Parameter" tab, where the pre-processing is continued by defining a parameter vector with initial values for the residual stress distribution through the film.

In the initial simulation step the cantilever is attached to a "box" on the left side and on the right side the nodes are fixed rigidly in all directions, shown in Fig. 2a. The mentioned "box" is simulating the untreated rest of the structure, which is connected to the cantilever to simulate slightly softer clamping conditions being closer to the experiment. The parameter vector containing the estimated residual stress field is applied as initial conditions and, after the nodes on the right side are released, the residual stresses inside the film relax, leading to a deflection of the micro-cantilever (see Fig. 2b). Subsequently, the FIB milling of the ILR-Area, highlighted on the FE model in Fig. 2c is performed in the simulation by deleting the corresponding element layers. At this point the Levenberg-Marquardt algorithm enters the calculation. The predefined parameter vector, containing the estimated residual stress distribution, is altered systematically with the goal to minimize the deviation between the simulated and experimental deflections during FIB milling. After the calculations are complete, the functions of the "Results" tab become available, where one has the possibility to display the resulting deflections and residual stresses in graphical form.

3. Results

In Fig. 4 the residual stress distribution through the Cu film calculated analytically and numerically is shown. Apart from some deviations, the residual stress runs through the Cu layers with a nearly constant value. This behavior is expected, as the sample went through thermo-mechanical treatment for a number of cycles during wafer curvature measurements. In this process most of the residual stresses caused by the fabrication process should have been relieved by plastic deformation. The

predominantly linear behavior of the deflection curve in Fig. 5 also confirms the observed residual stresses.

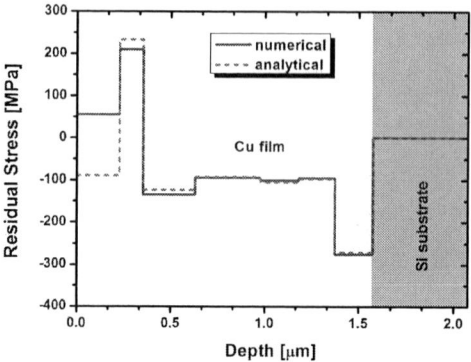

Figure 4: Numerically (blue solid line) and analytically (red dashed line) determined residual stress distributions through the Cu film.

A closer look at Fig. 5 explains the difference between the analytical and numerical values of the stress in the first removed layer. The deflections from experiment and simulation deviate. We can safely assume that the numerically determined stress will become of compressive nature if the simulated deflection approaches the experimental one. The matching of the stresses in the layer on top of the substrate is coincidental and merely a consequence of mismatching deflections. In order to minimize these deviations we have performed the calculations with different sets of optimization parameters. The system seems to be stuck in a local minimum, where it reaches the desired requirements to stop the optimization loop.

As shown in Fig. 6, we have also investigated the possibility of an orthotropic Si substrate with the directional parameters from Table 1. Comparing isotropic and orthotropic Si respectively, there are only minor deviations in the residual stress gradient in the Cu film, see Fig. 6.

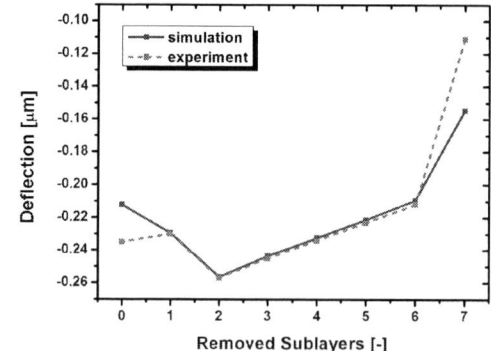

Figure 5: Comparison of the experimentally (red dashed line) and numerically (blue solid line) acquired cantilever deflections.

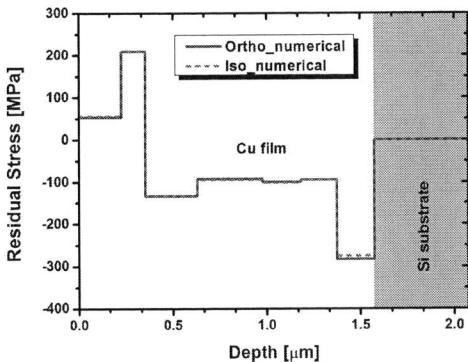

Figure 6: Numerical solution for the residual stress distribution through the Cu film for the case of orthotropic (blue solid line) and isotropic (red dashed line) Si substrate.

Also a calculation with a virtual compressive stress of 2000 MPa over the whole Cu film was performed to test the results with significantly higher stress values. The deviation between the deflections for isotropic and orthotropic material behavior varies around 0.5%, which is again negligible. Nevertheless, the calculations have to be performed for more samples to substantiate the findings above.

5. Summary and Outlook

In the presented work, we have investigated residual stresses occupying a thin film of Cu deposited on a Si wafer. Based on the ILR method, an approach has been developed where a FE simulation of an experiment is coupled with a least-squares optimization (Levenberg-Marquardt algorithm) in order to numerically evaluate the residual stress distribution through the Cu film. A neat and simple in-house developed GUI, feeding empirical data as input, supports the automatically performed calculations.

While the analytical calculation gives a fast estimate of the residual stress gradient the numerical implementation is much more powerful in terms of taking into account complex boundary conditions and material properties.

Accounting for orthotropic properties of the Si substrate does not give major changes in the residual stresses. However, since Si is the most common substrate of all, more calculations are required to draw general conclusions on the influence of the orthotropy on the stress distribution.

So far, we have investigated a Cu/Si micro-cantilever system without major flaws. In reality a composite structure will always contain imperfections like pores or cracks, which might act as initiators for crack propagation and/or delamination. Therefore, calculations with such conditions will be a future research direction, for which the present findings serve as a good starting point.

Acknowledgments

The authors thank H.P. Krückl and A. Drlicek for their helpful discussions and support. Financial support by the Austrian Federal Government (in particular from the Bundesministerium für Verkehr, Innovation und Technologie and the Bundesministerium für Wirtschaft, Familie und Jugend) and the Styrian Provincial Government, represented by steirische Forschungsförderungsgesellschaft mbH and by steirische Wirtschaftsförderungsgesellschaft mbH, within the research activities of the K2 Competence Centre on Integrated Research in Materials, Processing and Product Engineering, operated by Materials Center Leoben Forschung GmbH in the framework of the Austrian COMET Competence Centre Programme, is gratefully acknowledged. In addition partial financial support has been given by the FFG project 838841.

References

1. Takali, F. *et al*, "X-ray diffraction measurement of residual stress in epitaxial ZnO/α-Al$_2$O$_3$ thin film," Mechanics Research Communications, Vol. 38, Issue 3 (2011), pp. 186-191.

2. Hauk, V., Structural and Residual Stress Analysis by Nondestructive Methods, Elsevier (Amsterdam, 1997).

3. Freund, L. B. *et al*, Thin Film Materials: Stress, Defect Formation and Surface Evolution, Cambridge University Press (Cambridge U.K., 2003).

4. Kang, K. J. *et al*, "A method for in situ measurement of the residual stress in thin films by using the focused ion beam," Thin Solid Films, Vol. 443, (2011), pp. 71-77.

5. Sabaté, N. *et al*, "FIB-based technique for stress characterization on thin films for reliability purposes," Microelectronic Engineering, Vol. 84, (2007), pp. 1738-1787.

6. Korsunsky, A. M. *et al*, „Focused ion beam ring drilling for residual stress evaluation," Materials Letters, Vol. 63, (2009), pp. 1961-1963.

7. Song, X. *et al*, "Residual stress measurement in thin films at sub-micron scale using Focused Ion Beam milling and imaging,", Thin Solid Films, Vol. 520, (2012), pp. 2073-2076.

8. Korsunsky, A. M. *et al*, "Residual stress evaluation at the micrometer scale: Analysis of thin coatings by FIB milling and digital image correlation,", Surface & Coatings Technology, Vol. 205, (2005), pp. 2393-2403.

9. Sebastiani, M. *et al*, "Depth-resolved residual stress analysis of thin coatings by a new FIB-DIC method," Materials Science and engineering A, Vol. 528, (2011), pp. 7901-7908.

10. Krottenthaler, M. *et al*, "A simple method for residual stress measurements in thin films by means of focused ion beam milling and digital image correlation," Surface & Coatings Technology, Vol. 215, (2013), pp. 247-252.

11. Pan, B. *et al*, "Two-dimensional digital image correlation for in-plane displacement and strain measurement: a review," Measurement Science and Technology, Vol. 20, (2009), 17pp.

12. Massl, S. *et al*, "A direct method of determining complex depth profiles of residual stress in thin films on a nanoscale," Acta Materialia, Vol. 55, (2007), pp. 4835-4844.

13. Massl, S. *et al*, "Stress measurement in thin films with the ion beam layer removal method: Influence of experimental errors and parameters," Thin Solid Films, Vol. 516, (2008), pp. 8655-8662.

14. Massl, S. *et al*, "A new cantilever technique reveals spatial distributions of residual stresses in near-surface structures," Scripta Materialia, Vol. 59, (2008), pp. 503-506.

15. Kiener, D. *et al*, "FIB damage of Cu and possible consequences for miniaturized mechanical tests," Materials Science and Engineering A, Vol. 459, (2007), pp. 262-272.

16. Schöngrundner, R. *et al*, "Critical assessment of the determination of residual stress profiles in thin films by means of the ion beam layer removal method," Thin Solid Films, (2014) submitted.

17. Jiang, L. M. *et al*, "A modified layer-removal method for residual stress measurement in electrodeposited nickel films," Thin Solid Films, vol. 519, (2011), pp. 3249-3253.

18. Hopcroft, M. A. *et al*, "What is the Young's Modulus of Silicon?," Journal of Microelectromechanical Systems, vol. 19, No. 2 (2010), pp. 229-238.

19. Wortman, J. J. *et al*, "Young's modulus, shear modulus, and poisson's ratio in silicon and germanium," Journal of Applied Physics, Vol. 36, No. 2 (1965), pp. 153-156.

20. Matoy, K. *et al*, "A comparative micro-cantilever study of the mechanical behavior of silicon based passivation films," Thin Solid Films, Vol. 518, No. 1 (2009), pp. 247-256.

21. Bedell, S. W. *et al*, "Strain and lattice engineering for ge fet devices," Material Science in Semiconductor Processing, Vol. 9, (2005), pp. 165-169.

22. Huang, H. *et al*, "Tensile testing of free-standing Cu, Ag and Al thin films and Ag/Cu multilayers," Acta Materialia, Vol. 48, (2000), pp. 3261-3269.

23. Agnew, S. R. *et al*, "The influence of texture on the elastic properties of ultra-grain copper," Material Science and Engineering A, Vol. 242, (1998), pp. 174-180.

24. Sanders, P. G. *et al*, "Elastic and tensile behavior of nanocrystalline copper and palladium," Acta Materialia, Vol. 45, No. 10 (1997), pp. 4019-4025.

25. Levenberg, K. *et al*, "A method for the solution of certain problems in least squares," The Quarterly of Applied Mathematics, Vol. 2, (1944), pp. 164-168.

26. Marquardt, D. *et al*, "An algorithm for least-squares estimation of nonlinear parameters," Journal of Applied Mathematics, Vol. 11, (1963), pp. 431-441.

27. Simulia, Abaqus Documentation, Dassault Systems Simulia Austria GmbH, (2013), http://www.simulia.com/products/abaqus_fea.html.

Thermo-mechanical Analyses of Printed Board Assembly during Reflow Process for Warpage Prediction

Soonwan Chung, Seunghee Oh, Tackmo Lee, and Minyoung Park

Samsung Electronics Co., Ltd

Maetan dong 129, Samsung-ro, Yeongtong-gu, Suwon-si, Gyeonggi-do 443-742, Republic of Korea

E-mail: soon.chung@samsung.com, Phone: +82-31-200-5649, Fax: +82-31-200-2859

Abstract

One of the essential requirements for handheld devices such as smart phone, digital camer, and note-PC is the slim design to satisfy the customers' desires. PCB (Printed Circuit Board) should be also thinner for slim appearance. However, the PCB bending stiffness decreases as the PCB becomes thinner, which is one of the most difficult concerns to engineers. Especially, PCB deforms severely during reflow (soldering) process where the peak temperature goes up to 250°C. Therefore, it is necessary for thermo-mechanical quality/reliability engineers to predict PCB deformation at high temperature before PCB manufacturing. The purpose of this paper is to predict PBA (Printed Board Assembly) deformation based on thermo-mechanical finite element analysis by using the following capabilities. First, PCB is modelled in detail in order to obtain the interlayer stress caused by CTE or elastic modulus mismatch between each layer. Also, the contact boundary condition between PCB and rails in the reflower is considered because PCB deformation along the rails cannot be predicted in advance. Last, PCB multiple array placement can be controlled by the user to find the optimized PCB array for minimum PCB warpage. From the numerical results, it is seen that thermo-mechanical analyses of PBA based on detailed modeling can give the engineer PCB warpage prediction.

1. Introduction

According to the slimness of mobile products, hardware engineers try to diminish the PCB thickness from the early stage of product development. However, there are two major concerns to be solved so that slim PCB can be applied to mass production. One is the process quality issue. Since the bending rigidity of thin plate is theoretically proportional to the cube of plate thickness, slim PCB is apt to yield large deformation especially under high temperature. The warpage causes the interconnection defect between package and PCB during reflow process. The other is the reliability issue. As mentioned, slim PCB easily bends with the drop impact force as well as high temperature. Hence, the package on slim PCB should be concerned with drop reliability. The quality and reliability of slim PCB is not easily acquired, and should be studied concurrently at the begining of development.

In order to minimize PCB warpage, various PCB dielectric layer materials (e.g. FR-4) and dummy designs are applied. Most of PCB material companies are developing high heat-resistant dielectric materials of high Tg or low CTE. CMK is producing ZEROWARP®

material for package substrate, and it is composed high heat-resistant SR (Solder Resist) material.[1] Hitachi Chemical is producing high Tg glass epoxy multilayer material for package substrates with build-up construction.[2] Samsung Electro-Mechanics is developing high Tg LCP (Liquid Crystal Polymer)-based material for package substrate.[3] However, those high heat-resistant materials are relatively expensive, so the application to main PCB is still limited. On the other hand, many efforts are carried out to enhance drop reliability by changing structural design (e.g. package configuration, package type, and screw location)[4] or material design (e.g. solder ball joint, pad surface).[5]

In this paper, the authors address the importance of PBA warpage prediction by introducing the detailed modeling and numerical simulation of reflow process. If PCB warpage can be predicted in the design stage, time and cost caused by product re-design can be minimized. PCB deformation has been numerically analyzed under high temperature similar to reflow process and gravity effect.[6,7] However, PCB modeling is simplifed as the equivalent material properties along the thickness because the full modeling of each layer is a burden to cost and time. Nowadays, the PCB design corresponding to slimness requirement is so important and the accurate PCB warpage is needed to reduce the number of PCB re-design followed by PCB fabrication time. Therefore, the authors try to get the accurate PBA warpage by using the numerical capabilities such as detailed PCB and BGA modeling and contact constraint boundary condition between bottom surface of PCB and top surface of rail(support) in the thermo-mechanical analyses. PCB warpage from the numerical simulation suggested in this paper is verified by comparing with PCB contour measured by 3D full-field non-contact optical measurement technique, digital image correlation (DIC) technique.[8] Also, several numerical examples are shown to describe the PCB multiple array placement and the inclusion of BGA packages.

2. Pre-processing

In this paper, PBA thermo-mechanical simulation tool based on ANSYS® [9] software to calculate PCB warpage under reflow temperature profile. In this section, the numerical modeling of PCB and components(BGA, QFN, connector), the boundary & loading(gravity) condition, and some remarkable features of the numerical simulation tool developed in this study are described. For reference, the elastic material properties of PCB and components are only considered for simple calculation.

2.1 PCB modeling

PCB is composed of many internal layers such as Cu and insulating prepreg layers. The equivalent material properties along the thickness are often used for the computational efficiency. However, it cannot describe the interlayer stress caused by CTE or elastic modulus mismatch between each layer. In this paper, each internal layer is separately modeled by solid elements even though the number of degree of freedoms increase. Since the latest Cu trace width is reduced less than 100 , it is impossible to model the Cu traces directly. So, PCB is divided by rectangular sections for the in-plane modeling and the equivalent material properties for each section are used. The size of length and width of the setion can be selected from the beginning of PCB modeling according to the PCB size and accuracy level. The equivalent in-plane material properties in each layer are calculated by using Cu volume fraction as shown in eq.(1).

$$E_{eqv} = E_{cu} \times VF_{cu} + E_f \times VF_f \quad \ldots\ldots\ldots\ldots(1)$$

where E_{eqv} : Equivalent Young's modulus
E_{cu} : Young's modulus of Copper
E_f : Young's modulus of FR4
VF_{cu} : Volume fraction of Copper
VF_f : Volume fraction of FR4

The volume fraction is easily obtained by calculating the thermal conductivity in thickness direction. Fig.1 shows K_{zz} contour in ICEPAK® software. Actually, the modification of PCB e-CAD file is necessary before it is imported in ICEPAK in order to remove the text or duplicate line which can make errors in the finite element modeling.

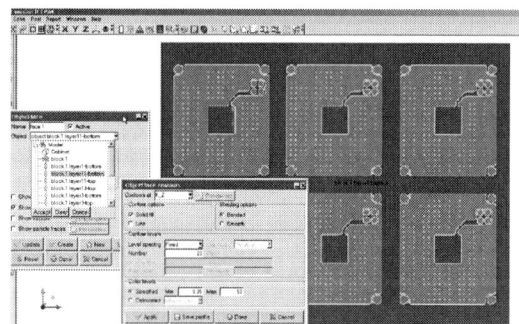

Figure 1. Contour of thermal conductivity (K_{zz})

The material properties of each layer are shown in Table 1. The material properties of Cu are independent of temperature, but those of FR4 are changed around glass transition temperature. In this study, CTE (Coefficient of Thermal Expansion) in the in-plane and out-of-plane direction is averaged respectively.

Table 1. Material properties of PCB internal layer

	Copper	FR-4
Elastic modulus [MPa]	68.9 10^3	6 10^3
Poisson's ratio	0.34	0.3
CTE(X,Y)[ppm/°C]	19	11.8
CTE(Z)[ppm/°C]	19	95.9
Density[kg/m³]	8890	1850

2.2 Component modeling

Some components which are relatively large can affect PCB warpage because those weights are not small and the soldering between component and PCB can generate a constraint force to PCB. As components, BGA, PoP, QFN, Connector, and Shield can are chosen and implemented in the simulation tool. If the outline dimension and mechanical properties are inputted (Fig.2), its finite element modeling can be generated. The solder ball or solder paste between component and PCB connects both after the temperature reaches peak temperature 250°C.

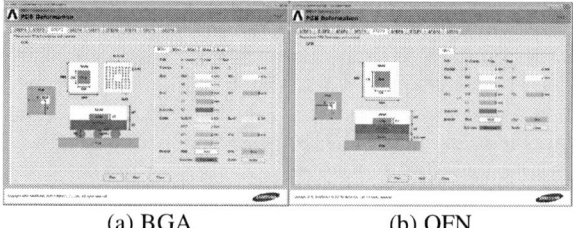

(a) BGA (b) QFN

Figure 2. Input interface for component modeling

2.3 Boundary condition

Generally, PCB is transported on top of two parallel rails during reflow process as shown in Fig.3. Since PCB deformation along the rails cannot be predicted in advance, the contact boundary condition between the bottom surface of PCB edges and the top surface of rails should be considered. In this paper, Penalty method and 3D 8-node Surface-to-Surface contact element are used for contact algorithm and contact element, respectively.

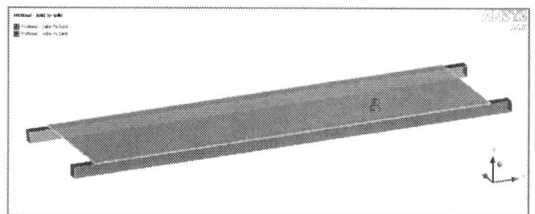

Figure 3. PCB on top of rails

2.4 PCB multiple array placement

Generally, PCB is manufactured in the way of mulitple array when PCB size is relatively small. It is important in the PCB global warpage how to place multiple PCB. One of the major capabilities of numerical

978-1-4799-4789-8/14 $31.00 © 2014 IEEE

simulation tool used in this study is that PCB array placement can be decided according to user preference. Fig.4 shows the procedure of PCB array placement. After one PCB is generated based on rectangle shape, the other PCB can be placed from the rotation, reverse, and rotation/reverse combination. This capability can be useful if an engineer wants to change PCB array after the previous result is not satisfactory.

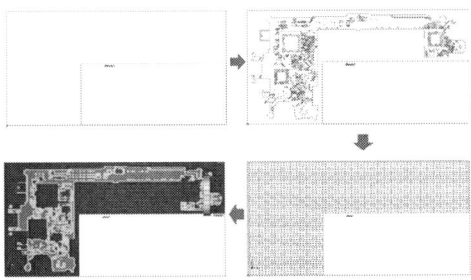

(a) First PCB generation

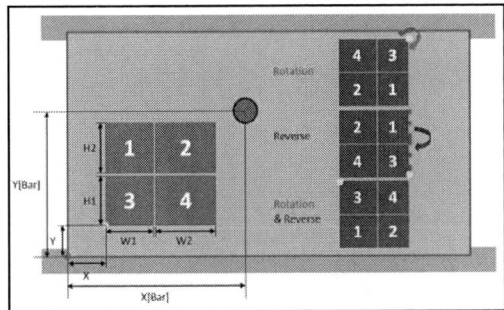

(b) Next PCB generation

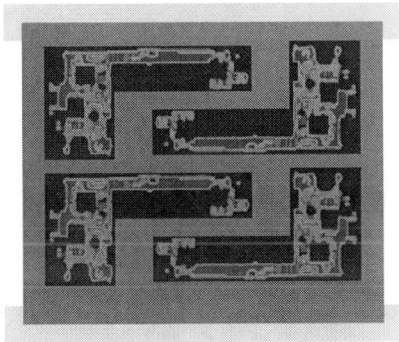

(c) Four array PCB generation

Figure 4. PCB multiple array placement

3. Numerical experiments

In this section, various PCBs are used to check the feasibility of numerical simulation tool. First, two kinds of PCB arrays are simulated for validation of simulation tool. Next, four PCBs and three BGA components in each PCB are simultaneously considered for more realistic PCB warpage simulation.

3.1 Numerical validation

The first example is four array PCB with eight Cu layers. Its dimension is 188.5mm 57.2mm 0.6mm As loading condition, temperature increase from 25 to 250°C and gravitational force are imposed at all the PCB. Fig.5(a) shows Cu traces of eight layers. The contours of PCB warpage and equivalent stress at 250°C are shown in Fig.5(b) and Fig.5(c). Since PCB array is alternatively arranged (top-bottom-top-bottom), warpage direction is changed, and maximum and minumum warpage are 0.215mm and -0.108mm, respectively. Therefore, the difference between maximum and minimum is 0.323mm. The maximum equivalent stress is about 270MPa. To validate the numerical results, PCB warpage is measured at 250°C by DIC technique as shown in Fig.6. The maximum and minimum warpage are 0.174mm and -0.183mm, respectively, so its difference is 0.357mm. Therefore, the numerical results are very close to the experimental results.

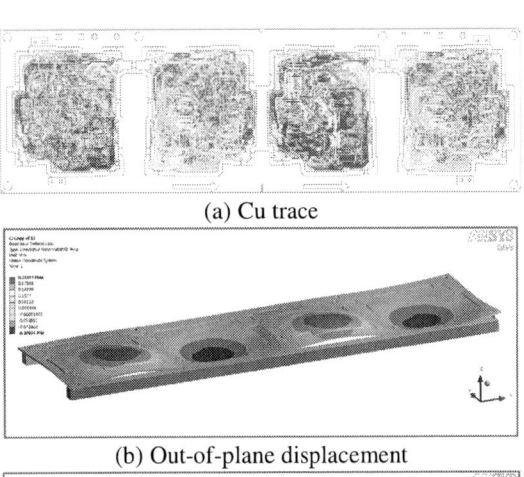

(a) Cu trace

(b) Out-of-plane displacement

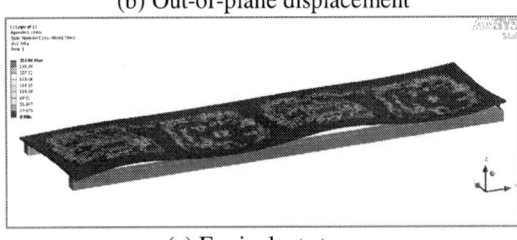

(c) Equivalent stress

Figure 5. Internal layer information and numerical results (PCB-1)

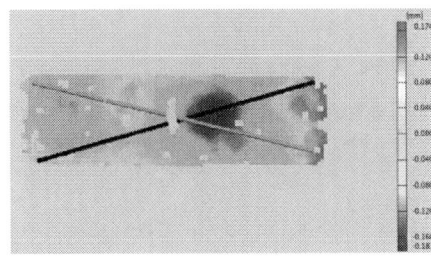

Figure 6. Out-of-plane displacement from experiment (PCB-1)

The second example is six array PCB with six Cu layers. Its dimension is 140mm　110mm　0.6mm As loading condition, temperature increase from 25 to 250°C and gravitational force are imposed at all the PCB. For boundary condition, PCB is constrained in the z-direction along the rails because the PCB configuraton is symmetric along the centerline. Fig.7(a) shows Cu traces of six layers. The contours of PCB warpage and equivalent stress at 250°C are shown in Fig.7(b) and Fig.7(c). Maximum and minumum warpage are 0.1mm and -0.68mm, respectively. Therefore, the difference between maximum and minimum is 0.78mm. PCB warpage contour at 250°C is shown in Fig.8. The maximum and minimum warpage are 0.4mm and -0.265 mm, respectively, so its difference is 0.665mm. It also shows that the numerical results can predict the experimental results within 20% numerical errors.

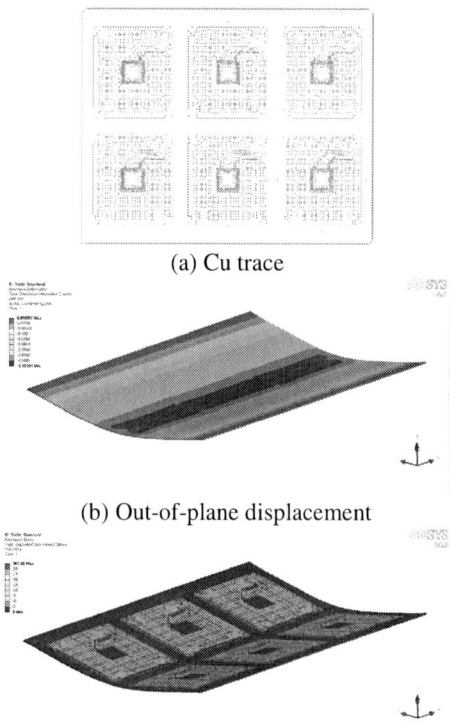

(a) Cu trace

(b) Out-of-plane displacement

(c) Equivalent stress

Figure 7. Internal layer information and numerical results (PCB-2)

Figure 8. Out-of-plane displacement from experiment (PCB-2)

3.2 PBA warpage prediction

In this section, PCB and BGA components are all included in the modeling, that is, PBA shown in Fig.4 is used for warpage simulation. PCB has four array and ten Cu layers. Its dimension is 190mm　145mm　0.65mm The numerical results are obtained from several steps. The first step is the PCB deflection caused by gravitational force as shown in Fig.9(a). The second and third steps are the PCB deflection at 250°C and room temperature after reflower.

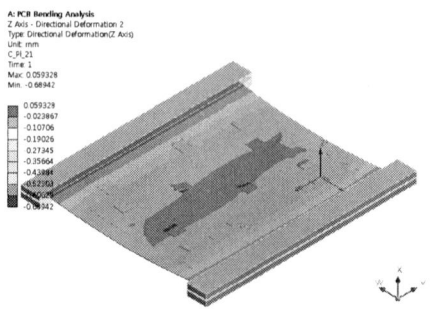

(a) 1st step (Initial state : Room Temp.)

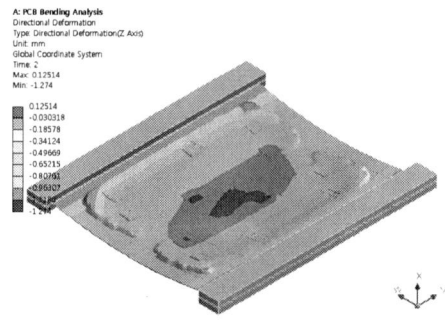

(b) 2nd step (250°C)

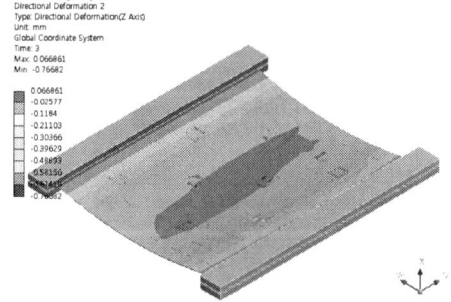

(c) 3rd step (back to Room Temp.)

Figure 9. Out-of-plane displacement contour (PCB-3)

Table 2. Out-of-plane displacement at each step (PCB-3)

Step	Max.	Min.	Max.-Min
1st (initial)	0.059	-0.689	0.748
2nd (high temp.)	0.125	-1.274	1.399
3rd (back to RT)	0.067	-0.767	0.834

From the displacement results, the transition of PCB warpage is natural in that PCB increases as temperature increases, and PCB decreases as temperature decreases. For reference, the experimental results are shown in Fig.10. The displacement along the black diagonal line shown in Fig.10(a) is plotted in Fig.10(b). The displacement contours of both Fig.9(a) and Fig.10(a) are very similar, and its transition from 1st step to 3rd step is also close each other. The difference of displacement at the first step is about 11.6%. However, it increases at the second and third step. This results show the shortcoming of the PCB warpage simulation tool. That is to say, the material properties used in this study are elastic and independent of temperature, so the plastic deformation of FR-4 cannot be reflected. Furthermore, the residual stress within internal layer generated at the PCB lamination process is ignored in this simulation.

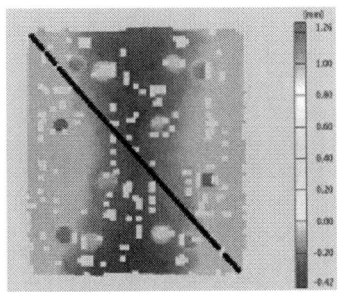

(a) Displacement contour

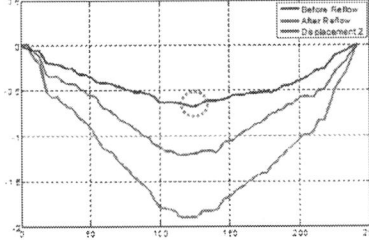

(b) Displacement along the diagonal line

Figure 10. Out-of-plane displacement (Experiment)

Table 3. Displacement comparison

Step	Max. – Min.		Difference
	Num.	Exp.	
1st (initial)	0.748	0.67	11.6%
2nd (high temp.)	1.399	1.89	26.0%
3rd (back to RT)	0.834	1.19	29.9%

4. Conclusions

In this paper, PCB and PBA warpages are predicted by using thermo-mechanical simulation tool. It has some key capabilities such as detailed modeling of PCB and components, contact boundary condition, and PCB

multiple array placement, etc. From numerical experiments, the numerical accuracy is validated. The authors have a plan to upgrade this simulation tool by including material properties as a function of temperature, temperature variation along the length for implementation of PCB transportation within the reflower.

Acknowledgments

The authors would like to acknowledge the valuable support of TSNE (Taesung S&E, Korea).

References

1. CMK Product information, "ZEROWARP", http://www.cmk-corp.com/product/new/zerowarp.html.

2. Hitachi Chemical Product information, "High Tg Glass Epoxy Multilayer Material", http://www.hitachi-chem.co.jp/english/products/bm/b02/006.html.

3. Samsung Electro-Mechanics Product information, "FC-CSP package substrate", http://www.samsungsem.com/product/printed_PackageSubstrate_FCCSP_en.jsp .

4. Soonwan Chung, Gyun Heo, Jae Kwak, Seunghee Oh, Yongwon Lee, Changsun Kang, and Tackmo Lee, "Development of PCB Design Guide and PCB Deformation Simulation Tool for Slim PCB Quality and Reliability," Proc 63rd Electronic Components and Technology Conference, Las Vegas, NV, May. 2013, pp.2157-2162.

5. Zhu, W.H., Xu, L., Pang, John H., Zhang, X.R., Poh, E., Sun, Y.F., Sun, Anthony Y.S., Wang, C.K., Tan, H.B., "Drop Reliability Study of PBGA Assemblies with SAC305, SAC105 and SAC105-Ni Solder Ball on Cu-OSP and ENIG Surface Finish," *Proc 58th Electronic Components and Technology Conf*, Lake Buena Vista, FL, May. 2008, pp. 1667-1672.

6. A.S. Halvi, W. Ahn, D. Agonafer, S. Novotny, "Simulation of PWB Warpage During Fabrication and Due to Reflow," in *Proc. IEEE Inter Society Conf. on Thermal Phenomena*, June 1-4, 2004, pp.674-678.

7. W. Sun, W.H. Zhu, K.S. Le, H.B. Tan, "Simulation Study on the Warpage Behavior and Board-level Temperature Cycling Reliability of PoP Potentially for High-speed Memory Packaging," in Proc. IEEE Int. Conf. on Electronic Packaging Technology & High Density Packaging (ICEPT-HDP), July 28-31, 2008, pp.1-8.

8. S.B. Park, R. Dhakal, R. Joshi, "Comparative Analysis of BGA Deformations and Strains using Digital Image Correlation and Moire Interferometry," in *Proc. Society for Experimental Mechanics*, Portland, OR, June 7-9, 2005.

9. K. Karimanal, A. Adhiya, K. Sahu, B. David, "ANSYS Multi-physics Solutions for LeadFree Electronics Manufacturing, 2009.

Design independent lifetime assessment method for PCBs under low cycle fatigue loading conditions

P. F. Fuchs[*], G. Pinter[**], T. Krivec[***]

[*] Polymer Competence Center Leoben GmbH, Roseggerstrasse 12, 8700 Leoben, Austria, peter.fuchs@pccl.at

[**] Material Science and Testing of Plastics, Department Polymer Engineering and Science, University of Leoben, Otto Gloeckel-Strasse 2, 8700 Leoben, Austria, gerald.pinter@unileoben.ac.at

[***] Austria Technologie & Systemtechnik AG, AT&S, Fabriksgasse 13, 8700 Leoben, Austria, t.krivec@ats.net

Abstract

In this work a method was worked on that assesses the lifetime of printed circuit boards (PCBs) under low cycle fatigue conditions. The method was based on finite element models and low cycle fatigue experiments. Verifying it, two significantly different PCB designs and several different PCB built ups were analyzed. Doing so, for the numerical part the individual layer materials were characterized, material models were defined and simulation models were built. A submodelling technique had to be applied in order to evaluate the local loading situation. For the experimental part corresponding PCBs were manufactured and a statistically relevant number of boards were tested in a board level cyclic bend test (BLCBT) while the critical connections were monitored. Based on the results of one PCB a correlation model describing the dependence of the cycles to failure on the local loading situation was formulated and used for the lifetime assessment of the other boards. A very good agreement between predicted and measured results could be shown.

1. Introduction

The industry demands on PCBs are constantly increasing. They should get smaller to integrate more functions while they should also perform more reliable (e.q. [1] and [2]). However, miniaturization is often contradicting the application lifetime. Thus, the PCB built ups have to be optimized to use the right materials at the right place in order to max out their potential.

Thereby a variety of different loads, depending on the PCB application, have to be taken into account. Common load cases are temperature, vibration, fatigue or impact. In this work the especially for handheld electronic products important impact performance has been focused on. It is a key issue for the related industry, as these devices are especially prone to be dropped whilst their reliability is important for the customer. The industry standard to test the impact behavior is an instrumented board level drop test (BLDT) [3]. An explicit correlation between the drop test performance and the low cyclic fatigue lifetime has been shown in a previous work [4].For the presented lifetime assessment the BLCBT was chosen to work with, as it is advantageous with respect to its idealization as a finite element model.

The use of finite element modelling in order to reduce the experimental testing effort in drop tests has been successfully evaluated by several authors (e.g. [5] [6] [7] [8] [9] [10] [11]). However, a main challenge still remains the actual lifetime assessment based on the simulation results. Different approaches have been reviewed by Wong et al. [2] and it was concluded, that a fatigue micro-damage model that accounts for the specific loading situation would be necessary for a robust failure model. In a previous work it was already successfully tried to formulate a correlation between the simulated local loading situation and the BLCBT lifetime for different PCB layer built ups [12]. However the loading situation and consequently the drop test performance also significantly depends on the overall PCB design. Thus in this work the focus was on the comparison and analysis of two different test vehicles.

2. Materials and PCBs

In this work two different PCB test vehicles (Design A and Design B) were analyzed. Their top view and schematic cross section is shown in Figure 1. Design A was based on the JEDEC test board [3] while Design B was a board design of the research industry partner. Both test vehicles were eight-layer PCBs with five (A) and two (B) mounted components respectively. The outer board dimensions were 132 x 76 x 1.064 mm (A) and 100 x 48 x 0.588 mm (B). The components used for both boards were ball grid array packages (type: 288 Ball BGA with 4 Rows of Balls, pitch: 500 μm, pad diameter: 350 μm and solder material: 96.5Sn/3.0A g/0.5Cu (SAC305)) with dimensions of 12 x 12 x 0.65 mm. While all copper layers had a thickness of 18 μm, the insulating layers differed with respect to their thickness and the applied material. The symmetric built up of the insulating layers of all tested PCBs is presented in Table 1. Thereby M1, M2, M3, M4 and M5 represent different epoxy resin matrix materials and the numbers 1500, 1080 and 1037 indicate the 'E' glass fabric woven reinforcements defined in the IPC specification [13]. Core stands for specific producer dependent fabric woven reinforcements. R1 and R2 are two fiber unreinforced epoxy resin layers.

978-1-4799-4789-8/14 $31.00 © 2014 IEEE

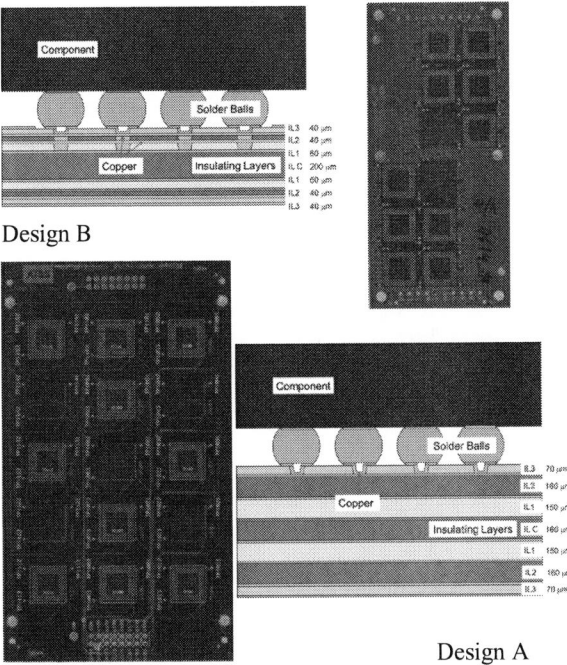

Figure 1: Top view and schematic cross section of the two different PCB designs analyzed in this work: Design A (bottom) and Design B (top).

Table 1: Overview of the insulating layer built up of all tested PCB designs.

Design	Insulating Layers			
Number	ILC	IL1	IL2	IL3
A1	M5 1500	M5 Core	M5 1500	R1
A2	M4 1500	M4 Core	M4 1500	R1
A3	M2 1500	M2 Core	M2 1500	R1
A4	M5 1500	M5 Core	M5 1500	M5 1080
B1	M1 Core	M1 1080	M1 1037	R2
B2	M1 Core	M1 1080	M1 1037	M1 1037
B3	M1 Core	M1 1080	M1 1037	R1
B4	M3 Core	M3 1080	M3 1037	R2
B5	M3 Core	M3 1080	M3 1037	M3 1037
B6	M6 Core	M6 1080	M6 1037	R2

The material models used in this work were determined by an extensive experimental material characterization based on the methods presented in a previous work [14]. Only the solder material behavior has been adopted from literature data [15]. Linear elastic orthotropic models have been defined for the fabric woven reinforced prepregs. For copper, the neat epoxy resin and the solder elastic plastic material models have been used. The prepreg matrix materials (M1 to M6) and the epoxy resins of the unreinforced layers (R1 and R2) are sorted according to their Young's modulus in descending order. The prepreg resin content depended on the applied fabric woven and was about 45 % for the 1500, 65 % for the 1080 and 70 % for the 1037 type.

3. Experiments

All boards listed in Table 1 were tested in a BLCBT. The test method and part of the results have already been presented in [12] and [4] and are summarized in this work. The BLCBT uses a three point bending fixture and applies a repeated sinusoidal bending load with 25 Hz. Two set-ups have been used to account for the different board sizes. For Design A and Design B a bearing distance of 105 and 90 mm has been used respectively. The support rollers and the fin had a diameter of 5mm. The set-up is shown in Figure 2 exemplarily for Design B.

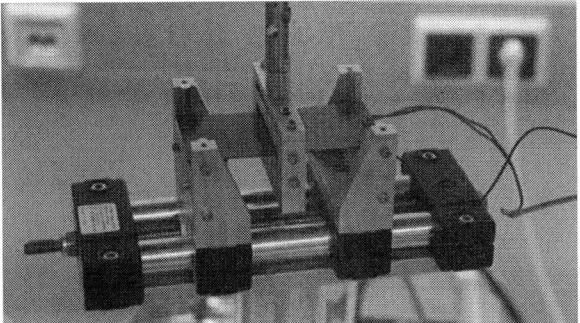

Figure 2: The board level cyclic bend test set-up for design B.

The failure detection methodology was like in the JEDEC drop test standard [3] using an event detector (256STD, Analysis Tech, Wakefield, US). Different deflection amplitude levels have been tested in order to measure times to failure for different local loading situations. An overview of all performed cyclic bend tests is given in Table 2. In order to avoid a lift-off of the boards from the fixture the lower deflection limit was set at 1mm for all boards. Consequently the defined amplitudes indicate a cyclic displacement between a lower limit of 1mm and an upper limit of 1 mm plus the given value. For Design A1 all defined amplitudes have been tested in order to have enough data to formulate the correlation between the local loading situation and the cycles to failure. For the other designs only chosen amplitudes have been analyzed, as the results were only needed to verify the correlation.

Table 2: Test matrix for the board level cyclic bend test.

Amplitude [mm]	1	1.75	2	2.75	3	3.5	4
Design Number	Measurement Repetitions [mm]						
A1	3	3	8	3	3	3	3
A2	-	-	8	-	-	-	3
A3	-	-	8	-	-	-	3
A4	3	3	8	3	3	3	3
B1	-	-	-	-	3	-	3
B2	-	-	-	-	3	-	3
B3	-	-	-	-	3	-	3
B4	-	-	-	-	3	-	3
B5	-	-	-	-	3	-	3
B6	-	-	-	-	3	-	3

978-1-4799-4789-8/14 $31.00 © 2014 IEEE

4. Simulation

All designs were analyzed in finite element simulations using the software Abaqus 6.12-1 (Daussault Systèmes Simulia Corp., Providence, RI, USA). The simulations were aiming in the prediction of the local loading situation at the different experimentally tested deflection amplitudes (see Table 2) in the BLCBT. In order to get detailed results a submodelling technique was applied. The bending load was applied on a global model and the results were used to define the boundary conditions of a more detailed local model of the chosen area of interest. Failures were known to be located close to, or in the solder balls, for which reason the solder ball carrying the highest loads in the global model was chosen to be analyzed in a submodel. In the global model shell elements were used for the board and the component in order to reduce the degrees of freedom. The different layers in the board were taken into account defining composite shell sections. The solder balls were modeled using solid elements to account for their free-form geometry. The mesh for the global model could be defined rather rough, as detailed results analysis was only planned for the submodels. Consequently the global model used element edge lengths between 0.07 and 1 mm while the submodel element edge lengths were between 0.02 and 0.001mm. All elements applied had a linear geometric order and used reduced integration. The global model and the submodel for Design B is shown in Figure 3. The global model of Design A was realized as quarter model as it had two symmetry planes which could be used. Otherwise the model was set-up analogous to Design B.

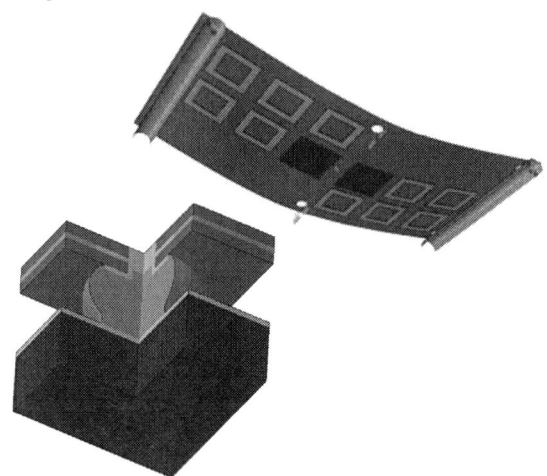

Figure 3: The simulation model for design B: local model (left) and global model (right)

5. Results and Lifetime Assessment

The results of the BLCBT experiments were summarized and presented in Figure 4. The cycles to failure of the tests listed in the matrix in Table 2 were plotted. The results of the tests performed with an amplitude of 1.75mm were not given in the figure, as the measured cycles to failure were over 50000 and thus out of the works focus on low cycle fatigue. For the 1mm amplitude tests no results were available as the tests were stopped after 200000 cycles and no failure has been detected at that point of time. The standard deviation for the PCB Design A1 is indicated in order to give an impression of the relatively large data scatter in the experimental cyclic lifetime determination. Nevertheless, as expected, a definite dependency of the cycles to failure from the applied amplitude could be observed. Furthermore, a significant difference between the analyzed designs was evident. All Design B PCBs except B2 performed better than the Design A boards indicating lower resulting local loads in Design B at the same global deflections.

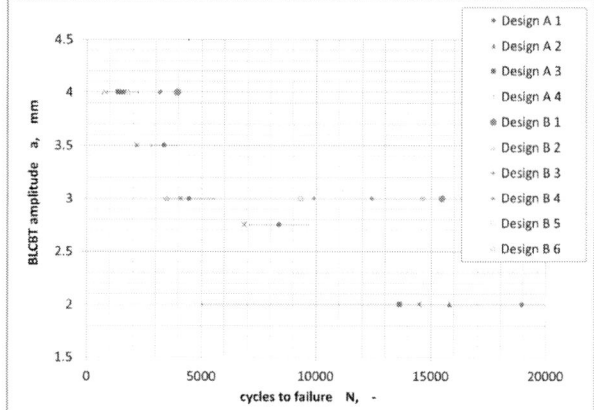

Figure 4: The experimental BLCBT results. The cycles to failures are shown for all tested amplitudes and PCB designs.

Failure analysis using light microscopy on cross sections of the solder balls showed why Design B2 showed a differing behavior. All designs but B2 failed due to a crack in the solder ball, while B2 failed in the outermost insulating layer. Figure 5 shows a comparisons of the failure modes for Design A1, B1 and B2. As the failure in the solder ball was the dominant mode for all but one design the further work was focused on it. The results of Design B2 will be taken into account in ongoing research.

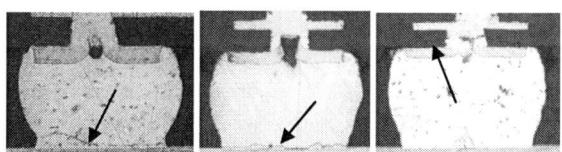

Figure 5: Cross sections of the failed region in PCB Design A1, B1 and B2 (from left to right).

The global simulation models results were used to identify the solder balls carrying the highest loads and to define the boundary conditions of the submodels. The solder ball showing the highest plastic equivalent strain (PEEQ) was then chosen for the detailed local analysis.

The PEEQ value was used as parameter as it is known to be widely and effectively used for the low cycle fatigue modeling of metals in general [16] and in particular for solder materials e.g. [17].

Using the submodels detailed analyses of the local strain distribution could be performed. In Figure 6 a cross section of the submodel of the Design A1 simulation is shown exemplarily. The PEEQ maxima were found to be in good agreement with the experimentally determined initial cracks (Figure 5). The submodel simulations were evaluated for all experimentally tested amplitudes and different board designs. Thus, local maximal PEEQ values could be attributed to every measured cyclic lifetime.

Figure 6: The submodel simulation results of Design A1 showing the plastic equivalent strain in the cross section.

The experimental and simulation results were finally used to generate a correlation plot. In Figure 7 the maximal local PEEQ values were plotted over the determined cycles to failure of the respective BLCBT. Based on this data it was tried to model the dependency of the cyclic lifetime on the local loading parameter. The Coffin-Manson equation [16]

$$N = C / (\varepsilon^p)^c$$

was used to describe a power-law based correlation. N refers to the number of cycles to failure and ε^p to the plastic strain and in this work to the PEEQ value respectively. C is the proportionality factor and c is the exponent. The model parameters were determined on the data of Design A using a least square fit. Thereby C and c yielded in values of 124.68 and 2.72. The resulting curve is indicated in the figure.

This so called characteristic failure curve was supposed to provide a general correlation between the local loading situation of the solder balls and the BLCBT lifetime. It should allow for the prediction of cycles to failures based on simulation results. The results of Design B were used to evaluate the approach. Thereby, the prediction quality of the characteristic failure curve could be analyzed. It can be seen in Figure 7 that the data of Design B fits well to the predicted trend. The discrepancies are far below the standard deviation of the test results approving the suitability of the methodology.

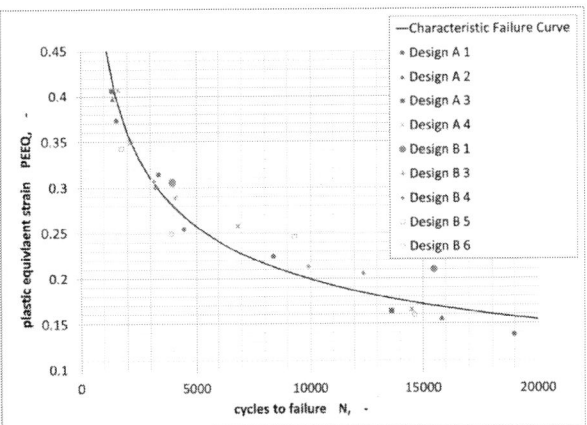

Figure 7: The plastic equivalent strain is plotted over the respective cycles to failure. The data is fitted by a characteristic failure curve.

6. Summary and Conclusion

In the work two significantly different PCB designs and several different PCB built ups have been analyzed. Their low cycle fatigue behavior was tested in a board level cyclic bend test. At the same time simulations of the experiments have been performed and using a submodelling technique the local loading situation was evaluated. The maximal plastic equivalent strain in the solder balls was determined and correlated to the experimental results. For the data of one design a so called characteristic failure curve based on the Coffin-Manson equation was fitted. The prediction quality of this curve was evaluated by the results of the second design. A good agreement could be shown. Despite the different geometry, the different component positions and the different copper structure the general trend of the results could be described very well by the determined correlation curve.

The work focused on the for this built ups dominant failure mode with a crack in the solder ball. In future work different failure modes, as failures in the outermost epoxy layers, will be taken into account. Therefore, different local loading parameters will have to be defined and additional characteristic failure curves will be generated. Based on these distinct curves it should be possible to not only assess the expected performance, but also the most likely failure mode.

Acknowledgments

The research work of this paper was performed at the Polymer Competence Center Leoben GmbH (PCCL, Austria) within the framework of the COMET-program of the Austrian Ministry of Traffic, Innovation and Technology with contributions by the University of Leoben and by the AT&S Austria Technologie & Systemtechnik Aktiengesellschaft. The PCCL is funded by the Austrian Government and the State Governments of Styria and Upper Austria.

References

1. Pradeep Lall, S.G.P.C.J.S.: Solder-Joint Reliability in Electronics under Shock and Vibration using Explicit Finite Element Sub-Modeling (2006)

2. Wong, E., Seah, S., Shim, V.: A review of board level solder joints for mobile applications. Microelectronics Reliability 48(11-12), 1747–1758 (2008). doi: 10.1016/j.microrel.2008.08.006

3. Jedec Solid State Technology Association: Board Level Drop Test Method of Components for Handheld Electronic Products(JESD22-B111) (2003)

4. Fuchs, P., Major, Z.: Cyclic bend tests for the reliability evaluation of printed circuit boards under dynamic loads. Frattura ed Integrità Strutturale(15), 64–73 (2011)

5. Amy, R.A., Aglietti, G.S., Richardson, G.: Accuracy of simplified printed circuit board finite element models. Microelectronics Reliability 50(1), 86–97 (2010). doi: 10.1016/j.microrel.2009.09.001

6. Dhiman, H.S., Xuejun Fan, Tiao Zhou: Modeling techniques for board level drop test for a wafer-level package. In: High Density Packaging (ICEPT-HDP), Shanghai, China, pp. 1–9. doi: 10.1109/ICEPT.2008.4607141

7. Le Coq, C., Tougui, A., Stempin, M.-P., Barreau, L.: Optimization for simulation of WL-CSP subjected to drop-test with plasticity behavior. Microelectronics Reliability 51(6), 1060–1068 (2011). doi: 10.1016/j.microrel.2011.03.011

8. Zhang, B., Ding, H., Sheng, X.: Modal analysis of board-level electronic package. Microelectronic Engineering 85(3), 610–620 (2008). doi: 10.1016/j.mee.2007.11.008

9. Tee, T.Y., Ng, H.S., Lim, C.T., Pek, E., Zhong, Z.: Impact life prediction modeling of TFBGA packages under board level drop test.

10. Tee, T.Y., Ng, H.S., Yap, D., Baraton, X., Zhong, Z.: Board level solder joint reliability modeling and testing of TFBGA packages for telecommunication applications. Microelectronics Reliability 43(7), 1117–1123 (2003). doi: 10.1016/S0026-2714(03)00127-6

11. Qu, X., Chen, Z., Qi, B., Lee, T., Wang, J.: Board level drop test and simulation of leaded and lead-free BGA-PCB assembly. Microelectronics Reliability 47(12), 2197–2204 (2007). doi: 10.1016/j.microrel.2006.10.017

12. Fuchs, P., Pinter, G., Major, Z.: PCB drop test lifetime assessment based on simulations and cyclic bend tests. Microelectronics Reliability(5), 774–781 (2013). doi: 10.1016/j.microrel.2013.01.001

13. IPC-4412A Amendment 1. Specification for Finished Fabric Woven from ''E'' Glass for Printed Boards (2008)

14. Fuchs, P., Pinter, G., Tonjec, M.: Determination of the orthotropic material properties of individual layers of printed circuit boards. Microelectronics Reliability 52(11), 2723–2730 (2012). doi: 10.1016/j.microrel.2012.04.019

15. Ma, H.: Characterization of lead-free solders of for electronic packaging. Dissertation (2007)

16. Halford, G.R.: Fatigue and Durability of Structural Materials. A S M International, Materials Park (2005)

17. Srinivas, V., Menon, S., Osterman, M., Pecht, M.G.: Modeling the rate-dependent durability of reduced-Ag SAC interconnects for area array packages under torsion loads. Journal of Electronic Materials 42(8), 2606–2614 (2013)

Determination of cyclic mechanical properties of thin copper layers for PCB applications

Klaus Fellner[1], Peter F. Fuchs[1], Thomas Antretter[2], Gerald Pinter[3], Ronald Schöngrundner[4]

[1]Polymer Competence Center Leoben GmbH, 8700 Leoben, Austria
[2]Institute of Mechanics, Montanuniversitaet Leoben, Austria
[3]Institute of Materials Science and Testing of Polymers, Montanuniversitaet Leoben, Austria
[4]Materials Center Leoben Forschung GmbH, 8700 Leoben, Austria

Abstract

The overall objective of this research work is the characterization of the mechanical behavior of Printed Circuit Boards (PCBs) under cyclic thermal loads. The conducting traces in PCBs are made from thin copper layers in an etching process. Hence, thin copper layers are characterized experimentally and subsequently cyclic material parameters are determined. The experimental characterization is conducted using cyclic tensile-compression tests at different temperatures and loading conditions. For these tests composite specimens made of five layers of copper and four layers of glass fiber reinforced epoxy resin are used. The obtained material response is modeled using the "Nonlinear isotropic/kinematic hardening model" built-in in the Finite Element Analysis-software Abaqus. For every loading case the optimal set of parameters is determined using an optimization procedure. Based on the known parameter sets of the individual loading cases the calibration of a "Nonlinear isotropic/kinematic hardening model" for all R-ratios and temperatures is attempted and the findings are discussed.

1. Introduction

In Printed Circuit Boards (PCBs) the conducting parts are made of copper while the dielectric parts are made of reinforced polymers, mostly glass fiber reinforced epoxy resins (FR4-prepregs). Since these materials show different coefficients of thermal expansion, stresses are induced when PCBs are thermally loaded. As PCBs are often subjected to cyclic thermal loads, the cyclic mechanical properties of the involved materials have to be known in order to be able to model the material response of the PCB correctly. An appropriate testing methodology for thin copper layers under cyclic tensile-compression loads was developed in an earlier paper [1]. In those cyclic tensile-compression tests force-strain curves of thin copper layers were obtained using a composite specimen consisting of layers of copper and FR4-prepreg. Out of those force-strain curves hysteresis loops were taken and used to determine the material input parameters for a "Nonlinear isotropic/kinematic hardening model" [2] which is available in the Finite Element Analysis (FEA)-software Abaqus (Abaqus 6.12, Simulia, Daussault Systèmes, Vélizy-Villacoublay, France). Current research in the field of PCBs places an emphasis on FEA-modelling [3-6] in order to save time and costs for experiments. The hardening properties of copper were determined in an optimization process based on the force-strain curves (hysteresis loops) obtained in the cyclic

tensile-compression tests. To this end a FEA model of the layered specimen using continuum shell elements in Abaqus was created. Using the built in "curve mapping algorithm" in the optimization software LS-Opt (LSTC, Livermore, CA, USA, 2011), the residuals between the measured and the simulated hysteresis loops were minimized. Abaqus was integrated as a user-defined solver in the LS-Opt optimization procedure. Finally the yield stress values were formulated as a function of the accumulated plastic strain.

2. Nonlinear isotropic/kinematic hardening model

This model is available in Abaqus and is based on the work of Lemaitre and Chaboche [7,8]. Detailed background information on this model can be found in the Abaqus Manual [9]. In this model the strain rate tensor $\dot{\boldsymbol{\varepsilon}}$ is additively decomposed into the elastic $\dot{\boldsymbol{\varepsilon}}^{el}$ and the plastic $\dot{\boldsymbol{\varepsilon}}^{pl}$ strain rate tensor. (Eq. 1)

$$\dot{\boldsymbol{\varepsilon}} = \dot{\boldsymbol{\varepsilon}}^{el} + \dot{\boldsymbol{\varepsilon}}^{pl} \tag{1}$$

The elastic region is assumed linear elastic (Eq. 2), where $\mathbf{D}$ stands for the fourth order elasticity tensor and $\boldsymbol{\sigma}$ for the stress tensor.

$$\boldsymbol{\sigma} = \mathbf{D} : \boldsymbol{\varepsilon} \tag{2}$$

The yield surface of this model is defined by (Eq. 3)

$$F = f(\boldsymbol{\sigma} - \boldsymbol{\alpha}) - \sigma^0 = 0, \tag{3}$$

Where σ^0 stands for the size of the yield surface and $f(\boldsymbol{\sigma} - \boldsymbol{\alpha})$ is defined for Mises plasticity as (Eq. 4)

$$f(\boldsymbol{\sigma} - \boldsymbol{\alpha}) = \sqrt{\frac{3}{2}(\boldsymbol{S} - \boldsymbol{\alpha}):(\boldsymbol{S} - \boldsymbol{\alpha})}. \tag{4}$$

In this equation $\boldsymbol{\alpha}$ represents the backstress tensor and $\mathbf{S}$ the deviatoric stress tensor. This model assumes associated plastic flow, so the rate of plastic flow $\dot{\boldsymbol{\varepsilon}}^{pl}$ is given by (Eq. 5)

$$\dot{\boldsymbol{\varepsilon}}^{pl} = \frac{\partial \mathrm{F}}{\partial \boldsymbol{\sigma}} \dot{\bar{\varepsilon}}^{pl} \tag{5}$$

where the equivalent plastic strain rate $\dot{\bar{\varepsilon}}^{pl}$ is given by (Eq. 6)

$$\dot{\bar{\varepsilon}}^{pl} = \sqrt{\frac{2}{3} \dot{\boldsymbol{\varepsilon}}^{pl} : \dot{\boldsymbol{\varepsilon}}^{pl}} \tag{6}$$

In this model the size of the yield surface σ^0 is a function of the accumulated equivalent plastic strain $\bar{\varepsilon}^{pl}$. In Abaqus, this function can be given in exponential or tabular form. The evolution law of the kth backstress $\boldsymbol{\alpha}_k$ is given by (Eq. 7)

$$\dot{\boldsymbol{\alpha}}_k = C_k \frac{1}{\sigma^0}(\boldsymbol{\sigma} - \boldsymbol{\alpha})\dot{\bar{\varepsilon}}^{pl} - \gamma_k \boldsymbol{\alpha}_k \dot{\bar{\varepsilon}}^{pl} + \frac{1}{C_k} \boldsymbol{\alpha}_k \frac{dC_k}{dT}. \tag{7}$$

The overall backstress $\boldsymbol{\alpha}$ is calculated as the sum of all backstresses $\boldsymbol{\alpha}_k$. C_k and γ_k are material parameters, $\frac{dC_k}{dT}$ is the rate of change of C_k with respect to temperature.

Hence, the aim of this research work is the experimental and numerical determination of the parameters C_k and γ_k as well as the identification of a suitable correlation between the size of the yield surface σ^0 and the equivalent plastic strain $(\overline{\varepsilon}^{pl})$ for a "Nonlinear isotropic/kinematic hardening model" of thin copper layers.

3. Experiments

Specimen manufacture

As copper layers used in PCBs have thicknesses as low as 18µm they cannot be tested under compression loads in a self-supporting way. Therefore, specimens consisting of five copper layers and four FR4-prepregs layers were manufactured in a compression molding process (a micrograph of the cross section and an image of the test specimen are shown in Figure 1).

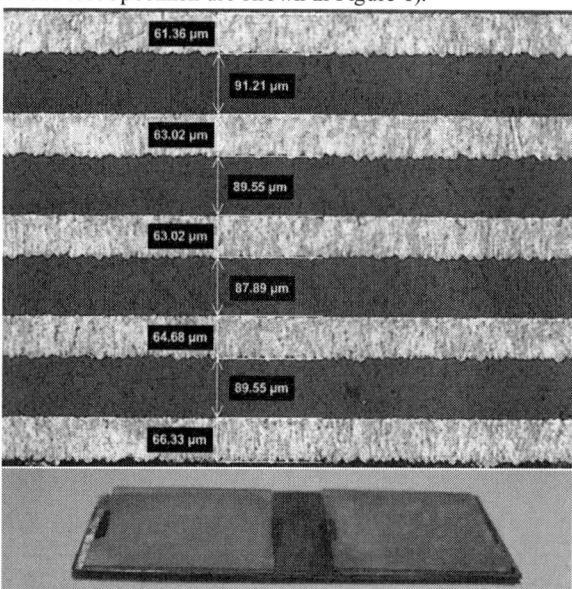

Figure 1: Upper figure shows a photo and a micrograph of the cross section of the layered specimen. [1]

Test setup

The equipment for the cyclic tensile-compression tests is shown in Figure 2. It was used in a MTS 810-22 testing machine for the experiments at room temperature (RT) and in a MTS 831 (MTS, Eden Prairie, MN, USA) testing machine for the experiments at higher temperatures (85°C, 145°C). The occurring strains were measured by digital image correlation using the Aramis-System (Gesellschaft für optische Messtechnik, GOM, Braunschweig, Germany), while the forces were recorded by the MTS sensors. The tests were carried out in a force-controlled manner, the waveform of the force was chosen as sinusoidal. The upper and lower limits of the force at RT are given in Table 1. The strain rate was kept constant for all tests by modifying the circular frequency. The

force limits were reduced at higher temperatures to account for the lower material strength. For the measurements at 85°C all values in Table 1 were multiplied by a factor of 0.9 and for the measurements at 145°C by a factor of 0.7. The gauge length was 5 mm. For each loading case at least three specimens were tested.

Figure 2: Testing device with specimen (red mark) in the climate chamber of the MTS 831 testing machine.

Table 1. Amplitude – Limits (UL=Upper Limit) at RT.

UL [N]	2000	1800	1600	1000
Lower Limits [N]	-2000	-2000	-2000	-2000
	-1800	-1800	-1800	
	-1600	-1600		
	-1000	-600		
	-600			
	-100			

Determination of the elastic properties

The elastic properties of the copper foil were measured dynamically by means of a micro-Dynamic Mechanical Analyzer (model µ-DMA RSAG2). Strip-shaped specimens with the dimensions 40 x 5 x 0.018 mm^3 were used. The specimens were loaded with a deformation amplitude of 0.02% and with a frequency of 1Hz. The considered temperature range was between room temperature and 150°C.

4. Parameter determination

Determination of the kinematic hardening behavior

Hysteresis loops obtained in the cyclic tensile-compression tests were used to determine the kinematic hardening parameters C_k and γ_k plus the current yield stress σ^0. For the sake of simplicity, for the time being only one backstress was used, the index k will thus be omitted henceforth in this paper. The parameter identification was accomplished using an optimization procedure. First a FEA-model of the test setup was created using Abaqus. The model takes advantage of the quarter-symmetry of the specimen and considers the influence of the fixture. Figure 3a shows the modelled region of the specimen, Figure 3b shows the FEA-model and the used boundary conditions to account for the symmetry and the fixture. Quadratic shell elements with reduced integration (Abaqus element type S8R, continuum shell formulation) are used. The loading direction displacement of the nodes

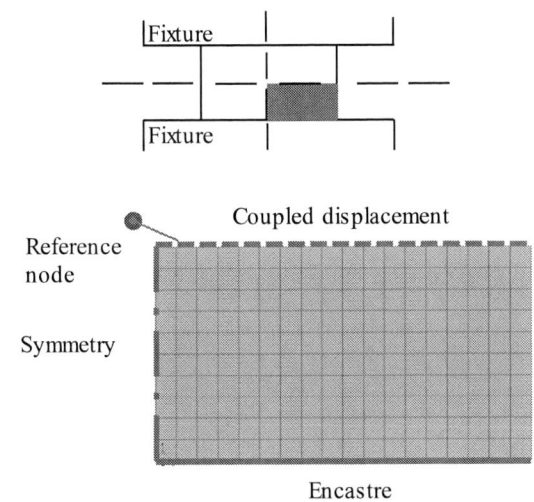

Figure 3: a) Modelled region of the specimen (red mark)
b) FEA-model of the specimen.

on the top of the model are kinematically coupled to a reference point, which guarantees that the symmetry line remains straight. The load is then applied to the reference node. The model was used to computationally determine the material response of certain sets of material parameters (C, γ and σ^0). The parameters were varied governed by an optimization algorithm. Figure 4 shows the optimization procedure by means of LS-Opt. Abaqus was used as a "user defined solver" in the LS-Opt optimization procedure. First, test sets of material parameters (C, γ and σ^0) were chosen based on full-factorial design of experiments. Moreover, also a D-optimal [10] design of experiments algorithm was used. Subsequently these sets of material parameters were entered into an Abaqus input file and the loading of the cyclic tensile-compression test was simulated. First, when required by the load case, the heating-up of the sample was simulated. This allows working out thermal strains and residual stresses due to the thermal mismatch of the constituents of the specimen. Afterwards it was subjected to the same cyclic loads as in the experiments. Since the material response of the model with constant yield stress σ^0 stabilizes very quickly, simulating only four cycles was found to be sufficient. To compare the simulated material response to the one obtained in the tests, a

"curve-mapping algorithm" [10] was employed. This algorithm calculates the area between two curves and uses it as a residual for optimization tasks. Based on these residuals, new sets of test parameters were chosen by LS-Opt. The material response of every set of parameters was plotted in order to be able to manually intervene in the selection of new material parameters.

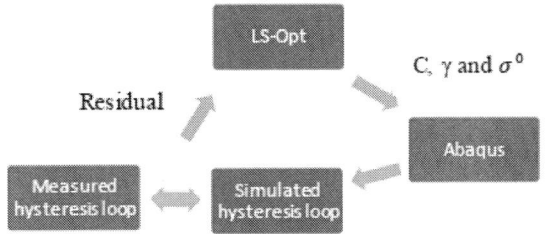

Figure 4: Optimization procedure for parameter determination

Determination of the isotropic hardening behavior

In the "Nonlinear isotropic/kinematic hardening model" the isotropic hardening component is represented by a relationship between yield stress σ^0 and equivalent plastic strain $\bar{\varepsilon}^{pl}$. Since this tabular relationship is not a-priori known a methodology has to be devised that allows to iteratively forecast the increment of equivalent plastic strain $\bar{\varepsilon}^{pl}$ pertaining to a given yield stress σ^0 within each cycle. As the experimentally determined current yield stress σ^0 depends on both the number of elapsed cycles and the loading situation, simulations of every loading situation using different values of the yield stress σ^0 were conducted. That way, a relationship between the yield stress and the equivalent plastic strain accumulated during each loading cycle could be obtained. In order to ensure, that these values are taken from a stabilized cycle, four loading cycles were simulated and the magnitude of the equivalent plastic strain increment that was added in the fourth loading cycle was chosen. This was done because the amount of equivalent plastic strain generated in the first loading cycle can significantly differ from the one obtained in the subsequent cycles. This was especially the case in asymmetric loading situations. By using a Matlab (Matlab R2013b, Natick, MA, USA) code these functions were fitted and analytical expressions were obtained. Figure 5 shows the procedure for the identification of the values of the equivalent plastic strain and explains how the sampling points of the tabular function were generated. It should be noted that the first entry of this function always has to be the initial yield stress and zero equivalent plastic strain. The sampling points were obtained step-by-step, because the amount of equivalent plastic strain $\bar{\varepsilon}^{pl}$ at cycle number N (see Figure 5) needs to be known for the determination of the amount of equivalent plastic strain $\bar{\varepsilon}^{pl}$ at cycle number M>N. Bearing in mind, that the values of the yield stress σ^0 are linearly interpolated between the sampling points, a Matlab script determines the suitable value of the equivalent plastic strain for a given cycle number M iteratively. This script basically adds up the equivalent plastic strain increments depending on the current yield stress. Here the analytical relationship between the yield stress and the equivalent plastic strain added during each loading cycle and the known amount of equivalent plastic strain $\overline{\varepsilon}^{pl}$ at cycle number N was used. Furthermore, the yield stresses σ^0 and the elapsed number of cycles were needed as input for this procedure. It was verified by simulating the cyclic tensile-compression tests using the obtained material parameters and comparing it to the experimental results.

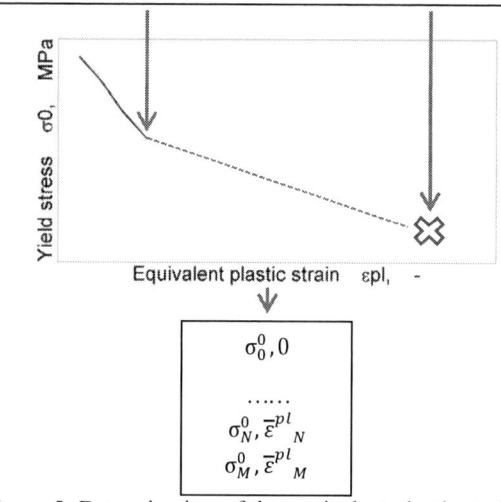

Figure 5: Determination of the equivalent plastic strain. The iteratively determined sampling point is marked.

5. Experimental results

In this section results of the cyclic tensile-compression tests are shown. The results of the tests at room temperature were described in an earlier paper [1]. The thermal and mechanical material properties of the FR4-prepreg were determined in [11]. These properties needed to be known in order to be able to extract the copper material behavior from the composite specimen behavior. Figures 6 and 7 show hysteresis loops obtained at testing temperatures 85°C and 145°C, at a R-ratio of -1. The strains shown in all following figures are caused by the mechanical loading. Thermal strains were not measured, because the recording of the heating-up was not possible with the ARAMIS-System. Hence, the digital image

correlation measurements were started when the specimens were heated through, but still free of mechanical load. In the following figures the number of elapsed cycles, the number of cycles to failure, N_f and the lower and upper limit of the sinusoidal loading function of the discussed specimens are given. As in [1], failure is defined as the point where the specimen is no longer able to transmit tensile loads. As the composite specimen shows many different failure modes, the number of cycles to failure scatters considerably. The curves show asymmetry with respect to the zero-strain line, although the loading is symmetric, because the first loading cycle starts in tensile

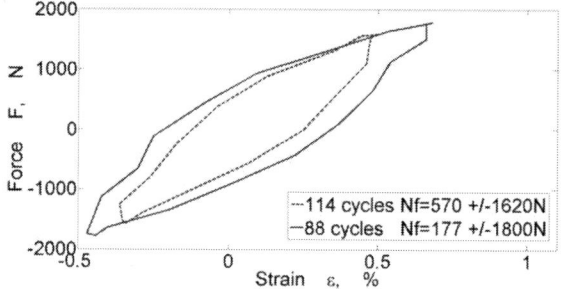

Figure 6: Stabilized force-strain curves, R= -1, testing temperature 85°C

direction which shifts the entire hysteresis loop to the right. Hence there residual tensile plastic strains develop. This effect was observed at all tested temperatures, but increases with temperature due to the lower material stiffness at higher temperatures. It also increases with higher loads, leading to higher plastic strains, as the

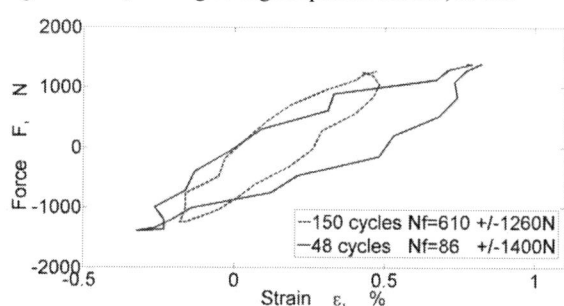

Figure 7: Stabilized force-strain curves, R= -1, testing temperature 145°C

different mean strains in the Figures 6 and 7 indicate. Figures 8 and 9 show the material response of composite

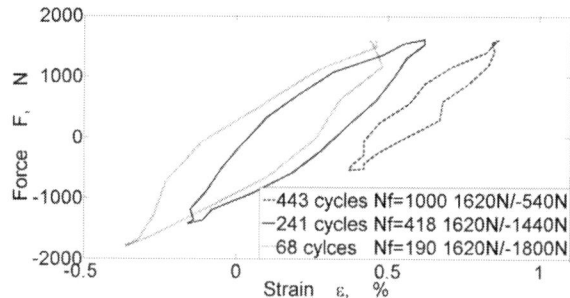

Figure 8: Stabilized force-strain curves, asymmetric loading, testing temperature 85°C

specimens subjected to asymmetric loading conditions. The force limits are given in the figures. Although the loading was reduced at higher temperatures keeping the R-ratios constant at all testing temperatures (see Table 1), the curves measured at 145°C show higher mean strains than the ones measured at 85°C. This effect can be explained with reduced yield stresses at higher temperatures and hence greater plastic strains created by

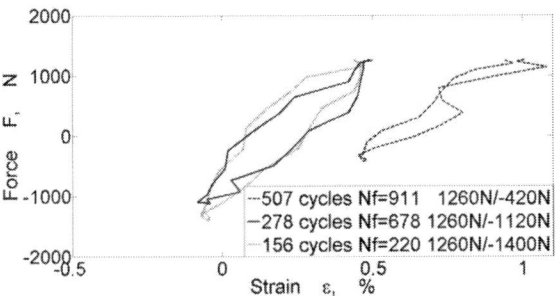

Figure 9: Stabilized force-strain curves, asymmetric loading, testing temperature 145°C

the first tensile loading step of the cyclic test. In general, the recording of smooth hysteresis loops at 145°C was more difficult than at 85°C or at room temperature, because the window pane of the climate chamber of the MTS 831 testing machine fogged up considerably. Therefore, at higher temperatures larger measuring tolerances were admitted in the optical strain measurements. All amplitudes listed in Table 1 could be tested successfully at all testing temperatures.

Elastic properties

The elastic properties are listed in Table 2 as a function of temperature. Since it is not possible to measure the Poisson's ratio with the DMA literature data [12] were used.

Table 2: Elastic properties of copper foil

T [°C]	Young's modulus E [GPa]	Poisson's ratio υ [-]
20	91.8	0.34
50	89.8	0.34
100	85.8	0.34
150	80.8	0.34

6. Results of the parameter determination

The results are discussed for the simulations of the measurements at room temperature and at 85°C.

Determination of the kinematic hardening behavior

The procedure was repeated for at least one stabilized cycle of every loading condition. One set of kinematic hardening parameters, suitable for all relevant loading conditions could be found. This set of parameters is given in Table 3.

Table 3: Kinematic hardening parameters

C [MPa]	γ [-]
250000	1900

For each inspected hysteresis loop the most suitable value of the yield stress σ^0 was determined. Figure 10 shows

two hysteresis loops, obtained at RT and the corresponding simulated material response, using the kinematic hardening parameters given in Table 3, but different yield stresses σ^0. Figure 11 shows two hysteresis loops at 85°C and the results of the corresponding simulations, which also used the parameters given in Table 3. The yield stresses σ^0 at 85°C were found to be lower than at RT. As the heating-up of the specimens could not be recorded with digital image correlation, only the strains caused by the mechanical load can directly be compared with the simulations. Nevertheless, the simulations must of course also reproduce the heating-up stage.

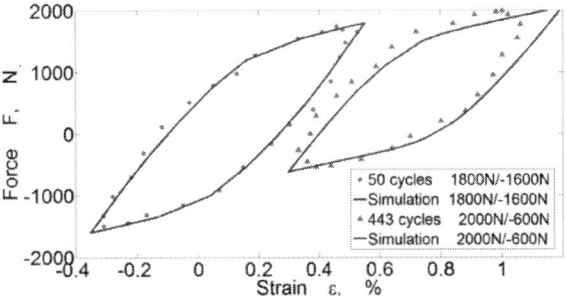

Figure 10: Hysteresis loops obtained under different loading conditions at RT and the corresponding fits.

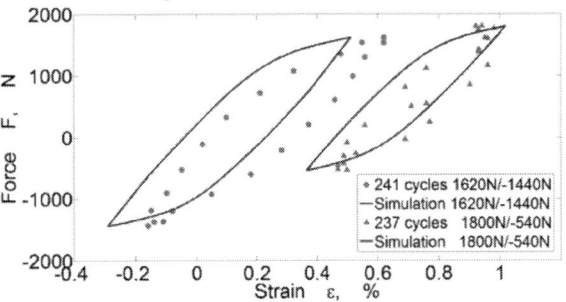

Figure 11: Hysteresis loops obtained under different loading conditions at 85°C and the corresponding fits.

Therefore, the heating-up was simulated, using the known coefficients of thermal expansion for the copper and for the FR4-prepreg from [11]. Then the mechanical loading was applied onto the model. For the figures in this section and also for the optimization process the thermal strains were subtracted from the total strains obtained in the simulations in order to ensure comparability and to account for the thermal loads. The simulated hysteresis loops in Figure 11 are shifted to the left compared to the measured results. This problem has been observed in several cases, but could not be solved with only one backstress used in the "Nonlinear isotropic/kinematic hardening model" (Eq.7). In further studies, different backstress evolution laws will be used to account for this problem. One possible solution could be the usage of a backstress evolution law including a threshold [13]. Furthermore, the agreement of the shapes of the hysteresis loops will improve using more than one backstress.

Isotropic hardening behavior

In Figure 12 the material response of a specimen subjected to sinusoidal loading (upper limit: 2000N, lower limit -600N) at room temperature is shown. The hysteresis loops obtained at different cycle numbers and the known yield stresses σ^0 at these cycle numbers were used to identify the tabular function, see Table 4. The amount of equivalent plastic strain $\bar{\varepsilon}^{pl}$ accumulated in each loading cycle was calculated.

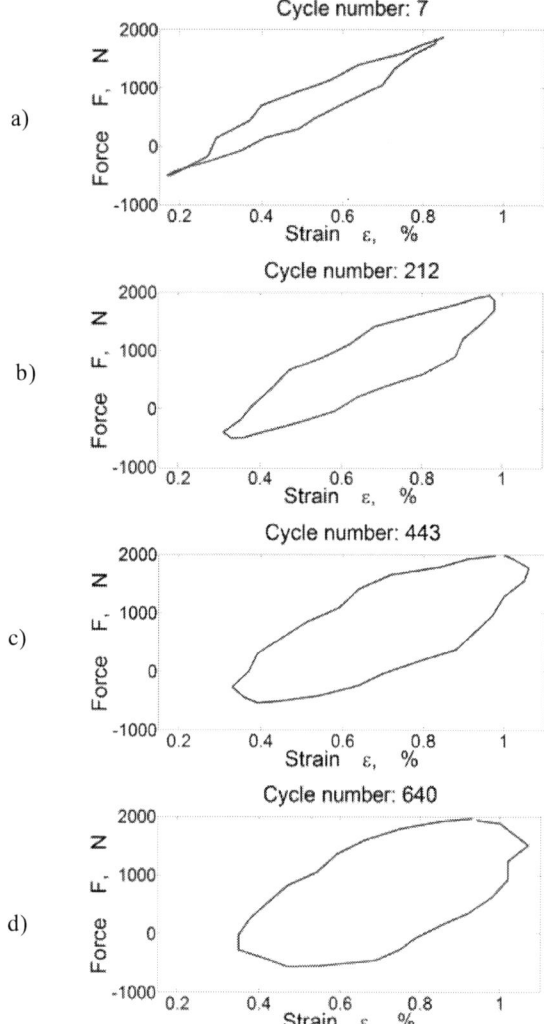

Figure 12: Hystersis loops obtained at different cycle numbers.

The yield stress obtained in the hysteresis loop shown in Figure 12a (cycle number 7) was also taken as initial yield stress, as the first cycles could not be measured properly due to the tuning of the testing machine. Since this research work focusses on lifetime assessment, no focus was put on the first cycles. As shown in Figure 12, the investigated copper grade shows cyclic softening. This is true for all tested strain ranges and temperatures. Comparing the isotropic hardening behaviors obtained at different strain ranges, they differ from each other. Hence it is not sufficient to model the isotropic hardening

Table 4: Isotropic hardening behavior. In the first column the corresponding hysteresis loop in Figure 12 is given.

	Yield stress σ^0 [MPa]	Equivalent plastic strain $\bar{\varepsilon}^{pl}$[-]
a)	105	0.0
a)	105	0.00984
b)	27	0.52033
c)	10	2.80569
d)	2	5.51740

behavior using a single relationship between yield stress σ^0 and equivalent plastic strain $\bar{\varepsilon}^{pl}$ as it is done in the "Nonlinear isotropic/kinematic hardening model". In further studies and extended model will be used which accounts for strain range memorization. [14-16]

7. Conclusion and Outlook

Cyclic material properties of thin copper layers could be obtained at different R-ratios, strain ranges and temperatures using cyclic tensile-compression tests. These tests were modelled in FEA-simulations in order to determine the material input parameters for a "Nonlinear isotropic/kinematic hardening model" of the thin copper layers in an optimization process. The model was found to be in good agreement with the measured results for most R-ratios. The model can be used for lifetime assessment of PCBs under thermal loads, when the loading situation at the most critical points can be estimated. This estimation is necessary, because the model does not account for plastic strain range memorization. Furthermore, it shows deficits in the correct reproduction of the kinematic shift of the hysteresis loops. The modelling of these effects will be the subject of near future investigations.

Acknowledgments

The authors wish to thank Stefan Hinterdorfer for support with the testing and Martin Krobath for practical explanations of the "Nonlinear isotropic/kinematic hardening model". The research work of this paper was performed at the Polymer Competence Center Leoben GmbH (PCCL, Austria) within the framework of the COMET-program of the Austrian Ministry of Traffic, Innovation and Technology with contributions by the University of Leoben and by the AT&S Austria Technologie & Systemtechnik Aktiengesellschaft. The PCCL is funded by the Austrian Government and the State Governments of Styria, Lower Austria and Upper Austria.

References

1. K. Fellner, P.F. Fuchs, G. Pinter, T. Antretter, T. Krivec, Method development for the cyclic characterization of thin copper layers for PCB applications, to be published in „Circuit World" 2014

2. Abaqus V6.12. Abaqus Analysis User's Manual. Providence, RI, USA: Abaqus Inc.; 2012.

3. Amy, Robin Alastair; Aglietti, Guglielmo S.; Richardson, Guy (2010): Accuracy of simplified printed circuit board finite element models. In: Microelectronics Reliability 50 (1), S. 86–97.

4. Fuchs, P.F; Pinter, G.; Major, Z. (2013): PCB drop test lifetime assessment based on simulations and cyclic bend tests. In: Microelectronics Reliability 53 (5), S. 774–781.

5. Li, L.; Kim, S.M; Song, S.H; Ku, T.W; Song, W.J; Kim, J. et al. (2008): Finite element modeling and simulation for bending analysis of multi-layer printed circuit boards using woven fiber composite. In: Journal of Materials Processing Technology 201 (1-3), S. 746–750.

6. Yang, Fan; Meguid, Shaker A. (2013): Efficient multi-level modeling technique for determining effective board drop reliability of PCB assembly. In: Microelectronics Reliability 53 (7), S. 975–984.

7. Lemaitre, J., and J.-L. Chaboche, Mechanics of Solid Materials, Cambridge University Press, 1990

8. Chaboche, J.L (2008): A review of some plasticity and viscoplasticity constitutive theories. In: International Journal of Plasticity 24 (10), S. 1642–1693

9. Abaqus V6.12. Abaqus Theory Manual. Providence, RI, USA: Abaqus Inc.; 2012.

10. LS-Opt 4.2, A design optimization and probabilistic analysis tool for the engineering analyst, LSTC, CA, USA, 2011

11. Fellner, K. (2012): Simulation of the "Single Via Thermal Cycle Test" - modelling and determination of the material input parameters. Master's Thesis. University of Leoben, Leoben. Werkstoffkunde und Prüfung der Kunststoffe.

12. H. M. Ledbetter and E. R. Naimon, "Elastic Porperties of Metals and Alloys. II. Copper," J.Phys.Chem., vol. 3, no. 4, pp. 897-935, 1974.

13. Chaboche, J.L., (1991): On some modifications of kinematic hardening to improve the description of ratchetting effects, In: International Journal of Plasticity 7 (7), S. 661–678

14. Chaboche, J.L., Dang-Van, K., Cordier, G., 1979. Modelization of the strain memory effect on the cyclic hardening of 316 stainless steel. In: SMIRT 5, Berlin.

15. Ohno, N., 1982. A constitutive model of cyclic plasticity with a non-hardening strain region. J. Appl. Mech. 49, 721.

16. Ohno, N., Kachi, Y., 1986. A constitutive model of cyclic plasticity for nonlinear hardening materials. J. Appl. Mech. 53, 395–403.

Drop Impact Simulations for Lifetime Assessment of PCB/BGA Assemblies regarding Pad Cratering

Grace L. Tsebo Simo[1], Hossein Shirangi[1], Matthias Nowottnick[2], Rainer Dudek[3],
Eberhard Kaulfersch[3], Sven Rzepka[3], Bernd Michel[3]

[1] Robert Bosch GmbH; Automotive Electronics, Engineering Assembly and Interconnect Technology (AE/EAI2),
Stuttgart, Germany; GraceLolita.TseboSimo@de.bosch.com
[2] University of Rostock; IEF/IGS, Institute of Electronic Appliances and Circuits; Rostock, Germany
[3] Fraunhofer ENAS, Micro Materials Center, Technologie Campus 3, D-09126 Chemnitz, Germany

Abstract

In automotive electronics, complex automotive functionalities are managed by car's computers such as electronic control units (ECU). Albeit extremely rare, accidental drop impacts may occur during transportation or mounting ECUs on automobiles, damaging the built-in printed circuit board (PCB)/ball grid array (BGA) package assembly. However, due to larger package dimensions together with heavy components such as capacitors mounted on the board surface, higher acceleration and stress levels can be achieved on ECU electronic components than on hand-held electronic devices during a drop impact. In such cases, the board level drop test methodology defined in the Joint Electron Device Engineering Council (JEDEC) standard needs to be modified in order to match the requirements in automotive applications. The experimental setup used in this study includes a test board clamped between two aluminum frames with the help of screws, in order to reproduce the real clamping conditions of the PCB in an ECU. Furthermore, the new board design allows mounting additional masses at the center of the PCB to take into account the effect of the mass of electronic components present in real ECUs. In this work, the mechanical behavior of PCB/BGA assemblies used in automotive applications when subjected to drop events is assessed. Numerical simulations of the board behavior are performed in order to analyze the transient structural response of the PCB and evaluate the local stresses on the board/joint interface responsible for pad cratering. By varying the loading conditions, different stress levels can be achieved on the PCB laminate directly under the solder joints and a stress-life curve for predicting the assembly lifetime is hence established.

1. Introduction

Among numerous packaging concepts, the BGA has established itself as one of the most popular mounting techniques for high input/output devices in the electronic industry. Its greatest advantage, compared for example to pin grid array (PGA), is to conciliate the requirements of producing a smaller package with the need for increasing the number of interconnection pins. Drop or shock events, resulting from mishandling during transportation or mounting of ECUs in automobiles, can considerably affect the built-in PCB/BGA assembly performance.

Indeed, they can induce damage on the solder joints or mechanical cracking of the PCB laminate directly under the joints, resulting from over-bending of the PCB. This possible assembly defect is a major reliability concern for companies manufacturing ECUs, especially since the introduction of lead-free soldering materials.

Most of the work about BGA Package reliability under drop impact conditions has been centered in the area of hand-held electronic products, since these are more prone to be accidentally dropped during their service life. A number of research teams analyzed the mechanical behavior of PCB/BGA assemblies under drop impact [1, 2, 3], with tests carried out under the guidance of the standard JEDEC methodology for consumer electronics. Based on energy or plastic strain criterions, a few models have been proposed for predicting the lifetime of solder joints [4]. However, validated lifetime models for PCB/BGA assemblies under various testing conditions and on a wide board bending frequency range has not been reported. Moreover, models available in the literature use for the most part plastic strain energy criterions, thus excluding the pad cratering failure mechanism.

This paper focuses on the mechanical response of the PCB during drop events and its correlation with the assembly reliability. It aims at establishing a model for predicting the lifetime of PCB/BGA assemblies under high-acceleration drop impact, independently from the testing or loading conditions. To that end, this work includes two principal steps.

The first step, namely the numerical simulation of the board behavior, aims at analyzing the transient structural response of the PCB and evaluating the local stresses on the board/joint interface responsible for pad cratering. Numerical modeling codes for BGA drop experiments are developed using the commercial software ANSYS Classic and the scripting language Ansys Parametric Design Language (APDL). Contrary to the standard input G method proposed by Tee et al. [2, 3] where the impact pulse measured on the test vehicle during the drop event was directly supplied as base excitation to the board, the simulation model includes all the components of the test vehicle with explicit contacts defined between them. The impact pulse monitored during the drop experiments is applied on the base of the lower frame as boundary condition. This allows simulating the effect of screws' tightening on the PCB dynamical response. Furthermore,

978-1-4799-4789-8/14 $31.00 © 2014 IEEE

the local stress distribution on the PCB laminate under the solder joints is assessed with the help of an accurate submodel.

The second step of this study includes the lifetime evaluation of the PCB/BGA assembly. To that end, the validated simulation model is employed to generate accurate Wöhler curves for lifetime evaluation of the assembly. For the first time, a global lifetime model for predicting the PCB/BGA assembly reliability regarding pad cratering is proposed in the form of S-N curve.

2. Experimental Setup

2.1. Test board

The test board used in this study consists of a quadratic four-layer PCB (100x100x1.6 mm) on which four packages were symmetrically mounted with respect to the PCB center. The BGA type used for the investigations is shown in Figure 1. It contains 292 solder balls with 0.8 mm pitch.

17x17 mm
Figure 1: Ball grid array

2.2. Drop testing machine

The drop tester used for the investigations is shown in Figure 2. The standard JEDEC board level drop test method for handheld electronic products is modified through a new experimental setup. This consists of a board assembly clamped between two aluminum frames. This mounting configuration aims at reproducing the real clamping conditions of the board in an ECU. The frames are installed on a base plate, which is fixed on a rigid drop table. All the connections are carried out with mounting screws, as explained in [5].

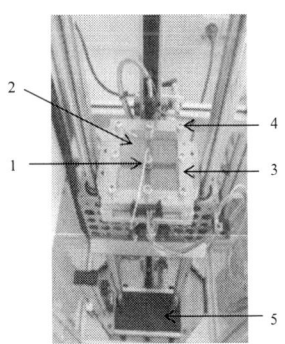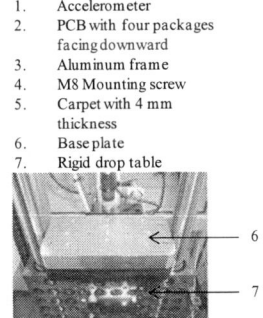

1. Accelerometer
2. PCB with four packages facing downward
3. Aluminum frame
4. M8 Mounting screw
5. Carpet with 4 mm thickness
6. Base plate
7. Rigid drop table

Figure 2: Drop tester [5]

The whole assembly is then dropped from a predefined height along guiding rods. Single-axis accelerometers allowing high level vibration measurements up to 5000G are installed on the base plate

or the PCB to monitor respectively the impact pulse or the output acceleration on the board center. Note that the accelerometers measure the acceleration on the axis corresponding to the drop direction.

2.3. Strain measurements

In order to experimentally validate the numerical model established to assess the mechanical behavior of the board, strain measurements were carried out at different locations on the PCB, with the help of encapsulated constantan gages with preattached ready-to-use cables. The corresponding strain and temperature ranges are ±3% and -50 to + 65 °C respectively. Figure 3 explains the position of each rosette on the PCB. The strain gauges measure the in-plane strains on the board during the drop event.

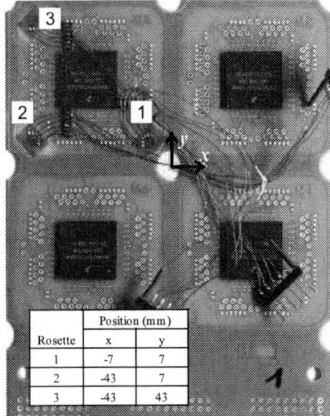

Rosette	Position (mm)	
	x	y
1	-7	7
2	-43	7
3	-43	43

Figure 3: Position of the rosettes

3. Fracture Mechanisms of the PCB/BGA assemblies in Drop Test: the pad cratering issue

As the test vehicle go into free fall, its gravitational potential energy is converted into kinetic energy and it moves with a downward velocity. During the impact on the carpet, an impact loading is created on the test vehicle, which results in shock waves propagating into the PCB. The board starts then vibrating. Due to the mismatch between the bending behavior of the PCB and the stiffer substrate together with the excessive curvature of the board, the solder joint experiences high stress concentrations, combining shear and peeling stresses. The stress level and the strain rate usually determine the failure mode. For small stress levels, solder fatigue caused by progressive damage accumulation during the loading cycles is the main failure mechanism. Since lead-free solder joints are strain rate dependent, high strain rates combined with high stress levels usually results in mechanical cracking of the brittle PCB laminate under the copper pad. Indeed, the higher the strain rate is, the greater the yield stress of the solder material gets. Figure 4 presents a typical pad crater after a drop test. It is a cohesive crack initiating at the corner of the copper pad and then propagating along the glass fibers.

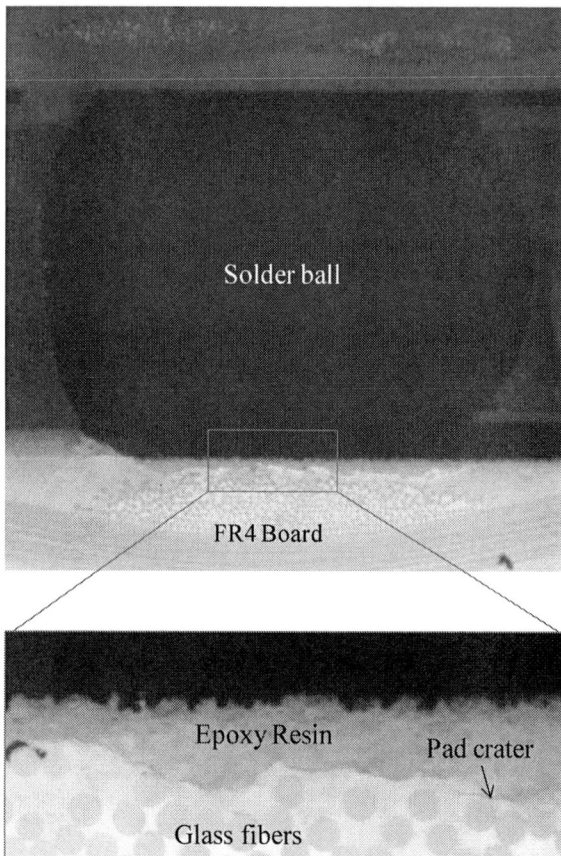

Solder ball

FR4 Board

Epoxy Resin

Pad crater

Glass fibers

Figure 4: Pad cratering

4. Numerical Analysis

4.1. Fundamentals of structural dynamics

Structural dynamics problems usually involve bodies subjected to time-dependent forces, thus combining spatial and time variables describing the state of the structure. By discretizing the space domain into finite elements, the complex time-space problem can be reduced to a time-dependent problem at some discrete points called nodes. Based on the equilibrium d'Alembert principle, the discretized linear equation of motion is then expressed as:

$$[M]\{\ddot{u}\}+[C]\{\dot{u}\}+[K]\{u\}=\{F(t)\} \quad (1)$$

$[M]$, $[C]$ and $[K]$ represent the structural mass, damping and stiffness matrices of the discretized structure. $\{\ddot{u}\}$, $\{\dot{u}\}$ and $\{u\}$ are respectively the vectors of nodal accelerations, velocities and displacements. The vector $\{F(t)\}$ is the time varying applied force.

The system is in dynamical equilibrium if this equation holds at any time t [6, 7, and 8].

4.2. Drop impact simulation

The drop test simulation was performed with the commercial finite element (FE) software ANSYS v14.5. The transient response of the PCB is obtained by solving the dynamical equation of motion (1). An implicit time integration scheme, namely the Newmark method is used. Indeed, implicit time integration schemes present the advantage to be unconditionally stable. The solution accuracy remains acceptable for large time steps.

4.2.1. Global model

The impact pulse on the base plate was monitored during the drop experiments. Therefore, it can be used as base excitation and directly input as boundary condition on a simplified simulation model. This is the so called input-G method. This modeling technique was developed by Tee et al. [2,3]. The impact acceleration pulse measured on the base plate during the drop impact is directly applied on the PCB as a boundary condition. Since the effect of friction of the guiding rods, the velocity before the drop impact and the characteristics of the carpet are already included in the monitored impact pulse, this method is of great efficiency and accuracy in comparison to free-fall drop simulation. However, the contact conditions between the mounting screws, the aluminum frames and the board are not considered through the standard input G method. To eliminate these uncertainties, a simulation model including the frames, the board and the mounting screws with explicit contacts defined between these components was elaborated. The impact pulse is applied on the base of the lower frame, not directly on the PCB.

In order to reduce the computational time, only ¼ of the test vehicle was modeled, since the geometry, loading and boundary conditions are symmetrical. Symmetry boundary conditions are applied on the symmetry planes. It should be noted that by applying the symmetry boundary conditions, the unsymmetrical vibration modes are ignored. Since the dominating mode responsible for BGA fracture is the PCB deflection in out-of-plane direction, the symmetry boundary conditions can be applied.

The finite element model included all components of the test vehicle such as the test board with one accelerometer, the frames and the mounting screws. An additional mass was modeled at the center of the PCB to take into account the effect of the mass of components such as capacitors present in real ECUs.

Solder balls were relatively fine meshed compared to the PCB and the interposer. Multipoint constraints were applied at the solder/PCB and the solder/interposer interfaces. The MPC method is an algorithmic strategy for generation of interface compatibility conditions defined in terms of either displacements or forces. Since the global model only aimed at modeling the dynamical response of the board, a more complicated numerical model with node to node connectivity between the solder balls and the PCB isn't needed. A detailed submodel is further developed to

978-1-4799-4789-8/14 $31.00 © 2014 IEEE

accurately model de stresses on the PCB and interconnection spheres.

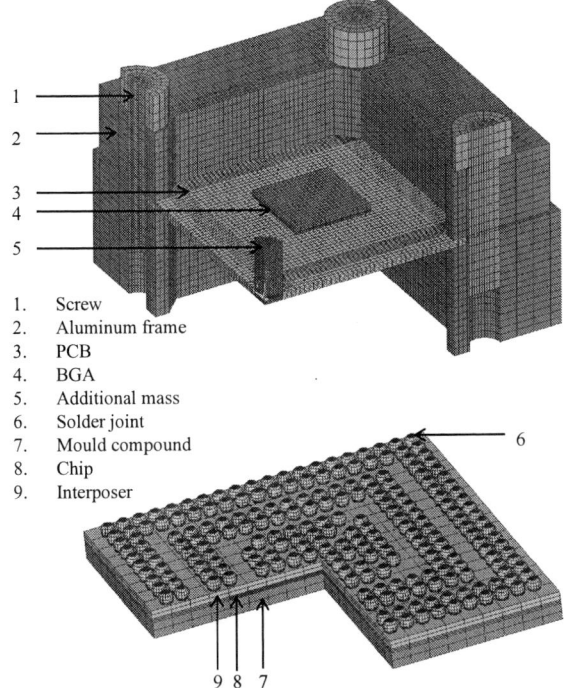

1. Screw
2. Aluminum frame
3. PCB
4. BGA
5. Additional mass
6. Solder joint
7. Mould compound
8. Chip
9. Interposer

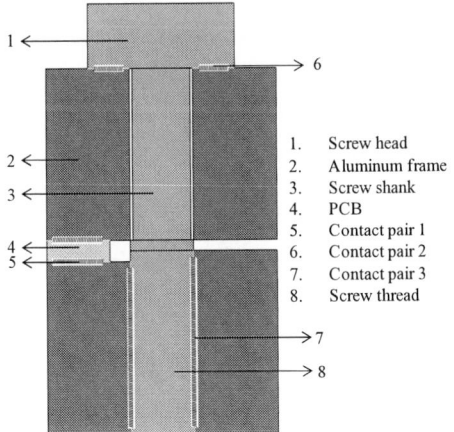

Figure 5: Simulation model

Furthermore the following contact pairs can be identified as shown in Figure 6:

- pair 1: nonlinear frictional contact between the frames and the PCB;
- pair 2: bonded contact between the screw head and the periphery of the mounting hole;
- pair 3: bonded contact between the frame and the screw thread.

Contact was modeled with the augmented Lagrange method, which combines the penalty and the Lagrange methods. Its main advantage resides in the less amount of penetration between the contact and the target surfaces, compared to pure penalty based approaches.

1. Screw head
2. Aluminum frame
3. Screw shank
4. PCB
5. Contact pair 1
6. Contact pair 2
7. Contact pair 3
8. Screw thread

Figure 6: Contact pairs

For the sake of accuracy, a pretension load corresponding to the mounting torque used by installing the screws on the frames is applied on the screw shank. The preload is created by inserting pretension elements into the bolt shank. Note that no preload is applied on the screw thread, since bonded contact between the frame and the thread could hinder the later from deforming properly.

Only linear elastic material models are included. Solder ball composition used in this study was Sn96.5Ag3.0Cu0.5 (SAC305). In reality, these solder alloys possess a viscoplastic material behavior. However, since the board bending frequency during a drop event is very high, we can assume that the solder joint undergo less plastic deformations.

The PCB and the interposer possess a layered structure with prepreg and core. They are modeled as orthotropic materials.

4.2.2. Submodel

Submodeling, also known as the cut-boundary displacement method, is based on the Saint-Venant principle. This principle allows the substitution of a set of boundary conditions by another statically equivalent set, and state that the stress distribution in a region reasonably distant from the applied boundary conditions remains the same in the two configurations.

The submodelling methodology as implemented in Ansys involves two separate numerical models: a global or coarse model representing the entire structure and a refined model focusing on a local region of the structure. The principle of the submodelling approach is as follows.

- First of all, the full model is used in order to determine the global displacements of the structure under external loading.
- Then, the global model displacement field is used to define boundary conditions for the refined submodel. The detailed stress state for local regions of the original model can hence be obtained.

Thus, the submodelling technique is a time saving approach in order to obtain local stress distributions on predefined regions of a complex structure.

The Submodel used in this study included all the components of a real BGA-ball as well as parts of the interposer and the PCB. Node to node connectivity is ensured between all the components and 3D solid elements are used for the meshing.

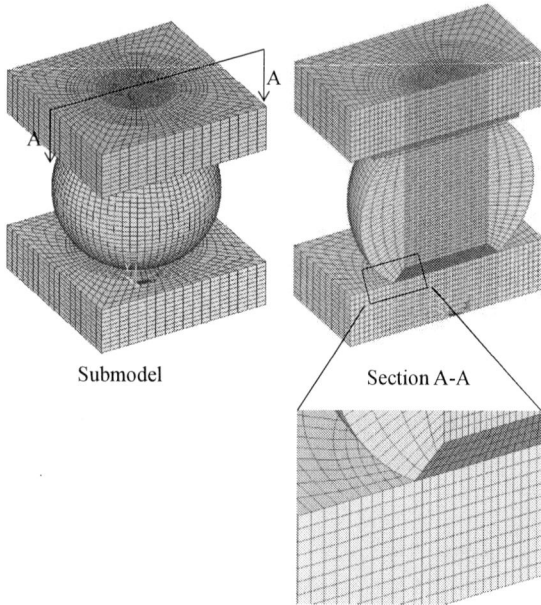

Submodel Section A-A

Figure 7: Submodel

4.3. Results and discussion

4.3.1. Modal analysis

Vibration modes are inherent properties of a structure, depending on its mass and stiffness. Each mode shape is associated to a particular frequency.

In order to validate the linear elastic material data used in this study, a modal analysis of the test vehicle is performed and the natural frequency and vibrations modes of the board are determined. The simulated values are then compared with experimental results obtained via a laser vibrometer.

Figure 8 shows the first two board vibration modes. A PCB without additional mass is considered. Since the modal analysis is linear, any nonlinearity such as contact isn't taken into account in the calculations. The stiffening effect of the torque on the dynamical behavior of the board is neglected.

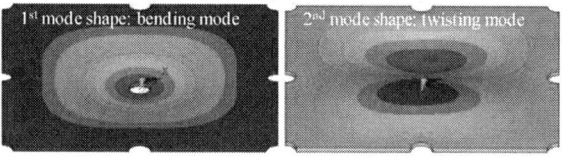

1st eigenfrequency:
–Simulation: 1066 Hz ~ Experiment: 1100 Hz
–Deviation simulation from experiment: 3%

Figure 8: 1st and 2nd mode shapes of the board

Simulation and experimental results match well. This is already a good indicator for the accuracy of the numerical model.

Furthermore, to understand the effect of the presence of an additional mass on the PCB dynamical behavior, modal analyses with various mass values ranging from 0

till 50 g are performed. In each case, the ratio between the second and the first eigenfrequency is calculated.

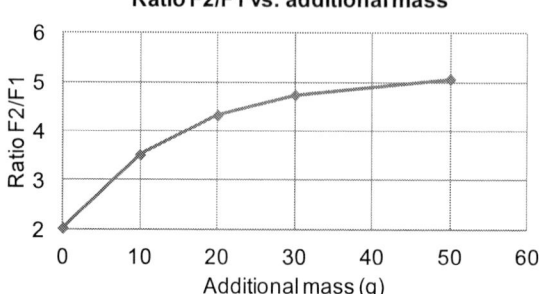

Figure 9: Effect of the additional mass on the bending behavior of the PCB

As can be seen on Figure 9, the bending mode becomes more dominant as the additional mass is increased. Indeed, an increasing frequency ratio denotes that the second natural frequency of the system, corresponding to the twisting mode, becomes farther and farther apart from the first one. Thus, the acceleration or strain curves on the PCB are expected to be smoother for high central masses.

4.3.2. Board level drop results and validation

The drop event simulation is performed by introducing the acceleration of the base plate during the drop impact as a boundary condition. The input acceleration is a half sine pulse characterized by two main parameters: its magnitude and its pulse duration. Table 1 shows the characteristics of the impact pulse depending on the drop height.

Table 1: characteristics of the impact pulse

Drop height (mm)	Peak acceleration (G)	Pulse duration (ms)
300	1000	0.425
500	2000	0.400
750	3400	0.274
1000	4400	0.233

Investigations were previously conducted to calibrate the appropriate Rayleigh damping parameters for modeling the dynamic response of the PCB.

Figure 10 summarizes the acceleration on the PCB center and the strain values at different locations on the board under different drop impact conditions. Experimental and numerical results are compared. The abbreviation R1 or R2 represents the number of the rosette whereas x or y represents the direction.

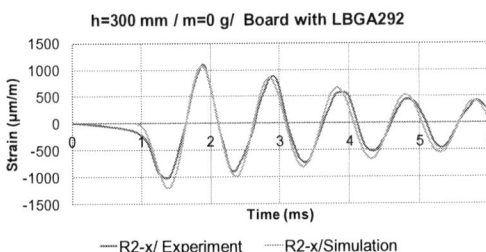

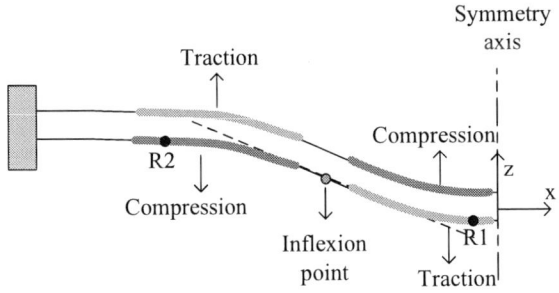

Figure 10: Acceleration on PCB center and in-plane strains on PCB

As can be observed in Figure 10, the experimental data match very well with the numerical results regardless of the testing conditions. The acceleration and strains on the test board attenuate progressively with the time. The bending strain on the rosette 2, near to the clamped support, is higher than the strain on the board center, where the PCB deflection is greatest. This is because of the bending moment, which is higher near the clamped support. Furthermore, the strains on the rosettes 1 and 2 are of opposite sign, indicating that the PCB curvature varies between these two points. To illustrate this, let us consider Figure 11, where the PCB deflection along the x axis is depicted.

Figure 11: Illustration of PCB deflection after the drop impact

In Figure 12, the out of plane displacement on the PCB center for the particular case of 500 mm drop height and 10 g central mass is depicted. The time history response of the assembly includes many bending cycles during which the bending direction switches from up to down and vice versa. Due to the PCB curvature in bending, the solder balls near the four corners of the BGA experience the highest shear and peeling stresses. However, depending on the board bending direction, the location of the joints under the highest stresses can switch from one corner to another.

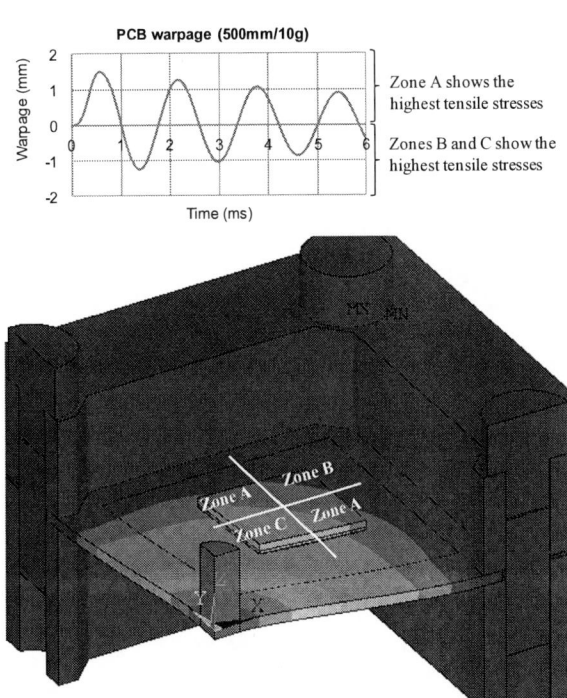

Figure 12: PCB warpage and zone classification

Using the validated global model and the elaborated submodel, the local stresses in the critical zones of the assembly can be accurately assessed.

4.3.3. Stress analysis[1]

For the local stress analysis, only the four corners balls of the BGA are considered, since they are subjected to the highest mechanical loading. It is found that the stresses on the Copper/PCB interface are higher than the stresses on the package side, due to the board deflection. In Figure 13, the peeling stresses on the PCB directly under the connection pads are depicted.

---

[1] In the remainder of this paper, the stresses are normalized to the peak stress values.

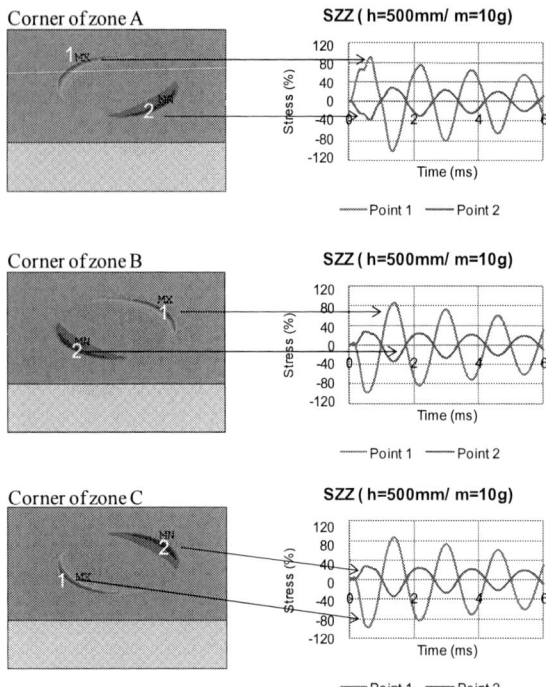

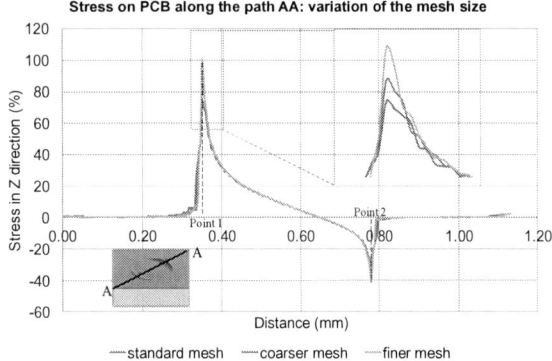

Figure 13: Normalized stresses on the PCB

It can be seen that the point 1, corresponding to the outside of the BGA, always shows the highest peeling stresses. It is therefore expected that cracks propagate from the outside to the inside.

Furthermore, let us consider the Figure 14, where the peeling stress along a predefined path on the PCB surface is depicted. Three different mesh sizes are used:

  − the initial mesh, as shown in Figure 7;
  − a coarser mesh, obtained by increasing the element size by approximately 33%;
  − a finer mesh, obtained by decreasing the element size by approximately 33% with regard to the standard mesh.

Figure 14: Normalized stresses on the PCB along the path AA - corner of zone C is considered

Stress singularities are observed at the points 1 and 2; the slope of the curves goes to infinity, regardless of the mesh size. In such cases, refining the finite element mesh will result in increasing stress values as the size of the elements get smaller. No mesh convergence can be reached.

To attenuate the effects due to stress singularities at the Copper/PCB interface, the stresses will be further evaluated on the 2nd row of elements, at a distance from 14 μm to the board surface.

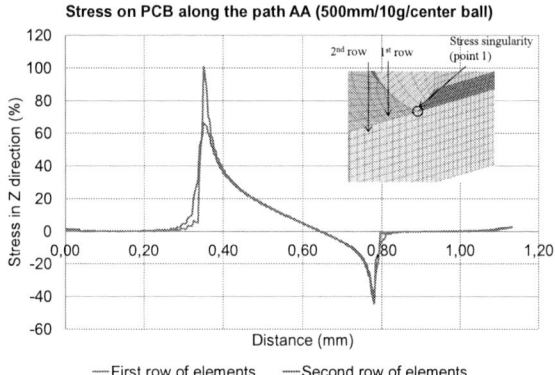

Figure 15: Normalized stresses on the PCB along the path AA: 1st vs. 2nd element row - corner of zone C

As can be observed in Figure 15, the stress results on the second and the first rows of elements are almost the same, except near the point 1, where stress intensification is observed.

In order to define an appropriate failure criterion for predicting pad cratering in the butter coat, the tensile, the shear and the 1st principal stresses on the PCB are assessed against the drop height and the additional mass on the board center.

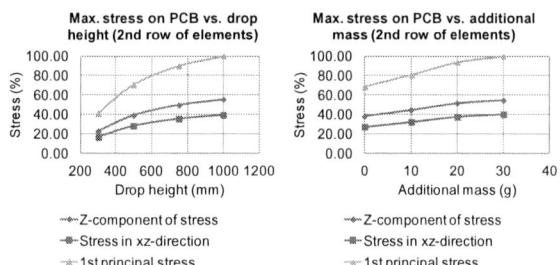

Figure 16: Comparison of different stress criteria: peak stresses are assessed on the 2nd row of elements - corner of zone A is considered

The higher the additional mass or the drop height is, the greater the peak stresses on the board gets. Indeed, the effective support excitation load [9] on the PCB center due to the impact pulse increases linearly with the additional mass, as well as greater drop heights lead to higher impact energies. Furthermore, it can be seen that peaks for components peel and shear stresses are close together. Therefore, assumptions about a dominant component of the stress tensor inducing the onset of pad cratering cannot be made. The first principal stress,

978-1-4799-4789-8/14 $31.00 © 2014 IEEE   134

combining the effects of the tensile and shear stresses, appears hence as a good indicator for predicting the assembly failure. Figure 17 shows the principal stress directions in the solder ball, the PCB and the interposer. As expected, the principal stress angle is greater than 0° and lesser than 90°, reflecting the combined effect of tensile loading in the drop direction and shear loading on the stress distribution in the submodel. Near the areas of high stress concentration on the PCB/solder and the solder/interposer interfaces, the principal stress angle is nearly 45°, indicating that the tensile and shear stresses are in the same order of magnitude. However, in the solder ball, the stress angle is greater than 45°, suggesting that the tensile stress becomes dominant.

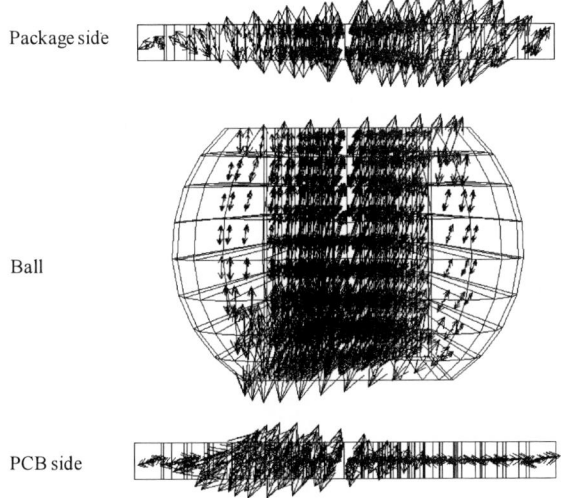

Figure 17: Trajectories of the 1st principal stress

In Figure 18, the effect of the drop height and the additional mass on the stress distribution on the PCB is assessed. On the first diagram, the drop height is varied while the additional mass is kept to a value of 10 g. The highest stresses are observed in the zone A. Moreover, as the drop height is increased, the peak 1st principal stress in the zone A becomes farther and farther apart from the peak stress in the zone C. To figure out this phenomenon, recall the Figure 12, were the zone classification was introduced. It was shown that the PCB bending direction influences the region of high stress concentrations: zone A shows the highest tensile stresses as the PCB bends up[2] while zone C shows the highest stress levels as the board bends down. However, after the drop impact, the first deflection orientation of the board is always the same as the drop direction. This implies that the zone A always experiences tensile stresses first, followed by the zone C as the board bending direction is reversed. Due to the damping, a fraction of the energy of the system is lost in each cycle of vibration. This implies that the PCB warpage gets smaller with the time and explains why the tensile stresses in the zone A, occurring directly after the

_____

[2] The bending direction refers here to the simulation model.

drop impact, are higher than the tensile stresses in the zone C, occurring while a portion of the energy of the system has already been dissipated.

On the second diagram of Figure 18, the central mass is varied from 0 to 30g while the drop height is kept to 500 mm. As can be observed, the stress values in the zones A and C get closer with an increasing additional mass. Indeed, the local curvature of the board near the Zone C gets higher for heavier extra masses, inducing stress intensification in this region.

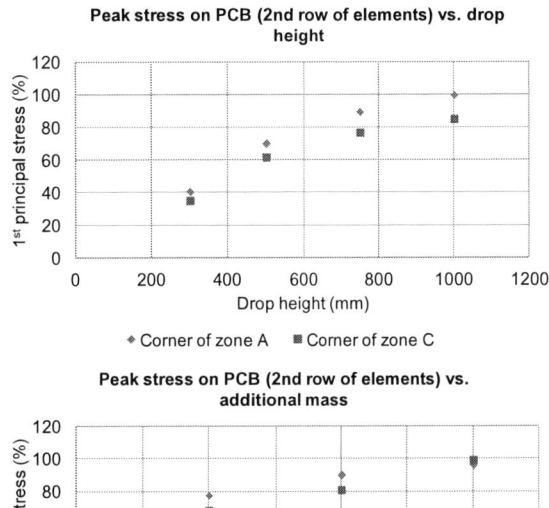

Figure 18: Comparison of normalized stress values in the zones A near the clamped support and the zone C near the board center

5. Lifetime Assessment of the PCB/BGA Assembly

The number of drops to electrical failure of each zone (A, B or C) of the assembly was experimentally assessed with the help of suitable daisy chain circuits. A resistance greater than or equal to 100Ω was considered as failure criterion. To ensure the statistical reliability of the results, two identical boards were drop tested for each loading condition. Hence, considering the quarter symmetry location of the BGAs on the board, the corresponding sample size was equal to 16 for the zone A, 8 for the zones B and C. Cross-sectional analyses of the assemblies after the drop tests revealed that pad cratering was the dominant failure mode [10]. According to the experimental data, components located in the zone A exhibits the shortest lifetime. Simulation results are hence in good agreement with experimental observations.

978-1-4799-4789-8/14 $31.00 © 2014 IEEE   135

Based on the stress results provided by the FE Simulation and the characteristic lifetime of the assembly obtained by fitting the experimental data to a Weibull distribution [10], a stress-life diagram for predicting the PCB/BGA assembly reliability is established. The number of drop to failure and the stresses are normalized to the highest lifetime and the maximum stress value, respectively.

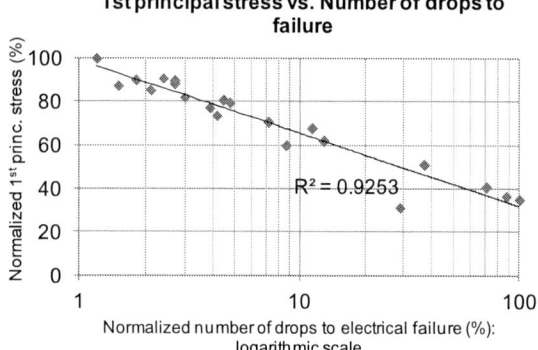

1st principal stress vs. Number of drops to failure

$R^2 = 0.9253$

◆ Zones A, B and C

Figure 19: Stress-Life curve

A linear correlation can be observed between the stress on the PCB laminate and the logarithm of the number of drops to failure. The S-N curve can be approximated by a straight line, obtained by using the least square method. From Figure 19, the corresponding regression coefficient is 0.96, indicating an excellent fit. Hence, the S-N relationship can be defined by an equation in the form:

$$S = a\log_x(N) + b \quad (3)$$

where a and b are the slope and the y-intercept of the curve with respect to an arbitrary base x, respectively[3].
Thus, for known stress amplitude, the cycles to failure can be extrapolated by using the following equation:

$$N = x^{(S-b)/a} \quad (4)$$

The number of drops to failure depends hence exponentially on the stress amplitude, and the constant $x^{1/a}$ can be seen as the corresponding growth rate.

Conclusions

This paper aimed at establishing a model for predicting the lifetime of PCB/BGA assemblies under high-acceleration drop impact regarding pad cratering. To that end, a numerical model has been implemented for predicting the real board behavior when dropped from a certain height in a predefined test vehicle. The input G method has been used, where the measured acceleration on the base table is applied to the model as a boundary condition. For better accuracy, complex contact conditions existing between the components of the test

[3] x is for example 10 in Figure 19

vehicle have also been modeled. The numerical model has been firstly validated against experimental measurements of acceleration on the PCB center and then against in-plane strains measurements at various locations on the board surface.

By varying the base excitation and the additional mass applied on the board center, different stress levels and loading rates could be achieved on the solder/board interface. Through a stress analysis of the assembly with the help of a suitable submodel, it was found that the zones of the package with the highest stresses depend on the PCB bending direction. A stress-life diagram for predicting the assembly reliability was finally established, by plotting the maximal stress on each zone against the corresponding experimental number of drops to failure. A linear relationship was found between the stress level on the PCB laminate and the logarithm of the number of cycles to failure. Finally, a model for lifetime assessment in form of an exponential law equation has been derived in order to predict the lifetime of the PCB/BGA assembly for known stress amplitudes.

References

1. S. Park, C. Shah, J. Kwak, C. Jang, and J. Pitarresi, "Transient dynamic simulation and full-field test validation for a slim-PCB of mobile phone under drop impact," in *Proc. 57th Electron. Compon. Technol. Conf. (ECTC)*, NV, USA, 2007, pp. 914–923.
2. J. Luan and T. Y. Tee, "Novel board level drop test simulation using implicit transient analysis with input-G method," in *Proc. 6th Electronics Packaging Technology Conf. (EPTC)*, Singapore, 2004, pp.671-677.
3. T. Y. Tee, J. Luan, E. Pek, C. T. Lim and Z. Zhong, "Novel numerical and experimental analysis of dynamic responses under board level drop test," in *Proc. 5th Thermal and Mechanical Simulation and Experiments in Microelectronics and Microsystems (EuroSimE)*, Belgium, 2004, pp.133-140.
4. L. Haohuan, Q. Xin, C. zhaoyi, W. Jiaji, T. Lee and H. Wang, "Lifetime Assessment of Solder Joints of BGA Package in Board Level Drop Test," in *Proc. 6th Electronic Packaging Technology Conf. (EPTC)*, Singapore, 2005, pp.1-5.
5. H. Shirangi, G. L. Tsebo Simo, Z. Wang, R. Unnikrishnan, T. Heinrich, "A Novel Drop Test Methodology for Highly Stressed Interconnects in Automotive Electronic Control Units," in *Proc. 64th Electronic Components and Technology Conf. (ECTC)*, FL, USA, 2014 (in press).
6. E. Wilson; "Dynamic Analysis by Numerical Integration: Normally, for earthquake loading, direct numerical integration is very slow," *Tech. Report, Computer & Structures*, Inc., 1998.
7. Y. K. Cheung, A. Y. T. Leung, "Finite Element Methods in Dynamics," *Kluwer Academic Publisher and Science Press*, 1991.

978-1-4799-4789-8/14 $31.00 © 2014 IEEE

8. ANSYS Structural Analysis Guide, Release 12.1, 2009.

9. C-L. Yeh, Y-S. Lai, "Support excitation scheme for transient analysis of JEDEC board-level drop test," *Microelectronics Reliability*, 2006, pp. 626-636.

10. Wang, Z.; "Experimental and Analytical Study of Fracture in Ball Grid Array Electronic Components under High Rate Impacts," *Institute for Microintegration (University of Stuttgart)*, Germany, 2013.

Predictive reliability using FEA simulations of power stacked ceramic capacitors for aeronautical applications

Warda Benhadjala[1,2], Bruno Levrier[1], Isabelle Bord-Majek[1], Laurent Béchou[1], Ephraim Suhir[3], Yves Ousten[1]

[1] IMS Laboratory, University of Bordeaux, UMR CNRS 5218, Talence, France
[2] CEA LETI, 17 rue des Martyrs, Grenoble, France
[3] University of California, Santa Cruz, California 95064, USA
warda.benhadjala@cea.fr, bruno.levrier@ims-bordeaux.fr, yves.ousten@ims-bordeaux.fr

Abstract

Viscoplastic finite-element simulation was used to predict reliability of solder joints in a high temperature 4-chips stacked capacitor mounted on a PCB under temperature cycling (-55°C to +125°C, 45min ramps/60min dwells). A three-dimensional (3D) model was built considering the materials properties of a commercial component. Capacitor materials were determined by using scanning electron microscopy (SEM) and energy dispersive x-ray spectroscopy (EDX). Thermomechanical properties, Anand parameters and Darveaux constant of the materials were incorporated into the simulation procedure to evaluate the mechanical strains and the variations in the plastic energy density for the high temperature solder joints of the 4-chips stacked capacitor. The number of cycles before the crack initiation in the solder joint and the number of cycles to failure have been calculated using Darveaux methodology. The obtained results showed that the maximum mechanical strains were localized at the bottom chip. It has been found that the number of cycles to failure exceeded 50,000.

1. Introduction

Technological evolution of aircraft and aeronautic systems is driven by energetic and environmental constraints that have strengthened in recent years in an attempt to reduce carbon dioxide emissions by 50%, the emission of nitrogen oxide and other greenhouse gases by 80% by the year 2020 [1]. To reach these objectives, one of the main areas of research in aeronautics electronics is being focused on the advent of the "all-electric plane". Owing to the advances in power electronics, the aircraft should be fully electric (out drive system), thereby reducing the fuel consumption, the maintenance costs and the devices carbon footprint.

Passive components, such as resistors, capacitors and inductors, are present in all the areas of electronics and correspond to 80% of the components of a whole circuit [2]. Improvement depends on the ability of passive components, especially capacitors, to reduce their volume and mass and improve their performance and reliability in aeronautical environment.

In recent years, the use of multilayered ceramic capacitors (MLCC) has been rapidly expanding in high reliability applications. Such capacitors have better performance than most other capacitor types in terms of capacitance density and resistance to harsh environments. Still, there was a strong incentive for improving their capacitance and operating voltage to meet the elevated aeronautical requirements. In this context, MLCC manufacturers proposed novel architectures for high temperature power ceramic capacitors, which consist in stacking and soldering several large MLCCs or "chips" (Figure 1). Electrical performances of these devices have been thereby enhanced significantly allowing for reaching capacitance values of 1300µF and nominal voltages above 500V [3]. Nonetheless, while many studies have been conducted on the evaluation of MLCC [4-6], high reliability of power stacked capacitors in operational conditions has not been yet demonstrated..

Figure 1: Photography of a 4 chips-stacked ceramic capacitor (chip dimensions: 1,90x3,50x0,27cm)

In the present work, reliability of a surface mounted 4-chips stacked ceramic capacitor has been evaluated by using the maximum principal stress in the capacitor's solder joint as a failure criterion. Indeed, among the various failure mechanisms occurring during ceramic capacitors thermal aging, the solder joints fatigue has been identified as one of the major reliability problems [7]. Viscoplastic finite-element analysis (FEA) method has been used to determine the solder joint fatigue life.

2. Structural Analyses

Material characteristics, such as, e.g., coefficient of thermal expansion (CTE) and elastic modulus are important parameters for conducting thermomechanical simulation. In order to improve the numerical model and obtain the required material data for finite element analyses, structural characterizations of the stacked-

978-1-4799-4789-8/14 $31.00 © 2014 IEEE

capacitor (Figure 1) were determined. Micro-sections have been prepared and observed by using scanning electron microscopy (SEM, JEOL JSM-6100) and energy dispersive x-ray spectroscopy (EDX, Oxford Inca X-act).

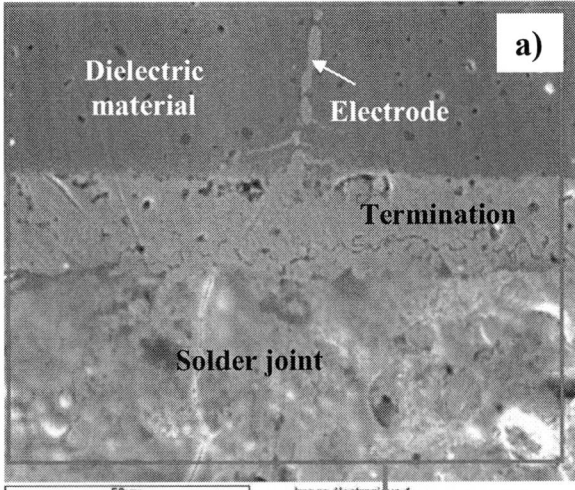

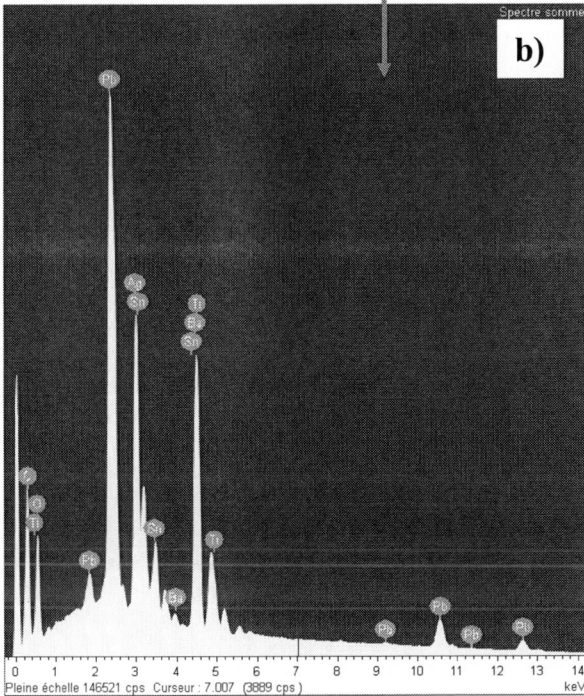

Figure 2 : (a) SEM image and (b) EDX spectrum of the analyzed 4-chips-stacked capacitor

The he EDX spectrum (Figure 2-a) indicates the presence of the elements Barium (Ba), Titanium (Ti) and Oxygen (O) in the dielectric layers, which means that they are composed by barium titanate (BaTiO$_3$) derived materials. More accurate quantitative analyses, carried out on the ceramics only, revealed a small amount of bismuth (Bi) (Table 1). Thus can be explained by the addition of a Bi-based dopant or adjuvant e.g. bismuth oxide (Bi$_2$O$_3$) during the fabrication process. Indeed, it has been demonstrated that the addition of such dopants enables one to reduce the sintering temperature and to improve the temperature stability of the BaTiO$_3$ dielectric constant [8,9]

Analyses have shown also that internal electrodes and terminations are mainly constituted by silver (Ag), and, to a lesser extent, by palladium (Pd). The low rate of palladium in AgPd alloys allows decreasing the MLCC costs. This modification generally requires lower sintering temperatures than in conventional MLCC manufacturing processes [10].

Element	Weight %
O	63,56
Ti	19,13
Ba	15,92
Bi	1,39

Table 1 : Distribution of elements in the dielectric

The lead-rich tin-lead (Sn-Pb) solder has been designed for high temperature applications [11].

3. Simulation methodology

3.1 Stacked capacitor model

A 3D model of a simplified stacked capacitor mounted on a PCB was set up using standard finite element software (ANSYS). To simplify the simulation process, only a quarter of the entire component was simulated without any loss of accuracy due to the inherent symmetry of the 3D model. Figure 3-a illustrates schematically the simplified capacitor structure used to build the model presented in Figure 3-b. The stacked-capacitor comprises four identical chips, C1 to C4, each characterized by a length of 38mm, a width of 19mm and a thickness of 2.8 mm. There is an 200µm high empty space between the chips. A 100µm-thick solder joint bonds the chips together and to the 300µm-thick connections. The inner electrodes and the termination have not been considered for the simulation. The component is fixed to the PCB and the modeled structure takes into account a 500µm gap between the PCB and the capacitor. The mesh (Figure 3-c) comprises 200 000 nodes and is made finer at the capacitor location.

978-1-4799-4789-8/14 $31.00 © 2014 IEEE   139

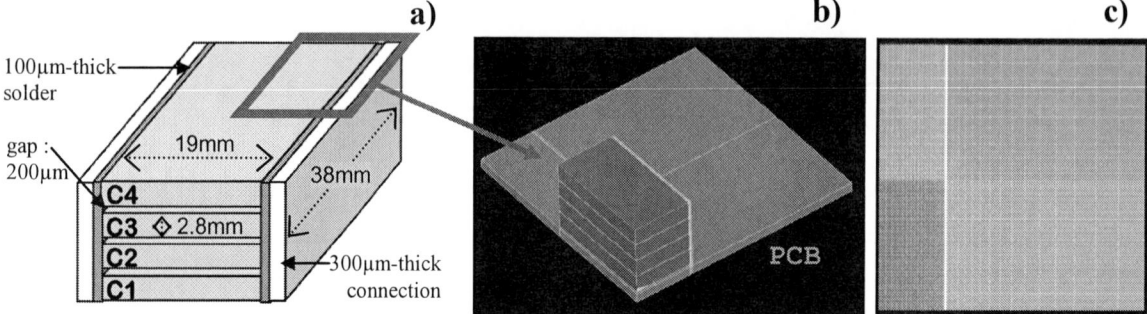

Figure 3 : a) Basic structure and b) finite element model of the 4-chips-stacked capacitor. c) Finite element mesh of the model (top view)

To prevent parasitic rotations during simulations, boundary constraints were established by blocking the node N1. Additional fixations, designated by FIXATION1, were obtained by blocking five nodes at the corner of the PCB surface. FIXATION1 mechanically holds the studied assembly to the stand of the electrodynamic shaker. In addition, symmetry conditions have been applied along the OX and OY axis. Figure 4 depicts the conditions applied.

3.2 Material properties

In the modeled structure, the materials considered for dielectric layers, solder, PCB and connections, have been selected to fit as much as possible to those determined during the structural analyses described above. The solder joint alloy, very rich in lead, was assimilated into the Pb97.5Sn2.5 traditionally used in power electronics.

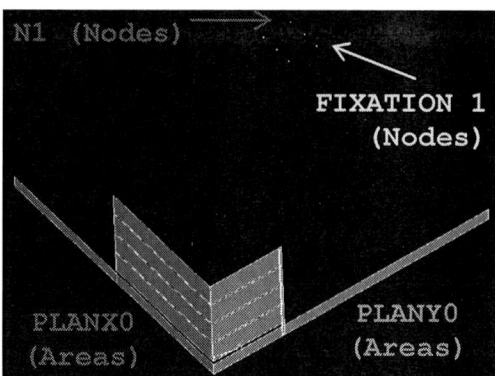

Figure 4 : Boundary conditions used for the simulation

Barium titanate was chosen as the dielectric material although the characterized capacitor ceramic contains also bismuth. Literature does not provide sufficient information on the thermomechanical properties of Bi_2O_3doped $BaTiO_3$. Nevertheless, it was shown that $BaTiO_3$ derived materials with a small amount of dopant may be mechanically approximated as pure $BaTiO_3$ [12]. For connections, we considered an iron/nickel alloy (FeNi) traditionally used for MLCCs [13].

Material properties required for the simulation, namely, density, elastic modulus (E), Poisson's ratio (ν), and the *CTE* (α) are given for different temperatures in Table 2 [14].

3.3 Viscoplastic finite-element analysis methodology

The life prediction methodology applied here is based on the correlation between the deformation energy and the crack growth rate in the solder joint. In this method, developed by Darveaux, the model utilizes finite element analysis to calculate the inelastic strain energy density accumulated per cycle during thermal cycling [15]. The strain energy density is then used with crack growth data to calculate the number of cycles to initiate a crack and the number of cycles to propagate crack through a joint. The fatigue life is calculated based on the joint length.

In this study, the Pb97.5Sn2.5 eutectic solder was modeled using VISCO 107 element model. VISCO 107 is based on Anand's unified viscoplasticity model and is available as a standard material model in ANSYS so that the user does not have to modify the source code. Anand's model is broken down into a flow equation and three evolution equations [15,16] :

Flow equation

$$\frac{d\varepsilon_p}{dt} = A \left[sinh\left(\xi \frac{\sigma}{s} \right) \right]^{1/m} \exp\left(-\frac{Q}{RT} \right)$$

Evolution equations

$$\frac{ds}{dt} = \left\{ h_0 |B|^a \frac{B}{|B|} \right\} \frac{d\varepsilon_p}{dt}; \quad a > 1$$

$$B = 1 - \frac{s}{s^*}$$

$$s^* = \hat{s} \left[\frac{d\varepsilon_p}{dt} \frac{1}{A} \exp\left(-\frac{Q}{RT} \right) \right]^n$$

where $d\varepsilon_p/dt$ is the inelastic strain rate, σ is the equivalent stress, s is the deformation resistance, s* represents a saturation value of s and T is the absolute temperature. Q/R, A, ξ, m, h_0, $\hat{s}$, a and n are Anand constants defined in Table 3 where Q and R are respectively the activation energy and the ideal gas constant.

978-1-4799-4789-8/14 $31.00 © 2014 IEEE   140

Material	Density (kg.m^{-3})	Property	Value						
			218K	298K	328K	348K	373K	473K	600K
FeNi (connections)	8250 @298K	E (GPa)	163	163	163	163	163		
		v	0,317	0,317	0,317	0,317	0,317		
		α (10^{-6}/K)	5,9	5,9	5,9	5,9	5,9		
Pb97.5Sn5 (solder joint)	10990 @512K	E (GPa)		15,7	14,5	14,0	13,6	10,7	0,8
		v		0,33			0,33	0,33	0,33
		α (10^{-6}/K)		28,7			28,7	28,7	28,7
BaTiO$_3$ (dielectric material)	5500 @298K	E (GPa)	107	107	107	107	107		
		v	0,33	0,33	0,33	0,33	0,33		
		α (10^{-6}/K)	5,0	5,0	5,0	5,0	5,0		
FR4 (PCB)	1800 @298K	E_x, E_y (GPa)	22,4	22,4	20,7		19,3		
		E_z (GPa)	1,6	1,6	1,2		1,0		
		v_{xy}	0,02	0,02	0,02		0,02		
		v_{yz}, v_{zx}	0,143	0,143	0,143		0,143		
		α_x (10^{-6}/K)	15,85	15,85	15,85		15,85		
		α_y (10^{-6}/K)	19,14	19,14	19,14		19,14		
		α_z (10^{-6}/K)	80,46	80,46	80,46		80,46		

Table 2 : Density (ρ), Elastic modulus (E), Poisson's ratio (v) and coefficient of thermal expansion (α) at various temperatures of the materials considered for the simulation [14]

The nine Anand's constants for Pb97.5Sn2.5 solder alloys have been published by Wang et al. and are listed in Table 3 [16].

The energy-based method proposed by Darveaux links the inelastic strain energy dissipation of solder joints to the fatigue life. This life consists of the life associated with crack initiation and the life associated with crack growth. The number of thermal cycles to crack initiation N_0 and the crack propagation rate per thermal cycle dL/dN are expressed by two simple power laws [15]:

$$N_0 = K_1 (\Delta W_{ave})^{K_2}$$

$$\frac{dL}{dN} = K_3 (\Delta W_{ave})^{K_4}$$

where L is the solder length. ΔW_{ave} is the average viscoplastic strain energy density accumulated per cycle as:

$$\Delta W_{ave} = \frac{\sum \Delta W . V}{\sum V}$$

where ΔW is the viscoplastic strain energy density accumulated per cycle of each element, V is the volume of each element, and K_1, K_2, K_3 and K_4 are the Darveaux crack growth correlation constants. These four constants depend on the material composition and other parameters, such as the FE model configuration, the solder constitutive model or the thickness of the element layers [13]. The crack growth correlation constants for Pb97.5Sn2.5 eutectic alloys were approximated by those determined for Sn37Pb alloys (see article by Xin Qu et al. [17]):

$$K_1 = 94.54 \text{ cycle/MPa K}_2$$

$$K_2 = -0.9334$$

$$K_3 = 0.005344 \text{ mm/(cycle MPa K}_4)$$

$$K_4 = 1.24$$

The solder joint fatigue life N_f, which corresponds to 63.2% population failure, is calculated by summing up the cycles to crack initiation with the number of cycles it takes for the crack to propagate across the solder joint length L:

$$N_f = N_0 + \frac{L}{\left(dL/dN \right)}$$

3.4 Thermal conditions

Three thermal cycles were simulated over a temperature range from -55°C to +125°C (i.e. 218 K to 398 K) with a ramp rate of 4°Cmin^{-1}. Each cycle consists of four load steps (ramp low, dwell low, ramp high and dwell high). The complete simulation incorporates twelve load steps. Publications that incorporate Darveaux's methodology have indicated the simulation of only two cycles (eight load steps). The simulation run times can be thus reduced by 30-35%, but the number of predicted thermal cycles to failure is reduced by approximately 5% [18].

Parameter (Unit)	Definition	Value
S_0 (MPa)	Initial value of deformation resistance	15.09
Q/R (°K)	Activation energy/ Boltzmann's constant	15583
A (s⁻¹)	Pre-exponential factor	3.25×10^{12}
ξ	Multiplier of stress	7
m	Strain rate sensitivity of stress	0.143
h_0 (MPa)	Hardening constant	1787.02
ŝ (Mpa)	Coefficient for deformation resistance saturation value	72.73
n	Strain rate sensitivity of saturation (deformation resistance) value	0.00437
a	Strain rate sensitivity of hardening	3.73

Table 3: Material parameters of viscoplastic Anand model for 97.5Pb2.5Sn [15].

4. Results

Figure 5-a and 5-b show respectively the von Mises stress distribution and the strain energy density in the solder joint of the 4 chips-stacked capacitor after three cycles. It can be seen that the highest stress remains below the yield stress (30MPa) and occurs at the C1 chip level. Thus any solder joint crack, if any, should begin at this location.

The obtained viscoplastic strain energy density accumulated per cycle (Figure 6) has been used to calculate the number of cycles to crack initiation N_0 equal to 740 cycles and the solder joint fatigue life N_f that achieved 55,515 cycles.

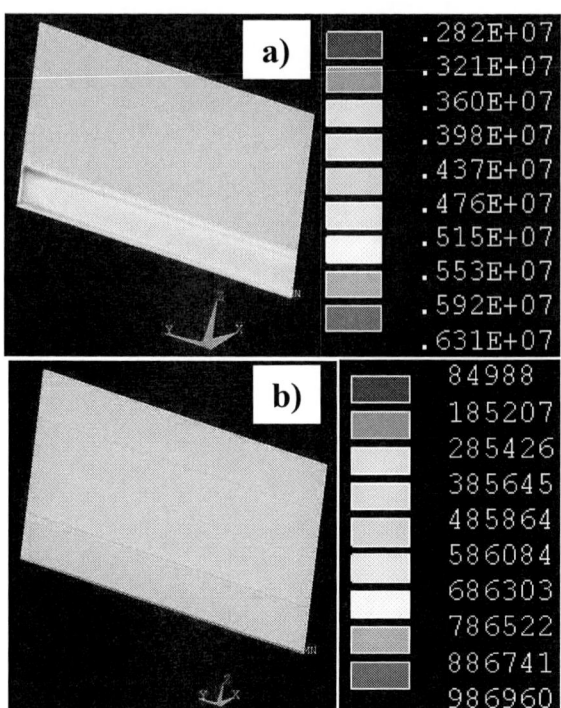

Figure 5 : a) Von Mises stress distribution (Pa) and b) strain energy density (J.m-3) in the solder joint of the 4 chips-stacked capacitor

Based on these results, the variation of the plastic energy density was also simulated in the solder volume of the chip C1 only, where it is most likely that a crack begins. While the number of thermal cycles to crack initiation remains approximately constant ($N_0 = 706$ cycles), the solder joint fatigue life decreases significantly to $N_f = 8289$ cycles due to the considerable reduction in the considered solder volume.

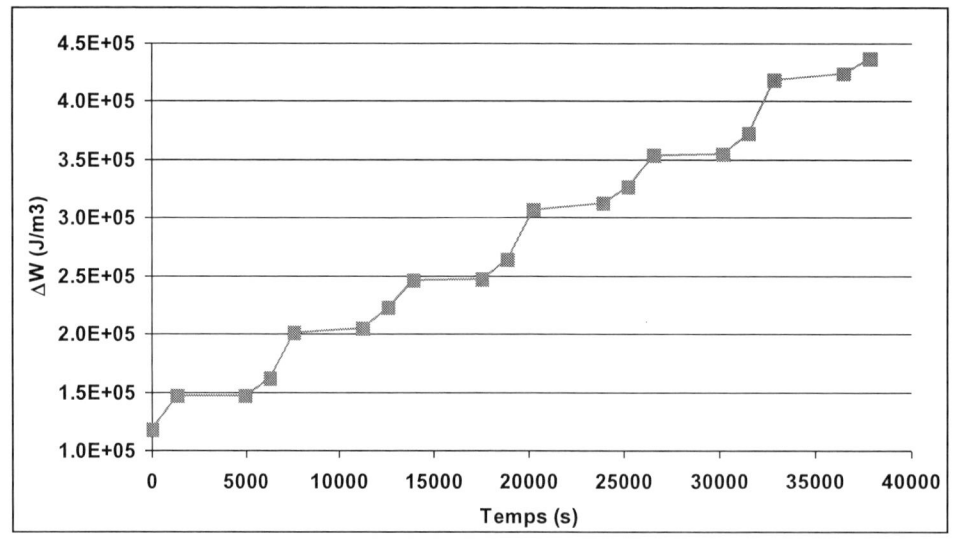

Figure 6 : Viscoplastic strain energy density accumulated per cycle

5. Conclusion

FEA have been carried out to predict the viscoplastic behavior of the solder joints in a high temperature 4-chips stacked capacitor mounted on a PCB under thermal stress (-55°C to +125°C, 45min ramps/60min dwells). The method uses the VISCO107 model based on Anand's viscoplastic constitutive law. Darveaux's crack growth rate model was applied to calculate solder joint fatigue life. At first, the analyses have been conducted in the entire solder volume. A fatigue life of 55,500 cycles, acceptable for aeronautical applications, has been determined. In a second step, the simulation has been performed in the solder volume of the chip C1 only, where the highest stress occur. While we observed that the number of thermal cycles to crack initiation is above 700 cycles in the both cases, the solder joint fatigue life decreases significantly cycles to 8,300 cycles. This strong decrease is due to the considerable reduction in the considered solder volume. Ongoing work includes accelerate ageing of 4-chips stacked capacitors under thermal stress to correlate the simulations with the experimental results.

Acknowledgments

The authors thank the DGAC for financial support (ISS2 project) and EURELNET Technology Transfer Unit for the simulations.

References

1. ALTEN, "Réalités et mutations du secteur aéronautique : le positionnement d'Alten ", ENTREPRISE EN VIE : http://www.entreprisesenvie. com/realites-et-mutations -du-secteur,119.html, 2011.

2. European Electronic Component Manufacturers Association (EECA) and European Passive Components Industry Association (EPCIA), "The Passive Component Industry in Europe - White Book", June 2002.

3. AVX, "Advanced Ceramic Capacitors for Power Supply, High Voltage and Tip & Ring Applications", version 13.2

4. Teverovsky A. "Thermal-Shock Testing and Fracturing of MLCCs Under Manual-Soldering Conditions", IEEE Transactions on Device and Materials Reliability, Vol. 12, pp. 413-419, 2012

5. Den Toonder, J.M.J. et al. "Residual Stresses in Multilayer Ceramic Capacitors: Measurement and Computation", Transactions of the ASME, Vol. 125, pp.506-511, 2003

6. Jin-Woo Park et al. "Thermo-Mechanical Stresses and Mechanical Reliability of Multilayer Ceramic Capacitors (MLCC)", J. Am. Ceram. Soc., Vol. 90, pp. 2151–2158, 2007.

7. Soon-Bok Lee et al., "A mechanistic model for fatigue life prediction of solder joint for electronic packages", Int. J. Fatigue, Vol. 19, No. 1 (1997), pp. 85-91.

8. Wuy S et al., "Effect of Bi_2O_3 Additive on the Microstructure and Dielectric Properties of $BaTiO_3$-Based Ceramics Sintered at Lower Temperature", Journal of Materials Sciences and. Technology, Vol. 26, pp.472-476, 2010.

9. Banjong D. et al, "Sintering Behavior and Microstructure of $BaTiO_3/Bi_2O_3$ Composite Ceramics", Proceedings of the 27th MST Annual Conference, Samui, Thailand, January 2010, pp.187

10. Lee, Y.-C. et al, "A Study of the Microstructure and Electric Properties of $Ba_2Ti_9O_{20}$-based Multilayer Ceramic Capacitors by Microwave Sintering", Procedia Engineering, Vol. 36, pp. 496- 450, 2012.

11. McCluskey F. P. et al., High Temperature Electronics, CRC Press, 1996.

12. Park, J. S. et al. ''Residual Stress Evolution in Multilayer Capacitors Corresponding to Layer Increase and Its Correlation to the Dielectric Permittivity,'' Journal of Applied Physics, Vol. 97, 094504-1, 2005.

13. EPCOS, "Multilayer ceramic capacitors", October 2006.

14. Bruno Levrier, Internal Report, 2011

15. Robert Darveaux, "Effect of Simulation Methodology on Solder Joint Crack Growth Correlation", Proceedings of 50th Electronic Components and Technology Conference (ECTC), Las Vegas, NV, May 2000, pp. 1048 – 1058.

16. Wang, G.Z. et al., "Applying Anand model to represent the viscoplastic deformation behaviour of solder alloys", Journal of electronic packaging, Vol. 123 (2001), pp. 247-253.

17. X. Qu et al, "Board level drop test and simulation of leaded and lead-free BGA-PCB assembly", Microelectronics Reliability, Vol. 47, pp. 2197-2204, 2007.

18. Bret A. Zahn, "Finite Element Based Solder Joint Fatigue Life Predictions for a Same Die Size - Stacked-Chip Scale - Ball Grid Array Package", Proc 27th Annual IEEE/SEMI International Electronics Manufacturing Technology Symposium, San Jose, CA, July 2002, pp. 274 – 284

An Investigation of the Tensile Deformation and Failure of an Epoxy/Cu Interface Using Coarse-Grained Molecular Dynamics Simulations

Shaorui Yang[1] and Jianmin Qu[1,2]

[1]Department Mechanical Engineering, [2]Department of Civil and Environmental Engineering
Northwestern University, Evanston, IL, 60208

Abstract

In this study, a coarse-grained model is developed to describe the interatomic interactions between a cross-linked epoxy and a copper substrate. Based on this model, the tensile deformation and failure of an epoxy/Cu bimaterial is studied. Attention is given to the microstructural evolution near the epoxy/Cu interface, and its effects on the overall stress-strain behavior of the bimaterial. It is found that, under uniaxial strain, plastic deformation in the epoxy/Cu bimaterial is localized to a thin layer of epoxy next to the epoxy/Cu interface. This discovery enables the definition of an interfacial zone. Consequently, a macroscopic cohesive zone model can be constructed at the continuum level where the separation of the interface is given by the stretching of the interfacial zone in its thickness direction.

1. Introduction

In our previous works [1, 2], we reported full-atomic molecular dynamics (MD) studies of tensile strength and deformation/failure mechanisms in an epoxy/Cu bimaterial. Based on the polymer consistent force field (PCFF), our MD simulations investigated the effects of strain rate, temperature, cross-link density and epoxy monomer functionality. Such full-atomic MD simulations provide useful insights regarding the atomistic interactions near the epoxy/Cu interface. However, we also found that the material volume that can be simulated by the full-atomic MD is rather small. Behavior of such small material volume may not be representative of the macroscopic behavior of the epoxy/Cu interface. More importantly, the PCFF used in the full-atomic MD is not capable of describing failure of atomic bonds in the epoxy, which severely limited its ability to simulate material failure.

In order to simulate the behavior of larger material volume, coarse-graining approaches have been developed. In such coarse-grained (CG) models, atoms are lumped into groups. Each group is called a bead. Once the interaction among the beads is described by a potential function, the beads can be treated as "super atoms" and MD simulations can be conducted for the ensemble of the beads. One advantage of such coarse-grained MD (CGMD) simulations is the reduced number of degrees of freedom, which enables the simulation of larger material volume. The other advantage is the larger mass of the beads which increases the integration timestep in simulating a physical process.

Most of the exiting CGMD simulations are for bulk materials. For polymer/metal interfaces, there is the additional challenge of coarse-graining the interfacial interactions between the polymer atoms and the metal substrate. There has been little work done in this regard. Site et al. [3] studied the adsorption of Bisphenol-A-polycarbonate (BPA-PC) on nickel (Ni) (111) surface using CGMD simulations. The BPA-PC versus Ni interaction in their coarse-grained model is described by a planar Lennard-Jones 10-4 potential, in which the parameters are obtained from quantum mechanical calculations. This coarse-grained model was later further developed to account for chain-end-orientations on the adsorption [4]. In another work, Farah et al. [5] used planar Lennard-Jones potential, but with a 9-3 functional form, to calculate the interactions between coarse-grained beads and the planar surface. However, this work is a parametric study on the effect of tuning the potential well depth. The potential was not parameterized rigorously. Wong et al. [6] studied the interaction energy and equilibrium distance between an epoxy molding compound and the copper substrate. It is not clear from their work how these quantities will be transferred to a coarse-grained MD simulation as potential parameters. The most relevant work to the current investigation is by Iwamoto [7-9], which establishes a meso-scale model for epoxy/Cu_2O interface. The Cu_2O substrate is coarse-grained in such a way that a 2×2 super cell is used as a bead. However, it is not clear from their work how the potential parameters describing the epoxy and Cu_2O interactions are derived.

2. Coarse-Grained Model Development

The epoxy molding compound of our interest is an EPN-BPA system, composed of Epoxy Phenol Novolac (EPN) as epoxy monomer and Bisphenol-A (BPA) as the cross-linking agent. Molecular structures of the monomers and the curing reaction mechanisms were discussed in [1, 2]. A coarse-grained (CG) model of this EPN-BPA system was developed in [10], which will be used in this study. For completeness, model is briefly outlined below.

The constitutive monomers (3mer and 4mer of EPN, and BPA) are represented as short chains of connected beads with identical mass M, Lennard-Jones potential-well depth ε and diameter σ. Beads interact with each other through the Lennard-Jones 12-6 potential, a harmonic angle bending potential, and a quartic bond stretching potential that allows smooth bond breakage. These potentials are listed below,

$$U_{LJ}(r) = 4\varepsilon \left[\left(\frac{\sigma}{r} \right)^{12} - \left(\frac{\sigma}{r} \right)^6 \right] , \tag{1}$$

$$U_b(r) = U_0 + k_4(r-r_c)^2(r-b_1-r_c)(r-b_2-r_c)\mathrm{H}(r_c-r)$$
$$+ 4\varepsilon \left[\left(\frac{\sigma}{r} \right)^{12} - \left(\frac{\sigma}{r} \right)^6 + \frac{1}{4} \right] \mathrm{H}(2^{1/6}\sigma-r)\mathrm{H}(r_c-r) , \tag{2}$$

$$U_a^{(i)}(\theta) = k_\theta^{(i)}(\theta - \theta_0^{(i)})^2, i=1,2,3 . \tag{3}$$

The optimized parameters in the above potentials were obtained in [10] .

To develop a CG model for the epoxy/Cu interface,

the first task is to group Cu atoms into beads. The concept here is to use a virtual face-centered-cubic crystalline structure, with appropriate lattice constant, to represent the CG Cu substrate. The lattice constant should not be too small otherwise the advantage of reducing the number of particles through coarse-graining is lost, nor should it be too large to prevent polymers from penetrating into the substrate, which is unphysical based on observations obtained from a previous full atomic study [11]. In the investigation by Stevens [12] the nearest-neighbor distance between the FCC substrate beads is chosen to be $1.204\,\sigma$ where σ is the Lennard-Jones diameter of the epoxy beads. This choice is adopted here, which corresponds to a lattice constant of $1.204\sqrt{2}\sigma = 7.463\text{Å}$.

In addition, to describe the interactions between the Cu beads, we link each Cu bead to its nearest neighbors on the face-center-cubic lattice site using a spring with a spring constant of $1000\varepsilon/\sigma^2 = 79.04\text{kcal/(mol·Å}^2)$. This is to maintain the crystalline structure of the substrate during the MD simulations.

Next, we construct our CG bimaterial potential based on the free energy surface of the epoxy/Cu system from an attached state to a detached state. This free energy surface will be computationally obtained by using a metadynamics method [13, 14]. The fundamental concept of the metadynamics method is to fill the system free energy well with a series of Gaussian potentials, and to track the filled Gaussian to reconstruct the free energy landscape [15]. The definition of the Gaussian potential depends on the selection of a set of collective variables, which are suitable to describe the physical process of interest. Let s_i, $i = 1, 2, \cdots, d$ be d number of collective variables. Then the Gaussian potential applied on the system at time t can be written as,

$$V(s,t) = \omega \sum_{\substack{t'=\tau_G, 2\tau_G, \cdots \\ t' < t}} \exp\left(-\sum_{i=1}^{d} \frac{[s_i(\mathbf{r}) - s_i(\mathbf{r}(t'))]^2}{2\delta_i^2}\right) \quad , \quad (4)$$

where s is a measure of the distance between the epoxy and the Cu substrate, δ_i is the Gaussian width of the i-th collective variable, ω and τ_G are the Gaussian height and the frequency at which the Gaussian potential is added, respectively. Once the Gaussian potential is added to the system, forces due to this potential will be exerted onto micro-coordinates, such as atom or bead positions. For example, let r_j be the system's j-th degree of freedom ($j \leq 3N$, N is the number of atoms or beads), then the force on this degree of freedom is

$$-\frac{\partial V(s,t)}{\partial r_j} = -\sum_{i=1}^{d} \frac{\partial V(s,t)}{\partial s_i} \frac{\partial s_i(\mathbf{r})}{\partial r_j} \quad . \quad (5)$$

The method has been applied in [15] to study the intrinsic strength between epoxy and silica.

In the current investigation, a 3mer-EPN, a 4mer-EPN and a BPA monomer are linked to form a "representative" molecule of the cross-linked epoxy. The representative molecule is attached to the Cu substrate as shown in Figure 1(a). The Cu substrate has

a surface area of $39.4 \times 38.7 \text{Å}^2$, and is periodic along both the in-plane directions. It contains 6 layers (111) atomic planes. The polymer consistent force-field (PCFF) is used to describe the full-atomic epoxy system. Between epoxy molecules and Cu substrate the Lennard-Jones (9-6) potential and the sixth-power mixing rule are applied to calculate the van der Waals forces [1]. After the model geometry and force-field parameters are defined, the model is equilibrated using the NVT MD simulation at 300K using the LAMMPS code [16].

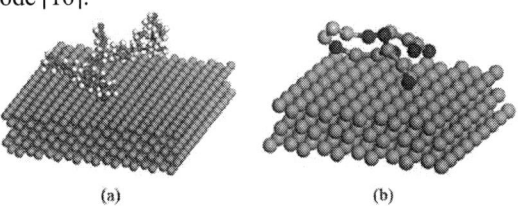

Figure 1. (a) full-atomic and (b) coarse-grained models of the representative molecule attached to the Cu substrate

To implement the metadynamics in MD simulations, the collective variable is chosen to be the distance s between the center of mass of the representative molecule and the Cu substrate surface. During our MD simulations, a Gaussian potential is added to the system periodically. Each time a Gaussian potential is added to the system, it adds forces to the system. Consequently, the distance s varies with increasing number of Gaussian potentials, but not necessarily monotonically. In other words, there is not a one-to-one correspondence between s and the number of Gaussian potentials. In fact, each value of s may be associated with many potentials depending on the total number of potentials added to the system. At a given time (i.e. after a certain number of Gaussian potentials are added to the system), metadynamics calculations can then identify which potentials are associated with a given distance s. Summation of these potentials gives the free energy associated with s at that particular given time. One may then plot the free energy versus s curve for any given time. As time increases (i.e., as more potentials are added to the system), the free energy versus s curves eventually converge to a single one, which is called the free energy surface of the epoxy/Cu system. In this work, the software package PLUMED [17] is used to conduct the metadynamics calculations. During the NVT MD simulations at 300K, a Gaussian potential of height $\omega = 0.005$ and width $\delta = 0.35$ is added at every 100 time steps. These choices of parameters were also used in [15] and found to be a good balance between accuracy and computational efficiency. Figure 2 shows the results of the metadynamics calculations. The convergence of the free energy surface is achieved at around 64ns. From the converged free energy surface (blue curve in Figure 2) the energy barrier between the attached and detached states is found to be approximately 262kcal/mol, and the equilibrium distance between the epoxy molecule's center of mass and the copper substrate surface is ~3.68Å.

978-1-4799-4789-8/14 $31.00 © 2014 IEEE

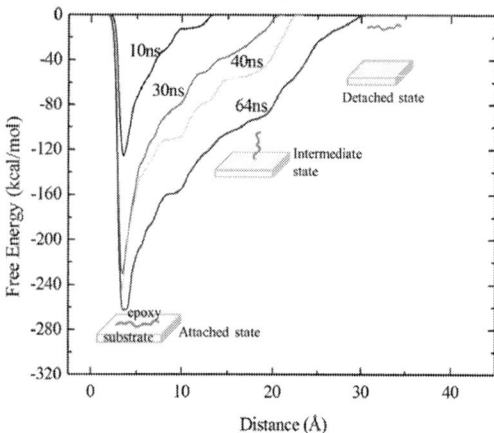

Figure 2. Free energy surface for the full-atomic model obtained from the metadynamics calculations.

In order to derive the CG model for the epoxy/Cu bimaterial, the Lennard-Jones 12-6 potential is used to describe the interactions between epoxy beads and the substrate beads. Let ε_{ew} and σ_{ew} be the two parameters of the Lennard-Jones potential function so that,

$$U_{ew} = 4\varepsilon_{ew}\left[\left(\frac{\sigma_{ew}}{r}\right)^{12} - \left(\frac{\sigma_{ew}}{r}\right)^{6}\right] . \qquad (6)$$

The task is then to determine the two parameters, ε_{ew} and σ_{ew}. To this end, we first construct the CG model for the full-atomic model of the epoxy molecule and Cu substrate system shown in Figure 1 (a). The corresponding CG model is shown in Figure 1(b). Next, we compute the free energy surface of the CG epoxy molecule and Cu substrate system by using the same steps used to compute the free energy surface of the full-atomic epoxy and Cu substrate system, see Figure 2. The free energy surface so computed will depend on the values of ε_{ew} and σ_{ew} used. We postulate that if the free energy surface of the CG model match that of the full-atomic model, the corresponding values of ε_{ew} and σ_{ew} will be the optimal values for the potential given in (6).

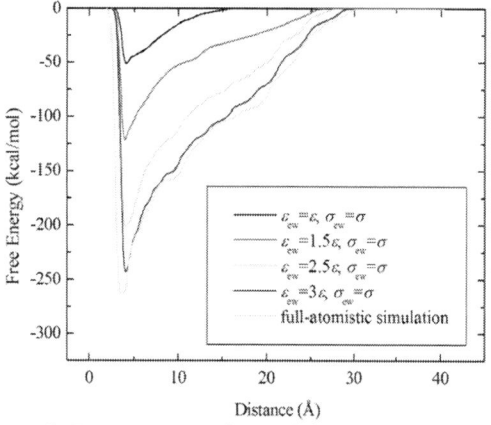

Figure 3. Free energy surfaces of the CG model computed using different CG Lennard-Jones parameters, compared to the free energy surface of the

corresponding full-atomic model.

Figure 3 plots the free energy surfaces using the CG bimaterial model with different Lennard-Jones potential parameters. A clear trend is the increase of energy barrier from the attached to the detached states with respect to the potential depth ε_{ew}. The equilibrium distance between the mass center of the epoxy molecule and the substrate surface maintains almost unchanged with changing ε_{ew}. Trial-and-error calculations shows that the free energy surface becomes very close to that obtained from the full-atomic model when $\sigma_{ew} = \sigma$ and $\varepsilon_{ew} = 3\varepsilon$, where σ and ε are the Lennard-Jones parameters for the CG epoxy, see (1) and **Error! Reference source not found.**. Thus this set of parameters will be used for the coarse-grained molecular dynamic (CGMD) simulations of the epoxy/Cu bimaterial at the coarse-grained level.

3. Coarse-Grained Study of Tensile Deformation of Bimaterial

Having the CG models for the bulk epoxy and the epoxy/Cu interface, we are ready to study the mechanical property and behavior of epoxy/Cu bimaterial at a length scale that is typically inaccessible by full-atomic simulations. The CG epxoy/Cu bimaterial model consists of a rectangular block for the epoxy siting on a Cu substrate containing 4 (111) atomic planes. Periodic boundary conditions are used on all the lateral surfaces. To build the simulation cell, beads representing epoxy and hardener monomers are randomly seeded inside the rectangular block. The stoichiometry ratio of 2:3:9 is used for the number of 3mer-EPN, 4mer-EPN, and BPA. The model has a total of 7,964,588 beads, corresponding to a physical volume of ~89×89×79nm^3 (at 300K).

After the beads are randomly seeded, the system is cross-linked at 500K using a timestep of 5fs, during which the top surface of the epoxy block and the bottom surface of the Cu substrate are kept traction free. In a dynamic cross-linking process we specify that if an EPN reactive bead is within a 5.7Å radius of a BPA reactive bead, there is a chance of 1% that this EPN bead will form a bond with the BPA. For every 10 MD steps, the system is checked for potential bond formations. In this way, it typically takes 400,000~500,000 MD steps to reach 90% conversion. Due to the periodic boundary condition in the lateral surfaces, the epoxy/Cu biamterial so constructed actually represents an infinite layer of epoxy attached to a Cu substrate. We note that epoxy layer interacts with the Cu substrate via the Lennard-Jones potential given by Eq. (6).

The tensile deformation of the bimaterial is simulated by prescribing a downward rigid-body velocity to the Cu substrate while fixing the very top layer (~1nm thick) of the epoxy block. The prescribed downward velocity corresponds to a strain rate of 10^8s^{-1} in the epoxy block. The temperature is kept constant at 300K using the Langevin thermostat [18] with the equation of motion being formulated as:

978-1-4799-4789-8/14 $31.00 © 2014 IEEE

$$m\ddot{\mathbf{r}}_i = \mathbf{f}_i - m\Gamma\dot{\mathbf{r}}_i + \mathbf{W}_i(t), \qquad (7)$$

where Γ is the damping constant and $\mathbf{W}_i$ is a random noise term. They satisfy the relation

$$\left\langle \mathbf{W}_i(t)\mathbf{W}_j(t')\right\rangle = 6k_B Tm\Gamma\delta_{ij}\delta(t-t') \quad , \qquad (8)$$

where k_B is the Boltzmann, T is the absolute temperature, m is the mass of the beads, δ_{ij} is the Kronecker delta, and $\delta(t)$ is the Dirac delta function. The last two terms on the right hand of (7) couple the system to a heat bath which maintains constant temperature. We found that using the Langevin thermostat is critical in keeping the system at a constant and uniform temperature.

4. Results and Discussions

The microstructure of the epoxy molding compound near the substrate surface is of great interest. Geometric and energetic constraints may cause structural inhomogeneity in the epoxy near the interface. Upon full equilibration at room temperature, the near-interface microstructure of the epoxy is analyzed. Figure 4 shows a zoomed-in view of the interfacial zone. To clearly demonstrate the microstructural features, the epoxy and BPA beads are not drawn to scale. A clear feature is that the epoxy beads tend to be densely packed into a few distinctive bands. The further away from the interface, the less distinctive the band structure becomes. Such structural feature is due to the rather planar geometry of the substrate surface, and the Lennard-Jones interactions between epoxy and substrate beads. *Ab initio* calculations [3] have shown that the benzene rings have a strong adsorption to metal surfaces and prefers to align itself in parallel with the Cu surface. In our previous full-atomic studies of the same epoxy/Cu bimaterial [11], densely packed benzene rings were also observed near the interface. These existing *ab initio* and full-atomic studies confirm that our coarse-grained model for the epoxy/Cu bimaterial is able to capture the structural characteristics of the epoxy/Cu interface at the atomistic level. More importantly, our CG model is able to capture the entire interfacial zone containing the heterogeneities. Due to computational limitations, most full-atomic studies are unable to capture the entire heterogeneous interfacial zone, thus can't reveal the layered microstructure near the interface.

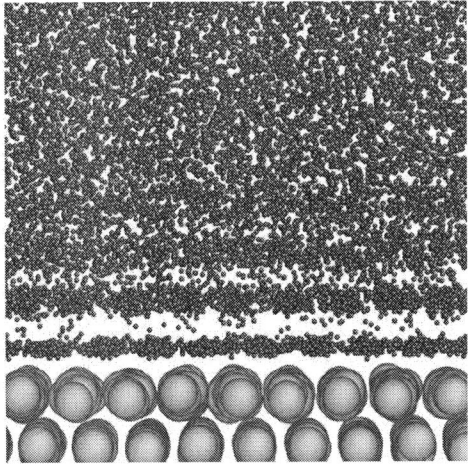

Figure 4. Zoomed-in view of the interfacial zone. Grey and red beads are the EPN and BPA beads, respectively. Yellow beads are the Cu beads.

To quantify the interfacial structural inhomogeneity observed in Figure 4, and to determine how far it extends into the bulk epoxy, we studied the variation of some physical properties of the epoxy with respect to the z (normal direction) coordinate. In Figure 5 (a) we plot the mass density of the epoxy along the z-direction, where $z = 0$ is the interface, and $z = 80$ nm is at the top of the epoxy block. It clearly shows an oscillatory behavior within the range of about 2nm from the interface, beyond which the mass density retains its bulk value, ~1.18g/cm^3. The oscillatory behavior of the density distribution agrees with the structural feature shown in Figure 4, i.e. a few distinctive layers of densely packed epoxy beads near the interface. Another quantity of interest is the projection on the z-axis of the average end-to-end distance of polymer network strands, $\left\langle z_{E-E}\right\rangle$, which is plotted in Figure 5 (b) as function of the z coordinate. We note that $\left\langle z_{E-E}\right\rangle$ indicates how "flat" a polymer strand is. If a strand is aligned parallel to the interface, then its two ends should be roughly on a same plane parallel to the interface. Thus, its projection onto the z-axis, $\left\langle z_{E-E}\right\rangle$ should be almost zero. If a strand is not flat, $\left\langle z_{E-E}\right\rangle$ should be representative of the size of the strand. Figure 5(b) shows the $\left\langle z_{E-E}\right\rangle$ versus z curve. It is seen that the strands near the interface are almost flat, and the flatness rapidly disappears away from the interface. Beyond about 2nm, the strands recover their bulk configuration. The results shown in Figure 5(a) and (b) are consistent in that they indicate the same range of structural inhomogeneity near the interface.

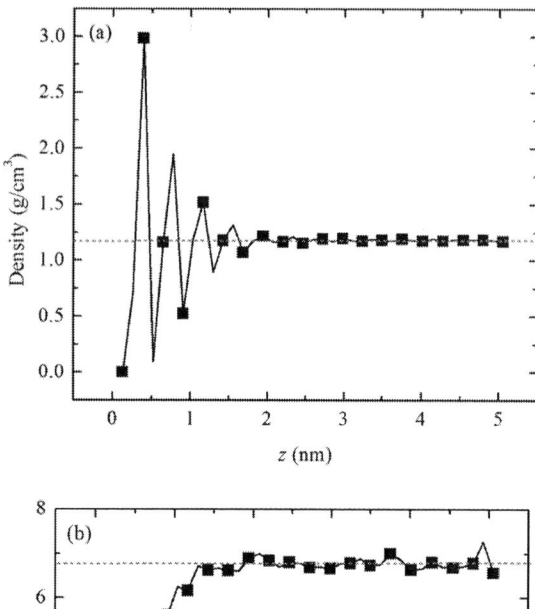

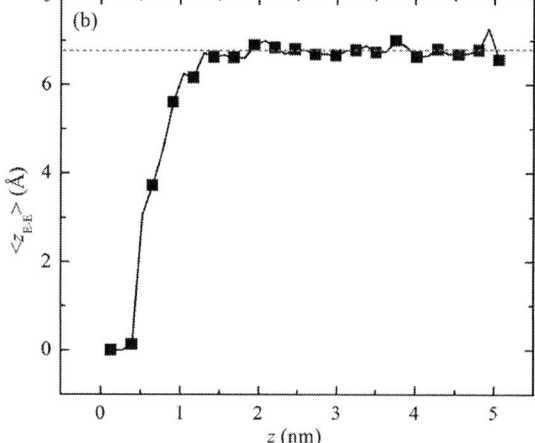

Figure 5. (a) mass density versus z; (b) Projection onto the z-axis of the end-to-end distance of network strands $\langle z_{E-E} \rangle$ versus z-axis. The dashed lines represent the bulk values.

By using the tensile simulation described in Sec. 3, the stress-strain curve is obtained and shown in Figure 6 for the strain rate of $10^8 s^{-1}$. The stress-strain behavior shows an elastic response until the stress reaches the yield strength of ~260MPa at the strain of ~5.4%. Beyond that, the bimaterial experiences a drastic relaxation of stress down to ~125MPa, which is due to the nucleation of cavities [10]. Such cavitation is caused by the tri-axial stress state with strong dilatational component. The tri-axial stress state is a consequence of the periodic boundary conditions prescribed in the lateral direction. In most practical applications, adhesives are indeed under such tri-axial state of stress due to the constraints by the adherends. At about 25% strain, the stress decreases drastically leading to the final interfacial failure.

In Figure 7, several snapshots of the deformed configuration are shown. It is seen that at 15% strain, cavities start to nucleate in the epoxy near the interface because of the relatively weaker force-field between the epoxy and the Cu substrate. The cavitation is also accompanied by extension of the polymer network strands from their coiled states to taut conformations.

Although not shown here, networks of polymer strands holding the inner wall of the cavities are observed inside the cavities [10].

After yielding and stress relaxation, a weak strain hardening takes place that extends to about 25% strain. This behavior is accompanied by expansion of the cavities. The second snapshot of

Figure 7 shows both longitudinally and laterally enlarged voids compared to the first snapshot. At the meantime, polymer strands are continuously transformed from their closed-packed to linear configurations. Upon reaching the ultimate failure at the strain of 25%, the epoxy within the interfacial zone is fully stretched. The strong covalent bonds between the polymer beads prevent further deformation within the epoxy. Therefore, the polymer strands are pulled off from the Cu substrate as shown in the snapshot at 30% strain in Figure 7. Such debonding eventually leads to a rather clean interfacial separation between the epoxy and the substrate as shown in the snapshot at 40% in Figure 7.

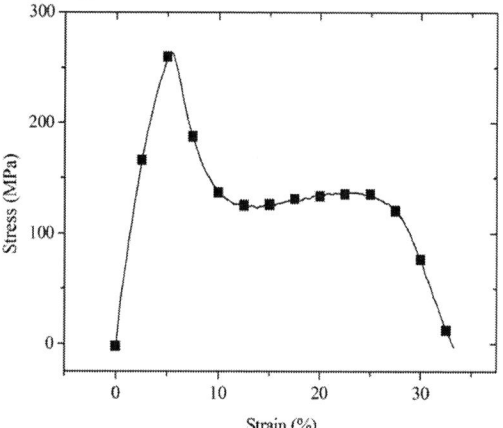

Figure 6. Stress-strain curves for the tensile simulation of epoxy/Cu bimaterial.

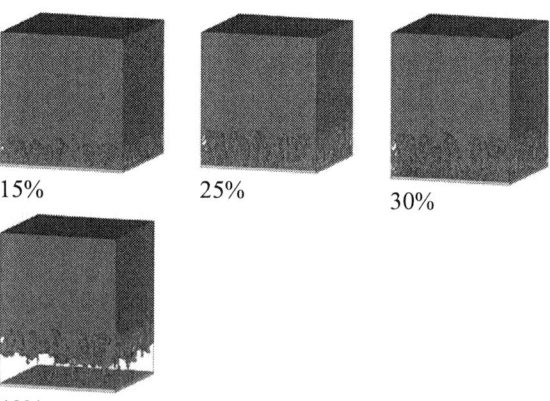

Figure 7 . Snapshots of the deformation process for the 90% cross-linked epoxy/Cu bimaterial under tension at the strain rate of $10^8 s^{-1}$

To investigate the deformation, the displacement profile within the epoxy is plotted in Figure 8(a) under three strain levels. The z-axis (horizontal) in Figure 8(a)

is the Lagrangian coordinate in that it indicates the location of the material particle in the undeformed state. For example, according to Figure 8(a), the displacement at $z = 0$ is about 4 nm under 5% of strain, while the displacement at z = 80 is zero. This means that the epoxy beads at the epoxy/Cu interface ($z = 0$) before the deformation has been displaced by 4 nm (downward) when the overall strain of the bimaterial is at 5%, while the epoxy beads originally located at the top of the epoxy block (z = 80 nm) remain fixed when the overall strain is 5%.

Several interesting observations can be made from Figure 8(a). First, the displacement profile before yielding (strain $\leq$ 5%) is nearly linear, which means that the strain (slop of the displacement profile) the epoxy is nearly a constant within the epoxy. Second, after yielding, the deformation in the epoxy is highly localized to a thin ($\sim$ 10 nm) zone next to the interface. The average strain inside the interfacial zone is approximately 115% when the overall strain is 15%, and approximately 185% when the overall strain is 25%. Outside this interfacial zone, the strain in the epoxy remains at about 2% everywhere at both 15% and 25% strain levels. In fact, although not shown, once yielding occurs, the strain outside the interfacial zone remains at ~2%, irrespective of the overall strain level. This suggests that the increased overall strain after yielding is all accommodated within the interfacial zone. Finally, the average strain outside the epoxy is higher before yielding than after yielding. This is caused by the cavitation induced softening of the epoxy inside the interfacial zone, which causes yielding and stress relaxation of the epoxy inside the interfacial zone.

We note that in our previous full-atomic studies [1] on the same epoxy/Cu bimaterial, such highly localized deformation was not observed. The reason is that the material volume used in the full-atomic investigation is less that the size of the interfacial zone, thus unable to capture the localized deformation. This demonstrates the necessity of accessing the correct length scale to fully understand the mechanical behavior of polymeric materials.

Further, Figure 8(a) also indicates that the deformed thickness of the interfacial zone increase with increasing applied load. This apparent increase of the interfacial zone thickness is entirely due to the increased deformation inside the interfacial zone. The total mass of the interfacial zone remains unchanged. This discovery motivates the use of a cohesive zone model in the macroscopic scale where the epoxy/Cu bimaterial is treated as a continuum. Thus, as the applied load increases, the initial thickness of the interfacial zone is stretched or elongated. Such elongation of the interfacial zone thickness can be defined as the separation between the Cu surface and the bulk epoxy in at the continuum scale. The relationship between this interfacial separation and the applied tensile stress can be easily obtained by keeping track of the displacement profile at each load increment. For example, it follows from Figure 8(a) that the separation is about 1.5 nm and 8.5 nm, respectively, when the applied overall strain is 15%

and 25%. From Figure 6, we find that the applied tensile stresses are 125MPa and 130MPa at 10% and 25% strain, respectively. Thus, one may conclude when the separation is 1.5 nm, the corresponding stress is 125MPa, and when the separation is 8.5 nm, the corresponding stress is 130MPa. Since the applied tensile stress is the same as the traction at the interface, the applied tensile stress versus separation relationship gives the desired traction versus separation law used in the cohesive zone model in continuum mechanics. The results are plotted in Figure 8(b). It is seen that the traction versus separation curve in Figure 8(b) shows the general features of traction versus separation laws for polymer/metal interfaces. The traction first increases almost linearly with increasing separation. Once reaching its maximum, the traction drops drastically, followed by a weak strain hardening until reaching the cohesive strength and critical separation.

We comment that the cohesive zone model is one of the most commonly used constitutive laws to describe the deformation and failure of materials interfaces [19, 20]. However, in most applications, the traction versus separation relationship is assumed *ad hoc* and calibrated via indirect experimental observations. Figure 8(b) is the first traction versus separation relationship for polymer/metal interfaces predicted based on the atomic microstructure at the interface.

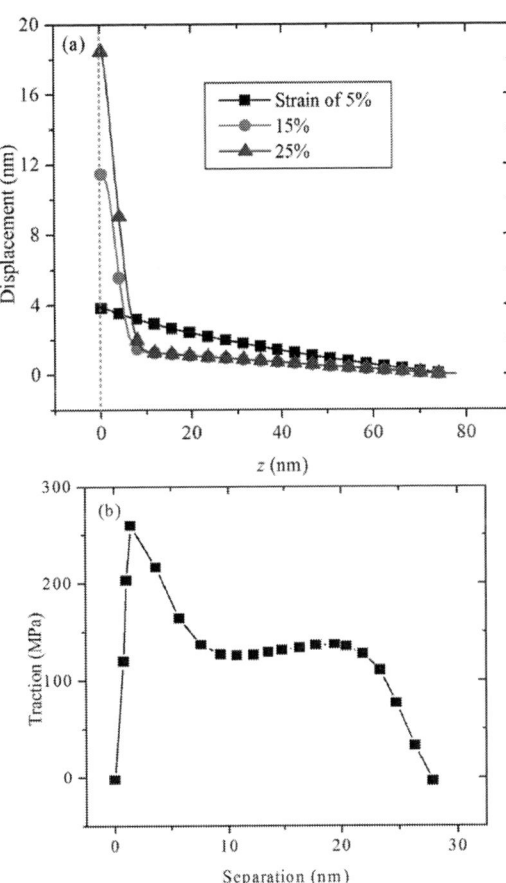

Figure 8. (a) Displacement profiles along the normal direction for the 90% cross-linked epoxy/Cu bimaterial under three overall strain levels. (b) The extracted

traction-separation relationship based on the definition of the interfacial zone.

During tensile deformation, the configuration of epoxy network strands also evolves. As shown in Figure 5(b), in the undeformed state, $\langle z_{E-E} \rangle$ has a uniform value of 6.7Å except for strands very close (< 2 nm) to the Cu substrate. Upon tensile deformation, it is expected that $\langle z_{E-E} \rangle$ for the near-surface strands should increase, since, according to Figure 7, strands near the interface are being pulled to align with the loading direction. This is confirmed by the results showing in Figure 9, which plots the $\langle z_{E-E} \rangle$ versus z curves under different overall strains. Before yielding, for example, at 5% overall strain, the $\langle z_{E-E} \rangle$ versus z curve is almost the same as that at 0% overall strain (undeformed configuration). After yielding, for example, at 15% strain, the $\langle z_{E-E} \rangle$ versus z curve profile shows a peak within the interfacial zone $0 < z < 10$ nm, and the height of the peak increases with increasing overall strain. This confirms our earlier expectations that the network strands within the interfacial zone are being pull to align with the direction of the applied load.

Although polymer strands are being continuously re-oriented towards the loading direction during the deformation, very few of them are broken. Even within the interfacial zone where the strain is highly localized, the number of broken bonds is less than 0.05% during the entire deformation process. At failure, the polymer strands are simply peeled off from the Cu substrate. In other words, the failure is dominantly adhesive at the epoxy/Cu interface rather than cohesive within the epoxy.

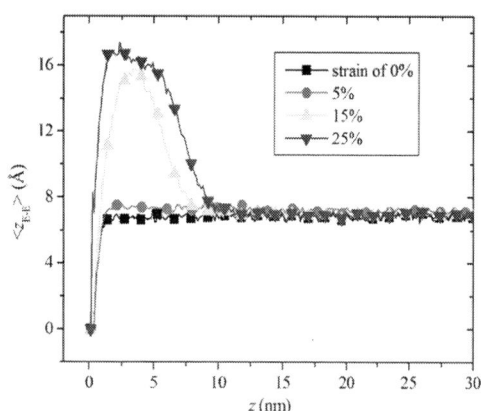

Figure 9. End-end distance (projection to the normal axis) of network strands along the normal direction at different overall strain levels.

5. Summary and Conclusion

In this article, a coarse-grained study of the interface between a cross-linked epoxy and a Cu substrate is presented. The bulk epoxy is modeled by a CG model developed in our earlier studies [10, 21]. To parameterize the CG potential describing the interfacial interactions at the epoxy/Cu interface, the

interfacial free energy landscape is first computed by the full-atomic MD simulations in conjunction with a metadynamics method. The same is then computed using the CGMD, in which the interfacial potential parameters are optimized until the interfacial free energy landscape matches that calculated from the full-atomic MD.

The newly developed CG model is then used to simulate the uniaxial strain deformation of an epoxy/Cu bimaterial. The simulation model is ~89×89×79nm³ and consists of nearly 8 million beads. Based on the simulation results, the following conclusions can be made.

For highly cross-linked epoxy (90% conversion), the initial deformation of the epoxy/Cu bimaterial is almost linearly elastic. Plastic deformation occurs via cavity nucleation and growth, and polymer chain stretching and breaking in the epoxy. After yielding, the plastic strain is highly localized in a thin layer of epoxy next to the epoxy/Cu interface. Further increase of the deformation only increases the plastic deformation within this layer of epoxy called interfacial zone, while the strain in the bulk epoxy remains unchanged with the increasing overall deformation. Consequently, thickness of the deformed interfacial zone increases with increasing overall deformation. By relating the traction to the change in the thickness of the deformed interfacial zone, a traction-separation relationship can be established from the CGMD simulations. This traction-separation relationship can then be used in the cohesive zone model in a continuum mechanics modeling of the epoxy/Cu interface.

Acknowledgements

The work was supported in part by a grant from NSF (CMMI-1200075).

References

1. Yang, S., F. Gao, and J. Qu, A molecular dynamics study of tensile strength between a highly-crosslinked epoxy molding compound and a copper substrate. Polymer, 2013. **54**(18): p. 5064-5074.
2. Yang, S. and J. Qu, Computing thermomechanical properties of crosslinked epoxy by molecular dynamic simulations. Polymer, 2012. **53**(21): p. 4806-4817.
3. Delle Site, L., et al., Polymers near Metal Surfaces: Selective Adsorption and Global Conformations. Physical Review Letters, 2002. **89**(15): p. 156103.
4. Abrams, C.F., L. Delle Site, and K. Kremer, Dual-resolution coarse-grained simulation of the bisphenol-A-polycarbonate/nickel interface. Physical Review E, 2003. **67**(2): p. 021807.
5. Farah, K., et al., Interphase Formation during Curing: Reactive Coarse Grained Molecular Dynamics Simulations. The Journal of Physical Chemistry C, 2011. **115**(33): p. 16451-16460.
6. Wong, C.K.Y., et al., Establishment of the coarse grained parameters for epoxy-copper interfacial separation. Journal of Applied Physics, 2012. **111**(9): p. 094906-7.

978-1-4799-4789-8/14 $31.00 © 2014 IEEE

7. Iwamoto, N., Developing the stress–strain curve to failure using mesoscale models parameterized from molecular models. Microelectronics Reliability, 2012. **52**(7): p. 1291-1299.

8. Iwamoto, N. Molecularly derived mesoscale modeling of an epoxy/Cu interface (part III*): Interface roughness. in Thermal, Mechanical and Multi-Physics Simulation and Experiments in Microelectronics and Microsystems (EuroSimE), 2012 13th International Conference on. 2012.

9. Iwamoto, N. Mechanical properties of an epoxy modeled using particle dynamics, as parameterized through molecular modeling. in Thermal, Mechanical & Multi-Physics Simulation, and Experiments in Microelectronics and Microsystems (EuroSimE), 2010 11th International Conference on. 2010.

10. Yang, S., Z. Cui, and J. Qu, A Coarse-Grained Model for Epoxy Molding Compound. Journal of Physical Chemistry B, 2014. **10.1021/jp409297t**.

11. Yang, S., F. Gao, and J. Qu, A molecular dynamics study of tensile strength between a highly-crosslinked epoxy molding compound and a copper substrate. Polymer, (0).

12. Stevens, M.J., Interfacial fracture between highly cross-linked polymer networks and a solid surface: Effect of interfacial bond density. Macromolecules, 2001. **34**(8): p. 2710-2718.

13. Laio, A. and F.L. Gervasio, Metadynamics: a method to simulate rare events and reconstruct the free energy in biophysics, chemistry and material science. Reports on Progress in Physics, 2008. **71**(12).

14. Laio, A. and M. Parrinello, Escaping free-energy minima. Proceedings of the National Academy of Sciences of the United States of America, 2002. **99**(20): p. 12562-12566.

15. Lau, D., O. Buyukozturk, and M.J. Buehler, Characterization of the intrinsic strength between epoxy and silica using a multiscale approach. Journal of Materials Research, 2012. **27**(14): p. 1787-1796.

16. Plimpton, S., Fast Parallel Algorithms for Short-Range Molecular-Dynamics. Journal of Computational Physics, 1995. **117**(1): p. 1-19.

17. Bonomi, M., et al., PLUMED: A portable plugin for free-energy calculations with molecular dynamics. Computer Physics Communications, 2009. **180**(10): p. 1961-1972.

18. Schneider, T. and E. Stoll, Molecular-dynamics study of a three-dimensional one-component model for distortive phase transitions. Physical Review B, 1978. **17**(3): p. 1302-1322.

19. Needleman, A., A Continuum Model for Void Nucleation by Inclusion Debonding. Journal of Applied Mechanics-Transactions of the Asme, 1987. **54**(3): p. 525-531.

20. Tvergaard, V. and J.W. Hutchinson, On the toughness of ductile adhesive joints. Journal of the Mechanics and Physics of Solids, 1996. **44**(5): p. 789-800.

21. Yang, S. and J. Qu, Coarse-Grained Molecular Dynamic Simulations of the Tensile Behavior of a Thermosetting Polymer. Physical Review E, 2014. **submitted**.

Methodology for supporting electronic system prototyping through semiautomatic component selection

ŚWIERCZYŃSKI R., URBAŃSKI K., WYMYSŁOWSKI A.
Wroclaw University of Technology
Janiszewskiego 11/17, 50-372 Wroclaw

Abstract

A new methodology and custom software tool is presented in this paper, which supports partial automation of designing and prototyping of electronic systems. The emphasis is put on component selection stage, which is realized in a substantially different way comparing to widely used parematic search engines. In presented approach besides of individual (component-level) parameters, also system-level constraints can be defined. As a result several sets of components fulfilling initial needs are automatically generated for further processing (i.e. system level modeling).

Together with the use of parameterized circuit templates and integration with external simulation tools like SPICE presented methodology speeds-up designing of electronic systems thanks to semi-automatic selection of system components from thousands of possible choices.

1. Introduction

Hybrid sensor systems are complex projects, where many different technologies can be used simultaneously in order to achieve required functionality. A MEMS pressure sensor with analog front-end and microcontroller with wireless communication peripherals mounted on flexible substrate is one of many examples. Component selection is only one of several steps during design process. During selection of components for such system not only electrical parameters of individual components are important, but also global (system-level) criteria must be taken into account. Co-design of such systems requires a special tool which could cover both analog and digital domain reflecting some technological constraints. Currently there are no universal tools available to carry out a whole design process (unlike tools for fully digital or analog designs). However, there are many computer programs available which might realize a single step of design process (e.g. simulation, optimization, schematic design and prototyping). Unfortunately, most of those programs does not offer standardized data exchange mechanism, which could be useful for a pipelining (the results of one program usually could be directly used as an input for other program). Additional aspect is that designing of sensor system requires using of electronic components which are available on market (such as operational amplifiers, analog-to-digital converters, digital-to-analog converters, microcontrollers, wireless modules). Paradoxically, large amount of available components instead of simplifying the design process makes it more complicated and time-consuming. Reference designs and parametric search engines delivered by manufacturers or distributors are commonly used to find relevant circuits, but obviously they are not optimized to find optimal set of components from various manufacturers.

The integration of analog and digital components on a single substrate or in a package requires more adequate methods and tools to design highly optimized mixed-signal circuits. Computer aided design (CAD) tools are becoming more complicated to fulfill increasing demands of design process. At the same time each step of design flow is becoming more and more separated from others. For the digital domain, this problem was solved years ago when hardware description languages (HDL) were introduced for describing digital behavior of very large scale integration (VLSI) systems. Due to high-level description of the digital system it is possible to develop system software even before the physical chip is produced. This trend is observable in embedded systems. It is obvious that similar tools are necessary to accelerate the design process of mixed-signal systems (ex. ASIC, SoC, SiP).

Based on authors' experience in designing of mixed-signal systems, in this paper a new methodology is presented for designing of sensor systems, with emphasis on improving component selection phase of this process.

2. Design Flow

Typical design process can be divided into several steps: concept and system specification, element selection and planning, system-level simulation (behavioral or physical), layout design, prototyping and finally measurements and verification. If the verification is not successful, the design process will be started again from component selection (Fig. 1).

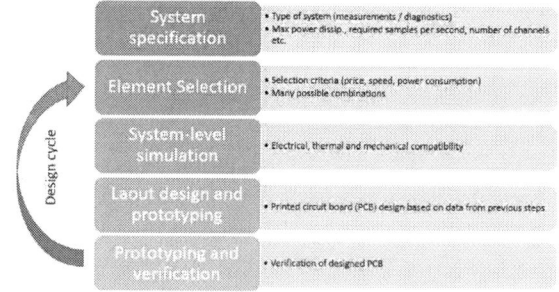

Fig. 1 Schematic diagram of design process.

The two first steps – system specification and element selection are critical for further development of the electronic system. Wrong specification (for example too narrow criteria) may cause significant increase in final price of a system. On the other hand, too wide or

978-1-4799-4789-8/14 $31.00 © 2014 IEEE

undefined input parameters may result in wrong operation of a final product. Furthermore, in order to reduce time-to-market it is essential to make those steps as short as possible. Electronic devices available on the market cover thousands of products (analog to digital converters – ADCs, microcontrollers – MCUs, sensors, amplifiers, current sources etc.) and this number is continuously rising. In order to successfully determine an appropriate set of elements a designer must take into account complex relationships between system elements and analyze large amount of components from different manufacturers. Obviously it is pointless to perform such task without special tools. Even though some manufacturers offer powerful search engines to find best matching elements, those tools often limited to only one database (only manufacturers' elements). Search engines available from distributors suffer from ability to process complex queries (at most several criteria from predefined list are available) As a solution to this problem a specialized software tool was developed – smart Search Engine for Electronic COMponents, further referenced as SEECOM. What makes this tool better than commonly used parametric search engines is its ability to follow several interactions between components, not only attributes of individual items. In relational database not only static values can be stored, it is also possible to store and use formulas describing complex characteristics of objects. However, often it is not possible to provide direct formula describing object or its properties. Instead, a model can be defined. Relational databases cannot use (process) such dynamic models directly and it is necessary to implement additional tool to bypass this limit. SEECOM allows partial integration of the database with simulation tools

like SPICE and makes component selection process faster due to its partial automation. Simulation tool or additional calculations are executed by control software automatically as they are needed. In addition, action scripts describing activity of the system can be defined and applied to determine required results under conditions similar to real operation.

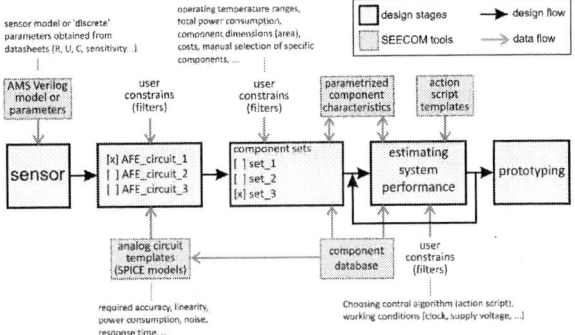

Fig. 3 Design and data flow in SEECOM.

SEECOM consists of two key components: relational SQL database and custom software for advanced data processing with built-in mechanisms for integration with external tools. The user is provided with an interface (fig. 3) used to enter input parameters and to observe results. Although SEECOM was initially designed to support initial steps of electronic system design – components selection, it will be also used as an interface to integrate and support every other step of the whole design process, described later in this paper.

The SQL database stores not only component data, but also information regarding component characteristics, action scripts and references to external software (simulation tools like SPICE, ANSYS, layout design tools like Eagle, KiCAD, Altium Designer). SEECOM does not directly connect to external applications. Instead, unified mechanism is implemented which allows using of custom plugins to interface with other software. Plugins can be realized as the locally executable code, but scripts for calculation software (MATLAB, ABAQS, …) running on other hosts are supported as well.

Design flow implemented in SEECOM (fig. 2) can identify five stages: (1) general circuit description, (2) searching for matching analog or digital front-ends, (3) generating of preliminary results (sets of component) based on user-specified criteria (4) estimation and simulation of system performance of selected sets, (5) choosing the optimal design and building a prototype. Similarly

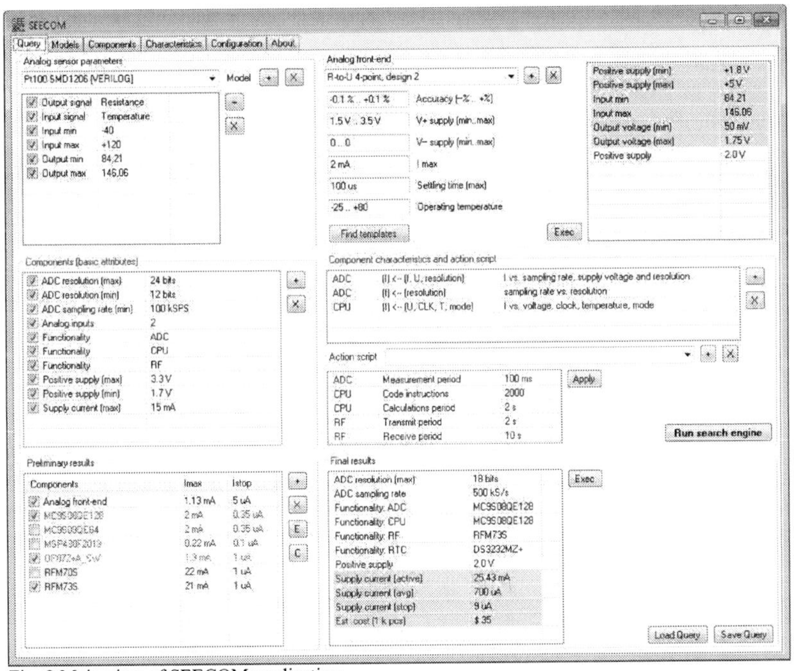

Fig. 2 Main view of SEECOM application.

978-1-4799-4789-8/14 $31.00 © 2014 IEEE

SEECOM is strongly focused on the first step of the whole design process which is component selection and system synthesis.

At the first stage user specifies most of the input parameters to begin the process of finding a matching circuit. This operation is visible in SEECOM query tab as selection of analog sensing element. The sensing element might be described as a 'black box' with some kind of output signal (voltage, current, capacitive, frequency). It can be selected from predefined set of elements (eg. Pt100) or circuit templates based on discrete components (Wheatstone bridge, RC network, current or voltage source). Alternatively sensor parameters can be obtained from the datasheet or reference manual. SEECOM is designed to connect with virtually any other software using custom plugins. It offers a possibility to acquire some specific model from other application upon sending request with optional parameters. Based on request a file containing model is created and then used by SEECOM. Eventually, some models might be downloaded from manufacturer's website.

3. Searching for matching front-ends

At this step the software is trying to find matching analog front-end circuits based on parameters defined in the first step and additional constraints (required accuracy, linearity, power consumption etc.). Circuits are stored in a database as parameterized templates. The sensor output is conditioned and processed to provide the corresponding measurement of the physical property. Because the full-scale output of most sensors are relatively small voltages, signal must be properly amplified and filtered with respect to the input parameters of ADC. It should be also taken into account that some ADCs are specifically designed for measurement systems thus on-chip amplification is possible through digital selection of gain. On the other hand, using specialized ADCs might cause significant increase of system costs. Every aspect should be taken into consideration during selection of conditioning elements. In order to calculate complicated relations between elements SEECOM stores set of parameters which describe devices stored in database (Fig. 4).

Data rate (total)	10000000	SPS	10 MSPS
Data rate per channel	10000000	SPS	10 MSPS
Effective number of bits	14.5	bits	14.5 bits
External clock frequency (max)	60000000	Hz	60 MHz
positive supply current (max)	0.25	A	250 mA
positive supply current (min)	0.22	A	220 mA
positive supply current (typ)	0.22	A	220 mA
idle current (max)	0.0008	A	800 uA
supply current per channel	0.22	A	220 mA
shutdown current	0.0008	A	800 uA
interface analog input	2		2
inteface parallel	1		1
External clock type (pulse)	1		1
Number of differential imput channels	1		1
Number of single ended input channels	0		0 p
Internal buffer	0		0 p
Is shutdown option	1		1
price 1k	21.35	$	21.35 $
Resolution	16	bits	16 bits
Number of simultaneous channels measurements	1		1
conversion time (min)	0.0000001	s	100 ns
operating temperature (max)	105	°C	105 °C

Fig. 4 Parameter definition of analog-to-digital converter (partial)

Currently it is not possible to synthesize a whole analog circuit using only predefined parameters. The first step of matching an analog front-end requires determining of a type of a sensor output. Typical sensors and their outputs are summarized in fig. 5. As the sensor outputs are generally categorized, there is no need to synthesize an analog circuit element by element. In [4,5,6,7] some typical solutions for certain sensors are shown and described. SEECOM fills-in circuit templates (net-list files) with values calculated from input parameters. In [8] a complete algorithm for such calculations is presented. After this step user might execute external SPICE simulation tool to verify suggested design from electrical point of view.

PROPERTY	SENSOR	ACTIVE/ PASSIVE	OUTPUT
Temperature	Thermocouple	Passive	Voltage
	Silicon	Active	Voltage/Current
	RTD	Active	Resistance
	Thermistor	Active	Resistance
Force / Pressure	Strain Gage	Active	Resistance
	Piezoelectric	Passive	Voltage
Acceleration	Accelerometer	Active	Capacitance
Position	LVDT	Active	AC Voltage
Light Intensity	Photodiode	Passive	Current

Fig. 5 Typical sensors and their outputs (summarized) [4].

4. Preliminary results generation and estimation of system performance

At the third stage user defines several constraints: system functionalities, operating temperature ranges, total power consumption, component dimensions (occupied area), costs, etc. Based on them the search engine returns several sets of components matching requirements. To reduce potentially large number of combinations it is possible to manually limit the set of processed components. In practice, the main purpose of the third stage is to reduce (filter-out) amount of result vectors processed by final stages. The best results are presented to the user. Since each set consists of specific circuits, and basic parameters of analog stage are already known, it is possible to estimate output values (system response) based on input parameters.

During the fourth step a few tasks should be done simultaneously to verify output of previous step. Firstly the proposed system should be verified against fulfilling all the specifications from the first and the second stage. When global parameters have been also specified (such as maximum total current consumption), an optional action script is executed to calculate the current for each component in certain working mode. For the first approximation, typical or maximum values from the datasheet will be enough to estimate current consumption. Additional external software can also be executed during this stage to perform system-level simulations in order to test signal integrity.

At this step it is often necessary to manually filter some of the proposed results. Typical example is entering stop or standby modes for a long periods of time (very low static current), with short activity periods (high active current). Knowing the duration of activity/stop modes and using previously calculated parameters, an average energy consumption can be estimated. In fact, it is not possible to choose the best MCU without taking action script into consideration. For applications requiring long inactivity periods and only simple calculations, probably the best choice is 8-bit or 16-bit MCU with very low static current consumption. When complex code needs to be executed, 32-bit MCU with excellent dynamic performance can be a better choice. Taking other system components into account requires manual selection of the results.

5. Conclusions

The improved methodology was presented which can be applied to reduce duration of component selection stage while designing of the electronic system. The improvement was achieved by automation of component selection process. The custom software tool (SEECOM) was implemented with additional database containing parameterized components and templates. Using presented approach a designer is able to define system-level initial constraints (for example, average current consumption, estimated dimensions of final device). Based on them, several sets of components with circuit templates are returned, ready to be processed during further design stages (modeling, verification, prototyping).

Acknowledgments

This work was performed in a frame of the "Partitioning and Modelling of System in Package (PARSIMO)" project; ENIAC-ED-13-08.

References

1. Wayne Current K., Considerations for an analog and mixed-signal computer-aided design tool, Mixed-Signal Design 2003, 23-25 Feb 2003, pp 15-20.
2. Ginés A.J., Peralías E., Reuda A., Madrid N.M., Seepold R., A Mixed-Signal Design Reuse Methodology Based on Parametric Behavioural Models with Non-Ideal Effects, Design Automation and Test in Europe Conference and Exhibition proceedings, 2002, pp. 310-314.
3. Bortolazzi, J.; Mueller-Glaser, K.D.; Towards computer aided specification of analog components, Custom Integrated Circuits Conference, Proceedings of the IEEE 1990
4. Kester W., Practical design techniques for sensor signal conditioning, Analog Devices Technical Reference Books, USA 1999, ISBN-0-916550-20-6
5. Kitchin C., Counts L. A designer's guide to instrumentation amplifiers, Analog Devices, USA 2006, G02678-15-9/06
6. Karki J., Signal Conditioning Wheatstone Resistive Bridge Sensors, Texas Instruments Application Report, Sep 1999
7. Regan T., Current Sense Circuit Collection, Linear Technology Application Note 105, Dec 2005
8. Bishop J., Pressure transducer-to-ADC application, Texas Instruments Application Report, Oct 2000

An accelerated method for characterization of bi-material interfaces in microelectronic packages under cyclic loading conditions

Emad A. Poshtan (1,2,3), Sven Rzepka (2), Bernd Michel (2), Christian Silber (1), Bernhard Wunderle (3)
1. Robert Bosch GmbH, Automotive Electronics, Reutlingen, Germany
2. Fraunhofer Institute ENAS, Dept. Micro Material Center, Chemnitz, Germany
3. Chemnitz University of Technology, Chemnitz, Germany
Email: emad.poshtan@de.bosch.com, Phone: +49(7121)35-37028

Abstract

In this paper, an accelerated and cost-effective characterization method for bi-material interfaces under cyclic loading using a Miniaturized Cyclic Mixed-mode Bending (MCMB) test setup is presented. The Modified Single Leg Bending (MSLB) samples are acquired directly from production-line Thin Quad Flat Package (TQFP) which provide a mixed-mode I + II loading condition. Under sub-critical cyclic loading, crack was found to occur at the polymer-metal interface. The crack length is measured using three methods: (i) in-situ measurements using microscope (ii) gray scale correlation method (iii) numerical method. The crack growth rate was found to have a power-law dependence on the strain Energy Release Rate (ERR) range. In addition influence of plasma cleaning on interfacial adhesion properties namely, crack initiation and propagation is discussed.

1. Introduction

Interfacial delamination is one of the most critical issues in the reliability of the IC (Integrated Circuit) packages in semiconductor industry. The package delamination arises in particular from the usage of a large number of bi-material interfaces and interaction of these material under thermo-mechanical load during processing (e.g. reflow), testing (e.g. moisture sensitivity level, thermal cycling) and operation (e.g. vibration).

The reliability of IC packages has been investigated using mixed-mode bending tests for crack initiation/propagation behavior of several types of specimens/loading conditions. Button shear tests [1] are applied for interface characterization of bi-material samples with limited range of Mode Mixity (MM) and so-called "handmade samples" (not from production-line). Four-point bending (4PB) tests were widely utilized [2], [3] as well as chisel bending tests [4] using a specific clamping mechanism for chip-MC interface characterization. Crack propagation in bulk material were investigated in mixed-mode condition by brasil-nut fixture [5]. The influence of moisture on the delamination behavior were also examined by [6], [7]. Advanced Mixed-mode bending (AMB) test recently were developed which can provide an enhanced mode mixity range with only couple of samples [8] under monotonic loading. Fatigue crack propagation along polymer-metal

interfaces in microelectronic packages was investigated using hand-made Double Cantilever Beam (DCB) samples under force-controlled loading condition by [9]. [10] investigated the Mixed-mode I + II fatigue/fracture characterization of hand-made composite bonded joints by Single-Leg Bending (SLB) test under subcritically force-controlled loading. They used an analytical equivalent crack method to overcome the difficulties associated to visual crack monitoring during the fatigue tests. [11] discussed the fatigue crack growth along interface between metal and ceramics submicron-thick films in inert environment using Four Point Bending (4PB) tests. They measured the crack growth by a numerical compliance method without the precise measurement of crack length (a). However to the best of the authors' knowledge, the cyclic crack propagation (sub-critical crack growth) under mixed mode thermo-mechanical condition for miniaturized production-line samples approx. with 10 mm length has not been fully investigated.

The formation of copper oxide and other contaminants has an effect on reliability of IC packages [12]. The oxidation can occur in assembly processes such as die attach curing and wire bonding that can accelerate the delamination growth. One of the typical methods to overcome this effect is the use of plasma cleaning as a pretreatment prior to coating process to increase package reliability. In this chemical process the surfaces are exposed to hydrogen-argon plasma gas mixtures to remove contaminants from metallic substrates. In [12] the effects of plasma process on adhesion strength between MC and copper leadframe using pulling tests under static loading is investigated. However the influence of plasma cleaning on Fatigue crack growth (FCG) is not discussed.

The thermo-mechanical behavior of TQFP package under Temperature Cycling (TC) namely influence of rate-dependent material on warpage were investigated in our work [13]. The evolution of ERR and MM during crack propagation between Molding compound (MC) and Copper Lead-Frame (LF) under TC test was discussed numerically in [14]. It was observed that the negative shear mode, mixed mode and positive shear mode are the dominating modes respectively in low temperature, mid-temperature and high temperature. The ERR values in negative mode (between 45 and 70 degrees) were considerably higher

978-1-4799-4789-8/14 $31.00 © 2014 IEEE

which shows a greater crack propagation probability in this mode range.

In this paper, the cyclic mixed-mode bending (CMMB) test set-up is presented. The aim for development of this apparatus is to provide an accelerated and cost-effective characterization method for bi-material interfaces under cyclic thermo-mechanical loading. CMMB aims to resemble the long-time delamination growth of IC packages under temperature cycling (TC) test. In addition the miniaturized and horizontal design of the testing machine facilitates the crack monitoring using SEM (scanning electron microscopy) or microscope during the fatigue tests.

The bi-material samples (MC/LF) are prepared directly from TQFP packages in production-line. Modified Single Leg Bending (MSLB) specimens have only one interface crack to facilitate the control of crack growth under cyclic loading and provide stable crack growth. In addition these samples can provide a good agreement between test and actual situation in terms of the critical mixed mode range. For more information you may refer to [14]. In this work monotonic loading and fatigue tests are carried out isothermally. Interfacial Sub-Critical Crack Growth (SCCG) is investigated. The evolution of Mode-Mixity (MM) and Strain Energy Release Rate (SERR) during crack propagation is investigated experimentally and numerically. In addition influence of plasma cleaning on crack in initiation and propagation is discussed. As an ongoing project the influence of temperature on SCCG will be discussed in the next paper.

2. Fatigue Crack Propagation (FCP)

Fatigue testing procedures are mainly classified in two groups namely: (i) High Cycle Fatigue (HCF) and (ii) Low Cycle Fatigue (LCF). In HCF load is applied in the linear domain which leads to macroscopically elastic deformation, although some small area can undergoes plastic deformation [15]. LCF occurs when the number of cycles is less than 10^3. Since the load applied is close to the yield strength, bulk plastic deformation can takes place [15]. Consequently plasticity of material should be taken into account [16]. Thus interfacial crack propagation in IC packages under TC (more than 10^3 cycles) is considered to be in the first group. Furthermore the slow rate of crack growth caused by low warpage of TQFP (50 μm) in each temperature cycle, provides the HCF situation.

HCF is classified in two criteria being: (i) infinite life design where Stress-Life methods are used [16] and (ii) damage tolerant design (DTD) or Fatigue Crack Growth (FCG) which is a Fracture Mechanics based method [17]. The FCG method correlates the rate of fatigue crack growth (da/dN) and the change of fracture parameter over the crack length (here G_{max}). In testing adhesives

and polymeric material, G_{max} is often used instead of the SERR amplitude ($\Delta G = G_{max} - G_{min}$) [16].

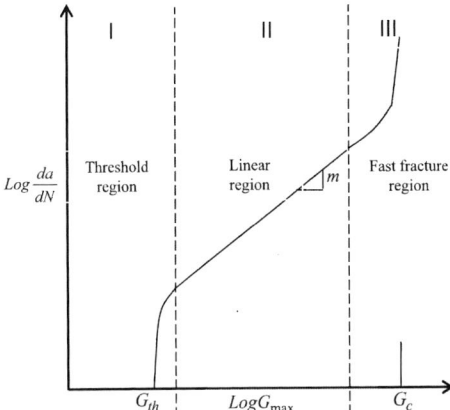

Figure 1.   Schematic fatigue crack growth curve in bi-logarithm scale.

Fig. 1 shows a schematic FCG curve in bi-logarithm scale which is typically observed in previous works [16], [11], [10], [9]. Region I or so-called threshold region is associated with threshold SERR (G_{th}), below which measurable crack growth does not occur. In region II, FCP curve is linear and fits with the Paris equation 1 [18].

$$\frac{da}{dN} = c(G_{max})^m \qquad (1)$$

,

where c and m are material constants. In region III unstable crack growth occurs as G_{max} approaches the critical SERR (G_c) in quasi-static loading. The ratio G_{th}/G_c indicates fatigue sensitivity. In many applications, region III does not significantly influence the total propagation life and can be ignored in the prediction of cycles to failure. The full sigmoidal shape of the curve can be described empirically by Eq. 2 [19] where p and q are additional material constants. In this paper the full sigmoidal equation is used to fit the SCCG results.

$$\frac{da}{dN} = c(G_{max})^m \frac{1 - (\frac{G_{th}}{G_{max}})^p}{1 - (\frac{G_{max}}{G_c})^q} \qquad (2)$$

3. An accelerated method for characterization of interfacial delamination under cyclic loading condition

Temperature cycling (TC) test is a common reliability testing method for semiconductor packages. During this test, IC packages are subjected to a typical temperature change from, i.e., -50C to 150C for a thousands of cycles which takes long time to perform. This justifies a need for an accelerated failure prediction method which can accelerate the qualification process.

A. Sample preparation

Knowing the thermo-mechanical behavior of the TQFP package under TC test, an accelerated interfacial characterization method under cyclic loading is proposed. This method is based on a specimen-centered approach. In other words, instead of adapting the test-specimens to a special testing machine, a universal test set-up is developed. The concept is also scalable, i.e. it has potential to work for small as well as small samples from different IC packages.

First, suitable test-specimens are prepared which are then loaded cyclically in a mixed-mode delamination test set-up. Then, in parallel, simulations have to be carried out to extract the critical fracture data, i.e. SERR as a function of MM. Finally, based on the simulation-experiment analysis a lifetime model can be established and applied to bi-material interfaces in heterogeneous system integration or to other packaging concepts and technologies used in microelectronics.

The main issues concerning bi-material samples made from packaging-relevant interface materials are: compatibility with real product, ability to resemble the actual working situation, efficient manufacturing (high number at low cost and in short time) and precrack-capability. The sample preparation process is illustrated in fig. 2. Two types of dummy packages (TQFP without chip) with and without plasma cleaning are acquired directly from production-line. In order to facilitate the crack propagation in one direction a tiny gap (0.5 mm) is inserted though the MC until the copper lead-frame using an automatic wafer dicing saw in step 1. In step 2 the TQFP is cut to stripes using similar saw. This sample shape of samples facilitates the in-situ interfacial crack tracking. The intactness of the samples were checked by SEM and SAM and no pre-delamination were observed.

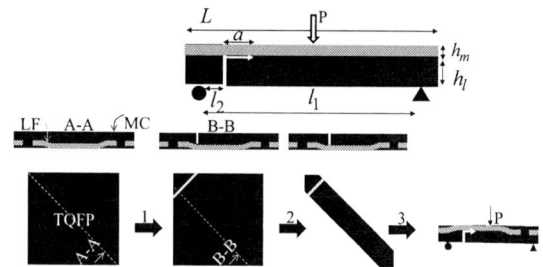

Figure 2. Sample preparation directly from production-line product (here TQFP 14x14).

The dimensions of the specimens are shown in tab. 1, where h_m and h_l are MC and LF thicknesses, respectively.

B. Design of the Miniaturized Cyclic Mixed-mode Bending (MCMB) apparatus

Based on the sample-centered approach a Miniaturized Cyclic Mixed-mode Bending apparatus is built around

Table 1. Specimens size in mm.

Specimen types	Width B	L	l_1	l_2	h_m	h_l
W and WO plasma	2	19	12	2	0.8	0.2

the sample. Using a piezomove flexure actuator combined with a closed loop control system, stable strain-controlled cyclic loading with a small amplitude (down to 5 μm) can be applied, as shown in fig.3. In addition not only critical crack growth, by monotonic loading, but also sub-critical crack growth, by cyclic loading, can be tested. The stiffness of the machine and clamps are checked by DIC (Digital Image Correlation). Two miniaturized positioning tables provide full access of actuator tip on sample. Compatible design of the apparatus allows combination with precision temperature forcing systems (heat stream) to investigate the influence of temperature on fatigue crack growth. Mixed-mode loading is applied on samples using a modified mixed-mode chisel setup. In this fixture a specimen is revolved relative to the load axis that leads to mixed-mode loading at the interfacial crack tip. Miniaturized and horizontal design of the testing machine facilitates the crack monitoring using SEM (scanning electron microscopy) or microscope during the fatigue tests fig. 4.

The displacement load δ and the force F at the loading point were continuously monitored during the tests with a strain-gauge sensor and a precious load cell, respectively. The humidity was kept constant during all tests.

C. Experimental approach

Crack growth is dependent on the control mode in constant amplitude fatigue tests [16]. It is observed that in displacement control, SERR decreases with crack length whereas in load control SERR increases. In displacement control a rapid initial crack propagation occurs, however, that the crack growth rate would decrease as the crack grows, and potentially stop before complete failure, if G_{th} is reached. This behavior allows us to generate the complete FCP curve from a single sample. Thus, in this work, displacement control was used to facilitates the determination of G_{th}.

Pre-cracking along the MC/LF interface was done as follows. The specimen was loaded monotonically under the constant displacement rate of 5 $\mu m/S$. After initiation of the interface crack, the specimen was unloaded.

Displacement controlled fatigue tests were performed with a loading ratio of $R = \delta_{min}/\delta_{max} = 0.1$ at 80% of the quasi-static crack propagation displacement. A testing frequency of 2 Hz with a sinusoidal waveform in constant displacement amplitude was applied.

Different approaches have been used to calculate the crack growth rate from the crack length as a function of cycles. Based on ASTM $E647 - 86a$ [20] the secant method in

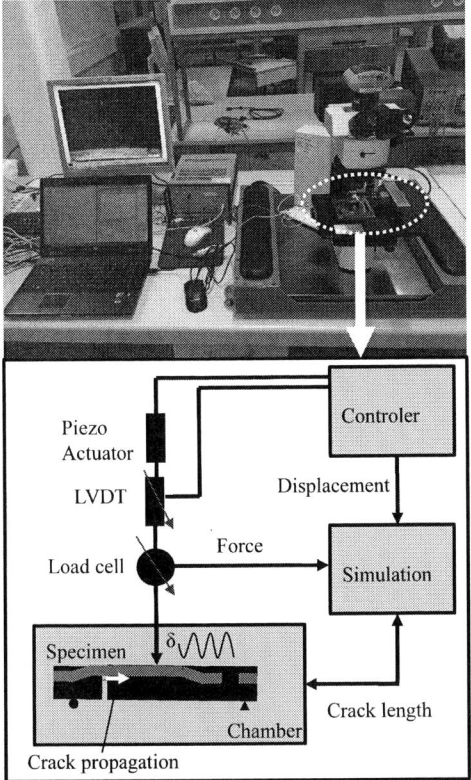

Figure 3. Experimental setup including the MCMB under microscope (top). Schematic view of Experimental-numerical analysis (down).

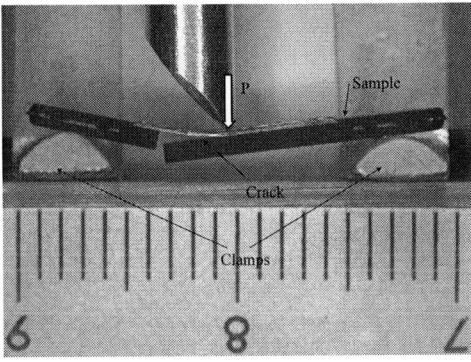

Figure 4. Side view of sample under MSLB test.

which the crack propagation rate is calculated from the slope between adjacent points using Eq. 3 and 4.

$$\left(\frac{da}{dN}\right)_{\bar{a}} = \frac{a_{i+1} - a_i}{N_{i+1} - N_i} \tag{3}$$

$$\bar{a} = \frac{a_{i+1} + a_i}{2}, \tag{4}$$

where i represents the i_{th} measurement performed dur-

ing the test in every 500 cycles.

D. Numerical approach

The FCG in MSLB specimens was numerically analyzed under the plane-stress condition by the commercial FEM code, ANSYS. The ERR and MM corresponding to each crack length and temperature is calculated using virtual crack closure technique (VCCT). The VCCT is based on Irwin's crack closure integral [21] and it was modified by [22] and [23]. According to Irwin, the work required to extend a crack by an infinitesimal distance is equal to the work required to close the crack to its original length. Thus, for homogeneous, isotropic, linear elastic materials, the components of strain energy release rate for mode I and mode II can be expressed as:

$$G_1 = \lim_{\Delta a \to 0} \int_0^{\Delta} \sigma_{yy}(\Delta a - r) \delta y(r) dr \tag{5}$$

$$G_2 = \lim_{\Delta a \to 0} \int_0^{\Delta} \sigma_{xy}(\Delta a - r) \delta x(r) dr \tag{6}$$

where Δa is a small crack extension; σ_{yy} and σ_{xy} are the normal and shear tractions, at a distance r ahead of the crack tip and δx and δy are the displacement jumps at a distance r behind the crack tip, along the x (sliding mode) and y (opening mode) directions, respectively. For a crack in homogeneous isotropic materials, the mode mixity based on strain energy release rate (SERR) can be express as:

$$\tan^2 \psi = \frac{G_2}{G_1} \tag{7}$$

The optimized viscoelastic material properties of the molding compound is applied in the numerical analysis. For more information regarding the optimization process and material model you can refer to [13].

Fig. 5 shows the SERR and MM during crack propagation under displacement control condition (R=0.1). The SERR values are normalized as $G_{norm} = \frac{G_{max}}{G_{max}|_{a \to 0}}$. The data are fitted to polynomial equations. These equations are used in section 4 to evaluate the G_{norm} values corresponding to each crack length. It is observed that shear dominated MM remains relatively constant while SERR decreases during crack propagation. Similar behavior is observed by [16], [24]. In this paper smiling bending of the samples , having LF on top of MC, is considered as the positive shear mode.

4. Results and discussion

A. Crack tracing

Three techniques were employed to determine the crack length during fatigue testing: (i) in-situ measurements using microscope (ii) gray scale correlation method (iii) using the numerical compliance method. The in-situ measurements were performed with a microscope,

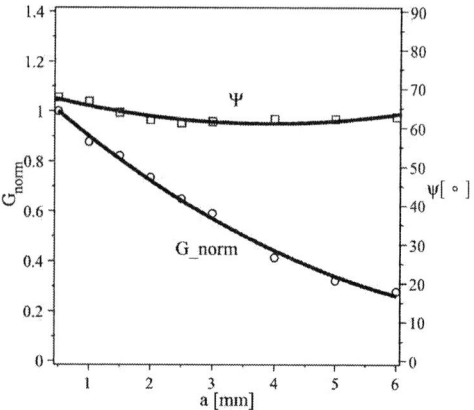

Figure 5. ERR and MM evolution during FCG under displacement control loading condition where a is the crack length.

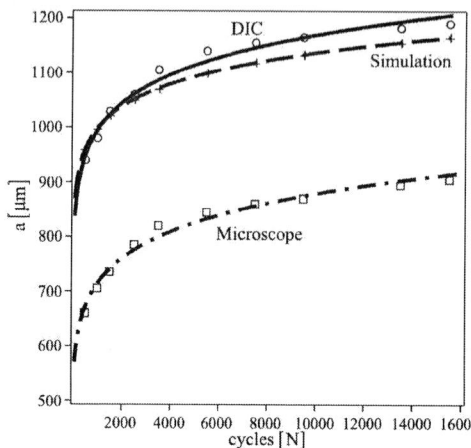

Figure 7. FCG comparison between simulation, microscope and DIC.

checking the crack length every 500-1000 cycles.

The digital cross correlation between gray scale images or DIC (Digital Image Correlation) is an optical field measuring method based on video or microscopic image acquisition and image processing. Using this method, relative displacement and strain distribution in the sample during crack propagation can be measured. In order to detect the crack tip, relative displacement of the copper LF to MC is measured. It is due to the fact that, unlike LF, MC is brittle and remains relatively bendless during crack propagation [25].

In numerical compliance method, the experimentally

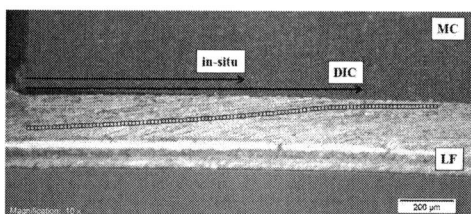

Figure 6. Crack tip detection comparison between microscope and DIC.

measured compliance during crack propagation is used as an input to simulation. Based on this input, crack length is evaluated numerically by an inverse fitting procedure.

Fig. 6 illustrates the crack between MC and LF after 14000 cycles. Relative displacement of the LF compared to MC at 14000 (blue) is compared to precrack condition (green). Here for easier detection of crack tip, only one row of the whole network is shown. The crack tip is consider to be at the meeting point of these two rows. It is observed that only the opened part of the crack can be detected by microscope.

In fig.7 the microscope and DIC crack tip detection methods are compared with simulation method. It is

observed that DIC method is in better agreement with simulation.

B. sub-critical crack growth

Fig.8 shows the relationships between fatigue crack growth rate and SERR in bi-logarithm scale for two types of samples being with and without plasma. The SERR values are normalized as using equations 5 and 9 where G_c^{WOP} and G_c^{WP} are the critical SERR for the Sample Types (ST). The regions I, II and III are clearly depicted in this graph. The data points are fitted to the full sigmoidal curve by eq. 2. The summary of key interfacial FCP parameters is shown in tab. 2 where fatigue sensitivity is calculated as $S_f = G_c - G_{th}$.

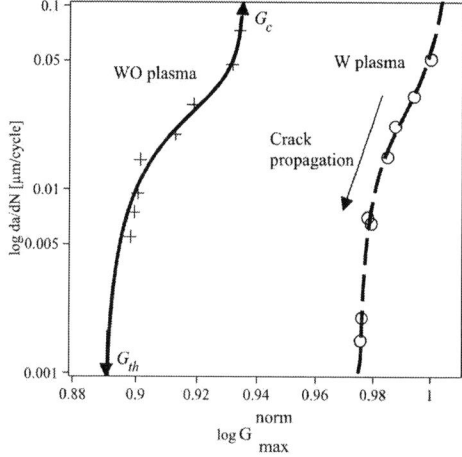

Figure 8. Interfacial SCCG in with and without plasma specimens.

978-1-4799-4789-8/14 $31.00 © 2014 IEEE

$$G_{max}^{norm} = \frac{G_{max}}{G_c^{ST}} * \eta \qquad (8)$$

$$\eta = \begin{cases} \dfrac{G_c^{WOP}}{G_c^{WP}} & ST:WOP \\[2ex] 1 & ST:WP \end{cases} \qquad (9)$$

where ST is the sample type which can be with or without plasma.

The specimen WO plasma shows a lower G_{th} compared to one with plasma. Thus, plasma treatment seems to increase the strength of the MC/LF interface in terms of initial FCG.

Table 2. Summary of key interfacial FCP parameters.

Specimen type	S_f	G_{th}^{norm}	G_c^{norm}
W plasma	0.03	0.97	1
WO plasma	0.05	0.89	0.93

The sample surface after tests was examined. Fig. 9 illustrates a perfect interfacial FCP without striations or MC particles (stick-slip behavior) is observed.

Figure 9. Surface of the specimen after test.

5. Conclusions

In this paper monotonic loading and fatigue tests are carried out using MCMB test set-up on the specimens directly from production-line. Sub-critical crack growth is observed along the surface between molding compound and copper lead-frame in the samples prepared from Thin Quad Flat Package (TQFP). The evolution of mode-mixity and strain energy release rate (SERR) during crack propagation is investigated experimentally and numerically. The FCG results show that plasma cleaning seems to increase the strength of the MC/LF interface in terms of initial FCG. As an ongoing project the influence of temperature will be discussed in the next paper.

6. Acknowledgments

I would like to thank Mrs. E. Noak and Mrs. B. Seiler from CWM company for DIC analysis using VEDDAC software, Dr. P. Gromala from Robert Bosch GmbH for his support in material modeling and STMicroelectronics for providing dummy TQFP packages. Further, fruitful discussions regarding simulation and experimental methods with Mr. M. Schulz from Jointlab Berlin is acknowledged.

References

[1] A. A O Tay, J. S. Phang, E. H. Wong, and R. Ranjan. A modified button-shear method for measuring fracture toughness of polymer-metal interfaces in ic packages. In *Electronic Components and Technology Conference, 2003. Proceedings. 53rd*, pages 1165–1169, 2003.

[2] PG Charalambides, J Lund, AG Evans, and RM McMeeking. A test specimen for determining the fracture resistance of bimaterial interfaces. *Journal of applied mechanics*, 56:77, 1989.

[3] Ines Hofinger, Matthias Oechsner, Hans-Achim Bahr, and Michael V Swain. Modified four-point bending specimen for determining the interface fracture energy for thin, brittle layers. *International Journal of Fracture*, 92(3):213–220, 1998.

[4] G Schlottig, I Maus, H Walter, KMB Jansen, H Pape, B Wunderle, and LJ Ernst. Interfacial fracture parameters of silicon-to-molding compound. In *Electronic Components and Technology Conference (ECTC), 2010 Proceedings 60th*, pages 1939–1945. IEEE, 2010.

[5] J-S Wang and Z Suo. Experimental determination of interfacial toughness curves using brazil-nut-sandwiches. *Acta Metallurgica et Materialia*, 38(7):1279–1290, 1990.

[6] Timothy P Ferguson and Jianmin Qu. Predictive model for adhesion loss of molding compounds from exposure to humid environments. In *Electronic Components and Technology Conference, 2006. Proceedings. 56th*, pages 7–pp. IEEE.

[7] MH Shirangi and B Michel. Mechanism of moisture diffusion, hygroscopic swelling, and adhesion degradation in epoxy molding compounds. In *Moisture Sensitivity of Plastic Packages of IC Devices*, pages 29–69. Springer, 2010.

[8] B. Wunderle, M. Schulz, J. Keller, I. Maus, H. Pape, and B. Michel. Advanced mixed-mode bending test: A rapid, inexpensive and accurate method for fracture-mechanical interface characterisation. In *Thermal and Thermomechanical Phenomena in Electronic Systems (ITherm), 2012 13th IEEE Intersociety Conference on*, pages 176–186, May 2012.

[9] John Guzek, Hamid Azimi, and Subra Suresh. Fatigue crack propagation along polymer-metal interfaces in microelectronic packages. *Components, Packaging, and Manufacturing Technology, Part A, IEEE Transactions on*, 20(4):496–504, 1997.

[10] MV Fernández, MFSF de Moura, LFM da Silva, and AT Marques. Mixed-mode i+ ii fatigue/fracture characterization of composite bonded joints using the single-leg bending test. *Composites Part A: Applied Science and Manufacturing*, 2012.

[11] Hiroyuki Hirakata, Masaya Kitazawa, and Takayuki Kitamura. Fatigue crack growth along interface between metal and ceramics submicron-thick films in inert environment. *Acta Materialia*, 54(1):89 – 97, 2006.

[12] JH Hsieh, LH Fong, S Yi, and G Metha. Plasma cleaning of copper leadframe with ar and ar/h2 gases. *Surface and Coatings Technology*, 112(1):245–249, 1999.

[13] E.A. Poshtan, S. Rzepka, B. Wunderle, C. Silber, T. von Bargen, and B. Michel. The effects of rate-dependent material properties and geometrical characteristics on thermo-mechanical behavior of tqfp package. *Electronics Packaging Technology Conference (EPTC), 2012 IEEE 14th*, pages 131 – 135, 2012.

[14] E. A. Poshtan, C. Silber, S. Rzepka, B. Michel, and B. Wunderle. An accelerated interfacial characterization method for microelectronic packages under automotive testing conditions - methodology and sample preparation. *Smart Systems Integration with Micro-and Nanotechnologies.*, 2013.

[15] A Manonukul and FPE Dunne. High–and low–cycle fatigue crack initiation using polycrystal plasticity. *Proceedings of the Royal Society of London. Series A: Mathematical, Physical and Engineering Sciences*, 460(2047):1881–1903, 2004.

[16] IA Ashcroft and SJ Shaw. Mode i fracture of epoxy bonded composite joints 2. fatigue loading. *International journal of adhesion and adhesives*, 22(2):151–167, 2002.

[17] A Pirondi and G Nicoletto. Fatigue crack growth in bonded dcb specimens. *Engineering fracture mechanics*, 71(4):859–871, 2004.

[18] PC Paris and F Erdogan. A critical analysis of crack propagation laws. *Journal of Basic Engineering*, 85:528, 1963.

[19] R. J. H. Wanhill H. L. Ewalds. *Fracture mechanics*. Edward Arnold, United Kingdom, 1984.

[20] E 647 standard test method for measurement of fatigue crack growth rates., 2008.

[21] Irwin GR. Analysis of stresses and strains near the end of a crack traversing a plate. *J Appl Mech*, 24:361364., 1957.

[22] E Fo Rybicki and MF Kanninen. A finite element calculation of stress intensity factors by a modified crack closure integral. *Engineering Fracture Mechanics*, 9(4):931–938, 1977.

[23] Ronald Krueger. The virtual crack closure technique: history, approach and applications. Technical report, DTIC Document, 2002.

[24] Marcello Manca, Amilcar Quispitupa, Christian Berggreen, and Leif A. Carlsson. Face/core debond fatigue crack growth characterization using the sandwich mixed mode bending specimen. *Composites Part A: Applied Science and Manufacturing*, 43(11):2120 – 2127, 2012.

[25] J. Auersperg, B. Seiler, E. Cadalen, R. Dudek, and B. Michel. Fracture mechanics based crack and delamination risk evaluation and rsm/doe concepts for advanced microelectronics applications. In *Thermal, Mechanical and Multi-Physics Simulation and Experiments in Micro-Electronics and Micro-Systems, 2005. EuroSimE 2005. Proceedings of the 6th International Conference on*, pages 197–200, April 2005.

978-1-4799-4789-8/14 $31.00 © 2014 IEEE

Simulation of stress distribution in assembled silicon dies and deflection of printed circuit boards

Katerina Macurova[a], Paul Angerer[a], Ronald Schöngrundner[a], Thomas Krivec[b], Mike Morianz[b], Thomas Antretter[c], Raul Bermejo[d], Martin Pletz[a,d], Michel Brizoux[e], Wilson Maia[e]

[a]Materials Center Leoben Forschung GmbH, Roseggerstraße 12, 8700 Leoben, Austria,

[b]AT&S AG, Fabriksgasse 13, 8700 Leoben, Austria,

[c]Institut für Mechanik, Montanuniversitaet Leoben, Peter-Tunner-Str. 5, 8700 Leoben, Austria,

[d]Institut für Struktur- und Funktionskeramik, Montanuniversitaet Leoben, Franz-Josef-Str. 18, 8700 Leoben, Austria,

[e]Thales Global Services, 18, avenue du Maréchal Juin, F-92360 Meudon-la-Forêt, France

e-mail: katerina.macurova@mcl.at

Abstract

The knowledge of thermally induced strains created during the assembly in Printed Circuit Boards (PCB) is an important issue for electronic packages. In the assembly process, a thin silicon-chip is attached onto a copper foil. The curing of the adhesive is followed by the cooling down of the assembled structure to room temperature. The different properties of the involved materials and the geometry of the structure induce stresses and deflection in the substrate, which can become critical for the further lamination process. In this work, the chip assembly process is investigated by means of a parametric FE analysis. The aim is to estimate the stress distribution in the silicon die and the deflection (warpage) of the entire architecture based on the assembly conditions. The key material properties (i.e. thermal expansion coefficient (CTE) and elastic constants) of all involved materials were determined as a function of the temperature (process relevant temperature up to 200°C) and used as input for the FE model. Special attention has been given to the determination of the volumetric shrinkage of the adhesive during the curing. The results predicted by the FE model are validated with experimental measurements using an X-ray diffraction method (Rocking-Curve-Technique), which enables the deflection of the attached silicon die to be determined. Good agreement between simulation and experiments is achieved.

1. Introduction

The ongoing trend of the electronic device miniaturization and increase of their functionality has established the chip embedding technology to be state-of-the art. The Embedded Component Packaging (ECP®) technology aims at integrating passive or active components in the inner layers of a multi-layer Printed Circuit Board (PCB), the backbone of electronic devices [1]. The study of stresses in the embedded components during embedding and deformations of the board are key issues in the PCB design. The force required to fracture brittle embedding components can be of the order of a few Newton. In previous works the failure of typical 2.0 x

2.0 x 0.12 mm^3 silicon dies was analyzed and the mechanical resistance (strength) was determined under biaxial bending [2,3,4]. The fracture loads for such a geometry ranged from 5 N to 20 N for a loading configuration that has some similarities with the loading of components during attachment. The increase of the PCBs complexity by embedding of large dies with hundreds of interconnections brings consequently higher requirements on the die package warpage.

An assembly process for making mechanical connections results in a thermal expansion (CTE) mismatch of the utilized materials in adhesively bonded packages [5,6]. The CTE mismatch during temperature change as well as the volumetric shrinkage during curing of the adhesive (cross linking of the molecular chains) at elevated temperature causes stress and leads to warpage of the packages [7]. In order to estimate the stresses and deflection in the board very accurately, a precise determination of all material properties is required.

In case of complex systems like microelectronic packaging structures, the use of the finite element method (FEM) enables the implementation of complex material models (e.g. elasto-plastic, phase transformations) as well as loading cases (e.g. adhesive curing process, cooling down, etc.) to analyze critical processes during the fabrication of printed circuit boards. The applicability of FEM to analyze the mechanical behavior of multilayer structures has been demonstrated [8,9]. In this regard, polymers exhibit time and temperature dependent viscoelastic material behavior [10,8]. The cure kinetics model has been incorporated [8,9]. Besides that, the plasticity of metal components represents another significant part of the analysis, which may have to be taken into account [11].

In order to validate the results from the FE simulation, non-destructive evaluation (NDE) techniques have been attempted aiming at estimating of the stresses in the parts involved as well as the deformation of the board. In the past decades, numerous optical methods for deformation measurements have been developed and employed as important engineering tools, e.g. Shadow Moiré method and Far Infrared Fizeau Interferometry [12]. The warpage measurement can be regarded as a cost-effective, quick and non-destructive method to validate and verify analytical and numerical results [13].

In this work, the chip assembly process is investigated by means of a parametric finite element analysis (FEA). The aim is twofold: (i) evaluate the stress development in the silicon die during chip assembly, and (ii) estimate the deflection (warpage) of the package based on the processing conditions. For the purpose of the numerical results validation, the X-ray diffraction method (Rocking-Curve-Technique) has been performed to measure the warpage of the silicon die. The validity of the used material models and the experimental results are discussed.

2. Die attachment process

The assembly process deals with the attachment of the silicon die onto a copper foil by an adhesive. This process consists of two main steps, (1) the flow of adhesive underneath the component during component placing, (2) the curing of the adhesive at elevated temperature followed by a cooling process. The first step has been investigated in a previous work [14]. In the following, the attachment process will be described and analyzed.

Figure 1a shows a cross-section of an assembly unit, which has been extracted from a board with uniformly distributed silicon dies (see Figure 1b). The investigated stacked die assembly represents a tri-layered structure, which consists of a silicon die (layer 1), a die attach adhesive (layer 2) and a copper foil (layer 3).

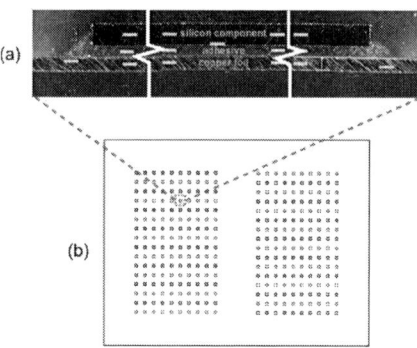

Figure 1: (a) The stacked die assembly in a cross-section, (b) The 24" x 18" copper panel with uniformly distributed chips

The different steps of the attachment process over temperature are illustrated in Figure 2. After the placement of the die onto the adhesive at room temperature (RT), the temperature is increased (step A-D), thus activating and speeding up the adhesive polymerization, i.e. the cross-linking in the polymer. During this process the volume of the adhesive decreases referred to as cure shrinkage. As a consequence, the adhesive polymer undergoes transition from liquid to the solid state. Mechanically relevant cure shrinkage starts

when the gel point of the cross-linking system is achieved. Thereby, the hardening of the adhesive develops continuously, leading to increasing force feedback effects on the silicon die and the copper foil. The polymerization is finished after certain time. Then, the package is subsequently cooled down to RT (step E).

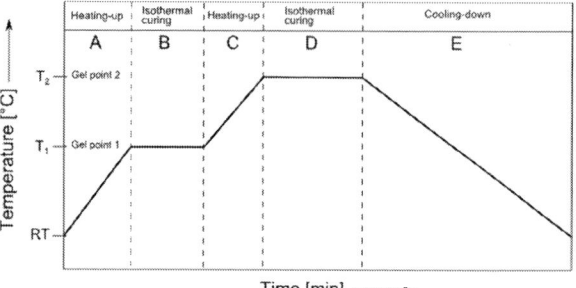

Figure 2: The attachment process temperature steps

3. Numerical modeling of the attachment process

In order to investigate the stress state and the warpage caused by thermal loading during the attachment process, a FEM model was developed. The analysis was conducted using the commercial finite element software ANSYS® [15].

To reduce the computational cost of the numerical analysis, only a quarter symmetry model with higher order elements was used, as shown in Figure 3.

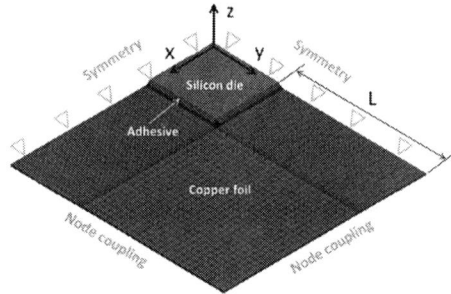

Figure 3: FE model of the stacked die assembly with BCs

The model incorporates the attachment assembly, capturing the silicon die (flat on both sides), the adhesive with its meniscus and the copper foil. The adhesive meniscus is formed during the chip attachment process [14], as can be observed in Figure 1a. All geometry parameters, such as height, length, thickness, etc., are taken from cross section measurements.

When assuming an adjacent die attached to the copper foil, the copper foil length L is set to extend one-half the distance between the modeled die and the adjacent one, as shown in Figure 3. A coupling of free edges of the copper foil was chosen; this condition represents the periodic assembly. The adhesive was perfectly bonded to the silicon die and to the copper foil by means of nodes sharing the interface. This decreases the complexity of the solution and eliminates contact nonlinearities.

978-1-4799-4789-8/14 $31.00 © 2014 IEEE

The dimensions of the modeled silicon die were 7 x 7 x 0.12 mm³. The adhesive bond line thickness (referred to as BLT) was 0.045 mm. The copper foil had the dimensions 22 x 22 x 0.07 mm³.

The adhesive curing is a transient process. However the cure kinetics is not investigated; instead a stepwise static approach was used in the numerical analysis. This methodology applies a uniform isothermal loading on the entire structure. For the case of isothermal curing steps, i.e. step B and D in Fig. 2, the curing process is simulated by a thermal shrinkage upon a temperature increase of $\Delta T = 1°C$ within the curing time, to account for the volume change of the curing adhesive. The adhesive polymer system is assumed to have completely reacted. Consequently, all calculations refer to a stress free (initial) state, which is defined by the reference temperature (T_{REF}). T_{REF} is set to a temperature where the mechanically relevant adhesive shrinkage starts. For the analysis, the heating-up process up to a curing temperature level T_1 (step A in Fig. 2) was not considered, because the stiffness of the uncured adhesive is negligible. The curing of the adhesive is assumed to occur progressively and homogenously in each step.

4. Material properties

In order to analyze the thermo-mechanical response of the system during the attachment process it is necessary to determine the key material properties of all involved materials as a function of the temperature.

Special attention is given to the derivation of the volumetric shrinkage of the polymer adhesive during its phase transformation, because the cure shrinkage is a significant part of the entire volume change.

ADHESIVE CURE SHRINKAGE

The cure shrinkage measurement is particularly challenging. In this work, we characterized the adhesive volumetric shrinkage by measuring the mechanically relevant shrinkage, using a modified rheology measurement approach [16]. The measured volumetric shrinkage ΔV is represented in Fig. 4 (gray dashed line). The curing temperature profile (black solid line) is also shown for illustrative purposes. During heating-up of the assembly from RT to T_1 (step A), no volumetric shrinkage occurs. An increase of the volumetric shrinkage was observed during the isothermal curing at T_1 (step B). At this temperature, the mechanically relevant adhesive shrinkage starts, reaching values of $\approx 1\%$. This temperature is set as the reference temperature, T_{REF}, for the stress calculations. At the following heating-up stage of the assembly to T_2, the volumetric shrinkage increases to $\approx 4\%$ (step C). During isothermal curing at T_2 the volume shrinkage decreases to $\approx 3\%$, observed at the end of the curing process (step D). Then again an increase in ΔV is observed during the cooling down process (step E).The final volumetric shrinkage of the adhesive ΔV_{final} is around 6% at RT.

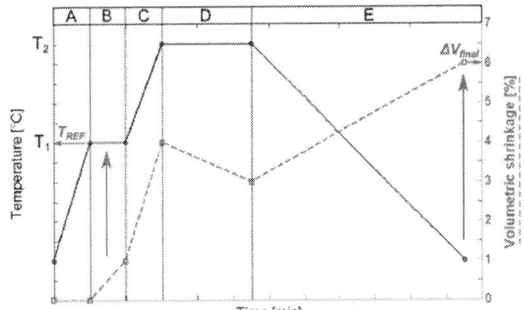

Figure 4: The measured adhesive volumetric shrinkage according to the curing temperatures (step B-E) over time.

ADHESIVE REACTIVE COEFFICIENT OF THERMAL EXPANSION

Since the adhesive volumetric shrinkage cannot be directly implemented into the numerical analysis, its amount is converted to an equivalent coefficient of thermal expansion, referred to as reactive CTE or $\alpha_{C,r}$. This approach is a modification of the work by Schmöller [17] and Böger [18].

The volumetric shrinkage measurement explained above provides the total shrinkage ΔV_{tot} at applied temperatures. This consists of (i) the thermal shrinkage ΔV_{therm}, caused by the adhesive CTE, and (ii) the chemical shrinkage ΔV_{chem}, related to the adhesive polymerization:

$$\Delta V_{tot} = \Delta V_{therm} + \Delta V_{chem} \qquad (1)$$

Although the individual contribution of the thermal and chemical shrinkage cannot be separated during curing, only the total shrinkage ΔV_{tot} is needed to estimate the coefficient of thermal expansion of the adhesive. The total volumetric shrinkage is equal to the reactive volumetric change ε_{vol}, which can be converted to the isotropic reactive length change ε_{len} as

$$\varepsilon_{len} = 1 - \sqrt[3]{1 - \varepsilon_{vol}} \ . \qquad (2)$$

According to [17] Eq. 2 can be simplified to

$$\varepsilon_{len} \approx \frac{\varepsilon_{vol}}{3} . \qquad (3)$$

The reactive length change, ε_{len} corresponds to the thermal strain, ε_{therm}. Thus the reactive thermal expansion coefficient $\alpha_{C,r}$ can be expressed as follows:

$$\alpha_{C,r} = \varepsilon_{therm} / \Delta T , \qquad (4)$$

where ΔT is the temperature change from the reference temperature T_{REF} to the given temperature (see Figure 5).

With regard to the solution methodology, $\alpha_{C,r}$ is derived for four temperature steps (from B to E). The steps B-D represent the adhesive curing, and step E the cooling-down to RT. In the isothermal curing steps, i.e. step B and D, the thermal loading is substituted by a slight temperature increase of $\Delta T = 1°C$ to account for the volume change of the curing adhesive. The calculated values of the reactive thermal expansion coefficient $\alpha_{C,r}$ for each temperature step are shown in Figure 5 (gray dashed line).

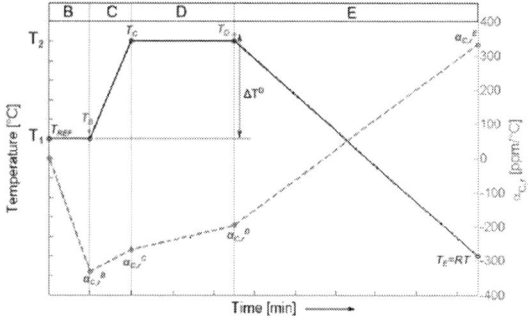

Figure 5: The reactive thermal expansion coefficient $\alpha_{C,r}$ according to the cure shrinkage.

ADHESIVE ELASTIC PROPERTIES

The adhesive elastic modulus was measured dynamically at relevant temperatures on fully cured rectangle-shaped adhesive specimens by using a micro-Dynamic Mechanical Analyzer (model μ-DMA RSAG2). The results are summarized in Table 1.

Table 1: Elastic properties of adhesive

T [°C]	Young's modulus E [GPa]	Poisson's ratio υ [-]
20	3.43	0.30
50	3.00	0.30
115	0.08	0.39
135	0.05	0.45

SINGLE CRYSTAL SILICON PROPERTIES

To determine the crystallographic orientation of the silicon die, an electron backscattering diffraction (EBSD) measurement was conducted. A ⟨100⟩-orientation was found. The orthotropic material properties for the determined crystallographic orientation were taken from the literature [19] and consequently used in the numerical analysis. Since the elastic properties of silicon are almost invariable over the range of the attachment process temperatures (20-130 °C) [20], only RT properties have been considered in the model.

Table 2: Approximate values of elasticity of standard silicon wafers in ⟨100⟩-orientation, at RT

Young's modulus [GPa]		Poisson's ratio υ [-]		Shear modulus [GPa]	
E_x	169	ν_{xy}	0.064	G_{xy}	50.9
E_y	169	ν_{yz}	0.36	G_{yz}	79.6
E_z	130	ν_{xz}	0.28	G_{xz}	79.6

The temperature dependent thermal expansion coefficient values were also taken from the literature [21], see Table 3.

Table 3: Single crystal silicon thermal expansion coefficient values

T [°C]	CTE [ppm/°C]
20	2.60
27	2.62
127	3.25

COPPER FOIL PROPERTIES

The copper foil elastic modulus was measured dynamically by using a micro-Dynamic Mechanical Analyzer (model μ-DMA RSAG2). Strip-shaped specimens with the dimensions 40 x 5 x 0.018 mm^3 were loaded with a deformation amplitude of 0.02% with a frequency of 1 Hz at temperatures between 20-150°C. The Poisson's ratio was taken from the literature [22]. Results are given in Table 4.

Table 4: Elastic properties of copper foil

T [°C]	Young's modulus E [GPa]	Poisson's ratio υ [-]
20	91.80	0.34
50	89.77	0.34
100	85.77	0.34
150	80.83	0.34

The temperature dependent thermal expansion coefficients for bulk copper were taken from the materials database provided by The National Institute of Standards and Technology (NIST) [23], see Table 5.

Table 5: Copper foil thermal expansion coefficient values

T [°C]	CTE [ppm/°C]
20	16.50
65	17.00
100	17.40
135	17.80

5. Results and discussion

The aim of the thermo-mechanical analysis was twofold: (1) evaluate the stress development in the silicon die during the attachment and (2) to estimate the deflection (warpage) of the stacked die package based on the processing conditions.

5.1. Stress field in the silicon die and the assembly

Several assumptions were adopted for the stress state evaluation, (a) the loading conditions lead to biaxial bending of the package, as $\sigma_x \cong \sigma_y$, (b) based on the package geometry there is a different stress distribution in the center, at the free edges and at the free corners of the package [6].

With regard to the curing process temperatures there is a change of the stress state in the assembly, and the package warpage. Particular attention is put on the evaluation of stresses in the temperature step E, since after cooling-down to RT the assembly is employed for further lamination. Deformations are exaggerated with a scale factor of 10 in all figures.

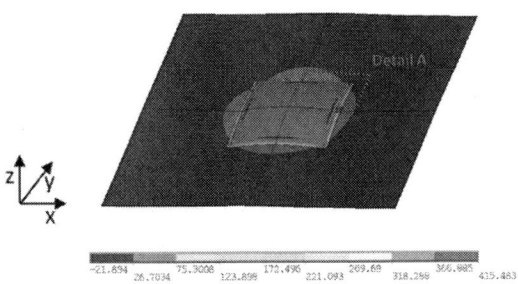

Figure 6(a): Maximum principal stress σ_I [MPa] in the assembly after the temperature step E, top view

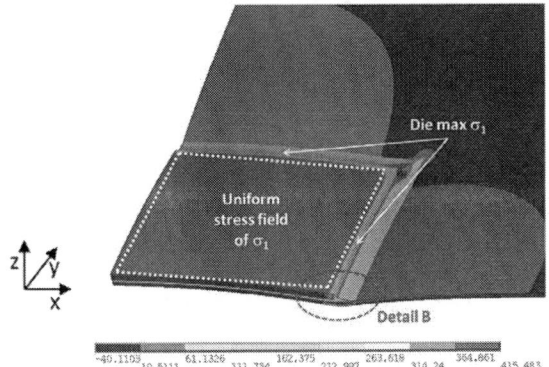

Figure 6(b): Detail A: Maximum principal stress σ_I [MPa]

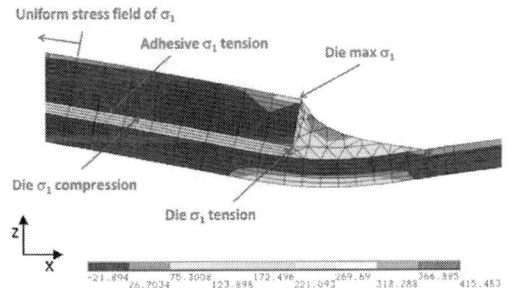

Figure 7: Detail B: The stress state of σ_I [MPa] in the silicon die, the adhesive bond line and the copper foil, at XZ cross-section

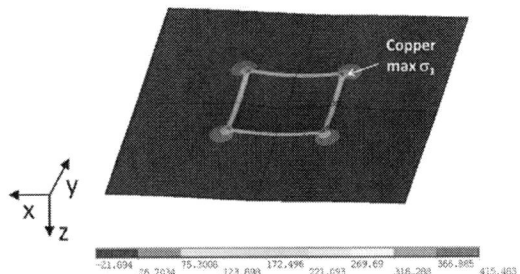

Figure 8: Maximum principal stress σ_I [MPa] in the assembly after the temperature step E, bottom view

The silicon die shows nearly uniform tensile stress field with a principal stress σ_I of app. 36 MPa located on the component top, as shown in Figure 6. Based on the package geometry and material combination the stresses increase in the vicinity of the edges and corners of the die. The maximal stress peak of σ_I is found to be close to the die edges and it is app. 236 MPa. The bottom of the die is under compression, σ_I is app. -0.2 MPa. The compressive mode changes to the tensile one near the die edges, σ_I is app. 62 MPa. The stress distribution close to the die edges is affected by the adhesive meniscus. For the numerical results presented here, the influence of numerical singularities (lie close to die edges) is not taken into consideration.

The biaxial strength (maximum failure stress) of the silicon die was determined experimentally using the ball-on-three-balls (B3B) test [2]. The maximum failure stress was found to be 3530 MPa. Based on the stress state comparison provided above one can conclude: The maximum stress level induced into the silicon die during cooling-down to RT is much lower than the biaxial strength and therefore not critical. No damage of the silicon die occurs in the attachment process.

In the bond line of the adhesive there is nearly uniform tensile stress σ_I of app. 95 MPa, as shown in Figure 7.

Underneath the silicon die, the warpage of the copper foil corresponds to biaxial bending. The maximal principal stresses σ_I in the copper are located below the adhesive meniscus, as shown in Figure 7. The stress maximum is app. 415 MPa, due to an applied linear elastic approach, and it occurs on the bottom of the copper foil, see Figure 8. The principal stress σ_I emerged in the copper foil during the temperature step E becomes very high, because only a linear elastic model of copper is used in the analysis. It should be noted that in fact the stress level of the copper foil is supposed to decrease due to the copper plasticity.

978-1-4799-4789-8/14 $31.00 © 2014 IEEE   167

5.2. Assembly warpage and experimental validation

In order to validate the accuracy of the used material models and the assumptions in the FE model the calculated warpage after cooling down to RT is compared with an experimentally measured warpage of the bonded die. In addition it is crucial to know the warpage of the package after attachment because the assembly is employed for lamination. Residual stresses, induced to the assembly during cooling, and the package warpage can cause problems during further lamination

Based on the numerical results, the curvature radius is derived from (x, y, z)-coordinates of 3 points lying on the top of the die. The curvature calculation is carried out in coordinate directions X and Y (in the XZ and YZ plane of symmetry), see Figure 9.

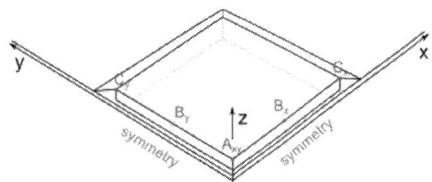

Figure 9: Points on the assembly geometry for the curvature radius evaluation

The determined values are compared with experimental measurements using an X-ray diffraction method (Rocking-Curve-Technique). This analysis technique can map major warpage features non-destructively even in fully encapsulated packaged chips [24].

The rocking curve experiments by X-ray diffraction of silicon dies were conducted on a D8 Discover diffractometer (Bruker AXS, Germany) in parallel beam geometry (40 kV, 35 mA, Cu Kα radiation). The beam diameter was reduced by means of a circular primary baffle with 0.3 mm diameter. During specimen scanning the detector was fixed at the Bragg angle 2Θ of the crystal plane of interest and the specimen was rotated about the ω-axis (perpendicular to the diffraction plane spanned by incident and diffracted beam). For this purpose a step size $\Delta\omega$ of 0.01° and a counting time of 1 s/step were used. The scanned ω-angle was limited to the region ± 2° near the maximum

The corresponding ω-angle refers to the direction of the lattice plane normal of the curved single crystal. The curvature K along one direction (X or Y) at one specific point (x, y) is equal to the reciprocal radius of an osculating circle in this point: $K = 1 / r$. The curvature radius was measured in two coordinate directions X and Y at RT.

The results, obtained from FEA and the rocking-curve measurement, are summed up in Table 6. A comparison shows that the numerical result is in a very good agreement with a difference of only 7% with the experimental result.

Table 6: The curvature radius of the warpage at XZ and YZ cross section

Evaluation technique	Curvature radius R [mm]	
	x-direction	y-direction
FEA	176	178
Rocking-Curve	165 ± 2	167 ± 2

6. Conclusions

With the hypothesis taken into account we showed that it is possible to predict the effect of the adhesive volumetric shrinkage during the attachment process on the assembly using a stepwise solution methodology described in the paper.

Based on FEA the maximum loading of the assembly during cooling-down to RT was demonstrated. It was proven that the maximum principal stress σ_I arising in the silicon die does not cause damage during attachment.

A comparison of the experimental measurements of the package warpage by X-ray diffraction technique shows a very good agreement with the numerical results.

The attachment process simulation is ongoing work. Future steps can be divided into four fields: The influence of numerical singularities on numerical results should be investigated. Furthermore, the level of interfacial stresses in the assembly is going to be determined. Interfaces represent a weak point of layered structures, since stresses arising there can cause delamination. Another open issue is the influence of copper plasticity on the copper stress state and the package warpage. Finally, since the assembly process is the first step in embedding silicon dies into PCBs, the next logical step is to study further lamination. The influence of the package warpage on the die loading will be evaluated during subsequent compression. The results of this study will permit to optimize process parameters and set of design rules.

Acknowledgments

Financial support by the Austrian Federal Government (in particular from the Bundesministerium für Verkehr, Innovation und Technologie and the Bundesministerium für Wirtschaft, Familie und Jugend) and the Styrian Provincial Government, represented by steirische Forschungsförderungsgesellschaft mbH and by steirische Wirtschaftsförderungsgesellschaft mbH, within the research activities of the K2 Competence Centre on Integrated Research in Materials, Processing and Product Engineering, operated by the Materials Center Leoben Forschung GmbH in the framework of the Austrian COMET Competence Centre Programme, is gratefully acknowledged.

References

1. Ostmann A, Manessis D, Stahr J, Beesley M, Cauwe M, De Baets J. "Industrial and technical aspects of chip embedding technology," in *Electronics System-Integration Technology Conference, 2008. ESTC 2008. 2nd*; Sept. 2008. p. 315-320.

2. Deluca M, Bermejo R, Pletz M, Supancic P, Danzer R. "Strength and fracture analysis of silicon-based components for embedding," *J. Eur. Ceram. Soc.*2011; 31(4): p. 549-558.

3. Deluca M, Bermejo R, Pletz M, Weßner M, Supancic P, Danzer R. "Influence of deposited metal structures on the failure mechanisms of silicon-based components," *J. Eur. Ceram. Soc.*2011; 32(16): p. 4371-4380.

4. Deluca M, Bermejo R, Pletz M, Morianz M, Stahr J, Supancic P, et al. "Local strength measurement technique for miniaturised silicon-based components," in *12th Int. Conf. on Thermal, Mechanical and Multi-Physics Simulation and Experiments in Microelectronics and Microsystems, EuroSimE*; 2011; Linz, Austria.

5. Timoshenko SP. "Analysis of bimetrial thermostats," *Journal of the Optical Society of America*Sept. 1925; 11.

6. Suhir, E. "Calculated thermally induced stresses in adhesively bonded and soldered assemblies," in *ISHM International Symposium on Microelectronics*; 1986; Atlanta, Georgia.

7. Yoo, H. Y.; Moon, B. H.; Kwank, J. S.; Kwank, Ch. W.; Lee, J. Y.; Borghard, T. J. Novel Die Attach Adhesive for Thin Quad Flat Package. Henkel.

8. Fatal T, Jansen KMB, de Vreugd J, Rzepka S. "Influence of Cure Dependency of molding Compound Properties on Warpage and Stress Distribution during and after the Encapsulation of Electronics Components," in *EuroSimE 2009*; 2009; Delft.

9. Boehme, B.; Jansen, K.M. B.; Rzepka, S.; Wolter, K. J. "Comprehensive material characterization of organic packaging materials," in *EurosimE 2009*; 2009; Delft.

10. Gromala P, Fisher S, Zoller T, Andreescu A, Duerr J, Rapp M, et al. "Internal stress strain measurement of large molded electronic control units," in *EuroSimE 2013*; 2013; Wroclaw.

11. Fu C, McDowell L, Ume IC. "A Finite Element Procedure of a Cyclic Thermoviscoplasticity Model for Solder and Copper Interconnects," *J. Electron. Packag.*1998; 120(1): p. 24-34.

12. Han B. "Thermal stresses in microelectronic subassemblies: Quantitative characterization using photomechanics methods," *Journal of Thermal Stresses*2003; 26: p. 583-613.

13. JESD22-B112A, Package Warpage Measurement of Surface-Mount Integrated Circuits at Elevated Temperature. 2005.

14. Macurova K, Kharicha A, Pletz M, Mataln M, Bermejo R, Schongrundner R, et al. "Multi-physics simulation of the component attachment within embedding process," in *EuroSimE 2013*; 2013; Wrocalw.

15. ANSYS® Academic Research. Release 14.5.

16. Haider M, Hubert P, Lassard L. "Cure shrinkage characterization and modeling of a polyester resin containing low profile additives," *Composites Part A: Applied Scinece and Manufacturing*2006; 38: p. 994-1009.

17. Schmöller G. Eingefrorene Spannungen:Konsequenzen für die Simulation am Beispiel von Klebverbindungen. München; 2006.

18. Böger T, Dilger K, Schmöller G. FE-Simulation der Klebstoffschwindung schwindung während des Aushärtevorgangs. München; 2001.

19. Hopcroft, M. A.; Nix, W. D.; Kenny, T. W. "What is the Young's Modulus of Silicon?," *Journal of Microelectromechanical Systems*2010; 19(2): p. 229-238.

20. http://www.ioffe.rssi.ru/SVA/NSM/Semicond/Si/mechanic.html#Elastic. [Online].

21. Masolin A, Bouchard PO, Martini R, Bernacki M. "Thermo-mechanical and fracture properties in single-crystal silicon," *J Mater Science*2013; 48: p. 979-988.

22. Ledbetter HM, Naimon ER. "Elastic Porperties of Metals and Alloys. II. Copper," *J.Phys.Chem.*1974; 3(4): p. 897-935.

23. http://www.itl.nist.gov/div898/handbook/pmd/section6/pmd641.htm. [Online].

24. Wong, C S; Benett, N; Allen, D; McNally, P.; Danilewsky, A. "A novel x-ray diffraction technique for analysis of die stress inside fully encapsulated packaged chips," in *Proc 4th Electronic System integration Technology Conference*; 2012; Amsterdam, Netherlands.

Acquisition Unit for In-Situ Stress Measurements in Smart Electronic Systems

Alicja Palczynska[a], Florian Pesth[a], Przemyslaw Jakub Gromala[a],
Tobias Melz[b], Dirk Mayer[b]
[a]Robert Bosch GmbH, Reliability Modeling and System Optimization (AE/EVR-MO)
Reutlingen, 72703, Germany
[b]Fraunhofer-Institut für Betriebsfestigkeit und Systemzuverlässigkeit LBF, Darmstadt, 64289, Germany

Abstract

Nowadays electronic systems used in automotive industry consolidate a variety of functionalities and features within one unit. Such smart systems consist of standard IC packaging, sensors and actuators. It is required that such systems will survive about 15 years of usage. To assure correct functionality prognostic and health monitoring (PHM) methods can be used. The potential benefits from implementation of these methods include among others reduction of maintenance costs and early warning of possible failure. For automotive industry, for which typical load history is not sufficiently known, this seems to be suitable approach. In today's reliability models some assumptions about the typical load history have to be made, which leads to inaccurate estimations on the remaining useful life. Though, in PHM approach one wants to make a prognosis of a failure so that maintenance can be done at the appropriate time. In this paper, the acquisition unit for in-situ measurements of internal stresses in a novel smart system module is developed. It allows recording the stresses from multiple piezoresistive sensors, at different locations on a surface of a chip, at the same time. As the measurements, that are planned, will be conducted when the smart system is assembled in the engine compartement, it is required that the acquisition unit is portable and immune to harsh conditions. The results of the measurements will be used to monitor the evolution of stress magnitudes in real work conditions.

1. Motivation

The objective of Prognostics and Health Monitoring (PHM) is to develop knowledge about the condition of a system to enable an assessment of its state-of-health which enables to schedule the maintenance action before the failure will occur. It is an interesting direction of research for automotive electronics, since the load history is different dependent on climate, frequency of usage and even driver. The potential benefits of application of PHM methods include warnings of system failures in advance, lower life-cycle cost (reduction of inspection costs, downtime etc.), increase of system availability (through extension of maintenance cycles) and knowledge of load history for future design, qualification and root cause failure analysis [1].

In this study the components of interest are large molded modules, which are used as ECU's in automotive applications. Their advantages are good hermeticity and resistance against harsh environment. However, the relatively large size of the module leads to high stresses at the integrated circuitry. In order to optimize the construction, it is needed to know the internal stress state within the module [2].

These kind of in-situ measurements were already made [2], however only in laboratory conditions. In order to do the field conditions measurements, a mobile acquisition unit has to be developed. Since the planned measurements would be conducted at several places on the Molded Control Unit (MCU), information must be gathered from several sensors at once. The scheme of a tested module is presented in Fig. 1.

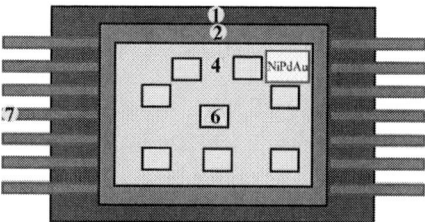

Fig. 1 **Scheme of the tested module, 1 - mold, 2 – ceramic substrate, 4 – LTCC, 6 – stress sensor, 7 - copper leads [2].**

Having this acquisition unit for field measurements, would also enable other experiments utilizing the same type of sensor. For instance, the simultaneous measurement with stress sensor and some other experimental method (like moiré interferometry) would be possible.

2. Design of Mobile Acquisition Unit

The sensor used in this study is a piezoresistive stress sensor, called IForce[3]. It has the advantage over other types of sensors (such as strain gages) that similarly built components are used in real products. In consequence, the obtained information can be interpreted as the stress state in actual devices. IForce has also advantages over classical, rosette-type stress sensors. Instead of typical Wheatstone bridge built on silicon resistors, which are relatively large, it is constructed of MOSFET transistors in a current mirror configuration. The fact that current mirrors are very sensitive on the changes of parameters of transistors was used. When stress acts on differently oriented channels of MOSFETs, it changes its resistivity differently. That's why IForce sensor has higher sensitivity and does not take as much place as other piezoresistive sensors [4]. Actually, in one package there is a whole matrix of sensors, what enables really high spatial resolution of measurement (Fig. 2).

978-1-4799-4789-8/14 $31.00 © 2014 IEEE

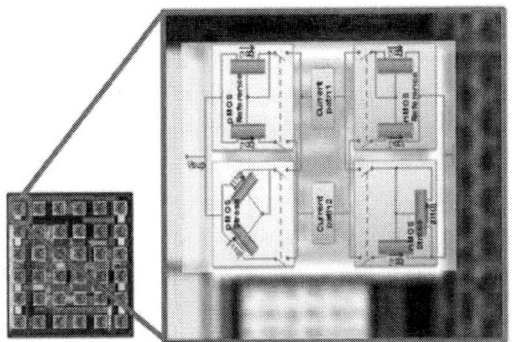

Fig. 2 Stress sensing cells at the top of the IForce [5]

Up to now, the data from mentioned sensors was gathered by a dedicated acquisition unit called "ChiRP Control Unit", developed by TZM. It consists of several functional units as shown in Fig. 3. Multifunction I/O (Input/Output) is a block which generates the signal to communicate with a sensor over power supply signal and which digitalizes the output of the sensor (input and output currents of current mirrors). Current Source and Waveform Amplification Unit provides a sufficient and stable amount of current to the chip, as well as amplifies the signal from the generator and from sensor. Programmable Control Logic (PLC) and PC have controlling functions – PLC is dedicated only to control the current source with waveform amplifier while PC controls the whole measurement process.

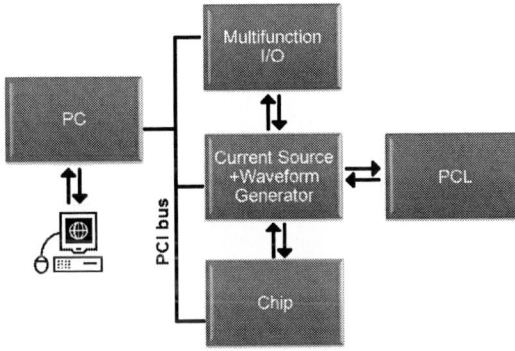

Fig. 3 Functional scheme of existing acquisition unit [6]

Additionally to the dedicated hardware, there is also software which allows for managing the process of gathering the data. It makes the procedure user-friendly, not only by enabling to set the whole measurement sequence (what measurements should be taken on which cells, in what order etc.) but also by processing and presenting the acquired data in a readable, easy to interpret way. An example of data gathered by this setup during thermal cycling in a laboratory conditions is presented in Fig. 4.

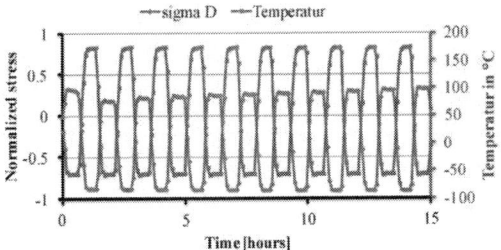

Fig. 4 Stress measurements during temperature cycling [2]

The major disadvantage of this system is lack of mobility. It is rather large (the size of a stationary PC station), requires some peripherals and a supply from common power line. Fig. 5 depicts the existing acquisition unit.
It can't be reasonably used outside the laboratory. It also allows measuring only one chip at a time and for the planned studies, it is required to handle between 6 and 9 sensors placed in different positions on a chip.

Fig. 5 A photograph of existing, commercial solution

The goal of current work is to minimize this system enough to be mobile, in order to conduct in-situ measurements in real working conditions, inside the vehicle. Control over whole process is taken by microcontroller discovery board STM32F4 and each other module of that system has to be build, in such a way, that assures the compatibility with it. It is also required that the whole physical system is supplied from the power supply existing in a car, that means from battery.

All inputs of the sensors are controlled from digital-to-analog converter (DAC). It includes the voltage generator which assures the power supply to the chip and communication with it, as well as driving of the current source. Outputs from the chip are digitalized by analog-to-digital converters (ADC). There are three signals that should be read from sensor – input and output currents of current mirror and the supply current, over which the data concerning the outgoing measurement are transmitted (like the address of measured cell, the phase of measurement etc.). Since all of these signals are of current

978-1-4799-4789-8/14 $31.00 © 2014 IEEE   171

modulation type, they have to be measured in a differential mode. The functional scheme of the developed solution is presented in Fig. 6.

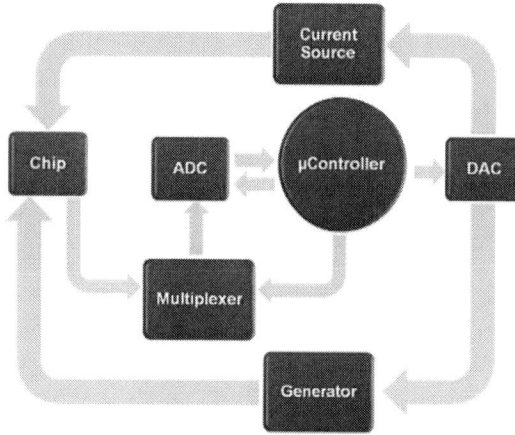

Fig. 6 The functional scheme of the developed acquisition unit

The prototype was built and tested. It is presented in Fig.7. The current source and voltage generator are both quite stable and accurate. The differences of actual and designed values of signals are just the result of scattered values of the resistors used, and do not exceed 2.5%. This is not crucial, because to calculate the stress these assumed values are not used. The only values that are needed for this are the currents in both branches of current mirror, so it is most important to measure it as accurately as possible. The validation of results obtained from the developed acquisition unit can be done using curves acquired with the established PC based system (like that in Fig. 4) as a reference.

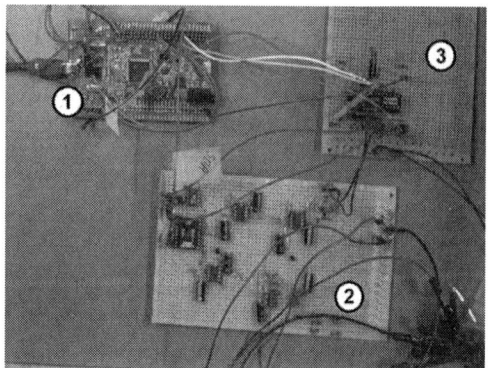

Fig. 7 The prototype of an acquisition unit, 1 – microcontroller, 2 – current source and generator, 3 – ADC.

The analog-to-digital converter has the 12-bit resolution. In existing prototype, the total error of conversion doesn't rise above ±2 Least Significant Bit (LSB). The conversion rate is about 50 samples/sec and it is not critical, because the planned measurements don't

concern rapidly changing phenomena. That's also the reason why the tests can be executed sequentially. Thus, in order to gather information from multiple sensors, a multiplexer can be used. However, collection of the data from multiple sensors during the long period of time requires the use of proper algorithms to store the data efficiently.

3. Algorithms

Managing data from a sensor network is related to numerous computational challenges especially in the context of storage and mining of information. Continuous flow of data over long periods of time requires the design of efficient techniques for processing which need to be executed in a step of the data collection, since it is not possible to store the entire dataset because of limited memory space. The existing algorithms for data mining are modified versions of clustering, regression, and anomaly detection techniques from other scientific fields [7]. There are several techniques to reduce data streams, such as sampling, Fourier Transform, histograms, wavelet transform, Ordered Overall Range (OOR) method or filters.

In [8] it is proposed to use a histogram approach to PHM. In this case the signal monitored in-situ is further processed to extract the relevant load parameters. The derivation of load parameters can be achieved using methods such as cycle counting algorithms (to find cycles in time-load signal) or Fast Fourier Transforms (to obtain information about the frequency content in a signal) [9]. These extracted parameters can then be stored in histograms, what leads to further data reduction. The preprocessed data are a good representation of usage history and can be used for remaining life prediction [8].

In [10] these reduction methods were actually implemented. The temperature and vibration acting on a Printed Circuit Board (PCB) placed under the hood of a car were monitored. For a data reduction method OOR was used, which converts an irregular history into a regular sequence of peaks and valleys. In this method, the smaller, least significant reversals can be eliminated by defining a screening level. The further reduction of data was obtained by using three-parameter rainflow algorithm. The classical rainflow method just counts the cycles and does not provide any additional information, like the mean load or cycle duration. That's why the modified method was used [10].

Another approach is presented in [11]. It is the wavelet decomposition for data mining in electronic assemblies. In this approach the signal is decomposed into a set of basis functions with two parameters – scale and translation. That method is used widely in the area of image processing to compress the data. Additionally to the reduction of the data, this method enables to calculate an energy signature for each sequence of wavelet packet. That value is a basis for the derivation of confidence values for health monitoring [11].

The algorithm based on filtering was applied for PHM of electronics in [12]. The method presented there is particle filtering which enables quantification of damage initiation, progression and Remaining Useful Life (RUL). This algorithm can be applied to noisy data sets. It is also recursive, which means that the update of results is made after each step and only one measurement is stored in a memory, which requires minimal data storage.

In a case of designed acquisition unit, definitely it is necessary to adopt some filtering method. The data will be collected over long period of time and the cycles of small amplitude can be omitted, because their contribution into accumulated stress is small. OOR seems to be suitable in this application, because it is quite straightforward to implement and recursive, so can be calculated in a real time. After filtering, reduction of data should be implemented. A reasonable approach for in-situ stress measurement would be rainflow algorithm, followed by histogram approach. This way, the cycles of defined amplitude and mean value would be counted and then their number stored in a bins of histogram. Obtained information would be a good representation of a load history and could be used for estimation of an accumulated damage and remaining useful life. The procedure of data mining is presented in Fig. 8.

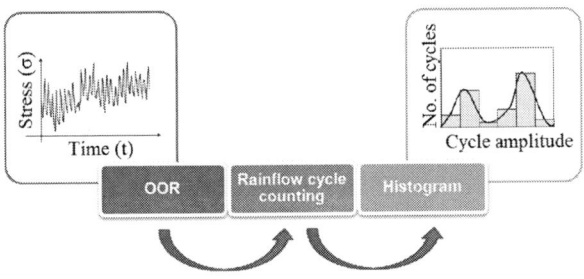

Fig. 8 Process of data reduction algorithm.

4. Summary

PHM methods draw more and more attention in all reliability studies. Now, it is possible to implement it not only in military or aircraft applications but also in automotive industry, as the cost of these systems are decreasing. The designed acquisition unit enables to implement these methods in stress state measurements of electronic devices dedicated to automotive applications. It allows gathering the data in-situ, in real work conditions of a vehicle. It would be also useful to lead other interesting studies that require a certain mobility of an experimental setup. Further development includes calibration of a measurement system, field tests and application of data reduction algorithms.

References

1. S.Rzepka et al., "EPoSS White Paper on Paper on Smart Systems Reliability", European Expert Workshop on Reliability of Smart Systems (EuWoRel), Berlin, Germany; September 16-17, 2013

2. Gromala, P.; Fischer, S.; Zoller, T.; Andreescu, A.; Duerr, J.; Rapp, M.; Wilde, J., "Internal stress state measurements of the large molded electronic control units," Thermal, Mechanical and Multi-Physics Simulation and Experiments in Microelectronics and Microsystems (EuroSimE), 2013 14th International Conference on , vol., no., pp.1,8, 14-17 April 2013

3. Kittel H. , Endler S., Osterwinter H. , Oesterle S., Schindler-Saetkow F, "Novel Stress Measurement System for Evaluation of Package Induced Stress", Smart Systems Integration 2008 - 2nd European Conference & Exhibition on Integration Issues of Miniaturized Systems - MOMS, MOEMS, ICS and Electronic Components, Barcelona April 2008

4. Robert Bosch GmbH, Abschlussbericht zum Verbundvorhaben iForceSens : Entwicklung eines integrierten Stressmesssystems zur Quantifizierung der 3D-Verformung von Sensorbauelementen in Abhängigkeit des Verpackungsprozesses, Technische Informationsbibliothek u. Universitätsbibliothek , Abstatt [u.a.] ; 2008

5. Zoller T., "Experimentelle Analyse von Eigenspannungen beim reaktiven Transfermolden von grossvolumigen Modulen", MSc Albert-Ludwigs-Universitat, Freiburg

6. IFORCESENS CHIRP CONTROL UNIT 30-004-A01 Instructions for Use, 2006 Steinbeis GmbH & Co. KG für Technologietransfer, TZ Mikroelektronik, Göppingen.

7. A. Appice, A. Ciampi, F. Fumarola, D. Malerba, "Data Mining Techniques in Sensor Networks", SpringerBriefs in Computer Science 2014

8. Vichare, N.M.; Pecht, M.G., "Prognostics and health management of electronics," Components and Packaging Technologies, IEEE Transactions on, vol.29, no.1, pp.222, 229, March 2006

9. N. Vichare, P. Rodgers, V. Eveloy, and M. G. Pecht, "Monitoring environment and usage of electronic products for health assessment and product design," in Proc. IEEE Workshop Accelerated Stress Testing Reliability (ASTR), Austin, TX, Oct. 2–5, 2005.

10. Ramakrishnan, A.; Pecht, M.G., "A life consumption monitoring methodology for electronic systems," Components and Packaging Technologies, IEEE Transactions on, vol.26, no.3, pp.625,634, Sept. 2003

11. Lall, P.; Choudhary, P.; Gupte, S.; Suhling, J.C., "Health Monitoring for Damage Initiation and Progression During Mechanical Shock in Electronic Assemblies," Components and Packaging Technologies, IEEE Transactions on , vol.31, no.1, pp.173,183, March 2008

12. Lall, P.; Lowe, R.; Goebel, K., "Particle filter models and phase sensitive detection for prognostication and health monitoring of leadfree electronics under shock and vibration," Electronic Components and Technology Conference (ECTC), 2011 IEEE 61st , vol., no., pp.1097,1109, May 31 2011-June 3 2011

Modelling of Non-stationary Processes in Optomechanical Thermal Microsensors

A. G. Kozlov
Omsk State University
Pr. Mira, 55a, Omsk, 644077, Russia
E-mail: kozlov1407@gmail.com; agk252@mail.ru

Abstract

An analytical method is developed to determine the frequency response of optomechanical thermal microsensors. The three types of the microsensors are considered: microsensor with the two supporting beams in adjacent corners of the plate; microsensor with the two supporting beams in opposite corners of the plate; microsensor with the four supporting beams. Taking into account the features of each type of the microsensors, in they structures, the domains of modelling are marked out. The domains are divided into the regions with homogeneous parameters. For each region the non-steady-state heat conduction equation is obtained that is solved by means of the time Fourier transform. The heat flux densities between the regions are determined using adjoint boundary conditions in the frequency domain. The analytical expression for the frequency responses of the microsensors is obtained. This method is applied to find the frequency responses, cutoff frequencies and time constants for the three types of optomechanical thermal microsensors. The dependencies of the sensitivity and the time constant on the length of the bi-material section of the beams for the microsensor with the two supporting beams in adjacent corners are obtained.

1. Introduction

Optomechanical thermal microsensors are a new type of thermal microsensors. The creation and investigation of these microsensors began twenty years ago and continue intensively in present time. Their work based on the following principles:

- use of bi-material effect in a supporting bridge cantilevers for displacement of a microplate;
- use of an optical readout;
- use of MEMS structures for increasing a temperature sensivity.

These principles allow one to exclude an electrical part from conversion and, respectively, to reduce electromagnetic effects on a microsensor.

Optomechanical thermal microsensors (bimaterial cantilevers) were first used as a chemical sensors to measure the heat generated in chemical reactions [1]. They were applied to measure IR radiation as IR detectors [2–4] or as scanning thermal imaging probes [5]. Up to now the question about the dynamic characteristics of optomechanical thermal microsensors remains open. Presently the investigations in this area are just starting [6, 7]. The dynamic characteristics depend on three subsystems of the microsensor: optical, thermal and mechanical. In most practical cases, it is enough to know the frequency response of a microsensor to determine its

dynamic characteristics. For a optomechanical thermal microsensor the frequency response can be presented as follows

$$S_{\omega} = S_{\omega}^{(\mathrm{op})} S_{\omega}^{(\mathrm{th})} S_{\omega}^{(\mathrm{mec})}, \qquad (1)$$

where S_{ω} is the frequency response of the microsensor; $S_{\omega}^{(\mathrm{op})}$, $S_{\omega}^{(\mathrm{th})}$, $S_{\omega}^{(\mathrm{mec})}$ are the frequency responses of the optical, thermal and mechanical subsystems of the microsensor, respectively. In the microsensor, thermal processes have the considerable lag in comparison with the optical and mechanical ones therefore the frequency response of the microsensors is determined by its thermal subsystem. To find the frequency response of the thermal subsystem it is necessary to consider the non-stationary thermal processes in the optomechanical thermal microsensor.

2. Method of modelling

The optomechanical thermal microsensors concern to thermal microsensors and are made with using surface micromachining. The basic elements of these microsensors are [8]:

1) the plate with absorbing layer, where the conversion of an optical radiation power into a heat power is occured;

2) the beams which support the plate over a substrate;

3) the bi-material sections which are the part of the beams and induce the deflection of the plate as result of the bi-material effect.

The optomechanical thermal microsensors under consideration are presented on Fig. 1.

To find the temperature distribution in the structure of the optomechanical thermal microsensor and then to determine its frequency response we will use the analytical method specifically proposed for these goals and taking into account the features of thermal microsensors [9]. This method was adapted to consider the non-stationary thermal processes in thermal microsensors [10]. The algorithm of the method for non-stationary case has the following sequence of the steps.

1. The thermally isolated structure (the plate and the sapporting beams) of a microsensor is divided into rectangular regions depending on the composition of layers and heat-generating conditions. Each region is replaced by an equivalent region with the homogeneous parameters: equivalent thickness, thermal conductivity, specific volume heat and thermal diffusivity. The values of the thickness, $d_{\mathrm{e}}^{(j)}$, and thermal conductivity, $\lambda_{\mathrm{e}}^{(j)}$, for the equivalent region j are determined as follows

---

978-1-4799-4789-8/14 $31.00 © 2014 IEEE   174

$$\lambda_e^{(j)} = \left(\sum_{i=1}^{n} k_i^{(j)} d_i^{(j)} \lambda_i^{(j)} \right) \Big/ d_e^{(j)} , \qquad (3)$$

where $d_i^{(j)}$ is the thickness of the layer i in the region j, n is the number of the layers in the region j, $k_i^{(j)}$ is the coefficient equal to the ratio of the total area of the layer i in the region j to the area of this region, $\lambda_i^{(j)}$ is the thermal conductivity of the layer i in the region j. The specific volume heat, $(\rho c)_e^{(j)}$ is determined as follows

$$(\rho c)_e^{(j)} = \left(\sum_{i=1}^{n} k_i^{(j)} d_i^{(j)} \rho_i^{(j)} c_i^{(j)} \right) \Big/ d_e^{(j)} , \qquad (4)$$

where $\rho_i^{(j)}$, $c_i^{(j)}$ are the density and the specific heat of the layer i in the region j, respectively.

Concerning the optomechanical thermal microsensors under consideration the division of their structure into regions has the following feature. To reduce the amount of computation in the structure of each microsensor the domain of modelling is marked out (Fig. 1). The size of this domain is determined by the symmetry of the microsensor structure (mirror symmetry or axial symmetry). Then, the domain of modelling is divided into regions (Fig. 2).

2. For each region, the heat exchange conditions with the environment by means of the heat transfer through gas medium and the radiant heat transfer, and with adjacent regions and the substrate by means of the thermal conduction and the initial condition are determined.

The basic ways of the heat exchange for each region are the heat transfer through the ambient air and the radiant transfer from the lower and upper surfaces and the thermal conduction transfer through the edges with the adjacent regions and the substrate. Consider these ways of the heat exchange in the regions in detail.

The total power of the surface heat losses per unit area of the region j ($N_t^{(j)}$) can be presented in the following form provided that the difference between the temperature of the region j (T_j) and the environment temperature (T_{en}) is small ($T_j - T_{en} \ll T_{en}$)

$$N_t^{(j)} = A_j \left(T_j - T_{en} \right), \qquad (5)$$

where

$$A_j = h_c + \frac{\lambda_a}{d_1^{(j)}} + 4\sigma \left(\varepsilon_l^{(j)} + \varepsilon_u^{(j)} \right) T_{en}^3 ; \qquad (6)$$

h_c is the convective coefficient for the given structure of the thermal microsensor; λ_a is the thermal conductivity of the ambient air, $d_1^{(j)}$ is the distance between the lower surface of the region j and the surface of the substrate; σ is the Stefan-Boltzmann constant, $\varepsilon_l^{(j)}$ and $\varepsilon_u^{(j)}$ are the emissivities of the lower and upper surfaces of the region j, respectively. The first term in Eq. (6) defines the heat dissipation from the top surface of the region by convection. The second term sets the heat dissipation

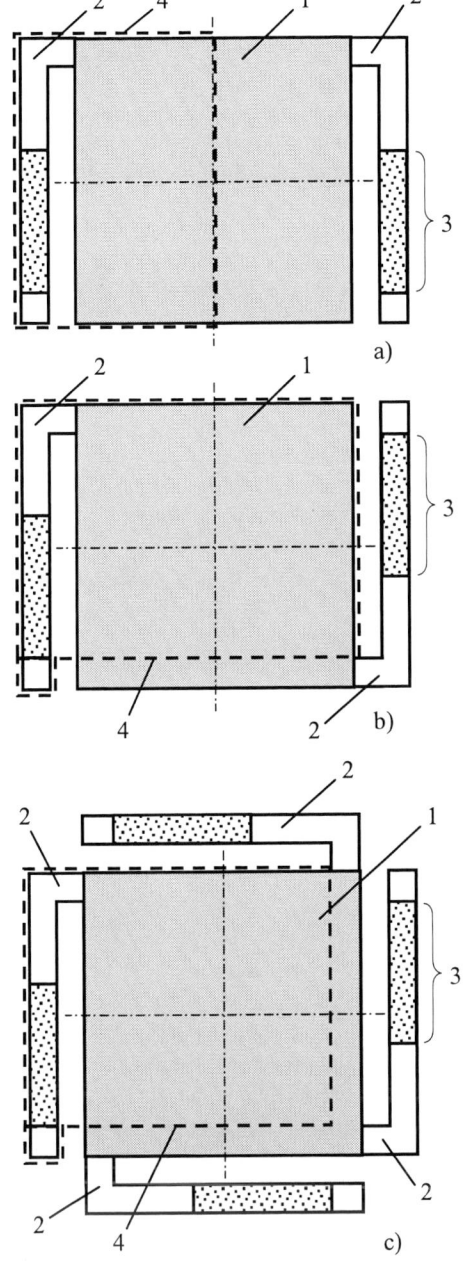

Figure 1: Optomechanical thermal microsensors based on the rectangular plate supported by: (a) two beams in adjacent corners; (b) two beams in opposite corners; (c) four beams; 1 – plate with absorbing layer; 2 – supporting beam; 3 – bi-material section of the supporting beam; 4 – domain of modelling

$$d_e^{(j)} = \sum_{i=1}^{n} k_i^{(j)} d_i^{(j)} ; \qquad (2)$$

from the lower surface of the region j by heat conduction through the air gap between this surface and the surface of the substrate. The third term defines the heat dissipation from the upper and lower surfaces of the region j by radiation.

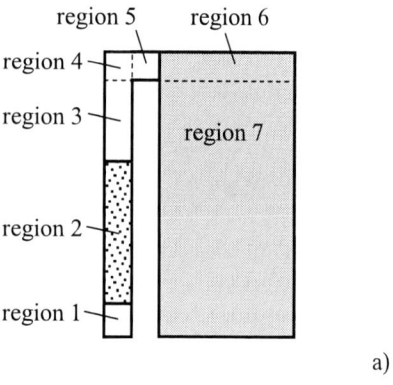

a)

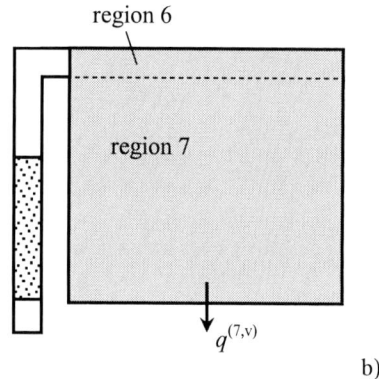

b)

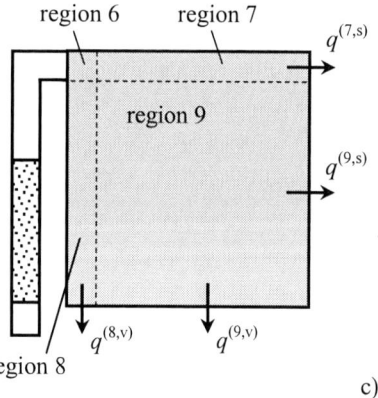

c)

Figure 2: Division of the domains of modelling into the regions for the microsensors based on the rectangular plate supported by: (a) two beams in adjacent corners; (b) two beams in opposite corners; (c) four beams; division of the beam into the region in the variants (b) and (c) is analogous to the variant (a)

Heat exchange conditions for each region with the adjacent ones and the support substrate are determined by the boundary conditions. This question was considered in detail in [9] where, depending on the combination of the boundary conditions, the types of regions were marked out.

The initial condition for each region determines the temperature distribution in the initial time. In the most practical cases, the initial condition for the regions can be presented as follows

$$T_j\big|_{t=0} = T_{en} .\qquad(7)$$

3. For each region, the non-stationary heat differential equation is defined and then this equation, the boundary conditions, and the initial condition are modified with using the time Fourier transform to exclude the time variable. The transformed equation is solved by the eigenfunction method. Its solution is presented with the help of eigenfunctions in the frequency domain. The heat flux densities between the regions included in the solution are presented as the sums of orthogonal functions with unknown weighting coefficients.

The common form of the non-stationary heat differential equation for the regions can be presented as follows

$$\frac{\partial T'_j}{\partial t} = a_e^{(j)} \frac{\partial^2 T'_j}{\partial x_j^2} + a_e^{(j)} \frac{\partial^2 T'_j}{\partial y_j^2} - p_j^2 T'_j - \varphi_j ,\qquad(8)$$

where

$$T'_j = T_j - T_{en} ,\qquad(9)$$

$$a_e^{(j)} = \lambda_e^{(j)} / (c\rho)_e^{(j)} ,\qquad(10)$$

$$p_j^2 = \left(A_j / d_e^{(j)} (c\rho)_e^{(j)} \right),\qquad(11)$$

$$\varphi_j = -q_j / d_e^{(j)} (c\rho)_e^{(j)} ,\qquad(12)$$

q_j is the power of heat generation per unit area in the region j.

Since the regions have the different boundary conditions, the solutions of the equation (8) for the various regions are different. In the structure of the optomechanical thermal microsensors the most of the regions does not have the boundaries with the substrate because consider a solution of this equation by the example of the region in which on all boundaries the Neumann boundary conditions are fulfilled. For the boundaries with other regions, the thickness ratio of these regions must be taken into account when formulating the Neumann boundary conditions. Therefore, in the right part of the Neumann boundary condition for such boundaries, it is necessary to use the product of the coefficient κ and the heat flux density q between the regions. The coefficient κ is equal to ratio between the thicknesses of the adjacent regions (for example, $\kappa^{(j,s)} = d_e^{(j)} / d_e^{(s)}$). The initial condition for the region occupied by the absorbing layer is the condition (7).

Using the time Fourier transform the non-stationary equation (8) with the initial condition (7) is transformed to the following forms

$$\frac{\partial^2 \widetilde{T}_j'}{\partial x_j^2} + \frac{\partial^2 \widetilde{T}_j'}{\partial y_j^2} - \left(\frac{p_j^2 + j\omega}{a_e^{(j)}}\right)T_j' = \frac{\widetilde{\varphi}_j}{a_e^{(j)}}, \qquad (13)$$

where

$$\widetilde{T}_j' = \int_{-\infty}^{\infty} T_j' \exp(-j\omega t)dt; \qquad (14)$$

$$\widetilde{\varphi}_j = \int_{-\infty}^{\infty} \varphi_j \exp(-j\omega t)dt. \qquad (15)$$

By analogy with the steady-state case [9], from the solution of the equation (13) with the boundary conditions one can obtain the expression for the temperature in the region j in the frequency domain. The expression for the temperature distribution in the regions with Neumann boundary conditions can be presented as follows

$$\widetilde{T}_j' = \widetilde{\varphi}_j\left(p_j^2 + j\omega\right)^{-1} + a_e^{(j)}\widetilde{D}_{0,0}^{(j)}\left[l_j b_j \lambda_e^{(j)}\left(p_j^2 + j\omega\right)\right]^{-1} +$$

$$\left(2/l_j b_j \lambda_e^{(j)}\right)\sum_{k=1}^{\infty}\widetilde{D}_{k,0}^{(j)}\left\{(k\pi/l_j)^2 + \left(p_j^2 + j\omega\right)/a_e^{(j)}\right\}^{-1}$$

$$\cos(k\pi x_j/l_j) + \left(2/l_j b_j \lambda_e^{(j)}\right)\sum_{m=1}^{\infty}\widetilde{D}_{0,m}^{(j)}\left\{(m\pi/b_j)^2 + \right.$$

$$\left[\left(p_j^2 + j\omega\right)/a_e^{(j)}\right]^{-1}\cos(m\pi y_j/b_j) +$$

$$\left(4/l_j b_j \lambda_e^{(j)}\right)\sum_{k=1}^{\infty}\sum_{m=1}^{\infty}\widetilde{D}_{k,m}^{(j)}\left\{(k\pi/l_j)^2 + (m\pi/b_j)^2 + \right.$$

$$\left[\left(p_j^2 + j\omega\right)/a_e^{(j)}\right]^{-1}\cos(k\pi x_j/l_j)\cos(m\pi y_j/b_j),$$

$$(16)$$

where

$$\widetilde{D}_{k,m}^{(j)} = -(-1)^k \kappa^{(j,s)}\widetilde{\delta}_m^{(j,s)} - (-1)^m \kappa^{(j,t)}\widetilde{\delta}_k^{(j,t)} + \kappa^{(j,u)}\widetilde{\delta}_m^{(j,u)} + \kappa^{(j,v)}\widetilde{\delta}_k^{(j,v)}. \qquad (17)$$

In the equation (16), the unknown quantities are the weighting coefficients, $\widetilde{\delta}_m^{(j,s)}$, $\widetilde{\delta}_k^{(j,t)}$, $\widetilde{\delta}_m^{(j,u)}$, and $\widetilde{\delta}_k^{(j,v)}$. The temperature distribution in the regions with other boundary conditions can be found in the analogous way.

4. The unknown weighting coefficients are determined using the adjoint boundary conditions in the frequency domain.

The determination of the unknown weighting coefficients is performed by the method similar to the method for the steady-state case [9]. The distinction of the non-stationary case lies in the dependence of the unknown weighting coefficients on the frequency. To find the unknown weighting coefficients it is necessary to consider the temperature equalities on all boundaries between the regions. From this consideration we obtain the generalized system of linear equations which is convenient to write in the matrix representation

$$\widetilde{M}\widetilde{\Lambda} = \widetilde{\Phi}, \qquad (18)$$

where $\widetilde{M}$ is the matrix of the coefficients; $\widetilde{\Lambda}$ is the vector of the unknown weighting coefficients; $\widetilde{\Phi}$ is the vector of the right parts.

For the variants of the optomechanical thermal microsensors with two supported beams (Figs. 1a and 1b)

the matrix $\widetilde{M}$ and the vectors $\widetilde{\Lambda}$ and $\widetilde{\Phi}$ can be written in block matrix and vector representation as follows

$$\widetilde{M} = \begin{bmatrix} \widetilde{A}_k^{(1,2)} & \widetilde{A}_k^{(2,3)} & 0 & 0 & 0 & 0 \\ \widetilde{B}_k^{(1,2)} & \widetilde{B}_k^{(2,3)} & \widetilde{B}_k^{(3,4)} & 0 & 0 & 0 \\ 0 & \widetilde{C}_k^{(2,3)} & \widetilde{C}_k^{(3,4)} & \widetilde{C}_m^{(4,5)} & 0 & 0 \\ 0 & 0 & \widetilde{D}_k^{(3,4)} & \widetilde{D}_m^{(4,5)} & \widetilde{D}_m^{(5,6)} & 0 \\ 0 & 0 & 0 & \widetilde{E}_m^{(4,5)} & \widetilde{E}_m^{(5,6)} & \widetilde{E}_k^{(6,7)} \\ 0 & 0 & 0 & 0 & \widetilde{F}_m^{(5,6)} & \widetilde{F}_k^{(6,7)} \end{bmatrix}; \qquad (19)$$

$$\widetilde{\Lambda} = \begin{bmatrix} \widetilde{\Lambda}_k^{(1,2)} \\ \widetilde{\Lambda}_k^{(2,3)} \\ \widetilde{\Lambda}_k^{(3,4)} \\ \widetilde{\Lambda}_m^{(4,5)} \\ \widetilde{\Lambda}_m^{(5,6)} \\ \widetilde{\Lambda}_k^{(6,7)} \end{bmatrix}; \qquad (20) \qquad \widetilde{\Phi} = \begin{bmatrix} 0 \\ 0 \\ 0 \\ 0 \\ \widetilde{\Phi}_m^{(5,6)} \\ \widetilde{\Phi}_k^{(6,7)} \end{bmatrix}. \qquad (21)$$

The designation of the submatrices and subvectors in Eqs. (19)-(21) is given in Ref. [9].

The difference between the two above mentioned variants is due to only the values of the coefficients in the submatrices $\widetilde{E}_k^{(6,7)}$ and $\widetilde{F}_k^{(6,7)}$. For these variants the domains of modelling are similar and the number of regions is identical.

For the variant of the optomechanical thermal microsensor with the four supported beams the matrix $\widetilde{M}$ and the vectors $\widetilde{\Lambda}$ and $\widetilde{\Phi}$ have the following dimensions: $\widetilde{M} = (13 \times 13)$; $\widetilde{\Lambda} = (13 \times 1)$; $\widetilde{\Phi} = (13 \times 1)$.

To determine the values of the unknown weighting coefficients in the frequency domain the solution of the system (17) must be conducted for every value of the frequency.

5. The inverse time Fourier transform is performed. In most practical cases, this transformation is required only for the temperature distribution in the regions with the absorbing layer and the bi-material section. To determine the frequency response such transformation is not necessary.

The frequency response of optomechanical thermal microsensors is defined as

$$S_\omega = \frac{\widetilde{z}}{\widetilde{P}}, \qquad (22)$$

where $\widetilde{z}$ is the vertical deflection of the plate in the frequency domain; $\widetilde{P}$ is the input power of a optical signal in the frequency domain. In the present model, the vertical deflection of the plate in the frequency domain can be found using the temperature distribution in the region 2 which contains the the bi-material section of the beam. Knowing this temperature distribution in the frequency domain one can find the bending of the beams and, respectively, the deflection of the plate. For this purpose the following differential equation for the bending of the beam is used [2]

978-1-4799-4789-8/14 $31.00 © 2014 IEEE   177

$$\frac{d^2\widetilde{z}}{d\widetilde{y}^2} = 6(\alpha_2 - \alpha_1)\frac{d_1^{(2)} + d_2^{(2)}}{\left[d_2^{(2)}\right]^2 K}\left[\widetilde{T}_2(y_2) - \widetilde{T}_2(y_2)\Big|_{y_2=0}\right],$$

(23)

where α_1 and α_2 are thermal expansion coefficients of the layers 1 and 2 in the region 2, respectively;

$$K = 4 + 6\left(\frac{d_1^{(2)}}{d_2^{(2)}}\right) + 4\left(\frac{d_1^{(2)}}{d_2^{(2)}}\right)^2 + \frac{E_1}{E_2}\left(\frac{d_1^{(2)}}{d_2^{(2)}}\right)^3 + \frac{E_2}{E_1}\left(\frac{d_2^{(2)}}{d_1^{(2)}}\right);$$

(24)

E_1 and E_2 are the Young's modules of the materials of the layers 1 and 2 in the region 2, respectively.

The decision of Eq. (23) allows one to find the deflection of the plate in the frequency domain.

3. Numerical results

The present model was used to calculate the frequency response for the three types of the optomechanical thermal microsensors (Fig. 1). For modelling, the following values of the parameters were used:

- dimensions of the plate: length - 50 μm; width - 50 μm; thickness - 0.5 μm;

- dimensions of the supporting beams: length along the plate - 50 μm; length of the section attached to the plate - 4 μm; width - 4 μm; thickness - 0.5 μm;

- material of the plate and the beams - silicon dioxide with the following properties: density of 2200.0 kg/m^3; specific heat capacity of 740.0 J/(kg·K); thermal conductivity of 1.2 W/(m·K); emissivity of 0.5; thermal expansion coefficient of $6.1 \cdot 10^{-6}$ K^{-1}; Young's modulus of 0.65 GPa;

- material of the absorbing layer - platinum with the following properties: density of 21450.0 kg/m^3; specific heat capacity of 140.0 J/(kg·K); thermal conductivity of 72 W/(m·K); emissivity of 0.9;

- dimensions of the bi-material section of the supporting beams: length - 34 μm; width - 4 μm;

- thickness of the layer 2 of the the bi-material section - 0.4 μm;

- material of the layer 2 of the the bi-material section – aluminium with the following properties: density of 2700.0 kg/m^3; specific heat capacity of 904.0 J/(kg·K); thermal conductivity of 236 W/(m·K); emissivity of 0.55; thermal expansion coefficient of $23.3 \cdot 10^{-6}$ K^{-1}; Young's modulus of 70.0 GPa.

The absorbing layer was assumed to occupy the all area of the plate on which it is disposed. The microsensors were believed to place in a package whose dimensions are larger than those of the microsensors. The distance between the lower surface of the plate and the substrate was assumed to be equal to 20 μm. The values of other parameters were chosen to be: the thermal conductivity of an ambient air – 0.026 W/(m K); the convective coefficient - 19.8 W/(m2 K); the environment temperature – 300 K; the power density of input radiation – 2000 W/m^2.

In considering the operation of the optomechanical thermal microsensors the temperature distribution in the bi-material section of the beams is important. This section

occupies the whole area of the region 2. The expression for the temperature distribution in the region 2 can be written as follows

$$\widetilde{T}_2' = a_e^{(2)}\widetilde{D}_{0,0}^{(2)}\left[l_2 b_2 \lambda_e^{(2)}\left(p_2^2 + j\omega\right)\right]^{-1} +$$

$$\left(2/l_2 b_2 \lambda_e^{(2)}\right)\sum_{k=1}^{\infty}\widetilde{D}_{k,0}^{(2)}\left\{(k\pi/l_2)^2 + \left(p_2^2 + j\omega\right)/a_e^{(2)}\right\}^{-1}$$

$$\cos(k\pi x_2/l_2) + \left(2/l_2 b_2 \lambda_e^{(2)}\right)\sum_{m=1}^{\infty}\widetilde{D}_{0,m}^{(2)}\left\{(m\pi/b_2)^2 + \right.$$

$$\left[\left(p_2^2 + j\omega\right)/a_e^{(2)}\right]\right\}^{-1}\cos(m\pi y_2/b_2) +$$

$$\left(4/l_2 b_2 \lambda_e^{(2)}\right)\sum_{k=1}^{\infty}\sum_{m=1}^{\infty}\widetilde{D}_{k,m}^{(2)}\left\{(k\pi/l_2)^2 + (m\pi/b_2)^2 + \right.$$

$$\left[\left(p_2^2 + j\omega\right)/a_e^{(2)}\right]\right\}^{-1}\cos(k\pi x_2/l_2)\cos(m\pi y_2/b_2),$$

(24)

where

$$\widetilde{D}_{k,m}^{(2)} = \kappa^{(1,2)}\widetilde{\delta}_k^{(1,2)} - (-1)^m \kappa^{(2,3)}\widetilde{\delta}_k^{(2,3)}.$$

(25)

Fig. 3 presents the temperature distribution in the region 2. As can be seen from Fig. 3 the temperature distribution varies along the y-direction and is constant along the x-direction. This fact indicates that for calculating the deflection of the plate we can use the Eq. (3).

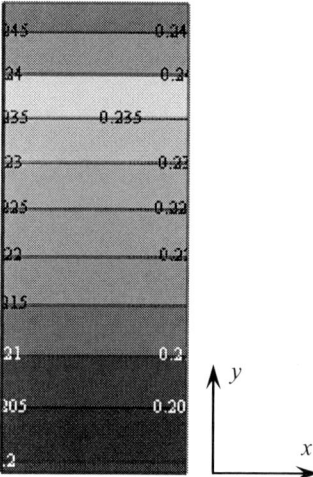

Figure 3: Temperature distribution in the region 2 (region with the bi-material section of the beam)

The frequency dependencies of the modules and arguments of the frequency response for the three types of the optomechanical thermal microsensors are shown in Fig. 4. As can be seen from the frequency dependencies of the modules (Fig. 4) among the microsensors the design variant with the plate supported by the two beams in adjacent corners has the most sensitivity. The variant with the plate supported by the four beams has the lowest sensitivity. The frequency dependencies of the argument for the three variants of the optomechanical thermal microsensors are almost identical. The differences between the frequency dependences of the arguments are presented in Fig. 5. Here we use the following notation: 1 - difference between the frequency dependences of the arguments of the microsensors with the two supporting beams in opposite corners and in adjacent corners; 2 - difference between the frequency dependences of the arguments of the microsensors with the four supporting beams and the two supporting beams in adjacent corners;

$|S_\omega|$, μm/W

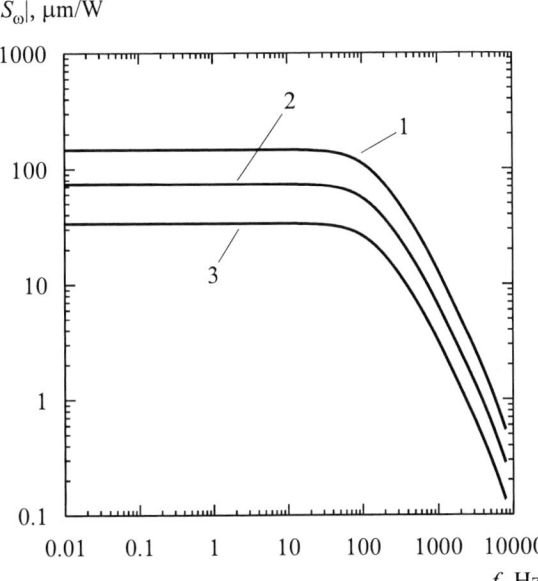

Arg(S_ω), degree

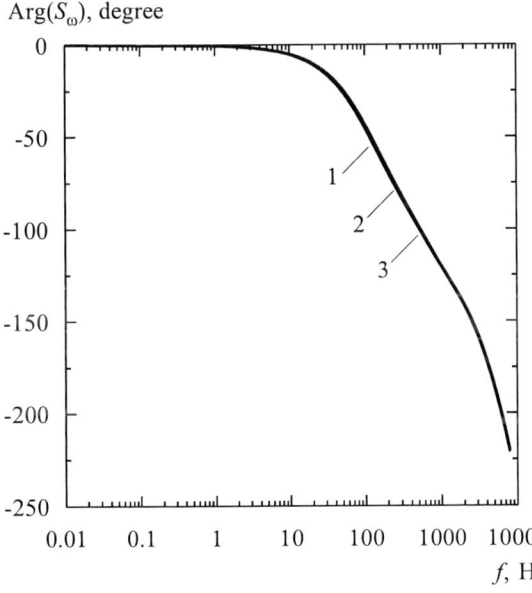

Figure 4: Modules and arguments of the frequency responses of the optomechanical thermal microsensors with: (1) two supporting beams in adjacent corners; (2) two supporting beams in opposite corners; (3) four supporting beams

ΔArg(S_ω), degree

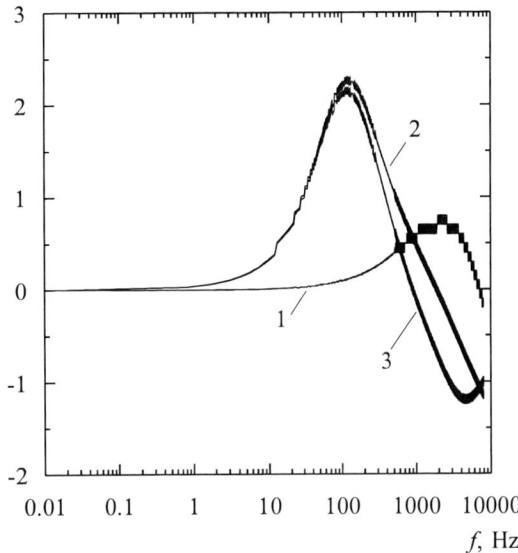

Figure 5: Differences between the arguments of the frequency responses of the optomechanical thermal microsensors

3 - difference between the frequency dependences of the arguments of the microsensors with the four supporting beams and the two supporting beams in oposite corners

Approximating the dynamics of the optomechanical thermal microsensors under investigation by the frequency response of first order inertia element allows one to estimate the relevant time constants for the microsensors. They are found using the frequency dependences of the argument. The time constant, τ, is equal to

$$\tau = \frac{1}{2\pi f_c}, \qquad (26)$$

where f_c is the cutoff frequency of the optomechanical thermal microsensor. The value of the cutoff frequency corresponds to the frequency on which the argument of the frequency response is equal to –45°. Based on the data presented on the Fig. 4 the following values of the cutoff frequency and the time constant for microsensors are obtained:

- the microsensor based on the rectangular plate supported by two beams in adjacent corners: f_c =99.4 Hz; τ =1.6 ms;

- the microsensor based on the rectangular plate supported by two beams in opposite corners: f_c =100.3 Hz; τ =1.59 ms;

- the microsensor based on the rectangular plate supported by four beams: f_c =106.6 Hz; τ =1.49 ms.

As can be seen from the present data the values of the cutoff frequency and the time constant for the various variants of the microsensors are close to each other, respectively. However, among them the microsensor based on the rectangular plate supported by four beams has the slightly better dynamic characteristics.

The present method was used to determine the dependences of the sensitivity and the time constant on the length of the bi-material section of the beams. These dependences for the microsensor with the plate supported by the two beams in adjacent corners are shown in Fig. 6. As can be seen from this figure the sensitivity increases and the time constant decreases with increasing the length of the bi-material section. This feature can be explained by the fact that the bi-material effect has a greater impact on the bending of the beam than the decrease of the temperature of the beam.

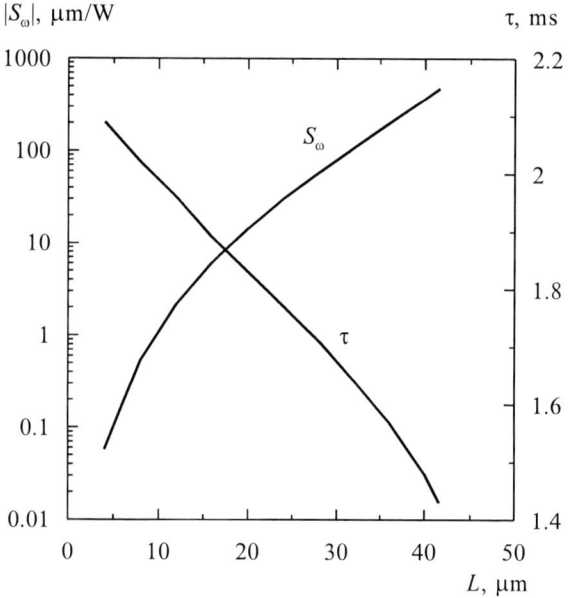

Figure 6: Modulus of the frequency response at ω=0 and time constant of the optomechanical thermal microsensor with the two supporting beams in adjacent corners as functions of the length of bi-material section of the beams

4. Conclusions

The present model allows one to exactly determine the dynamic characteristics of optomechanical thermal microsensors: frequency response, time constant, cutoff frequency. The method can also serve as the basis for creating analytical models allowing one to determine the response of the microsensors to the various input actions in the time domain and the noise equivalent power of the microsensors. The method can be used in the CAD systems of optomechanical thermal microsensors.

References

1. Gimzewski J. K., Gerber Ch., Meyer E., Schlittler R. R. "Observation of a Chemical Reaction Using a Micromechanical Sensor," *Chemical Physics Letters*, Vol. 217 (1994), pp. 589-594.
2. Barnes J. R., Stephenson R. J., Woodburn C. N., O'Shea S. J., Welland M. E., Rayment T., Gimzewski J. K., Gerber Ch. "A Femtojoule Calorimeter Using Micromechanical Sensors," *Rev. Sci. Instrum.*, Vol. 65 (1994), pp. 3793-3798.
3. Barnes J. R., Stephenson R. J., Welland M. E., Gerber Ch., Gimzewski J. K. "Photothermal Spectroscopy with Femtojoule Sensitivity Using a Micromechanical Device," *Nature*, Vol. 372 (1994), pp. 79-81.
4. Varesi J., Lai J., Perazzo T., Shi Z., Majumdar A. "Photothermal Measurements at Picowatt Resolution Using Uncooled Micro-optomechanical Sensors," *Appl. Phys. Lett.*, Vol. 71 (1997), pp. 306-308.
5. Majumdar A., Lai J., Chandrachood M., Nakabeppu O., Wu Y., Shi Z. "Thermal Imaging by Atomic Force Microscopy Using Thermocouple Cantilever Probes," *Rev. Sci. Instrum.*, Vol. 66 (1995), pp. 3584-3592.
6. Kwon B., Rosenberger M., Bhargava R., Cahill D. G., King W. P. "Dynamic Thermomechanical Response of Bimaterial Microcantilevers to Periodic Heating by Infrared Radiation," *Rev. Sci. Instrum.*, Vol. 83 (2012), 015003 (7p.).
7. Bijster R., de Vreugd J., Sadeghian H. "Dynamic Characterization of Bi-material Cantilevers," *Proc. 4th Int. Conf. on Sensor Device Technologies and Applications, SENSORDEVICES 2013*, Barcelona, Spain, August. 2013, pp. 1-8.
8. Miao Z., Zhang Q., Chen D., Guo Z., Dong F., Xiong Z., Wu X., Li C., Jiao B. "Uncooled IR imaging using optomechanical detectors," *Ultramicroscopy*, Vol. 107 (2007), pp. 610–616.
9. Kozlov A. G. "Analytical Modelling of Steady-state Temperature Distribution in Thermal Microsensors Using Fourier Method. Part 1. Theory," *Sensors Actuators A. Physical*, Vol. 101 (2002), pp. 283-298.
10. Kozlov A.G., Randjelović D., Djurić Z. "Analytical Modelling of Transient Processes in Thermal Microsensors," *Proc 12th. Int. Conf. on Thermal, Mechanical and Multiphysics Simulation and Experiments in Microelectronics and Microsystems, EuroSimE 2011*, Linz, Austria, April. 2011, pp. 1-7.

Multiphysics Modeling for Current Carrying Capability of a Power Package

Qiuxiao Qian, Yumin Liu and Yong Liu
Fairchild Semiconductor
South Portland, Maine 04106, USA

Abstract

In this paper, the impact of the lead frame design on the current carrying capability of the a power package is investigated. The coupled electrical-thermal and mechanical stress simulations are conducted, with the transient characteristics captured. The DoE simulations with regard to different lead frame design, different currents, and micro crack impact are studied to find the impact on current carrying capability. The simulation results show that the twisted Z shape lead design induces the highest stress level when the initial crack was induced by the assembly process, which could speed up the failure and reduce the current carrying capability.

1. Introduction

The design of electronic packages may change more or less to adapt for different environments and new applications. For a typical power electrical package, the design change includes the leadframe design, different silicon chips, different bond wire strategy, different material types, and etc. In this paper, the lead frame design change of a standard power package TO220 is investigated. In the high current application, thicker Al wires are required to be bonded from the source pad of silicon chip to the lead post. While in the original design, the lead post for source bond wires is relatively small. Therefore, there might be potential risks during the wire bonding process for the thicker Al wires. The major risk is the bond lift induced by the limited space for wire wedge bond cutting and the lead post clamping.

In order to eliminate the risk, the lead post for the source bond wires is designed with a wider geometry in the new layout. Accordingly, the middle lead area connecting the drain pad is shifted, and a twisted Z shape area is formed in order to produce a standard TO220 package. The twisted Z shape area of the middle lead may have potential risk after the assembly process. The stamping process during the leadframe manufacturing may have some impact on the twisted Z shape area of the middle lead, which may cause material property change, uneven thickness, and tiny micro-cracks in the local area. In this paper, the impact of the twisted Z shape area of the middle lead on the current carrying capability of the TO220 package is investigated. The coupled electrical-thermal and mechanical stress simulations are conducted, and the transient characteristics are also captured. The DoE simulations with regard to different lead frame designs, different currents, and micro cracks are completed to study the impact on current carrying capability. The simulation results show that the twisted Z shape lead induces the highest stress level when the initial crack was induced by the assembly process which

could speed up the failure and reduce the current carrying capability [1, 2, 3].

2. Lead Post Design Change of the TO220

Figure 1 shows the internal structural of the TO220 package with the original lead frame design. The source of power IGBT die is bonded to the right lead post through three 12 mil aluminum bond wires. The big diameter of bond wire helps to increase current carrying capability. But it makes the right lead over crowded. There is only one clamping point during wire bonding process. This is inadequate to fully clamp the lead post during wire bonding process, which may induce potential risks such as bond lift.

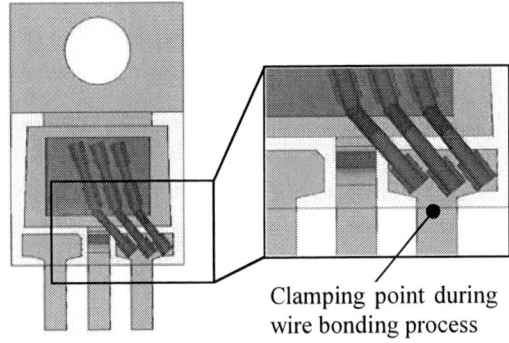

Figure 1: Original lead frame design for the TO220 package with 3 thick aluminum bond wires.

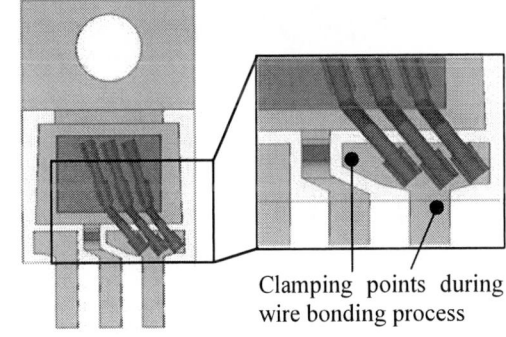

Figure 2: New lead frame design for the TO220 package with 3 thick aluminum bond wires.

In order to manufacture more robust products for high current applications, a new lead frame design is made for the TO220 package, as shown in Figure 2. Comparing

978-1-4799-4789-8/14 $31.00 © 2014 IEEE

with the original design, the right lead post is wider to allow a more space for clamping. Accordingly, the left lead is shortened and the middle lead is twisted to be "Z" shape. With the wider lead post for the source wires, two clamping points can be applied during the wire wedge bonding process. This can improve the wire bonding quality. However, the twisted "Z" shape area might be a weak point for the current carrying capability, which will be investigated in the following sections.

3. Electrical-Thermal Coupling Simulation

Coupled electrical-thermal transient simulations are carried out to study the current carrying capability of different lead frame designs. The 3-D FE model is shown in Figure 3. The TO220 package is mounted on a big metal cooling board. The metal cooling board is cooled by liquid in real test. The bottom surface of the cooling board is defined to be a constant temperature in simulation. There is 30μm thick thermal grease between the TO220 package and the metal cooling board. Figure 4 shows the internal structure of TO220 and the electrical constrain and loading. Silicon die is bonded on copper based lead frame with solder. The source of the switch is the top metal layer on silicon die. There are three 20mil aluminum wires bonded from die top metal to the right lead. The voltage of the right lead is defined to be "0". Different current loading is applied on the middle lead.

The material properties for the coupled electrical-thermal transient simulations are listed in Table 1 with electrical resistivity, thermal conductivity, specific heat and density.

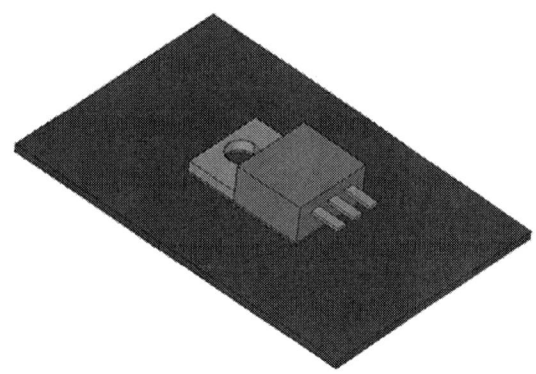

Figure 3: 3-D solid model for electrical-thermal coupling simulation.

The temperature distribution of the original straight middle lead design is shown in Figure 5. A 80A current is applied to the middle lead in this simulation. The metal cooling pad is 70 degree C which is correlated by the real test data. The max temperature of the package locates at the right lead. The reason is that the heat generated in the die, bond wire, middle lead and lead frame DAP might be

transferred to metal cooling board directly, but the right lead is far away from the metal cooling board.

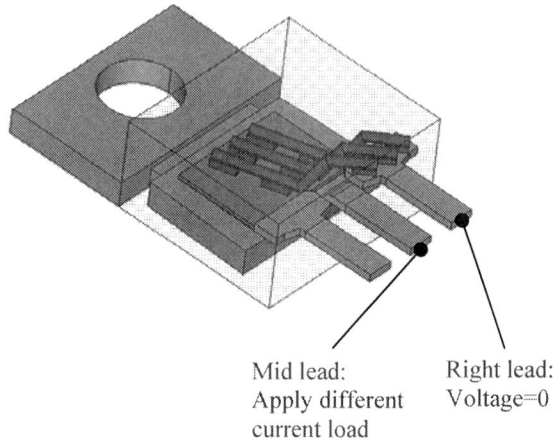

Mid lead:
Apply different current load

Right lead:
Voltage=0

Figure 4. 3D solid model of electrical thermal coupling simulation.

Table 1: Electrical & thermal properties for coupled electrical-thermal transient simulation

Mat	Electrical resistivity (Ω-m)	Density (Kg/m³)	Thermal conduct (W/m-C)	Specific heat (J/Kg-C)
lead frame	1.92e-8	8900	349	381.5
EMC	1e15	1850	0.71	836
solder	2.174e-7	11070	44	130
silicon	0.761e-3	2330	146	707
die top metal	2.72e-8	2700	210	900
Al wire	2.72e-8	2700	210	900
thermal grease	1e15	1850	1	836
metal plate	1.92e-8	8900	349	381.5

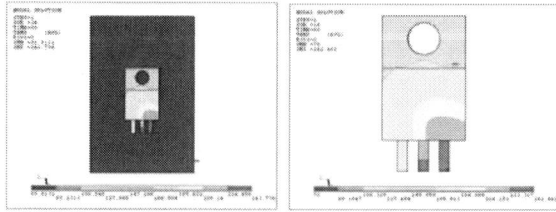

Figure 5. Temperature distribution of the original lead design (straight middle lead).

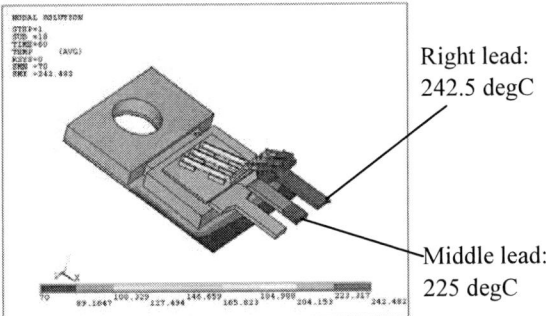

(a) Internal temperature distribution of the original lead design (straight middle lead).

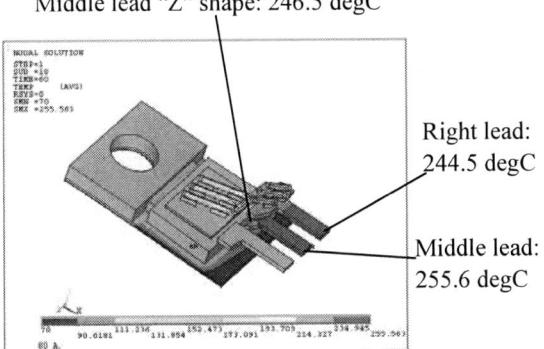

(b) Internal temperature distribution of the new lead design ("Z" shape twisted middle lead).

Figure 6: Temperature distribution of internal structures Of original and new LF designs.

The impact of different lead shape design is shown in Figure 6. Fig. 6 (a) shows the temperature distribution of the original straight middle lead design. The max temperature is 242.5 degree C, locating at the right lead tip. The temperature of middle lead is 225 degree C. Fig. 6 (b) shows the temperature distribution of new "Z" shape twisted middle lead design. The temperature of right lead is 244.5 degree C. The max temperature of the package locates at the middle lead tip which is 255.6 degree C. The temperature at the "Z" shape twisted area of the middle lead is 246.5 degree C.

Figure 7 shows the time history of temperature at "Z" shape twisted area of middle lead with different applied currents. Fig. 7 (a) shows the temperature curve with 30A current applied on the middle lead. The max temperature of "Z" shape twisted area of middle lead is a little higher than 90 degree C. Fig. 7 (b) shows the temperature curve with 40A current applied. The max temperature of "Z" shape twisted area is around 120 degree C. Fig. 8 (c) shows the temperature curve with 80A current applied. The max temperature of "Z" shape twisted area is more than 240 degree C. From the temperature vs time curves, it can be seen that the temperature increases rapidly within the first 3 seconds, and it becomes steady after around 20 seconds.

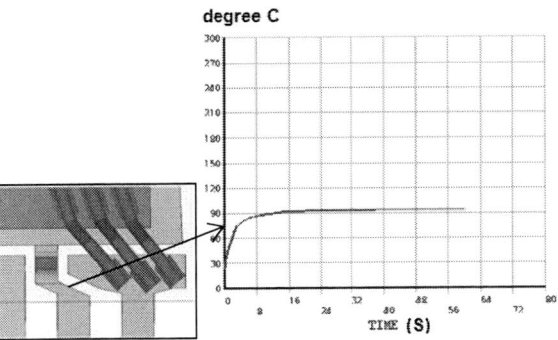

(a) Temperature vs time curve at "Z" shape twisted area of middle lead (30A)

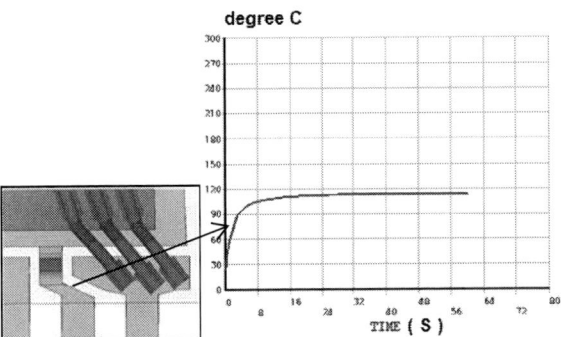

(b) Temperature vs time curve at "Z" shape twisted area of middle lead (40A)

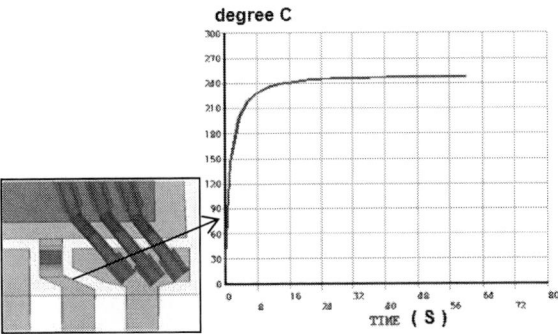

(c) Temperature vs time curve at "Z" shape twisted area of middle lead (80A)

Figure 7: Temperature curves of "Z" shape twisted area of middle lead with different currents applied.

The twisted "Z" shape middle lead design can share some space to the lead post for source bond wires. But the middle lead design may have potential risk. The stamping process during the leadframe manufacturing may have some impact on the twisted "Z" shape area of the middle lead. It may induce material property change, uneven thickness, and micro-crack in the local area. Fig. 8 shows SEM picture of "Z" shape twisted area of the middle lead. There are a lot of micro cracks in the "Z" shape position. The micro-cracks (or the micro voids) are induced during lead frame stamping process due to the local bending strain has exceeded its material elongation at the local twisted Z shape area. The electrical resistivity of the lead with micro cracks or voids will be increased.

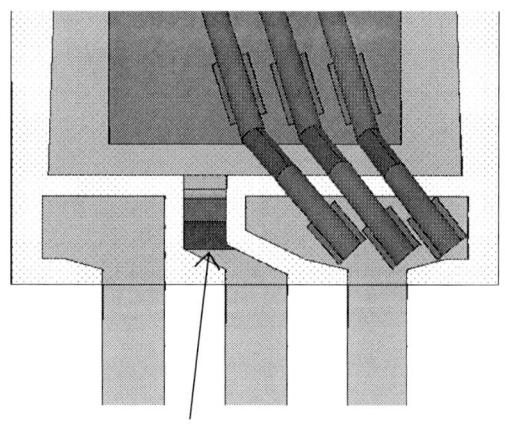

Figure 8. SEM picture of micro cracks in the optimized "Z" shape lead frame.

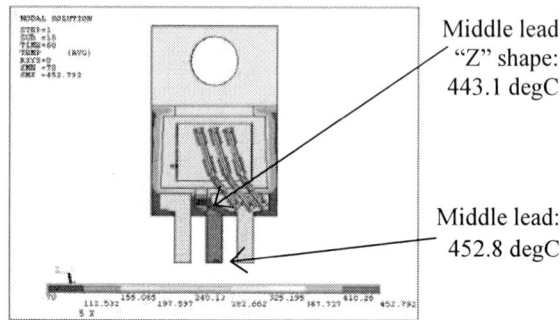

(a) Electrical resistance of "Z" shape lead part with micro cracks: $5 \times 1.92e-8\Omega$-m.

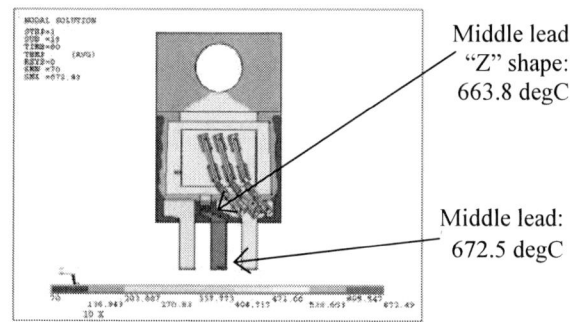

(b) Electrical resistance of "Z" shape lead part with micro cracks: $10 \times 1.92e-8\Omega$-m.

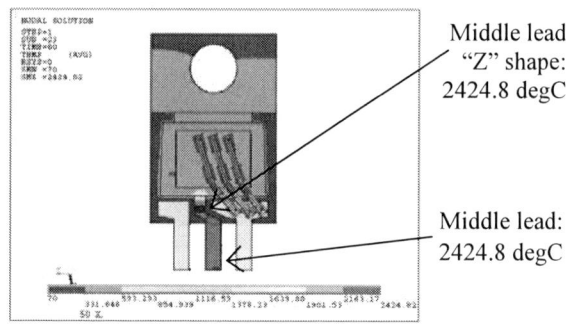

(c) Electrical resistance of "Z" shape lead part with micro cracks: $50 \times 1.92e-8\Omega$-m.

Figure 10: Temperature distribution with different electrical resistivity of "Z" shape twisted middle lead part with initial cracks.

Figure 9: Micro crack and higher resistivity area in "Z" shape twisted area of mid lead.

It's very difficult to simulate so many micro cracks in the lead in the coupled electrical-thermal simulation. But we can study its impact by changing different electrical resistivity in the lead with micro cracks. In Fig. 9 defined the range of the twisted "Z" shape part of the middle lead,

which has high dense micro cracks and therefore that range would have higher electrical resistivity.

Figure 10 shows the temperature contour of TO220's internal structure with different electrical resistance at "Z" shape twisted area with micro cracks. The electrical resistivity of other lead frame parts without micro cracks is 1.92e-8Ω-m as before. Fig. 10 (a) shows the temperature distribution when the electrical resistivity of "Z" shape lead part with micro cracks increases to be 5 times of the original value. The max temperature locates

978-1-4799-4789-8/14 $31.00 © 2014 IEEE

at middle lead tip. It's 452.8 degree C. The temperature at "Z" shape of middle lead is 443.1 degree C. Fig. 10 (b) shows the temperature distribution when the electrical resistivity of "Z" shape lead part with micro cracks increases to be 10 times of the original value. The max temperature locates at end of middle lead. It's 672.5 degree C. The temperature at "Z" shape of middle lead is 663.8 degree C. Fig. 10 (c) shows the temperature distribution when the electrical resistivity of "Z" shape lead part with micro cracks increased to be 50 times of the original value. The temperature of whole middle lead is 2424.8 degree C, which is much higher than the fuse temperature of lead frame.

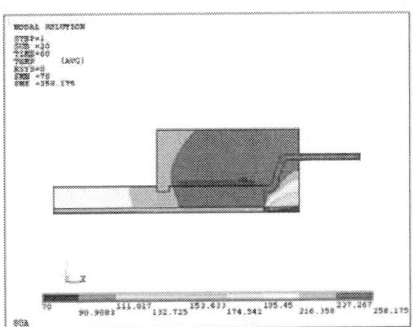

Figure 12: Finite element model of electrical thermal mechanical simulation (with initial crack in middle lead)

4. Electrical-Thermal-Mechanical Coupling Simulation

The finite element model of a 2-D TO220 package with initial cracks in lead frame is shown in Figure 11. Two cracks locate at the bent point of middle lead. Aluminum bond wires and the right lead are ignored in the 2-D model. The voltage of silicon die is defined to be "0". Current loading was applied on the middle lead. Metal cooling board is not considered in 2-D simulation. The bottom surface of the thermal grease is defined to be 70 degree C.

Figure 12 shows the temperature distribution of the 2-D electrical-thermal coupling simulation. The temperature data is then used as input for thermal-mechanical coupling simulation. Table 2 defines the material properties for the thermal-mechanical coupled simulation. The elastic modulus, poisson ratio and CTE of each material are listed in the Table. The lead frame is considered a bilinear material. Table 3 listed its yield stress and tangent modulus at different temperatures. Figure 13 shows the von-Mises stress of lead frame. The max stress appears at the initial crack tip of the middle lead. This would speed up the fuse failure of middle lead, and reduce the current carry capability.

Table 2: Material properties for thermal-mechanical coupling simulation.

Mat	Elastic modulus (Gpa)	Poisson ratio	CTE (ppm/C)
lead frame	118@25C 10@1000C	0.345	17.7
EMC	18	0.3	Alpha 1=16 Alpha 2=54 Tg=150C
solder	21.3	0.35	29.1
silicon	161	0.26	2.6
die top metal	68.9	0.33	20

Table 3: Yield stress and tangent modulus of lead frame at different temperature.

Temperature (°C)	25	250
Yield stress (MPa)	110	85
Tangent Modulus (MPa)	1100	850

Initial crack in middle lead

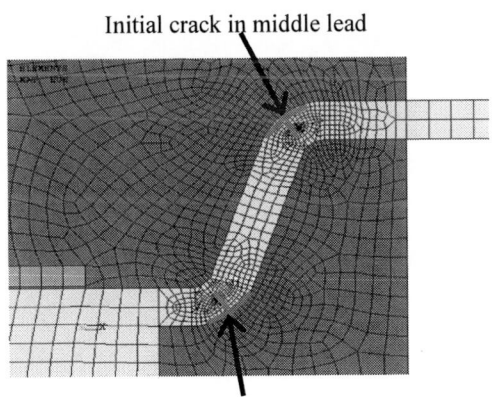

Initial cracks in middle lead

Figure 11: FE model of coupled electrical-thermal-mechanical simulation (with initial crack in middle lead)

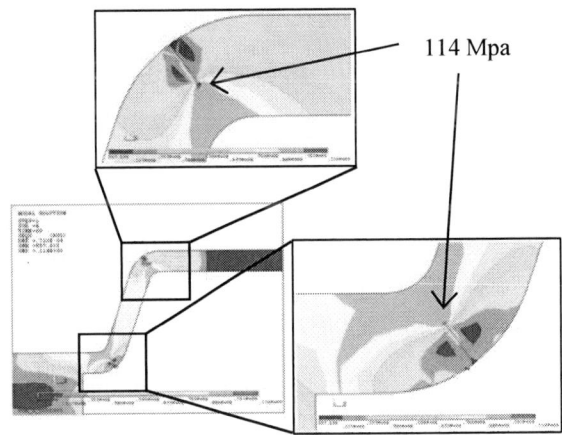

114 Mpa

Figure 13: Von Mises stress distribution.

5. Conclusions

In this paper, the impact of leadframe design with "Z" shape twisted middle lead on the current carrying capability of TO220 is studied. Modeling results shows that the max temperature of the original straight middle lead is 242.4 degree C when the input current in the middle lead is 80A. The max temperature of new design with twisted middle lead without initial cracks increases the temperature to be 256.5 degree C. In real application of lead frame manufacturing process, twisted lead would have the micro cracks or voids. Electrical-thermal simulation shows that the initial micro cracks in the lead, will induce much higher temperature as compared to the lead without the initial micro cracks. The coupled electrical, thermal-mechanical simulation results show that the initial micro cracks in middle lead have induced high stress concentration, which would combine with the high temperature of the center lead, speed up the lead fuse failure. As compared to the original design with straight center lead, the new design with twist Z shape center lead has lower current capability.

Acknowledgments

The authors wish to thank the support from Fairchild Semiconductor Reliability Group.

References

1. H. Wang, M. Liserre, and F. Blaabjerg, "Toward reliable power electronics: challenges, design tools, and opportunities," *IEEE Industrial Electronics Magazine,* Vol. 7, No. 2, pp. 17-26, Jun. 2013

2. Y.M. Liu, B. Newberry, Y. Liu and S. Martin, "Modeling characterization and reliability analysis of a power system in package," in Proc. *IEEE Electronic Components and Technology Conf. (ECTC)*, 2011, pp. 731-739.

3. Y.M. Liu, C.L. Wu, Y. Liu, D. Kinzer and O.S. Jeon, "Modeling for defects impact on electrical performance of power packages," in Proc. *IEEE Electronic Components and Technology Conf. (ECTC)*, 2010, pp. 403-410.

4. E.I. Almago, S.H. Paek, T.K. Lee, "Package Design Optimization for Electrial Performance of a Power Module using Finite Element Analysis," in Proc. *Electronic Packaging Technology Conference*, 2008, pp. 1023-1027.

5. Y.M. Liu, M.R.T. Carredo, Z.P. Hu, Y. Liu, T. Luk, S. Irving, "Effect of Wire Bonding and Die Layout on Electrical Performance of Power Packages," in Proc. *Thermal, Mechanical and Multi-Physics simulation and Experiments in Microeletronics and Microsystems (EuroSimE)*, 2009, pp. 1-6.

6. Y. Liu, S. Irving, M. RiouX, "Delamination Modeling for IC Package with Multiple Initial Cracks," in Proc. *IEEE Electronic Components and Technology Conf. (ECTC)*, 2003.

Comparison of Bondwire Life with Effective Strain Method and Cohesive Zone Method for a Power Package

Jicheng Zhang, Yangjian Xu

Fairchild –ZJUT Microelectronic Packaging Joint Lab, Zhejiang University
of Technology, Hangzhou 310032, China

Yong Liu

Fairchild Semiconductor, S. Portland, Maine, USA

Abstract

In this paper, the cohesive zone model was used to simulate the stiffness degradation of interfaces between dies and solder joints in a power module. Subsequently, the fatigue life of the solder joints was predicted by the effective strain method. In addition, the cohesive zone model, collabrating with a modified paris' law, was further utilized to investigate the reliability of a wirebond model. At the same time, the effect of the cohesive zone model parameters on the reliability was studied.

1. Introduction

The reliability of power modules is governed by the lifetime of wirebonds and solder joints in which the failure is easy to take place due to electrical, mechanical and thermal impacts[1]. Therefore, it is quite significant to develop a propriate method to predict the lives of the wirebonds or the solder joints in power modules. The effective strain method is widely used in the life prediction of solder joints since it can result in a more reasonable life estimation in most situations compared with the other methods. Reference [2] indicates that the mismatch of coefficient of thermal expansion (CTE) between solders and dies would cause the sepration of the interface during thermal cycling, which is prone to bring certain influence on the life prediction by using the regular methods. To more accurately evaluate the reliability of power modules, it needs a detailed investigation of the life prediction of solder balls in consideration of the interfacial separation or stiffness degradation.

The cohesive zone model (CZM), defines a traction-separation law at interfaces of two materials, which makes the facture and fatigue problem be effectively simulated through finite element. Arian Grams[3] utilized a cohesive zone model to simulate the fatigue crack extension and predict the life for an aluminum wire bond, which has shown that the CZM is a an effective method to simulate crack initiation and propagation of a wedge wire bond.

In our work, we used the CZM to simulate the interfacial stiffness degradation in a power module and then predict the fatigue life of solder joints and the bond wire by utilzing the effective strain method and the revised paris law. The effect of interfial stiffness

degradation on the reliability of power module is investigated.

2. Methods for simulation and life prediction

2.1 The effective strain method

The effective strain method in this paper is based on creep rupture, as given in Equation (1). it takes the creep strain as the theoretical basis for fatigue life prediction. The creep formula of Monkmann-Grant assumes the stress as constant during creep test. In case of varying stresses repeated in a cyclic form, an estimate of rupture time can be made by using a special form of time-fraction rule, given in Equation (2).

$$t_r = \frac{C}{\dot{\varepsilon}_{cr}} \tag{1}$$

$$N_f \left(\sum_{i=1}^{n} \frac{\Delta t_i}{t_{ri}} \right)_{one\ cycle} = 1 \tag{2}$$

Where $\dot{\varepsilon}_{cr}$ is the steady state creep strain rate for stress level σ_i; the constant C gives the "creep ductility" or the strain at the onset of failure; N_f is the number of repetitions or cycles to failure; n is the number of steps within a cycle; Δt_i is the time spent at stress level σ_i within a cycle; and t_{ri} is the rupture time for stress level σ_i.

$$N_f = \left(\frac{E_{cr}}{C} \right)_{one\ cycle} = 1 \tag{3}$$

Equations (1) and (2) can be simplified as equation (3). Equation (3) gives us the cycling creep life due to varying and repeated stresses for a single creep mechanism. The characteristic life of the effective strain method can be further written as equation (4) based on two creep mechanisms [4-5].

$$N_f = \left(\frac{E_1}{C_1} + \frac{E_2}{C_2} \right)^{-1}_{per\ cycle} \tag{4}$$

Where C_1, C_2 is the creep ductility of solder for these two creep mechanisms, which is determined by the authors[4] using the mean cycles to failure. E_1, E_2 is the accumulated creep strain for these two mechanisms, and were calculated by finite element simulation. C_1 and C_2 are taken as 50 and 15.87, respectively. Using these values, the life of the power module finally can be expressed as Equation (5).

978-1-4799-4789-8/14 $31.00 © 2014 IEEE

$$N_f = \left(0.02E_1 + 0.063E_2\right)^{-1} \qquad (5)$$

The post processing procedure requires a user defined output variables option to separate and store the corresponding creep strains for two damage mechanisms.

Equation (6) is the material model proposed by Wong et al[6]. The first item of the equation is used to describe the grain boundary sliding at the low stress, high temperature and low strain rate, while the second item is matrix creep at the high stress, low temperature, and high strain rate.

$$\dot{\varepsilon}_{cr} = B_1 \exp\left(\frac{-H}{kT}\right)\left(\frac{\sigma}{E}\right)^{n_1} + B_2 \exp\left(\frac{-H}{kT}\right)\left(\frac{\sigma}{E}\right)^{n_2} \qquad (6)$$

Where B_1, B_2 are constants; H is the activation energy; k is the universal gas constant; T is the absolute temperature; n_1 and n_2 are the indexes; E is the elastic modulus and its value is calculated by Equation (7) for a SAC solder in die attached model [5].

$$E = -0.088T + 56.024 \qquad (7)$$

Since Equation (6) is not a standard form of creep equation in ANSYS, a user subroutine was coded and complied based on the UPFS in Ansys. It needs to define the material creep parameters. Finally, by using the increment of creep strain obtained from each substep, we can calculate E_1 and E_2 according to the following Equations:

$$E_1 = \frac{\sum_{i=1}^{m}\left(\sum_{j=1}^{n} \varepsilon_{1ij} \cdot v_{ij}\right)}{\sum_{i=1}^{n} v_i} \qquad (8)$$

$$E_2 = \frac{\sum_{i=1}^{m}\left(\sum_{j=1}^{n} \varepsilon_{2ij} \cdot v_{ij}\right)}{\sum_{i=1}^{n} v_i} \qquad (9)$$

Where m is the number of elements to be analyzed; n is the number of substeps in last cycle; ε_{1ij}, ε_{2ij} represent the two types of creep strain; v_{ij} is the volume of the ith element in the jth substep.

By substituting E_1 and E_2 into Equation (5), we can acquire the life for each solder with which the critical solder can be determined.

2.2 The modified Paris law

For a low-cycle fatigue failure, Paris law is always a popular choice to predict its characteristic life. Hung et al [2] adapt the following paris' law to study the fatigue failure for a power module, in which the equivalent plastic strain fluctuation is taken as an evaluation index.

$$N_f = B_1 \Delta\varepsilon_{plastic}^{B_2} \qquad (10)$$

Where $\Delta\varepsilon_{plastic}$ represents the increment of equivalent plastic strain in the last cycle; B_1 and B_2 are two parameters for different materials, e.g., the values

B_1=113.18992 and B_2=-1.2233 are considered for Al wire.

2.3 Cohesive zone model

The CZM describes the interface fracture as a gradual damage process through the cohesive zone elements which are used as an interface layer between standard elements. With a defined relation of separation and traction, they can be used to constitute a crack at the interface between different materials.

In this study, the cohesive element formulation implemented in ANSYS® has been used. With the 3D 8-nodes interface elements (Fig. 1a), the relationship between traction and separation (Fig. 1b) can be defined through several parameters. Since the interface failure in electronic package is mainly dominated by shear slip, the maximum tangential traction T_{max} and the tangential displacement jump at the completion of debonding δ_t^c are the key CZM parameters.

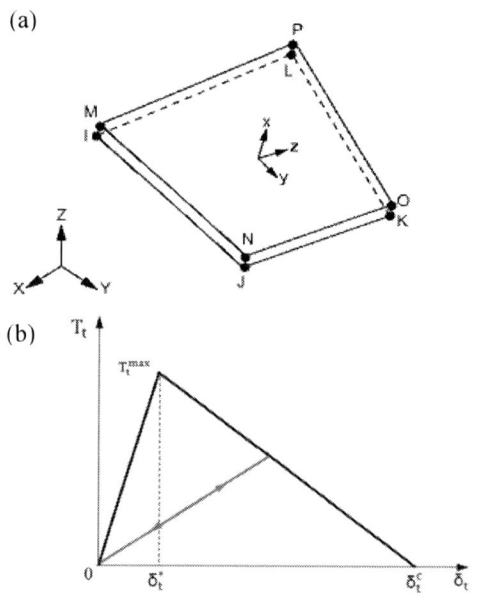

Fig. 1 Cohesive zone model: (a) cohesive zone element; (b) cohesive law

The relationship between tangential cohesive traction T_t and tangential displacement jump δ_t can be expressed as .

$$T_t = K_t \delta_t \left(1 - D_t\right) \qquad (11)$$

Where K_t is the tangential cohesive stiffness, D_t is the damage parameter associated with modedominated bilinear cohesive law, defined as in Equation (12).

$$\begin{cases} 0 & \delta_t^{max} \leq \delta_t^* \\ \left(\frac{\delta_t^{max} - \delta_t^*}{\delta_t^{max}}\right)\left(\frac{\delta_t^c}{\delta_t^c - \delta_t^*}\right) & \delta_t^* \leq \delta_t^{max} \leq \delta_t^c \\ 1 & \delta_t^{max} > \delta_t^c \end{cases} \qquad (12)$$

With this cohesive law, the crack at the interface can be well represented.

3. Life prediction for solder failure in the power module

3.1. FEM modeling of a power module

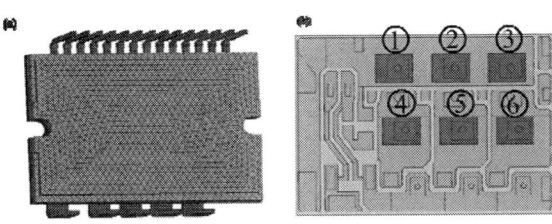

Fig. 2 FEA model: (a) the global model; (b) illustration of solder joint distribution

An automotive power module(APM)[6] from Fairchild, as shown in Fig. 2, was used to make an investigation here. The solder paste was characterized by a non-linear material constitutive model and the other materials were assumed elastic. Table 1 lists all the material property data used in the FEM simulationl and Table 2 lists the parameters adopted in Wong's model. In the simulation, the cohesive zone elements were set at the interfaces between solders and dies with the maximum tangential strength of 45Mpa and the maximum allowable tangential displacement jump of 0.005mm.

Table 1 Material Property Data for APM Model

Component	Material type	Material properties for APM model					
		Young's module (GPa)	Poisson's ratio	CTE (ppm/c)	Tg (°C)	Flexural strength(MPa)	Yield stress(MPa)
Die	Silicon	131	0.278	2.4	-	-	-
L/F	OFC	134.4	0.34	17.5	-	-	-
Solder	SAC 305	54.8	0.34	22	-	34.5	21.9
Ceramic	Al_2O_3	340	0.22	6.8	-	-	-
DBC copper	Copper	134.4	0.34	17.5	-	-	-
EMC	KCC5400GP	19.4(T=32°C) 18.0(T=100°C) 2.06(T=150°C) 0.78(T=200°C) 0.80(T=250°C)	0.3	9.1(T<Tg) 47.2(T>Tg)	143	-	-

Table 2 Parameters used in Wong's model

B_1(1/sec)	B_2(1/sec)	H(ev)	n_1	n_2
1.72e12	8.9e24	0.468	3	7

Thermal cycle loading was applied on the whole APM model according to the thermal cycling profile[7] shown in Fig. 3, firstly ramping to -45°C at the speed around 10.7°C/min, and keeping isothermal temperature for 15 min, and then ramping to 125°C, and keeping isothermal temperature for 15 min, finally ramping to -45°C at the speed around 10.7°C/min.

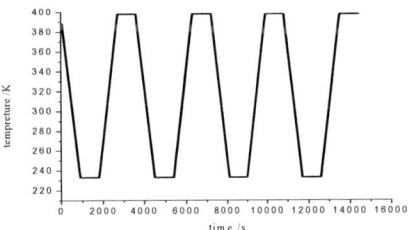

Fig.3 Thermal cycle profile

978-1-4799-4789-8/14 $31.00 © 2014 IEEE

3.2. Results and discussions

In order to investigate the influence of interface separation on life predction, both simulations have been made with and without considering the cohesive zone element (CZE). At the same time, their respective characteristic lives have been predicted and compared with each other. Through the FEM simulation under the same loading condition, the deformation contours in the last step are obtained for both cases, as shown in Fig. 4. The maximum displacement acquired from the model without embedding CZE is 0.02028 mm, while that from the model with CZE is 0.02026 mm. Therefore, the CZE at the interface can bring little influence on the results of the global model.

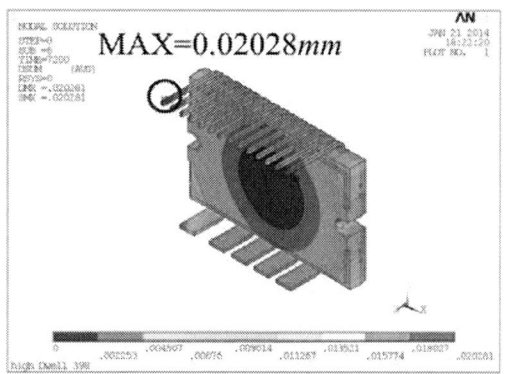

(a) Without the embedded CZE model

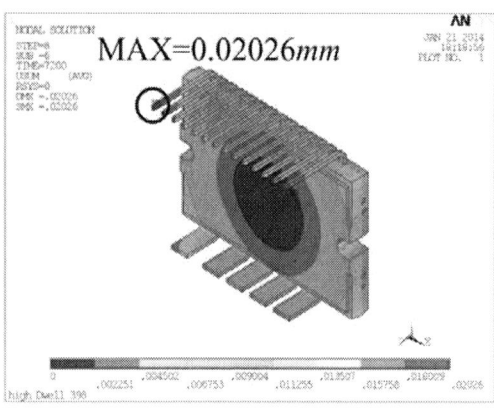

(b) With the embedded CZE model

Fig. 4 Deformation contours: (a) without embedded the CZE ; (b) with embedding the CZE.

Since the interface faliure of electronic packaging components is dominated by the mechanism of shear slip, the shear separation of CZE at the interfaces between solders and dies is a key index to characterize the interface failure. The XZ-plane shear separation distribution at the end of the last thermal cycle for all the CZEs is shown in Fig. 5, in which the maximum value (located at Solder 6) is 0.31E-6 mm. This value is much smaller than the max separation of 0.005 mm which is set for the CZM in the simulation. It means that the failure doesn't take place in all the CZEs at the present stage.

The history curve of XZ-plane shear separation is shown in Fig. 6. It changes along with the different thermal loading steps. During the first cycle, the maximum separation is 3.4E-4 mm, while it is 3.43E-4 mm for the second cycle. Hence, the interface separation becomes slightly larger with the increment of loading cycle, which matches with that we expected.

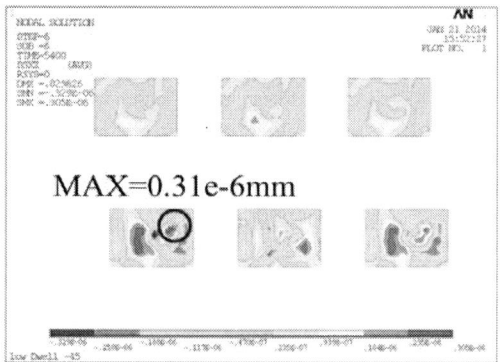

Fig. 5 the XZ-plane shear separation distribution at the interface

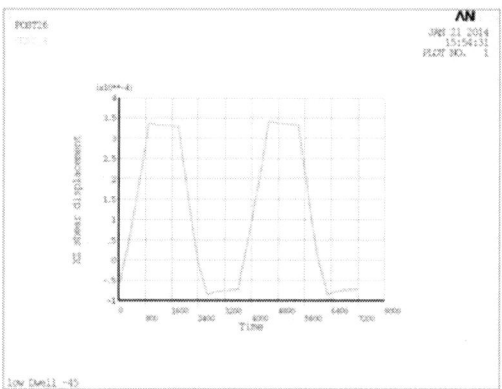

Fig. 6 the XZ-plane shear separation history curve

By calculating the accumulated creep strain during thermal cycling, we can finally obtain each solder's life by Equation (4). The characteristic fagitue lives of all the solder joints are shown in Fig. 7 in which the result from the CZE-embedded model is compared with that without considering CZE. It can be seen that Solder 4 has the shortest life for both kinds of models and the model with embedding CZE has a larger life than that without CZE for all the solder joints. Since embedding CZE at the interface will reduce the stress in solders, it will decrease their creep strain as well. It follows that the characteristic lives of all solder joints would be extended.

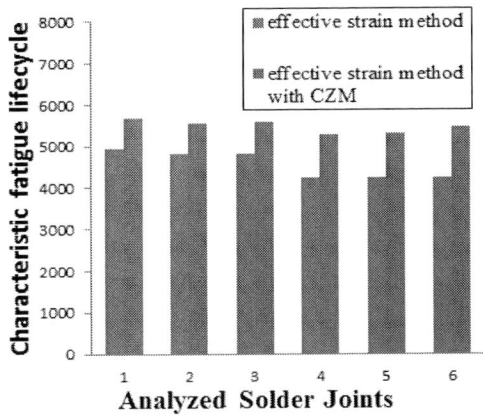

Fig. 7 Comparison of the solders' lives for two types of models (with and without embedding CZE)

4. Life prediction for wirebond failure in the power module

4.1. FEM modeling of power module

In order to further verify the effect of embedding CZE at the interfaces on the   life prediction model,   a wirebond power module is studied for the   life prediction based on Equation (10). In this wirebond power module, six dies are linked with leadframes through Al wires. The FE model is shown in Fig. 8.

Fig. 8 FEA model of a wirebond power module

Table 3 Material Property Data for Wirebond Model

Component	Material type	Material properties for wirebond model					
		Yong's module（GPa）	Poisson's ratio	CTE (ppm/C)	Reference temperature Tg(C)	Flexural strength (MPa)	Yield stress (MPa)
die	silicon	131	0.278	2.4	--		
L/F	OFC	134.4	0.34	17.5	--		
solder	SAC305	54.8	0.34	22	--	34.5	21.9
ceramic	Al2O3	340	0.22	6.8	--		
DBC copper	copper	134.4	0.34	17.5	--		
EMC	KCC5400GP	19.4(T=32) 18.0(T=100) 2.06(T=150) 0.78(T=200) 0.8(T=250)	0.3	9.1(T<Tg) 47.2(T>Tg)	143		
Al bondwire	Aluminum	70	0.35	23	--		

During thermal cycling, the Al wire would be damged due to undergoing an alternating change of elastic-plastic strain. Therefore, the Al wire is characterized by a bilinear elastic-plastic consititute model in our simulation, its yield stress is 30Mpa and tangent modulus is 216Mpa. The other material parameters are listed in Table 3. If the CZM model is considered, the CZEs are embedded at all the interfaces between Al wires and the other materials, as shown in Fig. 9, and the CZM parameters are shown in Table 4.

Table 4 CZM parameter

T_{max} （Mpa）	δ_t^c （mm）	α
45	0.005	0.5

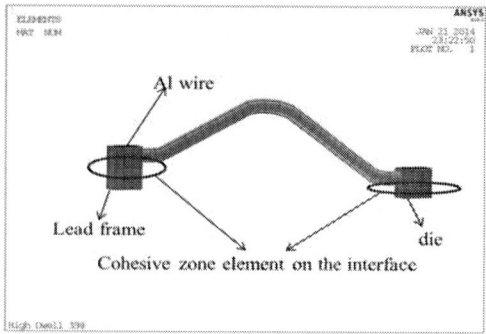

Fig. 9 Location of CZE in the wirebond structure

4.2 Results and discussions

The equivalent plastic strain distributions for both models (with and without embedding CZE) are illustrated in Fig. 10. The simulation results show that the max equivalent plastic strain of normal model is 0.041, while the max equivalent plastic strain is 0.071 for the CZE-embedded model. Note that the maximum equivalent plastic strain occurs at different wirebonds, but both are close to the interfaces. It verifies that the interfaces are prone to failure.

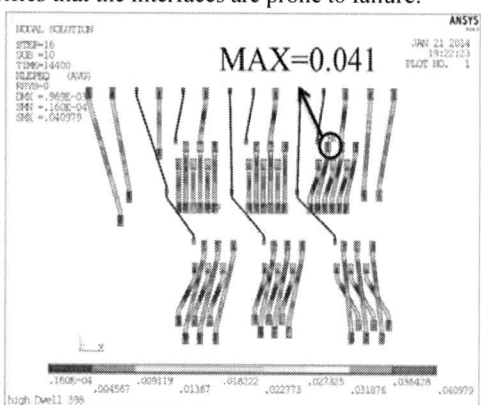

(a) Normal model

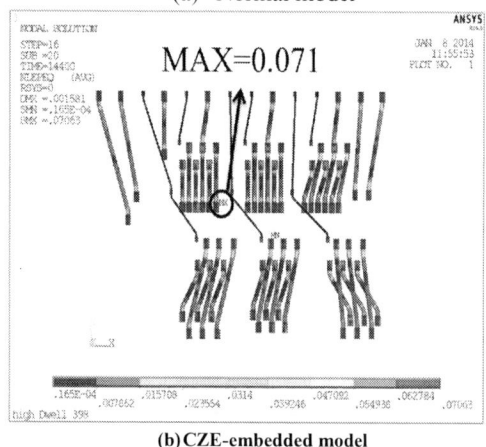

(b) CZE-embedded model

Fig. 10 The equivalent plastic strain distribution: (a) normal model with effective strain method, (b) CZE-embedded model.

The history of equivalent plastic strain for both method are illustrated in Fig. 11. The amplitude of equivalent plastic strain during the last cycle is used here. For the normal model, it's 0.0097; while for the CZE-embedded model, it's 0.016. By substituting the amplitude of equivalent plastic strain into Equation (10), we can obtain the fatigue characteristic lives of wirebonds: The life of the critical wire for normal model is 32854; and for the CZE-embedded model, it's 17812.

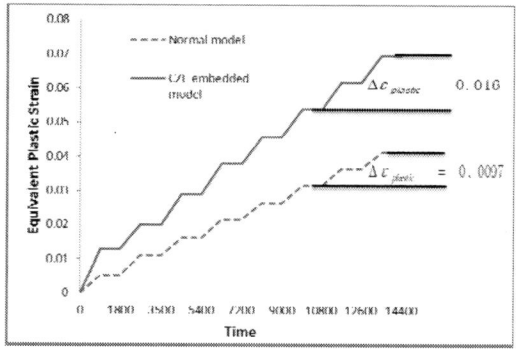

Fig. 11 The history of equivalent plastic strain

In order to study the influence of the shear strength T_{max} in CZM on the life prediction of wirebonds, we conducted a parameter study in terms of $T_{\max}$ and their fatigue lives are shown in Fig. 12. From this figure, we can see that the life is decreasing with the shear strength T_{max}. At the same time, the life for the model without embedding CZE is longer than the CZE-embedded model. The simulation result seems to conflict the first example while the solder joint life with CZM model is a little longer than the regular solder model withoutthe CZM. It might be caused by the reason that some local damages and failures have taken place at the interfaces in the present CZM wirebond model, which could induce local higher plastic strsin in the CZM model than the regular model without the CZM.

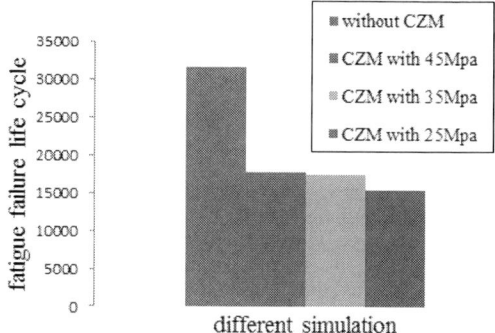

Fig. 12 The predicted fatigue lives for different cases

6. Conclusions

From the above results and discussions, we can get the following conclusions:

1) With the help of CZM, the stiffness degradation of the interfaces between different materials could be taken into consideration, which will affect the stress distribution apparently. If we consider the stiffness degradation, the simulation for the life evaluation will approach to a more real situation.

2) For the die attached model, the characteristic fatigure life increases when the interfaces are embedded with CZE, which is caused by the reason that the rigidness of the interface is slightly reduced due to the existence of cohesive zone, as compared to effective strain method. However, the life for the wirebond with CZE decreases since some local part of interfaces have encountered certain damage.

3) Parameter studies show that the characteristic life of wirebond is reduced as the tangential strength of CZM, T_{max} decreases with all other model parameters remain the same. This agrees with that we expected.

Acknowledgments

This work is supported by Fairchild Semiconductor Scholarship and National Nature Science Foundation of China (No. 51375448 and 51375447).

References

1. Bielen, Jeroen, J-J. Gommans, and Frank Theunis. "Prediction of high cycle fatigue in aluminum bond wires: A physics of failure approach combining experiments and multi-physics simulations." Thermal, Mechanical and Multiphysics Simulation and Experiments in Micro-Electronics and Micro-Systems, 2006. EuroSime 2006. 7th International Conference on. IEEE, 2006

2. Hung, T. Y. et al, "Thermal-mechanical behavior of the bonding wire for a power module subjected to the power cycling test," Microelectronics Reliability, Vol. 51, No. 9-11 (2011), pp. 1819-23

3. Grams, Arian, et al. "Simulation of an Aluminum Thick Wire Bond Fatigue Crack by Means of the Cohesive Zone Method." EuroSimE, 2013

4. Syed, A., "Solder Joint Life Prediction Model and Application to Ball Gid Array Design Optimization," 1996 proceedings of Experimental/Numerical Mechanics in Electronic Packaging, SEM, Vol. 1, pp. 136-144.

5. Liu, Yan, Yangjian Xu, and Yong Liu. "Reliability modeling analysis of a power module." Proc Thermal, Mechanical and Multi-Physics Simulation and Experiments in Microelectronics and Microsystems (EuroSimE), 2013 14th, 14-17 April 2013.

6. Wong B, Helling, D .E, and Clark, R .W. "A creep-rupture model for two-phase eutectic solder," IEEE CHMT, 1988, 11(3): 284-290.

Adhesion work analysis by molecular modelling and wetting angle measurement

Kamil Nouri Allaf, Dawid Jan Król, Artur Wymysłowski, Irena Zubel, Krzysztof Rola
Wroclaw University of Technology, Faculty of Microsystems Electronics and Photonics,
ul. Janiszewskiego 11/17, 50-372 Wrocław, Poland
e-mail: kamil.allaf@pwr.edu.pl

Abstract

The molecular modelling was be applied in order to calculate the work of adhesion between solutions of water, isopropyl alcohol (IPA) and silicon. The work of adhesion was calculated by using two computational methods based on experiment as well as numerical simulations. The method of wetting angle in case of the molecular modelling was used in order to compare simulation results with with the wetting angle experimental results. The main aim of the work was to make a comparison between the achieved values of the work of adhesion using simulations and the experiment. In comparison to the previously reported results [1], the current goal was to continue the research in order to improve and apply the developed method for surfactants, which are used in a process of silicon etching. Better understanding of adhesion phenomena can result in improvement of etching and the same can lead to the quality improvement of MEMS devices, as well as reduce their production cost. This work continues the previous research in that area where wetting angle and other methods where used to determine the adhesion work between IPA and silicon.

1. Introduction

Molecular modelling is one of the research methods based on computer techniques. It is also one of the computational chemistry methods, which allows calculations of dynamics and structure of any molecule. The molecular modelling uses the mathematical description of the dynamics and geometry of molecules using equation of the classical physics. In this model atoms are approximated by spheres and chemical bonds by springs. In the molecular modelling the quantum effects are not taken into account. For the description of particle's behaviour the force field, which is set of the potential energy and its parameters is used instead. It describes the force and types of the interaction between atoms. Other tools which are important in the molecular modelling is the so called molecular dynamics (MD) based on solving Newton equations and energy minimization of the molecules in order to predict the probable final structure of the molecule. In the molecular modelling methods it is important to use a number of assumptions as restriction of the area of simulation, determination of long-range interaction to cut-off distance and apply techniques of parallel computing in order to reduce significantly time of simulations. The molecular

modelling was used since the beginning mainly in case of chemistry and biology, but around nineties of the last century it has been implemented as well in other disciplines like for example nano- and micro-electronics. Modelling is very useful for determination of mechanical and some thermodynamic properties of a system of molecules. However, it cannot predict electrical or optical ones. The molecular modelling is also used when the current techniques like the finite element method in some type of calculation gives ambiguous results due to the scale of the analysis and there is an option for using new methods. The molecular modelling can be applied for searching the answers concerning phenomena which occur at the surface during anisotropic etching of silicon with the addition of a surfactant, which in fact is the main goal of the presented research. It can also be used to determine the phenomena occurring in a material when it is difficult or impossible to measure experimentally.

Anisotropic etching of silicon is one of the commonly used bulk micromachining technologies for fabricating MEMS and MOEMS (optical MEMS). The anisotropic etching process is performed in both organic and inorganic aqueous solutions. TMAH (tetramethylam-monium hydroxide) and KOH (potassium hydroxide) are the most popular etchants. TMAH solution is compatible with CMOS technology and has better etching selectivity but the KOH-based etchant is often preferred due to its lower cost and higher etch rate. In order to change the etching anisotropy and the surface morphology of the etched structures, surfactants or alcohols are added to the etch solution to modify parameters of the manufactured three-dimensional microstructures which are surface roughness, shape of convex corners and sidewalls inclination toward the substrate. Both surfactants and alcohols possess a hydrophilic as well as hydrophobic group. Such compounds are prone to adsorb at the liquid–gas interface by lowering the surface tension of the aqueous solutions. Similar adsorption is supposed to occur at the solid state–liquid interface. Therefore, the adsorption of surfactants or alcohols on the silicon surface is believed to be responsible for modifying the surface morphology and the etching anisotropy in the solutions containing additives. The KOH solution saturated with surfactants e.g. isopropyl alcohol or tert-butanol (figure 1), which is an organic tensioactive compound with larger molecules and higher boiling point than isopropanol could be successfully applied for shaping spatial structures in the wafers with non-standard crystallographic orientations [2, 10]. Surface tension of the solution with tert-butanol decreases more rapidly with the alcohol concentration than for isopropanol (figure 2). The more probable adsorption of the tert-butanol molecules than the isopropanol ones can be partly proved by measurements of surface tension of the solution with tert-butanol which decreases more rapidly with the alcohol concentration than for isopropanol (figure 2a). Thus the differences in values of surface excessmaxima (figure 2b) indicate that the tert-butanol molecules are adsorbed at the liquid–air interface more densely than the isopropanol ones when the solution is saturated. A similar difference in adsorption is believed to occur at the liquid–silicon crystal interface during etching in the KOH solutions saturated with alcohols.

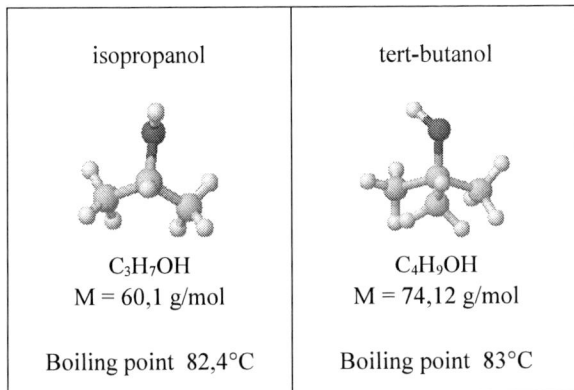

Figure 1. Molecules of isopropanol and tert-butanol surfactants.

The aim of the current work was to achieve the best fit of the molecular modelling simulation with experimental data in case of a solution of water and a surfactant on the silicon crystallographic surface (110) and (111) with different type of covering (silicon saturated with –H and –OH). Two surfactants were selected to perform the examination - the isopropyl alcohol and tert-butanol.

The research was divided into two steps, one for each surfactant. The first part of this research is presented in this paper in which isopropyl alcohol is used. The second part of the research will refer to the experiment of a solution of water with tert-butanol, which is still an ongoing research.

In this work, molecular dynamics simulations were applied in the calculation of the adhesion work. The adhesion phenomena occurs between two materials or surfaces that attract to each other through intermolecular forces. These forces are acting from several hundreds of picometers up to micrometers. The effect of this phenomena is that materials are connected to each other and a defined force is required to separate them. The work of adhesion is calculated using two computational methods and fundamental quantities

such as wetting angle and surface tension which are estimated and then the numerically evaluated values are compared with the experimental ones.

a)
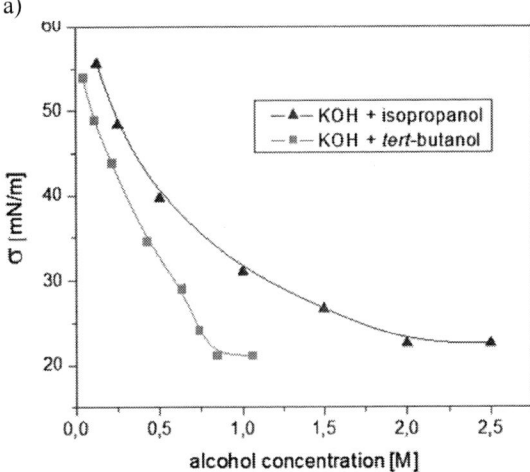

b)
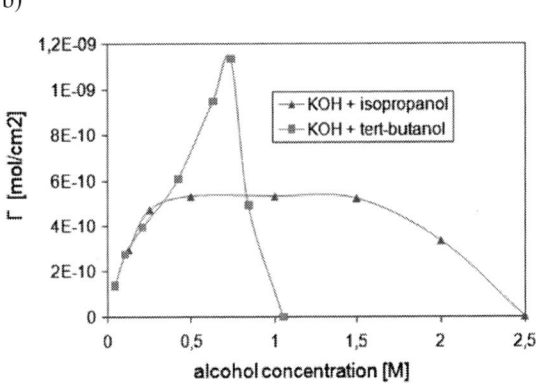

Figure 2. a) Surface tension of KOH solutions with isopropyl and tertiary-butyl alcohols addition versus concentration of alcohol, b) surface excess of the solutions versus concentration of alcohol calculated on the basis of the concentration dependence of the surface tension. [2]

2. Methodology

The work of adhesion of the surfactant solution on the surface of silicon was calculated using two computational methods. Results of these methods were compared with the wetting experiment. For that reason, the goniometer was used in order to determine the work of adhesion by measuring the wetting angle. Then the work of adhesion was calculated from Young-Dupre equation between surfactant solution and silicon (110) and (111) saturated with –H and –OH. The surface tension of the liquid was determined from simulations and then used to calculate the work of adhesion.

First method was based on the wetting angle experiment results which are then reproduced using molecular modelling and used in Young-Dupre formula. While in the second method the adoption of procedure described in [9] was used in order to determine the work of adhesion between the surfactant solution and silicon surface on the basis of surface tension values obtained in simulations. In contrast with the previous research the quality of molecular dynamics simulations were improved in this work and the accuracy of wetting angle measurement were also improved.

2.1. Wetting angle analysis
2.1.1. Experimental analysis and results
The wetting angle between surfactant solution (3% solution of IPA and water was prepared) and silicon surface was measured using the goniometer, as it was described in detail in the previous research [1]. The silicon wafer surface was cleaned first with Piranha solution and then etched in HF for around one minute (in order to obtain a clean surface saturated with –H) and washed in water for around 10 seconds. Then kept for 5 hours in deionized water to obtain saturation with –OH groups. Results of the wetting angle measurement as a function of time when droplet is placed on the silicon surface are shown in table 1.

Table 1. Wetting angle experimental results.

	Crystallographic plane	
	(110)	(111)
Silicon saturated with –H	66.4°	62.6°
Silicon saturated with –OH	59.1°	62.7°

It can be deduced from the above that the wetting angle is larger for plane (110) for silicon saturated with –H and for (111) for silicon saturated with –OH [4].

2.1.2. Numerical simulations
Accelrys Meterials Studio (version 6.1) software was used for molecular modelling. The numerical analysis was used by using two different approaches. In case of the first method, a droplet was simply placed on the surface while in the second case the droplet was placed at some distance from the surface. In fact, molecules are constantly moving, so the wetting angle is changing at the edge of the drop and the solid. Therefore a special MATLAB script [5] was applied and set accordingly for the analysis. Besides converting to Python scripting language, Isopropyl alcohol molecules were added to the calculation of single molecules and the droplet's centre of mass.

First, the droplet of liquid was placed on a solid. Then thermodynamics equilibrium was established between both substances. Simulations were performed by means of the Forcite module and Amorphous Cell

module, which provides solutions for dynamics simulations, molecular mechanics and visualisations of created materials (figure 3).

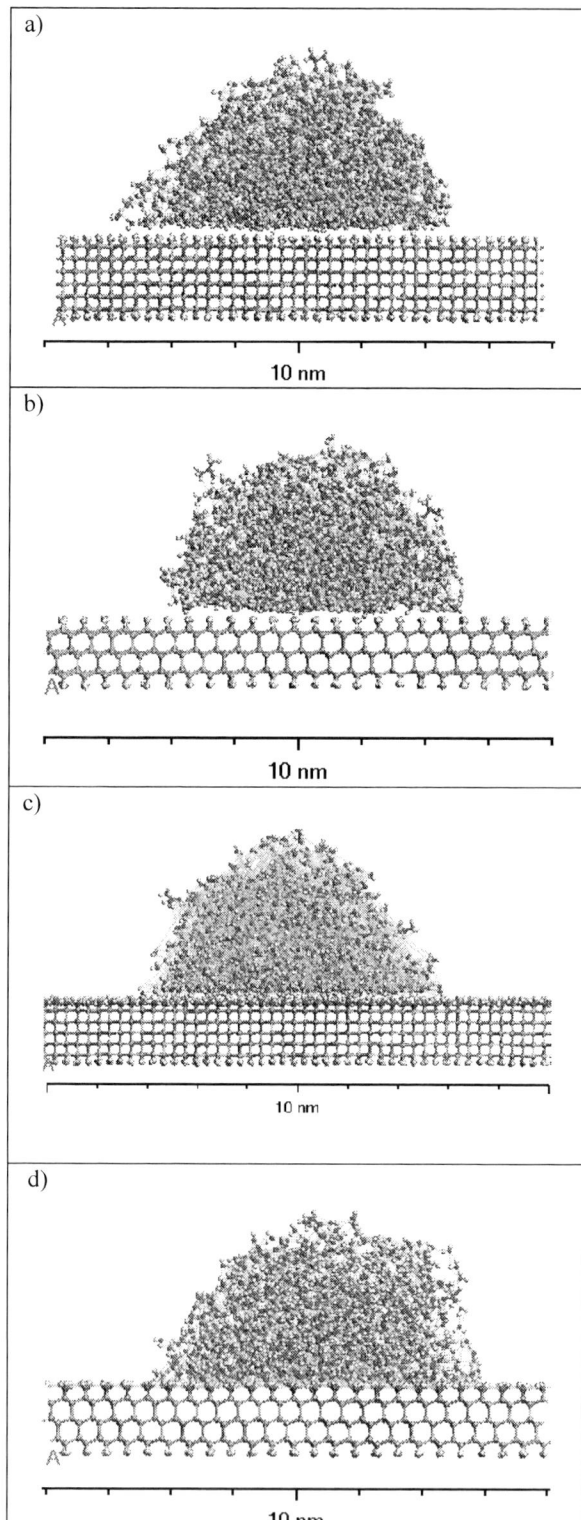

Figure 3. The droplet shape of IPA on silicon surface: a) Si (110) covered with –H, b) Si (111) covered with –H, c) Si(110) covered with –OH, d) Si (111) covered with -OH

There are several force fields that can be used in the simulations, which in fact has great impact on the final simulations' results. In this paper the COMPASS force field was used due to its capabilities for substances applied in the research [6]. Due to the low accuracy of wetting angle measurement in the previous research [1], the number of atoms in the droplet model was doubled in order to improve the quality of simulation results (all together around 2000 atoms).

The molecular modelling dynamics simulations were run with the time length of 20 ps and time resolution of 1 fs. Using NVT ensemble and the corresponding temperature of 293 K using Berendson thermostat with Decay constatnt 0.4 ps. In the final step the wetting angle was determined using a specially written script. The obtained numerical results are collected in table 2.

Table 2. Wetting angle values obtained from simulations.

	Crystallographic plane	
	(110)	(111)
Silicon saturated with –H	66.76°	64.91°
Silicon saturated with –OH	51.56°	53.38°

By comparing tables 1 and 2, it can be deduced that wetting angles values obtained during the simulation are comparable with the experimental ones.

2.1.3. Work of adhesion calculation

The work of adhesion was calculated from Young-Dupre equation between surfactant solution and silicon (2), which is given as follows [8]:

$$W_{adh} = \gamma_l (1 + cos\theta) \qquad (1)$$

where: γ_l - surface tension of the liquid solution,

θ - wetting angle.

The achieved values of work of adhesion using the experimental measurement of wetting angle and surface tension obtained from literature were used. The surface tension of the solution γ_l was obtained from [7] and was equal to 58 mN/m. Obtained results are shown in table 3.

Table 3. Work of adhesion values based on the wetting angle measurements

Silicon	Crystallographic plane

	(110)	(111)
saturated with –H	80.89 mN/m	82.60 mN/m
saturated with –OH	94.06 mN/m	92.60 mN/m

2.2. Work of adhesion calculation by surface tension analysis

The surface tension between the solution and silicon surface was calculated by using a procedure described in [9]. A bimaterial layered model was prepared for both silicon (110) and (111) crystallographic planes saturated with –H and –OH and the IPA solution. This model was used to prepare then models of separate materials by a principle of copying the complete model and separate materials by deleting one of them. The work of adhesion was derived by the following equation:

$$W_{adh} = \frac{E_1 + E_2 - E_{12}}{2A} \qquad (2)$$

In order to perform the examination three models were prepared: first one contained the IPA solution with layer of vacuum (with energy E_1), the second model contained the second material that is silicon with layer of vacuum (with energy E_2) and the third one contained both materials (energy E_{12}). Then E1 and E2 are summed and E12 is subtracted. The obtained result was divided by double horizontal cross-section area (A). Simulations were performed using the molecular dynamics option of Forcite module. Obtained values of energies were then used to calculate the work of adhesion which are shown in table 4.

Table 4. Work of adhesion calculations based on the surface tension analysis

Silicon	Crystallographic plane	
	(110)	(111)
saturated with –H	73.47 mN/m	81.92 mN/m
saturated with –OH	81.74 mN/m	72.44 mN/m

By comparing tables 3 and 4, it can be deduced that the work of adhesion values calculated according to the wetting angle experiment differ slightly from the values obtained from the simulation based on the surface tension analysis.

3. Conclusions

In the presented research the molecular modelling was applied for calculation of the work of adhesion. Two computational methods based on wetting angle and surface tension analysis were used and the corresponding numerical analysis results were compared with the experimental ones. In fact, the main aim of the current work was to achieve the best agreement between molecular modelling simulation results and experimental data in case of the solution of water and a surfactant on the silicon crystallographic surface (110) and (111) with different type of covering (silicon saturated with –H and –OH). The final results obtained from wetting angle measurement and during the numerical simulation were in a quite good agreement. In case of the work of adhesion calculation results based on the wetting angle experiment differed slightly from the values obtained from the simulation based on the surface tension analysis.

The same, the achieved results confirmed that the numerical analysis based on molecular modelling especially the molecular dynamics can be the valuable tool that can help to understand the physical phenomena at the nano scale and thus can be applied for improving performance of any electronic devices. It was also mentioned that the method of determining the surface energy can be applied for better understanding of adhesion interacting between any two materials. In contrast with the previous research the quality of molecular dynamics simulations were improved and the accuracy of wetting angle measurement were more accurate. The achieved outcomes of simulations suggest that the time of simulation compared to previous research can be reduced for the wetting angle determination and still with the acceptable accuracy. It is worth to notice that this result was possible to be achieved due to the improved Pyhton script, which currently is capable to evaluate the wetting angle values more accurately.

Moreover, another research is still ongoing that is focused on the adhesion work analysis using molecular modelling for another solution of water, which is a solution with tert-butanol instead of IPA as the examined so far. The anslysis will be continued in order to improve the developed method for surfactants, which are used in case of silicon etching. We expact that finally, the whole process of silicon etching can be improved e.g. in case of anisotropic processes.

Acknowledgments

Calculations have been carried out at Wroclaw Centre for Networking and Supercomputing (http://www.wcss.pl).

References

1. D.Krol, A.Wymyslowski, I.Zubel, K.Rola, Application of molecular modelling for analysis of a surface energy and its comparison with the experimental results based on wetting angle measurement, Thermal, Mechanical and Multi-Physics Simulation and Experiments in Microelectronics and Microsystems (EuroSimE), Wroclaw, 14th International Conference on 14-17 April 2013

2. K.P.Rola, I.Zubel, Investigation of Si(h k l) surfaces etched in KOH solutions saturated with tertiary-butyl alcohol, Journal of Micromechanics and Microengineering 21 (2011) 115026 (11pp)

3. D.R.Nieto, F.Santese, R.Toth, P.Posocco, S.Pricl, M.Fermeglia, Simple, Fast, and Accurate In silico Estimations of Contact Angle, Surface Tension, and Work of Adhesion of Water and Oil Nanodroplets on Amorphous Polypropylene Surfaces, ACS Applied Materials & Interfaces, 2012, 4 (6), PP 2855–2859.

4. I. Zubel, M. Kramkowska: The effect of the isopropyl on etching rate and roughness of (1 0 0) Si surface etched in KOH and TMAH solutions, Sensors and Actuators A: Physical Volume 93, Issue 2, 30 September 2001, Pages 138–147

5. J.LOUISE J. CRISCENTI AND JACQUELYN BRACCO, Molecular Modeling in Support of CO2 Sequestration and Enhanced Oil Recovery, Sandia National Labs January 2011.

6. Forcefield-Based Simulations http://northstar-www.dartmouth.edu/doc/insightII/ffbs/2_Forcefiel ds.html

7. Gonzalo Vhquez,* Estrella Alvarez, and Jose M. Navaza, Surface Tension of Alcohol + Water from 20 to 50 "C J. Chem. Eng. Data 1995,40, 611-614

8. Malcolm E. Schrader, Young-Dupre Revisited, 1995

9. O.Hölck, J.Bauer, O.Wittler, K.D.Lang, B.Michel, B.Wunderle, Experimental Contact Angle Determination and Characterisation of Interfacial Energies by Molecular Modelling of Chip to Epoxy Interfaces, 2011 Electronic Components and Technology Conference, 978-1-61284-498-5/11/$26.00 ©2011 IEEE

10. I. Zubel, K. Rola, J. Zalewska, Behavior of tensioactive compounds in the solutions for silicon anisotropic etching, Proc. of SPIE Vol. 8902 89022I-1

11. I. Zubel, M. Kramkowska, K. Rola, Silicon anisotropic etching in TMAH solutions containing alcohol and surfactant additives, Sensors and Actuators. A, 178, (2012) 126-135

Modeling and Simulation of a MEMS Thermal Actuator with Polysilicon Heater

Dries Dellaert, Jan Doutreloigne

Centre for Microsystems Technology (CMST), IMEC - Ghent University,
Technologiepark 914a, 9052 Gent, Belgium
Email: dries.dellaert@elis.ugent.be
Phone: +32-9-264-5369

Abstract

This paper describes the modeling and simulation of a MEMS thermal actuator which is heated by a polysilicon heater. The geometry of the structure was simplified to a one-dimensional structure in order to calculate the temperature profile and the resulting displacement. During the calculations also the stress distribution is obtained. Next to the modeling, the structure was simulated using the finite-element method. As the temperature profile depends on the displacement of the actuator, an iterative simulation approach was adopted. The structures were fabricated and their characteristics showed good agreement with modeled and simulated results.

1. Introduction

MEMS-switches for telecommunication require compact actuators, a low-resistivity signal path, good isolation and the actuators should be easily driven, preferably high impedant to minimize current losses. Guckel et al. [1] have demonstrated a thermal actuator where asymmetric heating is achieved by proper design of the actuator. This heatuator design manufactured in polysilicon [2] [3] is easily driven but cannot directly be used to route the signal, as both cold and hot arms are used to drive the actuator. The MEMSCAP DC switch [4] [5] solves this problem by using a silicon nitride tether as isolation structure, but suffers from a very low drive resistance. The bent-beam actuator [6] [7] has already been reported with polysilicon resistive heating underneath the beams [8], but these devices are not compact and can suffer from built-in stress, a disadvantage that is not present in the heatuator design if the beams are made equally long.

In the framework of designing a latching MEMS switch for uses in telecommunications, a thermal actuator was developed, which will be discussed in the following sections.

2. Actuator topology

The actuator developed in this work (see figure 1) consists out of two parallel electroplated nickel arms: a cold one and a hot one. These two arms are both connected to the substrate at one side and are connected to each other at the other side. Because of the temperature difference between both arms, the hot arm will expand more than the cold arm, causing the actuator to deflect sideways in the direction of the cold arm in the plane of the substrate. By transferring the thermal expansion in the x-direction to a

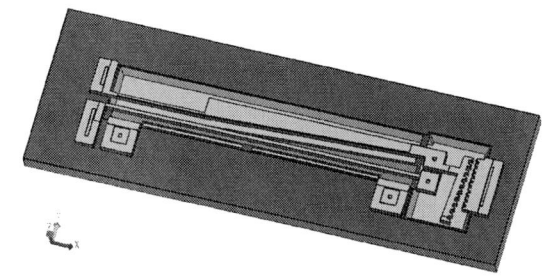

Figure 1. Topology of the proposed thermal actuator. (Upper nitride layer not drawn in order to show the poly heater.)

deflection in the y-direction, a displacement amplification is obtained. The cold arm is subdivided in a flex part, which is narrow, and a cold part, which is a lot wider. The cold arm will stay cooler because of the wider beam, which causes more heat loss to the substrate. The flex part is included to make the bending of the actuator possible.

The hot arm is heated by a polysilicon resistor, embedded in two layers of silicon nitride underneath the hot arm. By applying a current through the resistor it will heat up, causing the hot arm also to heat up. The poly heater is widened towards the end of the actuator so that the hot arm is still heated when the actuator is deflected. Because the heater is relatively close to the cold arm, this arm is also partly heated, which decreases the deflection. To minimize this parasitic effect, a cutout was made in the cold arm. Furthermore, the poly heater is anchored halfway its length to prevent it from sagging as its ends are anchored and the heater will also exhibit a thermal expansion. The cold and the hot arms are connected at their ends by means of a small nitride plate. This isolates both arms thermally, which increases the efficiency of the actuator. An other advantage is that both arms are also electrically isolated, which enables the two arms to carry different signals.

3. Analysis

In this section an electro-thermo-mechanical model is developed to calculate temperatures, displacements and stresses in the thermal actuator, which are compared to simulated results. In the modeling, the analysis is reduced to a one-dimensional one to simplify the calculations. For

978-1-4799-4789-8/14 $31.00 © 2014 IEEE

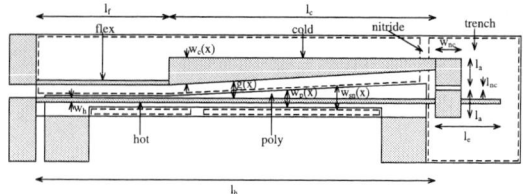

Figure 2. Geometry of the thermal actuator. ($l_h = 1000\mu m$.)

the following modeling, the geometrical parameters are defined in figure 2.

A. Electro-thermal analysis

The parts inside the actuator are relatively long compared to their cross sections, therefore a one-dimensional analysis will be performed. As heat conduction is the main heat transfer method in micro structures, convection and radiation will be neglected in this analysis [9] [10]. Using Fourier's law of heat conduction [11], a set of equations can be derived which describe the heat flow inside the structure. At every point along the longitudinal dimension, the sum of the the the net inwards heat flux and the generated heat inside that point should be equal to the total heat loss in that point to the substrate or to other parts in the structure. Because of the geometrical shape of the heater a simple analytical formulation is not possible. Therefore, a numeric approach is used here by discretizing the longitudinal edge in $N + 1$ points. With n the discretized longitudinal coordinate, the heat transfer equation inside the poly heater, the hot arm and the cold arm are given by:

$$
\begin{aligned}
&(2k_{sn}w_{sn}(n+1)t_{sn} + k_p w_p(n+1)t_p)\frac{T_p(n+1)}{\Delta} \\
&\quad - (2k_{sn}(w_{sn}(n+1)+w_{sn}(n))t_{sn} \\
&\quad + k_p(w_p(n+1)+w_p(n))t_p)\frac{T_p(n)}{\Delta} \\
&\quad + (2k_{sn}w_{sn}(n)t_{sn} + k_p w_p(n)t_p)\frac{T_p(n-1)}{\Delta} \\
&\quad + \frac{i^2\rho_p(1+\xi_p(T_p(n)-T_0))\Delta}{w_p(n)t_p} \\
&\quad - \frac{(T_p(n)-T_0)k_a(T_p(n),T_0)w_{sn}(n)S_p(n)\Delta}{t_{a2}} \\
&\quad - \frac{(T_p(n)-T_h(n))k_a(T_p(n),T_h(n))w_h S_h \Delta}{t_{a1}} = 0,
\end{aligned}
\tag{1}
$$

$$
\begin{aligned}
&k_n w_h t_n \frac{T_h(n+1)}{\Delta} - 2k_n w_h t_n \frac{T_h(n)}{\Delta} + k_n w_h t_n \frac{T_h(n-1)}{\Delta} \\
&\quad - \frac{(T_h(n)-T_p(n))k_a(T_h(n),T_p(n))w_h S_h \Delta}{t_{a1}} \\
&\quad - \frac{(T_h(n)-T_c(n))k_a(T_h(n),T_c(n))t_n S_{hc}(n)\Delta}{g(n)} = 0,
\end{aligned}
\tag{2}
$$

and

$$
\begin{aligned}
&k_n t_n w_c(n+1)\frac{T_c(n+1)}{\Delta} - k_n t_n(w_c(n+1)+w_c(n))\frac{T_c(n)}{\Delta} \\
&\quad + k_n t_n w_c(n)\frac{T_c(n-1)}{\Delta} \\
&\quad - \frac{(T_c(n)-T_0)k_a(T_c(n),T_0)w_c(n)S_c(n)\Delta}{t_{a2}} \\
&\quad - \frac{(T_c(n)-T_h(n))k_a(T_c(n),T_h(n))t_n S_{hc}(n)\Delta}{g(n)} = 0
\end{aligned}
\tag{3}
$$

respectively. With T the temperature, i the current through the heater, w and t the width and the thickness and k, ρ and ξ the thermal conductivity, the resistivity and the thermal coefficient of resistivity. The indices p, h, c and a indicate poly, hot, cold and air respectively; sn, n, $a1$ and $a2$ indicate silicon nitride, nickel, air between the heater and the hot arm and air between the heater and the substrate. T_0 is the substrate temperature, Δ is the length of a discrete element and $g(n)$ is the gap distance between the hot and the cold arm. The thermal conductivity of air $k_a(T_1, T_2)$ is calculated at the average temperature of T_1 and T_2. The poly heater loses not only heat to the substrate but also to the hot arm just above it. This outward heat flux is for the hot arm an inwards heat flux. Further, the hot arm loses also heat to the cold arm across the gap $g(n)$. The heat transfer between the different parts and the heat transfer between the substrate and the structure also take into account a shape factor S [12].

Above discretized differential equations need a set of boundary conditions to be solved. At the anchorage of the hot and the cold arm the temperature is equal to substrate temperature:

$$
T_h(0) = T_0 \tag{4}
$$

$$
T_c(0) = T_0. \tag{5}
$$

The endpoints of the poly heater are also equal to substrate temperature, but because of the connections to the heating element, the poly heater is pinned to substrate temperature over a distance of half the poly connection.

The final boundary conditions describe the heat flow at the endpoints of the hot and the cold arms. Because the two metal blocks, anchored to the silicon nitride, are relatively wide, they are assumed isothermal. The heat flow at the end of the hot or cold part is equal to the sum of the heat flow through the nitride part and the heat flow of the anchored part to the substrate. This results in following boundary conditions:

$$
\begin{aligned}
&-k_n w_c(N)t_n\left(\frac{T_c(N)}{\Delta} - \frac{T_c(N-1)}{\Delta}\right) \\
&\quad - \frac{(T_c(N)-T_0)k_a(T_c(N),T_0)w_{nc}\left(l_a + \frac{l_{nc}}{2}\right)S_e}{t_{a2}} \\
&\quad - (T_c(N)-T_h(N))\frac{k_{sn}w_{nc}2t_{sn}}{l_{nc}} = 0
\end{aligned}
\tag{6}
$$

978-1-4799-4789-8/14 $31.00 © 2014 IEEE

$$-k_n w_h t_n \left(\frac{T_h(N)}{\Delta} - \frac{T_h(N-1)}{\Delta} \right)$$

$$- \frac{(T_h(N) - T_0) k_a(T_h(N), T_0) w_{nc} \left(l_a + \frac{l_{nc}}{2} \right) S_e}{t_{a2}} \quad (7)$$

$$- (T_h(N) - T_c(N)) \frac{k_{sn} w_{nc} 2 t_{sn}}{l_{nc}} = 0,$$

with w_{nc} and l_{nc} the width and the length of the nitride connection between the two blocks, and l_a the length of the anchorage. S_e is a shape factor describing the heat flow of the two anchored blocks and the substrate.

Above equations made use of a shape factor to account for the excess heat flow caused by the two-dimensional fringing of the heat flow. In [12], an empirical formula has been proposed that describes the shape factor for heat flow between a line-shaped beam and an a large plane parallel to that beam. Using this formula, the shape factors for the poly heater and the cold arm become:

$$S_p(n) = \frac{2 t_{sn}}{w_{sn}(n)} \left(\frac{t_{a2}}{t_{sn}} + 1 \right) + 1 \quad (8)$$

and

$$S_c(n) = \frac{t_n}{w_c(n)} \left(\frac{2 t_{a2}}{t_n} + 1 \right) + 1. \quad (9)$$

For an estimation of the shape factor of the anchorage, the part is seen as a (short) line-shaped structure, connecting the cold and the hot arm. The shape factor here becomes

$$S_e = \frac{t_n}{w_{nc}} \left(\frac{2 t_{a2}}{t_n} + 1 \right) + 1. \quad (10)$$

To simplify the analysis, the shape factors S_{hc} and S_h are calculated for the situation in which the actuator is fully deflected. In this case, the cold arm is only heated by the hot arm and not by the poly heater, which is an approximation. The isothermal plane halfway the gap between the cold and the hot arm is parallel to the sidewalls of the beams, so the heat flow from the hot to the cold arm is the same as the situation in which the hot arm loses heat to a plane halfway the gap at a temperature which is the average of the two beams. A schematic representation of the heat transfer in this situation is depicted in figure 3. The heat transfer in this situation can be calculated by halving the temperature difference, halving the gap distance and the shape factor is then calculated with half the distance to the virtual plane. This results in:

$$S_{hc}(n) = \frac{w_h}{t_n} \left(\frac{g(n)}{w_h} + 1 \right) + 1. \quad (11)$$

S_h is also calculated as if the actuator is fully deflected (figure 4). In this situation the shape factor consists of half the shape factor like the situation of a large plane and half the vertical heat flow. S_h is thus given by:

$$S_h = \frac{1}{2} \left(\frac{t_n}{w_h} \left(\frac{2 t_{a1}}{t_n} + 1 \right) + 1 \right) + \frac{1}{2}. \quad (12)$$

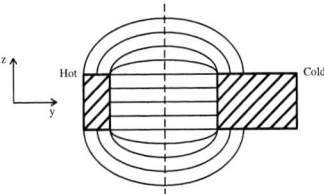

Figure 3. Estimation of the shape factor S_{hc} for the heat transfer between the hot and the cold arm.

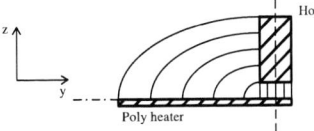

Figure 4. Estimation of the shape factor S_h for the heat transfer between the poly heater and the hot arm.

The difference equations together with the boundary conditions were put in a matrix description and were solved for the vectors T_p, T_h and T_c. As the thermal conductivity of air k_a is a function of temperature, the temperature profiles were calculated in an iterative way. Starting with an initial guess for the temperature, the difference equations together with the boundary conditions are solved with a matrix inversion. The newly found temperatures were then used as a new initial guess. The process was stopped once the difference between two consecutive solutions was small enough.

The resulting temperature profiles together with finite element simulations are shown in figure 5. The simulation was performed by enclosing the structure in a large volume of air ($1500 \mu m \times 500 \mu m \times 200 \mu m$, see figure 8) which boundaries are kept at room temperature. At the

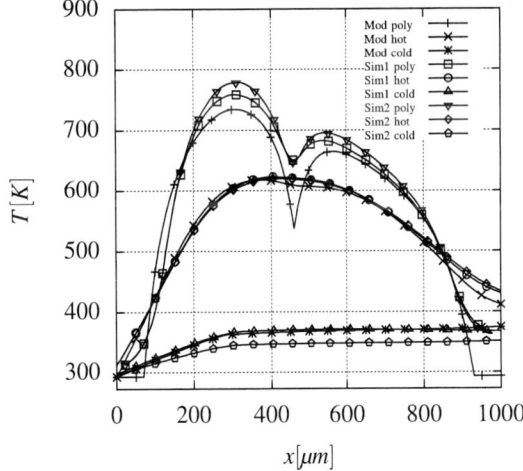

Figure 5. Modeled and simulated temperature profiles inside the actuator. Model: 11.4V, 10.4mA; Sim1: 11V, 10.4389mA; Sim2: 11V, 10.3359mA.

heater connections, a voltage was applied. Simulation *Sim*1 was performed on the undeflected structure, *Sim*2 was performed on the deflected structure which corresponds to the applied voltage (see section 3-C). As can be seen on the figure, the model makes a relatively good prediction of the temperature profile. The deviations can be addressed to the simple one-dimensional modeling and the shape factors, while the simulation takes into account the full three-dimensional temperature distribution.

B. Thermo-mechanical analysis

In the following section, the displacement and the stress in the structure will be calculated, making use of the temperature distribution in the structure. The geometry of the structure is simplified to the schematic in figure 6.

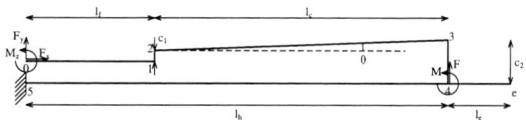

Figure 6.   Mechanical loads in the thermal actuator.

If the restrictions of fully anchoring of point 0 in figure 6 are removed, the point will translate a distance:

$$\delta = \int_{l_h} \alpha \left(T_h(x) - T_0 \right) dx - \int_{l_c + l_f} \alpha \left(T_c(x) - T_0 \right) dx. \quad (13)$$

Assuming that the TCE, α, is temperature independent, only the average temperature of the hot arm $\overline{T_h}$ and the average temperature of the cold arm $\overline{T_c}$ are needed.

To keep point 0 fully anchored at its original place, two reaction forces F_x and F_y, and a reaction moment M_z are needed. These reactions cause the following normal forces in the flex, cold and hot part:

$$
\begin{aligned}
N_f &= -F_x \\
N_c &= -F_x \cos\theta - F_y \sin\theta \\
N_h &= F_x,
\end{aligned}
\quad (14)
$$

and the following bending moments in the flex, cold and hot part:

$$
\begin{aligned}
M_f &= F_y s - M_z \\
M_c &= F_y l_f - M_z - F_x c_1 - F_x \sin\theta s + F_y \cos\theta s \\
M_h &= -F_y l_f + M_z + F_x c_1 + F_x \tan\theta l_c - F_y l_c \\
&\quad - F_x c_2 + F_y s + F s + M,
\end{aligned}
\quad (15)
$$

with s the distance variable in every separate part. To simplify the problem, the sections $[1-2]$, $[3-4]$ and $[4-e]$ (figure 6) are assumed rigid. Also dummy loads F and M are included for displacement calculations (see further).

The elongation calculated above can now be used to calculate the tip displacement. Using Castigliano's theorem [13] for a linear elastic structure, the following set of

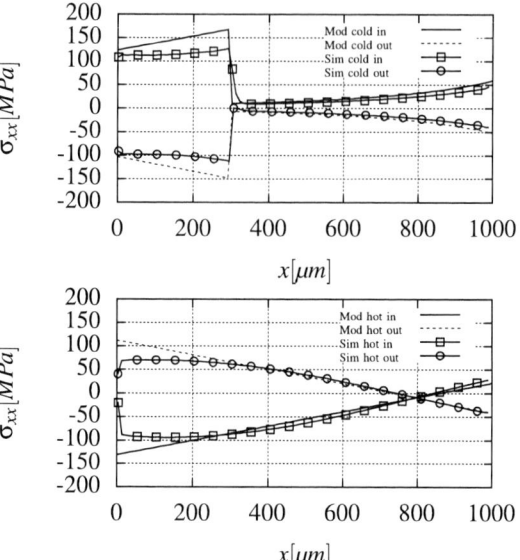

Figure 7.   Modeled and simulated σ_{xx} for an elongation of 2.25μm.

equations are obtained:

$$
\left. \frac{\partial U}{\partial F_x} \right|_{F=0, M=0} = -\delta \quad (16)
$$

$$
\left. \frac{\partial U}{\partial F_y} \right|_{F=0, M=0} = 0 \quad (17)
$$

$$
\left. \frac{\partial U}{\partial M_z} \right|_{F=0, M=0} = 0, \quad (18)
$$

which give a solution for the reaction forces F_x, F_y and M_z, with the strain energy U given by:

$$
\begin{aligned}
U =& \frac{1}{2} \int_0^{l_f} \frac{N_f^2}{EA_f} ds + \frac{1}{2} \int_0^{\frac{l_c}{\cos\theta}} \frac{N_c^2}{EA_c} ds \\
&+ \frac{1}{2} \int_0^{l_h} \frac{N_h^2}{EA_h} ds + \frac{1}{2} \int_0^{l_f} \frac{M_f^2}{EI_f} ds \\
&+ \frac{1}{2} \int_0^{\frac{l_c}{\cos\theta}} \frac{M_c^2}{EI_c} ds + \frac{1}{2} \int_0^{l_h} \frac{M_h^2}{EI_h} ds,
\end{aligned}
\quad (19)
$$

with E and A, elastic modulus and the cross-sectional area of the corresponding part respectively. The tensile residual stress in the nickel structure is not taken into account here, as both hot and cold arm are equally long. The tip displacement u_e is calculated as:

$$
u_e = \left. \frac{\partial U}{\partial F} \right|_{F=0, M=0} + l_e \left. \frac{\partial U}{\partial M} \right|_{F=0, M=0}. \quad (20)
$$

With the reaction forces and reaction moment known, the normal forces and bending moments in every section can now be calculated. Because of the long beams in the x-direction and the bending in the y-direction, the dominant stress will be the σ_{xx} component. For this geometry, σ_{xx}

results in:

$$\sigma_{xx} = -\frac{M_{bz}y'}{I_{zz}} + \frac{N}{A}, \qquad (21)$$

with M_{bz} the bending moment in the z-direction, N the normal force and I_{zz} and A the second moment of inertia and the area of the cross section. Figure 7 shows the resulting σ_{xx} stress for an elongation of $2.25\mu m$, which results in a tip displacement of $43\mu m$. As can be seen, the model can predict the stress very well. The deviation at the transitions between different parts is caused by three-dimensional effects, which were not included in the model.

C. Coupled electro-thermo-mechanical analysis

To calculate the deflection as function of the applied voltage, the analysis from section 3-A and section 3-B needs to be combined. In the model, the average temperature of the hot and the cold arm are calculated and transfered to the thermo-mechanical model. This can be done as the thermal coefficient of expansion is assumed to be temperature independent. A similar approach is used in the finite-element simulations: the average temperatures of the hot, cold and flex part are calculated, and these three temperatures are transfered to the thermo-mechanical simulation. Here the cold arm was split up in the flex and the cold part because the latter part is a lot wider.

A more accurate simulation, however, should also take into account that the thermal distribution depends on the deflection of the actuator. Indeed, in the deflected state the cold arm will be spaced further away from the poly heater, resulting in a lower temperature of this arm as can be seen in figure 5 and 8. Also, this arm will be closer to the edge of the trench, causing more heat loss to the substrate. As the temperature of the cold arm will decrease, the actuator will deflect further. To study the effect of this deflection dependence, an iterative approach is needed. In order not to rebuild the actuator 3-D model every time a new deflection is calculated, the actuator was drawn for a few deflections including the as-fabricated state. It should be noted that the deflected structures were actually rotated, pivoting at the anchorage of the flex part as seen on figure 8(b). Subsequently, all these structures were electro-thermally simulated (see figure 8) for a sweep of voltages and the average temperatures of the hot, cold and flex part were calculated, as well as the current through the heater. Starting from the temperatures of the as-fabricated structures, the deflection was simulated in an thermo-mechanical simulation. Next, this deflection was looked up in the list of drawn deflected structures and the interpolated values for the temperatures, using the electro-thermal simulations, were then used in the next thermo-mechanical simulation. The iteration process was stopped when consecutive temperatures were almost equal.

Figure 9 shows the electrical and the electro-mechanical characteristics of the actuator. Because of the lower temperature of the cold arm, the displacement will be higher

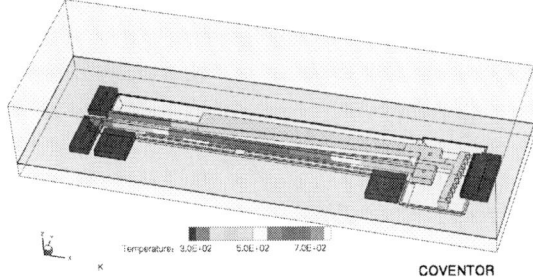

(a) Initial position.

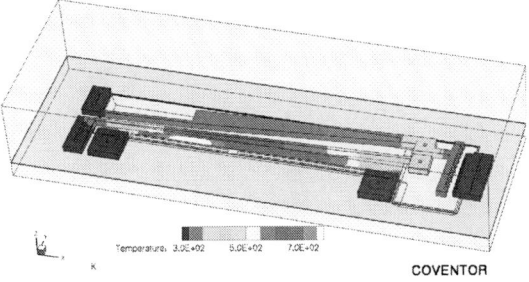

(b) Deflected position.

Figure 8. Temperature distribution in the actuator, driven by $11V$.

when using the method with iterations. From the figure it is also seen that the result of the model is actually closer to the simulated result without iterations (*Sim*1), despite that the model was calculated for the deflected state. This can be attributed to the simplistic modeling with shape factors. Additionally, these shape factors did not include the edges of the trench that caused extra cooling down of the cold arm.

4. Measurements

Figure 10 shows the modeled, simulated and measured characteristics of the thermal actuator. Good agreement is seen between the model, the simulation and the measured results. Displacements of $45\mu m$ are possible with an excitation of $12V$ and $11mA$. It is observed that the actual resistivity is slightly less than simulated. This could be due to process variations as the resistivity of the poly varies $\pm 13\%$ around its typical value [14]. Another cause may be an overestimation of the TCR in the simulations. It is also possible that the simulated temperature profile slightly overestimates the real temperature profile due to not perfectly known material properties.

Another observation on the displacement curves in figure 10 is that used actuators (not actuated for the first time) exhibit an offset as can be seen with *Meas*2 and *Meas*3. This is probably caused by plastic deformation of the beams in the actuator, accelerated by the high temperatures. Another possible cause may be friction, impeding the actuator to return to its initial position.

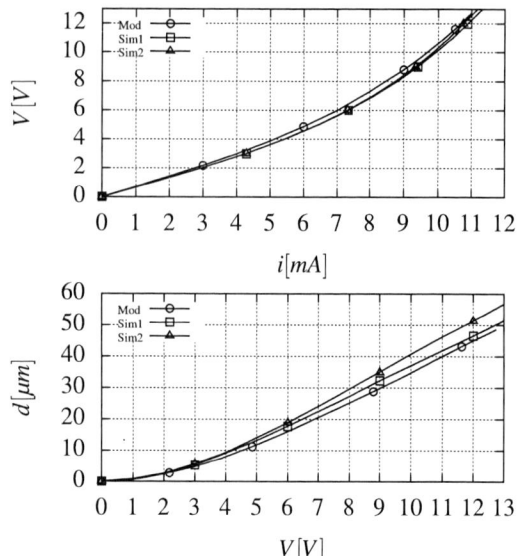

Figure 9. Modeled and simulated electro-thermo-mechanical characteristics.

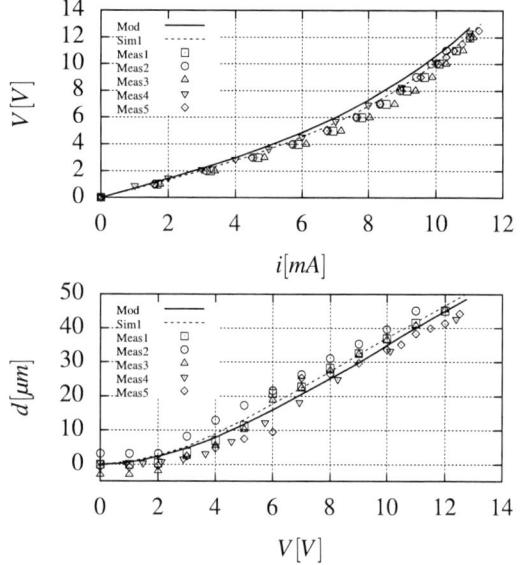

Figure 10. Measured, modeled and simulated electro-thermo-mechanical characteristics of the actuator.

5. Conclusions

The thermal actuator designed in this work exhibits a low drive current, which facilitates the implementation for system-in-package solutions. The beams, fabricated in nickel, make a low impedant signal path that can be used for signal routing.

For this actuator, an electro-thermo-mechanical model was developed, which could be used to quickly calculate temperatures, stresses and displacements, given the geometry of the thermal actuator. It was also indicated that the temperature profile depends on the deflection of the thermal actuator, which was investigated with an iterative simulation approach.

The modeled and simulated electrical and mechanical characteristics matched very well with the results measured on the fabricated devices.

6. Acknowledgment

This work is supported by a doctoral scholarship of the Special Research Fund (BOF) of Ghent University.

References

[1] H. Guckel, J. Klein, T. Christenson, K. Skrobis, M. Laudon, and E.G. Lovell. Thermo-magnetic metal flexure actuators. In *Solid-State Sensor and Actuator Workshop, 1992. 5th Technical Digest., IEEE*, pages 73 –75, jun 1992.

[2] J. H. Comtois and V. M. Bright. Applications for surface-micromachined polysilicon thermal actuators and arrays. *Sensors and Actuators A: Physical*, 58(1):19 – 25, 1997. Micromechanics Sections of Sensors and Actuators.

[3] J.H. Comtois, M.A. Michalicek, and C.C. Barron. Characterization of electrothermal actuators and arrays fabricated in a four-level, planarized surface-micromachined polycrystalline silicon process. In *Solid State Sensors and Actuators, 1997. TRANSDUCERS '97 Chicago., 1997 International Conference on*, volume 2, pages 769 –772 vol.2, jun 1997.

[4] V. Agrawal. A latching mems relay for dc and rf applications. In *Electrical Contacts, 2004. Proceedings of the 50th IEEE Holm Conference on Electrical Contacts and the 22nd International Conference on Electrical Contacts*, pages 222 – 225, sept. 2004.

[5] A. Maligno, D. Whalley, and V. Silberschmidt. Thermal fatigue life estimation and fracture mechanics studies of multilayered mems structures using a sub-domain approach. *World Journal of Mechanics*, 2(2):61 – 76, 2012.

[6] E.T. Enikov, S.S. Kedar, and K.V. Lazarov. Analytical model for analysis and design of v-shaped thermal microactuators. *Microelectromechanical Systems, Journal of*, 14(4):788 – 798, aug. 2005.

[7] D. Girbau, L. Pradell, A. Lazaro, and A. Nebot. Electrothermally actuated rf mems switches suspended on a low-resistivity substrate. *Microelectromechanical Systems, Journal of*, 16(5):1061 –1070, oct. 2007.

[8] M. Daneshmand, S. Fouladi, R.R. Mansour, M. Lisi, and T. Stajcer. Thermally-actuated latching rf mems switch. In *Microwave Symposium Digest, 2009. MTT '09. IEEE MTT-S International*, pages 1217 –1220, june 2009.

[9] R. Hickey, D. Sameoto, T. Hubbard, and M. Kujath. Time and frequency response of two-arm micromachined thermal actuators. *Journal of Micromechanics and Microengineering*, 13(1):40, 2003.

[10] O. Ozsun, B. E. Alaca, A. D. Yalcinkaya, M. Yilmaz, M. Zervas, and Y. Leblebici. On heat transfer at microscale with implications for microactuator design. *Journal of Micromechanics and Microengineering*, 19(4):045020, 2009.

[11] J. R. Welty, C. E. Wicks, R. E. Wilson, and G. Rorrer. *Fundamentals of Momentum, Heat, and Mass Transfer*. John Wiley & Sons, Inc., 2001.

[12] L. Lin and M. Chiao. Electrothermal responses of lineshape microstructures. *Sensors and Actuators A: Physical*, 55(1):35 – 41, 1996.

[13] N. Lobontiu and E. Garcia. *Mechanics of Microelectromechanical Systems*. Kluwer Academic Publishers, 2005.

[14] A. Cowen, R. Mahadevan, S. Johnson, and B. Hardy. Metalmumps design handbook, 2012.

Thermo-mechanical Stress of Underfilled 3D IC Packaging

Ming-Han Wang, Mei-Ling Wu

Department of Mechanical Engineering, National Sun Yat-Sen University, Kaohsiung, Taiwan
70 Lien-Hai Rd.
Kaohsiung, Taiwan (R.O.C)
E-mail address: m013020012@mail.nsysu.edu.tw
Tel.:+886-7-5254224

Abstract

In recent years, there has been a dramatic proliferation of research concerned with electronic products because of more various functions are integrate into the device and product's size has become smaller. As a result of these functional requirements, through silicon via (TSV) was investigated, this are getting considerable attentions not only from reducing the packaging size but also from shortening the interconnection's distance that can achieve the effect of enhancing signal transmission. TSVs are the vertical hole through the stacked IC, and they are also responsible for transferring signals between the ICs. Thus, they can improve the time delay of the signal transduction and allow better electrical performance than stacked ICs with wire bonding technology. However, a review of the literature indicates that electronic components will be affected easily by environmental factors such as humidity, pressure, and temperature. In general, the stacked ICs with TSV structure is easily affected by temperature changes than others factors since each material have different thermal expansion. In very recently, the stacked IC packaging has been primarily concerned with thermo-mechanical loadings than traditional single IC packaging, which leads some problems such as via cracking, die cracking and interfacial delamination and so on. The above problems not only affect the performance of the device but also lead the device fail. Hence, most of the studies [1-9] are focus on discussing thermal mechanical loading with simulation method. Some of them discuss the relationship between the TSV shape and the stresses [4, 5]. In addition, most of people just build local TSV structure to do their research [4-9]. Although it can save more time but it also increase the error percentage with real situation. And this paper build the three dimensional four layers stacked IC packaging model from Hsieh [1]'s paper which can more close to real situation. And setting the structure to be simulated from the temperature 150℃ to -50℃ which is as retreat temperature. This paper use ANSYS software which is based on finite element theory in order to reduce time used and save cost, as finite element simulation can provide results more quickly and cheaply than experiments. Moreover, the research mainly analyzes the maximum von-Mises stress in TSVs and micro-bumps. Besides, this paper will sort out geometries and material properties of underfill which will serious affect von-Mises stress value by Design of Experiments (DoE) analysis. Through the DoE analysis, the critical factors are selected as main design factors to reduce the von-Mises stresses. This study can provide the significant information to effectively design the products and increase the reliability. This information can also eliminate the testing time.

1. Introduction

System integration technology is significant because it has some advantages which are better performance than conventional technology and also because they satisfy recent requirements of electronic product requirements. Among today's electronic products, there are some common characteristics; faster signal transmission speeds, lighter weight and smaller size, and more input and output (I/O) counts. However, system integration technology is unlike the past chips which they only have a function, but it can provide multi-function as well as also satisfy the above conditions. And on of the most attention types integration technologies is system in packaging. SiP is widely used in most electronic products. Because SiP can stacked chips or packages in vertical direction which is called three dimensional package and it can also reduce the wiring (Figure 1) and make the faster signal transmission than horizontal package. These are the reasons why SiP has a pivotal position in the technology industry.

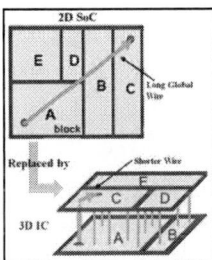

Figure 1: 2D SoC versus 3D IC [10]

However, there are lots of structures in SiP technology. It can be classified into two parts by through silicon via (TSV) structure. TSV is a hole through silicon and fill in metallic material, which it is responsible for transporting signals. Actually, the early SiP structures use wire bonding, flip chip, TAB and so on technology to stacked chips or packages. But above technologies still have some disadvantages to overcome until researchers invent TSV structure. TSV technology allows structures stacked more IC numbers and makes package thinner thickness. Besides, it can provide higher I/O density and better signal performance. Although the TSV is an

978-1-4799-4789-8/14 $31.00 © 2014 IEEE

advanced technology today, it can be affected by environmental factors which make it vulnerable to temperature, pressure and humidity. Moreover, SiP has serious heat dissipation problem and thermal stresses become an important issue. So this paper will focus on the four layers stacked ICs packaging with TSV structure which is affected by temperature fluctuations. The reason to cause thermal stress is the thermal expansion coefficient (CTE) mismatch. Since the packaging is composed of different materials, electronic products also exposed to environmental temperature fluctuations for a long time. Thus, thermal stress damage is a very important factor. In this research, TSVs and micro-bumps are the focus of the project since they are responsible for transferring signals of chips. This paper would contribution to reducing destruction of TSVs and micro-bumps.

More recently, with an increase in the use of the ANSYS software, which is based on the finite element method (FEM) simulations regarding the study of electronic packaging, we have seen a shift in the patterns of research. With the FEM as compared to the experimental method, there is more time-saving and money-saving. Consequently, the ANSYS software is adopted to analyze the thermal stress problem.

2. Preliminary Work

Research Process

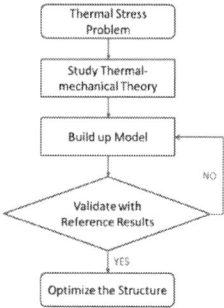

Figure 2: The flow chart of the study

Before starting the research, the research process must be planned, in order to effectively solve the problems and also to ensure the accuracy of the results. Therefore, the study is divided into the following sections: study thermal-mechanical theory, build up model, validate with theoretical results, and optimize the structure. Within those steps, building the model is the most important step because it will affect the accuracy and time consumption of the entire study.

After the completion of the model, the results will be analyzed until the results are converged. If the results are converge, it will be able to analyze the structure and provide the optimal design. Before the beginning of the research, we have to Figure out thermal-mechanical theory to know how thermal stress effects on stacked IC packaging under thermal-mechanical loading.

Literature Review

In our study, we also survey some papers about thermal-mechanical problem of TSV configuration and optimize underfill material. In thermal-mechanical problems, Dr. Hsieh are positively doing research about thermal problems in stacked IC packaging [1-3]. However, some people also study thermal stress problems about TSV configuration. Some people study the relationship between thermal stresses and via geometries. Then the reliability for TSV configuration can be improved by changing critical geometry factors [4-6].Besides, Dr. Hsieh also did some thermal stress studies on fcBGA structure by different underfills [12, 13]. Although underfill would cause larger peeling stress at the interface between chips and underfill, it can effectively stresses in thermal mechanical loadings.

Thermal-mechanical Theory

In order easily realize thermal stress theory, we show two-dimensional flip chip structure (Figure 3) which is proposed by Vandevelde [14]. As show in the Figure 3, component 1 is a chip, component 2 is a printed circuit board (PCB), and the parts between the chip and the PCB are solder joints.

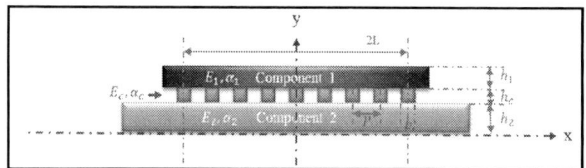

Figure 3: The schematic diagram of the flip chip package

Vandevelde analyzed the major factors for damage caused by thermal stress. Figure 4 shows the temperature difference results when external forces are applied to each component. In a temperature changing environment, each component is not only subjected to shear forces and axial forces, but is also affected by the bending moment. However, these external forces are due to the CTE mismatch as shown in Figure 5 that make each component distort differently.

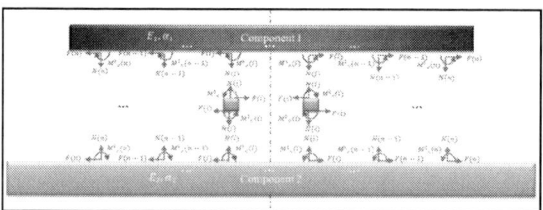

Figure 4: External forces applied to each component

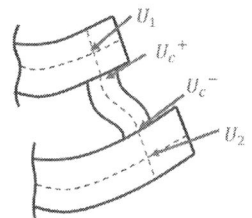

Figure 5: The distortion of the flip chip packaging

3. Methodology

Due to system integration technology develops rapidly. In addition, stacked IC packaging always exist thermal problems. Therefore, the research about stacked IC packaging is important. So we survey Dr. Hsieh's paper [1], and build up the same model. Then, validate with his results to prove our model can be trusted. After that, we analyze kinds of different underfill materials and provide the optimal underfill material characteristics by Taguchi method. Figure 6 is a 4-layer stacked IC packaging, which it is the same geometries and material properties with Dr. Hsieh's paper [1]. Table 1 and 2 are the geometric dimensions and material properties. There are copper via structures which connect with Sn/Ag/Cu micro-bumps and also responsible for transporting signals of 4-layer stacked ultra-thin dies. Besides, the underfill material is utilized to protect micro-bumps and reduce stresses and filled between chips. Silicon substrate is located beneath silicon dies and directly attached to testboard by Sn/Ag/Cu solder balls. Because of the structure is symmetry, the model can be replaced by quarter model. And it can also reduce analyzing time and provide the same accuracy with full model. There are 400 solder balls, 303 TSVs and 404 micro-bumps in the quarter section of this packaging. In this study, the ANSYS software is adopted. In Figure6, the coarse model are meshed with hexahedral elements by Solid185 element type. Therefore, the coarse model is applied the symmetrical boundary in symmetrical area and fixed in the bottom of symmetrical areas' interface. The loading condition is set from $150\,^{\circ}C$ to $-55\,^{\circ}C$.

In study process, the element size serious affects stress distributions in finite element analysis. Hence, the submodeling method (also known as "cut-boundary displacement method" or "specified boundary displacement method") has been applied. It is based on St. Venant's Principle and allows more accurate results in a particular region which is far enough away from the stress concentration region. In addition to this, it can also save more computing time which it don't have to analyze full model with finer mesh. Due to the micro-bumps and TSVs is the critical part in the submodel of packaging, so it is meshed finer than other parts. (Figure 6)

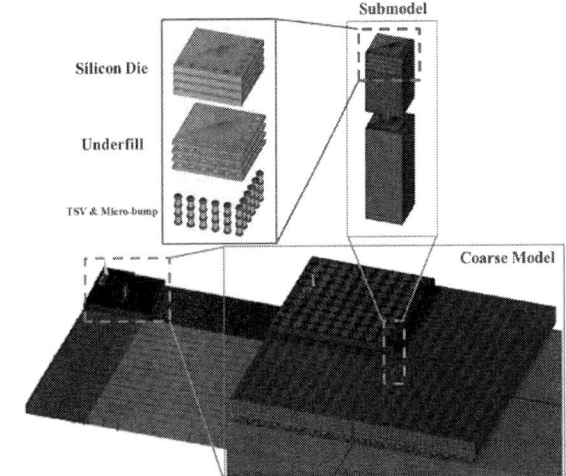

Figure 6: Finite element model of 4-layer stacked IC packaging

Table 1: Geometric dimensions for 4 layer stacked IC package (Unit: mm) [1]

Dimensions	Length	Width	Height
Silicon die	10.0	10.0	0.050
Substrate	20.0	20.0	0.725
Testboard	76.2	114.3	1.600
Dimensions	Diameter	Height	Pitch
Micro-Bump	0.05	0.025	0.1
Solder Ball	0.30	0.300	0.5
Copper Via	0.05	0.050	0.1

Table 2: Material properties for 4 layer stacked IC package [1, 11]

Material Properties	E (GPa)	v	CTE(ppm/K)
Silicon Die	162.5	0.29	3.2
Substrate	162.5	0.29	3.2
Testboard	22.0	0.28	18(X、Z) 75(Y)
Material Properties	E (GPa)	v	CTE(ppm/K)
Copper Via	120.5	0.34	17
Micro-Bump	74.84-0.08T	0.30	16.66+0.017T
Solder Ball	74.84-0.08T	0.30	16.66+0.017T
Underfill	6.9/0.15 (Tg=120 ℃)	0.33	29/100 (Tg=120 ℃)

4. Result and Discussion

Due to the reason of stress distribution is serious affected by element size. But displacement is converged that can be validated our model is accurate (Figure 7). Another reason why we have to make sure the displacement is correct, because the nodal displacement results in submodel of cut-boundary have to input from coarse model. After simulated, the von-Mises stress results (Figure 8) show the stress distribution of the submodel. From the results, it can be obvious observed that the maximum von-Mises stress always causes at the interface between two components. In submodel, the maximum von-Mises stress cause at the interface between underfill and chip. Because the CTE mismatch deviation between chip nd underfill are larger. If it exceed the attach strength, it would cause delamination phenomenon.

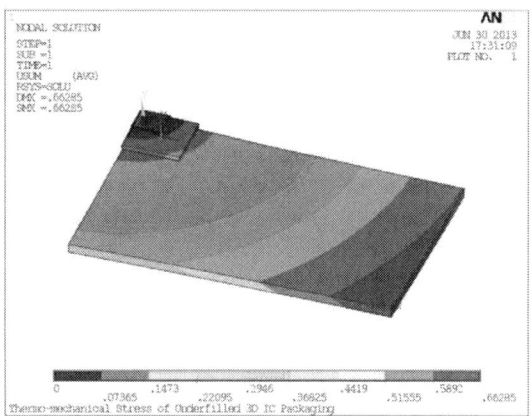

Figure 7: Displacement of coarse model

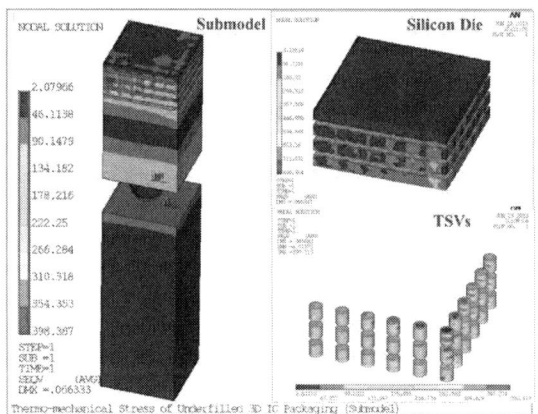

Figure 8: von-Mises stress of our submodel model

Consequently, the reliability of the packaging must be improved. So the material properties of different underfills is employed from Dr. Hsieh [12] (Table 3). Table 4 show maximum von-Mises stress results on each component by selected underfills. From the results, it can be discovered that underfill affect the von-Mises stress at micro-bump configuration obviously (the deviation between maximum and minimum value is larger than others), but the stresses of other components may not

affected apparently. Because underfill is filled to protect the micro-bumps. Meanwhile, it can also reduce thermal stress of the chips.

Table 3: Material properties of underfills [12]

Underfill	T_g ($^\circ C$)	E_1/E_2 (GPa)	ν	CTE α_1/α_2 (ppm/$^\circ$C)
Standard	120	6.9/0.15	0.33	29/110
A	120	6.9/0.15	0.35	29/110
B	88	11.0/0.24	0.35	31/95
C	100	10.7/0.09	0.35	30/110
D	122	7.9/0.20	0.35	26/107
E	104	10.3/0.30	0.35	22/88
F	104	8.1/0.09	0.35	27/109
G	110	9.9/0.13	0.35	27/90

Table 4: Maximum von-Mises stress of selected underfills (unit: MPa)

Underfill	Submodel	Silicon Die	Micro-bump (MPa)	TSV
Standard	407.027	808.036	1306.180	523.209
A	406.799	808.114	1278.740	529.387
B	466.611	772.670	984.745	519.401
C	448.920	773.679	1005.810	516.89
D	402.023	797.072	1212.710	516.888
E	393.918	775.792	1084.600	495.752
F	401.540	794.602	1189.090	518.001
G	401.666	777.807	1069.320	510.474

Therefore, we also analyze which critical underfill material factors serious affect von-Mises stresses in each component. However, E_2 and α_2 wouldn't affect results any more. Because underfill would transfer its material properties apparently after glass transition temperature (T_g), the case is simulated in retreat temperature.

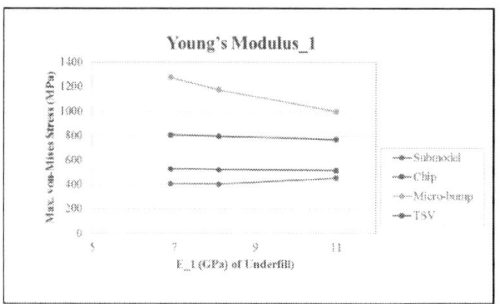

Figure 9: Max. von-Mises results Vs. E_1

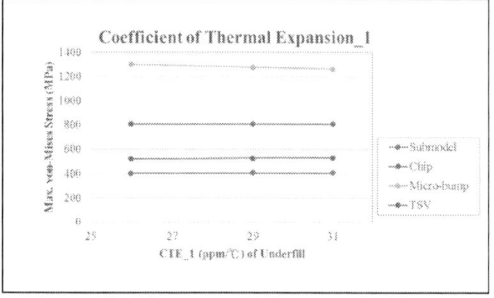

Figure 10: Max. von-Mises results Vs. α_1

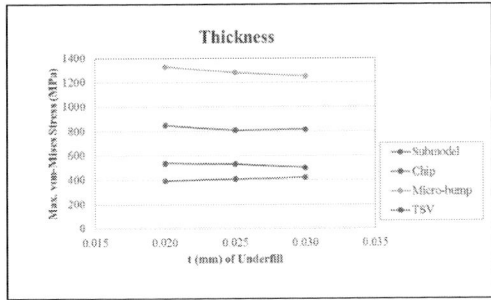

Figure 11: Max. von-Mises results Vs. thickness

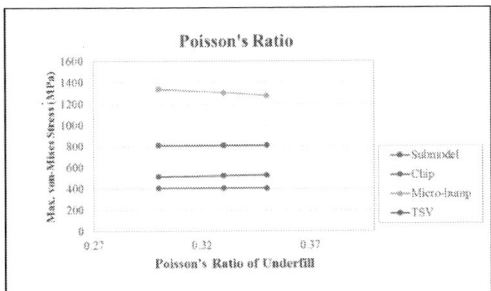

Figure 12: Max. von-Mises results Vs. v

Then, the results from Figure9 to Figure 12 are critical factors analysis of each component. Actually Coefficient of thermal expansion_1 doesn't affect the results very much due to the smooth curve. Young's modulus_1, thickness and Poisson's ratio are obvious affect maximum von-Mises stress of micro-bump. When the larger value, the von-Mises stress would be reduced. But it's not evident from other result. Thus we would adopt DoE method to get the best material properties of underfill.

5. Conclusions

The larger CTE mismatch deviation would cause the larger thermal stress, so the interface between underfill and chip would cause delamination phenomenon until it exceed the attach strength. E2 andα2 wouldn't affect results any more since the ambient temperature is below glass transition temperature (Tg) of underfill. Coefficient of thermal expansion_1 doesn't affect the results very much. Young's modulus_1, thickness and Poisson's ratio are obvious affect maximum von-Mises stress of micro-bump. When the larger value, the von-Mises stress would be reduced.

Acknowledgments

The authors would like to thank my advisor Dr. Wu and our members of MOEMS lab from National Sun Yat-Sen University, Kaohsiung, Taiwan, for their suggestions of simulation.

References

1. Hsieh, Ming-Che C., Yu, Chih-Kuang. (2008). Thermo-mechanical Simulations For 4-Layer Stacked IC Packages. International *Conference on Thermal, Mechanical and Multi-Physics Simulation and Experiments in Microelectronics and Micro-Systems. EuroSimE 2008.*
2. Hsieh, Ming-Che C., Yu, Chih-Kuang, Wu, Sheng-Tsai T. (2010). Thermo-mechanical simulative study for 3D vertical stacked IC packages with spacer structures. *Semiconductor Thermal Measurement and Management Symposium, 2010. SEMI-THERM 2010. 26th Annual IEEE.*
3. Hsieh, Ming-Che C., Lee, Wei Y. (2008). FEA modeling and DOE analysis for design optimization of 3D-WLP. *Electronics System-Integration Technology Conference, 2008. ESTC 2008. 2nd.*
4. SeMin Park, HanSur Bang, HeeSeon Bang, JaeSun You. (2012). Thermo-mechanical analysis of TSV and solder interconnects for different Cu pillar bump types. *Microelectronic Engineering, Volume 99, November 2012, Pages 38-42.*
5. Zhou, Jing, Yu, Daquan, He, Ran, WeiDai, Feng, Guo, Xueping, Song, Chongshen, Wang, Huijuan, Guidotti, Daniel, Cao, Liqiang Q., Wan, Lixi. (2011). Nonlinear thermal stress & strain analysis of through silicon vias with different structures and polymer filling. *Electronics Packaging Technology Conference (EPTC), 2011 IEEE 13th.*
6. E.J. Cheng, Y.-L. Shen. (2012). Thermal expansion behavior of through-silicon-via structures in three-dimensional microelectronic packaging. *Microelectronics Reliability, Volume 52, Issue 3, March 2012, Pages 534-540*
7. Vandevelde, Bart, Okoro, Chuckwudi A., Gonzalez, Mario, Swinnen, Bart, Beyne, Eric. (2008). Thermo-mechanics of 3D-wafer level and 3D stacked IC packaging technologies. *International Conference on Thermal, Mechanical and Multi-Physics Simulation and Experiments in Microelectronics and Micro-Systems, 2008. EuroSimE 2008.*
8. Liu, Xi, Chen, Qiao, Dixit, Pradeep, Chatterjee, Ritwik, Tummala, Rao R., Sitaraman, Suresh K. (2009). Failure mechanisms and optimum design for electroplated copper Through-Silicon Vias (TSV). *Electronic Components and Technology Conference, 2009. ECTC 2009. 59th.*
9. Guide, A. N. S. Y. S. (2012). ANSYS Release 14.0. *Swanson Analysis Systems, Houston.*
10. http://www.2cm.com.tw
11. Hsieh, M. C., Lee, C. C., & Hung, L. C. Comprehensive Thermo-Mechanical Stress Analyses and Underfill Selection of Large Die Flip Chip Ball Grid Array.
12. Hsieh, M. C., Lee, C. C., & Hung, L. C. Comprehensive Thermo-Mechanical Stress Analyses and Underfill Selection of Large Die Flip Chip Ball Grid Array.
13. Wang, G., Groothuis, S., Merrill, C., & Ho, P. S. (2004, June). Investigation of interfacial delamination for Cu/low k structures during flip-chip packaging. *In Thermal and Thermomechanical Phenomena in Electronic Systems, 2004. ITHERM'04. The Ninth Intersociety Conference on (Vol. 2, pp. 211-218). IEEE.*
14. Vandevelde, B., Christiaens, F., Beyne, E., Roggen, J., Peeters, J., Allaert, K., Vandepitte, D., Bergmans, J. (1998). Thermomechnical Models for Leadless Solder Interconnections in Flip Chip Assemblies. *IEEE Transactions on Components, Packaging, and Manufacturing Technology.*

Investigation Of Color Shift Of LEDs-Based lighting Products

S. Koh[1,2,3], H. Ye[2,4], M. Yazdan Mehr[2,4], J. Wei[2,3*], W.D. van Driel[2,5], L.B. Zhao[6], G.Q. Zhang[1,2,6]

[1]Beijing Research Center, Delft University of Technology, Beijing, China
[2]Delft University of Technology, EEMCS Faculty, Delft, the Netherlands
[3]State Key Laboratory of Solid State Lighting, Beijing, China
[4]Material innovation institute (M2i), Delft, the Netherlands
[5]Philips Lighting, Eindhoven, the Netherlands
[6]Institute of Semiconductor, Chinese Academy of Science, Beijing, China

*s.w.koh@tudelft.nl
Delft University of Technology/DIMES
2628 CT, Delft, The Netherland

Abstract

LED has the potential to revolutionize the lighting industry with their long lifetime and high efficacy. However, the long lifetime claimed by the manufacturers is often based solely on the estimated lumen depreciation and this is insufficient since the color shift of LEDs products is another important issue that need to be considered for LEDs products in order for them to perform their intended function. It has been shown through DOE's CALIPER program and GATEWAY program that color aging during LED products lifetime can be so extensive that it could be considered as a failure to the end-user. Nevertheless, although performance evaluation based on LED lumen depreciation had been well documented and established, literature on the performance evaluation based on color shift is very limited. This is because the mechanism for color shift is not well understood, nor the measurement / characterization methods. Therefore, an investigation of color aging over lifetime is urgently needed. Therefore, this paper will attempt to address the color shift of the plastic lenses and proposed a model for predicting color shift in LEDs lenses.

Introduction

According to International Energy Agency estimation [1], lighting application accounts for about 19% of world's total electricity usage. However, current artificial lighting technology, with the exception of solid-state lighting (SSL), is extremely inefficient (Figure 1). Hence, SSL, which is known for their low energy consumption and environment friendliness, has been recognized as the second revolution in the history of lighting. This is because switching to SSL will provide an annual global energy bill saving of €300 billion and a reduction of 1,000 MT of CO_2 emission. This is of especially strategic significance for the present high-energy demand of society and the current depletion of the world oil reserves. However, as shown in Figure 1, current artificial lighting technology, with the exception of Light Emitting Diodes (LEDs), is extremely inefficient [2]. Hence, LEDs has the potential to revolutionize the lighting industry, as they are known to have an efficacy of 150 lm/W. This is as compared to only about 15 lm/W for a conventional 60–100 W incandescent light bulb [3]. Furthermore, LEDs luminaires claimed to have a lifetime of more than 25,000 hours and this exceeds the life of nearly all the other light sources [4].

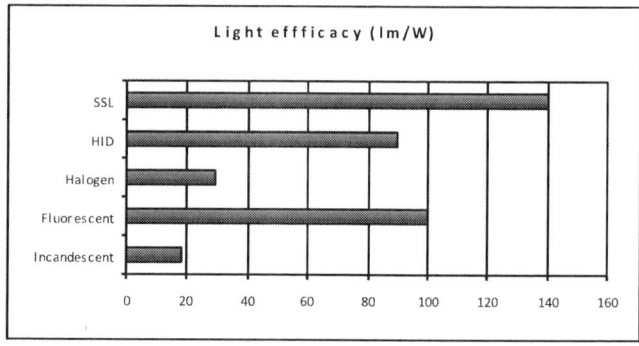

Figure 1: Lighting technology and their efficacy [2]

The long lifetime claimed by the manufacturers is often based on the estimated lumen maintenance [5,6]. Hence, in depth investigation on their lumen maintenance over time had been conducted and their degradation model had been well established [5-19]. The lumen depreciation over time had been found to follow an exponential relationship [5-14] as shown below.

$$L = \beta e^{\alpha t} \qquad (1)$$

In the above relation, L is the lumen maintenance at time t whereas α and β are the fitting constant [5].

$$\alpha = \alpha_0 e^{\frac{Q}{kT}} \qquad (2)$$

The relationship for lumen maintenance over time had been so established that industrial standards [20] and international standards [5,21] had been drafted based on the above relationship

However, this is insufficient since lumen depreciation is only one of the many possible failure modes. Focusing only on this single failure mode might mislead the industry and consumers since the proposed lumen maintenance testing procedure cannot sufficiently cover the real user profiles. In fact, DOE's CALIPER program [22] and Gateway program [23] had shown that color shift during LED products lifetime can be so extensive that it could be considered as a failure.

Nevertheless, literature on the performance evaluation based on color shift is very limited. As with any complex system, there may be a need to understand their failure mechanism of their critical sub-components in order to fully understand their reliability [24]. This paper will attempt to address the different failure mechanism of induces significant color shift in LEDs lamps.

Degradation mechanism

Yellowing of Polycarbonate plates

The discoloration and yellowing of polycarbonate lenses (PC) over time is one of the more common failure mechanisms, which will cause the reduction of LED light output [24] and also the shift of the color coordinates toward the yellowish color. In fact it was this shift toward yellowish color in the Smithsonian GATEWAY project that rendered the light to be considered to have loss their ability to perform its function at around 4000 hours although the color depreciation is only about 8%.

The rate of yellowing of these PC lenses had been found to be enhanced by prolonged exposure to short wavelength emission while being exposed to high temperature. This is because these conditions favor the formation of photo-Fries rearrangement and photo oxidation [25,26].

The photo-Fries rearrangement, as shown in Figure 2 consist of 3 main steps [25,26]:
1. The formation of two radicals,
2. Recombination,
3. Hydrogen abstraction.

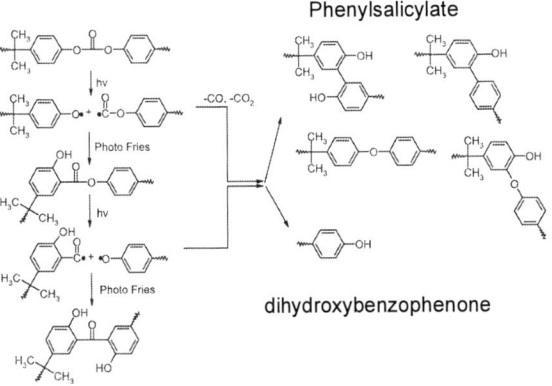

Figure 2: Photo-Fries rearrangement

Photo-oxidation, as shown in Figure 3, involves the following steps [24,25]:
1. The formation of free radicals
2. The reaction of free polymer radicals with oxygen
3. Rearrangements and chain scissions during branching

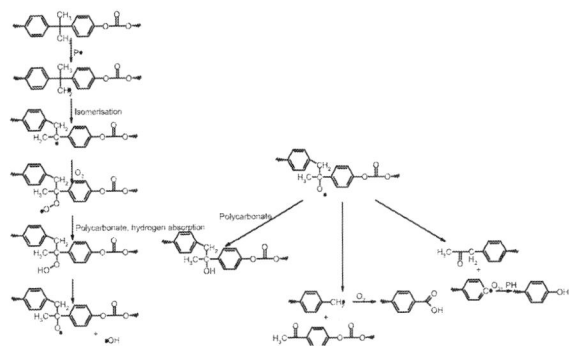

Figure 3: Photo oxidation

Despite these complex reactions, Koh et al [27] had demonstrated that the yellowing of the polycarbonate follow a linear relationship in a uv coordinates as shown in figure 3 and it is still dominated by a zero order reaction. Thus, their shift in color coordinates over time for different working temperature is as shown in eqn 3

$$\Delta \mathbf{uv} = \mathbf{A} + \mathbf{B} \exp\left(\frac{\mathbf{Ea}}{\mathbf{kT}}\right)\mathbf{t} \qquad (3)$$

In eqn (3), *A* is the incubation time in hours whereas *Ea* is the activation energy for yellowing in eV.

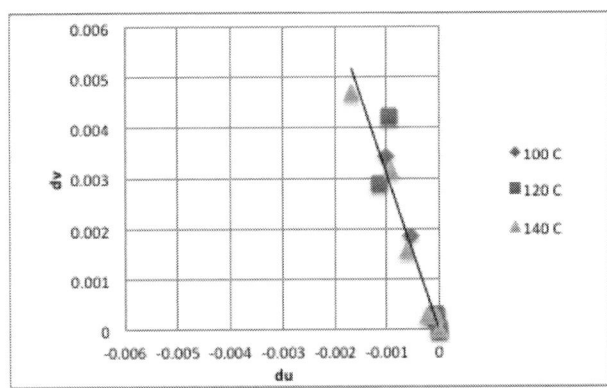

Figure 3: Color shift of BPA-PC plates for different conditions

Implementation

A LED lamp with remote phosphor is used in this research to study the feasibility of this test and also to formulate the maximum allowable acceleration factor.

For this particular lamp, a LED package with 6 LEDs is placed on a metal core PCB (MCPCB), which in turn is placed on a heat sink to dissipate heat away from the LEDs. A PC layer is then placed on the top of the LED package with a plastic support. A schematic drawing of the sample is shown in Figure 4.

978-1-4799-4789-8/14 $31.00 © 2014 IEEE

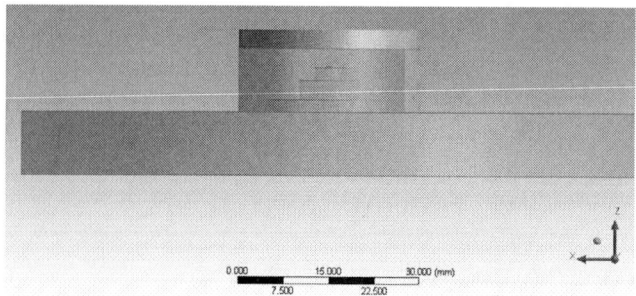

Figure 4: A schematic drawing of the sample

As shown in eqn 3, the rate of color shift is dependent on the temperature distribution of the LEDs lamps. Hence, a FEA simulation is used to analyze the temperature distribution of the LEDs lamps that is exposed to its normal working condition. The parameter used is listed in Table 1. Air convection and conduction is also modeled in the FEA simulation.

Table 1: Parameter used in the FEA simulation

Blue light		PC sample	
Power input	1.043 W	Diameter	50 mm
Optical power	0.034 W	Thickness	3 mm
Heat produced by LED	0.7 W	Efficiency of PC	80%*
Heat density in LED	2.43E9 W/m³	Heat produced by PC	0.0686 W
Number of LEDs	6	Heat density in PC	8.09E3 W/m³
Phosphor layer			
Efficiency of Phosphor layer	60 % (CT= 3000K)		
Heat produced by phosphor layer	0.1372 W		
Heat density in phosphor	1.62E4 W/m³		

Figure 5 shows the simulation result and it show that the temperature in the PC layer is around 40 °C. This temperature will be used during the subsequent analysis for the accelerated testing.

For the particular commercial PC material used in this LED lamps, its color shift had been found to be as follows:-

$$\Delta uv \approx 6.93e-6 * e^{\frac{-0.046}{k_bT}} * (t-500) \quad (4)$$

Given the plate's temperature is computed to be around 40°C, the color shift for this particular plate alone will be about 0.0065 CIE u'v'. Since one of the requirements for Department of Energy (DOE) energy star program [28] requires the LEDs lamp to have a color shift of less than 7 SDCM or the change in 0.007 CIE u'v' color space in 6000hrs, using this particular grade of PC materials will result in the lamps barely passing energy star's requirement.

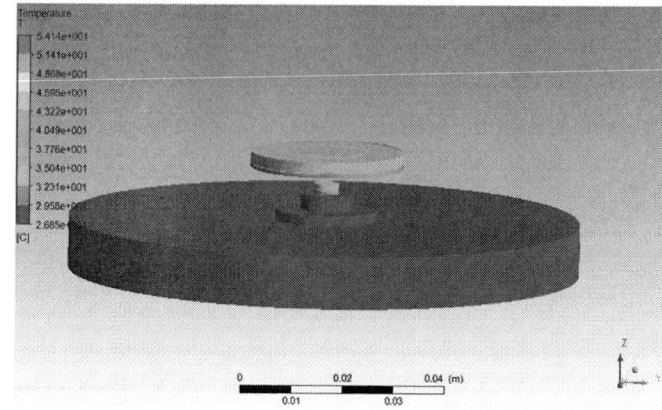

Figure 5: The temperature distribution of LED lamps as simulated by finite element analysis

Accelerated testing

The specified 6000 hours testing duration is much too long for the fast developing LED technology and industry, and it may cause unnecessary delay for new and good LED product creation and market introduction. There is a need to reduce the testing duration for color shift. Given that the transition temperature for most PC materials are about 150 °C and the uncertainty for u'v' measurement for most integrating sphere is about 0.001 or 0.002 du'v', a "logical" accelerated testing with the following conditions can be used in conjunction with eqn 3.

1. Du'v'=0.003
2. Temperature=140°C

Using eqn 3, the accelerated factor is computed to be around 5. Hence, we could postulated that the lamp using this polycarbonate material will not be able to pass the requirement for the energy star program with only 1200 hrs of testing as compared to 6000 hrs of testing for energy star method.

LEDs packages

Another common failure mechanisms for color shift are due to the failure of LEDs packages. A simulation tool, Light tool, is used to investigate the effect of these mechanisms that will affect the color shift for a 3000K LED lamps [29]. Figure 6 shows a summary of their results.

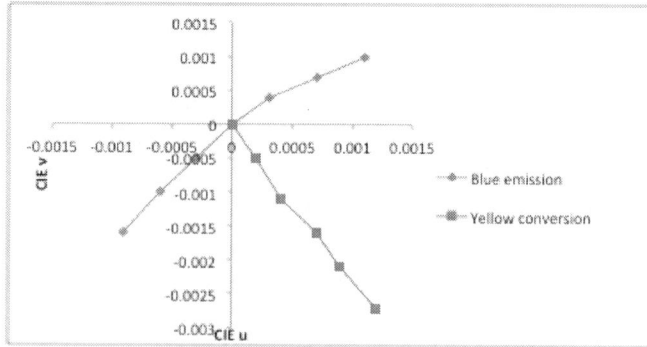

Figure 6: Effect of the degradation of blue LED and yellow conversion phosphor on color shift on LED lamps

Figure 6 shows that the magnitude of the color shift is linearly proportional to the degree of degradation. However, the direction of the shift is highly dependent on the different types of failure mechanisms. Nevertheless, the accelerated condition for other mechanism relating to LEDs can be deduced using the same methodology as outlines in the example using PC plates.

Conclusions

In this paper, some of the mechanism for color shift had been investigated. The yellowing and discoloration of these PC lenses is due to prolonged exposure to short wavelength emission at high working temperature, which resulted in photo-Fries rearrangement and photo oxidation. The effect of decreases in blue LED's lumen maintenance and the efficiency of yellow phosphor had also been investigated using simulation. Furthermore, using discoloration of PC lenses as an example, a methodology for accelerated testing for color shift had been proposed. For this particular grade of commercial PC, a acceleration factor of 5 had been achieved.

Acknowledgments

S. Koh and J. Wei will like to acknowledge the EMRP Researcher Grant ENG05 REG3 for their financial support whereas M. Yazdan Mehr will like to thank project number M71.9.10380 in the framework of the Research Program of the Materials innovation institute M2i for their financial support. Lastly, H. Ye will like to thank E63.9.10397 in the framework of the Research Program of the Materials innovation institute M2i for their financial support.

References

1. International Energy Agency. (2006). Light's Labours Lost – Fact Sheet. Available: http://www.iea.org/textbase/nppdf/free/2006/light_fact.pdf
2. T. Ribarich. (2010). "Build An Efficient HID Lamp Driver Circuit." Available: http://electronicdesign.com/article/design-solutions/Build-An-Efficient-HID-Lamp-Driver-Circuit.aspx
3. P. Mottier, LEDs for Lighting Applications: ISTE, 2009.
4. U. S. DOE., "LED Measurement Series: LED Luminaire Reliability". Available: http://apps1.eere.energy.gov/buildings/publications/pdfs/ssl/luminaire_reliability.pdf
5. Illuminating Engineering Society, TM-21-11 Standard. "Projecting Long Term Lumen Maintenance of LED Light Sources"; 2011
6. Philips lumileds, "Lifetime Behavior of LED Systems White Paper", 2010
7. Narendran, N. and Y. Gu, Life of LED-based white light sources. Journal of Display Technology, 2005. 1(1): p. 167.
8. Narendran, N., et al., Solid-state lighting: failure analysis of white LEDs. Journal of Crystal Growth, 2004. 268(3-4): p. 449-456.
9. Bürmen, M., F. Pernuš, and B. Likar, Accelerated estimation of spectral degradation of white GaN-based LEDs. Measurement Science and Technology, 2007. 18: p. 230-238.
10. Narendran, N., et al., Long-term Performance of White LEDs and Systems. 2007.
11. Narendran, N., et al. Performance characteristics of high-power light-emitting diodes. 2004.
12. Trevisanello, L., et al., Accelerated life test of high brightness light emitting diodes. IEEE Transactions on Device and Materials Reliability, 2008. 8(2): p. 304-311.
13. Nogueira, E., M. Vázquez, and N. Núñez, Evaluation of AlGaInP LEDs reliability based on accelerated tests. Microelectronics Reliability, 2009. 49(9-11): p. 1240-1243.
14. U.S Department of energy, Energy star program, "ENERGY STAR® Program Requirements or Integral LED Lamps ", 2013
15. M. Pecht and M.-H. Chang, "Failure Mechanisms and Reliability Issues in LEDs," in Solid State Lighting Reliability, ed: Springer, 2013, pp. 43-110.
16. M. Meneghini, L. Trevisanello, C. Sanna, G. Mura, M. Vanzi, G. Meneghesso, et al., "High temperature electro-optical degradation of InGaN/GaN HBLEDs," Microelectronics Reliability, vol. 47, pp. 1625-1629, 2007.
17. E. Jung, J. H. Ryu, C. H. Hong, and H. Kim, "Optical degradation of phosphor-converted white GaN-based light-emitting diodes under electro-thermal stress," Journal of The Electrochemical Society, vol. 158, pp. H132-H136, 2011.
18. G. Meneghesso, M. Meneghini, and E. Zanoni, "Recent results on the degradation of white LEDs for lighting," Journal of Physics D: Applied Physics, vol. 43, p. 354007, 2010.
19. M.-H. Chang, D. Das, P. Varde, and M. Pecht, "Light emitting diodes reliability review," Microelectronics Reliability, vol. 52, pp. 762-782, 2012.
20. ASSIST Recommends: LED Life Testing. Vol. 1-6, 2005. Lighting Research Center, Rensselaer Polytechnic Institute, Troy, NY, 2005.
21. CSAS-20, "Accelerated lumen depreciation life test for LED product", China Solid State Alliance Standards, Beijing, 2013
22. T Clark, "Color Shift", 2011 Solid-State Lighting Market Introduction Workshop Materials, 2011, Seattle, US
23. M Royer, R Tuttle, S Rosenfeld, N Miller, "Color Maintenance of LEDs in Laboratory and Field Applications," Report for Gateway demostration project, 2013
24. W. van Driel, et al. "LED system reliability", Proceedings 12th Int. Conf. on Thermal, Mechanical and Multiphysics Simulation and Experiments in Microelectronics and Microsystems, EuroSimE 2011.
25. Yazdan Mehr M, van Driel W.D, Jansen K.M.B, Deeben P, Boutelje M, Zhang G.Q. "Photodegradation of bisphenol A polycarbonate under blue light radiation and its effect on optical properties.", Journal of Optical Materials 2012, Vol. 35(3), Pages 504–508
26. M. Diepens, "Photodegradation and Stability of Bisphenol A Polycarbonate in Weathering Conditions," Thesis, Eindhoven University of Technology, 2009

978-1-4799-4789-8/14 $31.00 © 2014 IEEE

27. S. Koh, M. Yazdan Mehr, H. Ye, J. Wei, W.D. van Driel, G.Q. Zhang, "Color Shift Of Led-Based Products, ChinaSSL 2013, China, Beijing
28. U.S Department of energy, Energy star program, "ENERGY STAR® Program Requirements or Integral LED Lamps ", 2013
29. S. Koh "Degradation mechanism related to color shift of LEDs lamps", To be submitted

Crosstalk phenomena analysis using electromagnetic wave propagation by experimental an numerical simulation methods

Alicja Palczyńska[1], Artur Wymysłowski[1], Tomasz Bieniek[2], Grzegorz Janczyk[2], Daniel Pasquet[3], Thanh Vinh Dinh[3]

[1] Wroclaw University of Technology, Faculty of Microsystems Electronics and Photonics, Poland
[2] Institute of Electron Technology, Poland,
[3] LaMIPS, France
e-mail: artur.wymyslowski@pwr.wroc.pl

Abstract

Cross-talk (XT) is a phenomenon affecting signals propagated in electronic circuitry. In a general case if a signal is transmitted in one system it creates an undesired effect in another, which is due to the crosstalk. The crosstalk is usually caused by parasitic capacitances, inductances, or conductive coupling. In wireless communication, crosstalk is often denoted as a co-channel interference, and is related to adjacent-channel interference. The above is especially important for high frequency ranges, which are used in wireless RF applications. In integrated circuit design, crosstalk normally refers to a signal affecting another nearby signal. Usually the coupling is capacitive but other forms of coupling and effects on signal further away are sometimes important as well, especially in analogue and digital circuits. The main problem is how to prevent this phenomena in order to sustain the signal integrity. There are a wide variety of possible fixes, with the increased spacing, wire reordering and/or appropriate shielding. In fact it requires application of specific design rules during the prototyping stage. In the paper the experimental and numerical analysis of a crosstalk problem is presented. The experimental measurement were performed on manufactured test samples prepared on Si substrates with predefined configuration of aluminium lines / wires. Numerical modelling was based on finite element method (FEM) including both 2D and 3D simulations.

1. Introduction

In recent years the RF communication systems evolve rapidly. They can be used in portable wireless sensor systems to develop low-power systems based on tunable antennas, low noise oscillator or tunable filters. One of the area of study of interconnections in such a devices is electrical analysis of signal and power integrity and study of electromagnetic compatibility. Three basic approaches of that analysis are analytical, numerical and experimental methods. The topic of this paper is the study of electromagnetic cross-talk that can disturb the work of a device, which uses the RF. Additionally, this kind of study is interesting, because of the increasing data rate in digital electronics where influence of interconnects is important. The crosstalk may degrade the signal integrity, and thus it is important aspect in the design of every circuit. The severity of coupling depends upon the track geometry, the termination impedances and the types of signals. This phenomenon on transmission lines has spurred a lot of interest in recent years, and thus there are many aspects to cross-talk prediction and description [6].

There are a number of tools to evaluate the parameters of electrical network, such as impedance matrix, admittance matrix or scattering matrix. Impedance and admittance matrices are used to define the relationship between voltage and current in a network, while the scattering parameters links the input and output waves. In fact a four port network and corresponding matrices can be introduced (Fig.1).

Fig.1. General model of the investigated network

In case of impedance and admittance there are two types of matrices – denormalized and normalized, which are needed for transition to the scattering parameters. The denormalized one connects currents and voltages in a form [7]:

$$[V] = [Z][I]$$
$$[I] = [Y][V] \tag{1}$$

where: [V] – denormalized voltage, [I] – denormalized current [I], [Y] - matrices are of the forth order where coefficients Z_{ij} and Y_{ij} are respectively impedance and admittance between ports i and j. Additionally, there is a relation between [Z] and [Y] as follows:

$$[Z][Y] = 1 \tag{2}$$

The normalized impedance and admittance matrices connects the normalized values of currents and voltages. The current is normalized by dividing it by the square root of characteristic admittance of a transmission line and the voltage by dividing it by square root of its characteristic impedance and the normalized impedance and admittance matrices can be expressed in terms of denormalized ones:

$$[v] = [z][i]$$
$$[i] = [y][v] \tag{3}$$

where:

$$[z] = [\sqrt{Y_c}][Z][\sqrt{Y_c}]$$
$$[y] = [\sqrt{Z_c}][Y][\sqrt{Z_c}] \tag{4}$$

There is a relation [7]:

$$[\sqrt{Z_c}][\sqrt{Y_c}] = [1] \tag{5}$$

and the corresponding expression can also be given for a single element of each of the matrix [7].

2. Experimental analysis and results

2.1. Samples design and fabrication

Dedicated test structure addressing investigation of specific configurations of conducting strips in micro-scale was developed by Institute of Electron Technology from Warsaw, Poland (ITE). Based on the idea of the cross-talk, the corresponding test structures along with the dedicated mask sets were designed and fabricated (Fig. 2).

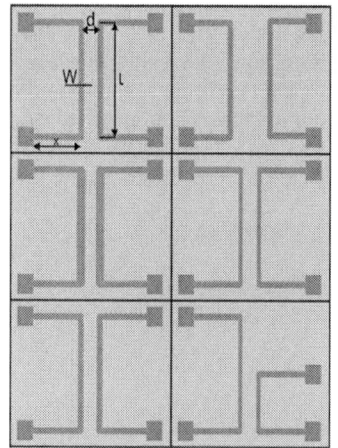

1st set:
W = const. = 30 um
d = 50, 75, 100, 200, 500 um
l = const = 10 mm
x = const = ?

2nd set:
W = 30,50,100 um
d = const. = 100 um
l = const = 10 mm
x = const = ?

3rd set:
W = const. = 30 um
d = const. = 100 um
l = 10; 5; 2,5; 1 mm
x = const = ?

Fig.2. The designed test structure configurations.

In fact the main challenge was to perform investigation on cross-talk patterns present in relatively long, neighbouring metal lines. A set of important parameters affecting the crosstalk was identified and: metal lines width, combination of different metal lines length, various spacing between metal lines, various length of metal lines, e.g. short line versus long lines. The resulting test structures contained 3 different measurement configurations (Fig. 3-6) :

- V1: the crosstalk effects between two long metal lines with different distance between them,
- V2: the cross-talk effects between two lines of different width,
- V3: the cross-talk effects between two lines of different length.

The first configuration (V1) was designed and optimized to measure cross-talk effects between two long metal strip lines fabricated in several layouts with varying inter-line distance (Fig.3). The geometry of the lines is the same: width is fixed to 30μm, length was fixed to 10400μm, whereas the distance between lines varies from 5μm, through 10μm, 20μm, 50μm, 75μm, 100μm, 200μm up to 500μm. Contact pads are squares: 200x200μm. The second configuration module (V2) was designed for cross-talk effect investigation in case of two lines (Fig.4). One line width was fixed to 30μm whereas the second line width varies from 3μm, through 5μm, 10μm up to 20μm. For each case there are two distances applied: 2μm and 5μm. The length of the strip was the same for all strip lines. It is the same as in case of setup V1: 10400μm. The third configuration module (V3) was designed to

investigate cross-talk issues in case when the first metal line strip was shorter than the second one (Fig.5). The designed test structure V3 incorporates setup with one line 10400μm length accompanied by the second line of 10400μm, 5400μm, 2900μm, 1400μm length. Distance between coupling lines was fixed to 100μm. Both lines width is 30μm.

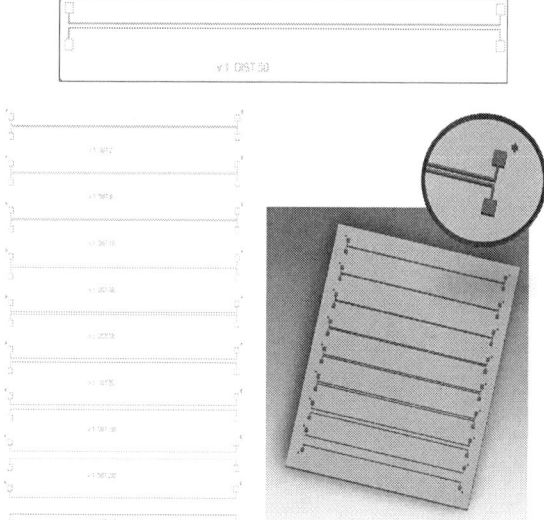

Fig.3. V1 test structures module – mask and 3D model.

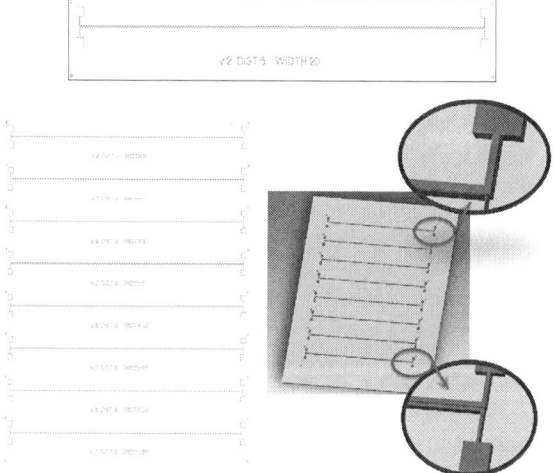

Fig.4. V2 test structures module – mask and 3D model.

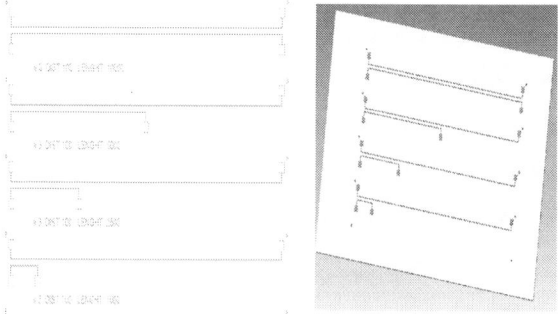

Fig.5. V3 test structures module – mask and 3D model.

978-1-4799-4789-8/14 $31.00 © 2014 IEEE

The final mask (Fig.6) and test structures (Fig.7) were successfully fabricated in ITE micro-fabrication facility by means and equipment of internal pilot CMOS&MEMS. Test structures have been fabricated on 4" N <111> (1-10Ωcm) silicon wafers.

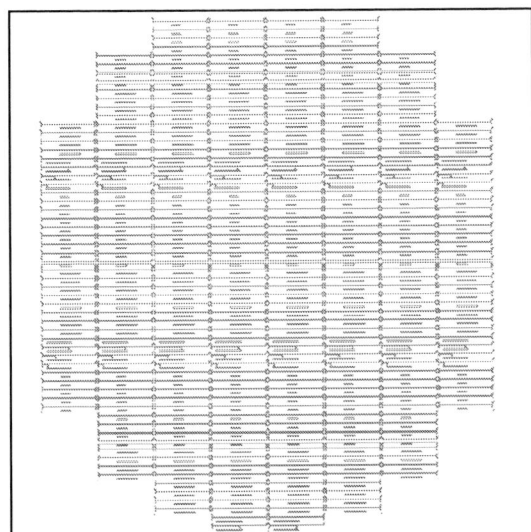

Fig. 6. Full mask set for developed test structures.

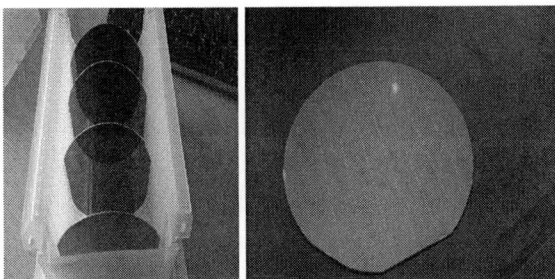

Fig.7. Wafers with fabricated test structures fabricated.

2.2. Impedance measurement of coupled lines and results

The impedance measurements was made on coupled lines from 20 Hz to 2 MHz and with LCR meter (Fig. 8).

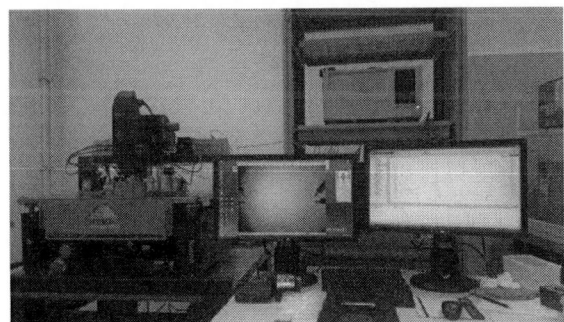

Fig.8. Measurement bench for the LCR meter.

The simplest electrical model of the test structures is given in figure 3. Z_A and Z_B are and inductance in series with a resistance while Z_C and Z_D are capacitances.

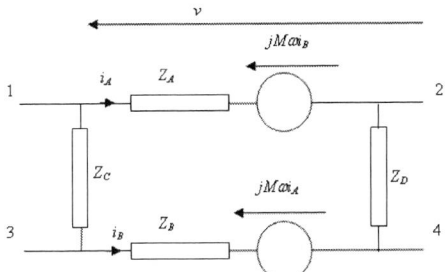

Fig.9. Proposed schematic with mutual inductances

The measured impedances between ports 1 and i are:

$$Z_{1-2} = \frac{Z_A\left(Z_B + Z_C + Z_D\right) + M^2\omega^2}{Z_A + Z_B + Z_C + Z_D - 2jM\omega}$$

$$Z_{1-3} = \frac{Z_C\left(Z_A + Z_B + Z_D - 2jM\omega\right)}{Z_A + Z_B + Z_C + Z_D - 2jM\omega} \tag{6}$$

$$Z_{1-4} = \frac{\left(Z_A + Z_D\right)\left(Z_B + Z_C\right) + M^2\omega^2}{Z_A + Z_B + Z_C + Z_D - 2jM\omega}$$

The expressions for the other impedances and admittances can be deducted by indices permutation. We keep the same model, but we directly replace the impedances by the corresponding elements (inductances or capacitances):

$$\begin{aligned} Z_A &= R_A + jL_A\omega \\ Z_B &= R_B + jL_B\omega \\ 1/Z_C &= jC_C\omega \\ 1/Z_D &= jC_D\omega \end{aligned} \tag{7}$$

If an expansion is done of ω to the first two terms, then:

$$\begin{aligned} Z_{1-2} &= R_A + j\omega\left(L_A - R_A^2 C_{CD}\right) \\ Y_{1-3} &= j\omega\left(C_C + C_D\right) + \omega^2 C_D^2\left(R_A + R_B\right) \\ Y_{1-4} &= j\omega\left(C_C + C_D\right) + \omega^2\left(C_C^2 R_B + C_D^2 R_A\right) \\ Y_{2-3} &= j\omega\left(C_C + C_D\right) + \omega^2\left(C_D^2 R_B + C_C^2 R_A\right) \\ Y_{2-4} &= j\omega\left(C_C + C_D\right) + \omega^2 C_C^2\left(R_A + R_B\right) \\ Z_{3-4} &= R_B + j\omega\left(L_B - R_B^2 C_{CD}\right) \end{aligned} \tag{8}$$

In the following figures 10a-d, the real and imaginary parts of the elements are shown against ω^2.

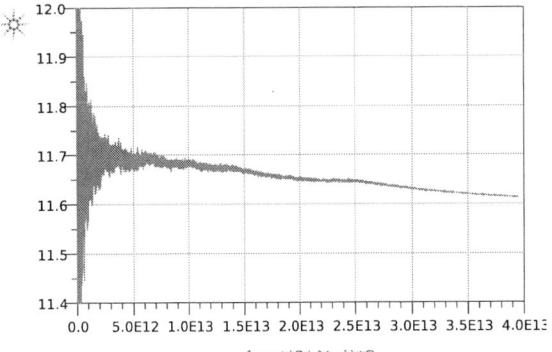

Fig.10a. $Re(Z_{1-2})$; $R_A = 11.7\ \Omega$

978-1-4799-4789-8/14 $31.00 © 2014 IEEE   217

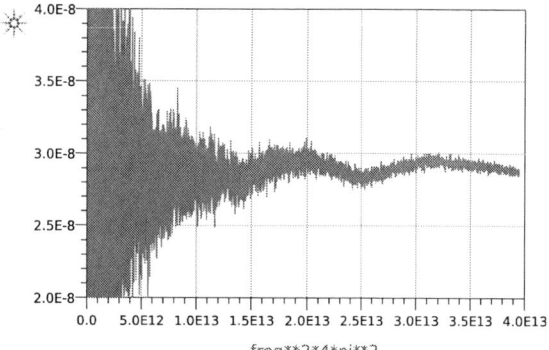

Fig.10b. $Im(Z_{1-2})/_o$; $L_A-R_A^2C_{CD}=28\ nH$

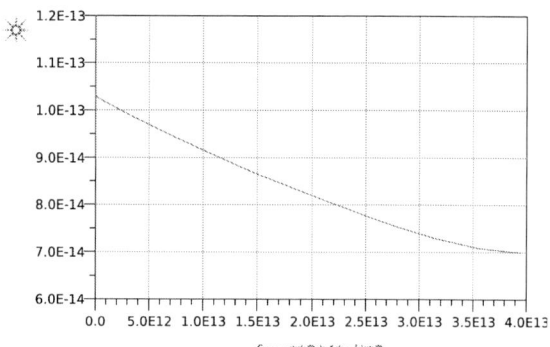

Fig.10c. $Im(Y_{1-3})/_o$; $C_C+C_D=103\ fF$;

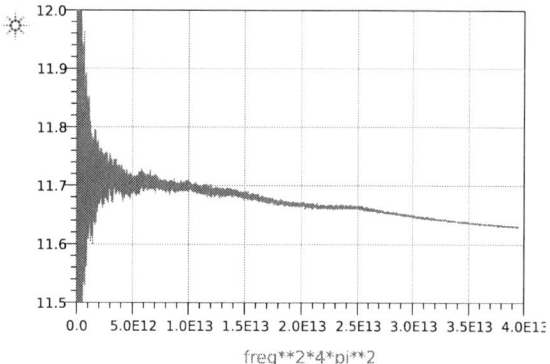

Fig.10d. $Re(Z_{3-4})$; $R_B=11.74\ \Omega$

The slopes of these curves can give further information if the expansion is performed up to the fourth term. In addition other measurements were used to complete the extraction, which allowed to obtain finally most of the elements: R_A=11.7 Ω; R_B=11.74 Ω; L_A=28 nH; L_B=23 nH; C_C+C_D=103 fF. The measurement accuracy did not allow to reach the mutual inductance M and to separate the capacitances C_C and C_D. To do this, the measurement should be made at higher frequencies and if possible in a four-port configuration with GSGSG probes.

3. Numerical analysis and results

The numerical analysis was done using FlexPDE package by PDE Solutions Inc, which is a script based software. There is no graphical interface or ready interfaces dedicated to solve a particular problem. The software requires knowledge on the both phenomena physics and FEM method.

In every script, there are four basic things to define [8]. At the beginning, the domain of interest have to be chosen. To do that, the coordinates system is selected, as well as the number of dimensions, then the shape of sub-domains is decided with different materials properties, by determining the corners, point-by-point. In 3D case, it is required to choose the layer in which each sub-domain is located. Then, there's a variables choice and equations statement. The two are closely linked since it must be just one equation for each variable in properly posed system. It has to be remembered, though, that the equation must be defined according to chosen coordinates system. When multiple variables are used, every equation has to be labelled with the name of variable it concerns. Next thing to do, is the definition the boundary conditions. There are two basic types of boundary conditions – Dirichlet's boundary conditions are denoted as value setting conditions, and Neumann's boundary conditions are referred as software natural conditions. They were to be defined on every edge in our domain, the internal ones as well as external. Last thing to define was the output. FlexPDE allows us to plot the solution over the whole domain or chosen sub-domain, obtain the values at specified point or integral over selected area. The problem can be presented in a form of a plot foa a specified time or an animation. Additionally, some parameters can be specified like mesh density, convergence error limit or some dependence between variables and material parameters.

3.1 Numerical simulations vs experimental results

For numerical modelling and experimental validation a number of criteria and test setups were taken into consideration (Fig.11-13) [5]. In the first setup V1 the dependence of cross-talk to distance of the path was investigated. For a whole series of the structures, the geometry of the lines was the same: width of path was 30 μm, the length was 10400 μm and the contact pads were squares 200 μm by 200 μm. However, the width between two path varied from 5 μm, through 10 μm, 20 μm, 50 μm, 75 μm, 100 μm, 200 μm up to 500 μm. The second experimental setup V2 was created to investigate the dependence of cross-talk to width of the path. The width of one of the paths was fixed at 30 μm and the second one was variable, and changes rom 3 μm through 5 μm, 10 μm up to 20 μm. There were created two series of that samples, one with 2 μm and the second with 5 μm distance between paths. The length of path was fixed, similarly as in the previous case, at 10400 μm. The third setup V3 was created to investigate the influence of length of path. One of the lines in this setup has length fixed at 10400 μm, and the second line's length varied from 1400 μm, through 2900 μm, 5400 μm up to 10400 μm. The distance between the paths was fixed at 100 μm, both lines had the same width of 30 μm.

978-1-4799-4789-8/14 $31.00 © 2014 IEEE   218

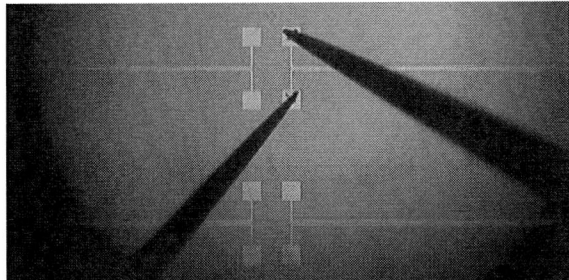

Fig.11. Samples of V1 series.

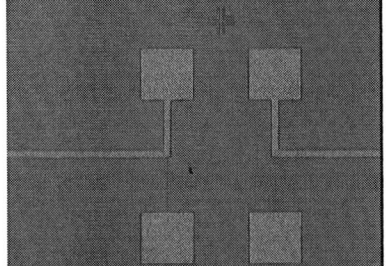

Fig.12. Samples of V2 series.

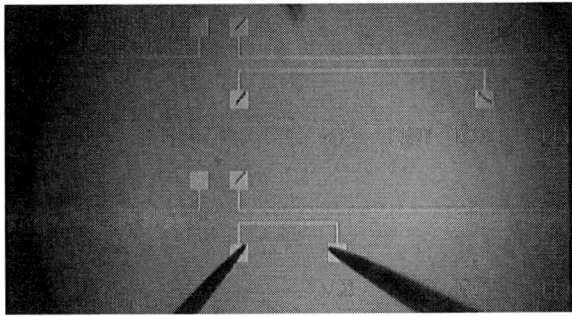

Fig.13. Samples of V3 series.

Impedance parameters of the test structures were measured in a range of frequencies from 20 Hz up to 2 MHz. In fact all parameters from 4 by 4 impedance matrix were recorded but for further considerations only some were taken into account. Figure 14 presents a numeration of pads, which was assumed for all setups.

Fig.14. Numeration of the pads.

The parameters of interest were as follows: 1 and 3, 1 and 4, 2 and 3, 2 and 4. In case of V1 the parameters were measured between pads 1 and 3, and between 1 and 4. Moreover, analyses was focused on an imaginable part of impedance since only the inductance and capacitance were later on simulated numerically. The results are easier to interpret, while it is presented as admittance. The measured spectra of experimental setups are given in figures 15-17.

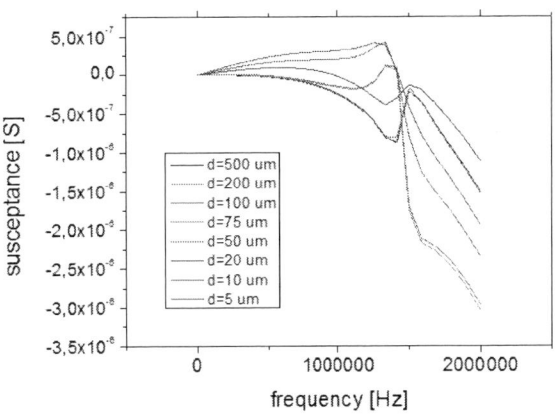

Fig.15. Susceptance of V1, measured between pads 1 and 4, with distance between pads as parameter.

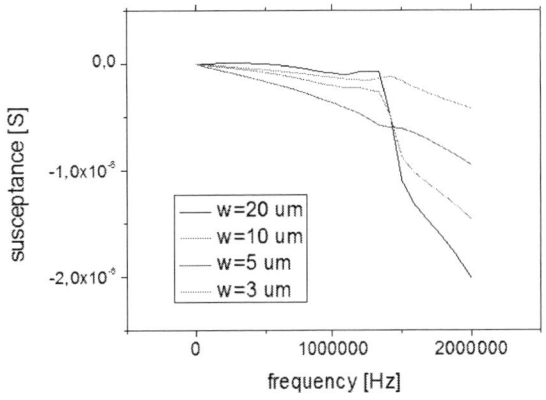

Fig.16. Susceptance of V2 with 5 μm, measured between pads 1 and 4, with width of path as a parameter.

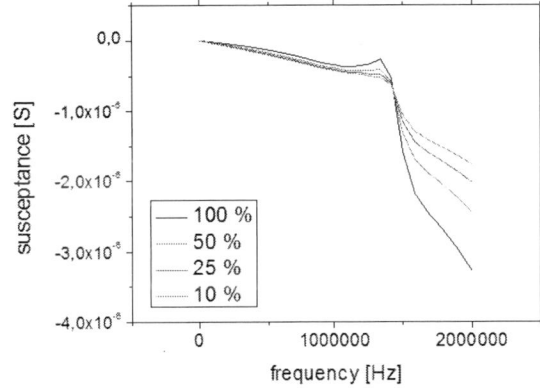

Fig.17. Susceptance of V3, measured between pads 1 and 4 with length of path as parameter (percent of original path).

In fact after the analysis of measurements results of all samples of V2 series with 2 μm distance between lines it turned out that they were shorted, so this part of experiment was ignored in further considerations. In fact in the spectra there can be seen two distinct regions of interest. For low frequencies, there is a major influence of capacitance on the impedance, then after some frequency (e.g. resonant frequency), there is a change of behaviour of that structure, which can be interpreted as a larger influence of inductive coupling. In order to cofirm that the phase of measured impedance was calculated. If G and B are known, then there are:

$$|Y| = \sqrt{B^2 + G^2}$$
$$\varphi = \arctan\left(\frac{B}{G}\right) \tag{9}$$

and it is given on figure 18. The achieved experimental results were then compared with the numerical simulations and two mechanisms of electromagnetic field coupling: capacitive and inductive effect were considered.

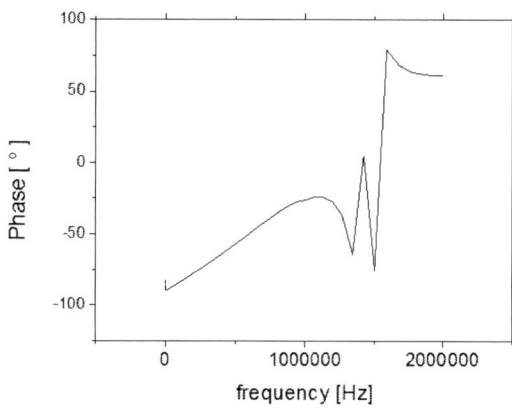

Fig.18. Phase calculated from measured spectrum, for V1 series, distance between path 10 µm.

3.1.1. Capacitive effect modelling

The capacitive effect is connected with existence of electrical field, which is described by the Gauss's Law. If the structure is treated as capacitor, then there is the same charge on both plates but of opposite signs. Thus, the whole charge in the area is equal to zero and the work of potential distribution can be conducted by an electrical field [9]:

$$\nabla V(x, y, z) = E \tag{10}$$

which can be replaced by Gauss's equation, which in fact was used for current simulations:

$$\nabla^2 V(x, y, z) = 0 \tag{11}$$

The capacitance of the structure can be calculated according to the formula [11]:

$$C = \frac{2 E_E}{\Delta V^2} \tag{12}$$

where: C–capacitance [F], E_E–energy of electrical field [J], ΔV–potential difference between plates [V], while the energy can be evaluated according to the formula [11]:

$$E_E = \varepsilon \frac{\nabla V \cdot \nabla V}{2} \tag{13}$$

The example of an evaluated distribution of electric potential is shown in figure 19.

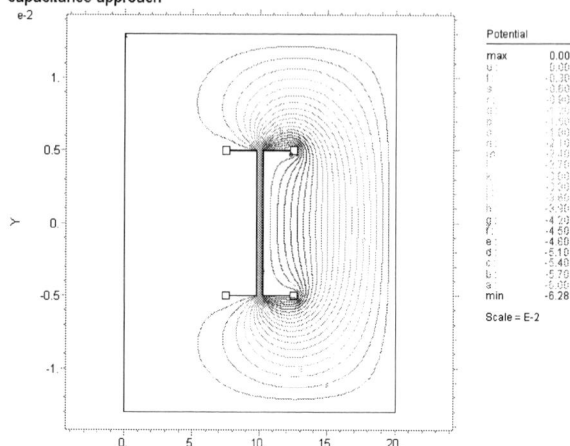

Fig.19. Distribution of the electric potential.

3.1.2. Inductive effect modelling

The inductive effect is connected with existence of magnetic field. In this case the Ampere's Law was used along with the potentials formulation. If the first term of that equation is treated as the current [12]:

$$J = \varepsilon \frac{\partial E}{\partial t} \tag{14}$$

then the Ampere's Law can be rewritten as:

$$\frac{1}{\mu} \nabla \times (\nabla \times A) = J \tag{15}$$

while taking into account the Coulomb gauge condition, the equation is given as:

$$-\frac{1}{\mu} \Delta A = J \tag{16}$$

On the other hand it is possible to extract the current by:
$$J = \sigma E \tag{17}$$
where due to the Faraday's law electric field is given as:

$$E = -\frac{\partial A}{\partial t} \tag{18}$$

Therefore the final equation that describes magnetic potential and that was used for simulations is given by:

$$\frac{1}{\mu} \Delta A - \sigma \frac{\partial A}{\partial t} + J_0 = 0 \tag{19}$$

The inductance in the structure was evaluated by the taking into acount energy stored in magnetic field [11]:

$$E_M = \frac{1}{\mu} \frac{B \cdot B}{2} \tag{20}$$

so the inductance can be calculated as:

$$L = \frac{2 E_M}{I^2} \tag{21}$$

The example of magnetic potential distribution over the domain is given in figure 20.

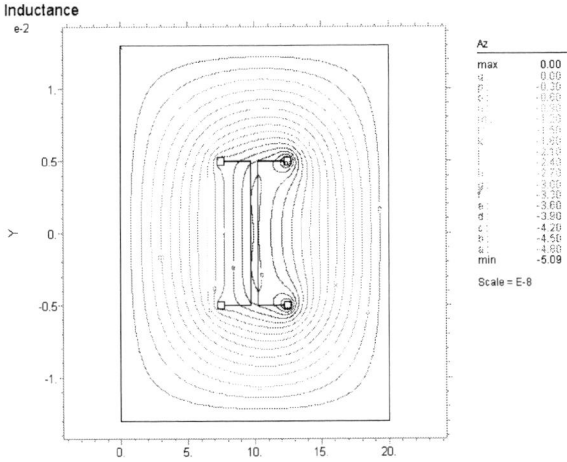

Fig.20. Distribution of the magnetic potential

3.1.3. The coupled field effect modelling

Two variables are used in this case, which is potential of magnetic field as well as potential of electrical field. The above is constituted by one equation [9]:

$$J = \sigma E \tag{22}$$

combining the distribution of electrical field in the domain and thus allowing capacitance calculation. Then the current can be introduced as the external one in the equation in terms of magnetic potential. Unfortunately, the computational cost was much bigger than that of two separates ones with the same results as for separate fields case. The only benefit was that the whole needed information can be obtained in one computational step.

3.2. Numerical simulation assumptions for 2D model

The 2D models were created and tested. For capacitive effect instead of calculating gradients and divergence in 3 dimensions, it was done in 2 dimensions. However, in case of an inductive effect, it was different due to the definition of magnetic potential. Finally, for capacitive effect an equation 11 was used while for inductive effect equation 19 [9]. Additionally the 2D solution required the modification of capacitance and inductance values, which were obtained per meter [9]:

$$\frac{C}{m} = \frac{2E_E}{\Delta V^2} \tag{23}$$

$$\frac{L}{m} = \frac{2E_M}{I^2} \tag{24}$$

where m stands for meters, while C and L in [F/m] and [H/m]. Additionally the results were multiplied by the height of the path (h=1.12 μm). Next step was to calculate the resultant susceptance of the structures. In order to do that an assumption was made of some equivalent circuit, which consisted of a serial connection of inductance and capacitance. The spectra of the circuit was simulated in Zview (Fig.21). For comparison, an experimental spectra for one of the test structures is given in figure 22.

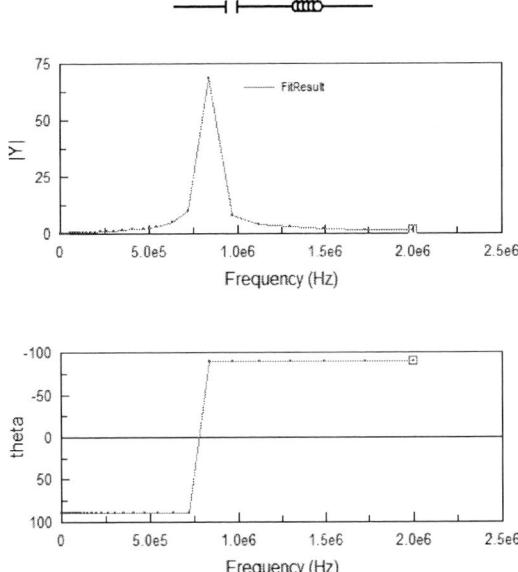

Fig.21. The simulated spectrum of inductor and capacitor connected in series, with C= 1 μm and L=100 nH.

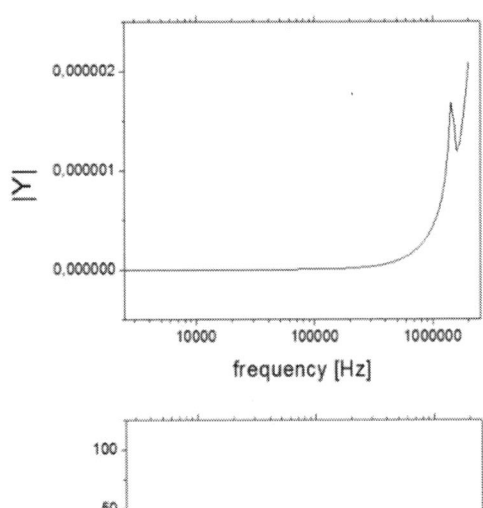

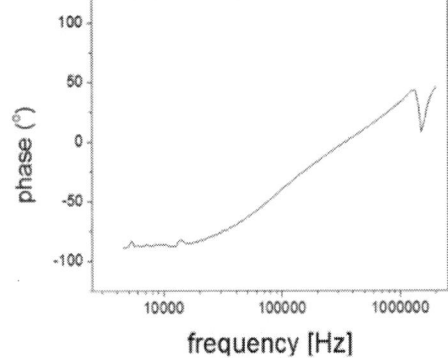

Fig.22. Admittance spectra, for V1 sample with 100 μm distance between lines.

As it can be seen in figure 21, the spectra of an equivalent circuit has similar general shape to the measured one, but with some additional loss effects. Nevertheless it can be concluded that the equivalent circuit can represent, to some extant, the real phenomena. Figure 22 presents one of the experimental representation of the spectra for one of the test samples. In case of a phase, the similar transition can be seen as in case of the equivalent circuit spectrum, while in case of an absolute value, it can be concluded that the resonance wasn't reached during the measurements.

3.3. Numerical simulation results of 2D model

Simulations were made at two different frequencies: 2000 Hz in low frequencies range and 2MHz in high frequencies range and a comparison with the experimental results was done. In case of V1 series (Fig.23-24), there is a good agreement of experiment and simulations. It's much better in case of low frequencies, Additionally, it can be seen in admittance spectra, that 2 MHz can be somewhere near resonance , so as any parasitic resistance isn't taken into considerations, it can be of different value.

V1 results, for small distances the difference between simulation and measurements is significant. For small frequencies the shape of plots is consistent.

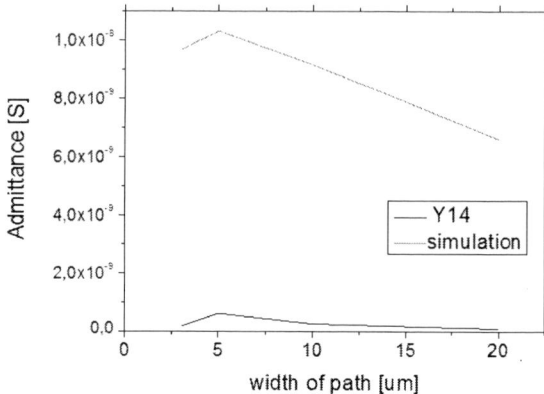

Fig.25. Results for V2 series, at frequency of 2000 Hz

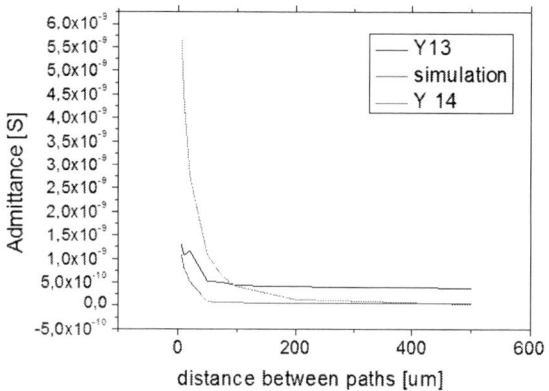

Fig.23. Results for V1 series, at frequency of 2000 Hz

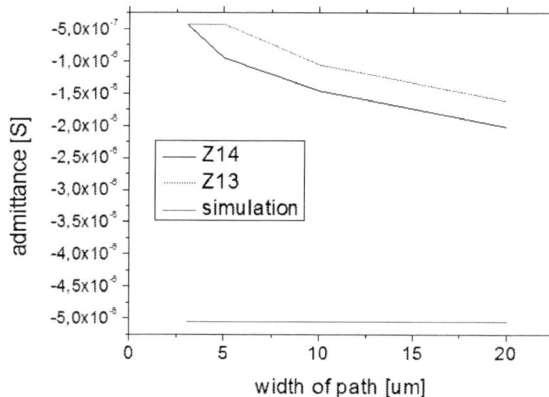

Fig.26. Results for V2 series, at frequency of 2 MHz

For V3 series (Fig.27-28), there is a good agreement between measurements and simulations and the trend is maintained. For high frequencies, it can be seen that the simulation results have smaller values.

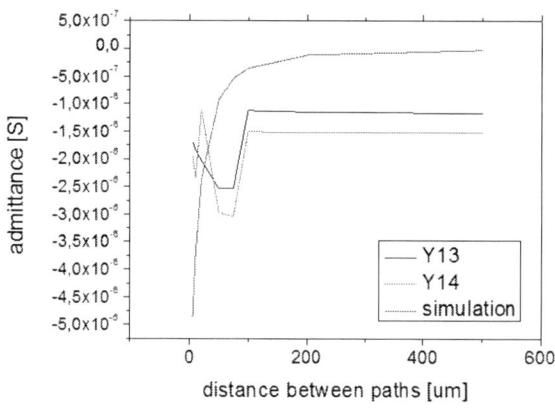

Fig.24. Results for V1 series, at frequency of 2 MHz

For V2 series (Fig.25-26), the accordance wasn't so good. The value range difference can be caused by the small distance between the paths, and as it can be seen in

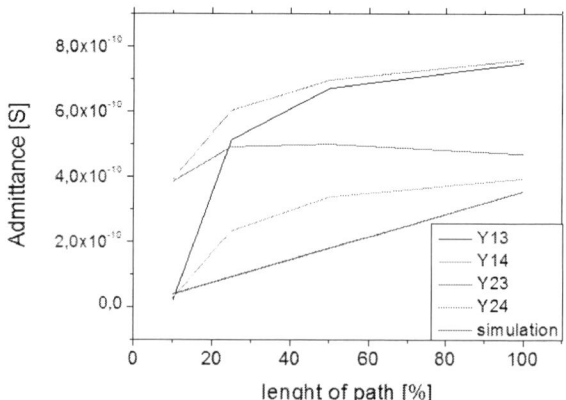

Fig.27. Results for V3 series, at frequency of 2000 Hz

978-1-4799-4789-8/14 $31.00 © 2014 IEEE

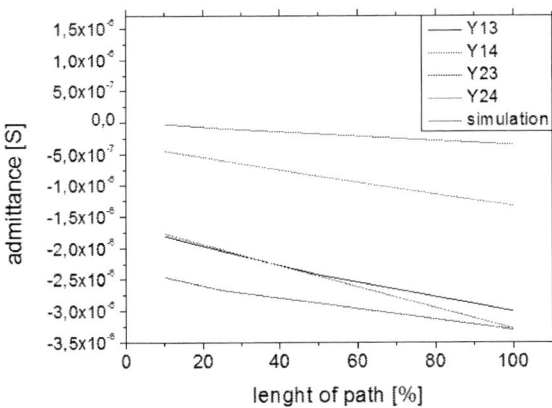

Fig.28. Results for V3 series, at frequency of 2 MHz

The 2D simulations are much less complicated and take less computational time. However, in this case, the substrate is not taken into account, so the results can be somewhat different from reality. The fringe field in z direction is also not take into account, so these results can be a bit smaller than in reality.

3.3. Numerical simulation results of 3D model

3D model is much more complex that the 2D one. As mentioned earlier, the thickness of aluminium paths is 1.12 μm, but also the thickness of silicon substrate is needed. It is typically 525 μm for 4 inch substrate, which was used to fabricate the test structures. Also in 3rd dimension, some space must be left above and under the substrate, in such a way, that the results wasn't impacted by the edge effects (the space under the substrate is not as vast as in the one above). Since the thickness has even smaller value that the width of path, it requires the use of smaller elements in space and the domain is larger, so the time of calculation grows proportionally. Still, the equations have to be adapted to the 3D case. For capacitive effect it's trivial. It is based on calculating the gradient and divergence in three dimensions. Again, it is not so easy for inductive effect modelling. An assumption is made that the external current is flowing only in x and y directions, so it is included only in equation for A_z. In the other directions component of current equals to zero [9]:

$$\frac{1}{\mu}\Delta A_x - \sigma\frac{\partial A_x}{\partial t} = 0$$

$$\frac{1}{\mu}\Delta A_y - \sigma\frac{\partial A_y}{\partial t} = 0 \qquad (25)$$

$$\frac{1}{\mu}\Delta A_z - \sigma\frac{\partial A_z}{\partial t} + J_0 = 0$$

The results of 3D modelling wasn't so much different from these obtained by 2D simulations (Fig.30). Additionally, as expected, due to the influence of a substrate, higher values were achieved, but in fact even higher that the measured parameters, as can be seen in figure 31.

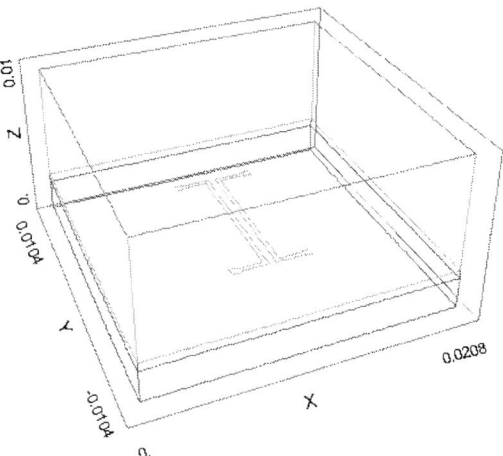

Fig.29. The 3D domain of calculations.

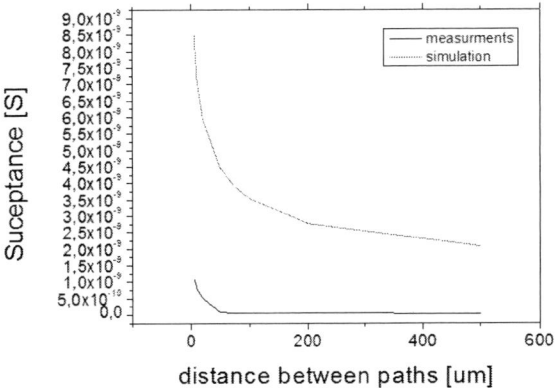

Fig.31. Results for 3D simulations and measurements

4. Conclusions

In this work, the experimental and numerical investigations are presented in case of the crosstalk problem in electronic circuits. In order to do that a special test structures were designed and manufactured. Then the structures were measured and simulated numerically. As a result, a good agreement between experimental and numerical results was achieved. In case of of numerical simulations it was shown that a wide range of frequencies can be analysed and 2D simulation results are similar to the corresponding 3D simulations, which were much more complicated and time consuming.

Acknowledgements

The above research was financed by PARSIMO project founded by JU ENIAC. Calculations have been carried out at Wroclaw Centre for Networking and Supercomputing (http://www.wcss.pl).

References

[1] P. Ciarlet, Notes de cours sur les equations de Maxwell, ENSTA 2012

[2] F.J. Savias, A gentle introduction to Finite Element Method, 2006

[3] P. Ciarlet, Lectures on The Finite Element Method, Tata Institute of Fundamental Research, 1975

[4] E.Suli, Finite Element Methods for Partial Differential Equations, University of Oxford, 2013

[5] Eliane Bécache,et al.,La Méthode des Eléments Finis. De la Théorie à la Pratique. II. Compléments, Les Presses de l'ENSTA, Collection Les Cours, 2010

[6] A. Wymyslowski, PARISMO project summary, Internal reports of WRUT, 2013

[7] Chen L.F. et al., Microwave Electronics - Measurement and Materials Characterization, 2004 John Wiley & Sons

[8] FlexPDE User Guide, www.pdesolutions.com

[9] G. Backstrom, Fields of Physics by Finite Element Analysis; Electricity, Magnetism, and Heat in 1D, 2D and 3D

[10] Comsol Model Gallery, comsol.com

[11] L.D. Landau, E.M. Lifszyc, Krótki kurs fizyki teoretycznej, Tom 1, Mechanika, Elektrodynamika, PWN 1978

[12] Gross, Paul W.; Kotiuga, P. Robert, Electromagnetic Theory and Computation – A Topological Approach, © 2004 Cambridge University Press

[13] J. Hirai et al., Study on Crosstalk in Inductive Transmission of Power and Information, IEEE Transactions on Industrial Electronics, Vol. 46, No. 6, December 1999

[14] A. Vittal et al., Crosstalk in VLSI Interconnection, IEEE Transactions on Computer-Aided Design of Integrated Circuits and Systems, Vol. 18, No. 12, December 1999

[15] T. Steinecke et al., EMI Modeling and Simulation in the IC Design Process, 17 the International Zurich Symposium on Electromagnetic Compatibility, 2006

[16] J.G. Yook et al., Characterization of High Frequency Interconnects Using Finite Difference Time Domain and Finite Element Method, IEEE Transactions on Microwave Theory and Techniques, Vol. 42, No. 9, September 1994

[17] Pasricha, Sudeep, On-Chip Communication Architectures - System on Chip Interconnect, 2008 Elsevier

[18] A.I. Mincer et al., Calculation of pad cross-talk in a thin-gap multiwire detector with pad redout, Nuclear Instruments and Methods in Physics Research A 404(1998), 41-50

[19] L. Lu, Computerized Optimization of Multi-Layered and/or Multi-Trace Microstrip-Lines for Reduced

Cross-Talk Interference, Computers Elect. Engineering Vol. 23, No. 3, pp.165-177, 1997

[20] P. Baine et al., Cross-talk suppression in SOI substrates, Solid State Electronics 49 (2005), 1461-1465

[21] H. Spieler, Low noise electronics in practical applications, Nuclear Instruments and Methods in Physics Research A 636(2011), S149-154

[22] D. N. Ladd et al., Spice Simulation Used to Characterize the Cross-Talk Reduction Effect of Additional Tracks Grounded with Vias on Printed Circuit Board, IEEE Transactions on Circuits and Systems II: Analog and Digital Signal Processing, Vol. 39, No. 6, June 1992

[23] M.L. Markel et al., Simulation Methods to Determine Cross-Talk in Three Dimensional Environment, ISBN 0-7803-1398-4/94/0000-0080, 1994 IEEE

[24] K. Hollaus et al., Simulation of Crosstalk on Printed Circuit Boards by FDTD, FEM and a Circuit Model, IEEE Transactions on Magnetics, Vol. 44, No. 6, June 2008

[25] T. Sasaki et al., Cross-Talk Simulation of a Multi-Track Magnetic Thin Film Head, IEEE Transactions on Magnetics, Vol. 25, No. 5, September 1989

[26] M. Chou et al., Efficient Reduced-Order Modeling for Transient Simulation of Three-dimensional Interconnect, ISBN 1063-6757/95, 1995 IEEE

[27] Z. Guoyan et al., The Simulation Analysis of Cross-Talk Behavior in SOI Mixed-Mode Integrated Circuits, ISBN 0-7803-6520-8/01, 2001 IEEE

[28] Y.S. Song et al., Suppression of the Mutual Coupling Between Two Adjacent Miniaturized Antennas Utilizing Printed Resonant Circuits, ISBN 978-1-4244-3647-7/09, 2009 IEEE

[29] L. Gal, On-Chip Cross Talk – the New Signal Integrity Challenge, IEEE 1995 Custom Integrated Circuits Conference

[30] Hirsch, Charles , Numerical Computation of Internal and External Flows – Fundamentals of Computational Fluid Dynamics (2nd Edition),© 2007 Elsevier

[31] K. Joardar, A Simple Approach to Modelling Cross-Talk in Integrated Circuits, IEEE Journal of Solid-State Circuits, Vol. 29, No. 10, October 1994

[32] D.S. Gao et al., Modeling and Simulation of Interconnection Delays and Crosstalks in High Speed Circuits, IEEE Transactions on Circuits and Systems, vol. 37, no.1, 1990

Size effect on the microbridges quality factor tested in free air space

Marius Pustan, Corina Birleanu, Florina Rusu, Cristian Dudescu, Ovidiu Belcin
Technical University of Cluj-Napoca, Department of Mechanical Systems Engineering,
Micro & Nano System Laboratory
Bd. Muncii 103-105, 400641 Cluj-Napoca, Romania
E-mail: Marius.Pustan@omt.utcluj.ro, Phone: +40 264 401665

Abstract

The influence of geometrical dimensions on the dynamical response of polysilicon microbridges is experimentally investigated and presented in this paper. The main goal is to determine the quality factor variation of microbridges if the samples length is modified. The quality factor is an important qualifier of mechanical micro-oscillators and provides information about the loss of energy during oscillations. The loss of energy in an oscillator structure depends on the sample geometry, the material properties and also is influenced by environmental conditions. In this paper the dynamic response of electrostatically actuated polysilicon microbridges fabricated in different geometrical dimensions is investigated. The geometrical dimensions have a significant influence on the frequency response of microbridges. Indeed, the oscillator stiffness, velocity and amplitude of oscillations, quality factor and the loss of energy coefficient are changed as a function of the samples dimensions. A high value of the quality factor and a low loss of energy coefficient are determined for microbridges with small length fabricated in the same width and the same thickness.

1. Introduction

The MEMS oscillators represent currently one of the important research areas of Microelectromechanical Systems (MEMS). The usual applications of MEMS oscillators are the radio-frequency electromechanical devices, the MEMS gyroscopes and the resonator sensors. Micro-switches, high-frequency filters, tunable capacitors and micromechanical inductors are currently RF-MEMS oscillators under intense development. These mechanical oscillators are essential components in communication circuits because they generally exhibit orders of magnitude higher quality factor Q than electrical components. Such devices can be designed to vibrate over a very wide frequency range (higher than 1GHz), making them ideal for ultra-stable oscillator and low filter functions for a wide range of transceiver types. Energy losses in high-frequency oscillators are critical to designers.

The Q-factor, a ratio of the energy stored in an oscillator to energy dissipated during each cycle, must be high for a micro-oscillator to be useful in a filter or frequency reference.

Different dissipation energy is presented in MEMS oscillators such as thermoelastic damping, anchor losses, and losses due to air flow (squeeze film damping). The relative dominance of different damping mechanisms depends on the operating conditions and dimensions of the oscillator structures. For most MEMS oscillators operating at ambient conditions, loss of energy due to air flow is the most dominant damping mechanism. The squeeze film damping is basically prevalent in systems where the air-gap thickness is insufficiently small compared to the dimensions of structure. For a given value of the air gap thickness, there are various factors that affect squeeze film damping such as the resonant frequency, pressure of surrounding medium, boundary conditions [1, 2, 3, 4].

In previous studies the changes in the dynamic response of MEMS oscillators has been investigated as a function of temperature [5]. Moreover, the position of the lower electrode has influence on the dynamical behavior of electrostatically actuated MEMS oscillators [6]. Changes in the environmental conditions influence the dynamic response of oscillators by modifying the structural damping of the material and the squeeze film coefficient. For instance airplane condition monitoring and satellite wireless communication applications require the operating of oscillators at temperatures as low as 50°C below zero. For the same input voltage, the oscillations amplitude and velocity of a MEMS oscillator increase if the temperature decreases. In the same way for high temperature operating conditions, the amplitude and velocity of oscillations decrease due to the heat energy dissipation called thermal damping, and the material heat softening called temperature relaxation [5].

The geometrical dimensions of oscillators also change the dynamic behavior of MEMS oscillators. Moreover, geometrical effect changes the loss of energy during oscillations. Experimental investigations of the dynamic response of MEMS oscillators fabricated in different geometrical dimensions are developed and discussed in this paper. The experimental investigations of the interactions effect between suspended plates and surrounding medium during oscillations is important to predict the quality factor and the loss of energy coefficient. The interaction between the surrounding medium and the oscillating plate is called squeeze film damping. The loss of energy based on this effect is frequency higher than internal dissipated energy.

The size effects on the quality factor and the loss of energy coefficient have been investigated by several researchers [4]. The size effect on quality factor determined based on the first vibration mode of microcantilevers tested in air is presented in [7] where take into consideration the geometrical effects, the existing theoretical models for damping analysis are

978-1-4799-4789-8/14 $31.00 © 2014 IEEE

modified. This study demonstrates that, the quality factor is proportional to the geometrical dimensions, and the experimental results were validated by numerical and analytical analysis. In the other study, the boundary effect on damping is analyzed considering an effective damping width of structure [8]. This study provides an analytical expression to compute the damping for rectangular plate with holes based on the Reynolds equation modification with the gas flow effect through the holes. The holes dimension effect on damping and stiffness contribution of gas film of a suspended oscillating plate is also studied and presented in [9].

The loss of energy coefficient Q^{-1} can be minimized if the oscillator operates in vacuum which requires expensive packaging. The other way to improve the quality factor and to decrease the loss of energy coefficient is to performed perforations to the oscillating plate. The holes decrease the loss of energy based on the air damping but reduce the device capacitance which is undesirable. The oscillator capacitance can be improved using smaller air-gap spacing. On the other hand, smaller air-gap spacing affects the pull-off position and increases the stiction risk. So, the oscillator can be designed with holes in order to decrease the loss of energy but without decreasing the system capacitance. Developing of new methods for the squeeze film analysis of MEMS oscillators is indispensable in the design of reliable MEMS devices.

In this paper, polysilicon microbridges with holes are investigated in order to determine the dimension effect on quality factor and on the loss of energy coefficient. During experimental tests, the same operating conditions of investigated MEMS oscillators are considered and the parameter that changes is the length of the samples. As a consequence, different responses of sample are obtained for the same input signal. MEMS oscillators are typically fabricated as a simple mechanical structure that vibrates in response to some actuation. Electrostatic actuation being the most common type of electromechanical energy conversion scheme in microelectromechanical systems, so the interest of this paper is focused on the dynamic analysis of MEMS oscillators composed of electrostatically actuated clamp-clamp beams.

The paper is organized as follow: (i) a theoretical approach of electrostatically actuated microbridges is presented in section 2; (ii) then, section 3 presents the experimental investigations performed in order to determine the effect of the samples length on the dynamic responses of MEMS oscillators and to estimate the change of the Q- factor and the loss of energy coefficient; (iii) in section 4, the experimental results are discussed and compared with numerical and analytical values; (iv) the significant insights of this research work are presented at the end of paper.

2. Theoretical approach

Figure 1 shows a schematic representation of investigated microbridge. When a DC voltage V_{DC} is applied between the lower electrode and vibrating structure, an electrostatic force is set up and the microbridge bends downwards and come to rest in a new position. To drive the oscillator at resonance, an AC harmonic load of amplitude V_{AC} vibrates the microbridge at the new deflected position.

If an voltage composed of DC and AC terms as

$$V(t) = V_{DC} + V_{AC}\cos(\omega t) \qquad (1)$$

is applied between microbridge electrodes, the electrostatic force applied on structure has a DC component as well as a harmonic component with the ω - frequency such as

$$F_e(t) = \frac{\varepsilon A V(t)^2}{2[g_0 - u_z(t)]^2} \qquad (2)$$

where ε is the permittivity of the free space, $A = w_e \times w$ - is the effective area of the capacitor (figure 1), w_e is the width of the lower electrode, w is the beam width, g_0 is the initial gap between flexible plate and substrate, and $u_z(t)$ is the displacement of mobile plate.

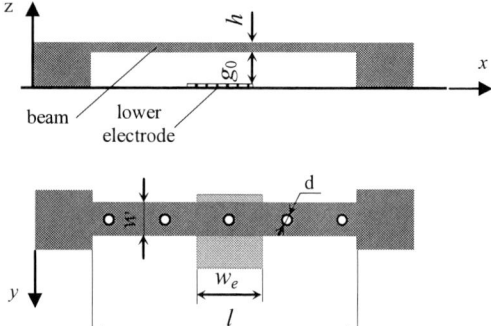

Figure 1: Schematic representation of a microbridge with holes under electrostatic actuation.

The dynamic response of microbridge shown in figure 1, subjected to a harmonic electrostatic force $F_e(t)$, is governed by the equation of motion

$$m\ddot{u}_z(t) + c\dot{u}_z(t) + k u_z(t) = F_e(t) \qquad (3)$$

where m is the effective mass of beam, c is the damping factor and k is the effective stiffness.

When only a DC voltage is applied across the microbridge, the static force balance equation, including the electrostatic force and the spring force is

$$\frac{192 E I_y}{l^3} u_z - \frac{\varepsilon A V_{DC}^2}{2(g_0 - u_z)^2} = 0 \qquad (4)$$

where u_z is the prestressed displacement of the midpoint of beam given by a DC signal, I_y is the cross-sectional moment of inertia about the bending axis, l is the length of the flexible part of beam, E is the material Young's modulus.

The motion energy of the investigated microbridge has two parts: one part is due to attractive electrostatic force and the other one is due to the mechanical stiffness of structure. The total energy stored in the micro-oscillator, can be written as

$$U_T = \frac{96EI_y}{l^3}u_z^2 - \frac{1}{2}\frac{\varepsilon A}{(g_0 - u_z)}V_{DC}^2 \qquad (5)$$

The equivalent stiffness k of microbridge at the prestressed position given by DC signal, obtained by linearizing the electric system around an equilibrium position $\tilde{u}_z$ is

$$k = \frac{192EI_y}{l^3} - \frac{\varepsilon A V_{DC}^2}{(g - \tilde{u}_z)^3} \qquad (6)$$

The resonant frequency of an electrostatically actuated microbridge can be computed as [10, 11]:

$$\omega_0 = \frac{1}{2\pi}\sqrt{\frac{\dfrac{192EI_y}{l^3} - \dfrac{\varepsilon A V_{DC}^2}{(g - \tilde{u}_z)^3}}{m_e}} \qquad (7)$$

where m_e, the equivalent mass of system which can be calculated as $m_e = 0.406 \times m$, m being the effective mass.

If the V_{DC} is zero the beam is not prestressed. The microbridge oscillates at initial position, the stiffness of beam is its mechanical stiffness and the resonant frequency is their natural frequency. Equations (6) and (7) become [11]

$$k = \frac{192EI_y}{l^3} \qquad (8)$$

$$\omega_0 = \frac{1}{2\pi}\sqrt{\frac{k}{m_e}} \qquad (9)$$

The dynamical response of system under DC and AC voltages is given by:

$$u_z(t) = \frac{u_z}{\sqrt{\left(1 - \left(\dfrac{\omega}{\omega_0}\right)^2\right)^2 + \left(2\xi\dfrac{\omega}{\omega_0}\right)^2}} \qquad (10)$$

where ξ is the damping ratio and ω_0 is the resonant frequency of beam given by equation (7).

The damping ratio ξ is any positive real number. For value of the damping ratio $0 \le \xi < 1$, the system has an oscillatory response. The system damping controls the amplitude of response when excited at resonance.

Usually, the response is plotted as a normalized quantity $u_z(t)/u_z$. When the driving frequency equals the resonant frequency $\omega = \omega_0$, the amplitude ratio reaches a maximum value.

At resonance, the amplitude ratio becomes

$$\frac{u_z(t)}{u_z} = \frac{1}{2\xi} \qquad (11)$$

An important qualifier of mechanical micro-oscillator is the quality factor Q. The quality factor is an expression of the cyclic energy loss in an oscillating system. In terms of energy, it is expressed as the energy stored E_s in oscillator divided by the energy dissipated E_d during one cycle of oscillation. At resonance, the quality factor can be expressed as [10, 11]

$$Q_r = \frac{m \cdot \omega}{c} \approx \frac{1}{2\xi} \qquad (12)$$

and the normalized response given by (11) is equal to Q_r. In equation (12), c is the damping factor equal by the sum between the damping due to intrinsic loses in oscillator and the viscous losses from the squeeze film. The damping factor for intrinsic losses in the structure depends on the material and geometry and are, therefore, constant for a given structure. In contrast, the damping constant due to the fluid flow depends on the structure geometry, resonant frequency and many other factors such as boundary conditions, medium pressure, inertial effect etc. In early 1917, analytical expressions for estimating the damping given by air between two circular plates for condenser transmitters were developed [12]. Several mathematical models exist today for the computation of the damping factor of the film for various geometry and different flow conditions.

The quality factor is also called sharpness at resonance, and it can be experimentally determined as

$$Q_r = \frac{\omega}{\Delta\omega} = \frac{\omega}{\omega_2 - \omega_1} \qquad (13)$$

where $\Delta\omega = \omega_2 - \omega_1$ is the frequency bandwidth corresponding to $0.707 u_z(t)_{max}$ on the frequency response experimental curves [6, 11].

3. Experimental investigations

The aim of the experimental tests is to find the influence of the sample length on the dynamic response of electrostatically actuated microbridges. For experimental tests a vibrometer analyzer is used. The samples are actuated using a white noise exciting signal with a DC voltage of 5V and the amplitude of driving current is 5V.

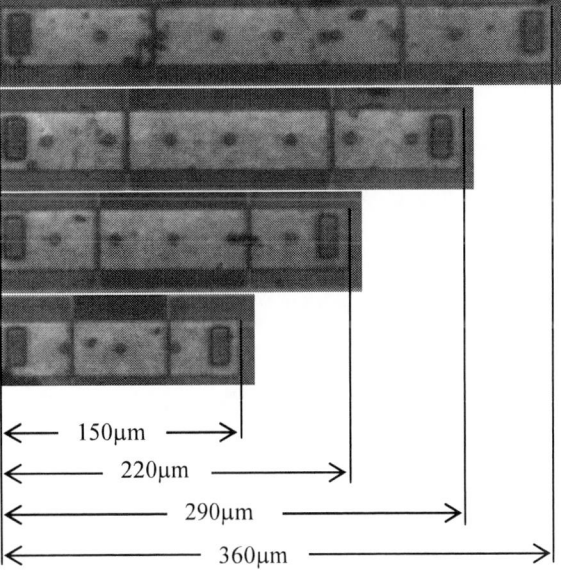

Figure 2: Polysilicon microbridges

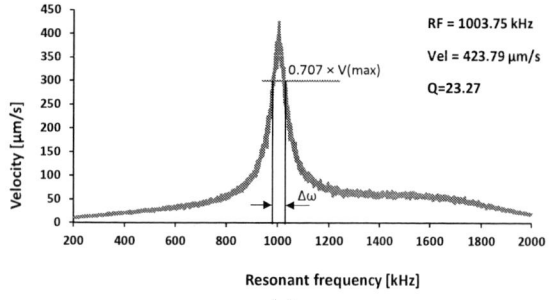

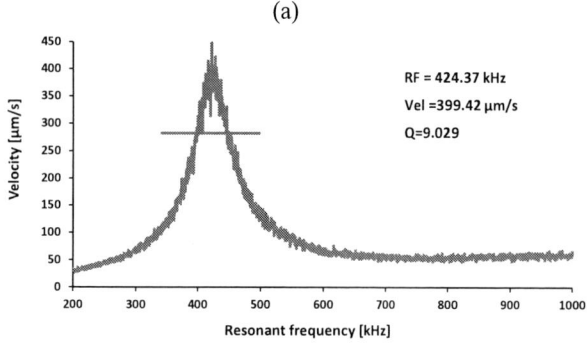

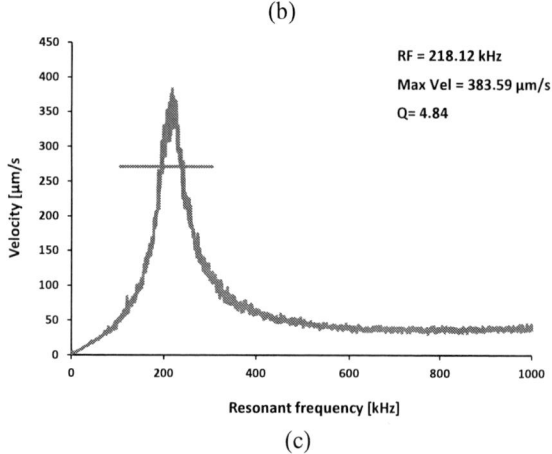

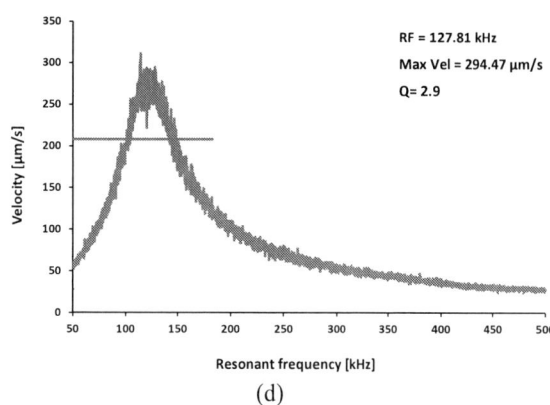

Figure 2 shows the polysilicon microbridges for experimental investigations. These are made with the same width w =30μm, and the same thickness t =1.9μm, with a gap between flexible plate and substrate of 2μm. The total lengths of samples including anchors are: 150μm, 220μm, 290μm, and 360μm. The lengths of flexible part of beam without anchor (used in numerical and analytical computation) are: 224μm, 194μm, 264μm, and 336μm. In order, to decrease the damping effect given by air the samples are manufactured with holes. The diameter of holes is equal to 3μm.

During experimental tests, the experimental curves of the resonant frequency are obtained. Figure 3 presents the experimental changes in the frequency response of investigated microbridges if the length of samples is modified. The tests are performed in ambient condition. The dynamic behaviors of microbridges are changed as a function of the samples lengths. The resonant frequency, velocity and amplitude of oscillations are modified if the sample lengths are changed. Moreover, the experimental quality factor decreases and the loss of energy coefficient increase if the length of sample increases, respectively.

4. Results and discussions

Polysilicon microbridges produced in different geometrical dimensions (different lengths) were experimentally investigated. The scope is to observe the changes in the frequency response and quality factor as a function of the samples lengths. All tests were performed under the same operating condition and the same impute voltage.

Experimental resonant frequency of investigated microbridges decreases from 1003.75 kHz to 127.81 kHz if the total lengths of suspended part of beam increases from 150μm to 360μm. Using equation (7) analytical resonant frequencies of investigated microbridges are computed. The lengths considered in computation are the distance between the samples anchors equal by: 224μm; 194μm; 264μm; 336μm. Figure 4 shows the experimental and theoretical resonant frequency of investigated microbridges as a function of the samples total length. Theoretical results of resonant frequency are in accordance with experimental values as it can be observed in figure 4.

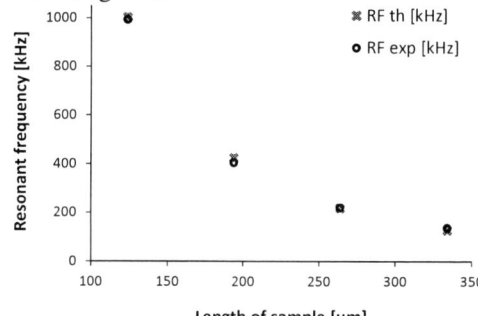

Figure 3: Frequency response experimental curves of investigated microbridges with the total lengths of: (a) 150 μm; (b) 220μm; (c) 290μm; (d) 360 μm.

Figure 4: Theoretical and experimental variation of resonant frequency as a function of the microbridges lengths.

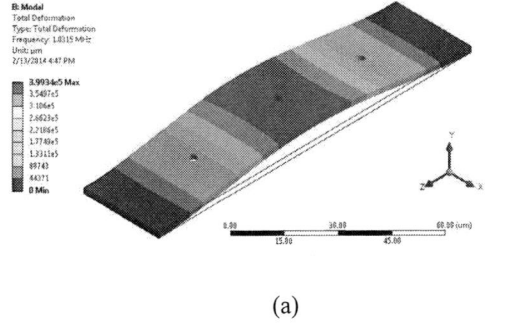

(a)

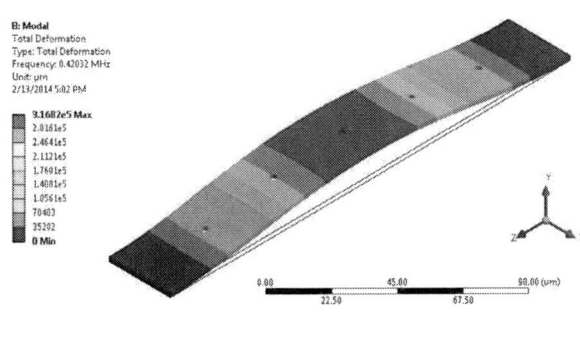

(b)

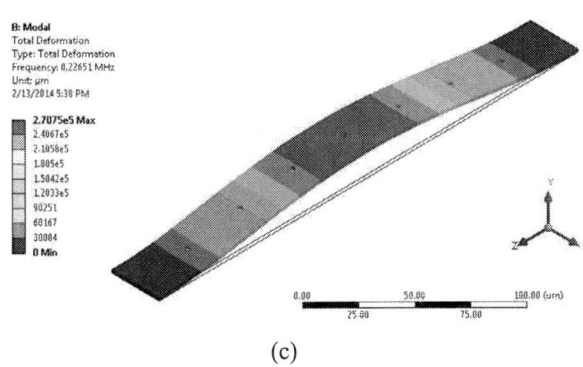

(c)

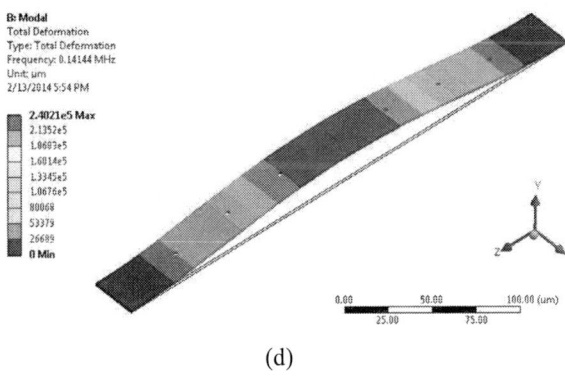

(d)

Figure 5: Modal analysis of the resonant frequency of microbridges with a total length of:
a) 150μm, (b) 220μm, (c) 290μm and (d) 360μm.

Finite element analysis of investigated microbridges is carried out using ANSYS/Workbench software. The scope of the finite element analysis (FEA) is to simulate the dynamical behavior of investigated microbridges and the changes in the frequency response as function of the sample lengths. The geometrical dimensions of microbridges used in FEA are the same with the dimensions of samples from experimental investigations and theoretical calculation. Material constants for polysilicon are taken from the literature: Young's modulus of 150GPa, Poisson's ratio of 0.22 and the material density of 2330 kg/m^3.

For the investigated microbridges, the Q-factor decreases if the length of samples increases respectively. The experimental loss coefficient (Q^{-1}) can be determined based on the quality factor. Loss of energy coefficient increases if the length of microbridges increases respectively. The experimental values of the Q- factor and the loss of energy Q^{-1}, for different lengths of micromembranes, are presented in table 1.

Table 1: Quality factor Q and the loss of energy coefficient Q^{-1} as a function of the samples lengths.

Length [μm]	Q	Q^{-1}
150	23.27	0.043
220	9.02	0.111
290	4.84	0.208
360	2.90	0.344

Analytically, the Q- factor can be expressed using equation (8) where the damping coefficient c due to squeeze - film c_{sq} and due to the loss through holes c_{holes} can be computed as

$$c = c_{sq} + c_{holes} \qquad (14)$$

The damping coefficient due to the squeeze-film is [13]

$$c_{sq} = \frac{16\sigma}{\pi^6} \frac{p_a \chi l^2}{\omega g_0} \sum_{m,n=odd} \frac{\frac{\Gamma^2}{\pi^2} + m^2\chi^2 + n^2}{(mn)^2 \left[\left(\frac{\Gamma^2}{\pi^2} + m^2\chi^2 + n^2 \right)^2 + \frac{\sigma^2}{\pi^4} \right]}$$

(15)

where: σ is the squeeze number that captures the compressibility effect, p_a is the air pressure, χ is the beam aspect ratio (χ = width/length), l – is the beam length, g_0 is the gap between microbridge and substrate, and Γ is a constant that captures the perforation effect.

The damping coefficient due to the loss through holes can be determined as [13]

$$c_{holes} = 8\pi\mu \left(\frac{h}{Q_{th}} + \Delta_E b \right) \cdot n \qquad (16)$$

where: μ is the dynamic viscosity of the environment, h is the beam thickness, Q_{th} is the flow rate factor which accounts for rarefaction effect in the flow through the parallel plates and through the holes, respectively (for slip

978-1-4799-4789-8/14 $31.00 © 2014 IEEE

flow regime), ΔE is the relative elongation of the hole length due to end effects, b is the holes radius, and n is the number of holes.

After numerical computation, the theoretical results of Q-factor are determined in the same range with the experimental values, as presented in figure 6.

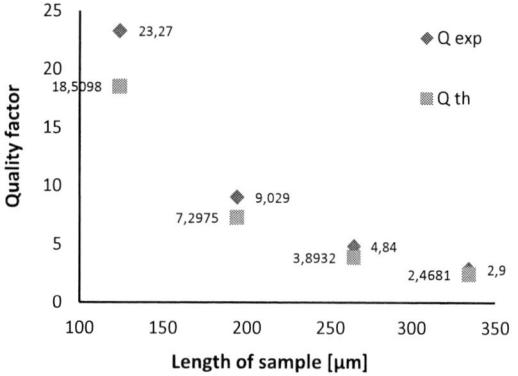

Figure 6: Theoretical and experimental variation of Q-factor as a function of the microbridges lengths.

Using the experimental values of Q-factors and based on equation (12), a shift in the damping ratio is observed as a function of the sample lengths. The damping ratio increases from 0.021 to 0.055, 0.103 and 0.172, respectively, if the length of sample increases from 150µm to 220µm, 290µm and 360µm. But, for each tested samples the damping ratios is $0 \leq \xi < 1$ which confirm that all microbridges have oscillatory responses under the exciting signal.

5. Conclusions

This paper presents the effect of geometrical dimensions on the dynamic response of polysilicon microbridges. Mechanical dynamic interaction between air presented into gap and the oscillating structures with holes are experimentally analyzed and presented in this paper. The samples length increases from 150µm to 360µm and the resonant frequency and quality factor is monitored. For the same operating conditions, the resonant frequency and quality factor decreases if the length of samples increasing, respectively. Moreover, if the quality factor decreases, the loss of energy increases, respectively. The experimental results of resonant frequency and Q-factor are compared with numerical and analytical results and these are in good agreement. The difference between them depends on the geometrical dimensions of samples, and on the real value of Young's modulus. In numerical simulation and analytical study, a theoretical value of Young's modulus (taken from literature) is considered.

Acknowledgments

This work was supported by a FP7 grant type MNT-ERA.NET no. 7-064/2012 "3-Scale modelling for robust-design of vibrating micro sensors".

References

1. Bao, M.H., Micro Mechanical Transducers: Pressure Sensors, Accelerometer and Gyroscopes, Elsevier B. V. (Netherland, 2000).
2. Bao, M., Yang, H., Sun, Y., French, P.J., "Squeeze-film air damping of thick hole plate, " *Sensors and Actuators A,* Vol. 108 (2003), pp. 212-217.
3. Homentcovschi, D., Miles, R.N., "Viscous damping of perforated planar micromechanicstructures, " *Sensors and Actuators A*, Vol. 119, No. 2 (2005), pp. 544-552.
4. Soma, A., De Pasquale, G., "Numerical and experimental comparison of MEMS suspended plates dynamic behavior under squeeze film damping effect," *Analog Integr Circ Sig Process*, Vol. 57, No. 3 (2008), pp.213-224.
5. Tadayon, M.A. *et al*, "Nonlinear Modeling and Simulation of Thermal Effects in Microcantilever Resonator Dynamic", *J. of Physics*, No. 34 (2006), pp. 89-94.
6. Pustan, M., Paquay, S., Rochus, V., Golinval J.C. "Effects of the electrode positions on the dynamical behaviour of electrostatically actuated MEMS resonators", *Proc 12th Int Conf on Thermal, Mechanic, Multi-Physics Simulation and Experiments in Microelectronics and Microsystems*, Linz, Austria, April. 2011, pp.1- 6.
7. Lee, J. H., Lee, S. T., Yao, C. M., Fang, W., "Comments on the size effect on the microcantilever quality factor in free air space". *J Micromech and Microeng*, Vol.17, No.1 (2006), pp.139–146.
8. Bao, M., Yang, H., Sun, Y., French, P. J., "Modified reynolds' equation and analytical analysis of squeeze-film air damping of perforated structures," *J. of Micromech and Micreng*, Vol.13, No.6 (2003), pp. 795-800.
9. Soma, A., Ballestra, A., Pennetta, A., Spinola, G., "Reduced order modelling of the squeeze film damping in mems," *Proc. ESDA Engineering Systems Design and Analysis,* Torino, Italy, July 2006, pp.1003-1014.
10. Lobontiu, N., Dynamics of Microelectromechanical System, Springer Science, (New York, 2007).
11. Pustan, M. *et al*, "*Dynamical behavior of smart MEMS in industrial applications*", in book Smart sensors and MEMS: Intelligent devices and microsystems for industrial applications, Edited by S Nihtianov and A L Estepa, in Electronic and Optical Materials No. 51, (Woodhead Publishing Ltd, 2003).
12. Crandall, I. B., "The air damped vibrating system: Theoretical calibration of the condenser transmitter," *Physics Review*, Vol. 11, No.6 (1917), pp. 449-460.
13. Pandey, A. K., Pratap, R., "A comparative study of analytical squeeze film damping models in rigid rectangular perforated MEMS structures with experimental results," *Microfluidics and Nanofluidics*, Vol. 4, No. 3 (2008), pp. 205-218.

FEM simulations for built-in reliability of innovative Liquid Crystal Polymer-based QFN packaging and Sn96.5Ag3Cu0.5 solder joint

Walide Chenniki*[a], Isabelle Bord-Majek[a], Bruno Levrier[a], Komkrisd Wongtimnoi[b],
Jean-Luc Diot[c], Yves Ousten[a]

[a] IMS Laboratory, University of Bordeaux, 351 Cours de la Libération, 33405 Talence Cedex, France
Tel: 05-40-00-24-29, e-mail: walide.chenniki@ims-bordeaux.fr
[b]INSA Lyon, 20 avenue Albert Einstein, 69621 Villeurbanne Cedex, France
[c]NovaPack SAS, 14 rue des Glairaux, 38120 Saint-Egrève, France

Abstract

In this study, Quad Flat No-lead (QFN) cavity package based on LCP and the reliability impact of the package geometry are investigated. A well-established model of Sn96.5Ag3Cu0.5 solder joint fatigue based on the Darveaux's methodology leading to strain energy density estimation is used. A dedicated Design of Experiments (DoE) is performed to assess the optimal thermo-mechanical properties of the LCP package leading to the maximum operating lifetime. A correlation between predicted lifetime results and optimal thermo-mechanical properties of the package is obtained depending on the geometry of the QFN under study.

1. Introduction

The development of a new generation of innovative thermoplastics for microelectronics and photonics applications is essential to propose cavity packages able to compete with ceramic packages in terms of weight, cost and design flexibility (transparent materials for optical components). Among plastic materials, thermoplastic resins such as Liquid Crystal Polymer (LCP) offer attractive properties including low gas permeation, high temperature resistance [1], thermal stability, and low dielectric constant (3.1 at 1 MHz and 2.8 at 10 GHz [2]).

In this study, Quad Flat No-lead (QFN) cavity package based on LCP and the reliability impact of the package geometry are investigated. A well-established model of Sn96.5Ag3Cu0.5 solder joint fatigue based on the Darveaux's methodology leading to strain energy density estimation is used [3]. This model allows to describe the creep behavior and the viscoplasticity of the solder alloy subjected to strain during temperature cycling sustained by the package mounted on board. According to this methodology and Finite Element Method (FEM) thermo-mechanical 3D simulations are performed using ANSYS software. The lifetime of the solder joint is then predicted.

In a previous study, a simulation model based on FEM was performed for QFN cavity packages [3]. A fatigue lifetime of 4510 cycles has been achieved. A difference of only 20% was observed between simulation and experimental results obtained with a cavity QFN package based on a commercial LCP (3744 cycles). This reasonable difference can be explained by the LCP mechanical properties considered as isotropic in the FEM analyses.

A dedicated Design of Experiments (DoE) is performed to assess the optimal thermo-mechanical properties of the LCP package leading to the maximum

operating lifetime. The main LCP mechanical properties are considered: Young's modulus and coefficient of thermal expansion (CTE). Each combination of parameters is simulated by FEM to obtain strain energy density per cycle of the solder joint. Finally, Taguchi methodology is used to post-process simulation results [4].

A correlation between predicted lifetime results and optimal thermo-mechanical properties of different QFN packages is obtained depending on the geometry of the QFN under study. The effect of LCP properties on the solder joint of QFN package is then established.

2. Finite Element Analysis fatigue model

Thermomechanical simulation is performed using Ansys Workbench version 14.0 [5]. Several solder joint fatigue lifetime models are described in the literature [6]. The most commonly used model to evaluate the solder joint fatigue lifetime is the methodology of Darveaux. This method is based on the calculation of the dissipated plastic energy density per cycle with the help of FEM simulation. In this model, the total fatigue lifetime consists in the calculation of the required time to initiate crack in the solder joint and the calculation of the crack growth rate.

2.1. Package geometry

Two QFN models are studied: QFN 7x7 – 32 leads and QFN 10x10 – 52 leads. The reliability of these two QFN packages mold with LCP is compared. ANSYS models and geometric parameters are shown respectively in Figure 1 and Table 1.

Due to the QFN package symmetry (see Figure 1) only ¼ of the assembly is modeled. Simulations are then performed on a "quadrant" section of the QFN to optimize the calculation time.

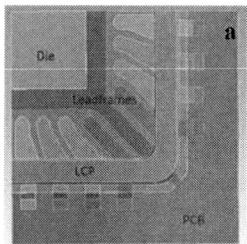 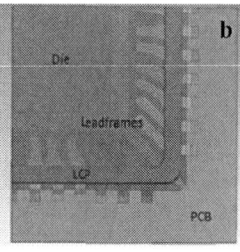

Figure 1: View of the model type "quadrant" applied to: QFN 7x7-32 leads (a) and QFN 10x10-52leads (b)

Parameters		Settings	
		QFN 7x7-32 leads	QFN 10x10-52 leads
Die Size		3 x 3 mm²	5.6 x 5.6 mm²
Die thickness		0.2 mm	0.2 mm
Package size		7 x 7 mm²	10 x 10 mm²
Leads width		0.3 mm	0.28 mm
Leads number		32	52
Pitch leads		0.65 mm	0.80 mm
Solder joint thickness		5 µm	5 µm

Table 1: Packages characteristics

A refined hexahedral meshing is performed in the solder joints as shown in Figure 2. A thinner meshing is performed on the solder joints that are the critical areas in terms of reliability.

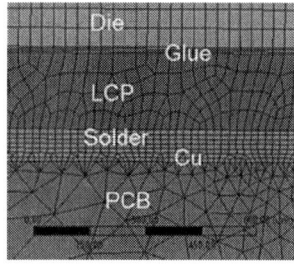

Figure 2: Junctions of solder joint with hexahedral fine mesh

2.2. Package material properties

The eutectic solder (Sn96.5Ag3Cu0.5) is a leadfree alloy. Anand's constitutive model is employed to take viscoplasticity and creep into account. Thus the solder joint fatigue can be calculated along the thermal cycles [7]. The constants of the constitutive model are shown in Table 2.

Parameters	Values
S_0 [MPa]	45.9
Q/R [K]	7460
A [1/sec]	$5.87.10^6$
ξ	2.0
m	0.0942
h_0 [MPa]	1375.98
s [MPa]	58.3
n	0.015
a	1.5

Table 2: Material parameters in Anand's viscoplastic model for Sn96.5Ag3Cu0.5

All other materials are assumed to be linear elastic and temperature independent (CTE and Young's modulus).

3. Solder joint fatigue model

Darveaux's model is the most used energy-based model for solder joint fatigue lifetime. The methodology is performed using the Anand's parameters to carry out the FEM calculation of the volume-averaged accumulated strain energy density (SED) per cycle. With the Darveaux's parameters, solder joint lifetime can be calculated using the following equations [8]:

$$N_0 = K_1 \Delta W_{ave}^{K_2} \quad (1)$$

$$\frac{da}{dN} = K_3 W_{ave}^{K_4} \quad (2)$$

$$\alpha = N_0 + \frac{a}{da/dN} \quad (3)$$

K_1, K_2, K_3 and K_4 are Darveaux's constants related to leadfree alloy material and geometry type [9]. N_0 is the number of cycles before crack initiation. da/dN is crack growth rate per cycle. ΔW_{ave} is the incremental inelastic energy per cycle in the solder joint volume. N is the total number of cycles before failure. a is the feature crack length, usually equals to pad diameter for BGA or pad length for plane solder joint. α is the characteristic lifetime when 63.2% of the population has failed.

4. Design of Experiments (DoE)

Several simulations have been performed using Design of Experiments (DoE) matrix. The DoE consists of several combinations of thermal and mechanical properties of the LCP. Each combination of parameters is then simulated by FEM. Finally the Taguchi methodology is used to post-processing of simulation results.

4.1. Parameters of DoE

The main LCP mechanical properties are considered: Young's modulus and CTE. Three different values of Young's modulus (5.7 GPa, 12.3 GPa and 20 GPa) and CTE (7.1 ppm/K, 10 ppm/K and 14ppm/K) are considered. The first set of values (5.7 GPa and 7.1

ppm/K) correspond to the experimental values obtained for commercial LCP (isotropic properties are considered in the simulations). The second set of values (12.3 GPa and 10 ppm/K) are the datasheets values. The latest set of values (20 GPa and 14ppm/K) are the extreme expected values for LCP. A thermal cycling is considered between -65°C and 145°C.

The experimental thermal cycles consist of a sequence of high and low temperature portions to trigger the initiation and propagation of thermo-mechanical fatigue cracks in solder joints. Figure 3 plots the experimental thermal cycling profile implemented in FEM simulations.

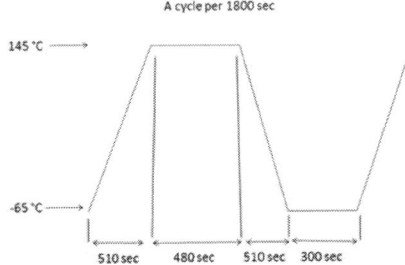

Figure 3: Thermal cycling profile

4.2. Simulation results

Different combinations of parameters are simulated. The result of the simulation is the strain energy density (SED) per cycle in the solder joint. The residual lifetime is then calculated using the Darveaux's methodology. Table 3 shows the obtained results.

Parameters		Results	
Young's modulus (GPa)	CTE (ppm)	α: characteristic lifetime (cycles)	
		QFN 7x7 – 32 leads	QFN 10x10 – 52 leads
A = 5.7	a = 7.1	8229	7288
A	b = 10	8699	7516
A	c = 14	9113	7780
B = 12.3	a	7523	6774
B	b	8218	7184
B	c	8943	7657
C = 20	a	6586	6139
C	b	7558	6721
C	c	8628	7463

Table 3: Simulations results

A L9 Taguchi table is first built considering all the parameters combinations and the simulation results.

Several L4 tables are then selected for the post-processing. The aim is to determine the interaction between each parameter and to extract the optimal value for CTE and Young's modulus. As an example Figure 4 presents the interaction thermal cycling/Young's modulus (a) and thermal cycling/CTE (b) for both QFN packages.

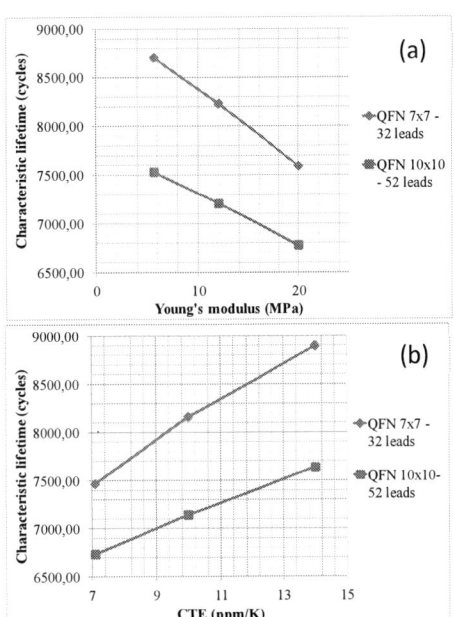

Figure 4: Young modulus (a) and CTE interaction (b)

Larger package size corresponds to a lower lifetime. This observation is in agreement with the literature [10].

According to Figure 4, the lifetime decreases with LCP Young's modulus and increases with CTE.

When the Young's modulus increases the LCP is more rigid. Then the risk of cracking of the solder increases.

When the CTE increases it tends towards the CTE of the other package materials decreasing the risk of cracking.

4.3. Optimum LCP mechanical properties

The interaction between the governing material and process parameters is then treated in the final analysis. Considering these interactions optimum values for CTE and Young's modulus of the LCP are assessed. The optimal parameters are determined graphically. The optimum are summarized in Table 4.

Parameters	Optimum		
	QFN 7x7- 32 leads	QFN 10x10- 52 leads	
Young modulus (GPa)	8.7	8.7	16
CTE (ppm/K)	8.7	8.7	12.3

Table 4: Optimal results values for CTE and Young's modulus of the LCP

Two optimum are identified using Taguchi methodology for both QFN packages: 8.7 GPa and 8.7 ppm/K. For the larger package two sets of values are determined. The second set 16 GPa and 12.3 ppm/K correspond to a mathematical solution not physically possible. In the datasheet the Young's Modulus of the

LCP is 12.7 GPa. The value of 16 GPa seems to be difficult to reach with such a polymer.

According to the simulations and the DoE, a QFN package with a maximum reliability is obtained with a value of 8.7 GPa for the Young's Modulus and 8.7 ppm/K for the CTE whatever the package size.

5. Conclusion

In order to assess the optimal thermo-mechanical properties of the LCP package resulting in the maximum operating lifetime, a special designed DoE has been performed. The main mechanical properties (Young's Modulus and CTE) have been considered.

The impact of each parameter on the fatigue lifetime has been investigated. The Taguchi methodology has been used to determine the optimal value for CTE and Young's Modulus of the LCP.

Whatever the package size, the Young's Modulus should tend towards 8.7 GPa and CTE towards 8.7 ppm/K. A maximal reliability is obtained using this set of values.

Acknowledgements

This work is done in the framework of MICROPLAST project, MINALOGIC and PLASTIPOLIS labeled project, sponsored by OSEO and FEDER.

References

[1] A.-V. Pham, « Packaging with Liquid Crystal Polymer », IEEE Microwave Magazine, vol. 12, n°. 5, p. 83–91, Aug. 2011

[2] E. C. Culbertson, « A new laminate material for high performance PCBs: liquid crystal polymer copper clad films », Electronic Components and Technology Conference, p. 520–523, May 1995

[3] W.Chenniki, I. Bord-Majek, J-L Diot, B. Levrier, E. Romain-Latu, V. Verriere, J. Vittu, Y. Ousten, « Thermo-mecanical simulation of cavity QFN package based on Liquid Cristal Polymer for reliability study », Minapad Conference 2012

[4] F. Salehuddin, I. Ahmad, F. A. Hamid, A. Zaharim, «Analyze and optimize the silicide thickness in 45nm CMOS technology using Taguchi method», ICSE Conference, June 2010

[5] Ansys, User's Manual, Swanson Analysis Systems, IV: 4.23-4.25, 1994

[6] W.W. Lee, L.T. Nguyen, G.D Selvaduray, « Solder joint fatigue models: review and applicability to chip scale packages », Microelectronics Reliability, vol. 40, n°.2, p. 231-244, Feb. 2000

[7] K. Mysore, G. Subbarayan, V. Gupta, et Ron Zhang, « Constitutive and Aging Behavior of Sn3.0Ag0.5Cu Solder Alloy », IEEE Transactions on Electronics Packaging Manufacturing, vol. 32, n°. 4, p. 221–232, Oct. 2009

[8] R. Darveaux, « Effect of simulation methodology on solder joint crack growth correlation », Proceedings of 50th Electronic Components & Technology Conference, p. 1048–1058, May 2000

[9] B. Levrier and Y. Ousten, « Eurelnet-Fiabilité-European Reliability Network », *Eurelnet* [Online]. Available: https://www.eurelnet.org/

[10] L. Li, «Reliability modeling and testing of advanced QFN packages», Electronic Components & Technology Conference, May 2013

Framework to Extract Cohesive Zone Parameters Using Double Cantilever Beam and Four-Point Bend Fracture Tests

Sathyanarayanan Raghavan*, Ilko Schmadlak**, George Leal***, and Suresh K. Sitaraman*
* The George W. Woodruff School of Mechanical Engineering
Georgia Institute of Technology; Atlanta; Georgia
** Freescale Halbleiter Deutschland GmbH, Munich, Germany
*** Freescale Semiconductor, Austin, Texas
suresh.sitaraman@me.getech.edu

Abstract

This paper focuses on extracting cohesive zone modeling (CZM) parameters for ultra low-*k* (ULK) interlayer dielectric (ILD) layers present in back end of line stack (BEOL) of flip-chip (FC) semiconductor devices. Unlike other fracture-mechanics based approaches, CZM can simulate crack initiation and propagation at several locations. However, additional parameters need to be determined to enable cohesive zone (CZ) elements to reliably predict the failure region. In this paper, we present a methodology, combing fracture experiments and finite-element (FE) simulations to extract CZM parameters under mixed-mode loading conditions. Using load vs. displacement data from the double cantilever beam (DCB) test and four-point bend test (FPBT) experiments, we have characterized all the parameters necessary for CZM. The developed cohesive zone failure criteria can then be applied to interfaces in BEOL stack models to identify the locations of crack initiation and propagation.

1. Assessing Under Bump Delamination

The BEOL structures of integrated circuits (IC) in FC packages are exposed to high thermo-mechanical stresses during lead-free solder reflow assembly. These stresses occur due to the mismatch in the coefficient of thermal expansion (CTE) between the silicon die and the organic substrate under the high thermal excursion from reflow temperature to room temperature. During such a cool down from reflow temperature, the solidified solder bumps couple the differential shrinkage of die and substrate, thus expose the underbump region of the die to high normal and shear stresses. Circular cracks under the FC solder bumps can occur as a triggered failure mode. These failures are often referred to as white-bumps.

To address such white-bump failures, microelectronics industry typically employs various design modifications in the BEOL stack and/or processing changes. In addition, fracture mechanics together with FE simulation is often used to determine the chances for crack propagation under reflow assembly and cool-down conditions, as described in detail in our previous publications [1, 2]. Although fracture mechanics is helpful in studying such failures, it invariably requires pre-existing cracks in locations where failures are likely to occur. The size of such pre-existing cracks is often chosen based on past experience. Also, when there are multiple locations where cracks could propagate, it is difficult to implement multiple crack fronts in fracture mechanics approach, as the stress fields from different cracks will influence the stress fields around proximal cracks.

2. Cohesive Zone Modeling Approach

Cohesive zone modeling approach can effectively overcome the aforementioned difficulties with fracture mechanics based approach. In CZM approach fracture is viewed as a progressive degradation of material along fictitious extension of crack tip. Under external loading, the interaction between the two fictitious surfaces, often called cohesive surfaces, is governed by traction-separation law. Thus, CZM requires additional parameters to be defined to fully characterize the interface of interest. In reality cracks propagate under mixed mode loading conditions, therefore in CZM the implementation of the fracture criteria is realized via separate traction-separation parameters for pure peel (mode I) and pure shear (mode II). Depending on the mode mixity, an effective combined criterion needs to be exceeded for delamination to propagate.

From the different traction-separation constitutive relations that have been developed for different application scenarios, the bilinear law [3, 4] has been found to best fit our study. The specific bilinear traction-separation approach used here is based on the model developed by Alfano and Cresfield [5] and the general form of bilinear traction-separation law is shown in Figure 1. In order to fully characterize the bilinear law, three independent parameters are required for each pure mode case, mode I (normal) and mode II (tangential) respectively. Therefore six independent parameters namely, maximum normal traction (T^n_{max}), critical normal separation (δ^n_c), ratio of separation at maximum normal traction (δ^{n*}) to δ^n_c (α^n), maximum tangential separation (T^t_{max}), critical tangential separation and ratio of separation at maximum tangential separation (δ^{t*}) to δ^t_c (α^t) are required to fully characterize bilinear law for mixed mode crack propagation scenarios.

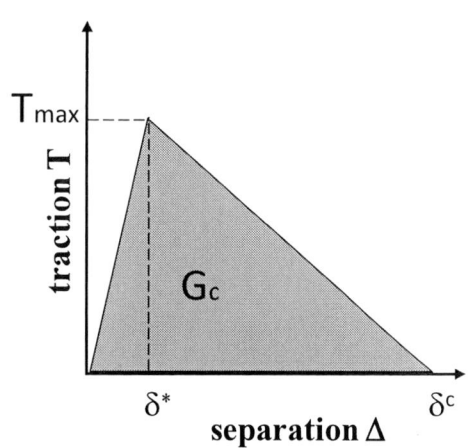

Figure 1: General bilinear traction separation curve

As shown in Figure 1, once the separation reaches the critical value δ_c, there is no more interaction between the cohesive surfaces. Therefore, the area under the curve represents the energy released when new, free surfaces are created or the critical strain energy release rate (ERR), G_c. In this work, we present a framework to extract the six independent parameters using a combination of experiments and finite-element simulations for the critical layer present in BEOL stack.

3. Fracture Tests to Extract CZ Parameters

In order to assess the strength of the failing interface, fracture test specimens of 65 mm length, 12 mm width and 0.78 mm thick are diced from a real product wafer. Each test sample consists of four dies separated by crack stop structures. As a first step, the solder bumps present in the test specimen are removed by fine polishing without damaging the BEOL stack underneath. The test specimen is then glued to a dummy sample of equal dimensions using an epoxy. The test sample sides are polished in order to increase the surface strength by reducing the number of micro cracks created during dicing. It should be pointed out that no starter crack is fabricated; the test samples are diced from real devices. Therefore, we begin our experiments with four-point bend test and use the crack generated by FPBT for further processing.

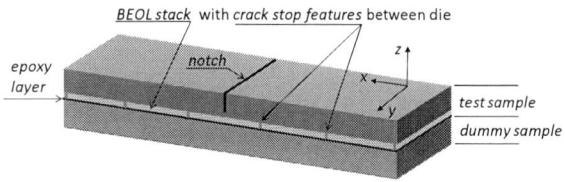

Figure 2: Schematic of four-point bend test specimen

A schematic of the FPBT sample is shown in Figure 2. A 700μm deep notch is made in the center of the test specimen by dicing, as illustrated. During the FPBT, as pointed out in Figure 3, crack starts from the notch tip and propagates into the BEOL stack. Once it hits the weakest layer, it starts propagating along the weakest layer perpendicular to notch, as shown in Figure 4 [6].

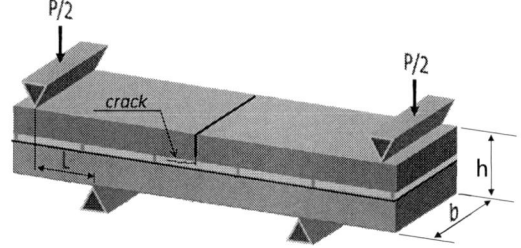

Figure 3: Schematic of four-point bend test

Figure 4: Crack propagating from notch tip into the BEOL stack

Since the moment between the two inner supports remains constant in FPBT, the force remains constant as the crack propagates. Figure 5 shows the measured load displacement curves for four-point bend testing. Focused ion beam (FIB) cross section of samples at various locations revealed that the cracks generated in the tests propagated through the same ultra low-k interlayer dielectric interface that had failed during flip-chip assembly process.

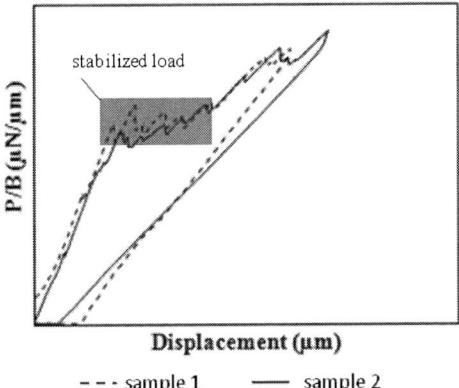

- - - sample 1   —— sample 2

Figure 5: Load displacement curves from four-point bend test

The critical energy release rate G_c for crack propagation can be obtained from the average stabilized load P while E and ν are the dummy specimen's Young's modulus and Poisson's ratio respectively [6].

$$G_c = \frac{21(1-\nu^2)P^2 L^2}{16Eb^2h^3} \qquad (1)$$

For symmetric sandwich specimen used in this work the mode mixity for crack propagation in FPBT is approximately 41° [7]. However, for characterizing CZ parameters, mode I critical energy release rate and mode II critical energy release rate are required. As mentioned earlier, no pre-fabricated crack is present in the sample, therefore, the crack generated by FPBT is used as starter crack for performing double cantilever beam test as shown schematically in Figure 6.

The DCB test is designed to apply a mode I load, allowing to extract G_{Ic} from it. In order to be able to apply the required peel load, the test samples are modified by attaching a bracket on each side of the pre-cracked specimen, as shown in Figure 6.

The mode I critical ERR under symmetric loading in a DCB test can be calculated by equation 2, derived from beam bending theory [8]

$$G_{Ic} = \frac{12(1-\nu^2)P^2 a^2}{Eb^2h^3} \qquad (2)$$

The calculation requires the length of the crack (a) to be known.

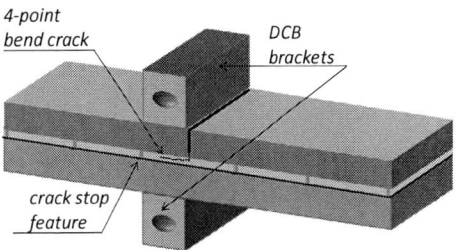

Figure 6: Schematic of DCB test specimen

Since optically measuring a crack of such small scale with the required accuracy can be a difficult task, the crack length is calculated using compliance calibration method. The pre-cracked specimen is first loaded until the crack begins to propagate. The tracked load-displacement curve indicates this event by a sudden drop in load, as shown in Figure 7. Subsequently, the reduction in stiffness was measured by several unloading and reloading steps until the sample fails as shown in the chart of Figure 7.

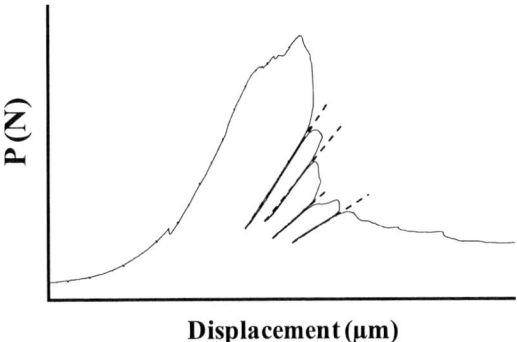

Figure 7: DCB load –displacement curve with compliance calibration

The decrease in stiffness is equivalent to the increase in compliance C, the reciprocal of the stiffness. The crack length is then determined from measured compliance value using equation 3 in which E_{xx} is the longitudinal elastic modulus and μ_{xy} the shear modulus in XZ plane (Fig.2) [9]. From the calculated crack length and peak load (P) G_{Ic} can be determined from equation 2,

$$C = \frac{8a^3}{E_{xx}Bh^3} + \frac{12a}{5Bh\mu_{xz}} \qquad (3)$$

The measured range of G_{Ic} from different samples was narrow which represented a good quality of the test results. The other two independent parameters required to define the mode I traction-separation curve, the maximum normal traction T^n_{max} as well as the separation ratio α^n are extracted by performing a finite-

978-1-4799-4789-8/14 $31.00 © 2014 IEEE   237

element simulation of DCB test with CZ elements placed along the critical layer determined from focused ion beam (FIB) cross sections of samples. Several simulations are performed by keeping α^n constant and varying T^n_{max}. In a next step, a sensitivity study on those two parameters is conducted to find the best fit for simulated load-displacement with experimental results. Figure 9 shows the resulting load displacement curves from three simulations and the experimentally measured curve respectively, in which a good match was found for $T^n_{max}{}^{**}$.

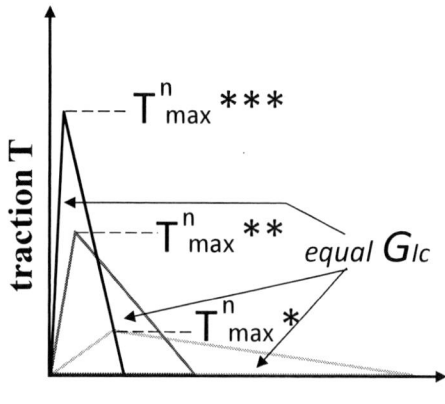

Figure 8: *Traction separation curves simulated to match simulation results with tests*

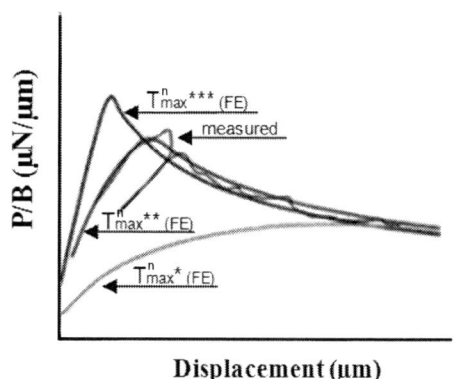

Figure 9: *Simulated and measured load displacement curves for DCB test*

The same approach is further used to determine the mode II traction separation parameters by performing FPBT simulations with CZ elements along the critical layer, keeping the mode I CZ parameters constant. From [10] it was assumed that:

$$G_{IIc} = 10 G_{Ic}.$$

Since the FPBT represents a mixed mode fracture configuration, the maximum shear traction T^t_{max} and α^t

are varied until the simulated load-displacement curve for the FPBT agrees with the measurements. Figure 10 shows the simulated test with crack advance in the interface. Figure 11 plots the simulated and measured load-displacement curves in comparison, revealing a fair agreement.

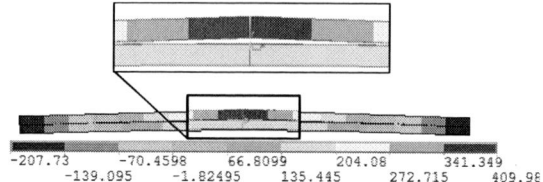

Figure 10: *Out of plane displacement contour plot of simulated four-point bend test*

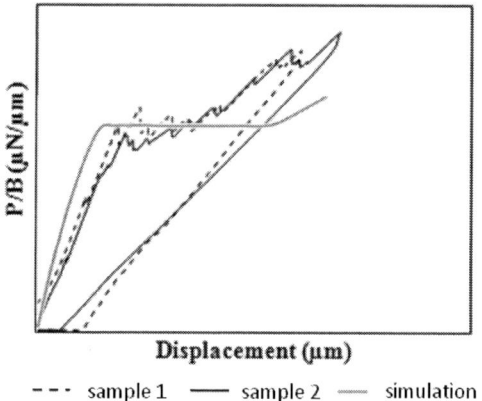

Figure 11: *Simulated and measured load displacement curves for FPBT*

4. Conclusion

The presented approach of measuring the critical energy release rate of ultra low-k layers of real back end of line structures with the four-point bend and the double cantilever beam fracture test, allows the identification of the mode I and mode II traction separation curves by the aid of finite element simulations. These represent a complete failure criterion for the measured interface for all possible mode mixity loading conditions. The found fracture parameters can be applied to simulation models which are able to predict fracture in the measured interlayer dielectric layers in good agreement with reality as shown in the direct comparison to the measurements.

It is planned to use the found failure criteria in more complex 3D models, simulating products under reflow conditions and conduct further validation with failure analysis findings. This further verification would be a large step towards the final aim which is the prediction of white bump risk in other products with increased accuracy and efficiency.

Acknowledgements

This work is supported by the Semiconductor Research Corporation. Industry liaisons or mentors for this project are: Torsten Hauck from Freescale and Chenzhou Lian from IBM. The authors are grateful for the valuable direction and suggestions from the industry mentors in carrying out this work.

References

1. Raghavan, S.; Schmadlak, I.; Leal, G.; Sitaraman, S., "Study of Chip-Package Interaction Parameters on Interlayer Dielectric Crack Propagation," *Device and Materials Reliability, IEEE Transactions*, Vol.PP, no.99, pp.1,1
   doi: 10.1109/TDMR.2013.2288255

2. Raghavan, S., Schmadlak, I., Leal, G., and Sitaraman, S., "Chip-Package Co-Design: Effect of Substrate Warpage on BEOL Reliability," *ASME 2013 International Mechanical Engineering Congress and Exposition (IMECE2013)*, San Diego, CA, November 2013, IMECE2013-65877.

3. G. T. Camacho and M. Ortiz, "Computational modelling of impact damage in brittle materials," *International Journal of Solids and Structures*, Vol. 33, pp. 2899-2938, Aug 1996.

4. P. H. Geubelle and J. S. Baylor, "Impact-induced delamination of composites: a 2D simulation," *Composites Part B-Engineering*, Vol. 29, pp. 589-602, 1998.

5. G. Alfano and M. A. Crisfield, "Finite element interface models for the delamination analysis of laminated composites: Mechanical and computational issues," *International Journal for Numerical Methods in Engineering*, Vol. 50, pp. 1701-1736, Mar 10 2001.

6. R. Dauskardt, M. Lane, Q. Ma, and N. Krishna, "Adhesion and debonding of multi-layer thin film structures," *Engineering Fracture Mechanics*, Vol. 61, pp. 141-162, Aug 1998.

7. P. G. Charalambides, H. C. Cao, J. Lund, and A. G. Evans, "Development of a Test Method for Measuring the Mixed-Mode Fracture-Resistance of Bimaterial Interfaces," *Mechanics of Materials*, Vol. 8, pp. 269-283, Feb 1990.

8. M. F. S. F. de Moura, J. J. L. Morais, and N. Dourado, "A new data reduction scheme for mode I wood fracture characterization using the double cantilever beam test," *Engineering Fracture Mechanics*, Vol. 75, pp. 3852-3865, Sep 2008.

9. J.J.L. Morais, M.F.S.F. de Moura, F.A.M. Pereira, J. Xavier, N. Dourado, M.I.R. Dias, J.M.T. Azevedo, "The double cantilever beam test applied to mode I fracture characterization of cortical bone tissue," *Journal of the Mechanical Behavior of Biomedical Materials*, Vol. 3, Issue 6, August 2010, pp. 446-453, ISSN 1751-6161,

10. X. Zhang, S. Im, R. Huang, and P. Ho, Chip-Package Interaction and Reliability Impact on Cu/Low-k Interconnects. (Norwood, MA: Artech House, 2008), ch. 2, pp. 23–59.

Measuring the Mechanical Relevant Shrinkage during In-Mold and Post-Mold Cure with the Stress Chip

F. Schindler-Saefkow[1,4], F. Rost[1,3], A. Rezaie-Adli[2],
K.M.B. Jansen[2], B. Wunderle[3], J. Keller[4], S. Rzepka[1], B. Michel[1]
[1] Fraunhofer ENAS, Micro Materials Center, Chemnitz, Germany;
[2] Delft University of Technology, Netherland;
[3] Technische Universität Chemnitz, Germany;
[4] AMIC GmbH, Berlin, Germany;
florian.schindler-saefkow@enas.fraunhofer.de

Abstract

The integration of smart systems into hybrid structures is one of the challenges addressed by the project MERGE – the cluster of excellence on Technologies for Multifunctional Lightweight Structures. As a first example, a sensor system is integrated that is able to explore the thermo-mechanical conditions these systems will typically be exposed to. After briefly describing the sensor system, the paper focuses on the results of the encapsulation step as part of the fabrication process mounting the sensor chip on the test board. The sensor system measures the mechanical stresses during and after transfer molding. In particular, the in-plane components on the chip surface were recorded with high accuracy [1, 2]. Based on these informations, material parameters have been deduced by combining experimental and simulation methods within a Design of Experiment (DoE) study.

During the encapsulation process, two sets of effects induce stress into the package simultaneously. On one hand, the coefficients of thermal expansion (CTE) lead to a thermal shrinkage of the materials during cooling from the curing to room temperature. On the other hand, the volume also decreases when the epoxy mold compound (EMC) is cured from its fluid into the final solid stage. This effect is called chemical cure shrinkage [3]. Separating both effects is really a challenge. The method shown in this paper allows quantifying the corresponding material parameters by combining the stress measurements with numerical parameter identification. Based on this method, the investigation on failure modes and reliability of the integrated smart systems can be improved.

Motivation

Measuring and simulating the cure shrinkage of epoxy materials (e.g. molding compounds) is one of the challenges of the last decade. The importance of knowing the chemical shrinkage will becomes clear by comparing the fracture toughness simulated with and without the residual stress induced by the curing process [6].

The comparison in Figure 1 shows that the simulation result is about 100% higher when the residual stresses are considered. Therefore, the magnitude of the residual stress needs to be determined. More general,

the relevant effects of cure kinetics need to be covered adequately by the constitutive models in order to reach the level of result accuracy that allows implementing virtual schemes based on numerical simulation in the design process for new electronics and smart cyber physical systems replacing tedious experimental tests and speeding up the development cycle.

Within this paper, we show a way of measuring the stress development during the molding and curing processes based on our stress sensing system [2]. The magnitude of the mechanically relevant shrinkage is determined and a parameter is introduced to model it quantitatively. Its value is identified by comparing the stress results from the experiment to those obtained by simulation and by minimizing the differences in an optimization step. This way, the simulation model is enabled to capture the residual stress within the package (chip, board and molding compound) exactly as it was measured in the experiment. The value of cure shrinkage obtained this way can subsequently be used in all further reliability simulations for this particular molding compound – as long a the molding and curing processes are not changed.

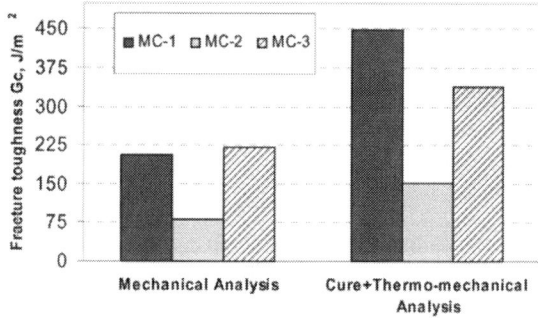

Figure 1 Comparison between fracture toughness for 3 EMC materials. Mechanical analysis contains only the critical load leading to the crack propagation. Cure+ thermo-mechanical analysis contains also the residual stresses arising from the manufacturing process [6].

The stress chip measurement system

The stress sensing system was developed by the publicly funded project 'iForceSence' (Germany) [2]. It provides readings of the in-plane components at the die surface. The difference σ_{Diff} between the two normal stress components σ_{xx} and σ_{yy} can readily be computed from the electrical currents I_{in} and I_{out} of orthogonally arranged transistors and the piezoresistive coefficient $\pi_{44}^{(p)}$ of the p-channel devices, equ. (1). The sum of the normal stresses σ_{sum} can be deduced from two stress states S_1, S_2 and the piezoresistive coefficients of the n-channel devices $\pi_{11}^{(n)}, \pi_{12}^{(n)}$ if the temperature is the same for booth stress states explored and the stress σ_{zz} is small, equ. (2). Then, equ. (3) yields the normal stress components [2, 4].

$$\sigma_{Diff} = \frac{2}{\pi_{44}^{(p)}}\left(\frac{I_{out} - I_{in}}{I_{out} + I_{in}}\right) \qquad (1)$$

$$\sigma_{sum} = \frac{2}{\pi_{11}^{(n)} + \pi_{12}^{(n)}}\left(1 - \frac{I_{in}^{s2} + I_{out}^{s2}}{I_{in}^{s1} + I_{out}^{s1}} - \pi_{12}^{(n)}\sigma_{zz}\right) \qquad (2)$$

$$\sigma_{xx} = \frac{\sigma_{Diff} + \sigma_{sum}}{2}; \quad \sigma_{yy} = \frac{\sigma_{Diff} - \sigma_{sum}}{2} \qquad (3)$$

On the sensor die, the individual cells are arranged as 2-D matrix with 273 µm pitch (Figure 2). Several chip configurations are available to represent different die sizes. Each sensor is accompanied by signal conditioning, data storage, and parallel/sequential conversion circuitry. The full system allows an in-situ determination of the stress states before, during and after the assembly steps, i.e., during fabrication and testing of new packages.

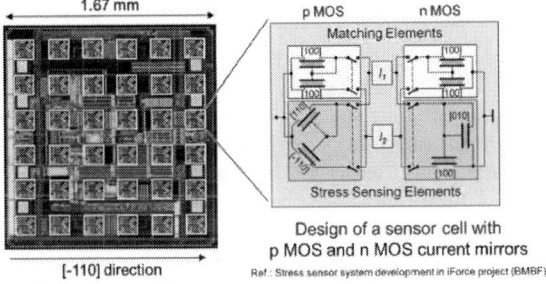

Figure 2 Stress chip (6x6 cells) and circuit block diagram of the measurement cells

The transfer mold process

The encapsulation process was studied in the Labs of Besi/Fico in Duiven, The Netherlands. The stress chips are assembled on a Chip on Board (CoB) setup. The flexible printed circuit (FPC) board has a 50 µm thin polyimide substrate. The length of the FPC strip allowed bringing the four contact pads of the stress chip outside of the mold machine as needed for in-situ measurement during molding. The FPC offers two places for the chip but just one chip per board was assembled.

Figure 3 CoB sample on flex-substrate after encapsulation

Figure 3 shows the sample after the encapsulation. The red quadrate indicates the position of the sensor chip (1.67 mm x 1.67 mm) surrounded by the epoxy mold compound of 75 mm x 75 mm.

Figure 4 presents the measured data of the stress chip during the encapsulation process. The mechanical stress is plotted in the black graph and the blue graph represents the temperature. The stress data results from the average value of the first row of stress cell at the top of chip (framed in Figure 4). In minute 5 -10 shows the reference state at room temperature (RT). By inserting the sample in the mold machine the temperature raised up to process temperature of 140 °C and a stress of 12 MPa was induced based on thermal expansion. At 15th minute the encapsulation was started by filling the mold with EMC. The temperature stayed constant and the stress went up to 76 MPa at 16th minute. The chemical curing of the EMC which is associated with volume shrinkage, took until the 31th minute. A change of stress of about 20 MPa was measured. After mold opening and ejection of the specimen both stress and temperature rapidly drop (minute 31-37).

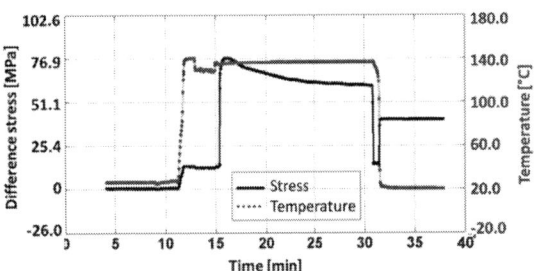

Figure 4 Transfer mold process: average difference stress of the top cell row difference stress and temperature vs. time

Furthermore the post mold cure process was measured also. There for the samples were heated up to 140°C three times in an oven. The recorded difference stress and temperature vs. time are plotted in Figure 5. Three thermal cycles of 40 min, 4h and 60 min at 140 °C were undertaken.

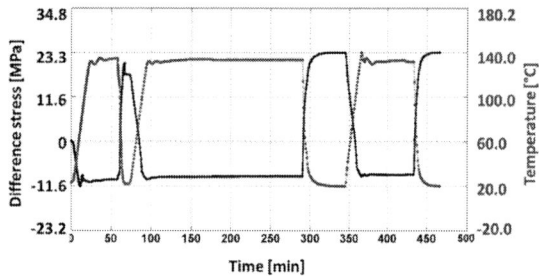

Figure 5 post mold cycle: average difference stress of the top cell row and temperature vs. time

FE-Simulation

The numerical simulation was prepared with ANSYS°14. The main reason was to estimate the behavior during the curing of the epoxy mold compound by determine the stress result on the chip surface and to compare it with the experimental results.

Therefore a 3D model was performed. Just one quarter of the geometry had to be modelled, because of symmetry orthogonal in xz and yz plain. In order to model the cure shrinkage we used the moisture expansion coefficient BETX in ANSYS°14. Similar to the coefficient of thermal expansion (CTE) the BETX describes the volumetric change by moisture expansion. The value was set to 0.8%, based on the datasheet of the EMC. For the reference state of the uncured EMC the saturation was set to 100% and reduced to 0%, which describes the state where the EMC is fully cured. That means the volume of EMC changed from uncured to the cured state by 0.8%.

Additional to the cure shrinkage boundary conditions a constant temperature of 140 °C was set as the process temperature during the encapsulation. Only in-plane displacement was allowed, because the specimen was fixed from the top and bottom mold during the encapsulation. The load step time was defined to 15000 sec which represents the in-mold and post-mold time at 140°C.

Component / material	Material law
Flex substrate / kapton	Linear elastic
Adhesive / epoxy	Linear elastic
Die, silicon	Linear elastic
Epoxy mold compound	Linear elastic

Table 1 Components and material laws of the simulation model

Table 1 lists all components respectively the materials with their applied material laws which were used in the FE-simulation. The material datas are based on data sheets. The viscoelastic behavior was measured by dynamic mechanical analysis (DMA) of the fully cured EMC at TU Delft (Figure 6).

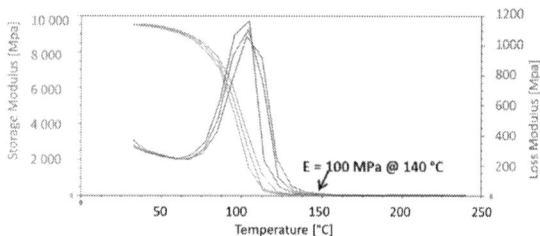

Figure 6 Epoxy mold compound, DMA measurements

Because of the isothermal conditions of 140 °C the EMC behaves in this temperature range linear elastic. Above the glass transition temperature of 110 °C the Young's module does not significant change as it is shown in the DMA curves in Figure 6.

Figure 7 displays on the left hand side the xz cross section of the model. The chip is surrounded by adhesive and EMC positioned on the flexible substrate. At the right hand side the z-displacement shows very clearly the decreasing of the EMC volume up to 10 μm.

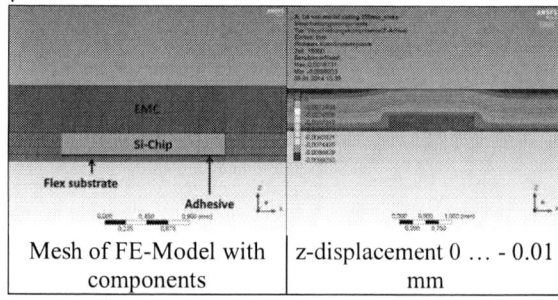

Mesh of FE-Model with components	z-displacement 0 … - 0.01 mm

Figure 7 XZ cross section of simulation model

Results of experiment and simulation

By determination the stress of mechanically relevant cure shrinkage and thermal shrinkage it has to been clarified at which point the package is in a thermo mechanical stabile state. To proof that the EMC is fully cured it is necessary to analyze the hysteresis of stress vs. temperature during the post mold cure temperature cycles (Figure 8). The plot shows the evolution over the three post mold temperature cycles of the difference stress from one cell of stress chip at the middle of the left edge. The first cycle had a high stress impact of 28 MPa. However, the stress impact between end of cycle no. 1 and cycle no. 2 amounts 10 MPa. The stress value of the last cycle no. 3 ends exactly at the level of the cycle before. That means that the package is in a stabile state up to a thermal load of 140 °C after two thermal cycles. Furthermore it is possible to say that till the end of the second cycle curing processes were still in progress.

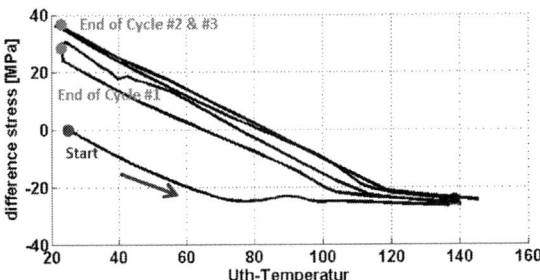

Figure 8 Result of experiment: difference stress vs. temperature of thermal cycle after encapsulation

By having a closer view on the evolution of the stress graph a change of the slope is distinguishable. Between 100 °C and 120 °C the stress gradient decreases which it is based on the glass transition temperature of EMC. This effect moves to higher temperature by the number of cycles. It is well known, that the material behaviour change with the state of curing [5]. This can be also monitored here with the stress chip.

Now the stress effects based on thermal shrinkage and mechanically relevant chemical shrinkage can be evaluated. As mentioned before the mechanical relevant chemical shrinkage has a stress impact till the 2nd post mold cycle. During the 3rd cycle only thermal shrinkage is working.

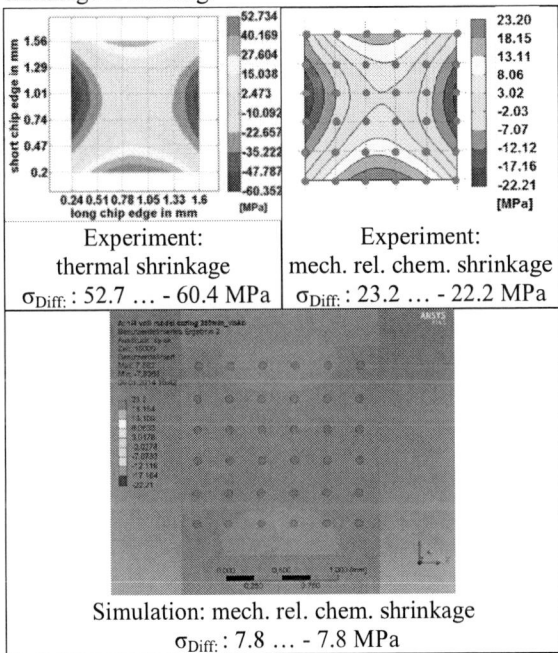

Figure 9 Difference stress induced by mechanically relevant cure shrinkage experiment and simulation

In Figure 9 the stress distribution of these effects are shown. The top left and top right stress plots are experimental results. The first one presents the stress impact of thermal shrinkage between RT to 140 °C during the 3rd cycle by reaching values from 52 ... - 60

MPa. In the second plot the induced stress caused of mech. rel. chemical shrinkage is shown. Therefore the stress states from the beginning of the in-mold-cure and the end of the post-mold-cure both at 140 °C are subtracted from each other. The difference stress ranged between 23.2 ...- 22.2 MPa. According that the bottom plot of Figure 9 shows the stress distribution from +/- 7.8 MPa of the FE-simulation by volumetric decreasing of EMC.

By a first comparison between the two experimental results it can be seen that the stress impact is more than 50 % higher by thermal shrinkage then by chemical shrinkage. The stress distribution gives in both cases qualitative the same pattern.

The meaning between the experimental result of cure shrinkage and the FEA results are explained in the following chapter.

Parameter identification

The parameter identification follows the method of adapting the FE-simulation to the experiment result by using Design of Experiment (DoE) analyses.

Therefore in Figure 9 (top right and bottom) the exact position of each measurement cell (mc) is pointed out by the red circles. Now a failure factor F calculates the sum of the difference between experiment and simulation at each measurement cell as is shown in equ. (4).

$$F = \sum_{k=1}^{n_{mc}} \left(\sigma_{k,sim} - \sigma_{k,exp} \right)^2 \quad (4)$$

OptiSLang 4.0 is employed to identify the correct value of BETX. A simple DoE is set up with the material parameter of cure shrinkage parameter (BETX) being the only factor. Its range is defined from 0.4% to 4%.

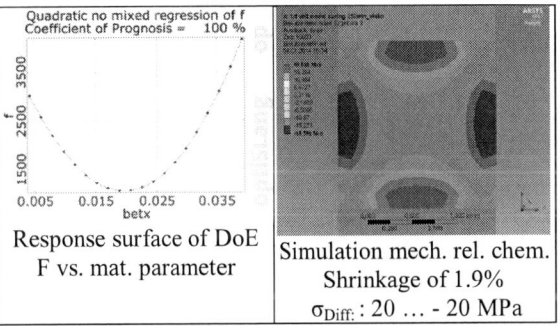

Figure 10 Result of DoE: optimized material parameter of EMC

The response surface as the result of the DoE analysis relates the input parameter, BETX, to the output parameter, F. It is shown in the left graph of Figure 10. The minimum of F yields the correct value of the material parameter BETX. Accordingly, the mechanically relevant cure shrinkage amounts to 1.96 %. The corresponding result of the FE simulation yields a stress distribution very similar to the experimental result. The computed range of difference stress now is between -20 MPa and 20 MPa (right plot of Figure 10). Hence, the parameter identification leads to a very close match between experiment and simulation. The remaining difference is as low as ±3 MPa dominantly attributed to experimental scatter.

Conclusion

The sensor chip was successfully used for the in-situ observation of the mechanical stresses occurring during transfer molding. Temperature and mechanical stresses could be monitored continuously. This is very interesting for the further investigations of hybrid components functionalized by smart systems within the project MERGE.

Furthermore, it was shown possible to detect material parameters of mechanically relevant chemical shrinkage by combining experimental tests, finite element analyses, and DoE routines. The actual time of curing and the changes in the material behavior during the curing could be found out. The stresses induced by different effects were successfully separated. In the particular example, the effect of cure shrinkage has 50% less impact on the stress then the thermal shrinkage from the molding temperature of 140°C to RT. The value of mechanically relevant shrinkage could be identified as 1.96 %, which is more than twice the value given in the datasheet of the material supplier. Here are some open questions that cannot be finally answered within this paper. Maybe the boundary conditions have more influence than expected; the material data's could be measured again or the data sheet information could to be verified differently.

The method allows measuring the residual stress of the packages and to transfer this information into simulation models. Likewise, this information leads to better approximation of reality and, hence, improves reliability investigations such as lifetime estimations, thermo mechanical assessments, and crack propagation studies.

A higher BETX value would be the result, if the FEM Model simulates with a visco-elastic material parameter. As the result the stress is dropping down with the same volume shrinkage. Two different BETX shrinkage parameter will be created (an elastic BETX and a visco-elastic BETX). FEM engineer decides which way to go considering the residual package stress in the next simulation step.

By showing the possibility of in-situ stress measurement during molding, other process steps could also be monitored and the influence of their process parameters can be investigated in the future. Hence, the stress measurement chip proofs their capability as health monitoring system for investigating the package reliability.

Acknowledgments

Part of this work was performed within the Federal Cluster of Excellence EXC 1075 "MERGE Technolo-gies for Multifunctional Lightweight Structures" supported by German Research Foundation (DFG). Financial support is gratefully acknowledged. The authors would like to thank H. Wensink from BESI Netherlands B.V. for leading the technical setup during the transfer molding experiments.

References

[1] F. Schindler-Saefkow et al. „Stress Chip Measurements of the Internal Package Stress for Process Characterization and Health Monitoring", EuroSimE, Cascais, Portugal, April 16-18 2012

[2] R.C.Jaeger et al., "CMOS Stress Sensors on (100) Silicon", IEEE Journal of Solid-State Circuits, Vol. 35, No. 1, January 2000

[3] M.F.Sousa et al., "Mechanically relevant chemical shrinkage of epoxy molding compounds", EuroSimE, Wroclaw, Poland, April 14-17 2013

[4] H. Kittel et al. "Novel stress measurement system for evaluation of package induced stress", SMART SYSTEMS INTEGRATION Barcelona, Spain, April 08-09, 2008

[5] Sadeghinia, et al., "Characterization and modeling the thermo-mechanical cure-dependent properties of epoxy molding compound", International Journal of Adhesion & Adhesives, October 13, 2011

[6] Shirangi M.H., et al.," Modeling Cure Shrinkage and Viscoelasticity to Enhance the Numerical Methods for Predicting Delamination in Semiconductor Packages", EuroSimE, 2009, Delft

CMOS STRESS SENSOR FOR 3D INTEGRATED CIRCUITS: THERMO-MECHANICAL EFFECTS OF THROUGH SILICON VIA (TSV) ON SURROUNDING SILICON

Komi Atchou EWUAME[1,2,a], Vincent FIORI[1,b], Karim INAL[2,c], Pierre-Olivier BOUCHARD[2,d], Sébastien GALLOIS-GARREIGNOT[1,e], Sylvain LIONTI[1,f], Clément TAVERNIER[1,g], Hervé JAOUEN[1,h]

[1]STMicroelectronics, 850 rue Jean Monnet, 38926 Crolles Cedex, France
[2]CEMEF, Mines ParisTech, rue Claude Daunesse CS 10207, 06904 Sophia Antipolis Cedex, France
[a]komiatchou.ewuame@st.com, [b]vincent.fiori@st.com, [c]karim.inal@mines-paristech.fr, [d]pierre-olivier.bouchard@mines-paristech.fr, [e]sebastien.gallois-garreignot@st.com, [f]sylvain.lionti@st.com, [g]clement.tavernier@st.com, [h]herve.jaouen@st.com

Abstract

This work aims at determining thermo-mechanical stresses induced by annealed copper filled Through Silicon Via (TSV) in single crystalline silicon by using MOS (Metal Oxide Semiconductor) rosette sensors. These sensors were specifically designed and embedded. Through the piezoresistive relations, the stress tensor was evaluated by carrying out electrical measurements on test vehicle. The MOS stress sensors would have been needed to be calibrated: first results of the calibration were obtained however, since they were still partial, they were not used to make the bridge from electric to mechanic quantities.
Experimental findings were based on the direct calculation of stresses from electrical measurements data and literature piezoresistive coefficients. In order to get only the TSV contribution and to suppress the manufacturing process variability contribution, an optimization calculation was needed. A finite element approach was also adopted to evaluate numerically the stresses induced by TSV. The stress values obtained from the optimization are in the range of the ones obtained by simulation in the sensor area. Thus, it can be stated that the methodology is relevant, and the results will be confirmed by extracting the true piezoresistive coefficients for the embedded MOS. Once calibration performed, the piezoresistive coefficients should enable getting more accurate stress values. At this stage, the quite good agreement between numerical and experimental results seems promising.

1. Introduction

The ever increasing improvement of semiconductor technologies related to 3D stacking processes, brings thermo-mechanical and residual stress issues to the forefront [1]. Numerous studies were published regarding stress fields distribution within 3D structures and the stress consequences on electrical devices were explored [2]. Among the different measurement methods allowing stress evaluation within devices, several in-situ MOS based stress sensors have been widely developed [3, 4].

In this paper, combined numerical and experimental investigations will be performed: calibration step will be carried out in order to determine the true relation between electrical current measurement and stress evaluation for these sensors. Then, a four point bending tool, which was specifically designed for sensors calibration will be presented. Finally, thanks to the evaluation of the piezoresistive coefficients and through the piezoresistivity relation [5], the link between drain current and stress will be established in order to assess specifically the TSV (Through Silicon Via) effects. TSVs are structures connecting multiple layers of stacked integrated circuits or 3D chips to deliver better performance, more functionalities and smaller packages of chips while consuming less power.
On the other hand, finite element models (FEM), focusing on the impact of TSV on the surrounding silicon will be presented. Numerical results will be compared to experimental measurements for correlation purposes.

978-1-4799-4789-8/14 $31.00 © 2014 IEEE

Comprehensive assessment of the induced stresses allows avoiding the priming of the first failure within products, the propagation of the failure in sub-layers. This also allows drawing design guidelines by preventing detrimental thermo-mechanical effects of TSVs on device performances, controlling the TSV process, optimizing products manufacturing and ensuring their quality and reliability.

2. Piezoresistive relations

Piezoresistivity describes the change in the electrical resistivity when mechanical stress is applied. This effect particularly occurs in semiconductor materials, making piezoresistive devices an attractive mean for stress sensoring in the microelectronic industry. In this part, the relations used for describing piezoresistivity are defined and its application is presented.

The MOS stress sensor (Fig. 1) is made up of four transistors in which channels are oriented in different crystallographic directions. As is always the case, MOS are made into the silicon, stress field is hence evaluated in a thin region of the chip, near the rosette. A tensor relation comes from the material properties and the "theory of piezoresistivity". According to this theory, the resistivity variation $\Delta\rho$, as well as the resistance (ΔR) and the drain current (ΔI_d) variations, is related to the applied stress by the relation (Eq.1) [5]:

$$\frac{\Delta R}{R} = -\frac{\Delta I_d}{I_d} = \frac{\Delta\rho}{\rho} = \pi\Delta\sigma \qquad (1)$$

With π the tensor of piezoresistive coefficients and $\Delta\sigma$ the applied mechanical stress tensor. The tensor $\frac{\Delta\rho}{\rho}$ (respectively the tensors $\frac{\Delta R}{R}$ and $\frac{\Delta I_d}{I_d}$) is built from scalar measurements of resistivity (respectively of resistance and drain current) in different directions.

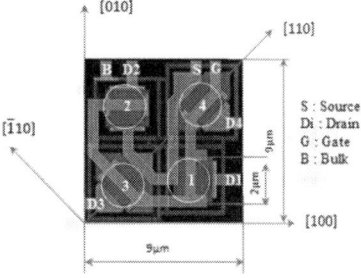

Figure 1- MOS rosette stress sensor

This relation is the same as the one used in microelectronics for evaluating stress in devices. Note that the resistance variation is often preferred for passive serpentines or doped active sensors ([3, 4]). In our transistor based ones, the drain current variation is rather used.

For the silicon cubic crystal, the three independent coefficients (π_{11}, π_{12} and π_{44}) are written in a shortened matrix (6x6) form, in ([100], [010], [001]) coordinate system (Eq. 2):

$$[\pi] = \begin{bmatrix} \pi_{11} & \pi_{12} & \pi_{12} & 0 & 0 & 0 \\ \pi_{12} & \pi_{11} & \pi_{12} & 0 & 0 & 0 \\ \pi_{12} & \pi_{12} & \pi_{11} & 0 & 0 & 0 \\ 0 & 0 & 0 & \pi_{44} & 0 & 0 \\ 0 & 0 & 0 & 0 & \pi_{44} & 0 \\ 0 & 0 & 0 & 0 & 0 & \pi_{44} \end{bmatrix} \qquad (2)$$

The knowledge of these coefficients is mandatory and their extraction process is called calibration.

3. Calibration

In the microelectronic industry, (001) substrate and two types of wafers are commonly used for chips manufacturing: the so-called rotated and not rotated wafers. For a rotated or <100> wafer, the x axis (respectively y axis) corresponds to the [100] (respectively [010]) direction (Fig. 2). For a not rotated or <110> wafer, the x axis (respectively y axis) corresponds to the [110] (respectively [$\bar{1}$10]) direction (Fig. 3).

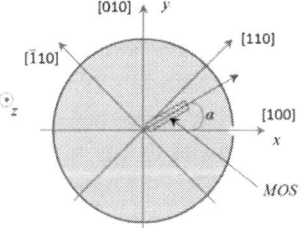

Figure 2- Schematic representation of <100> wafer

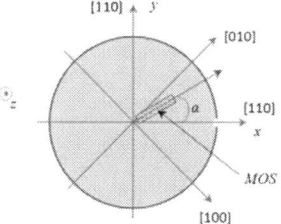

Figure 3- Schematic representation of <110> wafer

978-1-4799-4789-8/14 $31.00 © 2014 IEEE   246

α is the angle between the MOS channel and the x axis.

So, for a <110> wafer, the coordinate system transformation (i.e. rotation of 45° around the z axis) of $[\pi]$ leads to the following piezoresistive coefficients matrix (Eq. 3):

$$[\pi'] = [T] \, [\pi] \, [T]^{-1} \qquad (3)$$

With [T] (Eq. 4), the transformation matrix [6] and $[T]^{-1}$ its inverse;

$$[\pi'_{45°}] = \begin{bmatrix} \frac{1}{2}(\pi_{11}+\pi_{12}+\pi_{44}) & \frac{1}{2}(\pi_{11}+\pi_{12}-\pi_{44}) & \pi_{12} & 0 & 0 & 0 \\ \frac{1}{2}(\pi_{11}+\pi_{12}-\pi_{44}) & \frac{1}{2}(\pi_{11}+\pi_{12}+\pi_{44}) & \pi_{12} & 0 & 0 & 0 \\ \pi_{12} & \pi_{12} & \pi_{11} & 0 & 0 & 0 \\ 0 & 0 & 0 & \pi_{44} & 0 & 0 \\ 0 & 0 & 0 & 0 & \pi_{44} & 0 \\ 0 & 0 & 0 & 0 & 0 & \pi_{11}-\pi_{12} \end{bmatrix} \qquad (4)$$

To perform the calibration, a four-point bending machine (Fig. 4) is designed to apply a known uniform uniaxial stress into the sample.

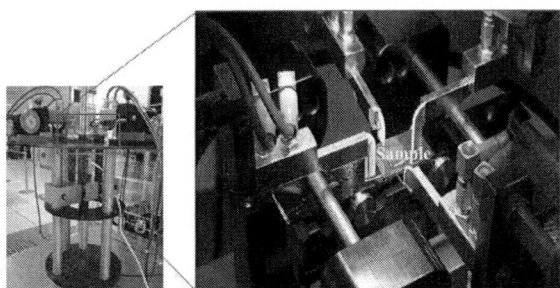

Figure 4- Pictures of the four point bending tool

<100> wafer

First, a sample is cut along the [100] direction. A uniform uniaxial stress ($\Delta\sigma_{L1}$) is applied in the [100] direction. Electrical measurements are performed on two MOS of the sensor (MOS1 and MOS2). The combination of Eq.1 and Eq.2 allows obtaining:

$$\pi_{12} = -\frac{\Delta I_{d2}}{I_{d2}^0}\frac{1}{\Delta\sigma_{L1}} \; ; \; \pi_{11} = -\frac{\Delta I_{d1}}{I_{d1}^0}\frac{1}{\Delta\sigma_{L1}} \qquad (5)$$

Then, another sample is cut at 45° (i.e. in the [110] direction). If a uniform uniaxial stress ($\Delta\sigma'_{L1}$) is applied in that direction, an electrical measurement on MOS3 allows obtaining the last piezoresistive coefficient π_{44} from relations Eq. 1 and Eq. 4:

$$-\frac{\Delta I_{d3}}{I_{d3}^0} = \frac{1}{2}(\pi_{11}+\pi_{12}+\pi_{44})\,\Delta\sigma'_{L1}$$

$[\pi']$ the new piezoresistive coefficients matrix (Eq. 4).

The calibration will be performed on the aforedescribed MOS rosette (Fig. 1).

Note that, in the following, I_{di} represents the drain current of the MOS i and I_{di}^0 its initial value; and the index of the drain current is related to the MOS index inside the rosette.

$$\Rightarrow \pi_{44} = -\frac{2}{\Delta\sigma'_{L1}}\frac{\Delta I_{d3}}{I_{d3}^0} + \frac{1}{\Delta\sigma_{L1}}\left(\frac{\Delta I_{d1}}{I_{d1}^0} + \frac{\Delta I_{d2}}{I_{d2}^0}\right) \qquad (6)$$

<110> wafer

The procedure is the same as the one described above. The difference here, is that the sample is first cut along the [110] direction, and then in the [100] direction. It is so needed to take care about the index of the MOS.

This calibration methodology is the same for n-MOS and p-MOS.

The calibration was started and the preliminary results obtained are plotted in Fig. 5.

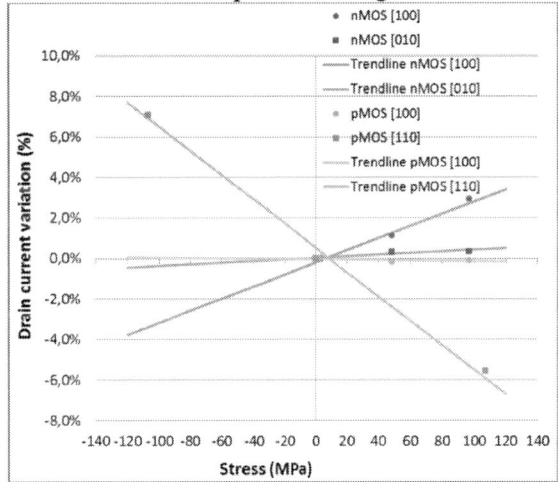

Figure 5- Preliminary results of calibration

One can notice that all the dots in Fig. 5 for each MOS type are almost aligned and the trend lines pass close to the axes origin. The slopes of these lines correspond to the piezoresistive coefficients. However, these results are still partial and the whole coefficients have not been determined yet. Until the complete measurement of the piezoresistive tensor, the coefficients extracted from previous works and related to similar MOS types are used in the following sections for stress evaluation.

4. Stress evaluation
It is then possible to evaluate stresses within a device thanks to electrical measurements.
In the literature ([3, 4] for example), vertical (σ_{zz}) and shear stresses (σ_{xy}) are sometimes not evaluated. In this section, the relations that allow evaluating these stresses are presented.
By combining Eq. 1, Eq. 2 and Eq. 5, one can write:

$$\begin{cases} a_1\sigma_{xx} + b_1\sigma_{yy} + c_1\sigma_{zz} + d_1\sigma_{xy} = -\frac{\Delta I_{d1}}{I_{d1}^0} & (\alpha = 0°) \\ a_2\sigma_{xx} + b_2\sigma_{yy} + c_2\sigma_{zz} + d_2\sigma_{xy} = -\frac{\Delta I_{d2}}{I_{d2}^0} & (\alpha = 90°) \\ a_3\sigma_{xx} + b_3\sigma_{yy} + c_3\sigma_{zz} + d_3\sigma_{xy} = -\frac{\Delta I_{d3}}{I_{d3}^0} & (\alpha = 45°) \\ a_4\sigma_{xx} + b_4\sigma_{yy} + c_4\sigma_{zz} + d_4\sigma_{xy} = -\frac{\Delta I_{d4}}{I_{d4}^0} & (\alpha = -45°) \end{cases} \quad (7)$$

With a_i, b_i, c_i and d_i the coefficients defined in appendix for each wafer type.
Again, these relations are unchanged for n-MOS and p-MOS.

5. Application to TSV induced stress

Test vehicle, description and sensors location
In this part, MOS rosette sensors, of both n and p types were used to evaluate the stress induced by the TSV. Two types of CMOS65 samples were manufactured: a wafer with TSVs and a one without TSVs. Note that designs are similar, and the presence of TSVs is managed by process routes: one sample followed the whole flow, whereas the other skipped the TSVs related steps. As shown in Fig. 6, the sensors were embedded into the wafers near the TSVs (i.e. at 12µm away from the center of the TSVs). Their size is given in Fig. 1. Electrical measurements from the two wafers were subtracted to evaluate the sole effect of the TSV.

Electrical measurements
Drain currents variations were measured on the aforementioned two wafers and the difference should correspond to the impact of TSV. The evaluation of stresses components requires four equations. As, there are four n-MOS and four p-MOS, two configurations were considered: two

p-MOS oriented at 0° and 90° and two n-MOS at 45° and -45° (config1), or two n-MOS at 0° and 90° and two p-MOS at 45° and -45° (config2). As an alternative to our own four points bending calibration, which is planned to be completed sooner, piezoresistive coefficients taken from the literature [5] were used for this study. Results of the direct calculation of stresses from Eq. 4, for the two configurations are presented in Table 1-b. However, preliminary internal studies showed that, independently to any stress variation, the current values of transistors are not the same between two similar wafers: this is related to the intrinsic variability of the CMOS process (Fig. 7-a).

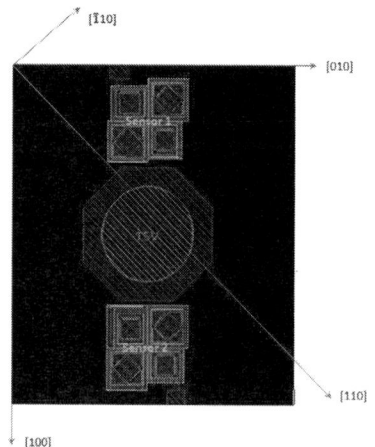

Figure 6- MOS rosette stress sensor and TSV location

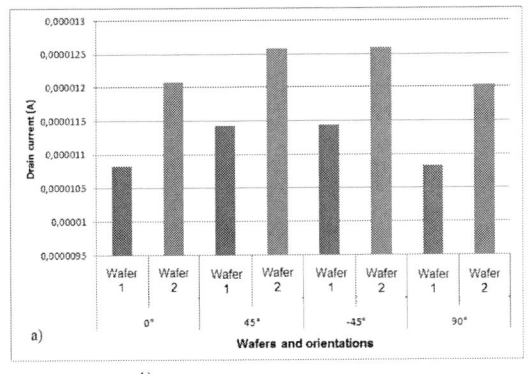

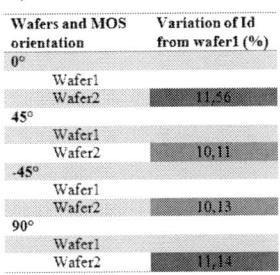

Figure 7- Representative diagram of process variability

Stress identification

So, in order to get only the TSV contribution, measurements were averaged and an optimization calculation was performed. Since the sensors provide a limited amount of equations and considering the whole unknowns of the system to solve, some assumptions had to be made: the drain currents for the process variability in [100] (0°) (respectively in [110] (45°)) direction are found to be the same as in [010] (90°) (respectively in [$\bar{1}$10] (-45°)) direction (Fig. 7-b). The simplified system is then obtained by adding the term $\frac{\Delta I_{dvi}}{I^0_{dvi}}$ to each line of Eq. 7. With ΔI_{dvi} the drain current variation related to process variability in the transistor number i of the rosette (see Fig. 1) and I^0_{dvi} its nominal value.

For this work, the so-called evolutionary optimization excel solver was used. This class of optimization method is particularly well adapted for avoiding local minima [7]. Based on the system of equations Eq. 7, a global cost function is built as the sum of the cost functions associated with each equation. By minimizing this global cost function, a set of stress values is found. The optimization initial conditions,

boundaries and the results are summarized in the following table (Table 1-a):

Table 1- a) Table summarizing the initial conditions, boundaries and results from optimization, b) direct calculation

a)	Initial values [MPa]	Lower limits [MPa]	Upper limits [MPa]	Obtained values [MPa]
σ_{xx}	75	-100	100	**66,08**
σ_{yy}	-75	-100	100	**-23,05**
σ_{zz}	-25	-100	100	**-13,04**
σ_{xy}	25	-50	50	**-5,19**
Cost function				0,01306

b)	Stress values [MPa] Config1	Stress values [MPa] Config2
σ_{xx}	**19**	**95**
σ_{yy}	**-51**	**6**
σ_{zz}	**-51**	**-109**
σ_{xy}	**0,3**	**-7**

The comparison of these results shows that the whole methodologies do not give similar stress fields, which is not suitable. Considering the direct approach (b), differences between configurations 1 and 2 would be attributed to piezoresistive coefficients inaccuracy and also to the fact that process variability is neglected. Hence, this method cannot be used as it and four point bending calibration is mandatory. As a consequence, it is needed to include the process variability and the measurements from the eight MOS in the equation system: the optimization method (a) should then provide a reliable estimation of the stress fields, which are:

σ_{xx} = 66,08 MPa ; σ_{yy} = -23,05 MPa ; σ_{zz} = -13,04 MPa ; σ_{xy} = -5,19 MPa.

TSV stress 3D simulation

This section deals with the numerical evaluation of stress fields induced by an isolated TSV in the silicon. Finite element simulations were performed with and without TSV using the ANSYS 14.5.R finite elements software. The copper TSV has a diameter of 10µm and a height of 80µm. The silicon oxide liner surrounding TSV has a thickness of 0.2µm and a height of 80µm. The model is 48µm long and 40µm wide. The PMD (Pre-Metal Dielectric), the low-k and the SiO$_2$ based interconnects have respectively thicknesses of 0.4µm, 1.67µm and 4.74µm.

978-1-4799-4789-8/14 $31.00 © 2014 IEEE

Due to geometrical symmetries, only one fourth of the stack was modeled (Fig. 8). Hexahedral quadratic elements (20 nodes) were used. Regarding boundary conditions, the node (x=0, y=0, z=0) was locked and the normal displacements of the areas A, B and the bottom base (z=0) were set to zero. The loading condition was a uniform thermal cooling from copper stress free temperature (i.e. 260°C) [8] to room temperature. No material property variation during cooling was also assumed. Materials were considered as isotropic and perfect adhesion was assumed at interfaces.

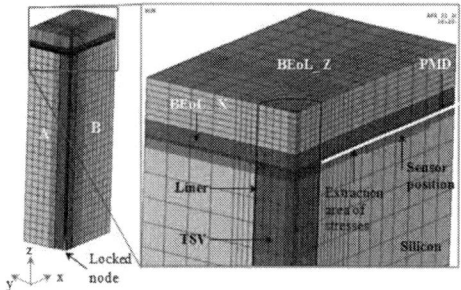

Figure 8- One fourth of the meshed model

Hence, simulations were performed under linear elastic hypothesis. The interconnects being patterned layers, analysis of the designed metal densities of the homogenized BEoL showed composition of 70% copper and 30% dielectrics. The stresses induced by TSV were obtained by subtracting stress fields from the model with TSV (Fig. 9) and those from the model without TSV. Stress variations were extracted at the top, middle and at the bottom of the models (Fig. 10).

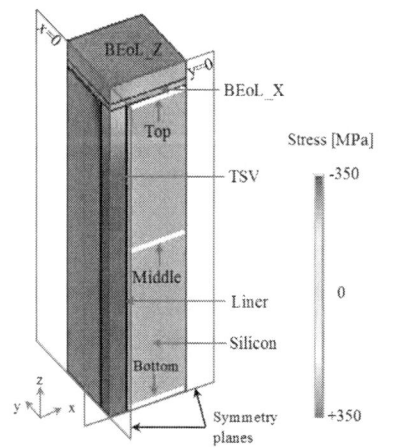

Figure 9- Stress σ_{xx} for one fourth of the simulated model with TSV

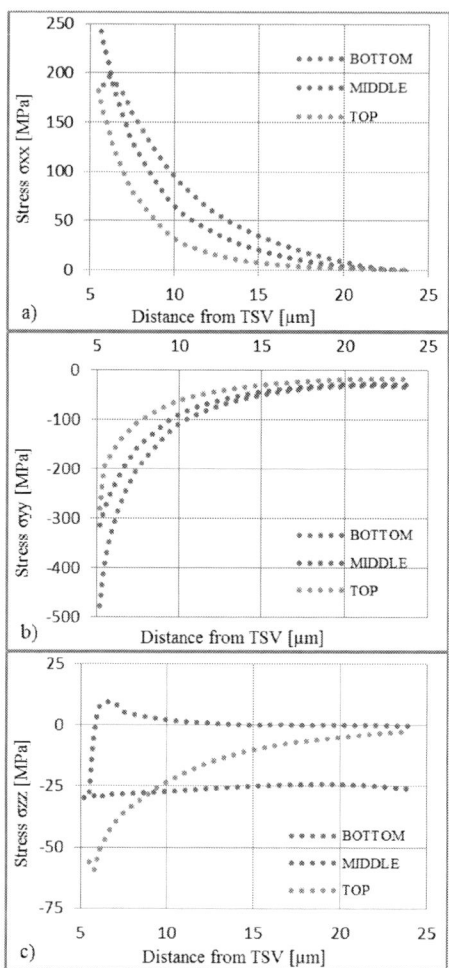

Figure 10- Stresses a)- σ_{xx}, b)- σ_{yy}, c)- σ_{zz} induced by TSV at the top, middle and at the bottom of the model

One can note that, in the silicon material, TSV induces tensile stresses (σ_{xx}) in the radial (x axis) direction, whereas in tangential (y axis) and vertical (z axis) directions, the stresses (σ_{yy}, σ_{zz}) are compressive.

Indeed, during the cooling, the model is contracted in the tangential direction (compressive stress), which by Poisson effect, generates a tensile stress in the perpendicular direction (radial direction). The surfaces of the model, except those to which are applied boundary conditions are free, so that from the center of the TSV to these surfaces the stress values decrease and tend to zero. This also occurs from the bottom to the top of the model as shown in Fig. 10 a), b).

A comparison of simulation results with the literature was limited to a qualitative analysis, since configurations used (material properties, model dimensions) are not the same.

One can mention the works of De Wolf and *al.* [2], and Li Yu and *al.* [9] in which, the radial (respectively tangential and vertical) stresses obtained by finite element simulation and micro Raman are tensile (respectively compressive). These stresses fall back to zero as one moves from the center of the TSV.

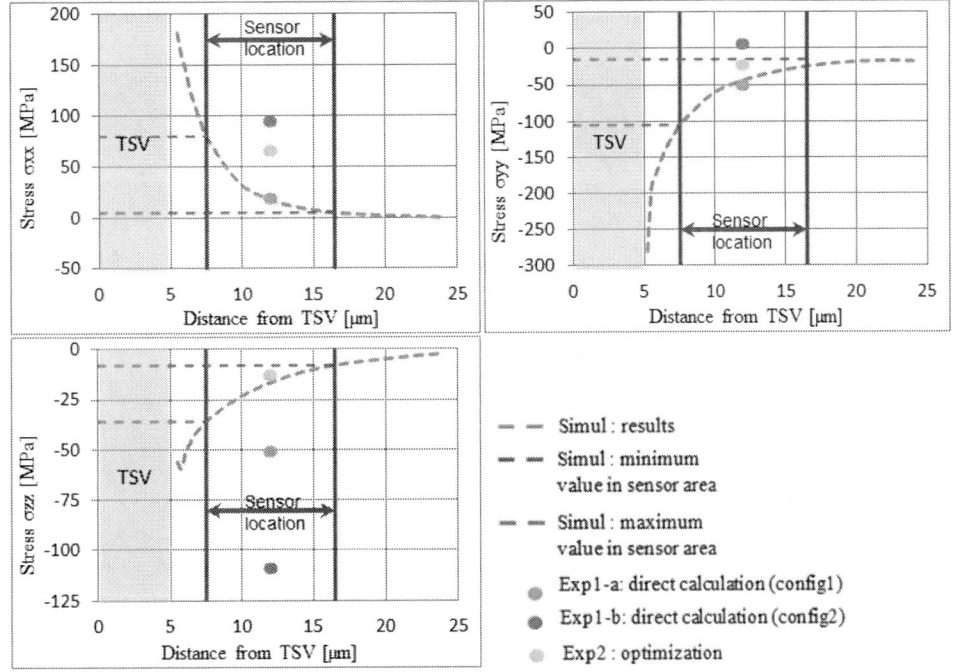

Figure 11- Stress induced by TSV at the top of the model, experimental and numerical results

According to the location of the sensor at the top of the model (from $x=7.5\mu m$ to $x=16.5\mu m$), the stress variation, i.e. the difference between maximum and minimum values, is quite large ($\Delta\sigma_{xx}=-80MPa$, $\Delta\sigma_{yy}=+80MPa$ and $\Delta\sigma_{zz}=+30MPa$). Hence, the comparison between sensor size and simulated stress fields shows that the accuracy of such sensor is likely to be weak. Averaged simulated stresses values and their variations in the sensor region are: $\bar{\sigma}_{xx}\approx30MPa\pm40$, $\bar{\sigma}_{yy}=-55MPa\pm40$ and $\bar{\sigma}_{zz}\approx-20MPa\pm15$.

According to Fig. 11, the stress values (except σ_{zz}) obtained from direct calculation for the first configuration are in the range of stresses obtained by simulation in the sensor area: $\sigma_{xx}\in[-10MPa, 70MPa]$, $\sigma_{yy}\in[-95MPa, -15MPa]$ and $\sigma_{zz}\in[-35MPa, -5MPa]$. But the stress values are out of the range for the second configuration. Furthermore, as mentioned before, this method cannot be used and these results must be dismissed. The technique employing the optimization procedure seems to be relevant by taking into account the process variability. However, it is difficult at this stage to confirm stress values calculated from experiments since the exact piezoresistive coefficients for the MOS sensors have not yet been calibrated.

6. Conclusion

In order to evaluate the stress induced by TSV in microelectronic devices, MOS rosette stress sensors were used, and the results are reported in this work. The calibration of the sensors was started and preliminary results were obtained. These results are partial and the piezoresistive coefficients were not fully determined, therefore they were not used for stress evaluation. In this work, a combined approach of experiment and simulation was carried out. Experimental investigations were based on the direct calculation of stresses from electrical measurements data and literature piezoresistive coefficients. This is in part related to process variability which leads to the need for using an optimization calculation in order to get only the TSV contribution. A finite element approach was also adopted to evaluate numerically the stresses induced by TSV. The stress values obtained from the optimization are in the range of the ones provided by simulation in the sensor area. Thus, it can be stated that the methodology is relevant, and at first, the results will be confirmed by extracting the true piezoresistive coefficients for the embedded MOS. At this stage however, the quite good agreement between numerical and experimental results seems promising. Pros and cons of such sensors allow determining a reliable work flow for stress probing. This should enable to provide further design and integration recommendations.

7. Appendix

$<100>$ wafer

$a_1 = \pi_{11}$; $a_2 = \pi_{12}$; $a_3 = \frac{1}{2}(\pi_{11} + \pi_{12})$

$a_4 = \frac{1}{2}(\pi_{11} + \pi_{12})$

$b_1 = \pi_{12}$; $b_2 = \pi_{11}$; $b_3 = \frac{1}{2}(\pi_{11} + \pi_{12})$

$b_4 = \frac{1}{2}(\pi_{11} + \pi_{12})$; $c_1 = c_2 = c_3 = c_4 = \pi_{12}$

$d_1 = 0$; $d_2 = 0$; $d_3 = -\pi_{44}$; $d_4 = \pi_{44}$

$<110>$ wafer

$a_1 = \frac{1}{2}(\pi_{11} + \pi_{12} + \pi_{44})$;

$a_2 = \frac{1}{2}(\pi_{11} + \pi_{12} - \pi_{44})$; $a_3 = \frac{1}{2}(\pi_{11} + \pi_{12})$

$a_4 = \frac{1}{2}(\pi_{11} + \pi_{12})$; $b_1 = \frac{1}{2}(\pi_{11} + \pi_{12} - \pi_{44})$

$b_2 = \frac{1}{2}(\pi_{11} + \pi_{12} + \pi_{44})$; $b_3 = \frac{1}{2}(\pi_{11} + \pi_{12})$

$b_4 = \frac{1}{2}(\pi_{11} + \pi_{12})$; $c_1 = c_2 = c_3 = c_4 = \pi_{12}$

$d_1 = 0$; $d_2 = 0$; $d_3 = -(\pi_{11} - \pi_{12})$; $d_4 = \pi_{11} - \pi_{12}$

References

1. International Technology Roadmap for Semiconductors (ITRS), 2010.
2. I. De Wolf, V. Simons, V. Cherman, R. Labie, B. Vandevelde, E. Beyne and IMEC, In-depth Raman Spectroscopy Analysis of Various Parameters Affecting the Mechanical Stress near the Surface and Bulk of Cu-TSVs, IEEE, 978-1-4673-1965-2, 2012.
3. K. Aditya, X. Zhang, Q. X. Zhang, M. C. Jong, G. Huang, L. W. S. Vincent, C. Lee, J. H. Lau, D. L. Kwong, R. R. Tummula and G. Meyer-Berg, Residual Stress Analysis in Thin Device Wafer Using Piezoresistive Stress Sensor, IEEE Transactions on Components Packaging and Manufacturing Technology, vol. 1, p. 6, 2011.
4. X. Zhang, K. Aditya, Q. Zhanga, Y. Onga, S. Hoa, C. Khonga, V. Kripesha, J. Laua, D.-L. Kwonga, V. Sundaramb, R. Tummulab and G. Meyer-Berg, Application of Piezoresistive Stress Sensors in Ultra Thin Device Handling and Characterization, Sensors and Actuators A156, pp. 2-7, 2009.
5. C. S. Smith, Piezoresistance effect in germanium and silicon, Phys. Rev., vol. 94, pp. 42-49, 1954.
6. J. C. Suhling, R. C. Jaeger, Silicon Piezoresistive Stress Sensors and Their Application in Electronic Packaging, IEEE Sensors, vol. 1, No. 1, pp. 14-30, 2001.
7. E. Roux, P-O. Bouchard, Kriging metamodel global optimization of clinching joining processes accounting for ductile damage, Journal of Materials Processsing Technology, 213(7):1038-1047, 2013.
8. M. Gregoire, Properties of Thin Film and Copper Interconnects. Thesis in Science and Engineering of Materials. Grenoble : National Polytechnic Institute of Grenoble, 248p, 2006.
9. L. Yu, W. Chang, K. Zuo, J. Wang, D. Yu, D. Boning, Methodology for Analysis of TSV Stress Induced Transistor Variation and Circuit Performance, IEEE Quality Electronic Design, pp. 216-222, 2012.

Structure Design and Reliability Assessment of Double-sided with Double-chip Stacking Packaging

Yen-Fu Su[1], Chun-Te Lin[2], Tzu-Ying Kuo[2], and Kuo-Ning Chiang[1]
[1]Advanced Micro-system Packaging and Nano-Mechanics Research Lab.
Dept. of Power Mechanical Engineering, National Tsing Hua University,
Hsinchu, Taiwan 300, R.O.C.
[2]Electronic and Optoelectronics Research Labs, Industrial Technology Research Institute (ITRI)
Phone: 886-3-5742925, Fax: 886-3-5745377 and E-mail: knchiang@pme.nthu.edu.tw

Abstract

Recently, consumer electronics demand has been geared towards lightweight, high efficiency, and small form factor devices. These characteristics can be accomplished by using three-dimensional (3D) integrated circuit (IC) technology. This study proposes a double-chip stacking structure in an embedded fan-out wafer level packaging (WLP) with double-sided interconnections. Regarding to the proposed structure, the finite element (FE) model is established to investigate the actual thermo-mechanical behavior during thermal cycling loading. The suitable geometry design of buffer layer and carrier can enhance the reliability of solder joint. The application of soft filler and passivation materials can increase the predicted fatigue life to more than 2,500 cycles. However, the high coefficient of thermal expansion (CTE) of soft filler material will induce significant deformation and excessive strain on the trace layer. Based on the above simulation methodology and life prediction model, the development of the proposed structure can be optimized within a feasible time. This effective methodology is necessary in the electronic packaging industry to reduce time-to-market and fabrication cost when a new packaging structure is being developed.

1. Introduction

In the past year, the development of the microelectronic industry was driven by Moore's law . However, physical limitations will occur in the complementary metal-oxide semiconductor (CMOS) process in the future. To achieve high performance, small form factor and lightweight application, the development of the microelectronic industry can be fulfilled by the more than Moore's technology which is executed by system-in-package (SiP) structures. Through silicon via (TSV) is the core technique of 3D-IC technology; it is utilized to interconnect stacked chips, thereby shortening signal transmission time, reducing power consumption, and solving signal delay issues can be accomplished. However, as chip size decreases and the number of components on the chip increases, the space for forming TSV seems quite limited; therefore, other approach such as fan-out type packaging using through molding via (TMV) technology could be an alternative solution for SiP.

In recent years, 3D-IC packaging has been rapidly adopted for SiP by using package-on-package (PoP)

stacking . However, the shrinkage of pitches and pads at the die side is significanty faster than that at the substrate side. This situation can be ragarded as an interconnection gap. Infineon has developed a fan-out wafer level BGA (WLB), called embedded WLB (eWLB), that can solve interconnection gap. Yew et al. developed the chip-on-metal WLCSP technology with fan-out capability. The reliability characteristic of this structure in the packaging level was described using FE analysis incorporated with factorial analysis to obtain sensitivity information of the packaging. It concluded that a thinner chip, superior trace scheme, and suitable stress buffer layer application could decrease the stress/strain value in the metal line.

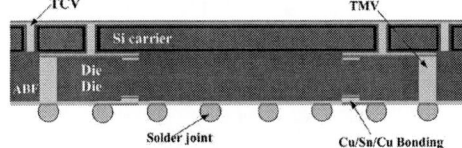

Figure 1: Schematic of the double-sided with double-chip packaging structure

As techniques in the microelectronic process progress, next-generation eWLB design will tend toward a double-sided eWLB with vertical interconnections, a multi-layer redistribution layer (RDL) eWLB, a large size eWLB, and multiple chips eWLB . Chiang et al. invented a structure, which first and second electronics devices are stacked and they are embedded in a chip carrier. There are conductive vias in the carrier to connect the first and second electronic devices. The present study proposes a double-chip stacking structure in an embedded fan-out wafer-level packaging with double-sided interconnections, as shown in Fig. 1. The double-sided RDL ensures that the PoP structure can be achieved. In addition, this stacking structure involves the development of heterogeneous integration technology, multi-functional and different-sized dies can be incorporated into a single packaging unit interconnected through lateral RDLs and vertical TMVs.

Previous study has analyzed the thermal performance of the double-sided with double-chip packaging structure through FE analysis because of issues of concerning the embedded structure and stacked chips. Besides, Chen et al. built up the FE model of twin die stacked package, and the emperical fatigue model was applied to predict the reliability of solder ball under a cyclic temperature loading. In this study, the reliability issue of interconnections remains in the proposed structure is observed. The FE analysis incorporated with the Coffin-

Manson relation for predicting fatigue life of solder joint which was validated through an experiment is adopted. Structure design and material selection are also studied through FE analysis.

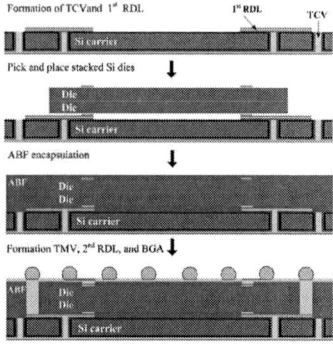

Figure 2: Manufacturing process

2. Process Flow for Double-sided with Double-chip Stacking Structure

The manufacturing process of the double-sided with double-chip packaging structure is shown in Fig. 2. First, through carrier vias (TCVs) are formed in the silicon carrier wafer through laser and plating processes. The 1st RDL is also made on the silicon carrier. The dies are thinned to almost 50 μm, and the back sides of the two thinned dies are bonded with an adhesive layer. The stacked dies are picked and placed on the silicon carrier wafer. The dies are then encapsulated by using Ajinomoto build-up film (ABF) to broaden the package area, thereby providing additional base material when proceeding with the 2nd RDL and TMV. TMVs are formed in the periphery of the ABF through laser drilling and Cu plating processes. The 2nd RDL is then formed. The electrical input/output (I/O) and TMVs are interconnected via RDLs; I/Os are thus re-routed from the dies to the ABF layer to ensure a relaxed pitch. After the bumping process, the wafer is diced into numerous packaging units, and then it is rotated 180° for packaging stacking or mounting on PCB. The thickness of the package unit is controlled to be less than 500 μm. A die size of 5 mm × 5 mm is selected to make the proposed structure. The exterior dimensions of the unit after encapsulation are 6.6 mm × 6.6 mm, and the pitch of the solder joint pad is 0.8 mm. 52 I/Os are present in one packaging unit. The pad opening measures 0.2 mm, and a SnAg lead-free solder is applied. The original pitch of the chip pad is 0.4 mm before being extended to 0.8 mm via the fan-out structure.

3. Solder Joint Thermal Fatigue Life Prediction

Electronic devices would be subjected to temperature loading during operation, and thermal stress is induced because of the coefficients of thermal expansion (CTE) mismatch among different materials. The Coffin-Manson relation based on plastic strain has been extensively used to assess the reliability of solder joint . The Coffin-Manson relation is described as following:

$$N_f = C(\Delta \varepsilon_{eq}^{pl})^{-\alpha} \qquad (1)$$

where N_f is the mean cycle to failure, C is the fatigue ductility coefficient, α is the fatigue ductility exponent , and $\Delta \varepsilon_{eq}^{pl}$ is the incremental equivalent plastic strain which means the solder joints in advance package, such as WLP and flip chip, would be subjected to each directional stress. In this research, the maximum equivalent plastic strain after one cycle is obtained through FE analysis and is substituted into the Coffin–Manson relation to predict the fatigue life of solder joints for reliability assessment.

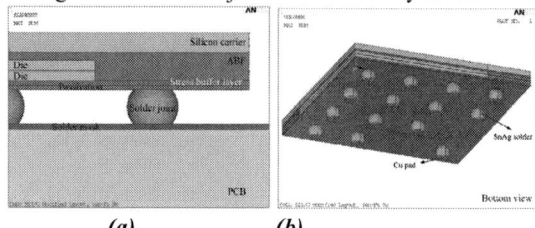

(a)   *(b)*

Figure 3: FE model (a) Cross-sectional view; (b) Arrangement of solder joints

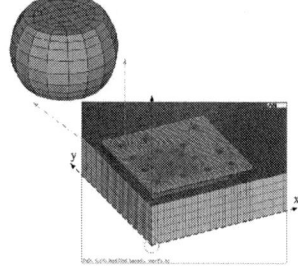

Figure 4: The mesh contour of the FE model

4. FE Modeling of Double-sided with Double-chip Stacking Structure

Before the proposed structure begins to be manufactured, the actual mechanical behavior during thermal cycling loading and reliability assessment should be analyzed. The simulation procedure and the suitable mesh density for the solder joint has been validated through traditional WLP and thermal cycling test. A 3D FE model of the double-sided with double-chip stacking structure is established by using commercial software ANSYS®. It is composed of stacked dies, adhesive film, ABF, stress buffer layer, copper pad, TMV, silicon carrier, solder joint and PCB, as show in Fig. 3(a). The RDLs are simplified as passivation layers to reduce the computation resource. The arrangement of solder joints is in a 6 × 6 array within the chip area, and the other solder joints are arranged in the peripheral of package, as illustrated in Fig. 3(b). A total of 8 TMVs are placed in the ABF region near the symmetric planes.

The mesh density in the stress critical region of solder joints, which is the contact surface of the die and PCB, is maintained as same as the validation model, as shown in Fig. 4. Given the symmetrical characteristic, the quarter FE model is developed. All materials are assumed to be isotropic and linearly elastic except for the SnAg solder, which is considered as a temperature-dependent and elastic-plastic material (Table 1). The proposed structure is subjected to thermal cycling loading with temperature ranging from -40 °C to 125 °C.

Table 1: Material property

Material	Young's modulus (GPa)	Poisson's ratio	CTE (ppm/°C)
Silicon	130	0.28	2.62
ABF	3.5	0.3	60
Cu pad	68.9	0.34	16.7
Solder mask	3.5	0.35	30
PCB	18.2	0.3	16
Polyimide	2.8	0.34	50
96.5Sn/3.5Ag	nonlinear	0.4	22.36

5. Failure Mechanism and Reliability Assessment

After one cycle simulation, the deformation of the proposed structure is shown in Fig. 5. As the temperature increases from 25 °C to 125 °C, the thermal stress is induced because of the CTE mismatch between different materials, and the package structure is warped upward, as shown in Fig. 5(a). Meanwhile, Fig. 5(b) indicates that the package structure is warped downward when the packaging cools down from 25 °C to -40 °C. Therefore, interconnection experiences significant nonlinear thermal stress/strain related to low-cycle fatigue.

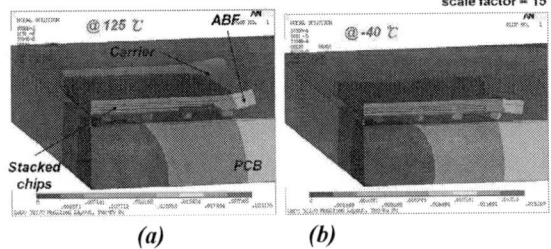

(a) *(b)*

Figure 5: Deformation during thermal cycling loading
(a) 125 °C; (b)-40 °C (scale factor =15)

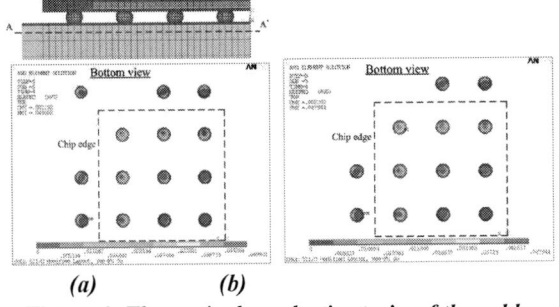

(a) *(b)*

Figure 6: The equivalent plastic strain of the solder joints (a) With dummy ball; (b) Without dummy ball

The equivalent plastic strain pattern of solder joints after one cycle is shown in Fig. 6. The maximum equivalent plastic strain occurs at the outmost solder joint where DNP is the largest. However, the outmost solder joints on the corner are used to design the bypass trace line for a daisy chain circuit. So, these solder joints can be regarded as dummy balls. As shown in Fig. 6(b), the maximum equivalent plastic strain on the solder joints, except dummy balls, which occurs on the corner of the

solder joints within the chip region after one cycle. After five thermal cycle computations, the incremental equivalent plastic strain of the solder joints converges to 4.7%. According to the Coffin–Manson relationship, the mean cycle-to-failure is predicted as approximately 200 cycles. Although the stress buffer layer can absorb the energy acting on the solder joints, the thickness and material property of the buffer layer affects the fatigue life. In the proposed structure, Young's modulus of ABF and lead-free solder are in the same order (GPa). The ABF produces significant deformation on the solder joints as a CTE mismatch between ABF and silicon carrier is induced. The strain induced by a CTE mismatch between the packaging and PCB is also performed on the solder joints. The thickness of the buffer layer is only 25 μm. Thus, the ability of releasing energy, which acts on solder joints under the stacked dies, is restricted.

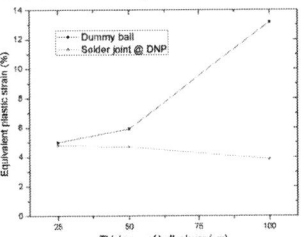

Figure 7: Equivalent plastic strain under different thickness of buffer layer

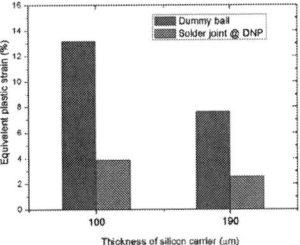

Figure 8: The effect of silicon carrier thickness (buffer layer = 100 μm)

6. Material Selection and Geometry Design

The baseline model of the double-sided with double-chip stacking structure is composed of a 25 μm-thick ABF stress buffer layer to release the energy. To clarify the mechanical behavior and improve the reliability of the solder joints, the geometric design and material selection are analyzed. The original thickness of the stress buffer layer is 25 μm and that of the ABF is 170 μm. The thickness of the stress buffer layer is changed to 50 μm and 100 μm, and the thickness of the entire ABF is also changed to 195 μm and 245 μm. The simulation results are shown in Fig. 7. It indicates that the thick stress buffer layer can effectively protect the solder joints in absorbing the stress/strain induced by a CTE mismatch between packaging and PCB. A 20% decreasing in equivalent plastic strain of solder joint is obtained as the thickness of the buffer layer increases from 25 μm to 100 μm. The mean cycles to failure is enhanced to approximately 300 cycles. However, the dummy ball is subjected to severe thermal stress, because the thick ABF causes a more significant CTE mismatch than the original model. For improving the reliability of the solder joints, the thickness of the silicon carrier is also increased from 100μm to 190 μm. The results present that 40% recuction in the equivalent plastic strain of the dummy ball is obtained, and the mean cycles to failure is increased to approximately 600 cycles, as shown in Fig. 8.

In addition, the material selection is also an important issue to improve the reliability of solder joints. Silicone (E=0.05 GPa, CTE=167 ppm/°C) replaces ABF as the filler material, and SINR (E=0.09 GPa, CTE=150 ppm/°C) replaces polyimide as the passivation layer in the following FE analysis. The maximum equivalent plastic strain occurs on the solder joints, which is under the TMV (Fig. 9), and the value after one cycle is effectively dropped to 1.13%. The relatively small equivalent plastic strain causes the predicted fatigue life of the solder joint to reach over 2,500 cycles. The main reason for the improvement is the application of a soft filler and passivation material. Although the soft filler material can be regarded as a stress buffer layer to protect the solder joints, the extreme expansion induces significant deformation on the trace layer, which crosses the transition region of the chip and filler material. The detailed FE model including trace pattern has to be established to evaluate the mechanical behavior and assess the reliability of trace line in the future.

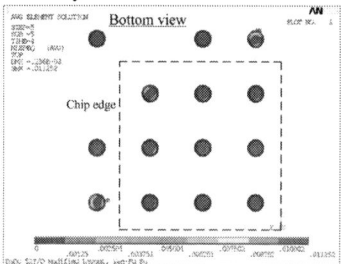

Figure 9: The effect of soft filler and passivation materials

7. Conclusion

This study proposes a double-chip stacking structure in an embedded fan-out wafer-level packaging with double-sided interconnections. The proposed packaging structure can be manufactured in batch process; its PoP structure can also be achieved through double-sided interconnections. Regarding to the reliability of the proposed structure, the FE model is established to investigate the actual thermo-mechanical behavior during thermal cycling loading. The reliability assessment of the baseline model in the predicted fatigue life of solder joint on the chip corner is only 200 cycles. The suitable geometric designs, such as the thickness of the stress buffer layer and silicon carrier, can effectively enhance the reliability. On the other hand, the application of soft filler and passivation materials can increase the predicted fatigue life of solder joint to more than 2,500 cycles. However, the high CTE of soft filler material will induce

significant deformation and strain on the trace layer which should be further studied. By this effective methodology, the geometry of the proposed structure is analyzed to enhance the reliability.

Acknowledgments

The authors would like to thank the National Science Council for providing financial support (NSC102-2221-E-007-038-MY3) and the members of the ITRI for supplying the test samples used in the study.

References

[1] G. E. Moore, "Cramming more components onto integrated circuits," *Electronics,* vol. 38,(1965), pp. 114-117

[2] G. Q. Zhang, M. Graef, and F. van Roosmalen, "The rationale and paradigm of "more than Moore"," *Proc 56th Electronic Components and Technology Conference*, San Diago, CA, USA, May 30-Jun. 2, 2006

[3] J. Zheng, Z. Zhang, Y. Chen, and J. Shi, "3D stacked package technology and its application prospects," *Proc International Conference on New Trends in Information and Service Science*, Beijing, China, Jun. 30-Jul. 2, 2009

[4] T. Meyer, G. Ofner, S. Bradl, M. Brunnbauer, and R. Hagen, "Embedded Wafer Level Ball Grid Array (eWLB)," *Proc 10th Electronics Packaging Technology Conference*, Singapore, Dec. 9-12, 2008

[5] M. C. Yew, C. A. Yuan, C. J. Wu, D. C. Hu, W. K. Yang, and K. N. Chiang, "Investigation of the Trace Line Failure Mechanism and Design of Flexible Wafer Level Packaging," *IEEE Transactions on Advanced Packaging,* vol. 32,(2009), pp. 390-398

[6] Y. Jin, X. Baraton, S. W. Yoon, Y. Lin, P. C. Marimuthu, V. P. Ganesh, T. Meyer, and A. Bahr, "Next generation eWLB (embedded wafer level BGA) packaging," *Proc 12th Electronics Packaging Technology Conference*, Singapore, Dec. 8-10, 2010

[7] C. Y. Chou, M. C. Yew, and K. N. Chiang, "Thin Stack Package Using Embedded-Type Chip Carrier," TW 395318, 2013.

[8] Y. F. Su, C. T. Lin, T. Y. Kuo, and K. N. Chiang, "Development of double-sided with double-chip stacking structure using panel level embedded wafer level packaging," *Proc 63rd Electronic Components and Technology Conference*, Las Vegas, NV, USA, May 27-31, 2013

[9] R. S. Chen, C. H. Huang, and Y. Z. Xie, "Application of optimal design on twin die stacked package by reliability indicator of average SED concept," *Journal of Mechanics,* vol. 28,(2012), pp. 135-142

[10] S. S. Manson, *Thermal stress and low-cycle fatigue*, New York, McGraw-Hill, 1966.

2D Micro-Chamber for DC Plasma working at Low Power

Veronique Rochus, Vladimir Samara , Bart Vereecke, Philippe Soussan, Bart Onsia and Xavier Rottenberg
IMEC, Belgium
Imec, Kapeldreef 75, B-3001 Leuven, Belgium
Veronique.Rochus@imec.be

Abstract

The plasma micro-chambers proposed in the literature make typically usage of relatively high RF power applied to cavities characterized by their 3D geometry, difficult to integrate on wafer. This work reports on the design, wafer-level fabrication and characterization of 2D DC plasma micro-chambers working at atmospheric pressure with noble and inert gases like helium and argon. The MEMS technology developed for this purpose allows the definition of small gaps in order to reduce the power consumption. The strike and sustain electrodes are made of Titanium Nitrite, material of choice for its hardness, and thus resistance to the ion bombardment, as well as his high melting point temperature, that allows the proximity and contact with high temperature plasmas. Measurements were performed, applying a high voltage to these electrodes, and measuring the relation between the voltage and the current when the plasma is ignited. Considering different gaps between the electrodes we can extract then the power consumed in the plasma and optimize the 2D micro-chamber.

1. Introduction

In the last few years, atmospheric microplasmas have gained increasing attention in the plasma community. Working at high pressure and sub-mm scale, these plasmas present non-thermal characteristics, high electron densities and non-Maxwellian electron energy distributions which make them complex to understand and study numerically [1]. However these properties make them appealing for numerous applications like the synthesis of nanomaterials, for chemical or bacterial decontamination, for the development of novel photonic devices, bio-medical applications, displays, radiation sources, micro-chemical analysis systems, gas analyzers, photo-detectors, microlasers, dynamic millimeter and microwave devices, microreactors, propulsion systems, aerodynamic flow control, and environmental applications [2-3].

The plasma micro-chambers proposed in the literature make typically usage of relatively high RF power applied to cavities characterized by their 3D geometry, difficult to integrate on wafer, with typical dimensions of the order of 100µm. RF plasmas are often used, for example for material processing, due the tendency of DC plasma to be more aggressive.

However, in comparison with DC plasmas, RF plasmas require more complex and more powerful generators. They are further difficult to confine, what makes these difficult to incorporate on wafer level.

The goal of our work is to fabricate a DC plasma micro-chamber working at atmospheric pressure with noble and inert gases like helium and argon in a 2D MEMS technology platform with small gap in order to reduce the power consumption in the plasma. In a first step, we evaluate the voltage required to ignite the plasma based on the analytical formulations of Paschen's law. Then, these values are compared to off-wafer strike and sustain measurements obtained using two macroscopic discs separated by a varying micro-scale gaps. Finally, we report and analyze strike and sustain measurements performed no-wafer in a MEMS technology implementing titanium nitride [4-5] electrodes isolated from a standard silicon substrate through a 5µm thick silicon oxide layer.

2. Paschen's law

A. Analytical Study

A plasma is a gas in which an important fraction of the atoms is ionized, so that the electrons and ions are separately free [6-7]. To generate a plasma, atoms or molecules of the constituent gas first have to be ionized. Once created (due to random processes as thermal collision or gamma rays) the initial ion-electron pair is accelerated by the applied electric field. The electrons collide with other gas molecules, creating new electrons and ions. When all the electrons reach the anode, the process is extinguished unless a secondary electron emitted by collision of the ions on the cathode restarts the process. This mechanism of free electron population enhancement though collisions and secondary emissions is referred to as avalanche (as seen in Figure 1: *Townsend Avalanche mechanism to ignite the plasma*) mathematically formulated by Paschen and Townsend.

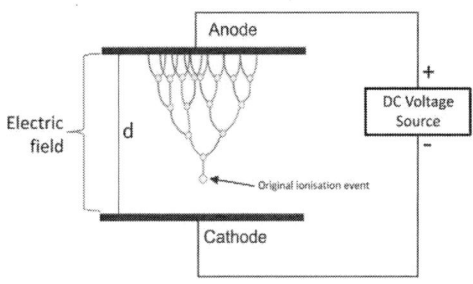

Figure 1: Townsend Avalanche mechanism to ignite the plasma [6].

In Paschen's law the breakdown voltage is computed by the expression [6-7]

$$V_b = \frac{Bpd}{\ln(pd) + \ln\left(\dfrac{A}{\ln(1/\gamma_i + 1)}\right)}$$

where Vb is the breakdown voltage, A and B are gas-dependent constants, p is the pressure, d is the distance between electrodes and γ is the secondary emission coefficient, dependent on the electrode material.

Using the parameters reported in table 1, in conjunction with a γ parameter equal to 0.01, we obtain the "Paschen curves" depicted in Figure 2. The computed minimum breakdown voltages at atmospheric pressure are about 142 V for helium and 160V for argon and are obtained for separation gaps respectively 55µm and 12µm.

Table 1 Gas parameters for Paschen's law

	A [Torr^{-1}cm^{-1}]	B[V.Torr^{-1}cm^{-1}]
Helium	3	34
Argon	14	180

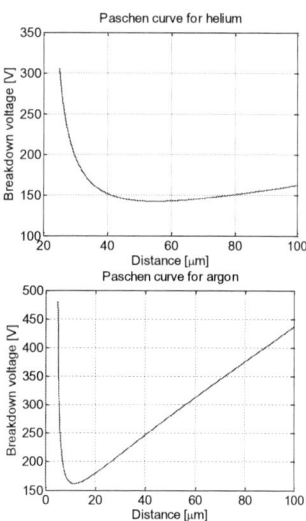

Figure 2: Computed Paschen curves for helium and argon at atmospheric pressure.

B. Measurements set-up

In order to confirm the actual strike conditions for atmospheric He and Ar plasmas, we experimentally evaluate these on a macro set-up. We evaluate Paschen curves using the manual probe station shown in Figure 3 and macroscopic aluminum parallel-plate electrodes separated by a gap manually tunable from 10µm to 500µm. Argon and helium are directly flushed between the discs at atmospheric pressure but the gas concentration is not accurately controlled.

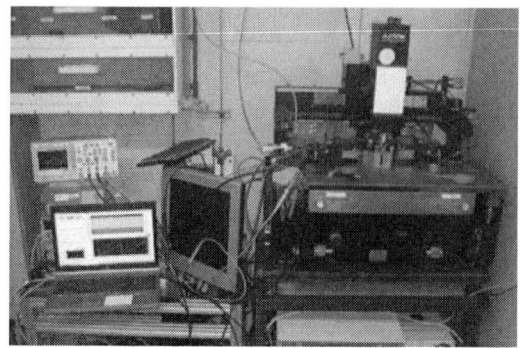

Figure 3: Macro set-up for off-wafer measurements

The electrical circuit used to ignite the plasma between the electrodes is illustrated in Figure 4. A resistor is used to limit the current passing through the circuit once the plasma is ignited.

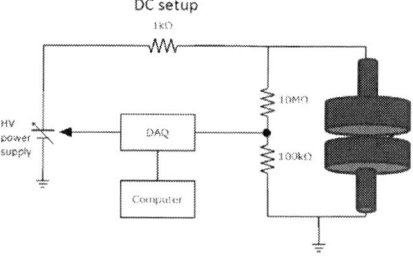

Figure 4: Electrical circuit implemented to study strike and sustain parameters in case of macro set-up.

C. Measurements results

The detection of plasma ignition and extraction of plasma parameters is in general a complex task. In this particular case, the strike event can be directly detected through the light emitted in the intra disc cavity as shown in Figure 5.

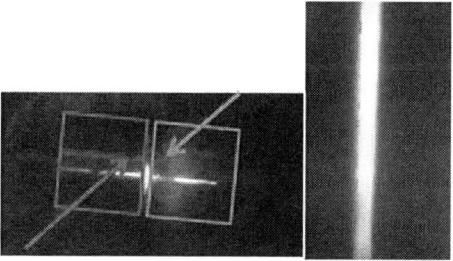

Figure 5: Plasma ignition between two macroscopic discs separated by a micro-gap.

The global shape of the Paschen curve obtained experimentally is plotted in Figure 6 for helium and Figure 7 for argon. The measured minimum breakdown/strike voltages at atmospheric pressure are about 350V for helium and 225V for argon and are obtained for separation gaps respectively 100μm and 50μm. The observed discrepancies between these measurements and the analytical results reported in the previous section are attributed mainly to the oxidation of the aluminum electrodes, the limited control of the atmospheric conditions and the non-ideal geometry of the electrodes. Indeed the measured flat response of the devices at small separation is for example known to be due to the non-perfect parallelism of the electrodes.

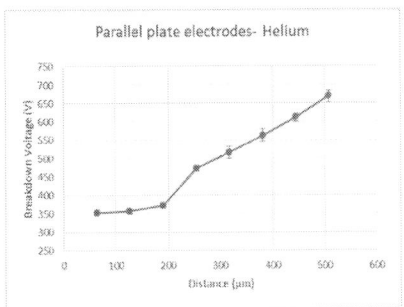

Figure 6: Breakdown voltage measured for two parallel discs at atmospheric pressure in helium.

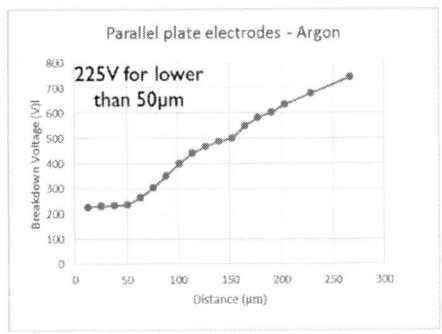

Figure 7: Breakdown voltage measured for two parallel discs at atmospheric pressure in argon.

From these measurements we can conclude that the dc voltage needed to ignite a stable plasma will be in the order of few hundreds volts. Indeed the value is strongly dependent on the geometry of the electrode and we cannot directly extrapolate our conclusions to the 2D MEMS structures.

3. Micro-chamber

A. Process description and layout

The process used to produce these micro-devices is quite simple, as sketched in the schematic cross-section of Figure 8. A silicon substrate is first coated with a 5μm thick layer of insulating silicon oxide. Then, a 100nm thick layer of titanium nitride is deposited and patterned to form the electrodes and the plasma gaps.

Figure 8: Cross section of a MEMS plasma micro-chamber.

To try and mimic a Paschen curve experimental evaluation, several micro-chambers geometries have been implemented as illustrated in Figure 9 where variations of gaps and electrode dimensions are shown.

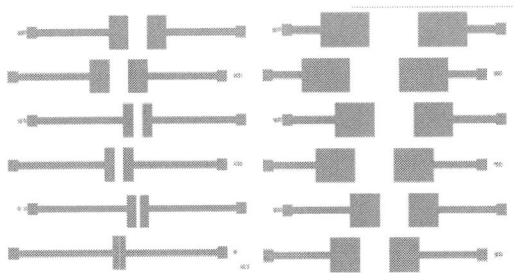

Figure 9: Layout of the devices with gap going from 7μm to 500μm.

B. Measurement Set-up Description

The measurement set-up is depicted in Figure 10. In the present case, the tests are performed in an environmental chamber where the pressure and gas composition can be accurately controlled. The wafer with plasma micro-chambers is placed in the test set-up. The air in the test-chamber is first pumped out. The chamber is then filled with the gas under test, ensuring a close to 100% purity of the test atmosphere. Then a high voltage is applied on the electrodes using a DC generator that can reach 1000V and is limited to a maximum of 2mA. A resistance of 560kΩ is placed between the generator and the electrode. Before ignition this resistor has a negligible effect on the voltage applied between the electrodes. However when the plasma is ignited, a current passes between the electrodes that needs to be limited, what is realized using the external series resistance. Finally a resistance of 1kΩ is used to estimate the current in the plasma by measuring the voltage drop on it, while the diodes protect the acquisition card from high voltage.

978-1-4799-4789-8/14 $31.00 © 2014 IEEE   259

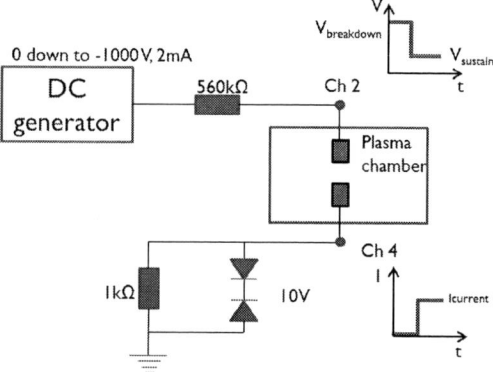

Figure 10: Electrical circuit implemented to study strike and sustain parameters in case of on-wafer measurements in environmental chamber.

The plasma ignition is detected by measuring a discontinuity in the current in order to have an accurate characterization of the plasma. We can define and measure then different characteristic parameters of the ignited plasma (as shown in Figure 11):

- The *breakdown voltage* (measured at Ch2) is the voltage applied between the electrodes at which the plasma appears.
- The *sustain voltage* (measured at Ch2) is the voltage measured between the electrodes when plasma is established between the electrodes.
- The *sustain current* (measured at Ch4) is the current passing through the plasma chamber.
- The *applied voltage* (measured by the generator) is the voltage applied by the generator at plasma ignition which is slightly lower than the breakdown voltage due to the resistance of 560Ω.

Using these parameters we can extract the I-V curve of the plasma to observe its nonlinear behavior and compute the power consumed by the plasma chamber itself.

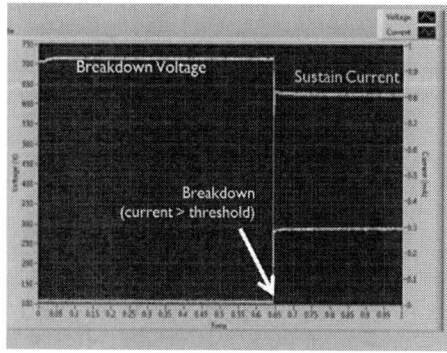

Figure 11: Plasma detection and plasma parameter extraction.

C. Measurements results in helium

The applied voltage, the breakdown voltage and the sustain voltage are plotted in Figure 12. We can observe that the breakdown voltage is slightly lower than the applied voltage due to the high resistance, but follows the same trend. The spread observed for the breakdown voltage (and then for applied voltage) is explained by the probability to create the first electron by ionization is due to random processes as thermal collision or gamma rays. The ignition will then have a certain stochastic behavior. The sustain voltage is much lower than the other two and is in the order of 200V. We can see in Figure 12 that it increases more or less linearly with the gap distance. The lower voltages are obtained for the 7μm gap devices around 180V.

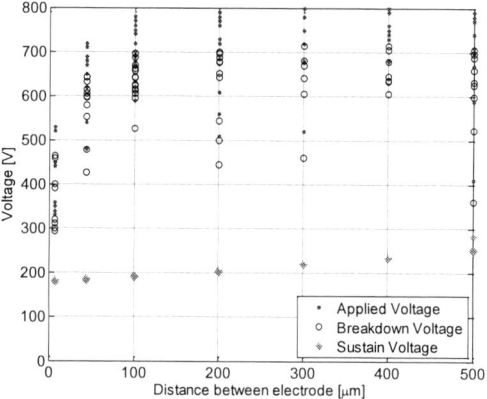

Figure 12: Applied voltage, breakdown voltage and sustain voltage measured for different gap dimensions.

By plotting the I-V curve of the plasma (sustain voltage versus sustain current), we can observe different behaviors (see Figure 13). First we clearly see that the sustain voltage increases when the gap increases. For gaps lower than 50μm, the I-V curve shows the same linear dependency and can be fitted by a linear approximation with a slope of 11Ω. For higher gap distances, the I-V curve becomes more nonlinear and these data can be fitted either by a quadratic curve or by a nonlinear resistance inversely proportional to the current square as proposed in [8]. The fitting curves plotted in Figure 13 uses this last approximation.

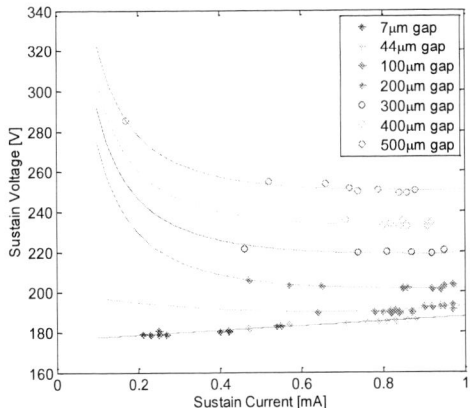

Figure 13: Measured I-V curve for on-wafer micro-chambers with different gaps in helium under atmospheric pressure.

This modification of trend between the large and small gap can be explained by the different regions that can be observed in DC plasma. In Figure 14 we can observe the main regions appearing in a glow discharge. Around the cathode we have the Townsend discharge or negative glow region where the electrons are repulsed from the cathode and create the avalanche behavior descripted in section 2A. The light emitted around the anode comes from the anode glow region. Between these two glowing regions, there is the positive column region which can be shrunk and finally disappears when the gap is reduced. In the measurements of gap lower that 50μm only the Townsend discharge and the negative glow are present and there is no space for a positive column. That explains the linear behavior of the I-V curve. For larger gap the positive column starts to be present, modifies the total resistance of the plasma and makes it more nonlinear.

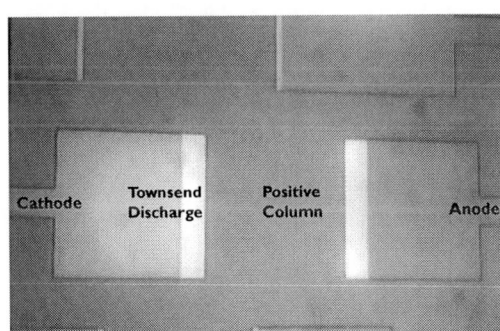

Figure 14: Picture of plasma between two electrodes where we can identify the different glow regions.

Finally we can compute the power consumed in the plasma chamber by multiplying the sustain voltage by the sustain current. As shown in *Figure 15*, the minimum power is obtained for the

7μm gap device for which we have a power varying between 37mW to 100mW. 40mW will then be the minimum power consumed by a stable plasma chamber. Note that it is possible to ignite the plasma at a relatively high voltage and then reduce the voltage keeping the plasma alive. This strategy will reduce the plasma power but degrades the stability of the plasma. Indeed if the plasma is weakened, it can extinguish and the voltage is not high enough to ignite it again.

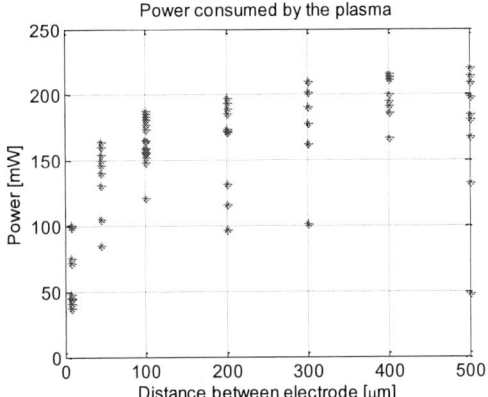

Figure 15: Power consumed in the plasma chamber depending on the gap between the electrodes.

D.   Measurements results in argon.

The same measurements are now performed in argon. *Figure 16* shows the applied voltage, the breakdown voltage and the sustain voltage for different gaps between the electrodes. We can see a trend but it is less clear than for the helium measurements. Indeed in argon we observed that the plasma ignition is more violent and needs really to be controlled by a low current in order to not directly create sparks.

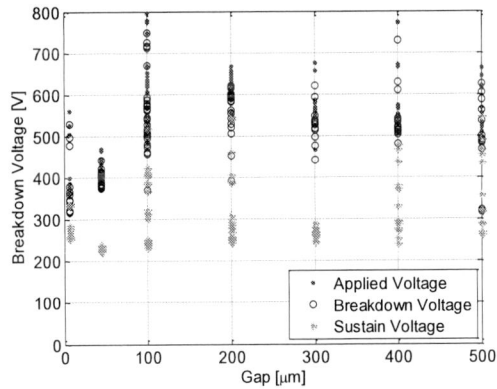

Figure 16: Applied voltage, breakdown voltage and sustain voltage in argon.

The I-V curves (sustain voltage versus sustain current) are plotted in Figure 17. In this case, we can also see that the 7μm and the 44μm gap devices show a positive linear resistance behavior but don't have the same slope. The physic in the Townsend discharge area for argon is more complex than for helium. For larger gaps all the IV measurements shows the same behavior which cannot be anymore fitted by a simple polynomial approximation. In that case the only way to fit the data is to use the nonlinear resistance inversely proportional the current square as proposed in paper [9].

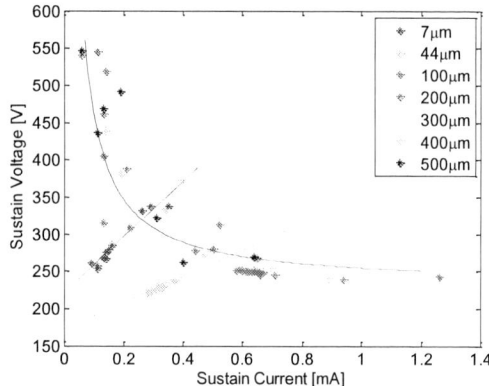

Figure 17: Measured I-V curve for on-wafer micro-chambers with different gaps in argon under atmospheric pressure.

Finally the power is computed for each gaps and plotted in *Figure 18*. The minimum power reached is about 25mW for the 7µm gap devices, but due to the strongly nonlinear behavior of the plasma we can also reach relatively low power at higher gap distance.

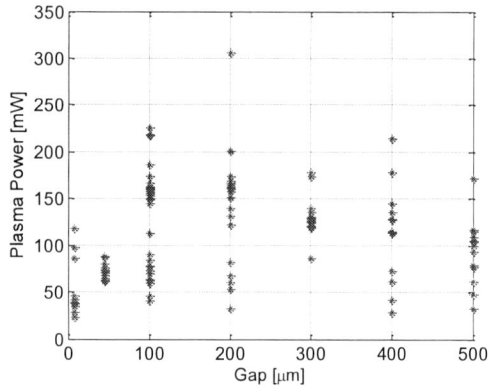

Figure 18: Power consumption in the plasma chamber in argon at atmospheric pressure.

4. Conclusions

This paper reports on the study of DC plasma in a 2D micro-chamber fabricated with MEMS technology. When the plasma ignites, the voltage applied on the plasma chamber switches from the breakdown voltage to the voltage needed to sustain the plasma. At this moment the current passing through the plasma is measured (about 0.3mA).

These measurements are performed on devices with a gap between 7µm to 500µm. We observe that the breakdown voltage decreases when the gap decreases, but due to the 2D geometry, the breakdown voltage varies a lot during successive measurements and requires at least 300V to ignite a stable plasma. The sustain current versus the sustain voltage is compared for the different gaps and two different behaviors are observed: at small gap, the plasma behaves as a positive resistance which is characteristic to the Townsend discharge. When the gap is larger than 100µm, a positive column appears after the Townsend discharge zone and the slope of the current-voltage characteristic of the plasma becomes negative and nonlinear. The minimum power consumption in the plasma chamber is about 37mW in helium and 25mW in argon. The 7µm-gap device seems then to be the best candidate. Concerning the gas, the argon shows a lower power consumption, but its plasma behavior is more stochastic than the helium plasma.

References

1. Microplasmas for nanomaterials synthesis, Davide Mariotti and R Mohan Sankaran, J. Phys. D: Appl. Phys. 43, 2010.

2. Microplasmas and applications, K H Becker K H Schoenbach and J G Eden, J. Phys. D: Appl. Phys, 2006

3. Microplasmas: Sources, Particle Kinetics, and Biomedical Applications, F. Iza, G. Jun Kim, S. Min Lee, J. Koo Lee, J. L. Walsh, Y. T. Zhang, M. G. Kong, Plasma Process. Polym. 2008, 5, 322–344

4. Titanium nitride electrodes for micro-gap discharge, Chung-Fon Hsieh, Shyankay Jou, Microelectronics Journal 37 (2006) 867–870.

5. Electrical resistivity of titanium nitride thin films prepared by ion beam-assisted deposition, K. Lala, A.K. Meikapa,*, S.K. Chattopadhyaya, S.K. Chatterjeea, M. Ghoshb, K. Babac, R. Hatadac, Physica B 307 (2001) 150–157

6. http://silas.psfc.mit.edu/introplasma/index.html

7. On the potential difference required for spark initiation in air, hydrogen, and carbon dioxide at different pressures, Friedrich Paschen (1889), Annalen der Physik 273 (5): 69–75

8. DC normal glow discharges in atmospheric pressure atomic and molecular gases, Plasma Source Sci. Technol. 17, 2008

9. A Simple Model of Spark Gap Discharge Phase, H. Habib, Eng. &Tech. Journal, Vol. 31, Part (A), No.9, 2013

The Shear Strength of nano-Ag Solders and the use of Ag Interconnects in the Design and Manufacture of SiGe-based Thermo-Electric Modules

Michael Edwards, Klas Brinkfeldt, Melina Da Silva, and Dag Andersson
Swerea IVF AB
Argongatan 30, SE-43153, Mölndal, Sweden
michael.edwards@swerea.se, Tel: +46 (0)31-706 6000

Abstract

Thermo-electric modules can be used to convert heat into electricity by utilizing the Seeback effect. It is now possible to buy BiTe thermo-electric modules that can operate up to temperatures of around 300°C. However, many applications, such as the harvesting of exhaust gas from large vehicles or gas turbine heat, may occur at higher temperatures Therefore, new materials and manufacturing processes need to be developed to produce packaged TEM that can operate at a maximum operating temperature of 650°C. Two critical areas in the manufacture of a SiGe TEM are the choice and strength of materials used to both solder the TE material to the rest of the module and the metal used for the interconnects. The interconnection material needs to be sufficiently strong to withstand large temperature fluctuations while maintaining a low contact resistance, as well as being compatible with the nano-Ag solder. Shear force tests of the sintered thermo electrical leg material showed that the joints are brittle when sintered to W metallized AlN substrates are used and ductile fracture behavior when sintered to Cu metallized AlN substrates using the NanoTach K nano silver paste. Almost all of the joints were found to be brittle when using the NachTach X nano silver paste. Shear testing of the solder joints showed that the X paste joints were variable in strength and stiffness, having a typical Young's modulus between 10 and 100 MPa at room temperature. The K paste joints were stiffer, but had a similar strength as compared to the X paste joints.

1. Introduction

Thermo-electric (TE) materials transform heat energy to electric power using the Seeback effect. They are considered to be a reliable method without moving parts and have been deployed as part of the power generation in space applications for several decades [1-3]. BiTe thermo-electric modules have been commercially available for some time [4] and they can operate up to temperatures of around 300°C. Thermo-electric materials are also under investigation as waste heat recovery devices in industrial and automotive applications [5-7], which sometimes require operation at higher temperatures.

In addition to the TE-material itself, the properties of the interface and electrode materials in a TEM are important. If the joining and current-carrying materials do not have high thermal and electrical conductivities, the efficiency is decreased due to resistive losses. Sintered silver particles have recently gained attention in the field of power electronics due to the high thermal and electrical conductivity of the material. The nanosilver paste used in this work has originally been developed for use as a die attach material in power applications, where these properties are important. The use of sintered silver particles as an alternative to other types of joining materials for the electrical connection between thermoelectric materials and the substrate in thermoelectric modules (TEM) has been investigated in [8].

In this work, further investigations of the shear strength of the sintered nano silver joints are made with the intent to create a more accurate TEM simulation model. The interfacial structure of a TE device has a critical role in its reliability, so the properties of the sintered nano-Ag need to be well understood if the TEM is to be modeled to a good degree of accuracy.

2. TEM Fabrication Process

The thermo-electric modules (TEMs) consist of two AlN substrates covered with either W or Cu metallizations, four SiGe-based legs (two n- and two p-type) of thermoelectric material and Ag interconnects. The basic schematic of a completed TEM is shown in figure 1.

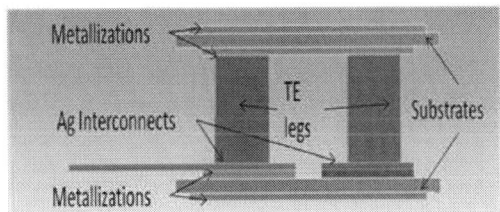

Figure 1: Schematic of AlN/Cu TEM with the Ag strip contacts.

The substrates are then connected to both the legs and interconnects (as shown in figure 2) using a nano-Ag paste, which acts like a brazing material. The completed TEM is then bonded together in a furnace at 310°C, which sinters the nano-Ag paste. The interconnection material needs to be sufficiently strong to withstand large temperature fluctuations while maintaining a low contact resistance, as well as being compatible with the nano-Ag solder. For this reason 300 μm thick Ag-strips were inserted onto the cold side of the device to act as

978-1-4799-4789-8/14 $31.00 © 2014 IEEE

interconnects, using the method shown in figure 2 (a). Figure 2 shows a basic schematic of the method used to manufacture TEMs in this project.

The process to make the connection involved screen-printing of the nano-Ag paste onto both hot and cold side substrate metallizations, the TE-legs were then attached to the hot and cold side metallizations, which was followed by a sintering process in a furnace at 310 °C with pressure applied at the peak temperature. The peak temperature and pressure was maintained for 20 minutes. The resultant joining interface is able to operate at much higher temperatures, which are theoretically near the melting temperature of Ag at Tm = 961 °C, if the joint itself is not required to withstand mechanical loads. The same sintering process was used for both types of nano-Ag paste described in the paper.

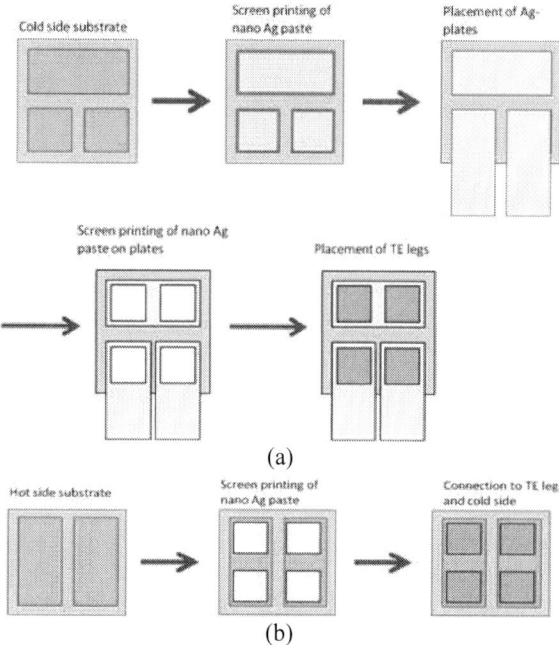

(a)

(b)

Figure 2: Schematic diagram of the TEM fabrication process. (a) Method for placing the Ag plates on the cold side of the TEM to act as interconnects and (b) shows the process used to connect the hot side substrate to the TEM.

3. Initial FE simulations of a TEM with the Ag interconnects

In the design process, thermo-mechanical FE simulations of the stresses and strains in the materials were used to select the materials used to build the TEMs, as well as enabling geometric optimization.

Model Geometry

The purpose of the model was to predict the thermo-mechanical behavior of a TEM consisting of four SiGe legs spread over a 20 x 20 mm^2 AlN substrate metalized with either W or Cu. It is important to understand the

thermo-mechanical behavior of the TEM when it is subjected to a thermal gradient of 600°C during operation. Figure 3 shows the geometry of the model and its constituent parts.

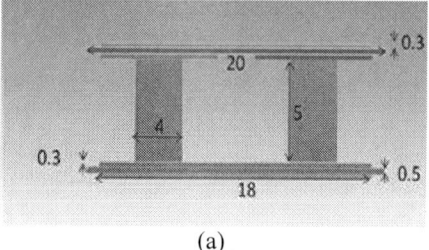

(a)

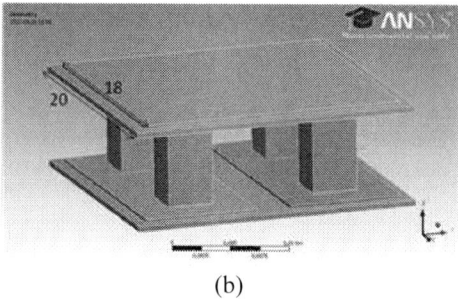

(b)

Figure 3: Schematic with the (a - b) key dimensions (mm) of the TEM labeled.The model is three-dimensional, with appropriate dimensions used for the TEM unit.

Table 1. Materials used for different parts of the TEM unit in the FE model.

Part	Material
Plate	Cu, W
Substrate	AlN, (alumina for shear tests)
Metallization	Cu, W
Joining material	nano Ag
Intercoonects	Ag
TE-leg	SiGe

Model Setup

The models described in this paper were drawn using NX 8.5. Once imported into Workbench, the thermal conditions shown in Figure 4 were applied to the model and it was solved in a thermal model solver.

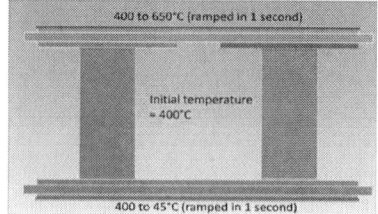

Figure 4: Thermal Loads applied to the FE model.

Once the thermal part of the model was solved, its results were exported into a mechanical model solver and the mechanical constraints were added. Figure 5 shows this model schematically.

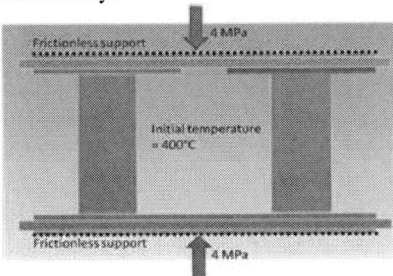

Figure 5: Mechanical loads applied to the model.

Integration with Ag contacts and interconnects

The first generation of fabricated TEMs in this work had connection issues, with the high resistance of the wires that connected the TEM and the wires becoming detached. High contact resistance results in high losses in the TEM output power, reducing the efficiency of the module.

A new connection concept consisted of placing 300 µm Ag strips under the TE legs on the cold side of the device to create a better power contact. A schematic of the module with the Ag strip contacts is shown in Figure 1. The additional sintered joints each have an assumed thickness ~10 µm, adding complexity to the model and increasing the amount of computational resources needed to solve a full TEM model. Therefore two different simulation approaches were developed to model the thermo-mechanical behavior of a TEM with the Ag connections:

• Model of entire TEM, neglecting the solder joints (model 1).
• A more detailed model of cold side of the module, which includes the solder joints (model 2).

The results from both models shown in tables 3 and 4 suggest that there will be an increase in the overall thermo-mechanical stress on the TEM models and the peak stress now occurs in the TE legs, rather than the AlN substrates. For model 1 this peak can be found where the TE legs meet the Ag interconnects for AlN/W TEMs and where the TE legs meet the hot side metallization for the AlN/Cu modules. The peak in stress can be found in the TE legs, just above the solder joint in model 2. These results indicate that the TE legs could fail if there are any pre-existing cracks in the material. Finally, these results show that when the Ag contact strips are added, the AlN/W modules appear perform similarly thermo-mechanically to the AlN/Cu modules. Therefore, it is possible that the choice of the best TEM substrate could be decided by which bonds best with the nano-Ag solder. Results of the simulations are shown in Figure 7.

Effects of Adding Pressure

In order to improve the reliability and bond strength of the nano-Ag sintered joints, the TEM is sintered together under 4 MPa of pressure and 10 MPa of pressure is applied to the module during operation. The results are shown in Table 5. Based on the simulations, the application of pressure appears to reduce the maximum amount of stress by 30 % through redistribution. It should also be noted that the model indicates that the AlN/Cu modules have the more favourable stress profile.

Table 2: The maximum stress, strain and deformation results for various material combination

Substrate	Stress intensity (MPa)	Plastic Strain on solders	Elastic Strain	Deformation (µm)
AlN/W	136.4	0.011	0.038	27.4
AlN/Cu	384.0	0.349	0.010	23.5

Table 3: Results from model 1, with the sintered joints neglected.

Substrate	Max deformation (µm)	Max Plastic strain	Max Principal stress (MPa)	Max Von Mises stress (MPa)
AlN/Cu	56.81	0.078	870.1	1103
AlN/W	97.61	0.020	1355.3	1233.1

Table 4: Results from model 2, using the properties of bulk Ag for the solder joints.

Substrate	Max deformation (µm)	Max Plastic strain	Max Principal stress (MPa)	Max Von Mises stress (MPa)
AlN/Cu	126.0	0.029	1026	914.7
AlN/W	40.9	0.008	910.2	1161.6

Table 5: Results from the application of pressure constraints.

Substrate	Max deformation (µm)	Max Plastic strain	Max Principal stress (MPa)	Max Von Mises stress (MPa)
AlN/Cu	111.85	0.015	695.64	657.26
AlN/W	91.93	0.018	620.86	831.11

4. Shear Testing of Nano-Ag joint used to connect the TE legs to the substrates

In order to test the strength of the nano-Ag solders at room temperature, some SiGe TE legs were attached to hot side substrates and were subjected to a shear stress test. Two different types of potential TEM substrates, AlN/W and AlN/Cu, were subjected to this test in order to determine the relative strength of the Nano-Ag bond to the substrate. The exact details of the substrates and metallizations are contained in table 6.

Table 6: Substrates and metallizations used in the TEMs.

Substrate	Metallization
AlN/Cu	300 μm Cu/0.1 μm Au
AlN/W	16 μm porous W/10 μm Ni/0.1 μm Au

Furthermore, two different grades of nano-Ag paste nanotech K and X pastes from NBE-Tech, Virginia were used also tested with each type of substrate. An image of the experimental setup used for the shear tests is shown in figure 6.

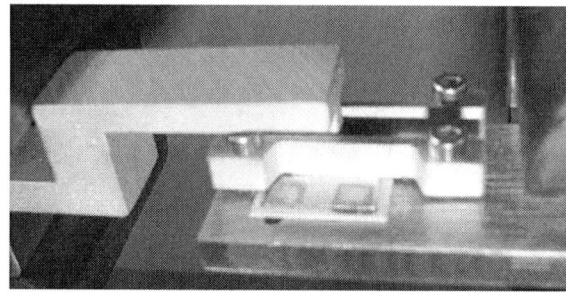

Figure 6: Experimental setup used for the shear test measurements.

Results from the NanoTach K Paste

Figure 7 shows shear diagrams for the AlN/Cu and AlN/W samples made using the naniTach paste. For the AlN/Cu substrates, only two N-type material samples were tested successfully. The tests indicate a brittle fracture mechanism for the AlN/W, whilst the AlN/Cu samples experience some ductile behavior prior to fracture. For both substrate types, the results were largely dispersed.

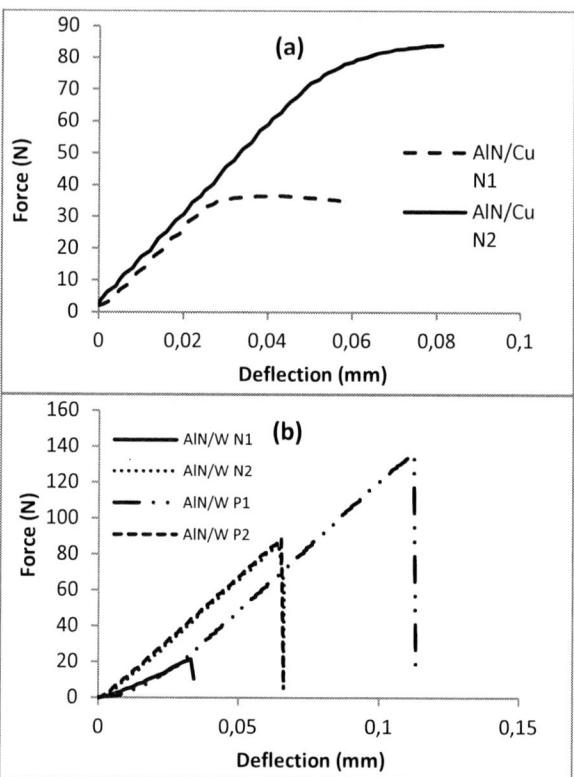

Figure 7: Shear stress as a function of the actual movement of the shear tool. a.) shows the results from the AlN/Cu substrates and b.) are the results from the AlN/W substrates.

An analysis of the fracture surfaces was done (Figure 8). It showed that the fracture surface contained mostly Ag for the AlN/Cu N2 sample while the AlN/W P1 sample fracture surface also included significant amounts of other elements. Au, Si, and Ge were easily identified in the spectrum.

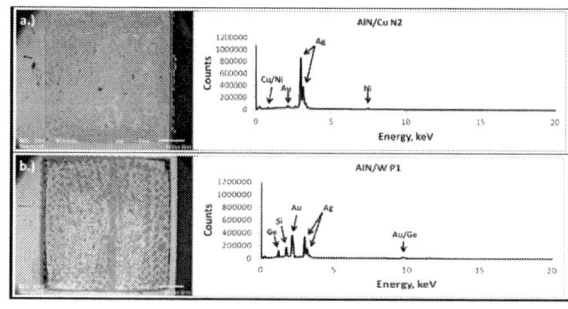

Figure 8: SEM image and EDX spectral response of the shear test fracture surfaces of a.) AlN/Cu N2 and b.) AlN/W P1 samples.

This suggests that the AlN/W substrate samples fail at the solder TE leg interface and the AlN/Cu substrates fail where at the interface between the substrate metallization and the solder.

Results from the NanoTach X Paste

Since NanoTach K paste has been discontinued by the manufacturer, the NanoTach X paste can be used as an alternative nano-Ag brazing joint. Preliminary test found that joining this material by following the manufacturer's instructions (i.e. with no pressure applied at the brazing step) lead to unsuitable joints for this application, so it was decided to use the same brazing method as for the Nantach K paste. Alumina was used rather than AlN with Cu metallizations because alumina and Cu DBC substrates are cheaper to purchase. AlN and alumina have similar elastic moduli and the purpose of the test was to test the strength of the solder joints, meaning that the difference between the substrate materials is minimal. The shear test results for the NanoTach X paste are shown in figure 9.

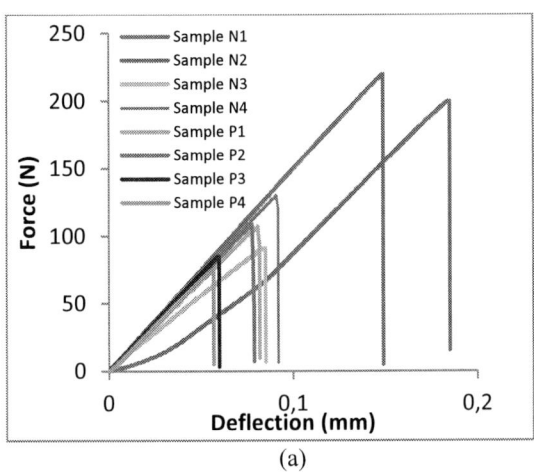

(a)

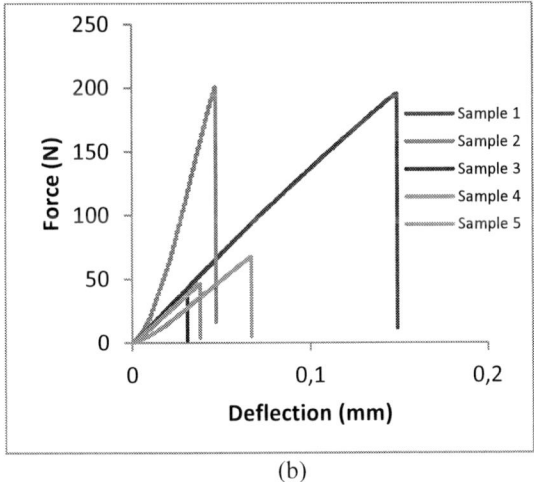

(b)

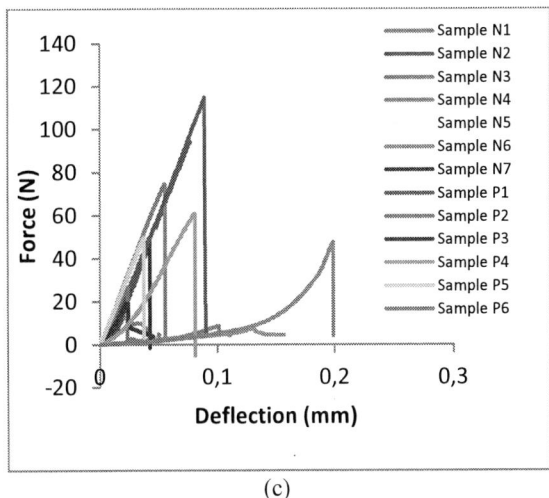

(c)

Figure 9: Force applied as a function of the actual movement of the shear tool where (a) shows the results from the AlN/W substrates, (b) are the results from the Alumina/Cu_Au substrates and (c) are the results from the Alumina/Cu_Ag substrates.

These results show that the sintered paste tends to produce a brittle solder joint that cracks at the yield point. This is no surprise because the solder joint is basically a sintered porous ceramic. There is some evidence of some non-linear deformation in the Alumina/Cu_Ag substrates, which indicates a less-common ductile failure mode similar to that observed with the AlN / Cu samples made using K paste.

It should also be noted that there is a lot of variability between the results of similar samples and this is the result of a few factors. Firstly, no two solder joints are identical, with the thickness, surface roughness and porosity of each solder joint being different. Previous research [9, 10] has shown that increasing the surface roughness weakens solder joints, results in the large variability seen in the fracture strength of the samples. The surface roughness, as well as other factors such as the even application of pressure during the sintering step, results in the thickness of the solder joint being variable and this introduces error into the strain measurement. Finally, variations in the porosity of the nano-Ag joint will both affect its elastic moduli and fracture strength, with both decreasing as porosity is increased.

5. SEM Imaging and EDS Analysis of the NanoTach Solder Joints

NanoTach K

In order to determine the quality of the NanoTach K solder joint, Scanning Electron Microscopy (SEM) and Energy Dispersive X-Ray Spectroscopy (EDS) analysis were performed on cross-sectioned TEMs. The specimens were firstly vacuum impregnated with epoxy prior to cutting, and subsequently mechanically ground/ polished. The SEM used for this analysis is a JEOL JSM-6610LV equipped with a Bruker XFlash EDS system. Selected images of these analyses for a TEM with an AlN/Cu substrate are shown in figure 10.

(a)

(b)

Figure 10: (a) SEM and (b) EDS images of the NanoTach K solder joint between a SiGe TE leg and a AlN/Cu substrate.

The SEM and EDS images in figure 10 show a consistent layer of Ag solder with a low porosity between the DBC substrate and the SiGe TE leg. This suggests that the ~20 μm solder joint will be sufficiently ductile to allow some plastic deformation, such as that seen in figure 7 (a). It should also be noted that the Cu metallization has been plated with Ni/Au layers to help the solder joint bond to it and a thin Au layer was also sputtered onto the TE leg for the same reason. This bond is likely to be stronger as a result.

NanoTach X

In order to determine the composition and appearance of the NanoTach solder joint, two additional samples were built, sintered and then cross-sectioned. One sample was a built on a AlN/W substrate and the other on a Alumina/Cu_Ag substrate. These samples were then prepared the same way as the

paste K sample SEM cross section and placed into the same SEM for analysis. Figure 11 contains images of this analysis.

The SEM images shown in figure 11 demonstrate that for both types of substrates the solder joint is porous and has a variable thickness between 20 and 40 μm, due to the surface roughnesses of both the substrate and the TE leg. This suggests that the solder joint will be both weaker and less stiff than if the joints consists of non porous Ag on a smooth surface. Furthermore, it explains why the strength of individual solder joints is variable and most of the failures are brittle. Cracks can propagate through the porous material or through the weak interfaces quickly, meaning that there is little chance for the joint to plastically deform. The images in figure 11 show evidence of cracking that probably occurred either during the sintering process or SEM sample preparation, demonstrating the propensity of the joint to crack. Therefore, all the evidence found so far suggests that the solder joints are weak and brittle. Additionally, using the material properties of bulk Ag for the solder joint appears to be inappropriate.

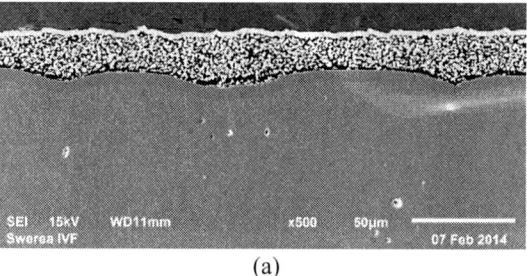

(a)

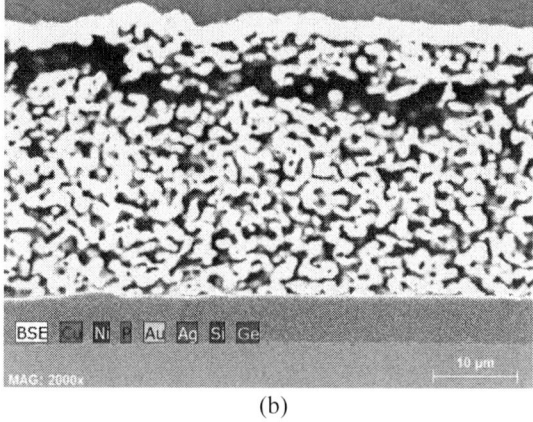

(b)

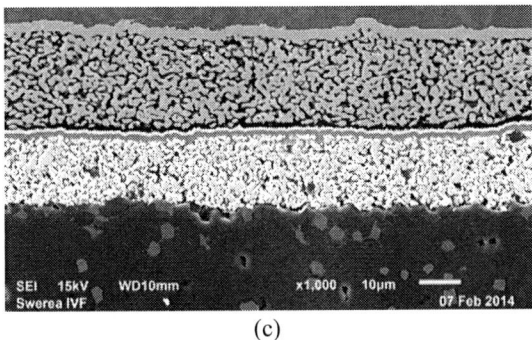

(c)

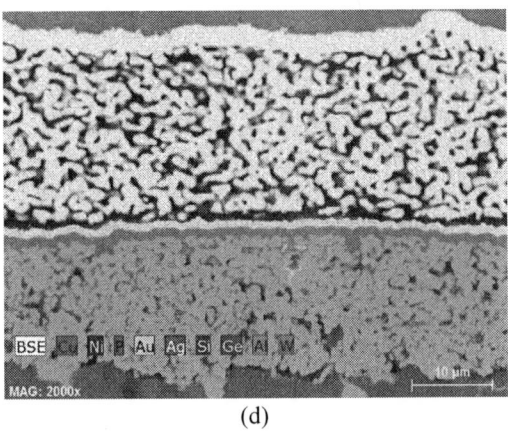

(d)

Figure 11: SEM and EDS images of the NanoTach X solder joint for the Alumina/Cu_Ag and AlN/W substrates, where (a) is an SEM image of the solder joint and (b) is an EDS material composition spectra for the Alumina/Cu_Ag substrate. With (c) being an SEM image and (d) being an EDS spectra for the solder joint on an AlN/W substrate.

6. FE Simulation of the Shear Stress Tests

Since the solder joints display linear stress/strain behavior until failure, it should be possible to FE simulate the shear tests. A FE model was firstly drawn using NX8.5 CAD software and then simulated in ANSYS. Figure 9 shows a diagram of the FE model of the shear test.

Table 7: The properties of nano-Ag solder used in the FE model

Solder thickness	30 μm
Poisson's ratio	0.37
Young's modulus	10, 25, 50, 75, 100, 400 MPa

The modulus of the solder joint will be fitted to the experimental data and these materials properties are contained in table 7. The materials properties for the metallizations, TE leg and substrate are well known and have been defined previously [11-20]. The

validated materials properties can then be used to simulate the behavior of the interconnects. The FEM data will help predict the behavior of the nano-Ag solder joint used to connect the module together.

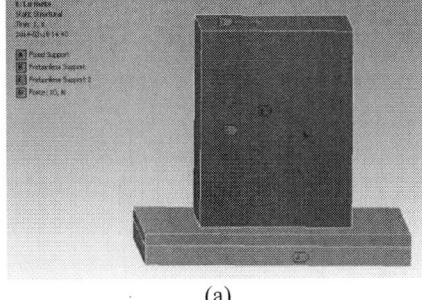

(a)

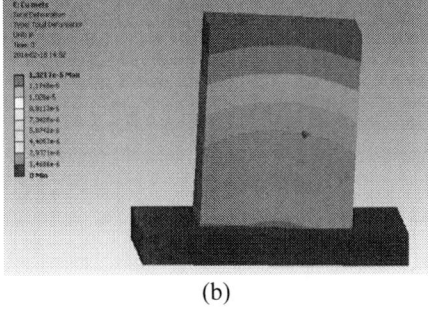

(b)

Figure 12: (a) is a schematic of the mechanical boundary conditions applied to the model and (b) shows a deformation measurement from the FE model used to simulate the behavior of the TEM with the nano-Ag solder joints.

A shear force, similar to that applied experimentally was applied to the model and its magnitude was increased linearly. The force and extension, which in this case corresponds to the deformation measured at the top of TE leg in the model (see figure 12), are plotted and compared to the experimental data. The results from the FE simulations and a select sample of the experimental data are shown in figure 13.

The plot in figure 13 shows that the elastic moduli of the X-paste solder joints, on which shear tests were performed, correspond to the FE simulations between 10 and 100 MPa, so highly variable. The K-paste solder joint has a much higher Young's modulus of around 400 MPa. The variability in stiffness is likely caused by variations in joint porosity and the surface roughness's of both the TE leg and solder joint. The surface roughness also means that the thickness of a solder joint varies across the interface, making it difficult to get a precise measurement of stiffness. However, these measurements are sufficient to validate the FE simulations and provide a suitable elastic modulus value for the solder joint in future simulations.

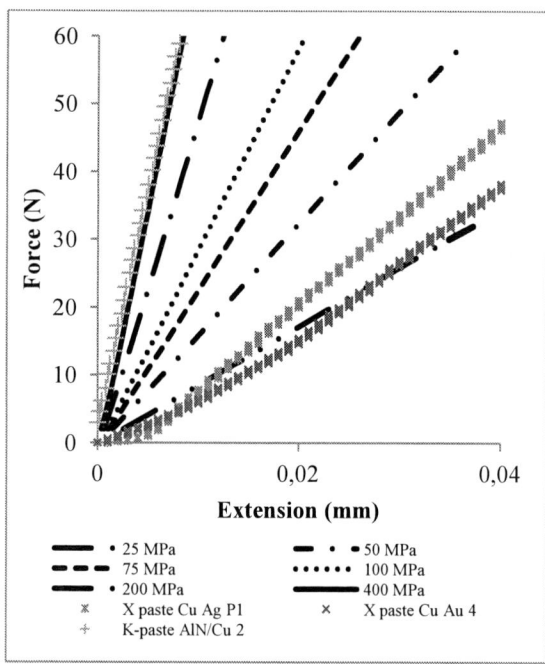

Extension (mm)

Legend:
- • 25 MPa
- • • 50 MPa
- • 75 MPa
- • • • • 100 MPa
- • 200 MPa
- 400 MPa
- ✕ X paste Cu Ag P1
- × X paste Cu Au 4
- + K-paste AlN/Cu 2

Figure 13: A plot comparing the selected experimental results with the FE simulations.

7. Discussion

The results shown in this paper prove that the X paste nano-Ag joint used to bond the TEMs is not as stiff of as strong as bulk Ag due to the porosity of the joint. The joint is highly porous, with a modulus of 10 to 100 MPa and very brittle. The solder joint produced using the K paste with the same materials is much stiffer and more ductile due to the much lower porosity. The brittleness is the direct result of the porosity since particles in a porous material can easily move apart, rather than slide across the atomic plane, which results in fractures rather than plastic deformation. Potential methods for strengthening the material would involve reducing the joint's porosity. Methods for reducing the porosity of the joints include increasing the sintering temperature and increasing the pressure applied to the joint whilst sintering. The potential risk of applying more pressure during sintering includes the fracturing of the substrates or TE legs. Increasing the sintering temperature will result in more CTE mismatch stress, which could also break the TEM. Therefore, to maximize the chances of the TEM having a long operating live, the sintering process needs further optimization and the solder joints need strengthening. It should be noted that he solder joints on the hot side of the module may be strengthened during operation at 650°C because this will reduce joint porosity.

One of the reasons for using the same screen printing and sintering processes for both types of paste was to directly compare to quality of the solder joints. The NanoTach K solder joint is a lot less porous and consequently stiffer than the NanoTach X joint, as shown in figure 13. However, the strength of both pastes, according to the shear tests appears to be similar for the AlN/Cu substrates. The NanoTach X paste appears to bond better to the W_Ni_Au metallizations than the K paste despite being porous and less stiff. The electrical performance for paste X for this application is unknown, but is likely to be inferior due to the higher porosity.

8. Conclusions

Further FE simulations and the manufacture of more TEMs are required to improve the paste X solder joint sinter process. Further FE models are required to simulate the module under operation with improved accuracy, now that the solder properties are better known. The results shown in this paper suggest that the problems are caused by its high porosity, which can be reduced by sintering at a higher temperature and with more pressure applied.

Acknowledgments
This work was supported by the FP7 EU-project NEAT under grant agreement NMP4-SL-2011-263440.

References

[1] J. F. Mondt and J. H. Ambrus, "Thermoelectric and thermionic conversion technology," 1984.

[2] G. L. Bennett, J. J. Lombardo, R. J. Hemler, G. Silverman, C. Whitmore, W. R. Amos, et al., "The General Purpose Heat Source Radioisotope Thermoelectric Generator: A Truly General Purpose Space RTG," in SPACE TECHNOLOGY AND APPLICATIONS INTERNATIONAL FORUM STAIF 2008: 12th Conference on Thermophysics Applications in Microgravity; 1st Symposium on Space Resource Utilization; 25th Symposium on Space Nuclear Power and Propulsion; 6th Conference on Human/Robotic Technology and the Vision for Space Exploration; 6th Symposium on Space Colonization; 5th Symposium on New Frontiers and Future Concept, 2008, pp. 663-671.

[3] V. Ravi, S. Firdosy, T. Caillat, E. Brandon, K. Van Der Walde, L. Maricic, et al., "Thermal expansion studies of selected high-temperature thermoelectric materials," Journal of Electronic Materials, vol. 38, pp. 1433-1442, 2009.

[4] A. S. Kushch, J. C. Bass, S. Ghamaty, and N. Eisner, "Thermoelectric development at Hi-Z technology," in Thermoelectrics, 2001. Proceedings ICT 2001. XX International Conference on, 2001, pp. 422-430.

[5] D. Crane, J. LaGrandeur, V. Jovovic, M. Ranalli, M. Adldinger, E. Poliquin, et al., "TEG On-Vehicle Performance and Model Validation and What It Means for Further TEG Development," Journal of Electronic Materials, vol. 42, pp. 1582-1591.

[6] H. Schock, G. Brereton, E. Case, J. D'Angelo, T. Hogan, M. Lyle, et al., "Prospects for Implementation of Thermoelectric Generators as Waste Heat Recovery Systems in Class 8 Truck Applications," Journal of Energy Resources Technology, vol. 135, p. 022001.

[7] G. Roy, E. Matagne, and P. Jacques, "A Global Design Approach for Large-Scale Thermoelectric Energy Harvesting Systems," Journal of Electronic Materials, vol. 42, pp. 1781-1788.

[8] K. Brinkfeldt, J. Simon, K. Romanjek, S. Noel, J. Räthel, M. Edwards, et al., "Sintered Nano-Ag as Joining Material for Thermoelectric Modules," in 11th European Conference on Thermoelectrics (ECT) 2013, Noordwijk, Netherlands, 2013.

[9] C. Buttay, A. Masson, J. Li, M. C. Johnson, M. Lazar, C. Raynaud, et al., "Die Attach of Power Devices Using Silver Sintering-Bonding Process Optimization and Characterization," in Proceedings of the High Temperature Electronics Network conference (HiTEN 2011), pp. 1-7.

[10] F. Sarvar, D. C. Whalley, and P. P. Conway, "Thermal interface materials-A review of the state of the art," in Electronics Systemintegration Technology Conference, 2006. 1st, 2006, pp. 1292-1302.

[11] S. Riffat and X. Ma, "Thermoelectrics: a review of present and potential applications," Applied Thermal Engineering, vol. 23, pp. 913-935, 2003.

[12] H. O. Pierson, "13 - Covalent Nitrides: Properties and General Characteristics," in Handbook of Refractory Carbides and Nitrides, ed Westwood, NJ: William Andrew Publishing, 1996, pp. 223-247.

[13] N. Scoville, C. Bajgar, J. Rolfe, J. P. Fleurial, and J. Vandersande, "Thermal conductivity reduction in SiGe alloys by the addition of nanophase particles," Nanostructured Materials, vol. 5, pp. 207-223, 2// 1995.

[14] J. P. Andrews, "The variation of Young's Modulus at high temperatures," Proceedings of the Physical Society of London, vol. 37, p. 169, 2002.

[15] J. G. Bai, K. D. Creehan, and H. A. Kuhn, "Inkjet printable nanosilver suspensions for enhanced sintering quality in rapid manufacturing," Nanotechnology, vol. 18, p. 185701, 2007.

[16] J. G. Bai, J. N. Calata, and G.-Q. Lu, "Processing and characterization of nanosilver pastes for die-attaching SiC devices," Electronics Packaging Manufacturing, IEEE Transactions on, vol. 30, pp. 241-245, 2007.

[17] Y. Mei, G.-Q. Lu, X. Chen, S. Luo, and D. Ibitayo, "Effect of Oxygen Partial Pressure on Silver Migration of Low-Temperature Sintered Nanosilver Die-Attach Material," Device and Materials Reliability, IEEE Transactions on, vol. 11, pp. 312-315, 2011.

[18] D. R. Smith and F. Fickett, "Low-temperature properties of silver," Journal of Research-National Institute of Standards and Technology, vol. 100, pp. 119-119, 1995.

[19] G. Skoro, C. Booth, and J. Back, "Tungsten behaviour at high temperature and high stress," in IPAC'10, Kyoto, Japan, 2010.

[20] E. Lassner and W.-D. Schubert, Tungsten: properties, chemistry, technology of the element, alloys, and chemical compounds: Springer, 1999.

Analysis of RF-MEMS Switches in Failure Mode:
Towards a More Robust Design

Thomas Kuenzig[1], Tatek Muschol[1], Jacopo Iannacci[2], Gabriele Schrag[1] and Gerhard Wachutka[1]

[1]Institute for Physics of Electrotechnology, Munich University of Technology, Munich, Germany
[2]Center for Materials and Microsystems (CMM), Fondazione Bruno Kessler (FBK), Trento, Italy
E-mail: kuenzig@tep.ei.tum.de

Abstract

We present comprehensive theoretical and experimental investigations on one of the most relevant failure mechanisms in RF-MEMS switches, namely electrically induced stiction. In particular, we analyze an RF-MEMS switch equipped with an embedded active thermal recovery appliance by deriving and applying a 3D, problem-adapted, coupled finite element (FE) model including all relevant mechanical, electrical, thermal, and fluidic effects. The accuracy and predictive power of the simulations is ensured by a dedicated calibration procedure based on highly accurate characterization techniques such as white light interferometry and laser Doppler vibrometry. Applying the calibrated model, we studied the switch operation during failure and recovery in all details and identified the most important design parameters affecting its reliability with a view to improving the recovery capability as well as optimizing the overall performance towards a more robust switch design.

1. Introduction

Compared to the solid-state switches commonly employed in the field of RF applications, MEMS relays offer several advantages, such as high isolation and low insertion loss [1]. However, their poor reliability still limits the use in commercial products. In particular, it is sticking of the movable parts during operation caused by micro-welding [2] (welding of ohmic contacts due to high local current densities) and dielectric charging [3] (trapping of charges in dielectric layers), which is still one of the major concerns. Producing stiffer switches, which are more robust against stiction due to the increased mechanical restoring force, however, would lead to higher actuation voltages and, thus, limit their application especially for mobile devices. This constitutes a typical trade-off situation between reliability and application field. One promising remedy to solve this problem is the active recovery appliance reported in [4], which can be activated in case of failure to restore the functionality of the micro-switch. Essential for the successful design of such a system and, thus, for its durability during failure, is a profound understanding of the operation and the underlying physical effects, which can be gained by detailed theoretical investigations and predictive simulations. To this end, we derived a physically based, coupled simulation model on the basis of the finite element (FE) method, which, on the one hand, allows for detailed insight in the device operation, but, on the other

hand, is fast enough for efficient, comprehensive design studies.

The paper is arranged as follows: In section 2 we introduce the investigated switch design, its working principle and the thermal recovery capability. The modeling concept and the workflow of the simulations are described in detail in section 3. Special emphasis is laid on the calibration and validation of the 3D FE model as the basis for reliable and predictive simulation, which is presented in section 4. The results obtained from the simulations are discussed in section 5, before we conclude our findings in section 6.

2. Device description and working principle

The investigated RF switch depicted in Fig. 1 belongs to the class of electrostatically actuated ohmic MEMS switches and is fabricated at the Fondazione Bruno Kessler.

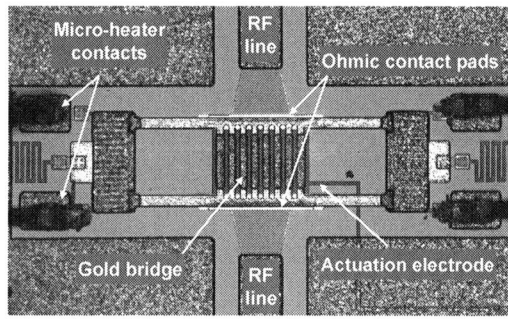

Figure 1: Microphotograph of the switch under investigation. The movable gold bridge can be pulled down towards the contact pads by applying a bias voltage between bridge and actuation electrode, thereby closing the RF signal line.

It consists of a gold bridge hinged by four beams at the anchors. The ridges of the bridge are stiffened by an additional gold layer. Applying a bias voltage between gold bridge and an actuation electrode buried in an oxide layer beneath, the switch can be pulled down onto underlying ohmic contact pads in order to close an RF signal line (see Fig. 1).

The before mentioned recovery appliance is realized by embedding resistive micro-heaters underneath the anchors as visualized in Fig 2. Activating one or both micro-heaters in case of stiction will increase the temperature inside the bridge. The resulting thermal expansion is supposed to exert forces to the membrane, to break up possible welding spots and to recover the switch

978-1-4799-4789-8/14 $31.00 © 2014 IEEE

from failures caused by micro-welding. Additionally, the temperature rise could be used to accelerate discharging processes to prevent stiction due to dielectric charging. In the following we focus on the failure scenario of micro-welding.

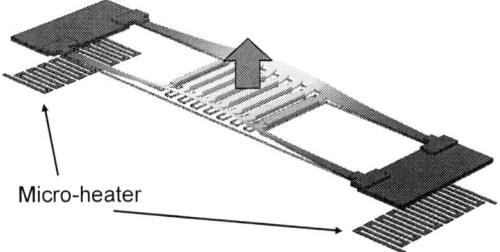

Figure 2: Schematic view of the embedded recovery appliance. Resistive micro-heaters buried underneath each anchor enable the heating-up of the device to restore the switch in case of stiction. Shear forces induced by the thermal expansion of the gold bridge are intended to break the welded contacts, and the temperature rise in the substrate speeds up the discharging of the dielectric layers.

3. Modeling and simulation

The detailed investigation of the switch in failure situations requires the consideration of several involved energy domains. To this end, we derived a physically based modeling scheme that covers all contributing mechanical (including mechanical contact), electrical, thermal, and fluidic effects, while, at the same time, it keeps the computational expense affordable. To achieve the latter, simplifying assumptions have to be made, since a model including all relevant effects on FE level would

not be manageable nor computable, especially in case of transient simulations. Thus, the thermo-mechanical part is modeled by a 3D FE approach, while other energy domains are included as analytical or semi-analytical models. The electrostatic forces, calculated using a plate capacitor approximation, and fluidic damping, approximated using a constant damping factor gained from measurements, are implemented as boundary loads in the mechanical domain and Joule heating is considered as body load in the thermal domain (detailed description see [5]).

Besides this, we divide the modeling process into two steps in order to tailor the simulation domains for each subproblem accordingly. The resulting simulation workflow is illustrated in Fig. 3. The first step simulates the heat propagation inside the whole device including the substrate, the micro-heaters and all oxide layers around them in a pure thermal analysis aiming at the temperature evolution inside the bridge. The desired temperatures are then extracted and applied as boundary conditions to the model in step 2, which is restricted only to the gold bridge and the anchor regions, since these are the relevant parts to be considered for the thermo-mechanical model applied to calculate the forces acting on welding spots. Step 2 itself is divided into three sub-steps. In step 2a the deformation of the switch in closed position is determined, in step 2b one welding spot is chosen and fixed at the closed position and the electrostatic force is switched off. The resulting force at the welded contact delivers the mechanical restoring force of the device. Finally, in step 2c, the thermal load is switched on to emulate the activation of the heater. The reaction force at the welding spot of step 2c consists of the mechanical restoring force and, in addition, the force caused by

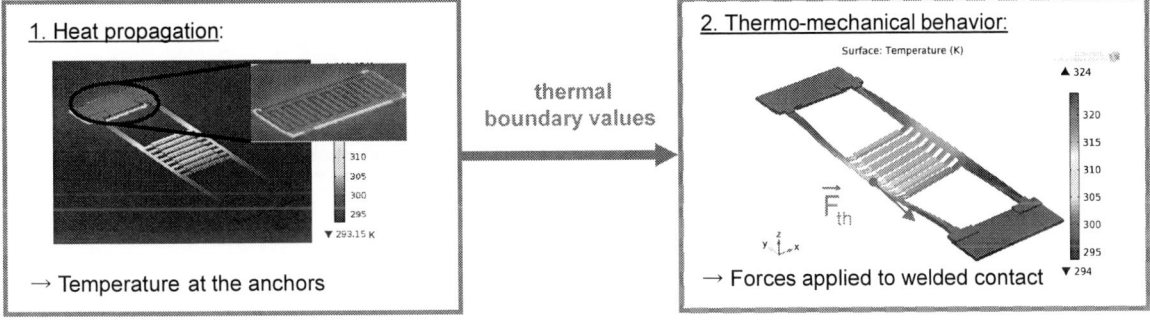

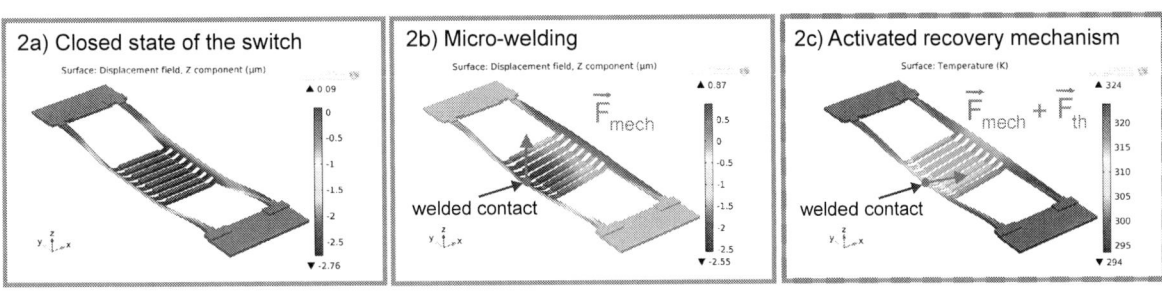

Figure 3: Workflow of the simulation approach. Boxes indicate the type of simulation (Thermal (red), mechanical (blue) and thermo-mechanical (red-blue))

978-1-4799-4789-8/14 $31.00 © 2014 IEEE

thermal expansion, the latter is – in the following – denoted as thermally induced force. The thermally induced force is now extracted by evaluating the reaction forces obtained from the thermo-mechanical FE model.

4. Model calibration and validation

Complex simulation models including several energy domains need a proper calibration and validation procedure in order to ensure accurate and meaningful results, from which trustable suggestions for the optimization of the device can be deduced. The dedicated calibration procedure applied in this work starts with the separate verification of the mechanical and thermal submodel, respectively, before connecting them to form a multi-energy domain coupled simulation model. For this purpose the switches were statically and dynamically characterized by optical measurements employing a white light interferometer and a laser Doppler vibrometer.

Special emphasis has to be laid on the calibration of the process-induced stress inside the gold bridge, since the process dependent stress values dominate the mechanical behavior of the switch. To this end, the mean stress of the gold layer is calibrated to the measured first eigenfrequency of the bridge.

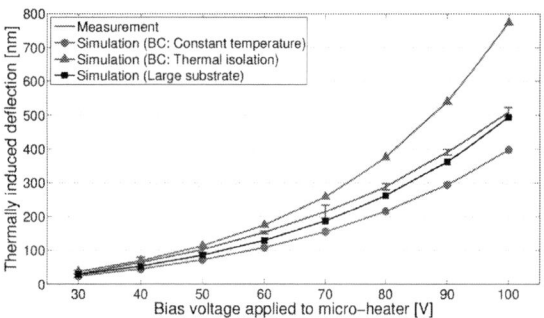

Figure 4: Thermally induced deflection of the gold bridge over bias voltages applied to the heater for different thermal models. The obtained simulation results are compared to measurements.

The heat propagation inside the device also has to be verified, which can hardly be assessed directly by measurements. Instead, it can be verified indirectly by recording the thermally-induced steady-state deflection of the bridge due to the activation of a micro-heater. More specifically, we heat up the switch in its rest position (without contact/stiction), evaluate the deformation in dynamic measurements and extract the steady-state deflection; the results are shown in Fig. 4. Additionally, we compared the deflections obtained from simulation using different lateral thermal boundary conditions for the substrate in Fig. 4 as well. The bottom of the substrate is hereby always modeled as ideal heat sink. Using thermally isolating lateral boundaries (triangles) the deflections are overestimated, using constant temperature (circles), which means ideal heat sink conditions, the bending caused by thermal expansion is underestimated. The most realistic reproduction of the heat propagation,

however, can be ensured by enlarging the lateral dimensions of the substrate until the choice of boundary conditions has only negligible impact on the simulation results, which means, the generated heat can spread into the substrate without being disturbed by boundary conditions. The resulting curve (rectangles) agrees very well with measurements, attesting the validity of the thermal model.

Furthermore, it revealed that the process-induced stress is not homogeneously distributed over the cross-section of the gold bridge, since chromium atoms from the seed layer (the gold bridge is fabricated by electroplating using a thin chromium deposit as seed layer) diffuse into the gold layer during process and produce additional stress leading to a vertical stress distribution. For the sake of simplicity and since this is a well-known approach to account for distributed stresses, the gold bridge was modeled by introducing two layers of different initial stress values. The usual way to further calibrate stress distributions inside microstructures is to measure the initial deformation in rest position. For the investigated switches, however, this approach exhibits some drawbacks, namely, for structures with high surface roughness, like the gold layers in this device, the bending curve cannot be measured with adequate accuracy and fully clamped devices only show small sensitivity in deformation due to changes in the distribution of process-induced stress.

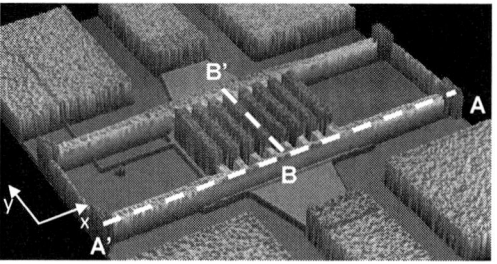

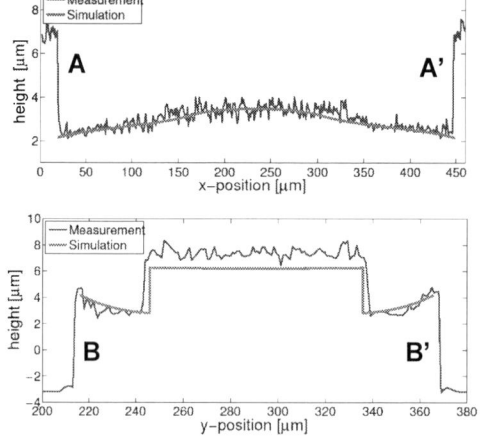

Figure 5: 3D-profile of the MEMS switch as recorded with a white light interferometer. The deformation in the rest position is used to validate the initial stress distribution inside the gold bridge.

In order to solve these difficulties an alternative method is used for calibration. Previous investigations have shown that the amplitude of the thermally-induced deflection, depicted in Fig. 4, is very sensitive to changes in the distribution of the process-induced stress, too, and, thus, better suited for calibration. Evaluating this steady-state deflection over a large range of bias voltages enables to calibrate both, the thermal model and the initial stress distribution using the same method. For low heating voltages the deflection amplitude is more affected by the stress distribution, since the difference between the different thermal models is small. For high heating voltages the simulated deflection shows larger dependence on the thermal boundary conditions, which enables the validation of initial stress distribution inside the bridge and, at the same time, to verify the thermal boundary conditions.

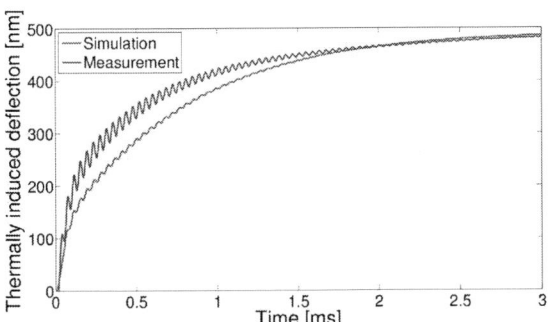

Figure 6: Thermal expansion of the gold bridge caused by a heating voltage of 100V and a current of 3.6 mA. Both, simulation and measurement show a superimposed mechanical oscillation at the resonance frequency of the gold bridge. The transient vertical deflection is used to validate the heat propagation inside the device.

The previously mentioned measurement of the deformation in rest position determined by white light interferometry can now be used for further validation of the obtained stress parameters (see Fig. 5). The excellent agreement between simulated and measured bending line confirms the success of the applied calibration strategy for the process-induced stress. In addition, the transient deflection curve of the bridge during heat-up is presented in Fig. 6. Both, simulation and measurement show the upward bending of the bridge due to thermal expansion being superimposed by mechanical oscillations at the first eigenfrequency of the gold bridge. The very good agreement of numerical and experimental results proves the validity of the thermo-mechanical simulations and demonstrates the high level of fidelity of the derived model.

5. Results and discussion

Applying the calibrated model we are now able to perform the simulation sequence described in section 3 aiming at the extraction of the desired forces acting on welding spots during failure. As a first outcome of the simulation we obtain the temperature distribution inside the switch and the substrate, which is shown in Fig. 7.

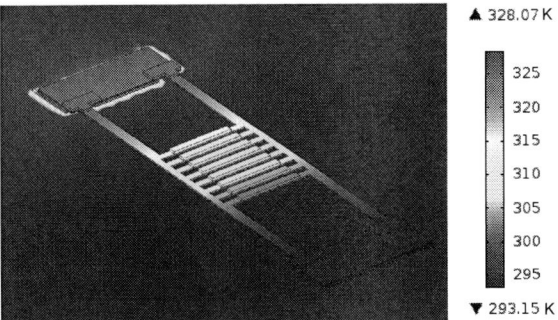

Figure 7: Temperature distribution inside the device as generated by the activated micro-heater.

The heat generated by the micro-heater leads to a significant temperature rise inside the respective anchor and the gold bridge, while the second anchor and most part of the substrate exhibit only small changes in temperature. This constitutes an excellent precondition for the exploitation of the thermo-mechanical expansion with respect to exert forces to a sticking bridge. The amount of heat flowing into the switch and therewith the achieved temperature rise strongly depends on the detailed design of the micro-heater and the surrounding oxide layers it is buried in. The impact of these geometrical features can be investigated and, hence, optimized applying the derived model, which is described explicitly in [6].

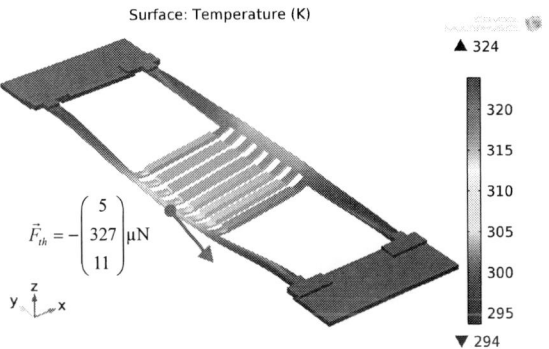

Figure 8: Thermo-mechanically deformed bride: Thermal expansion exerts shear forces on the welding spots, which can be extracted and evaluated from simulations.

As a second fundamental result, forces acting on potentially welded contacts can be calculated. An exemplary force exerted by thermal expansion on an emulated welding spot is visualized in Fig. 8. An interesting result is that the thermally induced force exhibits no significant contribution in the vertical (z-) direction, which would add to the mechanical restoring force. Instead, nearly the total amount of the thermal energy is transduced into a shear force (y-direction), which may break up the welded contacts in order to restore the functionality of the switch. It revealed that this shear force is in the same order of magnitude as the mechanical restoring force and as the strength of welded gold-to-gold contacts reported in literature [7], which are in the range of 100-400 μN. This constitutes a first promising result for the practical use of the demonstrated concept during the intended failure situations.

The simulated forces cannot only be used to assess the effectiveness of the restoring mechanism but, in the first place, to identify decisive design parameters for the improvement of the device towards a more robust design. Exemplary results of these investigations and their interpretations, which are new compared to [6] are presented in the following.

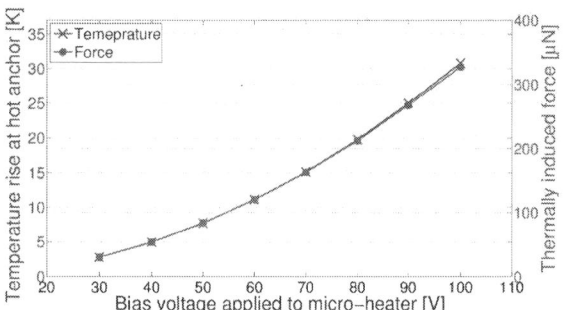

Figure 9: Temperature rise at hot anchor (crosses) and thermally induced force (circles) plotted versus bias voltage applied to the micro-heater. The force exerted on the welded contact grows likewise the temperature rise inside the anchor region.

Fig. 9 shows a plot of the contact forces (circles) versus voltage applied to the micro-heater. It is clearly visible that the force increases with applied heating voltage and there is no indication for saturation. Furthermore, the characteristic of the obtained forces turns out to be nearly congruent to that of the temperature rise (crosses) in the hot anchor. This indicates that the magnitude of the generated force solely depends on the temperature inside the bridge and the suspensions. This means, in return, that optimization of the exerted forces can be achieved by increasing the temperature at the hot anchor, which increases also the temperature inside the gold bridge. Ways to optimize the heat propagation aiming at maximizing the temperature and therewith the efficiency of the recovery capability have been already discussed in [6].

In addition, the correlation between generated heat and force exerted on welded spots is illustrated in Fig. 10. It shows a nearly linear dependence between exerted forces and thermal power, which is fed into the heater to generate the force. If we define the ratio between the magnitude of the generated force and the injected thermal power to be the efficiency of the mechanism, this means that the latter stays constant for different heating voltages.

Another interesting result affects possible pulse forms for the activation of the micro-heater and can be derived from the transient characteristics in Fig. 6. The recorded transient thermally induced upward deflection of the

bridge is a measure for the temperature rise inside the switch. As the superimposed mechanical oscillations show, the thermal time constants for heat propagation are much larger than those for the mechanical motion, which, consequently, impedes possible optimization by advanced control of the micro-heater, such as resonant pulsing to increase the achievable forces. Nevertheless, an optimum pulse width of about 1 ms can be identified from this curve as a rule of thumb that on one hand maximizes the exerted force while, at the same time minimizes the electrical energy, which has to be supplied. Accordingly, the resulting energies that have to be provided for the recovery of the bride are in the range of about 50-350 μJ for the investigated device.

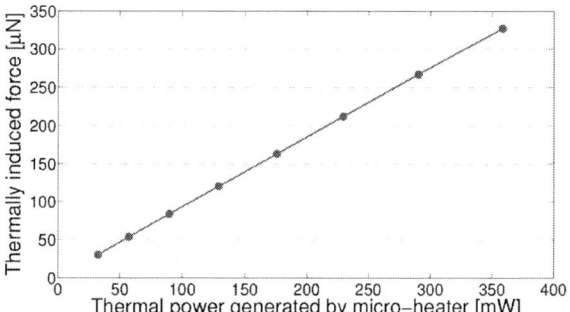

Figure 10: The force exerted on micro-welding spots exhibits an approximately linear correlation with the power converted in the micro-heater.

All in all, the presented results show already great promise for the improvement of RF-MEMS switches towards a higher robustness during operation and give rise to further design studies to optimize the switches accordingly. Furthermore, it could be demonstrated that with a physics-based and tailored simulation approach accompanied by a proper calibration procedure we are able to evaluate switches in failure situations on a reliable basis aiming at the optimization of the switch designs towards higher robustness, which can also be transferred to a broad variety of switching devices.

6. Conclusions

We presented theoretical and experimental investigations on an electrostatically actuated RF-MEMS switch equipped with an embedded recovery capability in the failure scenario of micro-welding. The introduced problem-adapted 3D FE model allows for considering all relevant mechanical, thermal, electrical and fluidic effects at acceptable computational cost. Using a sequential simulation approach we are able to extract forces on welded contacts exerted during failure and recovery. A dedicated calibration procedure to determine process-induced stress and its distribution as well as to verify the heat propagation has been applied based on static and dynamic measurements using a white-light interferometer and a laser-Doppler vibrometer. The careful and proper calibration strategy pays off with regard to a very good agreement of simulation and measurement, which, finally, ensures the accuracy and the predictive power of the derived model.

The obtained results offer detailed insights into the switch operation in failure situations, such as mechanical restoring forces during stiction, temperature evolution and exerted forces during recovery. Furthermore, ways to optimize the efficiency of the recovery capability are identified aiming at increasing the exerted forces and minimizing the required electrical power. The hereof gained profound understanding of the behavior of the MEMS switch in case of failure and recovery demonstrates the power of the applied tailored simulation model as a basis for the optimization of such devices towards a more robust design.

References

1. Rebeiz G M, RF MEMS - Theory, Design, and Technology, John Wiley & Sons, Inc., (New Jersey, 2003), p. 5.
2. Tazzoli A, Meneghesso G, "Acceleration of Microwelding on Ohmic RF-MEMS Switches", *J. of Microelectromechanical Systems*, Vol. 20, No. 3 (2011), pp. 552-554.
3. van Spengen W M, Puers R, Mertens R, *et al.*, "A comprehensive model to predict the charging and reliability of capacitive RF MEMS switches", *J. Micromech. Microeng.*, Vol. 14 (2004), pp. 514-521.
4. Iannacci J, Repchankova A, Faes A, *et al.*, "Enhancement of RF-MEMS switch reliability through an active anti-stiction heat-based mechanism", *J. Microelectronics Reliability*, Vol. 50 (2010), pp. 1599-1603.
5. Kuenzig T, Iannacci J, Schrag G, *et al.*, "Study of an active thermal recovery mechanism for an electrostatically actuated RF-MEMS switch", *Proc EuroSimE*, Cascais, Portugal, April. 2012, pp. 1–7.
6. Kuenzig T, Schrag G, Iannacci J, "Modeling and simulation of an active restoring mechanism for high reliability switches in RF-MEMS technology", *J. Microelectronics Reliability*, Vol. 52 (2012), pp. 2235-2239.
7. Iannacci J, Faes A, Repchankova A, *et al.*, "An active heat-based restoring mechanism for improving the reliability of RF-MEMS switches", *J. Microelectronics Reliability*, Vol. 51 (2011), pp. 1869-1873.

Thermo-mechanical simulation of plastic deformation during temperature cycling of bond wires for power electronic modules

A. Wright[1], A.Hutzler[1], A.Schletz[1], P. Pichler[1,2]

[1]Fraunhofer Institute for Integrated Systems and Device Technology IISB, Erlangen & Nürnberg, Germany
[2]Chair of Electron Devices, University of Erlangen-Nuremberg, Erlangen, Germany

Abstract

Modelling was undertaken to investigate the role of bond wire size on reliability in power electronic converters. Experiments have shown that thin 125 µm Al wires used in place of 375 µm Al wires alleviate bond wire lift-off and further outlast other sources of failure such as solder degradation in a power module. To investigate the role of bond-wire size on wire lift-off, the effective plastic strain was estimated through thermo-mechanical simulation. Three-dimensional models were constructed for the thin and thick bond wires, respectively. For the critical deformation of the aluminium bond wires during thermal cycling, a temperature-dependent bi-linear plasticity model was used. The effect of a difference in yield strength for the thin wires was also investigated. Maximum as well as volumetrically averaged values of the effective plastic strain showed significant differences between the thick and thin wires and wires with different yield strengths. The modelling results show higher effective plastic strain for the thick wires – supporting the experimental findings.

1. Introduction

Power electronic modules are currently limited in their lifetime by failure of internal bond wires [1]. After a number of temperature cycles, cumulative plastic deformation occurs at the interface between the bond wire and the bond pad as a result of inhomogeneous thermal expansion within the complete structure comprising the silicon device, solder, direct bonded copper (DBC) substrate, and Al bond wires. Due to this thermo-mechanical stress, the plastic deformation evolves into cracks and finally the lift-off of the bond wires. A current approach to extending lifetime is changing from aluminium to copper-based bonding wires [2–3], however this is still in development along with other alternatives. Active power cycling (cyclically switching the device on and off) performed at Fraunhofer IISB [4] has shown that module lifetime can be extended by decreasing the diameter of the bond wire from 375 µm to 125 µm while increasing the number of the wires to maintain the current density. This was successful to the extent that the modules failed by solder degradation instead of wire lift-off.

This study aims towards a better understanding—through thermo-mechanical modelling—of why the smaller wires are more resistant to lift-off. According to the Coffin-Manson model of fatigue crack-growth, the accumulated plastic deformation is key to predicting reliability. For this, the bottom of the bonded foot of the

Al wire is of principal interest as it is experimentally known that the cracks form just above the interface to the bond pad [5–7]. Models were constructed for both sizes of bond wire on a power electronic module substrate. These underwent a full temperature cycle, heating followed by cooling.

In addition to investigating the influence of the thickness of the bond wire, it was also desired to model the effects of different yield strengths for the thin wires. Tensile testing performed by Merkle et al. [8] showed a higher yield strength for thin aluminium wires which was explained by a different thermal treatment of the wires. As this could have an influence on the degree of cyclic damage incurred in the aluminium, simulations of the structures with thin wires were performed with both measured yield strengths. The effect of the yield strength was then evaluated through comparison of the effective plastic strain. This same comparison was then made between the thick and thin wires to evaluate the effect of the wire size on reliability.

2. Model Set-up and Material Parameters

To assess the plastic deformation of the bond wire during temperature cycling, the relevant parts of the power module were modelled. This included the bonded wire upon the bond pad, the silicon device, solder and DBC substrate as shown in Figure 1.

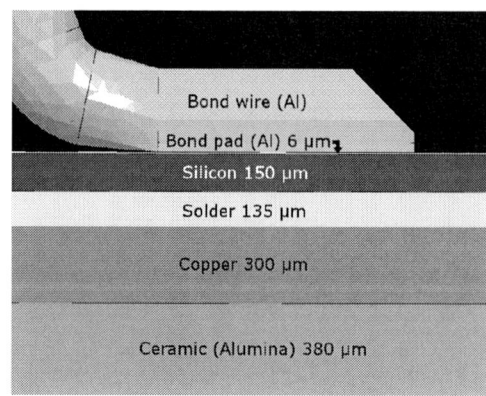

Figure 1: Structure of bond wire on DBC substrate. The bond pad (6 µm thick) corresponds to the thin white line above the silicon layer.

The models were created using the design modeller within the workbench in Ansys [9]. Further pre-processing was carried out in Ansys Mechanical and LS-PrePost with the model readied for execution in LS-Dyna [10]. Two three-dimensional models were created, one for each wire size. In order to have the correct bonded form

978-1-4799-4789-8/14 $31.00 © 2014 IEEE

of the wires in the simulation, the bonded 'feet' from both wire sizes were measured with a laser profilometer. The resultant geometry was then transferred into the model. For the loop geometry of the bonded wires, the models shown in Figure 2 and Figure 3 were aligned to cross sectional photographs taken of the bonded wires.

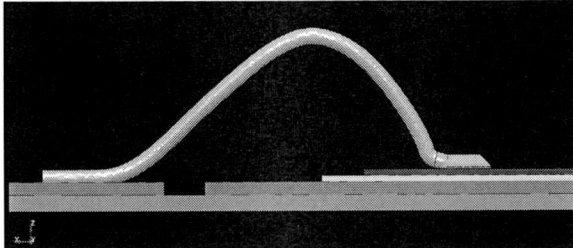

Figure 2: 375 µm diameter bond wire model (parts correspond to those shown in Figure 1), loop as per a cross section photo.

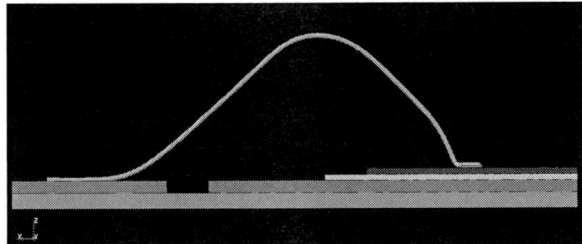

Figure 3: 125 µm diameter bond wire model (parts correspond to those shown in Figure 1), loop as per a cross section photo.

For reducing computational costs, only half of the geometry was simulated (with a no-pass boundary condition) assuming mirror symmetry as shown below in Figure 4.

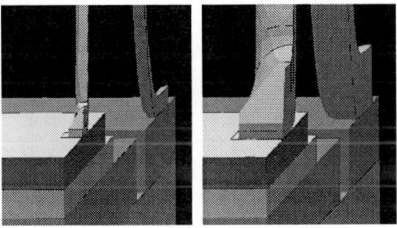

Figure 4: Half models of both wires were simulated for diameters of 375 µm and 125 µm – left and right respectively

Material parameter assignment began with the mechanical properties for the bond wires. Tensile testing performed by Merkel et al. [8] showed the 125 µm wires to have a higher yield point (42 MPa) than the thick wire (31 MPa). The reason given was that the thin wires were hardened by annealing to facilitate bonding. Their tests also provided the Young's modulus and the tangent modulus. Thermal softening results in a reduction of these parameters with increasing temperature. The temperature dependence of Young's modulus was adopted from the work of Köster [11], that of the yield point from the flow-

stress measurements taken by Ball [12], and that of the tangent modulus from Liu et al. [13]. These parameters were then used in the frame of a thermal bi-linear plasticity model.

Parameters for the ceramic substrate were obtained from the manufacturer Curamik GmbH and are listed in Table 1. In order to model the plastic response of the system, coefficients of thermal expansion were assigned to all model parts, see Table 1. As the primary focus of the simulation is the deformation of the aluminium, the other materials were modelled as elastic materials.

Table 1 Material parameters

Material	Density (g/cm³)	Young's Modulus (GPa)	Poisson's ratio	CTE (ppm/ K)	Yield point (MPa)	Tangent Modulus (MPa)
Al, 125 µm	2.7	64	0.35	23	31/42.3	45.7
Al, 375 µm	"	"	"	"	31	"
Copper	8.9	11	0.343	16		
Silicon	2.33	185	0.28	3		
Alumina	4	303	0.21	6.8		

Once the material parameters had been assigned in Ansys Mechanical, meshing of the structure was undertaken. The main area of interest in this work comprised the bottom of the bond foot, bond pad bond foot interface as well as the top of the silicon device. This area was assigned a finer mesh for higher solution precision as shown in Figure 5 for both wires.

Figure 5: Mesh overview for bond feet, (a) thick wire and (b) thin. A finer mesh was used for the area of interest comprising the interface between the bond pad foot and the top of the silicon.

With the mesh created, the model was transferred to LS-PrePost for simulating a thermal cycle from 25 °C to 155 °C and returning to 25 °C as shown in Figure 6. This time dependence was adopted from measurements of the temperature at the bond feet during active power cycles. For the sake of an easier comparison of the simulation results, a homogeneous distribution of the temperature was assumed within the simulated structure.

978-1-4799-4789-8/14 $31.00 © 2014 IEEE

With pre-processing complete, the model was transferred for calculation to the solver LS-Dyna [10]. As the real-life temperature cycle occurs over 30 seconds, strain-rate effects were ignored in the simulation. The simulation cycle could then proceed at a much faster rate.

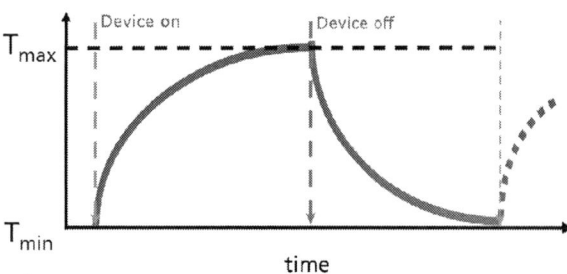

Figure 6: Simulated temperature profile adopted from measurements of active power cycles. Heating occurs rapidly from T_{min} (25 °C) to T_{max} (155 °C) followed by cooling. Cycle period is 30 seconds.

3. Results

The resultant effective plastic strain on the undersides of the bond feet can be seen in Figure 7 and Figure 8 for strains in the range from 0 to 3%. A higher degree of deformation can be seen on the thick wire compared to the thin wires. The temporal evolution of the effective plastic strain over the temperature cycle was then evaluated for a more detailed view of the maximum values. These values were obtained from the lower portions of the bond feet at the back (wire loop end) for the thick wire and the thin wires. The results of this, shown in Figure 9, show the highest plastic strain in the thick wire.

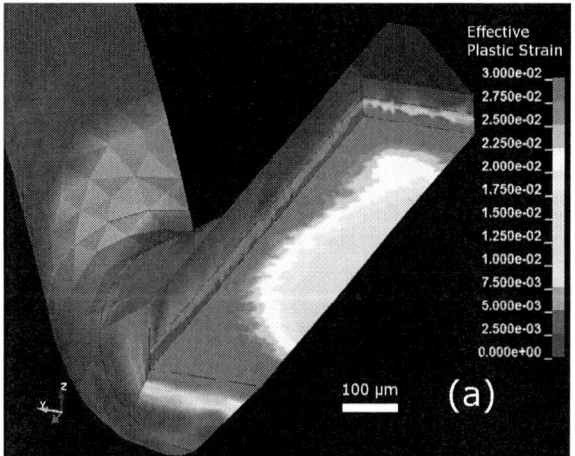

Figure 7: Plastic strain on the underside of the 375 μm wire's bond foot half-structure after a single temperature cycle from 25 °C to 155 °C and returning to 25 °C. Comparing this to the thin wires in Figure 8 a higher degree of plastic strain can be seen.

The maximum deformation is very similar for the two thin wires with the high-yield-strength wire resulting in a slightly higher value. However, viewing the strain distribution on the undersides of the thin-wire bond feet, a slightly higher degree of deformation on the low-yield wire can be seen.

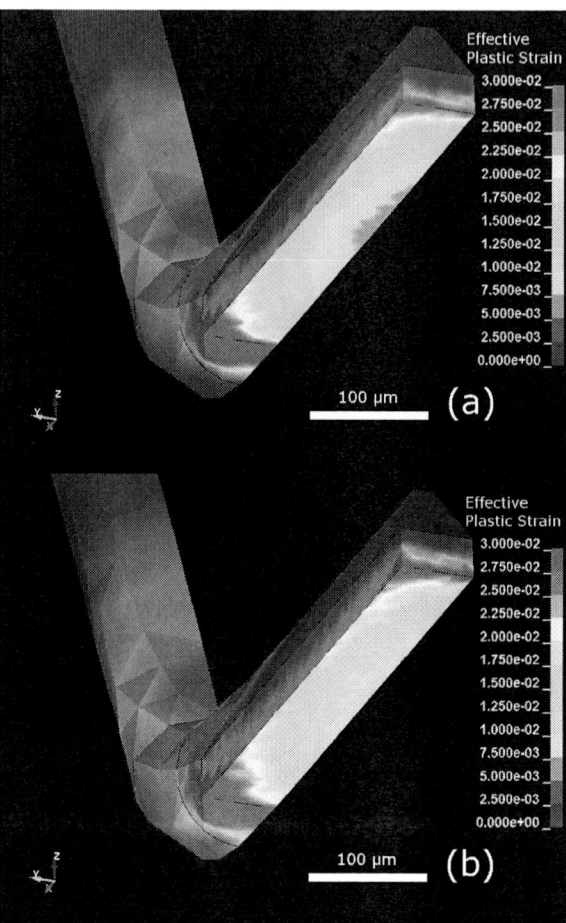

Figure 8: Plastic strain on the undersides of the bond feet half-structures for the 125 μm wires after a single temperature cycle from 25 °C to 155 °C and returning to 25 °C. A slightly higher degree of plastic strain can be seen for the wire with low yield strength (b) compared to that with high yield strength (a).

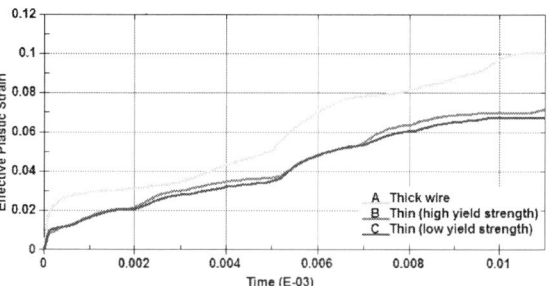

Figure 9: Maximum values of effective plastic strain, bottom 38 μm of the bond feet.

To evaluate the degree of plastic strain in the volume at the rear of the bond feet, comparisons of volume-averaged effective plastic strain were undertaken. A volume was taken from the underside the bond feet, the lowest 8 μm (see the highlighted area in Figure 10), where the maximum of the plastic strain was found. In this volume, the results of the individual finite elements were averaged.

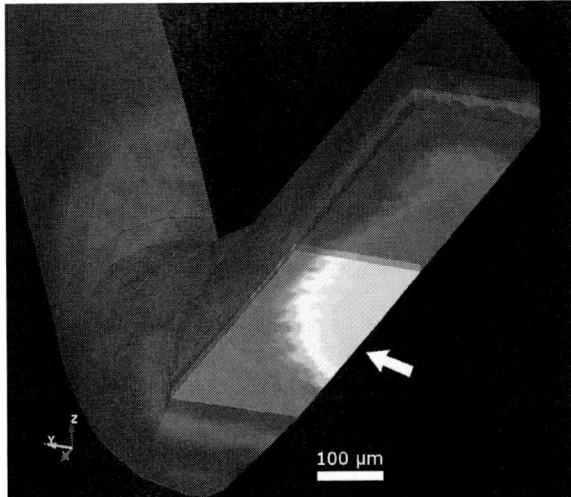

Figure 10: The volume taken for averaging (highlighted area indicated by arrow), half of the bottom 8 μm of the bond feet – where the maximum plastic strain was found.

In comparison to the maximum strain values which were less than 50% higher in the thick wire than in the thin ones, averaged strain values were even found to be twice as large as shown in Figure 11. This emphasizes the larger degree of deformation of the thick wire in general. The thin wires compared to each other reveal the high yield strength wire to have less overall plastic strain. With reliability analysis, this difference may reveal an advantage in reliability performance.

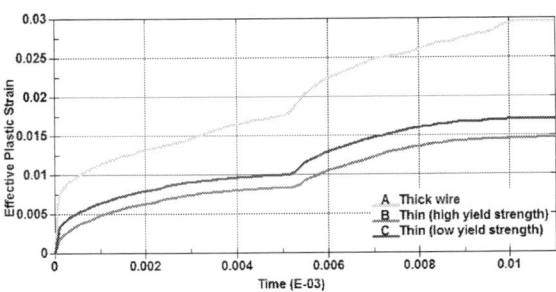

Figure 11: Volume averaged effective plastic strain around the strain maximum at the bond feet.

4. Conclusions

To investigate the role of bond wire size on wire lift-off, the effective plastic strain was estimated through thermo-mechanical simulation. Three-dimensional half models were constructed for the thin and thick bond wires. Plastic deformation in them was modelled via a temperature-dependent bi-linear plasticity model. Both high and low yield strengths were simulated for the thin wires to investigate its effect. Including the direct bonded copper substrate, solder and the silicon device, the effects of a heating and cooling thermal cycle was simulated with LS-Dyna.

Comparison of the maximum effective plastic strain showed higher deformation for the 375 μm wire compared to the 125 μm wires. Averaging showed the degree of plastic deformation in the thick wire to be double that of the thin wires. Additionally, the high yield strength of the thin wires showed a lower degree of deformation which may be beneficial for its reliability performance. Consequently, from the simulations, thin wires are expected to withstand a higher number of cycles than thick wires, in complete agreement with the experimental observation which showed a more than ten times higher lifetime.

References

[1] Lutz, J., Semiconductor power devices: *Physics, characteristics, reliability*, Springer (Heidelberg, New York, 2011), p. 392

[2] Schmidt, R., Scheuermann, U., and Milke E., "Al-clad Cu wire bonds multiply power cycling lifetime of advanced power modules," *PCIM Europe: International Exhibition and Conference for Power Electronics, Intelligent Motion, Renewable Energy and Energy Management,* Nürnberg, Germany, May. 2012, pp. 776–783.

[3] Rudzki, J., Osterwald, F., Becker, M., and Eisele, R., "Novel Cu-bond contacts on sintered metal buffer for power module with extended capabilities," *PCIM Europe: International Exhibition and Conference for Power Electronics, Intelligent Motion, Renewable Energy and Energy Management,* Nürnberg, Germany, May. 2012, pp. 784–791.

[4] Hutzler, A. and Schletz, A, "The easiest way to avoid bond-wire liftoffs," *IMAPS Workshop Wirebonding:,* San Jose, CA, USA, January. 2014.

[5] Goehre, J., Schneider-Ramelow, M., Geißler, U., and Lang, K.-D., "Interface degradation of Al heavy wire bonds on power semiconductors during active power cycling measured by the shear test," *6th International Conference on Integrated Power Electronics Systems: CIPS,* Nürnberg, Germany, March. 2010.

[6] Goehre, J., Geissler, U., Schneider-Ramelow, M., and Lang, K.-D, "Influence of bonding parameters on the reliability of heavy wire bonds on power semiconductors," *7th International Conference on Integrated Power Electronics Systems: CIPS,* Nürnberg, Germany, March. 2012.

[7] Ciappa, M., "Selected failure mechanisms of modern power modules," *Microelectronics Reliability*, Vol. 42, No. 4-5 (2002), pp. 653–667.

[8] Merkle, L., Sonner, M., and Petzold, M., "Lifetime prediction of thick aluminium wire bonds for

mechanical cyclic loads," *Microelectronics Reliability*, Vol. 54, No. 2 (2014), pp. 417–424.

[9]   Ansys Inc, <u>Workbench, version 15.</u>

[10]   Livermore Software Technology Corporation, <u>LS-Dyna, version 971 R7.0.0.</u>

[11]   Köster, W., "Die Temperaturabhängigkeit des Elastizitätsmoduls reiner Metalle," *Zeitschrift für Metallkunde*, Vol. 39, No. 1 (1948), pp. 1–9.

[12]   Ball, C. J., "The flow stress of polycrystalline aluminium," *Philosophical Magazine*, Vol. 2, No. 20 (1957), pp. 1011–1017.

[13]   Liu, D., Tsai, C., and Lzu, S., "Determination of temperature-dependent elasto-plastic properties of thin-film by MD nanoindentation simulations and an inverse GA/FEM computational scheme," *Computers, Materials and Continua*, Vol. 11, No. 2 (2009), pp. 147

Electrical Characteristics Evolution of the
Deep Trench Termination Diode Based on a Finite Elements Simulation Approach

F. Baccar, F. Le Henaff, L. Théolier, S. Azzopardi, E. Woirgard

IMS Laboratory, University of Bordeaux, 351 Cours de la Libération
33405 Talence Cedex – France
e-mail: fedia.baccar@ims-bordeaux.fr

Abstract

The main contribution of this work consists in showing the possibility to use the Cyclotene 4026-46 BCB (BenzoCycloButen) resin in thick layer to realize Deep Trench Termination (DT^2). The development of the DT^2 in power devices strongly depends on its reliability. 2D finite elements simulations were used to determinate the electrical characteristics after mechanical stresses created in the structure. It appears that void created inside the structure does not affect the structure's characteristic; however it changes when a quantity of charge was added at interface Silicon/BCB.

1. Introduction

Recently, deep trench structure has become one of the design options for semiconductor devices due to the advancement in the Micro Electro Mechanical Systems (MEMS) process technology [1]. The DT^2 was presented in 2009 [2] and then, several studies [1, 3] have proposed electrical improvements. Many studies demonstrate the possibility to use BCB passivation layers in Wafer Level Chip Scale Package (WLCSP) process [4] or use it in order to create moisture sensor [5], but no reliability study has been done on power devices.

The first aim of this work is to study the reliability of a 1200 Volts Deep Trench Termination diode. DT^2 structure is one of the candidates for future high voltage power devices edge termination design for reducing the chip area and improves the reverse voltage. Then, in a second step, it will be interesting to avoid long sample preparation for optical analyses and to propose simulations in order to explain ageing phenomena. Indeed, it takes a lot of time to prepare samples for optical observations, due to the care needed to go through coating and polishing. Moreover, the water used for polishing may inflate the BCB volume and defects may appear which will not be the result of ageing. We propose to use finite elements simulation, to justify the electrical variations that can be observed after passive thermal ageing. Several degradations have been implemented in the simulated structure such as creating voids and adding traps to the interface between BCB and silicon.

2. Presentation of the Deep trench Termination structure

The top view schematic of the 1200V structure is presented in Figure 1.

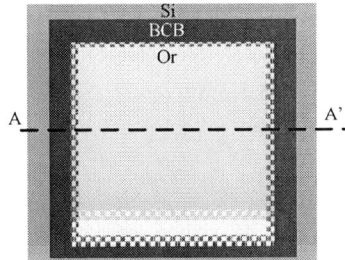

Figure 1: Top view of the schematic structure.

Figure 2 shows the 2D cross section structure of the DT^2 diode. The diode is composed by a 10^{14} cm^{-3} N$^-$ concentrations substrate with a P$^+$ implantation in top layer and a N$^+$ implantation in the bottom layer. A deep trench (105µm x 75µm) is filled by BCB in order to realize the termination. A field plate is needed to draw out the electrostatic potential in the trench.

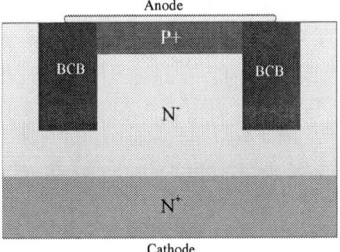

Figure 2: Cross section of the schematic structure.

3. Presentation of the limits of the optical observations

Figure 3 describes electrical reverse characteristics of two DT^2 diodes before and after 50 hours passive of thermal ageing (-40°C, +125°C).

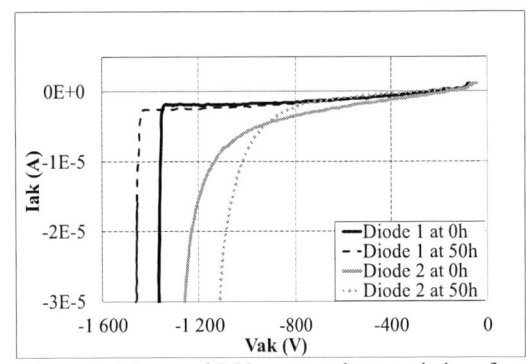

Figure 3: Measured I-V reverse characteristics after 50 hours of thermal ageing.

This figure exhibits the breakdown voltage variations after ageing. An increase or decrease of 100 volts can be observed. In order to explain these variations, samples are prepared for optical observations. Figure 4 presents optical observation of the real component, after 50 hours passive thermal ageing (-40°C, +125°C).

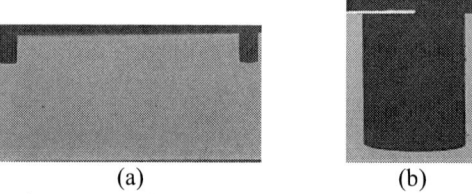

(a)	(b)

Figure 4: Optical observation (a) of the entire device and (b) a focus on the trench.

Optical observation does not highlight that peeling of the resin and silicon can explain these variations, and hence, more analyses are requested. Therefore, a simulation approach is proposed.

Thermal variations induce mechanical stress; so it is possible that a separation of materials along the Si/BCB interface and fixed charges appear.

ANSYS is used to localize the maximum stress in the structure after an increase in temperature and TCAD SENTAURUS is used to visualize the electrical characteristics after degradations created in the structure.

4. ANSYS simulations

First of all, using ANSYS Modeler software, a 3D structure of our chip composed of silicon, BCB and gold layers (see table 1 for material properties) was created. Due to the presence of two symmetries, the final structure is 1/4 of the initial structure, as shown in Figure 5. This simplification was used to reduce the simulation time. The major part of the structure was meshed into rectangle elements except for the area near the curved lines where regular trapezia elements were used. The model loading was an isothermal temperature applied to the entire structure in order to obtain the mechanical stress distribution in the structure.

Table 1: Material properties for FEM analysis.

	Density (Kg.cm⁻³)	CTE (10⁻⁶/°C)	Young's modulus (GPa)	Poisson's ratio
Gold	1930	14.2	79	0.44
BCB	1000	42	2.9	0.34
Silicone	2300	2.8	130	0.28

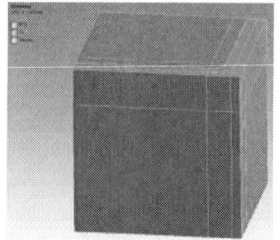

Figure 5: base model of the chip for ANSYS FEM analysis (green = silicon, grey = BCB and blue = gold).

Figure 6 and Figure 7 show the mechanical stress distribution in the BCB and silicon after the thermal load. A logarithmic scale was used to obtain a visual distribution of those mechanical stresses. As expected, the most stressed areas for both materials are located near the BCB/silicon interface on the right angle edge. Indeed, from a geometrical point of view, this geometrical singularity, shown in Figure 7, is the most complex part of the chip structure. Furthermore, the most stressed area of the gold layer is on top of this singularity and the interface BCB/silicon around the trench present high mechanical stress.

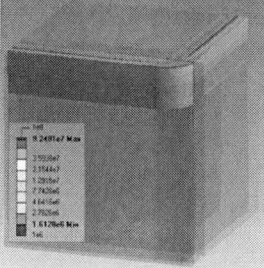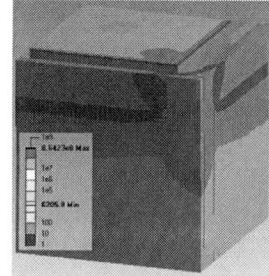

Figure 6: Mechanical stress distribution in the BCB (left) and the silicon (right).

Figure 7: Zoom of the BCB and the silicon parts where the most stressed areas are located.

FEM analysis has been useful to understand the behavior of the chip under thermal environment (125°C and isotherm cases) and areas under important mechanical stress have been located on the interface BCB/silicon).

978-1-4799-4789-8/14 $31.00 © 2014 IEEE

5. TCAD SENTAURUS simulations

The simulated structure is composed by two finite element cells, a base cell and the termination (Figure 8). In order to consider the real die, it is necessary to take into account the different surfaces for each cell. It is necessary to define the *Area Factor* for the base cell regardless of the termination one.

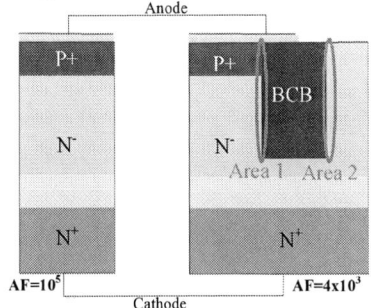

Figure 8: Base cell and termination.

Sentaurus Device is a single-device simulator, and a mixed-mode device and circuit simulator. A single-device command file is defined through the mesh, contacts, physical models, and solve command specifications [6]. For a multi-device simulation, the command file must include specifications of the mesh (*File section*), contacts (*Electrode section*), and physical models (*Physics section*) for each device. A circuit netlist must be defined to connect the devices (

Figure 9), and solve commands.

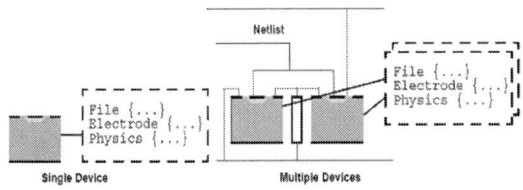

Figure 9: multi device simulation and its circuit netlist.

Sentaurus Device also provides a number of compact models for use in mixed-mode; in this work Spice has been implemented for defined the system of the circuit (Figure 10): D1, DT present respectively the central and elementary cell.

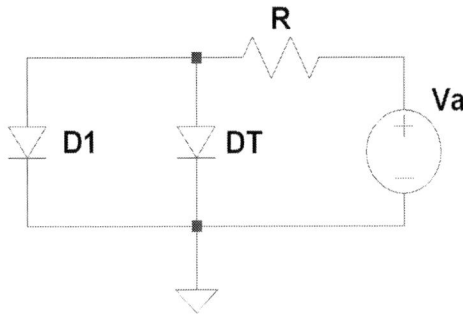

Figure 10: Circuit for SENTAURUS mixed mode.

5. A. Void created in the structure

5. A.1. Breakdown voltage variations

At the interface Si/BCB, more precisely in internal interface of the termination (Area 1) and external interface of the termination (Area 2), a void with different width (Wv) has been created to observe its impact on the structure.

Figure 11 presents the reverse characteristics after created the void in different parts of the device. A same observation can also be obtained in Area 1 and Area 2 (Figure 8).

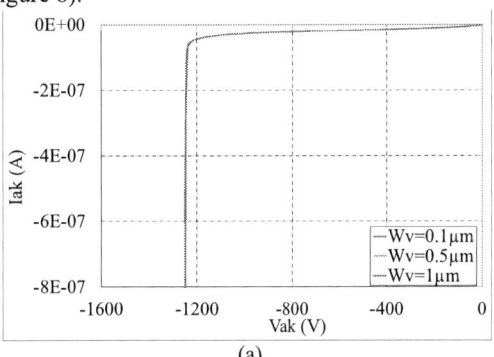

(a)

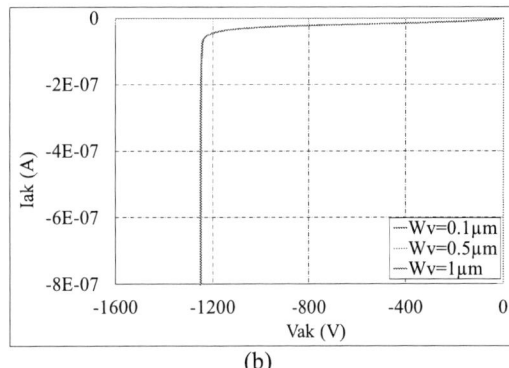

(b)

Figure 11: void width effect on the reverse characteristic in Area 1 (a) and Area 2 (b).

These figures exhibit that a separation at the interface does not affect the breakdown voltage. This is explained by the fact that the electric field is concentrated at the end of the field plate and does not make change in this region.

The insertion of a void in the structure at the interface Si/BCB modifies the distribution of potential lines due to the low permittivity of the void (Figure 12).

978-1-4799-4789-8/14 $31.00 © 2014 IEEE

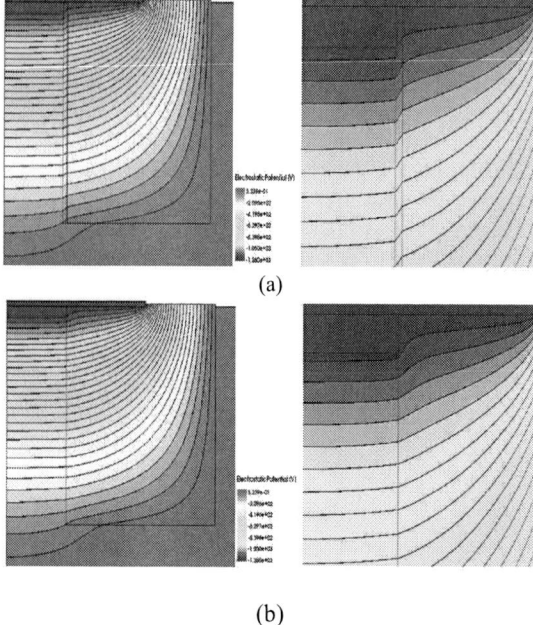

(a)

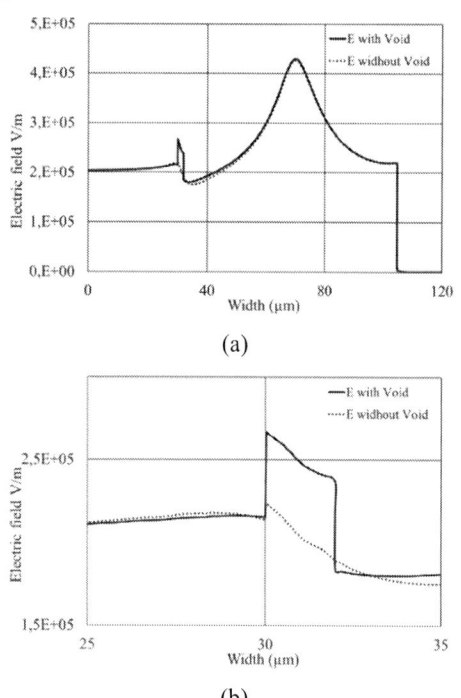

(b)

Figure 12: The distribution of potential lines in the structure and its zoom with (a) and without (b) void.

This change creates a local increase of the electric field, but remains very low and has no impact in the breakdown voltage of the structure.

Figure 13 presents the impact of the void created in the structure on electrical field along a horizontal cut-line.

(a)

(b)

Figure 13: Electrical filed along a horizontal cut-line (a) and its zoom (b).

It is interesting to observe that electrical field increases in the region near the void; however it has no effect in other regions, that confirms a separation at the interface does not affect the breakdown voltage.

5. A.2. Parasitic capacitance variation

When a reverse voltage is applied to a PN junction, the holes in the P-region are attracted to the anode terminal and electrons in the N-region are attracted to the cathode terminal. The resulting region contains almost no carrier, and is called the depletion region. The depletion region similarly acts to the dielectric of a capacitor, hence the creation of the capacity of the junction.

To check also the proper function of the structure, its capacitance is simulated. Figure 14 presents the capacitance of structure for different value of void.

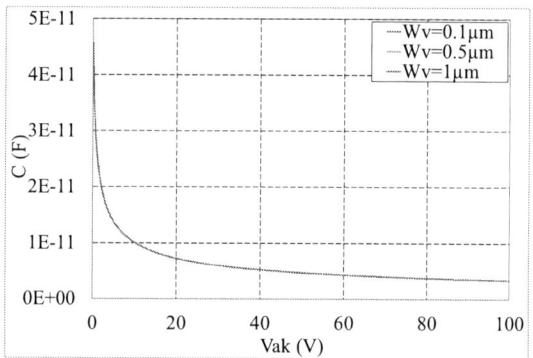

Figure 14: Effect of the void on the parasitic capacity (Area1).

Figure 14 illustrates that the void created at the interface has no effect on the parasitic capacity. All components have parasitic capacitances as illustrated in Figure 15.

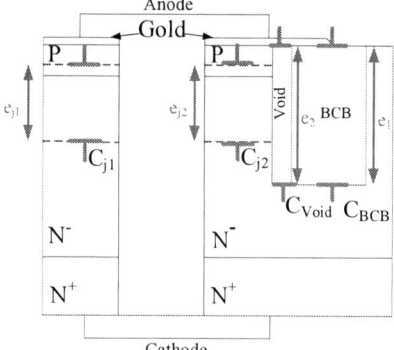

Figure 15: Parasitic capacitances of the structure.

Therefore, there are three capacitances: the junction capacitance associated to a depletion region, capacitance associated to the void, and to the BCB. The total capacitance is presented in Figure 16:

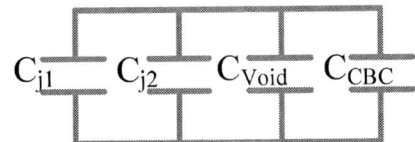

Figure 16: Total parasitic capacitance of the structure.

where C_{j1} and C_{j2} are the junction capacitances, C_{BCB} is the termination capacitance and the C_{Void} is the void capacitance.

The analytical expression of capacitance is given by equation (1)

$$C = \frac{\varepsilon_0 \varepsilon_r A}{e} \qquad (1)$$

Where ε_0 is the vacuum permittivity $(\varepsilon_0 = 8.854 \times 10^{-12} Fm^{-1})$, ε_r is the relative permittivity of material, e is the distance between the capacitor's plates, and A is the area of the capacitor's plate (Figure 17).

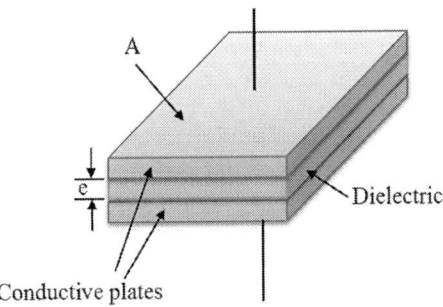

Figure 17: Capacity model.

The different values of capacitance at -1000V have been calculated in order to decrease the junction capacitances:

BCB Capacitance: $C_{BCB} = \dfrac{\varepsilon_0 \varepsilon_{BCB} A_1}{e_1}$

- ε_{BCB}=3.9
- A_1 =3 $10^5 \mu m^2$ } C_{BCB}=98nF
- e_1 =105μm

Void capacitance $C_{Void} = \dfrac{\varepsilon_0 \varepsilon_{Void} A_2}{e_2}$

- ε_{Void}=1
- A_2 =4 $10^2 \mu m^2$ } C_{Void}=33pF
- e_2 =105μm

Junction Capacitance D1 $C_{j1} = \dfrac{\varepsilon_0 \varepsilon_{Si} A_{j1}}{e_{j1}}$

- ε_{j1} =11.7
- A_{j1} =$10^6 \mu m^2$ } C_{j1}=1030nF
- e_{j1} =100μm

Junction capacitance DT $C_{j2} = \dfrac{\varepsilon_0 \varepsilon_{Si} A_{j2}}{e_{j2}}$

- ε_{j2} =11.7
- A_{j2} =12 $10^4 \mu m^2$ } C_{j2}=124nF
- e_{j2} =100μm

It can be noticed that: $C_{Void} \ll C_{BCB} \ll C_{j2} \ll C_{j1}$. This result confirms that a vacuum created in the structure has no effect on the electrical characteristic of diode.

5. B. Charge interface insertion

5. B.1. Breakdown voltage variations

Another type of degradation of the simulated structure is the insertion of charge in the interface Si/BCB. Figure 18 illustrates the I-V reverse characteristics with Si/BCB interface charge in Area 1 and Area 2.

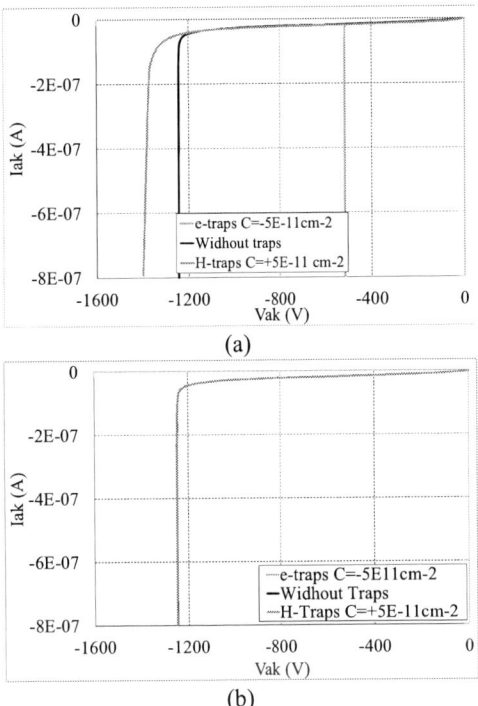

Figure 18: Traps Effect on the reverse characteristic in Area 1 (a) and Area 2 (b).

The breakdown voltage is modified by the insertion of charge interface in Area 1, but it is not modified in Area 2. Holes insertion creates a virtual P$^+$ vertical doping area and increases the breakdown voltage; while electrons insertion creates a virtual N$^+$ vertical doping area decreases the breakdown voltage. Holes and electrons in Area 2 have not effect in electrical characterization, indeed there is no depletion in this area but charges could have an influence on dynamic behavior.

5.B.2. Parasitic capacitance variation

Figure 19 points out the traps effect in the parasitic anode-cathode capacitance for the low value of reverse voltage.

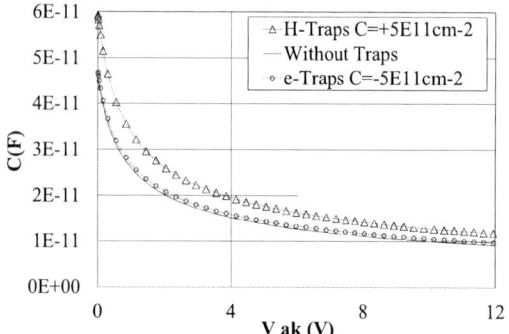

Figure 19: Effect of the traps on the parasitic capacitance (Area1) for the low value of reverse voltage.

Indeed holes insertion creates a virtual P^+ vertical doping area, so a new P^-N horizontal junction as expressed in Figure 20, and in this case the capacitance increases very fast, until the depletion region of the new junction is of maximum width.

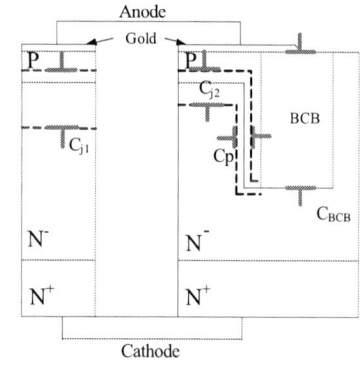

(a)

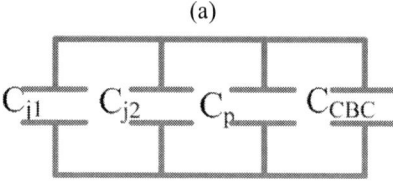

(b)

Figure 20: structure capacitance (a) and its equivalent circuit (b).

If it can be imagined that the diode capacitance can be likened to a parallel plate capacitor, then as the plate spacing (i.e. the depletion region width) increases, the capacitance should decrease. Thus, the junction has a variable capacitance that depends on the evolution of the voltage which is applied on it. It is shown that the holes insertion increases the breakdown voltage therefore increases the depletion region (Figure 22) which causes the decrease of its capacitance (Figure 21); while electrons insertion decreases the breakdown voltage, therefore decreases the depletion region which causes the increase of its capacitance.

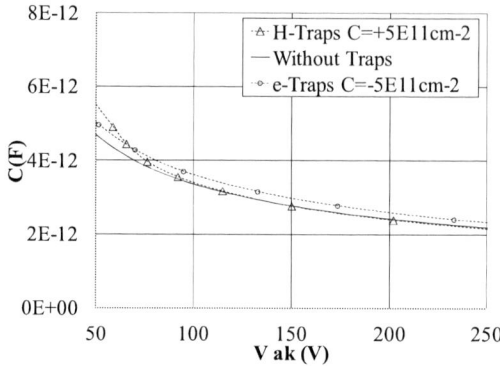

Figure 21: Effect of the traps on the parasitic capacity (Area1) for the higher value of BV.

It is observed that for negative charges (e-), the depletion region is smaller than for initial structure. This explains the capacitance rise for a same voltage. Furthermore, it seems the traps can be observed by a capacitance variation.

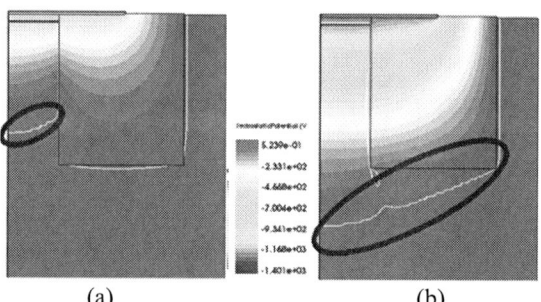

(a)   (b)

Figure 22: distribution of potential lines in the structure with e-traps (a) and without traps (b).

6. Conclusion

This paper presents the possibility to use the BCB resin in thick layer to realize deep trench termination.

It can be observed that reverse characteristics changing after 50 hours of power cycling. Optical microscopy could not permit to observed structural modification in order to explain the electrical variations. Thus, TCAD software is used to correlate structural modification with electrical characteristics in order to confirm the presence of the structural modification by combined characteristics.

Delamination on the Si/BCB interface has no impact on the proper function of the interface can not be observed by electrical variation. Moreover, the increasing or the decreasing of the breakdown voltage can be explained by the appearance of a quantity of charge at this interface.

Acknowledgments

This work was supported by the French National Research Agency (ANR) through SUPERSWITCH program (ANR 2011 BS09 033).

References

1. Kamibaba, R., Takahama, K., Omura, I., "Design of trench termination for High Voltage Device", *ISPSD 2010*, pp. 107-110, 2010.

2. Théolier, L., Mahfoz-kotb, H., Isoird, K., Morancho, F., *"A new junction technique: the Deep Trench Termination (DT²)"*, *ISPSD 2009*, pp. 176-179, 2009.

3. Seto, K., Kamibaba, R., Tsukuda, M., Omura, I., "Universal trench edge termination design", *ISPSD 2012*, pp. 161 - 164, 2012.

4. Lee, K.O., Yu, J., Kim, J.Y., Park, I.S., "Thermo-Mechanical Reliability of the Benzocyclebuten (BCB) film in a WLCSP process", *EMAP 2001*, pp. 84-87, 2001.

5. Tetelin A., Achen, A., Pouget, V., Pellet, C., Töpper, M., Lachaud, J.L., "Water solubility and diffusivity in BCB resins used in microelectronics packaging and sensor applications", *IMTC 2005*, pp. 792-796, 2005.

6. SENTAURUS TCAD Software, V12.06.

Molecular Dynamics Simulation of Adhesion Performance and Conformation Transition of SAM-modified Cu/Epoxy Interface under Electric Field

Stephen C.T. Kwok, Matthew M.F. Yuen
Department of Mechanical and Aerospace Engineering
The Hong Kong University of Science and Technology
Clear Water Bay, Kowloon, Hong Kong
kctaa@ust.hk, meymf@ust.hk
Phone: (852) 2358 8814 Fax: (852) 2358 8357

Abstract

This work aims to investigate the conformational transition of self-assembled monolayer (SAM) on copper (Cu) under influence of different electrical field strength through molecular dynamics (MD) simulation. The SAM was transited from the lying down to the standing-up conformation when electric field changed from positive to negative strength. With different SAM conformation on Cu, the performance in adhesion between Cu/Epoxy was also investigated. Epoxy model was incorporated for predicting the adhesive strength between SAM-modified Cu/Epoxy interface by using MD simulation. The results demonstrated SAM with standing up conformation and higher relative surface coverage gave rise to higher adhesive strength between Cu/Epoxy interface. This result provided further physical understanding for selecting the electric field during the SAM assembly process which aimed for the interfacial adhesion promotion application.

1. Introduction

Due to its high thermal and electrical conductivity, copper (Cu) still being the most common leadframe material used in electronics industry. Plastic packages with Cu as leadframe material typically use epoxy molding compounds (EMCs) as encapsulating materials since the latter have excellent insulating and mechanical properties. However, because of the weak Cu/epoxy interfacial adhesion, delamination is prone to happen under high moisture and high temperature environments. In order to tackle this problem, our previous report demonstrated an electrochemical assembly of thiol-based self-assembled monolayer (SAM) on Cu could improve Cu/epoxy interfacial adhesion by 20-fold with treatment time reduction by a factor of 32 [1] compared with passive immersion treatment method [2].

Apart from experimental work of studying interfacial adhesion between Cu/epoxy interface, Wong and coworkers [3] reported the use of MD model to study the Cu/epoxy interfacial separation process. Their work demonstrated the possibility of studying interfacial separation involving chemical bonding between the interfaces. He and coworkers [4] reported the use of MD model for predicting SAM-modified Cu/epoxy interfacial adhesive strength and the results provided a reasonable correlation with experimental work.

In seeing the effect of electrical field could reduce the preparation time of SAM on Cu, further work on conformational transition of SAM induced by electrical field and its effects towards adhesion between Cu/Epoxy interface is conducted by molecular dynamics (MD) simulations in this study.

The first part of this paper will focus on investigating the conformational transition of the SAM with various surface coverage values under the influence of electrical field. Next, molecular model of epoxy crosslink structure will be incorporated into the simulation. Epoxy model will be combined with the SAM modified Cu model with various surface coverage and conformation. The interfacial adhesion strength will be simulated in terms of interfacial energy for separating the SAM/Epoxy interface. Our results showed that SAM having lower surface coverage (3.06×10^{-10} mol/cm^2) will undergo more significant conformation change when compared to SAM having higher surface coverage value (10.2×10^{-10} mol/cm^2). SAM with higher surface coverage value will have a standing-up conformation, contributing to a much higher interfacial adhesive strength between Cu/Epoxy interface. This study provides a physical understanding of SAM under influence of electrical field. The results demonstrated that the conformation of SAM on Cu substrate can be tuned by applying different electrical field. With a proper conformation (standing-up conformation as in our case), the adhesive strength can be optimized through the application of proper electrical field strength.

2. MD Simulations

Molecular dynamics (MD) simulation was employed for studying the SAM conformational change under various electric field strengths and its adhesive strength between SAM-modified Cu/Epoxy interface. MD simulation was carried out by using commercial software package Materials Studio. (Accelrys Software). A Condensed-phase Optimized Molecular Potential for Atomistic Simulation Study (COMPASS) has been incorporated for the force field computation.

In the MD simulation, molecular model was built according to the chemical structure with Cu and 4-Hydroxythiophenol (4-HTP) layers. A unit cell was built with 2.55nm x 2.55nm which is periodic in x and y direction. Cu substrate was built with crystalline structure which was cleaved along the (0 0 1) plane. The 4-HTP layer was assigned on top of Cu layer with different numbers of 4-HTP molecules boned to Cu at a distance of 2.37Å along the vertical (z) direction. **Figure 1 (a)** shows the unit cell for the simulation. MD simulation was carried out to find out the optimized surface density of 4-HTP molecules on Cu substrate. It was carried out at 298k with the use of constant number of particles, constant volume and constant temperature (NVT) ensemble. The time interval for each iteration was 1fs with total 10ps computational time was simulated. The interfacial energy of Cu/4-HTP interfacial can be calculated by the energy difference according to equation (1):

$$E_{interfacial} = E_{total} - (E_{4-HTP} + E_{Cu}) \quad (1)$$

where E_{total} is the total energy of the whole system, E_{4-HTP} is the energy of 4-HTP SAM layer without Cu and E_{Cu} is the energy of Cu substrate without 4-HTP.

The interfacial energy represents a thermodynamic state. 4-HTP density on Cu having the lowest interfacial energy represents the most stable state that could be obtained at the simulated condition.

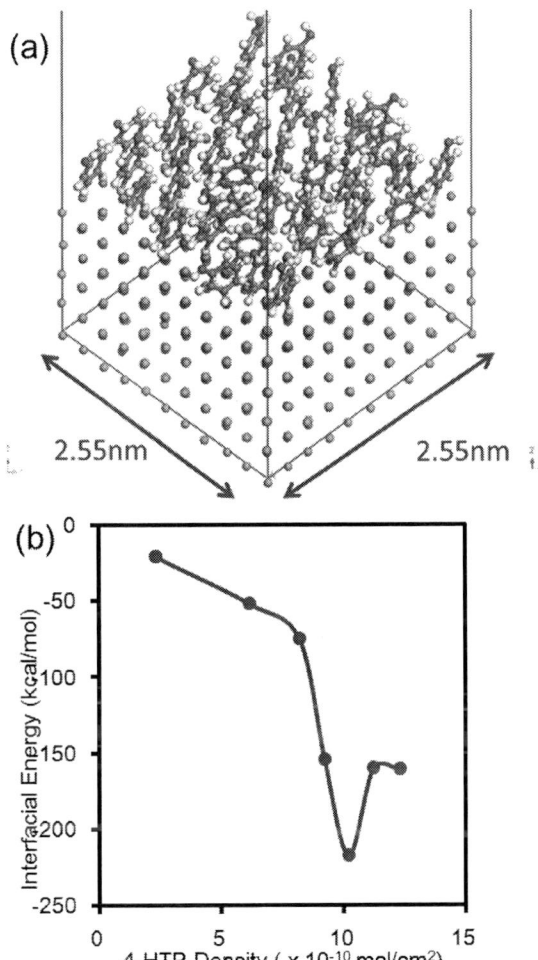

Figure 1. (a) Snapshot of 4-HTP configuration on Cu substrate after 10ps simulation time. Forty numbers of 4-HTP molecules in unit cell corresponds to 10.2 x 10^{-10} mol/cm^2 which results in lowest interfacial energy of -217.0 kcal/mol. (b) Interfacial Energy of Cu/4-HTP interface with different 4-HTP density.

For studying the electric field effect on 4-HTP layer, point charges were calculated and assigned to the 4-HTP molecules as well as its charged dipole hydroxyl group. An external electric field **E** was applied along the z axis. Under the electric field **E**, the force experienced by the atom i of charge q_i of 4-HTP molecules will be given by $\mathbf{F}=\mathbf{E}q_i$. The simulation time step was set to be 10ps. After that the trajectory file was output and the conformation of 4-HTP layer in terms of average oxygen height in hydroxyl group of 4-HTP molecules measured from Cu surface was calculated.

Epoxy layer was also modeled according to the molecular structure of Diglycidyl Ether of Bisphenol A (DGEBA) with 4, 4-diaminodiphenyl methane (MDA) as the curing agent. A similar crosslinking structure was simulated according to the procedures reported by Fan and coworkers [5]. After that the epoxy layer was combined with 4-HTP modified Cu model with various surface coverage and conformation from the electric field simulation. Then the simulation was carried out at 353k with the use of NVT ensemble. The time interval for each iteration was 1fs with total 10ps computational time was simulated. **Figure 2** demonstrates the simulation procedures in this study. In the simulation, we assume the separation of interface happened at the epoxy/4-HTP interface. As a result, the energy for separation can be calculated by the energy difference ΔE according to equation (2):

$$\Delta E = E_{total} - (E_{Cu+SAM} + E_{epoxy}) \quad (2)$$

where E_{total} is the total energy of the whole system, E_{Cu+SAM} is the energy of 4-HTP SAM and Cu layer without epoxy and E_{epoxy} is the energy of epoxy layer without 4-HTP and Cu.

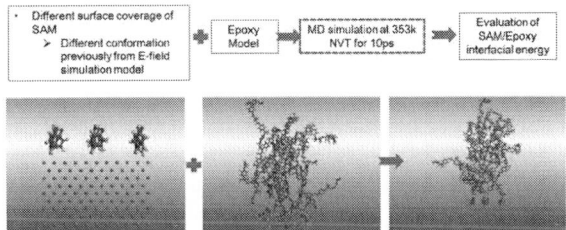

Figure 2. Simulation procedures of SAM model with different confromation and epoxy model for obtaining SAM/Epoxy interfacial energy.

3. Results and Discussion

3.1 Optimum 4-HTP coverage on Cu surface

Different numbers of 4-HTP molecules were put into the unit cell (2.55nm x 2.55nm) as described in *section 2*. The 4-HTP surface density ranged from 2.3 x 10^{-10} to 1.23 x 10^{-9} mol/cm^2, which was corresponding to 9 to 48 numbers of 4-HTP molecules in the unit cell. A series of MD simulations with condition described in *section 2* were carried out to find out the most stable density for a given configuration. Interfacial energies of 4-HTP/Cu interface were evaluated by equation (1) and the resulting

4-HTP density having lowest interfacial energy represent the most stable thermodynamic state. In **Figure 1(b)** the interfacial energies with various 4-HTP densities on Cu are shown. The interfacial energy dropped with increasing 4-HTP density, and the lowest energy state of 4-HTP density on Cu corresponds to 10.2×10^{-10} mol/cm^2. This configuration represented 40 numbers of 4-HTP molecules in the molecular model. With further increasing 4-HTP density, interfacial energy increased indicating they were thermodynamically less stable. We take this value to be 100% relative coverage of 4-HTP assembled on Cu.

3.2 Conformational transition of 4-HTP under different electric field strength

Different electric field strengths were applied to the unit cell containing various surface coverage of 4-HTP on Cu. The electric field ranged from -4V/Å to +4V/Å. After simulation time of 10ps, the trajectory files were output and we evaluated the average oxygen height of hydroxyl group in 4-HTP molecules. **Figure 3** shows the conformation transition of 4-HTP under different electric field strength with 30%-100% 4-HTP coverage. From the simulation results, it showed that 4-HTP having lower surface coverage (3.06×10^{-10} mol/cm^2) will undergo more significant conformation change when compared to higher surface coverage value (10.2×10^{-10} mol/cm^2). At -4V/Å, 4-HTP molecules showing a standing-up conformation while at -4V/Å 4-HTP molecules transited to a lying down conformation. However, the conformational change was not obvious at higher 4-HTP surface coverage. To have a more quantitative discussion, we calculated the average oxygen height of hydroxyl group in 4-HTP molecules measured from the surface of Cu layer. **Figure 4** shows the statistical measurement with three different 4-HTP surface coverage values. The average oxygen height of 30% coverage changed from 7.60Å at -4V/Å to 4.05Å at +4V/Å. In contrast, the average oxygen height of 100% coverage only has 0.03Å difference when the field changed from -4V/Å to +4V/Å. In the 30% surface coverage case, the negative field pulls the 4-HTP molecules in a standing-conformation while the positive field pushes them towards the copper surface.

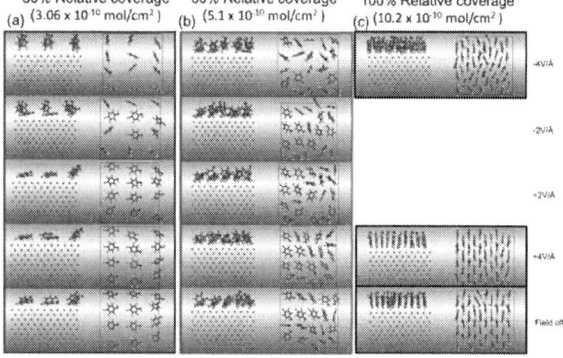

Figure 3. Conformation transition of 4-HTP on Cu under various electric field strength with relative coverage of (a) 3.06×10^{-10} (b) 5.1×10^{-10} and (c) 10.2×10^{-10} mol/cm^2.

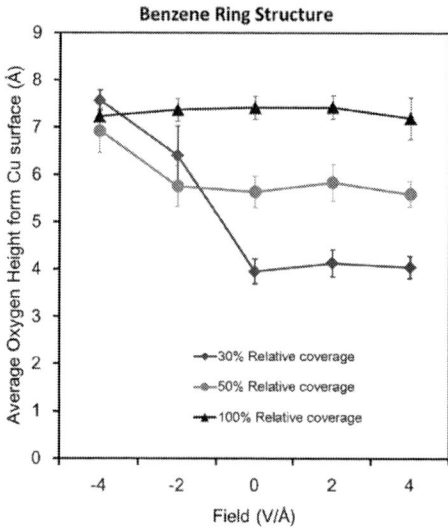

Figure 4. Average oxygen height in hydroxyl group of 4-HTP molecules measured from Cu surface after MD simulation under different electric field strength.

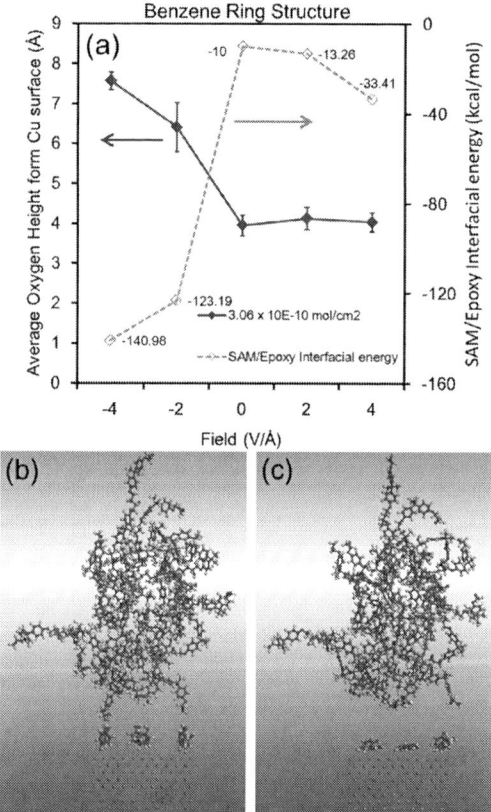

Figure 5. (a) Overlap plot of average oxygen height in 4-HTP molecule from Cu surface under different electric field strength and corresponding SAM/epoxy interfacial energy with 3.06×10^{-10} mol/cm^2 relative coverage. (b) Snapshot of SAM/epoxy model with 4-HTP conformation previously simulated at -4V/Å. (c) Snapshot of SAM/epoxy model with 4-HTP conformation previously simulated at 0V/Å.

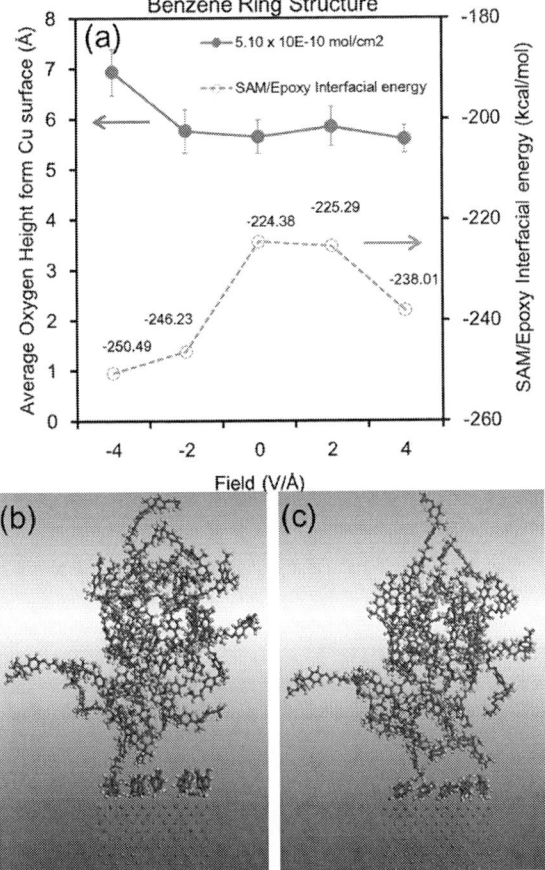

Figure 6. (a) Overlap plot of average oxygen height in 4-HTP molecule from Cu surface under different electric field strength and corresponding SAM/epoxy interfacial energy with 5.10×10^{-10} mol/cm^2 relative coverage. (b) Snapshot of SAM/epoxy model with 4-HTP conformation previously simulated at -4V/Å. (c) Snapshot of SAM/epoxy model with 4-HTP conformation previously simulated at 0V/Å.

3.3 Effect of adhesion performance of SAM modified Cu/Epoxy interface with different 4-HTP conformation

Epoxy model was further incorporated to previously simulated 4-HTP/Cu model with different 4-HTP conformation and surface coverage. The adhesive strength was simulated according to the procedures described in *section 2*. The simulated results are presented in **Figure 5-7** with 30%, 50% and 100% relative surface coverage respectively. From 30% to 100% surface coverage, the lowest SAM/Epoxy interfacial energy was changed from -140.98kcal/mol to -1445.19kcal/mol. This result demonstrated that with higher SAM surface coverage, the higher adhesive strength between Cu/Epoxy interface. This simulated result showed a similar trend as our previously published experimental work [1]. Both experimental and simulation work demonstrates this distinct adhesive strength improvement. Regarding the SAM conformational effect, we focused our discussion in

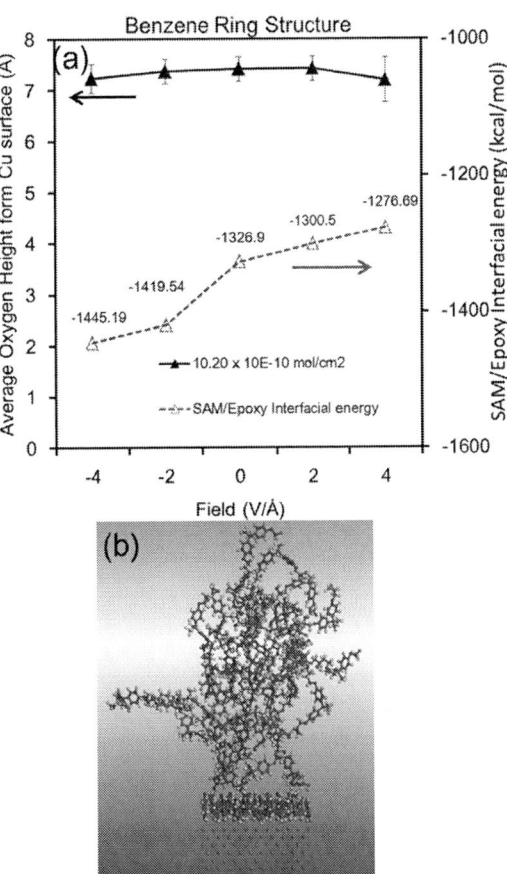

Figure 7. (a) Overlap plot of average oxygen height in 4-HTP molecule from Cu surface under different electric field strength and corresponding SAM/epoxy interfacial energy with 10.20×10^{-10} mol/cm^2 relative coverage. (b) Snapshot of SAM/epoxy model with 4-HTP conformation previously simulated at -4V/Å.

30% surface coverage case as the densely packed SAM did not undergo obvious conformational change. As in **Figure 5 (a)**, the SAM/Epoxy interfacial energy changed from -140.98kcal/mol with oxygen height 7.60Å to -33.41kcal/mol with oxygen height 4.00Å. There was a four-fold difference in interfacial energy under the same surface coverage value. This simulated result showed that in order to achieve an effective adhesion promotion effect, a standing-up conformation of SAM molecules is preferred. The simulated results were reasonable since for an epoxy ring to be covalently bonded to 4-HTP layer, the hydroxyl group from 4-HTP molecules were required to react with epoxy through reaction (3):

$$ROH + R' \triangle\!\!\!O \longrightarrow R' \overset{OH}{\underset{OR}{|}} \quad (3)$$

With a standing-up conformation (higher average oxygen height from Cu surface) of 4-HTP on Cu, more 4-HTP molecules on Cu can expose its end hydroxyl (OH) group which can then react with epoxy layer to form a covalent bond. In turn, this will result in a distinct adhesive strength improvement as a function of average oxygen height from Cu surface.

The interaction between the epoxy layer and the 4-HTP layer could also be visusalized in **Figure 5 (b) and (c)**. In **Figure 5 (b)**, the 4-HTP layer with a standing-up conformation (average oxygen height 7.60Å) showed a much closer interaction than the 4-HTP layer with a lying down conformation (average oxygen height 4.00Å) as in **Figure 5 (c)**. With increasing the surface coverage of the 4-HTP on Cu, the conformational transition of the 4-HTP under electric field became less obvious and the average oxygen height became higher. With a densely packed 4-HTP layer and a standing-up conformation, the simulated results showed a much closer interaction between the epoxy layer and the SAM layer as in **Figure 6 (b)**, **(c)** and in **Figure 7 (b)**.

4. Conclusions

In this work, the conformational transition of self-assembled monolayer (SAM) under influence of electric field was simulated by molecular dynamics (MD). In low surface coverage case, 4-Hydroxythiophenol (4-HTP) layer was transited from the lying down to the standing up conformation when electric field changed from positive to negative strength. However, the densely packed 4-HTP layer did not undergo obvious conformational change under the applied electric field. Epoxy model was incorporated into the simulation for evaluating the adhesion performance of SAM-modified Cu/Epoxy interface. Our results demonstrated SAM with a standing-up conformation and higher relative surface coverage gave rise to higher adhesive strength between Cu/Epoxy interface. To achieve an effective adhesion promotion effect, a standing-up conformation of SAM molecules is preferred especially in the low surface coverage case. The

current study provides physical understanding of tuning the adhesion performance of SAM-modified Cu/Epoxy interface by applying external electric field.

Acknowledgments

SCTK gratefully acknowledges HKUST project no. FSGRF12EG58 for financial support.

References

1. S.C.T. Kwok, F. Ciucci, M.M.F. Yuen, "Electrochemical Assembly of Thiol-based Monolayer on Copper for Epoxy-Cu Adhesion Improvement", *Electrochimica Acta*, 121 (2014) 57– 63.

2. C. Wong, S. Leung, H. Fan, M. Yuen, "Synergistic Toughening of Epoxy–Copper Interface Using a Thiol-Based Coupling Layer", *Journal of Adhesion Science and Technology* 25 (2011) 2081-2099.

3. C.K.Y. Wong, S.Y.Y. Leung, R.H. Poelma, K.M.B. Jansen, C.C.A. Yuan, W.D. van Driel and G. Zhang, 'Establishment of the coarse grained parameters for epoxy-copper interfacial separation', *Journal of Applied Physics* 111, (2012) 094906.

4. P. He, H. Fan and M.M.F. Yuen, 'Investigation of Benzenethiol (BT) Materials as Adhesion Promoter for Cu/Epoxy Interface Using Molecular Dynamic Simulation', *Proc. 12th Int. Conf. on Thermal, Mechanical and Multiphysics Simulation and Experiments in Microelectronics and Microsystems,* Lins, Austria, April 2011.

5. H. Fan and M.M.F. Yuen, 'Material properties of the cross-linked epoxy resin compound predicted by molecular dynamics simulation', *Polymer*, 48 (2007) 2174-2178.

Mutiphysics Study of RF/Microwave Planar Devices: Effect of the Input Signal Power

Miguel A. Sánchez-Soriano*, Michael Edwards[†], Yves Quéré*,
Dag Andersson[†], Stephane Cadiou* and Cédric Quendo*

* Lab-STICC, Université de Bretagne Occidentale, 6 Avenue Le Gorgeu, 29238 Brest, France
[†] Swerea IVF, Argongatan 30, SE-43153 Mölndal, Sweden
Email: m.sanchez.soriano@ieee.org, yves.quere@univ-brest.fr

Abstract

In this paper, the effects of the input signal power on microwave planar devices are studied in detail. In this context, a complete multiphysics study is performed, involving the electro-thermo-mechanical coupling in microwave components. For this study, a multiphysics simulator is used. As shown, for moderate input powers, the device transfer function can be altered, mainly in terms of an increase of losses and a frequency shift. Additionally, hot spots are to be appeared, whose location is related to the electromagnetic field distribution of the passive device under test. Guidelines are also provided to estimate the average power handling capability (APHC) of planar components. As an example, the multiphysics analysis of a microstrip coupled-line filter centered at 42 GHz is tackled taken into account different thermal and mechanical boundary conditions.

1. Introduction

RF/microwave planar devices implemented in microstrip, stripline, coplanar or the recent substrate integrated waveguide (SIW) technology are very important in the current communication systems, mainly due to their easy integration with other components/subsystems and their low cost (PCB techniques are used for their fabrication), weight and size [1]–[3]. Losses in microwave planar circuits can be important, and especially as the working frequency increases. In addition, there is currently an increasing tendency towards miniaturization that directly affects the thermal performance of circuits. The electrical properties of the dielectric (dielectric constant and loss tangent) can be strongly influenced by temperature. This causes a change in the RF behavior of the circuit, resulting in more losses in the circuit and a frequency shift in the response. For high demanding specifications, the frequency shift can be intolerable and require a redesign of the device.

Mechanical aspects such as thermal stress and expansion both in the conductor and dielectric are also important and need to be considered. Thermal expansion can have an important impact in the transfer function and can produce an additional frequency shift, which can be either in the same direction or opposite to the shift produced by the change of dielectric constant. The thermal stress can also limit the maximum input power that the circuit can withstand. All these factors make the electro-thermo-mechanical study in microwave planar devices a matter of great interest.

In this paper, the effect of the input signal power on the performance of RF/microwave planar devices is studied in detail. A complete electro-thermo-mechanical analysis (using Ansys Multiphysics) is undertaken on some conventional microwave planar devices, by way of illustration. Design guidelines are provided for the redesign of the devices in order to meet the specifications of a hypothetical communication system. The effect of the boundary and external conditions, such as convection, heat sink attachment, environment temperature are also discussed. Additionally, from this study, rules to estimate the maximum input power that a planar circuit can withstand are reported.

2. Electro-Thermal Analysis

Microwave planar devices basically have three loss mechanisms: ohmic, dielectric and radiation losses. The two former loss mechanisms are linearly proportional to the input power and generate heat in the circuit. Therefore, they can be defined as the internal heat sources in the structure. The latter does not generate any heat in the circuit and is thus neglected in this electro-thermal study. At a determined frequency, the planar circuit presents a well-defined electromagnetic distribution along the structure. The heat sources (metallic and dielectric losses) follow this electromagnetic pattern leading to a non-uniform heat flow generation. As a consequence, hot and cold areas appear in the circuit and a thermal profile is defined, where critical zones can be identified.

The first step is therefore to solve the Maxwell equations in the circuit at the frequency of interest in order to obtain the electromagnetic distribution. This first task can be done by means of a full wave simulator for complex structures or analytically for simpler ones by using transmission line theory, for example. The coupled electro-thermal governing equations both for the metal and dielectric substrate are

$$\rho_{d,i} C_{p,i} \left(\frac{\partial T}{\partial t} \right) = \nabla \left(k_i \nabla T \right) + q_i(\overline{x}, f, T) \qquad (1)$$

where p_d and C_p are the material density and specific heat, respectively, k is the thermal conductivity, T is the temperature and q is the heat source, which is a function of the position $\bar{x} = (x, y, z)$ in the circuit, frequency and temperature. p_d, C_p and k can be also temperature-dependent. The symbol i indicates where the equation is evaluated: metal or dielectric. The Maxwell equations solution is given implicitly in (1) through q_i. The electromagnetic fields reach the steady-state much sooner than the thermal field does, so that q_i is assumed to be time-independent in (1). For the case when an average or continuous wave (CW) is applied to the circuit, after a transient time the steady-state is also reached for the thermal field, thus, $\partial T / \partial t = 0$ in (1), leading to the Fourier's law of thermal conduction.

The heat contribution due to conductor losses is derived from the surface loss density as

$$q_c = \frac{R_s}{2} \int_S |\overrightarrow{H_t}| dS \qquad (2)$$

where $\overrightarrow{H_t}$ is the tangent magnetic field on the conductor, S denotes the surface of the conductor and R_s is the conductor surface resistance defined as

$$R_s = \sqrt{\frac{\pi f \mu_0}{\sigma}}$$

where μ_0 is the permeability of free space and σ is the metal's conductivity. Regarding the heat contribution caused by dielectric losses, it is derived from the volume loss density as

$$q_d = \frac{1}{2} \omega \epsilon_0 \epsilon_r'' \int_V |\overrightarrow{E}| dV \qquad (3)$$

where ω is the angular frequency, ϵ_0 is the permittivity of free space, ϵ_r'' is the imaginary part of the dielectric relative permittivity, $\overrightarrow{E}$ is the electric field and V denotes the volume of the dielectric. ϵ_r'' can be written as $\epsilon_r'' = \epsilon_r' \tan \delta$, where $\tan \delta$ is the dielectric loss tangent and ϵ_r' is the real part of the relative permittivity. The electro-thermal problem is completely defined by the contact/continuity conditions between conductor and dielectric substrate (this basically means continuity both for the heat flux and temperature at the interfaces) and with the Newton's law of cooling on the outside surfaces:

$$k_i \nabla T \overrightarrow{n} = h_j \left(T - T_{amb} \right) + \sigma_{rad} \epsilon_{rad,i} \left(T^4 - T_{amb}^4 \right) \quad (4)$$

where $\overrightarrow{n}$ is the normal to the surface j and T_{amb} is the ambient temperature of the surroundings. The first term of the right-hand side of (4) represents the heat loss by convection from the surface j, being h_j the convection coefficient of this surface, whereas the second term represents the heat loss by radiation, where σ_{rad} is the Stefan-Bolzmann constant and $\epsilon_{rad,j}$ is the infrared emissivity. The convection coefficient is typically in the range $h = 5 - 15$ W/m$^2 \cdot {}^o$C for natural convection. If a heat sink is attached at the bottom, an equivalent h_{eq} can be found ranging from 50-500 W/m^2, its value depending on the thermal resistance of the heat sink used [4].

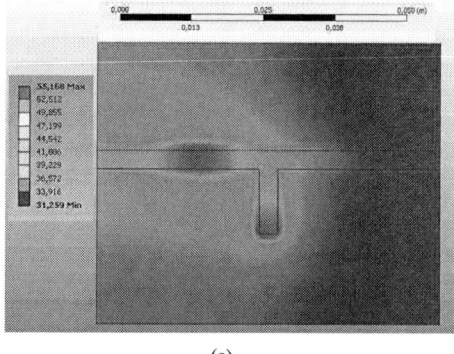

(a)

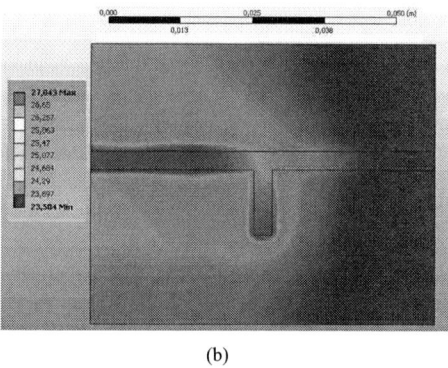

(b)

Figure 1. Thermal profile of a 1st order bandstop filter when a CW signal with an input power $P_{in} = 5$ W is applied to the circuit at its resonance frequency, $f_0 = 3.5$ GHz. For (a) $\tan \delta = 0.02$ and (b) $\tan \delta = 0.002$. Circuit designed on FR-4 substrate (substrate thickness $h = 1.5$ mm, conductor thickness $t = 32$ μm, relative permittivity $\epsilon_r = 4.4$, $K_d = 0.3$ W/m$\cdot {}^o$C, $K_c = 400$ W/m$\cdot {}^o$C). Natural convection is assumed on all the circuit surfaces with $h_{conv} = 10$ W/m$^2 \cdot {}^o$C. No heat transfer to the environment through radiation is considered in this example.

Some preliminary results can be obtained by directly inspecting the problem. The heat source due to conductor losses follow the magnetic field pattern of the circuit according to (2), whereas the heat source due to dielectric losses follow the circuit's electric field profile (see (3)). Generally for any microwave circuit, its electric and magnetic pattern at a determined frequency presents maximum and minimum values along the structure (standing waves), leading to maximum and minimum heat fluxes, and therefore, producing a non-uniform thermal profile on the circuit, where maximum and minimum temperatures are distinguished. As an exception, for when matched lines the electromagnetic pattern is uniform along the lines (no standing waves appear) leading to a generated uniform heat flux. For any microwave passive circuit, if one of the loss mechanisms is dominant over the other, it will define the circuit's thermal profile. As an example, Fig.1 shows the simulated steady-state thermal profile (using Ansys Multiphysics) of a first order bandstop filter analyzed at its center frequency ($f_0 = 3.5$ GHz) for two different situations (high and low dielectric $\tan \delta$). Such a structure at its resonance frequency presents a well-defined electromagnetic distribution with two standing waves: one in the open-ended stub and another between the input port and the cross junction between the feeding line and the open-ended stub where there is a virtual ground. Along the standing waves, the maximum values of electric and magnetic field are shifted a quarter-wavelength one respect to the other. Between the cross junction and the output port there is not a signal propagating and the electromagnetic fields are negligible in this region (leading to negligible q_c and q_d). For the higher $\tan \delta$ situation, dielectric losses are dominant over conductor losses, and as a consequence, there are hot spots around the maximum values of electric field, as seen in Fig. 1(a). For an average power handling capability (APHC) study of this device, the analysis should be focused on these points [5]. Furthermore, these point should be avoided to place lumped components. For the lower dielectric $\tan \delta$ case, conductor losses are slightly dominant over dielectric losses, and therefore, the circuit's thermal profile takes the shape of the combination of both electric and magnetic patterns. Obviously, for the second situation the resulting temperatures in the circuit are much lower than those for the first one because, even though q_c remains the same for both cases, q_d is around ten times lower for the second case.

3. Material Properties and their Dependence on Temperature

In the previous section it has been studied how the thermal profile of a microwave passive circuit can be computed when a microwave signal is applied to the circuit. Due to the dependence of material properties (both electrical and mechanical) with temperature, this generated thermal profile can provoke a change in the RF performance of the circuit, generally resulting in more losses in the circuit and in a frequency shift in its response as the input power increases. The increase of losses come from both the conductor and the dielectric substrate, since both σ and $\tan \delta$ generally increase with temperature, leading to an increase of q_c and q_d, respectively. Regarding the frequency shift in the response, it comes mainly from the variation of ϵ_r with temperature (which produces a change in the electrical dimensions of the circuit), but also from thermal expansion both for conductor and dielectric (variation of the physical dimensions of the circuit). The variation of ϵ_r with temperature ε_T depends on the kind of material used. For standard substrates employed in microwave applications, one can find a big difference among their associated ε_T. As an example [6],

- For PTFE/glass, $\varepsilon_T \sim$ -100 ppm/^{o}C.
- For FR-4, $\varepsilon_T \sim$ -200 ppm/^{o}C.
- For allumina, $\varepsilon_T \sim$ +200 ppm/^{o}C.
- For liquid crystal polymer (LCP), $\varepsilon_T \sim$ -20 ppm/^{o}C.

All provided values are for 10 GHz and for a temperature range 20-120 ^{o}C. The materials with positive ε_T will produce a negative frequency shift in the device response as temperature in the device increases, whereas there will be a positive frequency shift for negative-ε_T materials. From the previous list, the most stable material with respect to temperature is LCP. However, a mechanical analysis of the material is needed before this can be confirmed, since there are mechanical effects which can also modify the center frequency and limit the APHC.

A. Mechanical Properties and Static Structural Analysis

For the response frequency shift analysis a mechanical study should be also performed. The increase of temperature produces a thermal expansion both in the metal and dielectric part, quantified by the coefficient of thermal expansion (CTE). The CTE for dielectric is generally anisotropic, presenting its maximum value in z-direction. The CTE mismatch between the metal and dielectric as well as any mechanical constraints (e.g., circuit bottom layer fixed/stuck to the metal housing) produce thermal stresses in the circuit. The total deformation or strain in any part of the circuit is therefore due to the combination of thermal expansion and thermal stress in the circuit.

The material properties needed for the mechanical analysis are: CTE (in x, y and z-direction), Young's modulus (which represents the material stiffness) and Poisson's ratio. This means that a linear static structural analysis can be performed and the total deformation and stress in the circuit are solved. Material stress limits (such as the yield point) are also needed for an APHC study, as will be seen in the next section.

The deformation due to the temperature increase can produce a positive or negative frequency shift in the circuit response, depending on the circuit's layout and material properties. This shift can be in the same direction as that produced by ε_T, making the frequency shift more pronounced and therefore, a circuit redesign could be essential. Or on the contrary, the deformation effect could counteract the frequency shift due to ε_T.

4. Using a Multiphysics Simulator

The metal and dielectric electrical properties depend on temperature. It leads that electromagnetic fields will also be temperature-dependent making the resolution of (1) more complicated. The resolution of (1) can be faced by means of a closed-feedback process, which necessitates the use of a numerical simulator.

The resolution process can be summarized as follows: when the Maxwell equations are solved, at the beginning of the resolution process, the thermal profile is unknown. Thus, the computation procedure starts by solving the Maxwell equations for a given temperature (e.g., T_{amb}). After that, the thermal profile is obtained as Section 2 describes. At this point, the Maxwell equations are resolved taking into account the circuit's thermal profile along with the temperature-dependence of the materials. A new electromagnetic pattern is calculated leading to a new

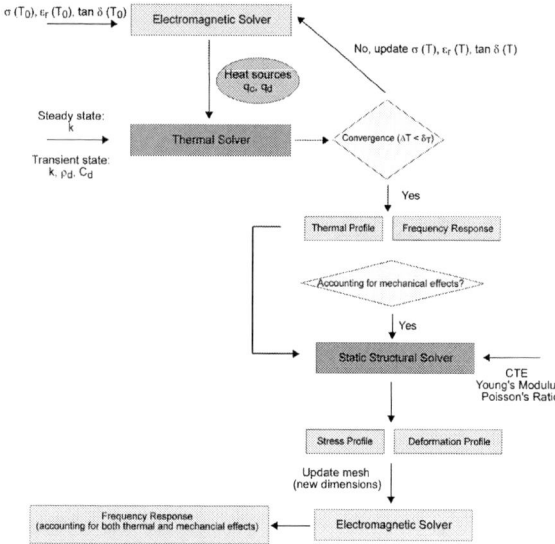

Figure 2. Flow chart of the procedure followed by the multiphysics simulator to perform an electro-thermo-mechanical simulation of a microwave device.

frequency response. The new thermal profile is computed and compared with the previous one. If the difference between both thermal profiles is above limits, then the process is repeated until the defined maximum differential temperature δ_T between the two latest computed profiles is achieved, meaning that the process has converged. The mechanical analysis can be also included in this process. Fig. 2 plots a flow chart of the procedure.

5. Average Power Handling Capability

Microstrip circuits may limit the maximum working input power of the communication system. Their maximum power handling capability can be determined by the heating in the materials which limits the APHC, and by the dielectric breakdown (related to the maximum peak voltage that the dielectric can withstand) limiting the peak power handling capability (PPHC). APHC value is generally lower than PPHC one, therefore, its study is of big importance. As P_{in} rises, the temperature in the circuit increases until the maximum temperature of operation is reached. The maximum temperature of operation can be defined as 1) that temperature where the electrical and mechanical performances of the circuit change (in general, the substrate glass transition temperature) or 2) that temperature which produces an excessive thermal stress in the circuit able to cause a failure. The minimum value of these two temperatures determines APHC ($= P_{in,max}$). The computation of the APHC must be focused, therefore, on the hottest regions of the circuit. In active devices, the hottest parts are around the active components, due to their high dissipation. However, in a passive circuit, the position of the hot regions are not obvious, and thus, a

Table 1. Mechanical Properties of Megtron 6

	Copper	Dielectric
CTE (x,y,z) (ppm/^{o}C)	17	16, 16, 45
Young's Modulus (GPa)	117	15
Poisson's Ratio	0.33	0.2
Yield Strength (MPa)	220	—

study as mentioned in the example of Section 2 must be done in order to determine their location.

For microstrip circuits with fixed temperature at ground (e.g., circuit inside a metal housing which equalizes the temperature on the ground plane), the heat flow mainly propagates along the cross section of the microstrip circuit, i.e., between the strip and the ground plane. Under this condition, at the hottest spots a gradient of temperature ΔT (^{o}C/W) between the strip and ground can be calculated as [1], [5], [7], [8]. ΔT is a characteristic property of the passive circuit and depends on the topology (electromagnetic performance) and on the material's thermal properties. The maximum temperature around the hottest spots can be predicted from

$$T_{max} = \Delta T \cdot P_{in} + T_{ref} \qquad (5)$$

where T_{ref} is the temperature at the circuit's ground, which depends on the input power, losses, and thermal boundary conditions [5]. As seen from (5), for a given P_{in} and knowing ΔT and T_{ref}, T_{max} in the circuit can be directly computed. Hence, it is possible to estimate if the circuit is working under the glass transition temperature for a given input power. Anyway, the APHC analysis is not completed and a structural analysis should be performed to compute the stress generated in the circuit, since, as previously mentioned, thermal stress can lead to a lower APHC limit.

6. Results and Discussion

As an example of analysis, a 3rd order hairpin coupled-line filter centered at 42 GHz is considered in this section. Fig. 3 shows the layout of the device. The filter is designed on Megtron 6 substrate from Panasonic, whose properties are the following [9]: $h = 0.2$ mm, $t = 18$ μm, $\epsilon_r = 3.61$ (at 40 GHz and at 22 ^{o}C), $\tan \delta = 0.006$ (at 40 GHz and at 22 ^{o}C), $K_d = 0.4$ W/m^{o}C, $K_c = 400$ W/m^{o}C. The electrical parameters are assumed to be linearly dependent with temperature (at 40 GHz) as follows: $\varepsilon_T = +50$ ppm/^{o}C, the variation of $\tan \delta$ is +2200 ppm/^{o}C and the variation of copper resistivity ($1/\sigma$) is +3400 ppm/^{o}C. Mechanical properties are listed in Table 1. The glass transition temperature T_g is 185 ^{o}C. The provided dielectric CTEs are for temperatures below T_g.

An electro-thermo-mechanical study has been performed of this device by using Ansys Multiphysics [10]. Three different load case studies are considered depending on the thermal and mechanical boundary conditions:

978-1-4799-4789-8/14 $31.00 © 2014 IEEE

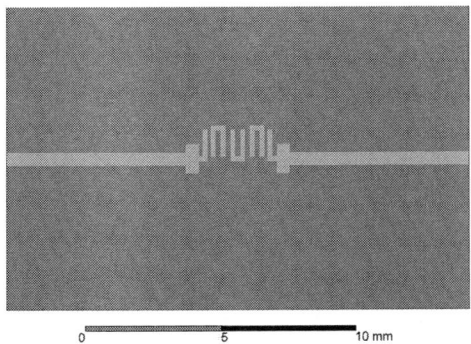

Figure 3. Layout of the hairpin coupled-line filter to be analyzed.

- Load Case 1: Fixed support at the bottom (ground) with a uniform temperature at the ground T_{ground}. In this case, convection and radiation heat transfer on the top layer are negligible since practically all of the generated heat flow goes from the line to the ground. This load case intends to simulate the situation where the circuit is enclosed in a metal housing with a heat sink attached at the bottom.
- Load Case 2: No mechanical constraints in the circuit. Cooling by natural convection on top and bottom layers, with $h_{conv} = 10$ W/m$^2 \cdot ^o$C.
- Load Case 3: Fixed support at the input/output ports. Cooling by natural convection on top and bottom layers, with $h_{conv} = 10$ W/m$^2 \cdot ^o$C.

Fig. 4 shows the simulated thermal profiles for load cases 1 and 2 at 42 GHz and for $P_{in} = 1$ W. Note that from a thermal point of view, load cases 2 and 3 are identical. At the center frequency, all resonators resonate, storing and dissipating energy. Thus, the hottest spots are located along them, as seen in this figure. For load case 1 the heat flow is better defined than for load case 2 due to the homogeneous temperature at the bottom [1], [5], which provides a clear heat flow path. In addition, temperatures are considerably lower for load case 1 since all generated heat is effectively delivered to the ground plane, providing an ideal cooling. This is not the situation for load case 2 where the cooling is by natural convection on both surfaces. Note also that in this example, for load case 1, T_{ground} equals T_{amb}, which leads to a very optimistic situation, since, generally, $T_{ground} > T_{amb}$.

A process as described in Section 4 has been carried out to evaluate the impact of the input power on the filter response. Fig. 5 plots the S-parameters of the filter for load case 3 under varied P_{in}. Load cases 2 and 3 are the cases where the thermo-mechanical impact in the frequency filter transfer function can be more important, since the temperatures are high for moderate input powers and there is not any important mechanical constraint limiting deformation. In this figure, the S-parameters in all the band of interest (30-50 GHz) have been computed

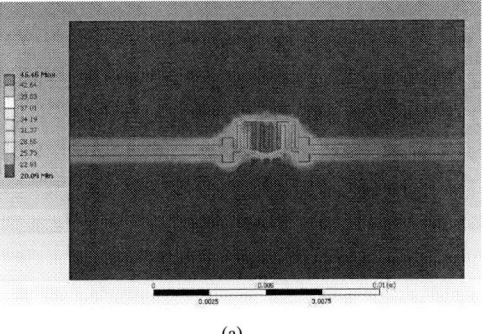

(a)

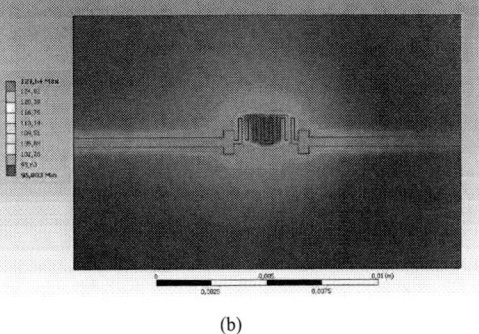

(b)

Figure 4. Thermal profiles of the filter under analysis at 42 GHz and $P_{in} = 1$ W. (a) Load case 1. (b) Load case 2. In this example, $T_{amb} = 22 \ ^o$C. For load case 1, $T_{ground} = T_{amb}$.

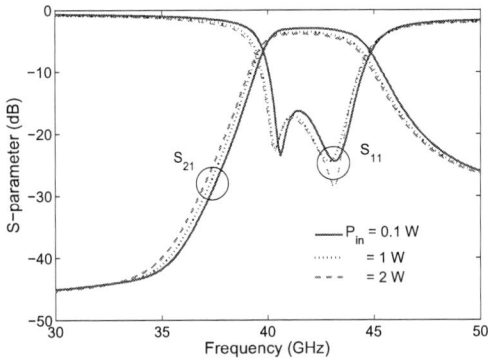

Figure 5. Filter responses under varied P_{in} for load case 3.

by using the thermal profile and mechanical deformation associated to each P_{in} at 42 GHz. This approximation works well within the passband bandwidth, since the thermal and deformation profiles do not change significantly along the passband. Out of band they can be quite different to those within passband, however their impact in the filter response is not very important. As Fig. 5 shows, increasing P_{in} leads to an increase of losses and a shift towards lower frequencies. For $P_{in} = 0.1$ W, the thermal and mechanical effects on the circuit performance are negligible. The frequency shift Δf between 0.1 and 2 W input power cases has been approximately 350 MHz. This

978-1-4799-4789-8/14 $31.00 © 2014 IEEE   299

frequency shift can be easily compensated by reducing the lengths of the resonators, for example. Δf could have been more important if a higher-ε_T material had been used instead. In such a case, the redesign could involve an additional modification of the coupling section between the resonators to keep the passband bandwidth.

A. Mechanical Analysis and APHC

A mechanical analysis has been performed for the three load cases under varied T_{ground} (for load case 1) and the temperature of the surroundings (for load cases 2 and 3) from 22 to 80 ^oC, and for different CW input signal powers (from 0.1 to 2 W) of 42 GHz. The analysis is focused on the maximum deformation suffered by the circuit, maximum equivalent (Von-Mises) stress and plastic deformation, with the latter indicating that the yield point is exceeded and therefore, some permanent deformation and a possible failure in the device.

For load case 1, deformation is negligible ($< 2 \ \mu m$) for any T_{ground} and P_{in}. The maximum stress occurs on the copper resonators, in particular, on the resonator's inside corners. The stress in the substrate for this load case does not seem to be significant. Fig. 6 shows the maximum equivalent stress and plastic strain on the circuit for copper. As seen in Fig.6(a), there is evidence of maximum stress convergence at $T_{ground} = 80 \ ^oC$ due to the copper yielding. Furthermore, as deduced from Fig. 6(b), for $T_{ground} > 60 \ ^oC$ plastic deformation occurs for any input power, defining a maximum working T_{ground}. For lower T_{ground}, $P_{in} = 2$ W can also produce plastic deformation.

Fig. 7 shows the maximum deformation and plastic strain for load case 2. The maximum deformation linearly increases with T_{amb}. For this load case, deformation can have an important impact on the filter response, especially for input powers higher than 1 W as seen in the previous subsection. Regarding copper plastic deformation, it only happens for $P_{in} = 2$ W case. However, temperatures in the circuit are very high even for $P_{in} = 1$ W (see Fig. 4)(b).

For load case 3, deformation is also important, and is mainly along the z-axis. The stress in both the conductor and dielectric is significant due to its particular fixed support, which makes the regions close to the input/output ports bend. In particular, the maximum stress peaks occur in the clamped input/output port corners rather than in the resonators (as for load case 1). Fig. 8 plots the maximum deformation and plastic strain (for the copper parts of the filter). This load case presents the highest stress impact of the three under-study load cases.

In view of these mechanical results, the APHC is limited by the mechanical stress for load cases 1 and 3 and by the glass transition temperature for load case 2. Therefore, to keep the circuit in safety ranges it would be recommended to work under $P_{in} < 1$ W and T_{ground}, $T_{amb} < 60 \ ^oC$ for the three load cases.

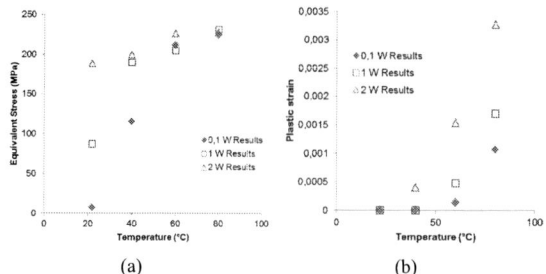

Figure 6. Equivalent (Von-Mises) stress (a) and plastic strain (b) as a function of T_{ground} under varied P_{in}, for load case 1.

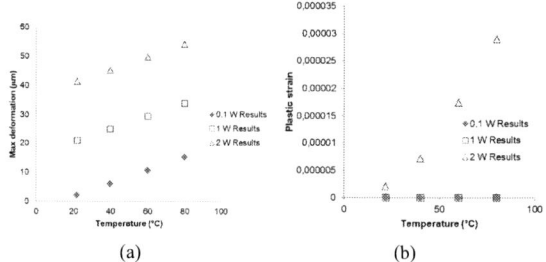

Figure 7. Maximum deformation (a) and plastic strain (b) as a function of the temperature in the circuit surroundings under varied P_{in}, for load case 2.

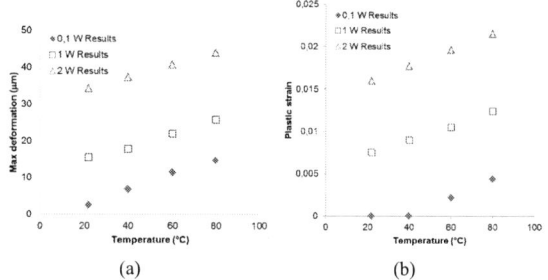

Figure 8. Maximum deformation (a) and plastic strain (b) as a function of the temperature in the circuit surroundings under varied P_{in}, for load case 3.

It should be mentioned that the APHC study should be performed particularly at frequencies where there are more losses rather than at the center frequency. In a bandpass filter, these frequencies are around the cutoff frequencies. Anyway, for this example, conservative limits have been given in the previous paragraph which should work for the entire passband bandwidth. Out of band, there is not any part of the circuit resonating, therefore, temperatures are lower than those in the passband and consequently, the APHC is higher.

7. Conclusions

In this work, a multiphysics study to evaluate the effects of the input signal power in microwave planar components has been presented. The electro-thermo-mechanical coupling in microwave devices has been analyzed in this study. It has been shown that the electromagnetic field distribution completely defines the thermal profile and, that the conventional statement of assuming a constant heat flux on the metal does not necessarily work for high frequency devices. In addition, due to the temperature dependence of materials, increasing the input power generally causes an increase of losses and a frequency shift in the device response. As a demonstration, a multiphysics analysis of a microstrip coupled-line filter centered at 42 GHz has been performed. The analysis has included different thermal and mechanical boundary conditions. A noticeable frequency shift has been occurred for two of the cases along with a considerable increment of losses. The stress in the copper metal has limited the average power handling capability for most of the load cases, defining a maximum input power of around 1 W to avoid any failure in the device.

8. Acknowledgements

This work has been supported by the Euripides European project MIDIMU-HD.

References

[1] K. C. Gupta, R. Garg, I. J. Bahl, and P. Bhartia, *Microstrip lines and slotlines*, 2nd ed. Artech house Boston, 1996.

[2] J. S. Hong and M. J. Lancaster, *Microstrip Filter for RF/Microwave Applications*. New York: John Wiley and Sons, inc., 2001.

[3] R. J. Cameron, C. M. Kudsia, and R. R. Mansour, *Microwave Filters for Communication Systems*. Hoboken, New Jersey.: John Wiley & Sons, Inc, 2007.

[4] Y. A. Cengel and A. J. Ghajar, *Heat and Mass Transfer: Fundamentals and Applications*, 4th ed. McGraw-Hill, 2011.

[5] M. Sanchez-Soriano, Y. Quere, V. L. Saux, C. Quendo, and S. Cadiou, "Average power handling capability of microstrip passive circuits considering metal housing and environment conditions," 2014, manuscript submitted for publication.

[6] D. Thompson, J. Papapolymerou, and M. Tentzeris, "High temperature dielectric stability of liquid crystal polymer at mm-wave frequencies," *IEEE Microwave and Wireless Components Letters*, vol. 15, no. 9, pp. 561–563, Sept 2005.

[7] W.-Y. Yin and X. Dong, "Wide-band characterization of average power handling capabilities of some microstrip interconnects on polyimide and polyimide/gaas substrates," *IEEE Transactions on Advanced Packaging*, vol. 28, no. 2, pp. 328–336, 2005.

[8] L.-S. Wu, X.-L. Zhou, W.-Y. Yin, M. Tang, and L. Zhou, "Characterization of average power handling capability of bandpass filters using planar half-wavelength microstrip resonators," *IEEE Microwave and Wireless Components Letters*, vol. 19, no. 11, pp. 686–688, 2009.

[9] Panasonic Corporation, Private communication and website, 2013.

[10] Ansys® Academic Research, 2013, Released 14.5.

Investigation of a Finned Baseplate Material and Thickness Variation for Thermal Performance of a SiC Power Module

Yafan Zhang[1,2,3], Ilja Belov[2], Mietek Bakowski[1], Jang-Kwon Lim[1], Peter Leisner[2,4] and Hans-Peter Nee[3]

[1]Acreo Swedish ICT AB
Electrum 236, Isafjordsgatan 22, SE-164 40 Kista, Sweden
E-mail: yafan.zhang{mietek.bakowski, jang-kwon.lim}@acreo.se, phone: +46 8 632 7700
[2] Jönköping University, School of Engineering, Mechanical Engineering - Materials and Manufacturing
PO Box 1026, SE-551 11 Jönköping, Sweden
E-mail: ilja.belov@jth.hj.se, phone: +46 36 101 686
[3] KTH School of Electrical Engineering, Electrical Energy Conversion
Teknikringen 33, SE-100 44 Stockholm, Sweden
E-mail: hansi@kth.se, phone: +46 8 790 6000
[4]SP Technical Research Institute of Sweden
PO Box 857 SE-501 15 Borås, Sweden / Copenhagen, Denmark
E-mail: peter.leisner@sp.se, phone: +46 10 516 5447

Abstract

A simplified transient computational fluid dynamics model of an automotive three-phase double-side liquid cooled silicon carbide power inverter, including pin-fin baseplates, has been developed and qualified for parametric studies. Effective heat transfer coefficients have been extracted from the detailed pin-fin baseplate model for two coolant volume flow rates 2 l/min and 6 l/min, at the coolant temperature 105 °C. The inverter model includes temperature dependent heat losses of SiC transistors and diodes, calculated for two driving cycles. Baseplate materials such as copper, aluminum-silicon carbide metal matrix composite, aluminium alloy 6061 as well as virtual materials have been evaluated in the parametric studies. Thermal conductivity, specific heat and density have been varied as well as thickness of the finned baseplates (1 to 3 mm). A trade-off between temperature of SiC chips and baseplate weight has been investigated by means of Pareto optimization.

The main results of the parametric studies include a weak dependence (1 to 3 °C) of the chip temperature on baseplate thickness. Furthermore, switching e.g. between copper and AlSiC results in 5 to 8 °C increase of the chip temperature, at 65 to 70 % baseplate weight reduction.

1. Introduction

Thermal management of power modules found e.g. in electric vehicle (EV) and hybrid electric vehicle (HEV) applications, is a challenging task that has to be solved in the design phase, often, by employing computer simulations. Heat losses in semiconductor chips are temperature dependent. Therefore, the results from semiconductor device simulations [1] have to be incorporated into a transient thermal model of the power module. Introduction of silicon carbide (SiC) transistors and diodes having higher power efficiency at higher current densities, and at higher operation temperatures than Si components, should lead to more compact power modules with a smaller thermal mass. The existing interconnect and packaging materials, however, require a

transition from single-side to double-side cooling [2, 3], for reliability reasons, in order to keep the operating temperatures and temperature cycles within limits defined by the reliability requirements. The weight is often also a design goal.

Furthermore, finned baseplates have become a feasible alternative to flat baseplates attached to a cooler [4]. The base of a pin-fin baseplate provides heat spreading and thermal mass. Therefore, physical properties of the baseplate material and the baseplate thickness are important design parameters.

Creation of a simplified representative model of a power module suitable for parametric studies has been of interest for design engineers [5, 6]. Thick copper layers, bus-bars and terminals contribute to the heat transfer path and have to be included in the model.

Even if finned-baseplate technologies for different industrial and automotive applications are readily available, a literature study reveals a lack of systematic comparisons between different baseplate materials and classifications thereof based on the thermal and weight criteria.

The paper presents a simulation methodology to assist a design engineer in quantifying the effect of different materials and thicknesses of finned baseplates on the chip temperatures in thermal transient and steady-state operation of a power module. First, attention is paid to the predictive accuracy of the computational fluid dynamics (CFD) model via evaluation of a reference pin-fin base plate using experimental data available in the literature [7, 8]. Second, a detailed CFD model of a double-side liquid cooled three-phase power inverter is created including temperature dependent heat losses of SiC transistors and diodes, computed from detailed finite-element (FE) semiconductor device simulations. Next, a simplified CFD model of a double-side cooled three-phase inverter is developed for parametric studies. Two modeling challenges are addressed: creating a representative compact model of the finned base plate and reduction of the three-phase inverter model to a single-phase model with symmetry boundary conditions.

978-1-4799-4789-8/14 $31.00 © 2014 IEEE

The physical properties of the pin-fin baseplate are varied to cover a set of typical pin-fin baseplate materials, such as copper (Cu), aluminium (Al), and aluminum-silicon carbide metal matrix composite (AlSiC). For reference purposes, comparisons to copper molybdenum (CuMo) and copper tungsten (CuW) materials are also provided. Results are presented for different baseplate thicknesses. The other varied parameters include driving cycle in terms of the worst operating points of the semiconductor devices, operation time (4 seconds and 20 seconds), and coolant flow rate 2 l/min and 6 l/min.

The results of the parametric study are quantified by means of Pareto optimization applied on the maximum temperature in the system and the weight of the finned baseplates, as conflicting performance criteria. Finally, cost aspects are briefly discussed.

2. Compact model of a pin-fin base plate

When evaluating a liquid cooled three-phase power inverter for HEV applications, the first challenge is to create a realistic compact model of the finned base plate, via extracting an effective heat transfer coefficient (EHTC) for the finned surface, which is assumed to be in contact with Ethylene Glycol based water solution. The surface heat transfer method is chosen for compact modeling. The reference copper pin-fin baseplate with staggered pin configuration is shown in Figure 1 [4, 7, 8].

Figure 1: Reference pin-fin baseplate

The diameter and height of the pin fin are 2.2 mm and 8.0 mm, respectively. In order to characterize the base plate, a junction-to-fluid thermal resistance of 0.19 K/W as reported in literature [7, 9] is taken as reference, as well as the packaging configuration with an integrated cooler [7], cooling channel geometry [4], and the coolant flow rate [8]. A detailed Flotherm CFD model shown in Figure 2, represents one third of the baseplate shown in Figure 1. The flow rate of the coolant is 10 l/min, and the distance between the edge pins and the aluminium (Al) channel walls is 1 mm. Each of the Si components (C1-C3) shown on the direct bonded copper (DBC) substrate approximates two transistor-diode pairs. For the mesh, shown in Figure 2, at 280 W power dissipation per component, and 70 °C fluid temperature, the maximum thermal resistance for component 2 is found to be 0.18 K/W, which agrees well with the reported value for the similar configuration.

The extracted EHTC for the pinned area has been found to be 16242 W/(m^{2}K). Additional simulations have shown that the extracted EHTC is nearly independent of the pin-fin structure length, i.e. modeling a one third of the baseplate can give representative EHTC values. The detailed model of the pin-fin baseplate is then replaced

with the EHTC applied to a flat surface of the baseplate. In order to compensate for both the conductive thermal resistance and the thermal mass of the pins, an effective material layer is introduced on the top of the base. The layer thickness is 2 mm, supplying the equivalent total volume of the pins. Physical properties of the effective fin layer are the following: thermal conductivity, $k_z = k_{copper}$, $k_x = k_y = 0$; specific heat, $c_p = c_{p,copper}$, and density, $\rho_{eff} = p \cdot \rho_{copper}$, where $p = N \cdot A_{pin} / A_b$, N is the number of pins on the base plate, A_{pin} is the cross-sectional area of a cylindrical pin, and A_b is the total area of the base plate surface.

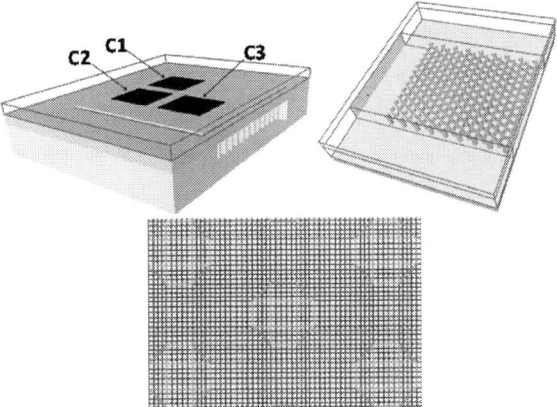

Figure 2: CFD model of the reference baseplate

The introduction of ρ_{eff} aims at moderating the time constant of the effective fin layer. The results of comparison between the transient detailed and compact models are presented in Figure 3. A good agreement between the simulation results justifies the usage of the EHTC extraction and compact modeling method for power module evaluation.

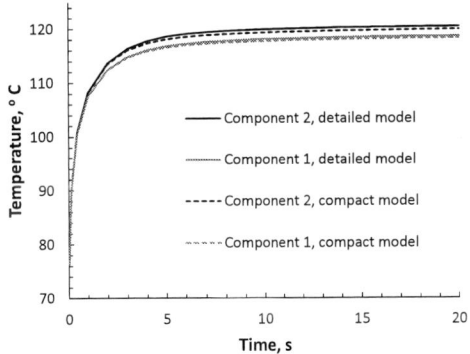

Figure 3: Comparison between detailed and compact models of the pin-fin baseplate

3. Temperature dependent heat losses

Transistor and diode temperature dependent heat losses in the following CFD model are obtained from FE TMA Medici simulations [1, 10]. Turn-on and turn-off switching losses for a diode and a metal-oxide semiconductor field-effect transistor (MOSFET) are obtained from switching simulations with a dc-link voltage of 600V and peak current value of 150A per chip.

The chip areas of the transistor and the diode are 0.81 cm^2 and 0.4 cm^2, respectively. Two heat loss profiles are developed which exemplify two extreme operating conditions of the inverter. One corresponds to the worst operating point of the transistor when the required voltage slope is 10 kV/µs (relatively fast switching). Such an operation condition of a MOSFET is in the following referred to as QM. It occurs e.g. when a vehicle is accelerating. The other heat loss profile represents the worst operating point of the diode when the required voltage slope is 5 kV/µs (relatively slow switching). Such an operating condition of a diode is in the following referred to as SD. It occurs e.g. when a vehicle is braking. Further details on conduction and switching loss calculation can be found in [11, 12]. The total heat losses per chip are plotted in Figure 4.

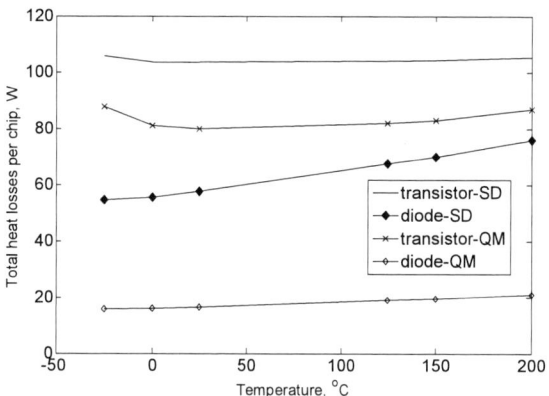

Figure 4: Temperature dependent heat losses of a SiC MOSFET and a diode

4. Detailed CFD model of a double-side cooled power module

A detailed simulation model of a three-phase inverter for HEV application has been developed. The electrical diagram realized in the thermal layout is shown in Figure 5. The target rating of the power inverter (in terms of dc-link voltage, rms output current, and switching frequency) is 600 V/300 A/20 kHz.

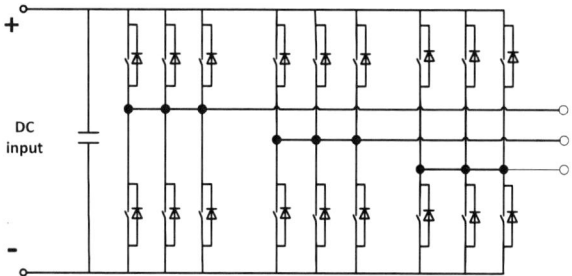

Figure 5: Electrical diagram of a three-phase inverter with SiC devices

The main thermal design principles have been adopted from [13], Figure 6. The solder-free sandwich design includes DBC power substrates (Al$_2$O$_3$: 0.3 mm, Cu: 0.3 mm) which reside on baseplates with thermal grease and silicone elastomer pads (thickness 0.5 mm) as thermal interface materials. The SiC transistors and diodes (thickness 300 µm) are assumed to be nano-Ag sintered. Molybdenum spacers (height 1 mm) are placed on the front side of the dies. Mo discs realize signal routing between the upper and lower power substrates. Even though the thermal grease layer forms a significant thermal resistance, large DBC substrate areas soldered to the finned baseplate might cause thermo-mechanical problems. In this study, the preference is given to the engineering solution with grease. The module consists of three sections: a double-side cooled power inverter, a dc-link capacitor, and a multi-channel driver board, as shown in Figure 7 and Figure 8.

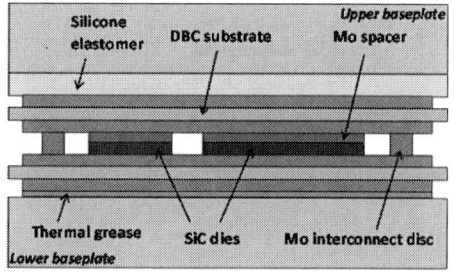

Figure 6: A double-side cooled transistor-diode pair; thermal interface material and interconnect/die attach on the die-copper, die-spacer and spacer-copper interfaces are shown

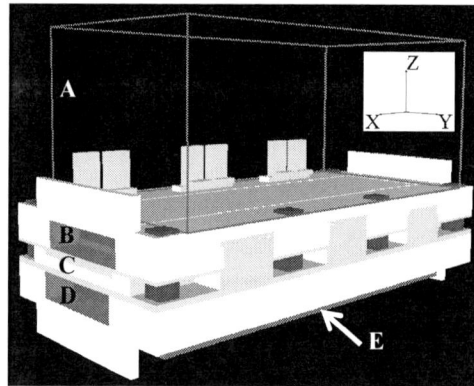

Figure 7: 3D view of the detailed model of the power inverter showing the placement of dc-link capacitor (A), cooling channels (B) and (D), inverter section (C), and a gate driver board (E)

Functional or reliability aspects of the power module are out of scope of the study. However, electrical layout constraints are taken into account, e.g. making the circuit as symmetrical as possible. The conductor length and inductive loop areas are minimized, e.g. via making a micro-strip configuration of dc-link capacitor terminals. The mechanical tolerances for cooling channel isolation, and space for mounting screws are considered. The power module dimensions without dc-link capacitor and driver board sections are 160×80×32 mm^3, including the assembled cooling channels.

978-1-4799-4789-8/14 $31.00 © 2014 IEEE

The power module is supposed to operate in an ambient air temperature of 125 °C corresponding to a harsh under-hood environment.

Cooling channels with Al alloy walls are designed for connecting in series to a coolant loop with a maximum coolant temperature of 105 °C. The cross-section of one cooling channel is 40×9 mm². The coolant flow rate is assumed to be either 2 l/min or 6 l/min to represent low and moderate flow rates, respectively. The initial temperature of the whole module is assumed to be 105 °C.

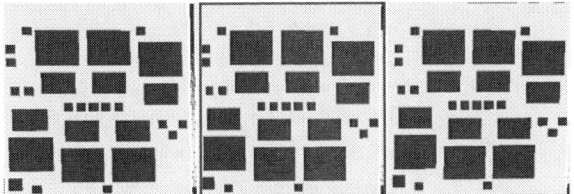

Figure 8: Top view of the three-phase inverter layout, without the upper substrate; phase B chosen for subsequent simplified modeling is framed

The pin dimensions and baseplate configuration is the same as described in Section 2 of the paper. The cooling channel width, however, is smaller. The EHTC of the finned area extracted according to the method described in Section 2, with the same mesh density, is equal to 15200 W/(m²K) for 6 l/min and about 12000 W/(m²K) for 2 l/min flow rate. The surface heat transfer method is applied for the finned baseplate modeling in the detailed model of the three-phase inverter.

Physical properties of materials present in the CFD model are summarized in Table 1 [14 - 17].

Table 1: Physical properties of materials

	k, W/(m·K)	ρ, kg/m³	c_p, J/(kg·K)
Silicone gel	0.18	1120	1600
Elastomer pad	6.5	1536	900
AlSiC[1] (detailed model)	$k_0 = 162$ $a = -0.12$ $T_{ref} = 100$	3010	741
AlSiC (simplified model)	156 (above 100°C)	3010	741
Copper	400	8930	385
Al 6061	180	2700	896
Molybdenum	138	10240	251
SiC[1]	$k_0 = 347$ $a = -0.9505$ $T_{ref} = 347$	3200	660
CuW	220	15000	211
CuMo	250	9500	323

[1] Temperature dependent property, $k = k_0 - a \cdot \left(T - T_{ref}\right)$.

The physical properties of the dc-link capacitor are taken from [13]. The nano-Ag layer and thermal grease are modeled as contact thermal resistances.

The studied baseplate materials include aluminium alloy 6061 (Al 6061), aluminum-silicon carbide metal matrix composite (AlSiC), copper tungsten (CuW), and

copper molybdenum (CuMo). The two later materials are taken merely for reference, since they are not appropriate for manufacturing pin-fin baseplates. Several types of AlSiC are available on the market; the choice of the material for this study is made on the basis of availability of the temperature dependent thermal conductivity for the wide temperature range provided in [14]. The CTE mismatch mitigation is out of scope of the present paper, only the thermal performance of the selected materials is studied.

The detailed CFD model contains 2.8 million grid cells. Open boundaries are set on the faces of the computational domain. Solid surfaces are represented as single radiant faces. Total source of heat, according to Figure 4, is applied over each rectangular cuboid volume representing a SiC diode or a transistor.

5. Simplified model

A simplified CFD model of a double-side cooled three-phase inverter, which would be suitable for parametric studies, is developed, Figure 9. The following assumptions and simplifications are made:

1. The dc-link capacitor, and the driver board are neglected.
2. Phase B of the inverter found in the middle is kept. The other identical phases are removed and symmetry boundary conditions are set to the domain faces which are normal to the X-axis.
3. The heat transfer coefficient 5 W/(m²K) is set on the other faces of the domain, which are in contact with solids.
4. The temperature dependence of the physical properties of the AlSiC baseplate is neglected, see Table 1.

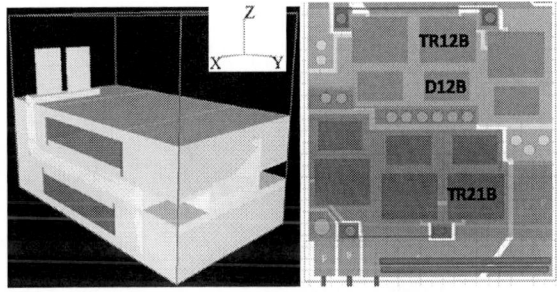

Figure 9: 3D view of the simplified model of a double-side liquid cooled power module, and electrical layout of phase B with indication of the monitored components

Three studies are conducted, where the baseplate material is either copper or AlSiC, the heat losses correspond to either the SD or QM conditions. The temperatures of the coldest, hottest and moderately hot components are monitored as shown in Figure 9.

It is evident from Figure 10, that the transition to the simplified model, containing nearly four times less grid cells, results in a percentage error in the chip temperature rise above the coolant temperature that is less than 1.5 %. Therefore, the results from the comparison between the models along with the results of the compact modeling of the baseplate in Section 2 qualify the developed simplified model for usage in parametric studies.

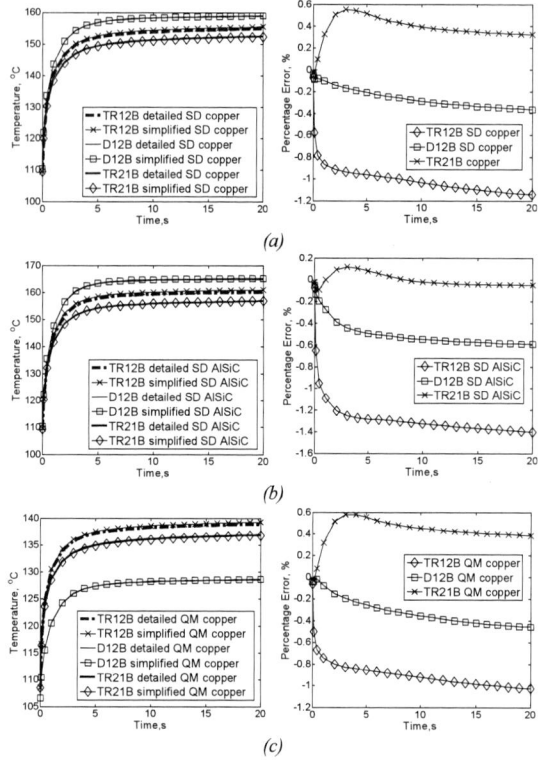

Figure 10: Evaluation of the simplified model for different baseplate materials and heat dissipations during the thermal transient operation; (a) copper, SD; (b) AlSiC, SD; (c) copper, QM

6. Parametric study setup

The purpose of the parametric study is to evaluate the influence of the thermal conductivity, density, specific heat and thickness of the baseplates on the maximum temperature of SiC components, in the SD (the worst case) operating conditions. The weight of the baseplate including the fin array is chosen as the second performance criterion.

A nearly steady state condition (20 s) and a well expressed transient operation (4 s) are studied. The baseplates with EHTCs corresponding to 2 l/min and 6 l/min coolant volume flow rates are evaluated. Each parametric study is performed for a fixed thickness of the baseplates and includes 400 simulations, generated with the Latin Hypercube design of experiment (DoE) method; each simulation requires less than five minutes on CPU i7-2630QM.

The physical properties of the pin-fin baseplate are varied to cover Cu, Al, AlSiC, CuMo and CuW materials, Table 2. Clearly, such a variation results in evaluation of non-existing or virtual materials. It would, however, demonstrate the hypothetical limits of baseplate thermal performance within the selected range of the properties. The thickness of the baseplates is varied as follows.

Upper baseplate / lower baseplate: 1 mm / 1 mm, 2 mm / 2 mm, 1 mm / 3 mm, and 3 mm / 3 mm.

Figure 11 clarifies on the cross-sectional baseplate structure. It shall be noticed that the upper plate facing the pad has a T-shape, with the lower part of that plate being 1.5 mm thick. This, as well as a relatively high thermal resistance of the elastomer pad is expected to mitigate the effect of the upper baseplate thickness variation. However, this T-shape design feature has been introduced to let a required space for dc-link capacitor terminals routed upwards, in the proposed power module design.

In order, to prove that the studied 400 designs approximate well the Pareto frontier in the temperature and weight criterion space, a cost function is defined and a simple weighted sum method has been employed to perform Pareto optimization [18].

Table 2: Baseplate physical property ranges

k, W/(m·K)	[150; 400]
ρ, kg/m³	[2685; 15000]
c$_p$, J/(kg·K)	[200; 960]

Figure 11: Geometry cross-section showing baseplate thickness variation in the parametric study

Cost function, CF, includes the temperature of the hottest component at SD conditions, i.e. the diode D12B, Figure 9, and the weight of the baseplate, including pins:

$$CF = w_1 \cdot W_b + w_2 \cdot T_{D12B} \quad (1)$$

$$w_1 = \mu_1 \cdot \theta_1 ; \quad w_2 = \mu_2 \cdot \theta_2 \quad (2)$$

$$\theta_1 = \frac{1}{W_{b\max} - W_{b\min}} \quad (3)$$

$$\theta_2 = \frac{1}{T_{D12B_{\max}} - T_{D12B_{\min}}} \quad (4)$$

where μ_1 and μ_2 are the weight factors ranging between 0 and 1; θ_1 and θ_2 are the normalization factors defined through maximal and minimal baseplate weight and temperature values, respectively, for the 400 simulated designs.

7. Parametric study results and discussion

Results of the parametric study for 20-second operation are shown in Figure 12. It reveals close-to-steady state operation of the module, where the positive temperature slopes are observed for each material, with respect to the baseplate thickness. Only highly conductive and relatively heavy copper baseplates still exhibit transient behavior. The maximum possible temperature difference 5-7 °C is determined when switching between copper and 68 % lighter AlSiC baseplates. Virtual materials with the highest thermal conductivity form the lower temperature boundary in Figure 12a. Experiments show that within the specified range of physical properties and thickness, calculated temperature variation is 8 °C. The maximum effect of thickness variation is 1 to 2 °C.

The results of the second parametric study are shown in Figure 13 and Figure 14, for 4-second operation. Here, a well expressed transient behavior takes place for most of the materials, except for lighter materials with low thermal conductivity (Al and AlSiC), which have nearly reached the steady state. The line in Figure 13b approaches Pareto frontier, which is found by response surface optimization (RSO) of the cost function defined by equations (1) - (4). The proximity of the circles in the figure to the straight line, indicates a good approximation of the Pareto frontier by DoE results. Therefore, only approximate DoE generated Pareto frontiers are shown in Figures 13a, 13c and 13d, and summarized in Figure 14 with the lines of different pattern and thickness.

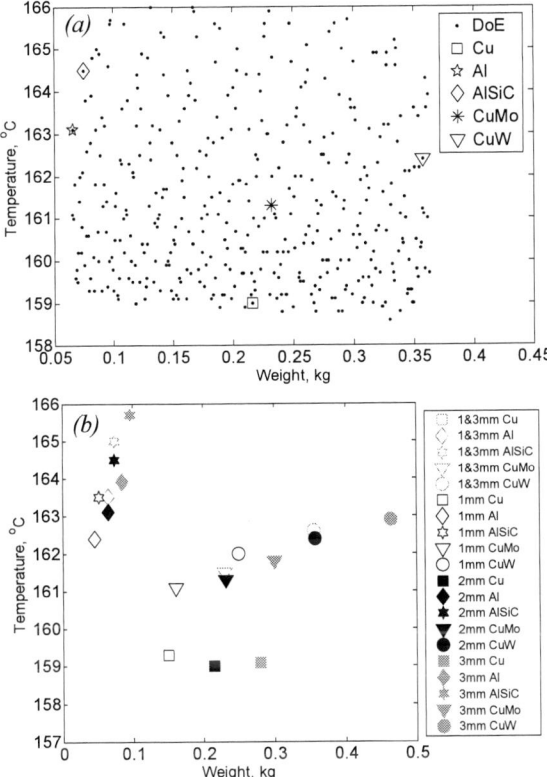

Figure 12: Parametric study results for 20 s operation at 6 l/min flow rate: (a) upper/lower baseplate thickness 2 mm / 2 mm; (b) overall results for real materials

It follows from the frontier locations in Figure 13 that the trade-off designs can deliver maximum 4 to 6 °C improvement at a fixed thickness of the baseplates, as a result of physical property variation. At that, the weight difference between the heaviest and the lightest baseplates is around 75 to 80 %. For thin baseplates the temperature decrease as a result of material variation is modest.

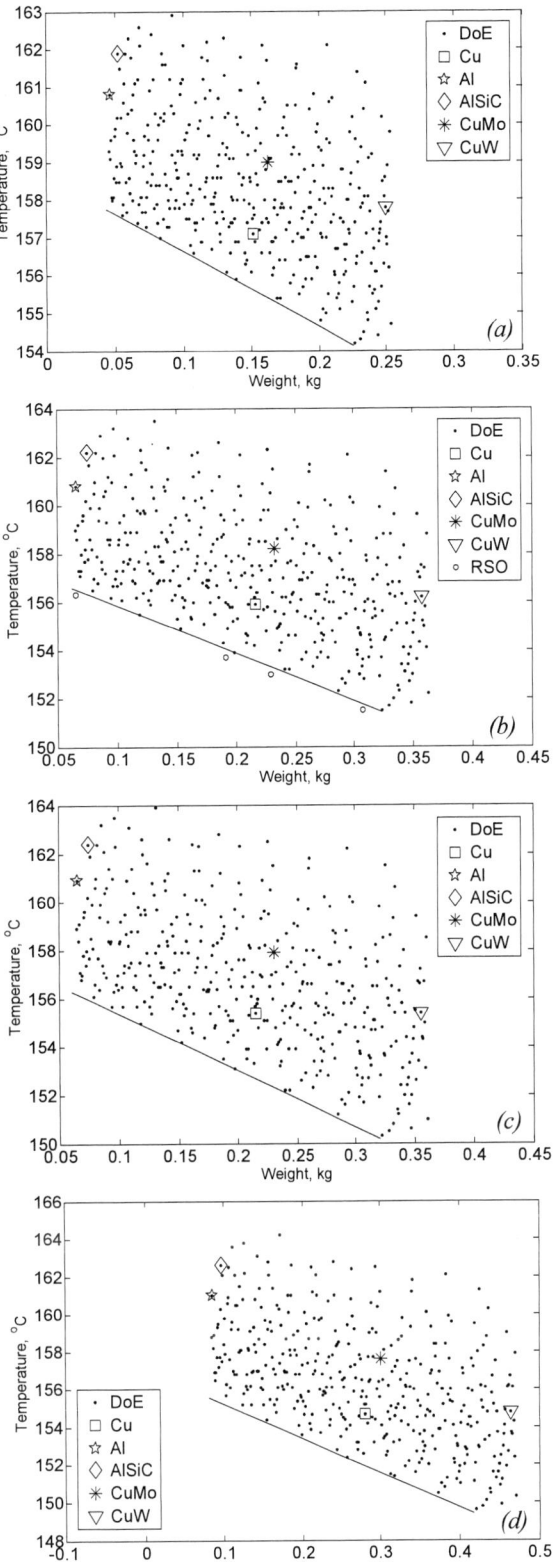

Figure 13: Parametric study results for 4 s operation at 6 l/min flow rate: (a) 1 mm / 1 mm; (b) 2 mm / 2 mm; (c) 1 mm / 3 mm; (d) 3 mm / 3 mm

Thickness variation leads to 1 to 3 °C temperature change. Experiments show that within the specified range of physical properties and thickness, the temperature span is 15 °C. Switching e.g. between copper and AlSiC results in 5 to 8 °C difference, at 65 to 70 % weight difference, depending on the baseplate thickness.

In the third parametric study the coolant volume flow rate is decreased to 2 l/min and this has resulted in a more expressed transient temperature behavior during 20 seconds, as expected, Figure 15.

effect on the temperature at the lower flow rate. At the more expressed transient behavior, Figure 15b, the effect is more significant. The lower flow rate results in 3 to 4 °C worse thermal performance. Such a modest effect can be related to high thermal resistance of the grease layer and the elastomer pad.

Finally, analysis has revealed the designs with similar physical properties of baseplates that form the frontiers, Table 3. In the table the baseplates with high densities (A and B) correspond to the designs with the large weight in the frontier plot, Figure 14. The baseplates with low densities (e.g. G and F) correspond to the designs with small weight and high temperature, on the frontiers. It can be seen from Table 3 that in case of the expressed transient operation (4 s), the trade-off designs have thermal conductivities approaching the one of copper, a specific heat closest to Al or AlSiC, and a significant density variation. There is a region in Figure 14 where the total baseplate weight is around 0.11 kg, and the designs have similar thermal performance, with the chip temperature around 156 °C. These designs are F (2 mm / 2 mm, and 1 mm / 3 mm) and G (3 mm / 3 mm baseplates).

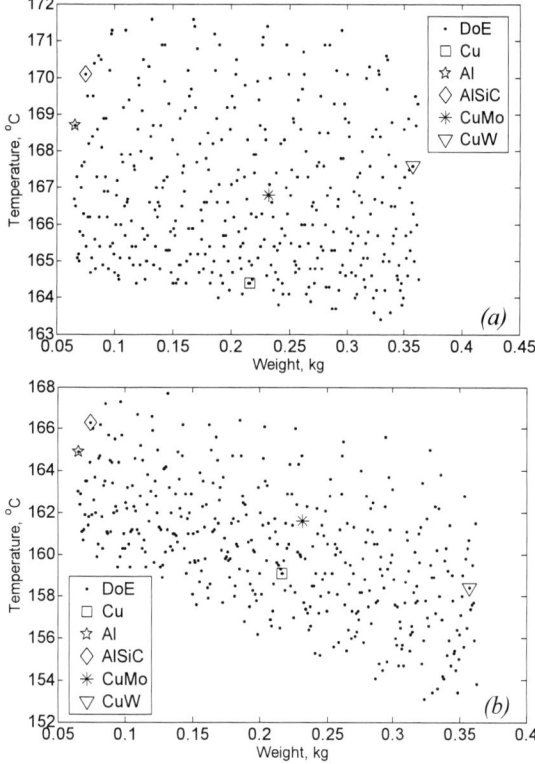

Figure 14: Summary of parametric study results for 4 s operation at 6 l/min flow rate

Figure 15: Parametric study results for 2 mm / 2 mm baseplates at 2 l/min flow rate: (a) 20 s, (b) 4 s

It is clear from Figure 15a and Figure 12a that switching between materials may have somewhat larger

Table 3: Physical properties of the baseplates representing the frontier points

	k, W/(m·K)	ρ, kg/m³	c_p, J/(kg·K)
A	368	13400	958
B	384	11900	909
C	396	9951	871
D	374	8196	939
E	378	6318	920
F	394	4902	869
G	382	3270	825

Table 4 summarizes the results for frontier designs with different baseplate thicknesses, revealing "the lightest" and "the coldest" designs as well as the designs with moderate performance with respect to both criteria.

Table 4: Thermal performance comparison

4-second operation		Design ref.	Weight, kg	Temperature, °C
1mm / 3 mm	coldest	A	0.32	150
	middle	E	0.15	154
	lightest	G	0.08	156
1 mm	coldest	A	0.23	154
	middle	E	0.11	157
	lightest	G	0.06	158
2 mm	coldest	A	0.32	152
	middle	E	0.15	155
	lightest	G	0.08	157
3 mm	coldest	A	0.42	150
	middle	E	0.20	154
	lightest	G	0.10	156

Among the realistic materials for pin-fin baseplates are Al (6063 or 6061), copper and AlSiC variants. The conducted study focuses on an automotive application, and the material selection process has to be supported with material and processing cost considerations. Briefly, copper is at least four times more expensive and three times heavier than Al alloys. AlSiC pin-fin baseplates produced with pressure infiltration casting would cost more than copper metal injection molded baseplates, while having density and thermal conductivity comparable to Al alloys. Assumption made in this study regarding the pin-array on the baseplate, which is considered the same for the mentioned materials, is aimed at future, when it could be possible to make AlSiC pin-fins as high and as dense as copper pins [19].

8. Conclusion

Simulation methodology is developed to quantify the effect of different physical properties and thickness of finned baseplates on the thermal performance and weight of a double-side cooled SiC power module for HEV applications. A simplified transient model of a three-phase inverter suitable for parametric studies is developed and evaluated for two driving cycles and temperature dependent heat losses of SiC components. Parametric studies have revealed a small, up to 3 °C, variation of the component temperature with the baseplate thickness (1 to 3 mm). Designs with the lowest temperature are found to have high thermal conductivity and specific heat, as expected. Investigated practical finned baseplate materials are Al, AlSiC and copper. Switching e.g. between copper and AlSiC results in 5 to 8 °C increase in temperature at 65 to 70 % weight reduction.

Finally, the presented simulation methodology, including the model simplification process, can be employed for other parametric studies involving e.g. power substrate material composition, thermal interface and die attach materials.

Acknowledgements

The financial support from Vinnova through the BIKT program and from Swedish Energy Agency and Vinnova through the SiC Power Center is acknowledged.

References

1. Lim, J-K. *et al*, "Comparison of Total Losses of 1.2kV SiC JFET and BJT in DC-DC Converter Including Gate Driver," Materials Science Forum, Vol. 679-680 (2011), pp. 649-652.
2. Ning, P.Q. *et al*, "Double-Sided Cooling Design for Novel Sandwich Module," *Proc. 28th IEEE Conf. APEC 2013*, Long Beach, CA, USA, Mar. 2013, pp. 616-621.
3. Charboneau, B.C. *et al*, "Double-Sided Liquid Cooling for Power Semiconductor Devices Using Embedded Power Packaging," *IEEE Trans. on Ind. App.*, Vol. 44, No. 5 (2008), pp. 1645-1655.
4. Liang, Z.H. *et al*, "HybridPACK2- Advanced Cooling Concept and Package Technology for Hybrid Electric Vehicles," *Proc. IEEE Conf. VPPC 2008*, Sept. 2008, Harbin, China.
5. Kojima, T. *et al*, "Novel RC Compact Thermal Model of HV Inverter Module for Electro-Thermal Coupling

Simulation," *Proc. IEEE PCC 2007*, Nagoya, Japan, Apr. 2007, pp. 1025-1029.
6. Mrad, S. *et al*, "A Compact Transient Electrothermal Model for Integrated Power Systems: Automotive Application," *Proc. 35th Annual IEEE IECON 2009*, Porto, Portugal, Nov. 2009, pp. 3755-3760.
7. Graovac, D., "Power Modules and Discretes for Hybrid- and Electric Vehicle Applications," Infineon Technologies, Nov.11, 2010.
8. Christmann, A., *et al*, "Reliability of Power Modules in Hybrid Vehicles," *Proc. PCIM Europe 2009*, Nuremberg, Germany, May 2009.
9. Hohlfeld, O., *et al*, "Direct cooled modules - integrated heat sinks," *Proc. 7th Int. Conf. CIPS 2012*, Nuremberg, Germany, Mar. 2012.
10. Gustafsson, U., *et al*, "Static and Dynamic Properties of 4.5kV MOSFETs in 4H and 6H SiC - Simulation Study," *Proc. ICSCRM 1995*, Kyoto, Japan, 1995, Inst. Phys. Conf. Ser. No 142: Ch. 4.
11. Sargos, F., "IGBT Power Electronics Teaching System Principle for Sizing Power Converters," Application Note AN-8005, Semikron, 2013.
12. Staudt, I., "3L NPC & TNPC Topology," Application Note AN-11001, Revision 4, Semikron, Sept. 2012.
13. Zhang, Y.F., *et al*, "Thermal Evaluation of a Liquid/Air Cooled Integrated Power Inverter for Hybrid Vehicle Applications," *Proc. 14th IEEE Int. Conf. EuroSimE 2013*, Wroclaw, Poland, Apr. 2013.
14. Occhionero, M. *et al*, "A New Substate for Electronics Packaging: Aluminum-Silicon Carbide (AlSiC) Composites," *Proc. 4th Annual Portable by Design Conference, Santa Clara*, CA, USA, Mar. 1997, pp. 398-403.
15. Aluminum 6061, www.hmwire.com, date of access: 2014.1.28.
16. Tungsten-Copper Heat Sinks, Torrey Hills Technologies, LLC, www.torreyhillstech.com, date of access: 2014.01.27
17. A Brief Introduction to CTE and CuW, CuMo Thermal Management Materials Introduction, Torrey Hills Technologies, LLC, www.torreyhillstech.com, date of access: 2014.01.27
18. Mausser, H. *et al*, "Normalization and Other Topics in Multi-Objective Optimization," *Proc. of Fields-MITACS Industrial Problems Workshop*, Toronto, Canada, Aug. 2006.
19. Sonuparlak, B., "IGBT Power Module for HEV and EV," *Proc. IEEE APEC 2011*, Fort Worth, TX, USA, Mar. 2011.

Thermal characteristics of SiC diode assembly to ceramic substrate

Kisiel R. [1], Guziewicz M. [2], Myśliwiec M. [1,2], Kraśniewski J. [3], Janke W. [3]

[1] Institute of Microelectronics and Optoelectronics, Warsaw University of Technology, Poland,
Koszykowa 75, 00662 Warszawa

[2] Institute of Electron Technology, Al. Lotników 32/46, Warsaw, Poland

[3] Department of Electronics and Computer Science, Koszalin University of Technology, Poland
Email: kisiel@imio.pw.edu.pl

Abstract

Silicon carbide (SiC) semiconductor diodes are studied for high power and high temperature system applications. Our packaging technology is developing to ensure a working temperature above 300°C for Schottky and PIN diodes. This work presents an investigation on the thermal properties of proposed assembly of SiC die into a ceramic package. Ag micro particles and sintering process were used for the assembly. It was found that thermal resistance of such package is dependent on assembly technology and it is near 3 K/W for Schottky diode in the temperature range from room temperature up to 200°C, and the thermal resistance is in the range 10.5 ÷ 8.7 K/W for PIN diode at temperature in the range from 20°C up to 300°C.

1. Introduction

Silicon carbide (SiC) devices are promising components for high power and high temperature system applications, which working temperature can exceed 300°C. Such devices need a special package, where materials and assembly technology should survive such critical temperature stresses. We propose a ceramic package with working temperatures as high as 350°C, Fig.1. [1]. The key question of such choice is about the way, how heat is dissipated from SiC structure to the ambient. The thermal conductance between SiC device and ceramic substrate generally describes the effectiveness of heat removal or in similar way, an efficiency of a spreading of the power taken from SiC structure away. A promising joint technology is using sintered Ag layer for this purpose [2]. Ag micro particle layers can be sintered at temperatures significantly lower than the melting temperature of Ag, typically 0.2-0.4 T_m. Once sintered the Ag joint will have a melting temperature similar to bulk Ag (961°C). This property permits to avoid the remelting problem in other steps of assembly processes of power devices. Researches from Samsung investigated the thermal resistance of LED package, where die was assembled into package using Ag paste, solder SAC 305 and Au/Sn eutectic bonding. They received the following thermal resistance results: 3.5°C/W for Au/Sn eutectic bonding, 11.5 ÷ 14.2 °C/W for Ag paste and 4.4 ÷ 4.6 °C/W for SAC solder [3]. Our proposal is using Ag micro-particles to assembly the SiC structure to DBC (Direct Bonded Copper) alumina substrate.

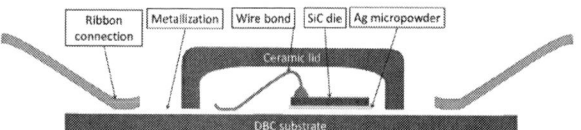

Fig.1 Conception of a ceramic package used in experiments [1]

To calculate the thermal resistance it is necessary to know SiC junction temperature, a dissipated power and case temperature. The junction temperature can be evaluated using optical methods (for example IR camera), physical contact methods (for example thermocouple) or electrical methods [4]. In the electrical method, the SiC chip can be used as a temperature sensor, because some of its electrical parameters are temperature sensitive and can be used as the thermo-sensitive electrical parameters (TSEPs). There is a linear relationship between the on-state voltage drop (V_F) across a p-n junction (corresponding to sufficiently low conduction current) and temperature. Once the calibration of TSEP was finished, the data can be used for junction temperature measurement. We applied this method for the junction temperature measurement. However it is worth to point out, that the chip temperature is not uniform, particularly when an active diode area is smaller than chip surface. So very important is to measure the junction temperature in steady-state condition.

2. Experiments, sample preparation

The test Schottky and PIN diodes were manufactured using as the base the 4H- SiC wafers from Cree Inc. The substrate was 0.368 mm thick 4H n-SiC(0001) wafer with resistivity of 0.02 Ωcm and with n-type epitaxial layers: the 1st layer - n= 1x10^18 cm^-3, the 2nd layer – n=9.89x10^15 cm^-3. Structure of PIN diode was formed as follow. Top epilayer was Al implanted to 150 nm depth through a mask and annealed at temperature of 1600°C for 20 min. to get p-type dopants on the level 2.5×10^19 cm^-3. Circular or rectangular mesas were formed using photolithography and RIE technique.

Before deposition of contact metallization the surface was cleaned in organic solvents (trichloroethylene, acetone, izopropanol) and sequentially etched in boiling NH_4OH-H_2O_2-$5H_2O$ and HCl-H_2O_2-$5H_2O$ solutions for 10 min., native oxide was removed using buffered HF immediately prior to the deposition of contact metallization. The 150 nm thick film of Ti was deposited by DC magnetron sputtering on bottom side. Ohmic contacts (o.c.) were formed by Rapid Thermal Annealing

978-1-4799-4789-8/14 $31.00 © 2014 IEEE

(RTA) in argon at temperature 1100°C (Ar, 3 min.). Electrical properties of ohmic contacts were characterized by linear current-voltage characteristic and specific contact resistance r_c of ~ 4×10^{-5} Ωcm^2 was calculated.

Multilayer metallization to p-SiC regions was formed by sputtering of Ti/Al/Ti (10/30/60 nm) and *lift off* technique. Next contact to p-SiC was formed by RTA annealing at 600°C (1 min.) to form TiAl alloy and following annealing at 1050°C (Ar, 2 min.) to form the ohmic contact. Metallization of the Ti(o.c.) contact was thickened by deposition of Ti/Au (100 nm/500 nm) and the TiAl(o.c.) contacts were thickened by (2 μm)Al layer.

In the case of Schottky diodes a similar ohmic contact on bottom side of the n-SiC wafer was formed at the beginning. The Schottky contact was formed from (50 nm)Ti/ (500 nm)Au bilayer on the top n-SiC epilayer where circular diode was patterned by using photolithography, subsequent DC sputtering of Ti and Au, and next *lift off* technique. No high temperature annealing was performed later on these diodes.

As a package substrate the DBC (Direct Bonding Copper) alumina initially covered by 200 μm layers of Cu from Curamik GmbH were used. The DBC plate was electroplated by (3 μm)Ni(P) and (1.5 μm)Au layers and cut on pieces size 10mm x 25 mm, where on the top of it the bonding pads were formed. Joining process between SiC die and DBC substrate was based on sintering with applying Ag micro particles, a flake shape (AX20LC, Amepox, Poland). To do this, the Ag particles were mixed with isopropyl alcohol and as a liquid was applied onto Au bond pad of DBC substrate. Next the Ag particles mixture was dried in air. The sintering process was performed under 10 MPa pressure in air at temperature 400°C during 40 min. The thickness of sintered layer is about of 20 μm. Next the electrical connections were made using wire bonding technology [5]. The electrical connections between Au bond pads onto SiC Schottky diode and Au pads onto DBC substrate were performed using Au wire diameter 50 μm and thermosonic technology. The electrical connections for PIN diode were made with 100 μm diameter Al wire bonded by using ultrasonic technology. Bond pads onto the PIN diode were formed by Al metallization and the pads onto DBC substrate were made with Ni layer. To seal the package, a ceramic lid was put onto DBC substrate and by using SG 705 dielectric paste (Heraeus) the package was encapsulated.

3. Thermal characteristic measurements

SiC diodes prepared according to description in previous Section have been investigated experimentally. The results of the measurements of electrical and thermal characteristics presented below have been obtained for Schottky diode (device no. 72) and PIN diode (device no. 115).

Current-voltage characteristics for wide range of ambient temperature up to 300°C were measured by a pulse method - to avoid the self-heating phenomenon, therefore the junction temperature was equal to the ambient temperature. Shapes of I-V characteristics are shown in Fig.2 (Schottky diode) and in Fig.3 (PIN diode) and particularly, the differences in the influence of temperature on both types of diodes, correspond to theoretical predictions. It is seen that both diodes work properly in the wide range of temperature. SiC Schottky diodes in the high-current, high-temperature region exhibit large values of the voltage drop that may lead to the device thermal instability [6].

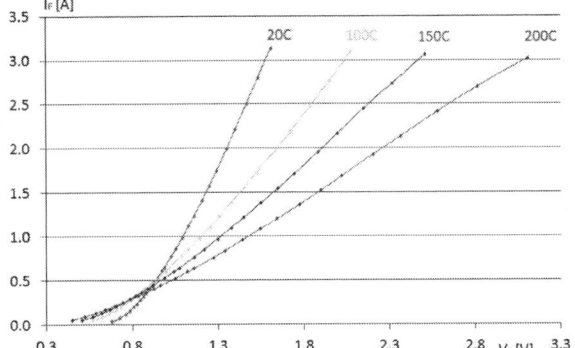

Fig.2 I-V characteristic of Schottky diode no. 72 at different ambient temperatures

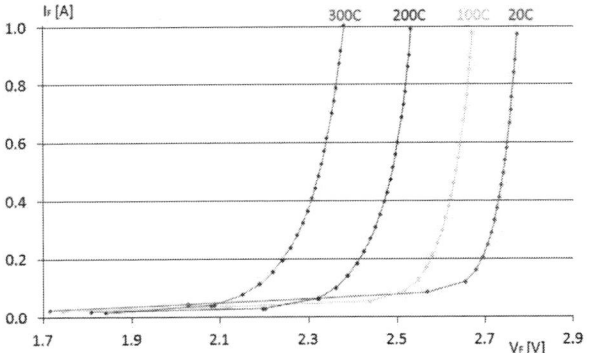

Fig.3 I-V characteristics of PIN diode no. 115 at different ambient temperatures.

For measurements of thermal resistance or impedance, the calibration curves have to be measured first. The voltage drop V_F, corresponding to low conduction current is chosen as temperature-sensitive parameter (that is typical for semiconductor device thermal measurements [7], [8], [9]).

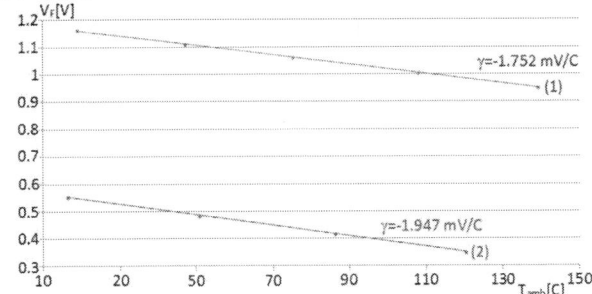

Fig.4 Calibration curves $V_F(T_{amb})$ of Schottky diode (2) no. 72 and PIN diode (1) no. 115 obtained for measurement current of I_M=1mA

978-1-4799-4789-8/14 $31.00 © 2014 IEEE

The calibration curves (i.e. dependences of V_F on T_{amb} at low current) measured for both diodes under test, are presented in Fig. 4. The slope values γ of $V_F(T_{amb})$ lines are shown in figures.

Experimental set-up was controlled by National Instruments PXI workstation equipped with fast AD/DA converter card, which allows an acquisition of samples at a rate of 10MHz. For measurements of junction-to-case thermal impedance $Z_{thj\text{-}c}(t)$, devices under test were mounted on aluminum A6405A profiled heatsink and placed inside of fan assisted temperature FEUTRON chamber 3213/15 allowing precise ambient temperature (T_{amb}) control. During the measurements, temperature of the heatsink was controlled by Hart Scientific 1522 Reference Thermometer with a thermistor probe, and it never exceeded the ambient temperature by more than 0.1K.

Exemplary measured curves of transient thermal impedance $Z_{thj\text{-}c}(t)$ of the investigated diodes for different ambient temperatures T_{amb} are presented in Figs 5 and 6:

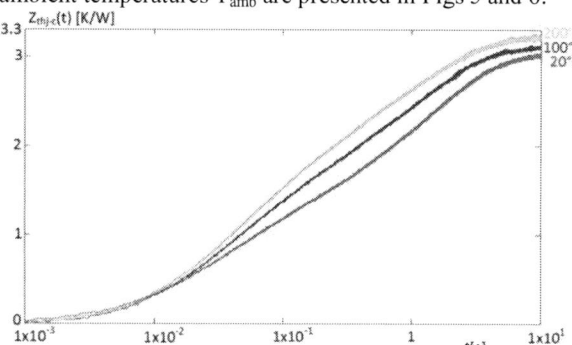

Fig.5 Transient thermal impedance curve of Schottky diode no. 72 for different ambient temperatures. Heating power P_H=4W

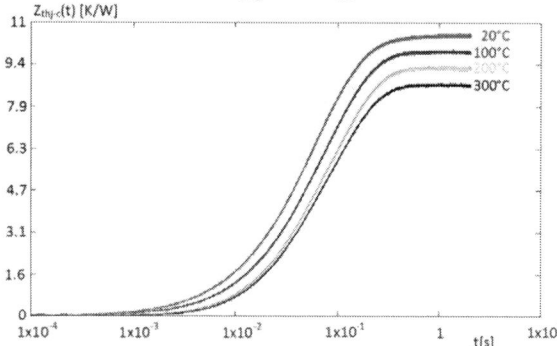

Fig.6 Transient thermal impedance curve of PIN diode no. 115 for different ambient temperatures. Heating power P_H=2W

Thermal resistance $R_{thj\text{-}c}$ values, corresponding to static conditions obtained from the above Figures are given in Table 1.

It is interesting, that for both investigated diodes, the differences in the dependencies of thermal resistance on ambient temperature, may be seen. It has been observed previously, that the influence of temperature on thermal resistance may be different for devices delivered by different manufactures [10].

Table 1. Temperature dependence of thermal resistance R_{th} for measured diodes

	Schottky no. 72	PIN no. 115
R_{th} measured after	10s	1s
R_{th} at T_{amb}= 20°C	3.02 K/W	10.49 K/W
R_{th} at T_{amb}= 100°C	3.12 K/W	9.90 K/W
R_{th} at T_{amb}= 200°C	3.22 K/W	9.32 K/W
R_{th} at T_{amb}= 300°C	-	8.71 K/W

In the presented case the differences can be also connected with the porosity of Ag layer used for assembly SiC diodes to DBC substrate. The Ag microparticles has a flake shape size from few up to 10µm and after sintering such layers can have different porosity. It is possible to see the differences in transient thermal impedance changes versus temperature for Schottky diode and PIN diode. It is worth to underline that the heat flow from junction to case goes through thick metal layers between SiC and ceramic (1.5 µm Au, 3 µm Ni and 200 µm Cu layer) and 200 µm Cu layer on the bottom of DBC ceramic. These metal layers have thickness comparable with alumina thickness 630µm.

The thermal resistance values obtained for assembly PIN diode into ceramic package are comparable with the results obtained by researches from Samsung were thermal resistance rang 11.5 ÷ 14.2 K/W were obtained for SiC diode assembled by Ag paste.

4. Conclusions

Our paper presents results of the study of thermal properties of two types of SiC diodes assembled into ceramic package. Thermal transient impedance as well as thermal resistance were measured in the ambient temperature range from 20°C up to 300°C. The SiC diodes were assembled onto DBC substrate using sintering process with applying flake shape Ag micro-particles. It was found that for Schottky diode the thermal resistance increases slightly from 3.02 K/W at room temperature up to 3.22 K/W at temperature of 200°C and for PIN diode thermal resistance decreases from 10.49 K/W at room temperature to 8.71 K/W at temperature of 300°C. Both diodes were assembled using similar technology. The difference in thermal resistance values can be explained by porosity of Ag sintered layer. Further investigations will be performed to explain these results.

Acknowledgments

The work was supported by The National Science Centre, NCN, Poland (Grant no. N N515 499240).

References

1. R. Kisiel, Z. Szczepański, P. Firek, M. Guziewicz, J. Grochowski, M. Myśliwiec, "Silver Micropowders as SiC Die Attach Material for High Temperature Applications", *Proc. of ISSE 2012*, Bad Aussee, Austria, (2012), pp. 144-148

2. Kim S. Siow "Are Sintered Silver Joints Ready for Use as Interconnect Material in Microelectronic Packaging?" *Journal of Electronic Materials*, (2014 TMS)

3. Hyun-Ho Kim, Sang-Hyun Choi, Sang-Hyun Shin, Young-Ki Lee, Seok-Moon Choi, Sung Yi, Thermal transient characteristics of die attach in high power LED PKG", *Microelectron. Reliab.* (EuroSimE) 48 (2008), pp. 445-454

4. Y.Avens, L. Dupont, Z. Khatir "Temperature Measurement of Power Semiconductor Devices by Thermo-Sensitive Electrical Parameters – A Review" *IEEE Transactions on Power Electronics*, vol.27, no6, (2012), pp.3081-3091

5. R. Kisiel, M. Guziewicz, Z.Szczepański, K. Król „An overview of Materials and Bonding Techniques for Inner Connections in SiC High Power and High Temperature Devices" *33rd International Spring Seminar on Electronics Technology ISSE2010*, Warszawa, (12-16 May 2010), p.128-132

6. W. Janke A. Hapka, The thermally induced limitations of SiC SBDs operation conductions, *Microelectronics Journal* 43 (2012), pp. 656-660

7. Department of defense MIL-STD-750D „Test method standard semiconductor devices", (28 Feb 1995)

8. EIA/JESD51 "Methodogy for the Thermal Measurement of Component Packages (Single Semiconductor Device)", Dec 1995 - Nov 2010

9. W. Janke, J. Kraśniewski "The investigations of transient thermal characteristics of microwave transistors", *Metrol. Meas. Syst.*, vol. XVI (2009), no.3, pp.433-442

10. W. Janke, A.Hapka, M. Oleksy, „Przejściowa impedancja termiczna diody Schottky'ego z węglika krzemu w szerokim zakresie temperatur", *Przegląd Elektrotechniczny* 11/2009, str. 203-205

Theoretical and Experimental Study of Thermal Management in High-Power AlInGaN LEDs

A.E. Chernyakov[1], A.L. Zakgeim[1], K.A. Bulashevich[2], S.Yu. Karpov[2], V.I. Smirnov[3], V.A. Sergeev[3]

[1]Submicron Heterostructures for Microelectronics Research & Engineering Center, RAS,
26 Polytekhnicheskaya Str., 194021 St. Petersburg, Russia
e-mail: chernyakov@mail.ioffe.ru
[2]STR Group – Soft-Impact, Ltd., P.O.Box 83, 194156 St. Petersburg, Russia
[3]Institute of Radio-engineering and Electronics of RAS, Ulyanovsk, Russia

Abstract

Current spreading in a high-power flip-chip light-emitting diode (LED) and its effect on the chip thermal resistance has been studied both theoretically and experimentally. Thermal resistances of various LED units have been determined by measuring the forward voltage relaxation under pulsed current excitation of the LED at varied duty cycle. The total thermal resistance of the chip is found to rise by ~20% while the LED operating current increasing from zero to 1 A.

The current density distribution in the LED active region predicted by coupled simulations of the current spreading and heat transfer agrees well with the measured near-field distribution of the light emission intensity. The observed rise in the thermal resistance is attributed to current crowding producing lateral non-uniformity in the temperature distribution inside the LED chip.

1. Introduction

The driving current and chip area of commercial high-power LEDs are continuously increased from year to year to provide ever higher output light flux [1]. This trend is accompanied by complication of the electrode geometry aimed at avoiding their shading effect on the light extraction efficiency [2] and wide use of the flip-chip mounting with the substrate removal after growth of LED structure. The new developments require more attention to pay to the thermal management of LEDs, commonly assessed in terms of the chip thermal resistance.

Conventionally, the thermal resistance of an LED chip is considered as the parameter independent of the driving current. This is valid, however, only at a uniform distribution of the heat generation rate inside the chip. The increase in the current leads normally to the current crowding, i. e. lateral non-uniformity of the vertical current density through the active region with the high-current regions formed next to the mesa edges. These regions provide dominant contribution to the heat generation thus forming a non-uniform lateral temperature distribution in the chip. As a result, the effective area of the heat release becomes smaller, and the heat removal from the chip to a submount is suppressed, as compared to the uniform case.

This work is aimed at detailed study of the current crowding effect on the LED thermal resistance and its variation with the driving current.

2. Experimental

Commercial Luxeon Rebel ES LED chip produced by Philips Lumileds with the area of 1340×1340 μm^2 was used in our study. To provide efficient heat sink, the chip was mounted with p-layer down on a ceramic submount and then on a metal-core printed circuit board (MCPCB). The schematic of the chip is shown in the inset of Fig. 2. The MCPCB was then mounted onto a radiator.

Thermal resistances of individual LED units were determined using the thermoelectric analogy. The heat transfer from the active region, through the ceramic substrate, and MCPCB was considered similarly to the current flow in an equivalent electrical circuit. There are two conventional equivalent circuits suggested by Foster and by Cauer (see Fig. 1). The Foster's circuit assumes parallel connection of resistors and capacitors, whereas in the Cauer's circuit all the capacitors are connected to a common bus. The capacitors in the circuits account for thermal capacities of the respective LED units and affect only transient characteristics of either heating or cooling.

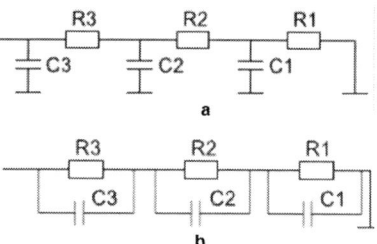

Figure 1: Cauer's (a) and Foster's (b) equivalent electric circuits used for thermal analysis.

Initially, the LED chip is driven by a test current of 5 mA. The temperature of the whole setup including LED, MCPCB and radiator is gradually increased, while the forward voltage corresponding to the test current is tabulated as a function of temperature. The linear voltage-temperature dependence with the coefficient of −1.733 mV/K is established by those measurements to be used further for evaluation of the p-n junction temperature. The test current is chosen in

978-1-4799-4789-8/14 $31.00 © 2014 IEEE

such a way, as to avoid the device self-heating. The temperature was measured at that with the accuracy better than 0.5 K.

Next, the thermal transient characteristics of the LED are studied by step-like switching on a constant driving current. While the LED is gradually heated, the heat generated in the chip is partly consumed by heating of the LED units (chip, submount, and MCPCB) and partly transferred to the radiator and then to the ambient air. The evolution of the p-n temperature during heating is measured by short pulses of the test current as described above. The post-processing analysis of the data within the structure-function approach allows extracting the components of the equivalent electrical circuit. Let C_Σ and R_Σ be the total thermal capacitance and total thermal resistance corresponding to the gap between the LED p-n junction and a certain horizontal plane inside the LED. Plotting the dC_Σ/dR_Σ derivative, i.e. the so called differential structure function, allows graphical identification of the peaks which indicate where the thermal properties of the LED are changed abruptly (Fig. 2). These peaks can be attributed to specific boundaries inside the LED while the distance between the peaks shows thermal resistance of the respective units of LED. More details on the measurement procedure can be found in [3] and references therein.

The above method is implemented in two different measurement equipment: (i) commercial Thermal Transient Tester T3Ster by MicRed, Ltd. and (ii) original equipment called LED Meter. They employ different heating modes and different procedures for the data analysis. T3Ster heats LED with a constant current and uses the Cauer's circuit for the data processing, while LED Meter utilizes a sequence of current pulses with of different duration and uses the Foster's circuit for processing.

Differential structure functions obtained with T3Ster at the driving currents of 100, 300, 700, and 900 mA are presented in Fig. 2. One can see that first two peaks attributed to the thermal resistances (i) from the p-n junction to the chip surface and (ii) from the chip surface to the submount are nearly the same for all the heating currents. Meanwhile, the third peak associated with the thermal resistance of the submount becomes shifted to higher resistances, as the current is increased, which indicates the growth of the respective thermal resistance with current. The increase of the total thermal resistance is also clearly seen from the shift of the last peak. The thermal resistance provided by MCPCB is estimated as 7 K/W, which is the most valuable contribution among all the LED units.

Variation of the total thermal resistance obtained experimentally with T3Ster and LED Meter are shown in Fig. 3, being in good agreement with each other. The thermal resistance is found to rise by ~20% from 10 K/W up to 12 K/W under increasing the driving current from 100 mA to 900 mA.

The T3Ster equipment is basically designed for electronic devices and its procedure of the data processing assumes that the electric power supplied to a device is completely converted into the heat. In LEDs, however, a certain fraction of the electric power is converted into the light and, therefore, does not contribute to the device heating. For consistency, all the results in this paper are reported by neglecting the light emission. So, the actual thermal resistance of the chip should be higher than that estimated from the above measurements by the factor of $1-\eta$ where η is the LED wall plug efficiency, which was about 30% with a slight variation over the studied current range.

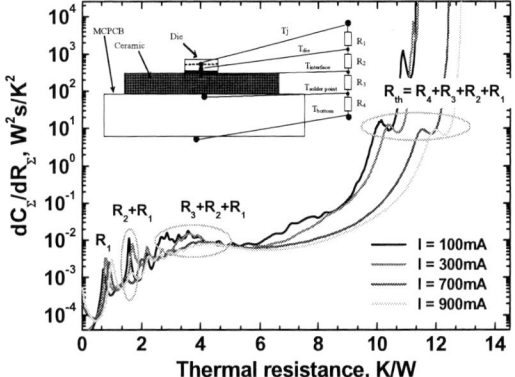

Figure 2: Differential structure functions obtained with T3Ster. Peaks indicate the sums of thermal resistances. Schematic of the chip and partial thermal resistances is given in the inset.

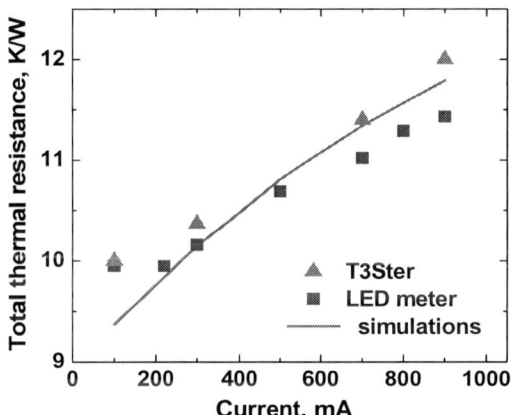

Figure 3: Total thermal resistance of the LED chip measured with T3Ster and LED Meter (symbols). Simulation results are shown by solid line.

3. Simulations

Coupled simulation of the current spreading in the LED chip and heat transfer in the LED chip and the submount was carried out by using the hybrid 1D/3D model [4]. The active region was considered as a thin layer with the Shockley's dependence of the current density j_{p-n} on the local p-n junction bias U_{p-n}:

$$j_{p-n} = j_s(T) \cdot \left[\exp\left(U_{p-n} / mkT\right) - 1\right] \quad , \qquad (1)$$

where m is the ideality factor and the dark current j_s is assumed to increase exponentially with temperature:

$$j_s(T) = j_0 \exp(-\hbar\omega / kT) \quad . \qquad (2)$$

Here $\hbar\omega$ is the photon energy and j_0 is the fitting parameter for the magnitude of the dark current which was adjusted to fit the experimental turn-on voltage. Internal quantum efficiency (IQE) dependence on the current density was approximated by the ABC-model with the recombination constants $A = 1\times10^7$ s^{-1}, $B = 2\times10^{-11}$ cm^3/s, and $C = 3\times10^{-30}$ cm^6/s.

Three-dimensional coupled problem of the current spreading and heat transfer was solved numerically using the commercial SimuLED package [5]. The sheet resistance of the n-GaN contact layer, 22 $\Omega/\square$, and the specific contact resistance of the contact to p-GaN, 10^{-3} $\Omega\cdot$cm^2, were chosen, respectively, as input parameters.

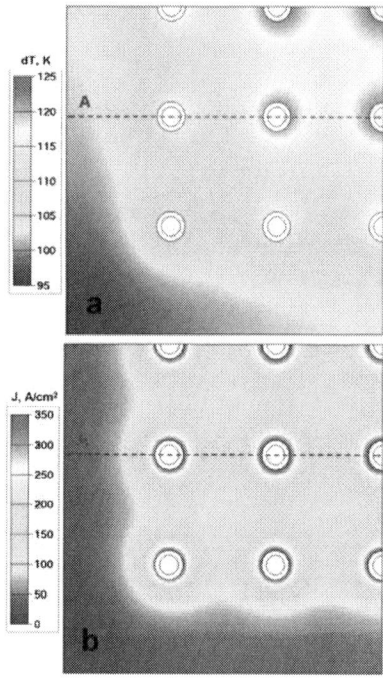

Figure 4: Simulated lateral distribution of the active region overheating (a) and vertical current density (b) at the driving current of 2A.

In the thermal simulations, we assumed the heat to be released through the bottom surface of the submount where the boundary conditions of the third kind for heat flux were used. The heat transfer coefficient was chosen to be 5×10^4 W/(m$^2\cdot$K) to fit the overall thermal resistance to the measured value. It is also more or less consistent with the measured thermal resistance of MCPCB, 7 K/W (see Sec.2). The submount thermal conductivity was assumed to be of 50 W/(m·K). The thermal resistance was calculated as a ratio of the

average overheating of the active region to the total electric input power to keep consistence with our experimental data. We would like to note that the submount should be included into the computational domain to enable lateral heat transfer. It is important for correct prediction of the lateral temperature distribution, as the bulk submount provide more intensive lateral heat transfer than the chip with removed substrate by itself, whose thickness is only ~3-5 μm.

Computed lateral distributions of the active region overheating, i.e. difference between the local and ambient temperature assumed to be of 300 K, and vertical current density are displayed in Fig. 4. Only a quarter of the whole chip is shown here due to the symmetry of the chip design. The empty circles in the figure represent n-mesas etched through the active region [6]. One can see the current density to localize substantially near the n-mesa edges, indicating considerable current crowding.

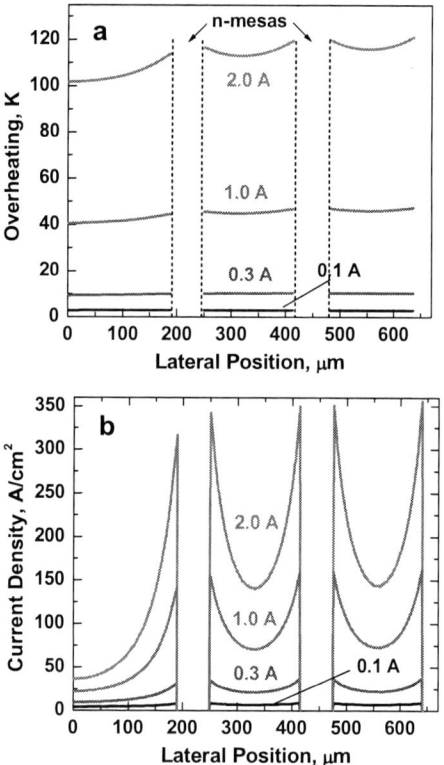

Figure 5: Lateral distribution of the active region overheating (a) and vertical current density (b) in the cross-section A of Fig. 4 simulated for various driving currents.

The active region overheating and current density in the cross-section A (see Fig.4) simulated for various driving currents is plotted in Fig. 5. It is seen the lateral current density non-uniformity to increase dramatically with current. At driving currents of 700 mA and higher, the maximum current density at the mesa edge is more than two times higher than the typical current density between the mesas. The current density near the external chip boundary is even lower.

Generally, the heat generation rate in the active region is roughly proportional to the current density with some deviation from the linearity coming from the lateral variation of the IQE and U_{p-n}. Nevertheless, the lateral distribution of the overheating is predicted to be much more uniform than that of the current density. Simulations show the lateral temperature variation not to exceed ~20% of the mean overheating at the driving current as high, as 2A. The latter fact is attributed to intensive lateral heat transfer in the submount, indicating the importance of its thermal properties for the overall thermal management of the LED chip.

4. Current Crowding

To access the current spreading pattern experimentally, the near-field distribution of electroluminescence (EL) intensity was monitored by using Mitutoyo optical microscope and Canon EOS 5D camera with 12 Mpxl matrix. The measured distribution of the near-field emission intensity at the driving current of 2A is shown in Fig. 6. The observed increase of the emission intensity next to the n-mesa edges is the evidence for considerable current crowding effect. Evolution of the emission intensity in the cross-section A with the driving current is presented in Fig. 7. It demonstrates clearly that the current non-uniformity becomes ever more pronounced at high driving currents.

5. Conclusions

Total thermal resistance of a high-power flip-chip LED and its constituents are determined from the measurements of the chip thermal transient characteristics during the device heating by a pulsed current. The total thermal resistance of the LED chip is found to grow by about 20% from 10 K/W up to 12 K/W while the driving current is increased from zero to 1A. The above values of the thermal resistance are obtained assuming complete conversion of the input electrical power to heat. The actual thermal resistance is higher by the factor of $1-\eta$ where $\eta \approx 0.3$ is the LED wall plug efficiency.

The variation of the thermal resistance is attributed to the effect of current crowding near the n-mesa edges and effective reduction of the area of intensive heat release from the chip into submount. This conclusion is supported by the coupled simulation of the current spreading and heat transfer in the LED. The existence of substantial current crowding is also confirmed by

direct observation of the near-field emission intensity pattern.

The results obtained point out the importance of the submount thermal properties for both reduction of the total thermal resistance of the chip and getting more uniform temperature distribution in the LED active region. A proper choice of the submount material seems to be rather critical for improvement of the overall thermal management in high-power LED chips.

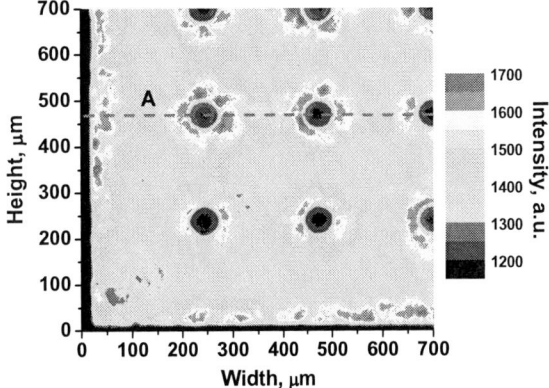

Figure 6: 2D near-field EL intensity distribution measured at 2A current. A quarter of the chip area is displayed here due to symmetry of its design.

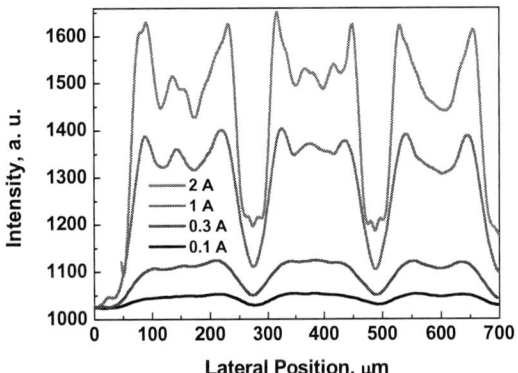

Figure 7: Near-field EL intensity distributions in the cross-section A of Fig. 6 measured at different driving currents.

Acknowledgments

The work was partly supported by Presidium of the RAS: Basic Research Program No. 24.

References

[1] M. R. Krames, O. B. Shchekin, R. Mueller-Mach, G. O. Mueller, Ling Zhou, G. Harbers, and M. G. Craford, "Status and Future of High-Power Light-Emitting Diodes for Solid-State Lighting," J. Display Technol., vol. 3, no. 2, p. 160, 2007.

[2] M. V. Bogdanov, K. A. Bulashevich, O. V. Khokhlev, I. Yu. Evstratov, M. S. Ramm, and S.

Yu. Karpov, "Current crowding effect on light extraction efficiency of thin-film LEDs," Phys. Status Solidi C, vol. 7, no. 7-8, p. 2124, 2010.

[3] M. Rencz and V. Székely, "Semiconductor Thermal Measurement and Management Symposium, 2004. Twentieth Annual IEEE," in Structure function evaluation of stacked dies, Budapest, 2004.

[4] M. V. Bogdanov, K. A. Bulashevich, I. Y. Evstratov, A. I. Zhmakin and S. Y. Karpov, "Coupled modeling of current spreading, thermal effects and light extraction in III-nitride light-emitting diodes," Semiconductor Science and Technology, vol. 23, no. 12, p. 10, 2008.

[5] [Online]. Available: www.str-soft.com.

[6] O. B. Shchekin, J. E. Epler, T. A. Trottier, T. Margalith, D. A. Steigerwald, M. O. Holcomb, P. S. Martin, and M. R. Krames, "High performance thin-film flip-chip InGaN-GaN light-emitting diodes," Appl. Phys. Lett. vol. 89, no. 7, p. 071109, 2006.

Characterizing and modelling the behaviour of an isotropic conductive adhesive in view of electronic assembly fatigue life studies under thermal cycling

S. Pin[1,2,3], M. Sartor[1], L. Michel[1], J. Parain[2], S. Dareys[3]

[1] Université de Toulouse ; ISAE, INSA ; Institut Clément Ader ; 10 av. Edouard Belin, F-31055 Toulouse, France,
[2] Thales Alenia Space, 26 Avenue Jean François Champollion, 31100 Toulouse, France
[3] CNES, Centre spatial de Toulouse, 18 avenue Edouard Belin, 31401 Toulouse Cedex 9, France

samuel.pin@isae.fr

Abstract

Isotropic conductive adhesives (ICA) are now widely used in industry for a panel of applications such as space electronics manufacturing. Epoxy based adhesives are compliant enough to reliably bond bi-materials with an important coefficient of thermal expansion (CTE) mismatch. However, they are still under investigation for the characterization of their fatigue behaviour. Engineers lack of life prediction tools, although they are available for metallic materials. It is all the more necessary when it comes to severe on-ground qualification tests used to accelerate the demonstration of the die attachment reliability.

This study deals with an epoxy based ICA with 80 wt% of silver flakes. The objectives are to characterize the glass/amorphous behaviour of the adhesive and to investigate its numerical implementation in ABAQUS for the on-ground test temperatures.

The adhesive viscoelasticity becomes more preponderant with increasing of the temperature. For the time being it has been characterized between 20°C and 50°C. Tensile and creeps tests have been performed in different conditions identification purposes. A viscoelastoplastic model called Two-layers has been fitted to experimental results with success to get the evolution of the material coefficients in function of temperature. The next step consists in the implementation of the complete adhesive behaviour in an ABAQUS 2D model of the microelectronic components in thermal cycling.

1. Introduction

The microelectronic domain for space applications still offers a panel of important challenges in terms of components scale, production efficiency and reliability. A die attachment technology has to be as compact as possible, as well as flexible for being used in different designs and has to be reliable for 18 years of mission duration. This makes the manufacturer to develop and improve methodologies like chip bonding with Isotropic Conductive Adhesive (ICA).

In comparison of soldering, epoxy based adhesives allow a better flexibility of designing thanks to their lower setting temperatures and their compliance regarding environmental loads.

Indeed, the stack of three materials with different properties leads to a Coefficient of Thermal Expansion (CTE) mismatch. Differential strains between the two adherents or with the adhesive itself result in a loading of the whole assembly to accommodate in function of the temperature similarly to laminates.

In orbit, the multiple eclipses and sun exposures of the satellite impose thermal cycling to the satellite. Onboard housing assures thermal control of the payload along the orbits. It does not avoid thermal cycling of the electronics that can experience variations around 30°C, their operating temperature. Thus, the reliability of any chip assembly is engaged and has to be qualified on-ground.

According to ESA specification [1], such assemblies have to endure at least 500 thermal cycles between -55°C and +125°C as presented in Figure 1.

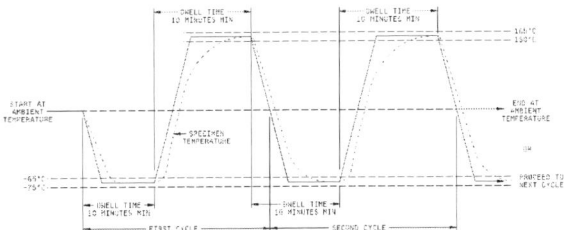

Figure 1 : ESA standard thermal cycle for any electronic component aboard

These tests are defined from an American standard [2] which imposes severe conditions of cycling on purpose. Indeed, the acceleration of fatigue accumulation is calculated in function of the range of testing temperatures. This is a direct legacy from fatigue of metallic materials.

Coffin and Manson have both assumed that the stress levels of the Low Cycles Fatigue domain (LCF) are high enough to implicate plastic strain in metallic materials [3]. Accumulating it throughout the cycles would lead to earlier rupture compared to High Cycles Fatigue (HCF) reaching over 10^6 cycles. They also explained that for this range of stress, the logarithmic law can be approximated with respect of the amount of accumulated plastic strain only. This is known as the Coffin-Manson law given in Eq. (1)

$$N_f = C \Delta \epsilon_p^{-b} \qquad (1)$$

where N_f is the number of cycles at failure, $\Delta \epsilon_p$ is the amount of plastic strain accumulated in one cycle and C and b are intrinsic material constants determined experimentally.

Further studies showed that this can be applied to thermal fatigue in the case of a large temperature range. Actually, thermal fatigue curves are a homothetic transformation of mechanical fatigue curves for the same average temperature. Eq. (2) is the fatigue law used for thermal cycling life prediction.

$$N_f = C' \Delta T^{-b'} \qquad (2)$$

The acceleration parameter can then be calculated from Eq. (2). Therefore, the qualification tests have to be severe to reach an equivalent fatigue state comparable to the system end of life with only a few cycles of high amplitude.

Norris and Landzberg have proposed another formulation for solder fatigue by considering creep behavior as shown by Eq. (3) from [4].

$$N_f = C f^n \Delta \epsilon_p exp \left(\frac{-E_a}{k T_{max}} \right) \qquad (3)$$

where f is the frequency of cycling, T_{max} is the temperature max, k is the Boltzmann constant and E_a is the activation energy defined experimentally as well as n.

The same strategy consisting in the modification of the Coffin-Manson fatigue law is adopted to fit with ICA bonding assemblies, the case of study. Some authors proposed novel formulations of a polymeric fatigue law as Sims et al. [5] for glass/epoxy composites and Rusanen et al. [6] for chip bonding. The first author presented a simple experimental law based on applied stress level for composites. Rusanen et al. come out with a law taking into account only creep strain range that is accumulated over the cycles. This inelastic strain is calculated by modelling the viscous effects with a Maxwell model without considering reversible and irreversible strain apart.

This study deals with the identification of what is meant to be accumulated during thermal cycling in the adhesive joint. The investigations start with the full characterization of the adhesive in its bulk configuration, scope of this paper. The main objective is to select a model that provides all the elements to describe the adhesive behavior. Then the material coefficients have to be identified at different temperatures using experimental results. The outputs of this work are meant to be implemented in a full 2D numerical model if a microelectronic chip available in ABAQUS to reproduce the mechanisms generating fatigue which are activated during thermal cycling.

2. Specimens preparation

To begin with the adhesive bulk characterization, a tensile test campaign has been carried out on "dog-bones" specimens. Their design is based on the standard ISO EN 527-2 but with smaller dimensions because of the cost of the studied commercial ICA. X-ray control of the specimens' health shows presence of porosities in the bulk on both intact and tested ones (cf. Figure 2).

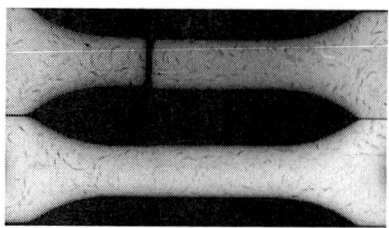

Figure 2 : X-ray inspection of the molded specimens

Porosities observed previously have to be removed. A solution is to mold a thinner volume so the adhesive is free to degas and the temperatures are more uniform across the section.

Thin plates of adhesive have been molded between silicone sheets. Adhesives bars are cut out from these plates with dimensions of 35 by 3 mm. Two thicknesses of 0.5 and 1 mm have been validated by X-ray scanning. This order of thickness makes previous tensile tests no longer possible. The new design of the specimens is thus adapted to a Dynamical Mechanical Analysis (DMA) facility which is made to characterize small polymeric specimens according to the standard ISO 6721. Health of the adhesive bars is largely improved (cf. Figure 3). Hence, the material characterization is performed from these specimens.

Figure 3 : Adhesives bars to be tested with a DMA facility

3. Experimental characterization

The material characterization using DMA facility begins with tensile tests at 0.1mm/min of loading rate. The dimensions of specimens allow reaching failure stress with the DMA low capacity of loading. The testing temperature is controlled within the DMA facility by means of a heater, liquid air and housing. Strain – stress curves of tensile tests at room temperatures are plotted in Figure 4.

978-1-4799-4789-8/14 $31.00 © 2014 IEEE

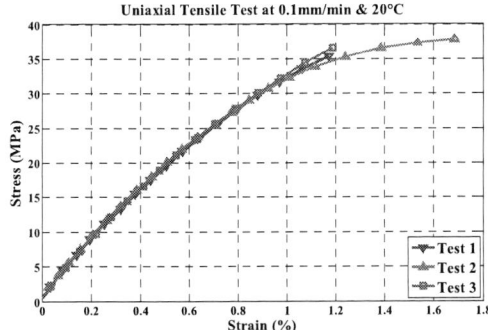

Figure 4 : DMA Tensile Tests at room temperature

The apparent Young's modulus of 5 GPa was measured for low strain levels to get a good approximation. Failure points are not well repeated but lie from 35 MPa to 40 MPa of stress and from 1.2% to 1.7% of strain. Additional tensile tests have been performed at low temperatures where the adhesive behaves more like an elastoplastic material with a negligible viscous aspect. Table 1 summarizes the measured properties showing the importance of test temperature.

Table 1 : Material properties (average values)

Temperature (°C)	-20	20	40	50	100
Young's modulus (GPa)	7	5	3	0.3	0.1
Failure stress (MPa)	50	40	25	15	3

Moreover, dependence to loading rate has been noticed. The adhesive gets this significant viscoelastic behavior from its epoxy matrix. The rest of the material characterization will thus be focused on viscous phenomena that may be involved in both elastic and plastic responses.

Creep tests have been chosen to deal with viscoelasticity according to [7]. They have been preferred to relaxation tests because it is more convenient with the DMA facility pneumatic actioner following the loading cell. The creep tests performed in this study consist in applying a constant stress level on a specimen during 10 min and release it under full monitoring during 10 min to 60 min. Figure 5 presents a 10 & 10 min test at 15 MPa of holding stress.

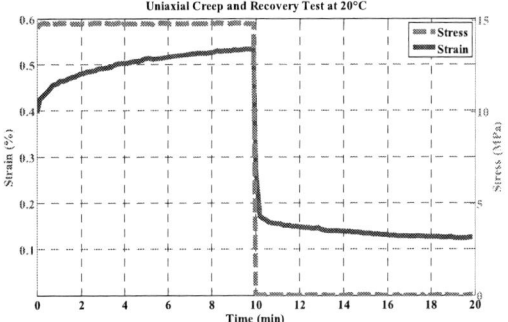

Figure 5 : Creep and Recovery test at 20°C with 10 min holding and releasing steps

Loading and unloading slopes have to be as short as possible to be approximately instantaneous compared to the characteristic time response of the material.

Multiple creep and recovery tests have been performed to fully characterize the viscoelastic behavior of the material. A first approximation of the viscoelastic components can be made with an asymptotic study of the strain – time curves for long enough duration. This becomes more difficult for higher temperatures as the characteristic time response of the material is too long to reach saturation in the creep step (10 min) as shown in Figure 6 with a testing temperature of 40°C.

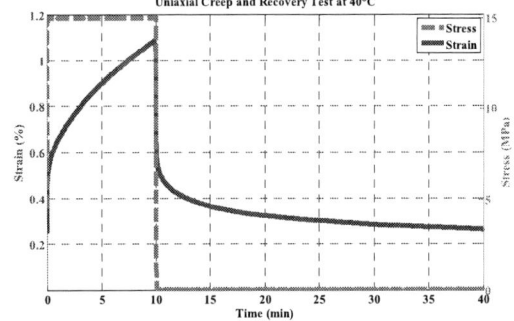

Figure 6 : Creep and Recovery test at 40°C showing long term characteristic time

The reader is invited to note that horizontal tangent is almost reached at the end of recovery in both tests with a non-zero value of strain. It is important to bear in mind that the ESA standard requires thermal shocks with at least 10 min of dwell time at temperature as shown in Figure 1. It means that if recovery must occur during a step, new loading step begins after 10 min without getting back to its initial state. In other words, both elastic recoverable strain and plastic strain can be gathered under inelastic residual strain which is accumulated over the cycles.

In this study, inelastic residual strain is considered with a plastic approach using hardening formulation. Plastic strain occurs when applied stress gets over the initial yield stress. To characterize the hardening law that fits, creep and recovery tests have been performed at different loading levels (cf. Figure 7).

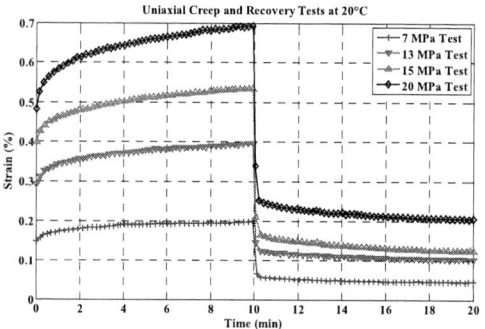

Figure 7 : Residual strain at the end of recovery is measured fo multitple loading levels

Measurement of the residual strain at the end of the recovery step allows establishing a hardening law in function of applied creep stress (cf. Figure 8).

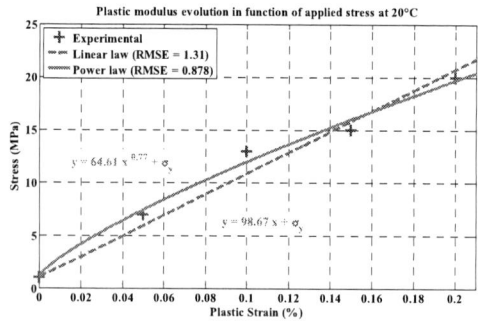

Figure 8 : The power law fits to the experimental residual strains at 20°C

At room temperature, the evolution of adhesive hardening can be taken linear or analogous to a power law like in Ramberg-Osgood hardening formulation. The starting point at zero plastic strain corresponds to the yield stress which is determined experimentally. There is a larger gap between the Root Mean Squares Errors of the approximating curves for 50°C tests as shown in Figure 9. It allows justifying the use of a power law to drive hardening in function of stress.

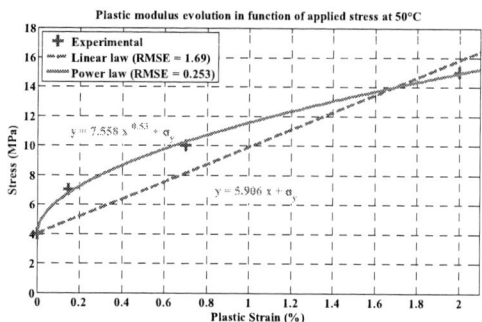

Figure 9 : The power law fits to the experimental residual strains at 50°C

A more effective way to investigate apparent plasticity is to perform different creep levels on a single specimen. This is called a Cyclic Creep and Recovery (CCR) test. They have been used by [8] to identify a nonlinear plastic law for a viscoelastic material. Figure 10 shows a succession of 3 creep and recovery steps at different stress levels.

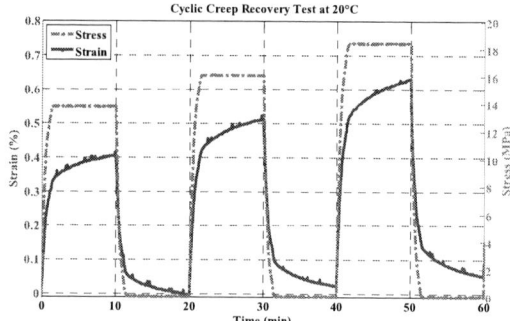

Figure 10 : Cyclic Creep Recovery tests help to identify the plastic behavior of a viscoelastic material

It allows finding in one run the evolution of plastic strain along with cycling at the condition that recovery is long enough to make residual strain appear. Indeed elastic strain, not yet recovered, accumulated over the cycles is a source of error if recovery steps duration is not long enough.

To investigate the time dependence of the inelastic domain, it is necessary to test the material under different holding durations. Creep steps have been thus performed successively on a single specimen similarly to a CCR test (cf. Figure 11).

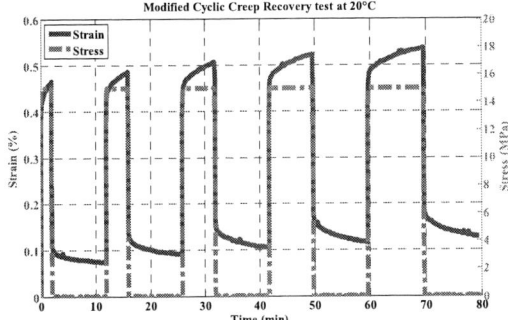

Figure 11 : The modified Cyclic Creep Recovery test allows characterizing viscoplasticity if it exists

For long enough recovery steps, residual strain keeps on increasing because of plastic strain flowing in function of time. It is still under investigation but it has to be considered for the full definition of the material model.

4. Results and numerical identification

The following section deals with material coefficients identification using numerical calculations fitting to experimental plots. In the previous section, model specifications have been defined regarding the experimental results.

A proper model has to be defined to perform identification and ABAQUS computation of the adhesive behavior. The most interesting viscoelastoplastic material available in ABAQUS is called the Two-layers model inspired by the work of Kichenin [9] (cf. Figure 12).

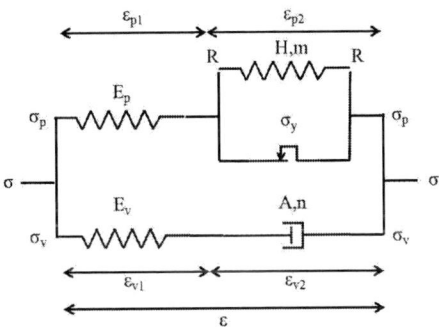

Figure 12 : Two-layers schematic representation

The damper has a double effect on the global behavior because it is placed in parallel of the "plastic branch" and in series of the "viscous branch". Therefore, its time delay is applied to both the elastic and plastic domains. The drawback of this "iso-strain" model is that only a single viscous law can be implemented. It means that plastic flow will behave exactly the same way in function of time that elastic flow.

A simulation code, based on the Two-layers model, has been developed using MATLAB. The identification by data fitting has been performed with the minimization of the least squares error between the numerical strains calculated on MATLAB and the experimental data.

The Two-layers model has got both the plastic and viscous "branches" laws based on the Von Mises criterion and an equivalent strain. For viscous flow, a Norton potential is formulated as follows:

$$\dot{\epsilon}_{v2}^{eq} = A(\sigma_v^{VM})^n \qquad (4)$$

The equivalent viscous strain rate is defined as:

$$\dot{\epsilon}_{v2}^{eq} = \sqrt{\frac{2}{3}\dot{\epsilon}_{v2} : \dot{\epsilon}_{v2}} \qquad (5)$$

The von Mises stress σ_v^{VM} in the viscous "branch" is:

$$\sigma_v^{VM} = \sqrt{\frac{3}{2}s_v : s_v} \qquad (6)$$

where s_v is the deviatoric stress relative to σ_v.

A Ramberg-Osgood plastic law has been chosen, considering the results presented in the previous section. Therefore, the equivalent plastic strain rate of the same form than (5) which corresponds to the plastic multiplier in the plastic "branch" is defined as follows:

$$\dot{\epsilon}_{p2}^{eq} = \left(\frac{\dot{R}}{H}\right)^{1/m} \qquad (7)$$

H and m are intrinsic material coefficients and R is the isotropic hardening variable defined in:

$$f_p = \sigma_p^{VM} - R - \sigma_y \qquad (8)$$

where f_p is the yield surface function.

A 1D calculation using an explicit method based on these criteria is reiterated for fitting. The optimization is realized with the *fmincon* function available in MATLAB and starts with a set of intervals from which each coefficient is taken. If a minimum is found, the iterations stop. Otherwise, if a bound is reached by a coefficient, the limits are readjusted and the optimization restarts from the last point.

Fitting a single curve from a creep and recovery test does not allow identifying hardening coefficients properly. Multiple stress levels are necessary to fully fit the hardening law. It means that several creep and recovery tests must be treated at the same time by the MATLAB script.

An insight of the fitting quality achieved with multiple creep and recovery tests for 20°C is given in Figure 13.

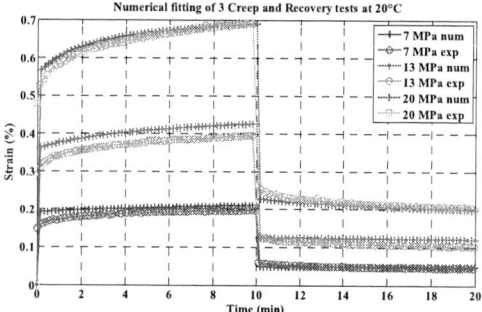

Figure 13 : Fitting insight of 3 creep and recovery tests at 20°C

The global behavior is fairly reproduced. Creep levels and residual strains are reached for every test. Progression of the creep flow and recovery in function of time is also accurately reproduced for tests at 50°C (cf. Figure 14).

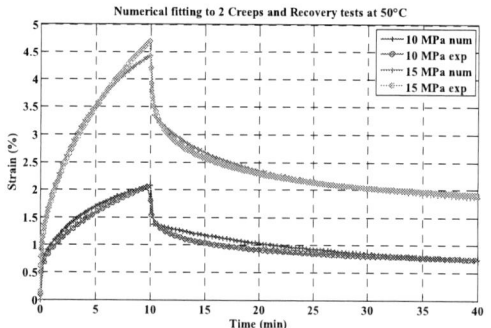

Figure 14 : Fitting insight of 3 creep and recovery tests at 50°C

The same precision is achieved for the other temperatures. To avoid fitting difficulties due to results dispersion, a second method consists in fitting numerical calculation to a CCR test for each temperature. It has been done at 50°C for comparison (cf. Figure 15).

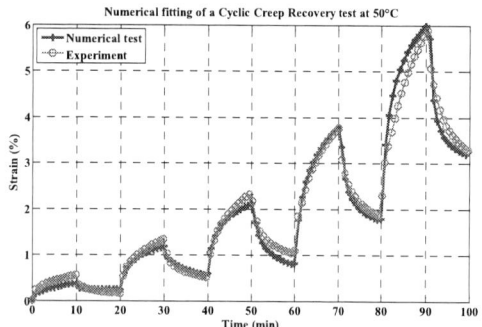

Figure 15 : Fitting insight of a Cyclic Creep Recovery test at 50°C

The resulting coefficients are close to the ones identified with multiple creep tests at 50°C with only 2%

of difference. The fitting process required less iterations and only one specimen.

Material coefficients are now available for the Two-layers model at 4 different temperatures. It is thus possible to get an approximated evolution of each as a function of temperature. Two linear laws deduced from the fitting process are plotted in Figure 16.

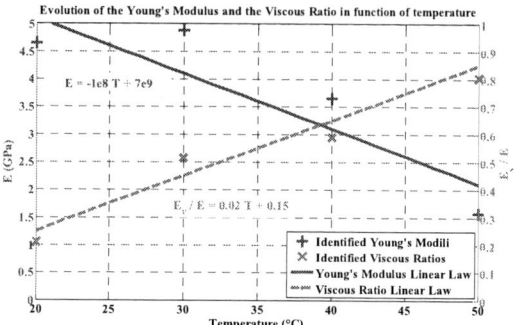

Figure 16 : Linear approximation of 2 material coefficients (E , E_v/E) in function of temperature

The viscous ratio is the part of the total instantaneous stiffness which is attributed to the "viscous branch". It is defined in the Two-layers of ABAQUS as the parameter f:

$$f = \frac{E_v}{E_v + E_p} = \frac{E_v}{E} \qquad (9)$$

The evolution trends of E and f show that the adhesive is softening as temperature increases. Moreover, the highest the temperature is, the more the remaining stiffness is carried by the "viscous branch". This means that the viscoelastic aspect becomes more predominant in contrary of the plastic flow that becomes negligible.

6. Conclusions

As shown in the first section, the understanding of fatigue and the development of life prediction tools for ICA used in space electronics is one of the most challenging issues in this domain. To fulfill the objectives of this study, a characterization of the adhesive material is realized in order to identify precisely the fatigue mechanisms to consider analytically or numerically.

The bulk properties of the adhesive are required to reproduce potential fatigue variables like plastic or creep strain. The tests performed on adhesive specimens highlights the presence of both.

The additional experiments made to characterize time dependence of the adhesive lead to the choice of the Two-layers model to simulate viscoelasticity and plasticity. Experiments also justify an isotropic hardening evolution based on Ramberg-Osgood power law. The resulting identification of material coefficient is satisfying. Numerical 1D calculation has been fitted to experimental plots with good accuracy. The evolution of 2 material coefficients (E and f) has been found to be linear in function of temperature.

At ends, material coefficients in function of temperature will allow calculation of the accumulation of residual inelastic and irreversible strain in a chip bonding over the thermal cycles. If will help to the formulation of a fatigue law that is meant to be experimentally validated or rejected.

References

[1] Capability approval programme for microwave hybrid integrated circuits (mhics) esa pss-01-612.

[2] Test method standard microcircuits - mil-std-883.

[3] S.S. Manson and T.J. Dolan. *Thermal stress and low cycle fatigue*, volume 33. 1966.

[4] K.C. Norris and A.H. Landzberg. Reliability of controlled collapse interconnections. *IBM Journal of Research and Development*, 13(3):266–271, 1969.

[5] D.F. Sims and V.H. Brogdon. Fatigue behavior of composites under different loading modes. *Fatigue of filamentary composite materials*, 636:185–205, 1977.

[6] O. Rusanen. Modelling of ica creep properties. PhD thesis, 2000.

[7] J.D. Ferry. *Viscoelastic properties of polymers*. John Wiley & Sons Inc, 1980.

[8] A. Launay, H. M. Mahaman , Y. Marco, I. Raoult, and F. Szmytka. Cyclic behaviour of short glass fibre reinforced polyamide: Experimental study and constitutive equations. *International Journal of Plasticity*, 27(8):1267–1293, 2011.

[9] J. Kichenin. *Comportement thermomécanique du polyéthylène. Application aux structures gazières*. PhD thesis, 1992.

Thermo-mechanical stress induced by CPI on 3D interposer package

Melina Lofrano and Mario Gonzalez
imec
Kapeldreef 75, 3001, Leuven, Belgium
Melina.Lofrano@imec.be, +32 16287925

Abstract

In this work Finite Element Modeling (FEM) is used to investigate the stress and deformation induced by chip package interaction on a 3D interposer package. A Design Of Experiments (DOE) was set up and parameters as epoxy mold compound (EMC) material properties, EMC thickness and die thickness were considered to study their effect on stress and package deformation. The results indicate that using a polymer that combines optimized mechanical properties with specific geometric dimensions makes it possible to reduce the package warpage and consequently the stress in the die. For this study, a multi-level sub modeling technique is used to access the stress distribution in the very small features in the packages, such as Cu μbumps. The stress analysis showed that by increasing the die thickness, the stress induced around μbumps increases. The results also showed that the EMC thickness has less impact on stress around μbumps than the die thickness.

1. Introduction

Nowadays heterogeneous systems such as logic and memory dies are built and packaged using separate technologies. In order to deliver high performance, low power consumption and low cost, it is desirable to integrate multiple dies, each using its own technology, in a single package [1]. The thermal performance and the mechanical stability of those systems is still one of the main concerns. The use of an intermediate carrier such as Silicon (Si) interposer between IC chips and substrate allows to improve the mechanical reliability of the interconnections and enables unmatched bandwidth, reduced power consumption and improves the heat dissipation [2]. However, as with all 3D technologies, it faces some challenges. The stacking of more chips in a thinner package stretches the performance envelope of all assembly process, materials and equipment [3].

During the packing process, where the stacked dies are assembled in a laminate substrate using underfilled Cu pillars and then overmolded at high temperature, is often observed high deformation and consequently stress at the die level when it returns to room temperature. These mechanical strain and stress are generated by the coefficient of thermal expansion (CTE) mismatch between the silicon die/interposer and the polymer based materials such as the epoxy mold

component (EMC), underfill and the substrate. These stresses are the main cause of failure.

In this work a design of experiments (DOE) has been set up to study the influence of EMC properties, EMC thickness and die thickness on the package deformation and stress of a 3D interposer package. Finite element modeling (FEM) was used to calculate the package warpage and with a multi-level sub modeling technique the stresses in the die due to the shrinkage of very small features, such as underfilled Cu μbumps and pillar, were obtained.

2. Design Of Experiments

In this work FEM is used to evaluate the Chip Package Interaction (CPI) effect on the package deformation. The 3D model consists of 2 identical dies of $8 \times 8 mm^2$ stacked on a $10 \times 20 mm^2$ Si interposer by underfilled Cu μbumps. The dies were placed at 2mm from each other. The μbumps consist of two Cu pads of 5μm thick and 25μm diameter with a 3μm thick Sn region in between, that after the reflow process is fully transformed in CuSn intermetallic. The stacked dies are flip-chip and mounted to a $25 \times 25 mm^2$ laminate substrate. The connection between 3D stack and substrate is made by underfilled Cu pillars. The flip chip Cu pillar is 50μm thick and has a diameter of 50μm. The package is overmoulded at 180°C where the whole package is considered stress free, Fig 1. The thickness of the laminate substrate was assumed 1.25mm and the thickness Si interposer was 100μm and they were kept constant for the entire DOE analysis. The deformations are obtained when the package is cooled down from overmould curing temperature to room temperature.

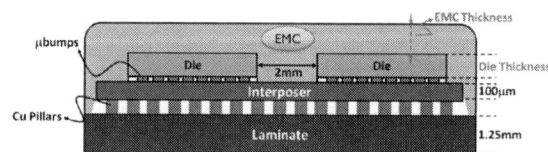

Fig 1: 3D interposer stack considered for DOE analysis

The DOE parameters are shown in red in the Fig. 1. Tables 1 and 2 summarize the mechanical material properties for 6 EMC's and the rest of the materials considered in this work. The range of the geometric parameter are:

- EMC thickness: $0 - 700\mu m$
- Die thickness: $100 - 500 \ \mu m$

978-1-4799-4789-8/14 $31.00 © 2014 IEEE

Table 1: EMC mechanical material properties

	Temperature of Glass Transition T_g [°C]	Elastic Modulus [MPa]	CTE [ppm/°C]
EMC1	134	23000 ($T<T_g$) 600 ($T>T_g$)	9 ($T<T_g$) 36 ($T>T_g$)
EMC2	120	22500 ($T<T_g$) 15 ($T>T_g$)	10 ($T<T_g$) 40 ($T>T_g$)
EMC3	130	25000 ($T<T_g$) 15 ($T>T_g$)	9 ($T<T_g$) 35 ($T>T_g$)
EMC4	135	26000 ($T<T_g$) 15 ($T>T_g$)	7 ($T<T_g$) 29 ($T>T_g$)
EMC5	150	25500 ($T<T_g$) 160 ($T>T_g$)	7 ($T<T_g$) 30 ($T>T_g$)
EMC6	175	20000 ($T<T_g$) 1100 ($T>T_g$)	12 ($T<T_g$) 36 ($T>T_g$)

Table2: Mechanical Material Properties

	Temperature of Glass Transition T_g [°C]	Elastic Modulus [MPa]	CTE [ppm/°C]	Poisson's ratio [-]
Si	--	169000	2.3	0.26
Die-to-Interposer Underfill	110	8384 ($T<T_g$) 148 ($T>T_g$)	31 ($T<T_g$) 116 ($T>T_g$)	0.3
Die-to-Laminate Underfill	120	11000	28 ($T<T_g$) 90 ($T>T_g$)	0.3
Laminate	>280	19300	16	0.19
Cu Pillar	--	117000	16.5	0.3
CuSn Intermetallic	--	131000	19	0.3

3. Results and Discussion of the DOE

In order to understand the impact of EMC mechanical properties on the package deformation and die stress a package configuration as shown in Fig. 2 is analyzed. For this configuration a die with 300µm and 500µm EMC thickness, measured from the top of the die, were simulated.

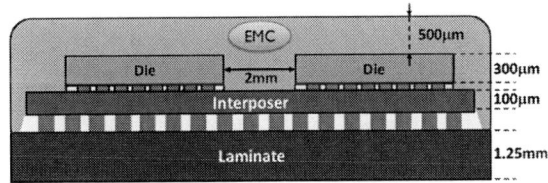

Fig. 2: 3D interposer package configuration for EMC mechanical properties analysis

The maximum out of plane deformation obtained for the 6 EMC's are shown in the Fig. 3.

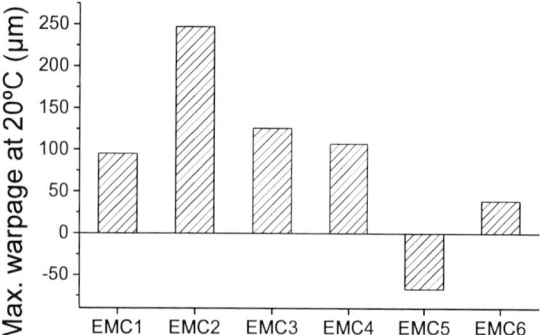

Fig. 3: 3D interposer maximum out of plane deformation at room temperature for 6 EMC's.

It can be seen from Fig. 3 that the package maximum out of plane deformation is higher for EMC2, 250μm, while EMC6 gives the lower out of plane deformation, 40μm. This deformation is mainly caused by the shrinkage of EMC. This is assumed to be stress free at the overmold cure temperature, 180°C. It means that when the package starts to cool, the temperature is above the glass transition T_g for all the EMC's and their CTE is very high. In the case of EMC6, which has a T_g close to the overmold temperature, T_g=175°C, this rapidly changes into the CTE below T_g, which is much lower than the CTE above T_g, thus the CTE mismatch between EMC6 and the Si die is reduced first. The EMC2 stays at CTE above T_g longer, T_g =120°C, resulting in a large CTE mismatch for a longer period and consequently a higher deformation. In Fig. 4 it is possible to see the time above and below T_g for all EMC's.

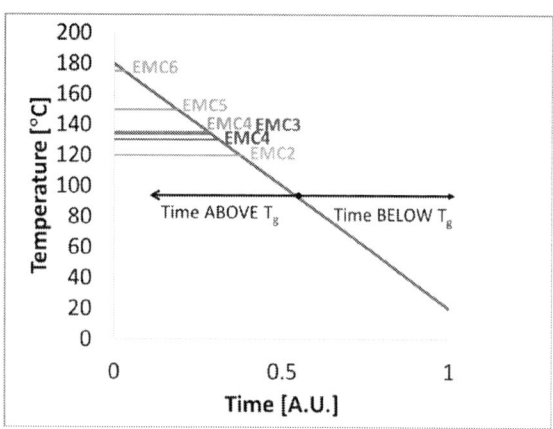

Fig. 4: Relative time on which EMC changes from CTE above T_g to CTE below T_g

Fig. 3 also shows that in the case of EMC5 the maximum out of plane deformation is negative, -67μm. Similar to EMC6, EMC5 has a high T_g. which means

that the majority of the shrinkage is determined by its CTE below T_g, 7ppm/°C. However the EMC5 CTE below T_g is much lower than the laminate substrate CTE, 16ppm/C. In this case the package stiffness is dominated by the laminate substrate that tends to bend downwards. The Fig. 5 shows the shape of the out of plane deformation for EMC01, EMC2, EMC5 and EMC6.

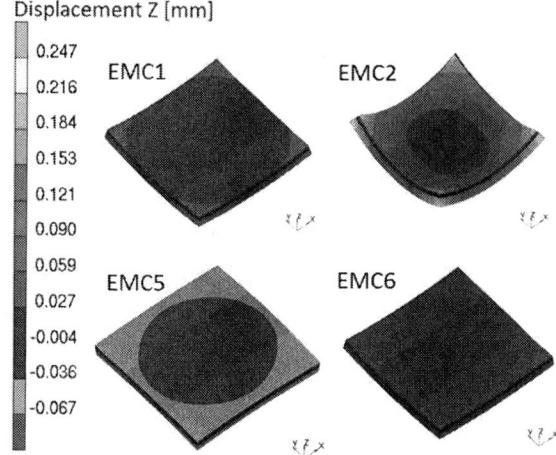

Fig. 5: 3D interposer out of plane deformation at room temperature

It can be seen that the shape of the out of plane deformation as well as its amplitude is highly dependent on the EMC mechanical material properties.

Two geometric DOE parameters have been considered: EMC thickness and die thickness. The EMC thickness is defined from the top of the die. It ranges from 0, meaning exposed die package, to 700μm EMC thickness, while the die thickness ranges from 100 to 500μm. Fig. 6. For this analysis the EMC was considered EMC1.

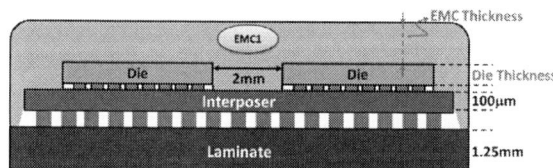

Fig. 6: 3D interposer package for DOE EMC thickness

When fixing die thickness of 100μm, it is possible to identify the influence of EMC thickness from 200 to 700μm. It is shown in Fig. 7 for EMC1.

978-1-4799-4789-8/14 $31.00 © 2014 IEEE

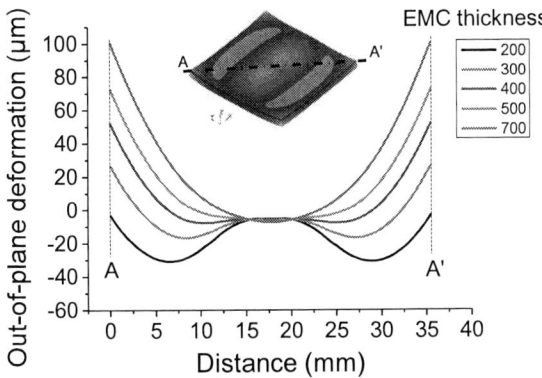

Fig. 7: Package out of plane deformation for different EMC thickness

It is observed that increasing the EMC thickness on top of the die increases the out of plane deformation of the package. This is due to the fact that when the EMC thickness increases there is more material shrinking what increases the deformation of the package.

Fig.8 shows the maximum out of plane deformation as function of die thickness for different EMC thickness. It is observed that the maximum out of plane deformation depends on both, die thickness and EMC thickness. When the stiffness of the EMC, dies and interposer equals the stiffness of the laminate an optimized flat package is obtained.

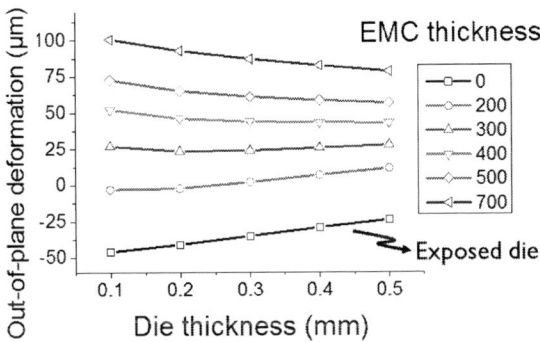

Fig. 8: Maximum out of plane deformation in function of die thickness for different EMC thickness.

3. Multi-level sub modeling technique: Results and discussions.

Modeling the package effects on the thermo-mechanical stress in the small features as such Cu μbumps is a challenge duo to big size difference. One way to deal with this problem is to use multilevel sub-modeling approach. The technique bridges the geometric gap between the global 3D interposer package (mm) and the local features such as Cu μbumps (μm). The technique consists of detailing the structure and refining the mesh at each sub-level. In this work only one level was needed to study the CPI

induced stress in the local μbump region on the dies. The Fig. 9 shows the detailed structure and its position on the global model.

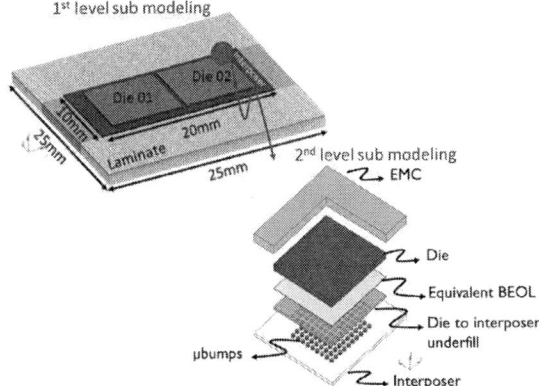

Fig. 9: Multi-level sub modeling: 1st level 3D interposer package, 2nd level detail of μbumps

For the multi-level sub modeling analysis only the geometric DOE parameters were taken into account. The EMC1 has been used in all analysis.

Fig. 10 shows one representation of the stress induced in the die by packaging. It can be observed that the Cu μbumps induces compressive stress on the Si die while in between the bumps it induces tensile stress. Due to overall package deformation, the stress is higher at the edges of the Si die and decreases towards the center.

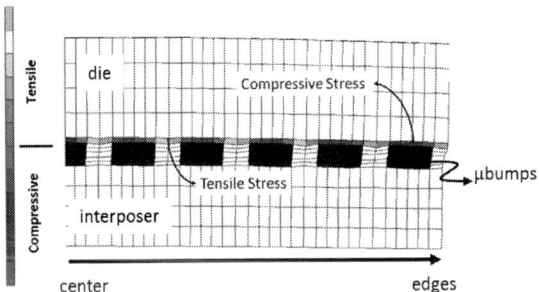

Fig. 10: Stress distribution on the Si die on top of μbumps array.

For a constant die thickness of 100μm, the stress at the μbumps region increases slightly with increasing of EMC thickness, Fig. 11. This increase represents a total of 12% from 200 to 700μm EMC thickness. The increasing of stress is caused by the combination of local Cu μbump and underfill shrinkage and the global package deformation. The global deformation introduces a twist deformation in the μbumps closer to the edges, [4]. It is also observed that EMC thickness does not induce a significant change in the tensile stress observed between the

978-1-4799-4789-8/14 $31.00 © 2014 IEEE

µbumps, this tensile stress is induced mainly by the local underfill shrinkage.

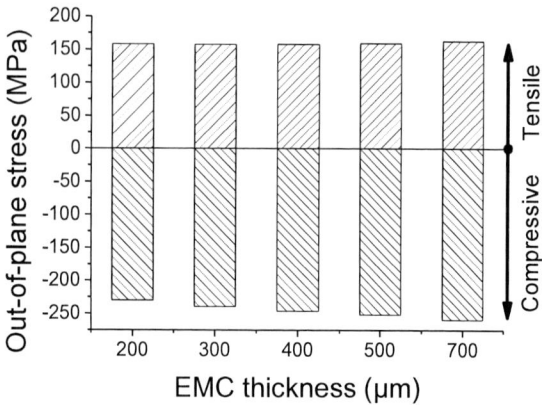

Fig. 11: Out of plane stress as function of EMC thickness for a die of 100µm thick

To identify the effect of the die thickness on the induced stress, a constant EMC thickness of 400µm from the top of the die has been considered. The results of the induced stress can be seen in Fig. 12. It is observed that the maximum out of plane stress increases considerably with an increasing die thickness. This effect is more accentuated in the die on top of µbumps regions where compressive stress is introduced. The increase varies from 250MPa for a 100µm die thickness to 550MPa for 500µm die thickness. It shows that the package stiffness has a direct impact on the local stress.

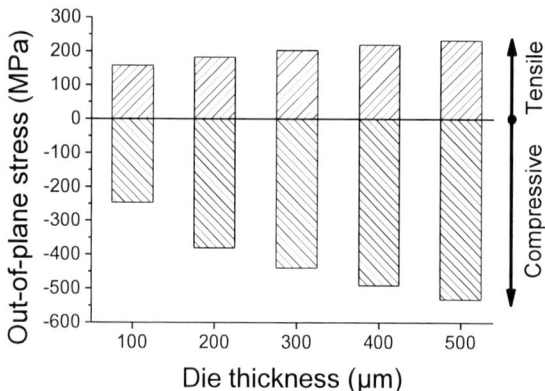

Fig. 12. Out of plane stress as function of die thickness for EMC thickness of 400µm

The deformation for a 100 and 500µm die thickness with 400µm EMC thickness can be seen in Fig. 13. The deformation is magnified by factor of 20.

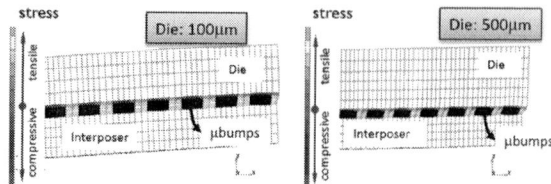

Fig. 13: Local stress distribution for a 100µm and 500µm die thickness with 400µm EMC thickness

4. Conclusions

In this work we used finite element modeling to study stress induced by CPI on 3D interposer package. A DOE has been set up to investigate the effect of EMC properties, EMC thickness and die thickness on the package deformation and stress. We have shown that the out of plane deformation is strongly dependent on the EMC properties. With an optimized EMC material properties and specific geometric configuration it is possible to obtain a flat package. The stress analysis showed that increasing die thickness the stress induced around µbumps increases suggesting that package stiffness has a direct impact on the local stress. The results also showed that the EMC thickness has less impact on local stress around µbumps than the die thickness.

References

1. Madden, L. Wu, E. Kim, N. Banijamali, B. Abugharbieh, K. Ramalingam, S. and Wu, X. "Advancing High Performance Heterogeneous Integration Through Die Stacking". Proceedings of the European Solid-State Device Research Conference (ESSDERC), 2012.
2. Matthias, T. Wimplinger, M. Pauzenberger, G. and Lindner, P. "New integration schemes for 2.5D interposer". International Symposium on VLSI Technology, Systems, and Applications (VLSI-TSA), 2013.
3. Karnezos, M. "3-D Packaging: Where All Technologies Come Together".. Int'l Electronics Manufacturing Technology Symposium, 2004
4. Lofrano, M. "A Multilevel sub-modeling approach to evaluate 3D IC packaging induced stress on hybrid interconnect structures", Microelectronic Engineering, 2013.

Theoretical and Experimental Investigations on Failure Mechanisms Occuring During Long-Term Cycling of Electrostatic Actuators

R. Behlert, T. Künzig, G. Schrag and G. Wachutka

Institute for Physics of Electrotechnology

Munich University of Technology

Arcisstr. 21, 80333 Munich

Abstract

We present an extensive study on dielectric charging effects, one of the major problems that limit the reliability of electrostatically actuated microdevices (such as the RF MEMS switches considered here) and, thus, their way into a broad commercial application. For the first time, we are able to provide quantitative statements on the amount of charge injected into the dielectric layers. They result from monitoring the long-term evolution of the switching voltages of the DUT recorded by a novel, on-purpose developed measurement setup, which enables also temperature-dependent investigations. Furthermore, the origin of the parasitic charges, their impact on the switching operation and measures to remove them from the dielectric layers could be identified.

1. Introduction

Electrostatic actuators constitute one of the basic component classes in today's microsystem design. They bear a huge potential in many application fields, e.g. the RF technology, where RF MEMS switches like the ones considered in this work are promising candidates for the use in high performance, reconfigurable RF passive components for wireless and telecommunication systems like tunable filters, reconfigurable impedance matching networks or high order switching matrices [1,2,3]. They provide several advantages like low manufacturing costs, low insertion losses, galvanic separation and fast switching behavior. Nevertheless, they could not find their way into broad industrial application so far, since they still suffer from reliability problems [4,5].

One major issue among others (see section 3) is charging of the dielectric layers during long-term operation, as well as the temperature dependence of the switching characteristics, which may cause drift in the operation voltage and, in the worst case, failure in the device function due to stiction of the movable mechanical parts. These effects have been already reported in literature [6], but a deeper understanding of the underlying mechanisms is still a matter of on-going research.

In this paper we present measurements and analysis on both reliability issues with the goal to obtain a better insight in the physics of dielectric charging, also with subject to temperature changes, to determine the total amount of accumulated parasitic charge for various actuation scenarios and operation conditions and, based on these findings, to derive new methods to avoid these effects.

After a short description of the device operation, different reliability issues and the hereof resulting failure

modes are described in detail in sections 3 and 4, followed by the electromechanical model and the measurement set-up in sections 5 and 6. Results and their interpretation as well as a short summary will conclude this work.

2. Device Description

In this work, two different types of ohmic RF MEMS switches have been investigated. The principal design is depicted in Fig. 1. They consist of a perforated rectangular gold membrane hinged by four cantilevers, which allow the membrane to move in vertical direction, thus acting as top electrode of the switch. A slightly smaller bottom electrode manufactured from polysilicon is located beneath the membrane. It is covered by a dielectric material (silicon dioxide) in order to avoid electrical shortening. The air gap between top-electrode and dielectric layer amounts to approximately 3 µm.

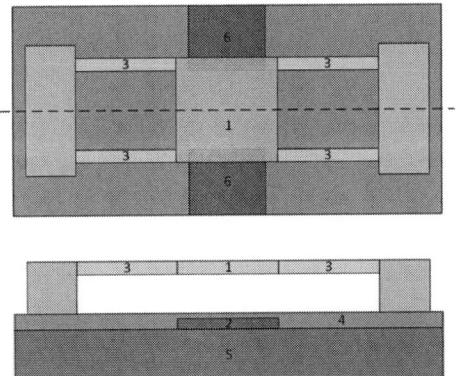

Fig. 1: Schematic top and cross-sectional view of an RF-MEMS-switch: 1) top electrode/membrane 2) bottom electrode 3) cantilevers 4) dielectric 5) substrate 6) CPW.

When the top electrode is actuated by applying an either positive or negative bias voltage to the bottom electrode, it deflects due to the attractive electrostatic force. Beyond a critical value, the so-called pull-in voltage, the electrostatic force exceeds the mechanical force, the membrane snaps towards the bottom electrode and the switch is in its closed state. Then the major part of the membrane is in flat contact with the dielectric layer, overlaps with the coplanar waveguide (CPW) and the RF signal can be conducted through the grounded top electrode.

When the bias voltage is decreased again and reaches the so-called pull-out voltage, the membrane jumps back to its rest position, the switch is in its open state and the RF signal path is interrupted. Due to the non-linear nature of the electromechanically coupled problem, the pull-out voltage is inherently lower than the pull-in voltage [6].

Since voltages of both polarities can be applied to actuate the switch, it is necessary to record four voltage values (two positive and two negative pull-voltages, respectively) in order to characterize the switching behavior to a sufficient extent.

3. Reliability Issues in Electrostatic MEMS Actuators

There are several reliabilty issues, which are known to potentially occur during long-term operation of electrostatic actuators like the RF MEMS switches investigated in this work.

One is the mechanical wear of the contact pads, which is due to the permanent bouncing of the top electrode during closure, especially becoming evident for soft contact materials like gold, which is used in the RF switches considered in this work. This effect impairs the transmission behavior, but can be avoided by applying optimized actuation pulses or by designing the switches as capacitive devices [7].

Another reliability issue, which will be discussed later on in this paper, is that the switching characteristics are altered by temperature changes. This effect is due to a strong dependence of the pull-voltages on the ambient temperature and necessitates to adjust the actuation voltages in order to ensure safe operation.

The most relevant failure mechanism occuring in electrostatic devices, however, is stiction of the movable top electrode to the underlying material. The two major effects causing this phenomenon are microwelding [8] and dielectric charging, the latter being discussed intensively in the following.

Dielectric charging implies that charges accumulate near the surface of the dielectric material, inducing an electric field, which, hence, changes the switching characteristics. Although the dielectric material is an insulator, a certain amount of intrinsic charge can be found inside the material. When an external electric field is applied between top and bottom electrode, those charges are polarized and accumulate at the interfaces to the air gap and the bottom electrode. A positive voltage at the bottom electrode therefore leads to positive charges at the surface of the silicon dioxide (see Fig. 2, left).

In contrast to polarization the second charging mechanism, the so-called injection, induces a negatively charged surface. When the switch is closed and the gold membrane touches the dielectric layer, the applied electric field is very high and charge carriers can be injected into traps inside the insulator. When the switch opens again these charges remain trapped having a very low mobility in lateral direction [9]. In this case a positive actuation voltage at the bottom electrode leads to negative charges at the surface of the dielectric layer (Fig. 2, right). The different failure modes induced through charging of the dielectric layers will be described in detail in the next section.

4. Failure Modes Induced by Dielectric Charging

Once the dielectric is charged, the switching characteristics change. The additional charges induce a parasitic electrical field which is superimposed on the external electrical field. The electrostatic force, pushing the membrane down, is therefore either amplified or reduced, depending on the polarity of the parasitic charge and the applied bias voltage. In consequence the four characteristic pull voltages drift to more positive or more negative values compared to the uncharged situation. This phenomenon is called shift and is observed and discussed by various authors [6,7,10]. As sketched in Fig. 3, the first failure mode happens, as soon as one of the pull-in voltages exceeds the applied bias voltage U_{act}. Then the switch will not close as it is supposed to, which can simply be "healed" by enlarging U_{act}.

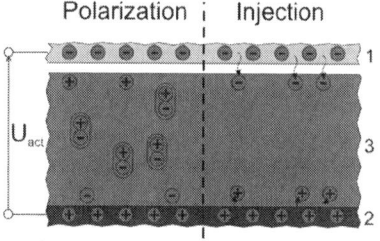

Fig. 2 : Polarization (left) and injection (right) of charge carriers from the electrodes (1) and (2) inside an insulator (3): Both charging mechanisms induce surface charges.

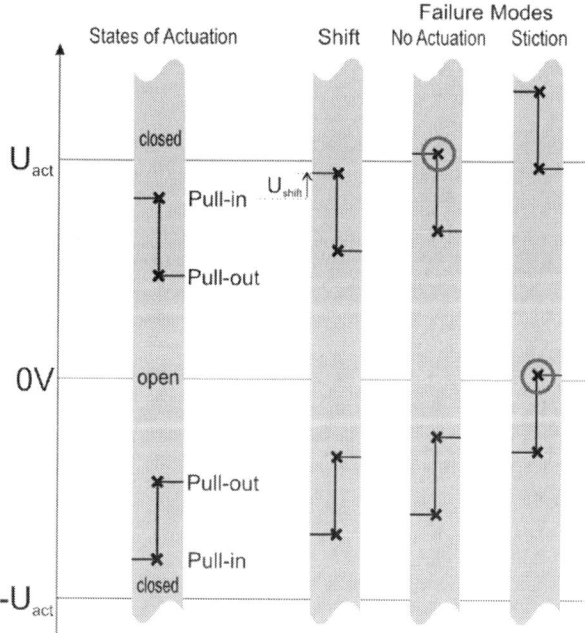

Fig. 3 : Different states of the switch during actuation: Depending on U_{act} and the previous state of the switch the device is supposed to be either in open (green) or closed (red) state. During unipolar actuation the pull voltages tend to shift, in the displayed example towards more positive values, until failure of the switch occurs.

978-1-4799-4789-8/14 $31.00 © 2014 IEEE

The second failure mode occurs, when one of the pull-out voltages crosses the 0V-line. Then the switch stays in a permanently closed position, even if the bias voltage is decreased in order to open the switch. First, the membrane starts to detach, but, due to its inertia, it is not able to follow the switching pulse quickly enough, before the voltage range for the closed states is reached again (see Fig. 3) and, finally, the membrane sticks to the counterelectrode.

Additionally to shifting, the so-called narrowing of the pull-voltages is observed, a phenomenon, which is also caused by dielectric charging. When the parasitic charges inside the insulator are distributed heterogeneously or even both polarities are existent, the total values of all pull voltages decrease, which results in a narrowing of the associated pull-out and pull-in voltages (negative and positive values, respectively) and, hence, in a diminution of the difference between the pull-out voltages (= "switch open" condition, see Fig. 3). When this window gets too narrow, the switch stays in its closed state (=stiction, failure). This effect has already been mentioned in literature [7], but not yet investigated in detail.

5. Electromechanical Model

The calculation of the amount of parasitic charge is based on an analytical model that includes basically the two governing forces: the electrostatic force and the mechanical spring force.

The electrostatic behavior is modeled by assuming two ideal capacitances C_{air} and C_ε with area A and d_{air} and d_ε being the thicknesses of the air gap and the dielectric layer, respectively. Fringing fields due to the perforations inside the membrane and the boundaries of the electrodes are neglected. When the switch is uncharged and an external voltage U_{act} is applied to the capacitances in series (see Fig. 4), the electrical field in air $E_{air,0}$ reads as:

$$E_{air,0} = \frac{U_{act}}{d_{eff} - \Delta} \quad \text{with} \quad d_{eff} = d_{air} + \frac{d_\varepsilon}{\varepsilon_R} \quad (1)$$

Here, Δ is the deflection of the top electrode and ε_R the permittivity of the dielectric material.

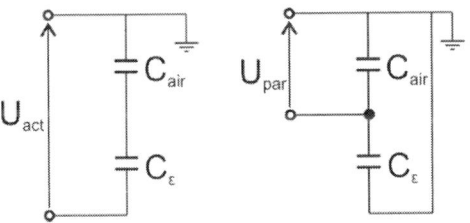

Fig. 4 : Equivalent circuit representation of the switch to calculate the electric field in air: uncharged switch with external actuation voltage U_{act} (left), non-actuated switch with parasitic charge inducing a voltage U_{par} across both capacitances (right).

A homogenously distributed parasitic charge density σ inside the dielectric layer can be interpreted as an extra voltage source U_{par}, which acts on the capacitances, that

are now in parallel (see Fig. 4). Hence, σ induces an electrical field $E_{air,\sigma}$:

$$E_{air,\sigma} = \frac{\sigma \cdot d_\varepsilon}{\varepsilon_0 \varepsilon_R (d_{eff} - \Delta)} \quad (2)$$

When the charged switch is actuated, both fields are superimposed to obtain the total electric field E_{air}

$$E_{air} = E_{air,0} + E_{air,\sigma} \quad (3)$$

The electrostatic force F_{el} acting on the top electrode can be calculated by

$$F_{el} = \frac{dW_{el}}{dz} = -\frac{1}{2}\varepsilon_0 A E_{air}^2 \quad (4)$$

F_{el} is the electrostatic force, which pulls the membrane down towards the dielectric layer. The mechanical spring force F_{mech} that pulls it back to its rest position is modeled as a simplified, linear spring-mass-system with spring stiffness k:

$$F_{mech} = k\Delta \quad (5)$$

For quasi-static motion, both forces are balanced and the relation between deflection Δ and actuation voltage U_{act} can be derived:

$$U(\Delta) = -\frac{\sigma \cdot d_\varepsilon}{\varepsilon_0 \varepsilon_R} \pm (d_{eff} - \Delta)\sqrt{\frac{2k}{\varepsilon_0 A}\Delta} \quad (6)$$

Fig. 5 shows a qualitative 3D plot of both forces. The equilibrium curve (6) is indicated by a red line, which is in detail also depicted in Fig. 6.

When the voltage is increased, the top electrode slowly deflects towards the counterelectrode following the red line in Fig. 5 and Fig. 6. When a certain voltage value is reached, the electrostatic force exceeds the mechanical force (see Fig. 5). Beyond this point (maximum value of the U-Δ-plot in Fig. 6), no stable equilibrium exists and the top electrode is accelerated towards the dielectric layer (=pull-in).

Electromechanical Model

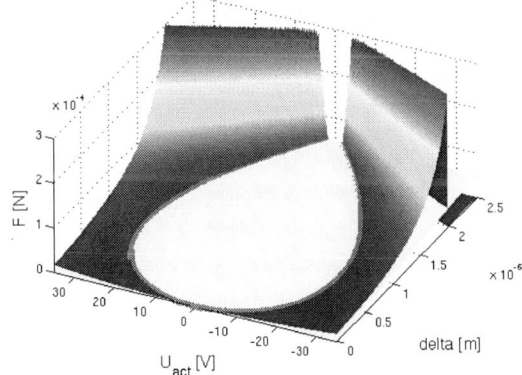

Fig. 5 : Plot of the two forces governing the operation of the switch: Electrostatic force (rainbow colors) and mechanical force (yellow). The equilibrium states are marked by the red line.

When the voltage is decreased until it reaches a value on the U-Δ-curve again, the top-electrode detaches from the dielectric and jumps to the equilibrium position which corresponds to this voltage value and, at the same time, exhibits a deflection no larger than one third of d_{eff}, which is the condition for stable equilibrium. This is why the pull-out point occurs at this specific voltage value, but at a smaller deflection, which can be seen in the U-Δ-curve of Fig. 6.

The pull voltages therefore can be expressed by

$$U_{pull-in} = -\frac{\sigma d_\varepsilon}{\varepsilon_0 \varepsilon_R} \pm d_{eff} \sqrt{\frac{8k}{27\varepsilon_0 A} d_{eff}} \quad (7)$$

$$U_{pull-out} = -\frac{\sigma d_\varepsilon}{\varepsilon_0 \varepsilon_R} \pm (d_{eff} - d_{air}) \sqrt{\frac{2k}{\varepsilon_0 A} d_{air}} \quad (8)$$

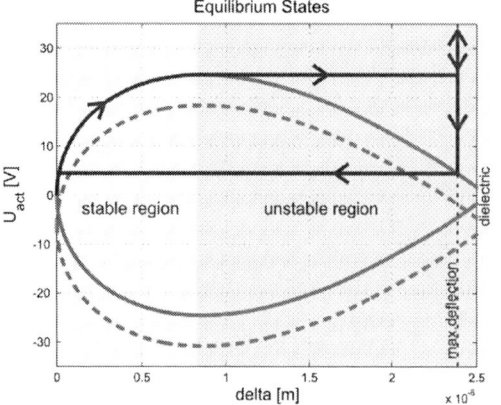

Fig. 6 : Equilibrium states of the electrostatic model: Pull-in and pull-out voltages can be extracted from these curves. In case of shift, the whole curve moves in vertical direction (dashed line).

The parameters needed as input for the above described electromechanical model are listed in Table 1. Accordingly, when they have successfully been extracted from measurements, we are able to describe the switching characteristics of the devices including also the shift due to dielectric charging through this model. As a result, the total amount of accumulated charges can be calculated by determining the shift of the pull voltages experimentally.

The above described electrostatic model has already been introduced by several authors in the past [6,10,11].

A	Area of the capacitance	e.g. 1,88e-8 m²
d_{air}	Air gap in rest position	appr. 2,5 μm
d_ε	Thickness dielectric	270 nm
k	Spring stiffness	appr. 20 N/m
ε_R	Permittivity	3,9

Table 1: Input parameters of the electromechanical model.

The narrowing effect, however, cannot be predicted by these equations. Hence, Rottenberg et al. [12]

developed a model, allowing for the incorporation of heterogeneous charge densities. This impacts the local electrostatic force and, thus, enables to model characteristics, that show narrowing of the pull-voltages.

6. Parameter Extraction and Measurement Set-up

The model derived in section 5 can only be applied to calculate the amount of the accumulated charge and to characterize the switches, if all relevant parameters denoted in Table 1 are known to a certain degree of accuracy. In this section we will describe the equipment and the measurement procedures, which have been used for their extraction.

The thickness of the air gap d_{air} has been determined using a Laser Doppler Vibrometer (LV). A triangular actuation voltage is applied (50 Hz), which opens and closes the switch continuously, but very slowly and the deflection of the membrane is recorded by the LV (Fig. 7). The difference between the deflection at 0V and that beyond the pull-in voltage then delivers the thickness of the air gap in the rest position of the switch. This value slightly differs from switch to switch but is in the range of 2,5 – 3 μm.

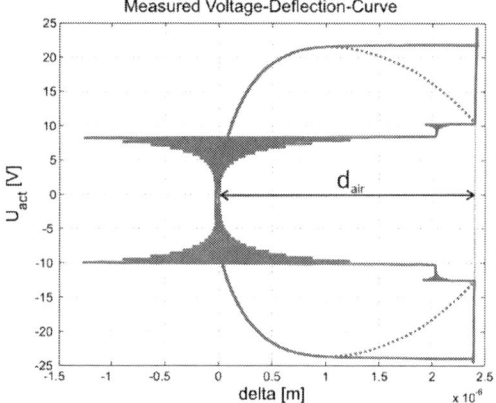

Fig. 7 : Voltage-deflection-curve of a slightly charged switch: The air gap is determined through the deflection occurring beyond the pull-in of the membrane.

The thickness of the dielectric material has to be taken from process data, because it cannot be determined, without destroying the device. It amounts to approximately 270 nm.

The effective Area A of the electrodes cannot be determined by White Light Interferometry (see Fig. 13), since the bottom electrode is not visible from the top. Therefore the exact layout of the device is imported from the design tool into an FEM software. Using the previously extracted thicknesses d_{air} and d_ε, the charge on the electrodes at a given voltage has been calculated, which delivers the capacitance of the whole setup, including also the perforations in the top electrode. From this value, the effective area of a corresponding ideal parallel plate capacitor can be determined, which amounts to 1,88e⁻⁸ m² for one type of the investigated switches.

The spring stiffness k of the membrane and the cantilevers is the most complicated parameter to measure. It cannot simply be calculated using the geometry and material values of the membrane, because all movable parts are mechanically pre-stressed from production. In order to include this parasitic effect an uncharged switch was activated by a triangular voltage and the deflection was recorded using the LV. Subsequently, we applied the electromechanical model to the deflection over voltage curve for $\sigma=0$ and extracted the unknown parameter k applying eq. (7). All effects like spring softening due to initially accumulated charges or prestress are then included, too. The value for k also varies for different switches and lies in the order of 20 N/m.

The most important measurement values, which have to be provided for the model are the pull-voltages themselves. To this end, a new on-purpose developed set-up was used, which enables automated measurements of electrostatically actuated devices during long-term operation. With this set-up it is possible to perform also temperature-dependent measurements from room temperature to about 100 °C. The switch, which is operated in series with a resistance, is actuated applying a low-frequency triangular voltage. When the pull-in voltage is exceeded, the capacitance of the switch jumps to very large values, inducing a current peak, respectively a voltage peak at the resistance, which is recorded by an USB-oscilloscope. When the switch opens again, the value of the capacitance jumps back to a smaller value, which can be seen as a discharging current through the resistance. Hence, within one triangular signal period the voltage at the resistance shows two larger peaks (=pull-ins) and two smaller peaks (=pull-outs). The locations of these peaks and, hence, the switching voltages, are then automatically detected by a peak-detection algorithm in LabView. The measurement setup is shown in Fig. 8.

Fig. 8: Pull-in measurement set-up: The waveform of the signal generator (1) is amplified (2) and applied to the DUT-resistance circuit (3). The USB-oscilloscope (4) measures the potential difference at the resistance and transmits the signal to the LabView software (5).

In order to investigate the dielectric charging phenomenon, different switching scenarios can be applied to the device. E.g., the device can be charged systematically by applying an unipolar rectangular actuation voltage for a certain time, thus emulating typical real world applications. Between these so-called stress periods a bipolar measurement signal is applied in order to release the accumulated charges and to deliver the non-deteriorated switching voltages.

The resulting pull-voltages over time curves allow for a comprehensive discussion of the long term behavior of the switches for constant as well as for varying temperature.

7. Measurement Results

Two types of switches were investigated in this work. In order to investigate the effects of dielectric charging the evolution of the pull-voltages over time was recorded during stressing (=charging) and relaxing (=discharging) of the switches.

Both designs show significant shifts of the pull-voltages U_{shift}, but the absolute value strongly depends on the switch type and the respective previous history, viz. how long and in which actuation mode it was used before. Typically, the relative value of U_{shift} amounts to 8-16% after one hour of actuation (see Fig. 9). At higher temperatures the switch accumulates more parasitic charge, and hence shows a larger shift within shorter time constants, but also discharges faster, when an alternating voltage is applied.

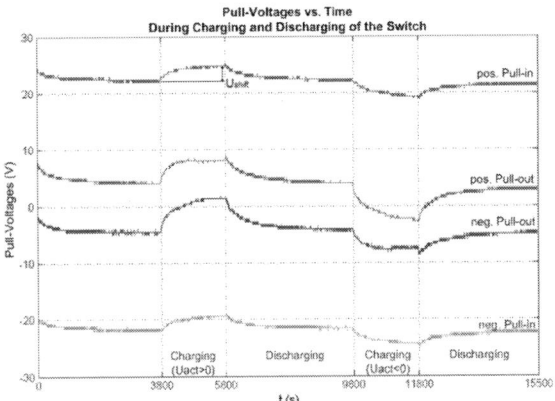

Fig. 9 : Measurement cycle with two charging and discharging periods: The characteristic four pull voltages are plotted against time.

For very long actuation times the narrowing effect could be observed, which manifests in the trend of all pull voltages to smaller absolut values (Fig. 10). It turned out that new switches that were never actuated before show a stronger narrowing effect than older ones.

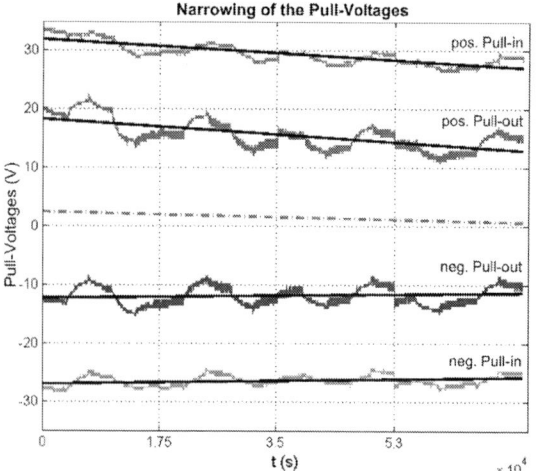

Fig. 10: Narrowing effect of the pull voltages during long-term-measurement.

The temperature dependent measurements revealed a strong dependency of the pull voltages from the temperature (see Fig. 11). It turned out that here the observed narrowing effect is not based on dielectric charging, like the narrowing described in section 4, but can be explained by spring softening inside the suspensions due to a decreasing Young's modulus inside the gold layer with temperature and, consequently, a smaller spring stiffness and lower pull-voltages. Fig. 12 shows the spring stiffness k calculated for temperatures from 20 °C to 60 °C for different actuation voltages, resulting in an averaged spring stiffness evolution (blue curve).

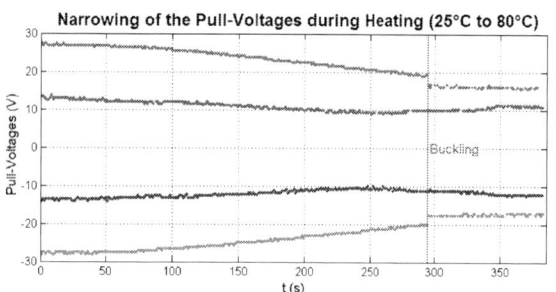

Fig. 11: Measurement of the pull voltages during heating: With increasing temperature the absolute values of the pull voltages decrease.

When a critical temperature is exceeded, so called buckling occurs. The membrane expands and is mechanically distorted due to the enhanced mechanical stress inside the structure. The characteristics of the switch change abruptly at this point, as can be seen in Fig. 11.

The amount of accumulated charge σ can be calculated applying eq. (7). Since the spring stiffness k can only be extracted for uncharged, new ("virgin") devices, we were only able to determine values of the parasitic charge σ for these switches. The resulting space charge densities of the investigated switches were in the

range of of $\pm$3e-8 C/cm² at room temperature and $\pm$8e-8 C/cm² for 80°C. This conforms well to the results reported in [13] for metal-insulator-metal structures.

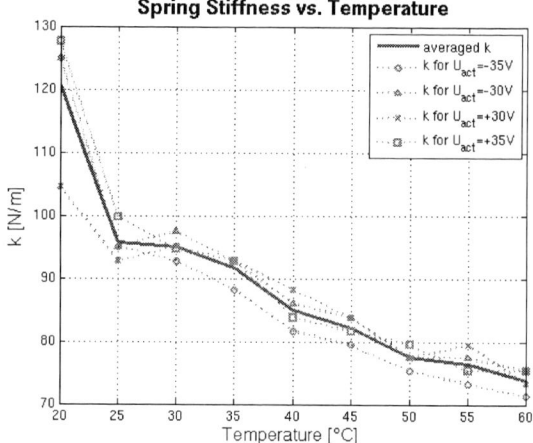

Fig. 12: Spring Softening due to higher temperatures: The spring stiffness is determined using eq. 7 and the methods, which are described in section 6.

Apart from the quantitative statements we could additionally confirm that a positive actuation voltage leads to accumulation of mainly negative charge carriers, when the top electrode is grounded. Hence, the switching voltages shift in positive direction. These findings prove that injection of carriers from the top electrode into the dielectric layer can be regarded as the main charging mechanism and not the polarization of the insulating material.

Furthermore a comprehensive discussion of the temperature influence is now possible. During the experiments it was observable, that stiction occurred much more frequently at higher temperatures than at room temperature. An example of a sticking membrane is shown in Fig. 13. This can be explained by two causes. On the one hand higher temperatures lead to stronger charging of the dielectric material, hence failure modes due to shift and narrowing are relevant after shorter actuation times. On the other hand this problem is reinforced by the temperature based narrowing, because failure modes like the crossing of the 0V-line happen faster, when the pull-voltage is already smaller without any parasitic charge being present.

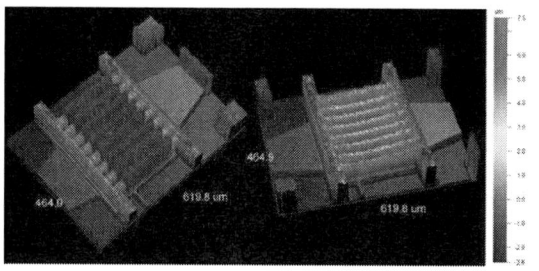

Fig. 13: White-Light-Interferometer image of a switch before (left) and after failure (right) caused by stiction.

8. Conclusions

An on-purpose developed measurement setup was designed to automatically measure pull voltages of electrostatic switches for various stress scenarios. The recorded pull-voltage over time evolutions were used to investigate the switching characteristics for long-time actuation at room temperature as well as at higher temperatures including also the heating-up phase. For the first time the accumulated charge inside the dielectric material was calculated for real RF MEMS switches using the proposed model. The necessary parameters could be extracted from laser vibrometric and white light interferometric measurements. Furthermore the mechanical behavior of the switch for rising temperatures was investigated. The obtained results give detailed insight into various failure mechanisms that occur during actuation of the switch and summarize the main reasons for the hampering access to a broader industrial application.

Applying bipolar actuation pulses for a certain time was identified as one successful measure to systematically discharge the switches. Though these discharging cycles showed to reduce the shift of the pull-voltages, they had no measurable influence on the narrowing effect.

One focus of ongoing work is the confirmation of the Frenkel-Poole effect as the main charging mechanism, which was proposed by several authors in the past, but is not yet verified [7,14].

Finally, the presented methods and models can easily be transferred and applied to other electrostatic devices, thus offering a valuable tool for design improvements towards devices, which exhibit a higher robustness against dielectric charging.

References

1. Malczewski, A. et al, "A family of MEMS tunable filters for advanced RF applications", Proc. of the International Microwave Symposium 2011, pp.1-4.
2. Domingue, F. et al, "A reconfigurable impedance matching network using dual-beam MEMS switches for an extended operating frequency range", Proc. of the Microwave Symposium 2010, pp 1552-1555.
3. Daneshmand, M. et al, "RF MEMS Satellite Switch Matrices", IEEE Microwave Magazine. 12-5 (2011) 92-109.
4. Wibbeler, J. et al, "Parasitic charging of dielectric surfaces in capacitive microelectromechanical systems (MEMS)", ELSEVIER Sensors and Actuators, Vol. A71 (1998), pp. 74-80.
5. Reid, J.R., "Simulation and Measurement of Dielectric Charging in Electrostatically Actuated Capacitive Microwave Switches", Proceedings of International Conference in Modeling and Simulation of Microsystems, 2002, p. 250.
6. van Spengen, W.M. et al, "A comprehensive model to predict the charging and reliability of capacitive RF MEMS switches", Journal of Micromechanics and Microengineering Vol. 14, No. 4, 2004, pp. 514 ff.
7. van Spengen, W.M., "Capacitive RF MEMS switch dielectric charging and reliability: a critical review with recommendations", Journal of Micromechanics and Microengineering, Vol. 22, No. 7, 2012.
8. Kuenzig, T. et al, "Modeling and simulation of an active restoring mechanism for high reliability switches in RF-MEMS technology", European Symposium on the Reliability of Electron Devices, Failure Physics und Analysis (ESREF),Vol. 52, No 9-10, 2012, pp. 2235-2239.
9. Papanioannou, G. et al, Advanced Microwave and Millimeter Wave Technologies: Semiconductor Devices, Circuits and Systems, Chapter 14, 2010.
10. Behlert, R., „Untersuchung der parasitären Ladungseinlagerung bei mikromechanischen Hochfrequenzschaltern", Technischer Report 2013/02, Lehrstuhl für technische Elektrophysik, Technische Universität Müenchen.
11. Mellé, S. et al., "Modeling of the dielectric charging kinetic for capacitive RF-MEMS", Proceedings of IEEE Microwave Symposium Digest, 2005.
12. Rottenberg, X. et al, "Analytical Model of the DC Actuation of Electrostatic MEMS Devices With Distributed Dielectric Charging and Nonplanar Electrodes", Journal of Micromechanics and Microengineering, Vol. 16, No. 5, 2007, pp. 1243 ff.
13. Yuan, X. et al, "Acceleration of Dielectric Charging in RF MEMS Capacitive Switches", IEEE Transactions on Device and Materials Reliability, Vol. 6, No. 4, 2006, pp. 556 ff.
14. Papanioannou, G. et al, "Dielectric charging mechanism in RF-MEMS capacitive switches", Proceeding of Microwave Integrated Circuit Conference, 2007.

A model in predicting color of LED packages with different phosphor layer dimensions

Cell K Y Wong*[1,3], Stanley Y Y Leung[1], Y J Xiong [1], Cadmus C.A. Yuan[2,3], and G Q Zhang[4]

[1] State Key Laboratory of Solid State Lighting, Changzhou, Jiangsu, 213161, China
[2] State Key Laboratory of Solid State Lighting, Haidian, Beijing, 100086, China
[3] Research and Development Center for Semiconductor Lighting, Institute of Semiconductors, Chinese Academy of Sciences, Haidian, Beijing, 100086, China
[4] Delft Institute of Microsystems and Nanoelectronics (DIMES), Delft University of Technology, Delft, the Netherlands
*cellwong@sklssl.org

Abstract

This paper describes a methodology in predicting color of the LED packages in different phosphor geometry for package free LED. Through resolving the peaks attributed from the LED and phosphor, the Phosphor Spectral Portion-Vol. ratio relations was developed. From the relations, the spectrum for a LED with known phosphor geometry can be constructed. Color of the LED can then be deduced. The method serves the geometric design of the phosphor layer for package free LEDs.

Introduction

With the progressive development of LEDs for illumination, quality of light has become key technology concern in recent years [1]. Package free LED [2-4], with miniaturized footprint that allows more space for optical and thermal design has been regarded as a promising technology for high quality products.

The design of package free LED is principally different from SMD LED since it does not involve the lead-frame. A common SMD LED whose geometry of the phosphor layer was dictated by the fixed die cavity structure on lead-frame (for example 5050, 3014, etc.), the design of the phosphor material is mainly focus on the phosphor materials concentration and their combination in order to fulfill the target specifications. Package free LED is simply constructed by a LED die, and the light emitting surfaces are covered with a phosphor layer material. The geometries of the phosphor layer is no longer limited by a cavity. Since the color conversion is dependent on both the phosphor chemistry and light scattering path, the design of the package free LED must simultaneously consider the phosphor concentration and the geometries of the phosphor layer.

Currently, the phosphor design are adopted from trial and accumulating experiences from the SMD packages. The common industry practice in producing LED light source within a target color range is by selecting phosphor candidates with the color co-ordinates in a CIE 1931 color space based on CIE x, y of the blue LEDs according to the color temperature requirement. Package are designed in terms of phosphor concentration, volume, etc. The performance will be adjusted according to engineer experience. The product development time depends on selection of phosphor and concentration with respective to silicone in trial and error experiments. The lack of a systematic analysis guideline lengthen the design cycle of package free LED.

This study aimed to develop a methodology in predicting the color of phosphor-converted package free LED as a function of phosphor geometry. The study investigates the impact of phosphor geometry on the phosphor peak in the spectrum of the LED packages which are excited by blue LEDs. The methodology in constructing the spectral power distribution (SPD) from the phosphor geometry and deducing the color of the package is established and demonstrated. This study helps in formulating the geometric design of the phosphor layer in a package free LED.

Spectral power distribution of phosphor converted LED

The color of a phosphor converted LED is dependent on the radiant flux from blue LED, conversion of the phosphor layer and radiation reflected by the substrate. Fig. 1 shows the impact factor on the color of a LED packages. The color conversion is dependent on the phosphor layer. The phosphor layer typically consist of transparent silicone filled with inorganic phosphor filler particles. During the LED operation, blue rays are emitted from the LED die. Part of the blue rays are absorbed by the phosphor particle and re-emit light rays of higher wavelength and rest of the blue rays are directly transmitted through the silicone. The mixed of these emitting light rays produce the white light. For the devices which consist of blue LEDs of the same type with identical spectral power distribution and peak wavelength, the behavior on phosphor excitation would be the same. The probability scattering of the blue rays by the phosphor particles is dependent on particle concentration and the length of travelling path. In a given set of materials and the phosphor concentration, the spectral radiant emission of the devices is a function of the geometry of the phosphor layer.

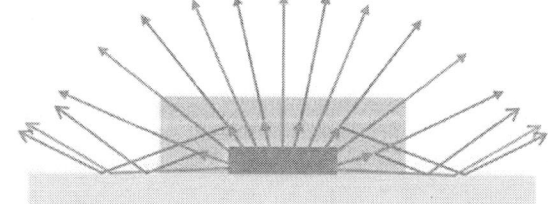

Fig. 1 Impacting factor on the color of a LED packages

The dependence of phosphor layer geometry on color is observed. Fig. 2a) shows a picture of two package free LED samples of assembled from the same type of blue LEDs with different phosphor layer thickness. A phosphor layer with smaller thickness was given in sample 2 which appeared to be more bluish compared to sample 1. The measured spectral power distributions of the corresponding samples are shown in Fig. 2b). In accordance, relatively higher radiant power of lower wavelength (blue radiation) was observed in the

978-1-4799-4789-8/14 $31.00 © 2014 IEEE

spectrum of sample 2. The effect of geometry is important for the phosphor layer design for package free LED.

The color quality of a package free LED can be revealed by the spectral power distribution of the radiant power. The spectrum represents the intensity of light irradiation as a function of different wavelength. In a phosphor-converted white LED, the combined blue radiation and phosphor radiation determines the shape and intensity of a spectrum. Based on the shape and intensity of a spectrum, information of luminous flux and color can be deduced.

a)

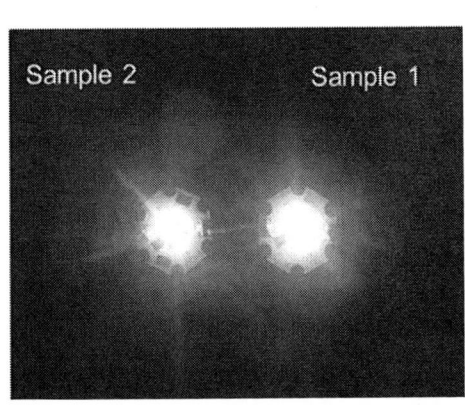

b)

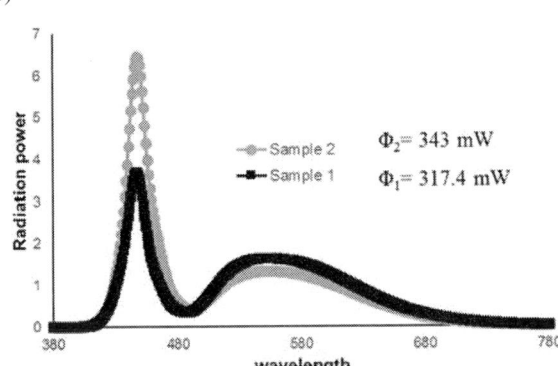

Fig. 2 a) color of two LED packages with different phosphor dimensions; b) the corresponding spectra of the package samples

A schematic spectral power distribution of a phosphor-converted white LED was given in Fig. 3. The spectrum consists of two peaks, a peak attributes to the blue LED and a yellow peak attributes to color conversion of the phosphor layer. The area of the peaks represents the intensity of light irradiated by the white LED. The total radiation power from the white LED (Φ_F) can be calculated from sum of the radiation power of the blue LED Φ_B and the radiation power from the phosphor Φ_{PHO}.

$$\Phi_F = \Phi_B + \Phi_{PHO} \qquad (1)$$

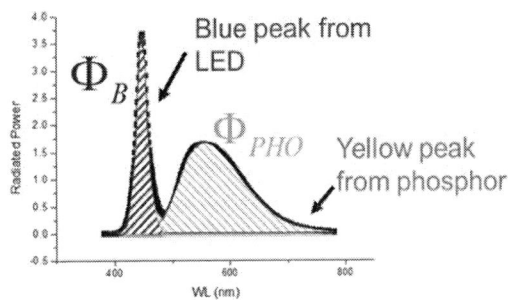

Fig. 3 Contribution of irradiation of phosphor-converted white LED

Fig. 4 illustrates the normalized spectra of Sample 1 and 2. The figure reveals that both the shape and position of the blue LED does not change with the two samples. Also, the position of the phosphor peaks does not change. The proportion of radiant flux from phosphor relative to radiant flux from LED die in Sample 2 is smaller in comparing to Sample 1, which is attributed from the smaller volume of phosphor layer. These findings implies that the area of the phosphor peak relates to the size of the phosphor layer.

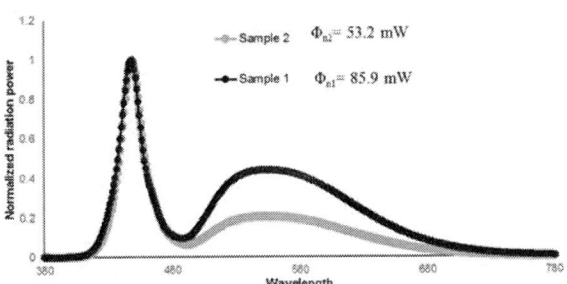

Fig. 4 Normalized spectra of Sample 1 and Sample 2

In the normalized spectra, the radiation power was normalized to the peak intensity along the wavelength. The radiation power of the phosphor in a spectrum can be calculated by equation (2):

$$\Phi_{PHO} = \Phi_B V \qquad (2)$$

where V is a factor related to the dimension of the phosphor layer.

The proportion of the radiation power emitted from phosphor layer (A_{PHO}) is formulated in equation (3):

$$A_{PHO} = \frac{\Phi_{PHO}}{\Phi_F} = \frac{\Phi_{PHO}}{\Phi_B + \Phi_{PHO}} \qquad (3)$$

by substituting equation (2) into equation (3),

$$A_{PHO} = \frac{\Phi_B V}{\Phi_B + \Phi_B V} \qquad (4)$$

further simplification of equation (4)

$$A_{PHO} = \frac{1}{1 + \dfrac{1}{V}} \qquad (5)$$

With a fixed materials system, Equation (5) revealed that A_{PHO} is a function of the dimension of the phosphor layer only. It is implied that if we are able to obtain A_{PHO} as a function of volume ratio, Φ_B and Φ_{PHO} can be deduced. Since with a fixed materials system, the shape functions of the emission spectrum from the LED die (blue) and the phosphor are independent of the power. Together with the total radiation power information, the overall spectrum can be predicted. The color of the package can be deduced from the spectral power distribution via the color matching function.

Methodology

Three types of white LED packages was prepared for the study. 2 different types of blue LEDs with 1W power was adopted for fabrication of the sample. The blue LEDs were bonded on ceramic substrates with silver surface. A layer of phosphor in different dimensions was covered on blue LEDs for color conversion. The phosphor layer were made from YAG phosphor which has been dispersed in silicone in a weight percent of 10. The detailed description of the 3 samples is given in Table 1.

Table 1 Configurations of the samples used in this study

Sample	LED type	Peak wavelength (nm)	Phosphor volume (x 10^8 um^3)
1	A	447	6.7
2	A	447	3.5
3	B	447	7.6

Determination of LED and phosphor spectrum

The spectral power distribution of these three samples has been measured by an 0.5m integration sphere (EVERFINE ASI-2-0.5m). The result is co-plotted in Fig. 5.

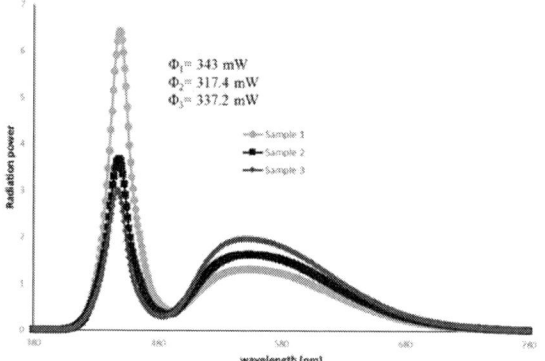

Fig. 5 Spectral power distribution of the three samples

By multiple peaks fitting of the measured spectrum, the radiation power of the blue radiation and the phosphor radiation can be resolved. Fig. 6 shows the result of the resolved peaks for Sample 1. In order to ensure the accuracy of the result, the sum of radiation power of the resolved peaks (Φ_T) has to be in reasonable matching with the total radiation power in the measurement (Φ_M). The resolved radiation power for the blue (Φ_B) and the phosphor peaks (Φ_{PHO}) given in Table x.

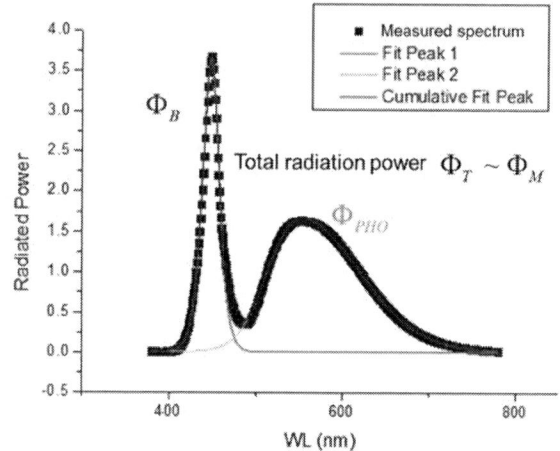

Fig. 6 Schematic of the ray-tracing model in color prediction

Phosphor Spectral Portion-volume ratio relations

For two devices in the different phosphor, the ratio of A_{PHO} is proportional to the ratio of the phosphor volume,

$$\frac{A_{PHO2}}{A_{PHO1}} = \frac{1 + \frac{1}{\frac{1}{V_2}}}{1 + \frac{1}{\frac{1}{V_1}}} = f\left(\frac{V_2}{V_1}\right) \tag{6}$$

From equation (6), A_{PHO} is in linear relationship with the phosphor volume. Therefore, a functional form can be assumed,

$$A_{PHO} = a + bV \tag{7}$$

where a and b are the fitting parameters. By linear fitting of equation (7) for the data points with white LED samples in different phosphor dimension, the relationship can be accomplished. The relations helps in predicting the Phosphor Spectral Portion in the spectrum of white LED with given phosphor dimension.

With the Phosphor Spectral Portion, the spectrum of the white LED can generated from SPD of a reference white LED with known blue LED SPD [$i_{B,r}\left(\lambda\right)$] and the phosphor SPD [$i_{PHO,r}\left(\lambda\right)$] whose $A_{B,r}$ and $A_{PHO,r}$ are known as illustrated in Fig. 7. The spectrum of the phosphor emission and the LED emission can be delineated from equation (8),

$$i_{PHO,S}\left(\lambda\right) = \frac{A_{PHO,S}}{A_{PHO,r}} i_{PHO,r}\left(\lambda\right)$$

$$i_{B,S}\left(\lambda\right) = \frac{1 - A_{PHO,S}}{1 - A_{PHO,r}} i_{B,r}\left(\lambda\right) \tag{8}$$

The full spectrum of the sample can be evaluated by summation as shown in equation (9),

$$i_S\left(\lambda\right) = i_{B,S}\left(\lambda\right) + i_{PHO,S}\left(\lambda\right) \tag{9}$$

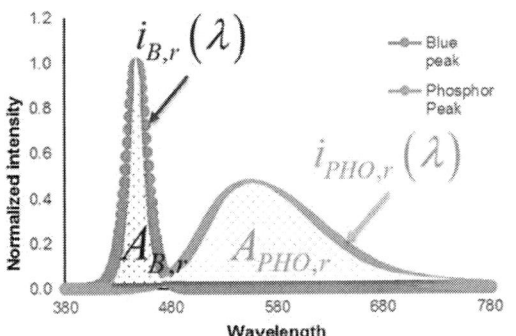

Fig. 7 SPD for the blue LED, $i_{PHO,S}(\lambda)$ and the phosphor, $i_{PHO,S}(\lambda)$

Spectrum construction

The spectrum, $I(\lambda)$ can be constructed from the normalized spectrum as given in equation (10),

$$I(\lambda) = \frac{\Phi_P}{\Phi_S} i_S(\lambda) \tag{10}$$

where Φ_P and Φ_s is the radiation power of the given sample and that of the normalized spectrum respectively. Correspondingly,

$$\Phi_P = \int_{380}^{780} I(\lambda)d\lambda \tag{11}$$

$$\Phi_S = \int_{380}^{780} i_S(\lambda)d\lambda \tag{12}$$

Fig. 8 illustrates the predicted spectrum of Sample 3 based on the developed methodology.

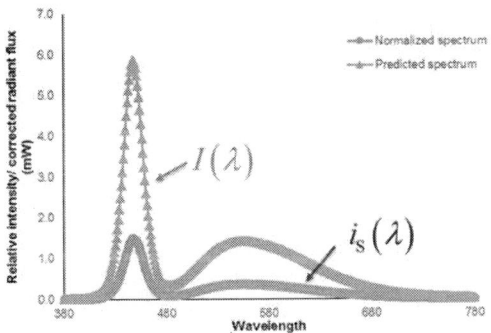

Fig. 8 The constructed spectrum $I(\lambda)$ of Sample 3

Benchmarking
Color determination and evaluation

The color co-ordinate of the sample in known dimension can be deduced with the known radiation power of the device together with the color matching function [5]. In order to evaluate the model accuracy, color parameters of the fabricated LED samples was measured. Six samples with the geometry of Sample 3 were examined. The electro-optical measurement was conducted by an 0.5m integration sphere (EVERFINE ASI-2-0.5m).

To further quantify the accuracy, the duv as defined in the the CIE1976 color space were calculated with the transformation equations (7) as below:

$$u' = \frac{4X}{X + 15Y + 3Z}$$
$$v' = \frac{9X}{X + 15Y + 3Z} \tag{13}$$

The color deviation is quantified by comparing with the predicted value CIE(u,v) with the measurement result CIE(u',v'). Distance (duv) between the two points indicates the color mismatch and is calculated by the equation (8).

$$duv = \sqrt{[(u' - u'_m)^2 + (v' - v'_m)^2]} \tag{14}$$

where CIE(u'_m, v'_m) refers to the color from measurement. Low duv implies little color mismatch between the model and experiment which implies good prediction.

Result

The resolved radiation power for the blue (Φ_B) and the phosphor peaks (Φ_{PHO}) and the Phosphor Spectral Portion are given in Table 2:

Table 2 The resolved blue and phosphor radiation power and Phosphor Spectral Portion of the samples

Sample	Model prediction				Measurement
	A_{PHO}	Φ_B	Φ_{PHO}	Φ_F	Φ_M
1	71.2	91.7	226.7	319	317
2	56.0	150.9	189.8	343	343
3	78.7	71.9	265.5	337	337

Fig. 9 plots the Phosphor Spectral Portion - Vol. ratio relations. The linear function is deduced as equation (15)

$$A_{PHO} = 36.717 + 35.89V \tag{15}$$

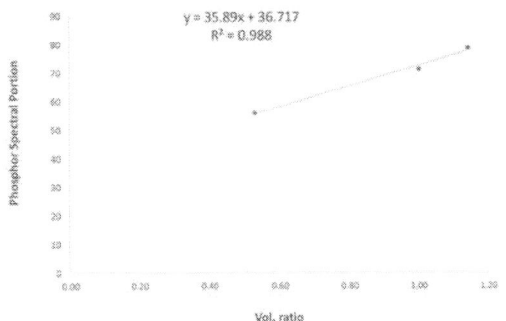

Fig. 9 Phosphor Spectral Portion- Volume ratio relations

For sample 3 whose Vol. ratio equals to 1.14, the A_{PHO} is given as 77.6. The portion of the phosphor peak in the predicted spectrum for sample 3 is 77.6% which implies the corresponding portion of the blue peak is 22.4%.

With the spectral portion data, the spectrum can be constructed from the SPD of the reference sample (Sample 1) as given by equation (4-6). Fig. 10 illustrates the spectral power distribution of sample 3.

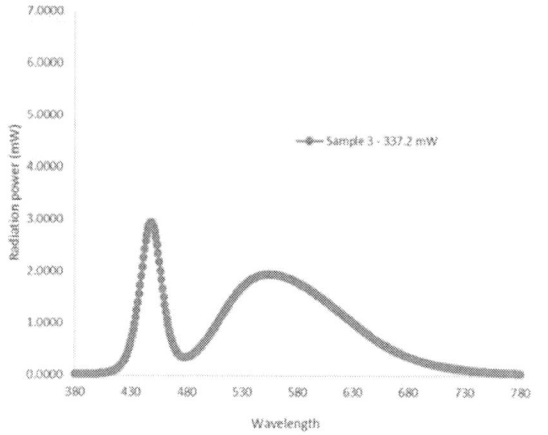

Fig. 10 The constructed spectrum of sample 3

The Tristimulus values XYZ in CIE 1931 color space of Sample 3 was deduced from the SPD in Fig. 10. The value of X, Y, Z is 154.7, 165.6 and 137.7 respectively. Color co-ordinate of the samples in the CIE 1976 (u', v') color space was deduced and listed in Table 3.

Table 3 summarizes the average data of the color co-ordinate of the measured samples and compared with the predicted result. The color difference (duv) between the predicted value and that of the measurement was 0.0018, which is within the tolerance of manufacturing variations.

Table 3 Comparison of the u', v' of Sample 3 based on prediction and measurement

	u'_m	v'_m	duv
Model prediction	0.2023	0.4883	
Measurement	0.2025±0.0003	0.4866±0.0049	0.0018

Conclusions

The methodology in predicting the color of package free LED with known phosphor geometry is established. The portion of radiation power emitting from the phosphor layer is linearly proportional to the volume of the layer. The Phosphor Spectral Portion- vol. ratio relations has been deduced. The spectrum of LED samples was constructed through the fitting parameter of the blue peak and the phosphor peak. The model was validated by comparing the color co-ordinate of a sample calculated from this model and the measured value with the color difference of 0.0018 was deduced, which is within the tolerance of manufacturing variations. The principle of the analyzing scheme established the guideline on the geometric design of the phosphor layer in a package free LED. The method can serve as a helpful tool for reducing design cycle in the new generation of package free LED.

Acknowledgement

The author would like to thank Gonqi Fan of State Key Laboratory of Solid State Lighting (Changzhou Base) for the sample measurement and discussion.

References

[1]   Y. Ohno, "Color rendering and luminous efficacy of white LED spectra," in *Optical Science and Technology, the SPIE 49th Annual Meeting*, 2004, pp. 88-98.

[2]   (2013, 16/2/2014). TSMC released package-free PoD LED light panel. Available: http://www.lednews.org/tsmc-released-package-free-pod-led-light-panel/

[3]   (2013, 16/2/2014). LED Manufacturers Under Pricing Pressure Rush into Package Free Chip. Available: http://en.ofweek.com/news/LED-Manufacturers-Under-Pricing-Pressure-Rush-into-Package-Free-Chip-1408

[4]   (2014, 16/2/2014). 2013 Review: Package Free LED Technology has became the highlight in 2013. Available: http://wellmaxgroup.com/package-free-technology-became-the-highlight/

[5]   A. Zukauskas, M. Shur, and R. Gaska, *Introduction to solid state lighting* vol. 122: John Wiley & Sons New York, 2002.

Design aspects for CPI robust BEOL

Mario Gonzalez, Luka Kljucar, Bart Vandevelde, Ingrid De Wolf and Zsolt Tokei

imec

Kapeldreef 75, 3001, Leuven, Belgium

e-mail: mario.gonzalez@imec.be

Abstract

In this paper we present our methodology to establish a quantitative comparison of the induced stresses at different locations of the package and their effect on the strength of the back-end-of-line (BEOL). A Chip Stack Package (CSP) with tight pitch and lead free solder joints is used as test vehicle. Different configurations of the interconnection between the solder balls and the BEOL, including a stiff passivation layer combined with a polyimide stress buffer layer with different thickness and openings are analyzed. It was found that the bending moment of the outermost solder joint induces high tensile stresses in the BEOL layer and this stress is reduced by increasing the thickness of the passivation layer. For this particular case, an optimal geometry of the stress buffer, in terms of thickness and open diameter is proposed. The stresses and energy release rate (ERR) induced on the BEOL is analyzed in a 2 metal layer configuration with different densities of via interconnections. The strength of the BEOL is improved when increasing both, the stiffness of the low-k material and the density of vias.

1. Introduction

The progress done towards the miniaturization and performance of integrated circuits demands the development of new packages and materials. A good example is the introduction of highly porous materials in the Back-End-of-line (BEOL) to reduce the k-value needed to prevent leakage between metal interconnections of the circuitry and minimize the time delay losses in the interconnections. Although these nano-porous materials improve the electrical performance and reliability of the circuit [1-4], it decreases at the same time the mechanical stability [4-7]. The thermal and residual stresses induced during the different packaging process steps might be catastrophic for different components of the package, including (but not limited) the BEOL. Thermo-mechanical deformation of the package can be directly transferred to the Cu/low-k interconnect, inducing large local stresses to drive interfacial crack formation and propagation.

In this paper we will study, with 3D Finite Element simulations, different approaches in order to reduce the risk of failures in the BEOL. In a first instance, for a given BEOL design, we focus on the passivation and stress buffer layer to reduce the risk of failures due to the high concentration of stresses in region surrounding the solder interconnections. The thickness of the passivation layer and/or the addition of a stress buffer layer made of polyimide are simulated.

In the second part of this work, we analyze the design and density of the metal interconnection and the material's properties of the low-k dielectric materials used in the BEOL. Their influence on the induced stress and the energy release rate is studied.

2. Influence of passivation layer

The model assembly used in this work is a Chip Scale Package (CSP) with a silicon die size of 6.8x6.8 mm^2 and a thickness of 200 µm. The chip is attached to a 200 µm thick printed circuit board (PCB), using standard lead free solder joint with a pitch of 100 µm. The standoff height and copper pad size was 45µm and 65µm respectively. A schematic of the test vehicle is shown in Figure 1.

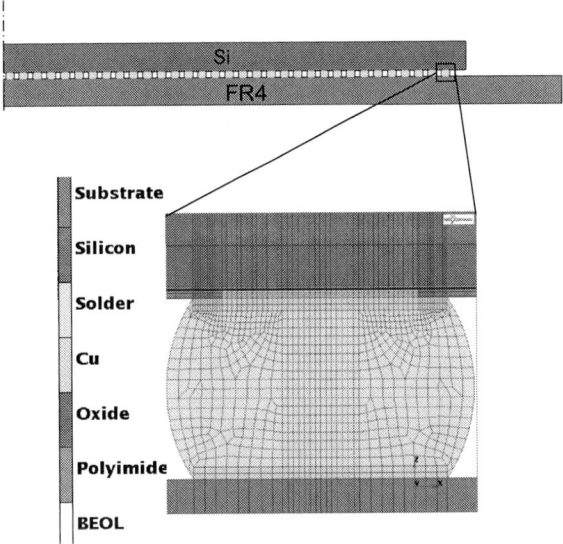

Figure 1. 3D finite element model of the CSP package including the baseline configuration of the solder joint

The underfill material was not included in the simulation to mimic the flip chip mass reflow assembly, where the maximum stress is induced to the BEOL. The stress and strain are calculated when cooling down the structure from 210°C to room temperature. Details of the BEOL were omitted in this model because its contribution to the global deformation is minimal given the relatively small thickness compared to the whole package. The equivalent properties of the BEOL layer were calculated using a Representative Volume Element (RVE). By applying three normal loads, 3 shear loads and 1 thermal load, orthotropic equivalent properties were obtained. A low-k material with a Young's modulus of 5 GPa was

978-1-4799-4789-8/14 $31.00 © 2014 IEEE

used. A description of the RVE and the method to calculate the equivalent properties were reported previously in [8].

The material properties used in this model are listed in Table 1. All materials except the solder joint were models as linear elastic. Solder joint was simulated as elasto-plastic material with Young's modulus and yield stress dependent on temperature.

Table 1. Material's properties

Material	Young's Modulus (GPa)		Poisson's ratio (-)	CTE (ppm/°K)
FR-4	20		0.22	12
Copper pad	117		0.3	16.7
Si	169		0.26	2.3
SnAgCu [9,10]	$52.71-67.14 \times 10^{-3} \cdot T - 5.87 \times 10^{-5} \cdot T^2$		0.4	17.60
BEOL	$E_1 = 99.87$ $E_2 = 101.63$ $E_3 = 46.41$	$G_{12} = 38.13$ $G_{23} = 32.51$ $G_{31} = 10.92$	$v_{12} = 0.2237$ $v_{23} = 0.1847$ $v_{31} = 0.0860$	$\alpha_1 = 3.45$ $\alpha_2 = 3.75$ $\alpha_3 = 10.94$
SiO2	71.4		0.17	0.68
PI	5		0.3	35

T = Temperature in °C

In order to reduce the thermo-mechanical stresses induced in the BEOL layers during the bonding step, several possibilities are analyzed. The first one consists in increasing the passivation thickness. By adding a stiff material between the copper pad and the BEOL, the pressure that the solder joint projects on the BEOL is distributed in a wider region, allowing to reduce the stress concentration in the edge of the solder joint. The edge of the solder joint is of particular interest because during the flip chip assembly, and because there is no underfill material at this step, the solder joints are subjected to a combined shear force and bending moment.

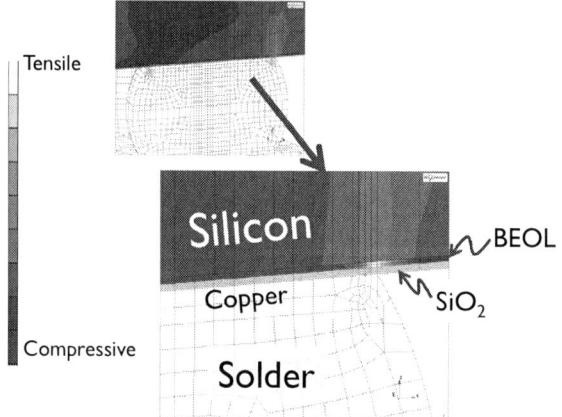

Figure 2. Influence of Oxide layer

In our experiment we use SiO_2 as passivation layer. The thickness of this layer was varied from 500 nm up to 3 μm. The maximum peel and shear stress in the equivalent BEOL layer is consider as failure criteria. Figure 2 shows a cross section of the outermost solder

joint showing the peel stress distribution in the equivalent BEOL and the silicon die. Due to the higher shrinkage of the PCB regarding the silicon die, the bending moment in the solder causes a high tensile stress in the outer edge of the solder joint and compressive stress in the inner side.

Figure 3 depicts the maximum peel and shear stress induced in the BEOL layer. The stresses in the BEOL is reduced as the thickness of the SiO_2 layer increases. A reduction of 36% and 48% for the out of plane and shear stress respectively is obtained while increasing the passivation thickness from 500 nm to 3 μm.

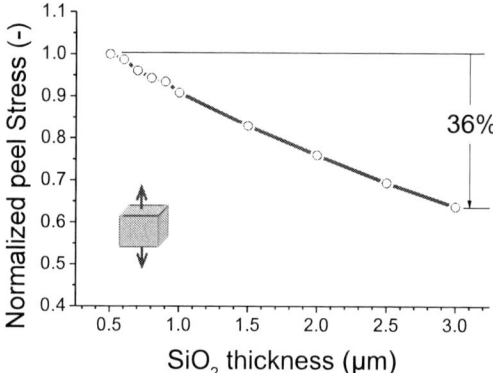

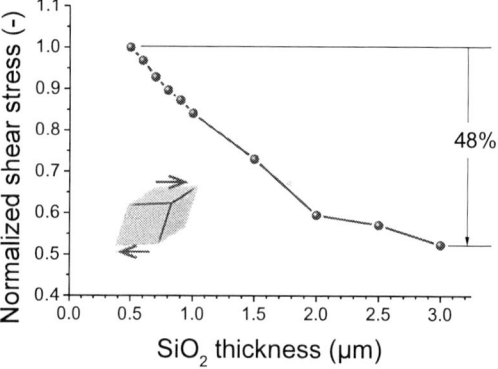

Figure 3. (a) Maximum induced peel stress in the BEOL located in the outer edge of the outermost solder joint. (b) Maximum induced shear stress in the BEOL

2. Stress buffer layer

In order to minimize the induced stresses due to the bending moment of the solder joint, we can either increase the standoff or reduce the solder joint diameter (not included in these simulations). One can reach similar effects by adding a stress buffer layer that decouples the peel force induced at the edge of the solder. The influence of this buffer layer is investigated. In the simulation, the opening of the stress buffer was kept constant to 50 μm and the thickness of the layer was varied from zero (no buffer layer) to 3 μm.

Due to the buffer layer, the stress concentration is shifted from the edge of the solder joint (point A in Figure

4a) to the edge of the opening of the buffer layer (point B in figure 4a). As the thickness of the buffer layer increases, the peel stress in the edge of the solder joints is reduced, however an increase in the stress in the edge of the buffer layer is observed. The optimum thickness is the point where the two regions have the same induced stress. For this particular case this point correspond to a thickness 0.8 μm.

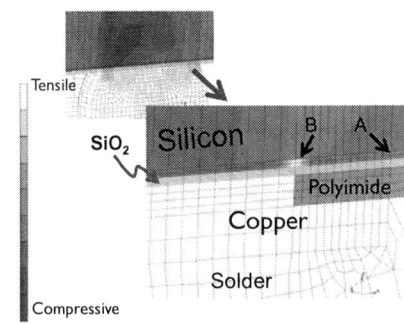

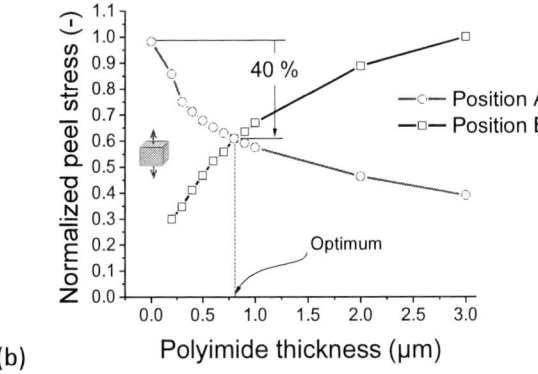

Figure 4. Polyimide as stress relief layer. (a) Finite Element Model; (b) Induced stress in the BEOL for a buffer layer opening of 50 μm and a passivation thickness of 500 nm.

The effect of the polyimide layer opening was also studied. The polyimide buffer layer decouples the peel stress induced by the bending moment of the solder joint. This effect is maximized by reducing the diameter of the buffer layer opening (increase the PI area between solder joint and BEOL). The peel stress induced in the edge of the solder joint is reduced, however, the induced stress in the edge of the buffer layer is increased. A balance between the induced stress in the edge of the bump and the edge of the buffer layer needs to be obtained.

Figure 5 depicts the variation of the peel stress in the BEOL layer for different buffer layer openings. An optimum value of 40 μm is obtained for this specific configuration.

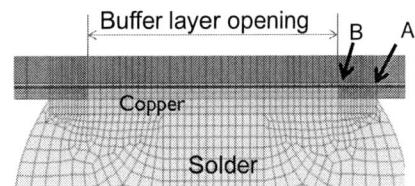

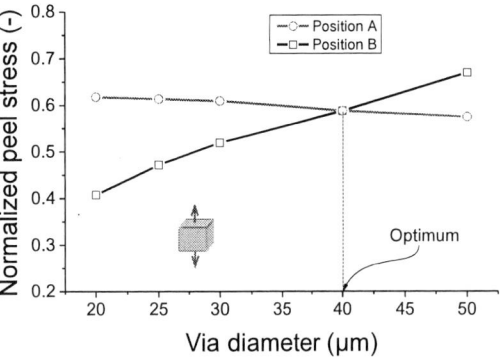

Figure 5. Induced peel stress in the BEOL layer. Location B indicated the edge of the PI buffer layer. This position change depending on the opening diameter. The passivation and PI thickness of 500nm and 1 μm respectively are used.

3. Study on the BEOL pattern structure

In order to understand the deformation of the BEOL layers, we built a detail FEM of the BEOL with 2 metal layers and 1 single via interconnection. The stress in the metal via and the ERR for a predefined crack are calculated for different densities of vias going from 0.51% up to 4%. The via density is simulated by increasing the length of the metal interconnections and keeping the geometry of the via constant. The metal lines and the ratio metal/low-k is also kept constant to 50%. The width of the metal lines is fixed to 45 nm and the cross section has an aspect ratio of 2 (thickness of metal lines equal to 90 nm). The initial crack is consider to be at the interface low-k/SiCN layer.

The FEM model is depicted in Figure 6. The bottom surface of the model is constrained in the vertical direction, while the top surface is free to move in all directions, but constrained to be parallel to the bottom surface. Periodic boundary conditions are applied to the other 4 faces, by linking the opposite faces between them.

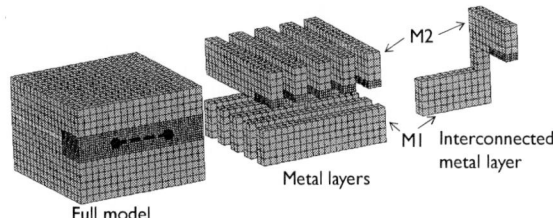

Figure 6. 3D FEM of the BEOL. A pre-crack is introduced in the interface low-k/SICN with surface of 225 nm x 45 nm.

In order to calculate the maximum stress in the metal via and the ERR at the crack tip, the simulation is done in two steps. In a first instance, without the presence of a crack, a normal load, equivalent to a pressure going from 100 to 500 MPa, is applied to the top surface of the RVE. The stress in the via is calculated by averaging the stress over the whole elements composing the via. In a second step, a crack is introduced in the model and the maximum ERR is calculated in the crack tip.

Figure 7 depicts an example of the calculated stress distribution and ERR for a via density of 2.77% and an applied stress of 100 MPa. The ERR was calculated using the Virtual Crack Closure Technique (VCCT). In all simulated cases, the crack surface was kept constant with a surface of 225 nm x 45 nm. This crack length was chosen to cover all the range of the via densities that were simulated.

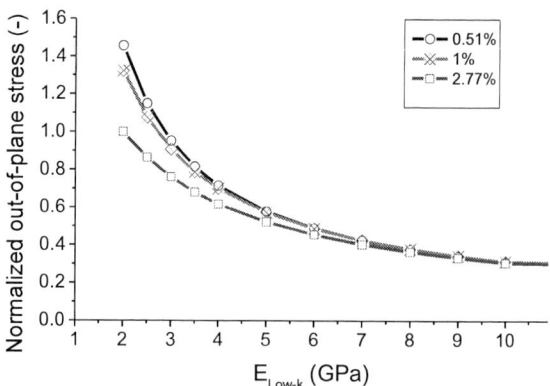

Figure 8. Variation of out-of-plane stress in the metal vias as function the Young's Modulus of the low-k materials and via densities.

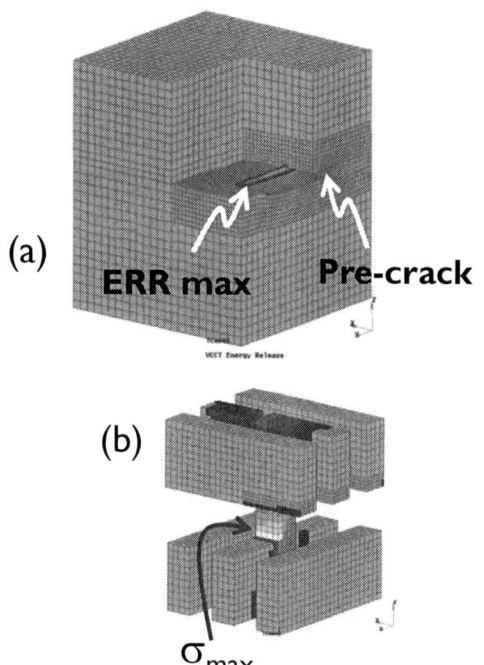

Figure 7. (a) Energy Release Rate; (b) Peel stress distribution in the via interconnection and metal lines.

Figure 8 shows the normalized average induced out-of-plane stress in the metal via for different via densities and for Young's Modulus of the low-k material varying from 2 GPa until 10 GPa. From these curves we can see that the induced stress in the via is not linearly related to the Young's Modulus of the low-k and it is dependent to the via density. We can observe from these curves that if we reduce the Young's modulus from 10 GPa (~low-k 2.5) to 4 GPa (~low-k 2.3), the induced stress increases 2.3 times for the case of 0.51% of via density and 2 times for the via density of 2.77%. Similar behavior has been observed for the ERR calculated in the second step of the model as depicted in Figure 9.

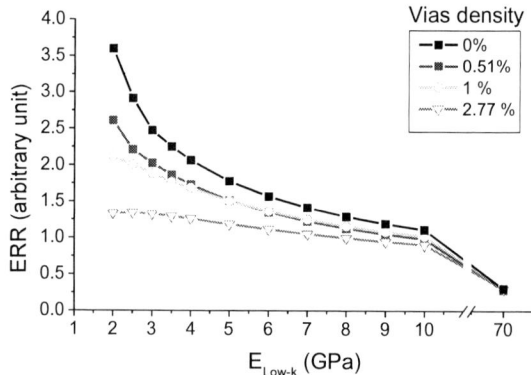

Figure 9. Variation of the ERR in the crack tip as function of Young's Modulus of low-k materials and via density.

We can conclude from these plots that if the technology requirements demands the use of ultra-low-k materials with Young's modulus of 4 GPa and below, the BEOL mechanical stability can be reinforced by increasing the via density.

4. Conclusions

Both the thick silicon oxide that helps to distribute the stress on a wider region and the soft PI layer that reduces the effective diameter of the solder joint plays and important role in reducing the risk of failure in the BEOL. Furthermore, the influence of Young's modulus of the low-k on the induced stress and/or energy release rate is reduced by increasing the via density.

A combination of design factors such as package type and dimensions, the inclusion of combined stiff and soft passivation layers, and a reinforced BEOL with high via densities, can increase the probability of having a robust and reliable package design.

References

[1] L. Zhao *et al.* "Ultra Low-k Materials: Challenges of Scaling". *ECS Trans.* 2010 33(12): 117-123; doi:10.1149/1.3501038

[2] R.J.O.M. Hoofman *et al.* "Challenges in the implementation of low-k dielectrics in the back-end of line". *Microelectronic Engineering* Volume 80, 17 June 2005, Pages 337–344.

[3] M. Pantouvaki *et al.* "Advanced Organic Polymers for the Aggressive Scaling of Low-k Materials". *Japanese Journal of Applied Physics* 50 (2011) 04DB01.

[4] K. Maex *et al. Journal of Applied Physics*, 93(11), 8793-8841 (2003).

[5] A. Urbanowicz, *et al.* "Fabrication of porogen residue free ultra low-k PECVD material by subsequent H2-afterglow plasma treatment and UV curing". *Advanced Metallization Conference: AMC.* 2009 Baltimore, MD. 13-15 October 2009.

[6] A. Urbanowicz *et al.*, "Improving mechanical robustness of ultralow-k SiOCH plasma enhanced chemical vapor deposition glasses by controlled porogen decomposition prior to UV-hardening". Journal of Applied Physics 107, 104122, 2010.

[7] M. Gonzalez, *et al.* "Modeling the substrate effects on nanoindentation mechanical property measurement". *Proc. of the 10th Int. Conf. Thermal, Mechanical and Multiphysic Simulation and Experiments in Micro-Electronic,* Delf, Netherlands, p. 486 (2009).

[8] M. Gonzalez *et al.*, ''Chip Package Interaction (CPI): Thermo Mechanical Challenges in 3D Technologies'', *Electronics Packaging Technology Conference (EPTC)*, pp. 547-551, 2012.

[9] A. Schubert, *et al.* "Reliability assessment of Flip Chip assemblies with lead-free solder joints". *Proc. of the 52nd Electronic Components and Technology Conference,* San Diego, CA, USA (2002) 1246-1255.

[10] J. H. Lau, *et al.* "Creep Analysis of Wafer Level Chip Scale Package (WLCSP) with 96.5Sn-3.5Ag and 100In Lead Free Solder Joints and Microvia Build-Up" *Printed Circuit Board. Journal of Electronic Materials*. Vol. 124, No. 2. (June 2002) 69-76.

Numerical simulations of piezoelectric MEMS energy harvesters

Giacomo Gafforelli[1], Raffaele Ardito[1], Alberto Corigliano[1], Carlo Valzasina [2], Francesco Procopio [2]

[1]Dept. of Civil and Environmental Engineering, Politecnico di Milano, Piazza Leonardo da Vinci 32, Milan, Italy
[2]STMicroelectronics, AMS Group, Via Tolomeo 1, 20010 Cornaredo, Italy
alberto.corigliano@polimi.it

Abstract

The application of piezoelectric materials in MEMS energy harvesters is continuously increasing, with the immediate corollary of a fundamental need for improved computational tools in order to optimize the performances at the design level. In this paper, a refined, yet simple model is proposed with the aim of providing fast and insightful solutions to the multi-physics problem of piezoelectric energy harvesting. The main objective is to retain a simple structural model (Euler-Bernoulli beam), with the inclusion of effects connected to the actual three-dimensional shape of the device. A thorough presentation of the analytical model is presented, along with its validation by comparison with the results of full 3D computations.

1. Introduction

Piezoelectric materials have been widely used in many technological applications at the macroscopic scale. In recent times, such materials have been entering in the field of Micro-Electro-Mechanical Systems (MEMS): in particular, the coupling between electrical and mechanical behavior is feasibly used in MEMS energy harvesters (which exploit the so-called "direct effect") and actuators (based on the "indirect effect"). Piezoelectric MEMS have been proven to be an attractive technology for harvesting small magnitudes of energy from ambient vibrations. This technology promises to eliminate the need for batteries or complex wiring in microsensors/microsystems, moving closer towards battery-less, autonomous sensors systems and networks which recover on-site the energy they need to fulfill their tasks.

At the present time, most of the micro piezoelectric harvesters reported in the literature are cantilever laminated beams and plates with thin films of lead zirconate titanate $Pb(Zr,Ti)O_3$ (PZT) on Si or SiNx substrate [1], [2]. The multi-physics simulation of piezoelectric effect can be obtained by considering that the structural members are represented by a laminate composite with piezoelectric and silicon layers [3], the active layer is then attached to an external load resistance which reproduces the circuitry employed for the power management. Numerous models have been reported, but the effect of 3D strain field on the coupling has not always been considered [4], [5]. In this paper a simple 1D model is built in order to simulate piezoelectric thin beams and plate harvesters. Starting from the fully coupled 3D constitutive equations of piezoelectricity, appropriate hypotheses are introduced to model strains and stresses so that the 1D model takes into account the 3D effects. It is worth noting that such a behavior is usually neglected if the mechanical response of the beam is considered, in view of a small difference with respect to standard analytical results. Conversely, in the presence of piezoelectric coupling, 3D effects involve a significant variation of the results in terms of electrical quantities. Consequently, the problem should be carefully studied in order to provide a reliable estimate of the energy production in a wide range of geometrical configurations.

The sectional behavior of the beam is studied through the Classical Lamination Theory (CLT, specifically modified in order to introduce the piezoelectric coupling) and a reduced order model is built through separation of time and shape variables and the introduction of a suitable shape function for the beam deformation. The resulting coupled equations are solved analytically in the frequency domain and numerically in the time domain by means of step-by-step integration (α-method). For the sake of comparison, simulations have been carried out also by means of fully 3D model built with a commercial code, coupled with user's routines in order to introduce the effect of the electrical circuit. The results show that the proposed model is in excellent agreement with the numerical outcomes, with substantial improvements with respect to the standard analytical solutions.

2. Geometry

The piezolaminated cantilever beam is presented in Figure 1, where L is the length, h the total thickness and b the width. The x_3-coordinate originates in the neutral axis and is directed downwards, x_1-coordinate lies along the beam axis while the x_2-coordinate originates in the middle of the beam, so that $-b/2 \leq x_2 \leq b/2$. A piezoelectric layer (for instance, made by PZT material) is placed on top of the beam substrate and is activated in d31-mode when the beam deflects: this means that the axial deformation of the layer causes an electric field in the vertical direction (along x_3-axis). To implement d31-mode, the polarization vector is opposite to x_3-axis and the electrodes span both the upper and the lower surfaces of PZT thin film. In this way, when the PZT layer is stretched along x_1-axis, the generated charge is collected by the electrodes.

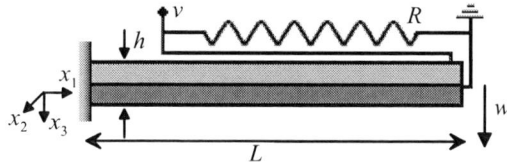

Figure 1: Cantilever piezoelectric harvester

The bottom electrode is grounded while the other is attached to an external load resistance (R) which

978-1-4799-4789-8/14 $31.00 © 2014 IEEE

represents an ideal external circuit employed for the management of the power generated by PZT.

3. Modelling

In this Section, the standard notation of piezoelectric analysis is adopted: stress component are denoted by the capital letter T and strain components by the capital letter S.

The beam final stack is not homogeneous since different deposited layers are employed. The mechanical response of the layered beam can be obtained by means of a number of theories, examined in [6]. Plane sections are considered to remain plane after deformation and the displacement vector results:

$$\mathbf{s} = \begin{bmatrix} -x_3\varphi_2(x_1) \\ s_2(x_1,x_2,x_3,\Lambda) \\ w_3(x_1)+\hat{s}_3(x_1,x_2,x_3,\Lambda) \end{bmatrix} \quad (1)$$

The vertical and the in-plane displacements are assumed to depend on parameter Λ which can be interpreted as *in-plane slenderness* of the beam defined as $\Lambda=L/b$. Functions s_2 and $\hat{s}_3$ are chosen trying to satisfy some specific hypotheses on the stress state. Let us denote with T_{ii}, S_{ii} the stress and the strain tensors, respectively. First, since the beam is considered thin, the vertical stress is assumed to be null, $T_{33}=0$. Second, the in-plane stress must be $T_{22}=0$ at $x_2 = \pm b/2$. Moreover, when $\Lambda\rightarrow0$ the beam can be considered as infinitely wide and the strain condition $S_{22}=0$ must be verified; on the other hand when $\Lambda\rightarrow\infty$ the beam can be imagined with null width and the condition $T_{22}=0$ must be verified. Considering the aforementioned hypotheses, the constitutive equations for the piezoelectric material reads [4]:

- when $\Lambda\rightarrow0$: $S_{22}=0$; $T_{33}=0$ ("Plane strain")

$$T_{11} = E\left(1-v^2\right)^{-1}S_{11} - \left(e_{31}-v\left(1-v\right)^{-1}e_{33}\right)E_3$$
$$T_{22} = Ev\left(1-v^2\right)^{-1}S_{11} - \left(e_{32}-v\left(1-v\right)^{-1}e_{33}\right)E_3$$
$$S_{33} = v\left(v-1\right)^{-1}S_{11} + E^{-1}\left(1-v\right)^{-1}\left(1-v+2v^2\right)e_{33}E_3 \quad (2)$$
$$D_3 = \left(e_{31}-v\left(1-v\right)^{-1}e_{33}\right)S_{11} +$$
$$+ \left(\varepsilon_{33}^S+\left(E\left(1-v\right)\right)^{-1}\left(1-v+2v^2\right)e_{33}^2\right)E_3$$

- when $\Lambda\rightarrow\infty$ or $x_2=\pm b/2$: $T_{22}=0$; $T_{33}=0$ ("Uniaxial Stress")

$$T_{11} = ES_{11}-\left(e_{31}-ve_{32}-ve_{33}\right)E_3$$
$$S_{22} = -vS_{11}+E^{-1}\left(1+v\right)\left(\left(1-v\right)e_{32}-ve_{33}\right)E_3$$
$$S_{33} = -vS_{11}+E^{-1}\left(1+v\right)\left(-ve_{32}+\left(1-v\right)e_{33}\right)E_3 \quad (3)$$
$$D_3 = \left(e_{31}-e_{32}v-e_{33}v\right)S_{11} +$$
$$+ \left(E^{-1}\left(1+v\right)\left(\left(e_{32}+e_{33}\right)^2\left(1-v\right)-2e_{32}e_{33}\right)+\varepsilon_{33}^S\right)E_3$$

where E_i, D_i are electric field and electric displacement field components, E is the Young's modulus and v is the Poisson's ratio, e_{ij} and ε_{ii} are the piezoelectric stress constant and the dielectric constant computed at constant mechanical stress. It is worth noting that, in view of the geometric features of the problem, only a subset of the tensor and vector components should be explicitly involved in the analysis.

By imposing $T_{33}=0$ one finds:

$$S_{33} = -v\left(1-v\right)^{-1}\left(S_{11}+S_{22}\right)+c_3 \quad (4)$$

where:

$$c_3 = E^{-1}\left(1-v\right)^{-1}\left(1+v\right)\left(1-2v\right)e_{33}E_3 \quad (5)$$

To obtain with a single model both eq. (2) and (3), S_{22} must be multiplied times a shape function, $f_\Lambda(x_2,\Lambda)$ that must be 1 when $\Lambda\rightarrow\infty$ or $x_2=\pm b/2$ and must be 0 when $\Lambda\rightarrow0$:

$$S_{22} = \left(-vS_{11}+c_2\right)f_\Lambda \quad (6)$$

where:

$$c_2 = E^{-1}\left(1+v\right)\left(\left(1-v\right)e_{32}-ve_{33}\right)E_3 \quad (7)$$

The function $f_\Lambda(x_2,\Lambda)$ is given by:

$$f_\Lambda\left(x_2,\Lambda\right) = \left(1-A_\Lambda\left(\Lambda\right)\right)\left(2|x_2|/b\right)^{B_\Lambda(\Lambda)}+A_\Lambda\left(\Lambda\right) \quad (8)$$

with $\xi = 2x_2/b$ and:

$$A_\Lambda\left(\Lambda\right) = \Lambda^{a_\Lambda}\left(\Lambda^{a_\Lambda}+b_\Lambda\right)^{-1} \quad (9)$$

$$B_\Lambda\left(\Lambda\right) = 1+\Lambda^{-1} \quad (10)$$

where $a_\Lambda = 0.8$ and $b_\Lambda = 2.2$ are used as fitting parameters.

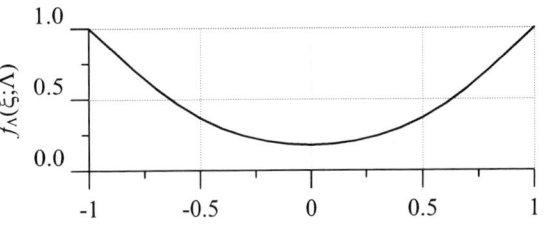

Figure 2: Shape function of in-plane strain

The stress and the electric displacement accordingly result:

$$T_{11} = E\left(1-v^2\right)^{-1}\left(1-v^2f_\Lambda\right)S_{11} +$$
$$-\left(e_{31}-ve_{32}f_\Lambda-v\left(1-v\right)^{-1}\left(1-vf_\Lambda\right)e_{33}\right)E_3$$
$$T_{22} = \left(1-f_\Lambda\right)\left(Ev\left(1-v^2\right)^{-1}S_{11}-\left(e_{32}-v\left(1-v\right)^{-1}e_{33}\right)E_3\right) \quad (11)$$
$$D_3 = \left(e_{31}-vf_\Lambda e_{32}-v\left(1-v\right)^{-1}\left(1-vf_\Lambda\right)e_{33}\right)S_{11} +$$
$$+ \left(\begin{array}{c} E^{-1}\left(1-v^2\right)\left(e_{32}-v\left(1-v\right)^{-1}e_{33}\right)^2 f_\Lambda + \\ + E^{-1}\left(1-v\right)^{-1}\left(1+v\right)\left(1-2v\right)e_{33}^2+\varepsilon_{33}^S \end{array}\right)E_3$$

Very thin beams are usually employed in MEMS devices, thus the Bernoulli hypothesis and CLT is adopted herein such that: $\varphi_2 = w_3'$. In this formula, $\bullet'$ means derivative with respect to the variable. The electric potential is constant on the electrodes, grounded on the bottom electrode and free to change on the upper one (v). According to the piezoelectric constitutive law, the electric field is proportional to strain which is linear across the piezoelectric layer thickness, thus the electric potential (ϕ) should be considered quadratic across the thickness of piezoelectric. However, as long as the

978-1-4799-4789-8/14 $31.00 © 2014 IEEE

piezoelectric layer is thin, a linear potential can be adopted:

$$\phi = x_3^* t_P^{-1} v \qquad (12)$$

where t_P is the piezoelectric layer thickness and x_3^* is the vertical upward coordinate in the piezoelectric layer (Figure 3).

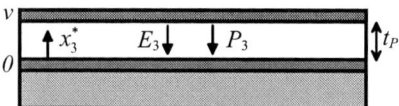

Figure 3: Polarization and electric potential between electrodes

4. Equation of motion

The equation of motion is computed using the dissipative form of Euler-Lagrange equations:

$$\frac{d}{dt}\left(\frac{\partial L}{\partial \dot{q}_i}\right) - \frac{\partial L}{\partial q_i} + \frac{\partial D}{\partial \dot{q}_i} = 0 \qquad (13)$$

where Δ is the Dissipation function and Λ is the Lagrangian function which is given by:

$$L = K - (E - W) \qquad (14)$$

K, E are the kinetic and internal energy and Ω is the external work.

The internal energy is given by:

$$E = \frac{1}{2}\int_V \left(T_{11}S_{11} + T_{22}S_{22} - D_3E_3\right)dV \qquad (15)$$

Given a stratification where $x_{3,k}$ is the coordinate of the k-layer and nl the total number of layers and since the T_{11}, T_{22}, S_{22} and D_3 are expressed as a function of S_{11} and E_3, eq.(11), the internal energy results:

$$E = \int_0^L \frac{b}{2}\sum_{k=1}^{nl}\int_{x_{3,k-1}}^{x_{3,k}}\left(\hat{E}_\Lambda^k S_{11}S_{11} - 2\hat{e}_\Lambda^k E_3 S_{11} - \hat{\varepsilon}_\Lambda^k E_3 E_3\right)dx_3 dx_1 \quad (16)$$

where

$$\hat{E}_\Lambda = \frac{1}{b}\int_{-b/2}^{+b/2} E\left(1-v^2\right)^{-1}\left(1-v^2\tilde{f}_\Lambda\right)dx_2 \qquad (17)$$

$$\hat{e}_\Lambda = \frac{1}{b}\int_{-b/2}^{+b/2}\left(e_{31} - e_{32}v\tilde{f}_\Lambda - v\left(1-v\right)^{-1}\left(1-v\tilde{f}_\Lambda\right)e_{33}\right)dx_2 \quad (18)$$

$$\hat{\varepsilon}_\Lambda = \frac{1}{b}\int_{-b/2}^{+b/2}\left(\begin{array}{c}E^{-1}\left(1-v^2\right)\left(e_{32} - v\left(1-v\right)^{-1}e_{33}\right)^2\tilde{f}_\Lambda + \\ +E^{-1}\left(1-v\right)^{-1}\left(1+v\right)\left(1-2v\right)e_{33}^2 + \varepsilon_{33}^S\end{array}\right)dx_2 \quad (19)$$

are averaged values of the elastic, piezoelectric and dielectric constants which also depend on the parameter Λ and:

$$\tilde{f}_\Lambda(x_2,\Lambda) = f_\Lambda(x_2,\Lambda)\left(2 - f_\Lambda(x_2,\Lambda)\right) \qquad (20)$$

It should be noticed that when $f_\Lambda = 1$ and $f_\Lambda = 0$ eqs. (17)-(19) reduce to the coefficients of eq. (2) and eq.(3).

Moreover, by means of compatibility and assuming the hypotheses in eq.(1) and (12), the strain and the electric field result:

$$S_{11} = s_1' = -x_3 w_3'' \qquad E_3 = -\phi' = t_P^{-1}v \qquad (21)$$

and the internal energy is written:

$$E = \int_0^L \frac{b}{2}\sum_{k=1}^{nl}\int_{x_{3,k-1}}^{x_{3,k}}\left(\hat{E}_\Lambda^k x_3^2 w_3''^2 - 2\hat{e}_\Lambda^k t_P^{-1}v w_3'' - \hat{\varepsilon}_\Lambda^k t_P^{-2}v^2\right)dx_3 dx_1 \quad (22)$$

By integrating on the thickness, one obtains the generalized internal stiffness $C_{\chi\chi}$ usually defined for the theory of laminates. In such a case the new constitutive law also contains, in the integrated constitutive equations, the generalized piezoelectric coupling coefficient $C_{\chi v}$. Moreover, the same procedure is adopted on the electrical part of the constitutive law obtaining the generalized internal capacity C_{vv}. [3].

$$C_{\chi\chi} = \sum_{k=1}^{nl}\int_{x_{3,k-1}}^{x_{3,k}}\hat{E}_\Lambda^k x_3^2 dx_3$$

$$C_{\chi v} = \sum_{k=1}^{nl}\int_{x_{3,k-1}}^{x_{3,k}} t_P^{-1}\hat{e}_\Lambda^k x_3 dx_3 \qquad C_{vv} = \sum_{k=1}^{nl}\int_{x_{3,k-1}}^{x_{3,k}} t_P^{-2}\hat{\varepsilon}_\Lambda^k dx_3 \quad (23)$$

and the internal energy finally results:

$$E = \int_0^L \left(bC_{\chi\chi}w_3''^2 - 2bC_{\chi v}v w_3'' - bC_{vv}v^2\right)dx_1 \qquad (24)$$

The external work is given by:

$$W = F_t w + \int_0^{L+b/2}\int_{-b/2} f_d w_3 dx_2 dx_1 - qv \qquad (25)$$

where F_t is a tip force at the free edge, f_d is a distributed force on the top surface and q the total charge on the top electrode.

The kinetic energy is given by:

$$K = \frac{1}{2}\left(m_t\dot{w}^2 + J_t\dot{w}'^2 + 2M_t\dot{w}\dot{w}' + \int_0^{L+b/2}\int_{-b/2}\dot{w}_3^2 m_d dx_2 dx_1\right)(26)$$

where m_t, J_t and M_t are the mass and the inertial and static moment of an eventual tip mass attached at the free edge. w and w' and are the vertical displacement and the rotation of the beam evaluated at the free edge; $\dot{\bullet}$ means derivation with respect to time. m_d is the distributed mass of the beam:

$$m_d = \sum_{k=1}^{nl}\int_{x_{3,k-1}}^{x_{3,k}}\rho^k dx_3 \qquad (27)$$

where ρ^k is the density of the k-th layer (the rotational inertia of the beam's cross-sections can be neglected as long as the beam is thin).

By means of separation of variables the mechanical unknowns reduces to the displacement at the free edge:

$$w_3(x_1) = \psi_w(x_1)w(t) \qquad (28)$$

where ψ_w is a shape function.

Substituting eq.(28) in eq.(24)-(26) and computing eq.(13) with a viscous dissipation function:

$$D = 1/2 c_M \dot{w}^2 \qquad (29)$$

978-1-4799-4789-8/14 $31.00 © 2014 IEEE 350

where c_M is the linear mechanical damping coefficient; the equation of motion is finally obtained:

$$m\ddot{w}+c_M\dot{w}+k_Lw-\Theta_{\chi v}v=F$$
$$k_Ev+\Theta_{\chi v}w=q \qquad (30)$$

The coefficients are computed by integrating the shape functions and the generalized constitutive coefficients on the area of the beam:

$$m=m_t+J_t\psi_w'^2(L)+2M_t\psi_w'(L)+\int_0^L bm_d\psi_w^2\,\mathrm{d}x_1$$

$$F=F_t+\int_0^L bf_d\psi_w\,\mathrm{d}x_1 \qquad k_L=\int_0^L bC_{xx}\psi_w''^2\,\mathrm{d}x_1 \qquad (31)$$

$$k_E=\int_0^L bC_{vv}\,\mathrm{d}x_1 \qquad \Theta_{\chi v}=\int_0^L bC_{\chi v}\psi_w''\,\mathrm{d}x_1$$

where m is the total inertial term; k_L is the linear elastic stiffness; k_E is the internal capacitance of PZT while $\Theta_{\chi v}$ is the coupling constant.

The electric charge collected by the electrodes is managed by an external circuitry which provides the power supply for the self-powered electronic device. Different schemes of circuitries are investigated in [7]. The harvester provides AC voltage and the simplest solution is the coupling with an external load resistance:

$$q=-R^{-1}v \qquad (32)$$

Eq.(31) and eq.(32) are usually studied separately by the commercial codes because they belong to two different physical domains which interact through electric potential and electric charge. The solution technique developed for commercial codes is explained in section **Errore. L'origine riferimento non è stata trovata.**. In case of the 1D code developed herein, the two physics are solved simultaneously.

5. Solution in the frequency domain

The oscillator frequency response to harmonic excitations is studied through Harmonic Balance Method [8] which provides an exact solution for linear systems. Harmonic balance mimics the spectrum analyzer in simply assuming that the response to a sinusoidal excitation is a sinusoid at the same frequency (ω). A sinusoidal couple of trial solution is substituted in the equations, the coefficients of same harmonics are equated and the Frequency Response Function (FRF) is computed:

$$Y=\left(\begin{array}{c}-\Omega_M^2+1+\kappa_{\chi v}^2\Omega_M^2\left(\Omega_M^2+\Omega_E^2\right)^{-1}+\\+\mathrm{i}\left(2\zeta_M+\kappa_{\chi v}^2\Omega_E\left(\Omega_M^2+\Omega_E^2\right)^{-1}\right)\Omega_M\end{array}\right)^{-1} \qquad (33)$$

where $Y=W/W_0$ is the dimensionless amplitude, W_0 is static displacement, $\Omega_M=\omega/\omega_r$ is the dimensionless excitation frequency, ω_r is the resonance angular frequency, $\Omega_E=1/Rk_E\omega_r$ is the dimensionless cut-off frequency of the circuit (Rk_E is the time constant of the RC circuit) and $\kappa_{\eta v}=\Theta_{\chi v}/(k_Ek_L)^{1/2}$ is the effective piezoelectric coupling coefficient. $\kappa_{\eta v}$ is a global measure of the degree of coupling which takes into account the coupling coefficient $\kappa_{33}=e_{33}/(E\varepsilon_{33}^S)^{1/2}$, and geometrical aspects. In case of a pure piezoelectric system, $\kappa_{\eta v}$ would

exactly coincide with κ_{33}. The equivalent stiffness and the damping ratio depend on the piezoelectric coupling, on the excitation frequency and on the load resistance:

$$k_{eq}=k_L\left(1+\kappa_{\chi v}^2\Omega_M^2\left(\Omega_M^2+\Omega_E^2\right)^{-1}\right)$$
$$\zeta_{eq}=\zeta_M+1/2\,\kappa_{\chi v}^2\Omega_E\left(\Omega_M^2+\Omega_E^2\right)^{-1} \qquad (34)$$

The power generation is the power dissipated by the equivalent electrical dashpot and results [9]:

$$P=F^2\omega_n\Big/m\kappa_{\chi v}^2\Omega_E\Omega_M^2\left(\Omega_M^2+\Omega_E^2\right)\times$$

$$\times\left(\begin{array}{c}\left(\left(\Omega_M^2+\Omega_E^2\right)\left(-\Omega_M^2+1\right)+\kappa_{\chi v}^2\Omega_M^2\right)^2\\+\left(\left(2\zeta_M\left(\Omega_M^2+\Omega_E^2\right)+\kappa_{\chi v}^2\Omega_E\right)\Omega_M\right)^2\end{array}\right)^{-1} \qquad (35)$$

The power can be maximized with respect to the electric load, or similarly to the cut-off frequency:

$$\Omega_E=\Omega_M\left(\frac{4\zeta_M^2\Omega_M^2+\left(1+\kappa_{\chi v}^2-\Omega_M^2\right)^2}{4\zeta_M^2\Omega_M^2+\left(1-\Omega_M^2\right)^2}\right)^{1/2} \qquad (36)$$

The FRF computed at the optimal load resistance shows two peaks that occur near the mechanical resonance frequency and near a second frequency called anti-resonance. Peaks would perfectly occur at these frequencies if mechanical damping had been neglected. The presence of two peaks is explained by the fact that the power ($P=V^2/R$) can be maximized both at high resistance (high voltage) and at low resistance (low voltage). It can be easily shown that in a general case the peaks are obtained when the injected electrical damping equals the mechanical damping of the system [10].

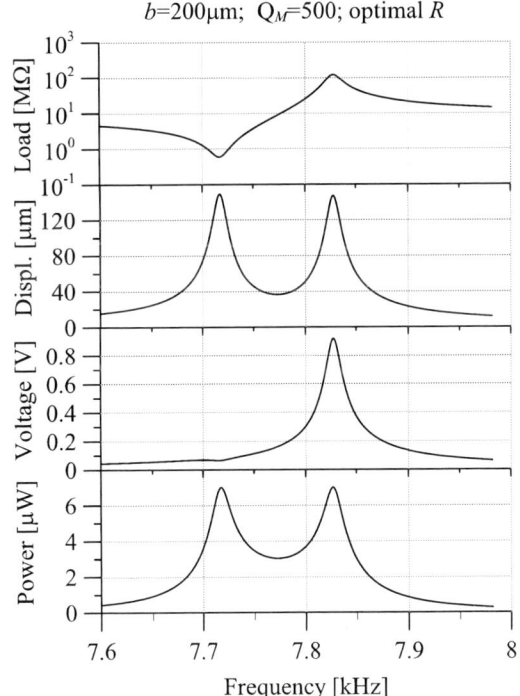

$b=200\mu m;\ Q_M=500;$ optimal R

Figure 4: Frequency response, b=200μm, Q_M=500, optimal R.

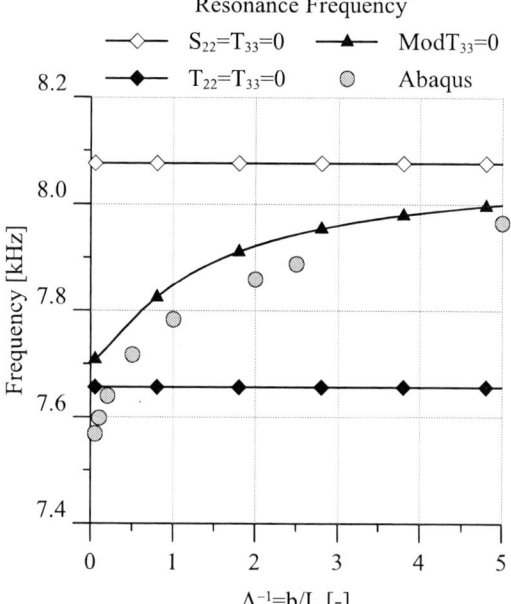

Figure 5: Resonance frequency

$$\mathbf{M} = \begin{bmatrix} m & 0 \\ 0 & 0 \end{bmatrix} \quad \mathbf{C} = \begin{bmatrix} c_M & 0 \\ \Theta_{\chi v} & k_E \end{bmatrix}$$

$$\mathbf{K} = \begin{bmatrix} k_L & -\Theta_{\chi v} \\ 0 & R^{-1} \end{bmatrix} \quad \mathbf{F} = \begin{bmatrix} F \\ 0 \end{bmatrix} \tag{39}$$

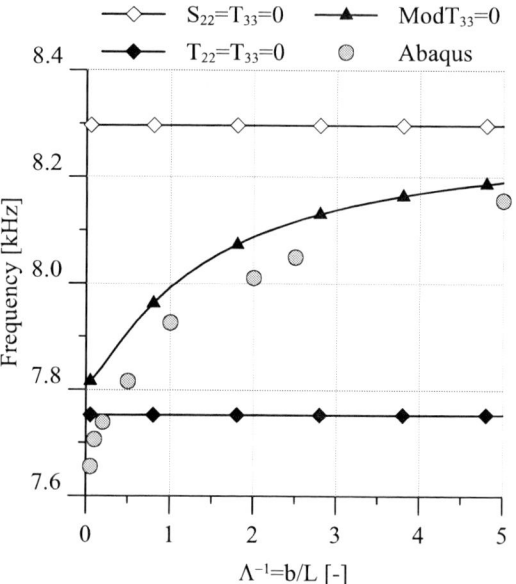

Figure 6: Anti-Resonance frequency

While the resonance frequency is purely a mechanical parameter, the anti-resonance frequency depends on the effective piezoelectric coupling coefficient and it can be used as a measure of the global coupling:

$$f_{ar} = f_r \sqrt{1 + \kappa_{\chi v}^2} \tag{37}$$

where f_r and f_{ar} are the resonance and anti-resonance frequency respectively.

Figure 5 shows that classical models cannot describe the dependence of resonance frequencies to the width of the device. On the other hand the modified model qualitatively reproduces the variation in frequency. It still does not perfectly match with numerical results because of a mismatch in the computation of the modal mass associated to the first vibrating mode.

Figure 6 shows that a good agreement is also obtained for the anti-resonance frequency. This result is profoundly different from the previous since anti-resonance measures the coupling parameters while resonance only mechanical ones. This means that the model can correctly describe the variation in the coupling due to different width ratio.

6. Solution in the time domain

Hughes α-method [11] has been chosen to solve eq.(30) and (32) for time dependent excitation. The equations are rearranged in matrix formulation in the variable $\mathbf{x}^T = \begin{bmatrix} w & v \end{bmatrix}$:

$$\mathbf{M}\ddot{\mathbf{x}} + \mathbf{C}\dot{\mathbf{x}} + \mathbf{K}\mathbf{x} = \mathbf{F} \tag{38}$$

where:

7. £D Finite Element model

To validate the model proposed in sections 2 and 3 a commercial code, ABAQUS, is employed. The code allows using piezoelectric constitutive equations in the classical structural mechanic environment. However, only 3D models can be built, this is an important limitation since piezoelectric elements are usually employed in laminated beams or plates. This results in a large numerical effort since a high number of elements must be employed to correctly reproduce the behavior of high aspect ratio components. Moreover, the solver of lumped electrical circuitry is not available in ABAQUS. For what concern the solution of piezoelectric behavior, the code solves an equation similar to eq. (30) built on the basis of 3D finite element provided with the additional voltage degree of freedom. The voltage is grounded on the bottom surface of piezoelectric material while is constrained to be uniform on the top surface, this to reproduce the presence of electrodes. The electric charge plays the role played in structural mechanics by the external force. The connection with the electric circuit has been built *ad hoc* through an external subroutine which interacts in a staggered procedure through the electric potential and the electric charge which act respectively as variable and load in the piezoelectric module and as load and variable in the electric module. The scheme of the adopted algorithm is shown in Figure 7. In this study the subroutine reduces to a discrete version of Ohm's law, but there is not any

978-1-4799-4789-8/14 $31.00 © 2014 IEEE   352

particular limitation on the equations that could be implemented.

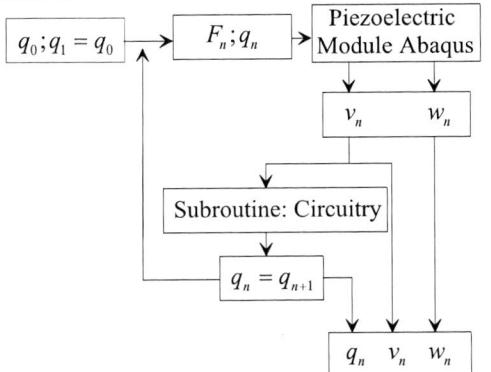

Figure 7: Algorithm for the resolution of the coupled piezoelectric-electric problem

The explicit algorithm can be considered correct as long as the variation of the charge in every step is small and the difference in the voltage computed with q_n and q_{n+1} is negligible. This is obtained for a small enough time step (Δ). In case of a pure resistive circuit, the value of time step that assures the stability of the algorithm depends on the external load resistance.

8. Analysis

Comparative analyses have been performed to validate the model. A 2 layers (2μm PZT on 6μm silicon substrate) cantilever beam has been chosen. The length is 1000μm, the width is parametric and varies from 50μm to 5000μm. Geometrical and material properties of the layers are given in Table I.

	ρ [g/cm^3]	E [GPa]	v [-]	e_{31} [N/mV]	e_{33} [N/mV]	$\varepsilon_{33\text{-}r}$ [g/cm^3]
PZT	7.70	100	0.30	-12	20	2000
SIL	2.33	148	0.33	0	0	0

Table I: Material properties

Open circuit static analyses have been performed with a tip load $F=1\mu N/\mu m$ at the beam free edge. Classical models have been compared to the modified model.

While the response of classical models is constant with respect to the width, the response of the modified model correctly describe the variation in stiffness (Figure 8) and more important, the variation of the upper electrode voltage and correspondingly of coupling coefficient (Figure 9). From the mechanical point of view the mean error can be reduced from 8% to less than 1%, while the error on the voltage reduces from 27% to less than 3%. Considering that the power will be the voltage squared, the reduction of the error in the power generation is expected to be even more significant.

Quasi static analyses have been performed; a smooth step load followed by a constant plateau has been considered. The dashed line in Figure 10 reproduces the applied force (right axis) vs. time. It is also shown that the

voltage gradually decreases to zero since the external load resistance dissipates energy reducing the charge on the electrodes. The decreasing speed depends on the value of resistance, the higher the resistance faster the voltage decreases.

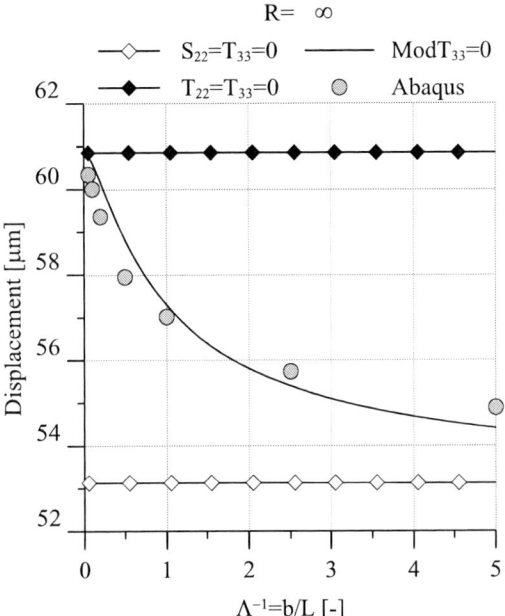

Figure 8: Open circuit Static Analysis, Displacement

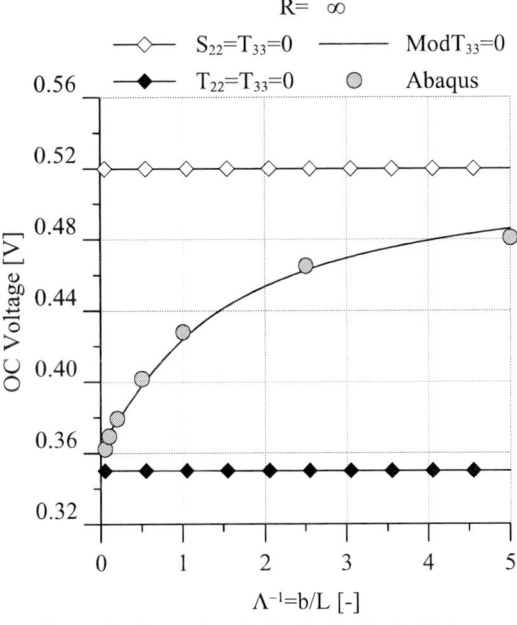

Figure 9: Open circuit Static Analysis, Voltage

Parametric analyses have been performed varying the load resistance with $b=200\mu m$ and $b=2000\mu m$, the results are shown in Figures 10-14. When $R\rightarrow\infty$ the solution should collapse to the open circuit solution of Figure 9, while when $R\rightarrow0$ short circuit condition is obtained and no power can be harvested. A peak power generation is

978-1-4799-4789-8/14 $31.00 © 2014 IEEE   353

obtained for an optimal value of load resistance which inversely depends on the capacity of the device. All models predict the same optimal load resistance as ABAQUS. Moreover, as expected, the voltage range does not significantly depend on the width of the device since both the coupling constant and the capacity depend linearly to the width. On the other hand the power generation increases (barely linearly) while increasing the width.

This same analysis has been performed for different values of width and the peak power per unit width has been collected in Figure 15. It is shown that the modified 1D model correctly predicts a cantilever type beam for all range of length-beam ratios that have been considered. On the other hand plane stress and plane strain models predict that the power generation is linear to the width and they always respectively under- and overestimate voltage and power. As shown by Figure 16 the mean error on the power generation can be reduced from about 30% to less than 5%. It can also be concluded that wider harvester not only recover more power, which is obvious, but also has a higher power density than narrower beams.

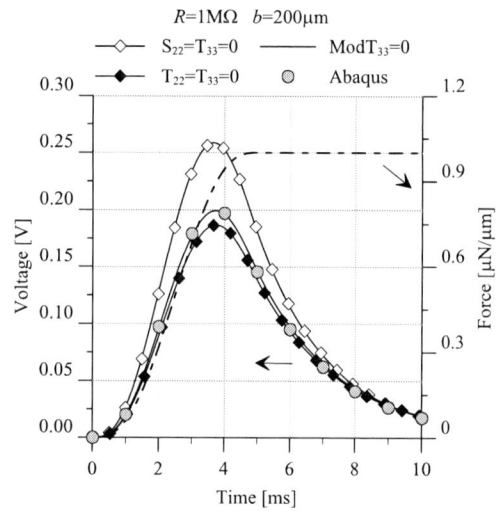

Figure 10: Quasi Static analysis, b=200 μm, R=1MΩ, Voltage and Force

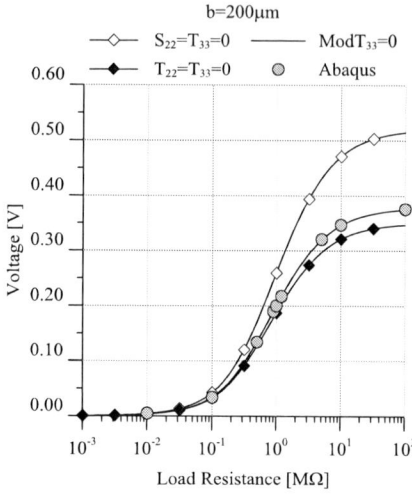

Figure 11: Quasi Static analysis, b=200 μm, Voltage

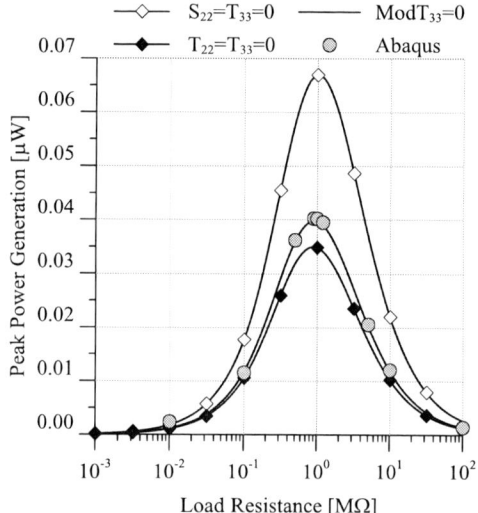

Figure 12: Quasi Static analysis, b=200 μm, Power

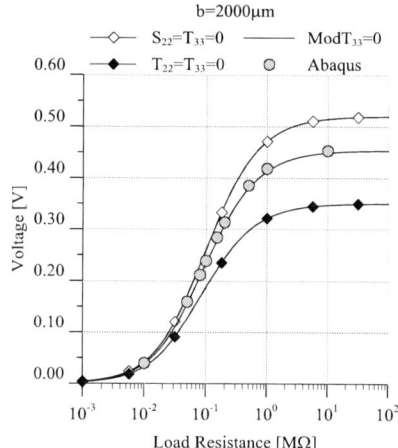

Figure 13: Quasi Static analysis, b=2000 μm, Voltage

978-1-4799-4789-8/14 $31.00 © 2014 IEEE

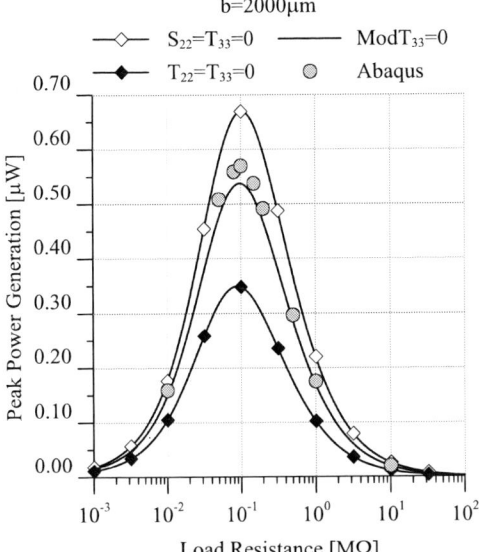

Figure 14: Quasi Static analysis, b=2000 μm, Power

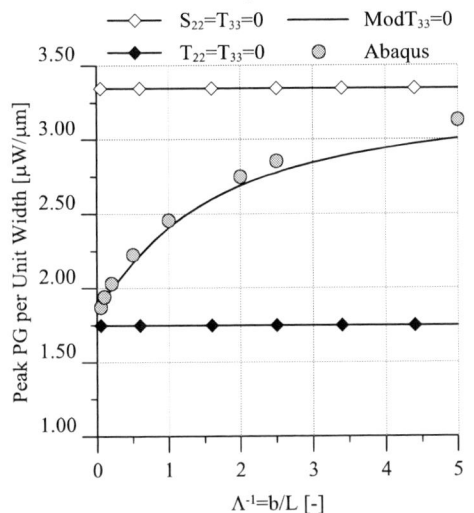

Figure 15: Peak Power generation at Optimal R

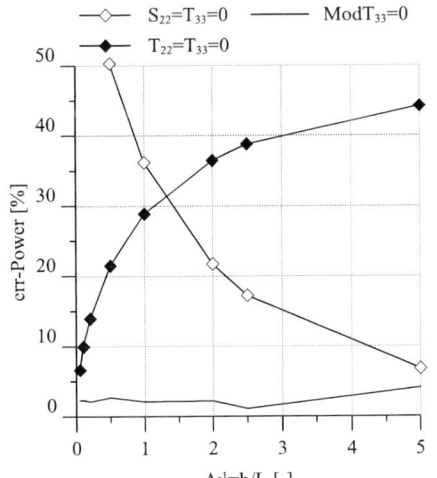

Figure 16: Error on Power Generation

9. Conclusion

A simple 1D model has been built to perform multi-physics simulation for MEMS piezoelectric harvesters. Suitable hypotheses on stresses and strains have been implemented to take into account of 3D effects on elastic and electric parameters. A mixed theory which includes the effect of the transverse strain has been considered such that boundary conditions and limit stress and strain configurations are recovered. The piezolaminated beam with modified elastic and electric constants has been studied through the Classical Lamination Theory (CLT) specifically modified to introduce the piezoelectric coupling and a lumped parameters model has been built through separation of variables. The resulting coupled equations have been analytically solved in the frequency domain and have been numerically solved in the time domain by means of the α-method.

The aforementioned model has been validated by comparison with fully 3D multi-physics simulation performed by a commercial code (ABAQUS) where an external subroutine has been added to simulate the coupling with an external load resistance. The comparison with the commercial codes shows that the 1D model is accurate enough to reproduce qualitatively and quantitatively the response of the harvester gaining in terms of model dimensions and computational time. The validated model can be used to perform simulations on MEMS scale piezoelectric energy harvesters.

Acknowledgment

The authors wish to thanks the ENIAC Joint Undertaking, Key Enabling Technology, project Lab4MEMS, grant n°325622, for partial funding of this research.

10. References

1. Roundy, S, and Wright, P.K, "A piezoelectric vibration based generator for wireless electronics", *Smart Materials and Structures,* Vol. 13, No. 5 (2004), pp. 1131-1142.

2. Jeon, Y.B, *et al,* "MEMS power generator with transverse mode thin film PZT", *Sensors and Actuators, A: Physical*, Vol. 122, No. 1 SPEC. ISS. (2005), pp. 16-22.

3. Ardito, R, *et al,* "On the application of piezolaminated composites to diaphragm micropumps", *Composite Structures*, Vol. 99 (2013), pp. 231-240.

4. Maurini, C, *et al,* "On a model of layered piezoelectric beams including transverse stress effect", *International Journal of Solids and Structures,* Vol. 41, No. 16-17 (2004), pp. 4473-4502.

5. Maurini, C, *et al,* "Extension of the Euler-Bernoulli model of piezoelectric laminates to include 3D effects via a mixed approach", *Computers and Structures,* Vol. 84, No. 22-23 (2006), pp. 1438-1458.

6. Ballhause, D, *et al,* "A unified formulation to assess multilayered theories for piezoelectric plates",

Computers and Structures, Vol. 83, No. 15-16 (2005), pp. 1217-1235.

7. Guyomar, D, *et al*, "Toward energy harvesting using active materials and conversion improvement by nonlinear processing", *IEEE transactions on ultrasonics, ferroelectrics, and frequency control,* Vol. 52, No. 4 (2005), pp. 584-594.

8. Worden, K and Tomlinson, G.R Nonlinearity in structural dynamics: Detection, Identification and Modelling, Institute of Physics Publishing (London, UK, 2000).

9. du Toit, N.E, "Design considerations for MEMS-scale piezoelectric mechanical vibration energy harvesters", *Integrated Ferroelectrics*, Vol. 71 (2005), pp. 121-160.

10. Renno, J.M, *et al*, "On the optimal energy harvesting from a vibration source", *Journal of Sound and Vibration,* Vol. 320, No. 1-2 (2009), pp. 386-405.

11. Hughes, T.J.R, The Finite Element Method, Prentice-Hall (Englewood Cliffs, 1987), pp.532.

Analytical tool for electro-thermal modelling of microbolometers

Piotr Zając[1], Cezary Maj[1], Michał Szermer[1], Mykhaylo Lobur[2] and Andrzej Napieralski[1]

1) Dept. of Microelectronics and Computer Science, Technical University of Lodz, Wolczanska street 221/223,
90-924 Lodz, POLAND
2) CAD Dept., Lviv Polytechnic National University, UKRAINE, Lviv, S. Bandery street 12
e-mail: pzajac@dmcs.pl

Abstract

In this paper we present a tool which incorporates an analytical model of a microbolometer. Within the tool the user can freely change the input parameters such as dimensions and material properties and immediately obtain output parameters such as responsivity, thermal time constant etc. Moreover, the tool can be used to compute the transient thermal response of the microbolometer for a given radiation power and bias current. The model was validated against the results obtained from ANSYS for several different devices and the maximal relative error in transient temperature response was found to be only 3%.

1. Introduction

Microbolometers are an example of Microelectro-mechanical Systems (MEMS), used to measure electromagnetic radiation. The operation principle is quite simple: a material with high temperature coefficient of resistance (TCR) is exposed to the incoming radiation, it heats up and therefore changes its resistance. Such resistance change can be then measured and the power of the incoming radiation can be calculated. Microbolometers are mostly used to measure infrared radiation in thermal imaging. Currently, many manufacturers offer thermal cameras based on microbolometer arrays [1-3]. Nevertheless, this field still needs research so that cameras can have a better resolution and work with higher frame rate. Therefore, a multi-domain simulation of phenomena occurring in microbolometers is necessary.

Finite Element Method (FEM) – based tools like ANSYS[8] and COMSOL[9] are commonly used for simulating such devices. However, very often during the early design phase the exact device dimensions, shape or even used materials are not yet known and it is the role of the designer to find the optimal/desired parameters. Therefore, it may be very time consuming to use FEM analysis at this stage, especially if the device geometry has to be repeatedly changed and transient analysis with many time points has to be used. Hence the need arises for simpler and, more importantly, faster model. Thanks to such a model, it is possible to rapidly sweep through a wide range of parameters, obtain preliminary results and, in general, significantly reduce the design space. Obviously, after finding the optimal/desired parameters using the simpler model, the obtained results should be validated using detailed FEM simulation.

In the paper, we thoroughly describe such a model and validate it against results obtained using FEM simulation. We perform the validation for various device dimensions and different materials in the electrical, thermal and coupled electro-thermal domain. The validation is based on comparing the transient thermal response obtained from Ansys and from our model. The presented comparisons show than the model correctly predicts the thermal behaviour of the device for a wide range of input parameters. Based on this validation, weak points of the model are identified and corrective measures are applied. The perspectives for model development are also given, with special emphasis put on the need for incorporating the mechanical domain into the model.

2. General microbolometer architecture

The description of the structure of a microbolometer may be found in literature [4-6], so only a short summary will be presented here. Various microbolometer implementations vary in shape, but a general device structure always remains the same: a square-shaped bridge suspended over the substrate, supported by relatively thin legs. Thin supports are necessary to thermally separate the bridge from the substrate. A microbolometer contains two specific type of materials: one active material with a high TCR, and an insulator. Usually the thin layer of active material is located between two membranes made from insulator. To maximize the resistance change of the thin active layer due to temperature, it is sometimes serpentine-shaped (see Fig. 1). The insulating membrane allows increasing the receptive area of the device and at the same time it does not change its electrical behaviour. It also strengthens the structure from the mechanical point of view.

The specific microbolometer structure which is analyzed in this paper is shown in Fig. 1. Although all dimensions are treated like input parameters in our work, for the model validation (see section 5) it was assumed that the general structure remains the same. Of course, the fact that the model was not validated for other types of structure does not mean that it will not work in such cases; simply, the model would need to be tuned to adequately describe the new device shape.

Let us now describe the structure and specify the parameters which will be later used as inputs to our model. First, the active material layer is assumed to have the same width and thickness in every point of the device. Other two parameters describing this layer are its total size and the size of its inner serpentine part (Fig. 1).

978-1-4799-4789-8/14 $31.00 © 2014 IEEE

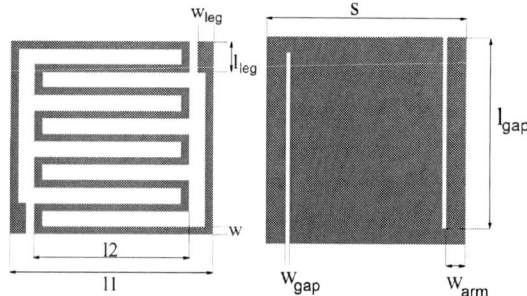

Fig. 1. Shape of the active layer (left) and the insulating membrane (right).

Note that it is also assumed that the number of serpentine turns stays the same. The necessary material properties of active layer are its electrical resistivity, thermal conductivity and TCR. Second, the insulating membrane is assumed to cover the entire structure and to have gaps along supporting arms. It is described by its total size, gap width and length and arm width (arm length is supposed to be equal to the total membrane size). The necessary material properties of insulating layer are its thermal conductivity, density and specific heat. All above-mentioned parameters and their symbols used later in this paper are summarized in Table 1.

Table 1. List of geometrical and material parameters

Name	Symbol
Active layer size	$l1$
Active layer inner size	$l2$
Active layer width	w
Active layer height	h
Leg length	l_{leg}
Leg width	w_{leg}
Leg height	h_{leg}
Membrane size	s
Top membrane height	h_t
Bottom membrane height	h_b
Gap length	l_{gap}
Gap width	w_{gap}
Arm width	w_{arm}
Electrical resistivity of active material	r
Thermal conductivity of active material	κ_a
Temperature coefficient of resistance of active material	α
Thermal conductivity of insulating material	κ_i
Density of insulating material	ρ
Specific heat of insulating material	c

3. Microbolometer model

The purpose of our model is to allow fast analysis in the early stage of the design process, allowing the designer to sweep across a wide range of parameters. Therefore, it was decided to use a simple RC equivalent circuit model [7] (see Fig. 2). Although such an approach constitutes a significant simplification with respect to

FEM-based simulation, it will be shown in section 5 that the resulting error is quite small. In exchange, the model has the advantage of allowing the designer to analytically calculate the temperature of the microbolometer at every time point. Let us now describe in detail how all output parameters of the microbolometer are obtained.

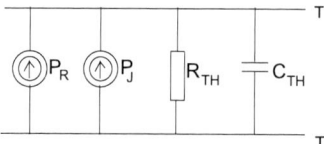

Fig. 2. Microbolometer model. P_R – radiation power, P_J – Joule heat power , R_{TH} – thermal resistance, C_{TH} – thermal capacity

Electrical domain.

We assume here that the resistivity of the insulator is much higher than that of the active material. Consequently, the bias current only flows through the active layer and only in this body Joule heat is generated. The main parameter in this domain is the resistance of the active layer. To obtain this parameter, it is first necessary to calculate the total length of active layer using appropriate dimensions from the Table 1. Once total length is calculated, the electrical resistance can be easily obtained using the well-known formula:

$$R = r\frac{l_a}{w_a h_a} \tag{1}$$

where r is the resistivity of the material, l_a is its length, w_a is its width and h_a is its height. From the electrical resistance and assuming the bias current is known, one can derive the Joule heat dissipated in the active layer using $P = I^2 R$ relationship. Thus, the first parameter (P_J) of the circuit presented in Fig. 2 is obtained.

Thermal domain.

The calculation of thermal parameters is more complex because both active and insulating layer has to be taken into consideration. Fortunately, there are some assumptions than can be made which greatly simplify the following analysis. Let us look at the results of the steady-state thermal analysis performed in ANSYS (Fig. 3).

Fig. 3. Steady-state thermal simulation of a microbolometer performed in ANSYS.

We can see that the temperature of the substrate is practically equal to the ambient temperature. The second observation is that the temperature of the membrane is distributed almost uniformly in the membrane. Therefore, the temperature drop only occurs along the supporting arms. In other words, the thermal resistance of the arms is much higher than the thermal resistance of the membrane and the substrate. Therefore, in the model, only the thermal resistance of arms is necessary. Note that each arm is composed of three layers: bottom membrane, active layer and top membrane.

$$R_{TH} = \frac{1}{\kappa} \frac{l_{arm}}{w_{arm} h_{arm}} \qquad (2)$$

In this case, the total resistance of three layers is obtained by first calculating the thermal resistance of each layer (see Eq. 2) and then deriving the total thermal resistance as a parallel connection of three individual resistances. The appropriate dimensions of the arms needed for Eq. 2 are calculated based on the parameters shown in Table 1.

When it comes to thermal capacity, necessary for transient simulation, it can be calculated using the formula Eq. 3:

$$C_{TH} = \rho c V \qquad (3)$$

To calculate the volume V in Eq. 3, we assume in our model that the volume of the active layer is significantly smaller than that of insulating membranes and is therefore neglected. Consequently, only the volume of membranes needs to be calculated. Again, we use the geometrical dimensions from the Table 1 to calculate the total membrane volume. Thus, we obtain two other parameters of the model, the thermal resistance R_{TH} and the thermal capacity C_{TH}.

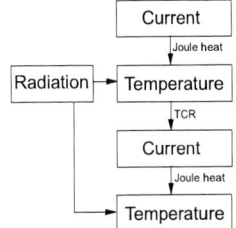

Fig. 4. Coupling between electrical and thermal domain based on one iteration step

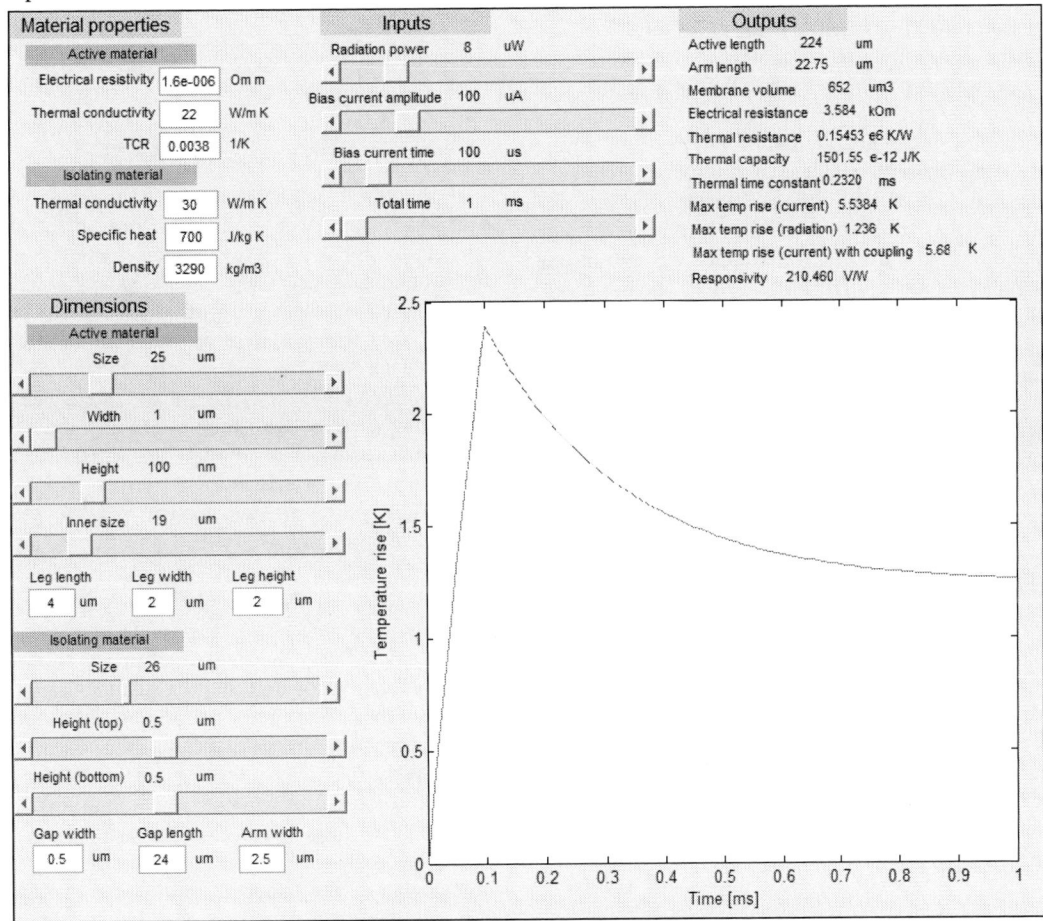

Fig. 5. A screenshot of the modelling tool developed in Matlab.

A careful reader may have noticed that so far in our model we ignored the coupling between thermal and electrical domain. Note that the thermal domain influences the electrical domain because the resistance of the active layer is a function of temperature. This effect in principle cannot be neglected because the materials used in microbolometers have relatively high TCR which means that the resistance can change significantly with temperature. Typically, in complex tools this problem is solved iteratively, however, after thorough analysis, it was discovered that in the case of a microbolometer, one iteration step already gives acceptable results. The reason is that the temperature change of the device is quite small, in the range of several degrees. Thus, such an approach (shown in Fig. 4) was implemented in our model. In short, we first calculate the resistance in room temperature, next the temperature change due to current and radiation is calculated. This new temperature is then used to calculate a new value of electrical resistance, which is in turn used to calculate the final temperature.

4. Modelling tool

The tool which uses the model described in the previous section was written in Matlab. The screenshot of the program is shown in Fig. 5. A user has the possibility to choose all material properties (like electrical resistivity, thermal conductivity, specific heat, TCR, etc.) and all geometrical dimensions of the device (total size of the device, the size of particular layers, the sizes of legs and arms, etc.). He can also define simulation inputs (bias current, current pulse time, radiation power, etc.). Based on these data, the tool calculates output parameters such as electrical resistance, thermal resistance, thermal capacity, thermal time constant, responsivity, etc. and, most importantly, shows the microbolometer's transient temperature response. Therefore, it is possible for example to quickly see what will be the maximum temperature reached by the microbolometer for a given radiation power, current pulse amplitude and current pulse time. Note that for the reason of convenience, sliders were added to the tool to control some inputs; since outputs and the graph are updated on-the-fly, a user can rapidly assess the influence of a parameter on output values just by moving the slider.

5. Validation of the model against FEM simulation

Since the model comprises many parameters, it would be difficult to compare the model with ANSYS for a wide range of all parameters. Therefore, it was decided to perform the comparison for several specific microbolometer structures. The transient temperature response was chosen as a method of comparison. We designed different geometries in ANSYS, used various material properties and simulation inputs (bias current and radiation power) and ran transient thermal simulations for each device. Then, the same devices were modeled in our tool and the results were compared. Although we performed the simulations for many structures, here we

present the analysis for five sample devices, whose parameters are given in Table 2.

Table 2. Parameters of five simulated devices

	Parameters
1	$l1$=25 μm, $l2$=19 μm, w=1 μm, h=100 nm, l_{leg}=4 μm, w_{leg}=2 μm, h_{leg}=2 μm, s=26 μm, h_f=0.5 μm, h_b=0.5 μm, l_{gap}=24 μm, w_{gap}=0.5 μm, w_{arm}=2.5 μm, r=1.6e-6 Ωm, κ_a=22 Wm^{-1}K^{-1}, α=0.0038 K^{-1}, κ_i=30 Wm^{-1}K^{-1}, ρ=3290 kg/m^3, c=700 Jkg^{-1}K^{-1} I_{bias}=100 μA, t_{pulse}=100 μs, P_R=8 μW
2	All dimensions from device 1 scaled 2x, material properties without change, I_{bias}=200 μA, t_{pulse}=200 μs, P_R=10 μW
3	$l1$=25.5 μm, $l2$=15.5 μm, w=1.5 μm, h=50 nm, l_{leg}=6 μm, w_{leg}=1.5 μm, h_{leg}=2 μm, s=26.5 μm, h_f=0.3 μm, h_b=0.3 μm, l_{gap}=24 μm, w_{gap}=2.5 μm, w_{arm}=2.5 μm, r=1.6e-6 Ωm, κ_a=22 Wm^{-1}K^{-1}, α=0.0038 K^{-1}, κ_i=30 Wm^{-1}K^{-1}, ρ=3290 kg/m^3, c=700 Jkg^{-1}K^{-1} I_{bias}=100 μA, t_{pulse}=100 μs, P_R=8 μW
4	All dimensions from device 3, r=3e-6 Ωm, κ_a=40 Wm^{-1}K^{-1}, α=0.01 K^{-1}, κ_i=15 Wm^{-1}K^{-1}, ρ=2000 kg/m^3, c=500 Jkg^{-1}K^{-1} I_{bias}=50 μA, t_{pulse}=100 μs, P_R=10 μW
5	All dimensions from device 3 scaled 1.5x, r=3e-6 Ωm, κ_a=40 Wm^{-1}K^{-1}, α=0.01 K^{-1}, κ_i=15 Wm^{-1}K^{-1}, ρ=2000 kg/m^3, c=500 Jkg^{-1}K^{-1} I_{bias}=150 μA, t_{pulse}=100 μs, P_R=15 μW

It should be emphasized that version 14 of ANSYS Workbench that we used does not directly support transient coupled electro-thermal simulation, so a workaround was necessary. The detailed description of our approach is however beyond the scope of this paper. In short, we performed iterative transient thermal analysis and for each step the inputs were calculated manually based on the outputs from the previous step. The figures 6 and 7 show the results of the performed simulations both in Matlab using our model and in ANSYS. The comparison was performed for a constant radiation power and a short pulse of bias current. On the y axis, the maximal temperature rise (with respect to the ambient) is shown.

It can be seen that the model is quite accurate: the maximal encountered error was equal to 3%, in the case of device 5. After a thorough analysis of all simulations results, it was discovered that the main source of error is probably the simplified calculation of the thermal resistance of supporting arms. Thus, if needed, the model could likely be improved by implementing a more complex thermal resistance modelling. A potential improvement might consist of using two thermal resistances with slightly different values, one resistance for calculating the temperature rise caused by the bias current and the second one for calculating the temperature rise from incoming radiation.

978-1-4799-4789-8/14 $31.00 © 2014 IEEE

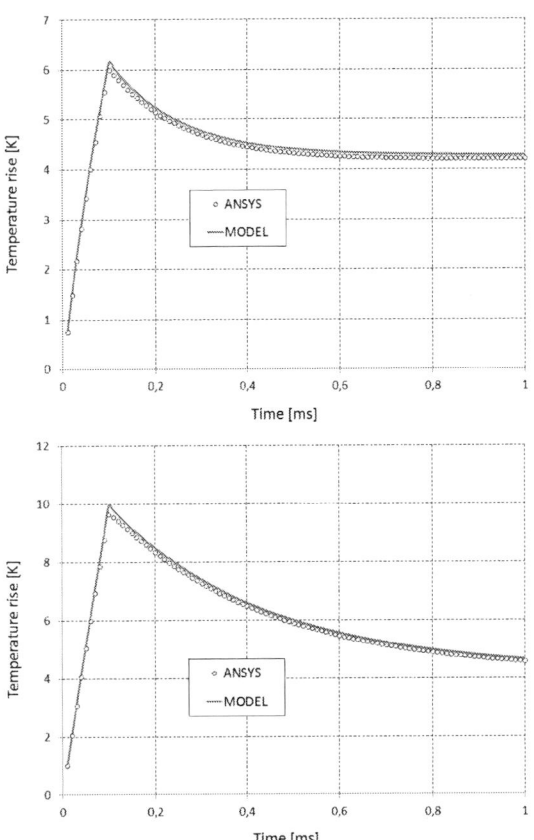

Fig. 7. Comparison of our model with ANSYS for devices 4 (top) and 5 (bottom)

In the future, the model will be expanded to include the mechanical domain which will allow the user to assess the structural integrity of the device under mechanical and thermal stress.

Acknowledgments

Research presented in the paper is supported by Marie Curie International Research Staff Exchange Scheme Fellowship within the 7th European Community Framework Programme - EduMEMS - Developing Multidomain MEMS Models for Educational Purposes, no. 269295.

References

1. Bhan R. K., Saxena R. S., Jalwania C. R., and Lomash S. K., "Uncooled Infrared Microbolometer Arrays and their Characterisation Techniques," *Def. Sci. J.*59, 580–590 (2009)
2. Matolin, D., Posch, C., Wohlgenannt, R., Maier, T., "A 64×64 pixel temporal contrast microbolometer infrared sensor", *IEEE International Symposium on Circuits and Systems*, pp.1644-1647, 18-21 May 2008
3. Wang H., Xinjian Y., Lai J., Li Y., "Fabricating microbolometer array on unplanar readout integrated circuit", *International Journal of Infrared and Millimeter Waves*, Vol. 26(5), pp.751–762

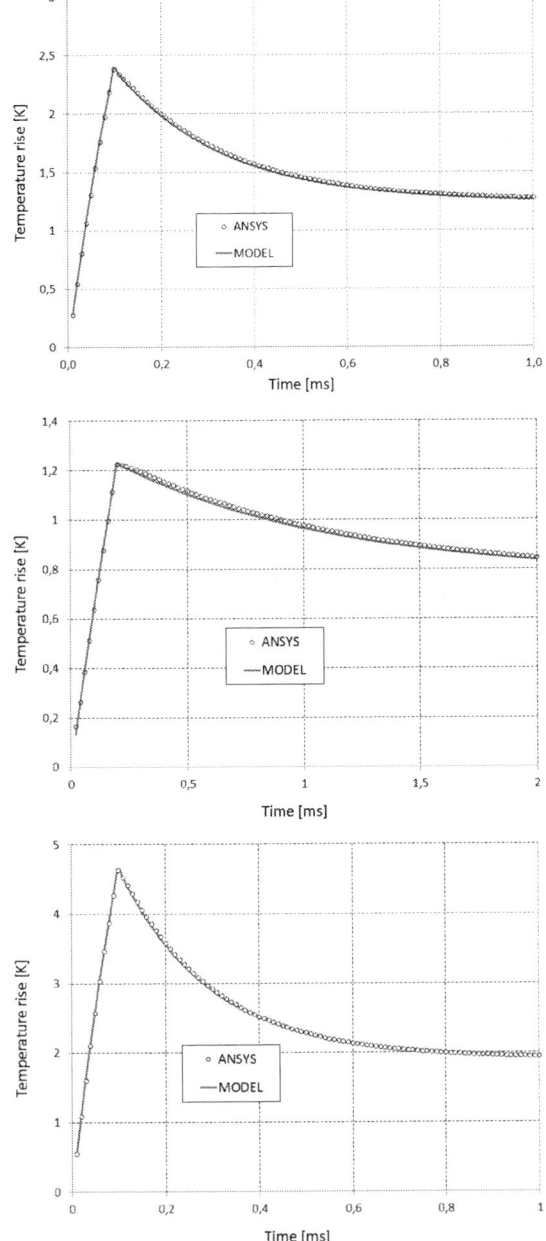

Fig. 6. Comparison of our model with ANSYS for devices 1 (top), 2 (middle) and 3 (bottom)

6. Conclusions and perspectives

It has been shown that despite its simplicity, our thermo-electric model of a microbolometer provides the results very similar to those obtained using relatively time-consuming FEM analysis. The disadvantage of the model is that its mathematical formulas depend on pre-defined shape of the device so if it is changed, the model requires a slight tuning. However, it may generally be expected that the RC equivalent circuit model will work correctly also for other types of microbolometer shapes.

4. Niklaus, F., Vieider, C., Jakobsen, H., "MEMS-based uncooled infrared bolometer arrays: a review" *Proceedings of the SPIE*, Vol. 6836, article id. 68360D (2008).

5. Woo-Bin Song and Talghader J. "Design and characterization of adaptive microbolometers", *Journal of Micromechanics and Microengineering*, Vol. 16, no. 5, 19 April 2006

6. Rogalski, A., "Infrared detectors: an overview", *Infrared Physics & Technology*, Vol. 43, Issues 3–5, June 2002, pp. 187-210

7. Vedel, C., Martin, J., Ouvrier Buffet, J., Tissot, J., Vilain, M. and Yon, J., "Amorphous silicon based uncooled microbolometer IRFPA", *Proceedings SPIE-Infrared Technology Applications XXV*, 1999, Vol. 3698, 276-83

8. ANSYS® Workbench™ 14, available at: http://www.ansys.com

9. COMSOL Multiphysics®, available at: http://www.comsol.com

Correlation of Activation Energy between LEDs and Luminaires in the Lumen Depreciation Test

Guangjun Lu[1,5], Cadmus Yuan[3,5], Xuejun Fan[4,5], G.Q. Zhang[2,3]

[1]Beijing Research Center, Delft University of Technology, Beijing, China
[2]Delft University of Technology, EEMCS Faculty, Delft, the Netherlands
[3]Institute of Semiconductors, Chinese Academy of Sciences, Haidian, Beijing, China
[4]Department of Mechanical Engineering, Lamar University, Beaumont, TX 77710, USA
[5]State Key Laboratory of Solid State Lighting, Beijing, China
xuejun.fan@lamar.edu, cayuan@gmail.com

Abstract

This paper investigated the correlation of activation energy between LED light source and LED luminaire in lumen depreciation test. Two nonlinear fitting methods were used to verify the stability of the different algorithms to extract the activation energy. The results show that the activation energy of LED luminaire is very close to the LED light source. Activation Energy is typically around 0.20eV and ranges from 0.18eV to 0.23eV for LED luminaire, compared to typically 0.17eV and ranging from 0.11eV to 0.26eV of LED light source.

Keywords: LED, Luminaire, Arrhenius Equation, Activation Energy, Lumen Depreciation Test.

Introduction

Unlike the traditional lighting, the lifetime of LEDs, LED modules and LED based lighting products is generally defined when the luminous flux falls to 70% (L70) of its initial value under natural operation conditions.[1-3]. With the application of the standard IES- LM-80-08 combined with TM-21, which is based on the proven assumption that the lumen decay follows the exponential decay model and the depreciation parameter obeys the Arrhenius equation, the lifetime of LED or LED modules can be predicted. However, many concerns exist. It is difficult to know how the lens or second optics impact the depreciation, how much the control circuit or driver contribute to the degradation , and what's impact of thermal design, and so on. Nevertheless, the activation energy Ea plays an important role in lumen depreciation prediction. This paper will investigate the correlation of Ea between LED light sources and LED lamps or luminaires based on the assumed exponential decay model and Arrhenius equation.

Method

Assume that both the LEDs and LED luminaires follow the exponential decay model as shown below in Equation (1) as follows.

$$\phi(t) = \beta e^{-\alpha t} \qquad (1)$$

Where $\phi(t)$ represents the normalized lumen output, t is the time, β is 1.0 theoretically, but varies around 1.0 when real experimental data is applied. α is depreciation parameter. According to Arrhenius function, α can be written as:

$$\alpha = A e^{-\frac{Ea}{k_b T}} \qquad (2)$$

Where A is the pre-factor, and Ea is activation energy, k_b is the Boltzmann constant (8.617385x10^{-5}eV/K) and T is the temperature in Kelvin.

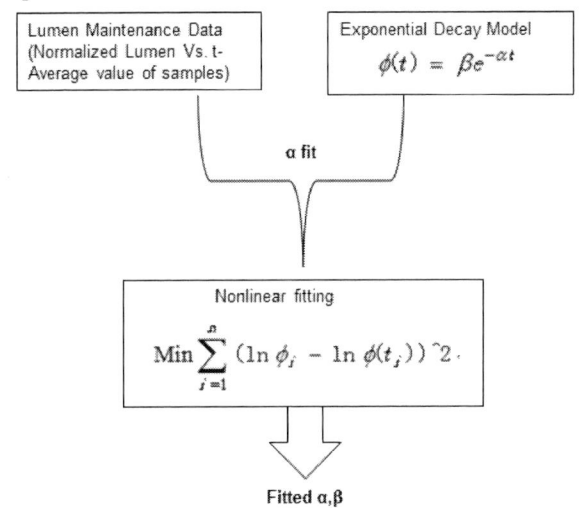

Fig.1 Depreciation parameter α fitting procedure

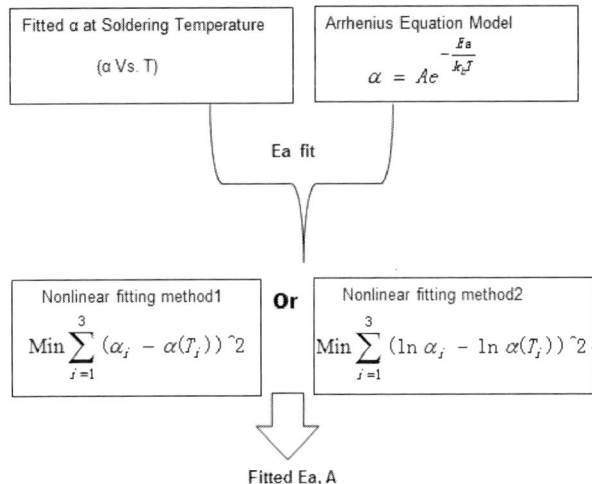

Fig.2 Activation energy Ea fitting procedure

Fig.1 and 2 show the fitting procedures for the depreciation parameter α and Activation Energy Ea respectively. Activation Energy Ea can be deducted through 2 steps via fitting method. First step is to obtain the α value with the

978-1-4799-4789-8/14 $31.00 © 2014 IEEE

available lumen depreciation data combined with exponential decay model by the nonlinear fitting method as shown in Fig.1 The second step is to use the fitted α value under different temperatures and the Arrhenius equation with a fitting method. There are two options for nonlinear fitting, as described in the Fig.2, in the second step. According to LM-80 and TM-21, the temperature used in the Arrhenius equation is soldering temperature. Here during the Ea fitting, the temperature used is also the soldering temperature for both LED light sources and LED luminaires.

While for the Ea fitting of LED light sources, the lumen depreciation data are from the available LM-80 database. For the Ea fitting of LED luminaires, lumen depreciation data from 4 brand luminaires were tested by authors. The soldering temperature is obtained through thermal couple during normal operating conditions. Luminaires tested are shown in Fig.3. Two down lights, one spot light (Par light) and one retrofit bulb are used.

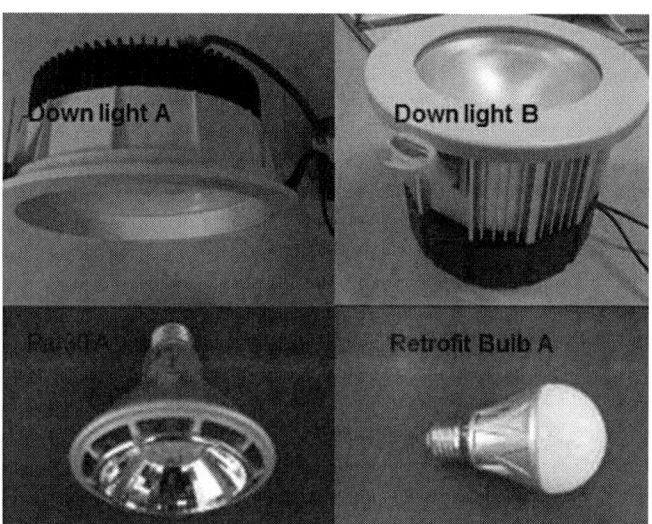

Fig.3 Pictures of Luminaires

Experiments

For each luminaire one sample was randomly chosen to do soldering temperature (Ts) measurement with thermal couples. Among three thermal couples used for the measurement, two (Ts,1 and Ts,2) were used for Ts measurement and one(Tr) was used for ambient temperature monitoring. Fig.4 shows the Ts measurement for the Retrofit bulb as an example. Table 1 shows the Ts measurement data for the 4 luminaires.

Ts used for the ambient temperature of 25deg.C is the average value of Ts readings at steady state. Since there is some difference (Td) between the ambient temperature Tr and the normal room temperature 25deg.C during measurement due to environmental temperature change, the adjustment of Ts was done in the Ea fitting. Take the measurement data of downlight A as an example, Tr reading is 26.5deg.C, the average of Ts is 59.8deg.C, and to take into consideration the temperature difference Td of 1.5deg.C as an adjustment, the actual Ts used for the ambient temperature of 25deg.C in the

Ea fitting is 58.3deg.C. Table 2 shows the actual Ts for Ea fitting.

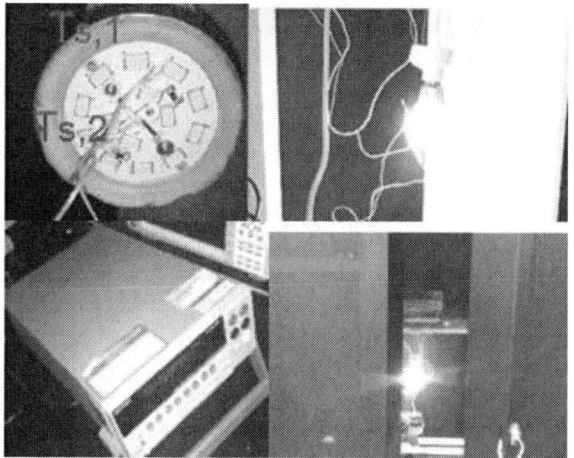

Fig.4 Ts measurement with thermal couples

Table1 Ts measurement data for luminaires (Unit:deg.C)

	DownlightA	DownlightB	Par30 A	Retrofit bulb A
Ts, 1	60.4	71.1	51.6	63.6
Ts. 2	59.1	73.0	/	63.7
Tr	26.5	25.9	25.6	19.1

Table 2 Ts for Ea fitting

Ta (deg. C)	Ts (deg. C)			
	DownlightA	DownlightB	Par30 A	Retrofit bulb A
25	58.3	71.2	51.0	69.6
55	88.3	101.2	81.0	99.6
85	118.3	131.2	111.0	129.6

Results &Discussion

Table 3 Ea fit results for Luminaires

Luminaires#	Fitting Method 1		Fitting Method 2	
	Ea	A	Ea	A
Downlight A	0.4134	13.79	0.2081	0.0289
Downlight B	0.4077	2.09	0.1958	0.0027
Par30	0.4193	5.92	0.1823	0.0043
Retrofit Bulb	0.4046	1.88	0.2285	0.0108

Table 4 Ea fit results for lighting sources (some of them)

LED Light Sources (only show part of them)	Fitting Method 2	
	Ea	A
	0.1797	0.0012
	0.2548	0.0287
	0.1493	0.0007
	0.2029	0.0062

Table3 shows the Ea fitting results for luminaires with 2 fitting methods. Ea fit results between two methods are quite different，around 0.2eV difference. Meanwhile, the pre-factor A is also quite different, 2~3 orders of magnitude difference.

Note that use the first nonlinear method, the Ea fit is unstable; the result depends much on the numerical precision and convergence condition, even the step length of the variables has much influence on the fitting results. So the Ea fit data with nonlinear method 1 shown in Table 3 is not reliable.

Hence, in the Ea fitting for the light sources, fitting method 1 was not used. Around 20 sets of LM-80 data were used, and some results were shown in Table 4.

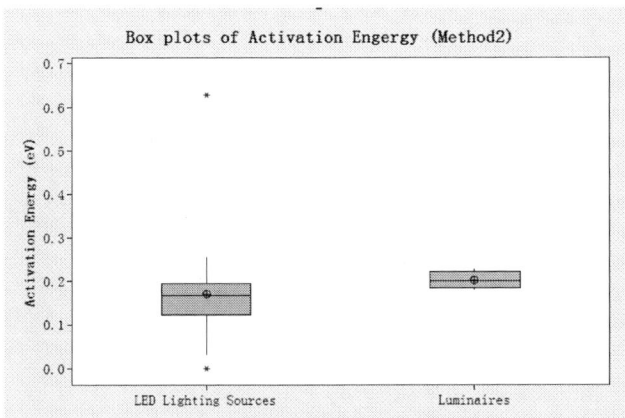

Fig.5 Box plots of Ea of Method 2

Fig.5 shows the box plots of Ea fitting results with nonlinear method 2 for both luminaires and light sources. Here all the Ea fitting results are included. Not considered the outliers, the Ea of LED light sources ranges from 0.11eV to 0.26eV, while the Ea of the Luminaires from 0.18eV to 0.23eV. Typical Ea value for the Luminaires is around 0.20eV, around 0.03eV higher than the light sources which have a typical value around 0.17eV.

Since the fitting method 2 is more robust and consistent with the linear fit result which set 1/T as a variable, Ea fit results with the fitting method2 is much more reliable. 0.03eV Ea difference between LED light sources and luminaires could be from a lot of factors as below

1) Measurement error. Including the lumen maintenance Data from LM-80 and luminaires measured with integrated sphere, temperature measurement with thermal couples.

2) Sample to sample variation. Different Ea values exist among different samples both for luminaire and light source

3) Contribution by the luminaire itself. Including the control circuit degradation during temperature strength loading, yellowing of lens or other optical parts, or possible interactions, and so on.

Conclusions

Two nonlinear fitting methods were used to verify the stability of the different algorithms to extract the activation energy. Correlation of activation energy between LED luminaires from some brand manufacturers and LED light sources from many brand suppliers were investigated. Results were shown the Ea value of LED luminaire is very close to the LED light source. Activation Energy is typically around 0.20eV and ranges from 0.18eV to 0.23eV for LED luminaire, compared to typically 0.17eV and ranging from 0.11eV to 0.26eV of LED light source.

Acknowledgement

Authors want to thank State Key Laboratory of Solid State Lighting (China) for the financial support of this project. Lu would like to give thanks to SKL colleagues including Yongqiao Qin, Min Jia, Hongyu Tang, Zenghui Fan and Heyuan Sun for measurement support and give thanks to Kai Lin for sample preparation and some lumen maintenance data support.

References

1. Van Driel W.D. and Fan X.J. Solid State Lighting Reliability Components to Systems, Springer, 2012, pp.413-426
2. Pradeep Lall, Junchao Wei et al, "L70 Life Prediction for Solid State Lighting Using Kalman Filter and Extended Kalman Filter Based Models", EC&TC 2013,pp.1454-1465
3. US DOE, "ENERGY STAR® Program Requirements for Solid State Lighting Luminaires", 2008, pp.4-4

Modelling the Lifetime of Aluminum Heavy Wire Bond Joints with a Crack Propagation Law

Arian Grams, Tobias Prewitz, Olaf Wittler, Stefan Schmitz, Andreas Middendorf, Klaus-Dieter Lang*)

Fraunhofer IZM, Gustav-Meyer-Allee 25, 13355 Berlin, Germany

Tel.: +49 (0)30 46403-138   Fax: +49 (0)30 46403-211

Email: arian.grams@izm.fraunhofer.de

*) Technische Universität Berlin, Microperipheric Technologies, TIB 4/2-1, Gustav-Meyer-Allee 25, 13355 Berlin, Germany

Abstract

In this study, an approach for enhanced lifetime modelling of wire bonds has been investigated. Numerical simulations of a 3D wire bond model have been used to acquire a suitable damage parameter. For the lifetime model, a modified Paris law for calculating crack growth per cycle has been employed, to consider the gradual area degradation due to thermo-mechanical loads.

With the knowledge of the crack propagation rates, the acquired lifetime model can easily be transferred to different wire bond geometries without repeating the experiments necessary to fit the crack growth parameters.

1. Introduction

In power electronics applications, the achievement of high reliability against active thermal cycles plays a major role during package development. A main limiting factor in standard technology is the degradation of aluminum wire bonds. Therefore, it is essential to quantify and improve their lifetime in order to be able to make use of this well established and cost efficient technology.

Wire bond joints show two main failure modes when subjected to thermal cycling: crack propagation in the heel area (heel crack) and in the bonding interface (lift-off). An example of the latter is shown in fig. 1.

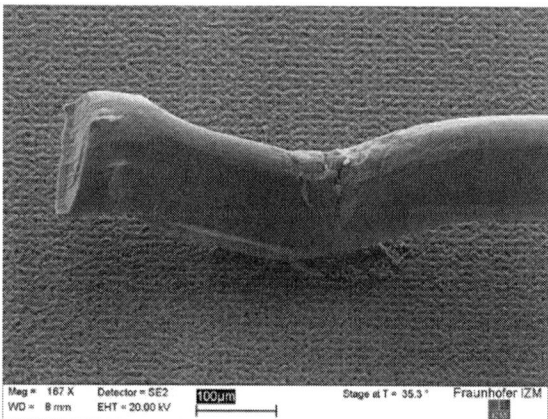

Fig. 1: Wire bond lift-off after 109,300 cycles at a temperature swing of 105 K [1].

Recent research revealed that silver-sintered chip-DBC interconnects show increased power cycling capability when compared to soldered interconnects. Furthermore, with optimized wire bonds the failure mode changes from wire bond heel crack to wire bond lift-off [2, 3].

Until today the lifetime prediction of a wire bond joint is not fully described. Although the reliability of wire bond joints has been a constant subject of research for the last three decades the thermo-mechanical simulation of the crack propagation in the bonding interface so far plays only a minor role. Most of these works use the Coffin-Manson approach to predict the joint lifetime [4-6]. There, the plastic strain per thermal cycle $\Delta \varepsilon_{pl}$ is correlated with the joint lifetime N_f by a factor C_1 and an exponent C_2 [7, 8].

$$N_f = C_1 \left(\Delta \varepsilon_{pl} \right)^{-C_2} \qquad (1)$$

The validity of these lifetime models is limited to the respectively investigated setup consisting of a specific wire diameter, bonding tool and bonding process parameters. Each of these parameters changes the resulting area size and the shape of the wire bond joint which leads to uncertainties in the lifetime prediction. In [9] this drawback was addressed by using finite element modelling and fracture mechanics to create a Paris law for the crack growth. This investigation was done for a 2D geometry and the results cannot be easily transferred to a real 3D wire bond joint. Recent research investigated the potential of Cohesive Zone Modelling of a 3D wire bond joint for the lifetime prediction of the lift-off failure mechanism [10]. The calculated shape of the crack growth is in very good agreement with those visible in shear-test pictures. But the unknown material parameters of the cohesive zone elements and the fact that the cohesive energy G_c is not a unique parameter for the wire bond joint failure prevents this method from being used for lifetime prediction.

In this paper a modified approach of the Paris law is proposed to predict the lifetime of a wire bond joint independently from its area size and shape which makes it usable for multiple wire diameters, bond tools and bond process parameters. By scaling the lifetime N_f of the wire bond joint with its area size A, the parameters C_1 and C_2 become independent from the wire bond joint geometry (see eq. 1). The damage criterion represents the averaged plastic strain per

978-1-4799-4789-8/14 $31.00 © 2014 IEEE

thermal cycle and is calculated as a weighted average value in the evaluation area using finite element modelling (see eq. 2 and 3). This approach has been proposed in [11] to compare the lifetime of BGAs with different solder ball diameters and its robustness and suitability was mainly investigated in [12]. The method of calculating $\Delta\varepsilon_{pl}$ has been further developed for large interfaces by [13].

$$\frac{dA}{dN} = C_1(\Delta\varepsilon_{pl})^{C_2} \qquad (2)$$

$$\Delta\varepsilon_{pl} = \frac{\sum_i \Delta\varepsilon_{pl,i} \cdot V_i}{\sum_i V_i} \qquad (3)$$

2. Geometry Model

To precisely acquire a damage parameter for the lifetime model, 3D simulations of the investigated wire bond wedge have been performed. For its large impact on the thermo-mechanical stresses, the wedge has been modelled accurately with the help of a deformation analysis as in [10]. Assuming the CTE-mismatch of wire and substrate is the driving force for crack propagation [14], the loop has not been modelled. The thickness of the silicon substrate is 175 µm and the thickness of the aluminum metallization is 5 µm.

According to [15], the crack is developing in the wire bond itself, close to the substrate metallization. For this model the crack is assumed to grow horizontally at a height of 5 µm above the substrate. As the damage parameter in eq. (2) depends on the current crack size, the wire bond interface has to be modelled in different stages of crack propagation in order to obtain $\Delta\varepsilon_{pl}$ values in the area of interest. In order to keep the modelling of different crack stages simple but accurate, a set of shear pictures has been evaluated. Shear pictures were taken after 10k, 25k, 50k and 100k active cycles, with the temperature in the area of the investigated bonds during cycling ranging from -20 °C to +100 °C. The pictures were evaluated regarding the occurring crack shapes by measuring the average crack lengths and positions. As a good approximation to the real crack shapes, the remaining shear plateau has been modelled elliptically. The position of the shear plateau and the characteristics of the ellipses have been adapted to average values from the shear pictures as in fig. 2 and eq. (4) to (8).

Looking at eq. (4) to (8), it can be seen that the smallest value of the length b is 147.5 µm (for $b_1 = 0$). For this value the crack length is zero.

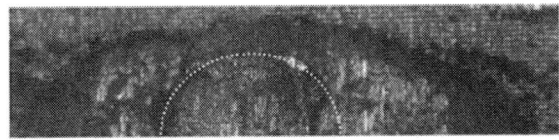
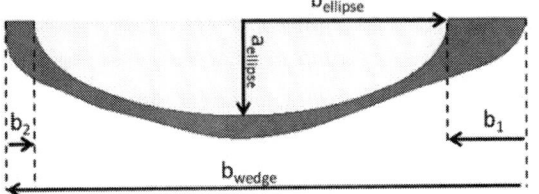

Fig. 2: Example of shear picture with elliptic contour on shear plateau (top) and geometric parameters for the generation of different crack states (bottom), top view

$$b_{wedge} = 1.09 \ mm \qquad (4)$$

$$\text{Ratio} \ ^{a_{ellipse}}\!/_{b_{ellipse}} = 0.83 = const. \qquad (5)$$

$$b_1 = 0.48 \cdot b_2 + 0.295 \qquad (6)$$

$$b = \frac{b_1 + b_2}{2} \qquad (7)$$

$$\text{crack length} \ b_{crack} = b - 147.5\mu m \qquad (8)$$

The crack itself was modeled as a 2 µm vertical cut through the wire bond, centered at a height of 5 µm above the substrate and starting at the outer edge of the ellipse, cp. fig. 3.

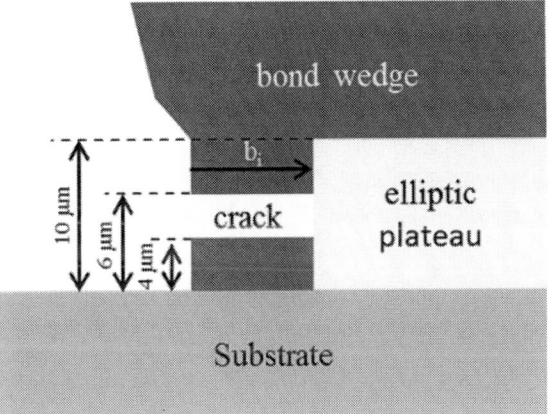

Fig. 3: Crack location in model, side view

978-1-4799-4789-8/14 $31.00 © 2014 IEEE   367

Looking at the available shear pictures, it was determined that the longitudinal crack length represents the crack area very well, cp. fig. 4.

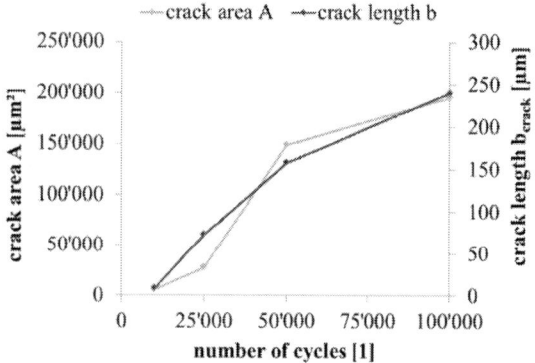

Fig. 4: Comparison of measured crack lengths and areas after active cycling

It can be seen, that the relative increase of crack length and crack area are closely corresponding to each other. Consequently the crack length b will be considered instead of the crack area A, because b is also the characteristic length for the geometry generation. Instead of the power law of eq. (2) then eq. (9) can be used for the lifetime model.

$$\frac{db}{dN} = C_1 (\Delta \varepsilon_{pl})^{C_2} \qquad (9)$$

The damage parameter $\Delta \varepsilon_{pl}$ depends on the current crack length b and has to be calculated for different crack sizes. The method of modelling the different crack stages presented here is fully parameterized. In total, 14 different crack stages have been modelled and calculated to obtain a $\Delta \varepsilon_{pl}(b)$-curve.

3. Boundary Conditions and Material Data

The setup of the simulation model and all calculations were performed with ANSYS® 15.0. Adequate boundary conditions have been applied to the simulation model to represent the thermo-mechanical load on the wire bond during active cycling.

First of all, to reduce computational effort a half model was used, so that no displacement in x-direction is allowed on the mirror plane of symmetry.

In order to account for the bending of the substrate itself, a separate simulation was used to calculate the substrate deformation during heating and cooling. Accordingly, deformations have been transferred as boundary conditions to the four top corners of the substrate of the actual investigated wire bond model. Tab. 1 lists the deformation on the four corner nodes, cp. fig. 5. The displacements between the two temperatures have been interpolated.

Fig. 5: Designation of displacements on four corner nodes of the substrate (see tab. 1)

Displacements in [µm]		P1	P2	P3	P4
at 100 °C	ux	0	0	0.218	0.209
	uy	0	0.838	0	0.833
	uz	0	-0.569	-0.069	-0.632
at -20 °C	ux	0	0	-0.115	-0.109
	uy	0	-0.545	0	-0.544
	uz	0	0.499	0.034	0.531

Tab. 1: Displacements at its corner nodes to consider the substrate bending

Finally, to take the temperature load into account, a uniform temperature has been applied to all bodies ranging from +100 °C to -20 °C. This assumes the temperature difference inside the modelled area to be small which is in alignment with findings by Goehre [16]. Tab. 2 lists material parameters at room temperature.

Material	Aluminum	Aluminum Metallization	Silicon
Youngs Modulus [GPa]	70 [17]	70	E_x: 169 [18] E_y: 169 [18] E_z: 130 [18]
Poisson Ratio	0.35 [19]	0.35	v_{xy}: 0.064 [18] v_{yz}: 0.36 [18] v_{xz}: 0.28 [18]
Shear Modulus [GPa]			G_{xy}: 50.9 [18] G_{yz}: 79.6 [18] G_{xz}: 79.6 [18]
Yield Strength [MPa]	30 [10]	140	-
Tangent Modulus [MPa]	216 [10]	216	-
CTE [10⁻⁶/K]	23 [19]	23	2.5 [20]

Tab. 2: Material data used for the simulations at room temperature

978-1-4799-4789-8/14 $31.00 © 2014 IEEE   368

Silicon is implemented as an orthotropic material model according to [18] with the silicon (100)-plane corresponding to the XY-plane of the model. The aluminum material model is elastic-plastic including kinematic hardening and the Young's modulus, Yield strength and Tangent modulus were measured using tensile tests. The material properties of the aluminum metallization are indicative. The actual properties are dependent on the processing parameters and on the resulting microstructure as well as the thickness of the metallization.

4. Evaluation Method

For the application of the modified Paris law a suitable damage parameter is necessary. Here, the accumulated plastic strain in the aluminum from cycle to cycle is considered, for it represents the damage accumulated in the material. Therefore, a number of thermal cycles are calculated and the difference in accumulated plastic strains from the last cycle to the one before is evaluated.

However, the value of $\Delta\varepsilon_{pl}$ still depends on the number of cycles calculated before evaluation, on the mesh size and on the area of averaging. Accordingly, preliminary studies have been performed to determine the best method of evaluation. It was found that the calculated strains on the exterior edge of the ellipses – where the crack front is located – strongly depend on the mesh size.

Fig. 6 shows the accumulated plastic strains along the longitudinal crack path representative for the damage to the interface. The values of the damage parameter are normalized. There, the number of horizontal element layers is varied, which was found to be the most crucial parameter.

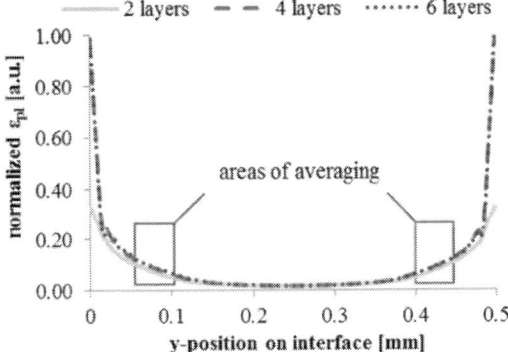

Fig. 6: Accumulated plastic strains along the interface area calculated for different mesh sizes

Farther than approximately 20 µm from the outer edges, the mesh dependency of the calculated strains could be neglected. Here, with some safety, the part of the crack path from 50 µm to 100 µm behind the crack front has been considered for the damage parameter.

The averaging is done over the described volume as in eq. (10).

$$\Delta\varepsilon_{pl} = \frac{\sum_{elem}\left(\varepsilon_{pl,elem,cyc4} - \varepsilon_{pl,elem,cyc3}\right) \cdot V_{elem}}{\sum_{elem} V_{elem}} \quad (10)$$

Of course, the distance between the averaging areas to the interface edge limits the minimum interface length, which can be used. So, for smaller interfaces a finer mesh might be necessary, so that averaging can be done in a smaller area and closer to the edge. A model with four horizontal layers and an element size of about 10 µm in the interface area was used for this investigation, so that a $\Delta\varepsilon_{pl}(b)$-curve could easily be calculated within one day.

To investigate the number of cycles necessary for the calculation, one model was evaluated for 20 thermal cycles. The results never showed oscillation after more than 3 cycles. Therefore, for all crack stages the accumulated plastic strains from the fourth to the third cycle have been considered for the damage parameter.

5. Lifetime Model

Using the simulation model and the developed evaluation method, a $\Delta\varepsilon_{pl}(b)$-curve has been recorded for 14 elliptic interface areas resulting from various values for the characteristic length b. The smallest length for b with the elliptic geometry model is 147.5 µm, cp. eq. (7) and (6).

Looking at the available shear pictures, the respective crack growth rates db/dN have been measured. As data is only available after certain numbers of cycles, between data sets the crack growth rates have been assumed to be constant. Fig. 7 shows the measured crack growth rates, as well as the calculated strains and a fit curve. It can be seen, that the calculated strain as well as the measured crack growth rates decrease with increasing crack length. As they share the same tendency, the data can be fitted well. The fit was generated using eq. (9) with appropriate values for C_1 and C_2 and the calculated $\Delta\varepsilon_{pl}(b)$-curve.

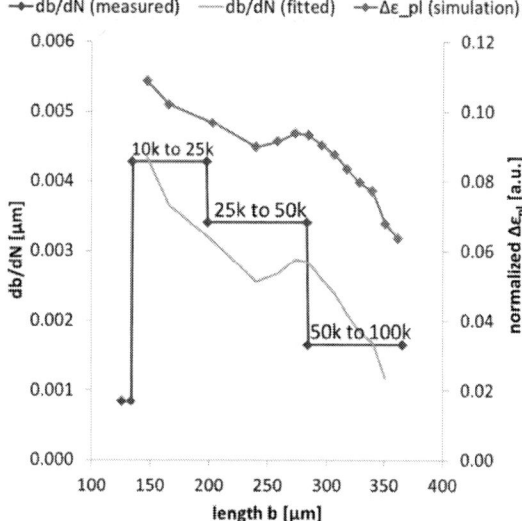

Fig. 7: Measured crack propagation rates *db/dN*, calculated damage parameters *Δε_pl* and fit for crack propagation rate using simulation results

The first experimental data point (from zero to 10k cycles) represents a very small crack propagation rate and cannot be covered with the same crack propagation law. There, crack initiation takes place so that the propagation rate is very slow.

To calculate a lifetime with the given $\Delta\varepsilon_{pl}(b)$-curve, the power law from eq. (9) has to be integrated as in eq. (11).

$$N = \int_{b_0}^{b_f} \frac{1}{C_1 \left[\Delta\varepsilon_{pl}(b)\right]^{c_2}} \, db \qquad (11)$$

The lower integral boundary b_0 can be chosen to be the initial value, i.e. for this geometry model 147.5 µm for $b_1 = 0$, cp. eq. (7) and (6). For the upper boundary b_f a failure criterion has to be defined. When the remaining interface area becomes smaller than the wire diameter, the thermal, electrical and mechanical performances of the interface are worse than the ones of the wire. So, for the investigated 400 µm wire, failure can be assumed for $b = 330$ µm for which the remaining interface area becomes smaller than the cross-sectional area of the wire. Because of the discontinuous data, the integral has been approximated by sums of i available data points as described in eq. (12).

$$N = \sum_i \left[0.5 \left(\frac{1}{C_1 \left[\Delta\varepsilon_{pl,i-1}(b)\right]^{c_2}} + \frac{1}{C_1 \left[\Delta\varepsilon_{pl,i}(b)\right]^{c_2}} \right) \cdot (b_i - b_{i-1}) \right] \qquad (12)$$

This way, the wire bond lifetime is calculated to be 64k cycles. Fig. 8 finally plots the measured and calculated crack lengths considering $b_1 = 0$ µm as zero crack length. The time for the initial crack to develop can be added to the lifetime if it is known.

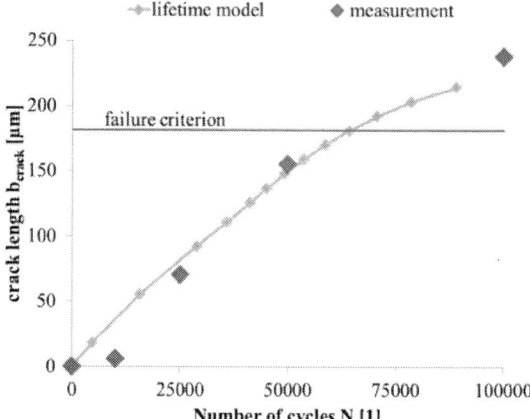

Fig. 8: Crack lengths during thermal cycling, once measured and once calculated using the lifetime model

6. Conclusions

The lifetime modelling approach in this work demonstrates how crack growth rate, due to thermal cycling, can be considered using a modified Paris law as an alternative to the more established Coffin-Manson law. It can be easily transferred to different geometries, so the logical next step would be to check how the same fit parameters can be applied to different geometries. Also, it has to be checked how the parameterized generation of cracked simulation models can work for alternative bond geometries.

The results for the current geometry showed, that the strains calculated by simulation present the same tendency as in the experiments, i.e. a decrease of crack growth rate over time. Therefore, the experimental data could be well fitted using the modified Paris law. Also, with a reasonable failure criterion the lifetime could be calculated in agree with the experimental results.

This method could also be valuable for considering damage accumulation. Therefore, the $\Delta\varepsilon_{pl}(b)$-curve could be recorded for relevant load levels and then different scenarios could simply be considered due to stepwise integration of eq. (12).

Acknowledgements

The results are based on research in the BMBF research activity "RoBE" (FKZ 16N11464) supported by the German Federal Ministry of Education and Research in the programme "IKT 2020". This support is gratefully acknowledged.

The authors would like to acknowledge Jens Göhre for sharing his experimental results.

References

[1] K.-D. Lang, J. Goehre and M. Schneider-Ramelow, "Interface Investigations and Modeling of Heavy Wire Bonds on Power Semiconductors for End of Life Determination," *10th Electronics Packaging Technology Conference*, 2008.

[2] R. Schmidt and U. Scheuermann, "Separating Failure Modes in Power Cycling Tests," *Integrated Power Electronics Systems (CIPS), 2012 7th International Conference on IEEE*, pp. 1-6, 2012.

[3] U. Scheuermann and R. Schmidt, "A New Lifetime Model for Advanced Power Modules with Sintered Chips and Optimized Al Wire Bonds," *Power Electronics, Intelligent Motion, Renewable Energy and Energy Management (PCIM)*, pp. 810-817, 2013.

[4] Y. Chen, X. Wu, I. Fedchenia, M. Gorbounov, V. Blasko and W. &. S. C. Veronesi, "A comprehensive analytical and experimental investigation of wire bond life for IGBT modules," *Applied Power Electronics Conference and Exposition (APEC), 2012 Twenty-Seventh Annual IEEE*, 2012.

[5] C. Hager, A. Stuck, Y. Tronel and R. &. F. W. Zehringer, "Comparison between finite-element and analytical calculations for the lifetime estimation of bond wires in IGBT modules," *Power Semiconductor Devices and ICs, 2000. ISPSD'00. Proceedings. 12th International Symposium on*, pp. 291-294, 2000.

[6] S. Ramminger, G. Mitic and P. &. W. G. Türkes, "Thermo-Mechanical Simulation of Wire Bonding Joints in Power Modules," *Proceedings of MSM conference*, pp. 483-486, 1999.

[7] L. Coffin, "A study of cyclic thermal stress on a ductile metal," *Trans Am Soc Mech Eng 76:931-50*, 1954.

[8] S. Manson, "Behaviour of materials under conditions of thermal stress," *NACA Report No.1170*, 1954.

[9] K. Sasaki, et. al., "Thermal and structural simulation techniques for estimating fatigue life of an IGBT module," *Power Semiconductor Devices and IC's, 2008. ISPSD'08. 20th International Symposium on IEEE*, pp. 181-184, 2008.

[10] A. Grams, et. al., "Simulation of an Aluminum Thick Wire Bond Fatigue Crack by Means of the Cohesive Zone Method," *EuroSimE*, 2013.

[11] J.-P. Clech, et. al., "A comprehensive surface mount reliability model covering several generations of packaging and assembly technology," *Components, Hybrids, and Manufacturing Technology, IEEE Transactions*, pp. 949-960, 1993.

[12] R. Darveaux, "Effect of simulation methodology on solder joint crack growth correlation," *Electronic Components & Technology Conference*, pp. 1048-1058, 2000.

[13] S. Deplanque, et.al., "Lifetime prediction of SnPb and SnAgCu solder joints of Chips on copper substrate based on crack propagation FE-analysis," *Thermal, Mechanical and Multiphysics Simulation and Experiments in Micro-Electronics and Micro-Systems*, pp. 1-8, 2006.

[14] T. Hung, S. Chiang, C. Huang, C. Lee and K. Chiang, "Thermal–mechanical behavior of the bonding wire for a power module subjected to the power cycling test," *Microelectronics Reliability 51*, p. 1819–1823, 2011.

[15] J. Goehre, M. Schneider-Ramelow, U. Geißler and K.-D. Lang, "Interface Degradation of Al Heavy Wire Bonds on Power Semiconductors during Active Power Cycling measured by the Shear Test," *CIPS*, p. Paper 3.4, March 2010.

[16] J. Goehre, U. Geißler, M. Schneider-Ramelow and K.-D. Lang, "Influence of Bonding Parameters on the Reliability of Heavy Wire Bonds on Power Semiconductors," *International Conference of Integrated Power Electronic (CIPS)*, 2012.

[17] Aluminium-Taschenbuch. Band 1: Grundlagen und Werkstoffe, 15. Aufl., Düsseldorf: Aluminium-Zentrale e.V., 1995.

[18] M. Hopcroft, W. D. Nix and T. W. Kenny, "What is the Young's Modulus of Silicon?," *Journal of Microelectromechanical Systems, VOL. 19*, pp. 229-238, April 2010.

[19] N. Fahey, Ultrasonic Determination of Elastic Constants at Room and Low Temperatures, Watertown, Massachusetts: Watertown Arsenal Laboratories, 1960.

[20] G. White, "Reference materials for thermal expansion: certified or not?," *Thermochimica Acta, 218*, pp. 83-99, 1993.

Thermo-mechanical Characterization of Passive Stress Sensors in Si interposer

Benjamin Vianne[a,b,c], Pierre Bar[a], Vincent Fiori[a], Sébastien Gallois-Garreignot[a],
Komi Atchou Ewuame[a], Pascal Chausse[b], Stéphanie Escoubas[c],
Nicolas Hotellier[a], Olivier Thomas[c]

[a] STMicroelectronics, 850 rue Jean Monnet, 38926 Crolles, France,
[b] CEA LETI, MINATEC Campus, 17 rue des Martyrs, 38054 Grenoble, France
[c] Aix-Marseille Université, CNRS, IM2NP UMR 7334,
Campus de Saint Jérôme, avenue Escadrille Normandie Niemen, 13397 Marseille Cedex 20, France
Email: benjamin.vianne@st.com

Abstract

Passive stress sensors have been integrated in a silicon interposer test vehicle to investigate thermo-mechanical stress in a typical 2.5D system. The present sensors are integrated in a rosette-shape consisting of eight oriented copper serpentines acting like strain gauges. An innovative design allows theoretically the calculation of a partial stress tensor, including three planar and one out-of-plane components. Electrical measurements at wafer level, combined to FIB/SEM cross-sections, revealed a strong impact of elaboration processes on the structures electrical characteristics. Numerical siumations using finite element analysis were built to evaluated the theoretical sensitivity of copper serpentine to mechanical strains. Finally a dedicated four-point bending tool coupled with a four-terminal resistance measurement setup was fabricated to extract experimentally the values of sensors sensitivity factors. Preliminary results depicted in this paper highlight a sensitivity to stress of distinctly oriented resistors. Several identified sources of data dispersion are inherent to the present measurement configuration and prevent a reliable calculation of strain gauges so-called "gauge factors".

1. Introduction

The evolution of packaging technologies follows basic rules laying on the need to have more I/Os, improved electrical performances and bandwidth, while keeping the path of Moore's law by reducing the silicon dimensions. As an example, the Flip-Chip (FC) technology has considerably increased the number of I/Os in smaller packages, bringing at the same time new issues in terms of costs and reliability. The growing gap between CMOS chips and classical Ball Grid Array (BGA) substrates is now driving the main microelectronics industry leaders into an economical non-sense. The always more complex chips made with the latest technology node at aggressive pitch are assembled on substrates which have to accumulate layers to bridge the dimensional scales, making the price of the substrate itself soaring compared to the silicon die cost. Besides, the introduction of fragile low-k dice

has generated strong thermo-mechanical issues regarding interactions with the package.

In response to standard packaging technologies, 2.5D passive silicon interposer technology is constantly catching attention with the window of opportunities it provides in the field of high wiring density, system partitioning, cost efficiency, low power and stress relief in fragile low-k dice [1], [2].

However, because of interposer's typical large size and small thickness due to TSV integration, thermo-mechanical reliability remains one of the most challenging issues concerning volume manufacturing which need to be investigated at multiple scales, from die-level to 2.5D system-level. Mechanical stresses induced by the Coefficients of Thermal Expansion (CTE) mismatch of materials (around 2.6×10^{-6} K^{-1} for Si, 17×10^{-6} K^{-1} for Cu, and 19×10^{-6} K^{-1} for BGA substrates) have proven to be detrimental in a thin-die integration flow [3]. As a matter of fact, failures in solder bumps regions are a common source of malfunction for flip-chip packages and become a critical concern in the case of 3D integrated devices. Assembly processes involving the stacking of several silicon dice become more and more complicated due to the chips' footprints and alternatives like thermo-compression can generate high mechanical strains in the interconnects level.

In this context, the need to assess thermo-mechanical stresses at die-level is relevant. In view of diverse stress characterization techniques which can be either time-consuming or destructive, the use of stress sensors offers the opportunity to have an easy access to in-situ stress in a silicon die, as well as being flexible enough to be operated at different time and environmental conditions. Stress sensors are especially a good option for measuring stress during an assembly step involving one or several thermal cycles.

As classical MOS-based stress sensors cannot be used in a passive silicon interposer which does not involve active devices, innovative passive stress sensors based on rosettes of serpentine resistors have been developed and embedded to quantify local strain states in interposer die

Figure 1. Microscope view of a passive stress sensor, as described in this paper.

(see Fig. 1). Their implementation in a copper interconnect level of interposer have been described in [4]. A first part of the paper will remind their physical principle. A second part will deal with the calibration of the sensors based on both electrical-mechanical coupling simulations and a dedicated four-points bending test bench, designed for measuring the sensors' response in presence of a known applied state of stress. Thanks to a specific design, these sensors will enable in-situ measurement of planar and vertical stress fields at critical steps during assembly and packaging flow as multiple DRAMs with logic die are stacked on silicon interposer. Preliminary results are depicted in this paper, and comparison between FEM, in-situ measurements and experimental observations will be carried out.

2. Passive stress sensors description

The complex and new issues introduced by 2.5D technology imply a deep understanding of thermo-mechanical stresses induced by specific assembly steps and interactions between package and the silicon die (e.g. copper pillars interactions during a bonding step). The following section will describe a method to measure strains by integrating rosettes of copper serpentines in a silicon interposer test vehicle. A further paragraph will deal with sensors calibration using a four-point bending dedicated test bench, allowing for coupled mechanical-electrical measurements of tested samples.

A. Strain gauge principle

The principle of quantifying mechanical strains by measuring the variation of an electrical resistance is widely known in literature [5]–[7]. At macro-scales, deformations of structural elements can be measured using passive resistors glued to the object using an appropriate adhesive. In microelectronics industry, piezoresistive stress sensors based on p-type and n-type silicon resistors have been used to monitor stress on chips associated with packaging steps [7]–[9] since they allow a direct measure of stress in active surface of silicon with a much better sensitivity

and a high spatial resolution. Other papers report the application of CMOS stress sensors based, resolving normal and shear stresses with resolution below 1 MPa, using temperature compensated cells. However, such sensors require active silicon surface and MOS-based sensors are to be dismissed in a passive silicon interposer integration flow. In this paper is presented an innovative integration of strain gauges in the Back-End-Of-Line (BEOL) level of a silicon interposer chip.

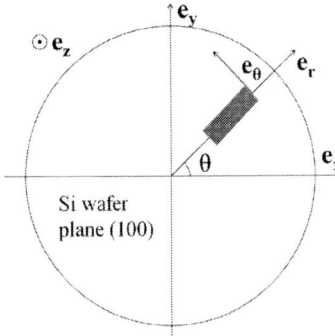

Figure 2. Schematic view of a passive stress sensor.

Following the strain gauge principle, a copper serpentine can be characterized by its geometrical features and its orientation with reference to the silicon surface (see Fig. 2). A linear model can describe the relationship between the relative variation of resistance value of a copper serpentine and mechanical strains in the device (see Eq. 1).

$$\frac{\Delta R}{R} = \frac{R(\sigma, T) - R(0,0)}{R(0,0)} \qquad (1)$$
$$= G_L \sigma_L + G_T \sigma_T + G_Z \sigma_Z + \alpha T$$

which gives in a cartesian local coordinate system :

$$\frac{\Delta R}{R} = (G_L \cos^2(\theta) + G_T \sin^2(\theta))\sigma_{xx}$$
$$+ (G_L \sin^2(\theta) + G_T \cos^2(\theta))\sigma_{yy}$$
$$+ G_Z \sigma_{zz} + (G_L - G_T)\sin(2\theta)\sigma_{xy} \qquad (2)$$
$$+ \alpha T$$

with G_L, G_T, G_Z respectively the sensitivity factors in longitudinal, transversal and out-of-plane direction of a serpentine, also known as gauge factors, α the thermal coefficient of resistance of electroplated copper.

B. Design of passive stress sensors in a test vehicle of silicon interposer

From Eq. 2, the resistance change of a single copper serpentine depends on four components of stress (σ_{xx}, σ_{yy}, σ_{zz}, and σ_{xy}). By assembling such strain gauges in rosette configuration and, by making the assumption that stress tensor does not vary in a sensor region, the calculation of these four components is achievable. As

978-1-4799-4789-8/14 $31.00 © 2014 IEEE

stated in [5], with identical design of resistors, it is only possible to measure three components of stress. Hence, by combining two different designs of serpentines (*i.e.* two different combinations of gauge factors) in a single sensor, it is possible to measure the partial stress tensor with four components.

By using four distinctly oriented serpentines ($\theta_1 = -45°$, $\theta_2 = 0°$, $\theta_3 = 45°$ and $\theta_4 = 90°$) with two different line lengths ("type I" and "type II" resistors) and assuming a constant temperature, we can write a linear system of four independent equations (Eq. 3) based on Eq. 2:

$$
\begin{cases}
\dfrac{\Delta R_1}{R_1} = \dfrac{G_L^{II} + G_T^{II}}{2}\sigma_{xx} + \dfrac{G_L^{II} + G_T^{II}}{2}\sigma_{yy} + G_Z^{II}\sigma_{zz} \\
\qquad + (G_T^{II} - G_L^{II})\sigma_{xy} \\[4pt]
\dfrac{\Delta R_2}{R_2} = G_L^{I}\sigma_{xx} + G_T^{I}\sigma_{yy} + G_Z^{I}\sigma_{zz} \\[4pt]
\dfrac{\Delta R_3}{R_3} = \dfrac{G_L^{II} + G_T^{II}}{2}\sigma_{xx} + \dfrac{G_L^{II} + G_T^{II}}{2}\sigma_{yy} + G_Z^{II}\sigma_{zz} \\
\qquad + (G_L^{II} - G_T^{II})\sigma_{xy} \\[4pt]
\dfrac{\Delta R_4}{R_4} = G_T^{I}\sigma_{xx} + G_L^{I}\sigma_{yy} + G_Z^{I}\sigma_{zz}
\end{cases}
\tag{3}
$$

with $(G_L^{I}; G_T^{I}; G_Z^{I})$ and $(G_L^{II}; G_T^{II}; G_Z^{II})$ the two combinations of gauge factors for type I and type II resistors respectively.

This system can be more simply written:

$$
\overline{\dfrac{\Delta R}{R}} = \overline{\overline{G}} \cdot \overline{\sigma}
\tag{4}
$$

The two different combinations of gauge factors make it invertible, allowing us to extract a partial stress field:

$$
\overline{\sigma} = \overline{\overline{G}}^{-1} \cdot \overline{\dfrac{\Delta R}{R}}
\tag{5}
$$

Based on this theoretical basis, passive stress sensors have been integrated in a silicon interposer test vehicle (see Fig. 3). They take the form of rosettes of eight serpentines arranged in four principal directions $\theta_1 = -45°$, $\theta_2 = 0°$, $\theta_3 = 45°$ and $\theta_4 = 90°$ (resistors are doubled for more statistics and to prospect possible stress gradients). The devices are elaborated using standard damascene process in a thick copper metallic level of BEOL and work on the same principle as macro-strain gauges, at the very difference that such devices are directly integrated close to silicon and not glued to the sample. "Type I" resistors (as described in previous paragraphs) are aligned in $\theta_2 = 0°$ and $\theta_4 = 90°$ directions , while "Type II" resistors are aligned in $\theta_1 = -45°$ and $\theta_3 = 45°$ directions. Details on geometrical values for each type of resistors are given in Tab. 1.

Table 1. Geometrical features of copper serpentines in BEOL level.

Type	Size (μm^2)	Space (μm)	Width (μm)
I	25×50	0.4	0.4
II	25×50	0.8	0.4

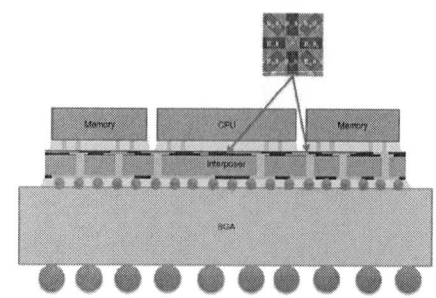

Figure 3. Schematic view of a 2.5D silicon interposer with passive stress sensors integrated on front-side.

A total of five stress sensors are dispensed at several locations on the test chip and intend to evaluate the influence of several parameters on the die stress like the copper pillars stress, the top dice positions or the distance to neutral point (see Fig. 4). Their placements were thought on the basis of 3D package-level simulations revealing several areas of interest (see Fig. 5) [10].

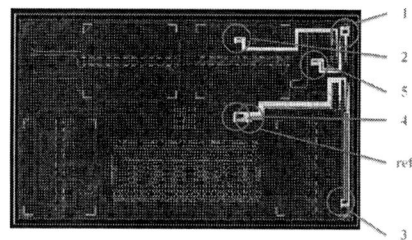

Figure 4. Design view of Si interposer test vehicle showing the locations of sensors. The die dimensions are about 20×30 mm^2

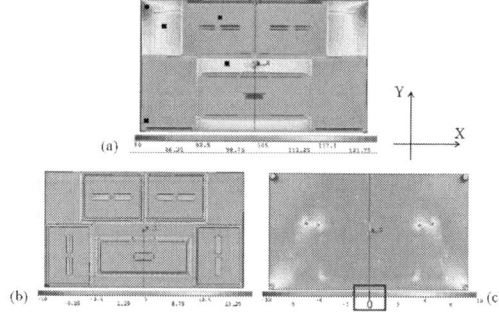

Figure 5. Top silicon stress components: (a) σ_{yy}, (b) σ_{zz}, and (c) σ_{xy}. The location of stress sensors is indicated with black squares in (a). σ_{xx} and σ_{yy} show similar and strong variations at die level (up to 124 MPa in corners) whereas the contributions of σ_{zz}, σ_{xy}, σ_{xz} and σ_{yz} remain weak (inferior to 20 MPa).

C. Electrical characterization

Electrical measurements have been performed at wafer-level on a total of 4950 devices, after realization of back-side interconnections and after debonding of the interposer wafer. Probings were done on thick circular copper pads at a temperature of 25°C, controlled by a chuck. Resistance values were recorded using four-terminal sensing technique to ensure a good precision. First results [4] showed large changes for each resistor at the scale of the wafer, no matter their orientation on the silicon wafer. Relative variations of nearly 20 % the center and the edge of the wafer are depicted on Fig. 6. These variations are mainly attributed to the Chemical Mechanical Polishing (CMP) step acting on the copper lines thickness and on the geometrical section of every resistor using a simple law (Eq. 6):

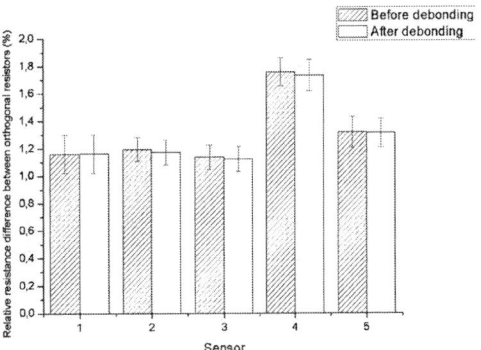

Figure 7. Relative differences between orthogonal type I resistors before and after debonding. Aligned resistors *i.e.* for $\theta = (0°; 180°)$ and $\theta = (-90°; 90°)$ have been averaged. Error bars correspond to standard deviation for all measured dice.

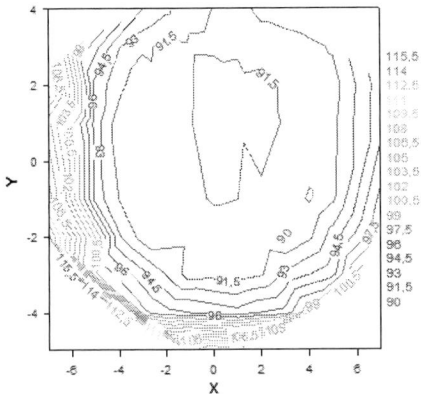

Figure 6. Wafer-level variations of a given resistor ($R(\theta = 0°)$ in sensor 1). All devices follow the same tendency. X and Y indicate arbitrary coordinates locating the measured dice on the wafer.

$$\frac{dR}{R} = \frac{d\rho}{\rho} + \frac{dL}{L} - \frac{dS}{S}$$
$$= \frac{d\rho}{\rho} + \frac{dL}{L} - \frac{dt}{t} - \frac{dw}{w} \qquad (6)$$

with ρ the resistivity of the conductor, L its length end S its section, made up from the line thickness t and width w.

At the sensor scale, local resistance change with orientation - especially between vertical and horizontal resistors - were also statistically reported [4], and confirmed with electrical measurements after debonding (see Fig. 7). Stress gradients can hardly be the root cause of resistance variations above 1 % between two close devices with the same design (see section 3-B) and several hypothesis were proposed to explain this phenomenon : (i) local morphological changes due to process drifts, (ii) microstructural effects in Cu or (iii) a piezoresistive effect in the Cu grains or in the TaN/Ta diffusion barrier. The first hypothesis being the most relevant, a physical characterization procedure has been carried out to check the morphological characteristics of several copper serpentines.

D. Morphological characterization

FIB/SEM cross-sections were performed on two orthogonal resistors in the same sensor and on two different dice

taken from the wafer edge and center (see Fig. 8). For each resistor, two cross-sections were performed (one site at the edge of a serpentine, another one at the center). Copper lines thickness and width were evaluated on SEM pictures and compared between structures.

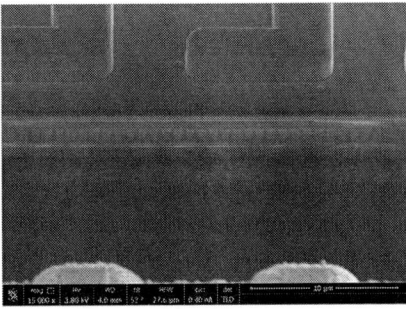

Figure 8. Enlarged view of a FIB cross-section performed in the center of a copper serpentine.

Pictures revealed combined etch and CMP non-uniformities with a demonstrated dependency on orientation. As stated in Eq. 6, this has a direct impact on resistors section and eventually on resistance value (see Fig. 9 and Fig. 10). If we consider that, at first order from Eq. 6, $\frac{dR}{R}$ is proportional to $-(\frac{dt}{t} + \frac{dw}{w})$, the relative percentage of variation of geometrical features (*e.g.* line width and thickness) matches the resistance variations previously stated (see Fig. 11). These observations explain then the sensitivity to orientation observed in wafer-level electrical characterizations. Even if the CMP, photolithography and dry etch process can be optimized to reduce the dispersion of line thickness and width, one can conclude that stress evaluation by comparing the single-step response of different devices from different dice is made impossible, since the expected sensitivities of strain gauges are in the same range as the process variability. However it is important to note that it does not prevent us from evaluating a stress history change on a single device.

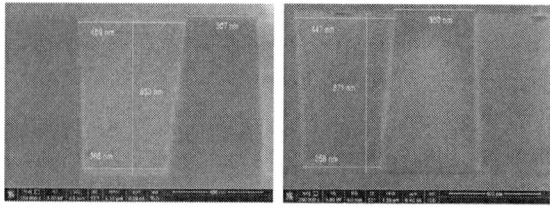

(a) Resistor oriented at θ=0°   (b) Resistor oriented at θ=90°

Figure 9.   Morphological features of two orthogonal resistors 9(a) and 9(b) of the same sensor showing anisotropical process non-uniformities.

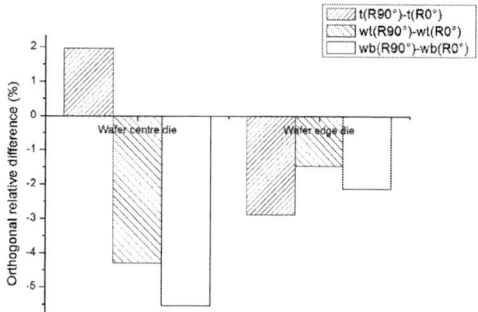

Figure 10.   Average orthogonal variations of copper line characteristics (thickness t, top width w_t and bottom width w_b).

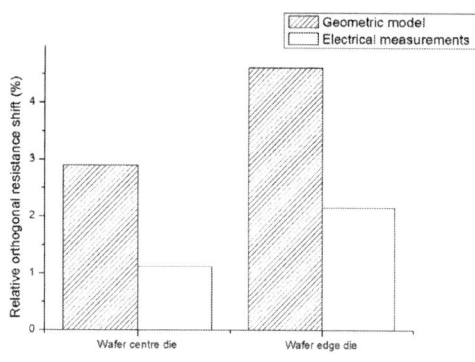

Figure 11.   Comparison of orthogonal resistance shifts calculated from a simple geometric law and observed on wafer-level characterization on the same two dice.

In the next sections, numerical and mechanical calibration of strain gauges will investigate the copper serpentine sensitivity to uniaxial stress and eventually lead to the extraction of gauge factors for each type of resistor.

3. Calibration of stress sensors

Coupled electrical-mechanical simulations using Ansys v14.0 have been built to bridge electrical measurements of sensors to stress in BEOL or in silicon. Their sensitivity factors to stress magnitudes in the principal directions of serpentines will be compared with the ones extracted from electrical measurements.

A. Methodology

The methodology from the concept to the realization and test of passive stress sensors can be described in Fig. 12:

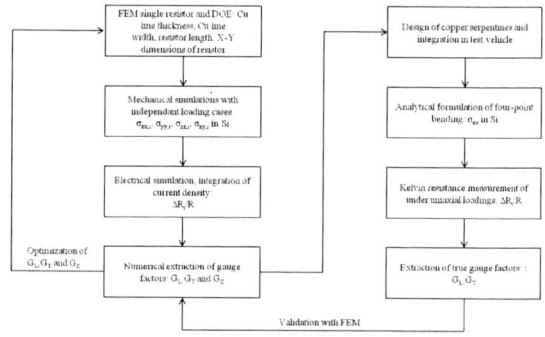

Figure 12.   Methodology of elaboration and test of passive stress sensors.

B. Numerical calibration procedure using FEM

Both designs of copper serpentines (type I and type II) have been modeled using material properties shown in Table 2. The model includes the surrounding oxide dielectric layers above and below the serpentine. Making the assumption that deformations are small enough to stay in elastic mode, all materials are considered elastic. Effect of anisotropy of silicon is taken into account by introducing orthotropic properties for (100) and (110) wafer (see Table 3). The other materials including dielectrics and copper are considered isotropic. 3D coupled-field solid elements with 20 nodes were used for electro-mechanical coupling simulations. One representation of type II resistor is shown in Fig. 13 showing both electrical and mechanical boundary conditions.

Table 2. Material properties at 20°C used in finite element modeling.

Materials	E (GPa)	ν	CTE ($10^{-6}.K^{-1}$)	ρ (Ω.m)
Cu [11]	128	0.33	17	2×10^{-8}
SiO$_2$ (USG)	75	0.25	0.5	1×10^{6}
Si [12]	See Table 3	0.28	2.6	2.3×10^{6}

A normalized potential is applied at the first extremity of the resistor while the other extremity of the resistor is connected to ground. The current is computed by integrating the current density along the copper serpentine. After the calculation of unstrained resistance (without

978-1-4799-4789-8/14 $31.00 © 2014 IEEE

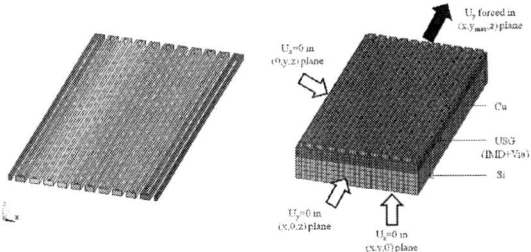

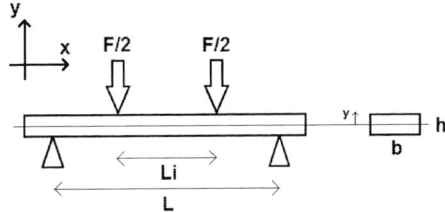

(a) Electrical boundary conditions in copper serpentine

(b) Mechanical boundary conditions for uniaxial loading in longitudinal direction of resistor (upper oxide layers are not shown for clarity)

Figure 13. Finite element model of a single copper serpentine corresponding to a "type II" resistor.

Table 3. Orthotropic properties of silicon (100) and (110) at 20°C [12].

Wafer	(100)	(110)
E_x (GPa)	130	169
E_y (GPa)	130	169
E_z (GPa)	130	130
ν_{xz}	0.278	0.362
ν_{yz}	0.278	0.362
ν_{xy}	0.278	0.064
G_{xz}	79.6	79.6
G_{yz}	79.6	79.6
G_{xy}	79.6	50.9

mechanical loading), three independent normalized displacements in the principal directions e_x, e_y and e_z, are applied to the model boundaries to generate a set of three independent equations. The three gauge factors, corresponding to an averaged stress in silicon, are finally extracted for both types of resistors. The details of the numerical calibration outputs are given in Table 4.

Table 4. Gauge factors extracted from FEM.

Resistor	Type I	Type II
G_L (GPa^{-1})	1.237^{-5}	1.223^{-5}
G_T (GPa^{-1})	-5.879^{-6}	-5.592^{-6}
G_Z (GPa^{-1})	-8.341^{-6}	-8.647^{-6}

C. Four-point bending apparatus

The theoretical gauge factors described in section 2-A, and numerically calculated in section 3-B need to be however experimentally determined. As described by Suhling [5], Chang [9] and Chen [13], a four-point bending fixture (see Fig. 14) is well adapted for such methodology. The principle lies in applying an isothermal and uniaxial loading on a rectangular strip.

The uniaxial compressive or tensile stress σ in the sensors areas between the two inner supports is determined using flexure formula from Bernouilli's beam theory [14]

Figure 14. Scheme representing a four-point bending setup.

(Eq. 7):

$$\sigma(y) = -\frac{My}{I_z} \qquad (7)$$

with M the bending moment, y the distance from the neutral axis and I the moment of inertia.

For a silicon strip of rectangular cross-section with thickness h and width b, the moment of inertia I_z is:

$$I_z = \frac{bh^3}{12} \qquad (8)$$

Finally, by fixing L and L_i - respectively the outer and inner span - one can get a relation between the applied load F and the maximum uniaxial stress σ (Eq. 9):

$$\sigma = -\frac{3F(L-L_i)}{2bh^2} \qquad (9)$$

D. Experimental calibration

The chip containing the stress sensors is glued to the silicon beam with strain gauges facing down the silicon support. The electrical measurement of integrated strain gauges is enabled by Kelvin measurements with four independant probes connected to Keithley 4200 parameter analyzer. The dedicated test bench shown in Fig. 15 and a scheme of the complete apparatus is given in Fig. 16. All tested resistors belong to the same sensor (No. 1 in Fig. 4).

Figure 15. Description of four-point bending set-up.

A forced voltage sweep was first applied on several structures to check for electrical setup influence on calculated resistance (i.e. the quotient $\frac{V_{measured}}{I_{measured}}$). All tested structures indicate the presence of typical self heating,

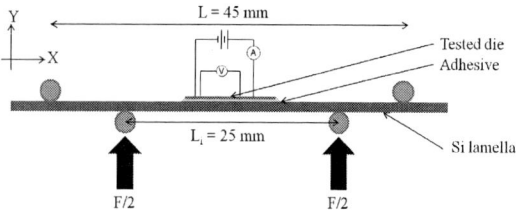

Figure 16. Electrical setup coupled with four-point point bending tool. The tested chip is glued to a silicon lamella.

nearly proportional to the square of the applied voltage (see Fig. 17). Nevertheless, the effect seems to be less pronounced at lower applied voltages - typically [50 mV - 200 mV] - justifying further measurements in this range. This implies also that comparison of resistance values must be made at a constant forced voltage bias to avoid an extra error due do temperature variation of copper resistivity.

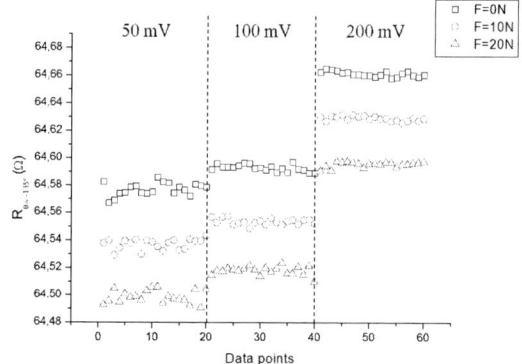

Figure 18. Example of recorded data for a single resistor. The self-heating effect shifts the resistance values depending on the applied voltage. The sensitivity to uniaxial stress is translated into a gap between the data series.

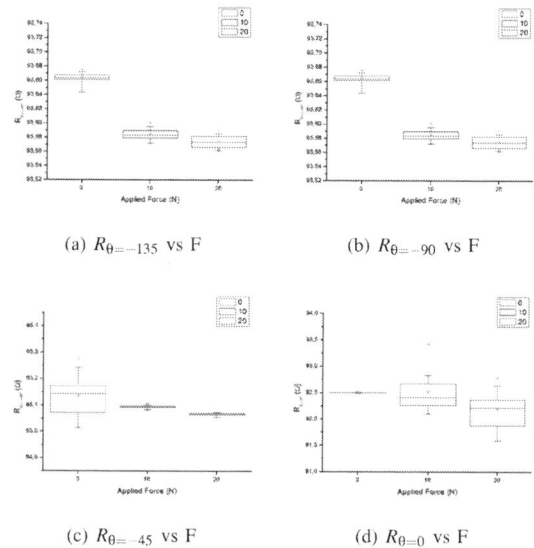

(a) $R_{\theta=-135}$ vs F   (b) $R_{\theta=-90}$ vs F

(c) $R_{\theta=-45}$ vs F   (d) $R_{\theta=0}$ vs F

Figure 19. Sensitivity to uniaxial stress of four distinctly oriented resistors in sensor No 1.

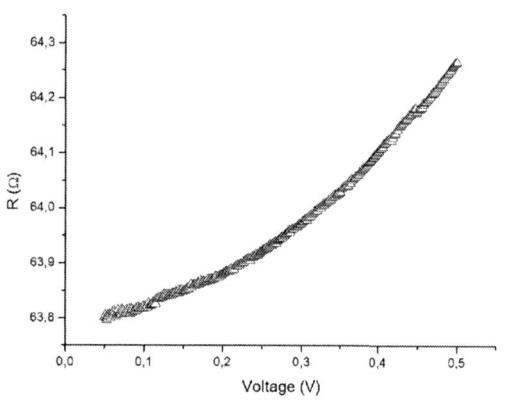

Figure 17. Illustration of self-heating phenomenon of a copper serpentine submitted to a voltage sweep.

Electrical measurements were then performed at three different applied voltages (50 mV, 100 mV and 200 mV) to account for self-heating effect. Three loading cases are applied using a piston related to a force cell: $F = 0$ N (no applied stress), $F = 10$ N and $F = 20$ N. Following Eq. 9, this corresponds to uniaxial stress in the longitudinal direction of the silicon lamella of about 0 MPa, 53 MPa and 107 MPa respectively in the region of sensors. Resistance values for each loading case and applied voltage correspond to the average of 20 data points (see Fig. 18). Sensitivity to stress is evaluated with the minimum bias voltage (i.e. 50 mV), assuming self-heating effect induces less error at this range. Results are depicted in Fig. 19.

While showing a sensitivity to stress of all copper serpentines, these preliminary results reflect the complexity of the experimental calibration task. Due to high dispersion on several data points (e.g. Fig. 19(c) and Fig. 19(d), the calculation of gauge factors is still out of reach at this stage. Several hypothesis explain the dispersion of some data: highlighted self-heating effect impacts directly the calculated value of resistance, while the adhesive supporting the die on the lamella in the chip-on-beam configuration probably introduce unwanted deformations due to its limited rigidity. A cured polymer could considerably reduce this measurement error in further attempts.

4. Conclusions

The weak sensitivity to mechanical strains, inherent to the design of passive sensors implies several conclusions and recommendations on the use of such stress measurement devices:

- process variabilities limits the potential calculation of partial stress tensor components to a stress history change on a single device; in other terms, stress calculation based on comparison of sensors responses from two different dice is not reliable;

- present configuration of experimental calibration apparatus seems to be prone to several inherent measurement errors inducing high dispersion on acquired data; a different sample setup will be investigated in further experiments.

5. Acknowledgements

Thank you to J.-F. Nowakowski, B. Rolland, S. Ricq, L .Clement and N. Bardos from STMicroelectronics Crolles for the technical support in metrology and characterization tools.

References

[1] J.U. Knickerbocker, P.S. Andry, E. Colgan, B. Dang, T. Dickson, X. Gu, C. Haymes, C. Jahnes, Y. Liu, J. Maria, R.J. Polastre, C.K. Tsang, L. Turlapati, B.C. Webb, L. Wiggins, and S.L. Wright. 2.5d and 3d technology challenges and test vehicle demonstrations. In *IEEE 62nd Electronic Components and Technology Conference (ECTC)*, pages 1068–1076, San Diego, CA, 2012.

[2] A. Rahman, Hong Shi, Zhe Li, D. Ibbotson, and S. Ramaswami. Design and manufacturing enablement for three-dimensional (3d) integrated circuits (ics). In *IEEE Custom Integrated Circuits Conference (CICC)*, pages 1–8, San Jose, CA, 2012.

[3] K. Murayama, M. Aizawa, K. Hara, M. Sunohara, K. Miyairi, K. Mori, J. Charbonnier, M. Assous, J.-P. Bally, G. Simon, and M. Higashi. Warpage control of silicon interposer for 2.5d package application. In *IEEE 63rd Electronic Components and Technology Conference (ECTC)*, pages 879–884, Las Vegas, NV, May 2013.

[4] B. Vianne, P. Bar, V. Fiori, S. Petitdidier, N. Chevrier, S. Gallois-Garreignot, A. Farcy, P. Chausse, S. Escoubas, N. Hotellier, and O. Thomas. Thermo-mechanical study of a 2.5d passive silicon interposer technology: Experimental, numerical and in-situ stress sensors developments. In *IEEE International 3D Systems Integration Conference (3DIC)*, pages 1–7, San Francisco, CA, 2013.

[5] J.C. Suhling and R.C. Jaeger. Silicon piezoresistive stress sensors and their application in electronic packaging. *IEEE Sensors Journal*, 1:14–30, 2001.

[6] P.-J. Tzeng, J.H. Lau, M.-J. Dai, S.-T. Wu, H.-C. Chien, Y.-L. Chao, C.-C. Chen, S.-C. Chen, C.-Y. Wu, C.-K. Lee, C.-J. Zhan, J.-C. Chen, Y.-F. Hsu, T.-K. Ku, and M.-J. Kao. Design, fabrication, and calibration of stress sensors embedded in a tsv interposer in a 300mm wafer. In *IEEE 62nd Electronic Components and Technology Conference (ECTC)*, pages 1731–1737, San Diego, CA, 2012.

[7] C. Liu. *Foundations of MEMS (2nd Edition)*. 2011.

[8] A.D. Trigg, Chai Tai Chong, Zhang Xiaowu, Chen Xian Tong, and Wai Leong Ching. Modular sensor chip design for package stress evaluation and reliability characterisation. *Microelectronics Reliability*, 52:1581–1585, 2012.

[9] Yu-Yao Chang, Hsien Chung, Ben-Je Lwo, and Kun-Fu Tseng. In situ stress stress and reliability monitoring on plastic packaging through piezoresistive stress sensor. In *IEEE Transactions on Components, Packaging and Manufacturing Technology*, pages 1358–1363, 2013.

[10] Ludovic Georgel. *Thermo-mechanical simulations of silicon interposer induced stress.* unpublished, 2012.

[11] R. Vayrette. *Analyse des contraintes mécaniques et de la résistivité des interconnexions de cuivre et des circuits intégrés : rôle de la microstructure et du confinement géométrique.* PhD thesis, Ecole Nationale Supérieure des Mines de Saint-Étienne, 2011.

[12] Anthony A. Kelly and Kevin M. Knowles. *Crystallography and Crystal Defects.* John Wiley and Sons, 2012.

[13] Yonggang Chen, Richard C. Jaeger, and Jeffrey C. Suhling. Cmos sensor arrays for high resolution die stress mapping in packaged integrated circuits. *IEEE Sensors Journal*, 13:2066–2076, 2013.

[14] Surya Patnaik and Dale Hopkins. *Strength of Materials: A New Unified Theory for the 21st Century.* Butterworth-Heinemann, 2003.

Molecular Dynamic Simulations of Maximum Pull-Out Forces of Embedded CNTs for Sensor Applications and Validating Nano Scale Experiments

Steffen Hartmann[1], Ole Hölck[1,2], Thomas Blaudeck[1], Sascha Hermann[1], Stefan E. Schulz[1,3],
Thomas Gessner[1,3], Bernhard Wunderle[1]

[1]Technische Universität Chemnitz
Reichenhainer Str. 70, 09126 Chemnitz, Germany
steffen.hartmann@etit.tu-chemnitz.de, sascha.hermann@zfm.tu-chemnitz.de,
thomas.blaudeck@zfm.tu-chemnitz.de, bernhard.wunderle@etit.tu-chemnitz.de
[2]Fraunhofer IZM Berlin
Gustav-Meyer-Allee 25, 13355 Berlin, Germany
ole.hoelck@izm.fraunhofer.de
[3]Fraunhofer ENAS Chemnitz
Technologie-Campus 3, 09126 Chemnitz, Germany
stefan.schulz@enas.fraunhofer.de, thomas.gessner@enas.fraunhofer.de

Abstract

We present investigations of pull-out tests on CNTs embedded in palladium by means of molecular dynamics (MD) and compare our results of maximum pull-out forces with values of nano scale in situ pull-out tests inside a scanning electron microscope (SEM). Our MD model allows the investigation of crucial influencing parameters on the interface behaviour, like CNT diameter, intrinsic CNT defects and functional groups. For the experiments we prepared simple specimens using silicon substrates and wafer level compliant technologies. We realised the nano scale experiment with a nanomanipulation system supporting an AFM cantilever with known stiffness as a force sensing element inside a SEM. Greyscale correlation has been used to evaluate the cantilever deflection. From simulations derived maximum pull-out forces are approximately 17 nN and depend on the existence of intrinsic defects or functional groups and weakly on temperature. Experimentally obtained maximum pull-out forces with values between $16 - 29$ nN are in good agreement with the computational predictions. Our results are of significant interest for the design and a failure-mechanistic treatment of future mechanical sensors with integrated single-walled CNTs showing high piezoresistive gauge factor or other nano scale systems incorporating CNT-metal interfaces.

1. Introduction

The design of future functional systems incorporating nano materials or molecular components (e.g. CNTs) will rely on the assessment of material data, which is required as modelling input. There are attempts to provide this data by means of atomistic simulations and simultaneously predict structure-property correlations [1], [2], but still validating experiments are required. For example, mechanical data like elastic modulus, elastic limit or other critical values have to be recovered based on small volumes and specimens, which will be integrated at the nano scale. The realisation of mechanical tests on small volumes and specimens may not be a straight forward task. In general, the typical engineering approach is to down scale classical methods, e.g. hardness test, bending test or tensile test. Several proposals and investigations have been made using nanoindentation [3], nanomanipulators [4] or micro/nano mechanical systems [5] (MEMS/NEMS).

A known question is, how the characterised material relates to the real device technology, a so called technology related specimen significance. This means: can one expect the same material behaviour, when comparing the material for characterisation and the material processed for the real device. Conservativley the answer is no in any case, because material properties are technology dependent. This may become negligible at the macro scale, but is an important issue when deriving material parameters at the micro and nano scale. In principle, one is bound to the constraint that the specimen or object of material characterisation has been exposed to the same technological processes like the material in the real device for application. With respect to nano material characterisation the trend to cope with the technology is to integrate the objective material, actuation and sensing components into one device, whereby the technological processes for the test device and the future functional device are the same or at least similar. For nano material characterisation in the field of MEMS/NEMS and electronic products silicon based MEMS devices are employed, which are designed to be made with wafer level technologies [6]. The consequence is a need to adjust the technology for this test device each time, when the technology for the future functional system has changed. This interaction between test device for material characterisation and future functional system with respect to technology adjustments is extremly time consuming and cost intensive.

978-1-4799-4789-8/14 $31.00 © 2014 IEEE

There are attempts to circumvent these issues by providing a test device, which may be used as a platform for different specimens [7]. In this case one needs to combine the test device and specimen after each fabrication. The challenge is then to merge a micro/nano scale specimen with a micro/nano scale test device, which again is depending on sophisticated manipulation tools and on the ability to obtain a thorough clamping of the specimen (e.g. with focus ion beam methods).

In contrast, the assessment of material properties at the nano scale is a complex task for its own and new methods may have to be invented and tested on applicability, practicability and adaptability. This is challenging, since the drive for material testing is in most cases bound to specific current research and development activity. Often the technology has to be explored as well. Yet, to explore possible characterisation methods for nano material characterisation, simple and low cost specimens are required, which can be rapidly manufactured for quick experiments. Also the test and specimen design should allow a fast adaption of numerical models, which are supporting the development of possible characterisation methods. Therefore an early interaction between material test and technology is needed to produce effectiv synergy with the benefit of a valuable mutual knowledge transfer for both disciplines.

In this study a novel micro-/macroscopic mechanical sensor system with integrated semiconducting single-walled carbon nanotubes (CNTs) [8] serves as the future functional system. CNTs are envisaged as a new sensing material due to their outstanding intrinsic piezoresistive gauge factor, which allows to detect forces and displacements at the nano or even pico scale [9], [10]. To this purpose individual single-walled CNTs should be suspended between a fixed frame and a movable mass to allow straining of the CNTs and a determination of their resistance. The mechanical support and the electrical connection are provided by metallic electrodes [11], [12] (e.g. palladium [13]). The material system to be characterised and explored is the interface between the CNT and its embedding metal. This is required, because during operation a mechanical load is acting on the CNT and its interface to the embedding metal. Due to unclarified bonding properties a detailed understanding of the behaviour of the stressed CNT inside its clamping is of fundamental importance for thermo-mechanical reliability predictions and failure-mechanistic treatment of future CNT devices.

A bonding strength of the CNT and its embedding material can be obtained with a pull-out test. To this purpose a CNT is with one end embedded into a matrix material and with the other end connected to a loading element. We developed a molecular dynamics (MD) model to predict such a pull-out test and reported results for CNT-gold interfaces [14]. Experimental pull-out tests have been performed for CNTs embedded in non-metallic ma-

trices [15], [16]. Thereby an AFM cantilever is employed as a force sensor inside a scanning electron microscope (SEM) [17]. The same approach has been used to debond multi-walled CNTs [18] or to determine the stress-strain relation of ropes of single-walled CNTs to recover the elastic modulus [17]. We adapt this procedure for the experimental validation of our MD simulations of CNT-palladium contacts.

2. Atomistic modelling

We used the LAMMPS code [19] for our computations. The cross-section of a representative MD-Model is shown in figure 1, where a (6,3) CNT is embedded in single crystalline palladium. We applied periodic boundary conditions perpendicular to the pull-out direction and integrated a vacuum slab in the simulation box to model the palladium surface.

For the self-interaction of carbon and palladium atoms we used the AIREBO [20] and EAM Potential [21], respectively. We assumed a van der Waals interaction between palladium and carbon atoms and approximated this with a Lennard-Jones potential style [22]. Parameters for the carbon-palladium interaction and for alcohol groups (*OH*-groups) attached to the CNT were taken from BioSym force fields [23], [24]. The pull-out procedure consists of repeated displacement steps (0.2 Å) of carbon atoms at the outer tip of the CNT and subsequent equilibration periods of 50 ps (timestep: 1 fs) under constant particle number, volume and temperature conditions at 1 K or 300 K. In effect the resulting force-displacement relations are obtained in a quasi-static way. Detailed informations about our MD modeling procedure can be found in [14] and [25].

For this study we chose uncapped CNTs of chirality (6,3), (9,6) and (17,3). The first two were chosen due to their semimetallic behaviour with probably high piezoresistiv gauge factor. The (17,3) CNT is a representative

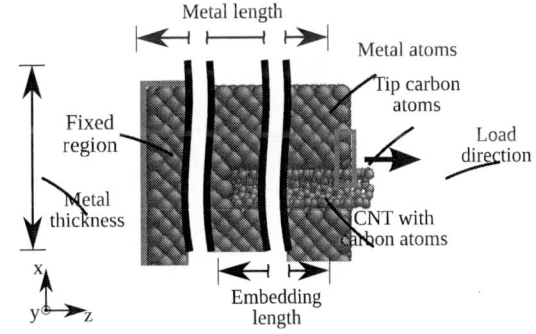

Figure 1. Representative geometry of a defect free pull-out test setup. Shown is the cross-section of a (6,3) CNT embedded in a single crystal palladium block with a <100> orientation with respect to the pull-out direction.

CNT type within the raw material for the experimental tests. Additionally we incorporated a 5-7 defect into the (6,3) CNT and attached four *OH*-groups. Since *OH*-groups can be unintentionally or intentionally attached to the CNT during processing [26], [27], we chose it as a functional group for our investigations. The embedding length of the CNT inside the palladium has been set to 50 Å or 100 Å.

We present a summary of the computational results in figure 3. We found a linear dependence of the maximum pull-out force as a function of the CNT diameter for uncapped CNTs. This has been also reported by Li et al. [28] for CNTs embedded in a ceramic matrix. When the diameter of an uncapped CNT is increased, the number of carbon atoms contributing to the interaction with the embedding material increases with a linear proportion. It has been also reported that for capped CNTs this linear dependency changes to a quadratic relation for weakly interacting interfacial atoms [28], because the number of carbon atoms on the spherical cap increases quadratically with the CNT diameter. Maximum pull-out forces of (6,3) CNTs with one 5-7 defect and with attached *OH*-groups are significantly increased. For *OH*-groups this effect can be explained with an interlocking mechanism. *OH*-groups act as anchors of the CNT in the matrix and add to the force necessary for the pull-out. Presumably an increased number of functional side groups or even longer side groups (e.g. carboxyl groups) can result in stresses high enough that a rupture of the CNT occurs upon mechanical loading.

In summary we can draw major conclusions for the pull-out behaviour of uncapped single-walled CNTs embedded in palladium from our MD simulations:

- For the investigated parameter space the failure mode is an interface fracture of the CNT and the palladium (see figure 2).
- The failure mechanism is a dry sliding detachment.
- The embedding length of defect free chiral CNTs has negligible influence on the pull-out force due to the incommensurate [29] interface structure of the CNT and the palladium.
- CNT diameter, 5-7 defects and functional OH-groups influence the maximum pull-out force.

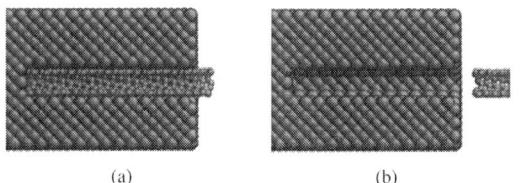

Figure 2. Cross-section geometries before (a) and after (b) the pull-out test for a (6,3) CNT. Clearly visible is a detachment fracture between the CNT and the palladium.

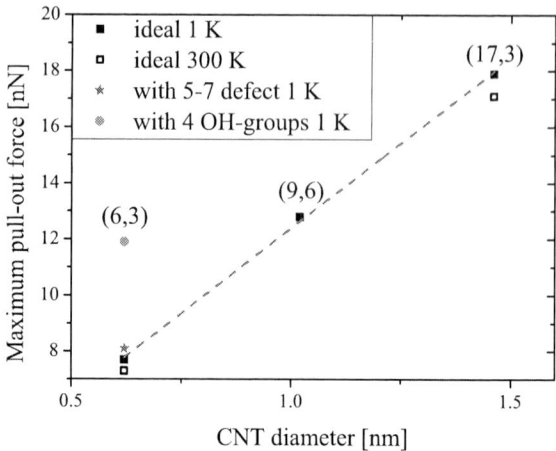

Figure 3. By means of MD simulation obtained maximum pull-out forces as a function of diameter for different uncapped CNT types at 1 K and 300 K. For the (6,3) CNT we also show results with intrinsic defects and with 4 OH-groups at 1 K. All results originate from models with a CNT embedding length of 5 nm.

3. Nano scale experiments

For the experimental pull-out test we adapted a procedure presented by Akita and Nakayama [18], Yamamoto et al. [15] and Tsuda et al. [16]. An AFM cantilever is mounted to a vacuum compatible nanomanipulation system and attached to a CNT to fulfil the load application. State of the art AFM cantilevers have stiffnesses between 0.01 N/m and 50 N/m and allow for a wide range of force resolution.

A standard SEM image contains 1024 x 768 pixels. With one image a cantilever of 100 μm length has to be monitored. This leads to an pixel resolution of ca. 100 nm. Thus, with a pursued force resolution of approximately 1 nN the cantilever stiffness has to be approximately 0.01 N/m.

We prepared specimens by employing standard silicon wafers, physical vapour deposition of palladium and deposition of an aqueous dispersion of CNTs. The CNT dispersion was prepared from solid material by sonication of a bucky paper sheet in deionized water at ca. 0.004 wt % nominal CNT concentration in the presence of an dispersing agent. A cleaned silicon wafer was supplied with a native oxide layer and a 200 nm thick aluminum undercoating layer obtained by sputtering. Consequently, a palladium layer of 50 nm thickness was deposited using standard electron beam deposition (EBD). After the metallization, the wafer was broken into chips of 15 mm x 20 mm in size. The deposition of the CNT dispersion was performed using a simple drop coating technique, placing droplets of ca. 50 μl CNT dispersion in the center of the hydrophilic palladium surface and storing the chips in an desiccator for 24 h in a fine vacuum. The samples were removed from the desiccator, flushed with a jet of

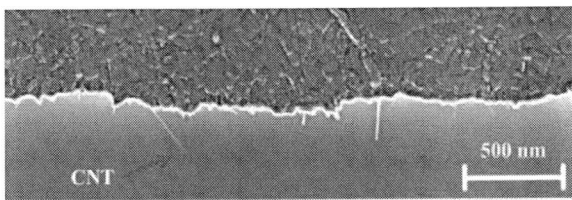

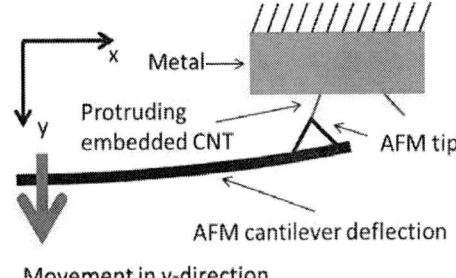

Figure 4. SEM image of fracture surface and free standing CNTs embedded in a metal matrix.

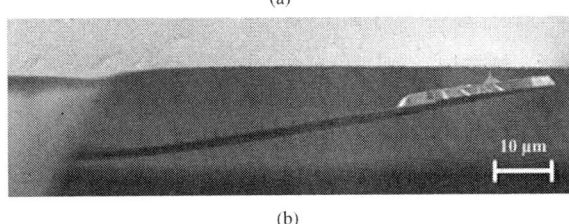

(b)

Figure 5. Sketch of experimental pull-out test in (a), and SEM image of its realisation in (b).

deionized water for 1 min and then purged in a nitrogen flow until they were dry. The CNT-coated samples were then covered with a 50 nm palladium layer by low power EBD evaporation. Prior to the pull-out test, the samples were broken in a way that the fracture was obtained along the diameter of the CNT deposition region. Figure 4 shows an SEM image of the fracture surface with free-standing CNTs.

One part of a fractured sample has been put on a SEM stage with a nanomanipulation system (Kammrath und Weiss GmbH). An AFM chip with a cantilever of given stiffness (0.0095 N/m and 10 % error) has been mounted on the nanomanipulator. By using the nanomanipulator with remote control the tip of the AFM cantilever has been moved to a protruding CNT until contact was made. CNTs were fixed to the AFM using electron beam induced depostion (EBID) [30]. To reduce a possible undesired change in properties of our sample we kept exposure times as short as possible. After two minutes of high dose beam exposure on the CNT and the AFM tip, we rectified the CNT and saved one SEM image (1024 x 768 pixels; pixel resolution ca. 100 nm) with a successive beam blank. Then we moved the nanomanipulator to put force onto the CNT, started a new scan and saved the new image. We repeated this procedure until we observed a failure. A sketch of the pull-out test and a SEM image of its realisation are displayed in figure 5. More details will be published elsewhere [31]

For each pull-out test exists a sequence of images. We determined the cantilever deflection d for each sequence of images by means of greyscale correlation [32]. From the deflection data we calculated with the given cantilever stiffness k the force on the cantilever. For the calculation of the tensile forces in the CNT we considered also a side movement of the AFM tip parallel to the fracture surface. The uncertainty ΔF of the determined force values depends on uncertainties of the cantilever stiffness Δk (given as 0.001 N/m) and of the beam deflection calculation Δd. We estimated Δd to be of twice the pixel resolution: 200 nm. Additional uncertainties originate from the underestimation of the cantilever length due to its coverage by the AFM chip. We calculated an additional force deviation of 10 % for each force value. We neglect total image drifts caused by thermal fluctuations due to the evaluation of

relative displacements. In summary we can estimate the total uncertainty for each force value to be $k\Delta d$ (2.0 nN in total) and an additional 20 % error.

Experimentally determined force-displacement relations are presented in figure 6. As a failure mode we observed a detachment of the the CNT and the AFM tip for #01, #02, #04 and #05. For #03 we found a detachment of the CNT and palladium matrix indicating a fracture between the CNT and the palladium matrix. Maximum pull-out forces vary between $16 - 29$ nN. The large scattering of the results can be attributed to an insufficient connection between the CNT and the AFM tip, the influence of scattering of CNT defects, unleveled load application or undesired EBID contamination occuring during the test. Contamination may reinforce the interface between the CNT and the palladium near the surface. Furthermore, we can not exclude for sure that CNT bundles have been pulled. By comparing the CNT length before and after the pull-out we could not estimate an additional CNT length after the failure for each test.

Comparing our results to published data from Yamamoto et al. [15] and Tsuda et al. [16] we find comparably small pull-out forces. Yamamoto et al. conducted in situ SEM pull-out experiments with AFM cantilevers of multi-walled CNTs emebdded in an alumina matrix. The failure mode observed was a "sword in sheath" detachment of outer and inner CNT shells. They observed no detachment between the multi-walled CNTs and the surrounding matrix. Maximum forces were determined to be of order 14.4 µN showing a strong interaction between large (diameter: 94 nm) multi-walled CNTs and the alumina matrix. In situ SEM pull-out experiments by Tsuda et al. of multi-walled CNTs embedded in PEEK

978-1-4799-4789-8/14 $31.00 © 2014 IEEE

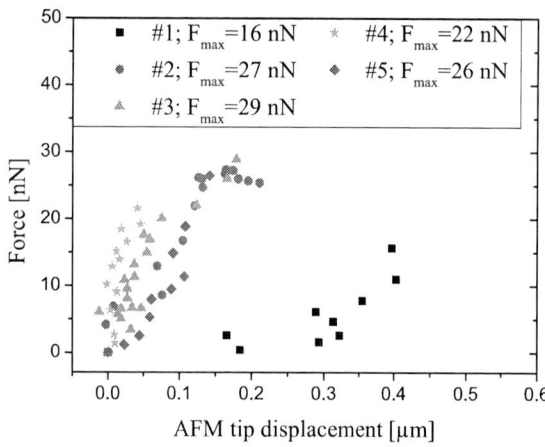

Figure 6. Experimental pull-out data for CNTs embedded in palladium.

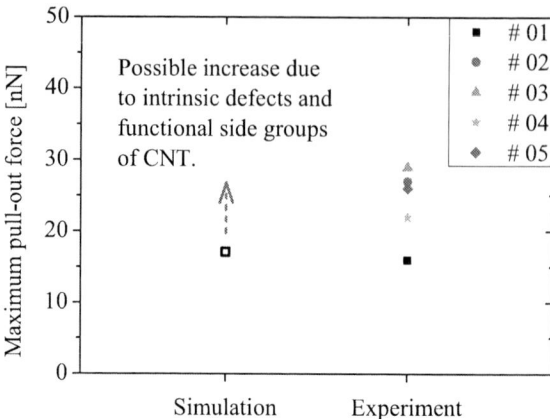

Figure 7. Pull-out force for a (17,3) CNT calculated by means of MD simulation at 300 K and the range of experimental values from the in situ pull-out tests. From experimental characterisation it is deduced that mostly (17,3) CNTs or CNTs of similar diameter are dispersed in the aqueous solution used for the sample preparation.

revealed pull-out forces of $5-15\,\mu N$ showing also a strong interaction.

4. Comparison of experimental and numerical results

Comparing the predicted maximum pull-out force for a CNT with a diameter of 1.46 nm and our experimental values of 16 nN to 29 nN we conclude a good agreement of the results (see figure 7). Still, from the conducted measurements we can not find with significance the same failure mode compared to the simulation, where we found an interface detachment. Comparing the resulting maximum forces of our simulations and experiments one has to consider the simplifications and specifications in the model and the properties of the real materials used in the experiment. MD simulations were conducted with uncapped CNTs only, but the experimental material may consists also of capped CNTs leading to higher pull-out forces. In our models we had a low defect density. However, we found a significant increase in pull-out forces of more than 50 % when attaching a small number of OH-groups to the (6,3) CNT. Higher amounts of defective CNTs in the experimental material can not be excluded as well and needs further analytical investigations. In fact, with the existence of defects the embedding length is becoming a relevant factor. Furthermore, we did not consider the effect of third party molecules in our simulations, although it may be very possible that there are residuals of the dispersing agent and H_2O at the interface between the CNTs and palladium of our samples. However, due to van der Waals interactions between CNTs and solution molecules we expect no reinforcing effect. A reinforcing effect leading to overestimations of the pull-out forces can be expected due to the contamination with hydrocarbon molecules during the pull-out test.

Assuming linear elastic material behaviour, an elastic modulus of ca. 970 GPa [33] and to have a CNT diameter of 1.46 nm we can estimate strains at maximum pull-out

forces to be ca. 1.2 % for the numerical results, 1.1 % for the lowest experimental maximum force value and 1.9 % for the largest experimental maximum force value. Although we could not observe a CNT rupture these strains being reasonable values for a covalently bonded material (see Yu et al. [17] reporting of experimentally derived maximum strains in the range between 1.1 % and 5.3 %).

5. Conclusion

Molecular dynamics simulations have been performed to predict maximum pull-out forces of uncapped single-walled CNTs embedded in palladium. Additionally, nano scale experiments by means of in situ SEM pull-out tests were conducted for validation. Experimental maximum pull-out forces are in the range of $16-29$ nN and can be compared to the value for a simulated (17,3) CNT that is approximately 17 nN. Comparing simulation and experiment we conclude a good aggreement in our data within the same order of magnitude. However, the simulated failure mode was always a CNT-Pd pull-out, but for the experimental failure mode we found mostly a CNT-AFM tip detachment with the evidence for a poor connection.

Ongoing MD computations will focus on larger functional groups (e.g. carboxyl groups) as well as third party molecules (surfactants and water) and their effect on the mechanical interface behaviour of CNT-metal contacts. Future activities will also involve experimental studies on CNTs embedded in other noble metals and carbide formers. To assess the possible correlation of functional side groups with maximum pull-out forces and associated failure modes analytical investigations on the CNT material are planned.

978-1-4799-4789-8/14 $31.00 © 2014 IEEE

6. Acknowledgements

This work has been founded by the VolkswagenStiftung within the project "Integration of dielectrophoretic deposited Carbon Nanotubes and their reliability in mechanical sensor systems" (I/85 086) and the Deutsche Forschungsgemeinschaft within the research unit FOR1713 "Sensoric Micro and Nano Systems". All MD simulations have been performed with the Chemnitzer Hochleistungs-Linux-Cluster (CHiC).

References

[1] B. Wunderle, E. Dermitzaki, O. Hölck, J. Bauer, H. Walter, Q. Shaik, K. Rätzke, F. Faupel, and B. Michel, "Molecular dynamics simulation and mechanical characterisation for the establishment of structure-property correlations for epoxy resins in microelectronics packaging applications," in *International Conference on Thermal, Mechanical and Multi-Physics Simulation and Experiments in Microelectronics and Microsystems (EuroSimE)*, (Delft), Apr. 2009.

[2] O. Hölck, E. Dermitzaki, B. Wunderle, J. Bauer, and B. Michel, "Basic thermo-mechanical property estimation of a 3D-crosslinked epoxy/SiO$_2$ interface using molecular modelling," *Microelectronics Reliability*, vol. 51, pp. 1027–1034, Mar. 2001.

[3] W. C. Oliver and G. M. Pharr, "An improved technique for determining hardness and elastic modulus using load and displacement sensing indentation experiments," *Journal of Materials Research*, vol. 7, pp. 1564–1583, Jan. 1992.

[4] J. E. Darnbrough, D. Liu, and F. P. E. J., "Micro-scale testing of ductile and brittle cantilever beam specimens in situ with a dual beam workstation," *Measurement Science and Technology*, vol. 24, p. 055010, Apr. 2013.

[5] Y. Zhu, A. Corigliano, and E. H. D., "A thermal actuator for nanoscale in situ microscopy testing: design and characterization," *Journal of Micromechanics and Microengineerung*, vol. 16, pp. 242–253, Jan. 2005.

[6] P. Meszmer, K. Hiller, S. Hartmann, A. Shaporin, D. May, R. D. Rodriguez, J. Arnold, G. Schondelmaier, J. Mehner, D. R. T. Zahn, and B. Wunderle, "Numerical characterization and experimental verifcation of an in-plane MEMS-actuator with thin-film aluminum heater," *accepted for publication in Microsystem Technologies*, 2014.

[7] S. Kumar, M. A. Haque, and H. Gao, "Notch insensitive fracture in nanoscale thin films," *Applied Physics Letters*, vol. 94, p. 253104, June 2009.

[8] S. Iijima, "Helical microtubules of graphitic carbon," *Nature*, vol. 354, pp. 56–58, Nov. 1991.

[9] M. A. Cullinan and M. L. Culpepper, "Carbon nanotubes as piezoresistive microelectromechanical sensors: Theory and experiment," *Physical Review B*, vol. 82, p. 115428, May 2010.

[10] C. Hierold, A. Jungen, C. Stampfer, and T. Helbling, "Nano electromechanical sensors based on carbon nanotubes," *Sensors and Actuators A*, vol. 136, pp. 51–61, May 2007.

[11] M. Chen, X. Song, Q. Lv, Z. Gan, and S. Liu, "Bonding of carbon nanotubes onto microelectrodes by localized induction heating," *Sensors and Actuators A*, vol. 170, pp. 202–206, June 2011.

[12] M. Muoth, T. Helbling, L. Durrer, S.-W. Lee, C. Roman, and C. Hierold, "Hysteresis-free operation of suspended carbon nanotube transistors," *Nature Nanotechnology*, vol. 5, pp. 589–592, Aug. 2010.

[13] M. P. Anantram and F. Leonard, "Physics of carbon nanotube electronic devices," *Reports on Progress in Physics*, vol. 69, pp. 507–561, Mar. 2006.

[14] S. Hartmann, O. Hölck, and B. Wunderle, "Molecular dynamics simulations for mechanical characterization of cnt/gold interface and its bonding strength," in *14th International Conference on Thermal, Mechanical and Multi-Physics Simulation and Experiments in Microelectronics and Microsystems (EuroSimE)*, (Wroclaw), Apr. 2013.

[15] G. Yamamoto, K. Shirasu, T. Hashida, T. Takagi, J. W. Suk, J. An, R. Piner, and R. S. Ruoff, "Nanotube fracture during the failure of carbon nanotube/alumina composites," *Carbon*, vol. 49, pp. 3709–3716, Apr. 2011.

[16] T. Tsuda, T. Ogasawara, F. Deng, and N. Takeda, "Direct measurements of interfacial shear strength of multi-walled carbon nanotube/peek composite using a nano-pullout method," *Composites Science and Technology*, vol. 71, pp. 1295–1300, Apr. 2011.

[17] M.-F. Yu, B. S. Files, S. Arepalli, and R. S. Ruoff, "Tensile loading of ropes of single wall carbon nanotubes and their mechanical properties," *Physical Review Letters*, vol. 84, pp. 5552–5555, Jan. 2000.

[18] S. Akita and Y. Nakayama, "Extraction of inner shell from multiwall carbon nanotubes for scanning probe microscope tip," *Japanese Journal of Applied Physics*, vol. 42, pp. 3933–3936, June 2003.

[19] S. Plimpton, "Fast parallel algorithms for short-range molecular dynamics," *Journal of Computational Physics*, vol. 117, pp. 1–19, Mar. 1995.

[20] S. J. Stuart, A. B. Tutein, and J. A. Harrison, "A reactive potential for hydrocarbons with intermolecular interactions," *Journal of Chemical Physics*, vol. 112, pp. 6472–6486, Jan. 2000.

[21] S. M. Foiles, M. I. Baskes, and M. S. Daw, "Embedded-atom-method functions for the fcc metals Cu, Ag, Au, Ni, Pd, Pt, and their alloys," *Physical Review B*, vol. 33, pp. 7983–7991, June 1986.

[22] G. C. Maitland, M. Rigby, E. B. Smith, and W. A. Wakeham, *Intermolecular Forces*. Oxford: Clarendon Press, 1981.

[23] *Biosym Technologies Inc.: CVFF forcefield file in new format, converted from original format file shipped with Discover 2.6.0 / InsightII 1.1.0 / Insight 2.6*, Sept. 1990.

[24] T. Halicioglu and G. M. Pound, "Calculation of potential energy parameters from crystalline state properties," *physica status solidi (a)*, vol. 30, pp. 619–623, Aug. 1975.

[25] S. Hartmann, O. Hölck, and B. Wunderle, "Mechanics of CNT-palladium interfaces for sensor applications simulated with molecular dynamics," *submitted for review to Procedia Materials Science*, 2014.

[26] N. Dementev, R. Ronca, and E. Borguet, "Oxygen-containing functionalities on the surface of multi-walled carbon nanotubes quantitatively determined by fluorescent labeling," *Applied Surface Science*, vol. 258, pp. 10185–10190, June 2012.

[27] P.-X. Hou, C. Liu, and H.-M. Cheng, "Purification of carbon nanotubes," *Carbon*, vol. 46, pp. 2003–2025, Sept. 2008.

[28] Y. Li, S. Liu, X. Han, L. Zhou, H. Ning, L. Wu, Alamusi, G. Yamamoto, G. Chang, T. Hashida, S. Atobe, and H. Fukunaga, "Pull-out simulations of a capped carbon nanotube in carbon nanotube-reinforced nanocomposites," *Journal of Applied Physics*, vol. 113, p. 144304, Apr. 2013.

[29] P. Bak, "Commensurate phases, incommensurate phases and the devil's staircase," *Reports on Progress in Physics*, vol. 45, pp. 587–629, June 1982.

[30] H. Nishijima, S. Kamo, S. Akita, and Y. Nakayama, "Carbon-nanotube tips for scanning probe microscopy: Preparation by a controlled process and observation of deoxyribonucleic acid," *Applied Physics Letters*, vol. 74, pp. 4061–4063, June 1999.

[31] S. Hartmann, O. Hölck, T. Blaudeck, S. Hermann, S. E. Schulz, T. Gessner, and B. Wunderle, "Quantitative in-situ sem pull-out experiments and molecular dynamics simulations of carbon nanotubes embedded in palladium," *submitted for review to Nanotechnology*, 2014.

[32] M. A. Sutton, W. J. Wolters, W. H. Peters, W. F. Ranson, and S. R. McNeill, "Determination of displacements using an improved digital correlation method," *Image and Vision Computing*, vol. 1, pp. 133–139, Aug. 1983.

[33] Y. Wu, M. Huang, F. Wang, X. M. H. Huang, S. Rosenblatt, L. Huang, H. Yan, S. P. OÂ´Brien, J. Hone, and T. F. Heinz, "Determination of the young's modulus of structurally defined carbon nanotubes," *Nano Letters*, vol. 8, pp. 4158–4161, Oct. 2008.

The underfill-microbump interaction mechanism in 3D ICs: impact and mitigation of induced stresses

[1,2]A. Ivankovic, [1]V. Cherman, [1]M. Gonzalez, [1]B. Vandevelde, [1,2]D. Vandepitte, [1]G. Beyer, [1]E. Beyne, [1,2]I. De Wolf

[1]imec, Kapeldreef 75, 3000 Leuven, Belgium
andrej.ivankovic@imec.be

[2]KU Leuven, Oude markt 13, 3000 Leuven, Belgium

Abstract

3D IC assembly processes are introducing new stress mechanisms not observed in 2D environments, which have significant effects on the performance of both BEOL and FEOL. This paper deals with the underfill-microbump interaction mechanism observed after 3D IC stacking and focuses on its scarcely explored impact on the FEOL.

FEOL stress sensors and finite element models are employed to analyze the interaction mechanism development on manufactured 2-tier stack test vehicles - memory on 130nm node logic die and 32nm node logic on logic dies. The logic dies vary from 25 to 50 nm in thickness with a thick memory die on top. 3D IC stacking stress reduction design guidelines are established for Si dies, underfill and microbumps such as die thickness, backside passivation, microbump diameter, pitch, height, and underfill Young's modulus, CTE and glass transition temperature. Furthermore, the equivalent zero stress stack bonding temperature and stress build up above underfill glass transition temperature is analyzed. Stress sensor evaluation methodology and stress impact on FEOL devices - planar and FinFETs is briefly discussed within the scope of the topic.

1. Introduction

The thermo-mechanical stress build-up during 3D IC assembly gains on importance with 3D technology evolution as it has the ability to directly affect both BEOL and FEOL performance introducing new stress mechanisms in comparison with a 2D environment [1-5]. The BEOL is usually first referred to as the weak spot after assembly, in particular due to introduction of low-k material. These new generations of electrically enhanced but mechanically weak dielectrics are already known to be prone to causing BEOL functional failures which leads to the topic of stress impact on the BEOL receiving significant attention. On the other hand, FEOL devices also exhibit unwanted sensitivity to stress in their surrounding, excluding the desired built-in device stress. With this in mind, 3D mainly focuses on immediate TSV stress impact [6-9]. However, a particular stress mechanism is observed after 3D IC stacking, the underfill-microbump thermo-mechanical interaction mechanism, which can pose a greater threat to FEOL devices [10-13,15] and is currently scarcely explored within 3D.

The main goal of this study is to provide design guidelines for the 3D IC stacking assembly phase in order to minimize impact on FEOL device performance and IC layout. The work will be presented in the following steps:

- Provide understanding on the origins of the underfill-microbump interaction mechanism
- Quantify the threat it poses to the FEOL and point out the importance of its study
- Provide concrete design guidelines for its mitigation based on assembled 3D stack prototypes and simulation projections

Stress sensor evaluation methodology and impact on FEOL devices of various technology nodes will be briefly discussed as much as it falls within the main topic. A broader overview on the same work package related to stress sensor evaluation currently published can be found here [11] and the impact of stress on various FEOL device types and in particular TSV stress impact here [6-9].

2. The underfill-microbump interaction mechanism

A 3D IC stack was produced consisting of a functional DRAM on logic structure, presented in *figure 1*. The 10.5 x 10.5 mm² logic die processed in 130 nm technology was thinned down to 25 μm and penetrated with 5 μm in diameter TSVs with an aspect ratio of 5:1. A 550 μm thick 8.5 x 7.5 mm² memory die was stacked on the logic die back-to-face by means of thermo-compression bonding at 250 °C. The gap between dies measured 13 μm consisting of disclosed underfill material and Cu-Sn based microbumps with 5 μm Cu pads on both sides and 3 μm of Sn in between.

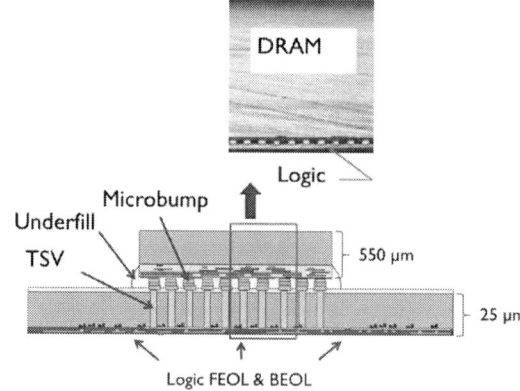

Figure 1 – First stack prototype, memory on logic

The microbumps, 30 µm in diameter, were separated at a distance of 250 µm. The image of the 3D stack above the illustration in *figure 1* emphasizes just how thin the logic die is compared to the memory in real scale. In further reference, the logic die will be considered as the top die with its active region being on its top side.

A 16 x 16 array of nFETs with channel width to length ratio 800 x 600 nm^2 was positioned above the microbumps on the further top side of the thin die in the logic FEOL, as indicated in *figure 1*. The transistor array covered an area of approximately 62 x 37 µm^2. With the microbump positioned in the center of the array, the full projected surface of the microbump was covered with nFETs plus an additional 16 µm from each lateral side. A reference array was copied at a distance far away from the microbump impact.

The currents of the nFETs were measured in saturation before and after stacking, with and without the presence of the underfill. *Figure 2* presents the measured currents from array row 6 passing close to the center of the underlying microbump, normalized to the values of the reference array.

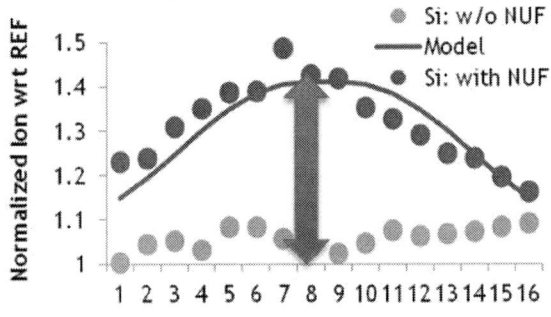

Figure 2 – Current shifts rising above 40% from underfill-microbump impact

It is revealed that the current of the nFETs above the microbumps in the presence of underfill shifted by more than 40%. Without the underfill, the nFETs exhibit negligible current shift. Following these results, it is indicative that the underfill ignites an impact on the FEOL that presents itself in full form in the vicinity of microbumps.

A finite element model is employed in MSC Marc [14] with the same stack geometry to further investigate this process. *Table 1* comprises the material properties used.

Table 1 – Stack material properties used in simulation

Material properties	Si	Cu	Cu-Sn	Underfill
Young's modulus [GPa]	169	117	108	T<Tg 2 - 8 T>Tg 0.1
CTE [ppm/°C]	2.6	16	19	T<Tg 30 - 80 T>Tg 100+

A span of underfill properties are noted due to a non-disclosure agreement for particular used underfills and as later a large group of underfills were simulated with values in this range. Underfill properties were obtained from DMA and TMA curves or alternatively official datasheets where not available. The current shift simulation results for the underfill used for the stack were added to the graph in *figure 2*, showing a good match.

The magnified displacements of the 3D stack obtained from the finite element model, presented in *figure 3* shed more light on the stress generation process happening in the background.

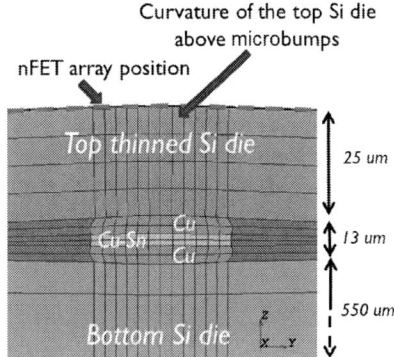

Figure 3 – Curvature of thinned Si observed above microbumps after stacking

A curvature of the thin top Si over the microbump is observed at room temperature after stacking. Looking into *table 1*, the underfill has by far the highest CTE of any material in its surrounding. The underfill-microbump interaction mechanism can be explained in the following manner. During cooling of the 3D stack from bonding temperature of 250 °C to room temperature, the underfill intensively shrinks more than surrounding materials and pulls the thin Si over the microbumps causing a local curvature in the microbump vicinity. The stretched and bent Si now exhibits mechanical stress and since Si is a fairly sensitive piezoresistive material, the mechanical stress reflects into mobility shift of current carriers, electrons and holes.

The historically established piezoresistance model, presented in *figure 4*, connects 6 independent mechanical stress components, σ_{xy}, to material resistivities, ρ_{xy}, through a linear combination involving a set of sensitivity factors, piezocoefficients, π_{xy}.

$$\frac{1}{\rho}\begin{bmatrix} \Delta\rho_{11} \\ \Delta\rho_{22} \\ \Delta\rho_{33} \\ \Delta\rho_{23} \\ \Delta\rho_{31} \\ \Delta\rho_{12} \end{bmatrix} = \begin{bmatrix} \Pi_{11} & \Pi_{12} & \Pi_{12} & 0 & 0 & 0 \\ \Pi_{12} & \Pi_{11} & \Pi_{12} & 0 & 0 & 0 \\ \Pi_{12} & \Pi_{12} & \Pi_{11} & 0 & 0 & 0 \\ 0 & 0 & 0 & \Pi_{44} & 0 & 0 \\ 0 & 0 & 0 & 0 & \Pi_{44} & 0 \\ 0 & 0 & 0 & 0 & 0 & \Pi_{44} \end{bmatrix}\begin{bmatrix} \sigma_{11} \\ \sigma_{22} \\ \sigma_{33} \\ \sigma_{23} \\ \sigma_{31} \\ \sigma_{12} \end{bmatrix}$$

$$\frac{\Delta\rho}{\rho} = -\frac{\Delta\mu}{\mu} = -\frac{\Delta I}{I}$$

Figure 4 – Piezoresistance model

The relative resistivity shift is in fact a consequence of the mobility shift which is directly related to current shift, as noted below the piezoresistance matrix in *figure 4*.

This is where the importance of investigating the sensitivity of FEOL devices to mechanical stress comes into place. Obtaining the sensitivity to mechanical stress in terms of the piezoresistance model means obtaining the piezocoefficients linking the measureable current shift of a device to mechanical stress. The created stress above the microbumps on the top side of the thin Si die consists of primarily in-plane components [10]. A customized 4-point bending setup, shown in *figure 5*, allows controllable application of in-plane stress to the die while simultaneously monitoring the currents of the FEOL devices.

Figure 5 – Customized 4-point bending setup for applying in-plane stress

A Si strip consisting of several dies is diced from the wafer and the particular module which comprises the devices of interest is wirebonded to a flexible substrate with conductive lines flowing through it. The final ''T'' shaped sample is placed into a socket on a PCB, as shown in *figure 5 a)*, leading eventually to SMUs needed for electrical measurements. Further on, the Si part of the sample is positioned in a 4-point bending setup as presented in *figure 5 b)* in order to apply uniaxial in-plane stress to the sample. The 4-point bending tool used consisted of a DTS delaminator with full computer motion control and data analysis. The Si part of the sample consists usually of 1 or 2 additional dies next to the wirebonded dies of interest.

The customized 4-point bending setup outputs a current to stress curve which slope indicates the piezocoefficient value related to in-plane stress. Along with vertical stress calibration, this is the first important step in stress characterization of a device. *Figure 6* shows

an example of FinFET vs. planar FET stress characterization curves, with obtained sensitivities summarized in *table 2*.

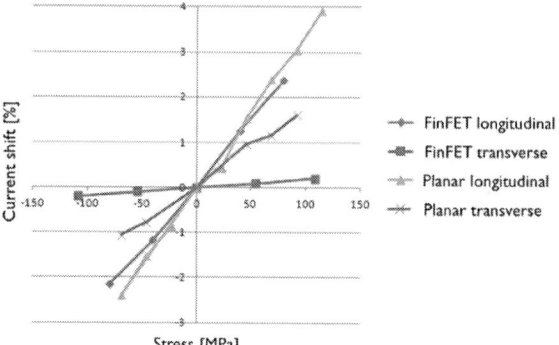

Figure 6 – FinFET vs. planar FET stress sensitivity

Table 2 – Stress sensitivity summary

Sensitivity [ppm/MPa]	Planar (n-type)	FinFET (n-type)
Longitudinal	331	300
Transverse	168	20

32 nm technology, 5 finger n-type FinFETs with a total width to length ratio of 500 x 900 nm^2 and 32 nm n-type planar FETs with width to length ratio 800 x 600 nm^2 were calibrated in [110] and [-110] direction with wafer surface being [001]. Longitudinal calibration refers to the stress being applied in the direction of the current and transverse to the stress being applied perpendicular to the current direction. All device currents were oriented in [110] direction. The in-plane stress calibration provided interesting results revealing similar sensitivity, or in other words piezocoefficients, in direction of the current and a distinctively lower sensitivity of the FinFET in the direction perpendicular to the current, as indicated in *table 2*. These findings give important information on device performance in a stressed surrounding and will have a direct impact on the layout which will be discussed in the following section.

In terms of stress sensors, for this particular back-to-face 3D stack where sensors are on the other Si side of the microbump and vertical stress is negligible, the 4-point bending setup provides a realistic environment for their evaluation and enables decisions to be made regarding their implementation. *Table 3* gives a sensitivity overview of 32 nm n- and p-type planar FETs considered for stress sensors in the [110]/[-110]/[001] reference frame. Longitudinal and transverse sensitivities were obtained in the saturation and linear transistor regime. The results point out 3 important things. Firstly, related to the overall sensitivities, the p-type FETs exhibit higher sensitivity. However, the two in-plane stress components, on the top Si side above the underlying microbumps are both tensile

[10,11] because the Si material is being stretched, which means they are of the same sign.

Table 3 – Planar FET stress sensitivities from the 32nm technology node

Technology node	FET type	Width/ Length (nm/nm)	Piezocoeff. type	Sensitivity Saturation (ppm/MPa)	Sensitivity Linear (ppm/MPa)
32 nm	N, long	500/900	longitudinal	-215	-248
	N, short	500/70	longitudinal	-100	-120
	N, long	500/900	transversal	-135	-155
	N, short	500/70	transversal	-51	-88
	P, long	500/900	longitudinal	406	457
	P, short	500/70	longitudinal	73	110
	P, long	500/900	transversal	-375	-440
	P, short	500/70	transversal	-165	-249

The p-type piezocoefficents are of the opposite sign. In a linear combination extracted from *figure 4*, presented in *eq. (1)*

$$(1) \qquad -\frac{\Delta I}{I} = \pi_l \sigma_l + \pi_t \sigma_t$$

the longitudinal component, $\pi_l \sigma_l$ and transverse component, $\pi_t \sigma_t$ will cancel each other leading to smaller current shifts. The n-type piezocoefficients are of the same sign and their stress components $\pi_l \sigma_l$ and $\pi_t \sigma_t$ will add up to enhance the final stress sensor current shift.

Therefore, if a transistor array is placed on the opposite side of the microbump as in *figure 1*, in terms of monitoring stress in 3D stacking, n-type FETs are preferable to p-type ones. Furthermore, long channel FETs exhibit higher sensitivity and they are more sensitive in the linear regime of operation than saturation.

3. Identifying the threat and impact of the underfill-microbump interaction

TSVs introduce distinguishable stress in Si and this matter, as one of 3D IC stacking foundations, is being thoroughly covered [6-9]. Solving issues related to TSV stress unfortunately does not solve issues of the 3D IC stacking for the FEOL. The underfill-microbump mechanism can induce larger stresses to the FEOL, as seen in *figure 2* if not addressed properly.

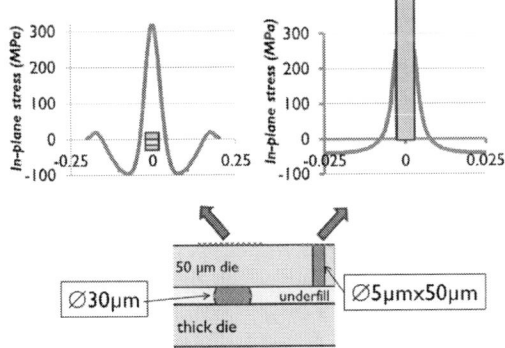

Figure 7 – Microbump and TSV stress impact areas

Furthermore, on top of the stress levels, the microbump impact area can be significantly larger than the TSV. Finite element based results presented in *figure 7*, derived from simulations verified by electrical measurements in *figure 2*, a 50 μm thinned top die on a thick die exhibits similar stress but a 10 times larger impact area around a microbump than a TSV.

This leads to an important effect of the underfill-microbump mechanism on the IC layout. TSVs require keep-out-zones, underfill-microbump impact can be significant and impact areas can be large, so how wide do microbump keep-out-zones have to be and how do they look like?

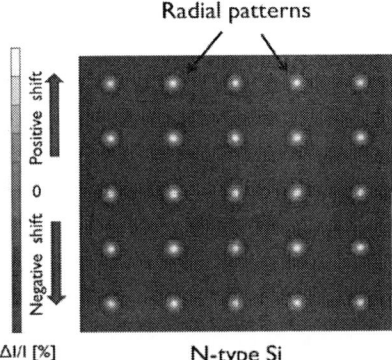

Figure 8 – N-type Si current shift patterns

Figure 8 and *figure 9* show current shift patterns on the top Si surface of a thin stacked die caused by interaction of underlying microbumps and underfill, for n-type Si and for p-type Si, respectively. The microbumps are 30 μm in diameter with a pitch of 200 μm in a total Si area simulated of 0.5x0.5 mm². The distance between the microbumps was chosen as such in order to prevent their stress patterns from overlapping.

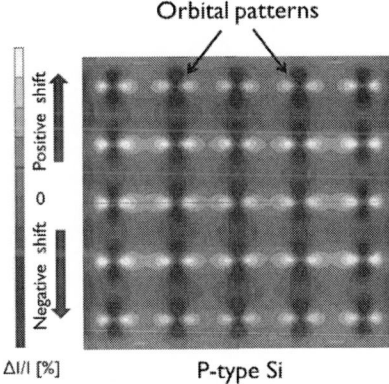

Figure 9 – P-type Si current shift patterns

N-type Si exhibits radial positive current shift patterns which gradually diminish from the microbump projected center and turn to mild negative shifts in between the microbumps. The positive current shift patterns are created by the tensile in-plane stresses on the Si surface from the interaction mechanism while the negative current shifts are a consequence of the less severe compressive stress caused by underfill alone.

P-type Si exhibits more complicated current shift patterns originating from the same stress patterns. The orbital current shift patterns in Si consist of lattices of positive and negative current shifts around the microbump projected centers. In contrast to n-type Si, the highest affected areas are not directly above the microbump but the areas in its immediate vicinity.

In practice, this means that n-type FEOL devices will prefer a radial keep-out-zone above the microbump and keep-out-zones for p-type devices will necessitate attention to 4 lattice regions, perhaps resulting in 4 rectangular regions. The keep-out-zones will be determined by the mechanical stress levels which are in direct correlation with underfill and microbump stack geometry and material and secondly, the sensitivity of FEOL devices. It is for this reason that determining stress sensitivity of FEOL devices, presented in the previous paragraph, is of highest importance as it has a direct relation to IC layout and further device operation.

4. Mitigation of induced stresses – observations and guidelines

Upon retrieval of electrical measurements from the first prototype stack with 25 µm thin top Si presented in *figure 2*, stress reduction after 3D stacking became a primary target. The finite element model added in *figure 2*, verified with these electrical measurements, was derived to find feasible ways of stress reduction and produce 3D stack design guidelines. The following paragraph will list all explored 3D stacking stress mitigation guidelines, summarize them by order of impact and present the results from the new stack prototype where some of the stress reduction proposals were already implemented.

The following geometrical stack properties were considered in the study (italicized implemented in new stack prototype):

- *Top die Si thickness*
- *Microbump pitch*
- *Microbump diameter*
- Standoff height
- Microbump arrays

The following material related stack properties were considered in the study:

- *Underfill choice (E, CTE, T_g)*
- Backside passivation

Figure 10 presents an illustration of the finite element model with indication of the investigated parameters and stress extraction position. In this study, the stress was extracted only from the thin Si side opposite to the microbumps. Stress in Si on the microbump side is considered to have similar in-plane stress values with addition of high vertical stress and is one of the focuses of further studies succeeding this paper [15]. *Table 4*

summarizes the initial values of stack parameters later varied.

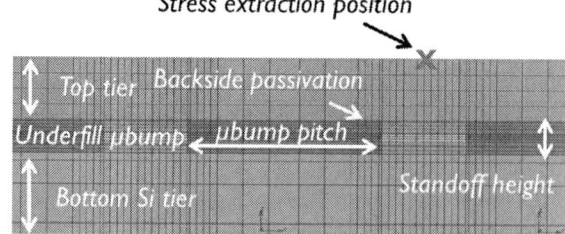

Figure 10 – Indication of explored stack parameters

Table 4 –Inital values of parameters

Initial stack parameters	Top Si thickness	µbump pitch	µbump diameter	Standoff height	UF choice	Backside passivation
	25 µm	200 µm	30 µm	13 µm	High stress UF	SiN

Figure 11 presents the effect of top Si thickness and microbump pitch. With a transition from top Si thickness of 25 µm to 50 µm, stress can already be reduced 3 times. Furthermore, placing the microbumps closer together will additionally lower stress levels.

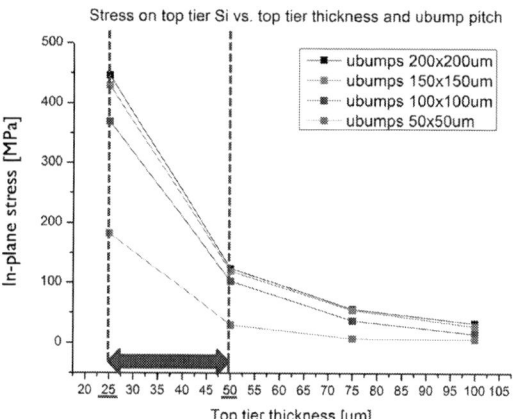

Figure 11 –Impact of Si thickness and microbump pitch

Microbumps with a 30 µm diameter on 200 µm pitch do not interact and act as standalone stressors. Placing them 4 times closer at a pitch of 50 µm decreases the stress levels 3 more times. Microbumps that are placed in sufficient vicinity one from the other prevent the underfill from bending the thin Si ultimately causing less stress.

Furthermore, increasing the size of smaller pitch microbump arrays will additionally benefit a larger Si area within the array. However, the Si on the edge of the microbump array is at risk as the array now acts as a platform over which edges the Si will eventually bend causing an edge effect. *Figure 12* shows a 5x5 microbump array stress pattern in thin on the opposite side of the underlying microbumps. The Si area above the outer microbumps of the array, over which the Si is most

978-1-4799-4789-8/14 $31.00 © 2014 IEEE   390

bent, exhibits a stress edge effect, while the Si area inside the microbump array benefits from the closer microbump configuration.

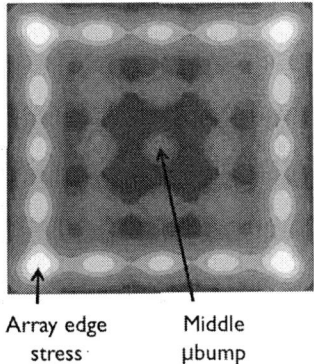

Array edge   Middle
stress   µbump

Figure 12 – Si stress edge effect

The effect of changing the microbump diameter is presented in *figure 13*. Decreasing the microbump diameter from 30 to 20 µm halves the stress. The effect of decreasing the microbump diameter is not straightforwardly comprehensible as a smaller surface would on first thought cause higher stress.

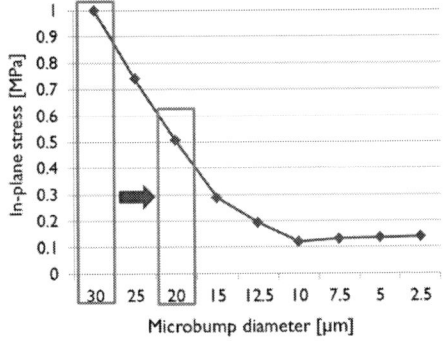

Figure 13 – Impact of microbump diameter

The standoff height between two dies can be considered as well. A smaller gap, inherently holds less underfill to pull on the Si. *Figure 14* presents the decrease of stress in Si with potential gap reduction.

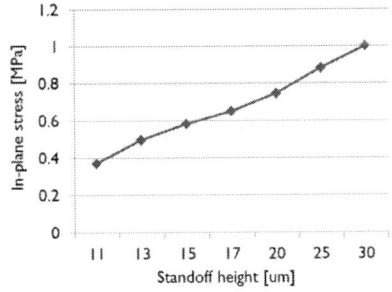

Figure 14 – Impact of stack standoff height

The underfill material is a crucial factor in stress reduction as it is in the core of the underfill-microbump

mechanism. A selection of underfill according to its thermo-mechanical properties, rather than processing feasibility alone, is of high importance. *Figure 15* shows a selection of simulated underfills that were available, with thermo-mechanical data obtained through their TMA and DMA curves or where not feasible, through their datasheets. The first column, *UF 6-A*, presents the underfill that was used in the first stack protoype from *figure 2*. Underfill *UF 2-D* causes 3 times less stress and is considered the best option from the thermo-mechanical standpoint. The choice of low stress underfill needs to be matched with processing feasibility in order to choose the optimal underfill. The CTE was previously reported to be the most influential underfill material parameter in its stress contribution shadowing also its Young's modulus [12].

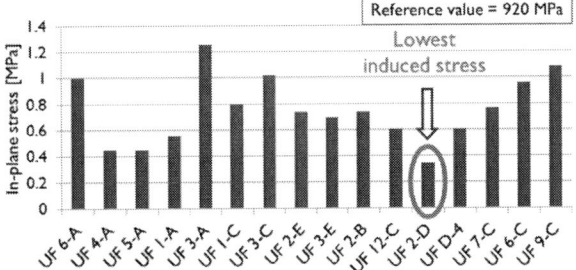

Figure 15 – Underfill selection vs. its stress impact

Here, the impact of the underfill glass transition temperature is added to the study, presented in *figure 16*. An indication of 5-10% stress reduction was present with 20 °C T_g increase.

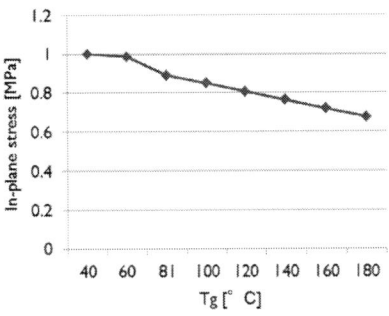

Figure 16 – Impact of underfill T_g

It is arguable whether the underfill is able to generate stress above its T_g and if processing temperatures above T_g should be taken into account at all in thermo-mechanical simulations. Several steps were taken in order to shed some light on this topic in the following paragraph.

In review of the aforementioned geometrical and material parameters than can be altered to mitigate stress, several options are available. *Table 5* orders the simulated parameters in order of impact and their trend necessary for stress reduction, with no. 1

representing the parameter contributing the most to stress reduction in Si.

Table 5 – Impact ordered stress reduction parameters

Order of impact	3D IC stack parameter	Parameter trend
1.	Si thickness	⬆
2.	UF choice	-
3.	Microbump diameter	⬇
4.	Microbump pitch	⬇
5.	Standoff height	⬇
6.	NxN array	⬆
7.	Backside passivation	⬆

A second 3D stack prototype was made taking into consideration the stress reduction guidelines. A 32 nm technology based thin die was stacked back to face to a full thickness die, as in the previous configuration in *figure 1*. The implemented changes from the first 3D stack were the following:

- Thin die increased from 25 µm to 50 µm
- Microbump diameter effectively 15 µm, with 30 µm top Cu pad, 15 µm bottom Cu pad and Cu-Sn region
- Low stress underfill used

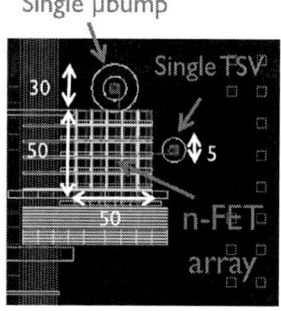

Figure 17 – Part of next generation stack IC layout

An n-FET array with 500 x 900 nm² channel areas, placed next to a single microbump was monitored. A single microbump was selected to enable direct comparison to the case in *figure 2*. *Figure 17* depicts the layout of the array and microbump of interest, with units in µm. Electrical measurement revealed a 4% maximal current shift which is drastically lower than the current shifts above 40% reported from the first stack in *figure 2*. The full profile of the measured n-FET array is presented in *figure 18*. Translated to stress, assuming equal in-plane stress components, transistor sensitivities from *table 3* and implementing in *eq. (1)*, the stress value equals to 114 MPa. Therefore, following the stress reduction guidelines, the new stack exhibits approximately 9 times lower stress, dropping from a reported 920 MPa in the first stack generation to 114 MPa in the next stack generation.

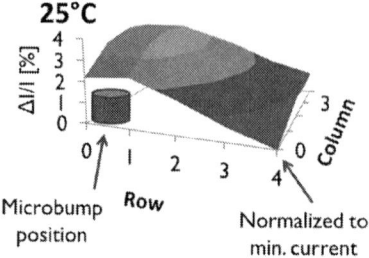

Figure 18 – Significantly decreased current shifts of next generation stack

5. 3D IC stacks at elevated temperatures – EZST

The following paragraphs discusses the stress buildup of the underfill-microbump mechanism above the underfill glass transition temperature. Usual discussions on this topic include questioning the underfill to be too soft above T_g to generate any stress and what equivalent zero stress temperature (EZST) should be considered for simulation purposes. In total 4 of the newer stacks from *figures 17* and *18* were measured at room and elevated temperatures – 25°C, 55°C and 85°C and compared with a finite element model ranging from room to bonding temperature. *Figure 19* summarizes these results.

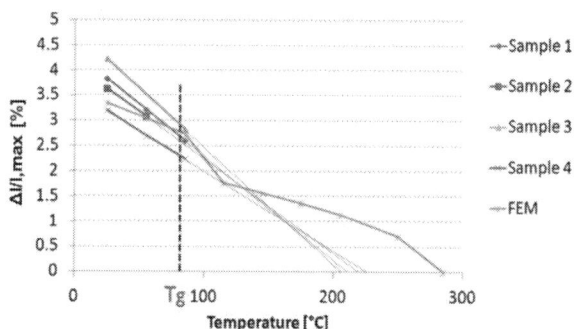

Figure 19 – In search of the EZST

All 4 stacks exhibited between 3 and 4% current shift at room temperature dropping by about 1% at the highest measured temperature of 85°C due to relaxation of stress at elevated temperatures. The glass transition temperature of the underfill was 81°C, exactly at the limit of the electrical measurement setup which means direct monitoring above T_g was not possible. The linearly projected EZST estimated just above 200°C. However, bonding temperature is done at 250°C and the underfill Young's modulus and CTE exhibit non-linear changes crossing the T_g which is why linear stress build up above T_g does not seem realistic. The finite element model considered cooling of the stack from a temperature above 250°C in order to take into account the cure shrinkage of the underfill before actual structure cooling starts. According to the finite element model, from bonding temperature the stress builds up slightly non-linearly to 1.5% current shift reaching the zone around T_g. Just before T_g, the changing underfill properties seem to cause a higher stress increase in Si and below T_g the stress linearly reaches a value above 3% current shift, within the range of the electrical measurements. The stress changes around T_g are sensitive to the underfills Young's modulus and CTE which are not always available from DMA and TMA curves. A slight change in Young's modulus or CTE changes the slope of the linear stress build up as well which can then better be matched to the slope of the electrical measurements. For this reason, obtaining precise underfill material properties is of high importance.

The finite element model clearly shows that stress build up above underfill T_g cannot be ignored and needs to be taken into account. The modeled current shift was matched with electrical measurements by taking an EZST slightly higher of the bonding temperature, added from calculations related to the underfill cure shrinkage. The stress build up above T_g equaled to more than 50 % of the total stress.

6. Conclusions

The underfill-microbump mechanism presents a potentially higher risk to FEOL devices than TSVs. During cooling of the 3D stack after die bonding, the underfill pulls the thinned Si die over underlying microbumps causing high stress in Si. In-situ electrical measurements on built 3D stacks with 25 μm thin top die showed transistor currents can shift above 40% due to stress effects on Si if the underfill-microbump interaction is not treated. Furthermore, the impact area of the microbump can be 10 times larger than the one of the TSV.

Understanding the stress patterns of this mechanism created in Si and evaluating the stress sensitivity of FEOL devices prior to usage is of high importance to defining underfill-microbump interaction related keep-out-zones. A modified 4-point bending setup is used for applying uniaxial in-plane stress in a controlled manner for evaluation of stress sensors and studying in-plane sensitivity of FEOL devices.

In-situ electrical measurements on 3D stacks and finite element modeling was employed to develop underfill-microbump stress mitigation guidelines. The following effects can have a distinctive impact in stress reduction: increasing the Si thickness, decreasing microbump pitch and diameter, grouping microbumps in arrays, decreasing die standoff height, choosing low stress underfill based on several of its material properties and tuning Si backside passivation. Stress build up above underfill T_g was explored and results point to more than 50% of stress generated already at temperatures higher than T_g.

Acknowledgments

The authors would like to thank core partners of the imec Industrial Affiliation Program (IIAP) on 3D Integration Technology.

References

1. E. Beyne, ''Electrical, Thermal and Mechanical Impact of 3D TSV and 3D Stacking Technology on Advanced CMOS Devices – Technology directions'', *3D Systems Integration Conference* (3D IC), pp. 1-6, 2012.

2. M. Gonzalez et al., ''Chip Package Interaction (CPI): Thermo Mechanical Challenges in 3D Technologies'', *Electronics Packaging Technology Conference* (EPTC), pp. 547-551, 2012.

3. G. Van der Plas et al., ''Design issues and considerations for low-cost 3-D TSV IC technology'', *Solid-State Circuits*, vol. 46, pp. 293–307 (2011).

4. J. Lin et al., ''High density 3D integration using CMOS foundry technologies for 28 nm node and beyond'' *Electron. Components Technol. Conf.* (ECTC), pp. 22–25 (2010).

5. L. Lin et al., ''Reliability characterization of Chip-on-Wafer-on-Substrate (CoWoS) 3D IC integration technology'' *Electron. Components Technol. Conf.* (ECTC), pp. 366–371, 2013.

6. V. Cherman et al., ''Impact of through silicon vias on front-end-of-line performance after thermal cycling and thermal storage'', *Int. Reliab. Phys. Symp.* (IRPS), 2B.3.1–2B.3.5, 2012.

7. W. Guo et al., ''Copper Through Si Via Induced Keep Out Zone for 10nm Node Bulk FinFET CMOS Technology'', *Int. Elec. Dev. Meet.* (IEDM), 12.8.1-12.8.4, 2013.

8. C. Yu et al., ''TSV process optimization for reduced device impact on 28nm CMOS'' *Symp. On VLSI Tech.* (VLSIT), pp. 138 – 139, 2011.

9. A. Mercha et al. ''Comprehensive analysis of the impact of single and arrays of through silicon vias induced stress on high-k / metal gate CMOS performance'' *Int. Elec. Dev. Meet.* (IEDM), pp. 2.2.1–2.2.4, 2010.

10. A. Ivankovic et al., ''Thermo-mechanical impact of underfill-microbump interaction in 3D stacked integrated circuits'', *Electronics Packaging Technology Conference* (EPTC), pp. 34-38 , 2012.

11. A. Ivankovic et al., ''FET arrays as CPI Sensors for 3D Stacking and Packaging Characterization'', *Int. Reliab. Phys. Symp.* (IRPS), pp. 2E.3.1 – 2E.3.9, 2012.

12. A. Ivankovic et al., ''Analysis of Microbump Induced Stress Effects in 3D Stacked IC Technologies'', *3D Systems Integration Conference* (3D IC), pp. 1-5, 2011.

13. B. Lemke, ''Experimental determination of stress distributions under electroless nickel bumps and correlation to numerical models'', *Sensors Journal*, vol. 11, pp. 2711–2717, 2011.

14. Marc 2010 User's Guide, MSC Software (2010)

15. V. Cherman et al., ''3D Stacking Induced Mechanical Stress Effects'', *Electron. Components Technol. Conf.* (ECTC), in press, 2014.

Characterization and modeling of the AuCuSn thin solder joint under thermal cycling

Tiphaine Pélisset[*†], Balamurugan Karunamurthy[*], Ralf Otremba[‡], Thomas Antretter[†]

[*] KAI - Kompetenzzentrum Automobil- und Industrieelektronik GmbH, Europastrasse 8, 9524 Villach, Autria
Email: tiphaine.pelisset@k-ai.at, Phone: +43 51777 19926
[†] Institute of Mechanics, Montanuniversitaet Leoben, Franz Josef Strasse 18, 8700 Leoben, Autria
[‡] Infineon Technologies AG, Am Campeon 1-12, 85579 Neubiberg, Germany

Abstract

Thin $Au_2Cu_6Sn_2$ solder joints have been identified to provide improved electrical and thermal device performance compared to thicker solder joints [1]. Both the material high thermal conductivity and the reduced joint thickness improve the thermal dissipation through the solder joint, making the thin $Au_2Cu_6Sn_2$ solder layer very attractive for high power devices. In this paper, we establish a material model for the $Au_2Cu_6Sn_2$ material in order to study solder thermal fatigue using finite element analysis.

1. Introduction

Accurately understanding and predicting failure modes in a package enables to shorten the product development cycle. The accurate prediction of the package behavior under operating conditions allows to identify the limitations of the packaging technology in terms of material and thickness. In this paper, the packaging technology under scrutiny is the soldering technique based on transient liquid phase bonding of the AuSn on copper substrate. One contributor to component failure is the degradation of the die attach and underlying metalization layer by crack formation or delamination. Delamination or cracks in the backside metalization leads to thermal failure of the device. These layers are thin films, observed to behave differently from their bulk counterpart, especially outside their elastic domain. Here we are using a well known technique, the wafer curvature measurement technique, to mechanically characterize the plastic behavior of the film, depending on temperature. An approach combining experiment and simulation enables the modeling of the film using continuum mechanical approaches. The so-identified material model is used to perform finite element simulations of the package.

In this paper, we first describe the methodology used to model the material behavior. Then we examine the behavior of the solder under thermal cycling.

2. Experimental characterization

The paper focuses on the solder system obtained by transient liquid phase bonding of Au-25wt%Sn and Cu [2], [3]. Four layers were sequentially sputtered onto a 220 μm-thick (100) silicon wafer: Al, Ti, NiV and Au-Sn. Then a 1000 nm-thick Cu layer was electrodeposited. Additional wafers were deposited with a single layer such

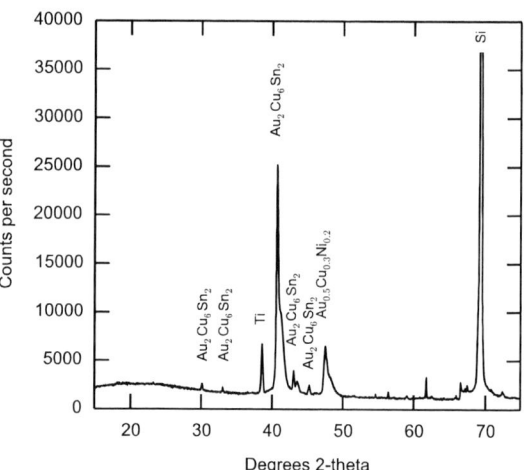

Figure 1: X-ray diffractogram of the backside metalization after 2 min annealing at 350 °C. The peak pattern allows to identify the $Au_2Cu_6Sn_2$, which is present in a polycrystalline state.

as Al, Ti or NiV. The wafers were cut into samples of $3\,cm \times 3\,cm$. The sample is then thermally cycled under N_2H to prevent copper oxidation. During the first cycle, the sample is heated from room temperature up to 350 °C, with a 2 min-hold time and naturally cooled down to 25 °C. During the following cycles, the sample is heated up to 450 °C.

During the phase transformation, the copper atoms diffuse into the Au-Sn layer to form another intermetallic compound. Qualitative X-ray analysis (XRD) shows that two phases are formed, consisting of the $Au_2Cu_6Sn_2$ phase, in a polycrystalline state and the $Au_{0.5}Cu_{0.3}Ni_{0.2}$ phase (Fig. 1). An additional TOF-SIMS analysis has confirmed that the $Au_{0.5}Cu_{0.3}Ni_{0.2}$ phase is only present in a small region close to the initial Au-Sn/NiV interface. This is in agreement with the work of Etschmaier et al.[4] who have identified the resulting intermetallic to be a homogeneous single phase $Au_2Cu_6Sn_2$.

Bilayer deformation arises from the differences in thermal expansion between the two materials. The induced-curvature κ was measured using a laser scanning technique [5]. The average biaxial film stress σ_f was cal-

978-1-4799-4789-8/14 $31.00 © 2014 IEEE

Table 1: Estimated fractional uncertainties for the thickness and the curvature.

Parameter	Fractional Uncertainty
h_{Si}	0.5 %
h_{Ti}	1.2 %
h_{NiV}	4.3 %
h_{Al}	3.1 %
h_{Cu}	5 %
h_{AuSn}	4.8 %
κ	0.1 %

Table 2: Young's modulus of the single metal phase and of the ternary phase.

Material	Young's modulus (GPa)
Au	78.0
Sn	49.9
Cu	129.8
at% weighted average	103.5
fit $Au_2Cu_6Sn_2$	135

culated from the substrate curvature by using Stoney's formula:

$$\sigma_f(T) = \frac{M_s(T) h_s^2}{6 h_f} [\kappa(T) - \kappa_0] \quad (1)$$

where the indices s, f, refer to the substrate and film respectively, h is the layer thickness, κ_0 the curvature of the bare substrate, $M_s = 1/(s_{11} + s_{12})$ the substrate biaxial modulus, s the compliance constants and T the temperature. The film is assumed to be initially stress-free at room temperature. The uncertainty on the curvature measurement was estimated at room temperature. A mirror was repeatedly mounted into the chamber and its curvature measured at $(25 \pm 2)\,°C$. The measurement was repeated N = 23 times. The standard deviation is calculated as $\sigma = \sqrt{\frac{1}{N-1} \sum (\kappa - \bar{\kappa})}$. The fractional uncertainties for the thickness are provided by in-line tool. Curvature and thickness fractional uncertainties are reported in the Table 1.

When multiple thin films are deposited onto a much thicker substrate, the total change of substrate curvature is a linear superposition of the curvature changes associated with the presence of each film. The force equilibrium in the multilayer system is given by:

$$\sigma_f h_f = \sigma_{Al} h_{Al} + \sigma_{Ti} h_{Ti} + \sigma_{NiV} h_{NiV} + \sigma_{AuCuSn} h_{AuCuSn} \quad (2)$$

It enables the calculation of the biaxial stress in the $Au_2Cu_6Sn_2$ film from the total biaxial stress of the multilayer sample and the stress of the individual layers.

3. Material parameters identification

A. Young's modulus estimate

Fig. 2 shows the stabilized stress-temperature cycle of the $Au_2Cu_6Sn_2$ phase. The hysteresis loop stabilizes at the third cycle. The film stress decreases linearly down to 150 °C. The film is under compressive stress at 450 °C.

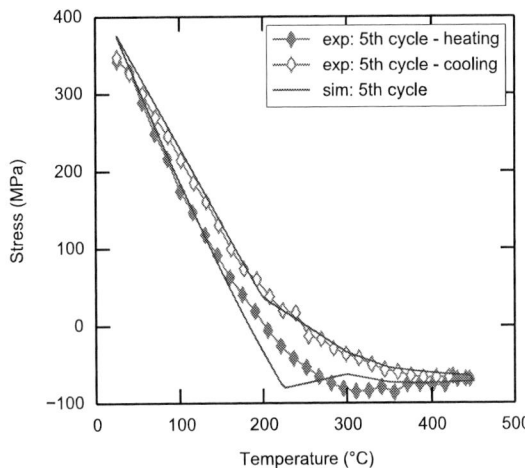

Figure 2: Stress-temperature cycle for the $Au_2Cu_6Sn_2$ film calculated from the stress-temperature data of the multilayer and each single layer. Not all the measurement points are displayed for readability.

The linear temperature dependence of the heating path corresponds to the elastic behavior of the material if the temperature dependence of the material properties is neglected. The slope of $\sigma(T)$ is given by:

$$\frac{d\sigma}{dT} = \frac{E_f}{1 - \nu_f} (\alpha_s - \alpha_f) (T - T_0) \quad (3)$$

where E_f is the film Young's modulus, ν_f its Poisson ratio, $\alpha_{s,f}$ the coefficient of thermal expansion of substrate and film, respectively, and T_0 the reference temperature. Three unknowns are present here: E_f, α_f and ν_f.

Fill [6] measured the thermal strain of a $Au_2Cu_6Sn_2$ sample over a temperature range from 25 °C to 250 °C. The linear coefficient of thermal expansion of the material was estimated $(15.8 \pm 1.3) \times 10^{-6}\,K^{-1}$. Considering the following thermal expansion for gold $(14.51 \times 10^{-6}\,K^{-1})$ [7], copper $(16.85 \times 10^{-6}\,K^{-1})$ [7] and tin $(15.8 \times 10^{-6}\,K^{-1})$ [8] and taking their weighted average $0.2\alpha_{Au} + 0.6\alpha_{Cu} + 0.2\alpha_{Sn}$ results in $15.96 \times 10^{-6}\,K^{-1}$ as an estimate. Metals usually have Poisson ratios around 0.3 so ν_f is set to 0.3. We assume a linear temperature dependence of the Young's modulus E_f. A fitting procedure is performed on the heating path with various assumptions concerning the end of the elastic domain. The smallest standard deviation from the aforementioned dependence was obtained by considering the material behaving elastically up to 200 °C. The Young's modulus at room temperature is (135 ± 10) GPa. The Young's modulus temperature dependence is $E(T) = 135[1 - 800 \times 10^{-6}(T - 25)]$ where E is given in GPa and T in °C.

Table 2 reports the bulk value of the Young's modulus for Au, Sn and Cu as well as the weighted average. The value obtained from the fitting procedure gives a larger estimate than the weighted average value.

B. Plastic behavior

The thermal loading of a bilayer is simulated by the finite element method. A 2D model is built consisting of the thickness direction only and a planar direction, as the deformation is uniform on the sample. Symmetry boundary conditions are applied on either side of the strip. [9]

The Chaboche model is used to capture the plastic behavior of the material. It is a flexible model which allows to consider non-linear kinematic hardening with several backstresses. Its parameters are the initial yield stress k, the initial hardening modulus C and the recovery parameter γ. With each additional backstress, a couple (C, γ) is added. Chaboche's model with one backstress and γ = 0 is equivalent to a linear kinematic hardening model. We evaluated successively linear kinematic hardening, non-linear kinematic hardening with one and then two backstresses. Two backstresses were found to better capture the plastic behavior.

Both the euclidian norm of the difference between experimental and the simulated $\sigma(T)$, $\sum(\sigma_{exp} - \sigma_{sim})^2$, and the maximum of the difference are used to evaluate the fit. Table 3 reports the two-backstress Chaboche parameters corresponding to the best fit reached: 63% of the points have a difference less than 10 MPa, 25% are in the range from 10 MPa to 20 MPa and 12% are in the range from 20 MPa to 50 MPa. The average of the difference $1/N \sum |\sigma_{exp} - \sigma_{sim}| = 10$ MPa, where N is the number of experimental (and simulated) points for the stabilized cycle.

4. Simulation and results

Thermal stresses arise in microelectronics devices during operation due to the mismatch of the coefficients of thermal expansion of the materials of the package. High levels of thermal stresses can result in device failure such as die cracking or delamination at interfaces. Subsequent

Table 3: Temperature dependent parameters of the Chaboche model with two backstresses.

Temperature (°C)	k (MPa)	C_1 (GPa)	γ_1	C_2 (GPa)	γ_2
25	400	5.0	10	5	1
200	68	5.0	10	5	1
300	15	5.5	10	5	1
350	10	6.5	10	5	1
500	1	7.0	10	1	1

Table 4: Convergence parameter value for various meshes.

Mesh	Element size (µm)	Convergence parameter value (MPa)
A	0.2	234
B	0.5	232
C	1	236
M	1	236

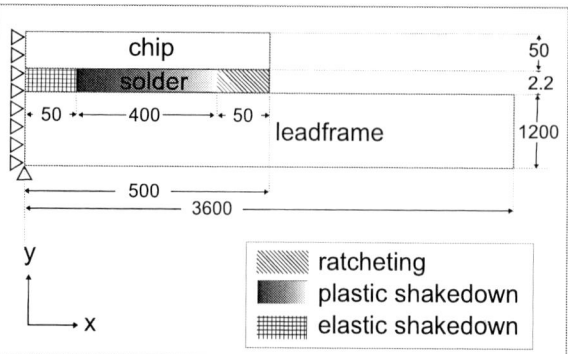

Figure 3: Model dimensions. The dimensions of the three-layer model are given in micrometers. The elastic shakedown, plastic shakedown and ratcheting zones are reported.

failure such as bond wire damage or electrical failure reduces the device lifetime. In order to enhance the reliability of the device, stress levels have to be accurately predicted to guide the choice of material and manufacturing processes. Parts of the discrete devices are electrically contacted from the backside through the leadframe as well as through the leads. The die-attach of such components is required to be electrically and thermally efficient to obtain a low-resistivity electrical contact and a large thermal dissipation. The mechanical integrity of the solder layer contributes to the device performance. The aim of this paragraph is to discuss the solder behavior and interfacial stresses arising in a microelectronics package.

A three-layer model is built with the typical dimensions of a TO-220 package. It contains three layers, the leadframe, a thick copper plate, on which a silicon carbide chip is soldered using the previously studied solder (Fig. 3). For now, no molding compound is considered. A 2D-model is used with plane strain condition and eight-node elements. Symmetry boundary conditions are applied along the centerline of the package. The layers are initially in a stress-free state. This assumption neglects the possible influence of residual stresses on the system behavior under thermal cycling. The three-layer system is cycled under uniform thermal loading from −55 °C to 150 °C.

A mesh convergence study is based on the von Mises stress value at the interface between leadframe and solder at 2.6 µm from the corner as the mesh convergence parameter, under a temperature rise of 100 °C. This point is chosen as being sufficiently far away from the singularity influence. The results of the convergence study are reported in Table 4. The element size refers to the refined region close to the corner. The convergence study is performed on a mesh with transition, well refined close to the sharp corner (meshes A,B,C). For comparison purpose, a fully mapped mesh (M) with uniform element size is compared to the transition mesh. The mesh C gives the same results as the mesh M. The influence of the

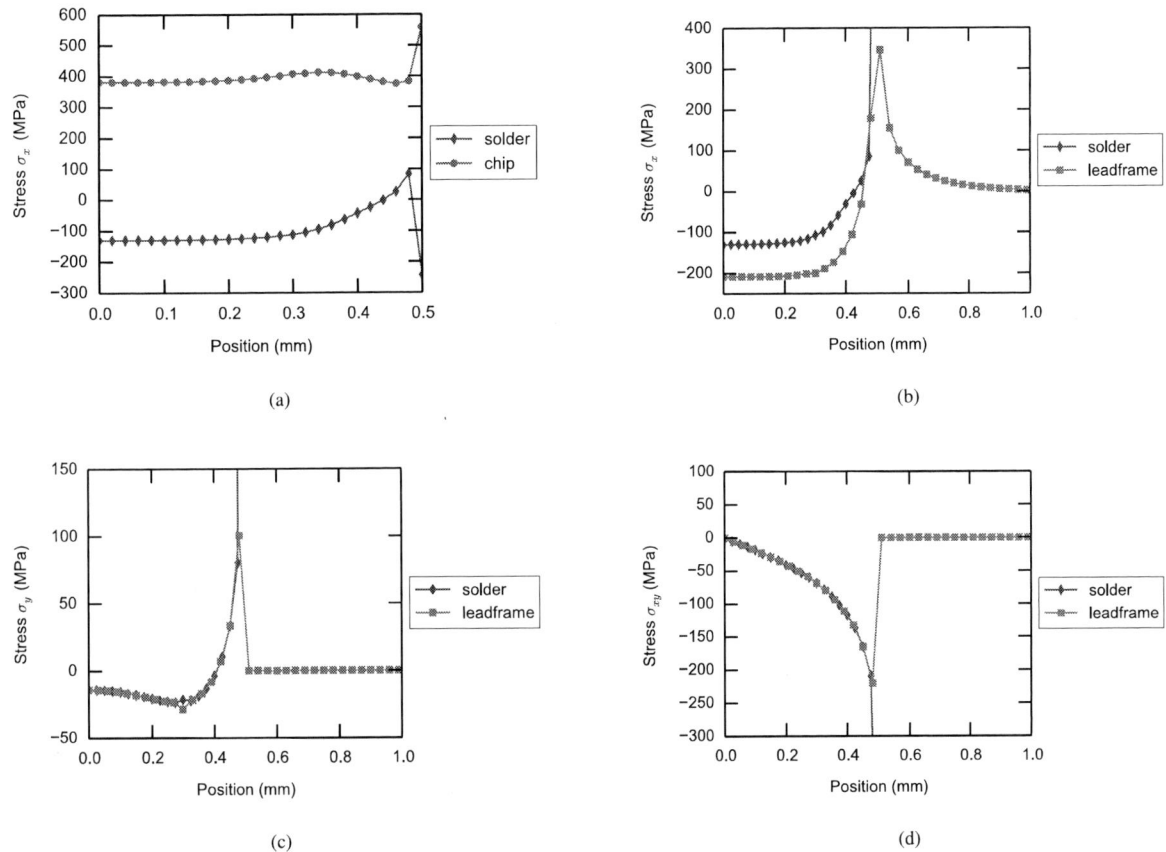

Figure 4: Interfacial stresses versus the x-position at 150 °C. The centerline of the chip is at x = 0, the corner of the chip at x = 0.5 mm and x > 0.5 mm is a position in the leadframe. (a) Solder/chip interfacial stress. The chip is in tensile state at high temperature while the solder is in compression, due to their relative CTE mismatch. (b) Solder/leadframe interfacial stress. Both solder and leadframe are in a compressive stress state at high temperature. (c) Solder/leadframe interfacial stress. The peeling stress is compressive below the chip and zero at the leadframe surface. (d) Solder/leadframe interfacial stress. The shear stress decreases slowly from the corner to the centerline, where it reaches zero, due to symmetry.

leadframe mesh on the stress close to the solder/leadframe interface is negligible. Further decreasing the size of the element in the neighborhood of the corner does not change significantly the convergence parameter value.

The interfacial stresses in the three-layer system during the first cycle at 150 °C are reported in Fig. 4. The stress values are interpolated from the integration points to the nodes of the interface. Fig. 4a shows the in-plane stresses at the solder/chip interface versus the position along the interface. The in-plane stress in the solder is compressive while it is tensile in the chip, due to their thermal expansion mismatch. The in-plane stress reaches a constant value away from the corner, stabilizing at the middle of the chip. The in-plane stress in the solder is uniform throughout its thickness (not shown here). The stresses at the solder/leadframe interface are reported in Fig. 4b, Fig. 4c and Fig. 4d. The σ_x component is

continuous along the interface, even though a singularity arises at the corner. The σ_y component is discontinuous due to the free boundary condition after the corner. At high temperature, both the leadframe and the solder are in compressive stress state below the chip. The peeling stress σ_y is compressive at the interface and zero at the free surface of the leadframe. The shear stress σ_{xy} at the solder/leadframe interface decreases slowly to reach zero at the centerline of the model. A very large stress singularity is observed in the simulation. Such stress singularity cannot exist physically. Once the stress level reaches the yield strength of the material, it yields and stress is redistributed to the neighboring region. The actual stress distribution is expected to display a smooth and gradual change along the interface. Usually, exact sharp corners are seldom observed. However, in microelectronics, the sawing procedure makes it possible to have such sharp

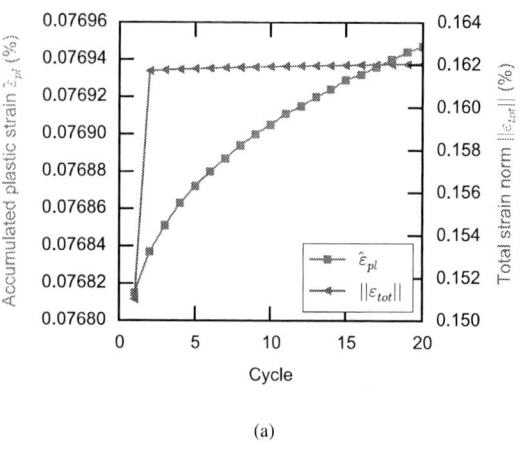

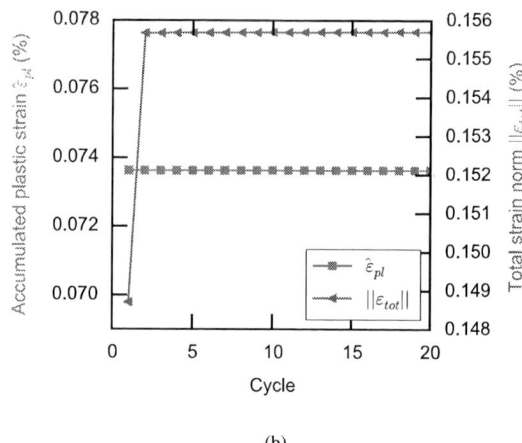

(a)   (b)

Figure 5: (a) Plastic shakedown; (b) Elastic shakedown

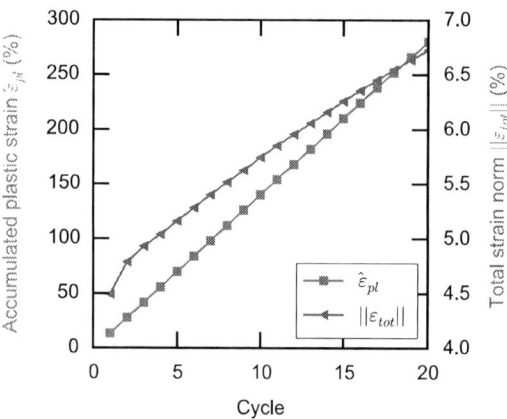

Figure 6: Ratcheting in a region of 50 μm close to the corner. The plastic strain accumulates infinitely and the mean of the norm of total strain increases

corner in the absence of chipping. Hence additional work is required to approach the treatment of the corner. Further work will rely on the use of cohesive zone element to study the interfacial delamination.

The second study point is the plastic behavior of the solder. The Chaboche model allows to simulate such effects as shakedown and ratcheting. Elastic shakedown consists of an initial plastic deformation, with plastic strain accumulation, soon followed by elastic further deformation. Plastic shakedown and ratcheting are characterized by the ever increasing accumulation of plastic strain. They can be distinguished by the norm of the total strain, which reaches a constant mean value in the case of plastic shakedown, while the mean value increases continuously for ratcheting. During the first thermal cycle, the solder

deforms plastically over the whole width of the chip, which is related to the simulation assumptions; twenty thermal cycles are simulated. The solder can be divided in three zones where elastic shakedown, plastic shakedown and ratcheting take place, shown in Fig. 3, Fig. 5 and Fig. 6. For readability, only the accumulated plastic strain value at the end of each cycle is plotted. The mean value of the norm of the total strain is calculated for each cycle and reported in the graphs. Ratcheting is occuring in a region up to 50 μm away from the corner (Fig. 6). Further away from the corner, it is not clear whether weak ratcheting or plastic shakedown will occur given the number of simulated cycles. Plastic shakedown (Fig. 5a) occurs in the region further away from the corner up to 50 μm from the centerline. Close to the centerline, elastic shakedown (Fig. 5b) takes place (Fig. 3). Ratcheting tends to decrease the failure life of a material [10]. Depending on the stress level, ductile failure caused by large ratcheting strain may occur in the zone with significant ratcheting behavior [11]. In the lower stress region, away from the corner, an undefined behavior between ratcheting and plastic shakedown will occur, which may lead to fatigue failure. The small region close to the centerline where elastic shakedown occurs may be damage free.

5. Conclusion

A methodology is presented to model the thin material layer used in the backside metalization of a chip-package. Under the assumption of isotropic material behavior, valid for small grain size material, the elastic modulus of the material has been determined. The parameters of a plastic material model can be fitted to the experimental curves. Here experiments have yet to be performed to determine the viscoplastic behavior of the material if relevant. Curvature measurement is an efficient technique to gain first insight into the thermomechanical behavior of an unknown material. The knowledge of the elasto-plastic behavior of a solder is a first step into further fatigue modeling of the solder joint. Interfacial stresses are mostly influenced by the singularity. Large plastic deformation of the solder is to be expected, leading to ductile failure of the solder close to the corner. Fatigue failure is to be expected in the region further away from the corner. To treat fatigue failure in the singularity region, cohesive zone element with cyclic behavior will be used in further work.

Acknowledgments

The authors gratefully acknowledge the support of Dr. Stefan Krivec (Infineon Technologies Austria AG) for the sample manufacturing and Prof. Erich Halwax and Michael Fugger (Vienna University of Technology) for the XRD measurements.

This work was jointly funded by the Austrian Research Promotion Agency (FFG, Project No. 831163) and the Carinthian Economic Promotion Fund (KWF, contract KWF-1521|22741|34186).

References

[1] M. Holz, J. Hilsenbeck, R. Otremba, A. Heinrich, P. Türkes, and R. Rupp, "SiC power devices: product improvement using diffusion soldering," *Materials Science Forum*, vol. 615-617, pp. 613–616, 2009.

[2] H.-J. Reichert, M. Deckers, and R. Zanner, "Method for producing a chip-substrate connection," US Patent 7 442 582, Oct 28, 2008.

[3] R. Otremba, "Connection structure semiconductor chip and electronic component including the connection structure and methods for producing the connection structure," US Patent 8 084 861, Dec 27, 2011.

[4] H. Etschmaier, J. Novák, H. Eder, and P. Hadley, "Reactions dynamics in diffusion soldered joints from the eutectic Au/Sn alloy on copper and silver substrate," *Intermetallics*, vol. 20, pp. 87–92, 2012.

[5] R. Huang, C. A. Taylor, S. Himmelsbach, H. Ceric, and T. Detzel, "Apparatus for measuring local stress of metallic films, using an array of parallel laser beams during rapid thermal processing," *Meas. Sci. Technol.*, vol. 21, pp. 1–9, 2010.

[6] M. Fill, "Die thermische Ausdehnung von miniaturisierten Merhlagen-Schichstrukturen," Master's thesis, Universität Wien, Fakultät der Physik, 2007.

[7] L. B. Freund and S. Suresh, *Thin Film Materials: Stress, Defect Formation and Surface Evolution*. Cambridge University Press, 2003.

[8] V. T. Deshpande and D. B. Sirdeshmukh, "Thermal Expansion of Tetragonal Tin," *Acta Crystallographica*, vol. 14, no. 4, p. 355, 1961.

[9] T. Pélisset, W. Heinz, B. Karunamurthy, and T. Antretter, "Constitutive modeling of thin films under thermal cycling," *European Microelectronics Packaging Conf.*, 2013.

[10] G. Kang, Y. Liu, J. Ding, and Q. Gao, "Uniaxial ratcheting and fatigue failure of tempered 42CrMo steel: Damage evolution and damage-coupled visco-plastic constitutive model," *International Journal of Plasticity*, vol. 25, no. 5, pp. 838–860, 2009.

[11] J. Peng, C.-Y. Zhou, Q. Dai, X.-H. He, and X. Yu, "Fatigue and ratcheting behaviors of CP-Ti at room temperature," *Materials Science and Engineering A*, vol. 590, pp. 329–337, 2014.

Challenges of viscoelastic characterization of low T_G epoxy based adhesives for Automotive Applications in DMA and relaxation experiments

I. Maus[1], H. Preu[1], M. Niessner[2], M. Fink[2], K.M.B. Jansen[3], R. Pantou[4,5]
B. Michel[4], B. Wunderle [4,5]

[1]Infineon Technologies AG, Wernerwerkstr. 2, 93049 Regensburg, Germany
[2]Infineon Technologies AG, Am Campeon 1-12, 85579 Neubiberg, Germany
[3]TU Delft, Industrial Design Engineering, The Netherlands
[3]Fraunhofer ENAS, Chemnitz Germany
[4]TU Chemnitz, Germany
Ingrid.maus@infineon.com

Abstract

Microelectronic devices integrate more and more diverse materials in order to miniaturize the packages. Epoxy resins enable smaller electronic devices with promising features [1]. The performance and the reliability of the product are highly dependent on the material behavior of the components and on their interaction under different loading situation.

Electrically conductive adhesives are widely used in semiconductor technology. The focus of this work is set on Isotropic Conductive Adhesives (ICA) with a high amount of electrically conductive filler particles.

The aim of this work is the material characterization of highly filled epoxy based die attach materials by dynamic mechanical analysis (DMA) and relaxation experiments in order to derive the elastic and a viscoelastic material model in a wide temperature range and the thermo mechanical analysis TMA measurement results to obtain the coefficient of thermal expansion CTE and information about the residual stresses in the adhesive material.

A comparison and discussion of measurement and simulation results from FEA with implemented material models obtained using DMA and relaxation experiment will be done. We use as an example to illustrate our method a measurement from two glue samples A and B.

The measurement of the epoxy based highly filled die attach material is a challenging topic. We show how to overcome the difficulties in measuring these materials.

1. Introduction

Epoxy-based adhesives are often used as thermal and electrical interfaces, which require excellent adhesion, heat transfer, and electrical properties. However due to the CTE-Mismatch in material properties, adhesive failure and delamination remain an integral part of the reliability issue since adhesive rupture or delamination may induce other electrical or mechanical failure mechanisms. It can be investigated in a virtual prototyping process using cost efficient simulation methods and saving qualification time.

Therefore a full characterization of the materials with a reliable material model for further simulation is a crucial topic.

We are facing challenges due to the inhomogeneous material properties and less reproducible measurement results compared e.g. with other epoxy based materials like molding compounds.

The aim of this study is to address these difficulties and establish a reliable method of measurement and characterization of the glue material. The experiments to illustrate the method were performed on samples from two glue materials both with a low glass transition temperature T_G around 50°C (glue A, B).

2. Measurement results

2.1 Thermo mechanical analysis

In a TA Instruments TMA machine Q400 the change in material length due to temperature ramping with 10°C per min from -60°C up to 260°C was measured.

To have reproducible results and to ensure that the samples are fully cured they are at least once heated above the glass transition temperature. Only the second heating run of the TMA measurement was used to evaluate the correct values for the coefficients of thermal expansion (CTE).

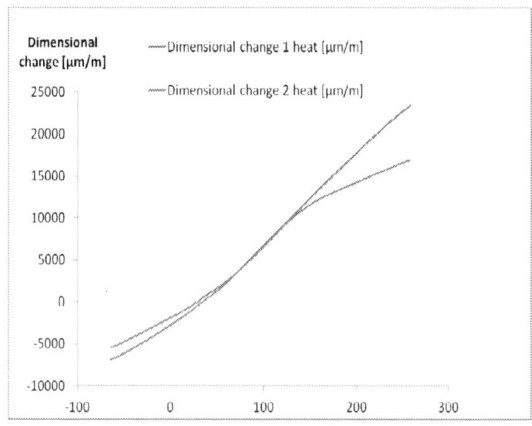

Figure 1 Thermo mechanical analysis TMA in the Q400

The CTE - calculated as the ratio of change in length and the temperature difference - below the glass transition temperature was found for the material A as 48 µm/(m°C) (i.e. ppm/K) and above the T_G the CTE is. 64 µm/(m°C).

The gray and red line in fig. 1 gives us the dimensional change with increasing temperature for the first and second heating, the gray and red dotted lines illustrate the derivative of the dimensional change over the temperature. In the most measured e.g. molding compounds there is a region above and below the glass transition temperature with a linear increase of the length of the sample with temperature, while a change in the slope indicates the location of T_G, which can be found as the intersection point of the slopes [2]. However, the glue doesn't show a clear linear region for the dimensional change above and below T_G; therefore for simulation in ANSYS® the whole curve Alpha from the second heating should be included.

2.2 Poisson's ratio

The Poisson's ratio was calculated using Digital Image Correlation DIC Veddac [3, 4] by measuring the transverse and longitudinal strain under an uniaxial tension in the Gabo Eplexor 500 test machine. The experiment was performed using a 15 mm length sample and applying a strain rate of 5% per minute. To monitor the deformation one image was every 0,15 s taken. The DIC method with the underlying mathematical algorithms for the software VEDDAC is based on the idea that images allow to record local unique object patterns which can be recognized again if the objects are stressed by thermal or mechanical loading. This is visualized in fig. 2.

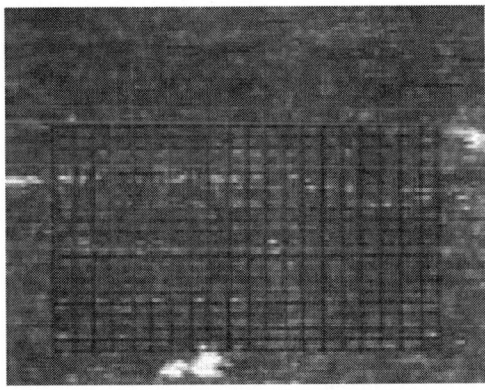

Figure 2: Deformation field on the sample surface before (blue grid) and after (red grid) loading obtained by DIC with VEDDAC in a longitudinal tensile experiment for the glue A (reference image in the background before loading)

The experimental results for the temperatures -10°C, 5°C, 25°C and 50°C are shown in fig. 2. The Poisson's ratio is around 0.24 for the temperature below the glass transition temperature T_G, which was found around 50°C. In the T_G-region the value increases from 0.24 to 0.29.

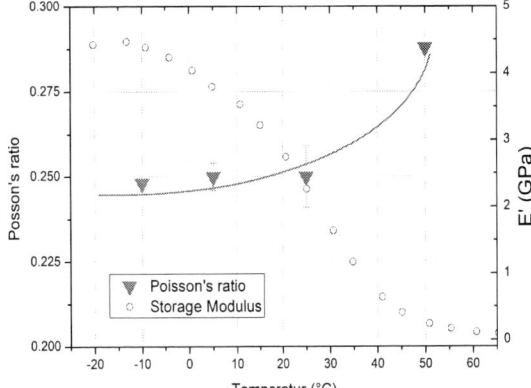

Figure 2: Poisson's ratio versus the temperature

After the transition from the glassy into the rubbery state the deformation after loading is too high and pattern recognition on the samples with DIC was not possible. The reason can be the sample geometry, the measurement was done on strips and the deformation in the clamping region is not negligible for the measurement around T_G. Anyhow in the fully rubbery state the Poisson's ration is supposed to be 0.5. However, future work is needed to prove this hypothesis for the glue material and to do the measurement with the DIC around T_G. This will be done on samples with a dog bone shape to increase the measurement accuracy by reducing the deformation in the clamping region.

2.3 Dynamic mechanical analysis

The viscoelastic material properties of the epoxy based glue die attach (DA) were determined in a dynamic mechanical analysis. The DMA device used is the mini tensile tester Q800 from TA Instruments. At frequency sweep measurements the material response of a dynamic (sinusoidal) load with oscillating amplitude of 20 µm at different frequencies (40, 20, 10, 3, 1Hz) is measured while the temperature is ramped continuously with 2°C/min from the starting temperature -60°C up to 260°C.

For the measurement setup the available temperature range of the chamber is between -150-600°C, the range of the applicable frequencies is between 0,01-200 Hz and the limitation for the maximum force is 18 N.

Several clamping configurations are available in order to perform the DMA experiment and measure the viscoelastic Young's modulus, e.g. the 3-Point-Bending-Test or the clamping for a tensile test [2, 6, 7]. The Young's modulus of an DA changes drastically over the temperature range. The glue becomes extremely soft for temperatures above T_G. The control loop of the TA Instrument Q800 in a 3-Point-Bending test reaches the limit of feasibility when in the measurement specimen such large deflection occurs; therefore the glue was measured in a tensile test.

The quality of the measurements is highly dependent on the sample preparation. On one hand the material is highly inhomogeneous due to the silver filler particle. On the other hand during the curing and polymerization process due to outgassing voids occur in the specimen.

To reduce the formation of voids the specimens were exposed to vacuum conditions for 4 min before the curing process. Afterwards the specimens are cured according to the specifications of the manufacturers at 125 °C for 45 min in a nitrogen-purged oven in a teflon cure cavity as shown in fig. 4. The geometry of the teflon cavity is 60x10x1.5 (mm).

Figure 3:
a. Teflon cavity to cure the glue A, B
b: Bulk Glue sample to perform DMA and relaxation experiment in a tensile test

Afterwards the samples are grinded to remove surface irregularities. Previous studies [6, 7, 8] already pointed out, that the stiffness of the sample – so the geometry of the sample – is crucial for correct measurements. Otherwise the compliance of the Q800 together with ist force limitations will cause remarkable deformation component and lead to a large error in the total response for the deformation.

Therefore a sample geometry optimization should be performed taking into account the effect of the compliance of the DMA machine. This will be shown later in more detail.

The sample geometry for the frequency sweep experiment as shown in fig. 3a is 25.2x4.9x1.2 (mm).

The storage modulus (reversible deformation energy), the loss modulus (irreversible deformation energy, heat) and the angular phase shift between stress and strain (tan delta) are monitored [9]. The storage modulus in the frequency sweep measurement is shown in fig. 3a.

a.)

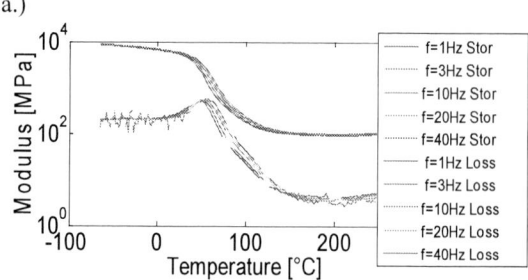

b.

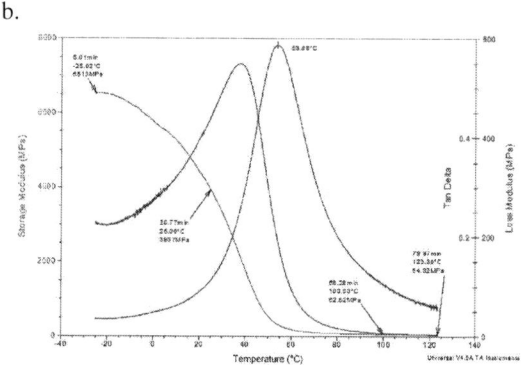

Figure 3:
a) Frequency sweep measurement: storage modulus and loss modulus from the frequency sweep measurement
b.) Standard DMA measurement performed with 1 Hz storage modulus, loss modulus and tan delta

The glass transition temperature was extracted from the peak of the tan delta value and it was around 50°C for glue A.

2.4 Relaxation experiment

Uniaxial stress relaxation experiments were performed in a TA Instruments Q800 testing machine at different temperatures. Since the Young's modulus of the epoxy based glue is strongly viscoelastic, the relaxation experiment is used in the time domain to extract the viscoelastic material properties. In the relaxation experiment a constant strain is held for 60 min and the force response is used to calculate the time dependent storage modulus. The stress required to hold this constant amount of deformation is recorded over time.

The first relaxation experiment was performed on the same sample geometry with the dimension 25.2x4.9x1.2 (mm) as was used for the DMA experiment assuming that the calculated strain from the standard DMA-(1 Hz) measurement would deliver

978-1-4799-4789-8/14 $31.00 © 2014 IEEE

relaxation experiment values from the linear viscoelastic range.

The relaxation experiment follows the procedure: first set the target temperature and hold it for 15 min until isothermal condition is achieved, in the second step the strain load of 0.015 is applied and holds for 60 min, this step is followed by a load free recovery time of 90 min. The results from these first relaxation experiments showed unexpected anomalous effects: By applying a certain strain level (ε=0,015) the force decreases at the beginning and afterwards increased again. The reason for this phenomenon was found to be due to a too small value for the applied strain in combination with a too stiff sample.

It is crucial to do the relaxation experiment with a strain value, which is small enough to be still in the linear viscoelastic range but still the value should be chosen big enough, that the Q800 compliance calibration does not affect the measured displacement. It means, even if the measurement with the Q800 gives some evidence, that a compliance correction will be done; the measured displacement includes both the sample displacement and the clamp displacement from the Q800.

The geometry of the samples was optimized regarding the requirements from the measurement setup. The frame stiffness of the Q800 is about 3E6 N/m; the clamp compliance is about 0.33 µm/N. The sample stiffness has to be noticeably smaller than the clamp stiffness of the Q800. Based on this fact the samples were cut and afterwards grinded to a 0.5 mm thickness. In order not to affect the measurement results the sample stiffness is a factor of 10 smaller then the clamp stiffness of the measurement setup. The new sample geometry with the dimension of 17,5x3.6x0.5 (mm) for the relaxation experiment can be seen in fig. 4.

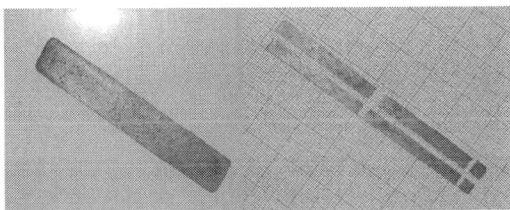

Figure 4: Sample preparation out of a bulk sample from the glue A

Several additional preliminary experiments were carried out to find out where the nonlinear region for the viscoelasticity/viscoplasticity starts. It is recommended to choose a value of about 80% of the strain value for which the nonlinear behavior of the sample response occurs [6].

In the result of the preliminary experiments the relaxation experiment was performed with the strain value of 0.15, which was a factor of 10 higher than in

the first experiments which show anomalous effects as hardening after relaxation as described before.

The T_G region was found about 50°C. In a range of 50 degrees above and below T_G no visible effects due to the relaxation are expected. For this reason the new relaxation experiments on the optimized sample geometry were carried out at following temperatures: -20,-10,0,10,20,30,40,50,60,70,80 and 125°C, around T_G with the temperature step of 10°C and in the glassy and the rubbery state only at one temperature. The relaxation modulus is calculated by the time dependent stress divided by the constant strain value. The experimental results are shown in fig. 5. At 125°C the sample stiffness in the rubbery state is too soft, so that the relaxation experiment failed. For the other temperature the relaxation doesn't show any anomalous behavior (like hardening). The sample shows, as expected in the T_G region the fastest relaxation.

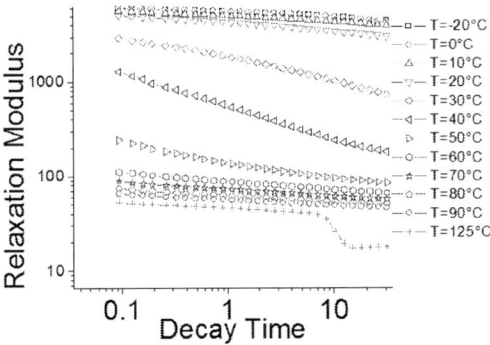

Figure 5: Relaxation modulus over the decay time

3 Material model for the simulation

The frequency sweep measurement was used to evaluate the viscoelastic material model for ANSYS for both material glue A and B. To evaluate the required material data a Matlab tool was written [10].

The material model extracted from the DMA was used to simulate the relaxation experiment and to compare the simulation results with the relaxation experiment measurement to test the quality of the material model.

In addition, the relaxation experiment for glue B was used to determine a master curve on the reduced time scale and to obtain the prony parameter and the shift curve for the material model.

For the characterization of the relaxation behavior the experiments in the frequency domain are preferred instead of characterizing the material based on relaxation experiments in the time domain. It takes less time to perform the experiments and to generate the needed experimental results for a full visco elastic characterization. Nevertheless a relaxation experiment is helpful to validate and optimize the material model for simulation done in a FEA.

The master curve for illustration reason was constructed using TTS (Time Temperature Superposition) [7] for the DA material A and is shown in fig.6. For rheologically simple materials the material response at one temperature can be related to another temperature response by a change in time scale. By shifting the curves (only horizontal shift was done) choosing as reference the curve at 50°C the master curve and the shifting values are obtained.

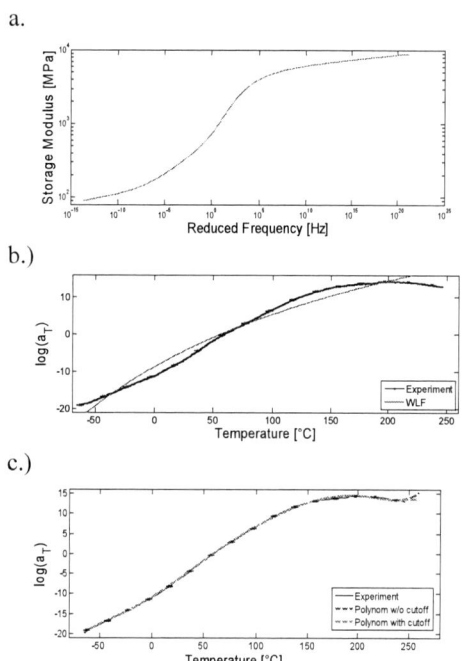

Figure 6:
a.) Master curve for glue A obtained from DMA results
b.) Shifting values used for the master curve above and the shift function using a WLF fit for glue A
c.) Shifting values used for the master curve above and the shift function using a polynomial fit for glue A

The shifting values were subsequently fitted with the WLF function, which is defined as follows

$$\log a_T = -\frac{C_1 \cdot (T - T_{ref})}{C_2 + (T - T_{ref})} \qquad \text{(eq. 1)}$$

with C1 and C2 as fitting constants, T_{ref} is the temperature of the non-shiftet part of the master curve.

For the glue A the fitting with WLF as shown in fig. 6.b was less accurate than fitting the curve with a polynomial fit function as shown in fig.6c. For the polynomial fit a degree of seven was chosen (s. blue curve polynomial fit w/o cut off) and for the values outside of the measurement range the values were additionally cut off, i.e. in the measurement range the shifting values were fitted with the polynomial function, above and below that range the last values were taken and used as a constant shift factor (polynomial fit with cut off). The polynomial function with the cut off was implemented as a User Function in ANSYS®; the normal polynomial function can be used without additional implementation of a user subroutine. The subsequent modeling of the DA in the relaxation experiment doesn't exceed the measurement range, therefore the polynomial function without a cut off was considered as sufficient. However, this is dependent on the application and has to be considered if the temperature ranges of the application exceed the measurement range.

The storage modulus in the frequency domain as a function of the angular frequency ω has the following form:

$$E'(\varpi) = E_0 \cdot \left(\alpha_\infty + \sum_{i=1}^{n} \alpha_i \cdot \frac{\tau_i^2 \cdot \varpi^2}{1 + \tau_i^2 \cdot \varpi^2} \right) \qquad \text{(eq. 2)}$$

With E_0 the storage modulus in the glassy state, the prony coefficient α_i defined as the quotient of Ei to E0 and the relaxation time τ_i.

The implementation of the viscoelastic data in ANSYS® requires both the prony series and the information how the series was constructed; i.e. the shifting based on the relation between time and temperature response of the material.

This prony series is obtained from a curve fitting procedure using eq. 2 in frequency domain or equation 3 in time domain by minimizing the error square of the experimental master curve.

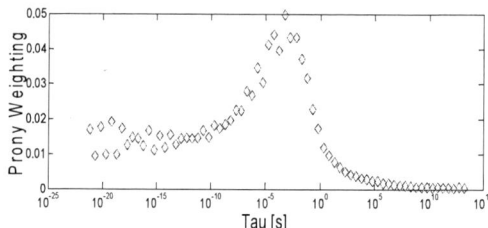

Figure 7: Weighted prony constants over the reduced time (glue A)

The weighted prony constants from the master curve in fig. 6a are plotted in fig. 7. The plot shows a smooth curve, which gives us some evidence about the quality of the measurement and the material model extracted out of it. However, further investigation is needed to optimize the values of the used prony constants for ANSYS and see the effect on the simulation results.

The results from the relaxation experiment are used to judge the quality of the results by comparing the simulation with the experimental results.

For the glue B in addition to the evaluation of the material model obtained from the frequency sweep measurement a second master curve and shifting function was extracted in the time domain from the relaxation experiment.

$$E'(t) = E_0 \cdot \left(\alpha_\infty + \sum_{i=1}^{n} \alpha_i \cdot e^{\frac{-t}{\tau_i}} \right) \qquad \text{(eq. 3)}$$

The relaxation modulus in time domain t can be expressed as shown in eq. 3 through a set of prony coefficients α_i and relaxation times τ_i.

The master curve and the required shifting values which were fitted with a WLF function are shown in fig. 8. A polynomial fit function of 7th degree without a cut off was also implemented.

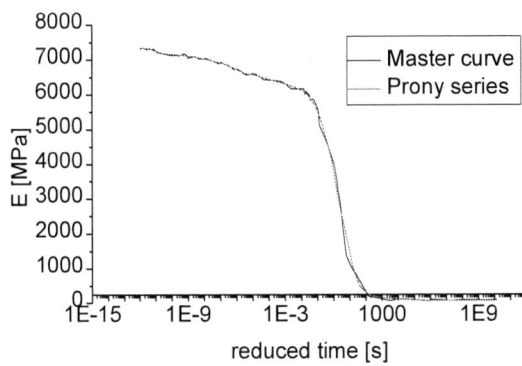

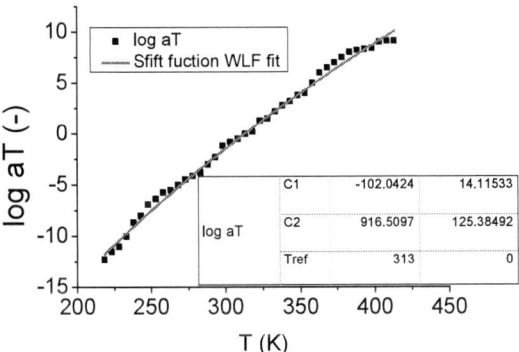

Figure 8:
a.) Master curve obtained from the relaxation experiment and the fitted prony series for ANSYS® for glue B
b.) Shifting values used for the master curve above and the shift function using a WLF fit for glue B

The focus of the work was set on the experimental part of the glue characterization. However simulation and experimental results for glue B for 2 temperatures in the T_G region and at one low and one high temperature will be discussed.

The simulation of the relaxation behavior was performed with the prony series from the frequency sweep experiment and WLF shift as well as using polynomial shift functions. Additionally the material data from the relaxation experiment were used.

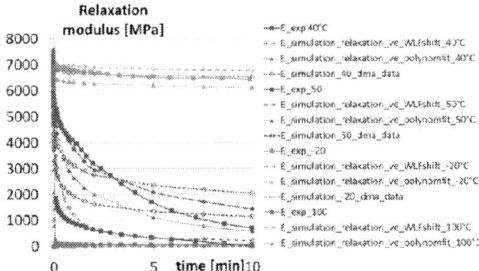

Figure 9: Relaxation modulus versus time for the DA B; comparison of the relaxation experiment and simulation with 3 material models

For the DA B a reasonable agreement between simulation and experimental results with all used material models can be observed. The agreement for sample B was good both in trend and magnitude, for sample A the trend was the same but the magnitude differed by a factor 1.5. In long term for the T_G region the agreement between the experimental data and the FEA simulations with the material model based on the relaxation experiment is better than for the material model based on the DMA measurement. The material model obtained from the frequency sweep measurement seems to give a stiffer response. The effect of using a polynomial fit function instead of a WLF-fit function to implement the time temperature superposition TTS is negligibly small.

Nevertheless there is still room to improve the agreement between experimental and simulation results. In the first step the effect of using different shift functions was investigated and considered to be small for the investigated temperature range. In the future additional improvement can be done by optimizing the prony series for ANSYS® toward a Gaussian distribution. Therefore further investigation is needed.

5. Summary and conclusion

We have established a method for measurement and characterization of highly filled epoxy based glue die attach materials and illustrated it in the case of two glue samples A and B, both with T_G near to the room temperature. Required material models for the FEA with *ANSYS®* were generated.

Since the application of the material in microelectronic industry is subjected to long term stresses and strain in normal usage in addition to the frequency sweep DMA experiment a viscoelastic measurement on the time scale as a relaxation experiment was performed.

Difficulties in the relaxation experiment were pointed out and it was shown how to overcome them i.e. by performing several preliminary experiments to find out the correct value for the applied load (strain value). The effect of the clamp stiffness (Q800 compliance) on the accuracy of the measurement results was discussed.

Different viscoelastic material models for one of the glues were investigated in a simulation model in *ANSYS*® in order to simulate the relaxation behavior for model verification. For the epoxy based glue die attach (sample B) a reasonable agreement between simulation and experimental results with all used material models can be observed. The prediction of the relaxation behavior at low and at high temperatures as well as in the most important temperature region around T_G is correctly predicted with all investigated material models. In long terms the prediction with the material model obtained from the frequency sweep measurement is less accurate than the prediction using the material model obtained from the relaxation experiment. However, the frequency sweep measurement method is preferred in the industry because of the much less measurement time needed to generate the data and the less effort and cost needed to perform the DMA measurement. The difficulties to perform a relaxation experiment for the glue in correct and reproducible way were pointed already out. When weighing up the costs required on the one hand and the benefits of the improvement of the results on the other hand the full characterization of the material in the frequency domain can be justified. However, the relaxation experiment is crucial for model verification.

The effect of using different shift function for the glue B can be considered as negligibly small, but further investigation is needed to improve the agreement between simulation and experimental results by optimizing the used material model for ANSYS.

Acknowledgments

The authors gratefully acknowledge for the support of the Analytical Department of IFX. Further we address special thanks to the Material Group of IFX especially Markus Fink and Philipp Mayinger for developing a Matlab tool for the data evaluation out of DMA measurements, to Joachim Mahler for his support and advice and Michael Bauer for his help with the sample materials for the measurement.

A part of the measurement was performed by the Fraunhofer Institute IZM and AMIC GmbH. Thanks to Hans Walter and Jürgen Keller to perform the Poisson's ratio measurement.

The study was funded by the BMBF in the European project EsiP.

References

1. Wunderle, B, "Thermo-mechanical reliability of flip-chip assemblies with heat spreader", PhD-Thesis, TU Berlin, 2003

2. Silva, K; Handbook of Adhesion Technology, Springer book, 2011

3. Keller, J; Vogel, D.; Gollhard A., "Characterization of Microcracks by Application of Digital Image Correlation to SPM Images ", *Proc. SPIE 5392, Testing, Reliability, and Application of Micro- and Nano-Material Systems II, 2005*

4. Keller, J, "Micro- and nanoscale characterization of polymeric materials by means of digital image correlation techniquesr", PhD-Thesis, TU Cottbus, 2005

5. B.; Michel, B., "Lifetime Modeling for Microsystems Integration – from Nano to Systems." Journal of Microsystem Technologies, Vol. 15, (2009), pp. 799-813.

6. Jansen, K.M.B., "Thermo mechanical modeling and characterization of polymers", course book, TU Delft, 2007

7. de Vreugd, J, "The effect of aging on molding compound properties ", PhD-Thesis, Delft University of Technology

8. Jansen, K.M.B; Hawryluk, M; Gromala, P., "Cure Dependent Characterization of Molding Compounds", *Proc 13th EuroSimE international thermal, mechanical & multi-physics simulation, and experiments in microelectronics and microsystems,* Lisbon, 2011

9. Unterhofer, K; Preu, H; Walter, J; Lorenz, G; Mack, W; Petzold, M "Material characterization to model linear viscoelastic behavior of thin organic polymer films in microelectronics," *Proc 13th EuroSimE international thermal, mechanical & multi-physics simulation, and experiments in microelectronics and microsystems,* Lisbon, 2011

10. Mayinger, P; "Linear viskoelastische Modellierung von Duroplasten für die Verpackung elektronischer Baugruppen", Masterthesis. TU Munich, 2013

Improvement of Freestanding CMOS-MEMS
Through Detailed Stress Analysis in Metallic Layers

S. Orellana[*‡], B. Arrazat[†], P. Fornara[‡], C. Rivero[‡], A. Di Giacomo[‡], S. Blayac[†], K. Inal[*], P. Montmitonnet[*]
[*]Mines ParisTech, CEMEF, UMR CNRS 7635. 1, rue Claude Daunesse, CS 10207, 06904 Sophia Antipolis Cedex, France
[†]Ecole Nationale Supérieure des Mines de Saint-Etienne, CMP, 880, route de Mimet, 13541 Gardanne, France
[‡]STMicroelectronics, TR&D, 190, avenue Célestin Coq, 13106 Rousset Cedex, France
[*‡] sebastian.orellana@mines-paristech.fr

Abstract

A freestanding cross-shaped structure designed as a planar rotation stress sensor [1], [2], [3] is manufactured using standard CMOS technology (Complementary Metal-Oxide-Semiconductor). The fabrication process induces thermal residual stresses which result in out-of-plane bending, which degrades the device reliability and precision. To control such movements, the design was studied under stress compensation using a bilayered aluminum (Al) / titanium nitride (TiN) structure. Likewise, a single layer of aluminum was studied, to determine a technological solution, with better compatibility.

Fabrication stresses have been measured using Stoney's formula based on bending of full-wafer coatings. The Finite Element Method (FEM) is used to model the effect of these stresses on the geometry after release, and the results are compared with measurements. For this purpose, a comb-shaped structure has been designed to relate residual stress in a freestanding Al-TiN bi-layered structure with its bending. Based on this, conservation or elimination of TiN layer is judged, so that the design remains planar after release. The model is then applied to the movement of the cross-shaped sensor after release, and a second optimization variable is studied for maximum sensitivity: the shape of the hinge between the two arms of the cross.

1. Introduction

The thin film material properties are one of the key concerns in Micro-Electro-Mechanical System (MEMS) devices, playing an important role for their robustness. The focus of the present paper is on the residual stress in thin metallic films. On the one hand, manufacturing heat treatments induce mechanical stresses that could bring about mechanical failures; the residual stress state in freestanding CMOS-MEMS may also cause bending of structures after release [4], or increase the probability of stiction of moving parts. It is therefore important to estimate stresses, and measurement by sensors inserted in the interconnections has been proposed [1]. On the other hand, residual stresses can be the driving force for micro-actuators; this is the basis of the CMOS-MEMS concept [5], an incentive to create metallic free-standing, mobile parts in a standard CMOS process [6].

The previous efforts to create a cross-shaped stress sensor conducted by Horsfall, Kasbari and Vayrette ([1], [2], [3])resulted in operational devices. Initially, the sensors are encapsulated in dielectrics. As a result of the deposition process, and in spite of annealing, the arms, the pointer and the beam, have a residual stress state. At the release stage, the underlying oxide is removed and residual stress is relaxed in the arms; if*e.g.* the expansion arms of cross-shaped structure were in tension, they shorten at release, pulling asymmetrically the pointer which will rotate clockwise (see Fig. 9). The angle of rotation is proportional to the initial stress, which confirms the residual stress sensor functionality.

But the out of plane deformation observed during HF release of industrial Al / TiN bilayers is still to be studied. The cross-shaped structure is not the most suitable design for measuring the deformation out of the plane. Also, the reliability of structure is a point to improve, finding the best sensitivity/reliability ratio.

This is the purpose of the present study, which consists of two stages: first, measure residual stresses in layered mobile structures; second, exploit the stress to manufacture mobile structure, measure their movement and relate it quantitatively with the state of stress. The FEM Comsol Multiphysics is used for this purpose.

The presentation of structures, their design, fabrication and characterization are presented in section 2. The configuration of FEM tool is described separately in subsection 2-D. The result section is split in two subsections to allow description of the two designs studied, both measurement and simulation.

2. Experimental

A. *Design and purpose of devices*

In order to study the residual stress stored in the metallic layer after their fabrication process, two structures were designed.

First, to measure the effects of residual stress in a composite metallic bilayer, mainly bending, we have designed the comb-shaped structure of Fig. 1a. It consists in a series of micro-beams, $1\mu m$ in width, 0.25 to $30\mu m$ in length, spaced by $1\mu m$ and 515 or $470nm$ in thickness. The materials of the bilayer have opposite stress states, tensile (+) for Aluminum (Al) [7], compressive (–) for titanium nitride (TiN) [8]: this generates bending in a freestanding film. The difference in stress state between Al and TiN bends the micro-beams downward. Measuring this curvature by interferometric profilometry will allow designing (with the help of FEM) the device such that stress compensation leads to almost straight combs.

For a better study of metallic film residual stress state, the cross-shaped structure (Fig. 1b) has been optimized. After ensuring that the structure is perfectly straight in the plane, a new design has been implemented. The hinge geometry has been varied, in order to improve the sensitivity/reliability of displacement in rotational applications. The Design of Experiment (DoE) is shown in Fig. 2. It is expected that the smaller the hinge, the more the pointer will rotate and thus, the higher the sensor sensitivity; on the other hand, it might make the sensor more fragile, less reliable. The compromise solution has to be studied.

A second improvement of the design consists in introducing a ruler on the left and right pads, to give more precision in the measurement of the rotational displacement (Fig. 1b)

technologies, located over the transistors and active components, serving the interconnections between them. In the silicon microelectronics industry, specifically in the BEOL part, the materials available are restricted to silicon, copper, aluminum, titanium, tungsten and their oxides and nitrides.

The metallic tri-layer Titanium (Ti) 7.5nm / Aluminum (AlCu) 470nm / Titanium Nitride 45nm is deposited by Physical Vapor Deposition (PVD) over Silicon Wafer. In fact, "titanium nitride" is a composition gradient layer, starting with pure Ti for adhesion purpose with progressively growing nitrogen proportion. It will however be called TiN in the following sections for simplicity. Similarly, as the Ti layer is 100 times thinner than Al layer, the tri-layer will be considered as bilayer Al/TiN. Then, 1.6μm of Silicon dioxide (SiO_2) is deposited by Chemical Vapor Deposition (CVD) using Tetraethyl-orthosilicate, TEOS. Then, the same metallic tri-layer is deposited on top (Fig. 3). Finally, photolithography is used to pattern parts of a thin film of metal. The last stage is the release of the structure, whereby the oxide is removed by an isotropic HF-Vapor etching during 7 min at 40°C.

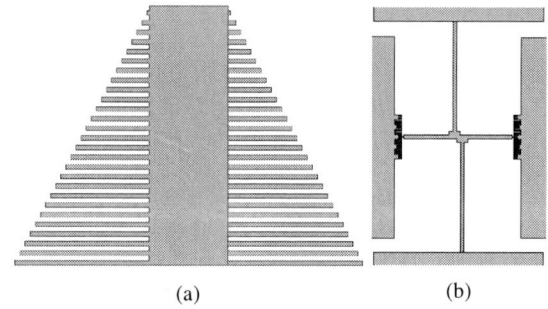

Figure 1: The two designs studied. a) The comb-shaped structure: out-of-plane-curvature measurement tool in a bi-metallic layer, b) Improved design of cross stress sensor for in-plane rotation study.

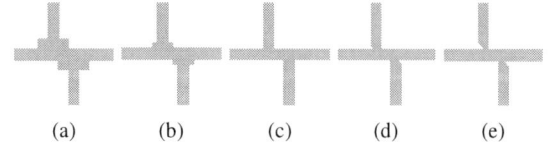

Figure 2: Different pivot designs, varying the hinge width: a) 3μm, b) 2μm, c) 1μm, d) 0.6μm, e) 0.3μm.

B. Device fabrication

Structures fabrication follows the CMOS process flow and the photolithography process, which consists in coating and patterning a specific material. The structures are placed in the Back End of Line (BEOL) of CMOS

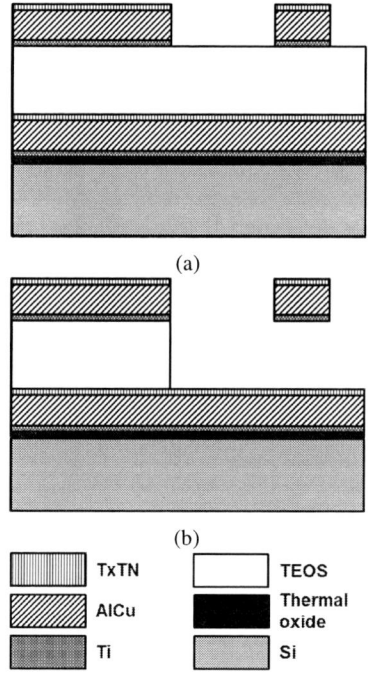

Figure 3: Technological stack used, a) before and b) after HF-Vapor etch of SiO_2.

C. Measurement procedure

The residual stresses (S_{film}) under the conditions of devices fabrication were deduced by measuring bending after deposition of each layer on full wafers. Using Stoney's formula with equi-biaxial stress assumption, a tension of about $+290MPa$ was found in Al and a compression of about $-1GPa$ in TiN.

After release of the cross-shaped structures, the pointer deviation is measured by means of Scanning Electron Microscope (SEM) observation. In the case of the comb-shaped structure, SEM observations are used to check complete etching of underlying dioxide. Released structures are observed in a SEM ultra 55 (Carl Zeiss), at a tension of $3kV$ to avoid charging electrically the metal surface, in the Secondary Electron (SE) mode to enhance relief.

The residual stress relaxation induces a bending which is measured by means of optical profilometer (Veeco-NT1100), in the Vertical Scanning Interferometry (VSI) mode. VSI mode creates a topographic map by analyzing interference between reflected light from the sample and a mirror at a known distance.

D. Configuring the FEM tool

The complete 3D structure is introduced in COMSOL Multiphysics $V4.3b$ FEM software. Structures bend (comb-shaped) or turn (cross-shaped) when their residual stress is relaxed as the release operation etches SiO_2 from under metallic pattern. The 3D steady-state simulations are performed by the "structural mechanics" option of COMSOL Multiphysics, using isotropic elastic behavior.

A linear elastic and isotropic Aluminum material chart was obtained from COMSOL materials library; Young's modulus is $70GPa$), Poisson's coefficient is 0.35. The TiN material had to be defined specifically, with Young's modulus $500GPa$ and Poisson's coefficient 0.25. In spite of the composition gradient, a homogeneous material has been assumed here.

The formulation used for structural analysis in COMSOL Multiphysics is a total Lagrangian one, referred to the material (initial) configuration. This means that the second Piola-Kirchhoff stress tensor is used together with the Green-Lagrange strain tensor:

$$\varepsilon = \frac{1}{2}[(\nabla \mathbf{u})^{\mathbf{T}} + \nabla \mathbf{u} + (\nabla \mathbf{u})^{\mathbf{T}} \cdot \nabla \mathbf{u}]$$

The Duhamel-Hooke's law relates the stress tensor to the strain tensor:

$$S - \mathbf{S_0} = \mathbf{C} : (\varepsilon - \varepsilon_0)$$

where $\mathbf{C}$ is the 4th order elasticity tensor, $\mathbf{S_0}$ and ε_0 are the initial stress and strain. For the boundary conditions, a fixed constraint ($\mathbf{u} = \mathbf{0}$) has been defined at the origin of arms, next to pads for cross-shaped structure, and in the residual oxide of central pad for comb-shaped structure.

Adaptive mesh option was activated in the FEM software, changing from 7835 to 197242 tetrahedral elements (3D) after 3 refinement steps (Fig. 4). The TiN layer, 10 times thinner than Al, is meshed in the "swept mode":

it copies the surface mesh pattern of the aluminum layer, creating triangular prism elements.

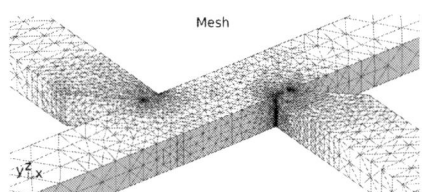

Figure 4: Adaptive tetrahedral mesh, refined in the hinge vicinity.

3. Results and Discussion

The two structures studied are presented through an experimental characterization and FEM analysis.

A. Bending of comb-shaped structures

1) Experimental: For a single Al layer, SEM images in tilted view (Fig. 5) show that the entire beams are correctly released. Prongs bend when their residual stress is relaxed as the release operation etches SiO_2 from under micro beams.

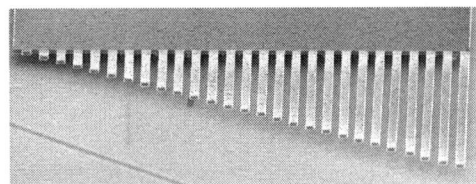

Figure 5: SEM images of Al combs after release.

As the SEM observation gives only qualitative information on the released shape, optical profilometry has been used to quantify the curvature of microbeams in the out of plane direction, as maps (Fig. 6) and as extracted profile (Fig. 7). Fig. 7a gives profiles of pure Al prongs, showing that 100% remain suspended; they even slightly bend upwards by $\approx 1\mu m$.

Moreover, all microbeams with Al/TiN layer are affected by negative (downward) bending (Fig. 7b). The movement increases with the microbeam length and reached a maximum displacement of $\approx 0.4\mu m$.

2) FEM analysis of combs structures: The displacement during the release stage has been modelled, using initial stress ($\mathbf{S_0}$) proportional to those measured on full wafers, *i.e.* $+290MPa$ for Al and $-1GPa$ for TiN. The stress in a line has been defined by Maniguet [9] as $(1 - \nu) \cdot S_{film}$, where ν is Poisson's coefficient and S_{film} is the biaxial stress of film.

When the constraint by SiO_2 is released, Al prongs bend upwards slightly ($0.1\mu m$). It is much smaller than the

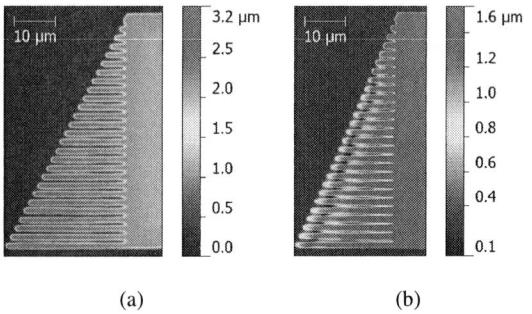

(a)   (b)

Figure 6: Profilometry image of a) Al layer and b) Al/TiN bilayer.

experimental measurement, which suggests that a stress gradient is present across the Al thickness.

As expected, addition of the TiN layer leads to downward bending, by *ca.* 1.5*μm*, which is somewhat larger than measurement, probably for the same reason.

Another reason might be that the boundary condition $\mathbf{u} = \mathbf{0}$ is imposed on the edge of the central pad. Yet oxide is etched away isotropically also from under the pad edges and extremities over a certain distance. The rigidity of the pad is therefore not infinite, and it can contribute to the out-of-plane displacement. In particular, the extremity of the pad supporting the longest prongs is itself a free cantilever; these prongs are not rigidly supported. When this is taken into account in the simulation, their bending is roughly doubled, closer to the measurement.

One can see clearly the influence of the TiN layer: thanks to its compressive stress, it compensates the tensile stress in aluminum.

The out-of-plane-curvature measured on comb-shaped structure will be used in the next section. It will be examined in particular if TiN is needed against bending or not, and what is the influence of its thickness on the device geometry. Then the effect of the hinge geometry will be addressed.

B. Optimization of stress sensor by design

1) Experimental: It has first been verified that Al (without TiN) cross-shaped structures have been released and remain suspended (Fig. 9).

Pointer deviation has been measured using SEM pictures, Fig. 10, thanks to the 'ruler' introduced in front of the arms tips.

The pointer deviation is drawn as a function of the hinge shape in Fig. 11 (blue triangles). It was considered a measurement error of 10%. The pink dots correspond to the simulation data, explained in the FEM analysis section.

As expected, keeping the same residual stress state, the smallest hinge gives the largest rotation. Furthermore, the rotation becomes smaller as the hinge stiffness increases.

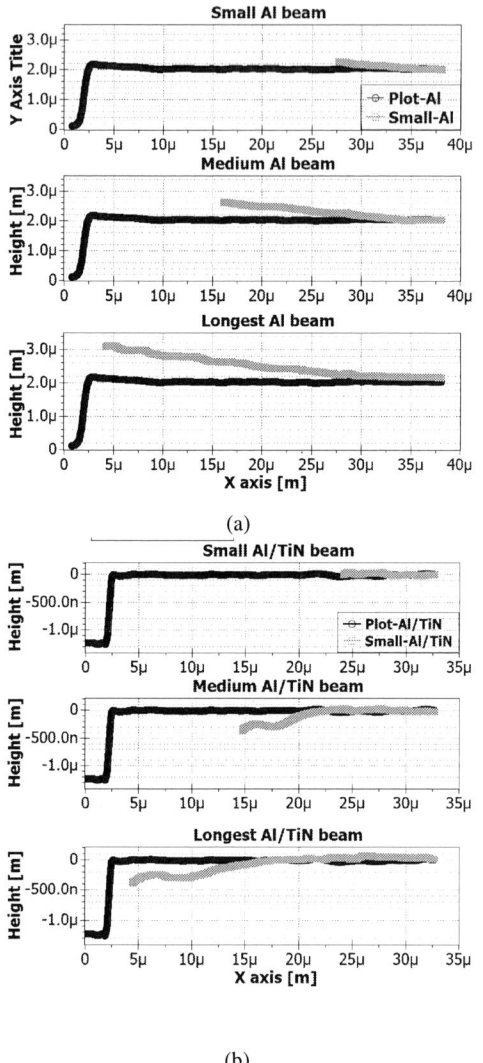

Figure 7: Experimental profiles of combs. Pink lines are taken along prongs; black lines are in the perpendicular direction, along the central pad. a) Al combs, b) Al/TiN combs.

As calculated by Vayrette's [3], rotations agree with experimental measurements.

The same structures with an Al/TiN layer were measured; the rotation angle decreases by 40 − 50% compared with Al layer. This difference is explained by Young's modulus of TiN, seven times stiffer ($\sim 500GPa$) than Aluminum ($70GPa$). Furthermore, contrary to the comb case, the Al / TiN bilayer induces bending of the mobile part, which is not present with the Al-only cross. This is modelled in next subsection

2) FEM analysis of cross-shaped stress sensor: In the FEM study, the residual stress is supposed uniaxial and

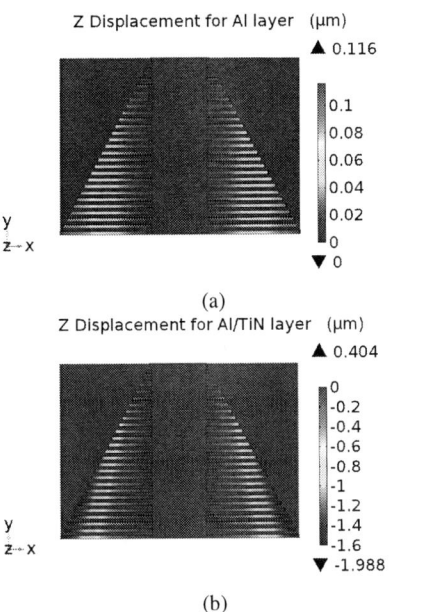

(a)

(b)

Figure 8: FEM Simulation of bending at release on the comb-shaped devices, a) Al layer and b) Al/TiN bilayer.

constant along the longitudinal direction of expansion arms. The simulation was performed, for all hinge shapes (see Fig. 11). Their results are close to experimental values, especially for largest hinge.

Fig. 12 shows the downward movement of the end of the mobile part for different TiN thickness, for the thinnest hinge. The only measured value (blue triangle) agrees with the simulations.

The out-of-plane displacement decreases with reduction of TiN thickness, canceling the 'z' displacement when TiN is completely removed. Furthermore, TiN stiffens the structure, limiting its rotation and sensitivity.

Relaxation of residual stress generates, through the elongation of the arms, the rotation of the mobile part. The simulation, as expected, obtains the displacement measured of the rotation (see figure 11).

Moreover, for the 'b5' hinge, Fig. 13 shows that a certain level of residual stress remains after release and rotation, which can be explained by a partial hindrance of the motion of the arms. Such a stress is not observed, after their release, in the completely free microbeams of comb-shaped structures. This stress map can be related to the experimental shape observation of Fig. 10. Bending is observed on horizontal arms, with maximum curvature (and stress) around $5\mu m$ from the hinge. $1\mu m$ away from the hinge, Fig. 13b shows a bending stress pattern, with surface stress $\approx \pm 180 MPa$; near the anchor pad (Fig. 13c), the S_{yy} stress component has fallen to half that value. At the very hinge, a very strong stress ($\pm 500 MPa$) testifies for the tendency to "close" the angle formed by the hinge

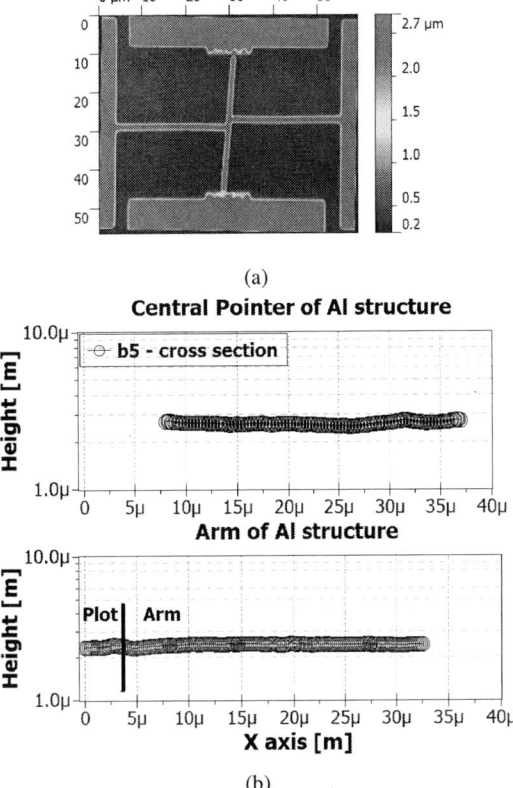

(a)

(b)

Figure 9: Al Cross-shaped, a) profilometry image, b) experimental profiles of Central pointer and Arm.

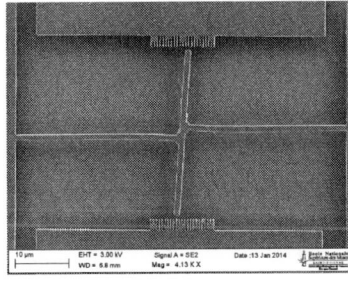

Figure 10: Cross-shaped stress sensor, rotation measured at the end of mobile part with this a 'ruler'.

(rotation in the sense shown by Fig. 10).

4. Conclusions

In order to optimize the cross-shaped stress sensor, a beam arrangement was designed to study the out-of-plane displacement. Comparison of Al and Al / TiN designs shows in the end that to reduce bending, using the materials and fabrication process available, the single Al layer is the most suitable for the mobile device in the

978-1-4799-4789-8/14 $31.00 © 2014 IEEE   411

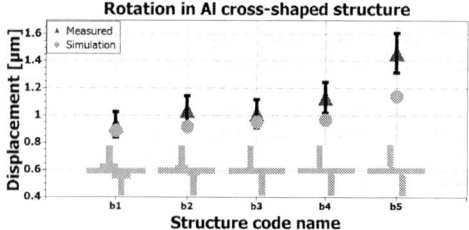

Figure 11: Rotation measured in the 5 hinge designs.

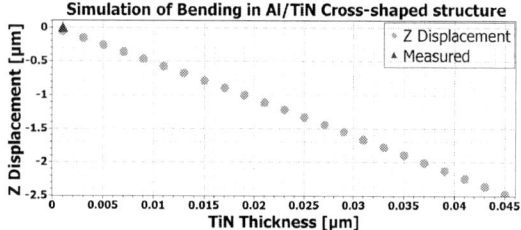

Figure 12: Numerical study of the influence of the TiN thickness over Al layer on bending of the cross-shaped structure.

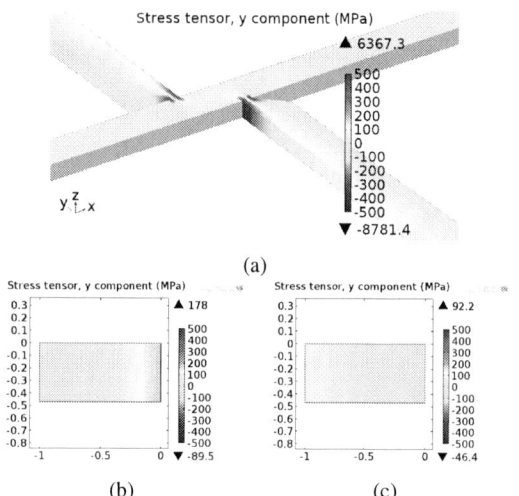

Figure 13: The residual stress in the 'y' axis - after release. Images; a) in the hinge vicinity, and cross section of arm b) $1\mu m$ from the hinge and c) $29\mu m$ from the hinge, near the anchor pad.

BEOL investigated here.

The technological modification was confirmed by implementation on cross-shaped structure, avoiding out-of-plane deformation.

A design work has also been developed, changing the cross hinge shape to increase the reliability of freestanding

applications. For the most flexible hinge, the displacement is maximal as expected, but the risk of material deformation is to be studied. That is why the 'b5' design, the smallest hinge, is more suitable for passive applications.

The design will be improved further in future work, *e.g.* by varying the separation distance of arms. Using simulation, the stress gradient should be studied to ensure correct interpretation of the experimental measurements. Another study could consist in relating this refined assessment of stress to grain microstructure orientation.

Finally, other applications of measurement can be developed with this passive sensor, *e.g.* thermo-mechanical properties [10].

5. Acknowledgements

The authors would like to thanks to Carole Vincent from STMicroelectronics, Rousset and Thierry Camilloni from CMP, Gardanne, for their support in HF release process.

SEM images and profilometry image were taken using equipment at CMP, Gardanne.

References

[1] A.B. Horsfall, J.M.M. dos Santos, S.M. Soare, N.G. Wright, A.G. O'Neill, S.J. Bull, A.J. Walton, A.M. Gundlach, and J.T.M. Stevenson. A novel sensor for the direct measurement of process induced residual stress in interconnects. In *33rd Conference on European Solid-State Device Research*, pages 115–118, Estoril, Portugal, 2003. IEEE.

[2] Moustafa Kasbari, Christian Rivero, Sylvain Blayac, Florian Cacho, Ola Bostrom, and Roland Fortunier. Direct Local Strain Measurement In Damascene Interconnects. *MRS Proceedings*, 990:0990–B07–06, February 2007.

[3] R. Vayrette, C. Rivero, B. Gros, S. Blayac, and K. Inal. Residual stress estimation in damascene copper interconnects using embedded sensors. *Microelectronic Engineering*, 87(3):412–415, March 2010.

[4] Daniel Ramos, Johann Mertens, Montserrat Calleja, and Javier Tamayo. Study of the origin of bending induced by bimetallic effect on microcantilever. *Sensors*, 7(9):1757–1765, September 2007.

[5] Raafat R. Mansour. RF MEMS-CMOS Device Integration: An Overview of the Potential for RF Researchers. *IEEE Microwave Magazine*, 14(1):39–56, January 2013.

[6] V. Burg, J. den Toonder, A. van Dijken, J. Hoefnagels, and M. Geers. Characterization Method for Mechanical Properties of Thin Freestanding Metal Films for RF-MEMS. In *7th. Int. Conf. on Thermal, Mechanical and Multiphysics Simulation and Experiments in Micro-Electronics and Micro-Systems*, pages 1–7, Como, Italy, 2006. IEEE.

[7] AB Horsfall, Kai Wang, J.M.M. Dos-Santos, S.M. Soare, S.J. Bull, N.G. Wright, A.G. O'Neill, J.G. Terry, A.J. Walton, A.M. Gundlach, and J.T.M. Stevenson. Dependence of Process Parameters on Stress Generation in Aluminum Thin Films. *IEEE Transactions on Device and Materials Reliability*, 4(3):482–487, September 2004.

[8] Tatsuya Matsue, Takao Hanabusa, and Yasukazu Ikeuchi. Residual stress and its thermal relaxation in TiN films. *Thin Solid Films*, 281-282:344–347, 1996.

[9] L Maniguet, M. Ignat, M. Dupeux, J.J. Bacmann, and Ph. Normandon. X-ray Determination and Analysis of Residual Stresses in Uniform Films and Patterned Lines of Tungsten. *MRS Proceedings*, 308:285, February 1993.

[10] B. Arrazat, S. Orellana, C. Rivero, P. Fornara, A. Di Giacomo, and K. Inal. From stress sensor towards back end of line embedded thermo-mechanical sensor. *Microelectronic Engineering. Unpublished*, December 2013.

978-1-4799-4789-8/14 $31.00 © 2014 IEEE

Electronic Control Package Model Calibration using Moiré Interferometry

Dae-Suk Kim[a], Bongtae Han[a], Arjun Yadur[b] and Przemyslaw Jakub Gromala[c]

[a]Mechanical Engineering Department, University of Maryland
College Park, MD 20742, USA
[b]Robert Bosch Engineering and Business Solutions Limited
Bangalore, 560095, India
[c]Robert Bosch GmbH, (AE/EDT3)
Reutlingen, 72703, Germany

ABSTRACT

Moiré interferometry is employed to test electronic control units, which are developed for automotive application utilizing transfer molding technology. The control units are subjected to a thermal cycle, and the two orthogonal in-plane displacement fields of the package cross-section are obtained at various temperatures. The results are used to verify and calibrate the complex 3-D finite element model of the units. The detailed experimental procedures including sample preparation are described. The modeling steps that lead to the calibrated finite element model are presented while comparing the experimental results with the numerical predictions.

1. INTRODUCTION

Electronic control units (ECU) were developed for automotive application utilizing transfer molding technology to endure harsh operating environmental conditions such as contaminants and moisture. It has been known that the interface between an aluminum bond and a copper lead inside a standard DPAK package in molded ECU is most vulnerable to the stresses caused by thermal cycles.

Numerical modeling can be used to analyze the reliability issues. However, the verification process is inevitable considering the complexity of the package as well as the nonlinear mechanical behavior of the key materials used in the package, including EMC and solders.

Moiré interferometry has been used effectively to verify numerical models due to its high sensitivity and high signal-to-ratio [1, 2]. This study focuses on verification and calibration of complex 3-D finite element models through moiré interferometry. The modeling steps that lead to the calibrated finite element model are presented after the experimental procedures are described.

2. ECU AND DPAK PACKAGE CONFIGURATION

Fig. 1a shows a molded ECU package. Red circles in **Fig. 1b** indicate DPAK packages investigated in this study. The inside DPAK packages are shown in **Fig. 2a**. Al bonds have diameters of 0.125 mm and 0.2 mm. After combined passive (-40 to 105 °C) and active (-40 to 85 °C) thermal cycles, it was observed that the Al bond was separated from the pads. The lifted Al bond after pull test is shown in **Fig. 2b**.

a)

b)

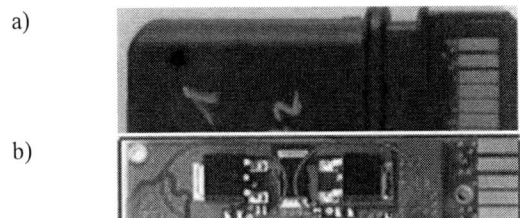

Fig. 1 (a) Molded and (b) unmolded ECU package

a) 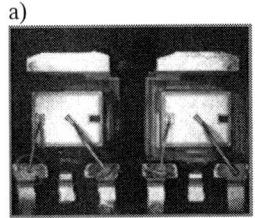   b)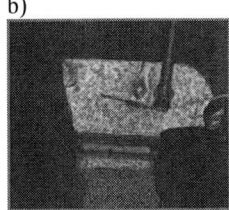

Fig. 2 (a) Inside DPAK packages and (b) the lifted Al bond after pull test

3. DESCRIPTION OF FEM MODEL

A numerical model simulating a moiré specimen was built using a commercially available FEA package (ANSYS workbench 14.5). **Fig. 3a** shows the model, where the critical side of interest containing Al bonds and Cu leads was exposed. The internal structures of the ECU package are shown in **Fig. 3b**.

a) 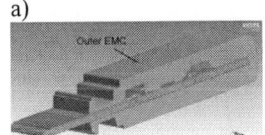   b)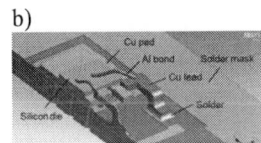

Fig. 3 3D model of moiré specimen (a) with and (b) without mold compund

The linear and nonlinear material properties used in the initial modelling are summarized in Table 1 and 2. All material properties were measured in-house using dynamic mechanical analyzer (DMA) and thermo-mechanical analyzer (TMA).

The FEM mesh of the 3-D model is shown in **Fig. 4**. Tetrahedron shape of mesh is dominant on the entire 3-D model because of the curved Al bonds and Cu vias. Contacts are applied between the surface of the unmolded ECU package and the surface of the inside outer EMC.

978-1-4799-4789-8/14 $31.00 © 2014 IEEE   413

Table 1 Material properties of chip and metals

Material	Young's modulus [MPa]	CTE [ppm/K]
Silicon die	168900	2.8e-6
Copper	125000	1,7E-05
Solder	32000	2,0E-05
Al bond	64000	25.3e-6
Ceramic	300000	8.0e-6

Table 2 Material properties of polymeric materials

Material	Young's modulus [MPa]	CTE [ppm/K] bellow T_g	CTE [ppm/K] above T_g	T_g [°C]
Prepreg	4700	X: 14E-6 Z: 16E-6 Y: 52E-6	X: 40E-6 Z: 40E-6 Y: 247E-6	149
Solder mask	4000	74E-6	140E-6	100
Outer EMC	25000	10E-6	39E-6	129
DPAK mold	16600	12E-6	44E-6	129

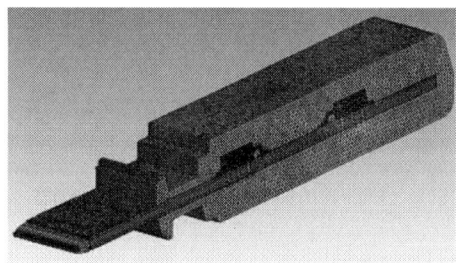

Fig. 4 FEM model of the package

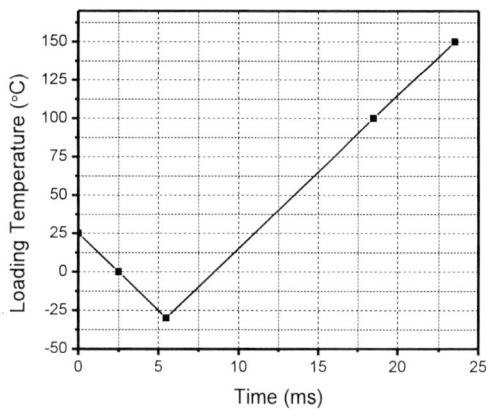

Fig. 5 Temperature loading conditions

The traction-free boundary condition was applied along the exposed surface (**Fig. 3a**). Uniform temperatures of 0, -30, 100, and 150 °C were applied to all nodes at a ramp of 10 °C/min (**Fig. 5**).

4. EXPERIMENTAL DATA OBTAINED FROM MOIRÉ INTERFEROMETRY

4.1. Experimental Technique: Real-Time Moiré interferometry

Moiré interferometry has been used extensively for thermal deformation analyses of electronic packaging and subassemblies. The principles of the technique are described in [3] and various applications can be found in [4, 5].

The real-time moiré setup used in this study is illustrated in **Fig. 6** [6, 7]. The specimen holder is not attached to the chamber. Instead, it is connected rigidly to the interferometer and it is essentially free from contact with the environmental chamber. Furthermore, the interferometer and the chamber are mounted on separate tables and thus the interferometer is mechanically isolated from the chamber. With this arrangement, the chamber vibrations are not transmitted to the specimen, and the moiré fringes can be documented while the chamber is being operated. Further details of the rod assembly and the temperature control can be found in Ref. [7].

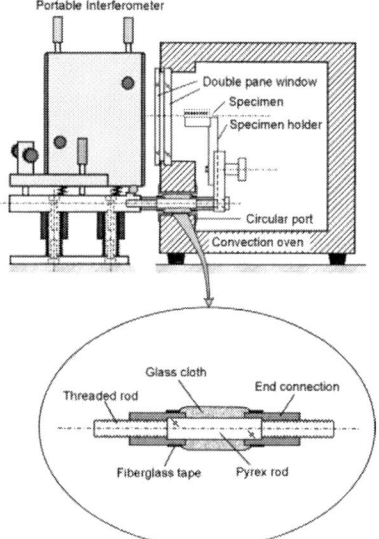

Fig. 6 Schematic of the moiré setup for real-time observation of thermal deformation

4.2. Specimen Preparation

The packages were ground to expose Al bond, Cu leads, chip, and DPAK mold. **Fig. 7a** shows the cross-section of the molded ECU package. The drag method was used to replicate the diffraction gratings (1200 lines per mm) on the cross section (**Fig. 7b**).

The specimen was then placed inside the thermal chamber and thermally-induced in-plane displacements of the ECU were measured at 0, -30, 100, and 150 °C.

a)

b)

Fig. 7 Cross section of (a) mold ECU package, (b) the replicated grating on mold ECU package

4.3. Experimental Results

Two zoom lenses were used to capture the whole field (**Fig. 8a**) and the zoomed-in field (**Fig. 8b**); the boundaries of the unmolded unit were added to the fringe patterns as references.

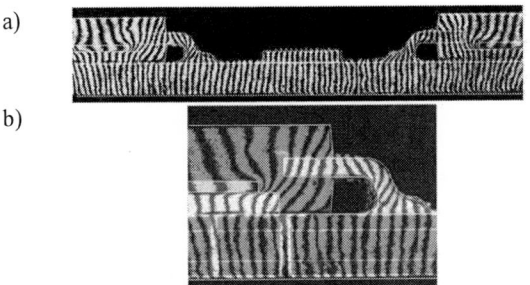

a)

b)

Fig. 8 U-field displacement contours of the unmolded EPM package obtained at 100 °C; (a) the whole field and (b) zoomed-in field.

Moiré interferometry fringes were analyzed to understand the effect of outer EMC. The phase shifting technique was implemented to obtain fractional fringe orders at each pixel and they were used to plot deformed configurations. The deformed configurations are shown in **Fig. 9**, where the relative deformation was magnified by 10 times. The red outlines and the yellow outlines indicate the geometries before and after deformation, respectively. The unmolded ECU produced larger bending deformation in the copper lead than the molded ECU.

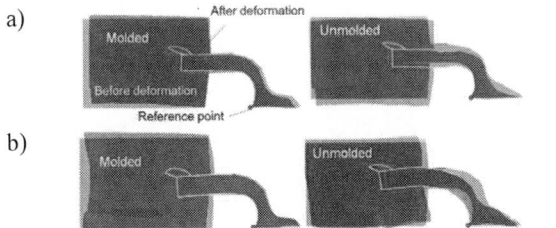

a)

b)

Fig. 9 Deformed configurations at (a) -30 °C and (b) 150 °C, obtained using phase shifting technique

The U displacements obtained at -30 °C and 150° C are shown in **Fig. 10** and **Fig. 11**. An interesting deformation was observed at 150 °C. The area marked "yellow" in the molded ECU (i.e., the DPAK mold) shows a deformation mode which is completely different from that of the unmolded ECU in **Fig. 10**. It is speculated that the DPAK mold was constrained by the chip more significantly at the high temperature due to its low modulus. The different deformation mode was not observed at the low temperature (**Fig. 11**) because of a high modulus of DPAK mold at the low temperature.

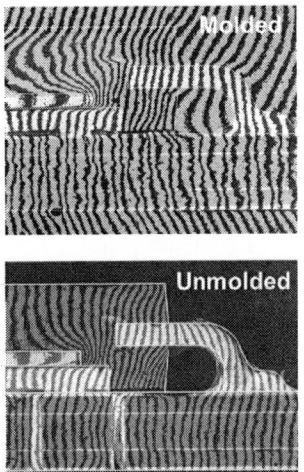

Fig. 10 U-field of the deformation of DPAK at 150 °C

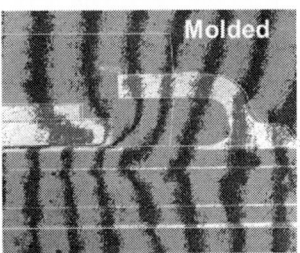

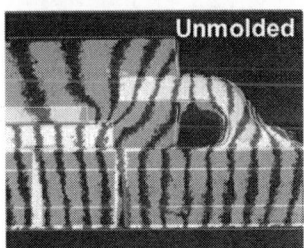

Fig. 11 U-field of the deformation of DPAK at -30 °C

The net bending displacements along the flat part of the copper lead were determined to show the deformation mode more clearly in **Fig. 12**. Displacements along AA' (the flat part of the copper lead) caused by heating from -30 °C to 150 °C were determined by subtracting the displacements at −30 °C from the displacements at 150 °C. For the molded ECU, the outer EMC constrained the DPAK mold and thus the net bending displacements along the flat part of the copper lead virtually diminished.

a) 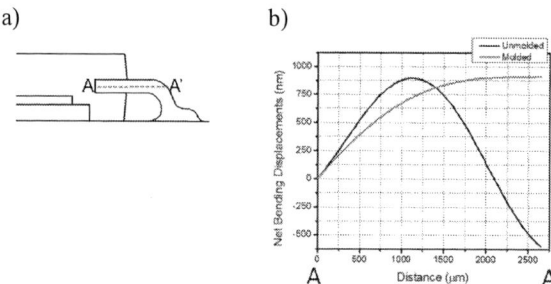 b)

Fig. 12 Deformed configurations of AA' caused by heating the packages from -30 °C to 150 °C.

5. VALIDATION OF NUMERICAL MODEL

The FEM model was validated using moiré displacement fields. **Fig. 13** shows an example, where the displacement contour interval of the FEM result was set equal to the moiré contour interval (417 nm/fringe order) for direct comparison. Correlation between the modeling and experimental results was used to verify the material properties used in the model.

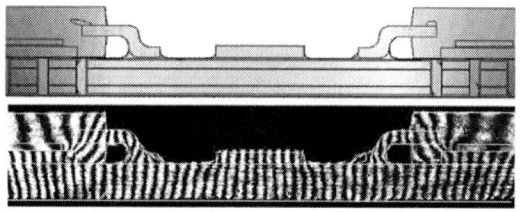

Fig. 13 Global U displacement field of the unmolded ECU obtained from the FEM (top) is compared with moiré fringes; the ECU is cooled from room to -30 °C.

Although the global fields show reasonable agreement, the zoomed-in fringe patterns show significant differences. An example of a zoomed-in fringe pattern is shown in **Fig. 14**, where the semi-transparent cross-section images are superimposed on the fringe patterns to assure accurate positioning of the regions of interest.

The comparison indicates that the U and V displacements of prepreg are approximately 10% larger than moiré displacement below T_g (circled areas in **Fig. 14**), while the V displacement of DPAK mold are 30% smaller than moiré displacement above T_g. In addition, the U and V displacements of the outer EMC are 30% larger than moiré displacement below T_g.

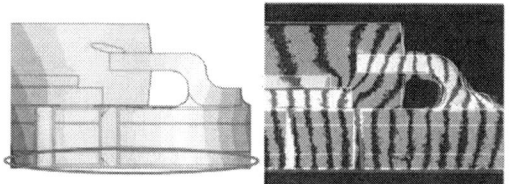

Fig. 14 Zoomed-in view of U displacement field obtained from the FEM (left) is compared with moiré fringes; the unmolded ECU is cooled from room to -30 °C.

To obtain the enhanced correlation, the CTE values were adjusted continuously until the difference of the

displacements between the FEM and moiré fringes became minimal. The CTE of prepreg was reduced by 10% below and above T_g in the X, Z, and Y direction. The CTE of the outer EMC was reduced by 28% below T_g and by 26% above T_g. Finally, the CTE of DPAK mold was increased by 10% above T_g.

a)
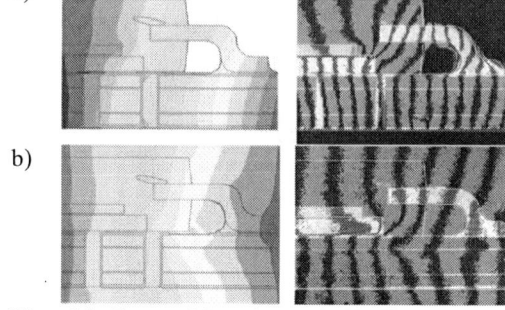
b)

Fig. 15 Zoomed-in view of U displacement field obtained from the FEM (left) after calibration; both (a) unmolded and (b) molded ECU package were cooled from room to -30 °C.

a)
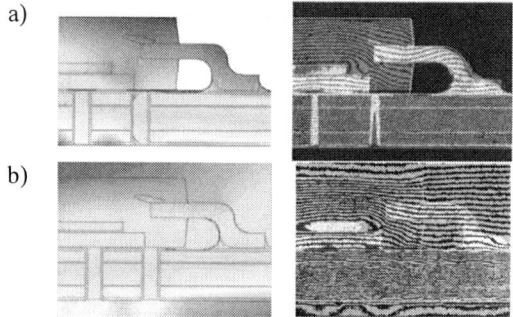
b)

Fig. 16 Zoomed-in view of V displacement field obtained from the FEM (left) after calibration; both (a) unmolded and (b) molded ECU package were heated from room to 150 °C.

The zoomed-in displacement obtained from the calibrated FEM results are shown in **Fig. 15** and **Fig. 16** for -30 °C and 150 °C, respectively. The difference between the experiment and numerical results is less than one micron. The calibrated FEM model will be further utilized to investigate the effect of the outer EMC on the critical interface of the Al bonds.

6. CONCLUSION

Deformations of electronic control units subjected to a thermal cycle were documented by moiré interferometry. The experimental results helped understand the package behavior subjected to the extreme thermal cycle condition. The results were further utilized to calibrate the 3-D FEM model. With the high sensitivity and high quality displacement fields provided by moiré interferometry, correlation within one micron was achieved at extreme temperatures (-30 °C and 150 °C). The calibrated model will be implemented extensively to investigate the effect of the outer EMC on the critical interface of the Al bonds and the results will be reported in the future publication.

978-1-4799-4789-8/14 $31.00 © 2014 IEEE   416

REFERENCES

[1] M. N. Variyam and B. Han, "DMDTM package model calibration using interferometry," *Texas Instruments Technical Journal,* (2001), pp. 1-9.

[2] B. Han, *et al.*, "Verification of numerical models used in microelectronics packaging design by interferometric displacement measurement methods," *Journal of Electronic Packaging,* Vol. 118, (1996), pp. 157-163.

[3] D. Post, *et al.*, High Sensitivity Moiré: Experimental Analysis for Mechanics and Materials, Springer-Verlag (New York, 1994)

[4] B. Han, "Thermal stresses in microelectronics subassemblies: Quantitative characterization using photomechanics methods," *Journal of Thermal Stresses,* Vol. 26, (2003), pp. 583-613.

[5] P. G. Ifju and B. Han, "Recent Applications of Moire Interferometry," *Experimental Mechanics,* Vol. 50, (2010), pp. 1129-1147.

[6] S. M. Cho, *et al.*, "Temperature dependent deformation analysis of ceramic ball grid array package assembly under accelerated thermal cycling condition," *Journal of Electronic Packaging,* Vol. 126, (2004), pp. 41-47.

[7] S. Cho and B. Han, "Observing real-time thermal deformations in electronic packaging," *Experimental Techniques,* Vol. 26, (2002), pp. 25-29.

A New Life-Test Equipment Designed for Medium-Duty Electromagnetic Contactors

Serkan BIYIK[1], Murat AYDIN[2]
Karadeniz Technical University
[1]Department of Metallurgical and Materials Engineering, 61080, Trabzon, Turkey
Tel: +90 462-377 36 41, E-mail: serkanbiyik@ktu.edu.tr
[2]Department of Mechanical Engineering, 61080, Trabzon, Turkey
Tel: +90 462-377 41 31, E-mail: maydin@ktu.edu.tr

Abstract

In this study, a new test equipment, which is capable of performing arc-erosion experiments of electrical contacts on a medium-duty contactor, was designed and manufactured. The current load value of the system was gradually adjustable up to 30 A via control buttons located on the device. The make/break counts (cycle) and delay times of electrical contacts were adjusted by an electronic flasher. The number of operations was readily adjusted to desired counts before tests, and was monitored along the whole test on the LCD display of up/down counter. Inductive loads consisting transformers were used to achieve real-life operating conditions. A contactor having pole number of 3, switching power of 37 kW, and switching current of 75 A was used to conduct arc-erosion experiments. The test conditions for this contactor were 220 V operating voltage, 50 Hz frequency, and 20 A switching current. The switching frequency and delay time were selected as 1,000 times per hour and 2.6 s, respectively. Arc-erosion experiments were carried out by using the contactor up to 40,000 operations. The contact surfaces were analyzed by using scanning electron microscopy. The chemical compositions nearby arc were determined by energy dispersive x-ray spectroscopy. The experimental results showed that the contact surfaces were reasonably affected by the arc-erosion, which leads to mass loss by material migration and/or evaporation. Arc-affected zones were enlarged with the increasing operation number. It was concluded that the stationary contacts underwent major erosion.

1. Introduction

Electromagnetic switches such as relays and contactors may be used for light, medium or heavy-duty applications depending upon their operating conditions [1]. One of the most important components of electromagnetic switches is electrical contact. Electrical contacts are responsible for controlling single or multiple electrical circuits via making or breaking them. The materials used to produce electrical contacts should have certain characteristics, such as good combination of electrical and thermal conductivities, wear performance, and resistance to erosion and welding. Otherwise, the contact erosion which is defined as the material loss at the contact surfaces, for example, due to material evaporation by an arc, will occur [2]. Below are required to minimize these unfavorable interactions, improve contact life and reduce the contact erosion to minimal level;

- high electrical conductivity to ease current flow,
- high thermal conductivity to dissipate arc heat away from contacts,
- high resistance to deteriorating effects of oxide, sulphide, and other compounds which cause insulation on surfaces,
- high melting point and low vapor pressure to limit arc erosion, material transfer and welding,
- high hardness to resist wear [3, 4].

In addition, the selection of contact materials should be optimized to provide acceptable safety, reliability, contact life and cost. The proper choice of a contact material for a specific application also depends on the characteristics of electrical circuit (current, voltage etc.), capacity, contact forces, number and frequency of make/break cycles. Furthermore, the type of supply (AC/DC), level (low/intermediate/high) and type of load (inductive, capacitive, resistive or a motor) play an important role in the selection of contact materials [5]. For example, the contact surfaces deteriorate with increasing operations due to the current flow and consequently the failure of the contact system is inevitable. Therefore, according to the type of application, examining the make/break operations of electrical contacts and controlling the formation of arc which damages the surfaces of both stationary and movable contacts are so important in an attempt to improve a new material for electrical contacts. For this reason, the properties of contact materials mentioned above should be examined and then the compatibility of the materials should be determined for any electrical application. On the other hand, there is not an ideal contact material which is capable of being used for all kind of applications and having all the features mentioned above [4]. Therefore, two or more materials having different properties should be combined to obtain a new material which has superior properties. Much work has previously been done in production and evaluation of various contact materials including some composites to investigate the electrical performance by using the arc-erosion testing equipment [6-14]. The common purpose of these studies is predicting the life span of electrical contacts which helps to eliminate or minimize the contact failures before happen and then to extend the contact life by improving arc-erosion performance or delimiting the negative effects of the arc [15-22]. However, the effect of different parameters such as current load, switching frequency and type of load on the arc-erosion performance of the contact materials has not been investigated in detail yet. Because, it has not been possible to fully investigate the arc-erosion performance of the contact materials by using conventional test rigs.

Here, unlike conventional test rigs, a new life-test equipment which capable of operating under a wide variety of current ranges (up to 30 A) with inductive loads, was designed and built. The arc deteriorations of electrical contacts of a contactor under a specified current load (20 A) were also examined by utilizing the test rig. The primary purpose of designing a new test rig is to investigate the arc-erosion performance of electrical contacts in detail and especially develop novel contact materials for electromagnetic switches such as relays and contactors.

2. Experimental

The new test rig operating at different current loads is designed and manufactured to perform arc-erosion experiments for various electromagnetic switches. The photograph of the test rig is shown in Fig. 1. The photograph of the aforementioned contactor was given in Fig. 2.

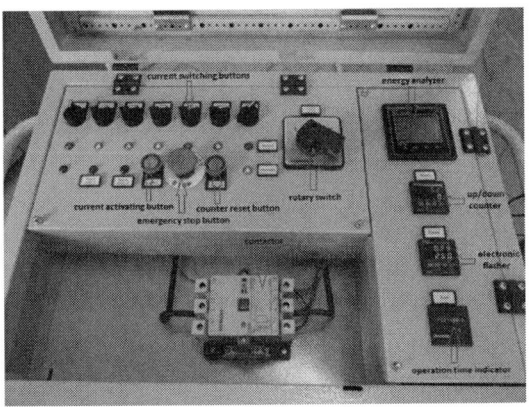

Fig. 1. Experimental setup which capable of performing arc-erosion tests up to 30 A.

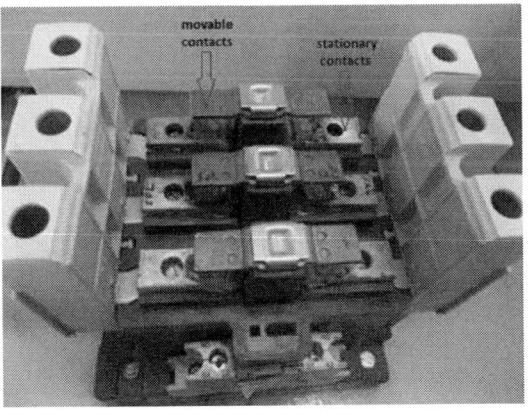

Fig. 2. The contactor.

Other components of the test rig are an electronic flasher, up/down counter, energy analyzer, operation time indicator, rotary switch, current activating and switching buttons, emergency stop button, transformers, circuit breakers and cooling system. Fig. 3 shows the transformers which are the source of inductive loads to achieve real-life operating conditions. In the present work, the current load value of the system was gradually adjusted to the specified level, namely 20 A, via control buttons located on the device. If necessary, this equipment may also be capable of operating at medium-duty current loads up to 30 A. From the functional point of view, the electronic flasher precisely allows to control the parameters (0-99.99 s - 0-9999 h), namely, make/break counts (cycle) and delay times of electrical contacts. The number of operations are readily adjusted to desired counts before tests and monitored along the whole test on the LCD screen of up/down counter located nearby electronic flasher. The test equipment also has an energy analyzer which capable of transferring data between the system and the computer. Computer monitoring allows controlling the characteristics related to the experiments (intensity of electric current, voltage and load etc.) throughout the whole life-test. Depending upon the surface deteriorations of electrical contacts in a contactor, variation of resistivity which blocks the current flow is determined by energy analyzer.

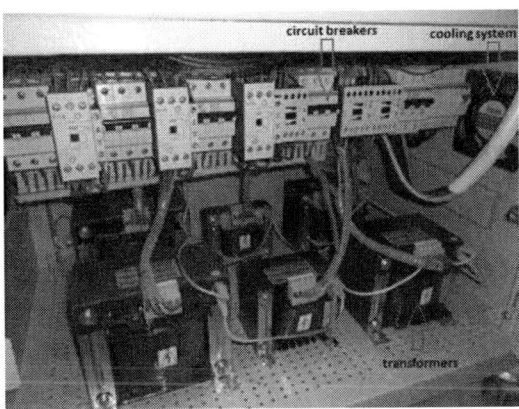

Fig. 3. Transformers which supply inductive loads to the contactor.

The test conditions were selected as 220 V operating voltage, 50 Hz frequency and 20 A switching current in the present work. The switching frequency and delay time were selected as 1,000/h and 2.6 s, respectively. Experimental conditions were given in detail in Table 1. Arc-erosion experiments were carried out on the contactor having pole number of 3, switching power of 37 kW, and contact current of 75 A, for each 5,000 operations up to 40,000. Mass loss determinations were performed after each 5,000 operations. Mass loss of the contacts was determined by a precision (10^{-5} g) weighing scale using KERN ABT 220-5DM model semi-micro analytical balance. In the sequel, the effects of the arc formed in contact surfaces were analyzed using scanning electron microscopy (SEM) on a Zeiss Evo LS 10 model microscope. The chemical composition of the tested electrical contacts was determined by EDS analysis of the SEM unit.

Table 1. Experimental conditions.

Contact Material	Silver alloy	
Circuit Conditions	AC 220V 20 A Inductive loads	
Frequency	50 Hz	
Number of Operations	5,000	10,000
	15,000	20,000
	25,000	30,000
	35,000	40,000
Switching Mode	AC	
Surrounding Gas	Air	
Switching Frequency	1,000 operations per hour	
Delay Time	2.6 s	

In line with the experiments, the effect of the electrical current on arc erosion was investigated and arc generated deteriorations with the increasing switching number were evaluated.

3. Results and discussion

The mass loss values obtained from the arc-erosion tests regarding aforementioned procedure are listed in Table 2. The values in this table were determined by using the measurements in each of 5,000 switching operations for both stationary and movable contacts tested at the current intensity of 20 A.

Table 2. Mass loss values of both stationary and movable contacts tested at 20 A.

Switching Number	Mass Loss of Movable Contacts (mg)	Mass Loss of Stationary Contacts (mg)	Overall Mass Loss (mg)
5,000	1.14	1.62	2.76
10,000	1.30	2.55	3.85
15,000	1.72	3.39	5.11
20,000	2.30	4.05	6.35
25,000	2.47	4.21	6.68
30,000	2.93	4.95	7.88
35,000	3.14	5.80	8.94
40,000	3.49	5.93	9.42

The curves showing the variation of mass loss with switching number for contacts which determined by using the data in Table 2 are given in Fig. 4. It can be seen from Table 2 and Fig. 4 that the mass loss in both contacts

increases with increasing number of switching operations. However, in certain values which especially correspond to 20,000 - 25,000 operations, no considerable change in mass loss occurs. These critical points last about no more than 5,000 operations and then, both contacts keep losing weights again in general. In addition, the mass loss in stationary contacts was found to be higher than that in movable contacts.

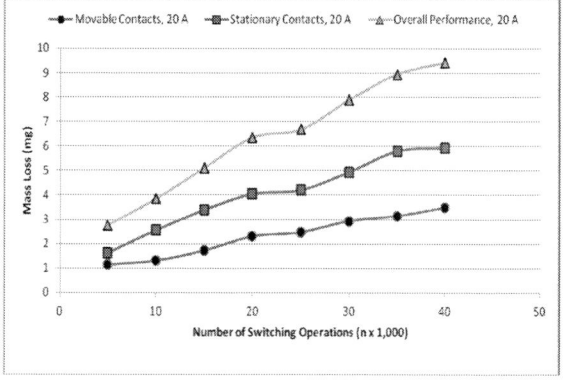

Fig. 4. Mass loss vs. switching number graph for both stationary and movable contacts tested at 20 A.

Depending upon the surface roughness of the materials, the real surface area of the intersection is much smaller than the area seemed on a macroscopical scale. In fact, the area of the intersection of contacts consists of small hills in considering microscopical scale and the real contact area between the contacts is the sum of these points. Therefore, the passage of current is being constricted through these small effective areas in the initial period of operation up to 5,000. Namely, the contact resistance is higher in the initial period. Therefore, the higher mass loss achieved in this period. In addition, this initial cycle may be thought as an adaptation stage for stationary and movable contacts due to surface roughness. The SEM micrographs obtained from the surfaces of both stationary and movable contacts after certain periods, namely per each 5,000 operation, of arc erosion experiments at 20 A given in Fig. 5 confirm this result. Fig. 5a and b show the mass losses of stationary and movable contacts in the first 5,000 operation. On the other hand, it was found that the mass loss in stationary contacts was higher than that in movable contacts, especially up to 10,000 operations. As seen in Figs. 5c and d, melting was dominant mechanism on the surfaces of stationary contacts and the reason why movable contacts remain less affected was related to material migration from stationary contacts to relatively cooler movable contacts. This may be attributed to the temperature differences between stationary and movable contacts, as movable contacts cool down as a consequence of their mechanical movement which subjected to air. In addition, some of the liquid metal penetrates to the holes of movable contacts in this step.

Stationary Contacts   Movable Contacts

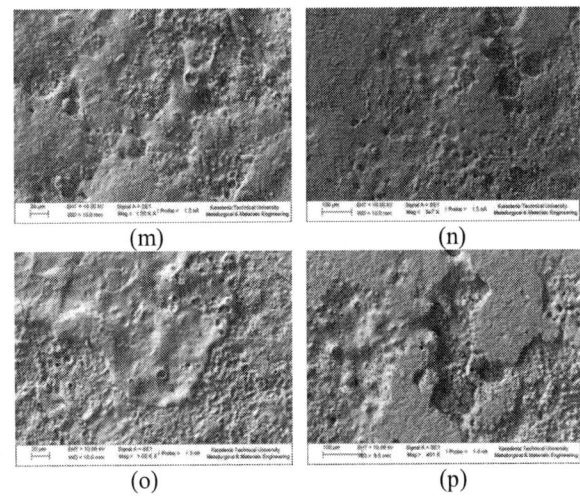

(m)   (n)

(o)   (p)

Fig. 5. continued.

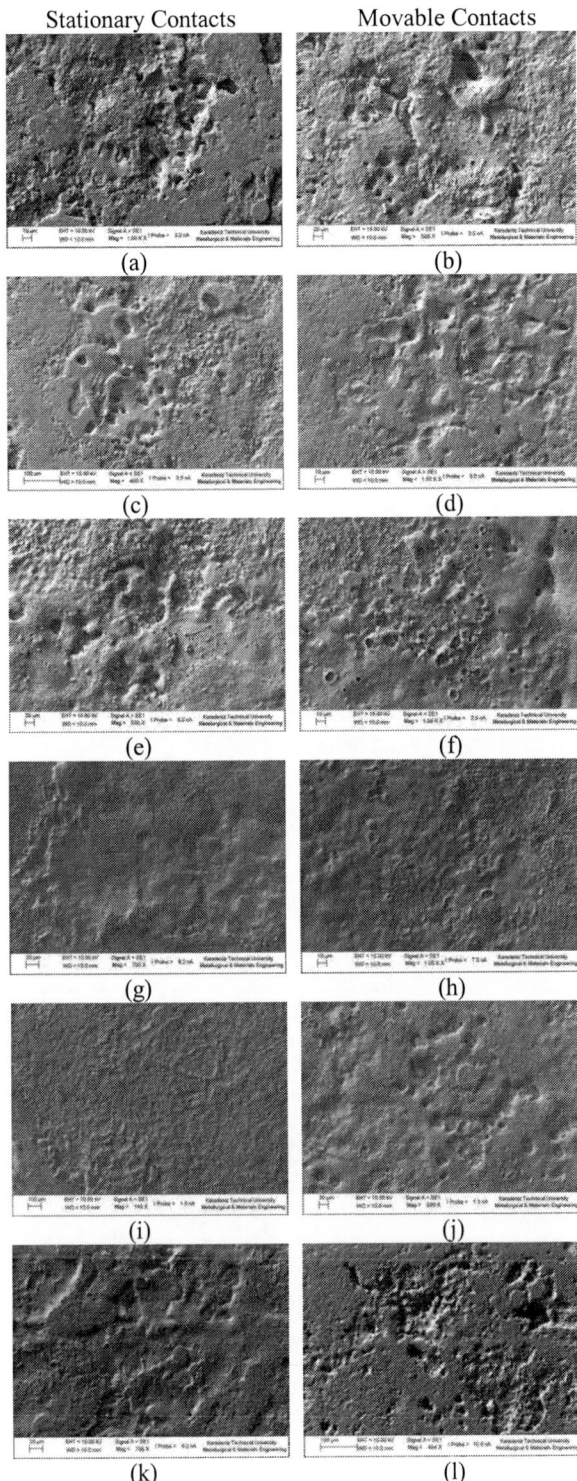

Fig. 5. SEM micrographs showing surfaces of contacts tested at 20 A after each 5,000 operation; (a, b): 5,000, (c, d):10,000, (e, f):15,000, (g, h):20,000, (i, j):25,000, (k, l):30,000, (m, n):35,000, (o, p):40,000.

With further material migration from stationary contacts to movable contacts, some craters were appeared on the surfaces of stationary contacts, while small hills and material outgrowth were observed on the opposite contacts for 15,000 operations as seen in Figs. 5e and f, respectively. It can be seen from Fig. 4 that, the graphs of stationary and movable contacts were almost parallel to each other between the operation numbers of 15,000 and 20,000. This is because, the mass-loss mechanisms by means of melting, evaporation and diminishing of the hills in both contacts are the same as seen in Figs. 5g and h. However, at the switching numbers between 20,000 and 25,000, very limited amount of the mass loss which is a desirable feature for arc-erosion performance, takes place in both contacts. This stage may also be defined as critical for both contacts as their mass loss remains nearly stable as seen in Fig. 4. The SEM micrographs obtained from the relatively smooth surfaces of both contacts at 20 A current intensity confirm this result as seen in Figs. 5i and j. The dominant melting mechanism was maintained as well and the cavities which can be easily seen in Fig. 5e, reduced to minimal level. Therefore, the points in the intersection of the contact area were increased and the distribution of the arc was enlarged from small points to large areas. This situation was also provided reducing mechanical forces per unit area on the contact surfaces which may be helpful to avoid or minimize the formation of microcracks which are detrimental for sustaining mechanical strength. Moreover, the effect of electrical cleaning due to sufficient load may be seen in operations between 20,000 and 25,000 for both contacts. This effect was brought the contact resistance back to lower levels and the minimal relative mass loss was achieved in this period. In addition, the surfaces are smooth, almost even as seen in Figs. 5i and j. After the critical point, the deteriorating effects of arc were reappeared by means of material outgrowth in stationary contacts and crater formations in the opposite contacts as seen in Figs. 5k and l, respectively. After 35,000

978-1-4799-4789-8/14 $31.00 © 2014 IEEE

operations, some of the melted materials were transferred from stationary contacts to movable contacts. Therefore, the rate of erosion and mass loss for stationary contacts was higher than the movable contacts (Fig. 4). Examining the SEM images, it is clear that the transferred material caused to occur hills on the surfaces of stationary contacts (Fig. 5m) while filling the pits on the movable contacts (Fig. 5n). In the last operation, the hills on the surfaces of stationary contacts diminished by hitting the movable contacts and this resulted damage in movable contact (Fig. 5p) while melt on the stationary contact (Fig. 5o). Consequently, the rate of mass loss for movable contacts was higher than the stationary contacts for operations between 35,000 and 40,000 as seen in Fig. 4. The crater on the surface of movable contact (Fig. 5p) is formed during the transformation of the hill (Fig. 5m) into a melting zone (Fig. 5o) on the stationary contact. The uneven continuous transfers eventually caused erosion on the surface of movable contacts. However, the electrical cleaning effect was partially seemed at current load range between 35,000 and 40,000 for stationary contacts and the surface topography was more stable than that before.

To show the effect of oxidation on the surfaces by means of arc-erosion performance, the contacts were examined in macroscopical scale and the photographs obtained from the surfaces were given in Fig. 6. In this figure, the contacts which were placed upper side of the frame are movable contacts while the lower side represents the stationary contacts. Here, attention must be paid for oxidation of the copper terminals due to the increasing arc temperatures. There are two main factors affecting the oxidation behavior of the contacts. Firstly, it should be noticed that the oxidation of the copper terminals due to the arc temperatures. Another important issue is the attachment quality of the contacts to the plates or terminals which were generally made of copper alloys. Several studies were conducted about the attachment quality and related issues of electrical contacts [23, 24].

Fig. 6. The electrical contact pairs of the contactor after the whole arc-erosion test: Movable contacts (upper), stationary contacts (lower).

Rivet, solder or braze type attachments may be utilized depending upon the type of application, load and current. The contacts should maintain their original shape in operation and this is possible with robust attachments of the contacts to the terminals. Apart from the attachment quality, the material of the terminal should allow to cool down of the contacts in operation. So, thermal conductivity of the terminals should be high enough to make this possible. For this reason, copper alloys are generally used as they have sufficient conductivity combining with mechanical strength. The oxidizing films formed on the terminals may spread over the contours of the contacts initially and propagate to the inner sides. In such case, the oxidizing films on the surface may disorder the electrical conductivity and cause the increase of contact resistance. Consequently, the control of the arc becomes more difficult and the rate of deterioration of the surfaces increases. In addition, ambient conditions are also effective on the oxidation behaviour of the contacts.

In order to better understand what lies beneath the formation of deteriorating mechanisms, such as an arc, craters, hills and melting zones, EDS analysis was performed. Below is the table showing the chemical compositions of the electrical contacts in arc-affected zones for switching numbers of 25,000 and 40,000.

Table 3. Chemical compositions of the contacts in arc-affected zones after 25,000 and 40,000 operations.

Chemical Composition (wt. %)	Type of the Contact and Switching Number			
	25,000 cycles		40,000 cycles	
	Stationary Contacts	Movable Contacts	Stationary Contacts	Movable Contacts
Ag	58.12	57.11	57.02	45.38
Cu	5.13	15.15	8.85	26.91
Sn	4.84	4.41	9.64	7.72
Fe	2.93	0.69	5.54	2.08
Bi	2.32	1.53	2.89	2.15
Cd	1.78	1.69	1.20	0.95
C	1.10	1.26	2.58	2.40
Al	0.29	0.03	0.25	0.06
Si	0.23	0.05	0.14	0.09
O	23.29	18.07	11.89	12.26

The primary elements for these contacts are silver, copper, tin and iron while the minor constituents are bismuth, cadmium, graphite, aluminum and silicon. Comparing the compositions of the arc-affective zones of the contacts obtained from two different switching numbers, the amount of silver content was decreased from 57.11 to 45.38 with the increasing number of operation as a consequence of melting and evaporation of silver which

978-1-4799-4789-8/14 $31.00 © 2014 IEEE   422

has low melting point. On the other hand, copper which has higher melting point than silver was replaced in the vicinity of regions lack of silver (Table 3).

Line scan analysis was also performed to observe the molten region. The chemical compositions of the liquid metal and the regions nearby it were determined by point analysis of EDS unit. The combined effect of line scan and point analysis of the stationary contact which were tested at 20 A after 40,000 operations was shown in Fig. 7.

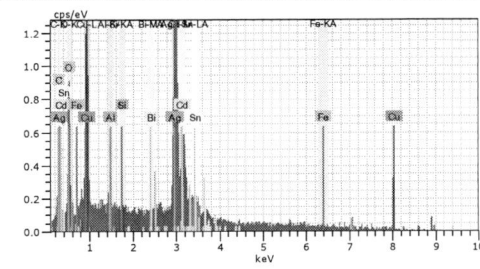

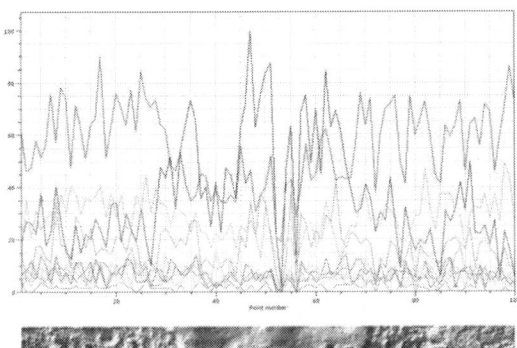

Fig. 7. The line scan and point analysis reports of stationary contact tested at 20 A after 40,000 operations.

The line scan analysis report (Fig. 7) showed that the major constituents of the molten region or alloy in liquid phase are copper (30.96%) and silver (28.53%). This region also contains iron (15.92%), tin (13.57%) and some minor additives which consist of bismuth, cadmium, graphite, aluminum and silicon. The line scan graph confirms that the regions having higher copper content were easily affected via arc by means of melting. This situation can be seen on Fig. 7 with the high intensity of blue colored copper peaks. However, the strength against melting and wear increased with increasing cadmium content as seen in the figure. The increasing frequency of the green colored cadmium peaks at the outside region of the melted zone confirms this result.

Apart from the line scan analysis, four different points were selected on the SEM micrograph which represents the chemical compositions in the vicinity of arc affected melting zone and results were given in Table 4. It can be seen from the table that zone A has minimum silver content whereas highest tin content among all zones. This table also shows that zone B and C have approximately the same silver content. In addition, zone B has higher iron content than that of zone C whereas zone C has higher copper content than zone B. As a consequence of the effect of the change in elemental distribution, zone B and C exhibit different appearance. Because locally melting region was more evident at zone C. Besides, cadmium content at zone B (2.89%) is higher than that at zone C (1.02%). In contrast, minimum cadmium content (0.40%) was determined at zone D. Therefore, this region was severely damaged by the deteriorating effects of arc, as well as zone A.

Table 4. Chemical compositions of different points of arc-affected zones after 40,000 cycles.

Chemical Composition (wt. %)	Zone			
	A	B	C	D
Ag	28.53	49.04	49.30	41.40
Cu	30.96	11.72	19.85	31.43
Sn	13.57	5.60	6.03	5.04
Fe	15.92	18.24	11.18	11.21
Bi	2.44	2.69	1.93	2.66
Cd	0.56	2.89	1.02	0.40
C	0.75	1.40	1.33	1.15
Al	0.83	0.78	0.32	0.55
Si	0.65	1.05	0.31	0.27
O	5.79	6.60	8.73	5.91

4. Conclusions

The following conclusions can be drawn for this investigation:

1. The experimental setup which capable of performing arc-erosion experiments for light and medium-duty contactors including test currents up to 30 A was designed and manufactured. The test rig is also available to investigate the effect of different test parameters such as current load and switching frequency on the arc-erosion performance of the electrical contacts under inductive loads.

2. The mass losses in both stationary and movable contacts were increased with the increasing switching number. However stationary contacts were exhibited higher mass loss than the movable contacts up to 40,000 operations.

978-1-4799-4789-8/14 $31.00 © 2014 IEEE

3. The effect of electrical cleaning was apparently seen in operations between 20,000 and 25,000 for the contacts tested at 20 A. This phenomena may be beneficial for the contact performance as it reduces the contact resistance back to the lower levels via breakdown of the non-conductive oxide films formed on the contact surfaces.

4. The chemical composition nearby arc was found to be effective on the surface deteriorations in both movable and stationary contacts.

5. The experimental setup designed and built for this work may be used to investigate the contact performances of various contactors having different contact materials and thereby the optimization of any contact material can be accomplished.

Acknowledgments

This work has been supported by the Scientific Research Fund of Karadeniz Technical University (KTUBAP, Project No: 1073).

References

1. Braunovic, M. *et al*, Electrical Contacts: Fundamentals, Applications and Technology, CRC Press, (2006).

2. Gurevich, V., Electric Relays: Principles and Applications, CRC Press, (2005).

3. Schwartz, M., Encyclopedia of Materials, Parts, and Finishes, CRC Press, (2002).

4. ASM Handbook vol. 2, Properties and Selection: Nonferrous Alloys and Special-Purpose Materials, ASM International, (1990).

5. ASM Handbook vol. 7, Powder Metal Technologies and Applications, ASM International, (1998).

6. Swingler, J., "Performance and arcing characteristics of Ag/Ni contact materials under DC resistive load conditions", *IET Science, Measurement and Technology*, 5, 2, (2011), pp. 37-45.

7. Wu, C. P., Yi, D. Q., Li, J., Xiao, L. R., Wang, B., Zheng, F., "Investigation on microstructure and performance of Ag/ZnO contact material", *Journal of Alloys and Compounds*, 457, (2008), pp. 565–570.

8. Cosovic, V., Talijan, N., Zivkovic, D., Minic, D., Zivkovic, Z., "Comparison of Properties of Silver-Metal Oxide Electrical Contact Materials", *Journal of Mining and Metallurgy*, Sect. B, 48, 1, (2012), pp. 131-141.

9. Mutzel, T., Niederreuther, R., "Contact Material Combinations for High Performance Switching Devices", *IEEE 58th Holm Conference on Electrical Contacts*, 2012, pp. 1-6.

10. Pons, F., Cherkaoui, M., Ilali, I., Dominiak, S. "Evolution of the AgCdO Contact Material Surface Microstructure with the Number of Arcs", *Journal of Electronic Materials*, 39, 4, (2010), pp. 456-463.

11. Wang, J., Liu, W., Li, D., Wang, Y., "The behavior and effect of CuO in Ag/SnO_2 materials", *Journal of Alloys and Compounds*, 588, (2014), pp. 378–383.

12. Chen, Z. K., Sawa, K., "Effect of Arc Behavior on Material Transfer: A Review", *IEEE Transactions on Components, Packaging, and Manufacturing Technology, Part A*, 21, 2, (1998), pp. 310-322.

13. Biyik, S., Aydin, M., "Investigation of the Compaction Behaviour of Ag8ZnO Composite Produced by Mechanical Alloying", *International Porous and Powder Materials Symposium and Exhibition*, Izmir, Turkey, 2013, pp. 145-149.

14. Franek, F., Neuhaus, A., Reichart, M., Schrank, C., "Mechatronical systems and experimental methods for investigations on tribology of electrical contacts", *Mechanical Systems and Signal Processing*, 22, (2008), pp. 1345–1355.

15. Zoz, H., Ren, H., Spath, N., "Improved $Ag-SnO_2$ Electrical Contact Material Produced by Mechanical Alloying", *Metall*, 53, (1999), pp. 423-428.

16. Wojtasik, K., Missol, W., "PM helps develop cadmium-free electrical contacts", *Metal Powder Report*, 59, 7, (2004), pp. 34-39.

17. Biyik, S., "Toz Metalurjisi Yontemiyle Gumus-Bor Oksit Esasli Kontak Malzemesi Uretimi ve Ozelliklerinin Incelenmesi", M.Sc. Thesis, Karadeniz Technical University, Trabzon, 2008.

18. Yang, X., Liang, S., Wang, X., Xiao, P., Fan, Z., "Effect of WC and CeO_2 on microstructure and properties of W–Cu electrical contact material", *International Journal of Refractory Metals & Hard Materials*, 28, (2010), pp. 305–311.

19. Wang, H., Wang, J., Wen, M., Zhao, J., Liu, G., "Preparation of $Ag/SnO_2+La_2O_3+Bi_2O_3$ Contact Material", *Proceedings of the 52nd IEEE Holm Conference on Electrical Contacts*, 2006, pp. 131-135.

20. Wingert, P., Bevington, R., Horn, G., "The Effects of Graphite Additions on the Performance of Silver-Nickel Contacts", *IEEE Transactions on Components, Hybrids, and Manufacturing Technology*, 14, 1, (1991), pp. 95-100.

21. Biyik, S., Arslan, F., "Investigation of Properties of Silver - Boric Oxide Based Contact Materials Produced by Powder Metallurgy", *6th International Powder Metallurgy Conference & Exhibition*, Ankara, Turkey, 2011, pp. 671-678.

22. Leung, C., Streicher, E., Fitzgerald, D., Cook, J., "High Current Erosion of Ag/SnO_2 Contacts and Evaluation of Indium Effects in Oxide Properties", *Proceedings of the 52nd IEEE Holm Conference on Electrical Contacts*, 2006, pp. 143-150.

23. Guojian, A., Horn, G., "Influence of Attachment Quality on the Performance of Arcing Electrical Contacts", *Proceedings of the 26th International Conference on Electrical Contacts*, 2012, pp. 398-401.

24. Behrens, V., Honig, T., Lutz, O., Schmitt, W., Spath, D., Worle, B., "Failure of Arcing Contacts in Low Voltage Switching Devices – Examples, Root Causes, Counter Measures", *Proceedings of the 56th IEEE Holm Conference on Electrical Contacts*, 2010, pp. 1-7.

Experimental Investigation and Interpretation of the real time, in situ Stress Measurement during Transfer Molding using the Piezoresistive Stress Chips

Ali R. Rezaie Adli [a], K.M.B. Jansen [a], Florian Schindler-Saefkow [b], Florian Rost [b],

[a] TU Delft, Delft, Netherlands

[b] Fraunhofer ENAS, Berlin, Germany

a.r.rezaieadli@tudelft.nl

Abstract

This paper describes a method used for experimental real-time monitoring of thermo-mechanical stress build-up during integrated circuit encapsulation. To detect the stress variations during molding, special stress measuring chips were employed. The working principle of the stress chip is based on the piezoresistive sensors embedded on the surface in a 6 by 6 matrix distribution. [1]

The tests were performed at several temperatures and initial conversion ranges. The stages of the molding were analyzed step by step and the shear stress and the normal stress difference distribution was investigated based on sensor locations and the orientation of the chip.

The chips are connected to a flexible polyimide board via four wire bonding at one side of the chip. The conductive tracks are extended to the edge of the board where the data acquisition system is soldered to connection pads. The material of the board ensures an elasticity inside the mold cavity and it is proven that it can withstand the weight and clamping pressure applied by the mold wall without causing any damage to the connections.

Experiment Design

The encapsulation process is transfer molding which is a highly reliable method for electronic component molding. The encapsulant is a fast curing, highly filled, thermosetting epoxy molding compound. The experiments were conducted at Besi-Fico facilities in Duiven, Netherlands.

The stress generated during the transfer molding is measured using a stress measuring chip which is developed by Bosch GmbH. The chip can measure the in-plane normal stress difference and in-plane shear stress values by the 36 piezo-resistive sensors embedded on a surface of 1.67 by 1.71mm. The details regarding the working principles of the sensors and the structure of the chip is not the scope of this project. However the detailed work can be found in H. Kittel's study [1]. A two layer flexible polyimide board was designed for the tests. The connection pads for the data transfer are positioned at the edge of the board which is extended to the outside of the mold cavity such that the real time data reading during the filling and cure was made possible. The die attach and the wire bonding of the chips to the board were performed at NXP in Nijmegen, Netherlands.

The board layout is given in figure 1. The dashed line represents the cavity where the encapsulation took place. The real mold geometry consists of four independent cavity sections where each one is equipped with two plungers that drive the epoxy melt via the gate and preheated runners, and finally through the four equidistant inlets into the mold cavity where the part of the board containing the stress sensing chip lies on one side. During the tests a dummy board is also placed at the other side of the mold for symmetrical distribution of pressure and speed adjustment.

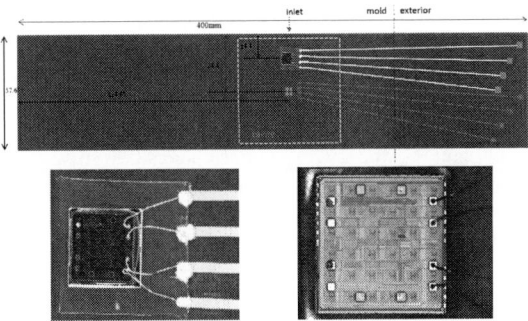

Fig 1. Board layout (top), the stress chip wire bonding (bottom left) and the chip sensor scheme (bottom right)

Stress Measurements

The tests were conducted at three isothermal mold temperatures of 120°C, 140°C and 160°C. The initial conversion level of the epoxy molding compound was also included as a test parameter and set either to zero or 20%. The calibrated normal stress difference and shear stress data collected from all the working sensors of the measurement conducted at 140° C are given in figure 2 and 3, respectively.

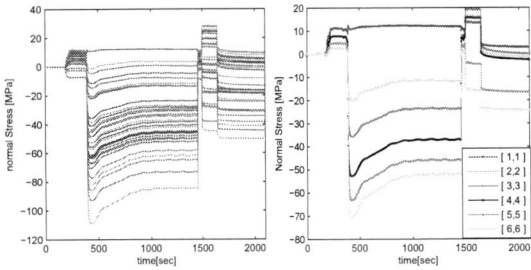

Fig 2. Normal stress difference data from the molding experiment conducted at 140° C. Left side is the data collected from all the sensors and the right side is the data from the selected range of sensors

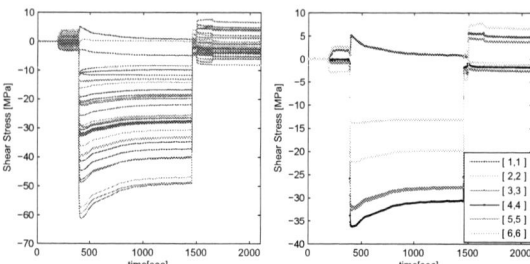

Fig 3. Shear stress data from the molding experiment conducted at 140° C. Left side is the data collected from all the sensors and the right side is the data from the selected range of sensors

Dealing with all of the sensors and understanding a pattern requires a lot of effort especially considering the fact that some sensors may deliver distorted results during certain parts of the encapsulation process. That is the reason to arrange some of the sensor data at the right hand side in order to offer a clearer visualization of the monitored stress with respect to the sensor location during the whole molding process. The sensors positions are indicated by the index numbers representing the row and column of the each sensor in a matrix arrangement. In these figures the diagonal line passing from the center of the chip is selected for the comparison of the stress values. The sixth row, for example, represents the sensors closest to the edge towards the inlet.

The tests started by inserting the board into the mold cavity which was already preheated to the test temperature. At this stage the cavity is epoxy-free. The shift in stress data given in figure 4 shows only the effect of thermal expansion of the polyimide flex-board, the chip and the adhesive used for die attach, altogether, on the sensors. The maximum detected shear stress value is 4 MPa while the highest normal stress difference recorded is around 12 MPa at the long edges of the chip.

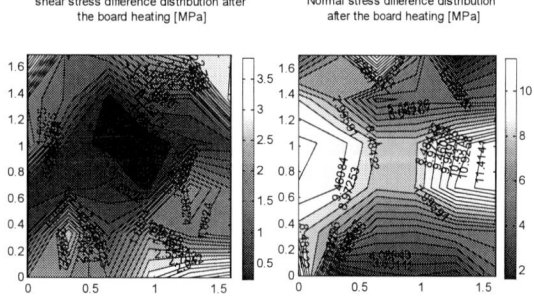

Fig 4. Stress difference when the board is inserted to the mold cavity (room temperature 24°C – mold temperature 140°C)

One of the driving forces of this project was the ability of the chip to collect data even during the filling stage of the molding, if the board is designed accordingly. The filling stage is particularly important

for the molding of the more vulnerable 3D die stacks which may not survive the viscous epoxy flow. During filling the resin exerts a drag force on the chip stack, proportional to the viscosity and the viscosity increases during filling due to the crosslink formation of the polymer.

The surface friction and so the drag force exposed on the chip surface is a function of the degree of reaction and the filling speed. Figure 5 shows the shear stress and the normal stress difference for the filling and packing stages independently. Even though the stress chip does not possess a fence type sensor and is designed specifically for the shrinkage induced stress measurements, there is still a considerable variation of stress data during the filling, detected by the sensors.

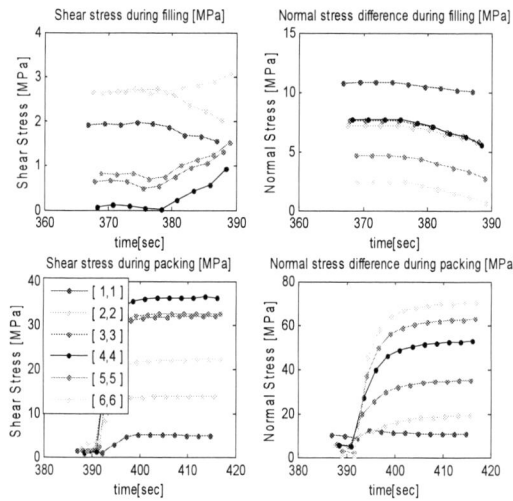

Fig 5. Filling phase shear stress and normal stress difference data at 140°C

The packing phase stress development is certainly affected by the plunger pressure increase at the end of filling in which the cavity is filled completely and in order to maintain a constant velocity the plunger requires some extra pressure. The packing phase takes only a brief period of time and the machine automatically switches to the cure stage as the pressure passes the 4 MPa boundary. After the packing stage, the densely packed thermoset in the cavity continues curing until it reaches close to full conversion at the end of cure phase.

The cure time was defined before the start of molding using a predetermined kinetic model of the epoxy molding compound. The conversion time graph of the thermosetting resin for the tested temperature range is shown in figure 7. The conversion model is based on Kamal- Sourour equation where the reaction rate is a function of conversion and isothermal temperature of the thermosetting polymer.

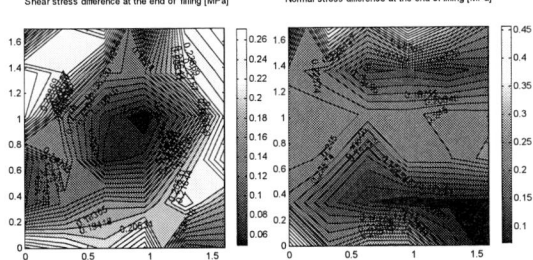

Fig 6. Shear stress (left) and the normal stress (right) difference distribution at the end of filling at 120° C

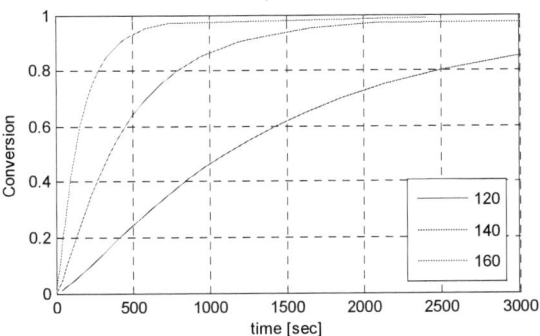

Fig 7. Conversion degree for the test temperatures

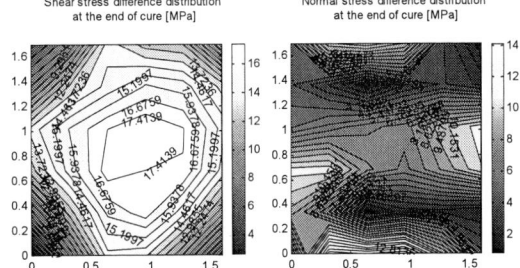

Fig 8. Shear stress (left) and the normal stress (right) difference distribution at the end of cure at 120° C

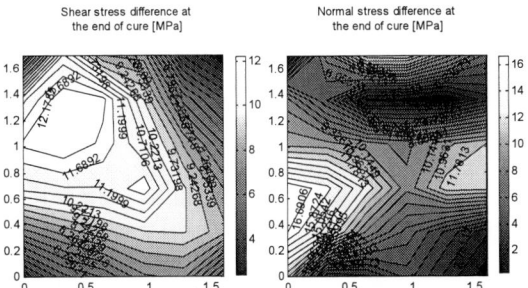

Fig 9. Shear stress (left) and the normal stress (right) difference distribution at the end of cure at 140° C and a %25 initial conversion

The packing and the end of cure phase can be clearly observed in figure 2 where the maximum variation in stress data are recorded between 400 and 1500 seconds. The portion between the two extremes is considered as the cure region where the thermosetting epoxy converts close to 100%. Another important issue about the start of the cure phase is the shift in shear and normal stress data during packing. The closer the sensor is to the inlet, the higher the monitored stress variation in packing. This phenomenon can be linked to the extra mold forced through the inlet to the cavity during packing. The orientation of the chips may differ for some of the experiments and for the specified test the chip is attached such that the inlet is at the bottom and so the flow is coming from the bottom to the top in the related figures. The orientation is the other way around for the 120° C measurement.

The difference in shear stress and normal stress difference data for the cure phase of the test performed at 120° C is given in figure 8.

The stress recorded at this stage is primarily due to the shrinkage of the epoxy resin which completes its crosslink formation till the end of cure. A circular distribution of the shear stress difference is observed with the origin at the center of the chip. The distribution of the normal stress difference presents a symmetrically scattered pattern. Almost the same pattern can be seen in figure 9 for the 140° C with a 25% of initial conversion. The initial conversion was achieved by loading the Epoxy molding compound to the chamber while the board was already placed in the cavity. The molding started when the specified conversion level was achieved based on the kinetic model.

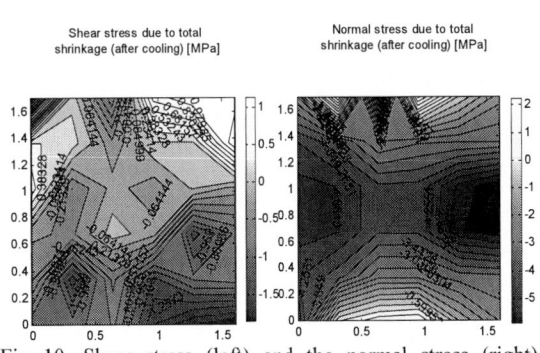

Fig 10. Shear stress (left) and the normal stress (right) difference distribution at the end of filling at 120° C

978-1-4799-4789-8/14 $31.00 © 2014 IEEE   427

The most direct outcome of the all measurements would be considered as the stress variation observed after the cure when the cooling starts. Figure 10 represents the shear stress and normal stress difference change between the start of molding and the end of cooling revealing the total shrinkage induced stress generated during molding.

Acknowledgements

We are particularly grateful to Henk Wensink and Hans Venrooij from Fico/Besi for supporting us during this project. We thank Chris Brouwer for assisting us during the measurements in Fico/Besi facilities in Duiven. We are sincerely thankful to Amar Mavinkurve from NXP for his feedback and support in chip bonding and die attach, and also we would like to thank Peter Koch who did all the chip connections manually.

References

[1] Hartmut Kittel, Novel stress Measurement system for evaluation of package induced stress, Robert Bosch GmbH, SMART SYSTEMS INTEGRATION Barcelona, Spain 2008

[2] Florian Schindler-Saefkow, Stress Chip Measurements of the internal Package Stress for Process Characterization and Health Monitoring, Fraunhofer ENAS, EurosimE, CasCais, Portugal, 2012

[3] Florian Schindler-Saefkow, Olaf Wittler, Thomas Schreier-Alt, Harmut Kittel, Bernd Michel, Package induced stress simulation and experimental verification, Integration issues of miniaturized systems- MOMS, MOEMS, ICS and electronic components (SSI), 2008, 2nd European conference and exhibition, 1-4

[4] Ricky Hardis, Julie L.P. Jessop, F. E. Peters, M. R. Kessler, Cure kinetics characterization and monitoring of an epoxy resin using DSC, Raman spectroscopy, and DEA, Journal of composites:Part A, 49 (2013) 100-108

[5] M. Khalil Abdullah, M. Z. Abdullah, M. A. Mujeebu and S. Kamaruddin, A study on the effect of epoxy molding compound (EMC) rheology during encapsulation of stacked chip scale packages (SCSP), Journal of reinforced plastics and composites, 2009, 28, 2527

[6] Thomas Schreier-Alt, C. Rebholz, Frank Ansorge, Simulation and experimental analysis of substrate overmolding, Microelectronics and Packaging Conference, 2009. EMPC 2009

B-Field Characterization and Equivalent Circuit Modeling of a Poly-SiGe-MEMS based Xylophone Bar Magnetometer

M. A. Farghaly[1,3], V. Rochus[2], X. Rottenberg[2], U. S. Mohammed[3] and H. A. C. Tilmans[2]

[1]KACST-Intel Consortium Center of Excellence in Nano-manufacturing Applications (CENA), Riyadh, Saudi Arabia

[2] IMEC, Leuven, Belgium

[3] Department of Electrical Engineering, Faculty of Engineering, Assiut University, Assiut, Egypt

Email: rochus@imec.be, Phone: +(32) 16 28 8534

Abstract

This paper reports a quantitative characterization of a poly-SiGe MEMS-based Xylophone Bar Magnetometer (XBM), thereby following a novel characterization method that is based on the measurement of the forward/backward transmission gains S_{21}/S_{21} of the XBM treated as a two-port network. More specifically, this was done through monitoring the absolute amplitude of the resonant peaks of S_{21} and S_{12} with changing magnetic induction B. Also, we present for the first time a novel equivalent circuit for the two-port XBM, modeling effectively the electro-magneto-mechanical behavior of the magnetometer. The experimental measurements showed that poly-SiGe XBM is capable of being a linear magnetic sensor in mT range with **a sensitivity** $0.1dB/mT$ with an excitation power $5dBm$ fed to the electrodynamic/electrostatic port and a biasing voltage $14V$ applied through the sense/drive capacitor.

1. Introduction

Global Positioning System (GPS) [1] is the most common way for navigation. However, it needs a direct line-of-sight with a satellite, which is not applicable inside high buildings. This could be crucial in case of firefighting or catastrophes. Thus, a backup aiding system is needed to do this function when GPS is either unreliable or unavailable. This backup system can be an Inertia Measurement Unit (IMU) that consists of three types of triaxial sensors (an accelerometer, a gyroscope and a magnetometer) and uses a dead reckoning technique to calculate current position and velocity relative to an initial reference frame (before losing line-of-sight with satellites).

Since these sensors are fabricated in different technologies, it is a real challenge to assemble these multi-sensors on a single a chip which leads to increasing cost and unavoidable errors due to axis misalignment. Amongst existing technologies, MEMS has the most promising to succeed in integration of theses three sensors. Besides, MEMS could be easily integrated with CMOS in a single chip, as offered by imec's SiGeMEMS platform technology [2]. The presented work is focusing on MEMS-based magnetometers whose readings provide calibration to lessen the inherited drift error in gyroscope readings [3].

This paper presents a preliminary characterization and equivalent circuit modeling of Xylophone Bar Magnetometer (XBM); A uni-axe magnetometer [4], [5]. Multiphysics simulations were reported in [6], thereby studying the effect of electro-thermal coupling on the mechanical performance of the XBM including different damping sources. Based on this study, the fabrication of prototype designs using imec's SiGeMEMS technology were reported and qualitative sensitivity measurements by moving a reference magnet closer or farther away from the XBM were presented [7]. The output was primarily monitored using a laser Doppler vibrometer for displacement measurements. The characterization reported here differs from previous work [7] in using a calibrated electromagnet that allowed us to do a quantitative measurements and further in monitoring and analyzing more precisely both the measured S_{21} and the S_{21} using a Vector Network Analyzer (VNA).

2. XBM's Dual Principle of Operation

The Xylophone Bar Magnetometer (XBM) is based on classical resonating free-free bar [7] [4] that is held at the nodes of its first fundamental transverse vibrational mode by supports, as shown in Fig. 1. These supports are located at 22.4% of the bar length of each end. The main concept of resonant XBM is to drive the bar by a force that its frequency matches the fundamental resonance frequency of the bar, which can be expressed as:

$$f_0 = \frac{1.029t}{L^2}\sqrt{\frac{E}{\rho}} \quad (1)$$

where L is the length of the bar, t is the thickness of the bar, E is Young's modulus and ρ is the density.

This causes the beam to deflect and the deflection in the vicinity of resonance frequency can be expressed as:

$$w(f) = \frac{w(f=0)}{\left[(1 - \frac{f^2}{f_0^2}) + j\frac{f}{Qf_0}\right]} \quad (2)$$

where $w(f=0)$ is the static deflection, f_0 is the natural resonance frequency and Q is the mechanical quality factor. Thus, the amplitude of vibration will be amplified by the mechanical quality factor Q [7] compared to its static deflection.

978-1-4799-4789-8/14 $31.00 © 2014 IEEE

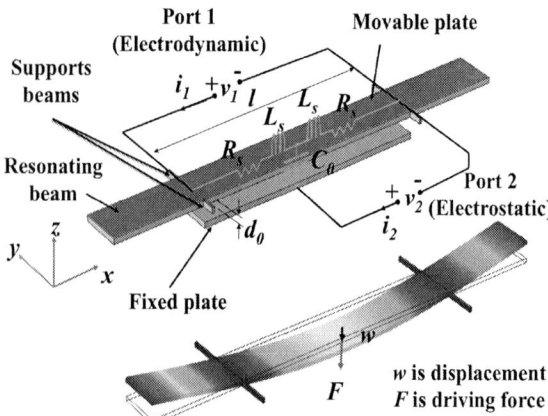

Figure 1: Schematic of the resonant XBM structure (plus an illustration of the fundamental flexural resonance). XBM has two ports (electrodynamic port 1 and electrostatic port 2) for driving and sensing, are clearly indicated. Presenting the beam with a T-network (indicated in light blue) modeling the direct electrical coupling through the beam between port 1 and port 2, as will be discussed.

Figure. 1 shows a typical structure of a capacitive poly-SiGe XBM. It consists of a movable structural poly-SiGe membrane ($4\mu m$ thick) and a fixed poly-SiGe sensing electrode ($80nm$ thick). They are separated by the gap d_0 and form the electrodes of a parallel plate capacitor C_0 for sensing/driving purposes. XBM has two electrical ports: An electrodynamic port 1 (between beam's supports for current) and an electrostatic port 2 (between one of the beam's supports and the sensing electrode). Resonant XBM may be operated in two ways; either electrodynamic-driving combined with electrostatic-sensing or electrostatic-driving combined with electrodynamic-sensing.

XBM may be driven electrodynamically (via port 1) using Lorentz force while the respective motion is electrostatically sensed using capacitive detection (via port 2). This is the most common way to operate resonant MEMS magnetic sensors. In such case, the sensors are known as resonant Lorentz force magnetic sensors. Figure 1 shows Lorentz force principle of operation is that a force $F = li_1B_y$, will act on an AC current i_1 (with frequency f_0 matches the bar's fundamental resonance frequency) carrying conductor with length l when it is placed in a magnetic field B_y. This force causes the beam to vibrate with an amplitude w directly proportional to the normal magnetic field component on current direction. This vibration is translated into a capacitance change, because the vibrating beam is one of two electrodes of a parallel plate capacitor. The capacitance change is directly proportional to magnetic field component B_y. This operating principle is fully described by forward transmission gain S_{21}.

Alternatively, an XBM may be electrostatically driven using capacitive excitation (via port 2) while the respective motion is electrodynamically sensed using electromagnetic induction (via port 1). This way is less common in literature, but it was used in [8]. As shown in Fig. 1, applying an AC input voltage v_2 (with frequency f_0 matches the bar's fundamental resonance frequency) between the capacitor electrodes causes an alternating electric force $F_e = \varepsilon_0 A_e V_b v_2 / d_0^2$ on the floating electrode, where ε_0 is the air permittivity, A_e is the electrodes area and V_b is the bias voltage. This causes the beam to vibrate around its static equilibrium position, cutting the magnetic field lines of B_y. In turn, a voltage $v_1 = B_y l \dot{w}$ is induced between nodal points of the beam, where $\dot{w}$ is the beam velocity in z-direction. This induced voltage is directionally proportional to the magnetic field component B_y. This operating principle is fully described by backward transmission gain S_{12}.

3. Equivalent Circuit Modeling

Equivalent circuit has been always an elegant tool in predicting the linear behavior of multi-physics systems as MEMS such that it compacts involved energy domains into one energy domain, the electrical domain, and save time that a FEM model take for convergence to a solution for just one explored design. Equivalent circuit model is developed starting from first law of thermodynamics (conversion of energy) and equations of equilibrium for energy domain, as described in [9], [10]. Beginning from first thermodynamic principle, an equivalent circuit of an ideal case of XBM (whose mechanical layer consists of two layers: metal layer and poly-SiGe layer; which involves no coupling between port 1 and port 2) could be represented by the following ABCD matrices [11]:

$$
\begin{bmatrix} v_2 \\ i_2 \end{bmatrix} = \begin{bmatrix} 1 & 1 \\ Y_{C_0} & 0 \end{bmatrix} \begin{bmatrix} 1/\Gamma & 0 \\ 0 & \Gamma \end{bmatrix} \begin{bmatrix} 1 & Z_{eq} \\ 0 & 1 \end{bmatrix}
$$
$$
\cdot \begin{bmatrix} 0 & 1/\Psi \\ \Psi & 0 \end{bmatrix} \begin{bmatrix} 1 & 2Z_c \\ 0 & 1 \end{bmatrix} \begin{bmatrix} v_1 \\ -i_1 \end{bmatrix} \tag{3}
$$

where $Y_{C_0} = j\omega C_0$ is the admittance of static capacitance at zero deflection, Γ is the electrical transduction factor, Z_{eq} is the equivalent mechanical impedance, Ψ is the magnetic transduction factor and $2Z_c = 2(R_s + j\omega L_s)$ is the electrical impedance of the metal layer.

This ideal case has no coupling between driving and sensing electrical ports. However, imec's SiGeMEMS platform provides only one layer (SiGe), as a mechanical layer, that is common between port 1 and port 2, as depicted in fig 1. This causes an inevitable direct electrical coupling between the electrical ports. This coupling can be modeled by presenting the beam as T-network, as shown in Fig. 1 and Fig. 2.

The Y-parameters for the equivalent circuit shown in

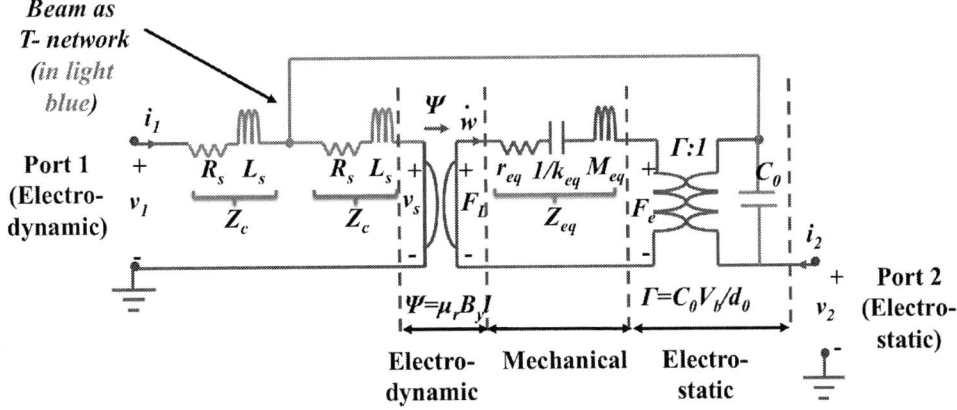

Figure 2: Equivalent circuit representation of XBM fabricated using imec's SIGeMEMS platform.

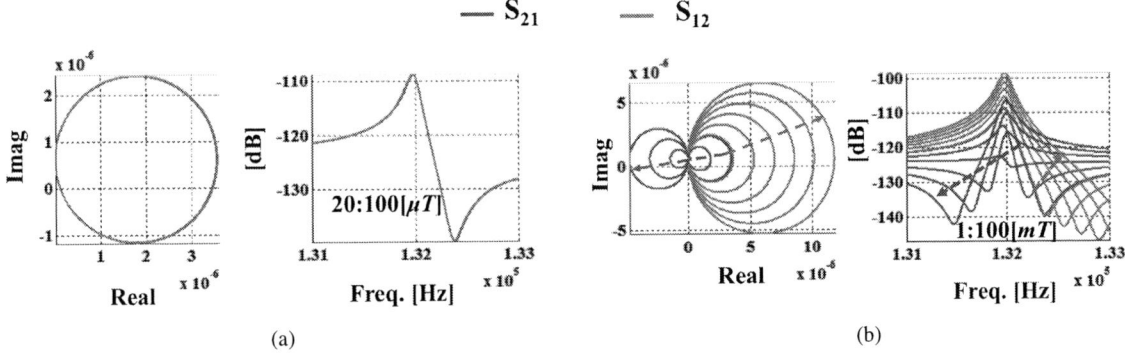

Figure 3: Polar and amplitude plots of S_{21} and S_{12} from equivalent circuit simulations of XBM two port network over magnetic field range (a) 20: 100[μT] range and in (b) 1:100 [mT] range, respectively. Dotted arrows show the direction of magnetic field B increase.

Fig. 2 are as follows:

$$Y_{11} = \frac{Z_{eq} + Z_c(\Gamma^2 + Y_{C_0} Z_{eq}) + Y_{C_0}\Psi^2}{Z_c^2(\Gamma^2 + Y_{C_0} Z_{eq}) + Z_c(Y_{C_0}\Psi^2 + 2Z_{eq}) + \Psi^2} \quad (4a)$$

$$Y_{12} = \frac{-(\Gamma\Psi + Z_c(\Gamma^2 + Y_{C_0} Z_{eq}) + Y_{C_0}\Psi^2)}{Z_c^2(\Gamma^2 + Y_{C_0} Z_{eq}) + Z_c(Y_{C_0}\Psi^2 + 2Z_{eq}) + \Psi^2} \quad (4b)$$

$$Y_{21} = \frac{-(-\Gamma\Psi + Z_c(\Gamma^2 + Y_{C_0} Z_{eq}) + Y_{C_0}\Psi^2)}{Z_c^2(\Gamma^2 + Y_{C_0} Z_{eq}) + Z_c(Y_{C_0}\Psi^2 + 2Z_{eq}) + \Psi^2} \quad (4c)$$

$$Y_{22} = \frac{2Z_c(\Gamma^2 + Y_{C_0} Z_{eq}) + Y_{C_0}\Psi^2}{Z_c^2(\Gamma^2 + Y_{C_0} Z_{eq}) + Z_c(Y_{C_0}\Psi^2 + 2Z_{eq}) + \Psi^2} \quad (4d)$$

It is obvious that numerators of Y_{12} and Y_{21} consist of two terms: $\Gamma\Psi$ (linearly dependent on magnetic field) and the term $Z_c(\Gamma^2 + Y_{C_0} Z_{eq}) + Y_{C_0}\Psi^2$ (represents a feedback effect due to the direct coupling between port 1 and port 2). The first term should dominate to still being able to get information about the ambient magnetic field. The second term causes degradation in the sensor performance that the sensor is only sensitive to magnetic field in mT range. Table 1 lists expressions and values of circuit elements for

equivalent circuit of XBM. Figure 3 shows simulations of the XBM equivalent circuit. As shown in Fig 3a, poly-SiGe XBM is insensitive to the change in magnetic field range 20:100 μT; the whole curves are nearly identical. However, S_{21} and S_{12} plots begin to respond the change in magnetic field range 1:100 mT.

4. B-Field Characterization

Magnetic field characterization was performed using an accurately calibrated magnet and a VNA. As shown in Fig. 4, the XBM and the electromagnet were placed inside a vacuum chamber. The VNA ports are connected to the electrical ports of XBM through leads that come out of the vacuum chamber. Bias-T is needed to couple RF and DC applied voltage to a single lead, especially for the electrostatic port and to keep their sources isolated.

A. Electromagnet description

For B-field characterization step, a homemade electromagnet was utilized to produce a magnetic field of induction B in mT range inside its gap. As shown in Fig.

978-1-4799-4789-8/14 $31.00 © 2014 IEEE

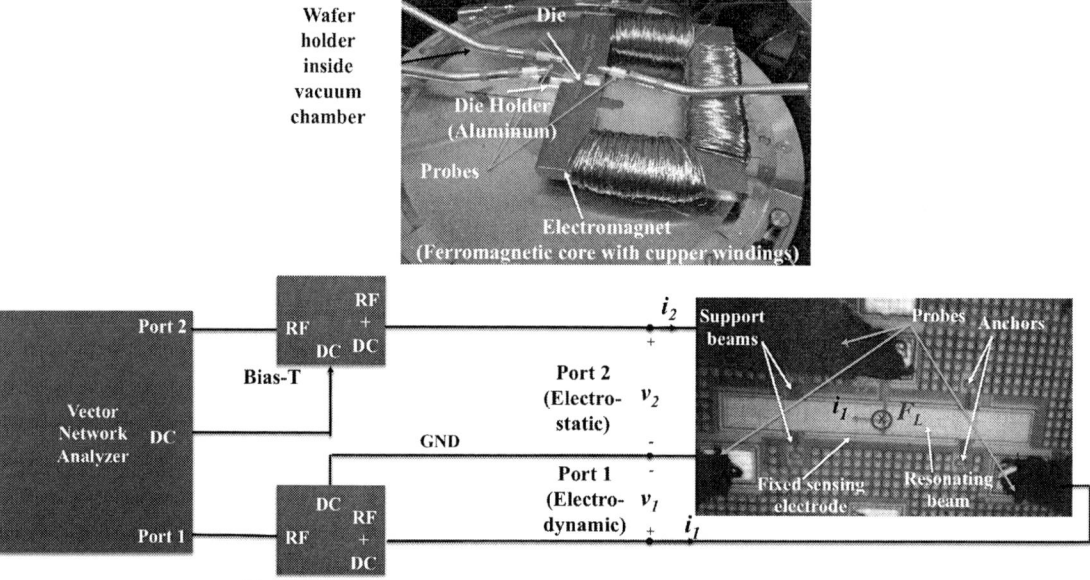

Figure 4: Typical measurements setup for a fabricated SiGe XBM (400 x 40 x 4μm) with support beams (30 x 2 x 4μm), mounted inside the gap of an electromagnet inside a vacuum chamber and a schematic diagram for external connection with VNA, showing XBM's electrical ports.

Table 1: Table contains values and expression of circuit elements for XBM as a lumped parameter system. These expressions are based on the following assumption that $w \ll d_0$. DC bias voltage $V_b = 20V$ and mechanical quality factor $Q = 1000$ for the listed values.

Variable	Expression (Lumped)	Value
R_s	$R_{support} + 0.5R_{beam}$	$35[\Omega]$
L_s	$\frac{1}{I^2} \int_V \frac{1}{2\mu} B^2 dV$	$0.13[nH]$
Ψ	$B_y l$	$220B_y[\mu NA^{-1}]$
ω_0	$2\pi f_0$	$829[krad/sec]$
L_{eq}	$M_{eq} = m = \rho_s Lwt$	$0.2915[nH]$
C_{eq}	$1/\omega_0^2 L_{eq}$	$5[mF]$
r_{eq}	$\omega_0 L_{eq}/Q$	$253.7[n\Omega]$
Γ	$C_0 V_b/d_0$	$0.173[\mu NV^{-1}]$
C_0	$\varepsilon_0 A_e/d_0$	$26[fF]$

4, this electromagnet has an air gap large enough to fit the sample. The core of the electromagnet is made out of iron (a high magnetic permeability μ material). Around the core, isolated copper wire was warped. Injecting a DC current through copper windings, creates a closed magnetic field loops through the core. This electromagnet could generate magnetic field that ranges from $\pm (8:110 mT)$ for a current $(0.1:2 A)$. Compared to a permanent magnet [7], using electromagnet is better because it is easier to manipulate the magnetic field strength B inside the air gap through controlling current and to ensure that the magnetic field direction is perpendicular to the current

passing through the XBM.

B. Two-Ports Electrical Measurements

Figure 5 shows typical measured and equivalent circuit's simulated plots of S_{21} and S_{12} for poly-SiGe XBM (400 x 40 x 4μm) with support beams (30 x 2 x 4μm). Measurements were done in vacuum at $5dBm$ power applied to the driving electrical port and a bias voltage $14V$ applied through the sense/drive capacitor C_0. Through usage of VNA, S_{12} and S_{21} were monitored for magnetic field range $\pm 8:110 mT$.

It is observed that the orientation and position of the resonance circles depends on the orientation of the magnetic field. The S_{12} and S_{21} readings are more pronounced for positive and negative magnetic field B, respectively. As shown in Fig. 5, simulations of the equivalent circuit, after subtracting parasitic vector from measurements and tuning the device parameters (e.g. R_s, M_{eq}, ... etc), display good agreement with the measured performance. The equivalent circuit parameters are as declared in Table 1 except values of Q, R_s, k_{eq}, r_{eq} and ω_0 that were tuned to minimize the difference between measured and simulated S_{21} and S_{12} for a set of 26 measurement data that were collected at different magnetic field values from $(-110:110mT)$. $Q = 2200$ is the average of the extracted quality factor for the all measured data. $R_s = 70\Omega$ which corresponds to high resistive SiGe routing lines (Sheet resistance= 20 Ω/sq) to the bondpads plus the intrinsic impedance of

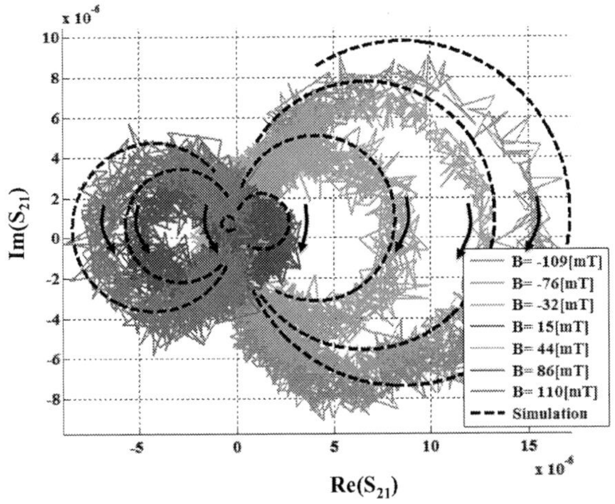

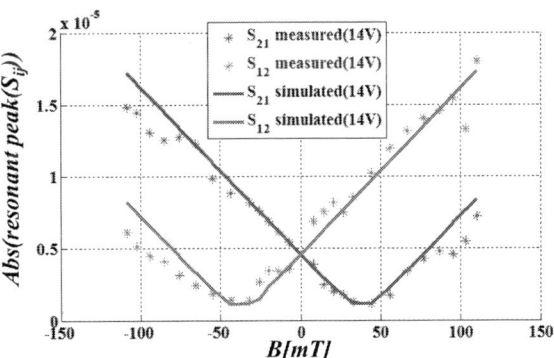

(a) S_{21} polar plots

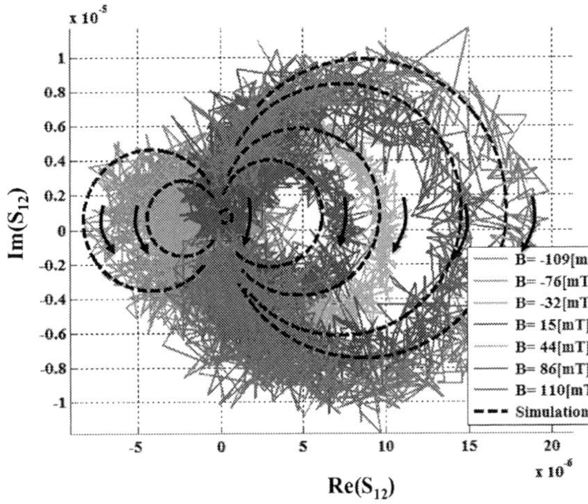

(b) S_{12} polar plots

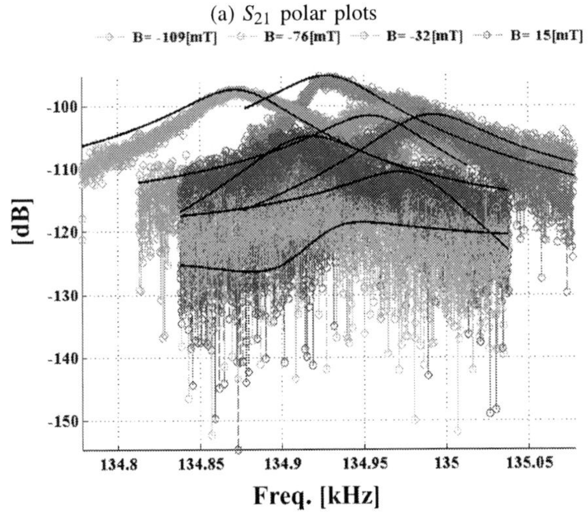

(c) S_{21} amplitude plots

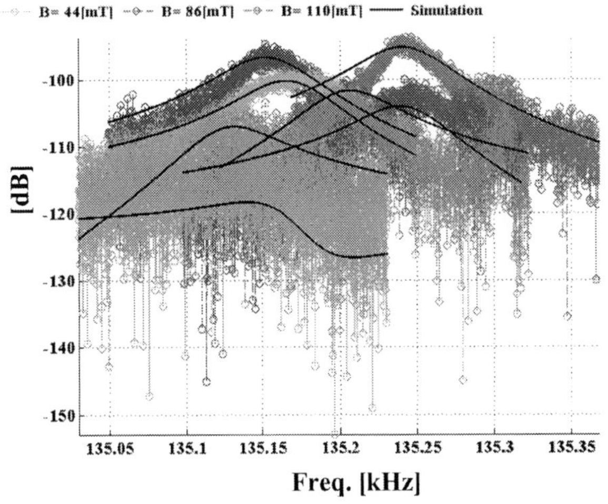

(d) S_{12} amplitude plots

Figure 5: Measured/ simulated polar and amplitude plots of S_{21} and S_{12} for positive and negative values magnetic field B. Arrows indicate the direction of increasing frequency.

the beam. Measurements were done consecutively and it was noticed that ω_0 changes a few hundreds of Hertz from one measurement to the next, as shown in Fig. 5c and Fig. 5d. The reason behind such behavior is probably due to the heat generated inside the beam, which causes an elongation in the beam length and in turn change the resonance frequency. Thus, a generic formula was used for $C_{eq} = 1/\omega_0^2 L_{eq}$ and $r_{eq} = \omega_0 L_{eq}/Q$ for each measurement.

Figure 6 shows the simulated and measured absolute amplitude of resonant peak of S_{12} and S_{21} as a function of the applied magnetic field B in the range from $(-110:110mT)$. The graph shows that the output characteristics of XBM as a magnetic sensor. XBM is applicable as a linear magnetic sensor for the use in the mT range with a sensitivity $0.1dB/mT$.

Figure 6: Plot of simulated and measured absolute resonant amplitude of S_{21} and S_{12} for magnetic field range $(-110:110\ mT)$ at bias voltage $14V$.

978-1-4799-4789-8/14 $31.00 © 2014 IEEE

5. Conclusions

In this paper, we demonstrate a quantitative characterization of a poly-SiGe-based MEMS XBM through monitoring the forward/backward transmission gains S_{21}/S_{21} of the XBM treated as a two-port network. The experimental measurements showed that poly-SiGe XBM is capable of being a linear magnetic sensor in mT range with a sensitivity $0.1dB/mT$. It is not sensitive for the targeted μT range because of inevitable electrical coupling between port 1 and port 2. Also, we present for the first time a novel equivalent circuit for the two-port XBM magnetometers, modeling effectively the measured electromagneto-mechanical behavior of poly-SiGe XBM.

References

[1] E. D. Kaplan, *Understanding GPS: Principle and applications.* ARTECH HOUSE, Boston, 1996.

[2] http://www.europractice-ic.com/MEMS_imec_SiGeMEMS.php.

[3] N. Ahmad, R. A. R. Ghazilla, N. M. Khairi, and V. Kasi, "Reviews on various inertial measurement unit (imu) sensor applications",," *International Journal of Signal Processing Systems*, vol. 1, no. 2, pp. 256–262, December 2013.

[4] D. Oursler, D. Wickenden, L. J. Zanetti, T. Kistenmacher, R. Givens, R. Osiander, J. Champion, and D. A. Lohr, "Development of johns hopkins xylophone bar magnetometer," *Johns Hopkins APL Technical digest*, vol. 20, no. 2, pp. 181–189, 1999.

[5] D. Wickenden, J. Champion, R. Osiander, R. Givens, J. Lamb, J. Miragliotta, D. Oursler, and T. Kistenmacher, "Micromachined polysilicon resonating xylophone bar magnetometer," *Acta Astronautica*, vol. 52, no. 2–6, pp. 421 – 425, 2003.

[6] H. Lamy, I. Niyonzima, P. Rochus, and V. Rochus, "A xylophone bar magnetometer for micro/pico satellites," *Acta Astronautica*, vol. 67, no. 7–8, pp. 793 – 809, 2010.

[7] V. Rochus, R. Jansen, H. A. C. Tilmans, X. Rottenberg, S. Ranvier, H. Lamy, C. Chen, and P. Rochus, "Poly-sige-based mems xylophone bar magnetometer," in *Sensors, 2012 IEEE*, 2012, pp. 1–4.

[8] K. Funk and W. Frey, "Magnetic field sensor having deformable conductor loop segment," US patent 6486665 B1, Nov. 2002.

[9] H. A. C. Tilmans, "Equivalent circuit representation of electromechanical transducers: I. lumped-parameter systems," *Journal of Micromechanics and Microengineering*, vol. 6, no. 1, p. 157, 1996.

[10] W. H. H. and M. J. R., *Electormechanical Dynamics Part I.* Wiely, 1968.

[11] M. A. Farghaly, "Modeling and design of mems-based magnetometers," MSc thesis, Assiut University, Egypt, December 2013.

Computationally Efficient and Stable Order Reduction Method for a Large-Scale Model of MEMS Piezoelectric Energy Harvester

M. Kudryavtsev[1], E. B. Rudnyi[2], T. Bechtold[1], J. G. Korvink[1]

[1]Institute for Microsystems Engineering, Freiburg University, Germany
[2]CADFEM GmbH, Munich, Germany
corresponding author: korvink@imtek.uni-freiburg.de

Abstract

In this work, we present a computationally efficient model order reduction technique for a large-scale multiport model of piezoelectric energy harvester. This novel technique generates stable reduced order models. The method combines model reduction based on Krylov subspaces and a Schur complement transformation of the resulting system. We demonstrate an excellent match between the full-scale and the reduced order model during transient and harmonic simulation.

1. Introduction

MEMS-based piezoelectric energy harvesters (see Fig.1) play the role of power suppliers for self-contained systems, such as remote monitoring sensors. The operational principle of one class of device is based on the conversion of ambient vibration energy to the electrical domain by exploiting a piezoelectric effect [1].

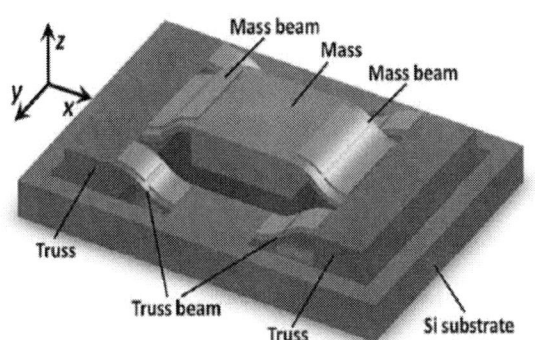

Figure 1: Schematic design of a piezoelectric energy harvester from [1].

The efficiency of the energy transformation highly depends on the harvester design as well as on the power circuit connected to it. To achieve the maximum power output, one needs to perform a system level-simulation, that is, the co-simulation of these two parts. Instead of over-simplifying piezoelectric models by replacing them with lumped elements (as in [2], for example), we employ nonlinear, multi-physical models, generated directly from finite element (FE) discretizations. The advantage is more realistic model behavior.

At the system-level, a piezoelectric energy harvester is represented by a block with input and output ports. The model presented in this work contains 3 input ports (one mechanical named ***displ*** and two electrical named ***el1_in***, ***el2_in***) and 5 output ports (three mechanical named ***centr***, ***north***, ***south*** and two electrical named ***el1_out***, ***el2_out***) as it is shown in Fig. 2.

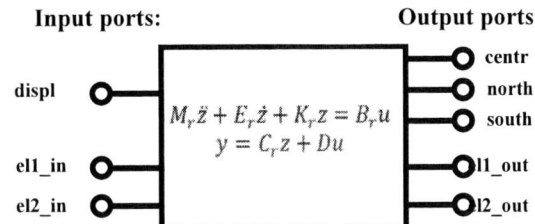

Figure 2: System level representation of energy harvester.

The relation between the input and output ports is described by a multiple-input multiple-output (MIMO) system Σ of second order linear ordinary differential equations (ODEs). Σ is generated from a FE discretization of the harvester model under the assumption of a loss-free piezoelectric material:

$$\Sigma: \begin{cases} \underbrace{\begin{bmatrix} [M] & [0] \\ [0] & [0] \end{bmatrix}}_{M} \cdot \begin{Bmatrix} \{\ddot{x}_1\} \\ \{\ddot{x}_2\} \end{Bmatrix} + \underbrace{\begin{bmatrix} [E] & [0] \\ [0] & [0] \end{bmatrix}}_{E} \cdot \begin{Bmatrix} \{\dot{x}_1\} \\ \{\dot{x}_2\} \end{Bmatrix} + \\ \underbrace{\begin{bmatrix} [K_{11}] & [K_{12}] \\ [K_{21}] & [K_{22}] \end{bmatrix}}_{K} \cdot \begin{Bmatrix} \{x_1\} \\ \{x_2\} \end{Bmatrix} = \underbrace{\begin{bmatrix} [B_1] \\ [B_2] \end{bmatrix}}_{B} \cdot \{u\} \\ \{y\} = \underbrace{[[C_1] \quad [C_2]]}_{C} \cdot \begin{Bmatrix} \{x_1\} \\ \{x_2\} \end{Bmatrix} \end{cases} \quad (1)$$

Where $[M]$ is a structural mass matrix, $[E]$ is a structural damping matrix, $[K_{11}]$ is a structural stiffness matrix, $[K_{12}]$ and $[K_{21}]$ are forward and reverse piezoelectric coupling matrices, $[K_{22}]$ is a dielectric conductivity matrix, $\{x_1\}$ and $\{x_2\}$ are parts of the state vector representing nodal displacements and nodal electrical potentials, $\{u\}$ is a vector of input port values, $[B_1]$ and $[B_2]$ are parts of a scattering matrix that translate the input vector into domain forces, $\{y\}$ is a vector of output port values, and $[C_1]$ and $[C_2]$ are parts of a gathering matrix which projects the system state vector onto the subspace of system outputs.

Because the original large-scale system (43748 equations) leads to excessively high computational cost, a model order reduction (MOR) technique has to be used.

2. Krylov Subspaces Based MOR

The methods of model reduction employing Krylov subspace techniques were shown to be efficient for generating stable reduced-order models of mechanical systems [4-6], and RLC circuits [7,8]. However, piezoelectric models combine mechanical and electrical domains and the state vector x of an initial large-scale system (1) consists of two parts of differing physical nature. The reduction methods do not specially take into consideration the multi-physical nature of a system, and do not preserve its structure.

In order to check the functionality and efficiency of different Krylov subspace based MOR methods, for the case of a multiport piezoelectric energy harvester model, we have tested following methods: first order block Arnoldi [8,9], SOAR [5,6], reduction of proportionally damped systems [4], and an MOR method which uses the superposition property of linear systems [7]. We have reduced the original large-scale model of 43748 degrees of freedom (DOF) down to 10 DOF (or 30 DOF with the method from [7]). This implies that the moments $\hat{m}_i$ of the transfer functions $\hat{H}$ associated with the reduced systems $\hat{\Sigma}$ match 10 moments m_i of the original transfer function H: $\hat{m}_i(0) = m_i(0)$; $i = 0, 1, 2, \ldots, k; k = 10$. In all the cases the transfer functions $\hat{H}$ displayed poles in $\mathbb{C}_+$, which means that the generated reduced models were instable [8].

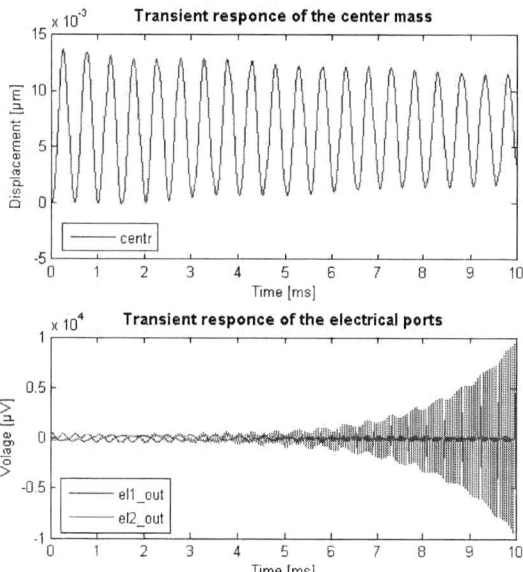

Figure 3: Transient response of the mechanical port *centr* and electrical ports *el1_out* and *el2_out* on the displacement step shows the instability of the reduced model.

Fig. 3 shows a transient response of the system, reduced using the SOAR algorithm from [6], on the step input at the *displ* port. It can be clearly seen that, while the response of the mechanical port *centr* is stable within the considered time segment, the behavior of the electrical ports *el1_out* and *el2_out* show and instability, and

therefore such model cannot be used to approximate the full-scale model.

3. Schur Complement Transformation

In [2], it was shown that, after a Schur complement transformation, which eliminates the electrical part x_2 of the state vector, the resulting full-scale system is composed of positive definite matrices and the reduced model, generated by a Krylov subspace based MOR, is stable. We will refer to this approach as "MOR after Schur". The usability of such a reduced model for the co-simulation with electrical power circuitry was demonstrated in [10]. The main disadvantage of such transformation, however, is that it increases the number of nonzero elements of the system stiffness matrix K (see Fig. 4). Consequently, the computational costs of matrix factorization, which is necessary for the MOR process, will rise. Table 1 shows that doubling the original system size results, after a Schur transformation, in an increase of the required RAM by 6.3, and an increase of the time for MOR by 81. Note that such calculations might be prohibitive for systems with more than 10^5 DOF, even when using computer clusters.

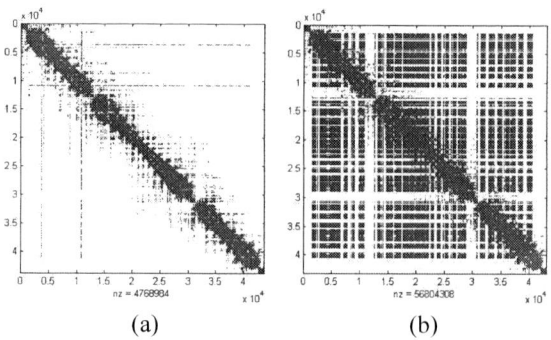

| | (a) | (b) |

Figure 4: For a system with 43748 DOFs, the initial system stiffness matrix K (a) has nz = 476894 nonzero elements, while after a Schur complement transformation (b) nz = 56804308.

Table 1: Comparison of the computation time and required RAM for systems derived from the same piezoelectric energy harvester model (all calculations were made using a 3.1 GHz processor with 8 GB RAM).

System size	43748 DOF	90663 DOF
Initial K matrix size	74 Mb	168 Mb
K matrix size after Schur transform	944 Mb	5.81 Gb
MOR time of the initial system	51 s	130 s
MOR time of the system after Schur transform	60 s	1 hour 21 min

In order to have the possibility to handle large-scale systems using average computers, which is necessary for many research and engineering tasks, computationally more efficient order reduction method is proposed in this work. It preserves the sparsity of original full-scale system and yet generates a stable reduced system as measured at the output.

4. Proposed Technique: Schur after MOR

As the first step, the original system (1) should be reduced using one of the second order Arnoldi based algorithms. As it was aforementioned, the matrices of the reduced system $\widehat{\Sigma}$ lose the structure of the full system Σ, therefore it is not possible to perform the Schur complement transformation explicitly. We propose to factorize the mass matrix $\widehat{M}$ of the system $\widehat{\Sigma}$ by means of an eigenvalue decomposition:

$$\widehat{M} = P\widetilde{M}P^{-1} \qquad (2)$$

where P is the matrix whose columns are eigenvectors of $\widehat{M}$, and $\widetilde{M}$ is a diagonal matrix of corresponding eigenvalues of $\widehat{M}$.

The reduced system $\widehat{\Sigma}$ is then projected onto the subspace colspan(P):

$$\widetilde{\Sigma}:\begin{cases} \underbrace{P^{-1}\widehat{M}P}_{\widetilde{M}}\ddot{\widetilde{x}} + \underbrace{P^{-1}\widehat{E}P}_{\widetilde{E}}\dot{\widetilde{x}} + \underbrace{P^{-1}\widehat{K}P}_{\widetilde{K}}\widetilde{x} = \underbrace{P^{-1}\widehat{B}}_{\widetilde{B}}u \\ \widetilde{y} = \underbrace{\widehat{C}P}_{\widetilde{C}}\widetilde{x} \end{cases} \qquad (3)$$

The resulting matrix $\widetilde{M}$ contains part $\widetilde{M}_2$ with eigenvalues which are close to zero and several orders of magnitude smaller than the others. As such eigenvalues do not have a significant influence on the mechanical behavior of the model, corresponding rows and columns of the system matrices $\widetilde{M}$ and $\widetilde{E}$ may be rounded to zero:

$$\widetilde{M} = \begin{bmatrix} [\widetilde{M}_1] & [0] \\ [0] & \underbrace{[\widetilde{M}_2]}_{\approx [0]} \end{bmatrix} \approx \begin{bmatrix} [\widetilde{M}_1] & [0] \\ [0] & [0] \end{bmatrix}$$

$$\widetilde{E} = \begin{bmatrix} [\widetilde{E}_{11}] & [\widetilde{E}_{12}] \\ [\widetilde{E}_{21}] & [\widetilde{E}_{22}] \end{bmatrix} \approx \begin{bmatrix} [\widetilde{E}_{11}] & [0] \\ [0] & [0] \end{bmatrix} \qquad (4)$$

We assume that the corresponding elements $\widetilde{x}_2$ of the state vector belong to the electrical domain. Exploiting this assumption, the reduced system $\widetilde{\Sigma}$ is transformed using a Schur complement transformation, as it was made for the full system in [2]:

$$\widehat{\Sigma}:\begin{cases} \widetilde{x}_2 = -\widetilde{K}_{22}^{-1}\widetilde{K}_{21}\widetilde{x}_1 + \widetilde{K}_{22}^{-1}\widetilde{B}_2 u \\ \widetilde{M}_1\ddot{\widetilde{x}}_1 + \widetilde{E}_{11}\dot{\widetilde{x}}_1 + (\widetilde{K}_{11} - \widetilde{K}_{12}\widetilde{K}_{22}^{-1}\widetilde{K}_{21})\widetilde{x}_1 = \\ \qquad (\widetilde{B}_1 - \widetilde{K}_{12}\widetilde{K}_{22}^{-1}\widetilde{B}_2)u \\ \widetilde{y} = (\widetilde{C}_1 - \widetilde{C}_2\widetilde{K}_{22}^{-1}\widetilde{K}_{21})\widetilde{x}_1 + (\widetilde{C}_2\widetilde{K}_{22}^{-1}\widetilde{B}_2)u \end{cases} \qquad (5)$$

5. Numerical Simulation results

In this section, an illustration of the proposed method applicability is given. A reduced model with 30 DOF was obtained from the full model (43748 DOF) using the MOR method from [7]. Then, two elements were cut from the state vector by means of a Schur transformation yielding the final system with 28 DOF.

Fig. 5-8 illustrate the excellent match of the reduced model with the full-scale model, and just the slight difference from the corresponding results for the MOR method from [2].

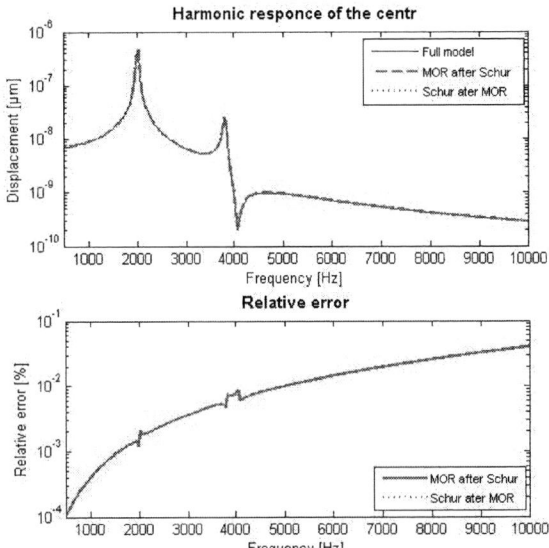

Figure 5: Harmonic response of the mechanical port **centr** *on the displacement excitation shows that the relative error of the reduced model doesn't exceed 0.04% over the entire frequency range.*

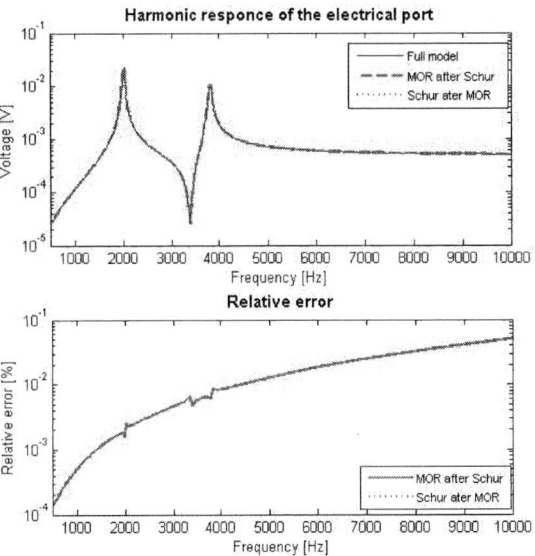

Figure 6: Harmonic response of the electrical port **el1_out** *on the displacement excitation shows that the relative error of the reduced model doesn't exceed 0.05% over the full frequency range.*

978-1-4799-4789-8/14 $31.00 © 2014 IEEE   437

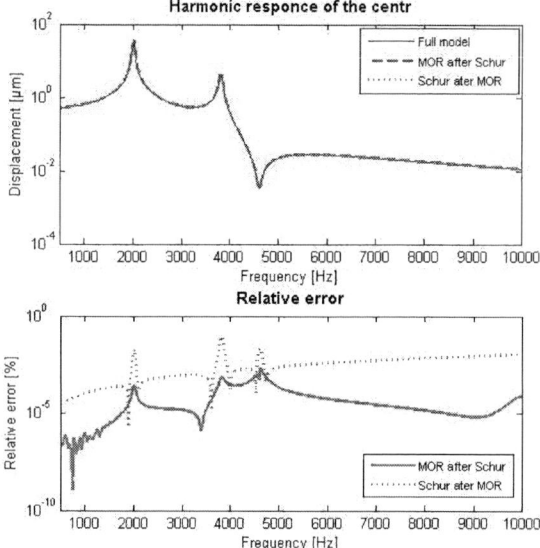

*Figure 7: Harmonic response of the mechanical port **centr** on the excitation of the electrical port **ell_in** shows that the relative error of the reduced model doesn't exceed 0.1% over the frequency range.*

As an error estimator for transient response (see Fig. 8), the root mean square error (RMSE) is used. The RMSE of the reduced model output $\hat{y}_t$ as compared with the full-scale system output y_t for times t is computed as:

$$RMSE = \sqrt{\frac{\sum_{t=1}^{n}(y_t - \hat{y}_t)^2}{n}} \qquad (6)$$

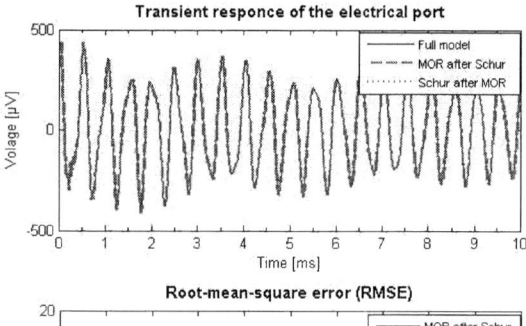

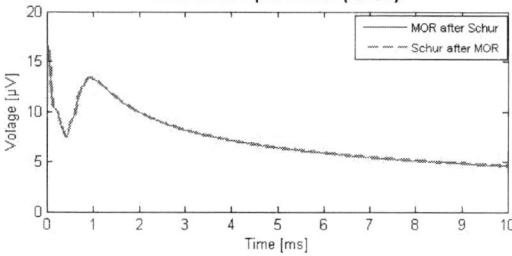

*Figure 8: Transient response of the electrical port **ell_out** on the displacement step shows that the root-mean-mean-square errors of both reduced models are almost the same.*

Please note that the resulting model's transfer function has poles solely in the left complex half-plane, and therefore the model is stable.

6. Conclusions and Outlook

It has been shown that the large-scale multiport electro-mechanical model of MEMS piezoelectric energy harvester can be efficiently reduced while preserving its stability. The efficiency is given by first reducing and then stabilizing the model.

Future work will focus on an alternative approach, to test and further develop the structure preserving reduction techniques, which may have advantages over existing MOR methods. An example of such a technique for large multi-port RLC circuits is presented in [11] and shows higher accuracy vs. the conventional PRIMA method.

References

1. Z. Wang et al., "A piezoelectric vibration harvester based on clamped-guided beams," *Proc. IEEE Micro Electro Mechanical Systems*, pp. 1201-1204, 2012.
2. Chien-Cheng Chen, Chi-Wei Kuo, Yao-Joe Yang, "Generating Passive Compact Models for Piezoelectric Devices," *Computer-Aided Design of Integrated Circuits and Systems*, IEEE Transactions on , vol.30, no.3, pp.464-467, March 2011.
3. T. Aftab, D. Hohlfeld, T. Bechtold, E. B Rudnyi and J. G. Korvink, "New modeling approach for micro energy harvesting system based on model order reduction enabling truly system level simulation," *Proc. Micromechanics and Microsystems Europe*, 2012.
4. R. Eid, B. Salimbahrami, B. Lohmann, E. B. Rudnyi, J. G. Korvink, "Parametric Order Reduction of Proportionally Damped Second-Order Systems," *Sensors and Materials*, v. 19, N 3, p. 149-164, 2007.
5. Z. Bai and Y. Su, "Dimension reduction of large-scale second-order dynamical systems via a second-order Arnoldi method," *SIAM J. Sci.Comput.*, 26(5):1692–1709, 2005.
6. Behnam Salimbahrami and Boris Lohmann, "Order reduction of large scale second-order systems using Krylov subspace methods," *Linear Algebra and its Applications*, v. 415, N 2-3, p. 385-405, 2006.
7. P. Benner, Lihong Feng, E. B. Rudnyi, "Using the Superposition Property for Model Reduction of Linear Systems with a Large Number of Inputs," *International Symposium on Mathematical Theory of Networks and Systems*, 2008.
8. Roland W. Freund, "Krylov-subspace methods for reduced-order modeling in circuit simulation," *J.Comput. Appl. Math.*, Vol. 123, 1-2, p. 395-421, 2000.
9. X.-D Do, and S.-G. Lee, "An efficient parallel SSHI rectifier for piezoelectric energy scavenging systems," *Advanced communication technology (ICACT)*, pp. 1394-1397, 2011.

10. F.Sayed, T. Aftab, D. Hohlfeld, T. Bechtold, J. G. Korvink, "Reduced Order Modeling Enables System Level Simulation of a MEMS Piezoelectric Energy Harvester with a Self-Supplied SSHI-Scheme," *Proc. EuroSimE 2013*, pp.1,6, 14-17, 2013.

11. Freund, R.W., "SPRIM: structure-preserving reduced-order interconnect macromodeling," *Proc. IEEE/ACM Int. Comput.-Aided Design Conf.*, pp.80 -87 2004.

Multiphysical modeling of nanosecond laser dicing on ultra-thin silicon wafers

G. Galasso[1,2*], M. Kaltenbacher[1], B. Karunamurthy[2], H. Eder[3], T. Polster[3]

[1]Vienna University of Technology, Wiedner Hauptstr. 6, A-1040, Vienna, Austria
[2]KAI Kompetenzzentrum GmbH, Europastr. 8, A-9524, Villach, Austria
[3]Infineon Technologies AG, Siemensstr. 2, A-9500, Villach, Austria
*email: germano.galasso@k-ai.at

Abstract

We propose an approach for the numerical modeling of a laser ablation (LA) process on silicon targets. The work is motivated by the increasing application of lasers in the separation of ultra-thin power semiconductors. In order to optimize the process, reduce the energy cost per laser pulse and minimize the extension of the thermally induced damage, a deeper insight into the mechanisms underlying laser dicing and a proper selection of laser settings are crucial. Numerical modeling is useful for understanding the tightly coupled physics involved in the interaction of laser with matter, as well as in the identification of the optimum laser configuration. With this aim, two numerical models have been prepared and combined. Initially, we set up a custom written one-dimensional hydrodynamic code which describes the main mechanisms triggered during LA, as vaporization and plasma formation. This first simulation allows to estimate the laser energy loss due to plasma absorption. The remaining available energy is used as input to perform a Finite Element transient thermal simulation on a three-dimensional geometry of the target. Here, an element deactivation technique is adopted to remove the vaporized elements from the computational mesh, therefore describing the geometry and the progressive formation of the ablated crater. The calculated crater geometries have been compared with experimental ones for two fluence values, showing reasonable agreement.

1. Introduction

A. Motivation

Since its discovery in 1960, the laser became a workhorse in the most disparate fields. Its unique properties, such as coherence, directionality and brightness, made it an ideal tool to be exploited in a wide range of applications. Laser is nowadays used in, among many others, microsurgery, precision machining, printing, communication, restoration of artworks. The semiconductor industry greatly benefits from lasers as well. They are applied in scribing, cleaning and finally, in chip's separation. The technological trend to move towards smaller scales, as well higher efficiency, are key aspects when producing reliable, robust ultra-thin silicon chips. One fundamental step during the fabrication of these devices is the die separation, where the chips are individually diced from

the wafer matrix. Currently, the two main technologies for chip dicing are mechanical and nanosecond (ns) laser cutting. While mechanical dicing remains the standard technology for dies that exceed 100 μm thickness, laser dicing showed advantages in the manufacturing of ultra-thin chips [1]. The reduced mechanical stability of a thin substrate, as well as the vibrations induced during mechanical dicing, often cause edge chipping and even shattering. Laser dicing, on the contrary, being a thermal process, exerts much smaller mechanical forces on the wafer and chipping is not observed [2]. Therefore, laser dicing becomes a serious alternative, although the breaking strength of the separated chips is reduced with respect to those mechanically diced. To better understand the effects of laser dicing on the chip's quality, a deeper insight into the process is needed. Certainly, the physics governing a LA process is far from being simple. When an intense laser beam hits a solid target, a long sequence of tightly interconnected processes is triggered. The high heat load deposited by one laser pulse causes melting and evaporation of the target within a very short time scale. The vapor expands on top of the irradiated target, interacting with the incoming laser beam. Experiments and theory showed that above a certain intensity threshold (about 10^8 W/cm^2 in case of metallic targets [3]) a laser-induced plasma is generated, which shields the target from the incoming light (plasma shielding phenomenon), affecting in turn the evolution of the entire process. Hence, plasma shielding has a significant impact on the dicing process: it results in a reduction of the ablation efficiency, uncontrollable material removal rate and ultimately, in the increase of the energy cost per pulse. Based on the above considerations the scope of this work is to provide a simplified, yet physically consistent, modeling strategy of the ns-LA process of silicon targets. In order to verify the model accuracy we compared the depth of craters induced by a single shot LA experiment to those calculated.

B. LA modeling: state of the art

With the increasing popularity of lasers, a considerable amount of scientific work has been done in the field. Surprisingly, there is still a profound disagreement about the actual mechanisms underlying ns-LA. In the last decades in fact, an uncountable amount of models flooded in the field of LA simulation. Nevertheless a common path way has not been established yet. The models, are

978-1-4799-4789-8/14 $31.00 © 2014 IEEE

often based on different assumptions and even contradicting interpretations. Models can range from simple semi-analytical one-dimensional [4], [5], [6], to multiphase and multidimensional hydrodynamic [7], [8], [9], [10] to kinetic ones [11]. Further, they may include or neglect [7], [9] the effect of a laser induced plasma developing on top of the target. Even the way plasma formation is treated and interpreted varies quite considerably. The hypothesis of local thermal equilibrium (LTE) is a quite popular assumption made in the description of plasma formation and it has been widely applied [12]. This implies that the various species composing the plasma posses, locally, the same energy. Given the very short time scales characterizing ns-LA, the LTE formulation is, at least, questionable in describing the initial stage of plasma formation. A more appropriate description should account for the collisional and radiative reactions governing the initial non-equilibrium formation of plasma species [3], [13], as done in this work. Moreover, Inverse Bremsstrahlung (IB) is commonly considered the only plasma absorption mechanism, while multi-photon ionization (MPI) is often neglected, while it plays a major role especially in case of ultra-violet (UV) laser wavelength [14], typically used in wafer dicing. Also the description of the target thermodynamics has been tackled at different levels of accuracy. The most common modeling procedure is to simply treat the target evaporation by means of the Clausius-Clapeyron law for ideal gases, regardless of the applied laser intensity[15], [13]. While this approach is justifiable in case that the critical temperature T_c is not reached, it has obviously little physical meaning when a high intensity pulse is applied and very high temperatures should be expected. This problem has been overcome in various ways: from considering the material as a transparent medium near T_c, which results into blocking the temperature at a maximum value not higher than the critical one [16], to equipping the model with appropriate multiphase equations of state (EOS) [17]. A thorough treatise and a critical analysis of the aforementioned aspects can be found in the works of Autrique et al [17], [18].

2. Model

A typical laser dicing process consists of several ns-pulses which are shot at high frequency f and displaced in space with a constant velocity v_{cut}. To maximize the spatial overlapping among pulses, the laser light is focused by a set of lenses, conferring to the beam the shape of an elongated ellipse along the cutting direction. This results in the formation of a deep crater, which can directly cut through a silicon wafer thinner than about 100 μm. The simulation of a real dicing process has to face two stringent and conflicting requirements. On one hand the processes involved in the single pulse interaction are described by a stiff set of partial and ordinary differential equations (PDEs and ODEs), which require tiny time steps to be solved. On the other hand, the simulation of the real

process presumes that long times are considered in a three-dimensional domain. Combining these aspects in one model would have required a far too high computational load. The modeling approach here proposed is as follows: Initially, we set up a one-dimensional hydrodynamic simulation based on the volumes of fluid (VOF) method, where the interplay between the target and the laser-induced plasma plume is analyzed for one single UV laser pulse of 10 ns full width at half maximum (FWHM) duration. The underlying model describes the bi-directional interaction of target and plasma domains and estimates the net amount of energy reaching the target after plasma absorption. Following this, a three-dimensional Finite Element (FE) commercial code (Ansys®) is used to perform a transient thermal analysis on a single pulse as well as on a train of moving and overlapping pulses, each carrying an averaged net amount of energy as determined by the preliminary hydrodynamic simulation.

A. 1D hydrodynamic model and laser induced plasma interaction

A self written 1D VOF model is used to describe the main phenomena characterizing the interaction of one pulse with a silicon target, namely: phase change of silicon in liquid and consequently into vapor, formation and evolution of a plasma plume induced by the laser, absorption of the laser through the plasma. The thermodynamics of the target is described by the energy equation, solved along the z-spatial coordinate in a Lagrangian formulation that accounts for the recession velocity v_{rec} of the crater due to material removal

$$\rho c_p^* \left(\frac{\partial T}{\partial t} - v_{rec} \frac{\partial T}{\partial z} \right) - \left(\frac{\partial}{\partial z} k \frac{\partial T}{\partial z} \right) = Q(t, z). \quad (1)$$

Here, ρ, c_p^* and k are the density, specific heat and thermal conductivity of the target respectively, while T is the unknown temperature to be determined. The source term on the right-hand side of the equation is due to the laser light absorption into the target, expressed via a Beer-Lambert law as [19]

$$Q(t, z) = (1 - R) \alpha I_{net}(t) \exp(-\alpha z). \quad (2)$$

In (2) R and α denote the reflectivity and absorption of silicon and $I_{net}(t)$ is the net irradiance reaching the target surface.

In order to account for the phase change from solid to liquid occurring at the melting temperature T_m, the latent heat of melting ΔH_m is included in the heat capacity formulation [14]

$$c_p^* = c_p + \Delta H_m \delta(T - T_m). \quad (3)$$

An appropriate δ-function is used to enforce the effect of ΔH_m only in a small temperature interval across T_m [20]. Equation (1) is solved by imposing ambient temperature T_{amb} =300 K as initial condition as well as boundary condition on the non irradiated surface, while the boundary condition on the irradiated surface at $z = 0$ is [15]

$$-k\frac{\partial T}{\partial z} = -\rho v_{rec}\Delta H_v\left(T_b\right),\qquad(4)$$

where ΔH_v is the latent heat of vaporization and T_b the boiling temperature of silicon. The evaporation process is supposed to evolve along the binodal vapor-branch of a Van Der Waals phase diagram for silicon, which can be easily retrieved by knowing the critical properties of the material.

Finally the crater velocity recession arises from two main material removal mechanisms: evaporation and volumetric mass removal [17]. The last mechanism occurs in case the target reaches critical temperature, where a clear distinction between liquid and vapor cannot be established. In this case the computational cells reaching critical temperature are removed from the target domain and shifted into the plume domain. The proper connection between target and plasma domains is provided by the Knudsen Layer (KL) relationships, which describe the transition of the vapor from a non equilibrium state, at the target side, to thermal and translational equilibrium at the plasma side. Via the KL boundary conditions both evaporation as well as condensation are accounted for, the last one occurring when the plume's pressure exceeds the vapor pressure at the target surface [21]. The set of Euler PDEs describes the plasma plume evolution in the presence of a background gas at 1atm, by applying the conservation of mass, momentum and energy to the mixture of plasma and background gas

$$\frac{\partial\rho}{\partial t} = -\frac{\partial(\rho v)}{\partial z}\qquad(5)$$

$$\frac{\partial\rho_v}{\partial t} = -\frac{\partial\left(\rho^{(Si)}v\right)}{\partial z}\qquad(6)$$

$$\frac{\partial(\rho v)}{\partial t} = -\frac{\partial}{\partial z}\left(\rho v^2 + P + \frac{\partial\tau}{\partial z}\right)\qquad(7)$$

$$\frac{\partial}{\partial t}\left[\rho\left(E + \frac{1}{2}v^2\right)\right] + \frac{\partial}{\partial z}\left[\rho\left(E + \frac{1}{2}v^2 + \frac{P}{\rho}\right)v\right]\qquad(8)$$
$$= -\frac{\partial}{\partial z}\left(\tau + \kappa\frac{\partial T}{\partial z}\right) + Q_{IB} + Q_{MPI} - Q_\varepsilon - Q_{pl}.$$

Here ρ is the total momentum density of the mixture (being the sum of the vapor density $\rho^{(Si)}$ and background gas density $\rho^{(b)}$), P the total pressure, ρv the total momentum density, ρE the total internal energy density, τ the viscosity and κ the conductivity of the plasma. The source terms at the right hand side of the energy equation represent the contributes given due to IB absorption Q_{IB}, MPI absorption Q_{MPI} and due to the radiation emitted

Table 1. Collisional (c) and radiative (r) processes and corresponding rate coefficients and atomic reactions

c	inv. Bremsst.	A_{ij}	$Si_m{}^k + e(\varepsilon) + h\nu_{laser} \rightarrow Si_m{}^k + e(\varepsilon')$
c	ei. ion./tb.rec.	γ_{ij}/δ_{ji}	$Si_m{}^k + e(\varepsilon) \leftrightarrow Si_n{}^{k+1} + e(\varepsilon') + e(\varepsilon'')$
c	ei. exc./deexc.	C_{ij}/D_{ji}	$Si_m{}^k + e(\varepsilon) \leftrightarrow Si_n{}^k + e(\varepsilon')$
r	phot. ion./rec.	β_{ij}/β_{ji}	$Si_m{}^k + n_{photon}h\nu_{laser} \leftrightarrow Si_0{}^{k+1} + e(\varepsilon)$
r	Bremsst.	E_{ji}	$Si_m{}^k + e(\varepsilon) \rightarrow Si_m{}^k + e(\varepsilon) + h\nu$

during IB Q_ε. Finally, Q_{pl} is included as a correction factor accounting for the spherical-like expansion of the plasma [22]. The proper closure of the Euler system is provided by the expressions of the pressure and specific internal energy, in the reasonable assumption that the plasma behaves as an ideal gas. The Euler equations have been solved by adopting a second order in space HLLC Riemann solver [22]. The temporal evolution of plasma's species is calculated by means of two approaches. At the beginning of the pulse, a set of ODEs rate equations is used to describe the initial formation and evolution of plasma species due to collisional and radiative processes. In this initial stage ionization of the background gas is neglected, and only the population of silicon atoms is described. This occurs mainly due to Inverse Bremsstrahlung absorption, Bremsstrahlung emission, Photon ionization and recombination, electron impact excitation and deexcitation, elctron impact ionization and three body recombination of atoms. In the most general form, a rate equation takes the following form [23]:

$$\frac{dn_i^{(Si)}}{dt} = -n_i^{(Si)}\sum_{j\neq i}^{N_L}W_{ij} + n_j^{(Si)}\sum_{j\neq i}^{N_L}W_{ji}\qquad 1 \leq i \leq N_L\quad(9)$$

where $n_i^{(Si)} = \rho^{(Si)}/m^{(Si)}$ is the number density of a specie at the i-th energy state (either an excited level n,m or a charge state k as shown in table 1), $n_j^{(Si)}$ the number density of a specie at a higher level j, $m^{(Si)}$ the mass of a silicon atom, N_L the maximum number of levels considered and W is the overall rate coefficient indicating the possible transitions towards a higher level (from i to j)

$$W_{ij} = n_e^{(Si)}C_{ij} + \beta_{ij} + n_e^{(Si)}\gamma_{ij} + A_{ij}I_0\qquad(10)$$

and towards a lower level (from j to i)

$$W_{ji} = n_e^{(Si)}D_{ji} + \beta_{ji} + n_e^{(Si)^2}\delta_{ji} + E_{ji}I_0\qquad(11)$$

with $n_e^{(Si)}$ being the number density of electrons composing the plume. The physical mechanism related to the various rate coefficients contained in (10) and (11) is summarized in table 1.

After the initial non-equilibrium plasma formation, the species temperature T tend to equilibrate. This indicates that an LTE condition is reached. At this point the various species population can be determined at each cell of

978-1-4799-4789-8/14 $31.00 © 2014 IEEE

the domain by employing the Boltzmann (evaluation of atomic distribution function) and Saha (ionization equilibrium) equations, the last one being [24]

$$\frac{n_e x_i^{(Si,b)}}{x_{i-1}^{(Si,b)}} = \left(\frac{2\pi m_e k_b T}{h^2}\right)^{3/2} \exp\left(-\frac{IP_i^{(Si,b)}}{k_b T}\right), \quad (12)$$

where m_e is the electron charge, k_b the Boltzmann constant, h the Planck constant, n_e the total electron number density of the mixture of silicon vapor and background gas, $x_i^{(Si,b)}$ the molar fraction of the ions with i-th charge number (belonging to the silicon vapor or to the background gas) and $IP_i^{(Si,b)}$ their ionization potential. Knowing the temporal evolution of the plume's species population it is possible to calculate, at each time step, the overall absorption coefficient of the plasma α_{pl}, as the summation of the absorption coefficient due to IB and due to MPI.

Finally, the net intensity I_{net} effectively deposited into the target after plasma absorption is

$$I_{net} = I_0 \left[1 - \exp\left(-\alpha_{pl} z_{pl}\right)\right], \quad (13)$$

with z_{pl} being the thickness of the plume and I_0 the nominal intensity, assumed to have a Gaussian temporal variation.

B. 3D Finite Element code and crater formation

A 3D FE model is used to perform a transient thermal analysis on a single laser pulse as well as on train of pulses (with frequency f=10 kHz), each characterized by a profile of the net intensity effectively reaching the target $I_{net}(t) \leq I_0(t)$ as determined by the preliminary hydrodynamic simulation. Additionally, the pulses are spatially displaced with a dicing velocity v_{cut}.

Equation (1) is now extended and solved into a three-dimensional domain. It is important to notice that for the laser wavelength under investigation (λ=355 nm) the optical absorption of silicon α is about 10^8 m^{-1}, meaning that the laser pulse is almost completely absorbed within a 10 nm layer from the irradiated surface. In this case, to properly describe a volumetric load $Q_\Omega(x,y,z,t)$, a mesh as fine as 1 nm would be needed along the target thickness. Considering the dimensions of a thin wafer, typically ranging from 20 μm to 100 μm, such a mesh would lead to an enormous computational load. On the other hand, in case of ns or longer pulses, the thermal diffusion is much higher than the optical absorption length.

Therefore, the above considerations justify the implementation of the heat load as a surface heat flux $Q_\Gamma(x,y,t)$, spatially moving along the cutting direction y with dicing speed v_{cut}, as expressed by the following relationship:

$$Q_\Gamma(x,y,t) = (1-R)\, 2I_{net}(t) \exp\left(\frac{x^2}{2\sigma_x^2} + \frac{(y-v_{cut}t)^2}{2\sigma_y^2}\right),$$
$$(14)$$

where the exponential term accounts for the spatial Gaussian intensity distribution over an elliptical footprint with semi-axes σ_x and σ_y.

Due to double symmetry of the elliptical laser footprint, the simulation of the single pulse LA has been performed on one-quarter of the domain, applying the same boundary conditions as described in the previous paragraph.

Finally, an element deactivation technique is adopted to remove the elements reaching critical temperature from the computational mesh, therefore describing the geometry and the progressive formation of the crater.

3. Results

A. 1D hydro-dynamic code

The results here reported refer to the numerical simulation of a single pulse LA process on a silicon sample. A UV (λ=355 nm) Nd:YAG laser is considered, characterized by a pulse irradiance of 1.5 GW/cm^2 and 10 ns FWHM duration.

As can be seen in figure 1(a) the Gaussian time dependence of the nominal pulse intensity I_0 (solid curve) is implemented. Since at the beginning of the heat pulse no plasma has yet developed, the laser radiation can entirely reach the target and the pulse energy be deposited into the material. Therefore, during this initial transient, the net intensity irradiating the target surface I_{net} (dotted curve) coincides with I_0. The energy deposited into the target will initiate melting and evaporation of the target's material. A vapor plume will start developing above the target surface. The plume dynamics can be easily understood by observing the evolution of the vapor parameter across the KL (Fig.1(b)). At around 5 ns a sudden jump of both vapor pressure ratio and Mach number occurs. The pressure ratio P_{KL}/P_S becomes lower than one indicating that evaporation started and the Mach number $M_{KL} = 1$ indicates that evaporation reaches almost immediately sonic speed.

A sonic vapor plume is therefore expanding above the target, simultaneously interacting with the incoming laser beam. The temporal evolution of the plume temperature and velocity is shown in figure 2(a)(b).

The energy carried by the laser heats up the vapor and a hot plasma starts to develop. At a certain time, here around 8 ns, the plasma is so ionized to become strongly absorbent, therefore shielding the target from the incoming light. This happens, as previously mentioned, mainly via IB and MPI. At this point the so called plasma breakdown occurred: the laser intensity reaching the target is now just a fraction of the nominal intensity. The reduced laser intensity I_{net} reaching the target will in turn slow the plume evaporation mechanism down. Simultaneously, the Mach number starts to drop below one, indicating indeed the deceleration of the vapor. The reduced evaporation rate results in a decrease of the plasma density (due to the plume advection), allowing

978-1-4799-4789-8/14 $31.00 © 2014 IEEE

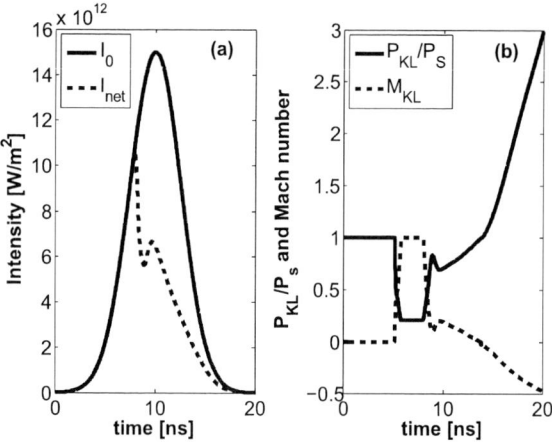

Figure 1. (a) Nominal intensity I_0 of one UV laser pulse with 10 ns FWHM duration (solid line) and net intensity I_{net} reaching the target after plasma absorption (dotted line). (b) Pressure ratio across the KL (solid line) and evaporation Mach number (dotted line).

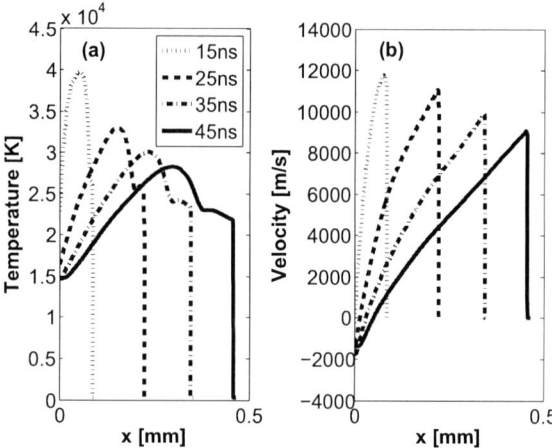

Figure 2. Plasma plume evolution at different times: (a) Temperature, (b) Velocity.

the laser radiation to reach the target again (notice the small re-bounce of intensity at about 10 ns in figure 1(a)). Finally, at later times (around 15 ns) condensation of the plume starts. As can be seen in figure 1(b), the pressure at the outer side of the KL becomes greater than the pressure at the target surface, the ratio P_{KL}/P_S becomes therefore greater than one and the Mach number negative.

B. 3D FE code

The 1D hydrodynamic simulation allowed to determine the temporal intensity profile $I_{net}(t)$ effectively reaching the target after plasma absorption. $I_{net}(t)$ has been used as the input intensity in the run of the 3D FE model. Here, the description of the crater formation has been achieved via a deactivation element technique. Due to the

double symmetry of the laser elliptical footprint, only one quarter of the domain is considered. The elements of the computational mesh are removed when their temperature reaches the critical one for silicon (around 7500 K [25]). In this way a crater is formed as shown in figure 3.

Figure 3. FE model: single pulse crater obtained via an element deactivation technique on one quarter of the target domain.

The results have been compared with real craters induced by UV single pulse LA for two fluence values (15 J/cm² and 25 J/cm²) and reasonable agreement between measured and calculated crater geometry has been found (fig. 4). Moreover, the thermal-transient simulation here considered, allows to qualitatively determine the extension of the heat affected zone (HAZ) around the crater as shown in figure 4.

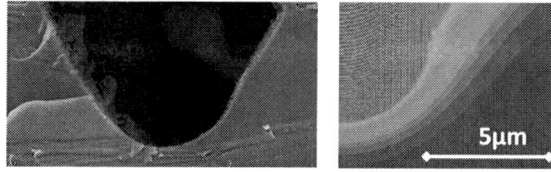

Figure 4. Single pulse LA of silicon at fluence 25 J/cm² : comparison between the sections of a real crater (left) and a calculated one (right, patterned area). The measured crater depth is 7 μm and the calculated one is 6.5 μm.

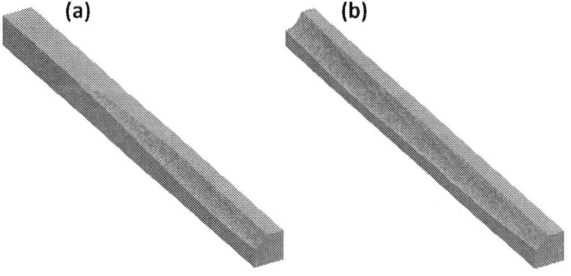

Figure 5. FE description of the crater formation due to a train of spatially moving and overlapping laser pulses: (a) 3 pluses, (b) 7 pulses.

This technique can also be applied to a train of pulses spatially moving with dicing speed v_{cut}. In this way the qualitative formation of the crater can be described. As an example, the technique has been applied on a train of pulses moving along a narrow silicon stripe, and the results are shown in figure 5.

4. Conclusions and outlook

In the manufacturing of power semiconductors, laser dicing represents nowadays a valid option for the separation of ultra-thin silicon chips. This work attempts to identify and model the main aspects underlying the interaction of a laser source with a silicon target, in a simplified yet physically self-consistent manner. For this purpose, two models are used: A self written 1D hydrodynamic code describes the cascade of interconnected processes underling LA: phase change, non-equilibrium plasma formation, plasma expansion and condensation, plasma absorption. This model allowed to estimate the amount of pulse energy effectively coupling to the silicon target. Following, a FE simulation has been conducted to describe the formation of a three dimensional crater in order to compare the numerical prediction with the experimental findings. The work has to be intended as a first step in setting up a general tool for the simulation of ns-LA processes. In the specific case, the model can be used as a complementary approach in the optimization of a laser dicing process. Several features still need to be further addressed. A better description of the target response near T_c would require a real multiphase EOS. Additional experimental tests are required in order to verify the repartition of laser energy between plasma and target. Finally, the numerical model proposed has been validated by a comparison of the calculated craters versus real ones, showing good agreement. A further verification of the model will require the comparison of the HAZ extension with single and multi-pulse experiments.

5. Acknowledgements

G. Galasso wants to express his gratitude to David Autrique for the numerous suggestions as well as for the provided help when preparing the 1D hydrodynamic code.

References

[1] DoHyung Kim, YoonJoo Kim, KyeongSool Seong, JaeKyu Song, BongChan Kim, ChanHa Hwang, and ChoonHeong Lee. Evaluation for uv laser dicing process and its reliability for various designs of stack chip scale package. In *Electronic Components and Technology Conference, 2009. ECTC 2009. 59th*, pages 1531–1536. IEEE, 2009.

[2] Jianhua Li, Hyeon Hwang, Eun-Chul Ahn, Qiang Chen, Pyoungwan Kim, Teakhoon Lee, Myeongkee Chung, and Taegyeong Chung. Laser dicing and subsequent die strength enhancement technologies for ultra-thin wafer. In *Electronic Components and Technology Conference, 2007. ECTC'07. Proceedings. 57th*, pages 761–766. IEEE, 2007.

[3] S Amoruso. Modeling of uv pulsed-laser ablation of metallic targets. *Applied Physics A*, 69(3):323–332, 1999.

[4] Dieter Bäuerle. *Laser processing and chemistry*, volume 3. Springer Berlin, 2000.

[5] P Solana, P Kapadia, JM Dowden, and PJ Marsden. An analytical model for the laser drilling of metals with absorption within the vapour. *Journal of Physics D: Applied Physics*, 32(8):942, 1999.

[6] V Schütz, U Stute, and A Horn. Thermodynamic investigations on the laser ablation rate of silicon over five fluence decades. 2013.

[7] Andreas Otto, Holger Koch, Karl-Heinz Leitz, and Michael Schmidt. Numerical simulations-a versatile approach for better understanding dynamics in laser material processing. *Physics Procedia*, 12:11–20, 2011.

[8] Kwan-Woo Park and Suck-Joo Na. Theoretical investigations on multiple-reflection and rayleigh absorption–emission–scattering effects in laser drilling. *Applied Surface Science*, 256(8):2392–2399, 2010.

[9] Anatoly Sotnikov, Harald Laux, and Bernd Stritzker. Experimental and numerical optimization of beam shapes for short-pulse ultraviolet laser cutting processing. *Physics Procedia*, 5:137–146, 2010.

[10] Sha Tao, Yun Zhou, Benxin Wu, and Yibo Gao. Infrared long nanosecond laser pulse ablation of silicon: Integrated two-dimensional modeling and time-resolved experimental study. *Applied Surface Science*, 258(19):7766–7773, 2012.

[11] AlexeyN. Volkov, GerardM. OâConnor, ThomasJ. Glynn, and GermanA. Lukyanov. Expansion of a laser plume from a silicon wafer in a wide range of ambient gas pressures. *Applied Physics A*, 92(4):927–932, 2008.

[12] Annemie Bogaerts, Zhaoyang Chen, Renaat Gijbels, and Akos Vertes. Laser ablation for analytical sampling: what can we learn from modeling? *Spectrochimica Acta Part B: Atomic Spectroscopy*, 58(11):1867–1893, 2003.

[13] Quanming Lu, Samuel S Mao, Xianglei Mao, and Richard E Russo. Theory analysis of wavelength dependence of laser-induced phase explosion of silicon. *Journal of Applied Physics*, 104(8):083301–083301, 2008.

[14] Robert Rozman, Igor Grabec, and Edvard Govekar. Influence of absorption mechanisms on laser-induced plasma plume. *Applied Surface Science*, 254(11):3295 – 3305, 2008.

[15] Mihai Stafe, Constantin Negutu, and Ion M Popescu. Theoretical determination of the ablation rate of metals in multiple-nanosecond laser pulses irradiation regime. *Applied surface science*, 253(15):6353–6358, 2007.

[16] Jong H Yoo, SH Jeong, R Greif, and RE Russo. Explosive change in crater properties during high power nanosecond laser ablation of silicon. *Journal of Applied physics*, 88(3):1638–1649, 2000.

[17] D Autrique, G Clair, D L'Hermite, V Alexiades, A Bogaerts, and B Rethfeld. The role of mass removal mechanisms in the onset of ns-laser induced plasma formation. *Journal of Applied Physics*, 114(2):023301, 2013.

[18] D Autrique, I Gornushkin, V Alexiades, Z Chen, A Bogaerts, and B Rethfeld. Revisiting the interplay between ablation, collisional, and radiative processes during ns-laser ablation. *Applied Physics Letters*, 103(17):174102–174102, 2013.

[19] Mihai Stafe. Theoretical photo-thermo-hydrodynamic approach to the laser ablation of metals. *Journal of Applied Physics*, 112(12):123112–123112, 2012.

[20] Wladimir Marine, Nadezhda M Bulgakova, Lionel Patrone, and Igor Ozerov. Insight into electronic mechanisms of nanosecond-laser ablation of silicon. *Journal of Applied Physics*, 103(9):094902, 2008.

[21] AV Gusarov and I Smurov. Near-surface laser–vapour coupling in nanosecond pulsed laser ablation. *Journal of Physics D: Applied Physics*, 36(23):2962, 2003.

[22] Eleuterio F Toro. *Riemann solvers and numerical methods for fluid dynamics: a practical introduction*. Springer, 2009.

[23] HK Chung, RW Lee, MH Chen, and Y Ralchenko. *The How To For FLYCHK*. NIST, 2008.

[24] B. Zel'dovich and Y. P. Raizer. *Physics of Shock Waves and High-Temperature Hydrodynamic Phenomena*, volume 1,2. Dover, New York, 2002.

[25] N. Honda and Y. Nagasaka. Vaporâliquid equilibria of silicon by the gibbs ensemble simulation. *International Journal of Thermophysics*, 20(3):837–846, 1999.

An Analytical Model for Thermal Failure Analysis of 3D IC Packaging

Jia-Shen Lan, and Mei-Ling Wu*
Department of Mechanical and Electro-mechanical Engineering
National Sun Yat-sen University
70 Lien-Hai Rd. Kaohsiung, Taiwan (R.O.C)
E-mail: meiling@mail.nsysu.edu.tw*

Abstract

The analytical model for temperature distribution in a multi-die stack with multiple heat sources is developed for calculating mean die temperature of a 3D IC package. The thermal resistance network model is set up based on heat dissipation paths from multi-die to ambient and is a composite of thermal spreading resistance and one-dimensional (1D) thermal resistance. Thermal spreading resistance comprises the majority of the thermal resistance when heat flows in the horizontal direction of the large plate. The present study investigates the role of determining temperature rise compared to thermal resistances intrinsic to the 3D technology, including thermal resistance of bonding layers and through-silicon-vias (TSVs). As the four thinner stacking chips in the 3D package are connected by TSVs and bumps, the Finite Element method (FEM) analysis is used to analyze the thermal management of the 3D Stacked IC package. The simulation model to obtain the multi-die temperature of 3D IC package was built up by ANSYS® APDL. The data comparison between the simulation and the analytical model showed that the analytical model is matched with the simulation model, demonstrating that the analytical model can be used to predict the thermal failure in 3D IC packages accurately. The main point in this paper is to use a simple concept and theoretical resistance network model to improve the thermal failure by redesigning the parameters or materials of the Printed Circuit Board (PCB).

1. Introduction

There are two ways to achieve through thermal management: One is prevention of thermal failure and the other is extending the useful life time of the electronic package. This paper will provide the analytical model to prevent the thermal failure. In the electronic package industry, power generation of microelectronic packages has conventionally been the main motivation for thermal management. The high operating temperature of electronic devices—caused by the combination of power and ambient conditions—is a serious concern. Temperature regulation is an important issue in electronic packaging because electronic devices may degrade permanently or even fail if not packaged properly. The purpose of the microelectronic cooling package is to dissipate heat, thus ensuring suitable operation and reliability. Recently, three-dimensional (3D) interconnection technology has been emphasized in high functionality devices, offering several electrical advantages such as increasing device density, bandwidth, and speed. While a 3D integration circuit (IC) provides some electrical benefits, it also causes significant challenges in thermal management. With higher temperature, transistors work slower due to the degradation of the carrier mobility. Some failures in thermal management include junction temperature, corrosion, electromigration failure, and increasing sub-threshold leakage. Therefore, this paper presents a way to predict the thermal failure for 3D IC packages.

Many reliability issues are currently associated with the 3D IC package, such as intermetallic compound (IMC) formation in micro-bumps, thermal stress in TSVs and micro-bumps, thermal management, known-good die (KGD), and the warpage of silicon chips [1]-[3]. However, most authors recognize thermal management as the critical issue, as it may cause the devices to degrade permanently or even fail. Some failures include junction temperature, corrosion, electromigration failure, and increasing sub-threshold leakage. Therefore, thermal management needs to be addressed effectively by preventing thermal failure or extending the useful life time of the material of electronic package. The thermal resistance network models are presented to predict the temperature of heat source and prevent thermal failure on microelectronic [4]-[7] and 3D IC packages [8], [9]. For multilevel VLSI interconnect lines incorporating via effect, the analytical models are presented to estimate temperature increase of multilevel VLSI with vias [10]-[13]. In addition, a simulation model is developed to analyze thermal behavior and temperature distribution of the 3D IC package [14], [15]. Therefore, this paper shows an effective method that can be applied to predict the thermal failure on the heat source. An analytical model of a 3D IC package with full structure is developed to estimate the temperature of stacked chips.

2. Methodology

In this paper, an analytical model is constructed, based on the calculation of thermal resistance for isotropic plate and the analysis of network model for the 3D IC package. The flowchart depicted in Fig. 1 shows the steps undertaken in this research, which was conducted in two stages, in order to discuss the thermal resistance of each element and analyze the heat path in thermal model.

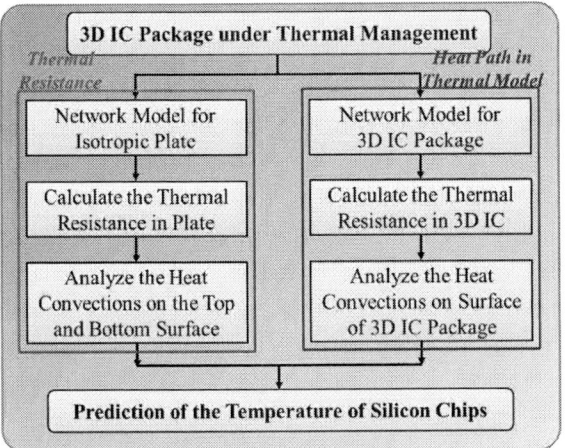

Figure 1: The Flowchart of Construting an Analytical Model for 3D IC Package

In thermal resistance, the one-dimensional thermal resistance and thermal spreading resistance are calculated by applying Laplace's equation, following the work by Muzychka et al. [16]-[18]. The thermal spreading resistance expresses the heat flow in the horizontal direction of an isotropic rectangular plate, while the one-dimensional thermal resistance expresses the heat flow from one side of the rectangular plate to the other. The general solution for two-dimensional thermal resistances is also presented, describing a constant heat source with conductive or convective cooling at the boundary.

The governing equation for the isotropic plate is Laplace's equation [16]:

$$\nabla^2 T = \frac{\partial^2 T}{\partial x^2} + \frac{\partial^2 T}{\partial y^2} + \frac{\partial^2 T}{\partial z^2} = 0 \qquad (1)$$

where
T is the temperature field at in the steady-state.

The boundary conditions for the isotropic plate are as follows:

$$\left.\frac{\partial T}{\partial z}\right|_{z=0} = -\frac{q}{k} \text{ (within the heat source area)} \qquad (2)$$

$$\left.\frac{\partial T}{\partial z}\right|_{z=0} = 0 \text{ (outside the heat source area)} \qquad (3)$$

$$\left.\frac{\partial T}{\partial z}\right|_{z=t} = -\frac{h}{k_1}[T(x,y,t) - T_f] \qquad (4)$$

$$\left.\frac{\partial T}{\partial x}\right|_{x=0,a} = 0 \qquad (5)$$

$$\left.\frac{\partial T}{\partial y}\right|_{y=0,b} = 0 \qquad (6)$$

where
q is the heat flux, k is the conductivity of isotropic plate, and T_f is the ambient temperature.

Therefore, $\theta(x, y, z)$ can be expressed as

$$
\begin{aligned}
\theta(x,y,z) = {}& A_0 + B_0 z \\
& + \sum_{m=1}^{\infty} \cos(\lambda z)[A_1 \cosh(\lambda z) + B_1 \sinh(\lambda z)] \\
& + \sum_{n=1}^{\infty} \cos(\delta z)[A_2 \cosh(\delta z) + B_2 \sinh(\delta z)] \\
& + \sum_{m=1}^{\infty}\sum_{n=1}^{\infty} \cos(\lambda x)\cos(\delta y)[A_3 \cosh(\beta z) + B_3 \sinh(\beta z)]
\end{aligned}
\qquad (7)
$$

where
$\lambda = m\pi/a$, $\delta = n\pi/b$, $\beta = \sqrt{\lambda^2 + \delta^2}$, a is the length of isotropic plate, and b is the width of isotropic plate.

For the case of a rectangular plate with convective cooling at the edge and on the bottom side, the solution is obtained by following the approach employed by Muzychka et al. [16]. The heat flux of the centroid and the heat convection of the edge and the bottom of isotropic plate are shown in Fig. 2.

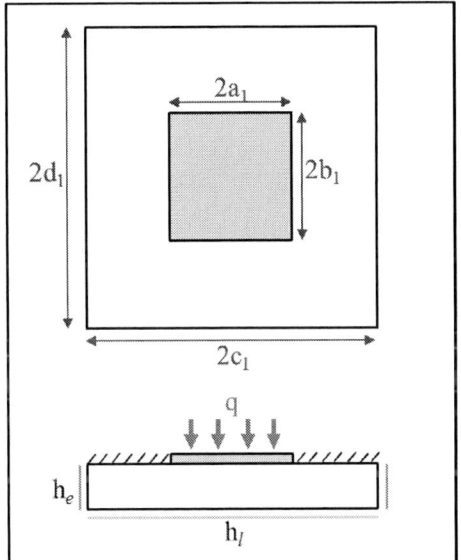

Figure 2: The Top View and Side View of Isotropic Plate with Heat Convection on the Edge and Bottom of Plate

The total resistance is calculated from the following expression:

$$R_t = \frac{c_1 d_1}{k_2 a_1^2 b_1^2} \sum_{m=1}^{\infty} \sum_{n=1}^{\infty} \frac{\sin^2(\delta_{xm}\frac{a_1}{c_1})\sin^2(\delta_{yn}\frac{b_1}{d_1})\phi_{mn}}{\delta_{xm}\delta_{ym}\beta_{mn}\left[\frac{\sin(2\delta_{xm})}{2}+\delta_{xm}\right]\left[\frac{\sin(2\delta_{yn})}{2}+\delta_{yn}\right]} \tag{8}$$

where

$$\delta_{xm}\tan(\delta_{xm}) = \frac{h_e c_1}{k_2}, \quad \delta_{yn}\tan(\delta_{yn}) = \frac{h_e d_1}{k_2},$$

$$\beta_{mn} = \sqrt{(\delta_{xm}/c_1)^2 + (\delta_{yn}/d_1)^2},$$

$$\phi_{mn} = \frac{\beta_{mn}t_1 + \left(\frac{h_l t_1}{k_2}\right)\tanh(\beta_{mn}t_1)}{\left(\frac{h_l t_1}{k_2}\right) + \beta_{mn}t_1 \tanh(\beta_{mn}t_1)},$$

a_1 and b_1 are the geometry of heat source, c_1 and d_1 are the geometry of plate, k_2 is thermal conductivity of plate, h_l is the bottom of heat convection, and h_e is the edge of heat convection.

For the case of a rectangular plate with convective cooling on the bottom side, and the adiabatic at the edge and on the top side, the solution is, once again, obtained using the work of Muzychka et al. [17]. The heat flux of the centroid and the heat convection of the bottom of the isotropic plate are shown in Fig. 3.

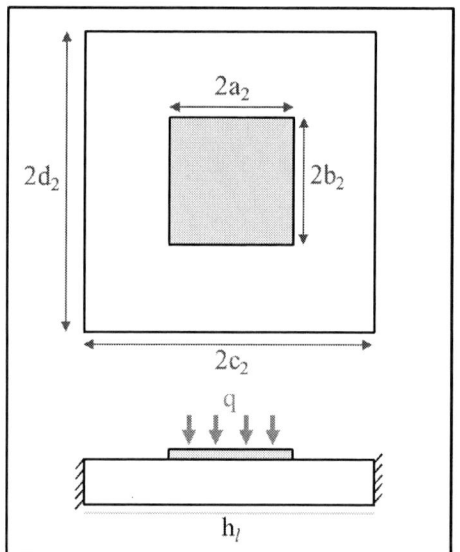

Figure 3: The Top View and Side View of Isotropic Plate with Heat Convection on the Bottom of Plate

The one dimensional thermal resistance and thermal spreading resistance are calculated from the following expression:

$$R_s = \frac{1}{2a_2^2 c_2 d_2 k_3} \sum_{m=1}^{\infty} \frac{\sin^2(a_2 \delta_m)}{\delta_m^3}\varphi(\delta_m)$$

$$+ \frac{1}{2b_2^2 c_2 d_2 k_3} \sum_{n=1}^{\infty} \frac{\sin^2(b_2 \lambda_n)}{\lambda_n^3}\varphi(\lambda_n) \tag{9}$$

$$+ \frac{1}{a_2^2 b_2^2 c_2 d_2 k_3} \sum_{m=1}^{\infty}\sum_{n=1}^{\infty} \frac{\sin^2(a_2 \delta_m)\sin^2(b_2 \lambda_n)}{\delta_m^2 \lambda_n^2 \beta_{mn}}\varphi(\beta_{mn})$$

$$R_{1D} = \frac{t_2}{4c_2 d_2 k_3} \tag{10}$$

$$R_t = R_{1D} + R_s \tag{11}$$

where

$$\delta_m = \frac{m\pi}{c_2}, \quad \lambda_n = \frac{n\pi}{d_2}, \quad \beta_{mn} = \sqrt{\delta_m^2 + \lambda_n^2},$$

$$\varphi(\xi) = \frac{(e^{2\xi t_2}+1)\xi - (1-e^{2\xi t_2})\left(\frac{h_l}{k_3}\right)}{(1-e^{2\xi t_2})\xi + (e^{2\xi t_2}+1)\left(\frac{h_l}{k_3}\right)},$$

a_2 and b_2 are the geometry of heat source, c_2 and d_2 are the geometry of plate, k_3 is thermal conductivity of plate, h_l is the bottom of heat convection.

In the heat path of the thermal model, the heat path of 3D IC package can be analyzed and designed based on the temperature distribution obtained in the simulation model. In the case study, the 3D IC package proposed by EOL/ITRI [14] was used, as shown in Fig. 4. The simulation model, referred to as the 3D vertical stacked IC package, is comprised of four layers of silicon chips, four layers of spacers, TSVs, bumps, substrate, solder ball, and PCB. The silicon chips are stacked tightly by bumps. The TSVs are surrounded the silicon chip and connect with bumps to transmit the signal from one chip to another. The material parameters of 3D IC package are listed in Table 1.

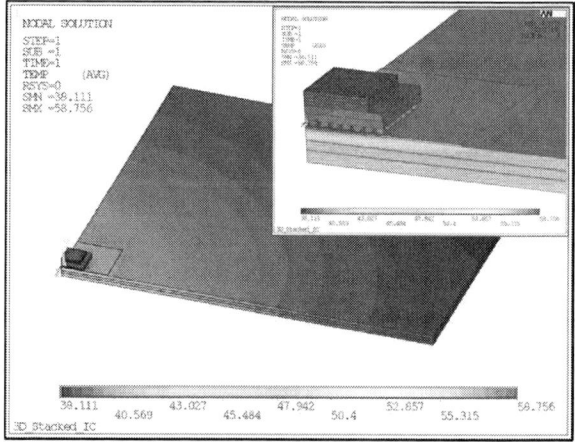

Figure 4: Temperature Distribution of 3D Stacked IC Package

The temperature distribution of the simulation output is used to analyze the heat path for 3D IC package. The heat generated by silicon chips dissipates to the ambient mainly via three mechanisms, namely the heat flow from silicon chips to the top and edge of stacked chips, the heat flow from silicon chips to the top of the substrate, and the heat flow from silicon chips to the top and bottom of PCB. Therefore, a thermal resistance network model of 3D IC package is designed by incorporating these main three heat paths. The thermal resistances of each element in the 3D IC package are calculated by applying Laplace's equation. An analytical model of 3D IC package, based on the calculation of thermal resistance and the analysis of the network model, is presented to estimate the mean temperature of silicon chips.

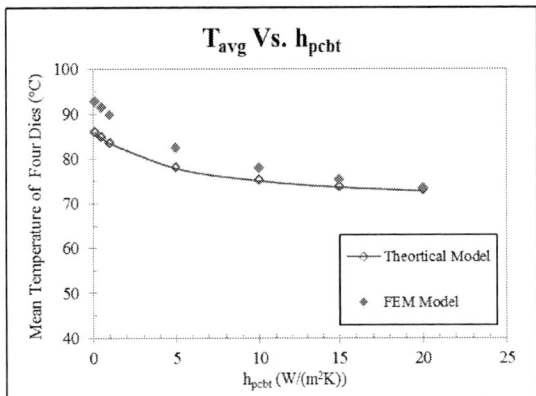

Figure 5: The Validation of Theoretical Model and FEM Model for the Mean Temperature of Four Chips with the Variable of h_{pcbt}

Table 1: Materials Properties of 3D IC Package [14]

Materials	k (W/mK)
Silicon Die	149
Silicon Substrate	149
Copper	385
Sn	63.2
ABF	0.2
FR-4	0.3
Effective Cu Trace Layer (80% Cu)	200 (x, y) 10 (z)
Solder Ball	33
Spacer	10

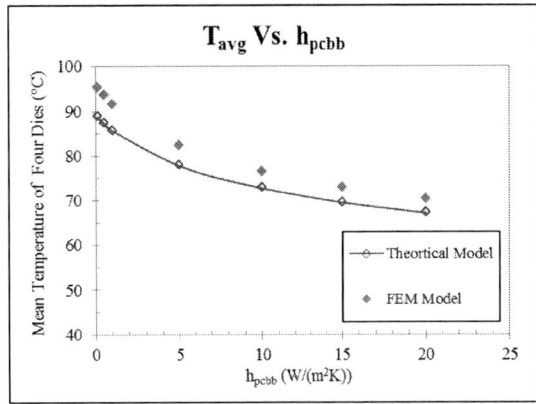

Figure 6: The Validation of Theoretical Model and FEM Model for the Mean Temperature of Four Chips with the Variable of h_{pcbb}

3. Result

The results are obtained by simulations and by applying the proposed model, as shown in Fig. 5-7. The difference between the results obtained in the simulations and the outputs of the proposed model is calculated by comparing the mean temperature of silicon chips. In most cases, the difference does not exceed 5% (shown in Fig. 5-7). However, the maximum difference of 7% is measured in the lower heat convection at the top and bottom of the PCB. Such a significant discrepancy in the findings stems from the difference in the model design and the simulation setup, whereby the former is aimed at two-dimensional structures, and the latter depicts a three-dimensional arrangement. The results indicate that the 3D IC package can be cooled down effectively by enhancing the heat convection at the top of the stacked silicon chips, as well as at the top and the bottom of the PCB.

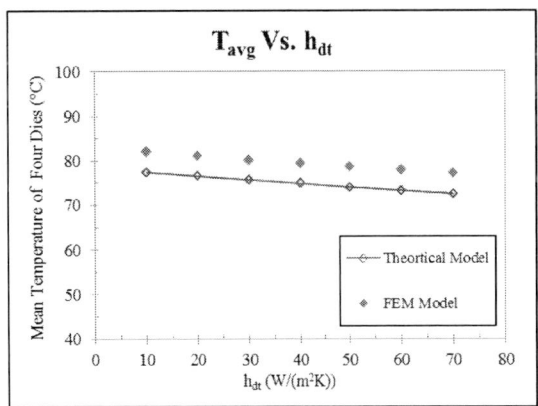

Figure 7: The Validation of Theoretical Model and FEM Model for the Mean Temperature of Four Chips with the Variable of h_{dt}

4. Conclusion

In this work, an analytical model of a 3D IC package was proposed, based on the calculation of thermal resistance and the analysis of the network model. The aim was to estimate the mean temperature of silicon chips and understand the heat convection paths of the 3D IC package. The results reveal these phenomena on the complete structure, including TSV and bump, and highlight the different thermal conductivity of material in 3D IC package. This paper provides a comprehensive model for the 3D IC package, thus improving the existing analytical approach to predicting the temperature of the heat source on the chip for the 3D IC package.

Reference

1. Lau, J. H., & Yue, T. G. (2012). Effects of TSVs (Through-silicon Vias) on Thermal Performances of 3D IC Integration System-in-package (SiP). *Microelectronics Reliability*. 52 (11). doi: 10.1016/j.microrel.2012.04.002.

2. Tu, K. N. (2011). Reliability Challenges in 3D IC Packaging Technology. *Microelectronics Reliability*. 51 (3). doi:10.1016/j.microrel.2010.09.031.

3. Cheng, E. J., & Shen, Y. L. (2012). Thermal Expansion Behavior of Through-silicon-via Structures in Three-dimensional Microelectronic Packaging. *Microelectronics Reliability*. 52 (3). doi: 10.1016/j.microrel.2011.11.001.

4. Luo, X., Mao, Z., Liu, J., & Liu, S. (2011). An Analytical Thermal Resistance Model for Calculating Mean Die Temperature of a Typical BGA Packaging. *Thermochimica Acta*. 512 (1). doi: 10.1016/j.tca.2010.10.009.

5. Ellison, G. N. (1992). Extensions of the Closed Form Method for Substrate Thermal Analyzers to Include Thermal Resistances from Source-to-substrate and Source-to-ambient. *Components, Hybrids, and Manufacturing Technology. IEEE Transactions on*. 15 (5). doi: 10.1109/33.180028.

6. Guan, D., Marz, M., & Liang, J. (2012). Analytical Solution of Thermal Spreading Resistance in Power Electronics. *Components, Packaging and Manufacturing Technology. IEEE Transactions on*. 2 (2). doi: 10.1109/TCPMT.2011.2162515.

7. Kim, L., & Shin, M. W. (2007). Thermal resistance measurement of LED package with Multichips. Components and Packaging Technologies. *IEEE Transactions on*. 30(4). doi: 10.1109/TCAPT.2007.906332.

8. Jain, A., Jones, R. E., Chatterjee, R., & Pozder, S. (2010). Analytical and Numerical Modeling of the Thermal Performance of Three-dimensional Integrated Circuits. *Components and Packaging Technologies. IEEE Transactions on*. 33 (1). doi: 10.1109/TCAPT.2009.2020916.

9. Choobineh, L., & Jain, A. (2012). Analytical Solution for Steady-State and Transient Temperature Fields in Vertically Stacked 3-D Integrated Circuits. *Components. Packaging and Manufacturing Technology. IEEE Transactions on*. 2 (12). doi: 10.1109/TCPMT.2012.2213820.

10. Huang, W., Ghosh, S., Velusamy, S., Sankaranarayanan, K., Skadron, K., & Stan, M. R. (2006). HotSpot: A Compact Thermal Modeling Methodology for Early-stage VLSI Design. *Very Large Scale Integration (VLSI) Systems. IEEE Transactions on*. 14 (5). doi: 10.1109/TVLSI.2006.876103.

11. Im, S., Srivastava, N., Banerjee, K., & Goodson, K. E. (2005). Scaling Analysis of Multilevel Interconnect Temperatures for High-performance ICs. *Electron Devices. IEEE Transactions on*. 52 (12). doi: 10.1109/TED.2005.859612.

12. Chiang, T. Y., Banerjee, K., & Saraswat, K. C. (2002). Analytical Thermal Model for Multilevel VLSI Interconnects Incorporating Via Effect. *IEEE Electron Device Letters*. 23 (1). doi: 10.1109/55.974803.

13. Wakil, J., Colgan, E. G., & Chen, S. (2011). Back-End-of-Line and Micro-C4 Thermal Resistance Contributions to 3-D Stack Packages. *Components, Packaging and Manufacturing Technology. IEEE Transactions on*. 1 (7). doi: 10.1109/TCPMT.2011.2109713.

14. Hsieh, M. C., Yu, C. K., & Wu, S. T. (2010). Thermo-mechanical Simulative Study for 3D Vertical Stacked IC Packages with Spacer Structures. *Proceedings of Semiconductor Thermal Measurement and Management Symposium 2010*, 21-25 Feburary. Santa Clara, CA.

15. Vaddina, K. R., Rahmani, A. M., Latif, K., Liljeberg, P., & Plosila, J. (2012). Thermal Modeling and Analysis of Advanced 3D Stacked Structures. *Procedia Engineering*. 30 (10). doi: 10.1016/j.proeng.2012.01.858.

16. Muzychka, Y. S., Yovanovich, M. M., & Culham, J. R. (2006). Influence of Geometry and Edge Cooling on Thermal Spreading Resistance. *Journal of Thermophysics and Heat Transfer*. 20 (2). doi: 10.2514/1.14807.

17. Muzychka, Y. S., Culham, J. R., & Yovanovich, M. M. (2003). Thermal Spreading Resistance of Eccentric Heat Sources on Rectangular Flux Channels. *Journal of Electronic packaging*. 125 (2). doi: 10.1115/1.1568125.

18. Yovanovich, M. M., Muzychka, Y. S., & Culham, J. R. (1999). Spreading Resistance of Isoflux Rectangles and Strips on Compound Flux Channels. *Journal of Thermophysics and Heat Transfer*. 13 (4). doi: 10.2514/2.6467.

Mechanical Analysis of Encapsulated Metal Interconnects under Transversal Load

B. Van Keymeulen[1], M. Gonzalez[2], F. Bossuyt[1], J. De Baets[1], and J. Vanfleteren[1]

[1] Centre for Microsystems Technology (CMST), imec and Ghent University, Technologiepark 914a, 9052 Gent, Belgium
[2] imec, Kapeldreef 75, 3001 Leuven, Belgium

Abstract

Novel insights regarding the ability of encapsulated metal interconnections to deform due to bending are presented. Encapsulated metal interconnections are used as electric conductor or measurement system within a wide range of applications fields, e.g. biomedical, wearable, textile applications. Nevertheless the mechanical analysis remains limited to reliability investigation of these configurations while deformability is as important for application fields where, so-called disappearing electronics is the main purpose.

An analysis based on the work needed to bend interconnections to a certain curvature will be used to compare different interconnection configurations with each other. The experimental as well as the simulation setup is based on PDMS encapsulated PI-enhanced Cu tracks. The results and conclusions are specific for this type of interconnections, but can be extended to a global conclusion about stretchable interconnections.

From the obtained insights it is proven that periodically meander-shaped interconnections need significant less work, up to more than 10 times less, to bend the interconnection to the same curvature compared to straight interconnection lines. Furthermore, the bendability of the straight interconnection lines is determined by the shape of the interconnection, where for meandered tracks the encapsulation will determine this factor, for an encapsulation thickness of maximum 1mm. It shows out, for the meander-shaped interconnection, that per increase of 250µm encapsulation thickness the work raises with a factor 2. For straight interconnection lines the work in function of the encapsulation thickness is limited to 20%/250µm. For encapsulations > 1mm, the encapsulation thickness will become the predominant factor which determines the deformability for both interconnection shapes.

1. Introduction

Encapsulated metal interconnections are widely investigated and used within applications where a certain degree of comfort is necessary from the user side of view or where deformability, one-time or more, is a necessary feature [1].

The traditional way to obtain electronic functionalities is to making use of conventional rigid PCBs. These kind of electronic circuits are not deformable from nature and don't accomplish the previous mentioned features regarding comfort and deformability.

Less traditional is the use of flexible circuit boards. These boards are bendable, which brings along a certain degree of comfort and deformability when one applies them in a bendable application, e.g. shoe sole.

Last years a novel approach has been invented, investigated and implemented, namely the stretchable circuit board technology. Here the interconnections are not only bendable, they are also stretchable and can be torsed without causing permanent damage under certain, limited, mechanical load conditions.

The deformability of the stretchable interconnections is widely investigated and reported in numerous papers by different research groups and institutes [2-8]. This gives us insights in design rules for reliable interconnections under, what will be called in this paper, longitudinal load conditions. Where one has now insights in developing reliable structures, one doesn't know how much work is needed to deform these kinds of interconnections. For this reason the stretchable encapsulated interconnection, made with meandered PI-supported copper tracks, will be compared with a flexible encapsulated interconnection, built up from the same material to obtain an objective comparison between both structures. Flexible interconnections are investigated because of the very simple design rules in comparison with meandered tracks and to obtain knowledge about the mechanical features of both configurations, which was not reported till now. Remark that 2D deformable structures can be very useful in certain devices, e.g. textile applications where the textile is typical not stretchable. If stretchable interconnections don't deliver additional advantages in such a case, flexible tracks can be used for their simplicity. It will be shown that unless the fact those stretchable as well as flexible interconnections are both 2D deformable, stretchable electronics delivers significant advantages regarding the ability to deform under a transversal load.

Because of the fact that flexible interconnections can't be stretched due to their straight nature, one other load case is needed instead of elongation to compare both kinds of structures. For this reason, and due to the fact that stretchable as well as flexible interconnections are bendable, a test environment will be used where the interconnections are bent to a certain curvature, while monitoring the delivered work to obtain this curvature.

The results of the numerical simulation are validated by practical experiments.

978-1-4799-4789-8/14 $31.00 © 2014 IEEE

Stretchable interconnections are not only having the ability to deform in longitudinal direction where flexible electronics don't have this ability. But, and this is the major observation of this investigation, stretchable electronics are more deformable, if they are bended, in comparison with flexible interconnections. From this it can be concluded that stretchable interconnections have advantages over flexible interconnections to implement in deformable but not necessarily extensible structures where comfort is a necessity, e.g. textile applications.

Furthermore, previous investigations [10] report that the thicker the encapsulation is, the more reliable a certain structure will behave. In this paper it is proven and validated that increasing the encapsulation thickness brings along a significant decrease in deformability. The flexural rigidity increases with a factor two per 250μm additional encapsulation thickness in case of meander shaped interconnections. This brings along a trade-off between reliability and deformability/comfort.

Unless the fact that the obtained results corresponds with PI-supported copper tracks, the conclusions can be transferred to other products serving as conductive material, encapsulation material and support material.

2. Technology description

This section summarizes the production method for the practical test vehicles used within this investigation. The production method for stretchable circuitry described in [7] will be used as starting point. Nevertheless some processing steps can be left out because only interconnections and not a whole circuitry is needed.

The used stack is a copper track supported with a layer of PI. The layer of PI has a certain offset in comparison with the copper track. The PI support layer is needed to enhance the life time of the metal interconnection under longitudinal load conditions as elongation, [9].

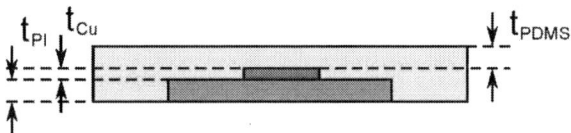

Figure 1: wanted stack of material within encapsulated PI-supported copper interconnection lines

For completeness the corresponding thicknesses for the different materials are summarized in *Table 1*.

Material	Thickness
PDMS	0μm-2000μm
Cu	18μm
PI	35μm

Table 1: material parameters in the height direction

The processing starts with a flexible foil built up of PI with Cu on top of it. The copper has a thickness of 18μm where the PI has a thickness of 35μm. This flexible foil needs to be processed following a conventional PCB manufacturing flow. The flexible foil is added to a rigid carrier to survive all the processing steps and to obtain an accurate product in the end. For this reason the flexible foil will be added to an FR4 board by means of a wax layer and vacuum lamination.

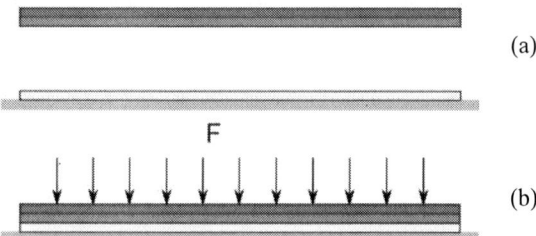

Figure 2: fix flexible foil to rigid carrier by wax layer

After adding the flexible foil on the FR4-board the structure undergoes a photolithography and etching process. We end up with a foil of PI with structured Cu at the locations where we want to have the interconnection lines.

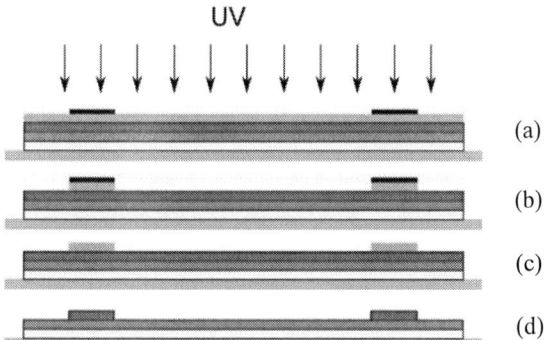

Figure 3: photolithography and etching

The PI layer will now be structured by means of a YAG-laser process. After cutting the PI, the residual PI is removed from the rigid board by means of manual peeling.

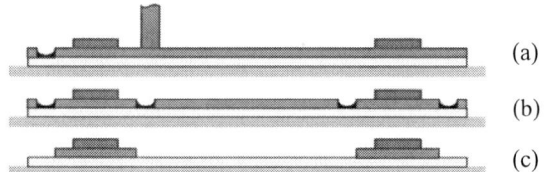

Figure 4: laser cutting and peeling

978-1-4799-4789-8/14 $31.00 © 2014 IEEE   452

The global structure will be covered with a layer of PDMS, Sylgard 186, by means of doctor blading. The whole specimen is cured for 6 hours on a temperature of 50°C. From now on the encapsulation is cured and the configurations can be cut out by hand or by laser ablation.

Figure 5: casting encapsulation and cutting structure

The resulting structures can be seen in *Figure 6* for the flexible as well as for the meandered metal interconnections.

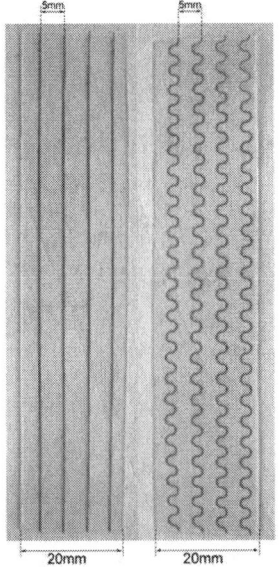

Figure 6: resulting samples (a) straight tracks (b) meandered tracks

3. Test vehicle description
A. Flexible interconnection
The outlines of a flexible interconnection are depicted in *Figure 7* as model for numerical analysis (a) and as practical test vehicle (b). It exists of a straight track of Cu with a width of 100μm. The PI has an offset of 100μm at both sides of the Cu track, resulting in a PI width of 300μm. Furthermore the test vehicle will exist of 4 parallel tracks with a distance of 5mm between the middle of the interconnect lines, as can be seen in *Figure 6*(a).

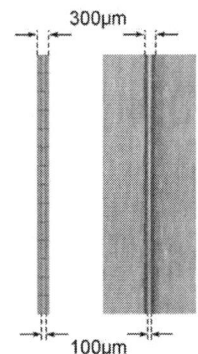

Figure 7: test vehicle for straight tracks (a) model (b) practical

B. Stretchable interconnection
The width of the copper as well as the offset of the PI support layer remains the same as for the straight tracks. The distance between the center of two meandered tracks remains also 5mm. The meander exists of two adjacent circles with an angle of 0° between the centers of two adjacent circles. Furthermore the R/W ratio of the copper tracks is 13. This makes the R=1,3mm.

For completeness a picture of the produced sample as well as the simulated design are depicted in *Figure 8*.

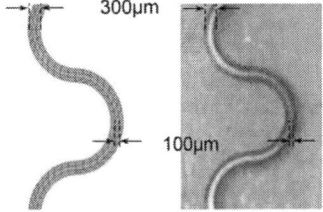

Figure 8: test vehicle for meandered tracks (a) model (b) practical

C. Summary
The dimensions of the model for numerical analysis are summarized in *Table 2*. The values are also indicated in *Figure 6*.

Parameter	Flexible interconnection	Stretchable interconnection
Length sample	108mm	108mm
Width sample	20mm	20mm
Width Cu track	0,1mm	0,1mm
Width PI track	0,3mm	0,3mm
Spacing tracks	5mm	5mm
Inner radius Cu	1,3mm	NA

Table 2: design parameters test vehicles

4. 3D numerical simulation

A. Setup

A schematic illustration of the simulation setup is shown in *Figure 9*. The right as well as the left downside of the encapsulated interconnection has a limited degree of freedom and can only move in the x-direction.

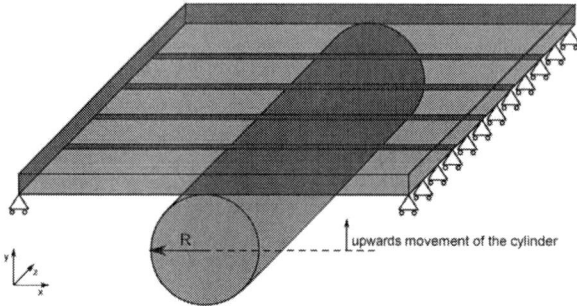

Figure 9: schematic overview simulation setup

A cylinder with a radius of 5mm will move upwards with a constant speed, 2,5mm/iteration, and will be the load for the interconnection. A total of 10 iterations, or an upwards movement of 25mm of the cylinder, is investigated.

The radius of the cylinder, *Figure 9* and *Figure 10(a)*, is rather small in comparison with the bending radius of the interconnection, see *Figure 10(b)* and *Figure 10(c)*. The reason behind this choice in simulation environment is due to the fact that we want to investigate the ability of the interconnection to bend. If a cylinder is chosen with a bigger radius, friction effects will have an influence on this test environment and the structure will be stretched where we want to neglect this property in this investigation. Furthermore, moving the cylinder too much in the upwards direction, relative to the length of the interconnection, would also bring friction and elongation effects along.

The used material parameters are based on [10] and are summarized in *Table 3*.

Material	Young's modulus	Poisson ratio	C_{10}
PDMS	NA	0.5	0,156
Cu	117 GPa	0,35	NA
PI	3,2 GPa	0,34	NA

Table 3: modelling parameters

The cylinder used to apply the load only touches the middle of the interconnection, which brings along a certain reaction force at the left and right side of the interconnection. This typical structure will now act as what is called in literature [11], a nonlinear stiffening spring.

Figure 10 shows the simulated behavior of the interconnection under bending. It can be seen that the higher the cylinder pushes the structure, the smaller its bending radius, or the higher the curvature, which is defined as $\kappa = \frac{1}{R}$.

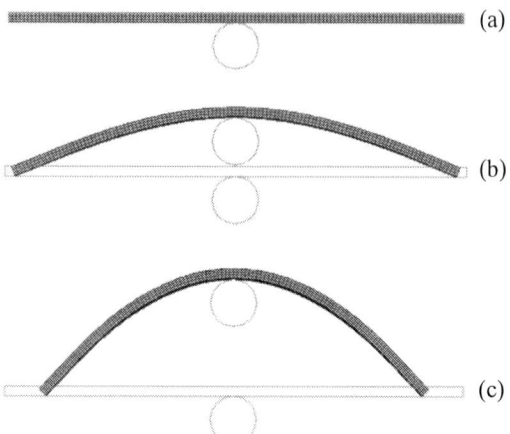

Figure 10: simulation model unbended (a), cylinder moved upwards 12,5mm (b), cylinder moved 25mm upwards (c).

The curvature of the bended interconnection changes along the interconnection, with a maximum curvature at the location where the cylinder touches the interconnection, *Figure 10* and Appendix A.

To compare the deformability of interconnections under transversal load the work needed to cause a certain degree of bending, within our investigation the maximum curvature, will be monitored.

B. Results

Figure 11 shows the work needed to bend the interconnection structure to a certain maximum curvature. This has been done for both interconnection shapes and for 6 different encapsulation thicknesses, these are 0µm, 250µm, 500µm, 750µm, 1000µm and 2000µm.

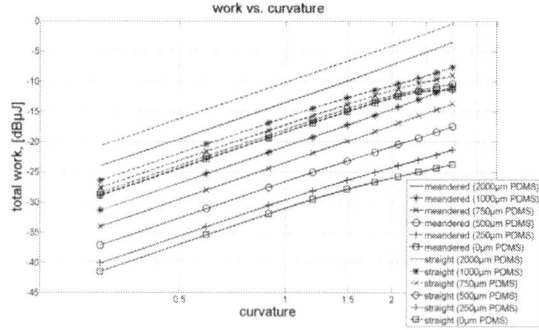

Figure 11: total work needed in function of the maximum curvature

In Appendix B it shows out that for small deflections the work to bend an interconnection has the following form:

$$10log_{10}W = 10log_{10}(EI) + 10log_{10}\left(\frac{3}{2l^3}\right)$$
$$+ 20log_{10}(y)$$

Where W, EI, l and y indicates respectively the work to bend the interconnection to a certain curvature, EI the flexural rigidity, l the half of the interconnection length and y the deflection of the interconnection.

The flexural rigidity, EI, depends on the encapsulation thickness and the shape of the interconnection. A difference in EI can be observed in *Figure 11* as an offset between two curves with different encapsulation thickness and/or interconnection shape. The second term is constant for all the different structures that have been tested due to the fact that the length of the interconnections is constant. The third term is the dependency of the deflection; this term will cause an increase of 20dB per decade of deflection for small deflections. In *Figure 11* an increase of work of about 20dB per decade curvature can be observed. This can be justified by the linear relationship between curvature and deflection for small displacements, see Appendix A.

Table 4 compares the work of the two investigated shapes with the same encapsulation thickness. It can be seen that the smaller the encapsulation is, the higher the influence of the interconnection shape becomes. The thicker the encapsulation thicknesses become the less influence the interconnection shapes will have, nevertheless at a rather thick encapsulation thickness of 2mm, a difference in work of a factor 2 can still be noticed between the different interconnection shapes.

Encapsulation thickness	ΔdB (straight – meandered)	Multiplication factor
0µm	12,54 dB	17,95
250µm	11,2 dB	13,18
500µm	8,68dB	7,38
750µm	6,44dB	4,41
1000µm	4,97dB	3,14
2000µm	3,38dB	2,18

Table 4: comparison of total work needed to bend encapsulated metal interconnects with identical encapsulation but different shape

The obtained results and equations from Appendix B make it possible to determine the flexural rigidity, EI, based on the numerical simulation and following equation:

$$EI = \frac{2Wl^3}{3y^2}$$

The results for the flexural rigidity based on numerical simulation are depicted in Table 5. Remark that the calculated EI is based on the work and deflection at the first iteration, y=2,5mm; because for this iteration the introduced errors due to simplified beam theory are negligible. It can be observed that for meander-shaped interconnections a significant difference in flexural rigidity is observed if the encapsulation thickness increases. For straight tracks this is not the case. The reason behind this behavior is that the initial stiffness of a straight interconnection without encapsulation is that high that the amount of encapsulation already needs to become very high if it needs to become predominant. Where for the stretchable interconnection a significant lower initial stiffness, of a factor 17, has been observed in comparison with the straight legs. This brings along that the stretchable interconnection its flexural rigidity is mostly determined by the encapsulation and only limited the shape.

Encapsulation thickness	Shape	Flexural rigidity µJm
0µm	Meander	6
250µm	Meander	8
500µm	Meander	16
750µm	Meander	33
1000µm	Meander	62
2000µm	Meander	337
0µm	Straight	108
250µm	Straight	111
500µm	Straight	121
750µm	Straight	147
1000µm	Straight	195
2000µm	Straight	734

Table 5: flexural rigidity in function of the shape and the encapsulation thickness

5. Lab experiment

To validate the simulated results the flexural rigidity of practical test vehicles with different encapsulation thicknesses and interconnection shapes are evaluated by means of a stiffness tester following the normalized test ASTM D1388. This test measures the overhanging length, this is the length at which the end of a test vehicle makes an angle of 41,5° with the horizontal, *Figure 12*.

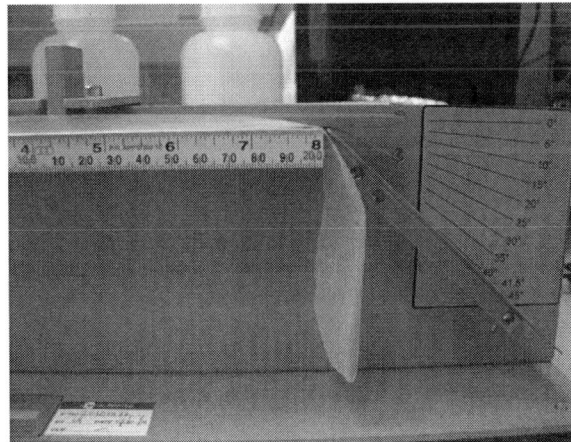

Figure 12: picture of the test environment

978-1-4799-4789-8/14 $31.00 © 2014 IEEE

To calculate the flexural rigidity of a test vehicle, the bending length c needs to be calculated from the overhang length o:

$$c = \frac{o}{2}$$

The bending length is the half of the overhang length by definition [12].

The flexural rigidity is obtained from the bending length by following formula:

$$EI = w.c^3$$

Where w is the weight of the sample per square meter and c is the bending length. Giving a flexural rigidity in µJm.

The measured overhang lengths, weights and thicknesses of the samples are indicated in Appendix C.

Next table gives the results for the different encapsulation thicknesses and interconnection shapes. The indicated values are calculated out of the averaged results over 3 samples per given structure. Furthermore, the overhang length per sample is measured 4 times, as described in ASTM D1388, and averaged.

A significant correlation between practical results can be observed, despite the minor differences in thickness between the practical and simulated samples.

Encapsulation thickness	Shape	Flexural rigidity µJm
344µm	Meander	9,7
456µm	Meander	17,9
630µm	Meander	32,1
280µm	Straight	85,7
470µm	Straight	124,9
630µm	Straight	147,9

Table 6: Flexural rigidity for the different tested test vehicles

Remark: the thickness of the samples has been measured using ASTM D3767 method. By using a normalized weight of 212g and a presser foot with a radius of 16mm.

6. Conclusions

Novel insights have been gained regarding the performance of stretchable and flexible metal interconnections to deform under a transversal load. These insights were gained from numerical simulation and verified by practical measurements.

It was proven in the preceding discussion that stretchable shapes have a significant comfort advantage over flexible shapes if they are bent. Up to more then 10 times less work is needed from the environment to bend a stretchable interconnection in comparison with a straight interconnection.

Furthermore an increase of 100% work per increase in thickness of 250µm was observed for the meandered interconnections. For the straight interconnections the shape of the interconnection is predominant for normal encapsulation thicknesses, 250µm-1000µm, and only a slight increase of max. 20% per additional 250µm encapsulation is perceived. If the encapsulation thickness increases significantly, >1mm, the shape of the interconnection will have less influence on the flexural rigidity of the structure in comparison with smaller encapsulation thicknesses, suggesting to use an encapsulation which is as thin as possible to obtain a comfortable interconnection.

The conclusions regarding encapsulation thicknesses and interconnections shapes are valid but not limited to this case of interconnections. Other elastomers, conductors, support layers can be considered as well as other processing techniques.

From these observations the claim that stretchable interconnections are in general more applicable for wearable, biomedical and textile applications is justified.

Acknowledgments

The authors would like to thank Sheila Dunphy and Steven Van Put for their practical support in the production of the test vehicles, Dow Corning for supplying Sylgard 186 and the European Commission for the financial support through the PASTA project (Grant Agreement 258724)

Appendix A: simplified beam theory

The reasoning about the behavior of the interconnection under bending can be better understood when an equivalent beam model is used. If one cuts the model of *Figure 10*(c) in the middle the equivalent schematic of *Figure 13*(a) is obtained. This schematic can be converted on its terms to one other schematic, *Figure 13*(b). If one now neglects the displacement in de x-direction, Δx, a model is obtained which is easier to analyze and widely discussed in the basic beam theory [11]. Remark that Δx is only small for rather small deflections, in this investigation this will be the case for the first 3-4 iterations as can be seen from the linearity in *Figure 11*.

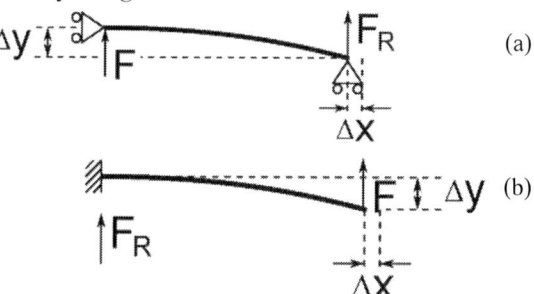

Figure 13: equivalent schematic of a half encapsulated interconnection

The curvature of a beam subjected to bending is given by:

$$\kappa = \frac{1}{R} = \frac{M}{EI}$$

The moment induced at a certain location x along the interconnection, see the equivalent schematic in *Figure 13*, is defined as:

$$M = F(x - l)$$

After introducing the last equation in the equation of the curvature, we obtain:

$$\kappa = \frac{1}{R} = \frac{F(x - l)}{EI}$$

With a maximum curvature at the location x=0:

$$\kappa_{max} = \frac{1}{R_{min}} = \frac{Fl}{EI}$$

Within the simulation model the location of x=0 in the equivalent model is the location where the cylinder touches the interconnection. From this point it is justified that the maximum curvature occurs at this location.

The applied force within the simulation environment is dependent on the deflection of the beam, y_{max}. The maximum deflection of a cantilever is given by:

$$y_{max} = \frac{Fl^3}{3EI}$$

Introducing the equation for the deflection in the last equation, we obtain:

$$\kappa_{max} = \frac{1}{R_{min}} = \frac{3y_{max}}{l^2}$$

Appendix B: determination of flexural rigidity

Work is defined in general by the following formula:

$$W = \int_C \vec{F} d\vec{r}$$

Where $\vec{F}$ is the force vector, in this case the force is only in the y-direction, and $\vec{r}$ is the displacement vector. The nodes where forces are applied are at the end of the interconnection, *Figure 13*(b). Furthermore, the displacement of these nodes of the interconnection moves in the x- as well as in the y-direction. To simplify previous equation the dot product of $\vec{F} \cdot d\vec{r}$ and due to the knowledge that there is only a force in the y-direction at the end of the interconnection, the equation for the dissipated work becomes:

$$W = \int_0^y F_l(y). dy$$

Where W is the total work, F_1 is the force applied by the load at the end of the cantilever. It is generally known from the beam theory that the displacement of a beam is dependent on the applied force or vise versa:

$$W = \int_0^y \frac{3EI}{l^3} y. dy$$

$$W = \frac{3EIy^2}{2l^3}$$

If the logarithm is taken of this equation we become:

$$10log_{10}W = 10log_{10}\left(\frac{3EI}{2l^3}. y^2\right)$$

$$10log_{10}W = 10log_{10}(EI) + 10log_{10}\left(\frac{3}{2l^3}\right) + 20log_{10}(y)$$

Appendix C: measurement results ASTM D1388

Encapsulation thickness	Shape	Weight [g/m^2]	Overhanging length [mm]
344µm	Meander	450	26
456µm	Meander	600	29
630µm	Meander	800	32
280µm	Straight	420	55
470µm	Straight	550	57
630µm	Straight	725	55

Table 7: Flexural rigidity for the different tested test vehicles

References

[1] T. Someya, *Stretchable Electronics*, 2012, Wiley, pp.1-484

[2] B. Balakrisnan, "Design of compliant meanders for applications in MEMS, actuators, and flexible electronics", Smart Materials and Structures, vol. 21, issue 7, Jul. 2012.

[3] C. Pang, "Recent advances in flexible sensors for wearable and implantable devices," Journal of Applied Polymer Science, Vol. 130, Issue 3, pp. 1429-1441, Nov. 2013.

[4] Yung-Yu Hsu, "Novel Strain Relief Design for Multilayer Thin Film Stretchable Interconnects," IEEE Transactions on Electron Devices, Vol. 60, Issue 7, Jul. 2013.

[5] L. Ming, "Design of two-dimensional horseshoe layout for stretchable electronics systems," Journal of Material Science, Vol. 48, Issue 24, Dec. 2013.

[6] A. Jahanshahi, "Stretchable circuits with horseshoe shaped conductors embedded in elastic polymers," Japanese Journal of Applied Physics, Vol. 52, Issue 5, May 2013.

[7] F. Bossuyt, "Stretchable electronics technology for large area applications: fabrication and mechanical characterization," IEEE Transactions on Components, Packaging and Manufacturing

Technology, Vol. 3, Issue 2, pp. 229-235, Feb. 2013.

[8] J. Vanfleteren, "Printed circuit board technology inspired stretchable circuits," MRS Bulletin, Vol. 37, Issue 3, pp. 254-260, Mar. 2012

[9] Yung-Yu Hsu, "Polyimide-Enhanced Stretchable Interconnects: Design, Fabrication and Characterization", IEEE Transactions on Electron Devices, Vol. 58, No. 8, Aug. 2011.

[10] Yung-Yu Hsu, "The effect of encapsulation on deformation behavior and failure mechanisms of stretchable interconnects," Thin Solid Films, 519(7), p. 2225-2234.

[11] J. Shigley, "Mechanical Engineering Design", International Edition, McGraw Hill, 1986, ISBN 0-07-100292-8

[12] P. Szablewski, "Numerical Analysis of Peirce's Cantilever Test for the Bending Rigidity of Textiles," Fibres and Textiles in Eastern Europe, Oct./Dec. 2003.

Design, Technology, Numerical Simulation and Optimization of Building Blocks of a Micro and Nano Scale Tensile Testing Platform with Focus on a Piezoresistive Force Sensor

Peter Meszmer[1], Karla Hiller[2], Daniel May[1], Steffen Hartmann[1], Alexey Shaporin[3],
Jan Mehner[3], Bernhard Wunderle[1]

[1]Technische Universität Chemnitz, Faculty for electrical engineering and information technologies,
Chair materials and reliability of microsystems, 09107 Chemnitz
[2]Technische Universität Chemnitz, Center for Microtechnologies ZfM, 09107 Chemnitz
[3]Technische Universität Chemnitz, Faculty for electrical engineering and information technologies,
Chair of Microsystems and Precision Engineering, 09107 Chemnitz

Email: peter.meszmer@etit.tu-chemnitz.de

Abstract

In this paper, building blocks of a MEMS tensile testing platform are presented. The building blocks include a thermo-mechanical MEMS actuator, driven by an aluminum thin-film heater on a thermal oxide for electrical insulation, a capacitive displacement sensor and a piezoresistive force sensor, capable of measuring forces on a nano-newton scale.

It is shown, that the presented building blocks fulfill the requirements for the use in a tensile loading stage for thermo-mechanical material characterization of one dimensional material samples on a micro- and nanoscopic scale under different environmental conditions, as varying temperatures, pressure, moisture. All components are realized in BDRIE technology, following a specimen centered approach.

In extension to previous presented actuators and sensors, the authors are aiming for high flexibility and full integratability of all components on the wafer-level and require for all building blocks the capability of electrical drive and electrical in situ readout, respectively.

1. Introduction

The thermo-mechanical reliability of microelectronic devices is based on an exact knowledge of the material and failure behavior and related parameters in the bulk and at the interfaces under given loading conditions such as temperature, moisture or vibration. The knowledge of this material and interface behavior forms the basis of a physics-of-failure based lifetime model used for predicting failures [1].

On the macro- and microscopic scale, methods for material characterization as tensile, shear and bending tests have been developed (cf. [2], [3]). For functional elements on the submicron or even nanoscopic scale, this procedures are, however, not feasible: First, these nano-functional elements cannot be clamped or fixed anymore without sacrificing reproducibility, integrity or meaningfulness of the results. Second, very often those elements

are mounted by self-assembling bottom-up processes, as e.g. dielectrophoretically deposited CNTs. As of this, a top down assembly would not test a realistic interface.

Therefore, new strategies have to be pursued and new testing methods to be developed. This may entail, that the loading mechanism or stage has to emerge around and after the nano-functional element to be tested is assembled. This signifies, that the processes required to create a loading stage have to be compatible with the process flow of heterogeneous integration of the nano-functional elements. Such a philosophy is called specimen centered approach.

Following this philosophy, this paper describes MEMS devices to form an universal micro-scale testing platform. Even on the microscopic scale, a tensile testing platform consists of three parts besides the specimen, which have to be integrated into a single chip: an actuator, a sensor to measure the force and a sensor to measure the displacement, which we call building blocks.

To achieve the given goals, a thermal actuator was designed, driven by a thin-film aluminum heater on top of an insulating SiO_2 layer. The displacement sensor is realized using the principle of differential capacitors and the force sensor is based on the piezoresistive effect.

The presented components are fabricated by Bonding and Deep Reactive Ion Etching (BDRIE) to ensure compatibility to the processes assembling the nano-functional elements of interest. Furthermore, we require for all building blocks the capability of electrical drive and electrical in situ readout, respectively, to ensure the usability of the testing platform in a wide range of applications and under various environmental conditions. The whole concept has been initially presented in [4].

2. Design Specifications of a MEMS Tensile Testing Platform

The design and the specifications of the components of the MEMS tensile testing platform, as shown in Figure 1, are based on the requirements of the samples in question. The range of specimens include, but is not limited to, microfluidic tubes [5], single wall carbon nanotubes (SWCNTs) [6] and other one dimensional materials as silicon, boron nitride and aluminum based nanowires, which are currently in the focus of research (cf. [7]–[10]).

In the here presented case, we restrict ourselves on the characterization of single wall carbon nanotubes (SWCNTs) with a length between 800 and 2000 nm. The SWCNTs have a modulus of elasticity of approximately $1\,TPa$ [11]. Considering a cross-section area of $1\,nm^2$, a force of $100\,nN$ can be estimated to achieve an elongation of 10%. As these considerations are very idealistic, we define minimum requirements for the components of the MEMS tensile testing platform as given in Table 1.

Displacement generated by actuator:	$> 1\,\mu m$
Resolution of force sensor:	$< 10\,nN$
Resolution of displacement sensor:	$< 10\,nm$

Table 1. Minimum requirements for the components of the MEMS tensile testing platform.

The designed MEMS tensile testing platform has to be capable to examine samples under different environmental conditions as varying temperature, moisture and pressure to account for later reliability testing under real life conditions. All building blocks have to be compatible with our already mentioned specimen centered approach, ensuring the full integration of specimen and building blocks.

Furthermore, we are aiming for a high flexibility of all building blocks to be able to adapt the components of the MEMS tensile testing platform to different samples as needed.

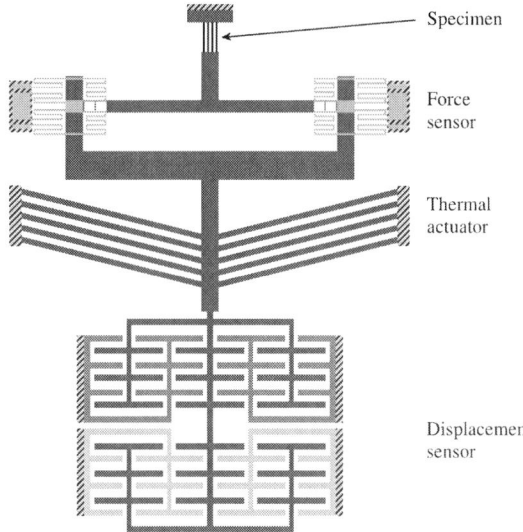

Figure 1. Concept of a MEMS tensile testing platform.

3. Components of a MEMS Tensile Testing Platform and Their Fabrication

The single MEMS building blocks are processed by Bonding and Deep Reactive Ion Etching (BDRIE) [12], [13] using equipment available at the Center for Microtechnologies Chemnitz (ZfM). The minimum structure width is given with $2\,\mu m$.

As to be developed and shown, the wafer-level processes allow a compatible processing of the nano-functional elements under research.

All components of the MEMS tensile testing platform are fabricated using a multi layered SOI-wafer. The technology flow starts with a substrate wafer, in which cavities are etched. Components above this cavities are able to move freely. To create a thin layer of silicon, which can be used to form the piezoresistive, $2\,\mu m$ thick parts of the force sensor, a SOI-wafer providing a $2\,\mu m$ active layer is used. The thin active layer is thereby located at the bottom side of the SOI-wafer. The layers are bonded onto the substrate wafer with the cavities and the handle wafer is later thinned down to an appropriate size to provide the needed thickness to form the displacement sensor. Then an additional oxide layer is deposited and patterned, followed by the metal layer structured to form electrical contacts and the heating meanders. Figure 2 depicts the different layers and their function. Finally, the actuator structure is etched into the silicon using deep reactive ion etching. Electrical contact to the piezoresistive layer is provided by insulated and metalized trenches through the insulation layers and the bulk silicon forming the displacement sensor.

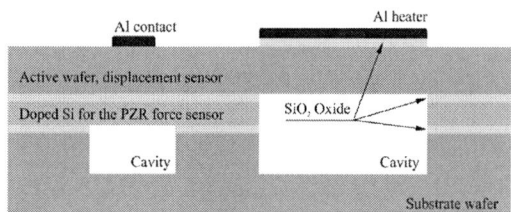

Figure 2. Cross-section depicting the layers and their function of the multi layered SOI-wafer used to fabricate the MEMS tensile testing platform (not to scale).

A. The Actuator

The actuation of the passive components of the MEMS tensile testing platform is done by means of a thermal actuator (cf. [14], [15]). The results presented in the following have been already published in [16] and we are going to summarize only the most important results.

As depicted in Figure 1 and visible in Figure 3, the thermal actuator (TA) can be described as a number of V-shaped arms, anchored at the outer rim of the actuator. At the tip of the V, all arms are connected by a shuttle. The arms are floating freely above a cavity as well as the

978-1-4799-4789-8/14 $31.00 © 2014 IEEE   460

Number of arms:		5 or 10
Length of arms:	l	$= 230\,\mu m$
Hight of arms:	h	$= 10\,\mu m$
Angle of arms:	α	$= 5$
Offset between two arms:	d	$= 15\,\mu m$
Thickness bulk silicon:	z_{Si}	$= 50\,\mu m$
Thickness SiO$_2$ insulator:	z_{SiO_2}	$= 300\,nm$
Thickness aluminum:	z_{Al}	$= 100\,nm$

Table 2. Parameters describing the thermal actuators used for the characterization and experimental verification of numerical finite element based simulations.

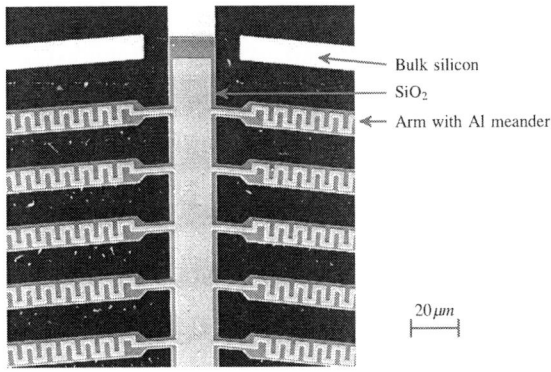

Figure 4. View on the connections between arms and shuttle with aluminum meander provided by a TESCAN PROXIMA SEM.

shuttle, which allows these parts to move laterally. The shuttle itself serves as connector to the remaining moving parts of the testing platform.

The actuator is electrically driven using contact pads located near the anchor points at the end of the V-shaped arms. Using these contact pads, a voltage difference is applied resulting in a current flow across the arms and the shuttle. The high current density causes joule heating, the resulting thermal expansion expands the arms and results in a movement of the shuttle.

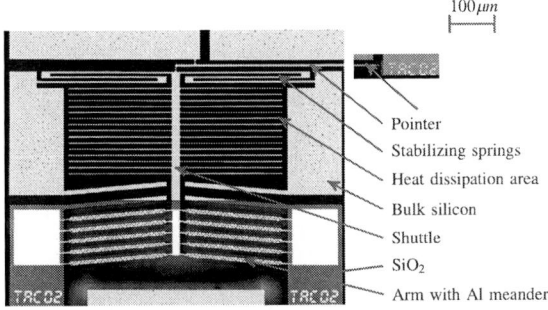

Figure 3. Low resolution overview of a TA provided by an optical microscope.

Due to the fact, that the thermal actuator can not be considered alone, but is physically connected to the remaining parts of the tensile testing platform, it has to be electrically insulated. The authors have chosen the path of vertical separation and insulation [16]. In such an approach, the heating of the bulk doped crystalline silicon (lighter gray with darker dots in Figure 3), forms the bottom layer and the base structure of the actuator, is done by an aluminum film (white) forming a meandering structure on top of an insulating silicon dioxide layer (medium gray in Figure 3). The bulk silicon houses the displacement sensor as well. Figure 4 provides a detailed view on the connections between the arms and the centered shuttle and makes the meandering structure visible.

The thermal actuators used for the characterization and experimental verification of numerical finite element based simulations can be described by the parameters shown in Table 2.

The numerical simulation of the thermal actuator is done my means of the commercial software package ANSYS 14.5, providing code for the coupled field analysis involving electrical, thermal and mechanical fields. The analysis is based on the SOLID226 element, providing twenty nodes with up to five degrees of freedom per node. The geometry is based on a cuboid.

Besides geometry and material parameters, additional input is given by the strength of the electric current across the contact pads. Furthermore, at the contact pads zero displacement as mechanical and room temperature as thermal boundary condition is applied, as well as the boundary conditions of heat dissipation by means of convection and radiation at the appropriate boundary elements. The ANSYS model of the five armed TA is shown in Figure 5. The detail on the left side depicts the two uppermost arms and their connection to the shuttle (right). Clearly visible are constrictions at the end of the arms, acting as joints and being thermally homogenizing. These constrictions are visible in Figure 4 as well. The whole TA can be described by the shown half-model, as the structure of the TA as well as the boundary conditions can be considered as symmetric.

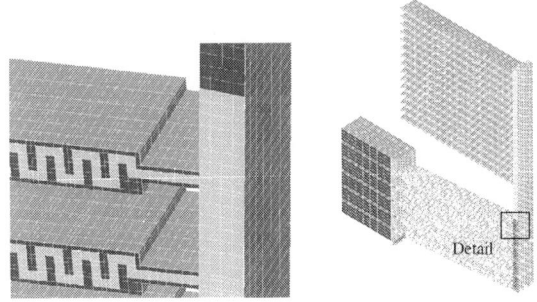

Figure 5. The elements of an ANSYS model. The area of the detail shown left is marked with a rectangle on the overview right. The bulk silicon is shown in gray, the SiO$_2$ insulator is black and the aluminum is shown in light gray.

978-1-4799-4789-8/14 $31.00 © 2014 IEEE

The generation of lateral in-plane displacement is the only task of the thermal actuator, based on thermal expansion of the arms, which moves the shuttle forward. Figure 6 plots the temperature dissipation on the surface of a five armed TA, driven by $I = 105\,mA$ ($I_s = 21\,mA$ per arm).

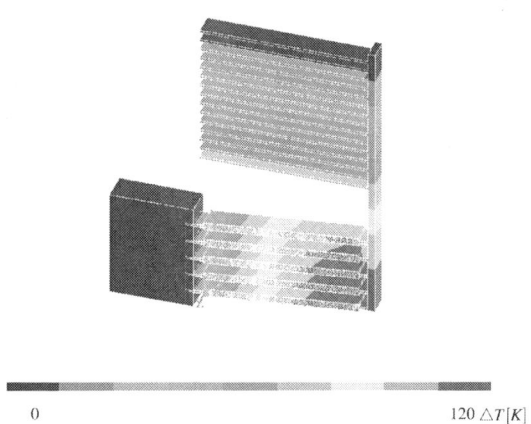

Figure 6. Results of finite element simulations of the temperature distribution on the surface of a five armed thermal actuator using a $I = 105\,mA$ ($I_s = 21\,mA$ per arm) drive.

Infrared imaging, as shown in Figure 7, provides a precise and accessible overview regarding the temperature distribution of a TA.

The measurements presented here and published in [16] are made by an Infratec ImageIR 8300 camera, providing a lateral resolution of $5\,\mu m$. The temperature resolution of the device is specified as less than $20\,mK$ at room temperature and decreases with increasing temperature. The emissivities of the relevant areas of a TA, summarized in Table 3, were experimentally determined and used for correlation of the IR measurement data.

Silicon:	0.93 - 0.97
Aluminum meander on top of SiO$_2$:	0.89

Table 3. Experimentally determined emissivities of the relevant areas of a thermal actuator.

Figure 8 depicts the temperature at two points on the lower and the upper end of the shuttle, respectively, using different drive currents. The high accordance of simulated data and experimentally obtained results has to be highlighted.

The lateral in-plane displacement can be measured using digital image correlation (DIC) [17]. The field of pixels, used for the calculation of the displacement is shown in Figure 9. Figure 10 presents the displacement measured using DIC on a ten armed TA with 5 inclined arms compared to data obtained by simulation. As one can observe, the correlation is very good indeed.

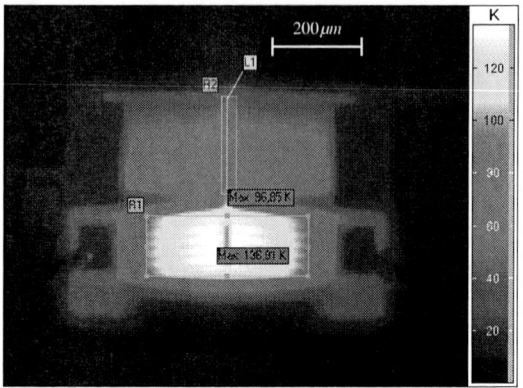

Figure 7. Infrared image of a five armed TA driven at $I = 105\,mA$ ($I_s = 21\,mA$ per arm).

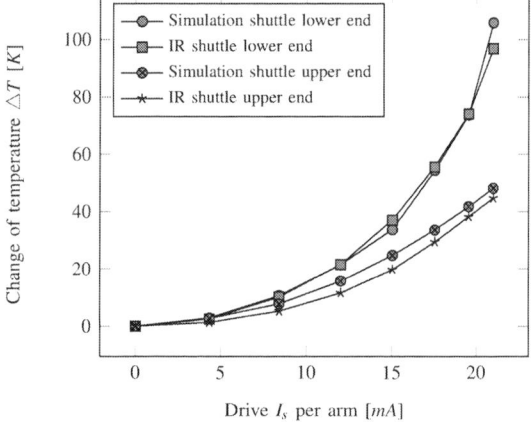

Figure 8. Temperature development measured on two points along the shuttle of a five armed TA based on a variation of drive currents. Data provided by infrared imaging and simulation.

Besides the data points obtained by simulation and DIC, Figure 10 depicts a quadratic reference curve as well. The shown reference is of the shape

$$\Delta y = a I_s^2, \qquad (1)$$

with the lateral displacement Δy, the drive per arm I_s and a scaling factor $a \approx 1.37\text{E-}3$. Δy can be approximated using the following chain:

$$I^2 R \sim Q = m c \Delta T, \qquad (2)$$

with the electric current I, electric resistivity R, generated heat Q, mass m and specific heat capacity c of the body in question. The extension $\Delta \ell$ of a beam caused by a temperature change ΔT is given by

$$\Delta \ell = \alpha_e \ell_0 \Delta T, \qquad (3)$$

with the linear thermal expansion coefficient α_e and the length ℓ_0 of the body in question, in the here examined case, the inclined arms of the thermal actuator. By combining (2) with (3) and using a small-angle approximation, we note

$$\Delta y \sim I^2. \quad (4)$$

The value $a \approx 1.37\text{E-}3$ of the scaling factor is determined using the least squares algorithm. The remaining terms of the obtained polynomial can be neglected and the norm of the residuals is given by 2.7E-4, quantifying the quality of the approximation.

However, the quadratic behavior indicates, that the main reason of heat dissipation is conduction and the effects of convection and radiation are insignificant.

The requirements presented in Section 2 include a displacement of the thermal actuator of more then $1\,\mu m$. It is shown in [16], that this can be achieved by using a TA designed with a small inclination angle of the arms as depicted in Figure 11 or by increasing the length of the arms compared to the actuators used to obtain the data presented in Figure 10. The results summarized here and extensively discussed in [16] indicate, that the described thermal actuator is capable of meeting the requirements formulated before. The successful concept of the presented actuator, heated by an aluminum meander on top of an insulator, capable of providing the specified travel range of more then $1\,\mu m$ if needed and electrically insulated against the other components of the tensile testing platform, fits seamlessly into the BDRIE technology process. The match regarding the latter points allows the full integration with the remaining components, as displacement and force sensor. Besides that, the flexible design in combination with numerical simulations, based on parametrized input files, allows the integration of an optimized actuator into a wide range of layouts and enables us to follow our specimen centered approach.

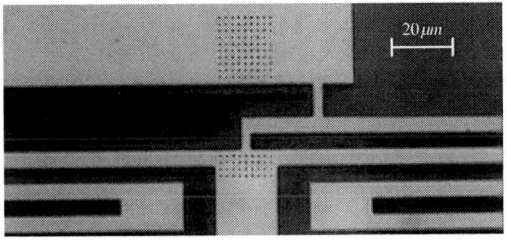

Figure 9. Field of pixels, used for the DIC based lateral displacement calculations.

B. The Displacement Sensor

The displacement sensor of the MEMS tensile testing platform has been implemented as a capacitive sensor. Following the principles presented in [18] and [19], the sensor is designed as a differential capacitor formed by one moving and two fixed parts, as depicted in Figure 12.

The displacement x of the movable part changes the capacitance C of a single cell of the capacitor formed

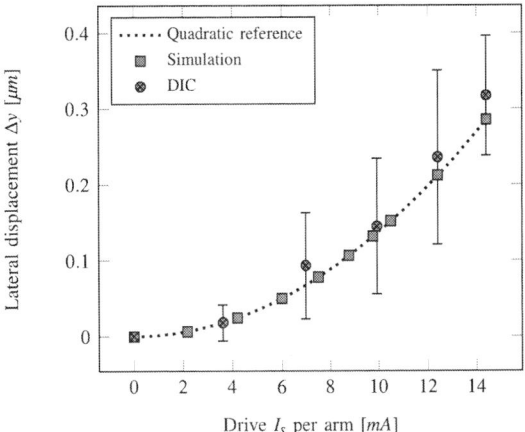

Figure 10. Lateral displacement of a ten armed TA with an arm angle of 5. Data provided by DIC and simulation.

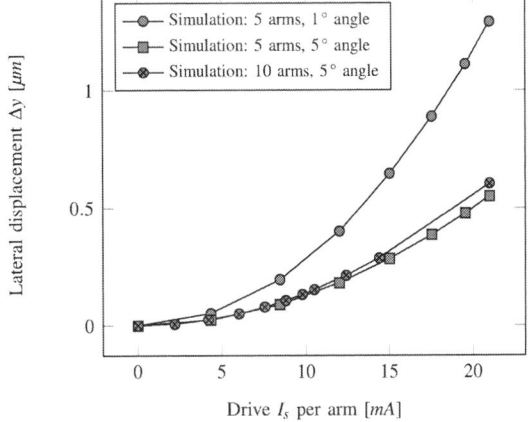

Figure 11. Expected lateral displacement of five and ten armed TAs with an arm angle of $1°$ or $5°$. Data provided by simulation.

by three electrodes by varying the distances d_0 and d_1 between the electrodes. The capacitance C is given by

$$C(x) = \varepsilon_0 \varepsilon_r \left(\frac{ht}{d_0 - x} + \frac{ht}{d_1 + x} \right), \quad (5)$$

with absolute and relative permittivity ε_0 and ε_r, respectively, the hight h of an electrode and the thickness t of the material forming the electrodes as shown in Figure 12. The choice of the design parameters d_0 and d_1 greatly affects the performance of the displacement sensor.

For technical reasons, a distance $d_0 = 2\,\mu m$ was chosen, as this represents the lower limit of the used lithography techniques. Choosing $d_0 = d_1$ would not allow any conclusions on the direction of movement and the sensitivity S in the operating point (neutral position) $x = 0$ would be zero, as the sensitivity at $x = 0$ is given by [19]:

$$S_{x=0} = \left. \frac{\partial C}{\partial x} \right|_{x=0} = \varepsilon_0 \varepsilon_r ht \left(\frac{1}{d_0^2} - \frac{1}{d_1^2} \right) = 0. \quad (6)$$

To increase the sensitivity of the sensor, the parameter d_1 has to be increased. The authors have chosen $d_1 = 5.5\,\mu m$, as this represents an acceptable compromise between needed space and sensitivity. With this configuration, one is able to reach approximately 80% of the sensitivity of a plate capacitor (c.f. [19]). Using the given parameters, the change of capacitance $\triangle C$ while varying the displacement x can be described by a nonlinear function.

By designing the sensor as a differential capacitor, $\triangle C$ can be linearized, while parasitic and base capacities could be reduced or - in a perfect setup - even be eliminated. The principle of a differential capacitor is realized using a mirrored geometry as illustrated in Figure 12. The two fixed parts and the common moving part form two independent capacitors. While the movable part is displaced in direction x, the distance d_0 is increased in the opening part of the sensor while simultaneously deceasing the distance d_0 by the factor $|x|$ in the closing part. The change of the capacitance is measured by an AC bridge.

The capacitance of a single cell can be controlled by the thickness t of the material forming the electrodes and their hight h. Besides that, the capacitance of the whole device can be increased by the usage of multiple cells as shown in 12. As choosing a higher thickness t has a positive effect on the capacitance of the sensor, shaping the sensor using the bulk silicon layer mentioned before, is advised. The bulk silicon layer has the role of stabilizing and supporting all structures. As of this it provides a thickness of at least $30\,\mu m$ and can therefore considered as optimal location for the sensor. Considering this, the capacitive displacement sensor offers high flexibility, as the target capacitance can be tuned by the choice of material thickness t, hight of the electrodes h, and the number of cells n in such a way, that an optimal base capacitance can be reached, which is determined by the evaluating electronics.

Due to the high number of adjustable parameters, the geometry of the displacement sensor can be adapted to the needs of the other components of the MEMS tensile testing platform to ensure the full integrability. The presented configuration provides a change of capacitance of $30\,fF$ per $10\,nm$, the evaluating electronics is optimized for a base capacitance of $C_0 = 10\,pF$ and provides a capacitive resolution of $4\,fF$ which yields a displacement resolution of approximately $1.4\,nm$.

Recapitulating the requirements given in Section 2, the capacitive displacement sensor fits into the described concept of a fully integrated MEMS tensile testing platform and fulfills the given requirements, offers electrical readout and complies with our specimen centered approach.

C. The Force Sensor

The MEMS force sensor is based on the piezoresistive effect (cf. [20]). Combining the piezoresistive effect and the Wheatstone bridge circuit yields a powerful base for

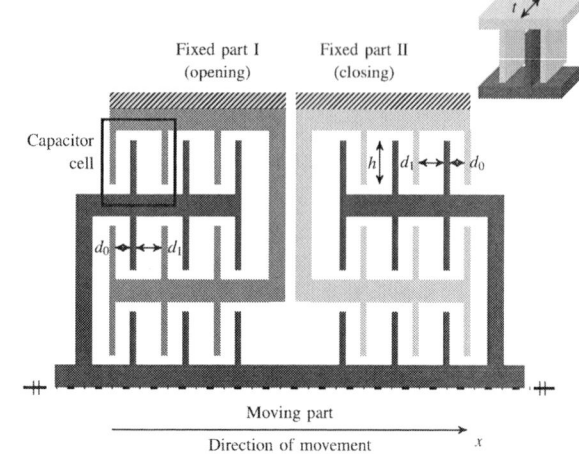

Figure 12. Top view of a simplified sector of a capacitative displacement sensor based on a differential capacitor. Perspective view of a single capacitor cell in the top right.

building a MEMS force sensor. During the development of the sensor, two main design variations have been studied.

Type I is formed by a bending beam, fixed to anchors on both sides and a moving shuttle in the center, providing connection to the thermal actuator discussed before and to a sample holder which is designed to fit to the sample in question. At the end of the arms of the beam, the piezoresistive structure, acting as a strain transducer, is embedded into the structure of the beam. This initial design, shown in Figure 13, closely follows the design presented in [21]. The main difference to the cited work is the fabrication using SOI-wafers, resulting in a simplified overall design and readout. In contrast to the design described in [21], the force sensor of a tensile testing platform has to be connected to an actuator and a displacement sensor. This creates a prestressing of the sensor, making it more complicated to perform a correct calibration.

As of this, a second design was studied and we are going to focus in this paper on the second type of the MEMS force sensor. The sensor of type II is formed by a bending beam, which is surrounded by an U-shaped carrier. At the end of the arms of the beam, the piezoresistive structure, acting as a strain transducer, is embedded into the structure of the beam, as in type I. The U-shaped carrier is connected to the thermal actuator and the bending beam is equipped with a connection to the sample holder mentioned before. The range of specimens include, but is not limited to, microfluidic tubes [5], single wall carbon nanotubes (SWCNTs) and other one dimensional materials as silicon, boron nitride and aluminum based nanowires. Beam, sensor and carrier are floating above a cavity, allowing the structure to move freely. Stabilizing springs to the left and the right of the arms of the U-shape ensure electrical drive and readout and provide the necessary stability. Figure 14 depicts the schematics of the structure. By removing or destroying the sample, the force sensor is free of outer loads and can thereby easily be calibrated.

To ensure electrical insulation of all components, the technique already discussed for the thermal actuator is used. The piezoactive layer is separated by insulating silicon dioxide layers from the substrate wafer and the layers above it, as depicted in Figure 2. Electrical contact is provided by insulated and metalized trenches through the insulation layers and the bulk silicon. This allows the full integration with the remaining components.

While the thermal actuator generates a displacement, the resistance of the sample in question results in a deformation of the bending beam of the sensor. The deformation strain results in compression and extension of accordingly designed areas in the beam.

These piezoresistive transducers are placed in the region of maximum stress and oriented in such a way, that they are in the pathway of electrical current and the change of electrical resistance can be measured. The magnified detail in Figure 14 depicts the piezoresistive transducers and the resulting deformation in the case of a tensile test. By reversing the operation direction of the actuator, a compression test can be undertaken.

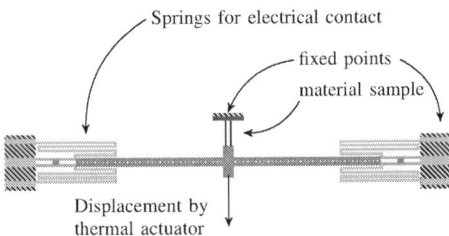

Figure 13.   Schematic representation of the force sensor type I.

The Wheatstone bridge circuit [22] is depicted in Figure 16 and the piezoresistive areas corresponding to the resistors R_1 to R_4 are shown in Figure 14. By applying a driving voltage U_A and measuring the bridge voltage U_E, the strain ε can be calculated by means of

$$\frac{U_A}{U_E} = |\varepsilon| k, \tag{7}$$

using the gage factor k. Monocrystalline silicon demonstrates a gage factor as high as 200, in comparison to metallic foil gages, which usually have a gage factor slightly over 2. This enables us to measure very small strains caused by forces on a nano-newton scale. Besides that, it has to be emphasized, that in the case of crystalline structures the gage factor is directional dependent and given by

$$k = \frac{\Delta \rho}{\rho \varepsilon} + 1 + 2\nu \tag{8}$$

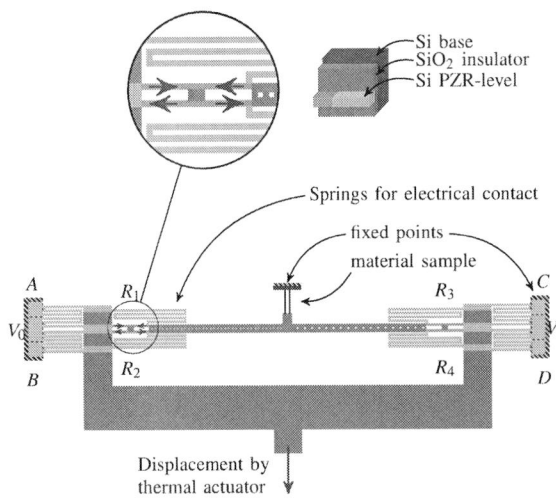

Figure 14.   Schematic representation of the force sensor type II. The magnified detail shows the piezoresistive area, the detail in the upper corner depicts the layers of the sensor.

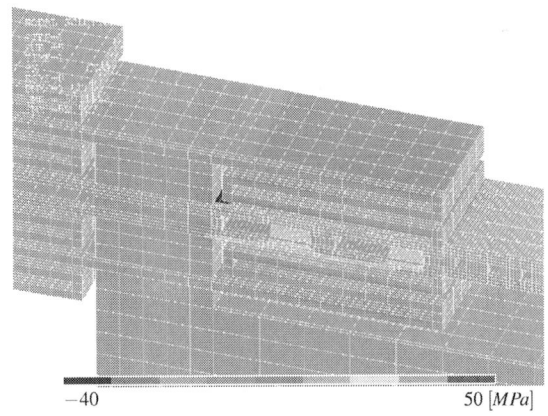

Figure 15.   Detail of a FE-simulation: Mechanical stress inside the piezoresistive elements under a displacement of $1\,\mu m$ and an incompressible sample.

using the Poisson's ratio ν and the directional dependent component

$$\frac{\Delta \rho_\omega}{\rho} = \sum_{\lambda=1}^{6} \pi_{\omega\lambda} \sigma_\lambda, \tag{9}$$

with current orientation ω, stress tensor σ and piezoresistance tensor π. As of this, the orientation of the sensor on the wafer is of importance.

The in this paper presented data is based on the following piezoresistance tensor π:

$$\pi = \begin{pmatrix} \pi_{11} & \pi_{12} & \pi_{12} & 0 & 0 & 0 \\ \pi_{12} & \pi_{11} & \pi_{12} & 0 & 0 & 0 \\ \pi_{12} & \pi_{12} & \pi_{11} & 0 & 0 & 0 \\ 0 & 0 & 0 & \pi_{44} & 0 & 0 \end{pmatrix},$$

with the three independent piezoresistive coefficients

$\pi_{11} = 6.6\text{E-}5\,MPa^{-1}$, $\pi_{12} = -1.1\text{E-}5\,MPa^{-1}$ and $\pi_{44} = 138.1\text{E-}5\,MPa^{-1}$ for p-type silicon [23].

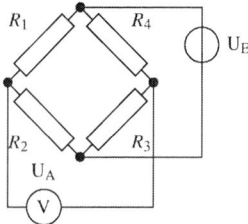

Figure 16. Wheatstone bridge circuit.

In the case of the presented force sensor, the driving voltage U_A is applied at the contact pads V_0 and V_1 as depicted in Figure 14. Using the additional pads A to D marked in Figure 14 as well, the full Wheatstone bridge circuit can be formed. The first arm of the bridge circuit is formed by short-circuiting the pads A and D, the second arm is created by short-circuiting the pads B and C. The bridge voltage U_E is then measured between the arms of the bridge.

The coupled field finite element simulations, as shown in Figure 15, are again based on commercial software package ANSYS 14.5 involving electrical, thermal and mechanical fields. As in the case of the thermal actuator, the analysis is based on the SOLID226 element, providing twenty nodes with up to five degrees of freedom per node.

The design of the force sensor can be described by a number of key parameters ranging from the thickness of the layers forming and supporting the sensor to geometrical parameters influencing the actual layout of the piezoresistive transducers and the surrounding structure. Figure 17 shows a selection of parameters and their influence on the bridge voltage U_E using absolute values based on a parameter variation of $\pm 10\%$. The presented values are based on finite element simulations. Figure 18 depicts the influence of the same selection of parameters on the bridge voltage U_E, but does not show absolute values and is based on a wider variation of parameters. The figure shows for all parameters a nearly linear or weak exponential influence on the bridge voltage U_E, making it easy to find the optimal parameter value of a single parameter at the outer edge of the definition area of the parameter, if the other parameters have been fixed due to outer limitations. The behavior becomes more clear if a single parameter is depicted across a wide variation of values, while all other parameters are fixed. Figure 19 draws such a scenario with varying beam length and varying length of the piezoresistive transducers.

For key parameters as shown in Table 4, a sensitivity of $3\frac{\mu V}{nN}$ using the sensor type I can be achieved using a drive of $12\,V$. Being able to calibrate the sensor of type II, goes along with a slight decrease in sensitivity doe to the surrounding U-shape. Using the parameters shown in Table 4 in combination with the sensor of type II, a sensitivity of $2.5\frac{\mu V}{nN}$ can be expected. As Figure 17 clearly depicts, the sensitivity can be increased by slightly increasing the driving voltage U_A as the connection between driving voltage U_A and bridge voltage U_E is linear.

Cross-section of piezoresistive elements:	$2 \times 2\,\mu m^2$
Length of piezoresistive elements:	$\sim 90\,\mu m$
Length of bending beams:	$\sim 670\,\mu m$

Table 4. Key parameters of one realization of the force sensor.

As discussed before, the MEMS tensile testing platform is fabricated using a BDRIE process and all parts of the platform a placed on different levels of the multi layered SOI-wafer. Figure 17 depicts clearly the insensitivity of the bridge voltage U_E of the sensor regarding the thickness of the insulating SiO_2 layer and the Si layer hosting the capacitative displacement sensor. This offers great flexibility regarding the design.

Furthermore, the piezoresistive force sensor provides a sensitivity as specified in Section 2, is electrically driven and offers electrical readout, fits in the concept of a MEMS tensile testing platform and therefore allows us to follow our specimen centric approach.

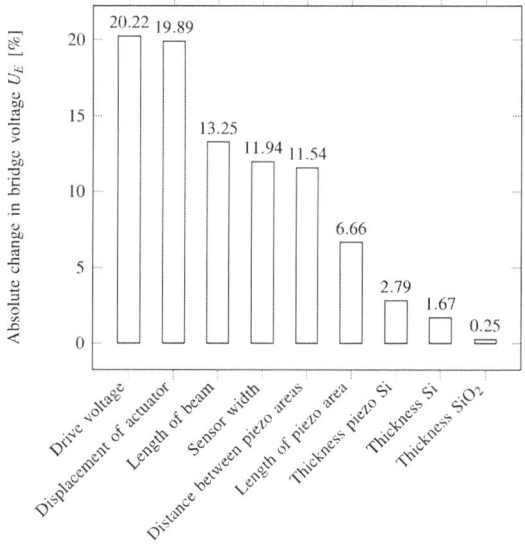

Figure 17. Sensitivity analysis of a set of selected parameters describing the force sensor type II based on finite element simulations regarding a reference set and parameter variations of $\pm 10\%$.

4. Conclusions

This paper focused on building blocks of a MEMS tensile testing platform, capable of thermo-mechanical material characterization on a micro- and nanoscopic scale under different environmental conditions, as varying temperatures, pressure, moisture.

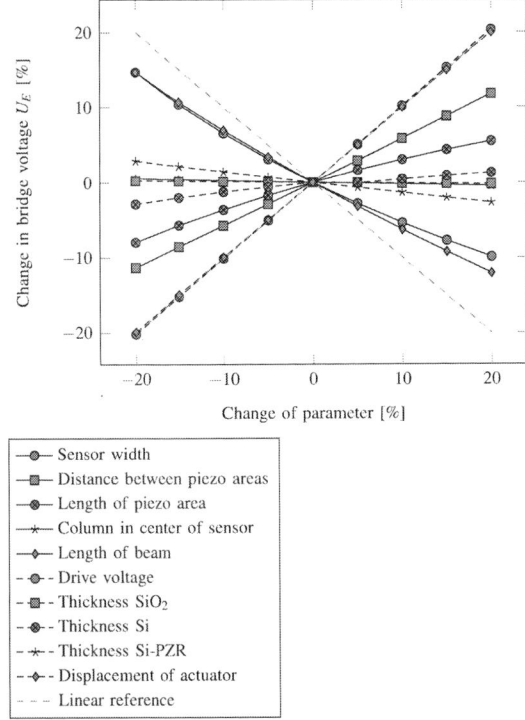

Figure 18. Sensitivity analysis of a set of selected parameters describing the force sensor type II based on finite element simulations regarding a reference set.

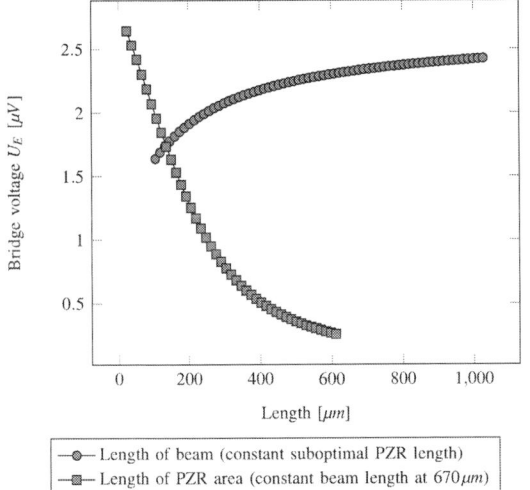

Figure 19. Development of the bridge voltage U_E of the force sensor type II under a load of $1\,nN$ while independently varying the length of the piezoresistive transducers and the length of the bending beam.

The goal of the integration of all components into the testing platform on the wafer-level is technologically challenging and can be considered as new approach.

The data presented in this paper shows the high flexibility of all components of the tensile testing platform. Furthermore, the shown results indicate, that all components are able to meet the requirements, formulated before.

The successful concept of an actuator, heated by an aluminum meander on top of an insulator, provides the specified travel range of more then $1\,\mu m$ if needed, is electrically insulated against the other components of the tensile testing platform and fits seamlessly into the BDRIE technology process.

The well known and in this paper recapitulated concept of a displacement sensor based on a differential capacitor, provides the capability of measuring displacements with a resolution below $10\,nm$, as specified before. The described flexibility of the layout of the sensors facilitates the overall layout of all components.

Finally, the presented piezoresistive force sensor, while defined on a microscopic scale, is capable of measuring forces on the nanoscopic scale. As the other components, it provides a high flexibility regarding outer parameters to achieve the needed sensitivity and shows insensitivity regarding geometric parameters needed to define the remaining parts of the MEMS tensile testing platform.

All components are electrically driven and provide electrical readout, enabling us to provide a test platform capable of tests under different environmental conditions. The flexible design in combination with numerical simulations, based on parametrized input files, allows the integration of optimized building blocks into a wide range of layouts and enables us to follow our specimen centered approach.

The integration of all components was successfully finalized shortly before the submission of this paper. A view on the tensile testing platform is shown in Figure 20 along with the electrical contacting using a multi contact wedge card, whose 15 needle tips are visible at the lower end of the picture. In the shown layout a holder for microfluidic tubes is attached to the force sensor, but is not visible doe to the size. The picture is provided by an optical microscope.

With samples on hand now, the authors are going to present simulated data in comparison to experimental data obtained by the MEMS tensile testing platform.

5. Acknowledgements

The authors wish to thank the Deutsche Forschungsgemeinschaft, DFG, for the financial support within the research unit 1713 "Sensoric Micro- and Nanosystems".

References

[1] B. Wunderle and B. Michel. Lifetime modeling for microsystems integration – from nano to systems. *J. of Microsystem Technologies*, 15(6):799 – 813, 2009.

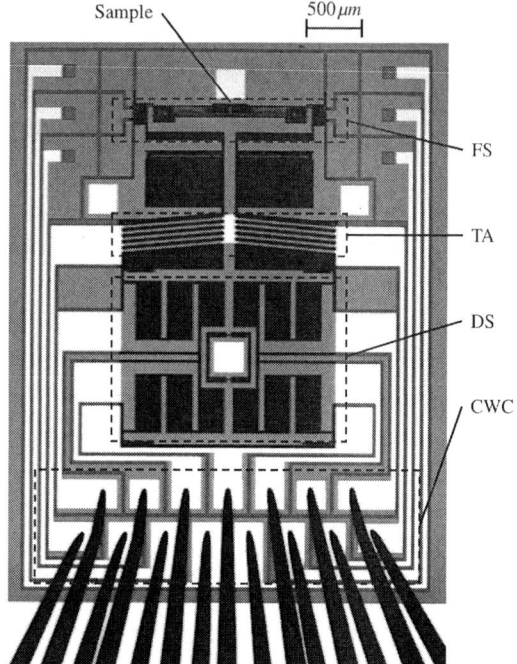

Figure 20. Integration of all components of the MEMS tensile testing platform. Visible is the force sensor (FS), the thermal actuator (TA), the displacement sensor (DS) and at the lower end of the picture the needle tips of a multi contact wedge card (CWC) used for electrical contacting. Picture provided by an optical microscope.

[2] K. Komvopoulos. Surface engineering and microtribology for microelectromechanical systems. *Wear*, 200(1–2):305–327, 1996.

[3] M.A. Haque and M.T.A. Saif. Microscale materials testing using mems actuators. *Journal of Microelectromechanical Systems*, 10(1):146–152, 2001.

[4] G. Schondelmaier, S. Hartmann, D. May, A. Shaporin, S. Voigt, R.D. Rodriguez, O.D. Gordan, D.R.T. Zahn, J. Mehner, K. Hiller, and B. Wunderle. Piezoresistive force sensor and thermal actuators usage as applications to nanosystems manipulation: Design, simulations, technology and experiments. *14th International Conference on Thermal, Mechanical and Multi-Physics Simulation and Experiments in Microelectronics and Microsystems (EuroSimE)*, pages 1–6, 2013.

[5] S. Böttner, S. Li, M.R. Jorgensen, and O.G. Schmidt. Vertically aligned rolled-up sio2 optical microcavities in add-drop configuration. *Appl. Phys. Lett.*, 102, 2013.

[6] S. Iijima. Helical microtubules of graphitic carbon. *Nature*, 354:56–58, 1991.

[7] C. Stampfer, T. Helbling, D. Obergfell, B. Schöberle, M. K. Tripp, A. Jungen, S. Roth, V.M. Bright, and C. Hierold. Fabrication of single-walled carbon-nanotube-based pressure sensors. *Nano Letters*, 6(2):233–237, 2006.

[8] C. Stampfer, A. Jungen, R. Linderman, D. Obergfell, S. Roth, and C. Hierold. Nano-electromechanical displacement sensing based on single-walled carbon nanotubes. *Nano Letters*, 6(7):1449–1453, 2006.

[9] Y. Mei, G. Huang, A.A. Solovev, E.B. Ureña, I. Mönch, F. Ding, T. Reindl, R.K.Y. Fu, P.K. Chu, and O.G. Schmidt. Versatile approach for integrative and functionalized tubes by strain engineering of nanomembranes on polymers. *Advanced Materials*, 20(21):4085–4090, 2008.

[10] Y. Mei, A.A. Solovev, S. Sanchez, and O.G. Schmidt. Rolled-up nanotech on polymers: From basic perception to self-propelled catalytic microengines. *Chemical Society Reviews*, 40(5):2109–2119, 2011.

[11] Y. Wu, M. Huang, F Wang, X. M. H. Huang, S. Rosenblatt, L. Huang, H. Yan, S. P. O´Brien, J. Hone, and T. F. Heinz. Determination of the young's modulus of structurally defined carbon nanotubes. *Nano Letters*, 8:4158–4161, 2008.

[12] K. Hiller, M. Kuechler, D. Billep, B. Schroeter, M. Dienel, D. Scheibner, and T. Gessner. Bonding and deep rie: a powerful combination for high-aspect-ratio sensors and actuators. *Proc. SPIE*, 5715:80–91, 2005.

[13] K. Hiller, S. Hahn, M. Küchler, D. Billep, R. Forke, T. Geßner, D. Köhler, S. Konietzka, and A. Pohle. Erweiterungen und anwendungen der bdrie-technologie zur herstellung hermetisch gekapselter sensoren mit hoher güte. In *Mikrosystemtechnik 2013 - Von Bauelementen zu Systemen*, 2013.

[14] J. Jonsmann, O. Sigmund, and S. Bouwstra. *Compliant electro-thermal microactuators*. 1999.

[15] Y. Zhu, A. Corigliano1, and H.D. Espinosa. A thermal actuator for nanoscale in situ microscopy testing: design and characterization. *Journal of Micromechanics and Microengineering*, 16(2):242–253, 2006.

[16] P. Meszmer, K. Hiller, S. Hartmann, A. Shaporin, D. May, R.D. Rodriguez, J. Arnold, G. Schondelmaier, J. Mehner, D.R.T. Zahn, and B. Wunderle. Numerical characterization and experimental verification of an in-plane MEMS-actuator with thin-film aluminum heater. *Microsystem Technologies*, 2014. Accepted for publication.

[17] J. Keller, D. Vogel, A. Schubert, and B. Michel. Displacement and strain field measurements from spm images. In B. Bhushan, H. Fuchs, and S. Hosaka, editors, *Applied Scanning Probe Methods I*. Springer, 2003.

[18] Y. Sun, B.J. Nelson, D.P. Potasek, and E. Enikov. A bulk microfabricated multi-axis capacitive cellular force sensor using transverse comb drives. *Journal of Micromechanics and Microengineering*, 12(6):832 – 840, 2002.

[19] M. Dienel. *Entwicklung und Analyse von Arrays mikromechanischer Beschleunigungssensoren*. PhD thesis, Technische Universität Chemnitz, 2009.

[20] M.H. Gnerlich. *Microelectromechanical Actuator and Sensor System for Measuring the Mechanical Compliance of Biological Cells*. PhD thesis, Lehigh University, 2012.

[21] M.H. Gnerlich, S.F. Perry, and S. Tatic-Lucic. A submersible piezoresistive mems lateral force sensor for a diagnostic biomechanics platform. *Sensors and Actuators A: Physical*, 188:111–119, 2012.

[22] K. Hoffmann. *Applying the Wheatstone Bridge Circuit*. HBM Publication, 2001.

[23] C. Liu. *Foundations of MEMS*. Prentice Hall, 2005.

Study on configuration design of interconnection in high power module

Li-Ling Liao[a,b], Tuan-Yu Hung[b], Chun-Kai Liu[a], Yen-Fu Su[b] and Kuo-Ning Chiang[b,*]

[a] Electronic and Optoelectronics Research Laboratories, Industrial Technology Research Institute.
[b] Dept. of Power Mechanical Engineering, National Tsing Hua University,
HsinChu, Taiwan 300, R.O.C.
Corresponding author: Tel: 886-3-5742925, Fax: 886-3-5745377,
E-mail:knchiang@pme.nthu.edu.tw

Abstract

In a high power module, a high electrical load may cause electromigration and induce the joule heating, subsequently raising the chip temperature. Temperature excursion in the IGBT chip may produce thermal stress, inducing failure in the metal wire and chip. Those failure modes degrade the reliability of a power module. This study elucidates the coupling behavior of a high power module through electro-thermal coupling analysis and thermo-mechanical analysis. Additionally, the configuration design of an aluminum wire is simulated and implemented. Related design parameters, e.g., wire arrangement position, wire number, wire bonding perimeter and wire height, are studied and the design trend of a metal wire is analysed and suggested as well.

Keyword: power module, IGBT chip, failure modes, electro-thermal coupling analysis and thermo-mechanical analysis.

1. Introduction

Owing to the increasing demand for electrical products with a high power density, the power module with high power and high operation temperature has received increasing attention [1]. A power module integrated with an insulated gate bipolar transistor (IGBT) chip has been widely applied in hybrid vehicles, DC/AC inverters and other electrical products. Despite a fast switching rate and low switching loss of the IGBT chip, the electrical load can induce the Joule heating, ultimately increasing the IGBT chip temperature. Thermal stress induced from electrical loading often results in failure in the metal wire or chip. Failure behaviors, e.g., heel cracking, fractures, wire lift-off and metallurgical damage of metal wire and IGBT chip, are detected during the reliability test [2-4]. Previous studies have analyzed the arrangement position of wire, based on electro-thermal simulation, to predict the temperature distribution; the maximum junction temperature is reduced as well [5]. The thermo-mechanical stress on metal wires has also been estimated by thermo-mechanical simulation during the current load [6]. A related investigation has examined the electro-thermal coupling effect, thermo-mechanical behavior [7] and life prediction of cycle fatigue in aluminum wires under a power cycling test [8]. According to their results, crack growth is visible in the interface between aluminum wires and aluminum-silicon during the thermal fatigue cycle test [3].

This study investigates the configuration design of interconnection in a high power module by using electro-thermal coupling and thermo-mechanical analyses. First, the current density of power module is predicted by using electro-thermal coupling analysis. The joule heating induced from electrical load significantly influences the temperature in the power module. The estimated temperature from the electro-thermal coupling analysis is regarded as temperature loading in the thermo-mechanical analysis. Additionally, mechanical behavior of the interconnection is assessed. Design parameters include the wire arrangement position, wire number, wire bonding perimeter and wire height are studied, and the design trend of aluminum wire is examined as well. Furthermore, FE analysis can assess the design trend [9,10] of interconnection efficiently and significantly reduce the development time of a high power module.

2. Power module, finite element mode and material properties

Figure 2 illustrates the established FE model based on a power module of 450 A and 1,200 V. Figure 1 shows the power module developed by the Industrial Technology Research Institute (Hsinchu, Taiwan). A constant temperature boundary condition was applied to below the bottom copper, which was measured below the center of the IGBT chip by a thermal meter. The heat transfer coefficient of the upper surface was set according to the following experimental equation of Ellison [11]:

$$h_c = 0.833 f \left(\frac{T_j - T_a}{L_{ch}} \right)^n$$

where L_{ch} denotes the characteristic length; f and n represent undetermined coefficients; T_j refers to the junction temperature; and T_a denotes the ambient temperature.

The electrical loading applied in region 1 is 70 A, and the output voltage set in region 2 is 0 V. Table 1 lists the material properties examined in the electro–thermal FE analysis. According to the data sheet of the ABB IGBT chip, electrical resistivity of the IGBT chip is temperature-dependent [12]. In this study, power loss of the IGBT chip is calculated and then incorporated in the simulation model. Table 2 lists the material properties examined in the thermo–mechanical FE analysis. To predict the mechanical behavior of the aluminum wire, the Young's modulus of the aluminum wire must be temperature-dependent.

978-1-4799-4789-8/14 $31.00 © 2014 IEEE

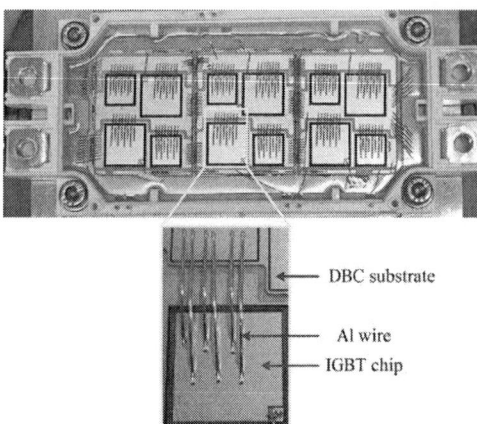

Fig. 1 450 A, 1200 V power module

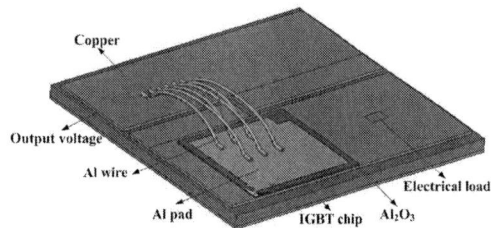

Fig. 2 3D finite element model of power module

Table 1 Material properties examined in the electro–thermal FE analysis

	Thermal conductively (W/mm*K)	Electrical resistivity (Ohm*mm)	Specific heat (J/Kg*K)	Density (Kg/mm³)
Aluminum	$237e^{-3}$	$2.65e^{-5}$	900	$2.7e^{-6}$
Silicon (IGBT)	$148e^{-3}$	Temp. dependent	700	$2.33e^{-6}$
Solder (SAC305)	$57e^{-3}$	$10.4e^{-5}$	230	$7.3e^{-6}$
Copper	$400e^{-3}$	$1.68e^{-5}$	380	$8.92e^{-6}$
Al₂O₃	$20e^{-3}$	$1e^{15}$	753	$3.96e^{-6}$

Table 2 Material properties examined in the thermo–mechanical FE analysis.

	Youn's modules (GPa)	Poisson ratio	CTE (ppm/°C)
Aluminum	Temp. dependent	0.33	24
Silicon (IGBT)	112.4	0.28	2.6
Solder (SAC305)	52.6	0.4	21.6
Copper	110	0.35	17
Al₂O₃	300	0.25	8

3. Electro-thermal coupling analysis and thermo-mechanical analysis

Electro-thermal coupling analysis was conducted using the established FE model under an electrical load of 70A. Figure 3 reveals that the predicted maximum current density is 2,050 A/mm². Maximum current density

occurred at the interface between an aluminum pad and aluminum wire; location also induced the current crowding effect. According to the literature, the predicted maximum current density may induce the behavior of the electromigration, resulting in crack formation and propagation during long-term operations [13].

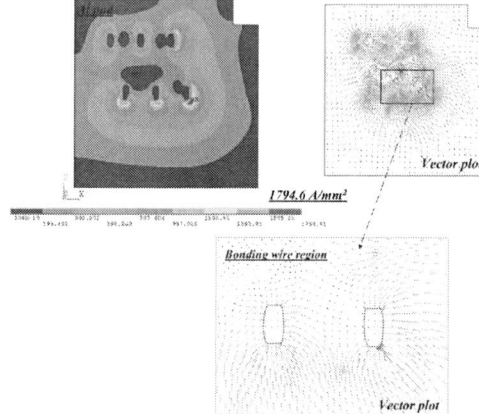

Fig. 3 Forecasted maximum current density and current crowding effect

A maximum temperature of 162.0 °C was occurred at the middle position of the aluminum wire (Fig. 4). Additionally, the simulatedtemperature of the IGBT chip is 105.9 °C. A high temperature in the IGBT chip may generate the coefficient of thermal expansion (CTE) mismatch with aluminum wires, resulting in wire lift-off from the interface between the aluminum pad and aluminum wires. The simulatedtemperature of IGBT chip was validated and compared with the measurement results [14]. Validation results indicate that the predicted chip temperature from the electro-thermal coupling simulation is reliable.

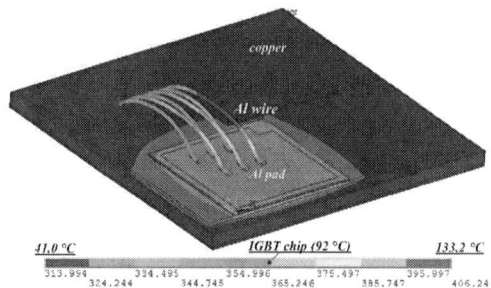

Fig. 4 Forecasted temperature of the FE model

The predicted temperature from the electro-thermal coupling analysis is viewed as the thermal loading and applied in thermo-mechanical simulation model to estimate the mechanical behavior of aluminum wire. The maximum von Mises stress of the aluminum wire is 15.5 MPa. Maximum von Mises stress occurs at the bottom side corner of the middle wire. This finding suggests that the aluminum wire may lead to the failure behavior of wire lift-off during a long-term cyclic operation.

Those failure modes degrade the reliability of a high power module and reduce its lifetime. This study attempts to reduce the maximum current density, chip temperature and thermal stress of the aluminum wire by using numerical simulation with the parametric analysis and acquire the design trends of aluminum wire.

4. Parametric analysis

4.1 Wire arrangement

This study first attempts to understand the differences between the linear arrangement and transposition arrangement in the maximum current density, wire/chip temperature and thermal stress of aluminum wire. The wire position of transposition arrangement (Fig. 3) can decrease in current density by 9.8 %. The maximum current density of the linear and transposition arrangement occurs entirely at the interface between the aluminum wire and pad. Following the electro-thermal coupling analysis, The wire/chip temperature of linear arrangement is higher than that of the transposition arrangement (Fig. 4) by about 4.5 and 1.8 °C, respectively. Owing to the slight differences of the chip temperature, the resulting von Misese stress of aluminum wire is unaffected The subsequent simulation uses the linear arrangement to understand the extent to which the wire configuration impacts electrical, thermal, and mechanical behaviors.

4.2 Wire number

This study also examines the extent to which the wire number affects electrical, thermal and mechanical behaviors. Figure 5 shows the electro-thermal coupling analysis, as estimated under different wire numbers. Analysis results indicate that each aluminum wire added can reduce the current density by approximately 16.6 % until the wire number increases to 8. Additionally, increasing the wire number can decrease the wire temperature, which is induced through the joule heating. However, the chip temperature decreases slightly. The behavior describes a situation in which the chip temperature is affected mainly by the cooling system, rather than the amount of wires. Owing to the slight decrease in the chip temperature, the subsequent thermal stress of aluminum wire remains declined (Fig. 6).

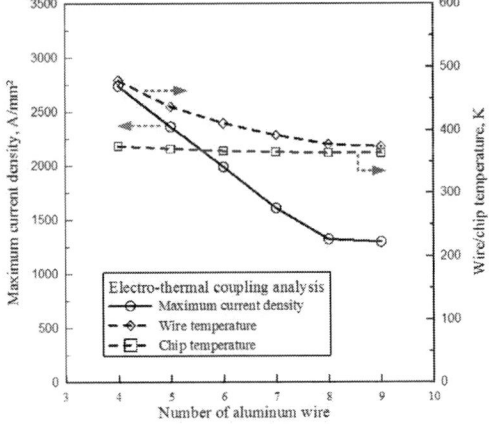

Figure 5 Electro-thermal coupling analysis, as estimated under different wire numbers

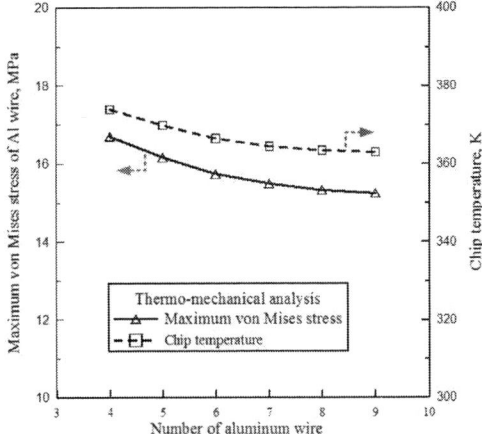

Figure 6 Thermo-mechanical analysis, as estimated under different wire numbers

4.3 Wire bonding perimeter

The ratio of wire bonding perimeter is defined to process numerical analysis. Figure 7 shows the electro-thermal coupling analysis predicted under different perimeter ratios. As the ratio of wire bonding perimeter reaches 1.1, the maximum current density changes negligibly. Additionally, increasing the ratio of the fixed wire bonding perimeter can reduce maximum current density by around 7.1 % before the 1.1 perimeter ratio. The Joule heat induced wire temperature is also reduced, which follows the increasing perimeter ratio. The chip temperature is obviously affected by a cooling system to maintain a constant, leading to a situation in which the thermal stress of aluminum wire remains unchanged continuously.

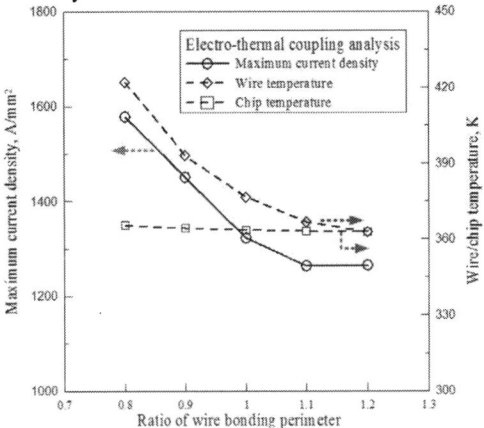

Figure 7 Electro-thermal coupling analysis predicted under different perimeter ratios

4.4 Wire height

Change in wire height can affect the angle between the bonding heel and aluminum pad, possibly influencing the wire stress. The ratio of wire height is defined to perform the numerical analysis. Figure 8 illustrates the electro-thermal coupling analysis under different wire height ratios. Height of the bonding wire negligibly

978-1-4799-4789-8/14 $31.00 © 2014 IEEE

impacts the maximum current density, which ranges from 1,325-1,313 A/mm^2. Although increasing the wire height can raise the wire temperature, the heat dissipation system still affects the chip temperature. As the chip temperature remains constant, change in wire height negligibly influences the thermal stress of aluminum wire.

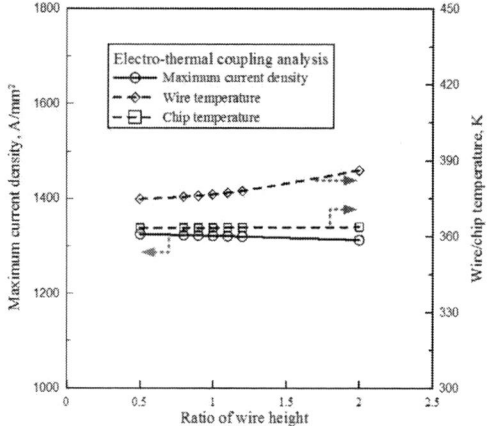

Figure 8 Electro-thermal coupling analysis estimated under different wire height ratios

5. Conclusions

This study elucidates the electro-thermo-mechanical coupling behavior of a high power module by using electro-thermal coupling and thermo-mechanical analyses. The configuration design of aluminum wire is also examined with respect to how to reduce the current density, wire/chip temperature and thermal stress. Analytical results indicate that the arrangement position of wire, wire number and wire bonding perimeter can reduce the current density, yet negligibly impact the chip temperature and thermal stress of wire. Additionally, the wire number and wire bonding perimeter added can also decrease the wire temperature. The wire height only changes the wire temperature, yet does not affect another behavior. Furthermore, the design trend of interconnection is realized, based on numerical analysis.

Acknowledgments

The authors would like to thank the National Science Council of Taiwan (NSC100-2221-E-007-014-MY3) for providing financial support for the current research.

References

[1] S. W. Yoon, M. D. Glover, H. A. Mantooth, and K. Shiozaki, "Reliable and repeatable bonding technology for high temperature automotive power modules for electrified vehicles," Journal of Micromechanics and Microengineering, vol. 23, 0150170, 2013.

[2] V. Smet, F. Forest, J. Huselstein, F. Richardeau, Z. Khatir, S. Lefebvre, and M. Berkani, "Ageing and Failure Modes of IGBT Modules in High-Temperature

Power Cycling," IEEE Trans. Ind. Electron., vol. 58, pp. 4931-4941, 2011.

[3] J. Onuki, M. Koizumi, and M. Suwa, "Reliability of thick Al wire bonds in IGBT modules for traction motor drives," IEEE Trans. Adv. Packag., vol. 23, pp. 108-112, 2000.

[4] W. S. Loh, M. Corfield, H. Lu, S. Hogg, T. Tilford, and C. M. Johnson, "Wire bond reliability for power electronic modules - effect of bonding temperature," presented at the 8th Thermal, Mechanical and Multi-Physics Simulation Experiments in Microelectronics and Micro-Systems, London, UK, 2007.

[5] M. Ishiko, M. Usui, T. Ohuchi, and M. Shirai, "Design concept for wire-bonding reliability improvement by optimizing position in power devices," Microelectron. J., vol. 37, pp. 262-268, 2006.

[6] H. Medjahed, P. E. Vidal, and B. Nogarede, "Thermo-mechanical stress of bonded wires used in high power modules with alternating and direct current modes," Microelectronics Reliability, vol. 52, pp. 1099-1104, 2012.

[7] T. Y. Hung, S. Y. Chiang, C. J. Huang, C. C. Lee, and K. N. Chiang, "Thermal–mechanical behavior of the bonding wire for a power module subjected to the power cycling test," Microelectronics Reliability, vol. 51, pp. 1819-1823, 2011.

[8] T. Y. Hung, L. L. Liao, C. C. Wang, W. H. Chi, and K. N. Chiang, "Life prediction of high cycle fatigue in aluminum bonding wires under power cycling test," IEEE Transactions on Device and Materials Reliability.

[9] R. S. Chen, C. H. Huang, and Y. Z. Xie, "Application of Optimal Design on Twin Die Stacked Package by Reliability Indicator of Average SED Concept," Journal of Mechanics, vol. 28, pp. 135-142, 2012.

[10] K. H. Tsai, C. L. Hwan, M. J. Lin, and Y. S. Huang, "Finite Element Based Point Stress Criterion for Predicting the Notched Strengths of Composite Plates," Journal of Mechanics, vol. 28, pp. 401-406, 2012.

[11] G. N. Ellison, Thermal Computations for Electronic Equipment, Van Nostrand Reinhole Company, New York, 1989.

[12] ABB, "5SMY 12M1280," ed, 2010.

[13] P. K. Tse and T. M. Lach, "Aluminum electromigration of 1-mil bond wire in octal inverter integrated circuits," in Proc. 45th Electronic Components and Technology Conference, Las Vegas, NV, USA, 1995

[14] L. L. Liao, T. Y. Hung, C. K. Liu, W. Li, M. J. Dai, and K. N. Chiang, "Electro-thermal finite element analysis and verification of power module with aluminum wire," Microelectronic Engineering, 2014, In Press.

2-Gb/s/pin DDR3 Memory Channel Design and Simulation for Carbon Reduction

Nansen Chen

MediaTek Inc.

No. 8, Dusing Rd. 1, Hsinchu Science Park, Hsinchu 300, Taiwan, R.O.C.

nansen.chen@mediatek.com

Abstract

The fabless semiconductor companies should take more responsibility for the global climate change when more and more consumer electronics are being produced. Under the trade-off between the signal quality of DDR3 memory and power consumption, the chip-package-board co-simulations were taken in the frequency domain up to 10 GHz and the time domain at 2 Gb/s to compare two types of channel designs. Results indicated that the proposed channel using the 2.5-layer PCB achieved the lower power consumption with acceptable eye diagrams of overlapping one DDR3 data byte that demonstrated 218-ps eye-aperture time, 1.47-V overshoot, and -0.05-V undershoot for the writing access, and 245-ps eye-aperture time, 1.83-V overshoot, and -0.25-V undershoot for the reading access. Moreover, there would be about 58 tons of carbon reduction per day if one third of global LCD TVs shipped each year use the 2.5-layer PCB design and the low-carbon DDR3 settings. Revising JEDEC Standard to implement two more weak drive strengths in the DDR3 SDRAM is recommended that is beneficial to reduce more power consumption in the consumer electronics.

1. Introduction

The necessary consumer electronics of home entertainment, such as liquid-crystal-display televisions (LCD TVs) and blu-ray disk (BD) players, with the high-speed memory and high data bandwidth connected to the internet will gain firsthand experience for accessing apps, video and more. The carbon footprint indicates how many grams or kilograms of CO_2 (or its equivalent) the product will produce over its life cycle, as shown in Fig. 1, and the use and materials acquisition stages are the first and the second domination for the carbon footprint of a LCD TV, respectively [1]. How people use the consumer electronics with less carbon emissions and still enjoy entertainment at home is an important subject for the fabless semiconductor companies. Particularly, the PCB is one of the major costs for consumer electronics. A 2-layer PCB has less manufacturing processes and cost than a 4-layer PCB, but pays the penalty of larger crosstalk and reflection due to the farther bottom ground layer. Many papers proposed the chip-package-board co-analysis to identify the channel effects on the signal quality or optimize the system performance of high-speed memory using high PCB layer count greater than four layers [2]–[4]. However, few papers presented to improve the signal quality of high-speed memory in the 2-layer PCB and analyze how to reduce power consumption with proper chip settings. This paper proposes a cost-effective low

layer count PCB, named 2.5-layer PCB, and evaluates if the full channel of chip-package-board can achieve DDR3 data accesses at 2 Gb/s. First, the electrical performance of two types of full channels in the frequency domain was compared using the full-wave electromagnetic solver to extract their S-parameters. Then the DDR3 eye diagrams of reading and writing one data byte were simulated with the signal and power integrity effects using the full channel S-parameters and chip SPICE netlists. Finally, the DDR3 settings with low power consumption were demonstrated and the total carbon reduction was also calculated.

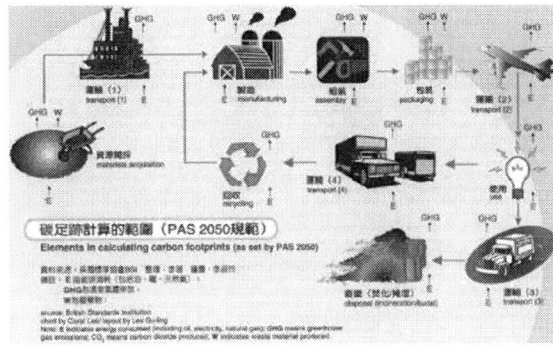

Figure 1. Elements in calculating carbon footprints [1].

2. PCB Stack-ups and Layout

In order to increase the data bandwidth in the LCD TV systems, the DDR3 controller is designed to drive two 16-bit DDR3 SDRAMs. Fig. 2 shows the PCB layout on the top layer. The routing topology is as follows.

a) DQ/DM/DQS/CLK/CS# nets: Dedicated routing between the controller and the memory on the top or bottom layer.

b) Other Addr./CMD nets: T-branch routing between the controller and the memories on the top or bottom layer with the reserved dual damping resistors (R_d) to reduce reflective signal.

c) Power supply: A low-cost low dropout (LDO) regulator to supply a 1.5 V I/O power and a 0.75 V reference power with a resistor divider referenced to the I/O power.

For high-speed digital signals, a neighboring solid reference plane is good for signal integrity. Therefore, a 4-layer PCB is easy to assign two solid reference planes in the second and the third layers of PCB, respectively, where the DDR3 traces are routed on both top and bottom layers. If reduction of the system cost or carbon emissions is a must, a 2-layer PCB would be a good candidate, but the DDR3 signal quality would be degraded due to lack of two metal layers for neighboring solid reference planes.

978-1-4799-4789-8/14 $31.00 © 2014 IEEE

Thus, a conductive metal layer, such as silver epoxy, copper epoxy, or graphite, is printed on the solder mask (a dielectric layer with 0.1 mm thickness) of DDR3 routing area in the 2-layer PCB, as illustrated in Fig. 3 and called as 2.5-layer PCB, where the conductive layer is connected to the ground or power net and becomes a neighboring solid reference plane for those high-speed digital signals. In the cause of cost issue, the graphite is selected instead of silver or copper epoxy, even though the conductivity of graphite (7×10^4 S/m) is far lower than that of copper (5.8×10^7 S/m).

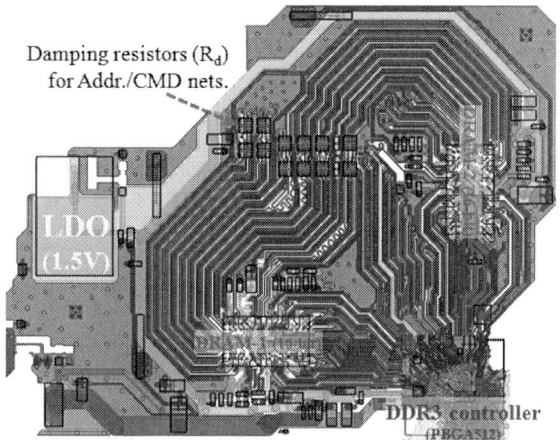

Figure 2. Layout of the DDR3 controller driving two 16-bit DDR3 SDRAMs.

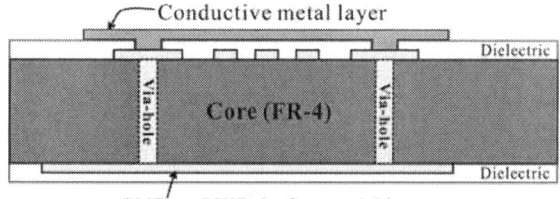

Figure 3. Cross-sectional view of 2.5-layer PCB with a conductive metal layer printed on the top layer.

3. System Performance

The full channel of DDR3 interface in the LCD TV system include a TV (controller) SoC encapsulated in the wirebonded plastic ball grid array (PBGA) package, a DDR3 DRAM encapsulated in the wirebonded fine-pitch ball grid array (FBGA) package, and a 2-layer PCB with the printed graphite layer. Due to different stack-ups of both packages, the channel S-parameters of DRAM package and controller package on the PCB were extracted by ANSYS HFSS and SIwave, respectively. Subsequently, the full channel of S-parameters was calculated from the controller I/O side to the DRAM I/O side by ANSYS Designer up to 10 GHz. The detailed simulation setups in SIwave, including material thickness, conductivity (σ), dielectric constant (D_k), and loss tangent (D_f), are stated in Table I.

TABLE I. DETAILED SIMULATION SETUPS IN SIWAVE

Type	Name	Thick.	Properties
Controller PBGA Package	Mold compound	1170 μm	$D_k = 3.6$, $D_f = 0.01$
	Solder mask (AUS308)	30 μm	$D_k = 4.3$, $D_f = 0.029$
	Layer-1 (Cu)	22 μm	$\sigma = 5.8 \times 10^7$ S/m
	PP (HL-832NXA)	100 μm	$D_k = 4.7$, $D_f = 0.01$
	Layer-2 (Cu)	18 μm	$\sigma = 5.8 \times 10^7$ S/m
	Core (HL-832NXA)	200 μm	$D_k = 4.7$, $D_f = 0.01$
	Layer-3 (Cu)	18 μm	$\sigma = 5.8 \times 10^7$ S/m
	PP (HL-832NXA)	100 μm	$D_k = 4.7$, $D_f = 0.01$
	Layer-4 (Cu)	22 μm	$\sigma = 5.8 \times 10^7$ S/m
	Solder mask (AUS308)	30 μm	$D_k = 4.3$, $D_f = 0.029$
	Air	250 μm	$D_k = 1.0006$
PCB	**Graphite layer**	100 μm	$\sigma = 7 \times 10^4$ S/m
	Solder mask	100 μm	$D_k = 3.4$, $D_f = 0.03$
	Top layer (Cu)	34 μm	$\sigma = 5.8 \times 10^7$ S/m
	Core (FR-4)	1473 μm	$D_k = 4.4$, $D_f = 0.02$
	Bottom layer (Cu)	34 μm	$\sigma = 5.8 \times 10^7$ S/m
	Solder mask	100 μm	$D_k = 3.4$, $D_f = 0.03$
	Graphite layer	100 μm	$\sigma = 7 \times 10^4$ S/m

A. Electrical performance in frequency domain

One data byte (Byte-0 including DQ0-7, DM0, and DQS0/DQS0#) and three power ports were simulated to extract the full channel S-parameters for two types of channel designs. Channel-1 is the full channel using the proposed 2.5-layer PCB printed with partial graphite layer. Channel-2 is the full channel using the traditional 2-layer PCB without any graphite layer. In Channel-2 design, both the top and bottom layers are routed with DDR3 signal traces. Thus, the nonideal return paths occur due to lack of a solid reference plane. Fig. 4 demonstrates S-parameters comparison for both channel designs. Specially, the comparison of channel S-parameters in magnitude for DQ1 net only is shown in Fig. 5. The maximum insertion loss slope (or rate) within 2 GHz is −1.54 dB/GHz for Channel-1 and −2.46 dB/GHz for Channel-2. In the same bandwidth, the maximum return loss is −8.11 dB for Channel-1 and −3.76 dB for Channel-2. Therefore, the graphite layer, although it is not a good conductor, is beneficial to reduce trace impedance, while the coupling is also depressed. With less energy loss in the frequency domain, better high-speed waveforms in the time-domain would be expected.

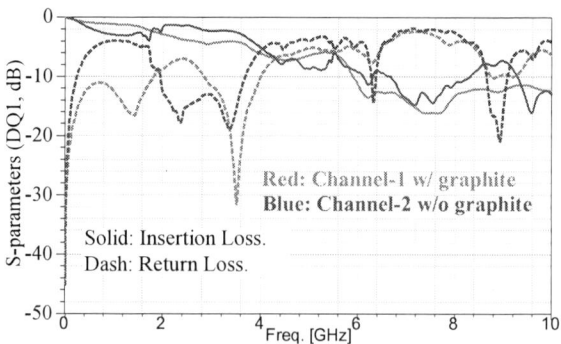

Figure 5. Comparison of DQ1 S-parameters in magnitude for full channel with and without graphite layer in PCB.

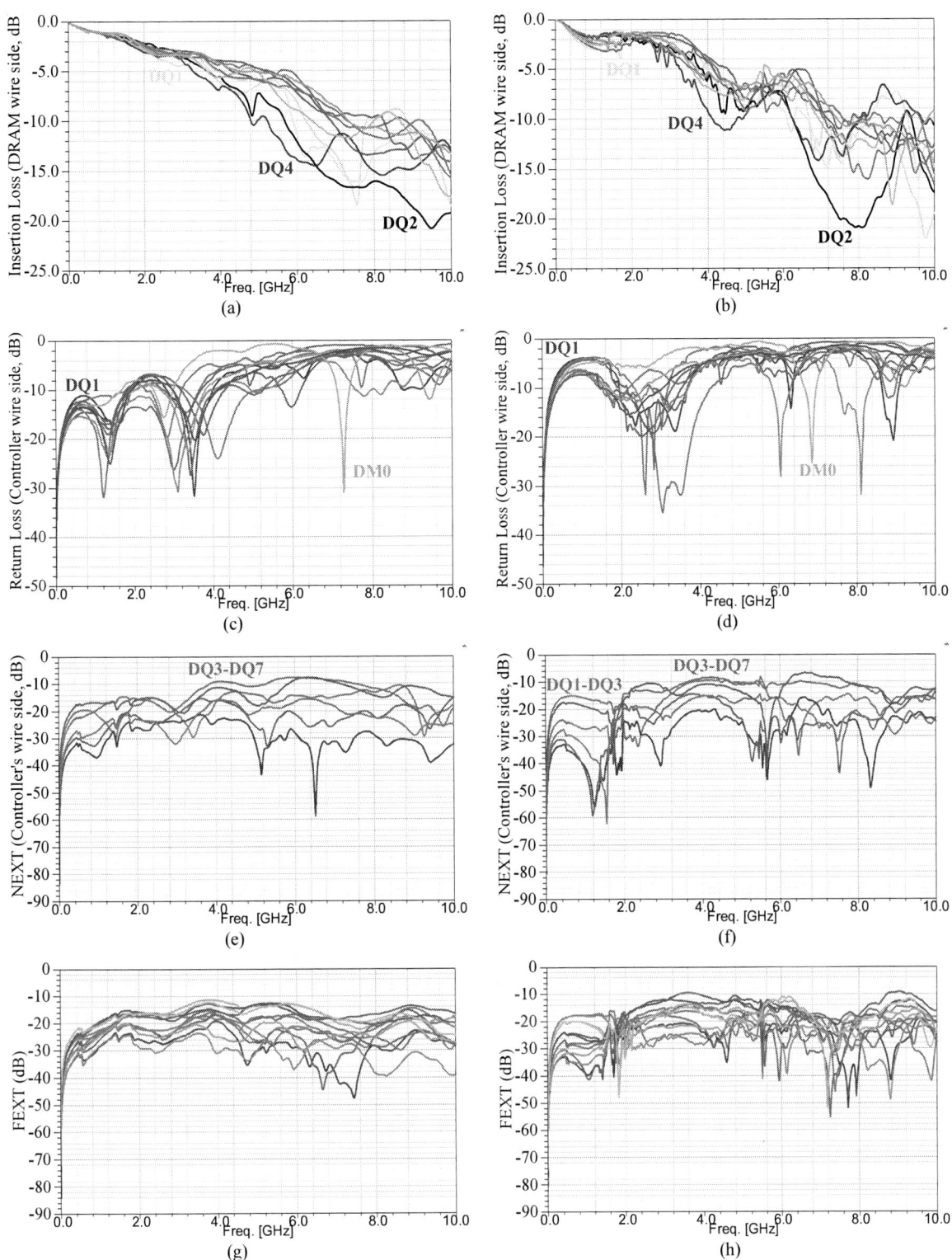

Figure 4. Comparison of full channel S-parameters in magnitude. (a) Insertion loss of Channel-1 with graphite layer. (b) Insertion loss of Channel-2 without graphite layer. (c) Return loss of Channel-1 with graphite layer. (d) Return loss of Channel-2 without graphite layer. (e) Near-end cross-talk of Channel-1 with graphite layer. (f) Near-end cross-talk of Channel-2 without graphite layer. (g) Far-end cross-talk of Channel-1 with graphite layer. (h) Far-end cross-talk of Channel-2 without graphite layer.

978-1-4799-4789-8/14 $31.00 © 2014 IEEE

B. Electrical performance in time domain

The full channel co-simulations from the DDR3 controller chip (I/O) pad to the DRAM chip (I/O) pad using the controller I/O netlists designed with the tsmc 40 nm silicon node, the channel S-parameters, and Micron DDR3 v68a netlists [5] were taken in Synopsys HSPICE for 220 ns. Both writing and reading data were at 2 Gb/s with pseudo random bit patterns (PRBP). The transient analysis with different on-die termination (ODT) and output drive impedance (ODI) settings were taken to evaluate the eye-diagrams and 1.5-V power consumption. Figs. 6 and 7 show the simulated eye-diagrams of reading and writing data for DRAM ODI = 48 Ω and ODT = 120 Ω, respectively and their corresponding voltage and timing margins are listed in Tables II and III. Although the JEDEC Standard only defines the DDR3 SDRAM specification, including features, functionalities, AC and DC characteristics, packages, and ball/signal assignments

[6], the margins for the voltages, including overshoot/undershoot, $V_{IH(dc)}$, and $V_{IL(dc)}$, and timings, including setup (tDS) and hold (tDH) times, measured on the controller side are also adopted with the same JEDEC Standard due to no global specification defined for the DRAM controller. Comparison of Channel-1 and Channel-2 with the voltage and timing margins, Channel-1 always has larger margins than Channel-2. For writing data at 2 Gb/s as summarized in Table III, Channel-1 conforms to all the JEDEC Spec, but Channel-2 does not. Fig. 8 shows the simulated eye-diagrams of reading and writing data for Channel-1 with ODT turned off. Under the control of proper drive strengths, the timing and voltage margins are still acceptable. That is why the graphite layer is helpful to reduce signal loss and coupling demonstrated in the frequency domain and preserve the high-frequency components for better eye-diagrams in the time-domain.

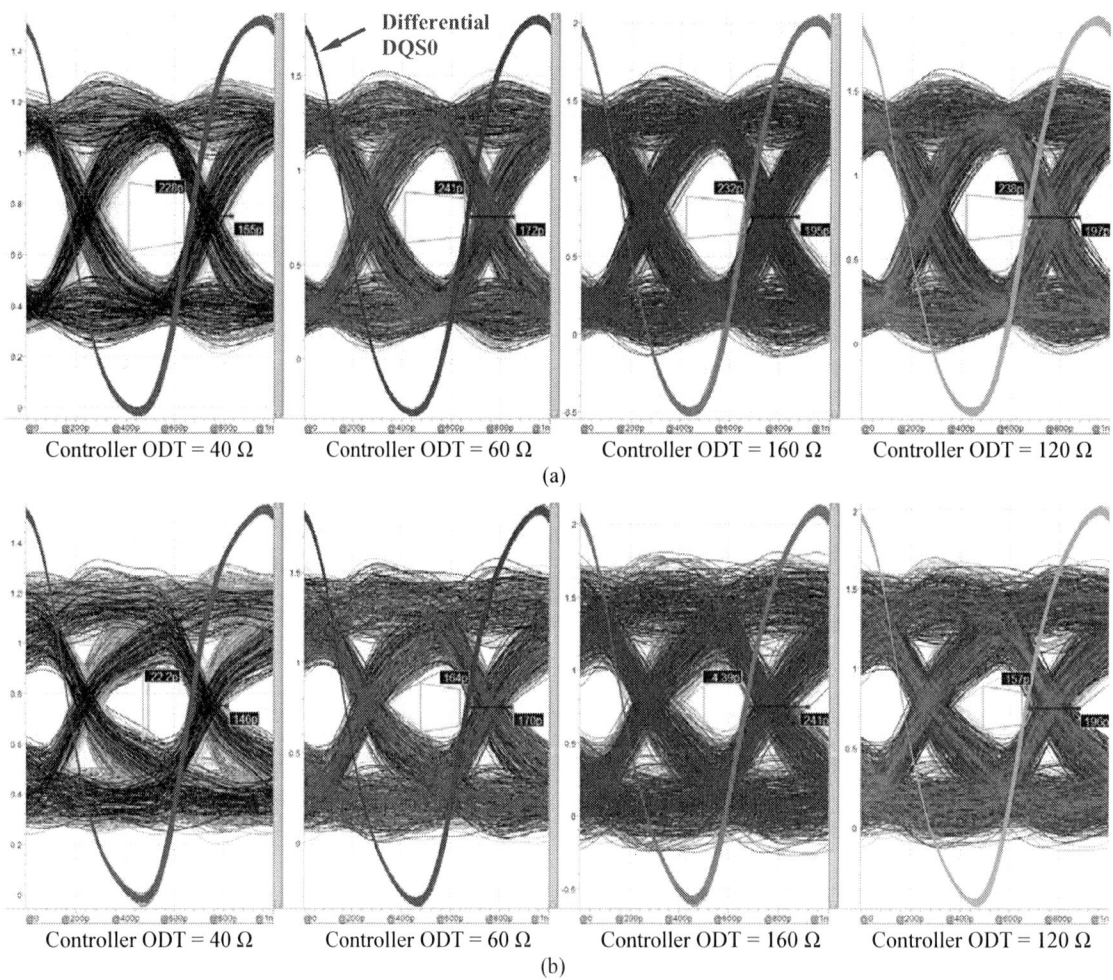

Figure 6. Simulated eye-diagrams of overlapping all Byte-0 signals on controller chip side for reading data at 2 Gb/s with DRAM ODI = 48 Ω. (a) Channel-1 with different controller ODTs. (b) Channel-2 with different controller ODTs.

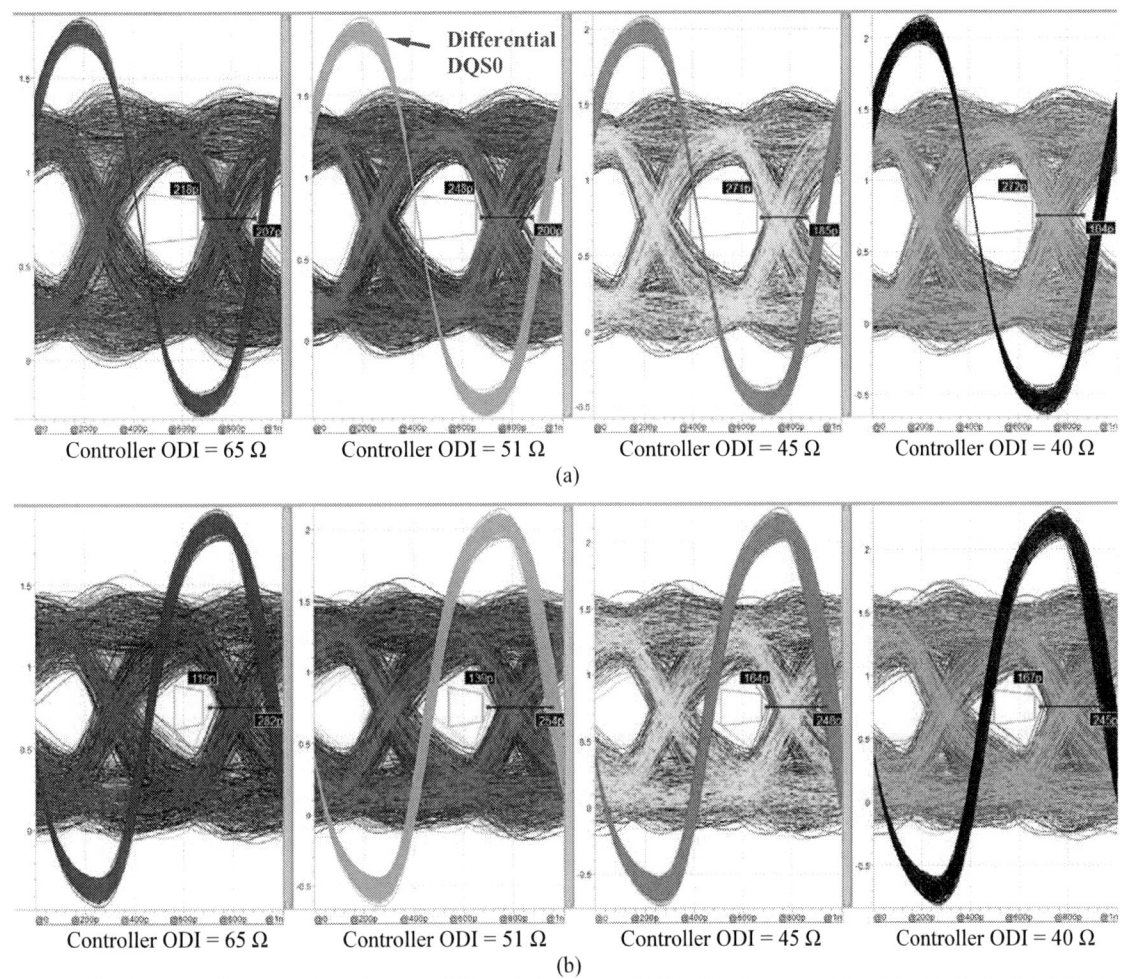

Figure 7. Simulated eye-diagrams of overlapping all Byte-0 signals on DRAM chip side for writing data at 2 Gb/s with DRAM ODT = 120 Ω. (a) Channel-1 with different controller ODIs. (b) Channel-2 with different controller ODIs.

TABLE II. Voltage and timing margins of reading data at 2 Gb/s on controller chip side with DDR3 DRAM ODI = 40 Ω and different on-die terminations (ODT)

ODT / Items	Channel-1 (w/ graphite)					Channel-2 (w/o graphite)					JEDEC SPEC (DDR3-2133)
	40 Ω	60 Ω	120 Ω	160 Ω	OFF	40 Ω	60 Ω	120 Ω	160 Ω	OFF	
$V_{IH(dc)}$ margin (mV)	170	>200	>200	>200	>200	58	101	100	92	NG	$V_{IH(dc)}$ = 850 mV
$V_{IL(dc)}$ margin (mV)	157	>200	>200	>200	>200	48	77	75	52	NG	$V_{IL(dc)}$ = 650 mV
Overshoot (V)	1.37	1.59	1.73	1.83	2.12	1.37	1.65	1.79	1.90	2.20	1.90 V
Undershoot (V)	0.16	-0.04	-0.16	-0.25	-0.62	0.16	-0.07	-0.22	-0.36	-0.76	-0.40 V
Skew (ps, V_{REF} to V_{REF})	151	179	173	174	232	156	172	205	260	NG	N/A
Horiz. eye-open (ps)	245	242	248	245	222	106	161	156	107	NG	tDS+tDH = 53 ps

Note: NG (No Go) means no timing or voltage margin defined in JEDEC spec.

TABLE III. Voltage and timing margins of writing data at 2 Gb/s on DRAM chip side with DDR3 controller ODI = 65 Ω and different on-die terminations (ODT)

ODT / Items	Channel-1 (w/ graphite)				Channel-2 (w/o graphite)				JEDEC SPEC (DDR3-2133)
	40 Ω	60 Ω	120 Ω	OFF	40 Ω	60 Ω	120 Ω	OFF	
$V_{IH(dc)}$ margin (mV)	103	141	180	98	NG	50	136	NG	$V_{IH(dc)}$ = 850 mV
$V_{IL(dc)}$ margin (mV)	140	158	167	92	NG	97	146	NG	$V_{IL(dc)}$ = 650 mV
Overshoot (V)	1.17	1.27	1.47	1.88	1.18	1.29	1.56	2.14	1.90 V
Undershoot (V)	0.25	0.14	-0.05	-0.40	0.25	0.14	-0.08	-0.63	-0.40 V
Skew (ps, V_{REF} to V_{REF})	167	178	207	313	195	222	282	NG	N/A
Horiz. eye-open (ps)	199	220	218	142	NG	109	119	NG	tDS+tDH = 53 ps

Note: NG (No Go) means no timing or voltage margin defined in JEDEC spec.

978-1-4799-4789-8/14 $31.00 © 2014 IEEE

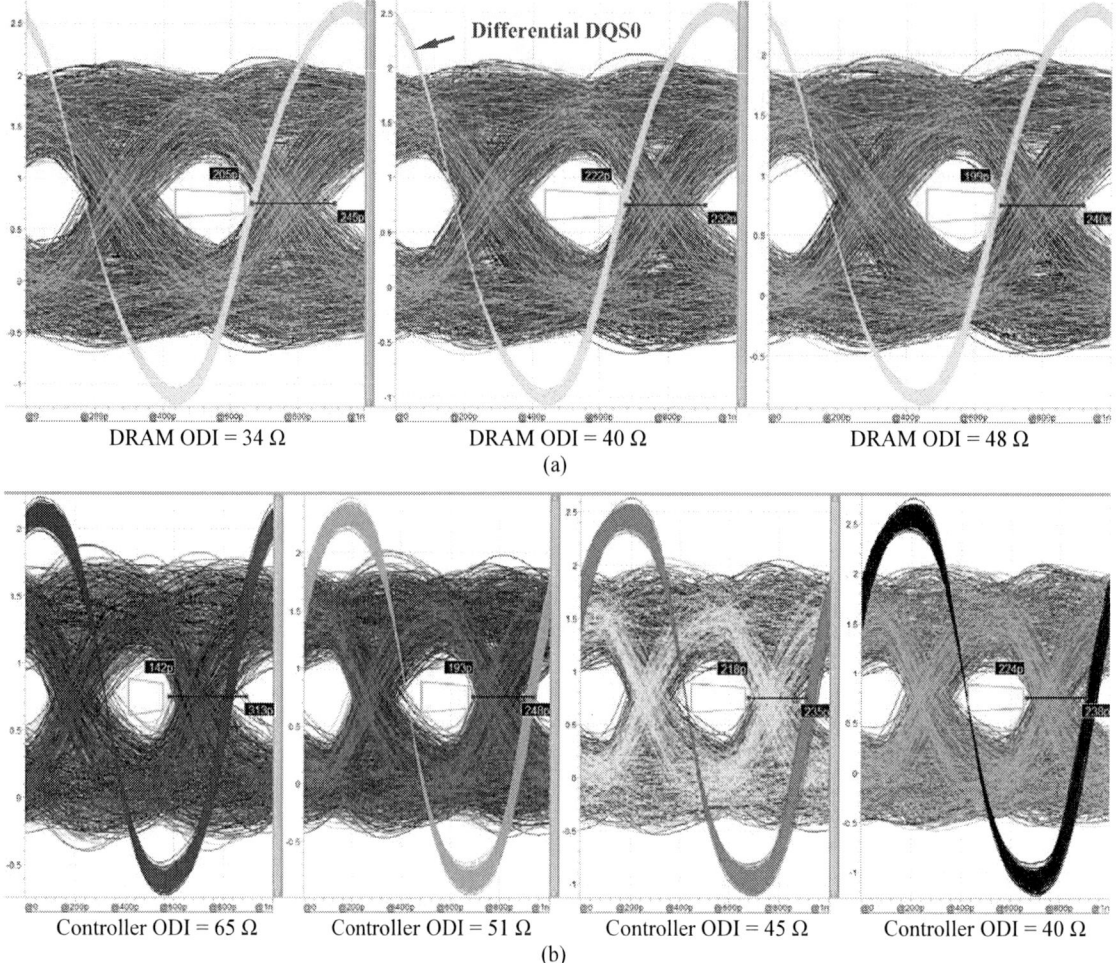

Figure 8. Simulated eye-diagrams of overlapping all Byte-0 signals on chip sides for Channel-1 with ODT turned off. (a) Reading data at 2 Gb/s with different DRAM ODIs. (b) Writing data at 2 Gb/s with different controller ODIs.

Fig. 9 also demonstrates the simulated eye-diagrams of Channel-1 for the CS# net with one clock cycle (1T at 1000 Mb/s) and overlapping some Addr./CMD signals with two clock cycles (2T at 500 Mb/s) including BA0-2, CAS#, RAS#, WE#, RESET#, CKE, and ODT. Because there is no ODT designed for Addr./CMD nets in the DDR3 DRAMs, the dual damping resistors (R_d) near the branch point of Addr./CMD traces in the PCB are reserved to reduce the reflective effects from both DRAMs. The results indicate that all the voltage and timing margins also conform to the JEDEC Spec.

C. Calculation of power consumption

If the drive strength is fixed from the transmitter, the bigger ODT enabled in the receiver will enlarge the eye-diagrams, but suffer the larger overshoot or undershoot, and vice versa. If the ODT is fixed in the receiver, the stronger drive strength from the transmitter will deliver more power and enlarge the eye-diagrams in the receiver, and vice versa. If the transmitter delivers the stronger drive strength, the receiver must use the smaller ODT to depress the overshoot and undershoot. Thus, such ODT and ODI settings will consume more power. Although

any ODT settings and ODI settings greater than 65 Ω in Channel-1 design have enough voltage and timing margins, the trade-off between the signal quality and power consumption is to select weaker drive strength from the transmitter and a bigger ODT enabled in the receiver. Tables IV and V summarize Channel-1 average current consumption from the 1.5 V regulator in 220 ns for reading and writing one data byte, respectively with different ODT and ODI settings. According to Table IV, the reading setting of ODT/ODI = 40/34 Ω consumes 41% more current than that of ODT/ODI = 160/48 Ω. From Table V, the writing setting of ODT/ODI = 40/40 Ω consumes 58% more current than that of ODT/ODI = OFF/65 Ω. Note that the DRAM ODI = 48 Ω (= 240Ω/5) is not defined in the JEDEC Spec, although there are two more ODI controls reserved in the More Register MR1 [6]. Revising the JEDEC standard to increase two ODI controls, such as 48 and 60 Ω, is recommended that will consume less power compared to the existing DDR3 DRAM design. The signal quality with weaker drive strength, i.e. larger ODI, is acceptable that was simulated and listed in Table III for the DRAM controller with ODI = 65 Ω.

978-1-4799-4789-8/14 $31.00 © 2014 IEEE

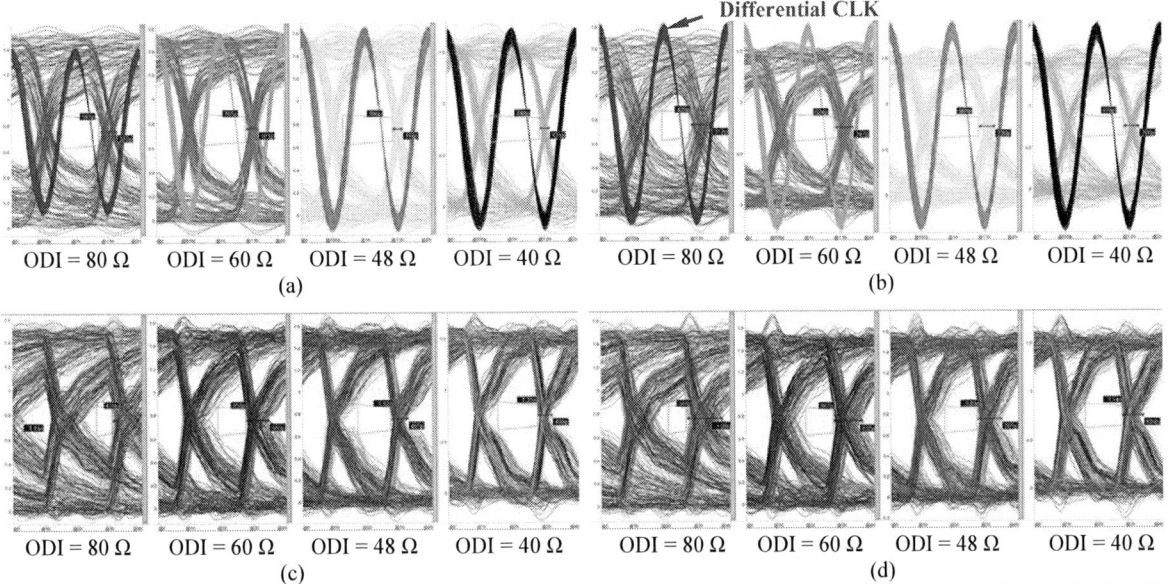

Figure 9.   Simulated eye-diagrams of overlapping Addr./CMD nets on DRAM chip sides for Channel-1 with dual R_d = 25 Ω in PCB. (a) CS# in DRAM-1 at 1000 Mb/s with different controller ODIs. (b) CS# in DRAM-2 at 1000 Mb/s with different controller ODIs. (c) 2T nets in DRAM-1 at 500 Mb/s with different controller ODIs. (d) 2T nets in DRAM-2 at 500 Mb/s with different controller ODIs.

TABLE IV. CHANNEL-1 AVERAGE CURRENT CONSUMPTION (CC_{AVG}) FOR READING ONE DATA BYTE (BYTE-0)

DRAM	Controller ODT and CC_{AVG} (mA)				
ODI	40 Ω	60 Ω	120 Ω	160 Ω	OFF
34 Ω	188	147	134	126	102
Decrement	0	-22%	-29%	-33%	-46%
40 Ω	177	135	122	114	91
Decrement	0	-24%	-31%	-36%	-49%
48 Ω	173	131	118	110	87
Decrement	0	-24%	-32%	-36%	-50%

TABLE V. CHANNEL-1 AVERAGE CURRENT CONSUMPTION (CC_{AVG}) FOR WRITING ONE DATA BYTE (BYTE-0)

Controller	DRAM ODT and CC_{AVG} (mA)			
ODI	40 Ω	60 Ω	120 Ω	OFF
40 Ω	212	180	147	118
Decrement	0	-15%	-31%	-44%
45 Ω	205	172	140	110
Decrement	0	-16%	-32%	-46%
51 Ω	198	165	133	103
Decrement	0	-17%	-33%	-48%
65 Ω	185	152	119	89
Decrement	0	-18%	-36%	-52%

The Channel-2 design is hard to achieve the acceptable voltage and timing margins if the ODT is turned off in the receiver. Therefore, the stronger drive strength is required to compensate the signal loss and the smaller ODT to depress overshoot and undershoot. Comparison of power consumption for both channel designs based on enough voltage and timing margins, Channel-2 consumes 15% more current or 20 mA for

reading one data byte and 50% more current or 89 mA for writing one data byte than Channel-1. The above comparison is calculated from the ODT/ODI settings of Channe-1 that are 160 Ω/48 Ω and OFF/65 Ω for reading and writing access, respectively. The ODT/ODI settings of Channe-2 are 60 Ω/40 Ω and 60 Ω/40 Ω for reading and writing access, respectively. Assume that a LCD TV is turned on five hours per day [1], Channel-1 with a 32-bit data access at 2 Gb/s consumes 1.635 watt-hours (Wh, 1.5 V I/O power) less energy than Channel-2. The global LCD TV shipments from 2011 exceed 200 million units [7]. If one third of TVs use the Channel-1 design, a total of power saving is 21.8 megawatts. That is equal to about 58 tons of carbon reduction *per day*, based on the emissions coefficient of 0.532 Kg CO_2/Kwh in Taiwan 2012.

4. Conclusions

The proposed cost-effective 2.5-layer PCB printed with a graphite layer in the DDR3 DRAM routing area can reduce signal coupling and loss effectively. The channel performances in both the frequency and time domains are correlated. This design in the LCD TVs or BD players would reduce carbon emissions significantly while preserve the signal quality with high data bandwidth. Revising JEDEC Standard to implement two more ODI controls with weaker drive strengths in the DRAM is recommended strongly that is beneficial to save more power compared to the existing DRAM design. Our future work would manufacture and verify the whole TV system to correlate the power savings with Channel-1 design.

Acknowledgments

The author would like to thank Dragon Chen of MediaTek Inc., Taiwan for his helpful discussion of the global LCD TV shipments.

References

1. C. Lee, "The Low-Carbon Future: Taiwanese Industry Goes Green," *Taiwan Panorama*, Vol. 35, No. 4 (2010), pp. 6-35.
2. H.-H. Chuang *et al.*, "Signal/Power Integrity Modeling of High-Speed Memory Modules Using Chip-Package-Board Coanalysis," *IEEE Trans. Electromagn. Compat.*,Vol. 52, No. 2 (2010), pp. 381-391.
3. J, Feng, B. Dhavale, J. Chandrasekhar, Y. Tretiakov, and D. Oh, "System level signal and power integrity analysis for 3200Mbps DDR4 interface," *Proc 63rd Electronic Components and Technology Conf*, Las Vegas, NV, May 2013, pp. 1081-1086.
4. G. R. Luevano, J. Shin, and T. Michalka, "Practical investigations of fiber weave effects on high-speed interfaces," *Proc 63rd Electronic Components and Technology Conf*, Las Vegas, NV, May 2013, pp. 2041-2045.
5. N. Chen, "Design of die-pad on exposed substrate (DOES) leadframe package for DDR3 interface applications," *Proc 13th Electronic Packaging Technology and High Density Packaging (ICEPT-HDP) Conf*, Guilin, China, Aug. 2012, pp. 567-574.
6. *DDR3 SDRAM Specification*, JEDEC Std. JESD79-3F, 2012.
7. T. Lo, "LCD TV market trends and outsourcing strategies for 2013," DIGITIMES Research. (2012, Dec.). [Online]. Available: http://www.digitimes.com/

System-level-model development of an SWCNT based piezoresistive sensor in VHDL-AMS

Vladimir Kolchuzhin[1], Jan Mehner[1], Erik Markert[2], Ulrich Heinkel[2], Christian Wagner[3], Jörg Schuster[5]
and Thomas Gessner[3,4,5]
Chemnitz University of Technology, Faculty for Electrical Engineering and Information Technologies,
[1]Chair of Microsystems and Precision Engineering, [2]Chair for Circuit and System Design,
[3]Center for Microtechnologies (ZfM), [4]Chair Microtechnology
Reichenhainer Str. 70, 09126 Chemnitz, Germany
[5]Fraunhofer Institute for Electronic Nano System (ENAS)
Technologie-Campus 3, 09126 Chemnitz, Germany
vladimir.kolchuzhin@etit.tu-chemnitz.de

Abstract

This article deals with the model development for a single walled carbon nanotube (SWCNT) piezoresistive sensor at system level design. The framework of VHDL-AMS is used for implementation and simulation, consisting of compact submodels that describe components performing heterogeneous functions. The SWCNT mechanical and electrical compact models presented in the article are based on the analytical model, lumped element model and the simulation results based on density functional theory (DFT). The macromodels of the MEMS are built using a reduced order modeling technique based on finite element simulations. This article presents and discusses the most important aspects of the development of system models and essential model parameters.

1. Introduction

Microelectromechanical systems, actuated by an electrostatic force, are widely used in various applications such as pressure sensors, microphones, switches, resonators, accelerometers, gyroscopes, RF-filters, tunable capacitors, and micromirrors. But the accuracy of geometry and therefore the characteristics (sensitivity, signal-to-noise ratio, and frequency range) of MEMS based sensors are limited by the technology tolerances.

Today, the nanoscale dimension is explored to look out for new sensing principles. One of them is the piezoresistive effect of carbon nanotubes, useful for a nano-scaled mechanical sensor. Several groups are working on this field by using either a mixture of CNTs [1, 2] or selected type of SWCNTs [3-5].

The NEMS/MEMS design is a very challenging multilevel and interdisciplinary task. The levels in design include synthesis of lumped elements, process sequence development and mask layout drawing, component and system simulations. Different algorithms and software tools are used for this purpose. Numerical simulations are used both as a design tool and for understanding complex device behavior. Commonly used numerical simulation methods (FDM, FEM and BEM) provide results for a given set of geometrical and material parameters. For CNTs, classical simulation approaches based on finite elements do not hold, because their electrical properties directly depend on the atomic positions of the carbon atoms. Therefore, quantum-mechanical simulations of

CNTs are done using density functional theory (DFT), which is a standard technique for simulations of systems consisting up to hundreds of atoms. As this number is small, periodic boundaries are used to give results for CNTs of finite length.

This article presents a system level model of an SWCNT based mechanical sensor and is subdivided into six chapters. Chapter 2 presents a short description of the CNT based mechanical sensor. In Chapter 3 the CNT compact model, its implementation using VHDL-AMS and the benchmarks are described. Chapter 4 presents the implementation process of MEMS macromodel using mode superposition technique. In Chapter 5 the developed models of CNTs are integrated into the MEMS model they are co-simulated. The paper closes with conclusions drawn from the performed studies and suggestions for future work.

2. Design concept of the CNT based mechanical sensor

The virtual prototype of the CNT based mechanical sensor consists of MEMS platform that is served to actuate a single-axis force. A schematic view on the sensor layout and interfaces is shown in Figure 1.

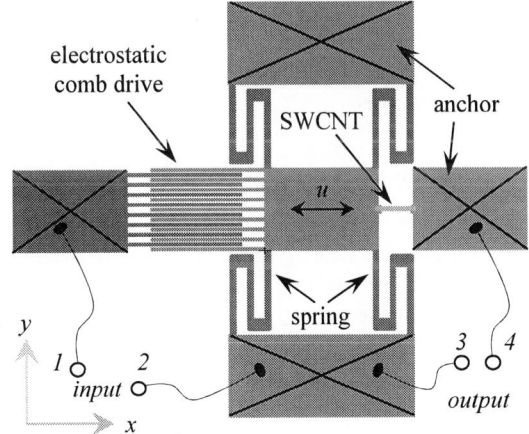

Figure 1: The 2D layout of a CNT based mechanical sensor

The single CNT is placed between the movable and fixed electrodes. The change in conductivity of the CNT,

978-1-4799-4789-8/14 $31.00 © 2014 IEEE

when strain is applied, is used to measure force and displacement.

It should be noted that the gauge factor of CNT decreases as the chiral angle increases. Zigzag tubes (n,0) have a maximum magnitude of gauge factor [1].

The length of the tube L_{CNT} is defined by minimal available air gap between the movable and the fixed electrodes (2µm in BDRIE process). In order to avoid damaging the CNT, a measurement is limited in the deformation range of 0 to +5 %. The electrostatic comb drive is used to set the operating point (OP) on the strain-displacement curve of CNT.

3. Compact models of piezoresistive effect for CNT

3.1. Analytical models

The change in electrical resistance of ideal CNT due to strain can be calculated by an empirical transport formula based on band-gap changes [6, 7]:

$$R = R_C + \frac{1}{|t|^2}\frac{h}{8e^2}\left[1 + \exp\left(\frac{E_{gap}(\varepsilon)}{kT}\right)\right]. \quad (3.1)$$

Here, R_C denotes the contact resistivity, which is neglected; t^2 is the transmission probability of the nanotube, usually taken as 1; h, e and k are physical constants; T denotes the temperature and E_{gap} is the bandgap of the nanotube. The bandgap of a strained CNT is linearly changing with the relative deformation. An analytical formula for CNTs with a large radius is given in [8] with a slight modification to fit DFT results:

$$E_{gap}(\varepsilon) = E_{gap}^0 + \mathrm{sgn}(2p+1)\cdot 3t_0\sqrt{2}(1+v)\varepsilon\cos(3\theta), \quad (3.2)$$

where p is a value determined by the chiral indices (n,m) of the nanotube in a way that $p = [n-m]_3 : p = 2 \mapsto p = -1$; $t_0 = 2.66$ eV is the hopping parameter; $v \approx 0.2$ stands for the Poisson's ratio; θ represents the chiral angle of the CNT and ε denotes the relative strain. Here, E_{gap}^0 is the bandgap of the unstrained state can be written for semiconducting CNTs as $E_{gap}^0 = 3a_0t_0/2r_0$, and for semimetallic as $E_{gap}^0 = t_0\cos(3\theta)\cdot(a_0/4r_0)^2$, where $a_0 = 1.42$ Å is the lattice constant of graphene and r_0 is the radius of the CNT. The maximum bandgap realized by straining CNTs is $E_{gap}^{max} = 3E_{gap}^0/2$. If CNTs are strained further, the band gap closes again. Such behavior can be understood by the underlying theory presented in [8, 16].

The available mechanical models for CNTs are plotted in Figure 2. The axial stiffness k_{CNT} of the SWCNT can be given as:

$$k_{CNT} = A\cdot E/L_{CNT}(0), \quad (3.3)$$

where E is the Young's modulus and A is the cross section area. The Van-der-Waals distance (6.3 Å) is the distance of two neighboring carbon sheets in graphite and is used to define the effective cross section area A^*, Figure 2b. For more details see [16]. The calculated stiffnesses for the CNTs are listed in Table 1.

Table 1: Overview of the calculated stiffnesses used in analytical models

CNT	radius r_0, Å	area, nm^2	area*, nm^2	stiffness k_{CNT}, N/m		
				rod	tube	DFT
(13,0)	5.092	0.815	2.016	0.41	1.00	0.54
(14,0)	5.484	0.945	2.171	0.47	1.09	0.58
(15,0)	5.876	1.085	2.326	0.54	1.16	0.62

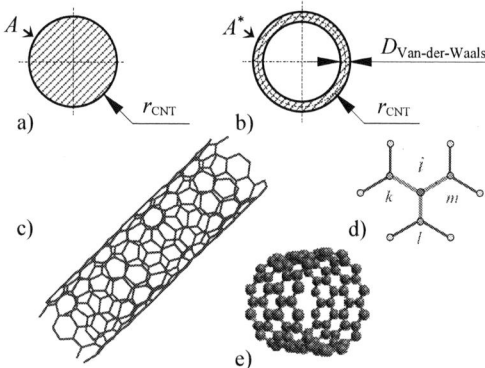

Figure 2: Overview of the mechanical models for CNT: a) rod; b) tube; c) FEM model using beam-like elements [9]; d) AFEM element for CNT taking into account second nearest-neighbor interactions [10]; d) the geometry of a (12,0)-nanotube and its unit cell for DFT simulation, two repetitions of that unit-cell are shown, $L_{cell}(n,0) = 4.26$ Å.

3.2. DFT modelling

DFT is a quantum mechanical theory, which solves the electronic structure problem in an efficient way. In contrast to the analytical model, where different approximations are used, it provides more reliable results and helps to confirm analytical models. In this special case, the analytical formula (3.2) has been corrected in comparison to the original publication [8]. Nevertheless, its computational cost is very high and only up to hundreds of atoms can be calculated.

For our simulations, we use Atomistix ToolKit (ATK) from QuantumWise [11, 12]. The detailed parameters of the calculations are shown in Table 2.

Table 2: Overview of the calculation parameters used in DFT calculations for the nanotubes

Basis set	Double zeta polarized [13]
k-points	1x1x20, Monkhorst-Pack grid [14]
Functional	LDA, Perdew and Wang [15]
Smearing	Fermi-Dirac, 300K = 25 meV
Max. forces	0.01 eV/Å = 0.01602 nN

As the amount of simulated atoms is very small, periodic boundary conditions have been applied to obtain the results for infinite CNTs. These results can be applied to ideal systems on the nano- and microscale, as the length of experimentally fabricated CNTs is long with respect to the electron wavelength. Thus, there will be no quantization effects in this direction in the experiments and infinite CNTs may be assumed for simulation. The results obtained by such kind of simulations are the mechanical and electrical properties of strained CNTs. For the mechanical part, the procedure is as following: the atoms of the CNTs are placed in a unit cell, which is periodic into all special directions, Figure 2e.

Then, a geometry optimization is performed to relax all internal forces. At each stretch step, the unit cell and the according atomic coordinates are scaled linearly, followed by another geometry relaxation. The total quantum-mechanical energy is saved and traced over the deformation range. The results for different CNTs are depicted in Figure 3, where DFT values and third order fits are shown.

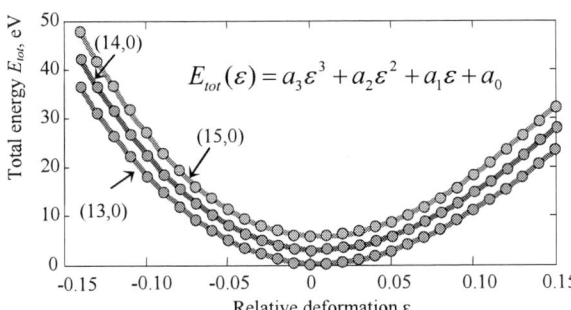

Figure 3: DFT values of the total energy of the (13,0), (14,0) and (15,0) CNTs traced over the relative deformation

A third order polynomial approximation describes the deformation over the whole range and the resulting stress-strain relation as well as the Young's modulus can be derived. The moduli obtained for the CNTs that are plotted in Figure 3 are shown in Table 3.

Table 3: Overview of the Young's moduli of the depicted carbon nanotubes

CNT	Young's modulus E, GPa
(13,0)	983.5 ± 1.1
(14,0)	985.6 ± 0.6
(15,0)	987.2 ± 0.3

Figure 4 shows the DFT values of the bandgap traced over the relative deformation. The $(n,0)$-CNTs with chiral angle $\theta=0°$ are the most sensitive ones. It is highlighted that different kinds of CNTs, which are geometrically very similar, show a strongly different behavior. For the (13,0) and (14,0) CNT, a band gap exists at zero strain. They differ in the gradient of the band gap at positive strain: for the (14,0) CNT, it decreases and for the (13,0) it increases.

The obtained results show only minor deviations of the DFT and the analytical model [16]. Thus, the simplified model can be used for sensor simulations for such kind of CNTs. These models are extremely helpful, to create compact models for lumped elements simulations for the sensor itself. DFT calculations take hours of time, whereas an analytical model is computed within a fraction of second. In contrast to DFT, analytical models are not that transferable, e.g. they break down at CNTs whose diameter is smaller than the Van-der-Waals distance. This is shown in [18]. Within fabrication, such small-diameter CNTs are rarely found and therefore, the analytical model is practically suitable.

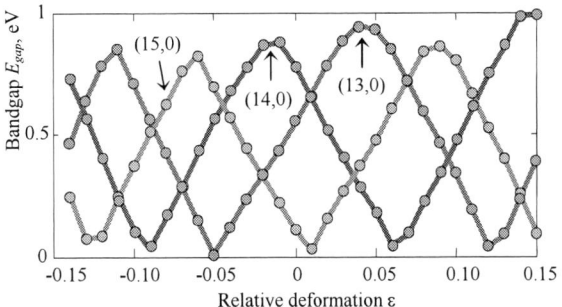

Figure 4: DFT values of the bandgap of the (13,0), (14,0) and (15,0) CNTs traced over the relative deformation

3.3. Implementation in VHDL-AMS

The VHDL-AMS model consists of an entity and an architecture sections which defines the terminals relating electrical and mechanical quantities based on the generalized Kirchhoff's laws. The entity of the SWCNT is presented as a four terminal device (two mechanical m_1 and m_2 terminals and two electrical e_1 and e_2 terminals) as shown in Figure 5. The architecture consists of the nonlinear mechanical spring unidirectional coupled to the strain dependent resistor.

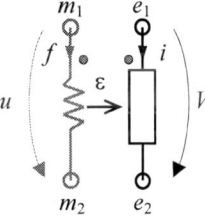

Figure 5: VHDL-AMS model of the single-walled CNT

The behavioral model of spring is defined by:
$f(t) = k_{CNT} u(t)$ in the linear case, and

$$f(t) = \frac{\partial E_{tot}(u)}{\partial u}$$ in the non-linear case,

where u is the displacement, f is the reaction force.

The next step is to define the behavioral model of the resistor. The current i flow through a resistor is defined by:

$$i(t) = V(t)/R_{CNT}(\varepsilon),$$

where $\varepsilon = u(t)/L_{CNT}(0)$ is the relative deformation of CNT, and V is the voltage applied across the CNT. The resistivity R_{CNT} of ideal CNT is calculated by an empirical transport formula (3.1), where the band gap of the system is of interest. The band gap is evaluated by DFT calculations as well as the analytical formula (3.2).

After development the VHDL-AMS code, the CNT-model is tested using the simulator *Simplorer*. It is necessary to specify the external circuitry (voltage source E1=1V, and controller units AM1, VM1), loads (mechanical force source F1 and controller units FM1, SM1) and solver parameters (time step size, total simulation time). The schematic of this measurement setup is depicted in Figure 6.

The simulation results are shown in Figure 7. Here time t is a formal parameter, used in order to define linearly changing input force F1, Figure 7a. The current I1 flowing through the CNT can be calibrated to get percentage change in applied strain, Figure 7d.

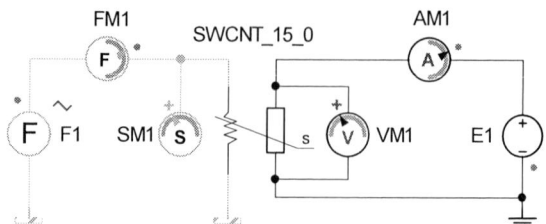

Figure 6: Testbench circuit for VHDL-AMS model of the single-walled CNT in Simplorer

The derivative results (deformation, bandgap and resistance of the (15,0) CNT) are depicted in Figure 8. It can be seen that for linearly changing input force, CNT shows nonlinear change in resistivity, Figure 8c. The resistance is also shown on a logarithmic scale (Figure 8d.), as it exponentially depends on the bandgap. The behavior of the (15,0) CNT differs, as its band gap is approximately zero in the original state and its resistivity increases in both tensile and compressive directions. After reaching the maximum resistivity, it decreases again. The reason for this awkward behavior is described in [8, 16].

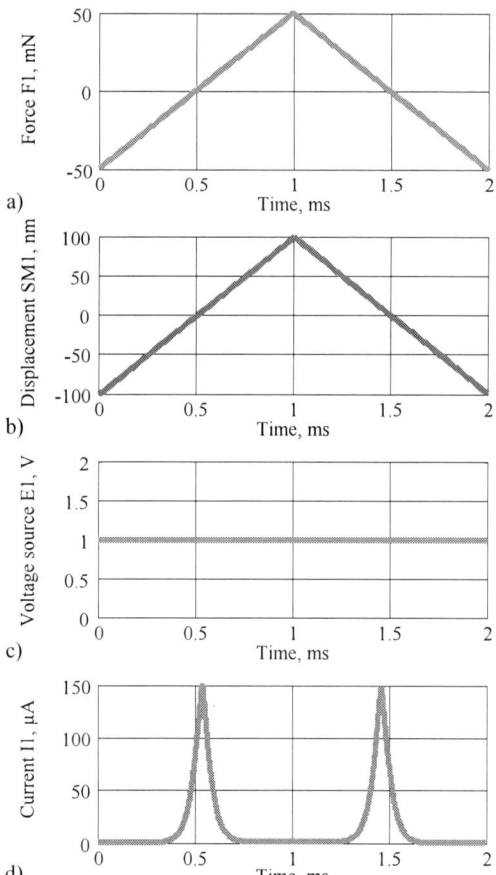

Figure 7: Results of an analysis for the semi-metallic (15,0) CNT in Simplorer

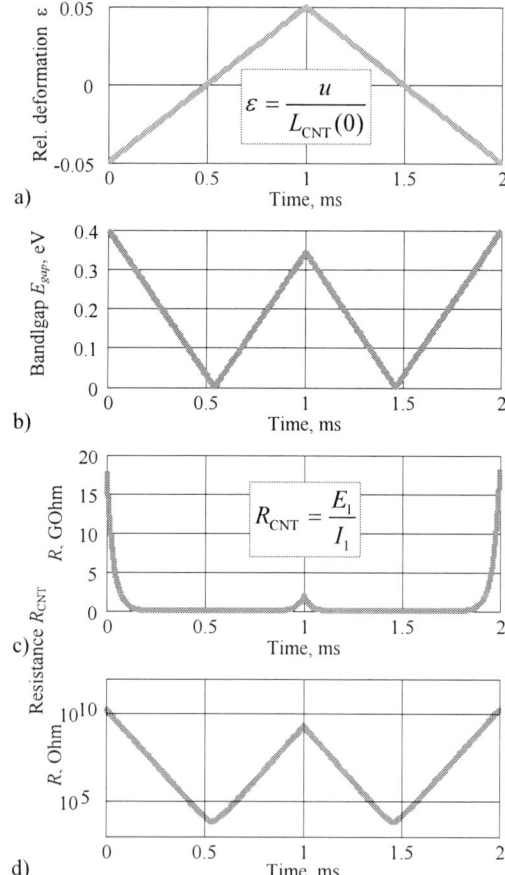

Figure 8: Derivites results of an analysis for the semi-metallic (15,0) CNT in Simplorer

4. MEMS platform

The presented macromodel of the MEMS platform is based on mode superposition technique [17]. The governing equation describing an electrostatically actuated structure in terms of modal coordinates q_i in the mechanical domain is defined by:

$$m_i \ddot{q}_i + d_i \dot{q}_i + f_i^r = f_i^e + f_i^m \qquad (4.1)$$

The modal masses m_i and modal damping constants d_i are calculated from the eigenfrequencies of the modes i and the entries of the modal stiffness matrix k_{ij}; $f_i^r = \dfrac{\partial E_{strain}}{\partial q_i}$ is the modal reaction force; f_i^m is the modal external force; E_{strain} is the strain energy;

$$f_i^e = \frac{1}{2} \sum_r \frac{\partial C_{kl}}{\partial q_i}(V_k - V_l)^2 \quad \text{is the modal electro-static}$$

force; r is the number of capacitances involved between the multiple electrodes. The capacitance C_{kl} between the electrodes k and l provides the electromechanical coupling.

In the electrical domain, the current I_j through the electrode j can be calculated from the stored charge:

$$I_j = \sum_r \left[\left(\sum_m \frac{\partial C_{kl}}{\partial q_m} \dot{q}_m \right)(V_k - V_l) + C_{kl}(\dot{V}_k - \dot{V}_l) \right] \cdot$$

$$\qquad (4.2)$$

The structural displacements u are calculated from modal amplitudes q_i by

$$u(x,y,z,t) = u_{ref} + \sum_{i=1}^{m} \phi_i(x,y,z) q_i(t), \qquad (4.3)$$

where u_{ref} are initial displacements, Φ are the eigenvectors involved in the reduced order model. The equations (4.1), (4.2) and (4.3) define the ROM macromodel, which fully describes the static, harmonic and dynamic nonlinear behavior of the flexible structure.

The strain energy, the mutual capacitances, the damping coefficients and the modal load forces are the parameters characterizing the coupled electromechanical system. The ROM macromodel is generated by numerical data sampling and subsequent fit algorithms. Each data point must be obtained by a set of separate FE run in the structural, and the electrostatic domains. At each point $(q_1, q_2, ..., q_m)$ the microstructure is displaced to a linear combination of m selected mode shapes in order to calculate the strain energy $E_{strain}(q_1, q_2, ..., q_m)$ in the structural domain. For example, in case of k modal amplitudes in each mode direction, the number of orthogonal sampling points would be k^m. At each point the $r = n(n-1)/2$ linear simulations are performed to compute lumped capacitances $C_r(q_1, q_2, ..., q_m)$ in the deformed electrostatic domain, where n is the number of conductors.

5. Co-simulation of the SWCNT and the MEMS platform

The created macromodel of the MEMS platform was integrated in a testbench of the complete sensor for verification. The equivalent circuit of the SWCNT based sensor for the first mode is shown in Figure 9.

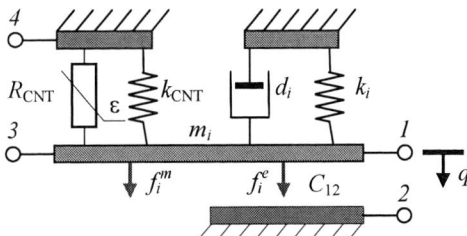

Figure 9: Equivalent circuit model of the SWCNT based sensor for the first mode with modal contribution factor 97%

A typical rectangular shaped comb drive design has a parabolic voltage-to-displacement relationship as shown in Figure 10, which is a function of the change in electrostatic force. Linearity of the electrostatic comb drive characteristics can be achieved by an optimization of the comb capacitors using FE analysis and optimization algorithms [18].

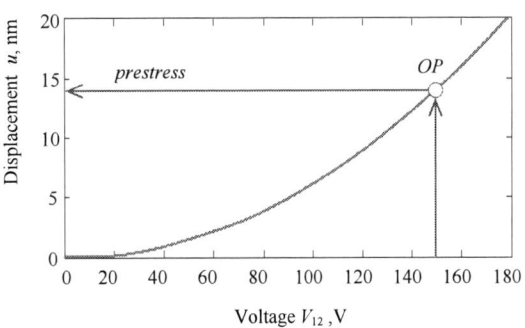

Figure 10: Voltage-to-displacement curve of the MEMS platform

The result of an analysis of the complete sensor in the electrical domain is shown in Figure 11.

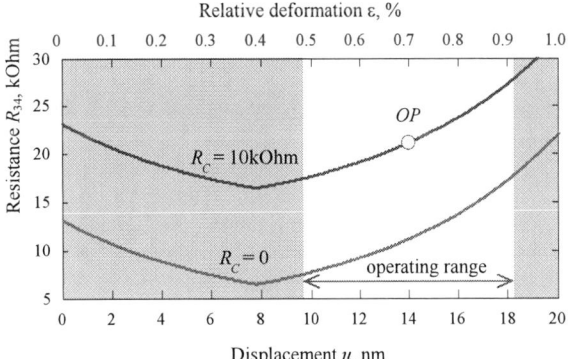

Figure 11: The output signal of the sensor with the (15,0) CNT

978-1-4799-4789-8/14 $31.00 © 2014 IEEE

Beyond this simple test case, one can simulate the CNT based mechanical sensor together with any other system environment or complex electronic circuit.

6. Conclusions

In the paper, the concept of CNT based MEMS strain gauge sensor is presented. The important aspects in developing the system models of CNTs, the parameters necessary in creating models are presented and discussed. Density function theory model is described which is used for quantum mechanical simulation of CNTs. DFT modelling is important because it deals with the properties of CNT that directly depend on the atomic positions of the carbon atoms in the tube. System level behavioral and structural abstract model of CNT strain gauge sensor is discussed using the hardware description language VHDL-AMS. The simulation results are shown. The library of developed NEMS/MEMS components is available for testing.

The presented sensor project is in concept development stage. Therefore a simple abstract model of a sensor is demonstrated in this paper. Future work consists of verification and finalizing the detailed specifications of the sensor, implementing precise and possibly every minimal electro-mechanical effects of the sensor in HDL models, considering other environmental effects on sensor such as temperature, humidity, pressure etc. Automation of model extraction, development SWCNT elements library of and co-simulation using different modelling tools are also the part of future work

Acknowledgments

This work has been done within the Research Unit 1713 which is funded by the German Research Association (DFG).

References

1. Cullinan, M. A., Culpepper, M. L., "Carbon nano-tubes as piezoresistive micro-electromechanical sensors: Theory and experiment," *Phys. Rev. B*, Vol. 82 (2010), 115428.

2. Mohamed, N. M., Kou, L. M., "Piezoresistive Effect of Aligned Multiwalled Carbon Nanotubes Array *J. of Applied Sciences*, Vol. 11 (2011), pp. 1386-1390.

3. Hierold, C., Helbling, T., Roman, C., Durrer, L., Jungen, A., Stampfer, C., "CNT based sensors," *Advances in Science and Technology*, Vol. 54 (2008), pp. 343-349.

4. Helbling, T., Roman, C., Durrer, L., Stampfer, C., Hierold, C., "Gauge Factor Tuning, Long-Term Stability, and Miniaturization of Nanoelectro-mechanical Carbon-Nanotube," *Sensors Electron Devices, IEEE Transactions on*, Vol. 58, No. 11 (2011), pp. 4053-4060.

5. Burg, B. R., Helbling, T., Hierold, C., and Poulikakos, D., "Piezoresistive pressure sensors with parallel

integration of individual single-walled carbon nanotubes," *J. Appl. Phys.*, AIP, Vol. 109 (2011), 064310.

6. Maiti, A., Svizhenko, A., and Anantram, M. P., "Electronic transport through carbon nanotubes: Effects of structural deformation and tube chirality," *Phys. Rev. Lett.*, Vol. 88, No. 12 (2002), p. 126805.

7. Minot, E. D., Yaish, Y., Sazonova, V., Park, J.-Y., Brink, M., McEuen, P. L., "Tuning Carbon Nanotube Band Gaps with Strain," *Phys. Rev. Lett.*, Vol. 90 (2003), p. 156401.

8. Yang, L., Han, J., "Electronic Structure of Deformed Carbon Nanotubes," *Phys. Rev. Lett.*, Vol. 85 (2000), pp. 154-157.

9. Bansal, R., Clark, J. V., "Lumped modeling of carbon nanotubes for M/NEMS simulation," *Microsyst. Technol.*, Vol. 18, No. 12 (2012), pp. 1963-1970.

10. Liu, B., Huang, Y., Jiang, H., Qu, S., Hwang, K.C., "The atomic-scale finite element method," *Comput. Methods Appl. Mech. Engrg.*, Vol. 193 (2004), pp. 1849-1864.

11. Atomistix ToolKit version 12.2, QuantumWise A/S, www.quantumwise.com

12. Soler, J. M., Artacho, E., Gale, J. D., García, A., Junquera, J., Ordejón, P., Sánchez-Portal, D., "The SIESTA method for ab initio order-N materials simulation," *J. Phys.: Condens. Matter*, Vol. 14, No. 11 (2002), pp. 2745-2779.

13. Abadir, G. B., Walus, K., Pulfrey, D. L., "Basis-set choice for DFT/NEGF simulations of carbon nanotubes," *J. Comput. Electronics*, Vol. 8, No. 1 (2009), pp. 1-9.

14. Monkhorst, H. J., Pack, J. D. ., "Special points for Brillouin-zone integrations," *Phys. Rev. B*, Vol. 13 (1976), pp. 5188-5192.

15. Perdew, J. P., Wang, Y., "Accurate and simple analytic representation of the electron-gas correlation energy," *Phys. Rev. B*, Vol. 45 (1992), pp. 13244-13249.

16. Wagner, C., Schuster, J., Geßner, T., "DFT investigations of the piezoresistive effect of carbon nanotubes for sensor application," *Phys. Stat. Sol. B*, Vol. 249, No. 12 (2012), pp. 2450-2453.

17. Schlegel, M., Bennini, F., Mehner, J., Herrmann, G., Mueller D., Doetzel, W., "Analyzing and Simulation of MEMS in VHDL-AMS Based on Reduced Order FE-Models," *IEEE Sensors Journal*, Vol. 5, No. 5 (2005), pp. 1019-1026.

18. Scheibner, D., Wibbeler, J., Mehner, J., Brämer, B., Gessner, T., Dötzel, W., "A Frequency Selective Silicon Vibration Sensor with Direct Electrostatic Stiffness Modulation," *Analog Integrated Circuits and Signal Processing*, Vol. 37 (2003), pp. 35-43.

FEM simulation and measurement validation of a cMUT cell

S.P. Mao[1,2,*], X. Rottenberg[1], V. Rochus[1], B. Nauwelaers[2] and H.A.C. Tilmans[1]

[1]IMEC v.z.w., Kapeldreef 75, 3001 Heverlee, Belgium

[2]Department of Electronic Engineering, Katholieke Universiteit Leuven, Leuven, Belgium

*mao@imec.be

Abstract

This paper presents Finite Element Modeling (FEM) simulation and experimental results of the acoustic and mechanical behavior of an isolated cMUT cell. cMUT cells fabricated in the SiGeMEMS technology of imec are used as test vehicles. Dynamic characteristics of a cMUT cell actuated by a pulse superimposed on a DC bias, such as resonance frequency, Q-factor and transient displacement, are studied using ANSYS FEM and Polytec vibrometer measurements. A good agreement is achieved between the FEM simulation results and optical measurements. The Rayleigh integral method is used to construct the spatial pressure field based on the transducer surface information obtained both from ANSYS and the optical measurements. The results are compared with hydrophone measurements, and good agreements are achieved. Acoustic measurements on a typical cMUT cell (measuring ~60μm on the side) in a fluorinert medium (FC-84) show a 7MHz center frequency, ~100% -6dB fractional bandwidth (FBW) and several kilo Pascal peak-to-peak pressure at millimeters scale. Our method provides a clear methodology to accurately predict the transient spatial pressure field of a cMUT cell, especially in the far field region, with little time cost.

1. Introduction

Capacitive micromachined ultrasound transducers (cMUTs) are very promising transducers for ultrasound applications. Comparing with the conventional piezoelectric ultrasonic transducers, cMUTs exhibit many unique features inherently. The flexural mode of the cMUT membrane reduces the mechanical impedance of the devices and improves the energy coupling with the ambient medium. Moreover, mircofabrication capabilities makes them easy for a low cost batch production with fine pitches. Besides, cMUTs also benefit from the underlying CMOS circuit, since the closer driving and read-out circuit means a better signal-noise-ratio. Electrical interconnection issue is also solved, which enables large scale 1D or more complicated 2D array configurations. Given above facts, cMUTs have been well recognized very suitable for the next generation ultrasonic diagnostic imaging [1]. In the past two decades, ultrasound imaging prototypes based on cMUTs have been reported by several groups [1-3].

However, the study of an individual cMUT cell, both in FEM (and analytical) modeling and in experiments, is often not addressed. In fact, the accurate prediction of output pressure and corresponding frequency response provides useful and important information for the design of ultrasonic transducers. Though the theory on acoustic radiators has been established and is well described for the traditional piezoelectric ultrasonic transducers relying on a thickness mode of vibration [4], the basic theory should also be applicable for cMUTs which operate in a flexural mode of vibration (of the membrane). However, cMUT cell requires a precise micromachining, has a much smaller aperture size and significant different aperture/lambda ratio. Moreover, the thin membrane enables cMUTs to have a much lower mechanical impedance than the conventional thickness mode transducers, which makes it more susceptible to loading by the acoustic medium [5].Therefore, the realistic dynamic response and corresponding spatial pressure field of a cMUT cell are still very complicated and not well examined, especially for the correlation study between theory and experiment. Numerous works on the modeling and simulation of cMUT have been done, such as the analytical model based equivalent circuit model [6-10] and FEM simulation [11-13]. The analytical model is fast but limited to some simple shapes such as circle and square, and approximations always have to be made for the modeling. FEM simulation on the other hand is very suitable for the complicated structures and has the highest accuracy but with a relative large computing time. To our best knowledge, the comparison of modeling and experiment is very scarce. There are some studies which combine FEM simulation and experiments, but lack either detailed comparison results [11] or a proper measurement correction method [12], [13]. Moreover, FEM simulations and experimental comparison of the transient response of a cMUT cell, especially with a clear methodology and strong conclusion, has not been found in the existing literature.

In this paper we use both FEM simulation and optical scan to obtain the transducer surface transient response, and then the results are utilized to construct the spatial pressure field by the Rayleigh integral method. Output pressure of the cMUT is also detected by the hydrophone, and the results of simulation and measurement are compared.

2. Device and technology build-up

An isolated square cMUT cell with rounded corners is chosen for the study. Isolating the cell has the advantage that any disturbing neighboring effects such as crosstalk, anchoring and so on are reduced to negligibly small levels. Figure 1(a) is the top view of the device; the diameter of the square membrane is 60μm, and there are two probe pads, the one closest to the device is connected to the bottom electrode and the other is connected to the top electrode. When the DC and AC voltage are applied onto the two pads separately or superimposed on one pad and make the other one grounding, the membrane will vibrate and the corresponding ultrasound wave will be generated in the load medium. In this paper the device is actuated in two mediums: air and a specific fluorinert: FC-84 (from 3M, USA).

A cross-section of the cMUT is shown in Fig. 1(b). The membrane is a composite of three different materials: a 1.5μm thick silicon oxide sealing layer, a 3.6μm thick SiGe structural layer and a 200nm thick SiC dielectric layer. The transducer gap defined by the SiGe bottom electrode and the SiGe membrane layer top electrode is composed of a 200nm (vacuum) gap surrounded by 200nm thick SiC dielectric layers. The cMUT cell is fabricated in imec's SiGeMEMS-based cMUT technology. Details about the processing can be found in one of our previous publications [14]. The main material properties used in this paper are listed in Table 1.

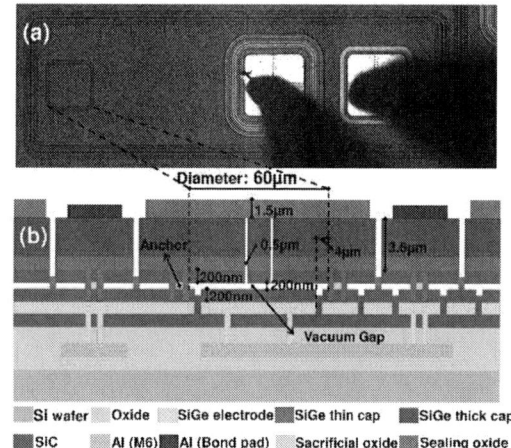

Figure 1: (a) Micrographs of the top view of the cMUT cell, (b) Cross-sectional view of the cMUT cell.

Table 1: Material properties as used in this paper.

Properties	SiO$_2$	SiGe	SiC	Air	FC-84
Young's modulus [GPa]	64	120	150	-	-
Poisson's ratio [-]	0.17	0.22	0.18	-	-
Density [kg/m^3]	2200	4775	3200	1.18	1730
Dielectric constant [-]	-	-	4.4	-	-
Speed of sound [m/s]	-	-	-	345	542

3. Modeling and Experiment

3.1. Finite element model

The commercial FEM software package ANSYS (ANSYS 13 non-commercial version, ANSYS Inc., USA) is used to simulate the cMUT cell. Figure 2 shows a sketch of the 3D FEM model of one quarter of the cMUT cell. Symmetrical boundary conditions are imposed on the two

cross sections in the top view (Fig. 2(a)). It should be noted that anti-symmetric harmonic modes are ignored when using a quarter model, however these anti-symmetric modes do not generate an effective output pressure for an individual cell [15]. A cross-sectional view of the FEM model is shown in Fig. 2(b); the cMUT sealing oxide layer, SiGe structural layer, SiC dielectric layer and the anchor are modeled using the SOLID186 element. The FLUID30 element is used to represent the FC-84 fluid medium, a spherical absorbing boundary FLUID130 is used to eliminate the reflection, and a TRANS126 element is used between the bottom and top electrode to represent the electrostatic transducer. In this model, no material internal stress is included. This is somewhat justified as it is found experimentally that the sealing oxide layer has a compressive stress which is almost equal to the tensile stress in the SiGe and SiC layer, and therefore the effect of the residual stress is cancelled. The anchoring is properly taken into account in the simulation. It is found that when using fixed edge conditions (i.e., without the anchor) the resonance frequency obtained from the FEM analysis is 20% higher than the situation with the anchor.

Two types of simulation are performed in this paper: (1) a pre-stressed harmonic simulation and (2) a transient simulation actuated by a pulse superimposed on a DC bias. The choice of the radius of the fluid medium is very crucial especially for the transient simulation. The minimum size is limited by 1/5 of the largest wavelength corresponding to the lowest operation frequency [16]. The maximum size is limited by the highest operation frequency, and a very fine mesh is required for the wave propagation and for the effective absorbance on the FLUID130 boundary at a high frequency. This results in a gigantic number of elements and a long computing time. Here, the radius of the fluid medium is set to 250μm.

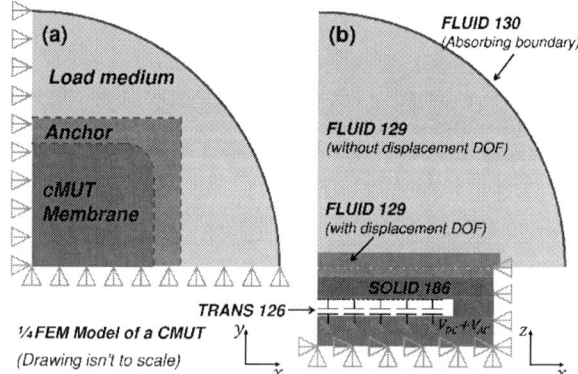

Figure 2: Illustration of the ANSYS FEM model: (a) Top view of the FEM model of a quarter of the cell; (b) Cross-section of the FEM model.

3.2. Measurements: Optical and acoustic

The dynamic behaviors of the cMUT cell are studied using the commercial Laser Doppler Vibrometer micro system analyzer (MSA-500, Polytec GmbH Waldbronn,

Germany). The experimental set-up is depicted in Fig. 3(a). The cMUT cell is emerged in FC-84 fluid which has a depth around 5mm. A +60V, 50ns duration pulse (with a rising and falling time smaller than 10nsec) is coming from a pulse generator (AVL-2-B-PN, Avtech Electrosystems Ltd., USA) and applied on the pad connected to the bottom electrode of the cMUT cell. A -160V DC bias which equals 60% of the theoretical pull-in voltage is applied to the other pad. Polarities of the pulse voltage and the DC bias are such that effective voltages add up. Measurements were performed in air and in FC-84. A correction is performed in the FC-84 measurement, and the measured displacement is divided by 1.261 which is the refractive index of FC-84 [17]. But the acousto-optic effect on the refractive index is not included, considering there is only several mega Pascal peak pressure variation during the measurement.

The experimental set-up used for measuring the sound pressure is shown in Fig. 3(b). A "tip type" calibrated hydrophone (HNP1000, Onda Inc., USA) is used for detecting the sound pressure at a certain distance away from the transducer surface. The hydrophone signal is amplified by the pre-amplifier and recorded by an oscilloscope (TDS3014B, Tektronix, Inc., USA). Furthermore, Labview (Labview, National Instruments, USA) is used to control the instruments and to collect and save the measured data of the oscilloscope.

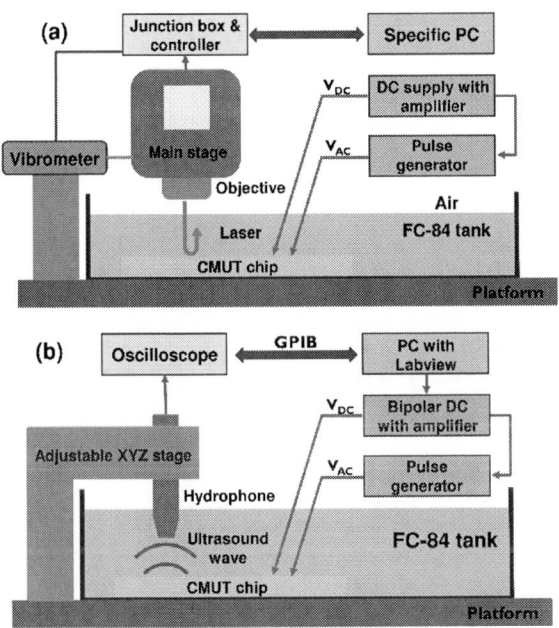

Figure 3: Schematic of the experimental set-ups: (a) Optical set-up for the dynamic characterization using a Polytec vibrometer system; (b) Acoustic set-up for the sound pressure measurements using a hydrophone.

4. Results and discussion

4.1. Dynamic (resonance) response

According to the acoustic theory [5], some fluid will vibrate together with the transducer as an amount of mass (called the radiation mass) attached to the transducer surface. The radiation mass will lower the membrane resonant frequency (or the center frequency in the pressure frequency response plot). In order to evaluate the effect of the fluid load, the frequency response characteristics of the cMUT cell is evaluated in both air and FC-84.

Results for air:

In ANSYS simulation the cMUT cell is initially stressed by a constant 160V DC bias, and then a pre-stressed harmonic simulation is performed. The displacement of the central point is evaluated and normalized to the maximum displacement (Fig. 4). In air, a resonance peak is simulated around 9.43 MHz, while the Polytec measurement gives a peak response at 9.5MHz. The center frequency mismatch between FEM simulation and measurement is negligibly small. However the difference in the observed Q-factor is significant. The Q-factor is characterized by the -3dB bandwidth of the resonance response relative to its center frequency. The Q-factor 496 obtained from ANSYS simulations is about two times larger than the measured Q-factor 232. This is considered reasonable, since besides the air damping losses there are no additional losses taken into account in the simulation. Therefore, the air damping is negligibly small and the material damping and/or the support or anchor (losses) damping are dominating. The results are summarized in Table 2.

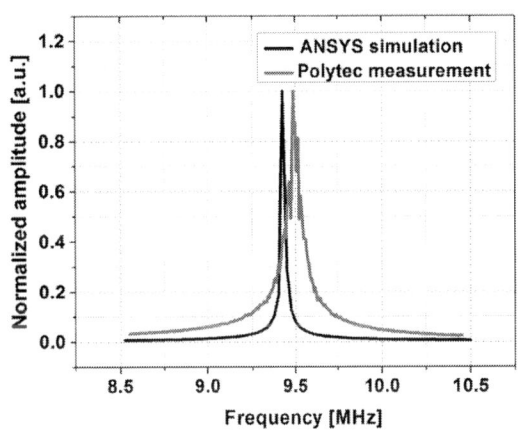

Figure 4: Resonance response of the cMUT cell in air.

Results for FC-84:

Similarly, a comparison is also done in FC-84 (Fig. 5); a polynomial fitting is used to reduce the noise in the fluid measurement. Both the simulation and the measurement show a clear shift in the resonance frequency compared to the measurements in air. The center frequency in FC-84 is around 5.5MHz (compared to 9.5MHz in air). This downshift is attributed to the significantly larger radiation mass of FC-84 as compared to air. The Q-factor observed in FC-84 is also largely reduced as compared to Q-factor

978-1-4799-4789-8/14 $31.00 © 2014 IEEE

in air. The Q-factor in FC-84 between simulation (Q-factor=1.2) and measurement (Q-factor=1.1) matches much better than the result in air due to the dominated fluid damping, and a Q-factor lower to one is obtained which illustrates the high bandwidth nature of the cMUT cell. The results are also summarized in Table 2.

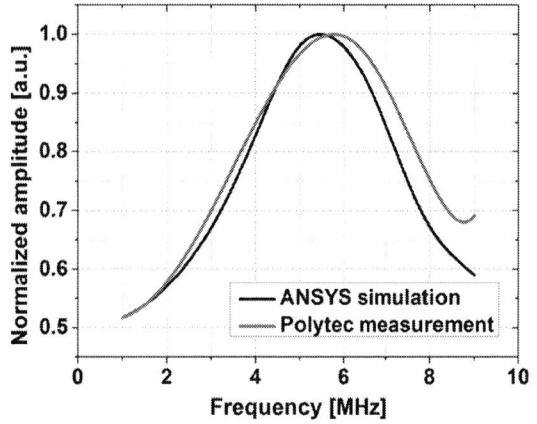

Figure 5: Resonance response of the cMUT cell in FC-84.

Table 2: Summary of resonance response results.

Medium	Properties	ANSYS	Polytec
Air	Center frequency [MHz]	9.43	9.50
	Q-factor	496	232
FC-84	Center frequency [MHz]	5.47	5.77
	Q-factor	1.2	1.1

4.2. Dynamic (displacement) response

A pre-stressed transient simulation in ANSYS firstly requires a transient simulation without the transient effect (ANSYS command: timint, off) and to apply a constant 160V DC bias only, and then a normal transient simulation with a pulse superimposed on the 160V DC bias is performed. Considering the delay in the Polytec measurement, 175ns time duration is set for the transient simulation without the transient effect which means the 60V, 50ns duration pulse is applied at 175ns time. The displacement of the central node is evaluated again. There is no static deflection in Polytec measurement, and the measurement displacement is adjusted in order to compare with the simulation.

Results for air:

Dynamic responses of the cMUT cell in air using ANSYS simulation and Polytec measurement are shown in Fig. 6. A good amplitude match is found at first several cycles. Due to the resonance frequency and Q-factor mismatch, the difference between simulation and measurement becomes significant after several cycles.

However in the measured case, because of the additional damping (material damping, anchor losses) besides the air damping, the response is expected to decay much faster than for the simulated cases.

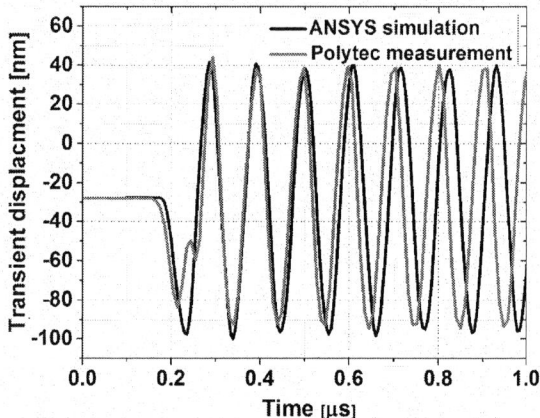

Figure 6: Transient displacement of the cMUT cell in air (Drive voltage: V_{DC}=-160V, V_{pulse}=+60V).

Results for FC-84:

A similar simulation and measurement is done in FC-84. Medium damping becomes significant when the device is emerged in FC-84, and the transient displacement signal is quickly damped both in the simulation and the measurement (Fig. 7).

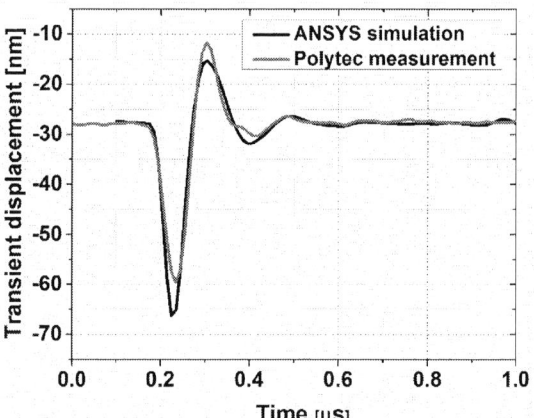

Figure 7: Transient displacement of the cMUT cell in FC-84 (Drive voltage: V_{DC}=-160V, V_{pulse}=+60V).

4.3. Sound pressure response in FC-84

4.3.1. Rayleigh integral method

Because of the finite dimensions (mm's) of the hydrophone tip, the acoustic sound pressure measurements can only be performed several millimeters away from the transducer surface. For the FEM analysis when using mesh sizes on the order of micron's, the region where because of the finite dimension of the hydrophone tip the pressure can be evaluated is limited to a few mm's distant from the transducer (diaphragm) surface, which is not available to be simulated. Furthermore, in the Polytec measurements only the displacement of the transducer surface can be recorded. In order to solve this problem, the frequency domain Rayleigh integral method ([18], [19]) is used to construct the spatial pressure field derived from the transducer surface velocity.

The general idea is to firstly mesh the transducer into the small rectangular element unit, acoustic path is calculated from the center point of the small unit to the spatial position, and pressure at each spatial position is a superposition of pressure from different units. According to the above transient simulation and measurement results, the response will be damped quickly and thus only a short time period simulation is needed and the simulation results are combined with a null vector at the end to represent a long time period signal. The transient displacement result is converted to the transient velocity, and then the transient velocity is transferred to the frequency domain by a Fast Fourier Transform (FFT). Spatial pressure in frequency domain is calculated by the Rayleigh integral based on the velocity. Finally the inverse Fast Fourier Transform (iFFT) is used to obtain the time domain pressure signal in the spatial space.

The great advantages of this frequency domain method are as follows: (1) Attenuation can be included in the spatial field construction, whereby the measured attenuation coefficient ($0.352f^{1.56}$ dB/cm/MHz^n) of fluorocarbon in reference [20] is used; (2) Frequency filter can be set, considering the hydrophone with a cut-off frequency above 20MHz and also the 24MHz measurement frequency span limit of the Polytec system, the frequency components higher than 25MHz are set to zero in the calculation of the pressure field.

4.3.2. Axial pressure amplitude

Figure 8 shows the axial pressure behavior of the cMUT cell. The transient pressure amplitude obtained by ANSYS or Polytec followed by the Rayleigh integral is plotted vs. distance from the transducer surface. Unlike the traditional piezoelectric ultrasonic transducer, there is no near field diffraction fluctuation and the far field transition

starts from the transducer surface. This is also expected because of the extremely small aperture/lambda ratio (<1) of the cMUT cell, much smaller than for conventional piezoelectric transducers. The effect of the attenuation (solid line in the figure) becomes pronounced in the millimeter scale, the region where the acoustic measurements are performed.

The results comparison is summarized in Table 3. A big difference at the first measurement point (at 3.24mm) between hydrophone measurement and ANSYS simulation is found. This is attributed to the near field transition (at around 3mm) of the hydrophone, which makes that measurements in this region or even closer to the cMUT cell are not "correct" any more, but are obscured by the hydrophone limitations. However, it is also found that the differences become large again as the displacement increases. This is probably because the attenuation in FC-84 is overestimated, making that the pressure decreases faster than it should. Moreover, the pressure measured by the hydrophone is converted using the calibration data in water while the measurement is actually performed in FC-84. This introduces another uncertainty in the hydrophone measurement data.

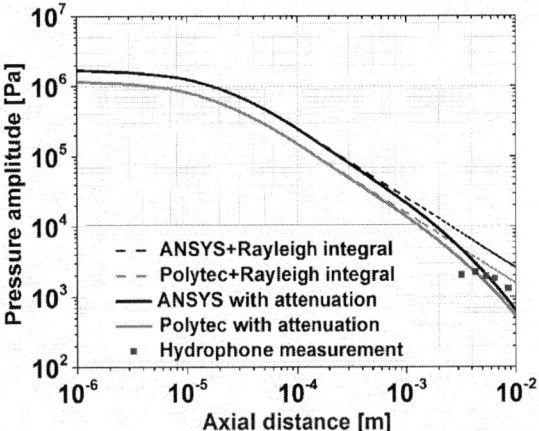

Figure 8: Axial sound pressure amplitude evaluated for several cases (Drive voltage: V_{DC}=-160V, V_{pulse}=+60V).

Table 3: Comparison of the pressure amplitude on the axial distance (with FC-84 attenuation). The %-error is always relative to the hydrophone measurements.

Distance [mm]	Hydrophone [kPa]	ANSYS [kPa]	%-error	Polytec [kPa]	%-error
3.24	2.037	4.660	128.8%	3.292	61.6%
4.32	2.230	2.989	34.0%	2.194	1.6%
5.40	1.950	2.063	5.8%	1.564	19.8%
6.48	1.820	1.490	18.1%	1.165	36.0%
8.64	1.291	0.867	32.8%	0.707	45.2%

Table 4: Comparison of the pulse response frequency spectrum (with FC-84 attenuation).

Properties	Hydrophone	ANSYS	%-error	Polytec	%-error
Center frequency [MHz]	7.366	6.50	11.8%	6.9	6.3%
-3dB FBW	57.15%	90.77%	58.8%	82.61%	44.6%
-6dB FBW	87.50%	129.23%	47.7%	105.79%	20.9%

4.3.3. Frequency spectrum in the far field

Figure 9 shows the simulated and measured normalized frequency spectra at a distance of 4.32mm from the surface of the cMUT cell. The acoustic measurement result is already corrected for by the sensitivity of hydrophone. However, the attenuation is not corrected here, since the effect of fluid attenuation is very big and we want to show the original measurement result. Due to the diffraction correction during the wave propagation [21], the center frequency in the far field is higher than the resonance frequency at the transducer surface. In fact, the center frequency can be as high as the resonance frequency without fluid load if there is no fluid attenuation. But, the attenuation coefficient is frequency dependent, which makes that the high frequency components decay faster than the low frequency ones. Finally, the center frequency comes to around 7 MHz as a result of the two effects and this is found for all three evaluation methods (see Table 4). A big mismatch in the high frequency region is found between ANSYS simulation on the one hand and Polytec measurements or hydrophone measurements on the other hand. This is probably because of some practical problems such as the slightly different pulse shape between simulation and measurement, or, any dielectric (SiC) charging of the device, or, the non-vacuum conditions of the cMUT cavity resulting in squeezed film damping.

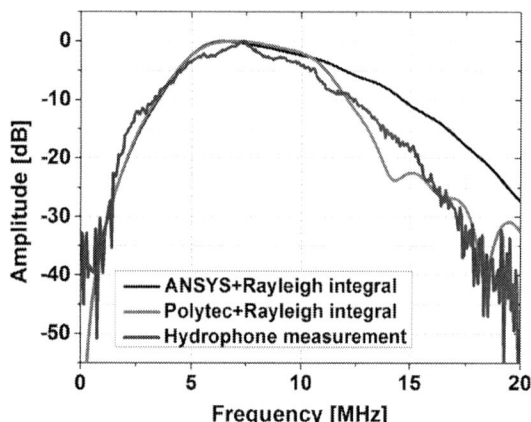

Figure 9: Comparison of the frequency spectrum at 4.32mm distance in FC-84 (Drive voltage: V_{DC}=-160V, V_{pulse}=+60V).

4.3.4. Transient pressure in the far field

Figure 10 shows a comparison between FEM, optical and acoustic measurements at a distance 4.32mm from the transducer surface in FC-84. Again, it is found, that the high frequency component of ANSYS simulation is larger than for the Polytec and hydrophone measurements, similar as seen in the frequency spectrum comparison. The pressure difference may be due to the uncertainty in the hydrophone sensitivity. Other effects like improper adjustment of the hydrophone position, the high voltage charging or possible membrane stresses may also affect the

results. All in all however, the results are already quite comparable within a small acceptable error.

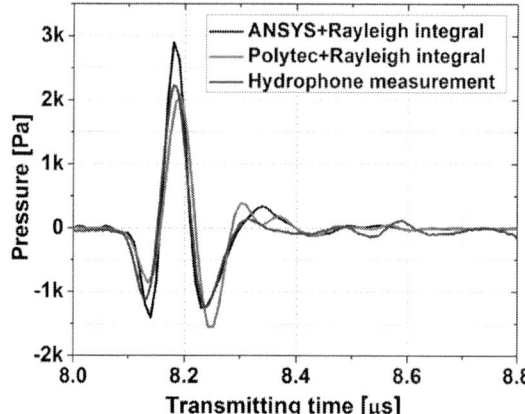

Figure 10: Comparison of the transient pressure at 4.32mm distance for FC-84 (Drive voltage: V_{DC}=-160V, V_{pulse}=+60V).

5. Conclusions

In this paper a clear correlation study between FEM simulation and experimental studies of an individual cMUT cell is presented, and a clear understanding of the cMUT cell behaviors is demonstrated. Good agreements are obtained between ANSYS simulation and Polytec measurement. The corresponding spatial pressure results also match well with the hydrophone measurement results. It proves that ANSYS combined with the Rayleigh integral method is a fast and effective pressure field simulation method. We believe it is very helpful for the design of complicated cMUT arrays for future imaging applications.

Acknowledgments

The authors should thanks Dr. Philippe Helin and Agnes Verbist for the cMUT processing, thanks Dr. Jeroen De Coster and Dr. Piotr Czarnecki for the friendly help during the cMUT test, and thanks Prof. Christ Glorieux, Dr. Jichuan Xiong and Liwang Liu from Department of Physics and Astronomy for the helpful discussion on the correction of the measurement.

References

[1] O. Oralkan, A. S. Ergun, J. A. Johnson, M. Karaman, U. Demirci, K. Kaviani, T. H. Lee, and B. T. Khuri-Yakub, "Capacitive Micromachined Ultrasonic Transducers: Next-Generation Arrays for Acoustic Imaging?" IEEE Transactions on Ultrasonics, Ferroelectrics, and Frequency Control, Vol. 49, No. 11 (2002), pp. 1596-1610.

[2] A. S. Savoia, G. Caliano, and M. Pappalardo, "A CMUT Probe for Medical Ultrasonography: From Microfabrication to System Integration," IEEE Transactions on Ultrasonics, Ferroelectrics, and Frequency Control, Vol. 59, No. 6 (2012), pp. 1127-1138.

[3] L. L. P. Wong, A. I. Chen, A. S. Logan, and J. T. W. Yeow, "An FPGA-Based Ultrasound Imaging System Using Capacitive Micromachined Ultrasonic Transducers," IEEE Transactions on Ultrasonics, Ferroelectrics, and Frequency Control, Vol. 59, No. 7 (2012), pp. 1513-1520.

[4] M. I. Gutiérrez, H. Calás, A. Ramos, A. Vera, and L. Leija, "Acoustic field modeling for physiotherapy ultrasound applicators by using approximated functions of measured non-uniform radiation distributions," Ultrasonics, 52 (2012), pp. 767–777.

[5] L. A. Kinsler, A. R. Frey, A. B. Coppens, and J. V. Sanders, Fundamentals of acoustics, Wiley, Jan. 2000.

[6] A. Lohfink and P. C. Eccardt, "Linear and Nonlinear Equivalent Circuit Modeling of CMUTs," IEEE Transactions on Ultrasonics, Ferroelectrics, and Frequency Control, Vol. 52, No. 12 (2005), pp. 2163-2172.

[7] I. O. Wygant, M. Kupnik, and B. T. Khuri Yakub, "Analytically calculating membrane displacement and the equivalent circuit model of a circular CMUT cell," 2008 IEEE International Ultrasonics Symposium Proceedings, Beijing, China, Nov 2008, pp. 2111-2114.

[8] H. Koymen, A. Atalar, E. Aydoğdu, C. Kocabaş, H. K. Oğuz, S. Olcum, A. Ozgurluk, and A. Unlugedik, "An Improved Lumped Element Nonlinear Circuit Model for a Circular CMUT Cell," IEEE Transactions on Ultrasonics, Ferroelectrics, and Frequency Control, Vol. 59, No. 8(2012), pp.1791-1799.

[9] A. Caronti, G. Caliano, A. Iula, and M. Pappalardo, "An Accurate Model for Capacitive Micromachined Ultrasonic Transducers," IEEE Transactions on Ultrasonics, Ferroelectrics, and Frequency Control, Vol. 49, No. 2 (2002), pp. 159-168.

[10] X. Rottenberg, A. Erismis, P. Czarnecki, Ph. Helin, A. Verbist and H. A. C. Tilmans, "Consistent Analytical Model for Single and Dual Thickness Capacitive Micromachined Ultrasound Transducers (cMUT)", 13th International Conference on EuroSimE, Apr. 2012, pp. 1–6.

[11] A. Caronti, G. Caliano, R. Carotenuto, A. Savoia, M. Pappalardo, E. Cianci, and V. Foglietti, "Capacitive micromachined ultrasonic transducer (CMUT) arrays for medical imaging," Microelectronics Journal, 37 (2006), pp. 770–777.

[12] B. Bayram, G. G. Yaralioglu, M. Kupnik, A. S. Ergun, O. Oralkan, A. Nikoozadeh, and B. T. Khuri-Yakub, "Dynamic Analysis of Capacitive Micromachined Ultrasonic Transducers," IEEE Transactions on Ultrasonics, Ferroelectrics, and Frequency Control, Vol. 52, No. 12(2005), pp. 2270-2275.

[13] G. G. Yaralioglu, A. S. Ergun, and B. T. Khuri-Yakub, "Finite-Element Analysis of Capacitive Micromachined Ultrasonic Transducers," IEEE Transactions on Ultrasonics, Ferroelectrics, and Frequency Control, Vol. 52, No. 12 (2005), pp. 2185-2198.

[14] Ph. Helin, P. Czarnecki, A. Verbist, G. Bryce, X. Rottenberg, and S. Severi, "Poly-SiGe-based CMUT array with high acoustical pressure", Proc. of MEMS 2012, Jan. 2012, pp. 305-308.

[15] Ch. H. Sherman and J. L. Butler, Transducers and Arrays for Underwater Sound, Springer, New York, 2007, pp.246-247.

[16] ANSYS 13 Help Documentation.

[17] Laser Vibrometer Measurements of Objects Immersed in Transparent Fluids, Polytec GmbH Laser Measurement Systems Application Note VIB-G-04, Nov 2005.

[18] W. N. Cobb, "Frequency domain method for the prediction of the ultrasonic field patterns of pulsed, focused radiators," The Journal of the Acoustical Society of America, vol. 75, issue 1 (1984), pp. 72-79.

[19] M. M. Goodsitt, E. L. Madsen, and J. A. Zagzebski, "Field patterns of pulsed, focused, ultrasonic radiators in attenuating and nonattenuating media," The Journal of the Acoustical Society of America, vol. 71, issue 2 (1982), pp. 318-329.

[20] E. M. Strohm, M. C. Kolios, "Sound velocity and attenuation measurements of perfluorocarbon liquids using photoacoustic methods," 2011 IEEE International Ultrasonics Symposium Proceedings, Oct 2011, Orlando, FL, USA, pp. 1-4.

[21] Th. L. Szabo, "Diagnostic Ultrasound Imaging: Inside Out," Elsevier Academic Press, 2004.

Microstructure Simulation of Grain Growth in Cu Through Silicon Via Using Phase-Field Modeling

Nabi Nabiollahi[1,2], Nele Moelans[2], Mario Gonzalez[1], Joke De Messemaeker[1], Christopher J. Wilson[1],
Kristof Croes[1], Eric Beyne[1] and Ingrid De Wolf[1,2]

[1] imec,
Kapeldreef 75, Leuven, Belgium
[2] Dept. of Metallurgy and Materials Engineering,
KULeuven, Leuven, Belgium
nabi.nabiollahi@imec.be

Abstract

In this paper, a time-efficient 3D phase-field model, for simulating grain growth in Through Silicon Via (TSV) is presented. This model is modified to model grain growth in the cylindrical shape of a TSV to capture the effect of temperature in its microstructure. The data generated from this simulation is used to explain large distribution of Cu pumping (i.e. non reversible thermal expansion of TSV). To achieve this, generated results must be used as an input in a Finite Element Model of a TSV structure to study the effect of grain growth and asymmetry in distribution of Cu pumping. Results generated from a sample FEM model with grain structure input confirms this capability.

1. Introduction

Through Silicon Vias (TSVs) are a key part of 3D System In Package (SIP) devices, enabling the vertical interconnection of stacked dies. Most often these are filled with Cu in polycrystalline form. Due to the large difference in coefficient of thermal expansion with Si, the exposure to high temperatures during subsequent processing steps causes irreversible extrusion of the Cu, referred to as 'Cu pumping'. This results in a relatively high tensile stress inside the Cu at room temperature. The distribution of both Cu pumping and Cu stress values shows a large spread over the TSVs of a single wafer [1-3]. For Cu pumping this spread is clearly correlated to variations in the Cu microstructure [4]. As potential reliability issues related to Cu pumping or Cu stress will first occur at the TSVs with the highest values for either, any model aiming to predict this behavior should include a statistical spread in addition to a median value. Therefore, variations in the Cu microstructure between TSVs and during exposure to the high BEOL processing temperatures (grain evolution [5]) must be taken into account.

Finite Element Models (FEM) for the study of reliability and failure mechanisms in TSVs encountered in literature assume isotropic Cu properties [6,7]. We are developing a finite element model for the thermo-mechanical behavior of Cu TSVs incorporating Cu microstructure, in order to capture the resulting variations and build further understanding of the role of Cu microstructure. This paper presents the first part, a time-efficient model for simulating grain growth inside the TSV using the phase-field method.

Phase-field modeling is widely used to simulate grain-growth in heterogeneous materials on a mesoscale. This method allows to simulate the evolution of the polycrystalline structure in a Cu TSV and when coupled with a FEM enables to include anisotropic properties as a function of grain orientation in the elasticity and plasticity models. In the phase field method, different order parameters are assigned to the different grain orientations and grain boundaries are described as diffuse transitions in the values of these order parameters. Moreover, differential equations are derived from kinetic and thermodynamic principles, based on the assumption that a reduction in bulk energy, interfacial energy or elastic energy, is the driving force for grain evolution [8, 9]. An important advantage of the phase-field method is that, thanks to the diffuse-interface description, there is no need to track the grain boundaries during microstructure evolution and therefore it is mathematically feasible to simulate the evolution of complex grain shapes and connected grain structures in 3D [8, 9].

In this work a Fast Fourier transform based method is used to solve the differential equations. The implementation was adapted to treat the cylindrical shape of the TSVs. The time steps in the simulations are related to a physical time and temperature through the thermodynamic and kinetic properties of the material. The resulting grain structures will be imported to a finite element code, where the effect of anisotropic elasticity in the different grains in a TSV will be analyzed, depending on the grain size. A description of the theoretical background of phase-field simulations with focus on grain growth and the modification implemented in phase-field method and a sample FEM to model grain growth in a cylindrical TSV are also included in this paper.

2. Phase-field method

Different phases of crystal orientations, structures, compositions in a heterogeneous material impose anisotropic physical properties. Phase-field method provides a way to model the evolution of the stated phases and enables us to analyze anisotropy in these materials. Although phase-field method is widely used in various applications including but not limited to solidification, multiphase systems and electromagnetism, in this section, the equations important for grain growth and crystal evolution are only covered. Readers are advised to refer to [8, 9] for in-depth review of phase-field method on modeling microstructural evolution.

All the grains in a Polycrystalline material are thermodynamically unstable and they continuously grow or shrink over time and under presence of physical load

978-1-4799-4789-8/14 $31.00 © 2014 IEEE

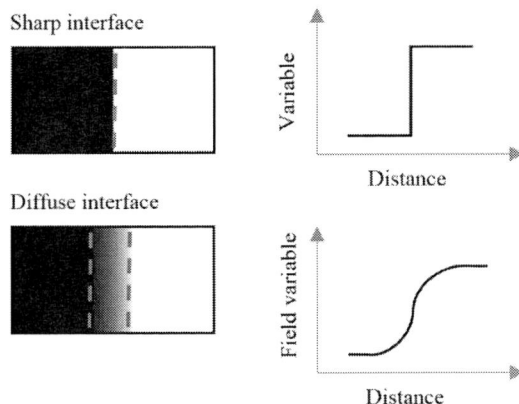

Figure 1: Sharp interface (Top) and diffuse interface (Bottom) representation, for defining grain boundaries

such as temperature, in order to reduce the total free energy.

According to equation 1 [9], the total free energy is calculated from sum of the bulk energy (which defines the composition of equilibrium phases), the interfacial energy and the elastic energy (interfacial and elastic energy, affect the equilibrium phase and define the shape of grain boundaries):

$$F = F_{bulk} + F_{int} + F_{el} \qquad (1)$$

Interfaces in a phase-field model are defined as diffuse-interface, i.e. in an interface, which is a narrow region, the properties change continuously. Alternatively, sharp-interface definition (used in other grain-growth methods such as Monte Carlo simulation), properties change are instant and sharp (figure 1). The field-variable in a domain, in phase-field model are multiplied by an order parameter, which is maximum or 1 within the domain and from near toward the interface, it declines, and equals to zero on the adjacent domain. The adjacent domains have a different order parameter with a same definition (on the reverse direction) [9].

Free energies are evaluated based on field variables. There are two main types of field variables: (a) composition, (b) non-conserved variables [9]. Composition variables (x_i), according to first law of thermodynamics, are conserved in the system. An example, concentration of phases, can change locally, however in the entire system sums up to a constant value (equation 2). In contrast, order parameters (η_k, non-conserved) do not necessarily obey the first law of thermodynamics. As an example grain orientation; during grain growth it is possible that a certain orientations shrink and disappear in a material because of growth of larger grains.

$$\sum_{i=1}^{c} x_i = 1 \qquad (2)$$

According to equation 3, free energy can be driven from field-variable (concentration, order parameter), and their gradient. In a homogeneous thermodynamic law free energy is independent of local variables, however in a heterogeneous material it varies locally, hence it should be explained by the concentrations and order parameters [9].

$$F(x_B, \eta_k) = \int_V \left[f_0(x_B, \eta_k) + \frac{\epsilon}{2}(\vec{\nabla} x_B)^2 + \sum_k \frac{\kappa_k}{2}(\vec{\nabla}\eta_k)^2 \right] d\vec{r} \qquad (3)$$

$f_0(x_B, \eta_k)$ is free energy density (J/m^3), where field-variables are constant (their gradient equals zero). $(\frac{\epsilon}{2}(\vec{\nabla} x_B)^2)$ and $(\frac{\kappa_k}{2}(\vec{\nabla}\eta_k)^2)$ are diffuse-interface characteristic of the domains. ϵ, k_k are positive coefficients and define interface thickness and energy. The microstructural evolution is preferred toward set of order parameters where minima(s) of the local free energies are achieved in subsequent time steps. In a single phase domain, free energy density is defined with only dependence on order parameter (η), as shown equation 4 [10]:

$$f_0(\eta_1, \eta_2, \dots) = \sum_{k=1}^{P} \left(-\frac{\alpha}{2}\eta_k^2 + \frac{\beta}{4}\eta_k^4 \right) + \gamma \sum_{i=1}^{p} \sum_{j \neq i}^{p} \eta_i^2 \eta_j^2 \qquad (4)$$

Finally, time-dependent Ginzburg-Landau equation is used to determine the order parameters (i.e. grain orientation in grain growth) at a newer time step based on free energy calculation and orientations at a previous time step [9,10].

$$\frac{\partial \eta_k(\vec{r}, t)}{\partial t} = -L_k \frac{\delta F}{\delta \eta_k(\vec{r}, t)}$$
$$= -L_k \left[\frac{\partial f_0}{\partial \eta_k} - \vec{\nabla}.\kappa_k \vec{\nabla}\eta_k \right] \qquad (5)$$

$$\frac{\partial \eta_k(\vec{r}, t)}{\partial t} = -L_k \left(-\alpha\eta_k + \beta\eta_k^3 + 2\gamma\eta_k \sum_{j \neq k}^{p} \eta_j^2 \right.$$
$$\left. - \kappa_k \nabla^2 \eta_k \right) \qquad (6)$$

Equation 6 is calculated for every orientation in the phase-field system, which can be calculation intensive when we deal with large number of order parameter. It is important to mention that there are also other equations used in other microstructural evolution models, which were not stated, interested readers can find more information in [9].

Aforementioned phase field differential equation can be solved using numerical methods such as finite differences. For numerical purpose, equation 6 can be discretized into equation 7 (for simplicity this equation is in 2D domain). This is defined under a uniform 2D or 3D space of equally spaced r by s grids as the grain growth domain. It is important to choose an optimal spacing size Δx, between grid points, fine enough to resolve interface profile detail, and large enough to avoid computational complexity and reduce memory usage. Equation 7 is defined for each orientation in the grid space, in time step n+1, calculated from its previous time step (n) and its adjacent grains. This equation can be solved using Newton-Raphson method. Using iterative techniques, large problems can also be optimally handled. This paper uses discrete Fourier transform for solving the equation in Fourier space. This method is advantageous for enabling use of a larger time step and avoiding separate calculation of gradients. In contrast, periodic boundary conditions are imposed to the domain and shall be treated when the structure requires a non-periodic boundary.

$$\frac{\eta_{r,s}^{n+1}-\eta_{r,s}^{n}}{\Delta t} = -L\left[\left(\frac{\partial f_0}{\partial \eta}\right)_{r,s}^{n} - \kappa\frac{\eta_{r-1,s}^{n}+\eta_{r,s-1}^{n}-4\eta_{r,s}^{n}+\eta_{r+1,s}^{n}+\eta_{r,s+1}^{n}}{(\Delta x)^2}\right] \quad (7)$$

In this equation $\eta_{r,s}^{n}$ are grain orientations in location r, s in space grid (2D) at time step n. Δx is the distance between neighboring grid points and κ and L as mentioned earlier are material constants and impact the interface energy and thickness [9].

Figure 2 shows an isotropic grain growth in 2D square domain using this method with a 2D Voronoi space of 50 grains as input. It is important to mention that for clarity sharp interface representation of grains are plotted through this paper. In the application of Cu pumping in TSVs, orientations at each time step are necessary in each elements in the finite element simulation, and sharp interface representation of grains are required to transfer these data.

3. Grain growth in TSV

The TSV shape studied here are high aspect ratio (>10) cylindrical trench inside Si substrate filled with electroplated Cu. The cylindrical TSV has a radius of r and height of h as illustrated in figure 3. The grain growth domain is the corresponding TSV cylinder depicted in

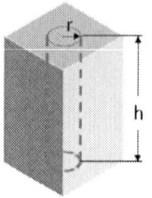

Figure 3: schema of a TSV inside Si, (defined geometry in phase-field simulation)

figure 3. Modifications are necessary in the numerical solution of phase-field method to achieve this, however for a discreet Fourier transform solution, equally spaced array of discrete points in input domain must be kept. The dimensions regarding the cuboid domain is consider 20% larger than the diameter of the TSV thus enough grid points should be considered not only for capturing the detail of grain boundaries, but also sufficient points to define nearly smooth circular edge of TSV's circular cross section (let's assume $a \times a \times b$ number of grid points). The program input requires p number of order parameter (grain orientations) to initialize a model of randomly distributed order parameters (η_k). The modified program also adds a dummy order parameter (η_{k+1}) to represent surrounding silicon. During the initialization phase, all points in the grid points are checked and when they are outside the circle, apart from what order parameter they may have, it is changed to a sharp η_{k+1}: ($\eta_{@x,y} = \eta_k$ if $(x - 0.5a)^2 + (y - 0.5\,a)^2 > (0.83a)^2$). It is important to assign a few layers on bottom of the cuboid to η_{k+1} as well, to treat the unwanted periodic boundary. (η_{k+1}) contributes in free energy calculations in the interfaces on the boundary of the TSV (Where two other grains also meet) and is calculated within the phase-field program; when new orientations are assigned, according to equation 7, it's order parameter is not considered in calculations. Hence (η_{k+1}) neither grows nor shrinks. It is important to mention that interface between this boundary is sharp from start of the simulation.

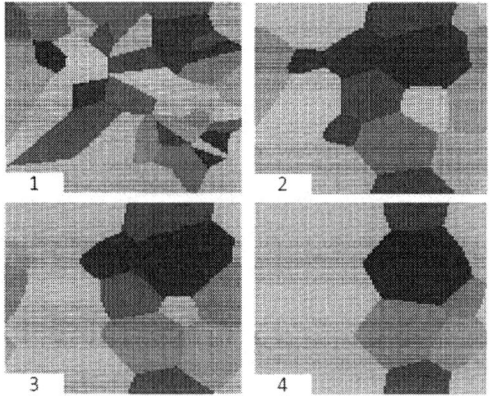

Figure 2: 2D isotropic grain growth in a periodic square domain using phase-field method described in this paper. The snapshots are captured at 1) 1st 2) 100th 3) 500th and 4) 1000th time step.

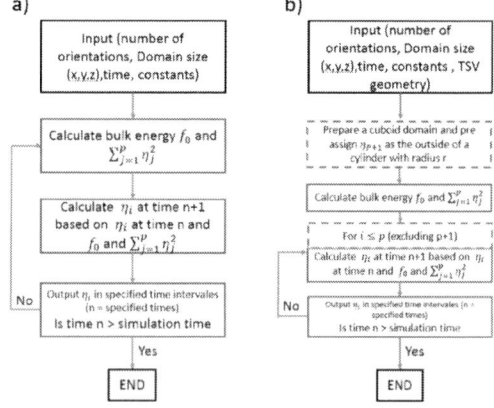

Figure 4: Flow chart of a) isotropic phase field model, b) modified model to include grain growth inside a cylindrical Through silicon via trench

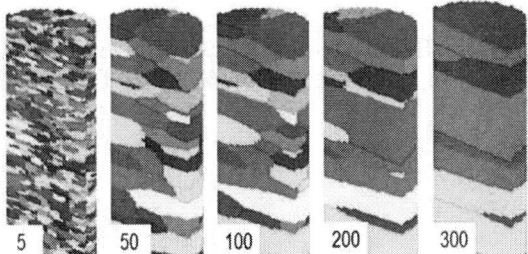

| 5 | 50 | 100 | 200 | 300 |

Figure 5: Cross section sharp interface image of microstructure evolution of TSV as obtained from a phase-field simulation (the numbers below each plot show the time steps of the simulation at which the snapshot was captured)

Figure 4, shows the general flowchart of modified and original phase-field program. The resulted η_i grains of equaly spaced grid points inside cylinder are transferred to Finite element model after converting to sharp interface representation. The grid points are divided to k number of sets containing each grid's (x, y, z) location and written to a file, to be read in a Finite Element model. Figure 5, shows 5 different snap shots of grains inside a cylinder in different time steps.

Figure 6, shows Cu pumping simulated in a sample FEM model based on the inputted grain structure from modified phase-field program. The FEM model has the same equally space grids used in phase-field model to enable one to one transfer of grain orientation data. After saving elements in sets of grains as done in input, the grains are assigned a random crystal orientation by rotating the material axis of each element, for depicting anisotropy. Results from single crystal model are also included in figure 6.b for direct comparison. In these models, Cu is assumed to have anisotropic elastic properties and isotropic plastic (temperature dependent Von-Mises yield criterion [11]) and thermal properties (table 1.). The thermal boundary condition in the simulation is temperature rise to 420°C and drop to the room temperature. As shown in figure 6, results strongly confirm the capability of this model to capture the impact of grain structures on asymmetric Cu pumping.

4. Conclusions

In this paper, we presented a time-efficient phase field model of grain growth, adapted to cylindrical shape of Cu TSV. Currently this model is being implemented in FEM to simulate the effect of anisotropy in Cu pumping in TSVs. The results of a sample FEM simulation presented in this paper also confirms that the grain structure is capable of capturing the asymmetry in Cu extrusion in TSVs.

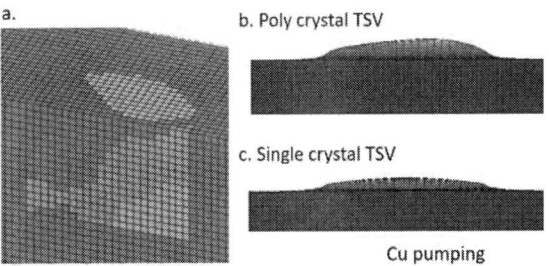

Figure 6: a. Cross section of a sample FEM model with input grain structure generated from phase-field model. b. Cu pumping in poly crystal TSV (a,) where asymmetric Cu extrusion is observed. c. same structure, with a single crystal TSV, where Cu pumping is symmetric.

References

1. De Wolf, Ingrid, et al. "Cu pumping in TSVs: Effect of pre-CMP thermal budget." Microelectronics Reliability 51.9 (2011): 1856-1859.

2. De Messemaeker, Joke, et al. "Impact of post-plating anneal and through-silicon via dimensions on Cu pumping.", ECTC, 2013

3. Cherman, V. O., et al. "Impact of Through Silicon Vias on Front-End-of-Line performance after thermal cycling and thermal storage." Reliability Physics Symposium (IRPS), 2012

4. De Messemaeker, Joke, et al. "Correlation between Cu microstructure and TSV Cu pumping", Submitted, ECTC, 2014

5. Ryu, S. K., et.al. "Characterization of thermal stresses in through-silicon vias for three-dimensional interconnects by bending beam technique. Applied Physics Letters, 100(4), 2012

6. Selvanayagam, Cheryl S., et al. "Nonlinear thermal stress/strain analyses of copper filled TSV (through silicon via) and their flip-chip microbumps." Advanced Packaging, 32.4 (2009): 720-728.

7. Liu, Xi, et al. "Failure mechanisms and optimum design for electroplated copper through-silicon vias (TSV)." ECTC, 2009..

8. Krill Iii, C. E., and L-Q. Chen. "Computer simulation of 3-D grain growth using a phase-field model." Acta Materialia 50.12 (2002): 3059-3075.

9. Moelans, Nele, et al. "An introduction to phase-field modeling of microstructure evolution." Calphad 32.2 (2008): 268-294.

10. Fan, Danan, and L-Q. Chen. "Computer simulation of grain growth using a continuum field model." Acta Material 45.2 (1997): 611-622.

11. Nabiollahi, N., et al. "Simulation of Cu pumping during TSV fabrication.", EuroSimE 2013.

12. Hertzberg, Richard W. "Deformation and Fracture Mechanics of Engineering Materials.", pp 14,

Material	Stiffness coeff.s (GPa)			CTE
	c_{11}	c_{12}	c_{44}	
Si	165.6	63.9	79.5	2.3×10^{-6}/°C
Cu	168.4	121.4	75.4	16.7×10^{-6}/°C

Table 1. Elastic material properties used in model [12]

Compact Thermal Modeling of Microbolometers

Janicki M., Zajac P., Szermer M., Napieralski A.

Department of Microelectronics and Computer Science, Lodz University of Technology,
Wolczanska 221/223, 90-924 Lodz, Poland
E-mail: janicki@dmcs.pl

Abstract

This paper presents an approach to dynamic thermal modeling of micromachined microbolometers. Firstly, all the important factors influencing temperature of infrared radiation sensing elements are identified in repeated numerical thermal simulations performed for a detailed device model, where temperature values were computed for time instants equidistantly spaced on the logarithmic time scale. Secondly, based on the simulation results all the important time constants contained in these dynamic thermal responses were properly identified what allowed the derivation of compact thermal models in the form RC equivalent circuits containing a limited number of stages. The resulting compact thermal models are suitable for the direct implementation in SPICE or any other multiphysics simulation environment.

1. Introduction

Various Micro-Electro-Mechanical Systems (MEMSs) are now commonly encountered in numerous everyday use machines and equipment, e.g. accelerometers present in cars and phones. Another MEMS device is a microbolometer which is used for sensing infrared radiation. Formerly, mainly because of their price, they were used predominantly in military applications, however now they became affordable and soon they could be even integrated as standard devices with portable devices allowing low-cost thermal imaging [1]-[2].

Considering that the operating principles of particular MEMS devices involve different electrical, mechanical, chemical and thermal phenomena, the simulation of these devices requires dedicated multiphysics tools. Such tools are offered various providers, e.g. by Ansys, Coventor or Comsol, but usually solutions are obtained after time consuming simulations of detailed system models what is not suitable for analyses of large sensor arrays.

Moreover, MEMS devices are usually integrated with electronic readout circuits, thus it would be advantageous to have the possibility of running simulations in standard circuit simulators, such as SPICE, using a compact model. This approach was already adopted in a few papers where compact model parameters were found based on complex geometrical calculations [3]-[4].

Thus, here we propose to derive compact models from device temperature responses. The next section is devoted to a brief description of microbolometer construction and its operating principles as well as the presentation of the particular structure used here for thermal simulations. The experimental part of the paper contains detailed analyses of simulation results and provides indications for further development of microbolometer models.

2. Microbolometers

Microbolometers are basically radiation sensors based on resistive principle in which radiant energy influences the resistance of the radiation sensing material. When the radiation absorption characteristics of the active material has its maximum in the infrared wave range, such sensors can be used to measure temperature. Since the amount of radiation is converted into the change of resistance the materials of choice should have quite high temperature coefficient of resistance but at the same time they should be compatible with MEMS technology. Therefore, typical materials used in practice are amorphous semiconductors, various metals and their oxides.

Except for the proper choice of active material, the device sensitivity can be increased by maximizing the temperature rise value for a given radiation intensity. This, as shown in Fig. 1, could be realized assuring high sensor thermal resistance manufacturing it as a very thin membrane suspended on cantilevers over semiconductor wafer surface. Additionally, this solution guarantees low thermal capacitance of the entire sensor structure and consequently quick response times. Moreover, when the membrane is suspended at the height of a few of microns, it forms with the substrate a resonant cavity reflecting radiation back from the wafer towards the sensor and consequently further increasing its sensitivity [5]-[6].

Microbolometer structures are usually etched on the semiconductor surface forming large matrices of sensors, known as infrared Focal Plane Arrays (FPAs), which are integrated in one piece of semiconductor substrate with their Readout Integrated Circuit (ROIC) and hermetically sealed in vacuum packages [7].

Taking into account that state-of-the-art FPAs contain hundreds of thousands of individual devices their detailed models cannot be effectively employed for simulations at the system level and then some appropriate reduced models have to be used. Furthermore, since the operation of microbolometers is highly temperature dependent their simulation requires accurate thermal models which would allow relatively fast but reliable determination of sensor key parameters, such as current or voltage responsivity.

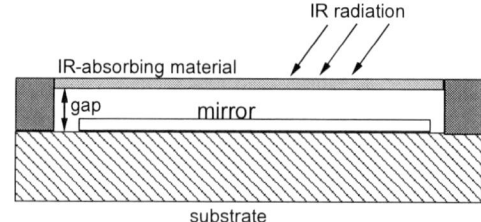

Figure 1: Simplified cross-section of a microbolometer.

978-1-4799-4789-8/14 $31.00 © 2014 IEEE

3. Experimental Results

For thermal simulations a sensor structure with 25 μm pitch was chosen. The membrane was suspended 2 μm above the wafer surface. The radiation absorbing element was a 70 nm thin winding titanium resistor sandwiched between 500 nm silicon nitride layers. The detailed model of this structure was created in the Ansys environment.

Initially, the goal of simulations was to determine the actual influence of individual factors affecting the sensing element temperature. Namely, three main heat generating factors were considered: incident infrared radiation, Joule heating by current flow in the sensing resistor and heating by the power dissipated in the ROIC electronics beneath the membrane.

During the simulations temperature values computed for logarithmically spaced time instants so as to capture all the thermal time constants contained in the dynamic responses. The results of these simulations are presented in Fig. 2 applying logarithmic time scale for both axes. Furthermore, the curves, in order to render possible their comparison, were normalized do as to obtain the same maximal temperature rise value.

As can be seen, when the heat source is located in the membrane, the solid and the dashed curves after 10 μs are identical. The difference in the beginning of the heating time might be attributed to the fact that the Joule heat is dissipated directly in the sensor whereas the incident radiation has to diffuse first through the top nitride layer.

When the membrane is heated from beneath by the ROIC, the heating curve looks entirely different reaching its final value only after a few milliseconds. This, in turn is caused by the fact that diffuses mainly to the substrate heating its thermal capacitance and only partly reaches the membrane.

More detailed analyses of these heating curves carried out applying a method similar to the multipoint moment matching technique described in [8, 9] allowed the precise determination of the thermal time constants present in the transient responses. The results of these analyses for each of the curves are given in Table 1 providing the values of thermal time constant and their corresponding thermal capacitances and resistances.

When the first two cases are considered, there exists one dominant time constant at 232 μs which correspond to the thermal resistance of 155 kK/W and the thermal capacitance of 1.53 nJ/K. Usually in literature, e.g. in [1] or [3], microbolometers are also characterized by a single thermal time constant reflecting the thermal capacitance of the membrane and the resistance of its supporting beams. These data obtained from Ansys simulations, also provided in the table, are very close to the ones obtained from the analyses of the heating curves.

Except for the dominant thermal time constants in the microsecond range there exist also time constants of the order of nanoseconds but their contributions to the total temperature rise are insignificant and could be neglected, even when analysing the Joule heating because the typical electrical pulse width during the sensor readout is around 100 μs.

The thermal parameters given in the table for the first two cases represent the RC values of the Foster ladder, which cannot have direct physical interpretation because capacitances are not directly connected to the thermal ground node, i.e. ambient temperature, but it facilitates the computation of the temperature response according to the following formula:

$$T(t) = P \sum_{i=0}^{n} R_{thi}\left[1 - \exp\left(-t/\tau_{thi}\right)\right] \qquad (1)$$

Each thermal time constant τ_{th} in the above equation is the product of thermal resistances R_{th} and capacitances C_{th} whereas P is the dissipated power.

When the influence of power dissipation in the ROIC is considered, the thermal model is different because heat only partially flows into the membrane, but the majority of it diffuses into the semiconductor substrate. Thus, the correct thermal model consists of two parallel RC ladder branches.

Table 1: Thermal parameters of reduced models.

Radiation heating		
Time constant (s)	Resistance (K/W)	Capacitance (J/K)
.2.27E-08	3.72E+01	6.11E-10
2.36E-04	1.55E+05	1.53E-09
Sensor bias current heating		
Time constant (s)	Resistance (K/W)	Capacitance (J/K)
3.06E-07	3.69E+03	8.29E-11
2.36E-04	1.55E+05	1.53E-09
ROIC electronics heating		
Time constant (s)	Resistance (K/W)	Capacitance (J/K)
1.49E-03	5.74E-03	2.59E-07
2.36E-04	1.55E+05	1.53E-09
Ansys		
Time constant (s)	Resistance (K/W)	Capacitance (J/K)
2.32E-04	1.55E+05	1.50E-09

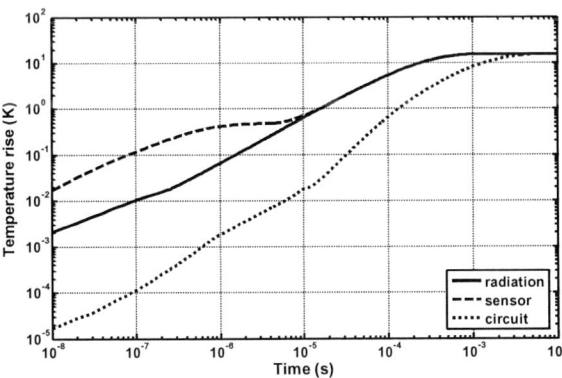

Figure 2: Normalized heating curves simulated with full distributed model.

Consequently, the thermal model parameter values given in the table correspond to two parallel RC stages. One of them represents the previously discussed dominant time constant related to the membrane whereas the other one reflects the presence of the large silicon substrate. Then, the formula to compute the time response of the system takes the following form:

$$T(t) = P \frac{R_{ths} R_{thm}}{R_{ths} + R_{thm}} \times$$
$$\times \left[1 - \exp\left(-t \middle/ \frac{(C_{ths} + C_{thm}) R_{ths} R_{thm}}{R_{ths} + R_{thm}} \right) \right] \quad (2)$$

where the indices s and m denote the substrate and the membrane respectively.

The value of the substrate thermal capacitance given in the table coincides very well with the one calculated theoretically for the volume of silicon considered in the simulation. However, it should be mentioned that in order to accelerate the simulations the structure included only the silicon volume under one sensor. If the total substrate volume were considered, the value of the capacitance would be much larger. Thus, depending on the cooling applied to the substrate, there might be cases when the consideration of more time constants is necessary.

The next stage in the numerical experiments was the computation of thermal responses based on the previously identified compact thermal models. For each of the cases considered here, the heating curves calculated from the compact models (dashed lines) were compared in Figs. 3-5 with the original Ansys simulations results (solid lines) obtained for the full distributed model.

The simulation accuracy with these extremely simple models was satisfactory and the simulation error with respect to the results from the detailed model did not exceed 10 % for all time instants. This is an important observation from the practical point of view because microbolometers are usually probed with short current pulses in order to avoid excessive sensor self-heating and increase its detectivity.

Another important benefit from using compact models is significantly shorter simulation time. The computation of results in Ansys took almost half an hour whereas the results from the compact model are available after just a few seconds. Besides, the preparation of the detailed model also requires significant amount of time, even for an experienced user.

4. Conclusions

This paper presented the analyses of microbolometer temperature responses to various factors influencing the temperature of the radiation sensing element. The results proved that in most practical cases the commonly used approximate thermal model consisting of just one thermal resistance and one thermal capacitance might be accurate enough.

However, the determination of exact model parameter values requires the knowledge of internal structure of the sensor and is subject to the availability of the detailed model. The main advantage of the approach adopted here is that basically the detailed model is not necessary for the generation of the compact model. Namely, though the analyses presented here were carried out on simulated data, the method allows the creation of a compact model also from measured dynamic responses. This solution can be also beneficial in cases when manufacturers do not want to reveal their confidential data.

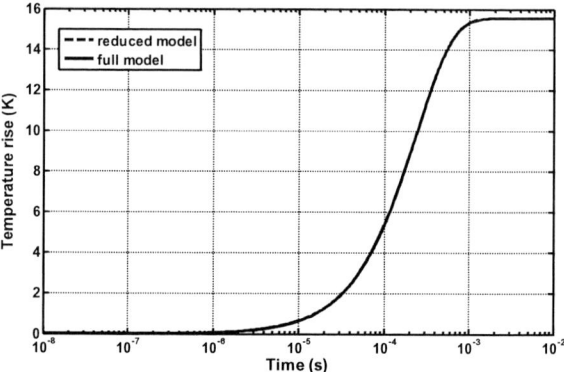

Figure 3: Simulated heating by radiation.

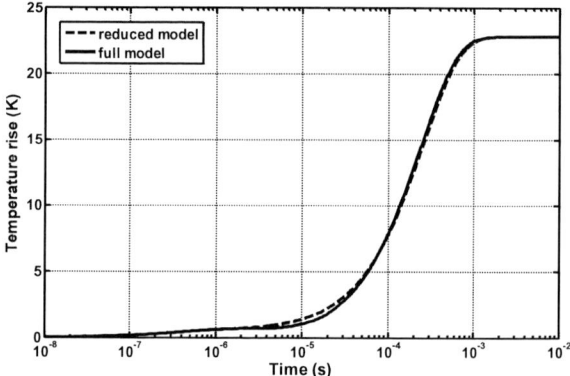

Figure 4: Simulated heating by sensor current.

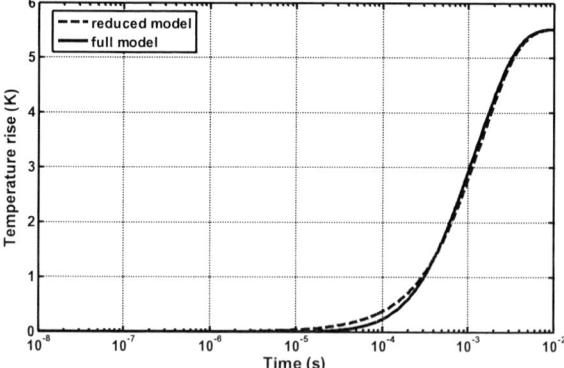

Figure 5: Simulated heating by ROIC.

Moreover, this behavioural approach to the generation of compact models based on the analysis of measured responses allows the determination of exact mathematical dependencies between qualities of various nature without any prior knowledge on investigated phenomena, what unfortunately often is the case in multiphysics simulations of MEMS devices.

Acknowledgments

The research presented in this paper was supported by the 7th EU Framework Programme project No 269295 'Developing Multidomain MEMS Models for Educational Purposes - EduMEMS'.

References

1. Rogalski, "Infrared Detectors: Status and Trends," *Progress in Quantum Electronics,* Vol. 27, (2003) pp. 59-210.

2. Klipstein, P., Mizrahi, U., Fraenkel, R., Shtrichman, "Status of Cooled and Uncooled Infrared Detectors at SCD," *Defence Science Journal,* Vol. 63 (2013), pp. 555-570.

3. Topaloglu, N., Nieva, P., Yavuz, M., Huissoon, J., "Modeling of Thermal Conductance in Uncooled Microbolometer Pixel," *Sensors and Actuators A*, Vol. 157 (2010), pp. 235-245.

4. Han, S., Chun, C., Han C., Park S., "Parameterized Simulation Program with IC Emphasis Modeling of Two-level Microbolometer," *Journal Electrical & Engineering Technology,* Vol. 6 (2011), pp. 270-274.

5. Moreno, M., Torres, A., Ambrosio, R. Kosarev, A., "Un-cooled Microbolometers with a-Ge_xSi_y Thermo-Sensing Films," *Bolometers,* Perera, U. (Ed.) (2012), InTech, DOI: 10.5772/32222.

6. Niklaus, F., Vieider, C., Jakobsen, H., "MEMS-based uncooled infrared bolometer arrays: a review," *Proc. SPIE*, Vol. 6836 (2008) 68360D.

7. Bhan, R. K., Saxena, R. S., Jalwania, C. R., Lomash, S. K., "Uncooled Infrared Microbolometer Arrays and Their Characterisation Techniques," *Defence Science Journal,* Vol. 59 (2009), pp. 580-590.

8. Codecasa, L., D'Amore, Maffezzoni, P, "Multipoint Moment Matching Reduction from Port Responses of Dynamic Thermal Networks," *IEEE T. Comp. Pack. Technologies,* Vol. 28 (2005) pp. 605-614.

9. Masana, F., "A Straightforward Analytical Method for Extraction Semiconductor Device Transient Thermal Parameters," *Microel. Reliability*, Vol. 47 (2007) pp. 2122-2128.

MedeA®: Atomistic Simulations for Designing and Testing Materials for Micro/Nano Electronics Systems

A. France-Lanord, D. Rigby, A. Mavromaras, V. Eyert*, P. Saxe, C. Freeman, and E. Wimmer

Materials Design SARL, Montrouge, France
Materials Design, Inc., Angel Fire, NM, USA
*Corresponding author: veyert@materialsdesign.com

Abstract

Results of atomic-scale simulations are presented including thermal conductivity, elastic moduli, diffusion, and adhesion. This type of simulations is most conveniently performed with the MedeA® computational environment, which comprises experimental structure databases together with building tools to construct models of complex solids, surfaces, and interfaces for both crystalline and amorphous systems. Central to MedeA® are state-of-the-art modules for the automated calculation of thermodynamic, structural, electronic, mechanical, vibrational, and transport properties combined with the corresponding graphical analysis and visualization tools. These capabilities are illustrated for both inorganic and organic materials. For Si-Ge alloys and amorphous-crystalline silicon superlattices we find a drastic reduction of the thermal conductivity compared with bulk crystalline Si. In addition, the Si-Ge alloys reveal a considerable sensitivity of their thermal conductivity to disorder. The second part of this study addresses properties of epoxy resin based thermosets, including their mechanical stiffness, thermal conductivity, and adhesion on alumina. In addition, we present calculated results for oxygen and water diffusivities in cross-linked epoxy systems and discuss factors influencing such diffusivities as, *e.g.*, mass effects or the concentration of residual hydroxyl groups in the polymer.

1. Introduction

The last decades have seen unprecedented progress of atomistic simulations in materials science and engineering. Nowadays, atomic-scale methods have evolved as an indispensable part of research both in industry and academia. In industrial research laboratories, the corresponding computational tools have found widespread application in solving complex engineering problems. At the same time, they are routinely used at universities and national laboratories to address fundamental questions of condensed matter research and to explore so far unknown materials with exciting properties.

The MedeA® computational environment of Materials Design has become a well-appreciated tool in this community. The MedeA® software offers a unique, comprehensive, and innovative software environment, which combines experimental structures and phase-diagrams with state-of-the-art computational procedures for property predictions for systems including alloys, semiconductors, ceramics, glasses, polymers, and fluids. Building on quantum mechanics, MedeA® facilitates the simulation of electronic structures and mechanical properties, as well as the thermal behavior for complex structures such as interfaces, heterostructures, grain boundaries, defect structures, and random alloys. Using high-performance molecular dynamics, MedeA® calculates transport properties such as diffusion coefficients and heat conductivity in amorphous oxides. Besides predicting fundamental properties of nanoelectric devices, the functionality of MedeA® addresses topics related to packaging, device assembly and component mounting. These include the prediction of thermal conductivity for semiconductors, metals, and oxidized metal heat sinks and epoxy thermoset encapsulation materials, as well as tools to investigate mechanical, adhesive and diffusive properties relevant to device performance and reliability.

In this paper, we demonstrate the capabilities of MedeA® with selected examples. In doing so, we focus especially on the thermal conductivity using forcefield methods as implemented in the software environment. The thermal conductivity is of high interest in different fields. In thermoelectrics, materials are sought with a high electrical conductivity combined with a low thermal conductivity as can be found in doped semiconductors with a high density of states near the band edges. In the present paper, we investigate the thermal conductivity of Si-Ge alloys and a-Si/c-Si superlattices and discuss the influence of defects, disorder, and amorphous structures. We find that in Si-Ge alloys the thermal conductivity can be reduced to 2 W/m/K for about 12% Ge as compared to 128 W/m/K of bulk crystalline silicon. However, layered systems are much more efficient than disordered materials in this respect.

The second part of the paper deals with epoxy resin based thermosets, which are widely used in semiconductor devices. In this context, a high thermal conductivity is desired in order to facilitate fast heat dissipation. Specifically, for the epoxy system consisting of a stoichiometric mixture of diglycidyl ether of bisphenol A (DGEBA) and meta phenylene diamine (m-PDA) cross-linked with degrees of cure up to ~80% we find good agreement with experimental data. Calculations of the elastic properties for the same thermoset reveal the sensitivity of the tensile modulus on the level of curing. Additionally, we address the adhesion of DGEBA-Jeffamine® on alumina. The simulations for the present model show that without direct chemical bonding between the organic and inorganic phases the work of separation is 0.15 Jm^{-2}. Finally, we study the diffusivities of oxygen and water, respectively, in the DGEBA-Jeffamine® epoxy systems. The calculations involve NVE molecular dynamics, which give access to mean-square displacements of the molecules. For DGEBA-Jeffamine® our results point to similar activation energies for oxygen and water and an increased diffusivity of oxygen as compared to water at ambient conditions.

2. Lattice Thermal Conductivity in Semiconductors

Lattice thermal conductivity is one of the fundamental transport properties of materials. Its knowledge is necessary, for instance, in evaluating semiconductor efficiency for thermoelectric device applications. As heat dissipation in microelectronics and nanoelectronics becomes more and more important, the ability to predict thermal conductivity is of crucial importance for modern science and technology.

There are currently two standard approaches for obtaining the lattice thermal conductivity, namely: (i) calculation of the phonon properties of the system, which are connected to the thermal conductivity through the Boltzmann transport equation (BTE) [1] and (ii) molecular dynamics (MD) simulations [2, 3] at both equilibrium (EMD) and non-equilibrium (NEMD).

Despite their inherent approximations, classical molecular dynamics simulations give very good results for thermal conductivity in agreement with experiments for various semiconductors around room temperature or higher [4, 5]. In particular, they allow computations for very large models of more than a million atoms representing complex geometries.

Quantum effects on the thermal transport are accounted for by using Boltzmann transport calculations, but this method suffers from several flaws. First, these calculations are much more expensive than MD: large or aperiodic systems cannot be simulated. Second, the relaxation time approximation used to solve the BTE drastically reduces the predictive power of the method since empirical parameters are needed. This is clearly opposed to the MD formalism, which is "classically exact", and does not rely on approximate theoretical expressions.

2.1 Thermal conductivity prediction using MedeA®

In the following we will sketch the basic principles of equilibrium molecular dynamics (EMD) and reverse non-equilibrium molecular dynamics (RNEMD) simulations for the calculation of the thermal conductivity. This methodology is implemented in Materials Design's computational environment MedeA®, which takes advantages of the LAMMPS [6, 7] molecular dynamics package, for both EMD and NEMD procedures. The capabilities of this approach are illustrated in the subsequent section.

2.1.1 Equilibrium Molecular Dynamics

At the equilibrium state, one can determine the thermal conductivity of a system using the Green-Kubo relation [8]:

$$\kappa = \frac{1}{3k_B V T^2} \int_0^\infty \langle J(t) J(0) \rangle dt \qquad (1)$$

where κ is the thermal conductivity tensor, k_B is the Boltzmann constant, V is the volume of the system, T its average temperature, and $J(t)$ the heat flux vector. Relatively long simulations - depending on the system size, and the species involved - in the NVE (constant number of atoms, volume, and energy) microcanonical ensemble are needed to obtain a well converged value of the heat flux, and therefore of its autocorrelation function. It is also necessary to model a sufficiently large domain, in order to avoid artificial size effects, due to the cutoff of the long wavelength phonons.

The major advantage of EMD simulations is that one can obtain the full thermal conductivity tensor in only one simulation. This is very important for systems showing important thermal anisotropy, like for instance Si-Ge superlattices. Equilibrium molecular dynamics have been used to determine the thermal properties of a wide range of semiconductors, including core-shell silicon nanowires [9], Si-Ge nanocomposites[10], and Bi$_2$Te$_3$ nano-wires [11].

2.1.2 Reverse Non-Equilibrium Molecular Dynamics

Transport phenomena intrinsically derive from the non-equilibrium state: for instance, when a temperature gradient is imposed on a system, the energy will flow until equilibration is achieved. This is the basis of non-equilibrium molecular dynamics simulations for thermal properties evaluation: in a way similar to experiments, one applies a temperature gradient, generally with the use of heat reservoirs, and evaluates the resulting heat flux. The thermal conductivity is then obtained using Fourier's law:

$$J = -\kappa \cdot \nabla T \qquad (2)$$

where κ is the thermal conductivity tensor, J is the heat flux vector, and ∇T is the temperature gradient.

The need of external thermostating can be suppressed by swapping the cause and the effect [12]. In reverse-NEMD (RNEMD) simulations, a heat flux is imposed by regularly exchanging the kinetic energies of hot and cold particles resulting in a temperature profile that can be determined once the steady state is reached. In this way, the total energy and total linear momentum are conserved; hence no external thermostating is needed.

The RNEMD method is implemented in practice by dividing the system in slabs following, by default, the z direction. The $z=0$ slab is defined as the cold slab, and the $z = L_z/2$ as the hot slab, L_z being the length of the simulation box. After every n time steps, the hottest atom in the cold slab and the coldest atom in the hot slab are identified and swapped: each of the components of their velocities are exchanged. Once the steady state is reached, the temperature profile is obtained by evaluating the mean temperature of each slab. The gradient in the z direction $\partial T/\partial z$ is then post-processed, by identifying the slope of the temperature profile. The corresponding diagonal component of the thermal conductivity tensor can then be calculated.

Generally speaking, NEMD procedures are well adapted for nanomaterials, and all kind of semiconductors. Simulations are usually less time-consuming than EMD ones, but one should pay attention to size effects, that are of greater importance. Moreover, only the diagonal component of the thermal conductivity tensor perpendicular to the slabs can be determined in this approach. Still, NEMD simulations are very well adapted to semiconductors, *e.g.,* in the form of superlattices [13–15].

2.2 Tailoring thermal properties using MedeA®: An example

In order to maximize the efficiency of thermoelectric devices, one aims at lowering the thermal conductivity of semiconductors without deteriorating their electronic transport properties. Molecular dynamics simulations allow predicting and quantifying the thermal conductivity and its changes on introducing structural distortions as, *e.g.,* defects and local disorder.

Here we will compare two ways to alter the thermal conductivity of bulk Si: by adjoining amorphous regions in between crystalline Si regions (a-Si/c-Si superlattices), and by replacing a certain number of Si atoms by Ge atoms. Both types of changes modify the phonon properties of the system, and it is possible to predict their impact using simple NEMD simulations.

With the help of MedeA® the following calculations can be very efficiently performed. To begin with, structural models are built using the corresponding tools implemented in MedeA®. Once these structural models are created, EMD or NEMD (depending on the structure – size, geometry, species) simulations using MedeA®-LAMMPS are carried out leading to the results shown in the figures below.

As a reference, the thermal conductivity of bulk silicon as computed using MedeA®'s Thermal Conductivity module amounts to 128 W/m/K, which compares excellently to the measured lattice thermal conductivity of 130 W/m/K (Ioffe Institute).

In this study, different systems were compared. First, the Random Substitution module of MedeA® was used to create $Si_{43}Ge$ by randomly replacing a Si atom by a Ge atom, this leading to a defect concentration of 2.27%. Second, superlattice models with the same Ge concentration were created by arranging the Ge atoms in layers, which are separated by 43 Si slabs. Finally,

for the latter arrangement $Si_{43}Ge_5$ superlattices were also considered.

RNEMD simulations were undertaken using MedeA® on these samples, leading to highly accurate results of the thermal conductivity. Finally, these results are compared to the ones of a recent study [16] on the thermal conductivity of a-Si/c-Si superlattices that was computed using NEMD with LAMMPS. Figure 1 shows three different samples, using MedeA®'s visualization tools.

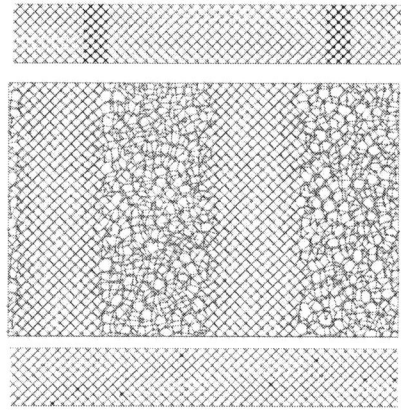

Figure 1: Three systems as generated from MedeA®'s visualization tools. Top: a layered $Si_{43}Ge_5$ sample. Middle: an amorphous/crystalline superlattice. Bottom: a Si system, with few random replacements of Ge atoms ($Si_{43}Ge$). Light color: Si, dark color: Ge.

Of course, these systems have different qualities and fields of application: The Si-Ge superlattice is one of the best material for high temperature thermoelectric applications, while a-Si/c-Si structures present a very high thermal conductivity anisotropy (the in-plane one is up to six times higher than the cross-plane one), very useful for heat spreading in optoelectronics [17] or heat shields, for instance.

Calculated temperature profiles of the $Si_{43}Ge$ random and layered structures as directly resulting from MedeA® are shown in Figure 2.

The results of the calculations are summarized in Figure 3, which shows the thermal conductivity of different systems as a function of the Ge concentration in Si-Ge and the fraction of amorphous Si in a-Si/c-Si, respectively. Given the thermal conductivity of crystalline Si of 128 W/m/K it is clear that the depression in a-Si/c-Si samples is due to the amorphous layers. Even with very thin layers, the thermal conductivity is reduced by a factor of 20 relative to the bulk value. Yet, according to Figure 3 the decrease of the thermal conductivity seems to saturate at about 25% of amorphous Si.

A similar drastic reduction of the thermal conductivity is observed for Si-Ge. On replacing only one out of 44 Si atoms by Ge the thermal conductivity drops by about a factor of 20 relative to bulk Si. Furthermore, the drop is even more drastic for the layered structure, which brings about further reduction to less than 4 W/m/K. Finally, for the layered systems it is found that the decrease in thermal conductivity is enhanced with increasing Ge concentration.

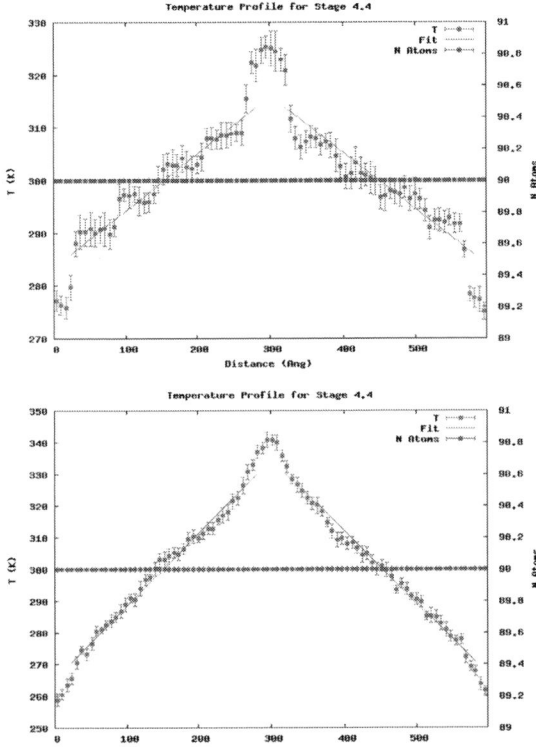

Figure 2: *Temperature profiles of layered (top) and random (bottom) $Si_{43}Ge$ structures. Circles and the dots making the horizontal line give the mean temperature and the number of atoms, respectively, in each slab. Thin lines are automatic fits to the mean temperature points, which are used to determine the temperature gradient.*

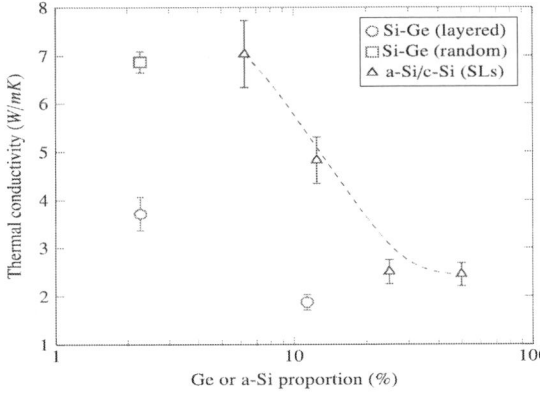

Figure 3: *Thermal conductivity as a function of the proportion of amorphous Si and Ge, respectively. Note that the two sets of calculations used different computation protocols. A quantitative comparison between the two series may thus contain systematic deviations.*

3. Thermo-mechanical Property Simulation of Epoxy-Based Thermosets

Epoxy resin based thermosets, in the form of epoxy molding compounds (EMCs) are widely used in a variety of electrical and electronic devices, ranging from use as insulating materials in high voltage transformers, cable terminations and associated equipment, to use as encapsulation materials in electronic components. At either extreme of the device scale, the essential requirement is that the epoxy acts as an insulator, while retaining the ability to dissipate excess heat generated within the device.

Epoxy thermosets themselves however suffer from the limitation that they typically possess relatively low thermal conductivity. Consequently filler consisting of electrically insulating but thermally conductive particulate material is often added to the EMCs used to fabricate the components. Often, this material consists of relatively high loadings of 50-80% silica micro- or nanoparticles bonded to the resin matrix using a silane-based coupling agent. However, the modest thermal conductivity of silica, coupled with a desire to exploit more effectively the properties of the cross-linked resin matrix has led to interest in using much lower loadings - up to 5% - of other fillers such as aluminum oxide, aluminum nitride, boron nitride, or even carbon nanotubes and nanoplatelets [18, 19].

From a device fabrication and durability perspective, the combination of cross-linked base resins, fillers, active electrical or electronic components and other parts such as connector leads presents interesting challenges, many of which are amenable to study at the atomistic level using quantum or classical mechanics simulation methods. Topics of interest include:

1. Effectiveness of the filler-coupling agent-epoxy interaction in enhancing, for example, mechanical properties of the device.

2. Absorption and aggregation of water after exposure to humid environments and the resulting effect on properties, including delamination and catastrophic failure (*e.g.* the 'popcorn' effect [20]).

3. Wetting and adhesion of resin and other device materials (*e.g.* Al_2O_3, semiconductor material, Au leads, *etc.*).

4. Thermal conductivity and thermal expansion of the cross-linked base resin and resin-filler systems.

In recent work, we have developed building, simulation and analysis tools within the MedeA® modeling environment suitable for conducting a variety of studies relevant to electrical and electronic devices and device packaging. Thus, for example, we have recently developed the MedeA® Thermoset Builder for construction of realistic models of epoxy and other thermosets. In the following sections we accordingly provide illustrations of applications of atomistic modeling to probe the behavior of specific regions within EMC encapsulated devices.

3.1 Interaction Between Cross-linked Epoxy and Alumina Surfaces

Initial studies of adhesion between epoxy and alumina interfaces have focused on the fully hydroxylated Al_2O_3 surface, viewed as representative of the extreme of the most common physical presentation of such surfaces. Separate layers of hydroxylated alumina, consisting of a 28.76Å × 33.21Å × 23.10Å slab of Al_2O_3, with initial structure of a hydroxylated unit cell first optimized using the MOPAC 2012 semiempirical quantum mechanics program, and cross-linked epoxy layer containing 40 diglycidyl ether of bisphenol A and 20 Jeffamine® D-230 cross-linker molecules (Figure 4) and commensurate a and b cell dimensions and thickness 30.30Å, were combined to form the composite interfacial system (Figure 5). The resulting structure was then subjected to constant volume, constant temperature (NVT) molecular dynamics simulation using the LAMMPS simulation program [7] combined with the PCFF+ forcefield, which has been developed since 2009 at Materials Design, based on the original PCFF class II forcefield [21] distributed with the LAMMPS software. Thorough NVT equilibration of the interface was followed by a lengthy production stage to sample configurations suitable for determining the ensemble average of the interfacial energy, which resulted in an interfacial energy of 0.15 J/m^2.

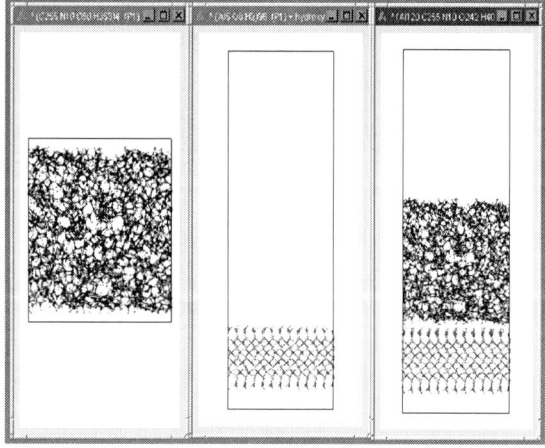

Figure 4: Diglycidyl ether of bisphenol A resin and polyoxypropylene diamine curing agent.

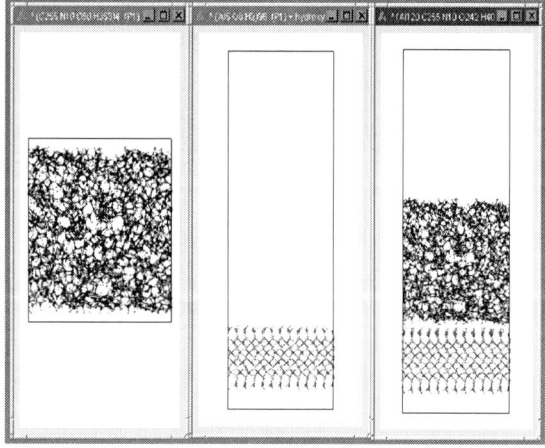

Figure 5: Assembled alumina-cross-linked epoxy interfacial model prior to equilibration.

3.2 Thermal Conductivity of Cross-linked Epoxy Base Resin

Thermal conductivity has been examined using a base epoxy system consisting of a stoichiometric mixture of diglycidyl ether of bisphenol A (DGEBA) and meta phenylene diamine (*m*PDA) cross-linked with degrees of cure up to ~80%, as studied experimentally by Kline [22], Krealing and Kline [23] and Cherkasova [24]. The simulations have employed the Müller-Plathe reverse non-equilibrium MD method outlined previously, using a periodic cell measuring 29.32Å × 29.32Å × 87.97Å. For systems of this size, simulations of duration 5-10 ns are normally required to achieve a precision better than ± 10%.

An example of a typical temperature profile is illustrated in Figure 6 for a system cured to slightly below 80%. The resulting thermal conductivity is calculated as 0.211 ± 0.024 W/m/K, which is slightly above the values in the range 0.176-0.188 W/m/K reported by Kline, but within the range 0.180-0.243 W/m/K measured between 298K and 363K by Cherkasova [24]. Further studies, probing the degree of cure in more detail and investigating the breadth of the distribution obtained with multiple cross-linked configurations, which our previous work has shown to be necessary when performing mechanical property studies [25], are accordingly desirable.

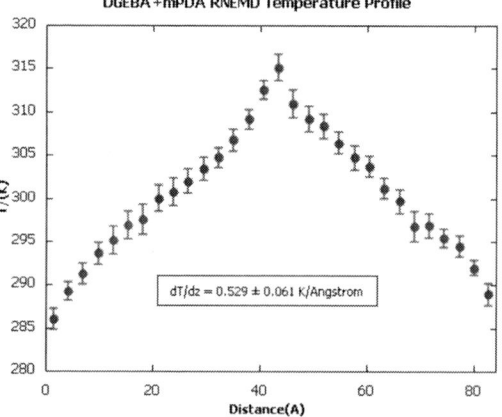

Figure 6: RNEMD temperature profile obtained for a sample of m-phenylene diamine cured DGEBA simulated at 298K.

3.3 Small Strain Elastic Constants of Cross-linked Epoxy

Small strain elastic constants for the DGEBA-mPDA system have been computed using multiple configurations of models prepared using 40 DGEBA molecules and 20 cross-linker molecules, for which the simulation cell sizes are slightly below 30 Å, containing ~2300 atoms. In almost all cases studied the degree of cure is close to 100%, as illustrated in Figure 7, which depicts the maximum bond strain of any bond during the 'curing' process applied by the MedeA® Thermoset Builder.

The tensile moduli for the batches of configurations analyzed – typically in groups of ~20 to permit optimal use of available computational resource – have been analyzed according to the Hill-Walpole method, which permits precise bounds definition when the moduli have been obtained for multiple small domains, as is the case whenever many independent atomistic realizations of a system have been simulated, as discussed at length for amorphous polymers by Suter and Eichinger [26], and applied in our previous work in this area [25].

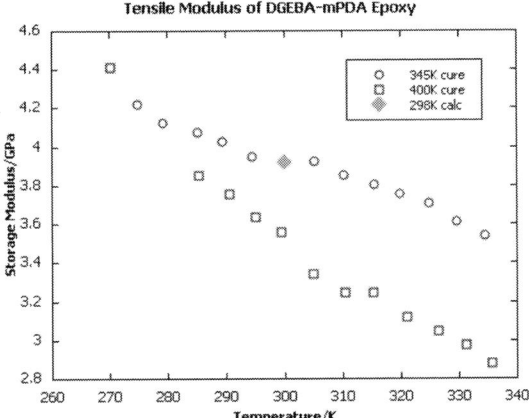

Figure 8: Comparison of calculated tensile moduli at 298K with experimental values obtained using different curing protocols and measured over a range of temperatures [23].

3.4 Oxygen and Water Transport in Cross-linked Epoxy

Barrier properties of cross-linked epoxies are of particular interest in coatings and adhesive applications, in which penetration of small molecules such as oxygen and water, followed by subsequent chemical reaction with metal substrates, can have significant effects on performance and integrity of the epoxy-substrate interface. Since the associated diffusion coefficients of even the smallest molecules – typically in the range $10^{-7} - 10^{-11}$ cm^2/s – are difficult to determine using atomistic molecular dynamics simulations with typical durations of perhaps a few nanoseconds, it is accordingly common to adopt an indirect method to study penetrant mobility. One preferred method involves computing diffusivities at elevated temperatures up to a few hundred degrees above the temperature of interest, followed by extrapolation to lower temperature with the assumption of Arrhenius behavior, which can be demonstrated experimentally in many systems. Thus one first performs a series of constant pressure (NPT) molecular dynamics simulations at a series of decreasing temperatures to estimate the system density, followed by a second, or accompanying, series of simulations in the microcanonical (NVE) ensemble at the calculated densities to monitor the mean square displacement, $\langle r^2 \rangle$ or MSD, of penetrant species as a function of time, from which the diffusivity D is obtained using the

standard Einstein relation $\langle r^2 \rangle = 6Dt$ (for diffusion in three dimensions).

A mean square displacement (MSD) plot is illustrated in Figure 9 for diffusion of oxygen in the DGEBA-Jeffamine® system discussed in section 3.1, at a temperature of 423K. Here it is observed that while the overall MSD often shows clear linear behavior, individual X, Y and Z components sometimes show evidence of the stepwise hopping that frequently dominate the diffusion mechanism.

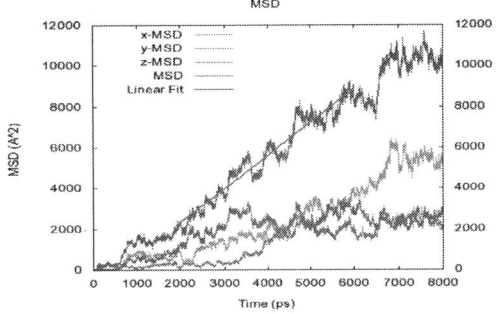

Figure 9: Mean square displacement plots for oxygen in cross-linked DGEBA-Jeffamine® epoxy system at 423K.

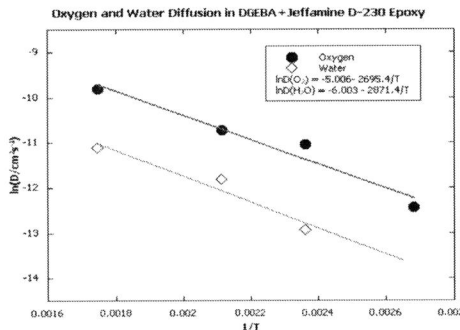

Figure 10: Arrhenius plot of oxygen and water diffusivities in cross-linked DGEBA-Jeffamine® epoxy systems. Densities at each temperature were determined from NPT molecular dynamics simulations.

The corresponding Arrhenius plots for oxygen and water diffusion in this epoxy system are illustrated in Figure 10, together with the least squares fits (inset). The slopes of the plots are observed to be similar for the two penetrants, suggesting similar activation energies for the diffusive process. Moreover, the results suggest that for this system the diffusivity of oxygen is higher than that of water by a factor of 4.5-5 under ambient conditions.

Experimentally, the relative diffusivities of oxygen and water in polymers can vary somewhat [27], with oxygen mobility sometimes exceeding that of water while in other systems the reverse applies. The situation in cross-linked epoxy systems has been discussed by Yarovsky and Evans [28], who applied

978-1-4799-4789-8/14 $31.00 © 2014 IEEE

molecular simulations to relatively small water-soluble epoxy models containing one of two phosphated epoxy components cross-linked using one of two different curing agents. These authors observed relative oxygen and water diffusivities in which either the oxygen or water was more mobile depending on the system, and argued that the slower oxygen diffusivities observed in two systems were mostly attributable to the higher mass of the oxygen, while the slower water diffusivity in a third system was attributable to the much higher concentration of residual hydroxyl groups remaining in the polymer after the cross-linking, with the hydrogen bonded interactions between bound hydroxyl and water reducing the overall water mobility. Such a situation is likely to apply to the DGEBA-Jeffamine® thermosets studied in this work, in which the cured systems contain two hydroxyl groups for each DGEBA moiety. Such interactions are also likely to lead to a higher solubility of water in the cross-linked epoxy, which is expected to increase the water *permeability* of the material in spite of the reduced water mobility. Consequently, small molecule solubility and its effect on permeability of the material as a whole is an ongoing topic of study.

4. Summary and Conclusions

As part of integrated computational materials engineering (ICME), atomistic simulations play an increasingly important role in the development of electronic systems to meet the challenges of higher performance, predictable reliability, lower cost, and environmental responsibility. Atomistic simulations provide understanding of key mechanisms governing the behavior of materials and they deliver quantitative materials property data, which can be used as input into finite element methods. In this context, the present work has demonstrated the application of MedeA®, a comprehensive state-of-the-art software environment for atomistic materials simulations, to the areas of thermal conductivity, elastic properties, adhesion, and diffusion.

As an example of an inorganic material, the effect of alloying and amorphization on the thermal conductivity of silicon has been demonstrated. Simulations revealed that only a few percent of Ge atoms reduce the thermal conductivity of crystalline silicon by a factor of about 20. The effect is even more dramatic if Ge atoms are ordered in the form of layers, where the thermal conductivity is reduced by a factor of 33 at 2% Ge and by about 60 at 12% Ge reaching values below 2 W/m/K. Also the introduction of amorphous Si into crystalline Si has a very pronounced effect on the thermal conductivity. Simulations showed that the thermal conductivity of pure silicon can be reduced from 128 W/m/K for a perfectly crystalline material to values as low as 4 W/m/K if 25% of silicon is amorphous. However, a further increase of the amorphous fraction does not lower the thermal conductivity below this value.

Calculations of the thermal conductivity of an epoxy thermoset were performed to show the capability of MedeA® to treat organic polymeric systems. Specifically, the epoxy system consisting of a stoichiometric mixture of diglycidyl ether of bisphenol A (DGEBA) and meta phenylene diamine (*m*PDA) cross-linked with degrees of cure up to ~80% has been investigated. The computed value of 0.211 ± 0.024 W/m/K is consistent with experimental data, thus showing the remarkable quantitative performance of current simulation technology.

For the same thermoset, namely DGEBA-*m*PDA calculations of the elastic properties have been performed, which show the sensitivity of the tensile modulus on the level of curing. An important aspect of reliable simulations of the materials properties for polymers is the ability to perform calculations on a series of different structural models in order to achieve statistically meaningful results. The MedeA® computational environment makes such large-scale computations readily doable with a minimum of human intervention, thus enabling the systematic screening of different materials under a range of various temperatures, pressures, and strains.

The adhesion of an epoxy on an inorganic surface has been demonstrated for the system of DGEBA-Jeffamine® on alumina. In this model, the surface is assumed to be covered by a layer of aluminum oxide with the dangling bonds on the surface saturated by hydroxyl group as they may occur in an ambient environment with moisture in the air. The simulations for the present model show that without direct chemical bonding between the organic and inorganic phases the work of separation is 0.15 J/m^2.

Finally, diffusivities of oxygen and water, respectively, in the DGEBA-Jeffamine® epoxy system were investigated. Molecular dynamics simulations of these molecules in the epoxy allow calculating their mean-square displacements, from which the corresponding diffusivities can be deduced. For DGEBA-Jeffamine® the simulations reveal similar activation energies for oxygen and water. Furthermore, the oxygen diffusivity is higher than that of water by a factor of 4.5-5 under ambient conditions.

In summary, the present work demonstrated the capabilities of state-of-the-art atomistic simulations as implemented in the MedeA® software environment in the computation of materials properties including thermal conductivity, elastic moduli, adhesion, and diffusion, thus providing insight at the atomistic level as well as quantitative materials property data. The examples discussed in this work represent only a fraction of the capabilities of MedeA®, which include the calculation of chemical reactions on surfaces as they occur in the processing and also corrosion of materials, predictions of electronic, optical, and magnetic properties for nanoscale structures, as well as the behavior of liquids and gases.

978-1-4799-4789-8/14 $31.00 © 2014 IEEE

Driven by the need to solve the tremendous challenges faced by the electronics industry and fuelled by the ever growing computing power, atomistic simulations are becoming increasingly valuable. In fact, leading electronics companies around the globe have made this type of approaches already an integral part of their R&D efforts. Leveraging the power and convenience of integrated software systems as implemented in MedeA®, the stage is set for exciting and rewarding applications of atomistic simulations as part of integrated computational materials engineering.

Acknowledgments

The authors would like to express their gratitude to the users of MedeA® for many fruitful discussions and stimulating suggestions. Furthermore, the authors thank their colleagues at Materials Design for all their support. A special thank goes to Jonathan Bell for his tireless work in keeping the computers and IT infrastructure of Materials Design, Inc. running smoothly and efficiently.

References

1. G. P. Srivastava. *The Physics of Phonons*. Taylor & Francis, 1990.

2. F. Müller-Plathe. *A simple nonequilibrium molecular dynamics method for calculating the thermal conductivity.* J. Chem. Phys., **106**, 6082 (1997).

3. P. Jund and R. Jullien. *Molecular-dynamics calculation of the thermal conductivity of vitreous silica.* Phys. Rev. B **59**, 13707 (1999).

4. S. Volz, J. B. Saulnier, G. Chen, and P. Beauchamp. *Computation of thermal conductivity of Si/Ge superlattices by molecular dynamics techniques.* Microelectronics Journal, **31**, 815 (2000).

5. J. R. Lukes, D. Y. Li, X. G. Liang, and C. L. Tien. *Molecular dynamics study of solid thin-film thermal conductivity.* J. Heat Transfer, **122**, 536 (2000).

6. S. Plimpton. *Fast parallel algorithms for short-range molecular dynamics.* J. Comp. Phys., **117**, 1 (1995).

7. S. Plimpton, R. Pollock, and M. Stevens. *Particle-mesh Ewald and rrespa for parallel molecular dynamics simulations.* Proceedings of the Eighth SIAM Conference on Parallel Processing for Scientific Computing, 1997.

8. K. Termentzidis, S. Merabia, P. Chantrenne, and P. Keblinski. *Cross-plane thermal conductivity of superlattices with rough interfaces using equilibrium and non-equilibrium molecular dynamics.* Int. J. Heat Mass Transfer, **54**, 2014 (2011).

9. Y. He and G. Galli. *Microscopic origin of the reduced thermal conductivity of silicon nanowires.* Phys. Rev. Lett. **108**, 215901 (2012).

10. X. Li and R. Yang. *Equilibrium molecular dynamics simulations for the thermal conductivity of Si/Ge nanocomposites.* J. Appl. Phys. **113**, 104306 (2013).

11. H. Y. Lv, H. J. Liu, J. Shi, X. F. Tang, and C. Uher. *Optimized thermoelectric performance of Bi2Te3 nanowires.* J. Mater. Chem. A **1**, 6831 (2013).

12. F. Müller-Plathe and D. Reith. *Cause and effect reversed in non-equilibrium molecular dynamics: an easy route to transport coefficients.* Comp. Theor. Polymer Science **9**, 203 (1999).

13. E. S. Landry and A. J. H. McGaughey. *Effect of interfacial species mixing on phonon transport in semiconductor superlattices.* Phys. Rev. B **79**, 075316 (2009).

14. K. Termentzidis, P. Chantrenne, and P. Keblinski. *Nonequilibrium molecular dynamics simulation of the in-plane thermal conductivity of superlattices with rough interfaces,* Phys. Rev. B **79**, 214307 (2009).

15. A. France-Lanord, E. Blandre, T. Albaret, S. Merabia, D. Lacroix, and K Termentzidis. *Atomistic amorphous/crystalline interface modelling for superlattices and core/shell nanowires.* J. Phys.: Cond. Matter **26**, 055011 (2014).

16. A. France-Lanord, S. Merabia, T. Albaret, D. Lacroix, and K. Termentzidis. *Thermal properties of amorphous/ crystalline silicon superlattices.* Subm. to Phys. Rev. B.

17. H.C. Nochetto, N.R. Jankowski, and A. Bar-Cohen. *GaN HEMT junction temperature dependence on diamond substrate anisotropy and thermal boundary resistance.* In: Compound Semiconductor Integrated Circuit Symposium (CSICS), 2012 IEEE, 1 (2012).

18. I. A. Tsekmes, R. Kochetov, P. H. F. Morshuis, and J. J. Smit, *Thermal Conductivity of Polymeric Composites: A Review,* Proc. IEEE International Conference on Solid Dielectrics, Bologna, Italy (2013), p. 678.

19. T. Zhou, X. Wang, P. Cheng, T. Wang, D. Xiang, and X. Wang, *Improving the Thermal Conductivity of Epoxy Resin by the Addition of a Mixture of Graphite Nanoplatelets and Silicon Carbide Microparticles,* eXPRESS Polymer Letters **7**, 585 (2013).

20. J. Zhao, C. Yang, Z. Huang, M. Wang, and C. Hu, *Mechanism of Adhesive Film Popcorn in Electronic Packaging,* in: Proc. 57th IEEE Electronic Components and Technology Conference, Reno, NV (2007), p. 487.

21. H. Sun, S. J. Mumby, J. R. Maple, A. T. and Hagler, *An Ab Initio CFF93 All-Atom Force Field for Polycarbonates,* J. Am. Chem. Soc, **116**, 2978 (1994).

22. D. E. Kline, *Thermal Conductivity Studies on Polymers,* J. Polym. Sci. **50**, 441 (1961).

23. R. P. Krealing and D. E. Kline, *Thermal Conductivity, Specific Heat and Dynamic Mechanical Behavior of Diglycidyl Ether of Bisphenol A Cured with m-Phenylenediamine,* J. Appl. Polym. Sci. **13**, 2411 (1969).

24. L. N. Cherkasova, Zh. Fiz. Chem., **33**, 1929 (1959).

25. D. Rigby, P. W. Saxe, C. M. Freeman, and B. Leblanc, *Computational Prediction of Mechanical Properties of Glassy Polymer Blends and Thermosets,* Proc. TMS 143rd Annual Meeting, San Diego, CA (2014), *in press.*

26. U. W. Suter and B. E. Eichinger, *Estimating Elastic Constants by Averaging over Simulated Structures,* Polymer **43**, 575 (2002).

27. S. Pauly, *Permeability and Diffusion Data,* in *Polymer Handbook,* J. Brandrup and E. H. Immergut (eds), 3rd Edition, John Wiley, New York (1989).

28. I. Yarovsky and E. Evans, *Computer Simulation of Structure and Properties of Crosslinked Polymers: Application to Epoxy Resins,* Polymer **43**, 963 (2002).

978-1-4799-4789-8/14 $31.00 © 2014 IEEE

FEM Stress Analysis of Various Solar Module
Concepts under Temperature Cycling Load

F. Kraemer, S. Wiese

Saarland University, Chair of Microintegration and Reliability, Saarbrucken, Germany
E-mail: f.kraemer@mx.uni-saarland.de

Abstract

The reliability of photovoltaic modules is essential for high electrical performance and a long operational lifetime. Both issues increase the profitability of photovoltaic electricity because these systems require only an initial installation invest. There are several aspects in a PV module which are able to reduce its profitability. An important aspect is the thermo-mechanical stress which is induced by day to night shifts at every day of operation. Since this stress obviously cannot be omitted the PV module set-up should reduce the resulting internal loads to a minimum. This paper analyzes the effects of the thermal induced stresses in three different module constructions.

The reliability of photovoltaic modules under thermo-mechanical loads is tested by the IEC standard 61215. This test method is reproduced in FEM simulations which are able to directly analyze the internal stresses of different PV modules. The investigation presented here applies a classic module assembly for H-patterned cells with a single front glass and a plastic back sheet which is the reference type. Another packaging type for H-patterned cells is the glass-glass module which replaces the back sheet by a second glass board. Finally there is a novel module type applying back contact solar cells. In this module type all electrical interconnections are supported by a substrate which is situated below the solar cells. This assembly is enclosed by a front glass and a plastic back sheet.

The different module assemblies are transferred to 3-D FE-models and subjected to temperature cycles. The mechanical analyses show that the solar cells are moved towards each other when temperatures decline and vice versa during temperature increase. This forced movement causes stresses and strains in the interconnection structures of the modules. The analyses reveal that those structures which are subjected to high mechanical loads are not supposed to cut the electrical interconnection because a failure may appear only in a small section of the according interconnection structure. Only in case of the glass-glass module the copper ribbons are subjected to high mechanical loads which may result in a complete cut of the series-connected solar cells.

1. Introduction

Long operational lifetimes of photovoltaic modules are beneficial for the solar industry. Longer lifetimes enable an increased profitability of PV modules or a reduced electricity price since PV systems require only an initial installation investment. Therefore reliability and the resulting lifetime is the crucial factor for these modules. In order to achieve a high reliability significant issues have to be identified in the design phase already. Among several aspects there is an annual derating of the modules power generation which is mostly caused by thermo-mechanical problems in the entire assembly. That is why new and current photovoltaic module assemblies have to be analyzed in terms of their mechanical behavior under varying temperatures.

This investigation is focused on the reliability of silicon based photovoltaic modules. At present there are three modules concepts applied in order to package the brittle silicon solar cells. These are the front glass back sheet assembly and a glass-glass module for H-patterned cells, respectively, as well as a back contact solar module. Each module concept has its individual cross-section which causes different thermo-mechanical loads on the photovoltaic cells and the interconnection structures between them.

2. Simulation models for PV modules

A silicon based solar module consists of about 60 individual cells. Nowadays there are two different PV cell types available which both require an electrical interconnection between them in order to conduct the generated electricity. First there is the H-patterned cell type which has a distinct division of the emitter contact on its front side and the base contact on the bottom side. Those cells are interconnected by a copper ribbon which connects the emitter of one cell with the base contact of the adjacent cell in order to form a series connection. Typically 10 to 12 cells are connected in series in order to form a cell string. These strings are put into a module and connected with each other. The packaging of the cell strings is done in two different ways as shown in figure 1. First there is the classic sandwich structure which consists of front glass, encapsulate sheet, cell string, encapsulate sheet and back sheet. Another option is the so called glass-glass module. In this module concept the back sheet is replaced by a second glass sheet. In this sandwich both glass sheets are roughly half as thick as the single front glass in the classic assembly. This way the glass-glass module has a symmetrical stack-up, which prevents the assembly from bowing due to different coefficients of thermal expansion. The classic module assembly has a lower weight which is advantageous on soft roofs.

Another idea of interconnecting individual cells is the application of back contact solar cells. This concept goes back to the 1970s which applies the so called metal-wrap-through technology [1-4]. In this technique the electrical connections are situated only on the back side of a solar cell. This layout is achieved by bypassing the emitter contacts to backside pads through metal filled holes in the silicon bulk. As shown in figure 2 this concept applies a flexible printed circuit board for the interconnection of the individual cells. Thus the copper ribbons of the tabber-stringer technology can be omitted. The remaining module stack-up is similar to the classic structure of the H-patterned cells applying a single front glass, encapsulate sheets and a back sheet. The advantage of the back contact module is an increased energy yield compared to the traditional assembly [4].

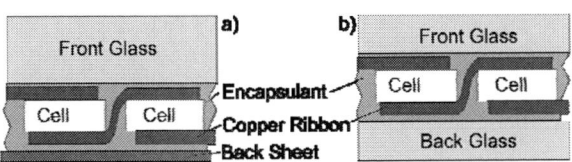

Figure 1: Schematic cross-sections for the packaging of H-patterned cells: a) classic front glass back sheet assembly; b) glass-glass module

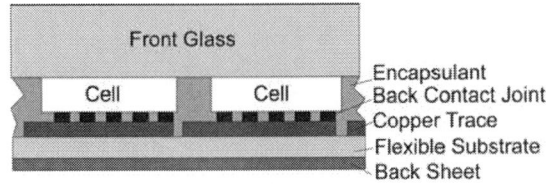

Figure 2: Schematic cross-section of a back contact solar module

All three module concepts are finally assembled in a vacuum lamination step. In this step the complete sandwich is heated and compressed in order to melt the encapsulate sheets, start the cross-linking in this polymer and release the generated gases out of the assembly. The sandwich is glued and the cell strings are encapsulated into a soft polymer. This production step is also the initiation of the module derating. It is the first thermo-mechanical stress to the module which is followed by temperature cycles resulting from day-night temperature shifts.

The investigation of the thermo-mechanical stresses and their effect on the module reliability is done by 3-D FEM simulations using the commercial software ANSYS. Critical module components are expected to be those who conduct the generated electricity. These components are the solder interconnections, the copper and the solar cells. Since there are rarely experiences with experimental failures which might be applicable for the validation of the FE models this analysis is limited to result comparisons between the models and material characteristics.

The FEM simulations presented here are executed in two steps applying the sub-modeling technique. This approach is necessary due to the large aspect-ratio between the module sizes of 1.66 x 1.0 m² to the small interconnection structures with a size of about 1.5 mm. Due to this approach the global model which represents the behavior of the complete module can be meshed in a coarse and effective way. The target of this step is the identification of the overall module deformation and the identification of highly stressed positions. The fine meshed sub-model is set at these high stress positions in order to analyze the mechanical stresses on the interconnection structures in the second simulation step.

The global models of all three module concepts are just one quarter of each complete module applying symmetry conditions at their inner edges. All models have the same size of 0.83 m in length and 0.5 m in width. These module quarters consist of arrays with 5 x 3 cells each. In order to achieve equal bending behaviors each module type has a thickness of 5.4 mm, which should be identical or at least close to productive modules. Although the global FE model consists only of a module quarter there is still a large aspect ratio to the electrical interconnection structures. That is why the thin layers of the solar cell strings and their electrical interconnections are concentrated into multilayer elements. Applying this method the global models for both H-patterned module concepts consist of about 207.000 elements while the back contact module type requires about 197.000 elements.

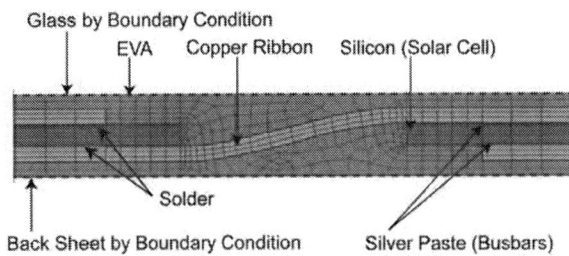

Figure 3: Cross-section through the sub-model cell interconnection for the H-patterned cells

The sub-models represent two adjacent solar cell quarters and the interconnection between them. Since the classic module and the glass-glass module both use H-patterned cells the sub-model is the same except its final position. The cross-section through the interconnection of two H-patterned cells is shown in figure 3. The sub-model represents each metallization layer which has a thickness of few micrometers each. Especially there is the silver paste which creates the solder contact with the copper ribbons and the back surface layer which is made of an aluminum paste. Furthermore the experimental solar cell has some tiny grid fingers on its front side but these structures are too small and uncritical in order to be represented in the model. The sub-model is surrounded by the EVA encapsulant up to the border to the front glass as well as the back glass and the back sheet, respectively.

978-1-4799-4789-8/14 $31.00 © 2014 IEEE   511

Thus the sub-model for the H-patterned solar cells requires about 434.000 hexahedral elements.

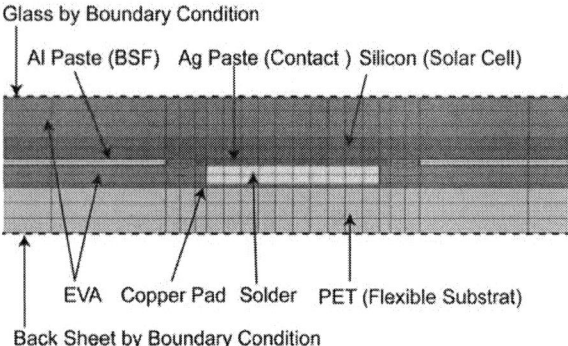

Figure 4: Cross-section of the FE sub-model for back contact solar cells; image of an emitter contact

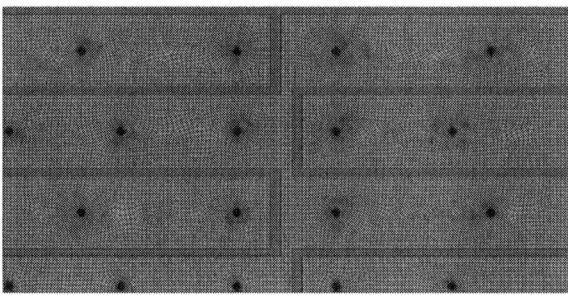

Figure 5: Bottom view on the FE mesh of the back contact module type. The copper interconnection structures are depicted in blue

The FE mesh of the sub-model for the back contact solar cells is shown in figures 4 and 5. Figure 4 shows a cross-section through a single back contact and the peripheral structure. The model setup is close to the sub-model for the H-patterned cells in order to create a similar mesh density. The sub-model represents both contact types. The emitter contact representing a front side contact is isolated from the surrounding metal layers as shown in figure 4. In case of a base contact the back surface field has a connection to the silver pad. Both silver contact pads have a diameter of 1.7 mm. Figure 5 shows the distribution of those contacts on both cell quarters by the dens mesh areas. The blue colored copper elements are arranged in a comb structure in order to achieve the serial connection between the adjacent cells. The total size of the back contact sub-model is equal to the size of the H-patterned version due to the equal experimental cell size. In total 239.000 hexahedral elements are required for the back contact sub-model.

Table 1 lists all material models which are applied in this investigation. Several material models have a temperature dependent behavior. That is why the table shows characteristic values at room temperature only. The dominating portion of the material models are taken from literature [5-8]. Silicon and glass are modeled independent from temperature since they do not vary their material behavior in the investigated range significantly. Due to their brittle characteristic there is no yield stress included either. The solder [9] and the copper [10] models are based on own measurements. The copper model applies a temperature dependent bilinear elastic-plastic behavior. An isotropic hardening behavior is suited to represent our measurement results. The solder behavior is represented by multi-linear curve of the time independent elastic-plastic material response. Furthermore the solder model accounts for the time dependent relaxation behavior due to creep. The ANSYS software offers a sinh-law which is found adequate to represent the experimental behavior.

Table 1: FE material models applied in this analysis and their characteristic values at room temperature

Material	Young's Modulus [GPa]	CTE [ppm/K]	Yield Stress [MPa]	Material model
Aluminum paste	6.0	11.9	28.3	Bi-linear temperature dep. elastic-plastic
Copper (initial)	85.7	17.0	95.1	Bi-linear temperature dep. elastic-plastic with isotropic hardening
EVA	0.012	15	-	Temperature dep. Young's Modulus
Silver paste	7.0	9.8	64.0	Bi-linear temperature dep. elastic-plastic
Silicon	130.0	2.6	-	elastic
Solder (SnPb)	32.0	25.2	48.0	Temperature dep. Young's Modulus & Yield stress; creep
PET	2.2	11.6	22.0	Bi-linear temperature dep. elastic-plastic
Tedlar	1.4	30.0	35.0	Bi-linear temperature dep. elastic-plastic
Glass	75.0	10.0	-	elastic

978-1-4799-4789-8/14 $31.00 © 2014 IEEE

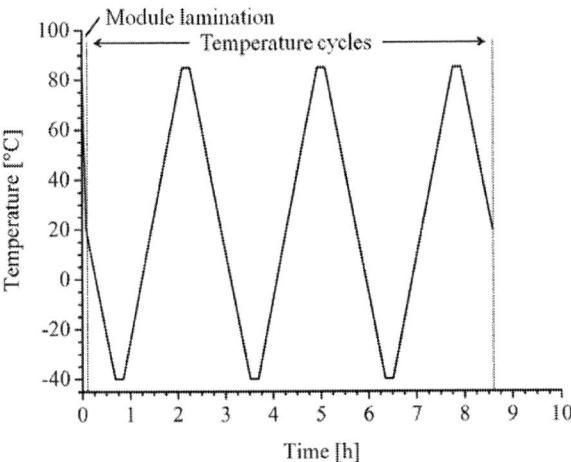

Figure 6: Temperature cycles applied in the FE analysis including initial cooling from lamination temperature

The thermo-mechanical stresses in the module assemblies are caused by temperature cycles according to the IEC standard 61215. This is the test standard for solar modules which targets the investigation of the long term reliability of PV panels. According to this test method the temperature varies between -40°C and 85°C. The proposed temperature cycles have slopes of 100 K/h and dwell times of 10 min at the maximum and minimum temperatures. The long temperature cycles are caused by the big mas of the module. The FEM simulations cover three of those temperature cycles which are necessary in order to evaluate results from steady state cycles. The simulations start with an initial cooling from the module lamination temperature. This cooling is expected to last five minutes. The module is expected to be stress-free at the beginning of this initial cooling step.

The global models are fixed at their outer edges along the thickness. This boundary condition represents an ideally stiff module frame which is not part of the geometrical models due to its delicate cross-section. Symmetry conditions are set at both inner edges of the global models in order virtually complete the entire solar panel.

3. FEM stress analysis

The FEM result analysis starts with the global model results. Due to their coarse meshing of global model the resulting module deformations are reliable results. Figures 7 – 9 show the maximum deformation of each module concept at -40°C which is the biggest temperature difference to the stress-free state. Based on the applied models and the boundary conditions the centers of the classic and the back contact module bend downwards. Due to the relations between coefficients of thermal expansion (CTE) and the geometrical conditions it has to be stated that only the silicon seems to interact with the thick front glass. The faster contraction of the glass than the silicon causes the observed deflection behavior.

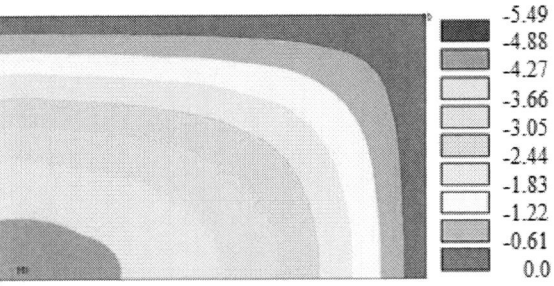

Figure 7: Deflection profile of the classic solar module at -40 °C

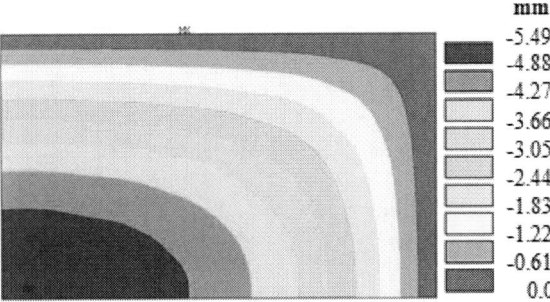

Figure 8: Deflection profile of the back contact module at -40 °C

Figure 9: Maximum deflection of the glass-glass module at -40°C

The biggest deformation of the classic solar module is 4.5 mm. In case of the back contact assembly the maximum module deflection reaches up to 5.5 mm which is 22 % higher than the classic module. The flexible substrate shifts the cells closer to the glass which reduces the thickness of the bending beam and enables bigger deformations. The glass-glass module does not bend due to the temperature changes which is not a surprise, see figure 9. As mentioned this module type has a symmetrical stack-up which is unable to cause any deflection. But still there are mismatches in the CTEs of the assembled materials. These differences cause stresses within the assembly which are investigated further with the sub-model.

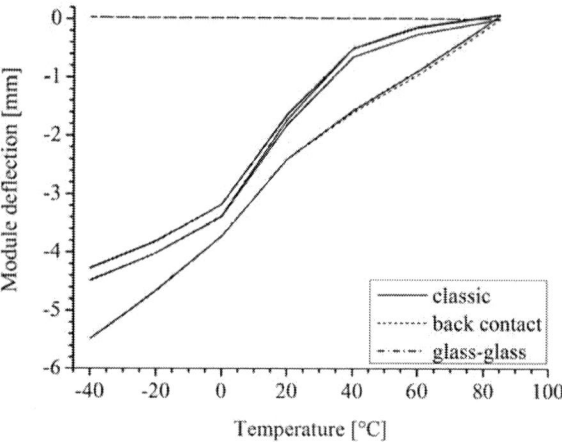

Figure 10: Deflections of the different module types over the temperature cycles

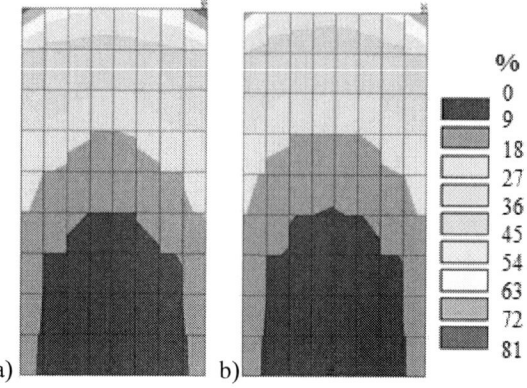

Figure 11: Accumulated creep strain in the solder layer between cell and copper ribbon. Creep strain at the end of the temperature course: a) glass-glass module; b) classic assembly

The temperature dependent deformations of all module types are condensed in figure 10. As shown in the previous figures there is a big difference between the glass-glass module and the glass-back-sheet assemblies. The glass-glass module shows no deformation within the complete temperature range. The other module types start to bend as soon as the temperature changes. For the back contact module there is an almost linear relation between temperature and deflection. This behavior proves the stiffened connection between glass and silicon. Due to this change there is almost no influence of the polymeric materials with their strong variation of the Young's modulus over temperature.

In case of the classic assembly there is another temperature dependent behavior. There is a change in the module deflection after the first temperature cycle proving that the model has to reach a steady-state condition. Furthermore there is no linear relation between temperature and deformation. At high temperatures there is little module deflection because the encapsulate material has a very low Young's modulus which causes a mechanical decoupling of silicon and glass. As soon as the EVA becomes stiffer the glass starts to work against the silicon causing a high deformation rate. When the EVA reaches the glass transition temperature at 0°C the deformation rate reduces again because the encapsulate material itself starts to work against the glass. Since the EVA has a higher CTE than the glass the direction of the deflection should be reversed but it its Young's modulus is still too small at the minimum temperature.

The deformation behavior presented for the global models is the input to the sub-models. Due to the smeared behavior in the multi-layer elements of the global model there is no reliable evaluation of stresses. That is why the variation of the cell gaps is taken as criterion for the most critical position in the photovoltaic module. The sub-models are put at the position with the biggest variation in the cell gap of each module type.

The result evaluations of the sub-models are focused on the electrical interconnection structures of the PV cells such as the solder layers and the copper ribbons and the copper sheets, respectively. Failures in these parts of the module will cause a reduced yield of the generated electricity.

The first sub-model result evaluations concern the stresses and strains in the solder. Figure 11 shows the accumulated creep strain of the solder layers in the H-patterned cell assemblies at the end of the simulated temperature course. The figure compares the results for the classic module assembly with the glass-glass module. In contrast to the negligible module deformation there is a remarkable solder creep in the glass-glass sub-model. However the basic creep behavior is similar for both module types. The highest creep strain appears at the corners of the solder layers. These high values reduce significantly towards the middle of the cell where only little creep strain is generated. The higher creep strain appears in the glass-glass assembly. At the corner elements the creep strain reaches up to 80 %. In case of the classic module the maximum creep strain is up to 70 %. The huge amount of accumulated creep strain might be overestimated due to the geometrical discontinuity at both corners. However there is more stress in the glass-glass module.

The creep strain distribution in the highest stressed solder pin of a back contact cell is shown in figure 12. The side view indicates that the solder pin is dominantly subjected to a shear load. This shear load is directed along the cell string. The analysis of all solder pins on a single cell shows that stress and strain increase for those pins which are closer to the cell edge. The highest creep strains appear in a diagonal direction through the solder pin, which is caused by the shear load. The highest strain appears at the bottom of the pin which is connected to the copper layer.

978-1-4799-4789-8/14 $31.00 © 2014 IEEE

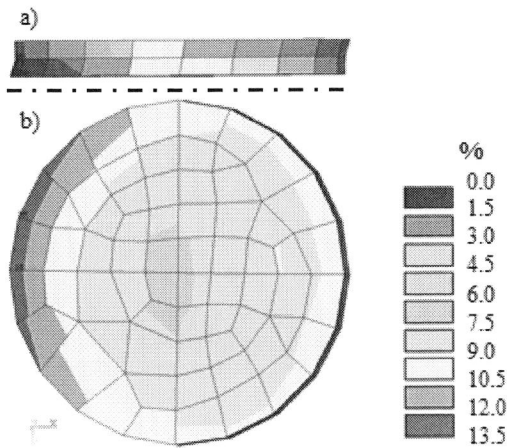

Figure 12: Highest accumulated creep strain in a solder pin at the end of the temperature cycles: a) side view; b) bottom view

The creep strain accumulated during the temperature cycles is much lower than at the edge of the H-patterned solder layers. The maximum strain reaches up to 13 %. The creep strain is distributed over the complete solder pin which becomes obvious due to the small scale. A low creep level is vital for this interconnection structure due to its small volume. That is why a complete failure of the electrical connection is more likely to appear than in case of the huge solder layers on the H-patterned cells. However the failure of a single solder pin may cause a loss of electrical power but it would not cause a complete cut of the power transmission in the module.

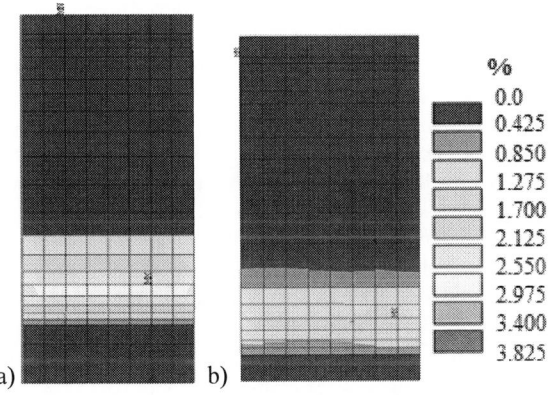

Figure 13: Top view of the distribution of plastic strain in the copper ribbon at the cell transition: a) ribbon in glass-glass module; b) classic module

Another critical part for the power transmission of a solar module is the copper ribbon and the copper sheet, respectively. Figure 13 shows the copper ribbons of the classic and the glass-glass module. The ribbon sections depicted in figure 13 show those parts which are located in the transition between two adjacent cells. In this area

the copper ribbons have to change between the top side of one cell to the back side of their neighbor. That is why the ribbons have a crimp which reduces their resistance against the cell movement and defines positions of the major deformation.

Figure 13 shows the accumulated plastic strain after the temperature cycles. The top view on the copper ribbon shows a single area of plastic deformation which is located on the inside of one curvature of the crimp. In the second curvature there is a similar distribution of the plastic strain. This strain distribution indicates that the copper crimp is dominantly compressed throughout the temperature course. The maximum plastic strain in the classic assembly reaches up to 1.4 %. In case of the glass-glass module there is again more plastic strain accumulated. The maximum plastic strain reaches up to 2.7 % which is almost the double plastic strain of the classic assembly. However the result of a single element should not be overestimated. Since the back side of a curvature deforms only elastically the difference in the complete cross-section of a copper ribbon is better compensated. But in general the glass-glass assembly is supposed to have a lower reliability.

The distribution of the accumulated plastic strain in the copper sheet of a back contact module is shown in figure 14. Plastic deformations in the copper appear only below the solder pins since these are the only positions where mechanical stress can be transferred from the silicon to the copper. The figure shows the position of highest plastic copper deformation which appears below the highest stressed solder pin. The shear load of the solder pins results in sickle shaped distributions of the plastic strain. Due to the single sickle below the solder the shear deformation of the pin appears only in one direction. The maximum plastic deformation is up to 3.8 % which is even higher than in the copper ribbon of the glass-glass module. This high plastic strain is tightly localized which may not cause complete cuts of the copper. A flaw in the copper may even help to reduce the stress in the entire interconnection.

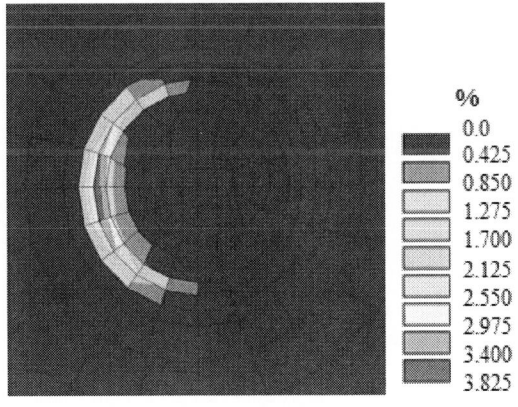

Figure 14: Top view on the plastic strain in the copper foil beneath the highest stressed solder pin at the end of the temperature cycles

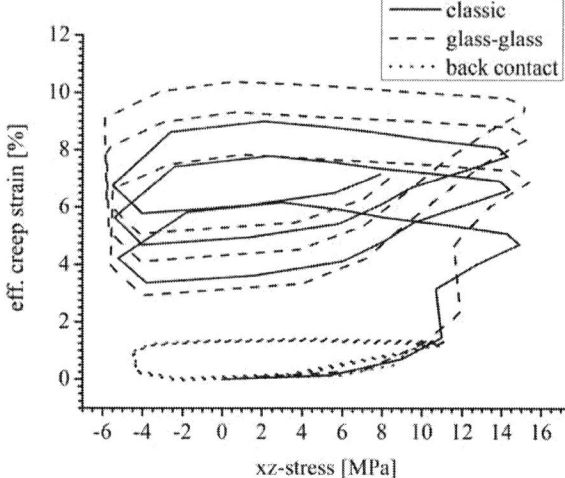

Figure 15: Hysteresis curves of the solder interconnections in all three PV module assemblies

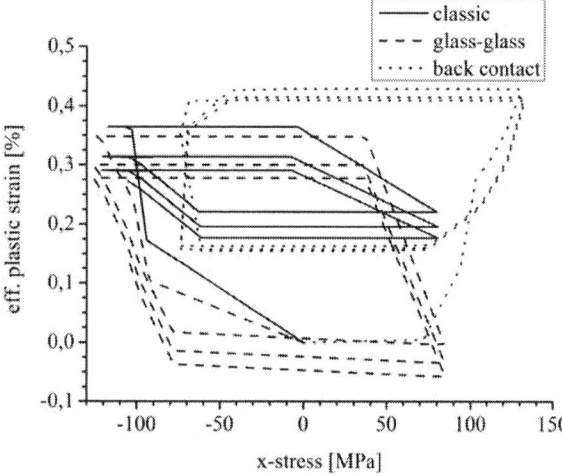

Figure 16: Resulting hysteresis curves in the copper interconnection structures of all investigated solar module types

Previous considerations revealed the critical parts of a solar module under temperature cycling conditions which are the interconnection structures of the PV cells. The assessment of the long term reliability of these structures has to include their stress-strain behavior within the temperature cycles. This assessment can be done best by hysteresis curves. Figure 15 shows hysteresis curves of the solder layers on H-patterned cells and in the solder pin of a back contact module. In case of the H-patterned cells a volume section close to the high stress parts is evaluated which has a similar volume as a single solder pin. The hysteresis curve presented here consists of the averaged x-z shear stress and the averaged effective creep strain. As indicated by the figures 11 and 12 the mechanical load of

the solder pin is much lower than in the solder layers. The hysteresis loops of the back contact solder are much smaller and they remain constant. In case of the solder layers on the H-patterned cells the glass-glass module creates higher loads on the investigated volume. Both curves require an initial temperature cycle in order to reach a steady-state deformation behavior.

Figure 16 shows the hysteresis loops of the copper interconnection structures. The data presented here are taken from the uppermost elements along the sickle in the copper sheet of the back contact structure. The mechanical load on the copper ribbons is evaluated in the innermost elements of one curvature in the crimp section. The hysteresis curves consist of the averaged tensile stress in x-direction and the averaged effective plastic strain. Again the biggest hysteresis loops are caused in the glass-glass assembly. The smallest hysteresis loops can be found in the classic module construction. In contrast to the solder behavior the copper hysteresis loops shrink after the initial temperature cycle. In case of the back contact module this shrinkage seems to be finished indicating a stable deformation progress after the three temperature cycles which are simulated here. As mentioned before a lifetime estimation of the sickle region in the copper sheet cannot be correlated with an electrical failure since this is just a small section of copper sheet.

4. Conclusions

The long term reliability of solar modules is crucial because it defines the cost of the generated electricity. Thus this paper presents a mechanical analysis of the interconnection structures in different solar module concepts. The analyses are executed by 3-D FEM simulations. This investigation includes three module types which are a classic solar module based on H-patterned PV cells with a single front glass and a plastic back sheet, a glass-glass module which is similar to the classic assembly but replaces the back sheet by a back glass in order to achieve a symmetrical stack up and finally a novel back contact module which applies metal-wrap-through solar cells. All module types are based on silicon solar cells which are no longer supposed to fail as soon as they have been laminated into the PV module. Thus this investigation is focused on the interconnection structures between the cells which have to withstand temperature cycles that appear in normal operation due to day to night temperature shifts.

The simulation results identify distinct module positions where high mechanical loads are created in the interconnection structures. In case of the solder high creep strains are accumulated within the simulated temperature cycles. These deformations can be found at the edges of the solder layers on H-patterned cells which are applied in the classic and the glass-glass module assemblies. Higher mechanical load of the solder appears in the glass-glass assembly although the module shows negligible deflection due to temperature variations. The high solder strains in both assemblies should cause early failures. But the deformations are very localized and the big remaining area is much less critical. The mechanical stress in the solder pins of the back contact cells is much lower. However a failing solder pin would have a bigger influence on the electrical performance of the module.

The stresses generated in the modules are sufficiently high in order to cause plastic deformations in the copper. In case of the back contact cells significant copper stress appears below the solder pins resulting in a sickle shaped deformation zone. For the H-patterned cells the plastic deformations are localized in the curvatures of the crimped copper ribbons in between adjacent cells. Again the highest mechanical loads appear in the glass-glass assembly. A complete crack of the copper ribbons cross-section is a serious reliability concern because it may cut a big portion of the modules power. The mechanical load in the copper sheet of a back contact module is in a similar range. But the considered sickle beneath the solder pin is just a section of the contact. Based on previous results a sickle crack may help to release stress in the back contacts and thus increase their reliability. The comparison of the three investigated solar module concepts indicates that the glass-glass assembly shows a serious potential to fail completely due to the high mechanical load in the copper crimps between adjacent cells.

References

1. M. Späth, et.al. "Solder version of 8 inch back-contacted solar cells", ECN-Report, 2005.

2. P.C. de Jong, et.al. "Single-step laminated full size PV modules with back-contacted MC-cells and conductive adhesives", Proceedings of 19th European Photovoltaic Solar Energy Conference, Paris, France, 2004, ECN-RX—04-067.

3. A.A. Mewe, et.al. "Reaching 16.4% module efficiency with back-contacted mc-Si solar cells", Proceedings of 24th European Photovoltaic Solar Energy Conference and Exhibition, Hamburg, Germany, September 21-25, 2009.

4. M. Späth, et.al. "A novel module assembly line using back contact solar cells", Proceedings of IEEE Photovoltaic Specialists Conference, San Diego, USA, May 11-15, 2008.

5. M.A. Salana, et.al. " On the Thermoelastic Analysis of Solar Cell Arrays and Related Material Properites", Technical Memorandum 33-753, National Aeronautics and Space Administration, California Institute of Technology, Pasadena, CA, February 15, 1976.

6. D.M. Kempe, et.al. "Acetic acid production and glass transition concerns with ethylene-vinyl acetate used in photovoltaic devices", Journal Solar Energy Materials and Solar Cells, Vol. 91 (2007), pp. 315-329.

7. C. Joubert, et.al. "Influence of the Crosslink Network Structure on Stress-Relaxation Behavior: Viscoelastic Modeling of the Compression Set Experiment", Journal of Polymer Science: Part B: Polymer Physics, Vol. 41 (2003), pp. 1779-1790.

8. 10. A.W. Czanderna, et.al. "Encapsulation of PV modules using ethylene vinyl acetate as a pottant: A critical review", Journal Solar Energy Materials and Solar Cells, Vol. 43 (1996), pp. 101-181.

9. S. Wiese, et.al. "Characterisation of constitutive behaviour of SnAg, SnAgCu and SnPb solder in flip chip joints". Sens Actuators A 2002; 99:188-93.

10. S. Wiese, et.al. "Mechanical Behaviour and Fatigue of Copper Ribbons used as Solar Cell Interconnectors", Proceedings of 11th International Conference on Thermal, Mechanical and Multi-Physics Simulation and Experiments in Microelectronics and Microsystems (EuroSimE), Bordeaux, France, April 26-28, pp. 1-5, 2010.

FEM Wire Bonding Simulation for Sensor Chip Applications

F. Kraemer, S. Wiese

Saarland University, Chair of Microintegration and Reliability

Campus A5 1, 66123 Saarbrucken, Germany

E-mail: f.kraemer@mx.uni-saarland.de

Abstract

The paper describes a three-dimensional dynamic finite element simulation of the wedge bonding process on a sensor chip design. The stiffness of the die bond connection has a strong influence on the sensitivity and accuracy of sensors. Soft die bonds are preferred in order to decouple the substrate stiffness from the flexibility of the sensor beams. However, a very soft die bond connection may cause trouble during the subsequent wire bonding process because the die is not sufficiently fixed to the substrate which may result in a partial absorption of the ultrasonic energy that is required for the welding of bond wire and pad metallization.

The FEM simulations presented in this paper were done by the commercial code LS-Dyna which is capable of high deformation speeds and high plastic deformations. The study investigates the effect of die adhesives with different Young's moduli on the resulting stresses and strains in the bond interface. Based on the simulation results contact forces are evaluated, too, in order to assess the induced ultrasonic energy which is essential for the contact formation. The simulation results show big differences in the resulting force peaks of up to 20 %. The bond pad stresses even differ by up to 70 %. The simulations prove a remarkable influence of the die adhesive stiffness on the mechanical contact loads during the wire bond process.

1. Introduction

The packaging of sensor chips is more and more challenging. Sensors are transducing elements which couple multiple energy domains. These interactions also define their operational behavior [1]. Novel sensors become smaller and lighter which increases their sensitivity for any external influence [2]. Furthermore the versatile sensitivity is challenging for the accurate determination of their accurate mechanical properties [3]. These developments lead to the target of a mechanical decoupling of the sensor chip from any external influences. First of all it leads to the further integration of analysis and transmission circuits on the sensor chip which omits stresses due to wafer to wafer bonding [4].

Another step towards the internal stress reduction is the mechanical decoupling of the sensor chip from the surrounding package. This can be achieved easily by the application of very soft adhesives in the die bonding process. However this trend is not uncritical for subsequent production steps such as the electrical interconnection of the sensor chips by wire bonds.

Especially the formation of wedge bonds requires the input of sufficient frictional energy in order to break the stiff aluminum-oxide layer on the bond wire and complete the micro welding process. Soft die adhesives may disturb the transmission of the mechanical energy and thus prevent the contact formation. This analysis investigates the influence of die adhesives with low Young's moduli on the wire bonding process by the transient finite-element simulations.

2. Finite element simulations

The effect of soft die adhesives on the wire bonding process is analyzed by a detailed 3-D FE model. The target of this investigation is a transient analysis of the stress distribution in a sensor chip assembly during wire bonding. That is why the whole process is transferred to a transient simulation code such as LS-Dyna.

Figure 1: Complete geometrical model including full-length bond capillary.

In this analysis the contact formation on the sensor chip is expected to be critical. Thus the 3-D FE model consists of the sensor chip, substrate, die adhesive, bond pad with nickel surface, bond wire and bond capillary, see figure 1. The sensor chip is modeled with its full-size. This enables the consideration of the real eigen-frequencies in the vibration system of die adhesive and chip which may become effective at a certain adhesive stiffness. The die is glued to the substrate on its entire surface. On top of the sensor chip there is a single copper bond pad with a size of 50 x 50 μm^2. The single pad is sufficient because the simulation cannot cover the formation of several bond interconnections at once and it reduces the number of elements. The modeled bond wire length represents the contact area and the section through the bond capillary. The bond wire has a diameter of 20 μm.

The wedge bond capillary is modeled in its full-size. Although the capillary has an almost rigid behavior its harmonic behavior during the ultrasonic oscillations is essential for a detailed analysis of the stresses and deformations created during the bond process [5]. An accurate vibration behavior requires the correct geometry, mass and material properties. The bond capillary has an irregular shape which is rather difficult for the mesh generation. The combination of v-notch and wire guidance disturbs any symmetry. That is why the final geometry has to be generated by multiple Boolean operations with assisting volumes. Finally, the volume of the notch is meshed by ten nodes tetrahedral elements. Due to this difficult mesh generation simplifications of the bond tool can still be found in literature [6].

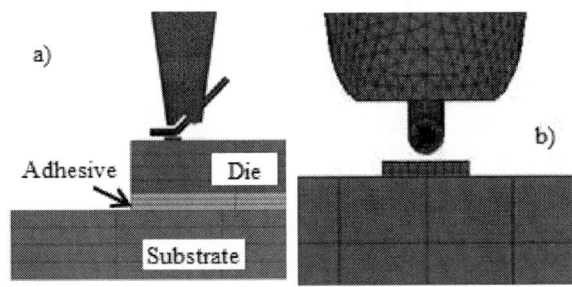

Figure 2: Detailed view of the pad area: a) side view on the complete stack-up of sensor chip, adhesive and substrate; b) front view on the section between chip and bond capillary.

Detailed pictures of the applied FE model are shown in figure 2. These images show the enormous size ratio between the bond capillary, the aluminum wire and the contact pad, respectively. These geometrical differences require different mesh densities which have to be achieved by transition areas. However, the geometrical model consists of 131.000 elements.

Furthermore, figure 2b shows the initial stand-off between bond pad and wire. Since the meshes of both sections cannot be put in an agreement contacts are required in order to define the interaction between the surfaces of both parts. The initiation of the contact

requires some time [7] which is overcome by an additional vertical movement of the capillary.

Table 1: Material behaviors and their properties in the FEM bond simulations.

Material	Mechanical behavior	Young's Modulus [GPa]	Yield stress [MPa]
Aluminum	elastic-plastic w/ isotropic hardening	67.0	145
Copper	elastic-plastic w/ isotropic hardening	95.0	150
Die adhesive	Elastic	variable	-
Nickel	Elastic	206.0	-
Silicon	isotropic elastic	169.0	-
Substrate	orthotropic elastic	21.0	-
Tungsten	Elastic-plastic w/ kinematic hardening	400.0	500

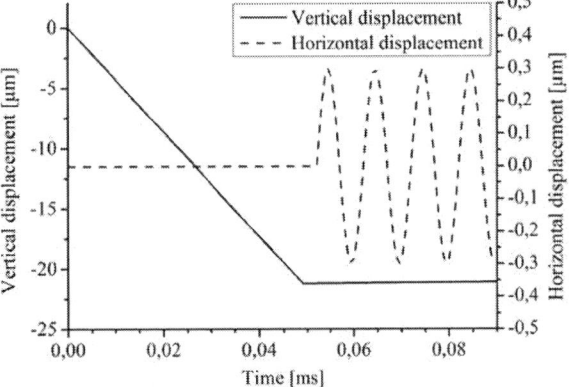

Figure 3: Displacement profile at the top of the bond capillary.

The material models applied in this investigation are condensed in table 1. The aluminum is modeled elastic-plastic because huge deformations are expected in the bond wire. The copper pad is not expected to show significant deformations however an elastic-plastic model is available [8] and it is applied. High stress and thus strong deformations may appear in the nickel surface but since there is no adequate value available so this model has to be simplified. Another simplification is used for the silicon, which is modeled isotropic elastic since no orientation of the silicon is known. The substrate has an orthotropic elastic material behavior. The tungsten in the bond capillary is also not expected to show remaining deformations however it has an elastic-plastic behavior in order to set a limit for the compressive stresses during the bond process. The die adhesive has an isotropic elastic material model. In this investigation its Young's modulus is modified from 1 MPa (soft) to 300 MPa (medium) and

1200 MPa (stiff) in order to investigate its effect on the resulting bond forces and deformations.

An irregular deformation of the geometrical model is prevented by a complete fixture of the bottom of the substrate. On the other side the bond capillary is able to move in any direction for the bond process. All capillary displacements are transferred through its top surface. The different stages of the capillary displacement are shown in figure 3. The simulation kicks-off with a vertical displacement in order to press the aluminum wire onto the bond pad with a certain force. After that the horizontal oscillations begin with a frequency of 100 kHz. The small horizontal amplitude is boosted by the resonant excitation of the complete bond capillary. The simulation covers up to 5 sinusoidal oscillations.

3. Results

Transient simulations result in a huge number of data depending on the chosen time-steps and the investigated time interval. The analysis of the bond process is split in two parts. High compressive stress is expected at the end of the vertical displacement of the bond capillary. Stress distributions and deformations are analyzed at this point in time. In the second part the ultrasonic movement is analyzed by the resulting forces acting on the contact between wire and bond pad.

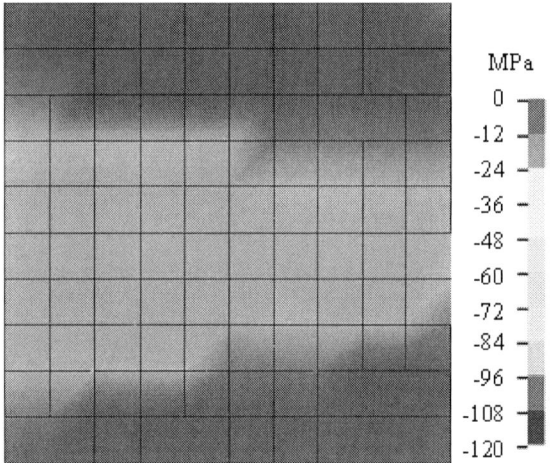

Figure 4: Distribution of 3rd principal stress in the copper pad at the end of the vertical capillary displacement. Model with soft die adhesive.

In the first step the stress on the surface of the copper pad is analyzed before the ultrasonic oscillations start. Figures 4 – 6 show the according distributions of the 3rd principal stress. Since the dominating load is compressive the highest stress values can be found for this criterion. The direction of the 3rd principal stress is acting through the pad towards the sensor chip. Low stress appears in case of the soft die adhesive. The significant stress is located in the middle of the pad where the bond wire is pressed onto the pad. The edges parallel to the bond

direction are almost stress-free. Elevated stresses can be found in case of the medium stiff die adhesive. Its Young's modulus is more than 2 orders of magnitude higher resulting in a stress increase of roughly 70 %. The highest in pad stress can be found for the stiffest die adhesive. The calculated stresses exceed 120 MPa which is almost double the value of the soft die adhesive.

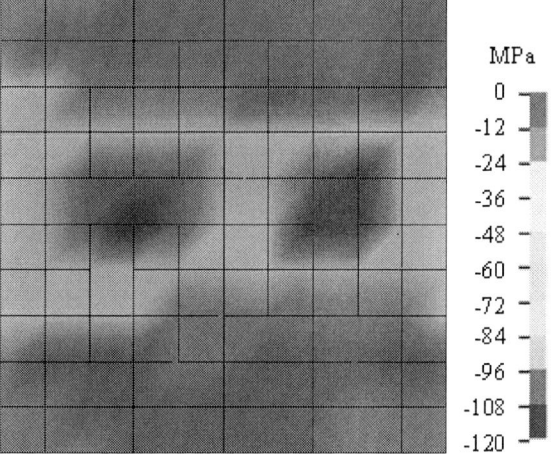

Figure 5: 3rd principal stress in the copper pad at the end of the vertical capillary displacement. Model with medium die adhesive stiffness.

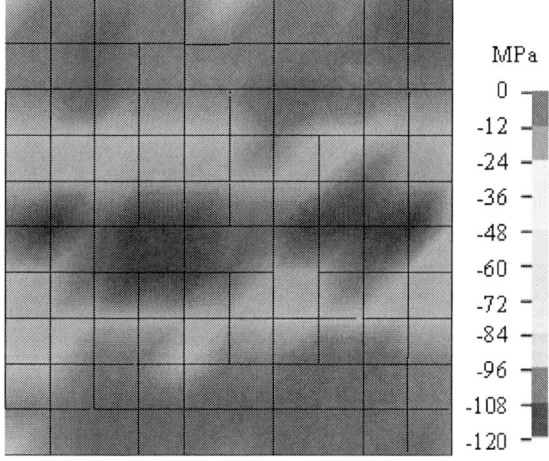

Figure 6: 3rd principal stress in the copper pad at the end of the vertical capillary displacement. Model with stiff die adhesive.

Further stress results were evaluated in the silicon, see figures 7 – 9. The figures show the section of the die which is located beneath the bond pad. This is the only section of the chip where reasonable stresses are present. Due to the compressive load the 3rd principal stress is chosen as result criterion again. Still the 3rd principal stress is directed towards the thickness of the stack-up. However the stress values in the silicon are much lower compared to the bond pad. In case of the assembly applying the soft die adhesive the maximum stress is far below 30 MPa which is less than half of the copper stress. Similar observations can be made in the other assemblies. According to the previous results the stress values in the silicon rise with the increased adhesive stiffness. But the silicon stress values do not increase as much as the copper stress. The difference of the peak stress values between the soft and the stiff assembly is just about 30 %. Thus the copper pad covers the majority of the compressive stress and distributes it. That is why uncritical stresses are created in the silicon chip in any case.

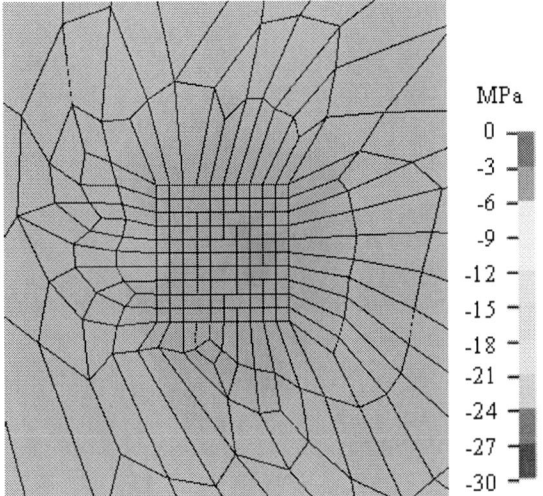

Figure 7:3rd principal stress in the sensor chip at the end of the vertical capillary displacement. Model with soft die adhesive.

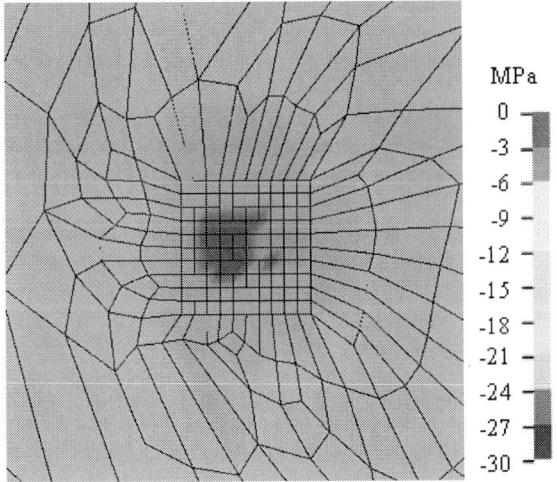

Figure 8: Distribution of 3rd principal stress in the silicon beneath the pad. Model with medium die adhesive stiffness.

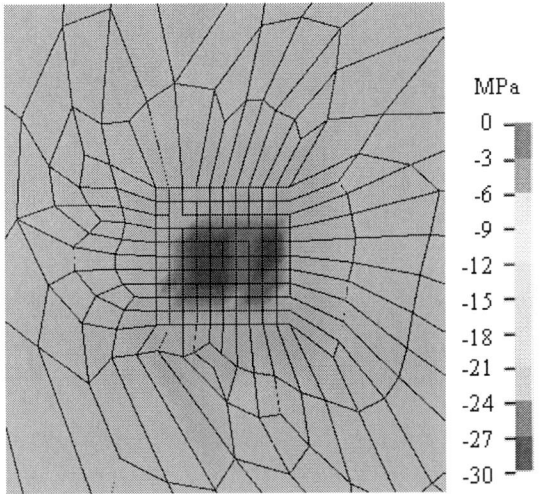

Figure 9: 3rd principal stress in the silicon at the end of the vertical capillary movement. Model with stiff die adhesive.

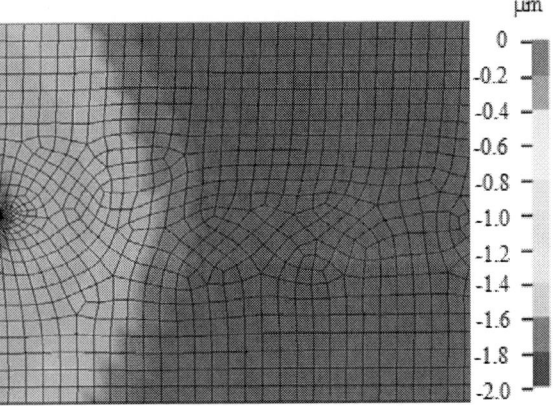

Figure 10: Compression of the soft die adhesive during the bond process.

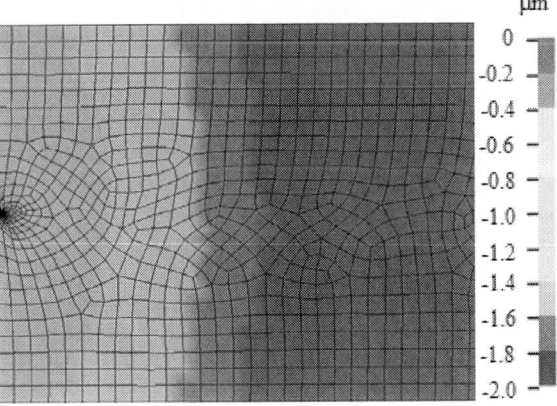

Figure 11: Distribution of penetration into the medium die adhesive.

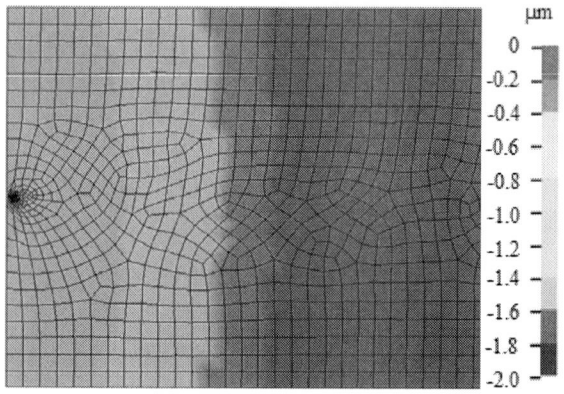

Figure 12: Penetration into the die adhesive with high Young's modulus.

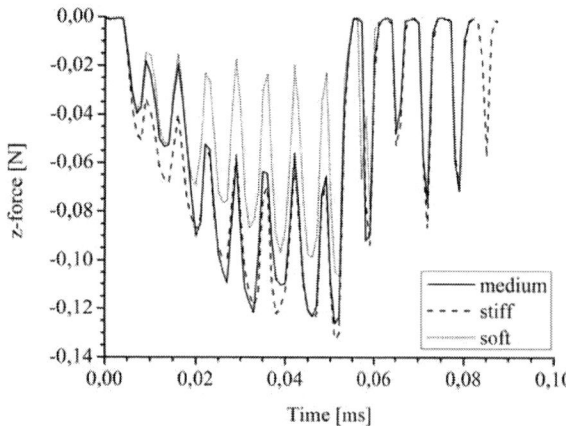

Figure 13: Time-dependent penetration force on the copper pad during bond process

Finally the penetration of the die adhesive is evaluated at the end of the compressive stage. This criterion shows the distribution of the local penetration over the adhesive area. Figures 10 – 12 show that the highest compression of the adhesive appears close to the bond pad. A reasonable compression of the die adhesive can be found up to one third of the chip length. As expected the highest compression can be found for the soft die adhesive. The maximum penetration is 1.9 μm which is almost 4 % of the initial height. The penetration depth reduces with increasing Young's modulus to 1.6 μm. The figures also show that the soft die adhesive has a lower ability to distribute the penetration over the gluing area. Noteworthy penetration can be found up to one quarter of the die length in a circular area around the pad. In case of the stiffer die adhesives the penetration is spread up to one third of the die length. The shape of the penetration is rectangular. The deeper and more localized adhesive penetration is also evident in the silicon which suggests higher stresses for the combination with the soft adhesive. But figure 7 proves that the mechanical resistance of the soft die adhesive is too low in order to create any stress.

As mentioned before the wire bond process is uncritical for the silicon even if it is die-bonded with a very soft adhesive.

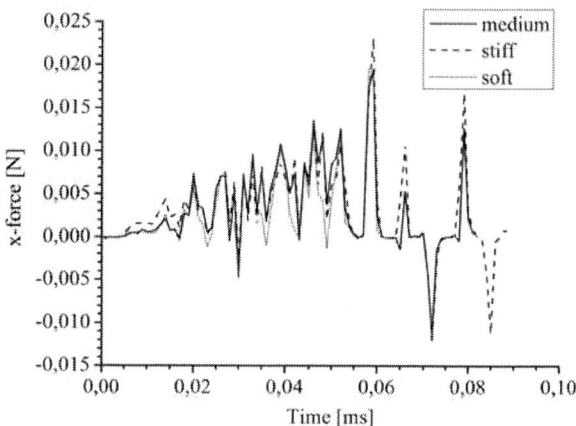

Figure 14: Time-dependent frictional force between wire and bond pad

The results discussed so far represent only a short moment within the entire wire bond process. The transient behavior of stresses is important for the assessment of the contact formation. This is done best at the interface between bond pad and aluminum wire. The simulation software provides different force signals at the according contact. A recalculation of these forces to stress values is omitted here due to the changing area of the compressed bond wire. The presented absolute force values are not representative because they are based on assumed friction values within the contact.

Figure 13 shows the development of the compressive z-force over the bond process. During the vertical displacement of the bond capillary the z-force increases in an oscillating way which is caused by the high displacement speed. The highest forces appear at the end of vertical displacement stage. Afterwards the z-forces decline within the ultrasonic vibration. In this stage the z-forces still appear in a vibrating shape with the peak forces at the turning points of the bond capillary. In general the calculated compression forces are in a reliable range [9]. The figure shows higher compressive forces for the assemblies with stiffer die adhesives. The maximum force peak appears for the stiff adhesive. The compressive force in the soft adhesive assembly is permanently lower. Its force peak is 15 % lower than for the medium adhesive and 20 % lower compared to the stiff adhesive model. These results confirm the previous findings at the end of the vertical capillary displacement.

The micro welding contact between the aluminum wire and the bond pad requires a sufficient energy in order to be created. This energy is created by the friction between both materials during the ultrasonic vibration. Thus a low frictional force may prevent the contact formation. That is why the force along the ultrasonic capillary displacement is evaluated, too. Figure 14 shows the resulting x-forces within the bond process. During the vertical displacement stage there are minor force peaks. Within the ultrasonic vibration the force peaks increase significantly. These force peaks appear short before the wire slides over the pad. Since there is no energy or force criterion which defines the sticking of the wire these force peaks appear periodically with the capillary movement. Again the force peaks increase with the stiffer die adhesive. The maximum x-force of the medium die adhesive assembly is 10 % higher and in case of the stiff die adhesive the maximum is 20 % higher compared to the soft assembly, respectively. These results cannot be rated exactly due to missing experimental data but it should be more critical.

4. Conclusions

This FEM analysis investigates the influence of different die adhesives with varying Young's moduli on the formation of a wedge bond contact. Sensor chip manufacturers require softer die adhesives in order to decouple the influence of the package stiffness on the sensing elements. However this may cause trouble for the formation of electrical interconnections to the sensor chips.

This transient FEM analyses cover the compression stage of the wire bond as well as the ultrasonic vibration for the micro welding. The simulation results show big differences between the resulting stresses in the bond pad and silicon chip according to the applied die adhesive stiffness. The difference in the maximum stress of the bond pad is up to 70 %. Similar remarkable differences can be found at the contact forces acting between aluminum wire and bond pad. At this interface the force peaks differ by 10 to 20 %. This difference may avoid the formation of a micro weld contact in case of critical process conditions. The simulation results prove the influence of the die bond adhesive on the electrical contact formation. Very soft adhesives may be unsuited for the entire packaging of sensor chips. The wedge bond process may have a small reliable process window.

References

1. T. Bechtold, et.al. "Enabling Technologies for System-Level Simulations of MEMS", Proceedings of the 14th International Conference on Thermal, Mechanical and Multi-Physics Simulation and Experiments in Microelectronics and Microsystems (EuroSimE), Wroclaw, Poland, April 15-17, pp. 1-6, 2013.

2. M. Zanaty, et.al. "Influence of nonlinear intermolecular forces on the harmonic behavior of NEM resonators", Proceedings of 14th EuroSimE, Wroclaw, Poland, April 15-17, pp. 1-7, 2013.

3. S.S. Mulay, et.al. "The fracture studies of polycrystalline silicon based MEMS", Proceedings of 14th EuroSimE, Wroclaw, Poland, April 15-17, pp. 1-9, 2013.

4. A. Ghisi, et.al. "A Multi-Scale Approach to Wafer to Wafer Metallic Bonding in MEMS", Proceedings of 14th EuroSimE, Wroclaw, Poland, April 15-17, pp. 1-6, 2013.

5. M. Mayer, et.al. "Ultrasonic Stresses in Thermosonic Ball Bonding", Proceedings of 12th EuroSimE, Linz, Austria, April 17-19, pp. 1-5, 2011.

6. Y. Ding, et.al. "Numerical analysis of ultrasonic wire bonding; Effects of bonding parameters on contact pressure and frictional energy", in Journal Mechanics and Materials, Vol. 38, pp. 11-24, 2006.

7. Livermore Software Technology Corporation "LS-DYNA Keyword User's Manual", May 2007, Version 971, www.lstc.com.

8. R. Meier, et.al. "Thermo-Mechanical Behaviour of Copper-Ribbon Materials", Proceedings of the 24th European PV Solar Energy Conference and Exhibition (EU PVSEC), Hamburg, Germany, Sept. 21-25, pp. 1-7, 2009.

9. A. Wright, et.al. "On the Thermo-Mechanical Modelling of a Ball Bonding Process with Ultrasonic Softening", Proceedings of 14th EuroSimE, Wroclaw, Poland, April 15-17, pp. 1-8, 2013.

Mechanical Stress Analysis in Photovoltaic Cells during the String-Ribbon Interconnection Process

F. Kraemer[1], J. Seib[2], E. Peter[2], S. Wiese[1]

[1]Saarland University, Chair of Microintegration and Reliability, Saarbrucken, Germany
[2]Robert Bosch GmbH, Corporate Sector Research (CR/APJ3), Stuttgart, Germany
E-mail: f.kraemer@mx.uni-saarland.de

Abstract

The paper analyzes the mechanical problems of interconnecting individual solar cells in order to create a photovoltaic module. Modern modules increase their produced electrical power steadily at a constant low voltage, which causes high currents through the interconnecting copper wires, also called copper ribbons. The resistance of the interconnections is crucial, because it has a significant influence on the total module efficiency. However, an increased cross-section of the copper ribbons leads to severe mechanical problems, because the thin silicon solar cells would tend to break more easily.

In this study the stresses created during the cell interconnection process are analyzed by 3-D FEM-simulations. These simulations are done by applying the commercial code ANSYS. The geometrical model consists of two adjacent cell quarters which are interconnected by one and a half copper ribbons. The geometrical model has a very fine mesh in critical cell sections in order to enable a result evaluation by path plots. The mechanical load is created by a temperature reduction from the solidification temperature of the lead-containing solder to room temperature.

The result evaluation by path plots highlights those cell sections, which are the most critical in the productive tabber-stringer process. These results cannot be found by contour plots since the path plots are able to identify positions with high stress gradients. Due to the dominating compressive load of the silicon after the cooling step it is difficult to find possible crack positions. Applying the path plots big stress difference can be found in very small sections which correlate well with experimentally observed failure positions. Now it is possible to understand the complex nature of failure formation in the silicon solar cells applying this result evaluation method.

1. Introduction

The reliability and the efficiency of photovoltaic modules are affected by distinct critical processes in the production. One of these critical process steps is the interconnection of single solar cells to cell strings, called tabber-stringer process. In this step copper ribbons are soldered simultaneously on the top and bottom side of a solar cell in order to create a serial connection between several cells. This process creates high mechanical stress in the silicon solar cells and the developments in the photovoltaic production will even increase this problem.

The target of cost-reductions of PV modules results in a steady thickness reduction of the silicon. At the same time there is the tendency to increase the cross-section of the copper ribbons in order to reduce the ohmic losses and compensate the higher cell efficiencies, respectively. Furthermore the photovoltaic industry has to change to real green products as the solder interconnection is still made by lead-containing solders. With these challenging developments it is necessary to understand the critical tabber-stringer process in order to be able to produce reliable products with a high yield. This paper is focused on the development of a deeper understanding of the mechanical stresses generated in this production step.

2. FEM simulations of the tabber-stringer process

The simulation results of previous investigations [1,2] have a drawback. The contour plots show high stress positions close to those areas where silicon cracks are observed in the production of PV cell strings. But the illustrated critical stress values have a compressive nature, which can be hardly correlated to the experimental failures. Based on those figures, it is difficult to forecast and understand those events.

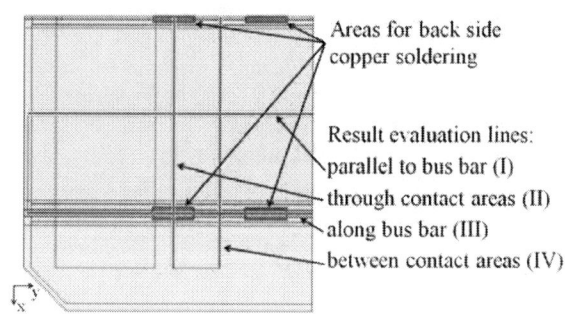

Figure 1: Area division in a quarter of a photovoltaic cell.

In order to achieve a better understanding of the mechanical problems which are created during the string-ribbon interconnection process, simulation results are evaluated in a more detailed way. Line or path plots are generated along interesting model parts additional to the stress and strain contour plots. Path plots extract results from all elements along a predefined path. In order to achieve reproducible and comparable results the FE model is modified in such a way, that a defined number of elements is created along the designated evaluation path.

This way a more difficult geometrical model has to be prepared for the successive meshing step, see figure 1. The figure shows four additional lines on the surface of a quarter of the solar cell. The line numbers II and III are of special interest since they are located in the critical sections of the cell.

The problems which arise in the tabber-stringer process can be explained best by the cross-section through the cell interconnection area shown in figure 2. The 165 µm thin solar cells are soldered with copper ribbons on top and bottom of the cell within a single step. Each copper ribbon has a thickness of 200 µm. The copper is soldered with lead-containing solder to the contact areas of the silicon, called bus bars. The solder layers have a thickness of just 10 µm. Due to the CTE mismatch the copper wants to compress the silicon on both sides, while the silicon keeps length of the copper ribbons almost constant. The mechanical stress can be hardly absorbed by the thin solder layers. That is why cell breakage is a typical failure after the tabber-stringer process.

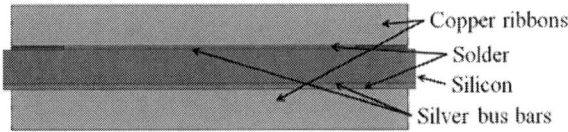

Figure 2: Cross-section through the interconnection area of a solar cell including the copper ribbons on top and bottom side.

Previous investigations and experimental observations [2-5] have shown that the bus bar areas are the most critical section of solar cells. This fact in combination with the path plot generation is the reason for a high mesh density in these areas. As mentioned before this mesh density is kept constant along all result evaluation lines. This model arrangement results in the FE net shown in figure 3. The complete model consists of two adjacent quarter cells and the copper ribbons in between them. The reason for the creation of both cells is to include possible interactions between both cells and the copper ribbon. The complete 3-D model consists of 218.000 hexahedral elements.

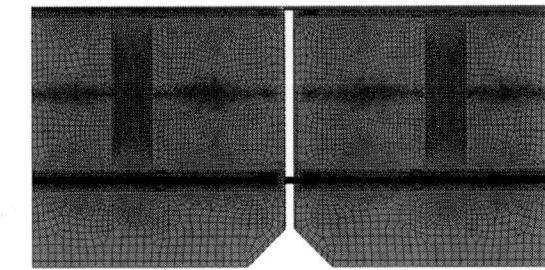

Figure 3: Geometrical model of the adjacent solar cell quarters and the interconnecting copper ribbons.

The basic load of the geometrical model is the cooling from the solidification temperature of the lead-containing solder at 183°C to the room temperature of 20°C. In modern tabber-stringer machines the layup of the copper ribbons and the peak heating of this assembly takes about 6 s. After that period the machines shift over to the next cell and repeat the soldering process. Meanwhile the already soldered cell strings are pushed to heating plates, which prevent the cells from a fast cooling. However, the resulting cooling profile is not very repeatable. Thus, an expected averaged cooling profile is taken for this investigation. This time-temperature-profile is shown in figure 4. The figure shows that the total cooling process is expected to take 60 s. In a further step the effect of an extended relaxation time on the stress distribution in the assembly is investigated by a dwell period of 72 h at room temperature. The results of this investigation will be discussed in the second part of the result evaluation.

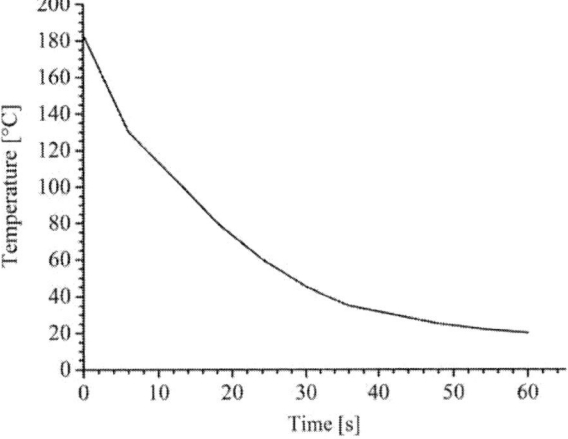

Figure 4: Cooling profile of solar cell strings after soldering of the copper ribbon. Time-temperature steps were taken as load of the FE model.

The limited number of material models required for this investigation is listed in table 1. The mechanical behavior of the solar cell is represented by the isotropic properties of multi-crystalline silicon. The more complex material models of copper and solder are based on own measurements, which were presented in [6] and [7], respectively. The mechanical behaviors of the aluminum paste and silver paste are assumed by the combination of glass, which is a main component of these pastes and the individual metal behaviors. Both materials are burned as thick-film pastes onto the solar cell in order to enable the conduction of the produced electricity. While the application of the silver paste is limited to the bus bar areas for the copper soldering, the aluminum paste covers the whole remaining back side of the PV cell. The measurement of their mechanical properties is difficult since the final behavior appears only in combination with silicon. However, their contribution to the final stress values in the silicon should be small since the pastes are expected to have little mechanical influence and they are just few micro-meters thin.

Table 1: Material models applied in FE-analysis and characteristic values at room temperature.

Material	Young's Modulus [GPa]	CTE [ppm/K]	Yield Stress [MPa]	Material model
Aluminum paste	6.0	11.9	28.3	Bi-linear temperature dep. elastic-plastic
Copper (initial)	85.7	17.0	95.1	Bi-linear temperature dep. elastic-plastic with isotropic hardening
Silver paste	7.0	9.8	64.0	Bi-linear temperature dep. elastic-plastic
Silicon	130.0	2.6	-	elastic
Solder (SnPb)	32.0	25.2	48.0	Temperature dep. Young's Modulus & Yield stress; creep

The geometry model is held down on top and bottom of the soldered cell-copper-stack-up. This assumption represents the fixtures of the soldering rods in the tabber-stringer machine. Symmetry conditions are introduced at the inner surfaces of the cell quarters.

3. FEM stress analysis after cooling

The initial analyses of the simulation results apply contour plots in order to get an overview of the deformations and stress distribution. Figure 5 shows the out-of-plane deformations of the model. Due to the boundary conditions there is little deformation between the outer and the central bus bar. The biggest deformation appears at the middle of the free cell edges. This deformation is caused by the CTE-mismatch of the aluminum paste and the silicon. Since the aluminum paste contracts faster than the silicon, the setup bends downwards. The amount of deformation of about 0.4 mm is in a reasonable range.

Figure 6 shows a contour plot of the 3rd principal stress in the silicon cell at the end of the cooling process. Due to the faster thermal shrinkage of the copper the dominating load of the silicon is contraction. The evaluation of the principal stresses proves this observation. While for the 1st principal stress values are calculated which are uncritical for silicon, high values are calculated for the 3rd principal stress. The contour plot

supports the assumption that high stresses appear at the interface between silicon and copper since both have a high Young's modulus and similar geometrical dimensions. Thus notable stress can be found along the bus bars, while the remaining solar cell is almost stress-free. The biggest compressive stress appears when the copper ribbons are connected to the silicon on top and bottom. This condition is met only at distinct areas, see figure 1. However, experimental silicon cracks are rarely initiated next to these back side contact areas. More failures appear at the beginning of the front side bus bar.

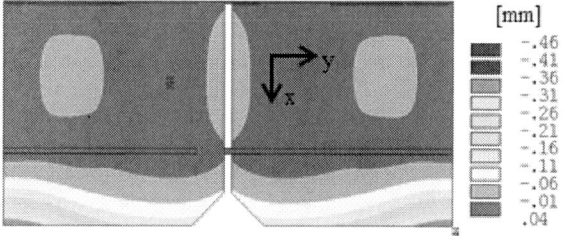

Figure 5: Contour plot of the out-of-plane deformation of the solar cell assembly after cooling to room temperature.

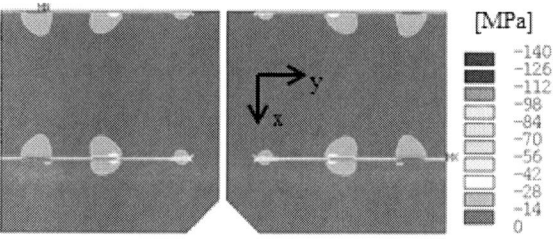

Figure 6: Contour plot of the 3rd principal stress in the silicon cells after cooling. Areas beneath the back side solder regions show highest stress.

Path plots of the lines parallel (I) and along the bus bar (III), see figure 1, are shown in figure 7. The figure shows the component of the x-stress along both paths, which is acting perpendicular to the direction of the paths. As discussed for the contour plot in figure 6, high stress appears along the bus bar, while little stress can be found in the remaining PV cell. Due to the high stress values along the bus bar, results are analyzed at the top and bottom side of PV cell. These results show a stress gradient through the cell thickness. The top side is stressed more than the bottom side. The reason for this effect is the complete connection of the copper ribbon to the continuous front bus bar along the cell length. The values of the x-stress path plots meet the values of the contour plot in figure 6. Thus, the orientation of the 3rd principal stress is along the x-direction at this section of the solar cell. However the path plot indicates that the areas around the back side copper contacts are the most

978-1-4799-4789-8/14 $31.00 © 2014 IEEE   526

critical, which does not meet the experimental findings. This criterion is not suited for a crack prognosis.

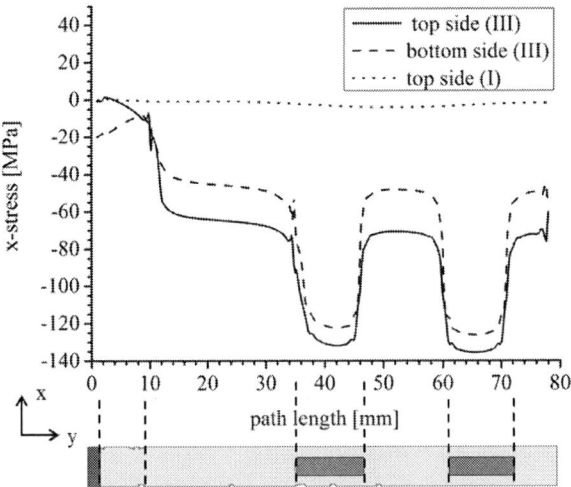

Figure 7: Path plot of the x-stress on the top and bottom silicon surface along the bus bar line, see fig. 1. There is high compressive stress as long as the copper is connected to the front side. Values even increase at areas of back side copper connection.

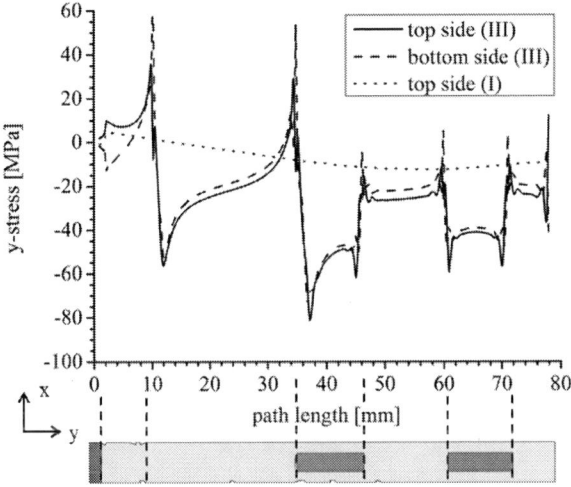

Figure 8: Path plot of the y-stress on the top and bottom silicon surface along the bus bar line. High stress gradients appear at the edge of contact areas.

Figure 8 shows the y-stress component along the result evaluation paths (I) and (III). Again there is almost no stress at the path parallel to bus bar (I). The stress increases slightly towards the middle of the cell but this value is uncritical. Beneath the bus bar there is notable stress. The stress values do not reach the values of the component in x-direction. But this result criterion shows some positive stress peaks. The comparison between the stress peaks and their position on the cell indicates that

these peaks appear at every change of the geometrical profile. These peaks are also very sharp resulting in high stress gradients. The biggest differences appear at about 8 mm and 35 mm, respectively. At 8 mm there is the beginning of the front side bus bar and thus this is where the connection between copper ribbon and silicon starts. At 35 mm there is the edge of the first back side contact area. So this is where the silicon is subjected to the mechanical load of both copper ribbons. At these positions the total stress difference reaches about 110 MPa to 130 MPa on a length of 2 mm. The higher stress gradients appear on the bottom side of the cell. Due to the big stress differences at both positions a certain crack risk has to be estimated. A final judgment seems to be impossible with these data. A further analysis applying linear elastic fracture mechanics should give more insight to the actual crack risk in this assembly, but this is not part of this investigation.

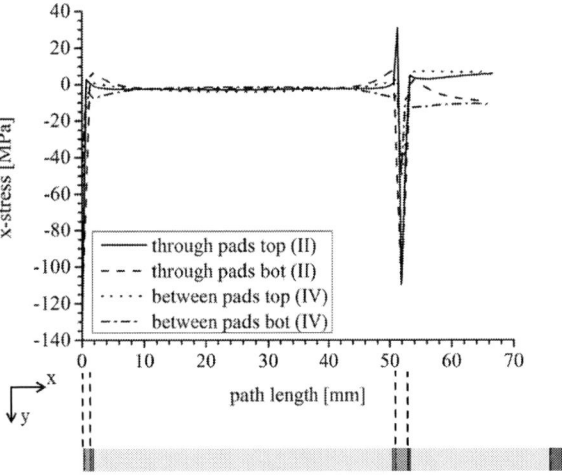

Figure 9: Path plot of the x-stress on top of the silicon cell through the contact areas, see figure 1. Stress appears only at the bus bars.

Further result evaluation is done at the remaining lines called through contact areas (II) and between contact areas (IV), see figure 1. The corresponding x-stress components are depicted in figure 9. This figure proves the observations of the path plots in figures 7 and 8. Little stress can be found between the bus bars, since the silicon is only interacting with the thin and soft aluminum back side paste. High stress values appear beneath both bus bars. As shown before the maximum stress value depends on the position along the bus bar. Thus higher stress peaks can be found on the path through contact areas (II). Both paths show a small influence of the high stressed bus bar areas on the surrounding silicon. This effect vanishes within few millimeters. Furthermore it becomes obvious that the resulting stress gradients are smaller than what was found in figure 8. This result corresponds to the experimental observation that silicon cracks do not appear at any position along the bus bars.

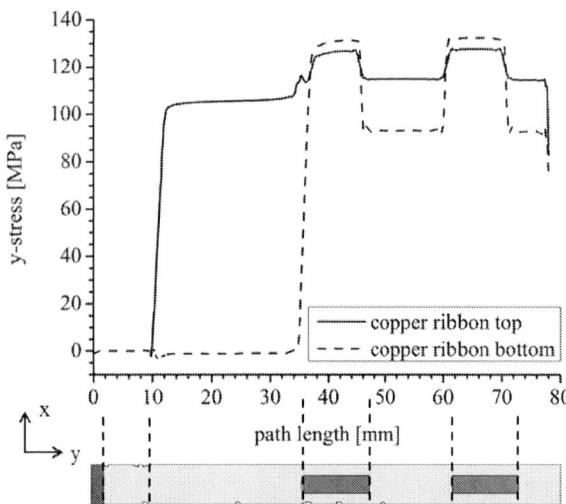

Figure 10: Path plot of the tensile stress in the copper ribbons on top and bottom of a solar cell.

Figure 10 shows path plot of the tensile stress on the surfaces of the top side and bottom side copper ribbons. Stress appears only as long as the copper is connected to the PV cell. Thus the top side copper ribbon is stress-free up to 8 mm and the bottom side copper ribbon is unloaded up to the edge of the first back contact area at 35 mm. After this the copper stress reaches the level of plastic deformation. These plastic deformations reduce the stress occurring in the silicon. Therefore copper ribbons with reduced yield stress are preferred. Basically, a high tensile stress is generated in the copper at this single tabber-stringer process step but the ribbons are not supposed to fail.

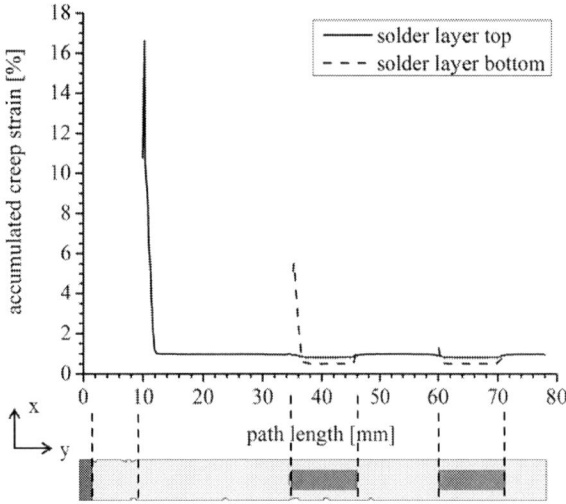

Figure 11: Path plots of the accumulated creep strain in the solder layers on top and bottom of the PV cell.

The path plots of the accumulated creep strain in the solder layers on the top side bus bar and in the back side contact areas are shown in figure 11. Very high creep strain is calculated at the edge of the front side bus bar. This may be caused by the big strain difference between silicon and copper that is suddenly bonded to each other at this position. The similar but much smaller effect can be found at the edge of the outermost back contact area. Short after this the local creep strain values decline to about 1 % on the top side and about 0.5 % in the back contact areas. Both results are far below any failure strain. Even the high creep strains at the edges of the interconnection zones are no concern. The strong localization of this effect may lead to local solder cracks but there is a long remaining length of the solder layer which is able to ensure the electrical connection between the solar cell and the copper ribbon.

The result evaluation presented so far has identified a risk of cracks in the brittle silicon cell. Different changes in the assembly, such as smaller copper ribbons or thicker solder layers, are suited to reduce the initial stresses right after the cooling step. However, this would increase the price of a solar module. That is why the effect of a defined relaxation is investigated here in order to reduce failures in following solar module production steps.

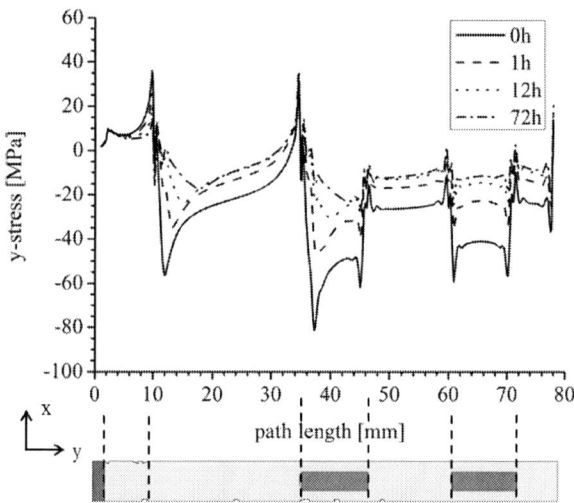

Figure 12: Path plot of the x-stress on top of the silicon solar cell at the end of the cooling process and after several hours of relaxation.

The stress relaxation investigated here applies a dwell time of up to 72 h at 20°C. The relaxation effect on the y-stress component along the bus bar path (III) is shown in figure 12. The y-stress seemed to be suited to predict silicon cracks at the edge of the front side bus bar. The figure shows a clear relaxation of the stress within the initial 12 h. Even within the first hour there is a significant reduction in stress of 30 to 50 %. The path plot indicates that this result is achieved by a better distribution of the stress. The result shows that a dwell time may help to reduce the risk of silicon cracks during and after the tabber-stringer process.

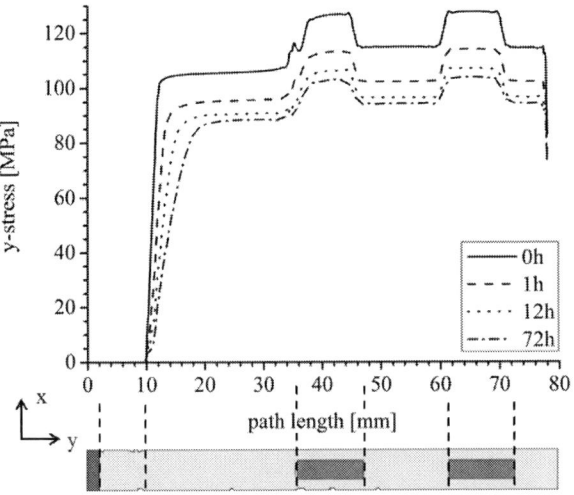

Figure 13: Path plots of the tensile stress in the top side copper ribbon at the end of the cooling process and after several hours of relaxation.

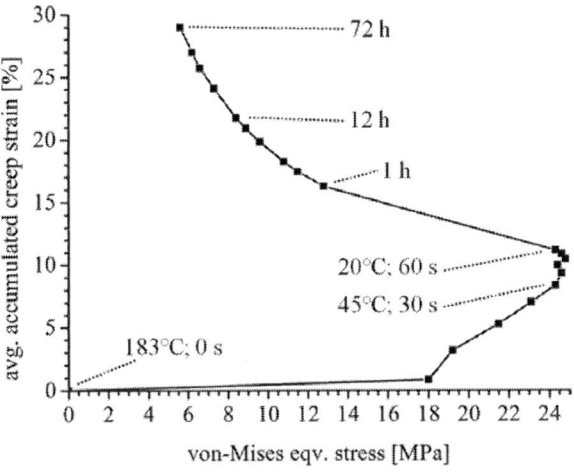

Figure 14: Averaged accumulated creep strain and von-Mises equivalent stress in the solder during cooling process and after 3 days of relaxation.

Figure 13 shows that the relaxation time also affects the stress in the copper ribbons. The figure depicts the path plot of the tensile stress only in a top side copper ribbon. As mentioned before the biggest effect appears in the first hour. After 12 h the further relaxation is almost negligible. In case of the copper the stress relaxation appears to be smaller than for the silicon. This is the case for the stress peaks since they are not as sharp as in the silicon. In the plateaus the stress relaxation is similar. Again the path plot highlights the better distribution of stress after the relaxation period.

The stress relaxation shown in the figures 12 and 13 has to be generated somewhere. This is done by the solder. Figure 14 shows the accumulated creep strain. The values presented in this graph are the averaged values of the solder layer above the top side bus bar tip. The graph supports the findings of the previous observations. It can be seen that the relaxation already starts at the end of the cooling phase. Due to small temperature reduction at the end of the cooling phase, see figure 4, the solder is able to compensate the expected stress increase. After the cooling there is an initial fast relaxation of stress which reduces over time. The creep-rate also reduces with the smaller stress. The averaged accumulated creep strain presented here is relatively high. Based on these simulation results it can be expected that first cracks may appear at the very edge of the solder layer. But this effect is hard to recognize in the regular production since there are still more than 130 mm of intact solder connection.

4. Conclusions

The paper analyzes the mechanical problems, which are created during the interconnection of individual solar cells in order to build a photovoltaic module. This tabber-stringer process is crucial because it is able to cause obvious cracks in the silicon solar cells and it initiates cracks, respectively, which cause reliability problems and power loss in the regular service.

The FE simulation results show high compressive stress in the silicon due to the higher coefficient of thermal expansion of the copper ribbons. These stresses are focused on the areas below the copper interconnections. Such results have already been known before but they do hardly explain the creation of cracks in the silicon. Applying path plots the previous results can be confirmed. Furthermore this result evaluation technique highlights the formation of stress gradients in the silicon cell. These gradients appear along the silicon thickness and at sharp geometrical edges of the assembly. The biggest gradients can be found at the edge of the front side bus bar and at the edge of the outermost back contact area. The edge of the front side bus bar is an interesting position because experimental cracks are often found here.

Furthermore the effect of an extended stress relaxation time is analyzed here. A dwell time of up to 72 h shows a significant reduction of the stress peaks of up to 50 %. Thereby the biggest stress reduction takes place within the first hour. The path plots show a wider distribution of stress after the relaxation period. According to the stress values failures should be prevented after the relaxation.

References

1. Wiese, S.; Kraemer, F.; Betzl, N.; Wald, D.: „Interconnection Technologies for Photovoltaic Modules – Analysis of technological and mechanical Problems". In: Proceedings of 11th International Conference on Thermal, Mechanical and Multi-Physics Simulation and Experiments in Microelectronics and Microsystems, 2010. p. 1 – 6.

978-1-4799-4789-8/14 $31.00 © 2014 IEEE

2. Kraemer, F.; Wiese, S.; Peter, E.; Seib, J.: „Analysis of Mechanical Problems of String-Ribbon Interconnections in Photovoltaic Modules". In: Proceedings of 27th European Photovoltaic Solar Energy Conference and Exhibition, 2012. p. 1162 – 1164.

3. Koentges, M.; Kajari-Schroeder, S.; Kunze, I.; Jahn, U.: „Crack Statistic of Crystalline Silicon Photovoltaic Modules". In: Proceedings of 26th European Photovoltaic Solar Energy Conference and Exhibition, 2011. p. 3290 – 3294.

4. Pingel, S.; Zemen, Y.; Frank, O.; Geipel, T.; Berghold, J.: „Mechanical Stability of Solar Cells within Solar Panels". In: Proceedings of 24th European Photovoltaic Solar Energy Conference and Exhibition, 2009. p. 3459 – 3463.

5. Dietrich, S.; Pander, M.; Ebert, M.: „Mechanical Challenges of PV-Modules and its Embedded Cells – Experiment and Finite Element Analysis". In: Proceedings of 24th European Photovoltaic Solar Energy Conference and Exhibition, 2009. p. 3427 – 3431.

6. Meier, R.; Kraemer, F.; Wiese, S.; Wolter, K.-J.; Bagdahn, J.: „Reliability of copper-ribbons in photovoltaic modules under thermo-mechanical loading". In: Conference Record of the IEEE Photovoltaic Specialists Conference, Art. no. 5614220; 2010. p. 1283-1288.

7. Wiese, S.; Feustel, F.; Meusel, E.: „Characterisation of constitutive behaviour of SnAg, SnAgCu and SnPb solder in flip chip joints". Sens Actuators A 2002; 99:188-93.

Analytical stress model for tin based solder material

M. Guyenot, A. Fix

ROBERT BOSCH GmbH, Schwieberdingen, Germany

Michael.Guyenot@de.bosch.com, Andreas.Fix@de.bosch.com

Abstract

In recent years the activity to predict lifetime focuses more and more on finite element analysis (FEA) using up growing computer power. The description of the material properties especially viscoplastic behavior have to be well known to bring FEA calculations in a good agreement with experimental results concerning solder joints [1, 2]. Typically, creeping of solder materials is specified by extensive mathematics modeling. New research activities concentrate on viscoplastic modeling according to Chaboche [3, 4].

This paper shows a new methodology for lifetime prediction based on stress measurements of tin solder specimens performed by Fraunhofer Society IKTS [5]. Characteristic hysteresis curves of the stress-strain behavior of tin based solder alloy were measured at different temperatures to enable the calculation of elastic and plastic elongation depending on time and temperature. This new approach gets by on complex mathematics equations and FEA. A stress based model combined with lifetime facts offers a rapid lifetime prediction of solder materials. Transferability to solder joints is shown by means of the component CR1206 and its reliability due to thermal cycling experiments. The adjusted analytic calculated results are evaluated by the experimental thermal cycling results and will be transformed to a calculation model for thermal cycle test procedure.

Introduction

Lifetime prediction of solder joints either use external load conditions (testing parameter) or strain respectively stress. Typical testing parameters for automotive applications are temperature differences ΔT, temperature gradients, dwell times t_h, peak temperatures T_{peak} and mean temperatures T_{mean}. The materials responses on these tests are in either case time dependent strains $\varepsilon(t,T)$ and stresses $\sigma(t,T)$. These values cannot be determined in the majority of cases by using analytical models. Therefore, FEA will be applied to get representative results for an experimental setup. One of widely-used failure criteria from FEA are accumulated creep strain $\Delta\varepsilon_{cr,acc}$ and the creep energy density $\Delta W_{cr,acc}$

Both testing parameters and materials responses slip into lifetime modeling. Coffin-Manson relations [7] describe the Wöhler (S/N) curve for low cycle fatigue as it can be seen in equation (1). Plastic strain amplitude $\Delta\varepsilon_{pl}$ correlates with the mean time to failure (MTTF) of a solder joint. $MTTF$ can be understood as the number of cycles to failure.

$$\frac{\Delta\varepsilon_{pl}}{2} = \varepsilon'_f \cdot (MTTF)^c \text{ or } MTTF = \left(\frac{\Delta\varepsilon_{pl}}{2\cdot\varepsilon'_f}\right)^{\frac{1}{c}} \quad (1)$$

with ε'_f = fatigue coefficient, c = Coffin-Manson factor

This correlation was adjusted to predict lifetime for solder joints in the past. Coffin-Manson like equations were used to estimate the end of life at a given time for variable damage parameters. Equation (2) shows the relationship between $MTTF$ and typical damage parameters.

$$MTTF = (\Delta Z)^{\frac{1}{c}} \text{ with } Z = \Delta T, \Delta\varepsilon_{cr,acc}, \Delta W_{cr,acc} \quad (2)$$

An extended equation [equ. 2] using thermal swings as well as dwell time and the thermal impact through by activation temperature was developed by Norris Landzberg / Engelmaier [8, 9].

$$MTTF = (\Delta T)^{k_1} \cdot (t_h)^{k_2} \cdot e^{\frac{k_3}{T_{peak}}} \quad (3)$$

with k_1, k_2 and k_3 as specific coefficients

A big disadvantage of the Z-Values from equation (2) is the necessity to get data from FEA or to use testing parameters. On the one hand FEA need time for modeling the geometry, for defining boundary conditions and

978-1-4799-4789-8/14 $31.00 © 2014 IEEE

for determining load conditions. On the other hand information about the strength of solder joints will be lost if testing parameters are only used. Therefore, the challenge is to derive the strength of a solder joint without using FEA. A strength based model was developed using experimental data from creep tests. For solder materials especially solder joints, creep strain is used to describe materials behaviour during test conditions.

The new approach of the present work focuses on mechanical stress as a damage parameter for solder material. Plastic stress amplitude (load range) is used to describe lifetime of solder joints in a S/N methodology. The stress will be separated in elastic (time independent) and plastic (creep) part for modelling. Therefore, it is necessary to know stress σ, strain ε and Young´s modules E of the solder material as a function of time and temperature. Figure 1 shows relationship between elastic and plastic stress and the S/N methodology and equation (4) describes the mathematical relationship.

$$\sigma = E \cdot \varepsilon \qquad (4)$$
$$\sigma = E \cdot (\varepsilon_{elast} + \varepsilon_{plast}) = E \cdot \varepsilon_{elast} + E \cdot \varepsilon_{plast}$$
$$\sigma = \sigma_{elast} + \sigma_{plast}$$

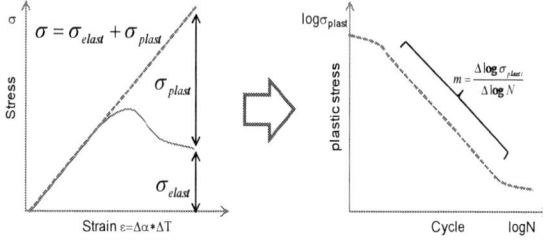

Figure 1: Schematic stress strain curve for linear elongation with separated plastic and elastic stress component (left), S/N curve for acceleration (right)

Experimental setting and methodology

The database for the present work bases upon collaboration between Fraunhofer Society IKTS and Robert Bosch GmbH. Metasch already presented in the past the advanced experimental procedure for determining the properties of a tin based solder alloy under cyclic load and isothermal conditions [5]. Hence, stress-diagrams were generated using this experimental setup. The applied strain rate at every strain stage has been varied in the range of 10^{-3} to 10^{-6} per second. At the end of every strain stage a time-limited relaxation experiment is performed, in which the specimen's length is kept constant, while the stress evolution is recorded. The observed cyclic stress-strain hysteresis was measured for different temperatures. An example for presented experimental data is shown in figure 2. The test procedure was the following: Each specimen was homogenized by cycling load and then elongated 1% for a comparable microstructure between the different samples. Then, the elongation of 1% was hold and the stress relaxation was determined as a function of time. The specimens were tested at different temperatures (25°C, 75°C, 125°C and 150°C).

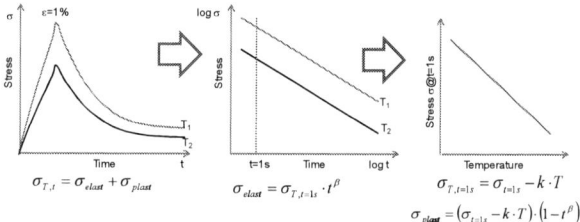

Figure 2: Systematic for relaxation fitting based measurement (left curve)

A double logarithm plot of the relaxation data shows linear dependency and could be fitted as a function of time and inertial stress $\sigma_{T,t=1s}$, which can be seen in figure 2 (middle). These initial stress $\sigma_{T,t=1s}$ could be understood as the stress level after one second of load for each temperature level. This inertial stress exhibits also a dependency from temperature. Figure 2 shows a linear relation between initial stress and temperature. Hysteresis curves for the bulk material was used to get temperature dependency of Young´s modules. The slope of the Hook's line was plotted for each temperature. The diagram reveals a linear temperature dependence which can be seen in Figure 3. That way it is possible to calculate occurant stress as a function of temperature and time analytically. For low cycle fatigue only plastic deformation is responsible for damaging and cracking. Therefore, plastic stress should be used for calculation stress level and has to combined with the results from thermal cycling experiments.

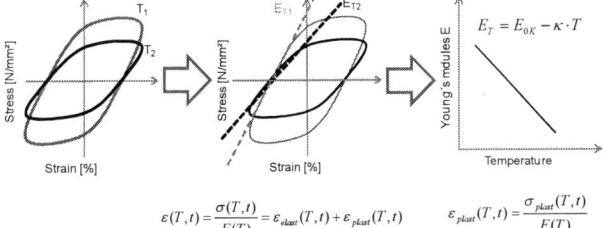

$$\varepsilon(T,t) = \frac{\sigma(T,t)}{E(T)} = \varepsilon_{elast}(T,t) + \varepsilon_{plast}(T,t) \qquad \varepsilon_{plast}(T,t) = \frac{\sigma_{plast}(T,t)}{E(T)}$$

Figure 3: Systematic for fitting Young`s modules based measurement (left curve)

Fitting methodology for material data

Starting with the measured stress-strain-curves it was possible to find out the relaxation coefficients. Figure 4 shows the test reading and double logarithm plot. It is possible to postulate that the relaxation is only time dependent and the stress level is temperature dependent. For this reason it is possible to calculate the elastic stress as a function of time and inertial temperature dependent stress level.

By using the data from figure 2 (middle) the best fit line could be described by the equation

$$\sigma_{elast} = \sigma_{T,t=1\sec} \cdot t^{\beta} = \sigma_{T,t=1\sec} \cdot t^{-0,11} \qquad (5)$$

with *t* = time, σ_{elast} =elastic stress rest, β = time exponent, $\sigma_{T,t=1s}$ = initial temperature depended stress one second after load relieving

The exponent β describes linear slope of relaxation and $\sigma_{T,t=1s}$ is the temperature depended stress factor. In the next step the temperature dependent stress $\sigma_{T,t=1s}$ for the solder material will be calculated for the time 1 second after load. This $\sigma_{T,t=1s}$ is the initial temperature depended stress one second after load relieving. Because soft solder materials show high creep rate, it is necessary to define a time for start point of stress parameter. Hence, it is also possible to compare different test velocities.

For determining the $\sigma_{T,t=1s}$, the data from fig. 4 was used. The inertial stress level $\sigma_{T,t=1s}$ was plotted as function of temperature. The plotted result in fig. 5 shows for $\sigma_{T,t=1s}$ a linear slope to the temperature. The extrapolation of this fit for $\sigma_{T,t=1s}$ cross the stress free level near by

melting temperature of the soft solder material at roughly 500 K.

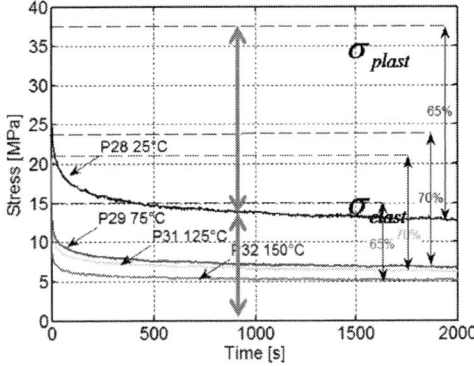

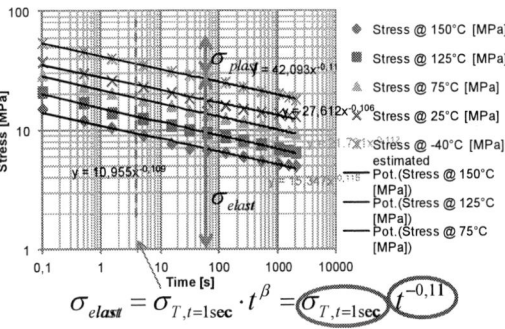

$$\sigma_{elast} = \sigma_{T,t=1\sec} \cdot t^{\beta} = \sigma_{T,t=1\sec} \cdot t^{-0,11}$$

Figure 4: Measured relaxation from tin based solder material for different temperature (top) and double logarithm plot of relaxation (down) including the best fit line with function

The temperature dependent stress one second behind the load shows a linear slope and can be described by following equation. The coefficient $\sigma_{T,t=1s}$ is the missing link from temperature to stress.

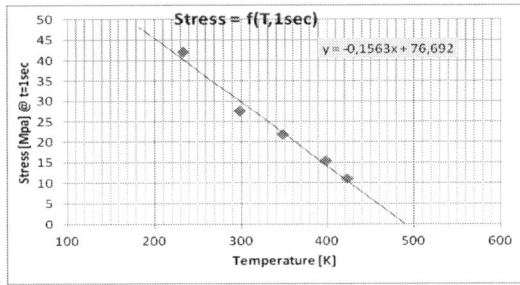

Figure 5: Temperature dependent stress one second after load

$$\sigma_{T,t=1s} = \sigma_{t=1s} - k \cdot T$$

$$\sigma_{T,t=1s} = 77\,MPa - 0{,}158\,\frac{MPa}{K} \cdot T \qquad (6)$$

with $\sigma_{t=1s}$ = material stress 0 K, T = temperature and k = specific temperature coefficient

978-1-4799-4789-8/14 $31.00 © 2014 IEEE

Through this relation (6) and equation (5) it is now possible to calculate the elastic stress according to time and temperature.

$$\sigma_{elast} = \sigma_{T,t=1sec} \cdot t^{\beta} \tag{7}$$

$$\sigma_{elast} = \left(\sigma_{t=1s} - k \cdot T\right) \cdot t^{-\beta} = \left(77\,MPa - 0{,}158\frac{MPa}{K} \cdot T\right) \cdot t^{-0{,}11}$$

Beside to the temperature dependences of stress also the Young´s modules are changing with temperature. For estimate the exact Young´s modules there are also material measurements at different temperatures used. The Fraunhofer Institute IKTS conducted the measurement. They used the same equipment and samples as used for figure 2. They measured the stress strain hysteresis at different temperatures. The measurement for one temperature is represented in figure 6. For these exemplarily temperature the stippled line illustrates the slope of stress strain. This slope is the Young´s modules and is speeded as function of temperature in figure 7. These Young's modules were fitted for the measured temperature and for -40°C estimated. The slope of Young´s modules depends on temperature indicated linear correlation between temperature and Young´s modules. These correlation could estimated by the following equation.

$$E_T = E_{0K} - \kappa \cdot T = 48.000\,MPa - 95{,}2 \cdot T \tag{8}$$

with E_{0K} = Young´s modules at zero degrees, κ = specific temperature coefficient and T = temperature.

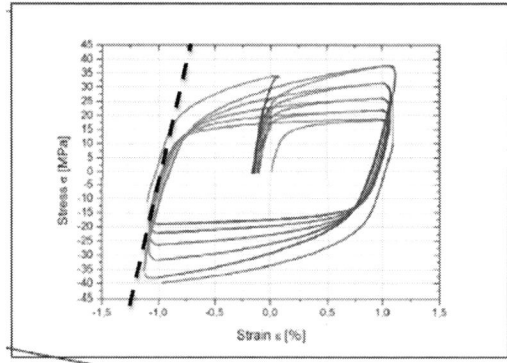

Figure 6: Measured hysteresis of stress strain for one temperature, stipples line Young`s modules

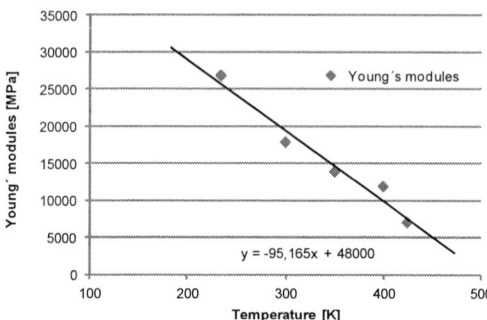

Figure 7: Young`s modules as function of temperature

At these step we have to summarise:

- Elastic and plastic stress of solder materials show a temperature and time dependence and could be described analytically

- Stress relaxation is only triggered by time

- Stress level is controlled by temperature

- Young´s modules has a linear temperature dependence

- An analytically calculation of stress through thermal cycling for solder joints could be possible

Analytical thermal cycle calculation

With these new material data it is now possible to calculate temperature dependent stress and relaxation for solder materials. For evaluation of these materials laws there are thermal cycle tests with ceramic component in size CR1206 in combination with FR4 printed circuit board (PCB) are calculated. Based on equations (4) the stress in solder material could be described by equation (9). In this calculation the elongation is substituted by the difference of thermal expansion coefficient ($\Delta\alpha$) and the temperature difference caused by thermal cycling (ΔT).

$$\sigma = \sigma_{elast} + \sigma_{plast}$$
$$= \varepsilon \cdot E_T = \Delta\alpha \cdot \Delta T \cdot E_T \tag{9}$$

Through these combination it is possible to bring mechanical and thermo-mechanical parameters into the stress model. For calculation the plastic stress the equation (9) has to be converted and combined with

equation   (5)
$$\sigma_{plast} = \sigma - \sigma_{elast}$$
$$= \Delta\alpha \cdot \Delta T \cdot E_T - \sigma_{T,t=1\sec} \cdot t^{\beta}$$
(10)

By integration of equation (8) into equation (10) the plastic stress can be described by

$$\sigma_{plast} = \Delta\alpha \cdot \Delta T \cdot (E_{0K} - \kappa \cdot T) - ((\sigma_{t=1s} - k \cdot T) \cdot t^{\beta})$$
(11)

In this calculation ΔT is ΔT_{test} less then stress free temperature gap $\Delta T_{\sigma=0}$. The effective thermal swing is responsible for thermal stress

$$\Delta T = \Delta T_{test} - \Delta T_{\sigma=0} \qquad (12)$$

The component CR1206 has a coefficient of thermal expansion $\alpha \sim 6$ ppm/K beside to a PCB with $\alpha \sim 16$ ppm/K. The difference of the thermal expansion coefficient between CR1206 and PCB is approximately $\Delta\alpha \sim 10$ ppm/K. It is giving by the thermo-mechanic of the component and system. The input from test procedure is the thermal swing ΔT, temperature T and the dwell time t. Through this setup parameter the stress and relaxation can be calculated by the following equation.

$$\sigma_{plast} = 10\frac{ppm}{K} \cdot \Delta T \cdot \left(48.000MPa - 95,2\frac{MPa}{K} \cdot T\right) - \left(77MPa - 0,158\frac{MPa}{K} \cdot T\right) \cdot t^{-0,11}$$
(13)

There were two different thermal cycle for the component CR1206 calculated with equation (13). The first thermal cycle was calculated for thermal shock in the temperature range from -40°C up to +85°C with each 30min dwell time and the second thermal shock for the temperature from -40°C up to +125°C. Figure 8 presented the temperature profile and the calculated stress. In the comparison of this thermal cycle there are a few interesting results to discuss:

In first cooling from 20°C to -40°C both diagrams show pressure up to -15MPa and little relaxation during dwell time. For the first heating up to 85°C the material get stress up to 6 MPa. Because the stress level is too low there is only a low relaxation in material. Beside thermal swing up to +125°C there are a higher stress level and more relaxation to observe.

The cooling from maximum temperature down to -40°C makes significant pressure. There are up to -27 MPa estimated for the

85°C thermal cycle version and up to -32 MPa for 125°C version estimated. At minus temperature both stresses show a high relaxation. Because the starting point of the stress level is different and the temperature swings for both thermal cycles are different and there are two stress level reached. The calculation shows a stress free temperature nearby 85°C. Because of these reasons the observed stresses nearly zero. For the thermal swing up to +125°C there are only small stresses estimated.

At the third cooling period both thermal cycling coming more and more in confident stress level. For reaching confidences it is necessary to calculate between three cycles. This effect is also well known from creep based simulations [10].

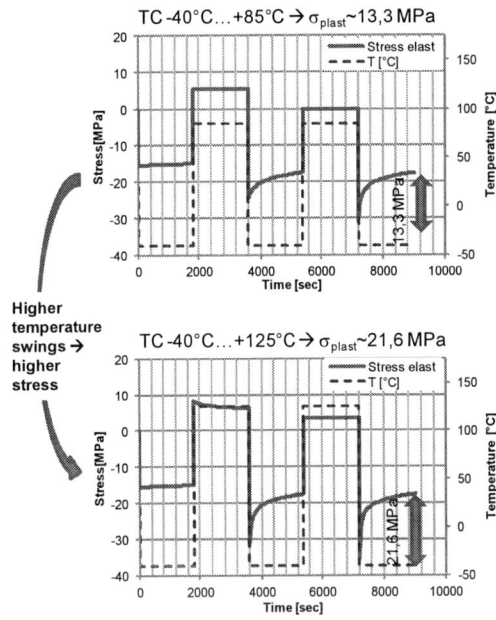

Figure 8: Calculate thermal cycle (-40°C...+85°C) and (-40°C...+125°C) with 30 minutes well times

Based on these computed results it is obviously that the maximum temperature has not significant influences on stress. The maximum stress is always at the minimum temperature. The main damaging factor is always the cooling period. Regarding stress behaviour, stresses could be estimated for -40°C up to +85°C with 13.3 MPa and for the thermal cycle from -40°C up to +125°C with 21.6 MPa.

978-1-4799-4789-8/14 $31.00 © 2014 IEEE   535

The main elements about the calculations of thermal cycle are:

- Higher temperature swings are responsible for higher stresses.
- The plastic stress is dominating after cooling down.

Combination of stress model with thermal cycle results

This chapter will fit the stress model on thermal cycling results for the SMD component CR1206. Through the lead free developing in the last years, a lot of thermal cycling test was done in community. As an example German government public funded projects LiVe, nanoPAL or InnoLot are mentioned. Beside to this works there were also a lot industrial exterminations of the new materials. This data pool is basic for the correlation between stress and number of thermal cycling capability. The stress was calculated by equ. (13) and plotted in correlation to experimental results (figure 9).

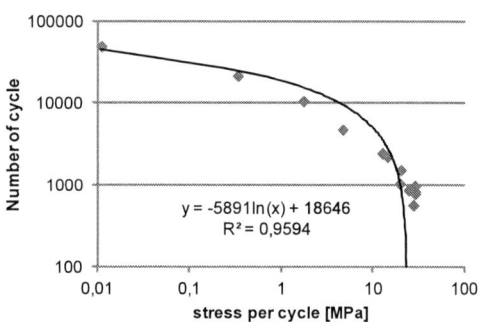

Figure 9: S/N curve, stress calculated for CR1206, number of cycle coming from test results

The lifetime prediction and/or description of thermal cycling capability is dependent on the calculated stresses (Figure 9). The best fit line isn't optimal, but could be used for calculation the thermal cycling capability and could explain by the equation 14.

$$N = a - c \log \sigma_{plast}$$
$$= a - c \cdot \log \left(\Delta \alpha \cdot (\Delta T) \cdot (E_{0K} - \kappa \cdot T_{\min}) - ((\sigma_{t=1s} - k \cdot T_{\min}) \cdot t^{\beta}) \right)$$
$$(14)$$

with N=Number of cycle and $N_{\sigma=1}$ = number of cycle at 1 MPa stress

By using the correct material parameter it is possible to estimate the thermal cycle capability of solders materials.

$$N = 18646 - 5891 \cdot \log \left(10 \frac{ppm}{K} \cdot \Delta T_{eff} \cdot \left(48.000 MPa - 95,2 \frac{MPa}{K} \cdot T_{\min} \right) - \left(77MPa - 0,158 \frac{MPa}{K} \cdot T_{\min} \right) \cdot t^{-0,11} \right)$$
$$(15)$$

It is now possible to calculate lifetime of thermal cycle capability based on measured material data through this methodology.

Conclusion

This work revels an analytical stress methodology for calculation life time of tin based solder materials. This new stress model should be used for easily calculation of solder joint capabilities. The model based strictly on material data and it was evaluated with experiments for the component CR1206. Future works will transfer the stress model to other components and solder materials, etc. This stress model was not developed to substitute works for a Chaboche like description of solder joint behavior but it is an alternative method for fast estimations even in the case of long term test with field loads.

In comparison to Coffin-Manson or to Norris-Landzberg this new model based on material measurements allows to implementing also dwell time. The model is more physical and could explain the material degradation more clearly through these factors.

Future work will transfer the material results to other component with different sizes, stiffness and internal buildup. The principal methodology should also be useable for other solder materials.

Acknowledgments

The authors would like to thank all colleagues for discussion and critical feedback. Special thank to A. Kabakchiev, F. Dörfler, S. Haag, B. Métais from the BOSCH research group CR/APJ3 and to colleagues from Fraunhofer IKTS R. Metasch, M. Röllig and to the colleagues from Fraunhofer ENAS R. Dudek and S. Rzepka.

978-1-4799-4789-8/14 $31.00 © 2014 IEEE

References

[1] A. Schubert, R. Dudek, E. Auerswald, A. Gollbardt, B. Michel, and H. Reichl. Fatigue life models of SnAgCu and SnPb solder joints evaluated by experiments and simulations. In 53th Electronic Components and Technology Conference ECTC, pages 603–610, New Orleans, USA, May 2003.

[2] S. Deplanque, W. Nuechter, M. Spraul, B. Wunderle, R. Dudek und B. Michel, „Relevance of primary creep in thermo-mechanical cycling for life-time prediction in Sn-based solders," in ˙s 6th. lnr. Confi on Thermal, Mechanical ond Multiphysics Simulation nnd Experiments in Micro-Electronics and Micm-System, EuroSimE, 2005.

[3] J. Chaboche, „A review of some plasticity and viscoplasticity constitutive theories," International Journal of Plasticity, Nr. Vol. 24, pp. pp. 1642-1693, 2008

[4] J. Chaboche, „Constitutive Equations for Cyclic Plasticity and Cyclic Viscoplasticity," International Journal of Plasticity, Nr. Vol. 5, pp. pp. 247-302, 1989.

[5] Metasch, R.; Roellig, M.; Kabakchiev, A.; Metais, B.; Ratchev, R.; Wolter, K.-J.; Experimental investigation of the visco-plastic mechanical properties of a Sn-based solder alloy for material modelling in Finite Element calculations of automotive electronics; EuroSimE2014, Ghent 2014

[6] Kabakchiev, A.; Metais, B.; Ratchev, R.; Guyenot, M., Buhl, P.; Hossfeld, M.; Schuler, X.; Metasch, R.; Roellig, R.; Description of the thermo-mechanical properties of a Sn-based solder alloy by a unified viscoplastic material model for Finite Element calculations ; EuroSimE2014, Ghent 2014

[7] Manson, S. S., Thermal Stress and Low-Cycle Fatigue, McGraw-Hill Book Company, New York, 1966.

[8] Norris,K.C.; Landzberg, A.H.: Reliability of Controlled Collapse Interconnections, IBM Journal of Research and Development, May 1969, pp.266-271

[9] Engelmaier, W., Turbini, L.J.: Design for Reliability in Advanced Electronic Packaging-A White Paper on Surface Mount and Related Technologies for the Surface Mount Council, Mendham/Atlanta, 1995

[10] Spraul, M.; Lebensdauerprognosen für gelötete Baugruppen mit Zinn-Blei und Zinn-Silber-Kupfer Lot für Temperaturwechselprüfung, dissertation, TU Berlin, 2007

A CRACK ANALYSIS MODEL FOR SILICON BASED SOLAR CELLS

Joseph Al Ahmar, Steffen Wiese
University of Saarland
Saarbrücken, Germany
E-Mail: j.ahmar@mx.uni-saarland.de

ABSTRACT:

Cracks in silicon layers of PV cells are a major problem for the photovoltaic industry and have an impact on the reliability of the PV technology. Especially considering the desire of developing and producing thinner and larger modules, which makes the cells more vulnerable to cracking.

In this work an analysis of the performance of a PV module under different load types has been done. The numerical evaluation of the stress fields is done using the finite element method. Critical regions containing micro cracks in silicon have been modelled using a special meshing technique. Finally, to answer the question if induced stresses on cracked cells lead to a complete separation of cell parts, the theory of linear elastic fracture mechanics is applied and the resulting stress intensity factors (SIF's) under different load conditions are calculated.

1. INTRODUCTION

Several steps during production induce thermal and mechanical stresses in the cell which can generate micro-cracks, especially by thermal treatments while printing the silver paste or creating grid and also during soldering process. The differential thermal expansion of copper and silicon especially at high temperatures, leads to the formation of micro cracks. Cracks in the module can also occur due to several mechanical loads during transport and installation.

Once the module is installed, it is regularly exposed to thermal and mechanical environmental loads such as heat, wind and snow. Due to repeated stresses on cracked cell regions, a crack growth could take place and lead to an isolation of cell parts and so cause a major reduction in the electrical performance of the module.

In this work an array of H-patterned cells is considered. A finite element model is generated using the commercial software ANSYS. Thermal and mechanical load types are applied in the executed simulations. Different techniques provided by the postprocessor are applied to extract the fracture parameters for a given crack configuration.

2. 2D-MODEL OF SILICON SOLAR CELL

For the following analysis we consider a 2D-Module of silicon solar cells. The cross-section on Figure 1 shows one PV cell with its different layers and the corresponding dimensions are presented in table 1.

Copper Ribbons provide the electrical interconnections between the cells. A solder layer joins silicon and Copper Ribbons. The Back sheet is a Tedlar foil which protects the module especially against humidity. The front of the module is covered by a glass plate that protects the internal parts. Between the front Glass and the Back sheet, the solar cells are encapsulated in ethylene vinyl acetate (EVA). Table 2 shows a detailed summary about the applied materials for the geometrical model. The values for the presented characteristic properties are given for room temperature. In this approach we assume an isotropic linear elastic material behavior for our investigation about cracks in silicon layer. As a brittle material, silicon shows linear elastic properties right up to the point of fracture. Thus, the theory of linear elastic fracture mechanics can be applied. We assume also a polycrystalline behavior for the silicon. Most of the applied materials show negligible changes for the temperature range used in the simulation. The elastic plastic material response of solder is illustrated by a multi-linear model. The time dependent part is described through a creep model in ANSYS.

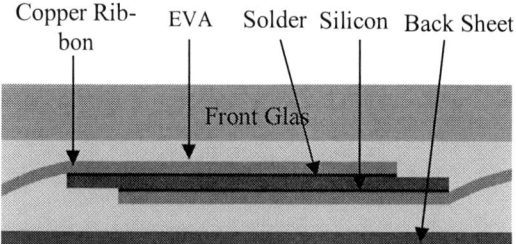

Figure 1: Schematic of a Photovoltaic Cell (Cross Section View).

Table 1: Solar cell dimension

	Length[mm]	Width[mm]
Si	156	0.18
Cu	147	0.2
Glass	156	4
Solder	147	0.025
Back sheet	156	0.3

Table 2: Summary of material models for FE-analysis with characteristic values at room temperature.

Material	Young's Modulus [GPa]	CTE [ppm/K]	Yield Stress [MPa]	Material model
Copper (initial)	85.7	17.0	95.1	Bi-linear temperature dep. elastic-plastic with isotropic hardening
Silicon	130.0	2.6	-	Linear-elastic
Glass	75.0	10.0	-	Linear-elastic
Tedlar	1.4	30.0	35.0	Bi-linear temperature dep. elastic-plastic
EVA	0.011	15		temperature dep. elastic
Solder	32.0	25.2	48.0	Temperature dep. Young's Modulus & Yield stress; creep

3. THEORETICAL APPROACH: FRACTURE MECHANICS

The fracture mechanics theory is applied to analyze structures with preexisting cracks. It characterizes the fracture behavior in structural parameters. Based on theory of linear elasticity, the concept of linear elastic fracture mechanics (LEFM) was derived. It is a simplified concept that is applicable as long as the plastic region around crack tip is vanishingly small compared to the rest of the material. In order to be able to perform a life prediction we must determine the onset of crack propagation and estimate critical load and crack dimensions.

In [1] an analysis is done to calculate the stress field around the crack tip under LEFM assumptions. As shown on figure 2, we assume a polar coordinate axis with origin at the crack tip the following expression is given

$$\lim_{r \to 0} \sigma_{ij}^n = \frac{K_n}{\sqrt{2\pi r}} f_{ij}^n(\theta) \qquad (1)$$

with σ_{ij} is the stress distribution, f_{ij} a dimensionless function of θ, K and the subscript n representing the SIF and the corresponding mode of loading, respectively. There are three different basic modes of crack surface displacement. Mode I describes a load, in which the crack surface displacements are

perpendicular to the crack plane. Mode II and III correspond to shearing displacements in the plane of the crack.

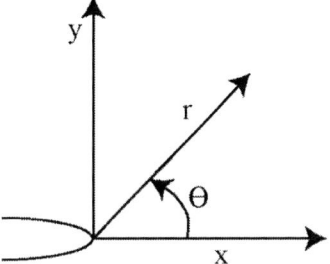

Figure 2: Crack tip coordinates

As the expression (1) shows, the fields near crack tip are determined by the SIF's. They define the amplitude of the crack tip singularity and the crack tip conditions and control the local deformation fields. Thus, knowing the value of the SIF for a given load provides criteria for the crack extension. The SIF should be compared to the material characteristic resistance to fracture K_C for the corresponding crack surface displacement mode, which describes the fracture toughness of the material. Thus, crack will grow under an applied load if

$$K \geq K_C. \qquad (2)$$

Another fracture criteria can be given using the energy release rate G. Crack will grow if

$$G \geq G_C, \qquad (3)$$

where G_c represents the energie requiered for a crack to propagate. In the case of a small plastic zone comparing to the crack dimension, G is related to K through the following formula

$$G = \frac{K_I^2}{E}, \qquad (4)$$

where E defines the elastic modulus of the material.

In this work two different extraction techniques are used to estimate the SIF's from the results of the finite element analysis using the postprocessor in ANSYS. The first one is the crack opening displacement [2]. It is a direct approach and uses the theoretical asymptotic displacement fields in combination with nodal displacement solution from the finite element method and applies it to calculate the SIF'S using the following relations for $\theta = 180°$ along the crack line and for plane strain assumption

978-1-4799-4789-8/14 $31.00 © 2014 IEEE

$$K_I = \frac{\mu\sqrt{2\pi}\,v}{\sqrt{r}(2-2\mu)},$$

$$K_{II} = \frac{\mu\sqrt{2\pi}\,u}{\sqrt{r}(2-2\mu)}, \tag{5}$$

$$K_{III} = \frac{\mu\sqrt{\pi}\,w}{\sqrt{2r}}$$

with u, v and w the x, y, and z nodal displacements, and v the Poisson's ratio. An improvement in this approach can be reached by using quarter-point-tip elements [3]. It is based on using standard quadratic elements and moving the mid-side node from the center position to the quarter position in the crack tip direction. Also one edge of the element should be collapsed at the crack tip. This approach is illustrated on figure 3.

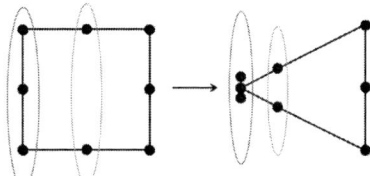

Figure 3: Quarter-point element

The second technique applied for the estimation of the SIF's is based on the calculation of the J-integral [4]. It is an energy approach and the calculated value J describes a fracture mechanics parameter and is under linear elastic assumptions equivalent to the energy release rate. Using the crack tip coordinate system x_1 and x_2 the J-integral is defined as

$$J = \lim_{\Gamma \to 0} \int_{\Gamma} [Wn_1 - \sigma_{ij}\frac{\partial u_i}{\partial x_1}n_j]d\Gamma \tag{6}$$

where w is the strain energy density, σ_{ij} the stress tensor, u_i the displacement field, and n_j the unit outward normal to the contour of integration.

4. THREE-POINT BENDING BEAM WITH INITIAL CRACK

Before applying the presented methods on the finite element model of solar cells, we investigate their accuracy through a finite element simulation of a three-point bending response of a silicon beam specimen. Figure 4 shows a typical geometry of Mode I fracture testing of beam. According to [5], the analytical formula for the calculation of the Mode I stress intensity factor is expressed through

$$K_I = \frac{P}{\sqrt{w}}f(\frac{a}{w}) \tag{7}$$

where l, w, a and P represent length, width, crack length and tensile force, respectively. The geometry factor $f(\frac{a}{w})$ is calculated through:

$$f\left(\frac{a}{w}\right) = \frac{3 \cdot \frac{S}{w} \cdot \sqrt{\frac{a}{w}}}{2 \cdot \left(1 + 2 \cdot \left(\frac{a}{w}\right)\right) \cdot \left(1 - \left(\frac{a}{w}\right)\right)^{\frac{3}{2}}} \tag{8}$$
$$\cdot \,[1.99 - \frac{a}{w} \cdot (1 - \frac{a}{w})(2.15 - 3.93 \cdot \left(\frac{a}{w}\right)$$
$$+ \, 2.7 \cdot \left(\frac{a}{w}\right)^2)]$$

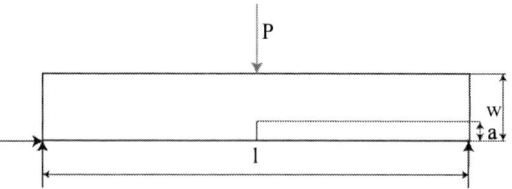

Figure 4: 2D Model of a three-point bending beam specimen.

Figure 5 shows the results of the Mode I stress intensity factor for different crack lengths. The values for K_I obtained from the finite element result through calculating the J-integral and also through applying the crack opening displacement method with quarter-point elements shows a good correspondence with the analytical results. This means that the method used can also be applied for other more complex geometry and crack configuration.

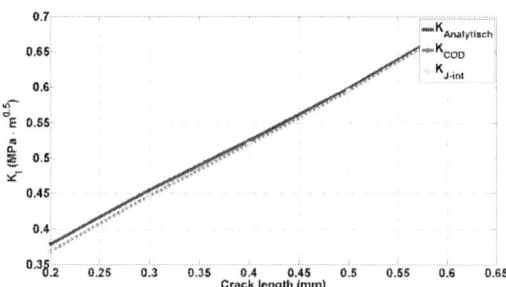

Figure 5: Mode I stress intensity factor for the three-point bending beam.

978-1-4799-4789-8/14 $31.00 © 2014 IEEE 540

5. SIMULATION AND RESULTS OF AN ARRAY OF THREE SOLAR CELLS WITH INITIAL CRACK

In this section we investigate the fracture mechanical properties of a 2D-model of an array of H-patterned cells. Figure 6 presents the schematic of the considered 3-cell array model. Cracks are initiated in silicon layer and the resulting stress distributions and stress intensity factors are estimated. The finite element mesh is done in ANSYS using quadratic 8-node elements PLANE183 with quarter point displacement on the crack tip. On Figure 7 and figure 8 the generated finite element meshes for the cell array and around crack tip are illustrated.

Figure 6: Schematic of an array of 3 cells (Cross Section View).

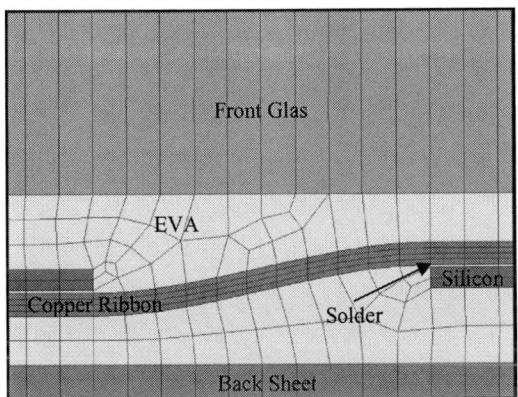

Figure 7: FE mesh of connected solar cells

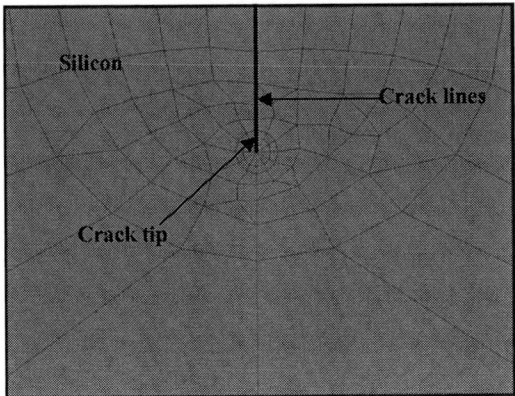

Figure 8: FE mesh around crack lines in silicon

A temperature cycling test (TCT) according to IEC 61215 is done. The temperature values are between -40°C and 85°C as shown on figure 9. In the first step the TCT is applied on cell array without initiated cracks. The stress distribution in the structure is estimated. Once the regions with highest stress values are located, cracks can be included to the finite element mesh there. The crack lines start from the silicon-solder interface and go straight through the silicon material.

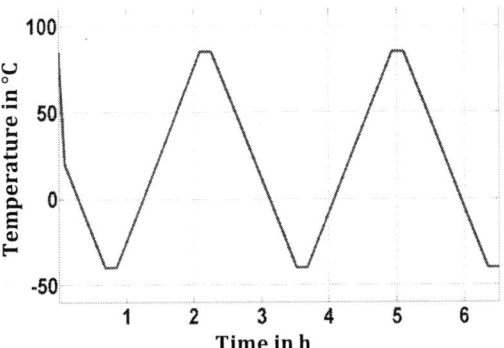

Figure 9: Temperature profile for Simulation

The TCT is now applied to the solar array with the initiated crack. Different crack lenghtes are applied and the resulting stress intensity factors for a 20 μm crack are shown on figure 10. The considered critical fracture toughness used in this work is 0.94 and corresponds to the polycrystalline silicon. The two different techniques used showed a good correspondence and the estimated fracture parameter exceed the fracture toughness of silicon for low temperatures between $-15°C$ and $-40°C$ which means that initiated crack would be able to propagate through the silicon layer and could lead to a cell breakage.

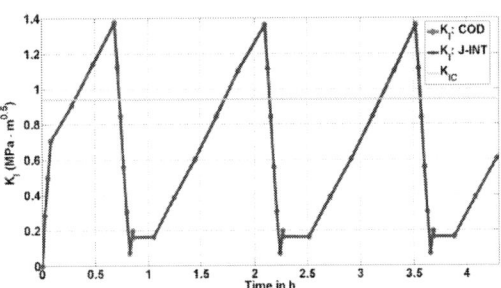

Figure 10: Results of mode I stress intensity factor during TCT (crack length = 20 μm).

In the next analysis a vertical uniform load of $10^{-3} \frac{N}{mm}$ is applied on the cell array. This corresponds to the load resulting from snow mass over the module. In simulation the load has been progressively increased up to the maximal value and the cells reach a temperature of $-10°C$ for a period of up to one week. Figure 11 represents the values of the SIF's estimated for a crack length range between 20 μm and 130 μm. Both applied methods show a good correspondence in the calculated SIF's. The snow load, in contrast to the TCT, does not lead to fracture parameters that exceed the fracture toughness of silicon for the considered crack geometry.

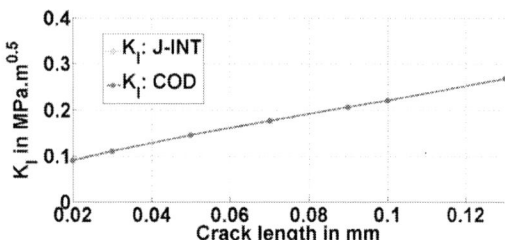

Figure 11: Results of mode I stress intensity factor for vertical uniform load.

6. CONCLUSION

This paper handle with the fracture mechanical behavior of silicon based solar cells. A Finite element analysis is performed and tools for extracting fracture parameters are presented. The TCT is done on PV cell arrays with initiated cracks in silicon layer and the resulting values of stress intensity factors are estimated. The fracture toughness of silicon is exceeded in the temperature range between $-15°C$ and $-40°C$ for a crack length value of 20 μm. The second analysis considers the thermo mechanical influence of a snow load on cracked cell array. The calculated fracture parameters show no overrun of the silicon fracture toughness value for the simulation time period in the considered crack geometry and orientation.The accuracy of the extraction technique applied has been validated through comparing the results of the energy approach (J-integral) to the values obtained from the direct approach related to the crack opening displacement technique.

REFERENCES:

[1] M.L. Williams: "On the Stress Distribution at the Base of a Stationary Crack." Journal of applied Mechanics, Vol. 24, 1957, pp. 361-364.

[2] P. C. Paris and G. C. Sih: "Stress analysis of cracks. In Fracture Toughness Testing and its Applications", ASTM STP 381, pp. 30-83, Philadelphia, 1965. ASTM (American Society for Testing and Materials).

[3] R. S. Barsoum: "On the Use of Isoparametric Finite Elements in Linear Fracture Mechanics," International Journal for Numerical Methods in Engineering, Vol. 10, pp. 25-37, 1976.

[4] Rice, J.: "A path independent integral and the approximate analysis of strain concentration by notches and cracks". Journal of Applied Mechanics, 35:379-386,1968.

[5] Tada, H., Paris,P., and Irwin, G.R., The Stress Analysis of Cracks Handbook, 3rd edition, ASME Press, New York, 2000.

[6] C.P. Chen and M.H. Leipold: "Fracture Toughness of Silicon", Am. Ceram. Bull., 59:469-472, 1980.

Effect of laminar air flow on probe burn for spring probe

Baha Zafer
İstanbul University - Dept. of Mechanical Engineering
İstanbul, Türkiye / baha.zafer@istanbul.edu.tr

Bahadır Tunaboylu
İstanbul Şehir University - Dept. of Industrial Engineering,
İstanbul, Türkiye / btunaboylu@sehir.edu.tr

Abstract

This paper investigates transient heat transfer between a heated spring probe and its air environment. A continuum finite volume simulation is used to analyze of heat flow within and from the resistively heated probe to its environment. Experimental results are conducted for spring probe with laminar air flow and without air flow. The numerical results and experimental results are compared and very good agreement is observed.

1. Introduction

The wafer contact is becoming an important issue because of power delivery to the chip through the probe from the tester generating conductive thermal effects due to Joule heating. Test system performance is dependent on the ability of probes to carry required test current to the chip. Besides the contact force and the test dc/pulsed currents, the material properties and geometry of the probe card needles also play important roles in wafer level testing [1].

In this paper, the 3D transient thermal conduction equation with Joule heating as a source term is used to compute temperature distribution on the probe body [2] and heat convection is computed between heated probe body and its environment. A special design spring beam was selected as the probing technology for this study. A computational model is used for the probes for predicting temperature rise along a probe body and heat loss in the case of convection under laminar flow situation in order to investigate probe burn effects.

2. Probe Burn Phenomenona

A probe will burn during wafer test when its probe current carrying ability (I_{max}) is exceeded. Probes are usually the weakest link in the test architecture and therefore first to burn during the test due to the constriction resistance. Therefore, in many applications it is important to accurately predict how much current can be handled by a cantilever probe tip or a spring probe body for designers. Test system performance is dependent on the ability of probes to carry required test current to the chip. Increasing the wafer probe current beyond its limit can drastically reduce life span by deforming or burning the probe tip.

The governing equation for transient heat conduction is

$$\frac{\partial}{\partial t}(\rho h) + \nabla \cdot (\vec{v} \rho h) = \nabla \cdot (k \nabla T + q_r) + S_h \qquad (1)$$

where ρ is the density and k is the thermal conductivity. h is the sensible enthalpy. The terms on the right-hand side of (1) are the heat flux due to conduction; q_r is radiation heat flux and volumetric heat sources, S_h, respectively. The second term on the left hand side of (1) represents the convective energy transfer and $\vec{v}$ is the motion fluid particles.

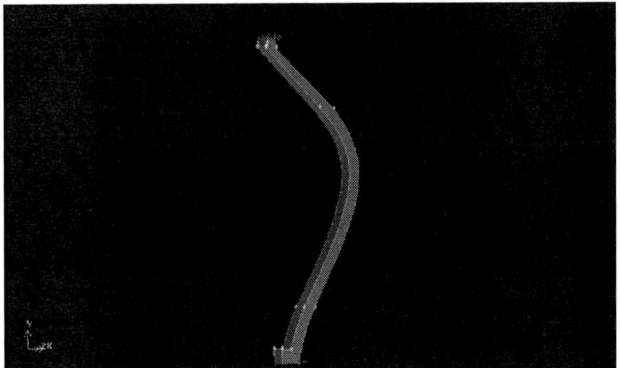

Figure 1: Solid geometrical model of spring probe.

When an electrical current passes through a cantilever probe during wafer level test, it heats the probe through a process known as Joule (or I^2R) heating [3-5]. With a sufficient increase in current flow through the wafer probe, the heat generated will exceed the natural heat dissipation and the probe temperature will increase. Eventually, the temperature of the probe reaches a critical value at which mechanical strength diminishes and plastic deformation begins. This decreases the contact force, which adversely affects electrical contact resistance. This, in turn, can generate unreliable test readings and wafer yield loss. The maximum allowable loss of contact force due to current passing through a deflected probe is 10%, as a generally accepted criterion in the wafer test industry [2].

3. Numerical Analysis

Because of the complications in modelling the test process, many earlier studies mainly depended on experiments to investigate the probe burn. However, with the development of the computer performance and advances in modelling packages, more researchers investigate the micro-mechanical systems by the numerical methods.

The heat conduction within spring probe and heat convection according to laminar flow is modelled by using shadow wall options of the commercial finite volume software. This transient conjugate heat transfer is used to investigate the effect of convection at the probe body.

978-1-4799-4789-8/14 $31.00 © 2014 IEEE

The thermal conduction effects between the spring/cantilever probe and wafer pad is modeled by using finite volume analysis. Heat flux values are computed corresponding to current values (i.e. Joule heating) which are imposed from tester according to following formulation:

$$S_h = \frac{V_b^2}{\rho_e L^2} = \left(\frac{I}{A_{tip}}\right)^2 \rho_e$$

where V_b is applied electrical potential and L is length of the cantilever probe. A_{tip} is the tip area of the wafer probe and electrical resistivity ρ_e is an inherent property of a material to resist the flow of electrical current through it by means of producing electrical resistance. Electrical resistivity is independent of the material geometry and depends on the temperature, given as;

$$\rho_e(T) = \rho_{e,o}\left[1 + \eta(T - T_o)\right]$$

where η is the temperature coefficient of resistivity and $\rho_{e,o}$ is the electrical resistivity defined at reference temperature T_0. We assumed that the adiabatic condition exists at the base and the temperature of the base is unchanged. The material properties remain unchanged with variable temperature along the spring probe body.

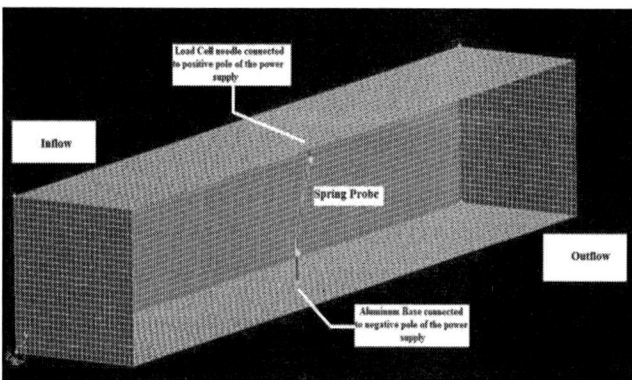

Figure 2: Computational domain for spring probe with laminar air flow.

Computational software does not provide the solution for joule heating, so the user defined function (UDF) is implemented for Joule heating as a source term. A three dimensional spring wafer probe model and computational domain of laminar flow is designed by a CAD program and it is used as an input for meshing. The spring probe body is meshed in the preprocessor program Gambit by using Hex/Wedge type elements with interval size 0.005 as shown in Figure 2. The mesh is exported to computational toolbox along with the physical properties of the probe body material and the initial conditions specified. Different cases are solved using transient solution. Time dependent solution was completed when it reached the steady-state condition.

4. Discussion and Results

The experimental setup provides a measurement for the CCC based on the mechanical degradation of probe as shown general view in Figure 3. It does not, however, provide any direct data about the temperature distribution along the spring probe body. Therefore, a different kind of approach is needed to observe heat effects on the body.

At this stage, a numerical procedure was carried out to investigate the temperature distribution with laminar flow and without air flow. In this method, the current is applied by the tester and it goes through the probe body and reaches to the wafer die, is an input. Applied current is used to link the electrical effect to the thermal effect by using Joule heating.

The temperature along probe body was observed and then compared with the melting point of the body material in the case of without laminar flow by solving heat transfer equation [5]. In this approach, if the temperature distribution at any point exceeds the melting point of the probe material then the applied current is defined as the CCC similar to the procedure in the experimental method.

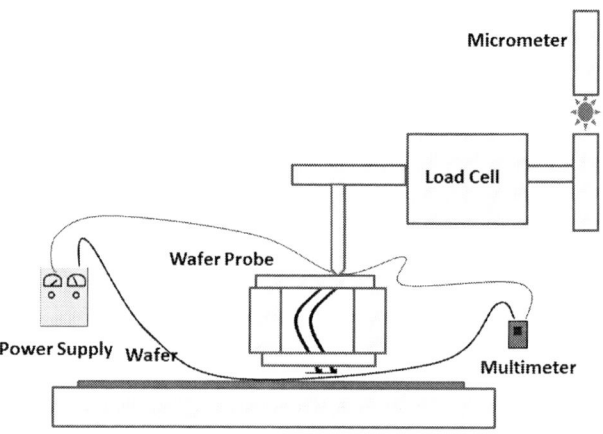

Figure 3: General view of experimental setup for spring probe.

Three dimensional transient conjugate heat transfer equation will be used in order to compare conducted experimental results with numerical solution. The computational domain is meshed with hexagonal structure and shown in Figure 2. The laminar flow is conducted in the Plexiglas canal which cover spring probe. The experimental results for spring probe without and with laminar flow are given in Figure 4. It is observed from experimental result, the current carrying capacity ($I_{\%10}$) is increased in the case of laminar air flow.

In order to compare the performance of numerical results, a different experimental set up was designed. Two different situations are set up, one without laminar flow and second with laminar flow, the experiments were conducted separately.

978-1-4799-4789-8/14 $31.00 © 2014 IEEE   544

The initial overdrive (OD) was applied for each case at different values, 3.4 gram for without laminar case and 3.6 gram for with laminar case are applied as shown in the experimental results in Figure 4.

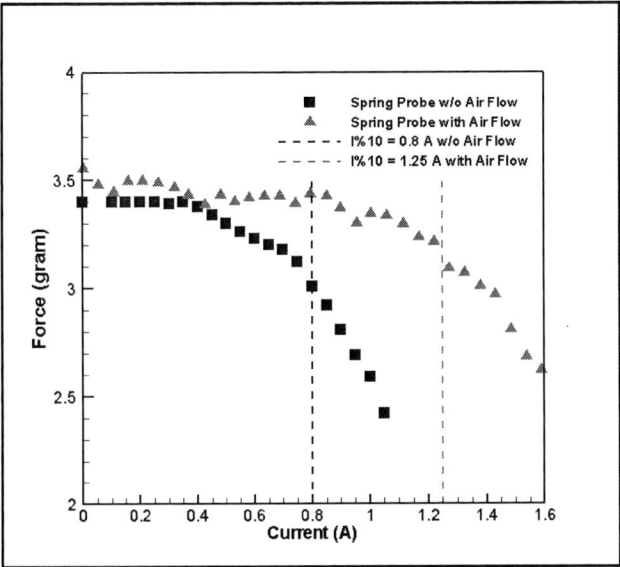

Figure 4: Experimental results for the Spring Probe [Base radius 2.5 mil] with air flow and without air flow. Observed current carrying capacity is $I_{\%10}=0.8$ amp. for w/o laminar flow case and for with laminar flow case is $I_{\%10}=1.25$ amp.

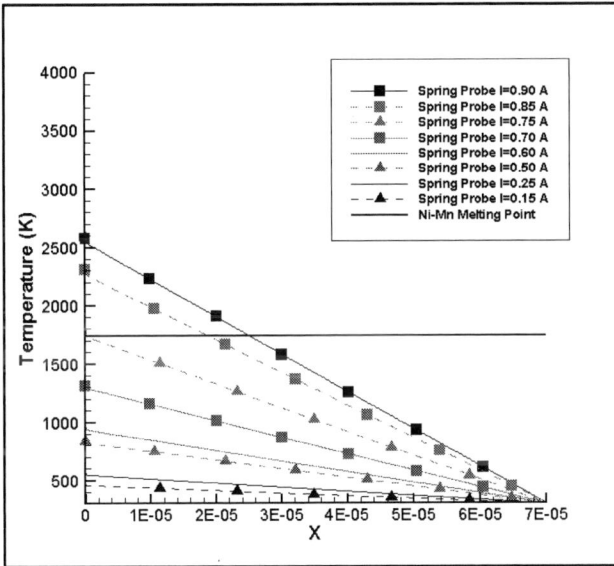

Figure 5: Numerical results for the Spring Probe [Base radius 2.5 mil] with laminar air flow. The temperature distribution along the spring probe from the tip to the epoxy base, $I_{max}=0.75$ amp.

This paper develops computational mechanics simulation framework for the special design spring wafer probe in order to investigate the probe burn phenomenon. The numerical results are computed and the temperature rise is observed along the probe body. In this case, the epoxy wall temperature, T_a is taken an ambient temperature and source term is computed using Joule heating effects. The experimental results are shown in Figure 4 and the converged numerical results are shown in Figure 5-6 for without and with laminar flow case respectively. In both cases, the nickel-alloy (Ni-Mn) is chosen for the spring probes.

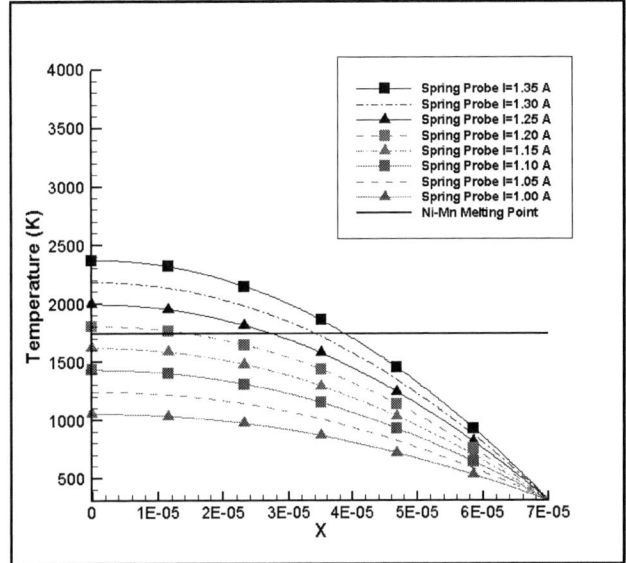

Figure 6: Numerical results for the Spring Probe [Base radius 2.5 mil] without laminar air flow. The temperature distribution along the spring probe from the tip to the epoxy base, 1.15 amp $<I_{max}<$ 1.20 amp.

The effect of the laminar flow on the temperature distribution along spring probe body is investigated. As the laminar air flow is conducted, the temperature rises characteristic is changed at the tip region and current carrying capacity (I_{ccc}) is reasonably increases. In the case of laminar air flow, as shown Figure 6, the current carrying capability has reached 1.25 amp. in experimental results and between 1.15 $<I_{max}<$ 1.20 amp. in numerical results. The fundamental results is that the laminar air flow is increase current carrying capacity nearly 60 % for spring probe and enlarged operational flexibility of the wafer probe tester.

As a conclusion, the main focus of this study was on identifying and evaluation of the state of the art numerical computations using Joule heating for the case of spring probe burn for wafer test in order to identify areas of future improvement in power delivery, thermal cycling and finally the life span of a probe. The results indicated that laminar air flow is increases the probe current carrying capability. Reasonably good agreement is observed between experimental and computational results.

ACKNOWLEDGMENT

This research was supported by a Marie Curie International Reintegration Grant within the 7th European Community Framework Program under grant # 271545. We also thank SV Probe Inc. R&D members for the discussions and contributions in this study.

REFERENCES

[1] M. Allison, 2005, "Wafer Probe Acquires a New Importance in Testing," Chip Scale Review, May-June Issue, p. 45-49.

[2] B. Tunaboylu, E. Caughey, 2003, "Vertical Probe Development for Copper Probing Challenges," Semiconductor Wafer Test Workshop, June.

[3] A. Mertol, 1995, "Estimation of aluminum and gold bond wire fusing current and fusing wire," IEEE Transactions on Components, Packaging and Manufacturing Technology, vol. 18 (1), p. 210-214.

[4] R. Holm, Electrical Contacts: Theory and Applications, 4th ed., Springer and Verlag, 2000.

[5] Shabany, Y., Heat Transfer : Thermal Management of Electronics, CRC Press, 2010.

[6] B. Zafer, M. H. Vishkasougheh and B. Tunaboylu, 2013, "Wafer test probe burn modeling and characterization" 14th IEEE International Conference on Thermal, Mechanical & Multiphysics Simulation and Experiments in Micro/Nano-Electronics and Microsystems (EuroSimE 2013).

Thermal management of electrical overload cases using thermo-electric modules and phase change buffer techniques: Simulation, technology and testing

M. Springborn[1,4], B. Wunderle[1], D. May[1], R. Mrossko[4], C.-A. Manier[2], H. Oppermann[2],
M. Abo Ras[3], R. Mitova[5]

1 Chemnitz University of Technology, Chemnitz, Germany
2 Fraunhofer Institute Reliability and Microintegration, IZM, Berlin, Germany
3 Berliner Nanotest & Design GmbH, Berlin, Germany
4 AMIC GmbH, Berlin, Germany
5 Schneider Electric, Grenoble, France

Abstract

In this paper, a new and dedicated phase change cooling concept is discussed, following the goal to buffer periodic overload operations of electric power modules and thus keeping junction temperatures constant. A top-mounted latent heat storage material (LHSM) buffer is applied, to soak the overload heat in conjunction with thermo-electric coolers (TECs), which control the temperature and phase change process. The cooling system is going to be realized within an IGBT converter module, undergoing repeated overload situations.

The concept features double-sided cooling and assembling as well as new materials and joining technologies such as transient liquid phase bonding/soldering and silver sintering. One-dimensional equivalent circuit estimations and transient electro-thermal FE simulations were used to calculate the cooling performance, extract optimization guidelines and discuss potential difficulties of the concept. Furthermore, the simulations are successively refined and brought into agreement with various test stands and characterization methods of reduced complexity.

1. Introduction

Dealing with the increasing performance and reliability requirements of today's power electronic, new materials and integration concepts are one of the main focuses of the ongoing packaging research. In this context, the thermal system design is mainly determined by the challenge of dissipating the increasing power loss (e.g. of electrical power modules) while reducing form factor, weight and cost. As the heat removal benefits from lower thermal resistances and also higher operation temperatures (thinking of convection as the final heat removal or heat sink), novel high-temperature joining processes (i.e. TLPS/TLPB and sinter silver) and double-sided cooling assembling are part of the ongoing technological research and reliability investigation.

Providing a workaround on the heat removal problem, several concepts and material technologies have been attempted and also successfully applied. Heat pipes for examples (commonly embedded into today's CPU coolers or notebooks) are of high interest, providing possibilities to enhance heat spreading or transfer. Techniques like Pulsating- and Loop Heat Pipes are still under investigation, aiming for improvement of the condensation/evaporation areas within, as well as new shapes and smaller scale, which widens embedding possibilities right up to approaches of DCB substrate integration [12 – 21]. Thermo-electric cooling (TEC) provides possibilities to intelligently cool hot spots [4, 22–29], though bringing the huge disadvantage of additional heat loss, that has to be drained by the post-TEC heat removal path (i.e. the heat sink). Another possibility is the use of latent heat storage materials (LHSM) that cool down hot spots of limited transient thermal loading [4, 6-9]. A specific cooling ability however is mainly determined by the LHSM's melting temperature not necessarily lying within the needed range of desired transient cooling temperatures. This is a main drawback here, and complicates a useful application. Wide melting temperature ranges, chemical stability due to repeated melting/freezing cycles, low thermal conductivities (even lower, when molten) and other aspects like toxicity and flammability are problems that have to be dealt with. A great review and summary of considerable applications and latent heat storage materials is given by Sharma and Sagara in 2005 [5].

Figure 1: Concept of TEC- and heat pipe assisted latent heat storage cooling within an IGBT power module

A potential and highly sophisticated application of **these techniques combined** is the focus of this work. The aim is to realize a new, dedicated and double-sided

978-1-4799-4789-8/14 $31.00 © 2014 IEEE   547

transient cooling system within a 4kW IGBT converter module, providing constant junction temperatures during a 60-second thermal overload situation of 150% power loss (114W in total). As seen in figure 1, a latent heat buffer, whose phase change transition is controlled or initiated by means of Bi_2Te_3 based TE cooling/heating will be applied onto the top-side of the power module. This buffer module will feature active cooling during the overload situation, while remaining passive (or in a recovery state) during normal operation. A conventional (convective) cooling system is applied at the bottom-side of the power converter and will feature the removal of power loss during the normal operation phase.

For the latent heat material an eutectic bismuth-tin solder will be used. Despite the more commonly application of paraffin waxes, this will bring the advantage of a much higher conductivity, while providing comparable values of latent heat densities, though being very heavy and costly. The thin melting temperature range, a melting point of $T_{Melt} = 138°C$ (slightly above desired junction temperatures of $T_{J,OL} = 130°C$) and a thermal conductivity of about 2 magnitudes higher as compared to other wax based LHSMs makes this solder compound a predestinated option. For a further enhancement of the thermal buffer's conductivity, it has been found that heat pipes are the only viable solution to conduct the thermal heating power into the LHSM. Finally, thermo-electric cooling will provide the temperature control to close the temperature gap of T_{Melt} minus $T_{J,OL}$, while the additional TEC's heat loss will be soaked by the LHSM buffer.

Accepting the very high thermal and technological complexity of the system, only a combination of these have shown to provide the desired cooling performance, canceling out the physical limits (of LHSM and TEC cooling) mentioned above. System and also cost efficiency, however is not a goal of this demonstrator providing a solution for problematic thermal transients, which are lacking an alternative thermal cooling concept.

New high-temperature joining technologies (i.e. TLPB, TLPS and sinter silver) and double-sided DCB assembling are the key enabling technologies, to make this thermal buffer concept work, also demanding for reliability and lifetime investigations referred to the thermal interfaces besides performance drifts of the thermally active components (i.e. LHSM [5] and TECs [32]).

2. One-dimensional modeling of a TEC assisted thermal buffer

In order to perform first estimations of the thermal system, simplified and generalized equivalent circuit calculations have been done. As the TE coolers will control the temperatures and melting process, dimensioning (number of TE legs/modules) and needed driving current are of major importance for the system design. These are calculated as well as derived values like the total heating power that has to be stored by the

thermal buffer, thus defining the needed volume of the LHSM.

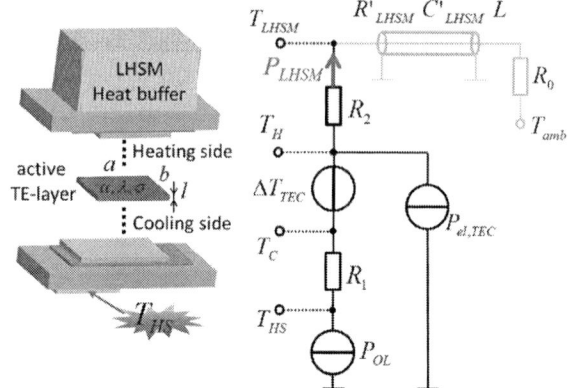

Figure 2: Thermal equivalent circuit of a transient TEC-assisted latent heat cooling system.

The TEC-LHSM cooling layer stack and the thermal equivalent circuit are shown in figure 2 and the meaning of the symbols is found in table 1. As the thermal buffer system has the task to soak the thermal overload power of the hot spot, while keeping the temperature at constant ($T_{HS,normal} = 125°C$) or at least below a maximum of 130°C ($\Delta T_{HS,OL} = 5K$), P_{OL} and $T_{HS,OL}$ are fix boundary conditions of the system. To calculate the required TEC current, the transmission line alike transient behavior of the LHSM buffer is ignored, assuming a perfect thermal conductivity ($R'_{LHSM} = 0$). Also, as a factor of safety, the additional bottom side heat flow based on the allowed $\Delta T_{HS,OL} = 5K$ will not be taken into consideration. In theory, this would reduce the overload power that has to be buffered by at least 5 W, depending on the performance of the bottom side cooling.

a, b, l, N_{TEC}	*Dimensions and number of the effective TE layers*
α, λ, σ	*Seebeck coefficient, thermal conductivity and el. resistivity of the TE layer*
T_C, T_H, ΔT_{TEC}, $P_{el,TEC}$	*TEC's cold and hot side temperature, temperature difference and power loss of the TEC*
T_{HS}, P_{OL}	*Hot spot temperature and TEC-related overload power, that has to be soaked*
R_1	*Cumulative thermal resistance of the layer structure between hot spot and TEC*
R_2	*Cumulative thermal resistance of the layer structure between TEC and LHSM buffer*
R'_{LHSM}, C'_{LHSM}	*thermal Resistance and capacitance per unit length of the LHSM buffer*
L, R_0, T_{LHSM}, T_{melt}	*length, thermal buffer heat sink resistance, LHSM and melting temperature*
	* Values referred to one TE layer or block $$C_X = C_{X,TOTAL} / N_{TEC}$$ $$P_{OL} = P_{OL,TOTAL} / N_{TEC}$$

Table 1: Symbols and description

As can be derived from thermo-electric theory [27,28,30,31], the net cooling ΔT_{TEC} of the TE layer is given as

$$\Delta T_{TEC} = T_H - T_C = \frac{\rho l^2 I^2 + 2ablP_{OL} - 2abl\alpha I(T_C + \Delta T_{TEC})}{2ab(-\alpha lI - ab\lambda)}$$

$$= \frac{\rho l^2 I^2 + 2ablP_{OL} - 2abl\alpha I T_C}{-2abab\lambda}$$

(1)

With the temperatures of the LHSM and hot spot (T_{LHSM} and $T_{HS,OL}$) given (i.e. LHSM's melting temperature and hot spot overload temperature, which is desired during overload operation) the following relations of the network can be stated.

$$T_C = T_{HS,OL} - R_1 P_{OL}$$
$$T_H = T_{LHSM} + R_2(P_{OL} + P_{el,TEC})$$
$$P_{el,TEC} = UI = \frac{\rho l}{ab}I^2 + \alpha I \Delta T_{TEC}$$

(2)

Putting (2) into (1) leads to the following expression (3), whose solution is the TEC current, which is needed to perform the desired cooling.

$$I^3 + AI^2 + BI + C = 0$$
$$A = \frac{-2abR_2\left(\sigma\lambda + \alpha^2\left(T_{HS,OL} - R_1 P_{OL}\right)\right) - \sigma l}{\alpha R_2 \sigma l}$$
$$B = \frac{2ab\left(P_{OL}(R_2 - R_1) + T_{HS,OL}\right)}{R_2 \sigma l}$$
$$C = \frac{2a^2b^2\lambda\left(T_{HS,OL} - T_{LHSM} - P_{OL}(R_2 + R_1)\right)}{\alpha R_2 \sigma l^2} - \frac{2abP_{OL}}{\alpha R_2 \sigma l}$$

(3)

Referred to the system's coefficient of performance an optimal number of TE couples (or TEC modules, when using the material parameters of an effective TEC model) exists, minimizing the TEC's heat loss. To reduce the needed LHSM buffer volume, this should be taken into account, if possible.

For the IGBT module the thermal resistances R_1 and R_2 are approximated by the thermal resistance series of the double-sided assembly structure as there are (from LHSM to IGBT/Diode, see figure 3):

- o Copper socket for the LHSM's casing
- o TECs (in parallel)
  - TIM layer
  - Upper bordering substrate
  - (Active TEC layer)
  - Lower bordering substrate
  - Solder layer

- o DCB substrate, AlN and copper
- o IGBT and Diode
  - TLPS layer, copper interposer and TLPB layer
  - Die (Diode and IGBT)

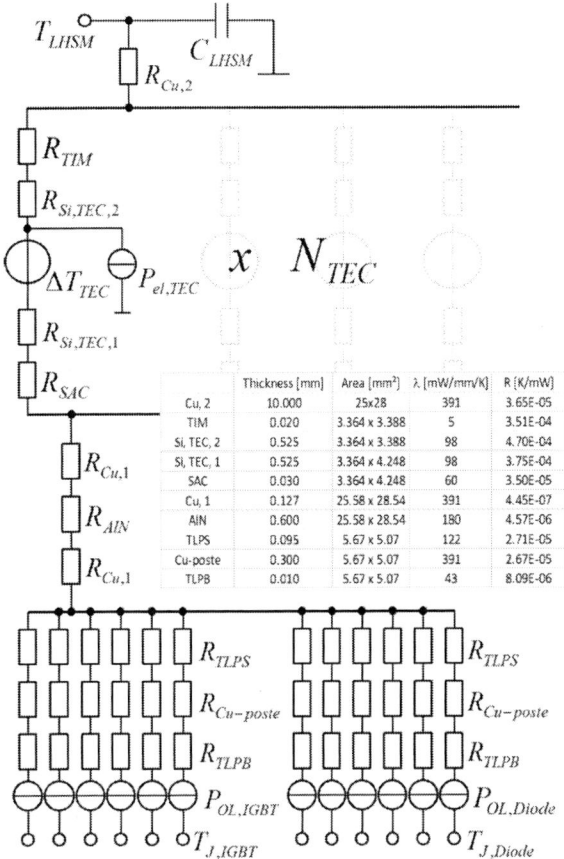

	Thickness [mm]	Area [mm²]	λ [mW/mm/K]	R [K/mW]
Cu, 2	10.000	25x28	391	3.65E-05
TIM	0.020	3.364 x 3.388	5	3.51E-04
Si, TEC, 2	0.525	3.364 x 3.388	98	4.70E-04
Si, TEC, 1	0.525	3.364 x 4.248	98	3.75E-04
SAC	0.030	3.364 x 4.248	60	3.50E-05
Cu, 1	0.127	25.58 x 28.54	391	4.45E-07
AlN	0.600	25.58 x 28.54	180	4.57E-06
TLPS	0.095	5.67 x 5.07	122	2.71E-05
Cu-poste	0.300	5.67 x 5.07	391	2.67E-05
TLPB	0.010	5.67 x 5.07	43	8.09E-06

Figure 3: Thermal resistance structure of the TEC-LHSM buffer on top of the power converter

The whole resistance stack is now simplified to the TEC-related thermal resistances R_1 and R_2 (see figure 2), neglecting spreading resistances and merging diodes and IGBTs by using the fact, that $P_{OL,Diode}$ is about 14.35% of $P_{OL,IGBT}$.

$$R_1 = R_{Si,TEC,1} + R_{SAC} + N_{TEC}\left(2R_{Cu,1} + R_{AlN} + \frac{R_{TLPS} + R_{Cu-poste} + R_{TLPB}}{6 \cdot 1.1435}\right)$$
$$R_2 = R_{Si,TEC,2} + R_{TIM} + N_{TEC}R_{Cu,2}$$

(4)

The TEC-related overload power P_{OL} (thermal power each TEC has to pump) is given in equation 5.

$$P_{OL} = \frac{P_{OL,tot}}{N_{TEC}} = \frac{6(P_{OL,Diode} + P_{OL,IGBT})}{N_{TEC}} = \frac{6.861 \cdot P_{OL,IGBT}}{N_{TEC}}$$

(5)

For the active TEC layer, the use of thin film TE modules was considered. The effective one-TE-block model data has been derived from comparing simulations (optimization routines) with experiments and data sheet values (see section 3).

As for the overload operation and for a LHSM of eutectic $Bi_{58}Sn_{42}$, the TECs will have to maintain a $\Delta T_{LHSM\text{-}HS}$ of at least 8 K ($T_{melt} = 138°C$, $T_{HS,max} = 130°C$) while pumping the heat load of $P_{OL,tot} = 43.02$ W. As shown in figure 4, a minimum number of ten and an optimal number of 38 thin film TE coolers are obtained, leading to TEC currents of $I_{TEC} = 0.9$ A and 0.265 A and a total heat power (converter overload + TEC's heat loss) of $P_{LHSM} = P_{OL,tot} + P_{el,TEC} = 120$ W and 69.1 W respectively.

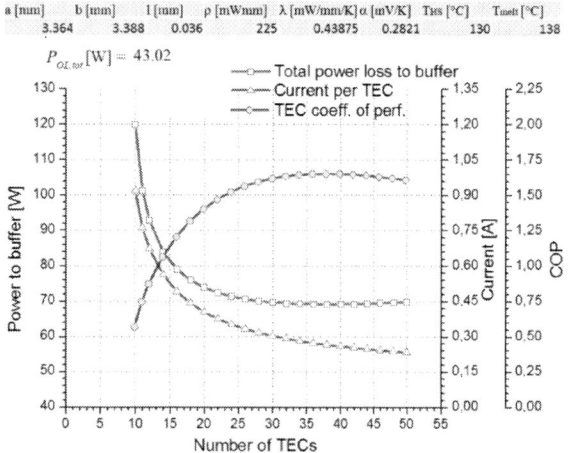

Figure 4: System COP and power loss as a function of the number of thin film TECs.

Due to the availability of the thin film modules, a small number of 12 TECs @ 0.67A was chosen for the TE simulations, leading to $P_{LHSM} = 93W$ and the heating energy of $Q_{OL} = 5.58kJ$ which has to be soaked by the LHSM during the overload of 60 s. For the $Bi_{58}Sn_{42}$ LHSM (latent heat density of 383 J/cm³) a volume of around 14.4 cm³ would be sufficient to cover this thermal capacity demand. In comparison to this, a block of aluminum would need to be more than 150 times larger (~2230cm³) considering a temperature rise of $\Delta T_{Al\text{-}block} = 1$ K.

3. Transient FEM and effective thermo-electric TEC modeling, validation measurements and material data

To reduce the complexity of the buffer's FE model, the TE coolers are modeled, using effective material data and only one single TE layer emulating the thermo-electric behavior of the complete TEC structure: numerous thermo-couples, metallization layers and thermal interfaces. This method has been described in [1, 2] and is depicted in figure 5 for the considered thin film TEC.

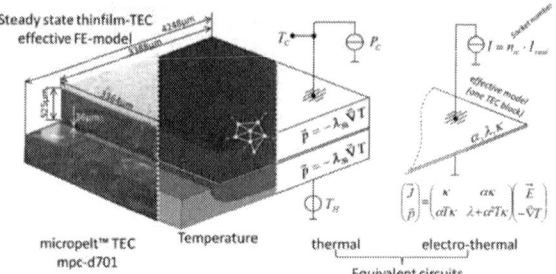

Figure 5: Field-coupled TE-simulation, principle of effective FE-model (thin film TEC) [2]

To confirm and also refine the data, including relation to samples as well as joining process degradation, TEC testing with a testing stand of reduced complexity were carried out.

Reduced complexity testing

In order to qualify the joining technology and to refine the simulation models by measurements on module level, a reduced complexity test device has been realized.

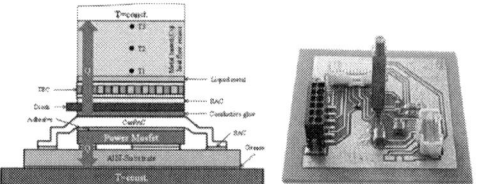

Figure 6: RC demonstrator schematic and assembled module

The main focus for the R.C. tester is on the testing of the transient TEC cooling of a thermal overload situation. The demonstrator consists of a heat source (MOSFET transistor), one TE module and a metal-based heat flow sensor on top. Several temperature sensors, including a diode in between the MOSFET and TEC and the heat flow sensor measure the thermal state of the system. The whole MOSFET-Diode-TEC-sensor stack is put in between two temperature controlled liquid-based heat sinks. Both the power dissipation in the transistor and the pumping power of the TE cooler can be controlled. Static operation as well as transients can be realized for power dissipation, heat pumping and data recording. See figure 6 for a conceptual sketch and the built RC device (exclusive the top and bottom heat sinks).

TEC validation and measurements

The R.C. demonstrator was used to validate the material data of the thin film TECs, characterized by the TEC testing device, described in [2]. Already in very good agreement, we could provide further refinement of these effective TEC data and a sample related characterization of the processed TECs.

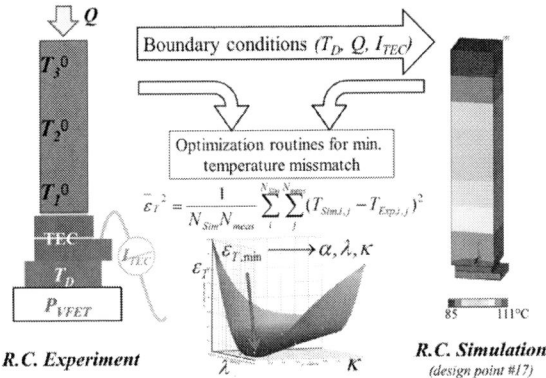

Figure 7: Principle of the TEC model optimization using the R.C. demonstrator

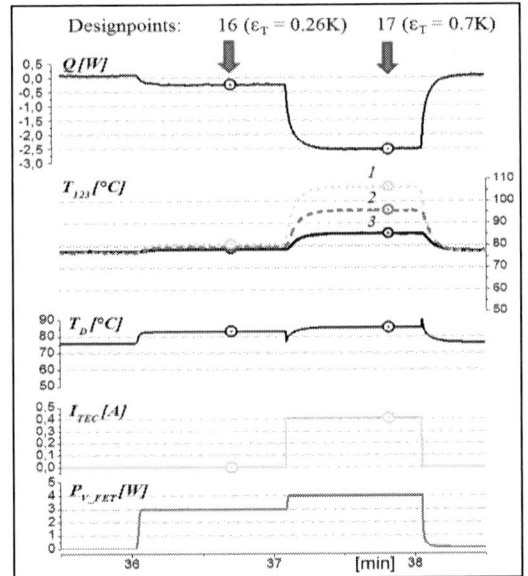

Figure 8: Simulation (circles) versus experiment (lines) plot (except of 2/24 design points)

Simulations are in very good agreement to experiments

Figure 7 shows the principle of the TEC data optimization, using the measured temperature and power data of the R.C. demo experiment as boundary conditions for an electro-thermal steady-state simulation. The temperature data of the heat flow sensor are used to fit the TEC-model performance by means of least square optimization. An excerpt of the simulation's design points is depicted in figure 8, using steady state data conditions within the transient measurements. The effective model shows excellent agreement between simulation and measurement. Since only minor scaling factors (for λ, ρ and α of the TE layer) had to be (re)applied, sample variation and drifts due to the soldering process can be stated to be of low influence.

As the thermal capacity of the very thin (~36μm) active layer of the TECs is only of minor impact to the transient TEC behavior, only static characterizations have

been done, also to keep the optimization process simple. Instead, literature data for density and specific heat of the bordering silicon substrates are used, that are of course well known. The optimized steady–state thermo-electric data that was found is listed in table 2.

Temperature [°C]	3	25	33	60	80	120
Th. conductivity [mW/mm/K]	0.394	0.398	0.402	0.411	0.422	0.422
El. Resistivity [μΩm]	216	216	216	216	216	216
Seebeck coeff. [μV/K]	285.4	287.3	288.1	290.6	293.8	293.8

Table 2: Refined temperature-dependent electro-thermal data of a R.C. demo TEC mpc-d701

Note, as this is an effective model, the data differs from the Bi_2Te_3 values given in the literature, depending on the number of TE couples and filling factor of the TE structure.

Transient thermal material data

The thermal and thermo-electric material data that is used in the simulations is given in table 3. Particle or volume percentage analyses have been carried out on the TLPB's and TLPS's micro sections to determine effective material data of these interfaces. The data of the heat pipes has been approximated using the given typical thermal resistance within the datasheet and are also discussed in [36].

Since most of the LHSM's literature lacks of exact $c_p(T)$ distribution data, a simplified "peak type" $c_p(T)$ model has been applied in analogy to the enthalpy based model description in [37].

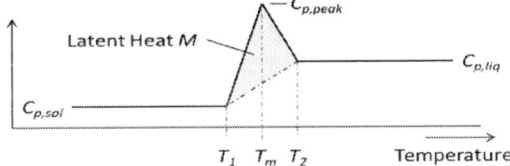

Figure 9: Specific heat of LHSM using a simplified "peak type" $c_p(T)$ distribution

The specific heat's peak value is defined by the latent heat M, the solid's and liquid's specific heat ($c_{p,sol}$ and $c_{p,liq}$) and the melting temperature range T_2-T_1 (equation 6).

$$c_{p,peak} = \frac{2M}{T_2 - T_1} + \frac{(c_{p,sol} + c_{p,liq})}{2} \qquad T_m = \frac{T_1 + T_2}{2} \qquad (6)$$

	Th. Cond. [mW/mm/K]	Density [t/mm³]	Specific heat [mJ/t/K]
Si	148	2.33E-9	712E6
TLPB (eff.)	43	8.53E-9	302E6
TLPS (eff.)	122.1	8.42E-9	310.7E6
Cu	398	8.92E-9	385E6
Al	237	2.7E-9	897E6
AlN	285	3.26E-9	740E6
HeatP. (eff.)	10k (100k) [36]	-	-
TIM TEC-LHSM	5	-	-
$Bi_{58}Sn_{42}$	19 (sol.) ~15 (liq.) [34, 35]	8.56E-9 [34]	c_{sol}= 167E6 c_{liq}= 201E6 T_1= 137°C T_2= 139°C M= 44.8E9 mJ/t [34]

Table 3: Thermal material data @ RT
Unity system: mm-t-s-A-K

4. Thermo-electric transient FE simulations: Thermal buffer concepts and feasibility study

The requirement to the TEC assisted thermal buffer is to provide a near-zero (referred to overload power) thermal resistance and literally soak all the OL heat, inducing no temperature increase. In order to develop guidelines and design rules, as well as to show the feasibility of the cooling concept, electro-thermal transient finite element simulations have been performed.

Compared to the 1D estimation, described in section 2 (and stating only minor effects of the heat capacity and in-plane heat spreading within the layers of the top-sided assembly), first simulations using a perfect conducting LHSM buffer have shown an excellent agreement with the one-dimensional model. However, applying realistic material data, 3 thermal difficulties could have been revealed:

> ➢ Insufficient diffusivity of the thermal LHSM buffer, during phase change, which leads to a non-uniform melting process and thus unwanted and considerable temperature increase of T_H and T_{HS} (i.e. the TEC's heating side and IGBT junction temperature) → ΔT_1

> ➢ "Back-swapping" of the stored energy, since there is no TEC current applied during the relaxation period (I_{TEC} = 0A) to hold the temperature difference T_{LHSM} -T_{HS}. As a consequence, the LHSM buffer heats up the IGBT. → ΔT_2

> ➢ The relaxation time or performance of the LHSM's top-sided cooling has to be dimensioned well, so that the thermal system can fully reset to a defined pre-overload situation. Otherwise, the stored energy of the system rises with each overload cycle and no transient cooling can be realized anymore. → ΔT_3

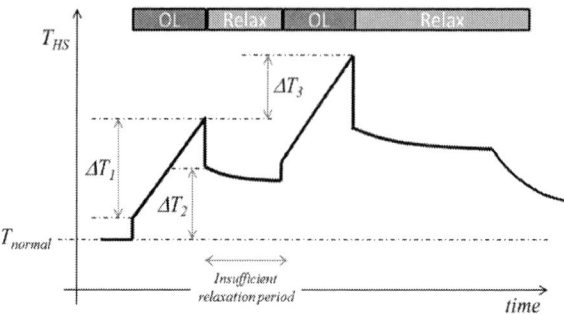

Figure 10: IGBT temperature increase, due to problematic thermal effects of the cooling concept

The different effects and impacts onto the IGBT temperature are depicted in figure 10.

To address the problem of diffusivity, which is the most critical, several LHSM buffer concepts with copper/aluminum foam- and wrap-type conductivity enhancements also in combination with heat pipes have been studied and assessed by the value of ΔT_1. The different concepts that have been considered as well as the final design, which will be described in detail below, are shown in figure 11. The effective material models and ΔT_1 results can be found in table 4.

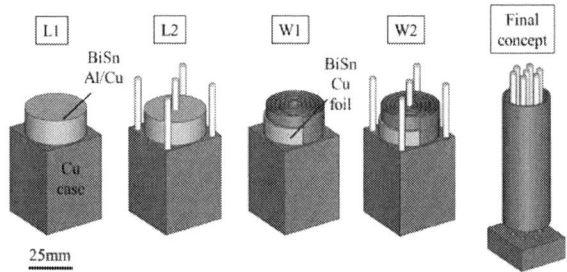

Figure 11: LHSM buffer concepts using heat pipes and copper/aluminum conductivity enhancing structures
L1 – foam/sponge type LHSM compound
L2 – L1 with heat pipes put into the case
W1 – wrap of copper/$Bi_{58}Sn_{42}$ foil as LHSM comp.
W2 – W1 with heat pipes put into the case

As can be seen in table 4, an effective diffusivity enhancement by the depicted concepts could only be sufficiently satisfied with a reactor type LHSM module. The conductivity gain of copper or aluminum based compounds is much too small, though decrease the LHSM volume. However, heat pipes provide a conducting thermal path, which is orders of magnitude higher compared to the LHSM. So these seem to be the only way to successfully address the problem of low thermal diffusion.

Concepts	LHSM compound (%Vol.)			LHSM Th. cond. λ [mW/mm/K]	Dens. ρ [t/mm³]	Spec. heat. C_p [mJ/t/K]	ΔT₁ [K]
	Bi Sn	Al	Cu				
L1	100	0	0	19 (sol.) ~15 (liq) #	8.56 E-9 #	c_{sol}= 167E6 c_{liq}= 201E6 T_1= 137°C T_2= 139°C M= 44.8E9 mJ/t *	14
L1	65	35	0	95 (sol.) 93 (liq.) +	6.51 E-9 ##	c_{sol}= 423E6 c_{liq}= 445E6 T_1= 137°C T_2= 139°C M= 28.8E9 mJ/t **	11
L1	65	0	35	152 (sol.) 149 (liq.) ++	8.69 E-9 ###	c_{sol}= 243E6 c_{liq}= 265E6 T_1= 137°C T_2= 139°C M= 28.8E9 mJ/t ***	10
L1	100	-	-	∞ #		*	0
fiction 20xλ_real	65	-	-	380 (sol.) 500 (liq.)	##	**	2.5
W1	65	35	0	φ,z: 95 (sol.) 93 (liq.) r: 28 (sol.) 22.3 (liq.) +++	##	**	11
W1	65	0	35	φ,z: 152 (sol.) 149 (liq.) r: 28.5 (sol.) 22.6 (liq.) ++++	###	***	10
L2	65	35	0	+	##	**	10
L2	65	0	35	++	###	***	9.5
W2	65	35	0	+++	##	**	10
W2	65	0	35	++++	###	***	9.5
final	85	No effective material model					3

Table 4: Lumped, wrap and "reactor type" model design testing, assessed by the temperature rise ΔT_1 during overload, T_{sol} = RT, T_{liq} > 138°C

Since the limiting factor (when using heat pipes), referring to the heat diffusion into the LHSM is the radial heat path from heat pipe to LHSM, only a dense arrangement of heat pipes put directly into the LHSM material has led to satisfying results.

"Reactor" type thermal buffer simulation

As shown in figure 12, the following solutions have been found, addressing the problematic thermal effects, described above:

> Diffusion enhancement through seven d3x50mm heat pipes (~18W power capability each) within the $Bi_{58}Sn_{42}$ LHSM buffer (figure 12, upper left, quarter model, capped hidden view)
> A TEC's relaxation phase de-ramping operation from 70mA to 0A during 9 minutes, according to the equivalent circuit calculations for P_{OL} = 0W (equation 3)

> And a relaxation period of 19minutes, to provide sufficient time to reset the thermal system.
> Note: Since thermal data of the bottom-side cooled EasyPim module is already available, and to reduce calculation time and the simulation's complexity, effective convection coefficient parameters have been applied underneath the IGBTs/Diodes six-pack, emulating the bottom-side cooling.

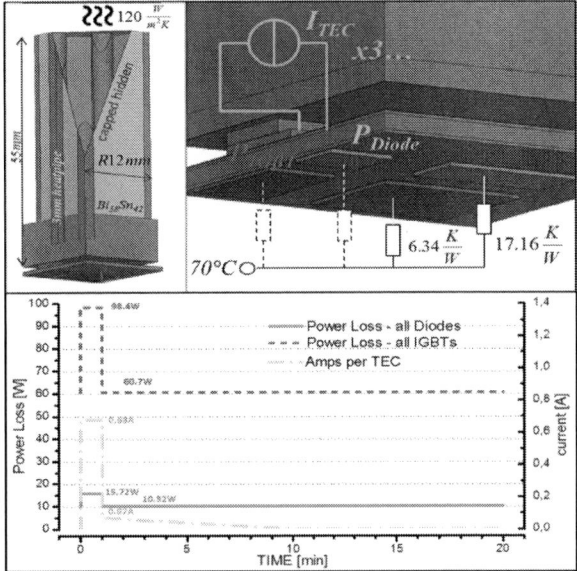

Figure 12: ¼ transient electro-thermal FE model and BC of the thermal buffer system for the IGBT power device.

Figure 13 shows that, in comparison to no-TEC-LHSM cooling and L1 (35% Al) the junction temperature can indeed be maintained at around 130°C. Only a small increase (to 132°C) of the junction temperature from normal to overload operation has to be accepted, stating that the problem is solved by this buffer architecture.

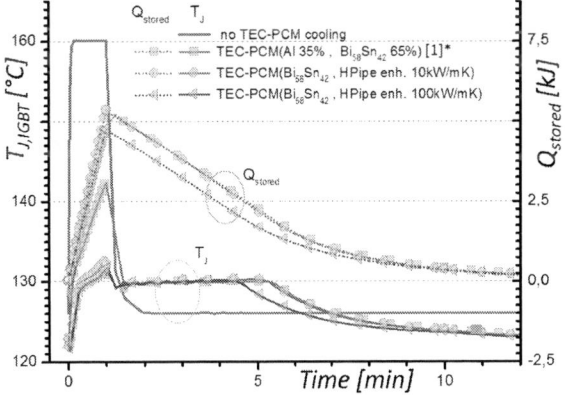

Figure 13: Simulation results: IGBT Junction temperature and stored thermal energy
* [1] with TEC post-cooling

The limiting diffusion factor as mentioned before is the radial heat path from heat pipe to LHSM, which can be stated because of the small difference between simulations of 10kW/m/K vs. 100kW/m/K heat pipe performance (gain saturation). This can also be seen in the temperature profile during the overload situation (figure 14), showing the superior vertical thermal heat conduction caused by the heat pipes.

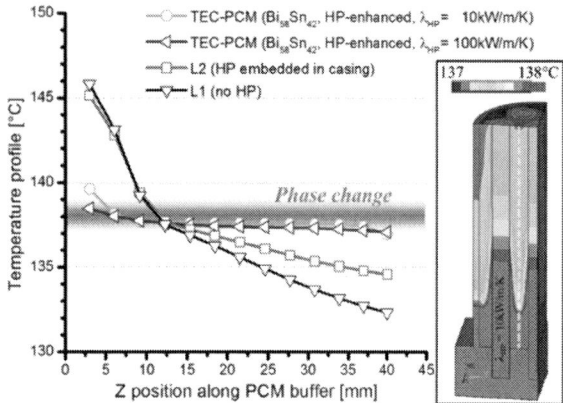

Figure 14: Temperature profiles of the buffer concepts, middle of heat buffer (lumped models) or in between of two heat pipes (reactor type concept) at $t_{OL} = 30s$

Referring to the energy balance or stored thermal energy, a thermal system reset can be stated at around 12 minutes.

Summarizing the design rules

➤ An optimal number of TE couples should be used to minimize the heat loss caused by the TECs, and thus power, that has to be buffered and conducted into the LHSM. This can be calculated as shown in section 2 and dependents on the loading conditions and thermal network.

➤ A sufficient LHSM volume is needed to buffer all the heat energy, including the TEC's heat loss, depending on the overload power and duty cycle of the power module. For a LHSM of Bi58Sn42 the latent heat density is 383J/cm³. To buffer the 5.5kJ of thermal overload energy, at least 15cm³ needs to be considered, excluding the volume of the heat pipes that are essential to thermally access the LHSM.

➤ A thermal diffusivity enhancement within the heat buffer has to be applied. The use of heat pipes is mandatory here to provide a sufficient vertical conduction. Note that heat pipes lose their functionality at higher power values, due to the dry out effect. Here a number of 7 heat pipes were considered, capable of 18W heat conduction each (~135% of the 92W power load into the buffer). In that context, the radial thermal conductivity from heat pipe to LHSM was found to be the

limiting factor of the thermal diffusion problem. So the performance of the heat pipe itself is only of minor influence, due to a gain saturation of λ_{HP} (shown in figure 13). Further improvement measures should aim for a radial mesh or in-plane fin structure to increase the heat diffusion.

➤ Post-cooling (see equation 3, $P_{OL} = 0W$) and a sufficient time of thermal recovery has to be applied.

With the feasibility of this concept proven, this system is going to be built in combination with test stands of reduced complexity, validating and also successively refining the simulations by means of material data and models.

5. Novel high-temperature double-side joining technologies and process sequences

Technological overview

Compared to conventional power module buildup (power die soldered on, and wire-bonded to substrate), double sided cooling presents the advantage of offering two solid thermal paths with good thermal conductivity for higher efficiency in thermal extraction and dissipation of energy losses. As such it widens the scope of power applications towards higher power density of components. Different assembly technologies can be applied for building double-sided cooling. They all rely in the achievement of interconnections targeting high temperature applications.

Transient liquid phase bonding (TLPB) is one of the alternative technologies. Unlike established soldering used for first level die bonding of vertical power components like IGBTs and diodes, the solder material cannot remelt when operation temperature exceeds the soldering temperature. Effectively during joining, the thin solder layer melts and fully transforms into intermetallic compounds, shifting the temperature of liquid phase formation towards higher values. Due to the thinness of the solder layer, the reaction is completed within a short time. After bonding the bondline consists in thermally stable phases, and no remelting of the joining material can take place. The bondline is maintained also very thin, enabling better heat transfer.

A second alternative is sintering of silver powders. Powder is mixed with organics to form a paste which is afterwards applied by stencil printing on the substrate. Subvariations in processes are considered, ie. with or without pressure application during sintering, to obtain a good bonding of the die with the substrate. During heating up, firing of silver paste takes place, organics evaporate followed by densification of silver particles. Sintering occurs by atomic diffusion engendering a bulk silver bond with low or no porosity. As a consequence, the achieved bondline has good heat transfer properties due to high conductivity of the bond material and is thermally stable. Moreover, silver sintered interconnects depicts better behavior under thermo-cycling fatigue than conventional SAC solders.

Transient liquid phase soldering (TLPS) represents the third alternative. Contrary to sinter silver sintering, the transient liquid phase soldering (TLPS) is a fully new developed joining technology based on leadfree composition. Similar to TLPB, this solder based interconnection technology relies on phase transformation. Contrary to TLPB which necessitates electroplating of the solder, TLPS is based on powder technology. Low and high melting powders are mixed together, typically Sn-based and copper powders. The resulting paste is applied by stencil printing and under the influence of heat temperature, the whole interconnect joint is fully transformed into intermetallics [33].

One of the potential drawback compared to silver sintering concerns the thermal efficiency of the interconnect. Also, the TLPS layer is far thicker than the TLPB bondline since the process is based on powder (figure 15). Intermetallics depict lower thermal conductivities than silver, probably resulting in a poorer thermal path, ie. a higher thermal resistance of the TLPS join compared to the other technologies.

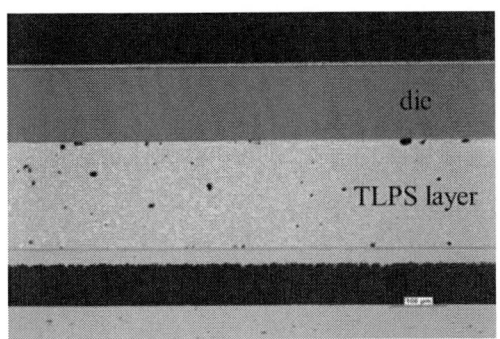

Figure 15: Micro section of a TLPS interconnect

Power module Assembly

The combination of the described technologies requires wafer level back end processes. First, active wafers must be post processed to obtain adequate surface finishes. Wafer post-processing includes temporary wafer bonding for safer handling and electroplating of copper and tin on the electrical contacts. After dicing, diodes and IGBTs (voltage class 1200 V) are mounted per flip-chip on substrate, typically Direct-Copper bond substrates (DCB). The gate and Emitter of the IGBTs and the anode of the diodes are connected to the substrate. The achieved TLPB bondline consists in a Cu6Sn5 intermetallic layer between two Cu3Sn intermetallics layers formed by interdiffusion of copper and tin. The thickness is in the range of some micrometers. Subsequently an underfilling is done to protect and isolate the connections. To build the inverter module, top-bottom electrical feedtrough are required. Copper elements are added and also joined using the TLPB approach. Afterwards the top DCB

substrate is mounted on the backside of the components using the two others technologies previously described, ie. TLPS or sintering. The resulting half bridge module is shown in figure 16. Here TLPS has been used to contact the devices with the second DCB substrates. As it can be seen in figure 16. TECs can further be mounted on the top layer of the top DCB.

Figure 16: Double-sided cooling power module (double half bridge inverter module) after assembly

6. Conclusion and outlook

Despite the fact, that thermo-electric coolers always contribute a considerable amount of power loss in addition to the devices heat loss, in combination with a latent heat storage buffer and designed properly, we have shown that this thermal buffer concept opens up new possibilities to meet problematic thermal transients and hot spot cooling in situations, where classical solutions fail.

As this thermal cooling concept is completely new, transient electro-thermal simulations have been performed to show the thermal feasibility and function of various buffer concepts. In that way, guidelines and thermal design rules could be given, to bypass limits, such as low the thermal diffusivity of the thermal buffer.

Simulation strategies, effective thermal modeling of a material phase change/melting and electro-thermal modeling of (thin film) TE coolers have been described and verified by experimental setups of reduced complexity. The requirements of the considered IGBT power module (~5K temperature increase of the IGBT junction during a 150% overload operation) could be successfully satisfied by a reactor type thermal buffer, using a LHSM of $Bi_{58}Sn_{42}$ in conjunction with thin film TE coolers and standard heat pipes. While the usability for common heat removal problems will have to be shown, transient low duty-cycle power modules have shown to be potential applications for such an advanced cooling device.

Furthermore, technological aspects of the double-sided DCB assembly have been discussed. The targeted high-temperature and –conducting joining technologies (TLPS/TLPB and sinter silver) as well as low tolerances during assembly and hierarchy of the soldering process are the key, to make this complex thermal system work.

The lifetime and reliability of the system will have to be considered as well, focusing mainly on the thermo-mechanical stability of the thermal interfaces as well as the temperature and lifetime degradation of the TE coolers and LHSM (e.g. ZT or melting temperature drift).

Acknowledgments

The authors gratefully acknowledge their funding by the EU FP 7 Integrated Project "Smartpower", Grant #288801.

Also we would like to thank our co-workers Dr. Wilhelm Maurer from Infineon Technologies and Dr. Tristan Caroff from CEA Liten for scientific discussions and Tobias Xhonneux from Taipro Engineering for help to build up the RC demonstrator.

References

1. B. Wunderle, C.-A. Manier, M. A. Ras, M. Springborn, D. May, H. Oppermann, M. Toepper, R. Mrossko, T. Xhonneux, T. Caroff, W. Maurer, und R. Mitova, "Double-sided cooling and thermo-electrical management of power transients for silicon chips on DCB-substrates for converter applications: Design, technology and test",19th International Workshop on Thermal Investigations of ICs and Systems (THERMINIC), 2013, S. 253–261.

2. B. Wunderle, M. Abo Ras, M. Springborn, D. May, J. Kleff, H. Oppermann, M. Topper, T. Caroff, R. Schacht, and R. Mitova. "Modelling and characterisa-tion of smart power devices". 18th International Workshop on Thermal Investigations of ICs and Systems (THERMINIC), 2012.

3. Caroff, T.; Mitova, R.; Wunderle, B.; Simon, J., "Transient cooling of power electronic devices using thermoelectric coolers coupled with phase change materials", Thermal Investigations of ICs and Systems (THERMINIC), 2013 19th International Workshop on , vol., no., pp.262,267, 25-27 Sept. 2013

4. X. Wang, J. Xu, F. Zhang et al., "Phase change materials at the cold/hot sides of thermoelectric cooler for temperature control". Int. Conf. on Smart Mat. and Nanotech. in Eng. Harbin, 2007

5. S.D. Sharma, K. Sagara, "Latent heat storage materials and systems: A review", International Journal of Green Energy, 2: 1–56, 2005

6. Murphy, E.; Michie, C.; White, H.; Johnstone, W., "Phase change materials for passive cooling of active optical components," Avionics, Fiber- Optics and Photonics Technology Conference (AVFOP), 2012 IEEE , vol., no., pp.11,12, 11-13 Sept. 2012

7. Chamarthy, P.; Utturkar, Y., "Theoretical evaluation and experimental investigation of microencapsulated phase change materials (MLHSM) in electronics cooling applications", Semiconductor Thermal Measurement and Management Symposium, 2009. SEMI-THERM 2009. 25th Annual IEEE , vol., no., pp.239,245, 15-19 March 2009

8. Faraji, M.; El Qarnia, H., "Cooling management of a protruding electronic components by using a phase change material heat sink," Electronics, Circuits and Systems, 2007. ICECS 2007. 14th IEEE International Conference on , vol., no., pp.174,177, 11-14 Dec. 2007

9. Faraji, M.; El Qarnia, H., "Cooling management of a protruding electronic components by using a phase change material heat sink," Electronics, Circuits and Systems, 2007. ICECS 2007. 14th IEEE International Conference on , vol., no., pp.174,177, 11-14 Dec. 2007

10. J. Torresola, "Solification properties of certain waxes and paraffins", M.Sc. Thesis Massachusetts institute of technology, 1998.

11. G. Meyer, "Thermal properties of micro-crystalline waxes in dependence on the degree of decoiling", SOFW-Journal, 135, pp. 42-50, 2009

12. Kim, J.; Golliher, E., "Steady state model of a micro loop heat pipe," Semiconductor Thermal Measurement and Management, 2002. Eighteenth Annual IEEE Symposium , vol., no., pp.137,144, 12-14 March 2002

13. Kammuang-lue, N.; Sakulchangsatjatai, P.; Terdtoon, P., "Effect of working fluids on thermal characteristic of a closed-loop pulsating heat pipe heat exchanger: A case of three heat dissipating devices," Electronics Packaging Technology Conference (EPTC), 2012 IEEE 14th , vol., no., pp.142,147, 5-7 Dec. 2012

14. Lin, Z.R.; Lin, W.Z.; Zhang, L.W.; Wang, S.F., "An experimental study on applying miniature loop heat pipes for laptop PC cooling," Semiconductor Thermal Measurement and Management Symposium (SEMI-THERM), 2013 29th Annual IEEE , vol., no., pp.154,158, 17-21 March 2013

15. Peters, T.B.; McCarthy, M.; Allison, J.; Dominguez-Espinosa, F.A.; Jenicek, D.; Kariya, H.A.; Staats, Wayne L.; Brisson, J.G.; Lang, Jeffrey H.; Wang, E.N., "Design of an Integrated Loop Heat Pipe Air-Cooled Heat Exchanger for High Performance Electronics," Components, Packaging and Manufacturing Technology, IEEE Transactions on , vol.2, no.10, pp.1637,1648, Oct. 2012

16. Ji Li; Daming Wang; Peterson, G.P., "A Compact Loop Heat Pipe With Flat Square Evaporator for High Power Chip Cooling," Components, Packaging and Manufacturing Technology, IEEE Transactions on , vol.1, no.4, pp.519,527, April 2011

17. Ivanova, M.; Avenas, Y.; Schaeffer, C.; Dezord, J.B.; Schulz-Harder, J., "Heat Pipe Integrated in Direct Bonded Copper (DBC) Technology for the Cooling of Power Electronics Packaging," Power Electronics Specialists Conference, 2005. PESC '05. IEEE 36th , vol., no., pp.1750,1755, 16-16 June 2005

18. Aoki, H.; Shioya, N.; Ikeda, M.; Kimura, Y., "Development of ultra thin plate-type heat pipe with less than 1 mm thickness," Semiconductor Thermal Measurement and Management Symposium, 2010. SEMI-THERM 2010. 26th Annual IEEE , vol., no., pp.217,222, 21-25 Feb. 2010

19. Elnaggar, M.H.A.; Abdullah, M.Z.; Munusamy, S.R.R., "Experimental and Numerical Studies of Finned L-Shape Heat Pipe for Notebook-PC Cooling," Components, Packaging and Manufacturing Technology, IEEE Transactions on , vol.3, no.6, pp.978,988, June 2013

20. S. Mahjoub, and A. Mahtabroshan, "Numerical simiulation of a conventional heat pipe," in Proc. World Acad. Sci., Eng. Technol. Conf., vol. 29. 2008, pp. 117–122.

21. J. C. Wang, "Novel thermal resistance network analysis of heat sink with embedded heat pipes", Jordan J. Mech. Ind. Eng., vol. 2, no. 1, pp. 23–30, 2008.

22. G.J. Snyder, M. Soto, R. Alley, D. Koester and B. Conner. "Hot Spot Cooling using Embedded Thermoelectric Coolers". Proc. 22nd Semitherm Conf. 2006.

23. Russel, M.K.; Ewing, D.; Ching, C.Y., "Numerical and Experimental Study of a Hybrid Thermoelectric Cooler Thermal Management System for Electronic Cooling," Components, Packaging and Manufacturing Technology, IEEE Transactions on , vol.2, no.10, pp.1608,1616, Oct. 2012

24. Bar-Cohen and P. Wang. "On-chip Hot Spot Remediation with Miniaturized Thermoelectric Coolers". Microgravity Sci. Technol. 21 (Suppl 1). pp. 351–359, 2009

25. F. Schindler-Saefkow, M. Jägle, B. Wunderle, H. Böttner and J. Nurnus. "Thermal-Electric Simulation of Micropeltier Elements for Hot Spot Cooling". Micromaterials and nanomaterials 6 (2007), S.218, ISSN: 1619-2486, 2007.

26. Z. Yuanyuan, Y. Jianlin, "Design optimization of thermoelectric cooling systems for applications in electronic devices". International journal of refrigeration, 35, 1139-1144, 2012

27. W. Seifert, M. Veltzen, C. Strümpel, W. Heiliger, and E. Müller, "Onedimensional modeling of a peltier element," in Proc. 20th Int. Conf. Thermoelectrics (ICT), Beijing, China, Jun. 2001.

28. M. Hodes, "On one-dimensional analysis of thermoelectric modules (TEMs)", IEEE Trans. Comp. Packag. Technol., vol. 28, no. 2, pp.218–229, Jun. 2005.

29. M. Labudovic and J.Li, "Modeling of TE Cooling of Pump Lasers")", IEEE Trans. Comp. Packag. Technol., vol. 27, no. 4, pp.724–730, Dec. 2004.

30. E.E. Antonova, D.C. Looman. "Finite elements for thermo-electric device analysis in ANSYS". ICT 2005, pp.215-218.

31. M. Jaegle. "Simulating Thermoelectric Effects with Finite Element Analysis using COMSOL". Poster Contribution at the ECT 2007.

32. Park, W. et al. "Effect of thermal cycling on commercial thermoelectric modules". IEEE Intersociety Conference on Thermal and Thermomechanical Phenomena in Electronic Systems (ITHERM) 2012, May 30 - June 1, San Diego, CA

33. Ehrhardt C., Hutter H., Oppermann H., Lang K. D.,. "Transient Liquid Phase Soldering – An emerging joining technique for power electronic devices", PCIM Europe 2013 , 14 – 16 May 2013, Nuremberg, pp. 591-599

34. Indium Corp. Indalloy® 281 Bi-Sn Solder Alloy, @ http://www.matweb.com

35. R. E. Krzhizhanovskii, N. P. Sidorova, I. A. Bogdanova. "Experimental investigation of the electrical resistivity of some molten bismuth-tin binary alloys and of the thermal conductivity of bismuth, tin, and a eutectic bismuth-tin alloy", Journal of engineering physics, January 1974, Volume 26, Issue 1, pp 33-36

36. Geoffery Thyrum, Ellen Cruse, "A simplified technique for modeling heat pipe assisted heat sinks", Advanced Packaging (2001), pp. 23-27 or @ http://electroiq.com

37. E.E.U. Haque, P.R.Hampson, "Modelling phase change in a 3D thermal transient analysis", Int. Journal of Multiphysics, 2013

Simulation and Measurement of Pressure Dependent Q-factors in NEMS Resonators

J. Manz, G. Schrag, G. Wachutka
Institute for Physics of Electrotechnology
Munich University of Technology
Arcisstr. 21, 80333 Munich

Abstract

The fluidic damping and the hereof resulting Q-factor of various mechanical resonators with gaps of nanometer size between suspended part and substrate was theoretically and experimentally determined. These investigations have been carried out in the pressure regime from atmospheric pressure down to about 3 Pa.

The air flow in the nanogap between the movable part of the device and the substrate was modeled by extending the mixed level model presented in [1, 2] to the slip flow and molecular dynamical regime. The pressure-dependent measurements were carried out using a Laser-Doppler vibrometer. The extracted Q factors conform very well with those expected from theory in every pressure regime. This is a noticeable result, because even at normal pressure the range of validity for a continuum-theoretical description is reached for the nanometer feature sizes considered in this work.

1. Introduction

The damping dynamics of mechanical transducers are characterized by the Q-factor, which is a measure of the energy dissipated in the device. The most prominent damping mechanisms in MEMS/NEMS structures are fluidic damping, anchor losses, thermoelastic damping, and surface damping [3].

Thermoelastic damping describes the energy loss originating from an irreversible heat flow due to the bending in the moving structure and is – for simple device geometries - well understood by the Zener Model [4, 5]. In the devices investigated in this work thermoelastic damping has no significant influence on the operation, thus it can be neglected. Anchor loss occurs when an acoustic bulk or surface wave is emitted from the moving structures and radiated into substrate at the anchor regions. The phenomenon surface damping comprises the wide field of all damping mechanisms which are related to the surface to volume ratio of the structure and is focus of current research. In this work we focus on the fluidic damping due to the airflow in the gap between the moving structure of the device and the substrate.

Since technological progress leads to continuously decreasing feature sizes and, in consequence, to smaller devices (Nano Electro Mechanical Systems, NEMS) the limit of standard continuum-theoretical description is often reached already for normal ambient conditions and the predictive determination of the Q-factor becomes a challenge. We discuss the fluidic damping in the moderate to high Knudsen regime, where the dimensions of the device are of the same order of magnitude or even much larger than the mean free path of the air molecules,

and we investigate the validity of an existing, but modified mixed-level approach [1, 2] for this regime. In order to validate the derived models, test structures of NEMS sensors of different geometries and perforation grade with air gaps up to 280 nm underneath the suspended membrane have been modeled and optically characterized by Laser-Doppler vibrometry. A vacuum chamber enables pressure variations and, hence, the extraction of Q-factors from room pressure down to about 3 Pa.

In section 2 the theoretical approach and the applied methodology is summarized, followed by the device description and the measurement setup in section 3 and the discussion of the results in section 4. Finally, a short summary and an outlook on work in progress conclude this paper.

2. Theoretical Approach and Modeling Methodology

The viscous damping occurring in the considered structures is dominated by the so-called squeeze film effect, which occurs, when a fluid is squeezed out of a small gap underneath a vertically moving plate. Hence, this phenomenon depends strongly on the time and spatial variation of the gap height and, thus, on the deflection of the moving structure.

Squeeze film damping can be described by the well-known two-dimensional Reynolds equation [6], a simplification of the Navier-Stokes equation, which holds under the assumption of laminar fluid flows and lateral device dimensions, which are much larger than the thickness of the fluid film. This equation is then evaluated by transforming it into a distributed finite network according to the mixed-level approach as described in [1, 2]. An on-purpose written conversion program provides this distributed flux-conserving network description of the Reynolds equation by taking the topological information from a finite element discretization of the device. In this Kirchhoffian network, the above mentioned Reynolds equation and the mass continuity equation constitute balance equations where the mass flow acts as generalized flux variable and the pressure as generalized force variable. Pressure drops at the boundaries and tentative perforations in the structure, which are not accounted for by the Reynolds equation, are considered by adding fluidic boundary resistors and a dedicated "hole-model", respectively [7]. The fluidic network is coded in a hardware description language and numerical simulations can be carried out using a standard system simulator. Converting the FE discretization into a finite network in combination with applying lumped elements for parts of the structure (so-called "mixed-level" approach [1, 2] reduces the degrees of freedom so drastically, that simulations can be carried out in the order of minutes.

However, it has to be kept in mind that the above discussed model is based on continuum theory, which is only valid, if the mean free path of the air molecules is much smaller than a characteristic length of the structure. This corresponds to Knudsen numbers (= ratio of mean free path to characteristic feature size) of below 0.001. For the NEMS structures considered in this work and described more detailed in the following section, the Knudsen numbers range from 0.1 to 10 000, depending on the ambient pressure. So they operate far beyond the continuum-theoretical regime.

In order to account for these gas rarefaction effects, correction factors for the specific elements of the Kirchhoffian network were introduced, which were taken from literature (overview given in [2]). For the two-dimensional Reynolds flow we apply a correction factor taken from [8] which approximates a calculation of the pressure-dependent mass flow underneath a square plate by Fukui [9, 10]. The effects at the boundaries and the holes are taken into account by a fitting formula from [11], which is based on the work of Sharipov [12]. Both, Fukui and Sharipov evolve their work from the Boltzmann equation.

Hence, we describe the fluidic effects in our test structures through a model, which has been derived on the basis of continuums theory although they operate in the transition to the free molecular regime. At the first glance, this looks counterintuitive. However, the introduced correction factors are based on formula derived from the Boltzmann equation and, hence, the aim of our experiments (see section 4) will be to investigate to which extent the above presented simulation technique is applicable to structures with very small gaps and at low pressure.

3. Device Description and Measurement Setup

The following section describes the design, the mechanical properties and the mode shapes of the tested structures. We measured two types of cantilevers (Fig. 2) and one perforated paddle-like structure, further-on denoted as "paddle" (Fig. 3).

The structures are fabricated on the basis of a silicon – silicon-germanium – silicon stack, where the silicon-germanium layer serves as sacrificial layer in a selective etch process [13]. Due to the almost similar lattice constants of silicon and silicon-germanium all layers can be grown in a monocrystalline structure which provides a good mechanical stability due to the absence of intrinsic stress. Fig. 1 shows an exemplary etch profile. The red rectangle visualizes the corresponding cross section of a 1 µm wide cantilever. The average thickness of the structure is 160 nm and the average height of the nanogap amounts to 280 nm.

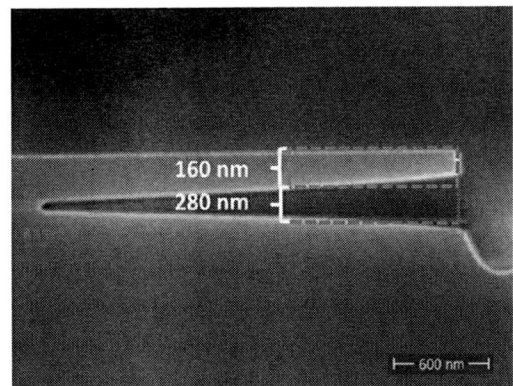

Fig. 1: Microphotographs of an etch profile of the nanogap. The red rectangle visualizes the corresponding cross section of the cantilever. The average thickness of the structure is 160 nm and the average height of the nanogap is 280 nm.

Both cantilevers exhibit a length of 15 µm and widths of 1 µm and 2 µm, respectively. Fig. 4 shows the simulated and measured mode shape of its first eigenmode.

The perforated paddle is 24 µm wide and is connected to the anchors via two clamps. In the first eigenmode the paddles tilts around the axis through the two clamps like a teeter-totter; in the higher eigenmodes the paddle area is deformed itself, see Fig. 5, in which FEM simulations of the first four mode shapes are shown. These mode shapes are needed to simulate the time-varying gap underneath the movable structure and, hence, are fed into the mixed-level model to calculate the pressure-dependent damping and, hereof, the Q factors of the respective eigenmodes. These are then extracted also from measurements for comparison with the simulated data.

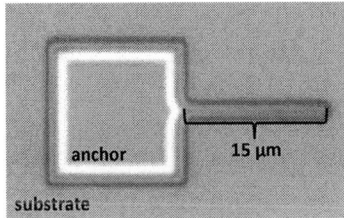

Fig. 2: Microphotograph of the cantilever.

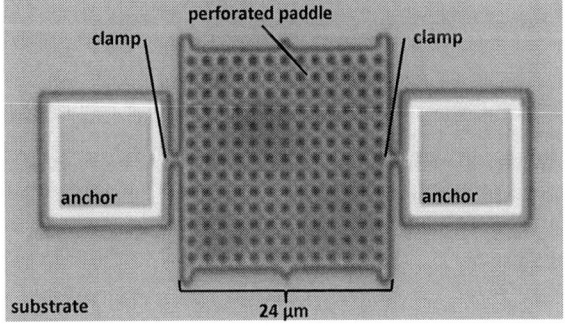

Fig. 3: Microphotograph of the perforated paddle.

 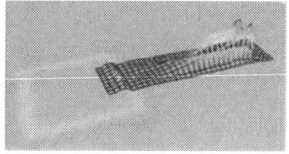

Fig. 4: Simulated and measured mode shape of the first eigenfrequency of a cantilever.

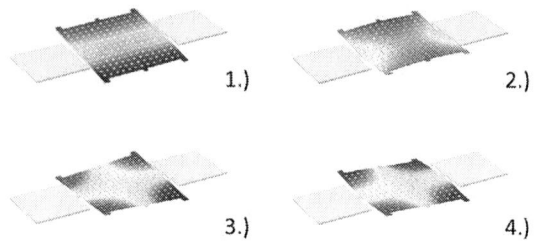

Fig. 5: Simulated mode shapes of the first four eigenfrequencies of the paddle.

In order to control the pressure during these measurements the structures were mounted in a vacuum chamber and characterized by a Laser-Doppler vibrometer, see Fig. 6.

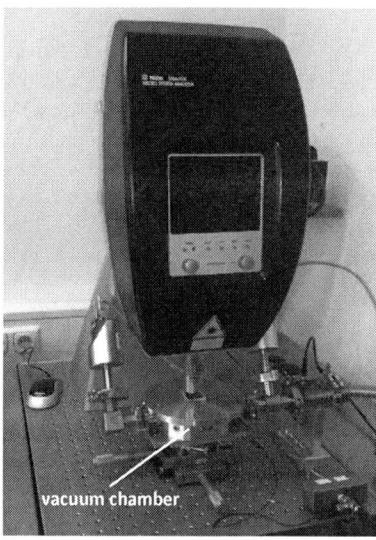

Fig. 6: Measurement Setup: Laser-Doppler vibrometer with vacuum chamber.

Since the structures are prototypes, they are not mounted on a chip carrier or electronically contacted. However, due to the high resolution of the vibrometer it is possible to perform the measurements only with the actuation from the ambient noise in the laboratory. The noise level in the considered frequency range is sufficiently flat and leads to displacement amplitudes in the range of nanometers.

The vibrometer provides a frequency spectrum like the exemplary for a cantilever depicted in Fig. 7. Here, the velocity and not the displacement is plotted on the y-

axis as this is the quantity measured by the laser Doppler vibrometer. The red dots denote the measured signal and the black line an interpolation of the measurement; the horizontal blue line indicates the -3dB bandwidth of the resonance peak, from which the Q-factor is determined.

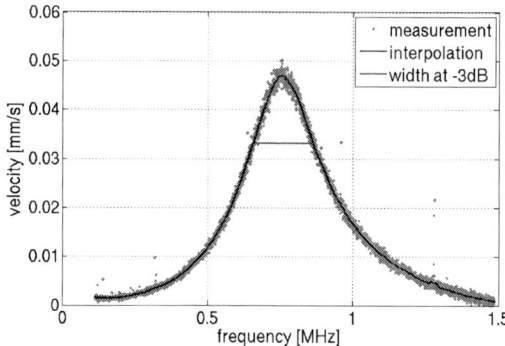

Fig. 7: Exemplary frequency spectrum of the cantilever at room pressure. The Q-factor is calculated from the width of the peak at -3dB.

For every structure and varying pressure, such a frequency spectrum is taken and evaluated, which leads to the double logarithmic pressure-dependent plots in Figs. 8, 9 and 10 which will be discussed in the next section.

4. Measurement Results and Comparison to Theory

In this section we discuss the results of the pressure-dependent measurements of the Q-factor and compare them to the simulations obtained with the mixed-level model described in section 2. The measurements are carried out in the pressure range from atmospheric pressure down to 3 Pa.

The double logarithmic plot in Fig. 8 shows the Q-factors of the first eigenmode of the 1 μm wide cantilever with an eigenfrequency of 0.85±0.06 MHz. The red line represents the simulation and the two blue lines the measured data of two equivalent cantilevers with the same geometrical dimensions on the same chip. Fig. 9 shows the Q-factors of the first eigenmode of the 2 μm wide cantilever with an eigenfrequency of 0.93±0.02 MHz.

The simulations show already a very good agreement with the measured data, especially for pressure values between 10^2 Pa to 10^4 Pa. At higher pressure, the noise level in the measurements increases and, in addition, the resonance peak in the frequency spectrum gets broader, so that measurements in the regime of 10^5 Pa to 10^4 Pa become extremely difficult. This leads to larger measurement errors and could explain the slightly higher deviations from the simulated values. At very low pressures the measured Q-factors reach a plateau, which is due to pressure-independent damping factors, like thermoelastic damping, anchor loss and surface damping (already mentioned in section 1). Since our simulations take only fluidic damping into account, these damping losses are not considered here.

Estimations carried out applying the Zener model [4, 5] to the investigated cantilevers revealed that the impact of thermoelastic damping is several orders of magnitude too small to explain the measured Q-factors. Surface damping, however, has an intricate impact on the Q-factor, which is – to our knowledge – not well understood and quantitatively very elusive. Hence, we assume anchor loss to be the relevant mechanism to cause the Q-factors of a few thousand in the low pressure regime, where fluidic damping becomes negligible. The focus of future work will therefore be to quantify this contribution of anchor loss by dedicated simulations, which is still a matter of research.

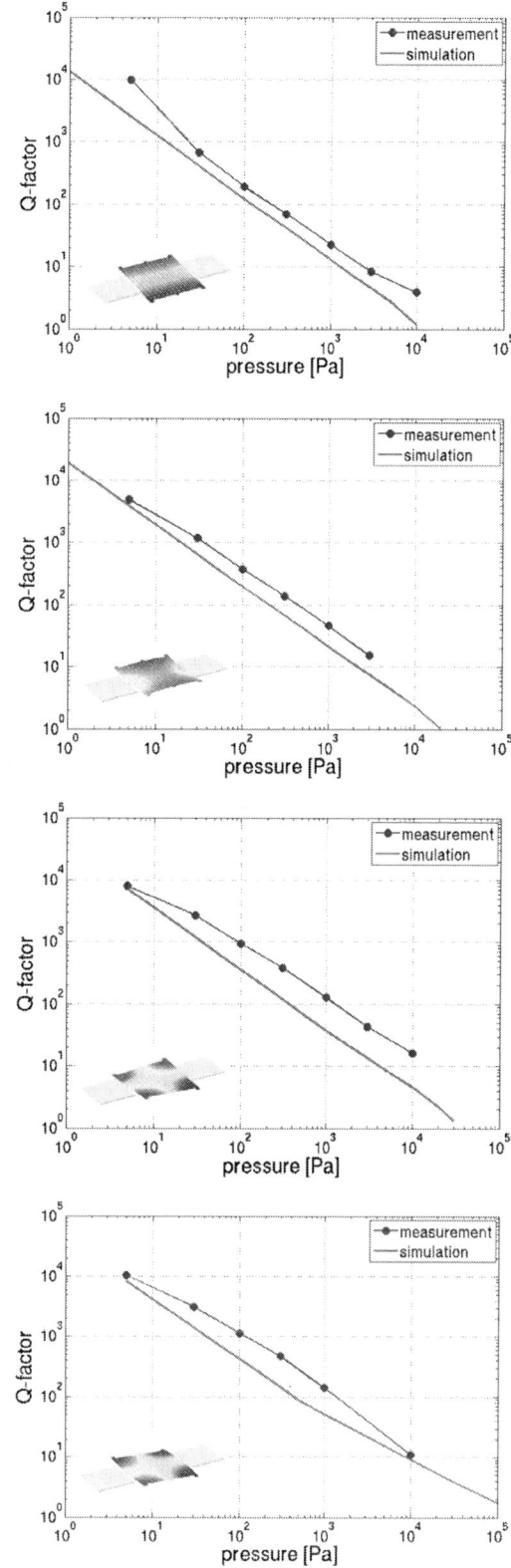

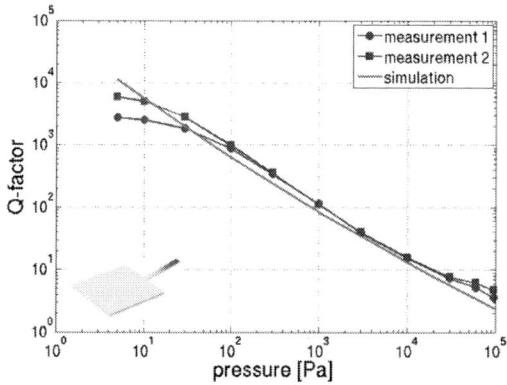

Fig. 8: Q-factor of the first eigenmode of the 1 µm wide cantilever; comparison between measurement and simulation. Two different cantilevers with the same geometrical dimensions located on the same chip have been characterized (measurement 1 and 2, resp.).

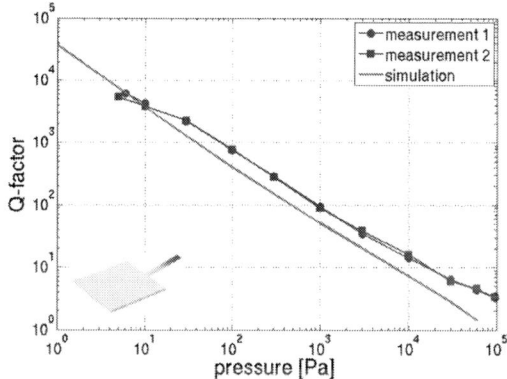

Fig. 9: Q-factor of the first eigenmode of the 2 µm wide cantilever; comparison between measurement and simulation. Two different cantilevers with the same geometrical dimensions located on the same chip have been characterized (measurement 1 and 2, resp.).

The results for the paddle are depicted in Fig. 10. We measured the pressure-dependent Q-factors of the first, second, third and fourth eigenmode of the paddle at 0.35 MHz, at 0.70 MHz, at 1.47 MHz and at 1.80 MHz, respectively, and extracted the Q-factors.

Fig. 10: Q-factors extracted from the first four eigenmodes of the paddle; comparison between measured and simulated data. At room pressure the system is over damped due to the small air gap underneath the structure.

978-1-4799-4789-8/14 $31.00 © 2014 IEEE   561

For all four frequencies we were not able to obtain analyzable data in the high pressure regime from 10^5 Pa to 10^4 Pa. This is not only due to the already mentioned large noise level at high pressure, but also and, in this case prevalently, because the system is overdamped in this regime and exhibits very low Q factors. In general, the simulation underestimates the damping, so in the simulation we obtain also Q-factors even up to atmospheric pressure for the fourth eigenmode. A second point to note is that no clear plateau can be observed in the pressure regime below 10 Pa in contrast to the cantilevers investigated before. We suppose that a similar plateau could be observed for even lower pressure values, which would entail the improvement of our vacuum chamber and is therefore matter of future work.

To summarize, we obtained already a very good accordance between measurement and simulation, although it is not yet clarified to which extent the correction factors are valid. By trend, the simulation model delivers more accurate results for the cantilevers. This is plausible, since the model of the cantilevers contains solely the Reynolds equation and the fluidic resistances accounting for the additional pressure drops at the boundaries while for the paddle further and more complicated models accounting for the perforations are added. Hence, this model constitutes an intricate interplay of various effects, which necessitates further detailed investigations in order to explore the area of validity of our modular approach, especially for devices and operation conditions in the rarefied gas regime.

5. Conclusions

We extracted pressure-dependent Q-factors for test devices with nanogaps underneath the movable membrane and compared them to simulated values obtained by a modular mixed-level model. In order to account for the large Knudsen numbers originating from the small feature size and/or the low ambient pressure this model was extended by correction factors taken from literature.

First results obtained with this model agree quite well with the measured data, even though there is quite a scatter in the published correction factors, which is not clarified to its full extent so far. In particular, there is still a lack of systematic experimental data in this regime applicable to MEMS and NEMS devices.

Further systematic theoretical and experimental investigations (test structures with varying geometrical dimensions and smaller air gaps down to less than 100 nm) are currently on their way in order to test the range of validity of our model and to figure out the impact of other damping mechanisms, especially in the low pressure region, where fluidic damping becomes negligible.

Acknowledgements

The authors wish to express their gratitude to Fraunhofer EMFT and in special to Mr. Martin Heigl who was in charge of the fabrication of the sophisticated NEMS structures.

References

1. Schrag, G. et al, System-level Modeling of MEMS Using Generalized Kirchhoffian Networks – Basic Principles in System-level Modeling of MEMS, Wiley (2013), pp.19-52.
2. Niessner, M. et al, Mixed-Level Approach for the Modeling of Distributed Effects in Microsystems in System-level Modeling of MEMS, Wiley (2013), pp. 163-187.
3. Weinberg, M., "Energy Loss in MEMS Resonators and the Impact on Inertial and RF Devices", *Proc. Transducers*, Denver, CO, June 2009, pp. 688-695.
4. Zener, C., "Internal Friction in Solids. I. Theory of Internal Friction in Reeds", *Physical Review*, Vol. 52, (1937), pp. 230-235.
5. Zener, C., "Internal Friction in Solids. II. General Theory of Thermoelastic Internal Friction", *Physical Review*, Vol. 53, (1938), pp. 90-99.
6. Hamrock, B. J. et al, Fundamentals of Fluid Film Lubrication, McGraw-Hill (1994).
7. Schrag, G., "Accurate System-level Damping Model for Highly Perforated Micromechanical Devices", *Sensors and Actuators A*, Vol. 111, (2004), pp. 222-228.
8. Veijola, T., "The Influence of Gas Surface Interaction on Gas-film Damping in a Silicon Accelerometer", *Sensors and Actuators A*, Vol. 66, (1998), pp. 83-92.
9. Fukui, S., "Analysis of a Ultra-Thin Gas Film Lubrication Based on Linearized Boltzmann Equation: First Report – Derivation of a Generalized Lubrication Equation Including Thermal Creep Flow", *ASME Journal of Tribology*, Vol. 110, (1988), pp. 253-261.
10. Fukui, S., "A Database for Interpolation of Poisseuille Flow Rates for High Knudsen Number Lubrication Problems", *ASME Journal of Tribology*, Vol. 112, (1990), pp. 78-83.
11. Veijola, T., "End Effects of Rare Gas Flow in Short Channels and in Squeezed Film Dampers", *Nanotech*, Vol. 1, (2002), pp. 104-107.
12. Sharipov, F., "Data on Internal Rarified Gas Flows", *J. Phys. Chem. Ref. Data*, Vol. 27, (1998), pp. 657-706.
13. Borel, S., "Isotropic Etching of SiGe Alloys with High Selectivity to Similar Materials", *Microelectronic Engineering*, Vol. 73-74, (2004), pp. 301-305.

Electro-Thermal Characterization of Through-Silicon Vias:

Aida Todri-Sanial
CNRS - LIRMM/University of Montpellier 2, France
Email: aida.todri@lirmm.fr

Abstract

Through-Silicon Vias (TSVs) are the vertial vias that enable three-dimensional integration by providing shorter, faster and denser interconnects. In this work, we investigate their thermal properties and show that TSVs used for power and ground connections can suffer from high thermal dissipations, which can lead to reliability and timing errors. Due to nature of current flow on 3D ICs (i.e. from package to each tier), we show that the TSVs near the package tier endure high current flows and high temperatures which eventually lead to Joule heating and electromigration (EM) phenomena. Such analyses bring forth the importance of power- and thermal-aware TSV placement.

1. Introduction

On-going miniuarization and advancements in manu-facturing and process technologies are making transitors faster and also allowing large die size for embedding more functionality and computing power. Ironically, interconnects are not getting faster with scaling and their power consumption is becoming a limit for high speed and band-width data transfer both on- and off-chip. Such limitations have motivated the industry to look into new materials and methodologies. 3D integration is a promising new manufacturing technology that offers a viable solution to the interconnect problem by utilizing Through-Silicon-Vias (TSVs) for integrating multi-layer circuits. As a prospect, 3D integration offers shorter, faster, and high bandwidth interconnects while enabling integration of many computational cores for high-performance comput-ing and heterogenous circuits for smart system integration.

The advantages on high performance and integration factor also come with a high cost on power density and elevated heat dissipation. Due to stacking many circuit layers of different switching activity and power demand, results in high power density and current flow. Due to the structure of the 3D stack, power (ground) voltage is delivered from package to first layer's global power network then through the TSVs is connected to the global power networks of the rest of layers. Figure 1 illustrates this concept. As each layer can be a die of different functionality, power demand and technology, it is repre-sented by a power network of different topology which can be connected from TSVs through a voltage regulator (VRM) in case if the voltage level needed for each layer is different. Due to such structure, the TSVs closer to the package experience the largest current flow which is

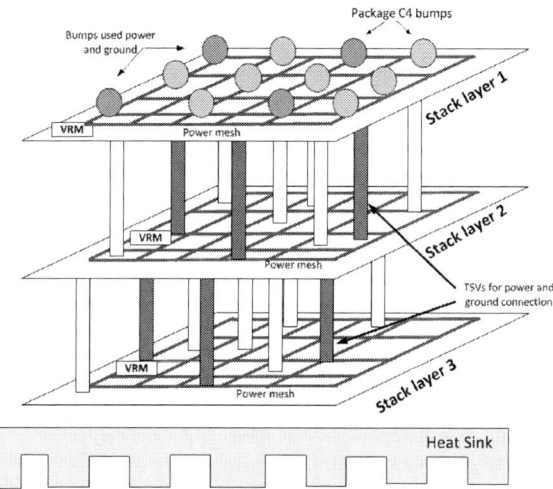

Figure 1: Illustration of 3D power delivery network for a three layer circuit using TSVs.

further distributed to each circuit layer. Another unique property of current flow on power delivery networks and power TSVs is that current is unidirectional as it flows from package to each circuit layer. Large amount of currents flowing unidirectionally for a long period of time can lead to Joule heating and electromigration (EM) phenomena. Elevated temperature levels along with large current demands, will impact the parasitics of TSVs (i.e. resistance changes with temperature) and further worsen the performance of TSVs and exacerbate the power supply noise.

Similarly, as circuit layer next to the heat sinks benefits from immediate cooling, whereas the rest of the layers can suffer from high thermal gradients due to long and increased heat conduction paths through the TSVs. As a result, 3D circuits would have higher power density and temperature levels compared to their 2D counterparts. Eventually, these effects would impact circuit perfor-mance, induce logical errors, degrade circuit reliability and reduce circuit lifetime, thus annihilating the very advantages of 3D integration.

There are several works in the literature that have looked into the TSV behaivor in 3D stacked ICs. These works can be grouped mainly into two groups, where the first group of papers focused on electrical modeling and

978-1-4799-4789-8/14 $31.00 © 2014 IEEE

characterization, and the second group on thermal modeling and characterization of TSVs [1], [2], [3], [4]. Some works utilize mathematical models based on finite-element based method or utilizing modeling and simulations tools such as COMSOL and ANSYS for performing detailed thermal simulations. There are fewer works that have looked into electro-thermal modeling and analysis [5], [6]. Complementary to these research efforts, this work aims to investigate the electro-thermal behaivor of TSVs based on their stacking order on the 3D systems.

In this work, we exploit electrical-thermal duality principle [6], [7] to represent a thermal network for performing fast thermal analysis. We consider two heat path flows on the 3D stack system, the main heat path flow is through the heat sink and secondary heat flow is through the package. Such study not only will help to understand the thermal distribution on each stack layer and TSVs impact, but also explore and identify potential knobs that can lead to power- and thermal-aware design methods for 3D stacked ICs. Contribution of this paper are:

- Investigate the impact of primary and secondary heat path flows and their impact on 3D stack temperature distribution.
- Electro-thermal analysis of TSVs on each tier
- Electromigration issues and occurence on TSVs

This paper is organized as follows. Section I describes the models utilized in this work. Section II provides the TSV thermal analysis while using electro-thermal duality principle. Experimental results are shown in Section III and in Section IV we conclude the paper.

2. Electro-Thermal Modeling

As shown in Figure 1, supply voltage is delivered from package bumps to the global power delivery (power mesh) of first stacked layer then distributed to other stacked layer's global power delivery through the TSVs. Depending on the technology, design style, and functionality each layer may have different power consumption and switching frequencies. They also might have different granularity power meshes with non-uniform grids. Despite these differences, power delivery networks are designed to deliver reliably the supply voltage with minimum voltage drop.

Several works have reported on the extraction of the parasitics of TSVs, micro-bumps (metallic bonds made of copper or copper tin alloy) and packge C4 bumps that provide the connections to the global power delivery network [1], [8]. Some authors have provided the analytical expressions of resistance, capacitance and inductance of TSV [1], [2], [3]. Authors in [9] have shown that the sidewall dielectric thickness is a critical parameter for TSV capacitance. TSV resistance is computed as [1], [2], [3], [9]:

$$R_{tsv} = \frac{\rho l_{tsv}}{\pi r_{tsv}^2} \quad (1)$$

where r_{tsv} is the radius of TSV, l_{tsv} is the length of the TSV, and ρ is the resistivity of the conducting material. Inductance of the TSV can be derived as in [10]:

$$L_{tsv} = \frac{\mu_0 l_{tsv}}{2\pi} [ln(\frac{2l_{tsv}}{r_{tsv}} - \frac{3}{4})] \quad (2)$$

where μ is the permeability of free space. The capacitance due to the oxide liner between silicon substrate and TSV, can be computed as in [10]:

$$C_{ox} = \frac{4\varepsilon_{si} t_{si}(r_{tsv} - t_{ox})}{t_{ox}} \quad (3)$$

where t_{si} and t_{ox} are the silicon and oxide liner thicknesses, and dielectric constance of silicon is $\varepsilon_{si} = 11.9\varepsilon_0$ where $\varepsilon_0 = 8.854 x 10^{-12} F/m$. TSV capacitance due to silicon substrate can be obtained as [10]:

$$C_{si} = \frac{\varepsilon_{si} \pi l_{tsv}}{ln\left(\frac{p}{2r_{tsv}} + \sqrt{\frac{p}{2r_{tsv}}^2 - 1}\right)} \quad (4)$$

where p is the TSV pitch. Overall, the TSV capacitance can be derived as:

$$C_{tsv} = \left(\frac{1}{C_{ox}} + \frac{1}{C_{si}}\right)^{-1} \quad (5)$$

For system-level analysis, both electrical and thermal models of TSVs are important. Electrical and thermal properties of TSVs and power meshes are closely coupled. Electrical current flowing through the TSVs or power grid branches result in Joule heating, hence in rise of temperature. Such temperature rise causes changes on the parasitics values of TSV and power grid branches (i.e. resistance and capacitance). Other studies have shown that the temperature rise in TSVs leads to mechanical stress due to unequal thermal expansion coefficient between the TSV filler material (i.e. copper) and silicon. The transistors around the TSVs may suffer from such stress and degrade their performance, which is why a keep out zone from TSVs is imposed as a design rule. The thermal resistance of TSV is based on thermal models that are derived in [1], [5].

Similarly, electro-thermal models are needed for the power delivery network. We use the one-dimensional (1-D) heat diffusion equation to avoid the expensive computations required by other simulators (i.e. FEM based), can be described as follow [11]:

$$\frac{d^2 T}{dx^2} = -\frac{Q^*_{eff}}{k_m} \quad (6)$$

where k_m is the thermal conductivity of the metal, T is the temperature of metal branch, x is the length, and Q^*_{eff} is the effective volumetric heat generated. The major source of heat generation is the power consumption of the circuit due its dynamic and static activity. Additionally, the power dissipated in interconnects lines (i.e. power grid branches) are also a source of heat generation. The power dissipation on metal line next for a segment $\triangle x$ can be expressed as:

$$P_{dis} = I_{rms}^2 \triangle R_E(x) \qquad (7)$$

where I_{rms} is the root mean square of the current flowing through the metal line, R_E is the electrical resistance which has a linear relationship with temperature and can be expressed as:

$$R_E(x) = R_0(1 + \beta T(x)) \qquad (8)$$

where R_0 represents the nominal resistance at room temperature of the metal line (i.e. power grid branch). Such equation allows to derive the resistance at any location x of the metal line as a function of temperature. This would mean partitioning the metal line into small segments of $\triangle x$ for computing the thermal model of metal lines [12]. However, this would result into a large thermal network which would be computationally expensive to simulate. Instead, authors at [12] exploited the *characteristic thermal length*, L_H to derive a lumped thermal model. The L_H represent the metal line length between any two vias which serve as boundary conditions for solving the heat equation [12]. It has been shown that the within the length, L_H, the metal line's self-heating (or Joule heating) will be dispersed through the vias to the next metal line rather than the dielectric layer. Thus, a lumped thermal model for the power meshes can be build as long as the grid branches lengths is within the thermal characteristic length, L_H. This not only reduces the size of the power thermal network but also improves the computational runtime of the electro-thermal analysis. Table 1 shows the various elements and their corresponding parasitics.

Table 1: Elements and their parasitics.

Element	Geometries	Parasitics
Silicon substrate		ρ=100Ω-cm
TSV	Diameter:2μm	R=43mΩ,
	Height:20μm	C=40fF
	Pitch=60μm	
μBumps	Diameter:5μm	R=40mΩ,
	Height:5μm	C=0.4fF
C4 Bumps	Diameter:100μm,	R=10mΩ,
	Pitch:200μm	C=9.5fF
Power tracks (Cu)	Mesh:1100μmx1100μm,	R=45mΩ
	Granularity=12x15	
Package (next to Tier 1)	Thermal resistance	20K/W
Heat Sink (next to Tier 3)	Thermal resistance	2K/W
Power density	Uniformly distributed	1.5μW/μm^2
Frequency		3.2GHz
Vdd		1V
TSVs for VDD		20
TSVs for GND		20

3. TSV Electro-Thermal Analysis

In this section, we perform various analyses for understanding both the electrical and thermal behavior of TSVs. We perform three case studies to investigate TSV, (i) impact of thermal analysis on worst case voltage drop on a 3D stacked system, (ii) the impact of package thermal

models as the secondary heat path flow, and (3) the reliability of TSVs in each stack, where depending on their location some TSVs can me more prone to reliability issues.

A. Impact of Thermal Analysis

Here, we perform electrical analysis only and derive the worst-case voltage drop on each circuit tier of the 3D stack. Figure 2 shows the worst case voltage drop map for each tier. Tier 1 being the closest to the package pins, has the shortest current path, thus the least amount of voltage drop. Whereas, Tier 3 is the furthest from the package pins, and current flows through long paths resulting in the largest amount of voltage drop. As temperature effects are not considered, the voltage drop is measured solely due to the impact of switching activity of the circuit. A maximum voltage drop of 165mV is obtained in Tier 3 for power delivery networks and 106mV is obtained on the ground delivery network.

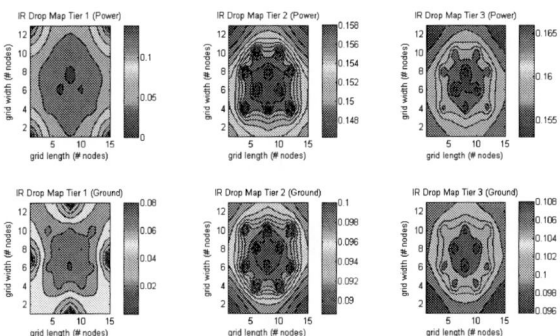

Figure 2: Voltage drop map for each tier without considering the temperature effect Here, only electrical analysis is performed.

B. Impact of Package Thermal Model

Here, we perform electro-thermal simulations for obtaining the worst case voltage drop and temperature in each tier. Generated heat from the switching circuits can be dispersed both laterally and vertically on the 3D stack through the TSVs. The primary heat path flow is through the heat sink which provides immediate cooling for the nearby Tier 3. A secondary heat path flow also exists through the package. To show the importance of the heat flow through the package, we perform electro-thermal simulations with and without the thermal model of the package. Results are shown in Figures 3 and 4. Figure 3a shows the voltage drop map on each tier for both power and ground networks. Figure 3b shows the temperature map for each tier on both power and ground networks, respectively. Similarly, Figure 4 shows the voltage and temperature map of each tier including the thermal model of the package.

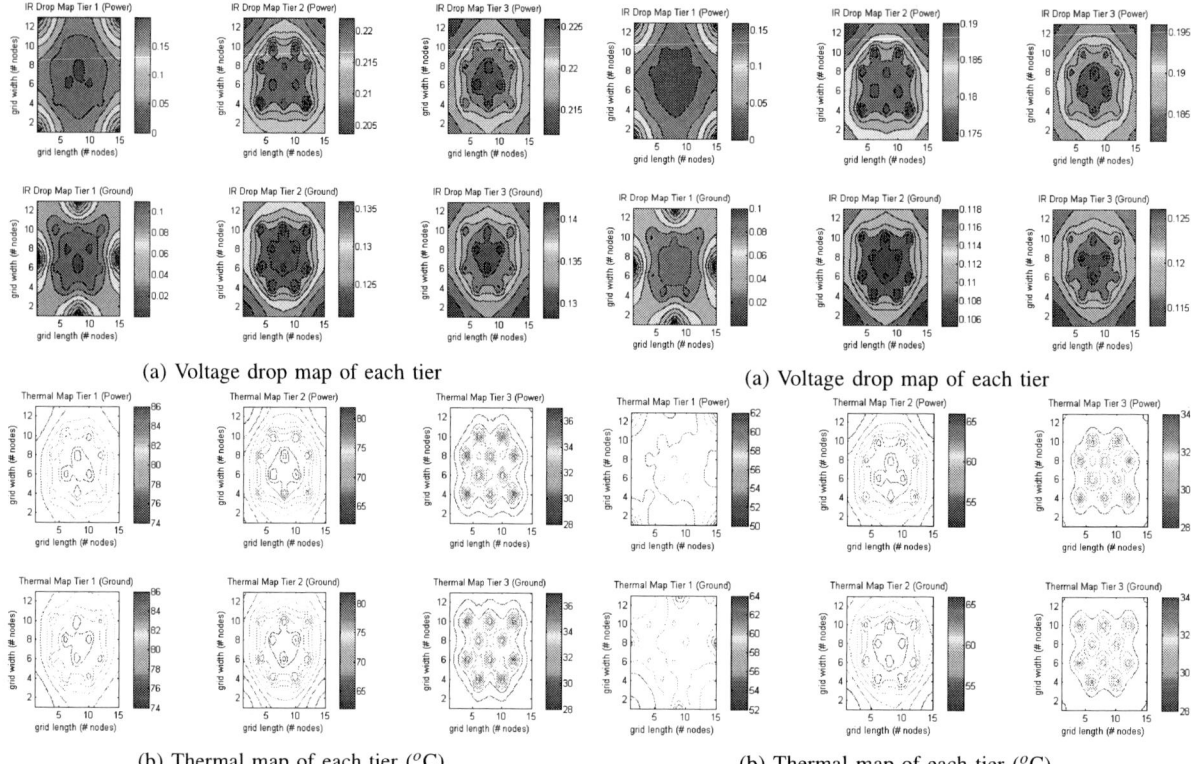

(a) Voltage drop map of each tier   (a) Voltage drop map of each tier

(b) Thermal map of each tier (^oC)   (b) Thermal map of each tier (^oC)

Figure 3: Electro-thermal simulation without considering the package model as secondary heat path flow.

Figure 4: Electro-thermal simulation with the package model as secondary heat path flow.

We observe a difference between the amount of voltage drop and temperature on each tier due to the package effect. Overall, we observe that the worst case voltage drop is reduced when the package thermal model is included. Similarly, the temparature on each tier is reduced when the effect of package thermal model is included. This is because, some of the heat can be dispersed through the package which ultimately reduces the overall temperature. The reduction in temperature also prompts reduction in voltage drop. Additionally, we note that performing electro-thermal simulation helps to determine voltage drop more accurately in comparison to electrical-only simulation which might underestimate voltage drop. Figure 5 shows the comparison of voltage drop for each tier when electrical-only and electro-thermal simulations with and without the package thermal model are performed. Figure 6 shows the worst case temperature difference between each tier when the secondary heat path of the package is considered.

C. TSV Reliability

In this section, we perform electro-thermal simulations for investigating the reliability of TSVs. In this work, we make the assumption that the TSVs are of uniform dimensions and uniformly distributed on each tier. There

are 20 TSVs on the three tier stack where 10 TSVs are between Tier 1 and 2, and 10 TSVs between Tier 2 and 3 for VDD and GND connections, respectively. Due to the nature of the current flow on a 3D stack, TSVs that are the closest to the package tend to have large amount of currents flowing through them. Such current is then distributed to the next tiers. Such concept is also illustrated in Figure 7. For example, for three tier stack, the TSVs between Tier 1 and 2 would have large amount of currents flowing through them which is later distributed to Tier 3.

In Figure 8, we show the amount of currents that flow on each TSVs. It is shown that the TSVs of Tier1-Tier2 have currents such as 70mA flowing through them, whereas, TSVs between Tier2-Tier3 have currents flowing in the range of 25mA. In Figure 9, we show the temperature levels obtained for each TSV on the stack. Due to proximity to heat sink located next to Tier 3, TSVs between Tier2-Tier3 have temperature levels in the range of 29^oC, whereas TSVs between Tier1-Tier2 have elevated temperature reaching up to 62^oC.

These results indicate that the TSVs located next to the package not only suffer from large amount of current flows but also of elevated temperatures. Both high current

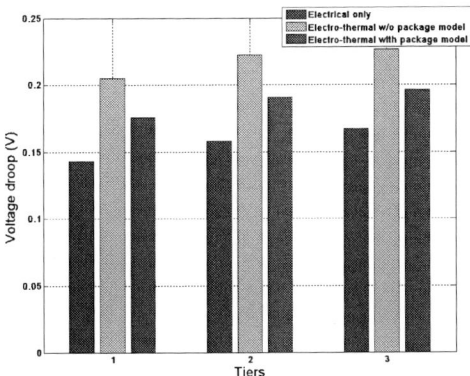

Figure 5: Worst case voltage drop on each tier when (i) electrical only simulations are performed while ignoring temperature effects, (ii) electro-thermal simulations are performed while ignoring package as secondary heat path, and (iii) electro-thermal simulations while taking into account the package as the secondary heat path flow.

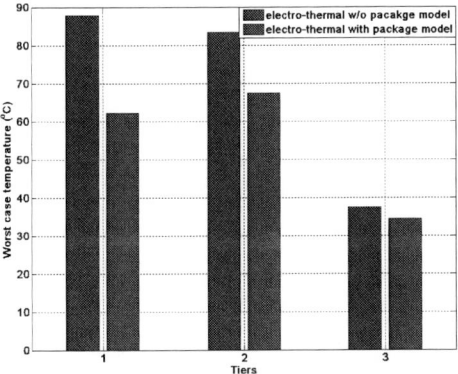

Figure 6: Worst case temperature on each tier when (i) electro-thermal simulations are performed while ignoring package as secondary heat path, and (ii) electro-thermal simulations while taking into account the package as the secondary heat path flow.

demand and high temperatures are the primary factors of Joule heating that eventually leads to electromigrations (EM) issues. Unique to power delivery networks, is that current flow is unidirectional. Such large amount of unidirectional currents in high temperature environment running for long period of time, can prompt to formation of hillocks and voids on metal lines, microbumps and TSVs [13].

Electromigration is the mechanism of metal-ion migration due to high current density stress and temperatures [14]. The currents flowing on 3D power delivery networks and power/ground TSVs are unidirectional currents which over long period of time would lead to failures or breaks on the metal line. An empirical formula was developed

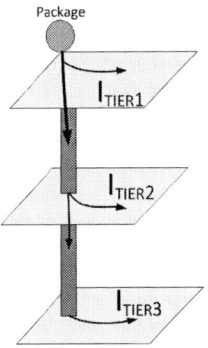

Figure 7: Illustration of current flow on a three tier stack where the TSVs near the package tend to have large current flows to distribute to the rest of the tiers.

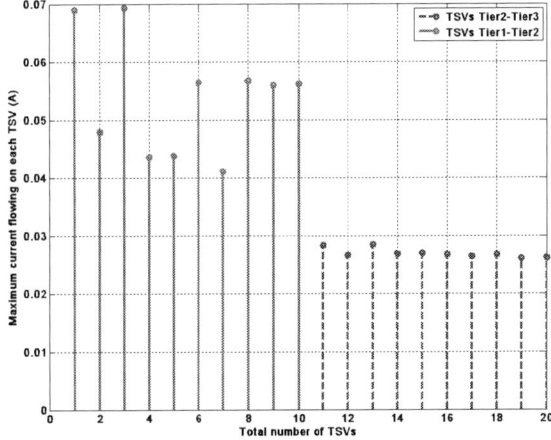

Figure 8: Current flowing through each TSV. It is observed that TSVs nearby the package (Tier 1) have the largest amount of current flow is further distributed to each tier.

by [14] to compute the mean time to failure (MTTF) of metal wires due to electromigration:

$$MTTF = \frac{C}{J^n} exp\left(\frac{E_a}{kT}\right) \qquad (9)$$

where C is a constant based on cross-sectional area of the interconnect, J is the current density, E_a is the activation energy for electromigration mechanism, k is the Boltzmann's constant, T is the temperature and n is an empirical constant. We have computed the $MTTF$ for each TSV on the 3D stack based on the temperatures and current flows obtained from the electro-thermal analyses shown in Figures 8 and 9. Figure 10 shows the ratioed $MTTF$ for each TSV with respect to the TSV with the highest $MTTF$ which is located next to the heat sink (low temperature and less current flow). Results indicate that the TSVs between Tier1-2 have a reduced $MTTF$ by 85% reaching as low as $0.15 * MTTF$ for TSV1 and TSV3. It is noted that

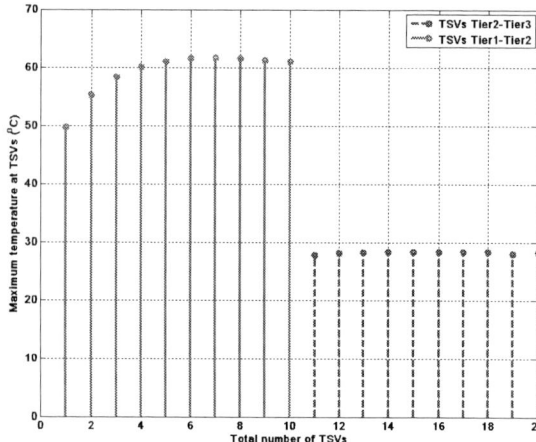

Figure 9: Worst case temperatures measured for each TSV. It is observed the TSVs furthest (Tier 1) from the heat sink (Tier 3) have the highest temperature levels.

TSVs depending on their stacking order can suffer from large current flows and high thermal gradients which may lead to reduced reliability and lifetime. These results also indicate the need to investigate the size, placement and distribution of TSVs for power and thermal integrity in 3D stacked ICs.

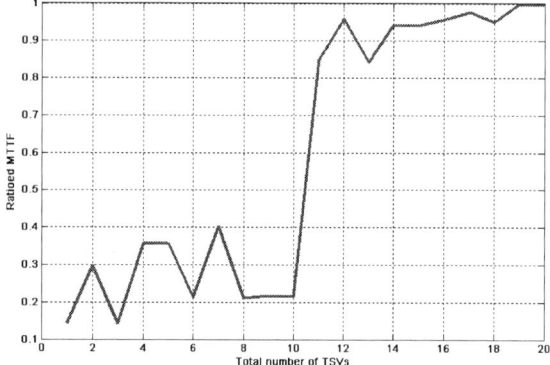

Figure 10: Ratioed MTTF for each TSV with respect to the TSV with the highest $MTTF$ (with low temperature and current flow).

4. Conclusions

In this work, we investigate electro-thermal behaivor of TSVs and show that TSVs used for power and ground connections can suffer from high thermal dissipations

which can lead to reliability and timing errors. Due to nature of current flow on 3D ICs (i.e. from package to each tier), we show that the TSVs near the package tier endure high current flows and high temperatures which eventually lead to Joule heating and electromigration (EM) phenomena.

5. Acknowledgements

This work has been funded by European project Master-3D of Catrene Program, grant number CT312.

References

[1] G. Katti, M. Stucchi, D. Velenis, B. Soree, K. De Meyer, W. Dehaene, "Temperature-Dependent Modeling and Characterization of Through-Silicon Via Capacitance," IEEE Transactions on Electron Device Letters, vol. 32, no.4, pp. 563-565, Apr. 2011.

[2] A. Jain, R.E. Jones, R. Chatterjee, S. Pozder, "Analytical and Numerical Modeling of the Thermal Performance of Three-Dimensional Integrated Circuits," IEEE Transactions on Components and Packaging Technologies, vol.33, no.1, pp.56-63, March 2010.

[3] H. Oprins, V. Cherman, M. Stucchi, B. Vandevelde, G.Van der Plas, P. Marchal, E. Beyne, "Steady state and transient thermal analysis of hot spots in 3D stacked ICs using dedicated test chips," IEEE Semiconductor Thermal Measurement and Management Symposium (SEMI-THERM), pp.131-137, March 2011.

[4] S. Swarup, Sh. X-D. Tan, Z. Liu, "Thermal Characterization of TSV based 3D Stacked ICs," IEEE EPEPS, pp.335-338, 2012.

[5] W. -S. Zhao, X.-P. Wang, W.-Y. Yin, "Electrothermal Effects in High Density TSV Arrays," Progress in Electromagnetic Research, vol. 115, pp. 223-242, 2011.

[6] A. Todri, S. Kundu, P. Girard, A. Bosio, L. Dilillo, A. Virazel, "A Study of Tapered 3-D TSVs for Power and Thermal Integrity," IEEE Transactions on Very Large Scale Integration (VLSI) Systems, vol.21, no.2, pp.306-319, 2013.

[7] A. Todri-Sanial, S. Kundu, P. Girard, A. Bosio, L. Dilillo, A. Virazel, "Globally Constrained Locally Optimized 3D Power Delivery Networks," IEEE Transaction on VLSI, doi: 10.1109/TVLSI.2013.2283800.

[8] G.B. Kromann, "Thermal Modeling and Experimental Characterization of the C4/Surface-Mount-Array Interconnect Technologies," IEEE Transactions on Components, Packaging, and Manufacturing Technology, vol.18, no.1, pp.87-93, Mar. 1995.

[9] I. Savidis, S. Alam, A. Jain, S. Pozder, R. Jones, R. Chaterjee, "Electrical Modeling and Characterization of Through-Silicon Vias (TSVs) for 3-D Integrated Circuits,"Microelectronics Journal vol. 41, pp.9-16, 2010.

[10] G. Katti, M. Stucchi, K. De Meyer, W. Dehaene, "Electrical Modeling and Characterization of Through Silicon Via for Three-Dimensional ICs," in IEEE Transactions on Electron Devices, vol.57, no.1, pp.256-262, 2010.

[11] M.N. Ozisik, Heat Transfer: A Basic Approach. New York, NY: McGraw-Hill, 1985.

[12] Y. Zhong, M.D.F. Wong, "Thermal-Area IR Drop Analysis in Large Power Grid," IEEE International Symposium on Quality Electronic Design (ISQED), pp. 194-199, 2008.

[13] Th. Frank, S. Moreau, C. Chappaz, L. Arnaud, P. Leduc, A. Thuaire, L. Anghel, "Electromigration Behavior of 3D-IC TSV Interconnects," IEEE Elec. Comp. Tech. Conf. (ECTC) pp. 326-330, 2012.

[14] Blech, "Electromigration in Thin Aluminium films on Titanium Nitride", in Journal of Applied Physics, Vol. 47, pp.1203-1208, 1976.

Experimental investigation of the visco-plastic mechanical properties of a Sn-based solder alloy for material modelling in Finite Element calculations of automotive electronics

R. Metasch, M. Roellig, A. Kabakchiev*, B. Metais*, R. Ratchev*, K. Meier**, K.-J. Wolter**

Fraunhofer Institute for Ceramic Technology and Systems, Material Diagnostics, Dresden, Germany
* Robert Bosch GmbH, Postfach 300240, 70442 Stuttgart, Germany
** Technische Universität Dresden, Electronics Packaging Laboratory, Germany
Tel.: +49-351-88815-581 Email: rene.metasch@ikts-md.fraunhofer.de

Abstract

Here, we present an advanced experimental procedure for determining the properties of a SnAg3.5 solder alloy in the strain range of primary creep under cyclic load and isothermal conditions. The challenge in this experiment is the accurate high-resolution measurement of sample elongation used for a closed-loop control, as well as avoiding the influence of sensor and specimen clamping. We realized reproducible strain rate control within a total specimen elongation of 60 μm. The tensile-compression experiment comprises strain rate variation for three strain amplitudes with integrated relaxation stages followed by a measurement of cyclic fatigue. The strain rate at every strain stage was varied in the range of 1E-3 to 1E-6 per second. At the end of every strain stage a time-limited relaxation experiment is performed, where the specimen's length is kept constant, while the stress evolution is recorded. Finally, the specimen is subjected to cyclic fatigue until a drop of 50 % of the initial materials strength is reached. The total procedure is performed in a temperature range from -40 to 150 °C.

We prove the capability of common creep models to map the observed cyclic stress-strain hysteresis as well as stress dependency on strain rate. The results reveal substantial limitations of common stationary creep models and strongly suggest the application of advanced visco-plastic material models for an accurate description of the solder alloy properties. The experimental data presented can be used for the calibration of unified visco-plastic constitutive models initially proposed by Chaboché et al. and further extended during the past two decades.

1. Motivation

The time-dependent material behaviour of high temperature solder alloys used in automotive electronic devices was addressed by the usage of different phenomenological material models. The state of the art models employed in the design of electronic products by Finite Element (FE) simulations are the Power Law, the Double Power Law and the Garofalo approach combined with the Arrhenius Law. These models describe the steady state creep rate behaviour of solder alloys as a function of mechanical stress and temperature. Steady state creep can be determined by force-controlled experiments under isothermal conditions easily realized by loading a solder specimen

by different weights [1]. Numerous experimental works reported by now suggest that the specimen strain at which steady state creep rate is reached is not constant, but strongly depends on the applied load [2], [3]. Furthermore, absolute sample strain needed to reach stationary creep is not negligible and, as a result, in most applications inelastic deformation in solder joints turns out to happen within the primary creep region. Thus, the non-linear evolution of deformation rate with time and load magnitude which is characteristic for the primary creep region has to be considered. However, the modelling of the solder alloy deformation by secondary creep is usually calibrated based on uni-directionally loaded tensile tests. Due to this approach, non-linear strain accumulation over time, material hardening and intrinsic degradation processes cannot be taken into account. These can take place in solder joints under operation, since they undergo a cyclic mechanical load during temperature variation, where stress evolution is governed by a complex non-linear dependency on deformation rate, strain accumulation, time and temperature.

2. Derivation of the test conditions

The soldering process is still the most widely used connection technology for realizing the electrical functionality of modern automotive electronic devices. During a solder re-melting and solidification process electrical components are connected to a circuit printed board by the formation of solder joints. Sn-based solder alloys which are used as joint material have a melting point of approximately 217 °C to 221 °C.

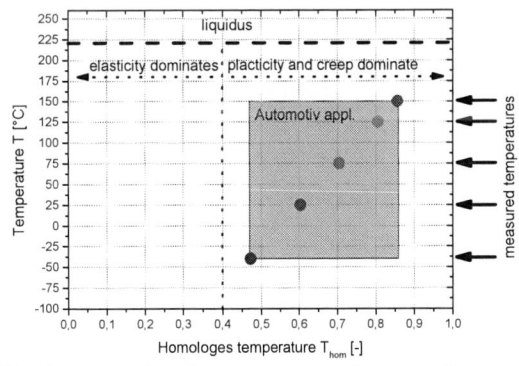

Fig. 1: Operational temperature of automotive application in terms of the homologous temperature of SnAg3.5 (T_m = 221°C)

The reliability of solder joints is determined by the operational temperature range, the surrounding materials and the material behaviour per se. This operational temperature range of automotive application is located between -40 °C and 150 °C. Under these thermal conditions the homologous temperature of SnAg3.5 is always over 0.4. Consequently, time- and temperature-dependent inelastic processes like plasticity and creep govern the deformation behaviour of the solder joint (see Fig. 1).

Changes of temperature lead to thermally induced mismatch between the different materials in an electronic device. As a consequence, solder joints deform depending on the stiffness of the neighbouring components and substrate material. In order to estimate the relevant strain amplitudes in automotive applications the maximal thermal strain was determined based on the expected mismatch between a typical resistor ceramic component and an FR4 printed circuit board (PCB). The resulting thermal strains are located well below 1 % strain (see Fig. 2).

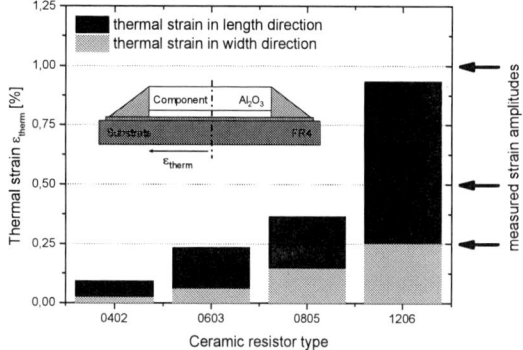

Fig. 2: Estimation of the strain amplitudes in terms of thermal mismatch between different ceramic resistors and a substrate material FR4.

Assuming different temperature rates occurring under operation of automotive electronic devices, the resulting strain rates in solder joints can be estimated (see Fig. 3).

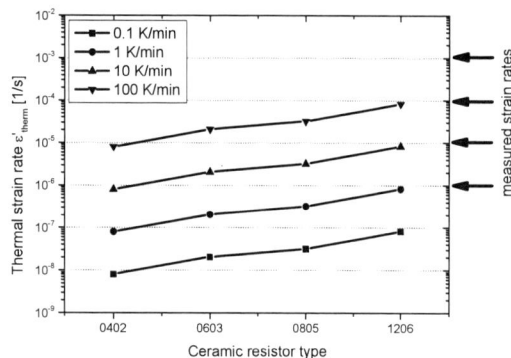

Fig. 3: Estimation of strain rates for different resistor component types.

For temperature gradients between 0.1 K/min and 100 K/min deformation rates can vary in a range between 1E-8 1/s and 1E-4 1/s. During special lab qualification tests even harsher thermal gradients are often performed, thus the upper limit of strain rate can be an order of magnitude higher. According to the estimation described above, the experimental procedure was adapted to obtain cyclic deformations in the strain rate regime between 1E-6 1/s and 1E-3 1/s. The experiment is designed to investigate the mechanical properties in the initial as-casted state of the material. Strain rates of 1E-7 down to 1E-8 1/s at the relevant strain amplitudes result in a substantial rise of total test time. In order to avoid the influence of creep degradation effects within the experiment the strain rates below 1E-6 1/s were not integrated in the measurement procedure (see Tab. 1).

Tab. 1: Time calculation example for strain rate experiments

Strain rate	Displacement	Total time (approx.)
1E-3 1/s	±1 % (120 μm)	20 s
1E-4 1/s	±1 % (120 μm)	3 min
1E-5 1/s	±1 % (120 μm)	33 min
1E-6 1/s	±1 % (120 μm)	6 h
1E-7 1/s	±1 % (120 μm)	56 h
1E-8 1/s	±1 % (120 μm)	23 d

At the end of every strain amplitude stage a measurement of stress relaxation in tension and compression was performed. For this, the maximal specimen's elongation in each of the three stages was kept constant while stress evolution was recorded.

Finally, the specimens were submitted to a low cycle fatigue (LCF) experiment until a stress drop of 50 % of the initial stress is reached.

In summary, the isothermal experimental procedure includes the following measurement types and parameters:

- 3 strain amplitude stages: 0.25 %, 0.5 % and 1 %,
- 4 strain rate variations: 1E-6 1/s, 1E-5 1/s, 1E-4 1/s and 1E-3 1/s,
- 6 stress relaxation measurements and
- Cyclic fatigue at a constant strain rate of 1E-3 1/s up to a stress drop of 50 %.

3. Test setup and experimental procedure

The key prerequisite to realize reproducible experimental conditions is the direct measurement of sample displacement, which avoids any influence through the specimen's clamping. Further important features are a closed loop control system realizing constant displacement speed as well as stable isothermal conditions in the temperature chamber and the laboratory.

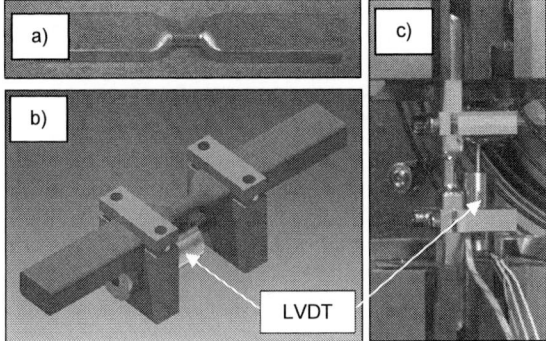

Fig. 4: Tensile-compression specimen (a) and non-contact measurement system for the displacement b) CAD view, c) real view

The inner region of the specimen has a constant diameter of 4 mm and a length of 6 mm. The contactless measurement system is clamped out of the region where the specimen is deformed. The shape and the measurement concept are designed to avoid any influence through the measurement system on the solder test region (see Fig. 4).

In order to realize comparable test results under different temperatures and materials a strain rate control is of utmost importance. It is necessary to obtain constant deformation speed during the load. For cyclic experiments in which only strain amplitude is tracked without strain rate control, the time-strain evolution changes due to the non-linear resistance of the material (s. Fig. 5).

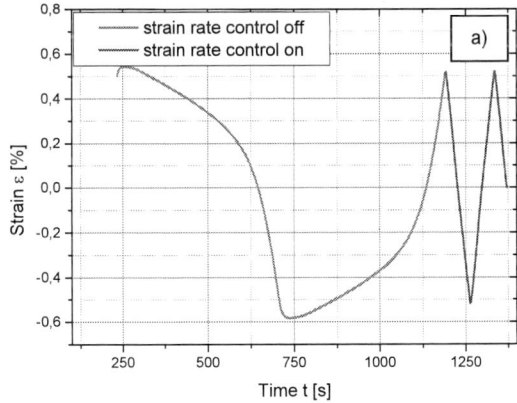

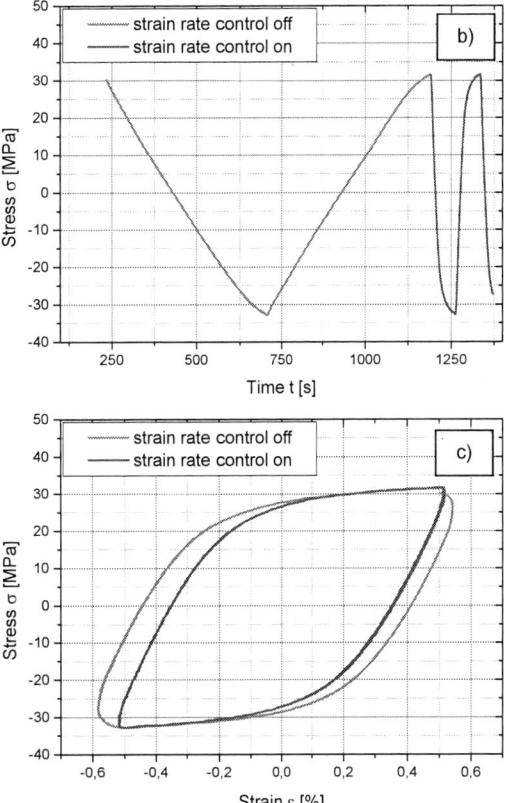

Fig. 5: Cyclic tension-compression tests with and without closed loop strain rate control a) strain-time, b) stress-time and c) stress-strain evolution.

In the absence of strain rate control, a nonlinear deformation speed of the specimen results, because the deformation behaviour changes depending on the load and the deformation mechanism. At the beginning after each force reversal, the deformation progress is rather slow and increases with increasing absolute stress. In the case of a closed loop control strain progress is linear in time, which is independent on the material behaviour and the temperature.

The full experimental progress with all experimental variations has a total time of approximately 3 days and can be seen in Fig. 6. The subsequent LCF-experiment takes total measurement time of up to 5 days.

The total procedure described above was implemented in two different material testing machines. The first system (see Fig. 7) is a completely self-developed setup, called μTester presented in 2009 [3]. For the present study, it was upgraded by the closed loop strain rate control as well as an optical inspection system for recording images triggered by specific events during the test process.

978-1-4799-4789-8/14 $31.00 © 2014 IEEE

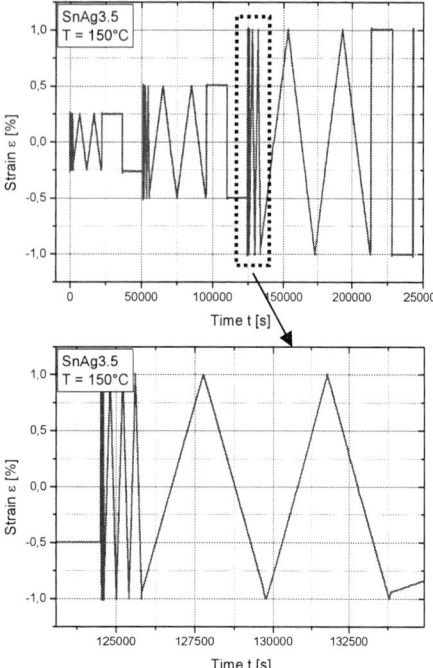

Fig. 6: Full experimental strain progress (without cyclic fatigue)

For low temperature experiments a commercial system from Zwick&Roell was used (see Fig. 8). A challenging issue was to obtain stabile test conditions below room temperature. For this, a series of modifications were performed. We implemented the displacement sensor concept developed in the home-made µTester setup. Further, a local heating of the rods near the penetrations in the temperature chamber was necessary in order to avoid freezing, which influences the stability of the force measurement. The temperature control of the force sensor was adapted, to avoid any signal drifts during the experiments, since the force sensor is initially mounted too close in to the temperature chamber.

Fig. 8: Test setup for temperatures below room temperatures with Zwick&Roell Z250 (a) and the specimen clamping in the temperature chamber (b)

4. Results

The following chapter presents selected results, which are extracted from the total multi-level test-procedure at five different temperatures. The complete experimental data is presented in the attachment.

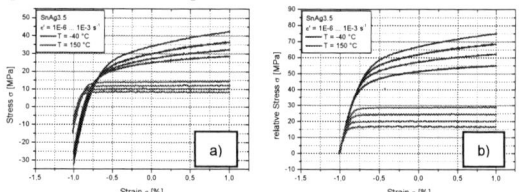

Fig. 9: Strain rate dependent stress-strain progress a) absolute b) relative

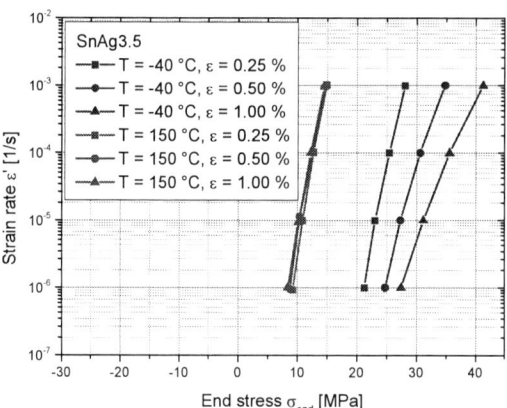

Fig. 10: End stress vs. strain rate for the three amplitude-levels measured at -40°C and 150°C

Fig. 9, shows a comparison of stress dependency on strain rate of SnAg3.5 for two different temperatures. At 150 °C the solder reaches stationary stress level after approximately 0.25 % of strain. For a temperature of -40 °C a significant continuous increase of stress during the total deformation is observed. This indicates that primary creep accompanied by hardening effects dominates the deformation behaviour at lower

Fig. 7: Test setup (left) for temperatures over room temperatures and optical inspection system (right) during the experiments

978-1-4799-4789-8/14 $31.00 © 2014 IEEE

temperature. This relation is further evident because of the evaluation of the stresses reached at the maximal amplitudes within the cyclic tests at different strain rates (see Fig. 10).

In the total temperature range studied, a continuous evolution of stress-strain behaviour of the solder is observed (s. Fig. 11).

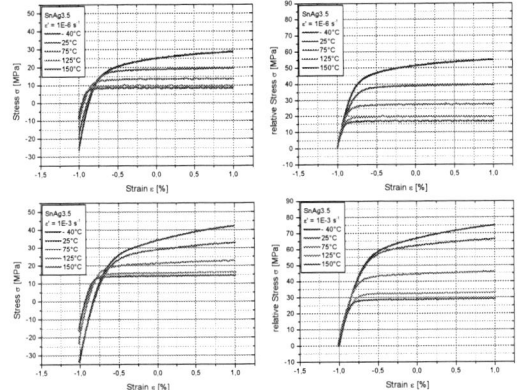

Fig. 11: Temperature dependent stress-strain progress (top: the lowest strain rate, bottom: the highest strain rate, left: absolute, right: relative)

The extraction of the stress slopes gives an overview of the conditions for which the solder behaviour shows stationary stress evolution (see Fig. 12). It is evident that only at temperature over 125 °C and strain rates below 1E-4 1/s stationary conditions exist. This is a strong indication of the importance of primary creep processes at the thermal loads common for automotive applications.

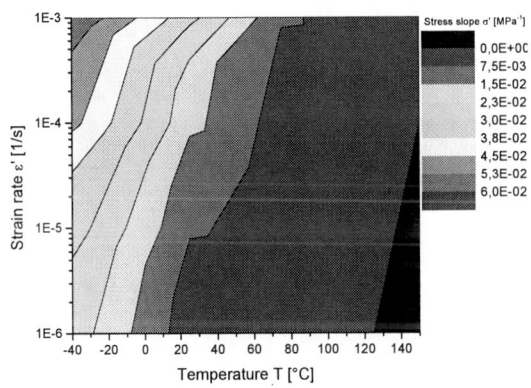

Fig. 12: Temperature and strain rate dependent stress slope

A comparison of stress-time evolution for these three strain amplitudes is shown in Fig. 13. There is a good correspondence between the stress levels obtained within different cycles of the experimental program. Thus, a modification of the mechanical properties of the material due to softening and degradation processes during the measurement

procedure can be excluded. At the highest strain rates a small increasing of the stress progress can be identified, which suggests a slight cyclic hardening during the whole experiment.

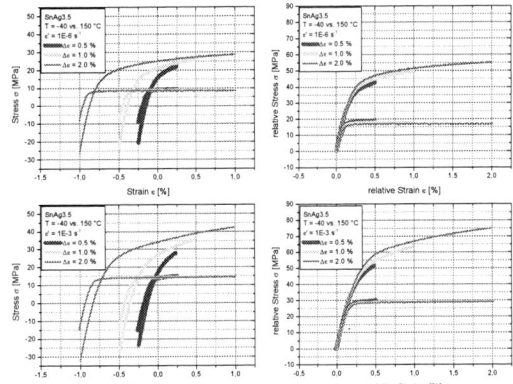

Fig. 13: Strain amplitude dependent stress progress (top: the lowest strain rate, bottom: the highest strain rate, left: absolute, right: relative)

The constant strain control for stress-relaxation stabilizes after approx. 1 s experimental time. Up to a time period of 1h stress evolution is comparable for all temperatures. However, at -40 °C stress relaxation is distinctly slower. For time periods above 1 h the curves tend to a stationary stress level.

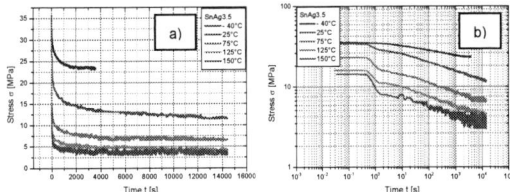

Fig. 14: Relaxation progress after for 1 % strain with a strain rate of 1E-3 1/s in tensile direction a) absolute b) relative

5. FE-Modelling of the experiment

A finite element calculation of the experiment was performed using a Garofalo material model. This model is describes stationary creep rate behaviour as a function of stress and temperature. This model can not represent the non-linear stress evolution at a given strain rate in the primary creep region, since hardening effects are not taken into account. Material model parameters were determined based on the stationary stress region for the measured strain rates at 150°C. A comparison of calculations and measurements for three different temperatures was performed. At 150 °C the model fits very well the measured stress progress. This temperature is representative for stationary creep behaviour. At 25 °C the model fits the measured maximal stresses, but fails to map the non-linear material behaviour influenced by primary creep. At -40 °C the model does not fit the measured stresses, since stationary creep is not reached within the strain amplitudes in the experiment (see Fig. 15).

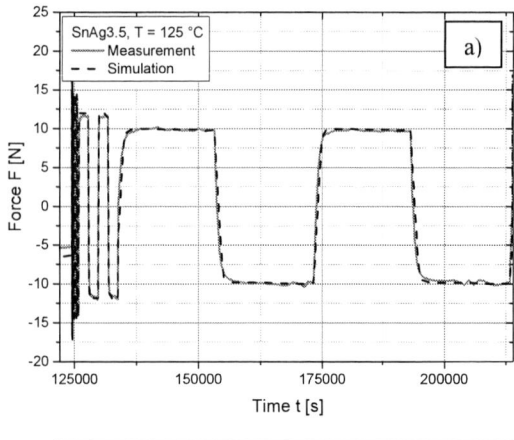

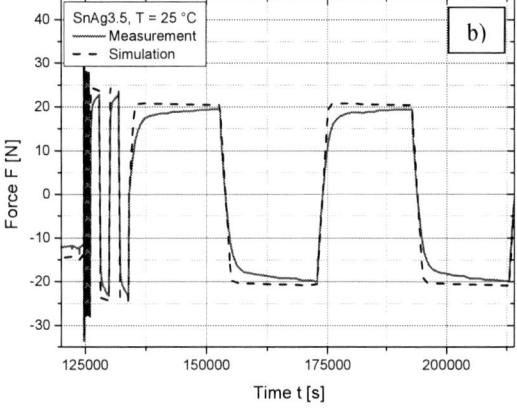

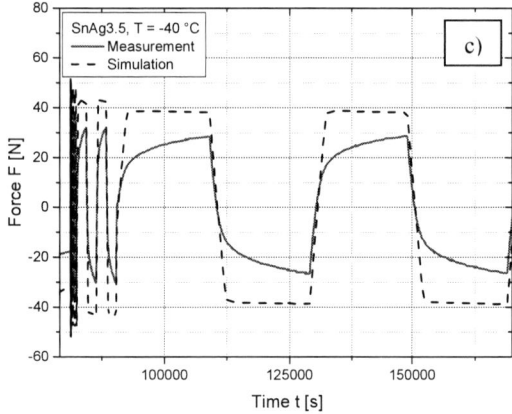

Fig. 15: Measured stress vs. simulated progresses a) 125 °C, b) 25 °C c) -40 °C

The determined model parameters which are used for this comparison are shown in Tab. 2.

Tab. 2: Adapted Sinh-model parameter for SnAg3.5

	A [1/s]	σ_0 [MPa]	n [-]	Q [kJ/mol]
Sinh	1.7E+10	22.3	11.3	93.4

In this case the typical strain rate versus stress plots of the material model is seen in Fig. 16.

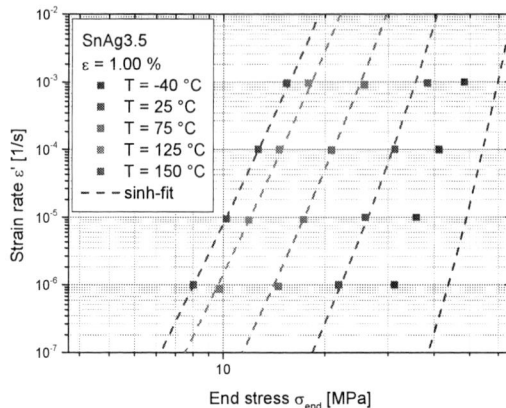

Fig. 16: Adapted Sinh-fit to model the secondary / stationary creep behaviour

The calculations were performed by modelling the specimen geometry, which is located within the mounting points of the displacement sensor (see Fig. 17). The length between the outer areas is 18 mm. In the simulation, the displacement progress from the experiments is used as a displacement condition applied on the outer faces of the sample geometry.

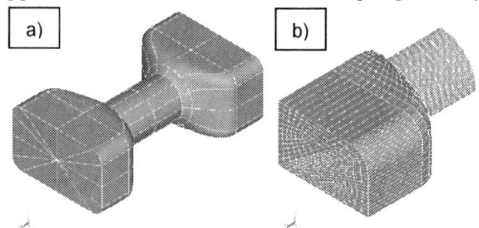

Fig. 17: FE-geometry model of the tensile and compression specimen a) whole model, b) reduced model with symmetric conditions)

For the comparisons the averaged stresses of the inner volume of the geometry model are evaluated and compared with the experiment (see Fig. 18 and Fig. 19).

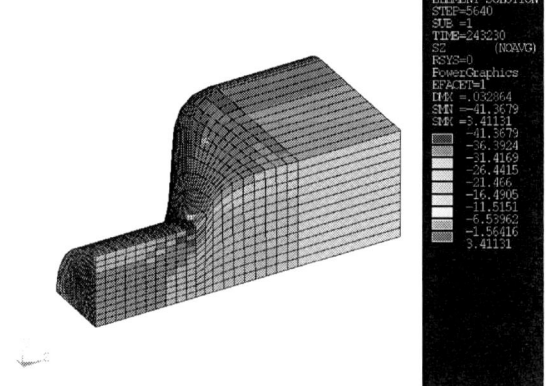

Fig. 18: Stress distribution example, tensile component in load direction

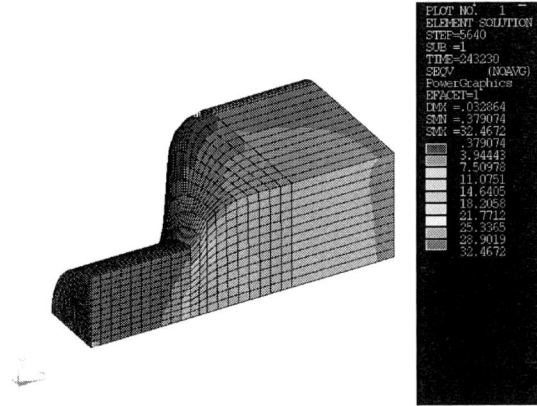

Fig. 19: Stress distribution example, equivalent stress by Mises

A suitable mechanical property for a comparison of simulation and experiment is the accumulated plastic strain energy density, since it is a commonly used value in life-time prediction models of a solder joints. The results are presented in Fig. 20.

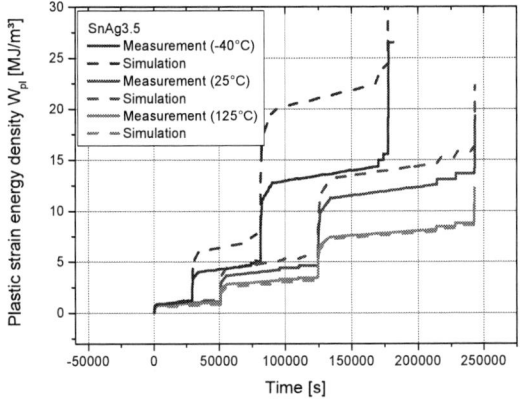

Fig. 20: Comparison of the plastic strain energy density progress during the experiment and the FE-simulation

At 125 °C the simulation results match the experimental results very well. The average difference is under 1 % over the whole progress. However, at lower temperatures, the difference between the simulation and the measurement substantially increase with decreasing temperature and increasing strain amplitude. At 25 °C the average differences increase to approx. 15 %, whereas at -40 °C the difference is approx. 40 %.

6. Summary

The present study shows the extensive investigation of the complex solder deformation behaviour through the combination of different experimental types. The combination of strain rate variations, relaxation and fatigue experiments reveal the complex non-linear material behaviour in the initial state without an influence of degradation processes. The advantage of the combined test procedure is that the time-depended material behaviour can be characterized in a wide load-parameter range on a single specimen. In this way, the influence of deviations between the properties of different samples does not impair the consistent investigation of the material specific stress-strain rate dependencies. The total measurement procedure at a given temperature takes approx. 4 days. Thus, a characterization of a solder alloy at 5 different temperatures with 5 repeats can be completed within 4 months. Furthermore, the experimental program designed for the present study provides an extensive database suitable for parameter calibration of various phenomenological constitutive models. The application of a stationary creep model for finite element simulation reveals significant difference between the measured and calculated stress evolution. With decreasing temperatures primary creep and hardening effects become increasingly dominant and govern the material behaviour. The results show that stationary creep behaviour is only relevant for temperatures above 125 °C, when automotive applications are considered. Finally, the present results suggest, that lifetime prediction techniques for electronic devices based on FE-calculations have to be extended by the usage of advanced visco-plastic material models [4]. These may be capable of an accurate description of the complex solder material behaviour at the relevant thermal and mechanical loading conditions.

References

[1] S. Wiese, M. Roellig, M. Mueller, S. Rzepka, K. Nocke, C. Luhmann, F. Kraemer, K. Meier, K.-J. Wolter. The Influence of Size and Composition on the Creep of SnAgCu Solder Joints. *Electronics Systemintegration Technology Conference ESTC*, pages: 912 – 925, Publication Year: 2006

[2] S. Deplanque, W. Nuechter, M. Spraul, B. Wunderle, R. Dudek, B. Michel. Relevance of primary creep in thermo-mechanical cycling for life-time prediction in sn-based solders. *Conference on Thermal, Mechanical ond Multiphysics Simulation and Experiments in Micro-Electronics and Micm-System, EuroSimE*, pages 71–78, Publication Year: 2005.

[3] R. Metasch, J.C. Boareto, M. Roellig, S. Wiese, K.-J. Wolter. Primary and tertiary creep properties of eutectic SnAg3.8Cu0.7 in bulk specimens. *Thermal, Mechanical and Multi-Physics simulation and Experiments in Microelectronics and Microsystems, EuroSimE*. Publication Year: 2009

[4] A. Kabakchiev, B. Métais, R. Ratchev, M. Guyenot, P. Buhl, M. Hossfeld, X. Schuler, R. Metasch, M. Roellig. Description of the thermo-mechanical properties of a Sn-based solder alloy by a unified viscoplastic material model for Finite Element calculations. *Thermal, Mechanical and Multi-Physics simulation and Experiments in Microelectronics and Microsystems, EuroSimE.* Publication Year: 2014

Attachments

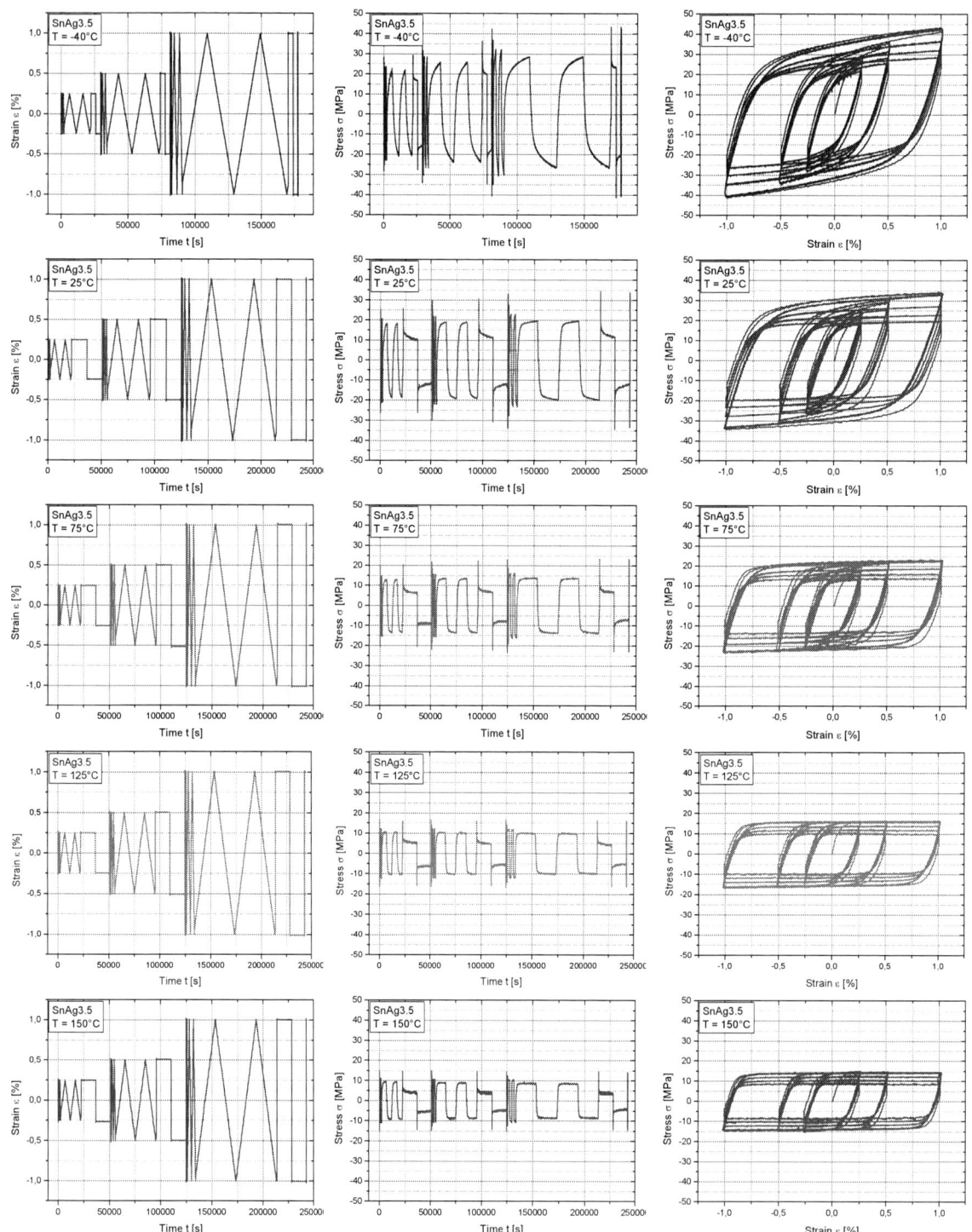

Analysis of mechanical properties of thermal cycled Cu Plated-Through Holes (PTH)

H. Walter[1], A.Kaltwasser[2], M. Broll[3], S.Huber[1], O. Wittler[1], K.-D. Lang[1,3]

[1] Fraunhofer IZM, Environmental and Reliability Engineering (ERE), Berlin
[2] Hochschule für Technik und Wirtschaft (HTW) University of Applied Siences, Berlin, Germany
[3] Technical University, Berlin, Germany

Fraunhofer IZM, Dept. Environmental and Reliability Engineering
D-13355 Berlin, Gustav-Meyer-Allee 25

Email: hans.walter@izm.fraunhofer.de, phone +49 (0) 30 46403-184

Abstract

The aim of thermo-mechanical reliability assessment in microelectronic packages is life time prediction under different thermal and mechanical induced stress loads. The analysis of long time stability of thermally loaded Plated-Through-Holes (PTH) in Printed Circuit Board (PCB) also requires an accurate determination of material data. This leads to application of different test and measurement methods, which are allowed to measure mechanical materials properties at micro- and/or nanostructural scale. This paper focuses on application of instrumented nanoindentation measurement technique for analysis of mechanical properties of microelectronic relevant electroplating copper. Nanoindentation method has been widely used for characterization of mechanical behaviour of devices in small volume (especially for PTH) and determined typically elastic mechanical properties (reduced modulus and hardness). In combination of modified Finite-Element (FE) simulation models and nanoindentation test results elastic and plastic material properties of copper in small scale were obtained. It was dimensionless functions for determination of presentable stresses developed, which allows to indicate the stress-strain curve of bulk materials. It is a precondition to implementation of this function that the indentation depth is out of indentation size effect. The presentation of calculated stress-strain curves by using of dimensionless function and the influence of thermal cycling of material behaviour of PTH are subject of this paper.

Keywords: nanoindentation, electroplated Copper, PTH, Printed circuited device

Introduction

Printed circuit boards (PCB) include a large number of electronic devices which are commonly surface mounted and additional components in form of active layers within each PCB. The knowledge of deformation and materials behaviour is important for understanding the reliability problems of microsystem applications.

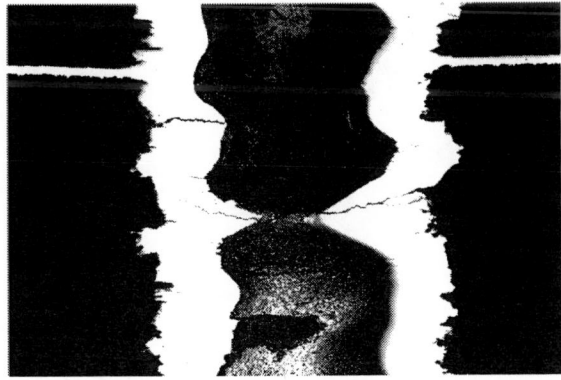

Fig. 1   Cracks in PTH via wall as a results of thermal cycling

For electrolessly deposited copper which use for Plated through holes (PTH) mechanical damages caused by thermal stresses and could be the reason for cracks in the wall.

Fig. 1 shows cracks in the PTH wall as a result of thermal cycling.

During the reliability analysis of electronic devices the local material behaviour of PTH and composite material is often needed. Therefore, an instrumented nanoindentation test is an established test method for characterization of mechanical properties at micro structural scale and/or nano scale. Because this test method is a useful tool for determining the mechanical heterogeneity of microstructure and for investigating the local mechanical properties of miniaturized electronic and engineering. In literature, numerous methods are presented for matching the results of finite element simulation with the experimental indentation data to get the elastic-plastic properties of thin metal layers [1-3].

Few experimentally studies have addressed the relationship between load-indentation behaviour and true stress-strain properties. Dowhan et. al. [2] presented a numerical optimization method for extraction of elastic-

978-1-4799-4789-8/14 $31.00 © 2014 IEEE

plastic properties of thin metal films based on nanoindentation results. Additionally, the 3-level, full factorial design of experiments (DOE) process was applied.

Also, Wittler et.al.[3] studied the Ramberg-Osgood relationship for extracting the elastic plastic behaviour of thin metal layers.

$$\varepsilon = \frac{\sigma}{E} + K\left(\frac{\sigma}{E}\right)^n \qquad (1)$$

where the total strain ε is a function of stress σ. This relation depends on the Young's Modulus E as well as the constants K and n, which describe the plastic hardening of the material. First results show different possible solutions for one indentation curve and independence of film thickness and indentation depth.

Extended models of Dao et. al, Bucaille et.al and Ogasawara et.al appear more suitable for evaluation of valuable results of elastic-plastic properties of metals by matching the load-displacement curves of nanoindentation and Finite-Elemente (FE-) simulation tools [4-8]. The usage of true stress-strain curve (Ramberg-Osgood-Relation) is also the basic for applying these models. But, crystalline metals are showing an indentation size effect (ISE). That means an increasing of hardness leads to reduction of indentation depth. However, both this behaviour and pile-up and sink-in at the side of indents lead to error of modulus and hardness measurements. These effects can be described and accounted for elastic-plastic properties by finite element simulation models based on experimental data of the indentation process.

Analysis of nanoindentation characteristics of Cu Plated-Through-Hole (PTH) vias

Printed circuit boards including the plated through holes were carefully cut into pieces and embedded into epoxy resin. The plated through holes were grinded perpendicular to their interconnection higth until the middle of their diameter. For the last grinding steps abrasive paper with high grid size were used to reduce the depth of plastic deformation. After grinding, several polishing steps including vibration polishing with colloidal silica followed. In a final step, the sample surface was etched to remove remaining influence of the mechanical preparation.

For nanoindentation experiments, Berkovich intender geometry with representative strain for determination of true stress-strain behaviour was used. First results for analysing load/displacement curves of electroplated copper (PTH) have showed the matching of experimental and simulation test data has provided valuable results on the elastic/plastic behaviour.

The indentation experiments for the determination of young modulus were carried out with a Berkovich indenter tip up to a displacement of 500nm

Fig. 2. Referring to the method of Oliver and Pharr [9], the Young modulus of the initial state of the electroplated copper internal layer was determined.

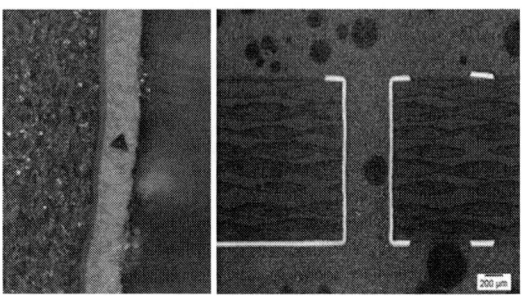

Fig. 2: *Indentation test with Berkovich tip on PTH via wall*

To derivate the macroscopic plastic material properties, the displacement was increased up to 3µm. The maximal displacement was set to this value to avoid any plastic deformation of the surrounding layers. Every PTH was tested several times and the results were averaged (*Fig. 3*).

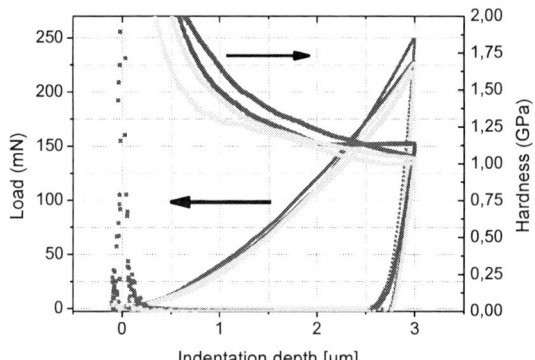

Fig. 3 *Load-indentation depth-curves obtained on PTH via (initial state)*

For analysis of the elastic-plastic properties of copper, the extended model of Ogasawara et. al., a Berkovich-Tip geometry was calculating using:

$$\frac{W_t}{h_{max}^3 \sigma_{0.0115}} = -0.020821 \left[\ln\left(\frac{E^{**}}{\sigma_{0.0115}}\right)\right]^3 + 2.5602 \left[\ln\left(\frac{E^{**}}{\sigma_{0.0115}}\right)\right]^2 - 3.7040 \left[\ln\left(\frac{E^{**}}{\sigma_{0.0115}}\right)\right] + 2.7725$$

$$\frac{S}{2h_{max}E^{**}} = (-0.04783n^2 + 0.04667n - 0.01906) \left[\ln\left(\frac{E^{**}}{\sigma_{0.0115}}\right)\right]^3 + (0.6455n^2 - 0.6325n$$
$$+ 0.2239) \left[\ln\left(\frac{E^{**}}{\sigma_{0.0115}}\right)\right]^2 + (-2.298n^2 + 2.025n - 0.4512) \left[\ln\left(\frac{E^{**}}{\sigma_{0.0115}}\right)\right] + (2.050n^2$$
$$- 1.502n + 2.109)$$

For calculation of the representative stresses of copper by using methods of Ogasavara, the stress-strain characteristics of electroplated cu was investigated by tensile test (*Fig. 4*).

Fig. 5 shows the 3D-simulation of the nanoindentation under consideration of the real sample geometry. The indenter tip is a three sided pyramid equivalent to a Berkovich-tip. The model consists of the 30µm thick electroplated copper layer, FR-4 substrate and molding mass from the sample preparation.

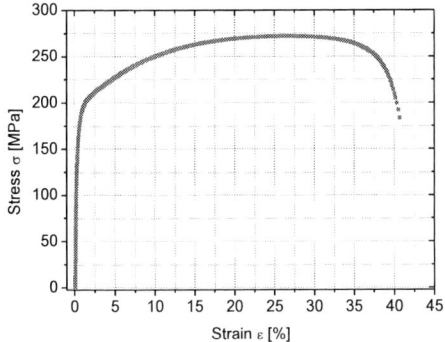

Fig. 4 *Stress-strain characteristics of Cu (electroplated)by tensile test.*

To include the influence of the tip blunting on the force-displacement curve, the experimental indenter area function is also implemented in the tip simulation model. Due to symmetric real sample geometry, the simulation model can be simplified as a cuboid.

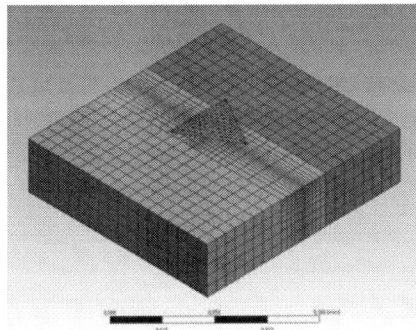

Fig. 5 *3D- model for simulation of load-indentation behaviour with Berkovich tip of embedded Cu-Sample*

Mechanical material data of copper and epoxy are required as the input for the simulation model and presented in Table 1.

Table 1 Material data of copper and epoxy @ room temperature (own measurements)

	Cu (electroplated)	Embedded Epoxy
Modulus (GPa)	128	2,6
Yield stress (MPa)	150	50
Elongation at Break (%)	~30	~ 3
Max. Stress (MPa)	260	50

However simulated hardness-displacement curves (H-h) extracted from different plastic material properties within a range of (yield stress "a" 160-290MPa; "b" 190-240, "c" 0,15-0,96) indicate an influence of adjacent layers for displacements greater than 1,5µm (**Fig. 6**). Further the curvature of hardness-depth curve seems to be similar for all tested combination of plastic material properties. The experimental hardness around 1,5 µm displacement is still changing and therefore, indicates the presence of the indentation size effect. Because of this the macroscopic hardness should be approximated below the experimental hardness value at 1,5µm displacement .

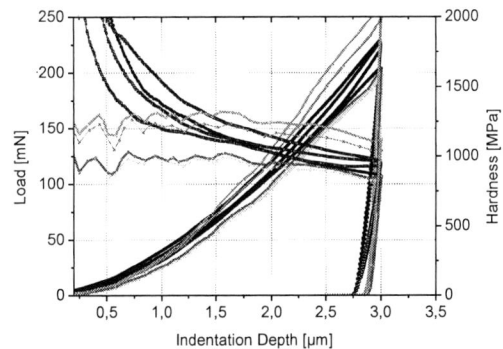

Fig. 6 *Adjustment of the experimental and simulated load-indentation-depth curve with respect to the approximated macroscopic hardness on electroplated copper.*

The results in [10,11] show that it is possible to determine the macroscopic true stress-strain curve from indentation measurements if the displacement is out of the indentation size effect (ISE). However in our case this boundary condition can`t be fulfilled without any plastic deformation of the adjacent layers. Because of this the experimental displacement lies within the ISE. Additionally we have to expect an overlapping effect of ISE and adjacent layers on the load – displacement curve. Therefore, the experimental macroscopic hardness of every sample was approximated with the help of experimental and simulated data. From the experimental hardness-displacement curve, the effect of ISE and influence of adjacent layers can't be clearly separated.

The adjacent FR4 and molding compound have lower hardness and young modulus than the copper layer and therefore decrease the measured hardness. However from the experimental load - depth curve can be seen that the hardness at 3µm displacement seems to get constant despite the influence from adjacent layers. That indicates a decrease of ISE influence. Due to this we approximate the macroscopic hardness of on plated through hole (PTH) copper as the experimental hardness at maximal displacement.

The yield-stresses of the copper layer were derived by adjusting the curvature of the experimental and simulated loading curve (*Fig. 7*).

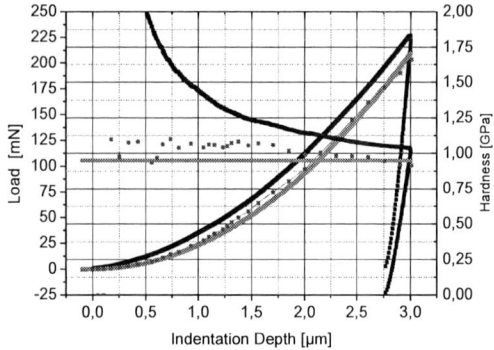

Fig. 7 *Experimental and simulated load-displacement curves of PTH-copper*

After accordance of simulated and experimental curvature the stress-strain data can be obtained from the simulation model according of modified Ramberg-Osgood equation:

$$\sigma_{pl} = \sigma_y + \left(\left[E \cdot \frac{\varepsilon_{pl}}{a}\right] \cdot \sigma_y^{(n-1)}\right)^{\frac{1}{n}} \qquad (2)$$

where a, n material parameter (See table 2)

Table 2 Used material parameters of PTH-vias (initial State) for simulation of stress strain curves

Meas.	Yield Stress (MPa)	a	n	Hardness (GPa)
1	167	7	2,36	0,95
2	220	5,04	4,5	0,95
3	243	6,29	5,75	0,95

The experiment has been matched with Simulation model in order to achieve a best fit of experimental and simulative curves. However different stress-strain curves can lead to nearly the same loading curve. Only one representative stress-strain value can be obtained at the intersection of all used stress-strain curves. This intersection is shown in *Fig. 8*.

Influence of thermal cycling of mechanical properties of printed through hole (PTH)

Finally, the effects of thermal cycling of materials properties of PCB included PTH) are studied. The thermal cycling test was performed to check the reliability of printed cuircuit board (PCB) with several PTHs. The following temperature profile of Minus 55°C up to + 125°C were used. The ramp rates of 25 K/min and dwell time of 10 min are also the parameters for thermal cycling analysis.

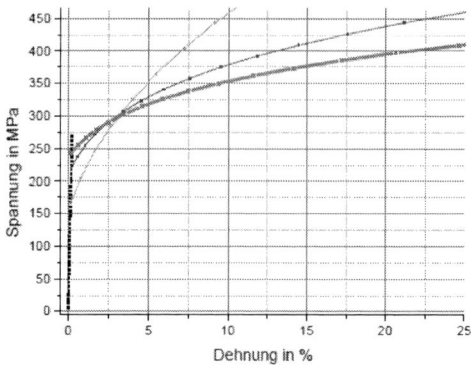

Fig. 8 *Stress-strain curve determined by nanoindentation for PTH (initial State)*

A plated through hole was considered to be failed when the electric resistivity changes more than 5% during the temperature cycle test. The increase of electrical resistivity in PTH is a massive damage by formation of cracks as a result of thermal cycling (further information in [12]) (*Fig. 9*) .

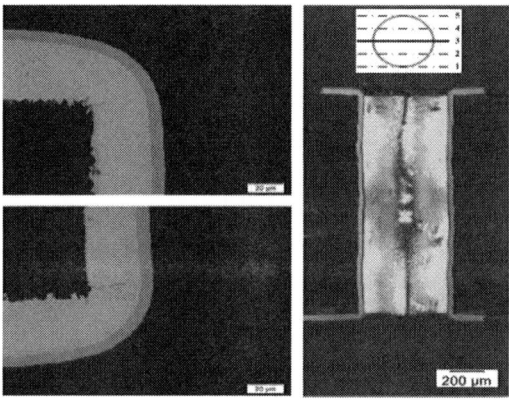

Fig. 9 *Cracks in PTh vias – in the corner during thermal cycling*

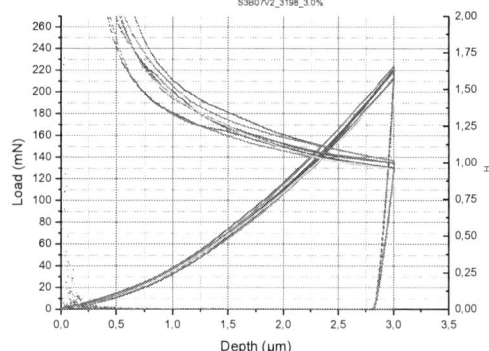

Fig. 10 *Load-indentation depth curves for thermal cycled PTHs*

Such failures are generally associated with changing of structure and material properties. The load-indentation depth curves are depicted in Fig. 10. for thermal cycled PTH –vias after 3200 cycles. As can be seen in Fig.1 the indentation modulus increases with increase of thermal cylces (Fig. 11) while the hardness decrease (Fig. 12).

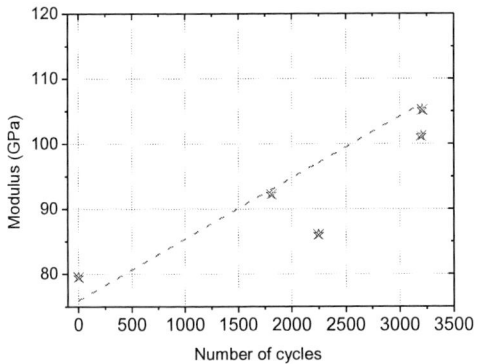

Fig. 11 *Increase of modulus while increase of thermal cycles*

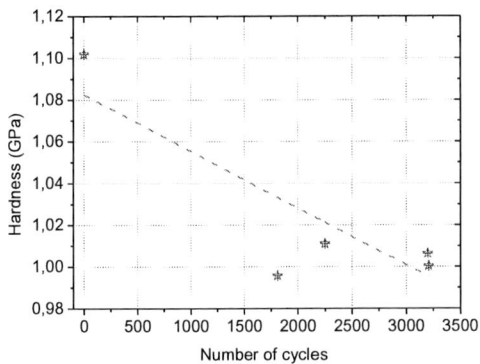

Fig. 12 *Decrease of Hardness while increase of thermal cycles*

The experimental data of cycled PTHs has been matched with the simulation model in order to achieve a best fit of experimental and simulation curves with parameter in Table 3. The coefficients for modulus, yield stress and material parameter a and b indicates Table 3

Table 3 Used material parameters of cycled PTH-vias for simulation of stress strain curves

Meas.	Yield Stress (MPa)	a	N	Hardness (GPa)
1	206	4,1	6,67	0,81
2	205	4,17	6,5	0,82
3	203	4,03	6,84	0,82

For this reason, several nanoindentation experiments were performed in conjunction with Electron Backscattered Diffraction (EBSD) technique. EBSD was used to analyze the influence of thermal cycling on the microstructural evolution of electroplated copper.

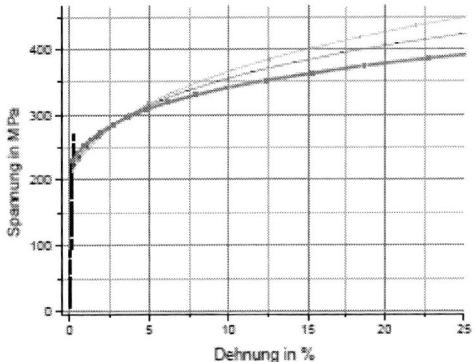

Fig. 13 *Inverse method for calculation of simulated stress strain curves*

This technique based on the detection of backscattered electrons in the depth on the sample surfaces of 50nm [1]. Backscattered electrons build so called Kikuchi pattern on the phosphor screen which can be related to the local crystal orientation. The three coordinate axes of the sample system consist of the normal direction (ND), transversal direction (TD) and rolling direction (RD) which is in accordance to the direction of the copper deposition direction (TD) and PTH depth (RD).

Fig. 14 and *Fig. 15* show the color coded inverse pole figure map with respect to the TD direction of the initial and the stressed state. Both inverse pole figure maps are representative for three other plated through holes with nearly same background.

Fig. 14 *Inverse pole figure map with respect to the TD of the initial state*

EBSD maps showed that during the thermal cycling the crystal orientation of electroplates copper is changed and indicating the occurrence of plastic deformation.

Fig. 15 *Inverse pole figure map with respect to the TD after thermal cycling*

All of this obtained information is needed for an adequate finite-element modelling of PCB/ PTH and for evaluation of reliability tests. The nanoindentated mechanical properties and plastic deformation mechanism of Cu are analysed to clarify the grain size effect on the mechanical properties of materials.

After thermal cycling, the PTH show a modified material behaviour. The hardness increased and the modulus decreased due to grain boundary sliding and/or of grain rotation and decreasing grain size (*Fig. 16*). Also the strength of copper increased with decreasing the grain size following the Hall-Petch relation (Equation):

$$\sigma_y = \sigma_0 + \sqrt{kd} \qquad (3)$$

Where σ_y, d are yield strength and grain size.

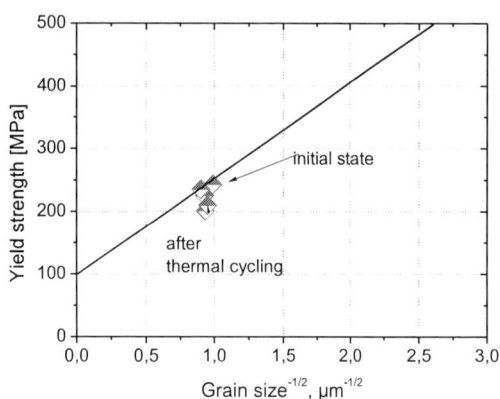

Fig. 16 *Yield strength as a function of grain size of Cu*

Analysis of hardness and modulus by nanoindentation measurements technique provides quantitative evaluation for representative mechanical properties of Cu under influence of their grain size, depend on thermal cycling. The sliding of grain boundary and grain rotation was observed due to temperature induced loading of PTH.

Cu in plated through hole with different grain size, may exhibit different mechanical behaviour.

Summary

In this study, the mechanical properties of electroplated Cu were measured by nanoindentation tests. Extended model of Ogasawara was used for evaluation of valuable results of elastic-plastic properties of metals by matching load-displacement curves of nanoindentation and Finite-Elemente (FE-) simulation tools. Thereby, the combination of nanoindentation test data and FE-Simulation results is possible to determine the elastic-plastic properties of PTH, depend on environmental conditions. First results of analyses show that thermal cycling leads to increasing the modulus and decreasing the hardness. It was also observed the yield strength decreased.

Acknowledgments

The authors would like to thank D. Blankenburg from University of Zwickau for thermal cycling tests of PCB. The work was supported by ZVEI, Germany. This funding was gratefully appreciated.

References

1. Dowhan, Ł. et. al. "Extraction of elastic–plastic material properties of thin films through nanoindentaion technique with support of numerical methods" *Microelectronics Reliability, 51, (2011), 1046–1053*

2. Dowhań,Ł.et.al., "Application of Numerical Optimization Algorithms Used to Investigation of Thin Films in Nanoindentation Test", *11th. Int. Conf. on Thermal, Mechanical and Multiphysics Simulation and Experiments in Micro-Electronics and Micro-Systems, EuroSimE 2010*

3. Wittler, O. et.al. ; "Mechanical characterisation of thin metal layers by modelling of the nanoindentation experiment", *2nd Electronics Systemintegration Technology Conferences, ESTC 2008*

4. Dao, M., et.al.; "Computational modeling of the forward and reverse problems in instrumented sharp indentation", *Acta Materialia, 49, (2001), 3899-3918*

5. Bucaille, J.L. et.al.;"Determination of plastic properties of metals by instrumented indentation using different sharp indenters", *Acta Materialia, 51, (2003), 1663–1678*

6. Ogasawara, N. et.al.; "Measuring the plastic properties of bulk materials by single indentation test", *Scripta Materialia, 54, (2006), pp. 65–70*

7. Bucaille, J.L. et.al.;"Thin A new technique to determine the elastoplastic properties of thin metallic films using sharp indenters", *Thin Solid Films, 447 – 448, (2004), pp. 239–245*

8. Ogasawara, N.;"Comments on "Extracting the plastic properties of metal materials from microindentation tests: Experimental comparison of recently published methods" by B. Guelorget, et al. [J. Mater. Res. 22, 1512 (2007)]: The correct methods of analyzing experimental data and reverse analysis of indentation tests, *Journal of materials research, 23, 03, (2008), 598-608*

9. Oliver, W.C.; et.al. "An improved technique for determining hardness and elastic modulus using load and displacement sensing indentation experiments", *Journal of Materials Research, 7, (1992), 1564-1583*

10. Kaltwasser, A.; "Experimentelle Methoden zur Bewertung der lokalen Werkstoff-Eigenschaften im Bereich der Durchkontaktierungen (German*), Master Thesis, HTW-Berlin, University of Applied Sciences, 2013*

11. Broll, M. et.al.; „Indentation zur Ermittlung elastisch-plastischer Werkstoffeigenschaftenvon metallischen Mikrostrukturen" (German), *7. DVS/GMM EBL-Fachtagung, Fellbach 2014*

12. Blankenburg,D.: „Entwicklung und Evaluierung eines in-situ-Messsystems für statistische Lebensdauer-betrachtungen von Durchkontaktierungen in Leiterplatten" (German), Master Thesis WH-Zwickau, *University of Applied Sciences, 2013*

Failure mode analysis and optimization of assembled high temperature pressure sensors

Roderich Zeiser[1], Suleman Ayub[1], Michael Berndt[1], Jens Müller[2] and Jürgen Wilde[1]

[1]University of Freiburg - IMTEK, [2]University of Ilmenau

Email: roderich.zeiser@imtek.uni-freiburg.de, Phone: +0049761-2037295

Abstract

Thermal-mechanical stresses are a dominant factor limiting the reliability of sensor-systems in harsh automotive environments. Strains and stresses and their effect on the performance and reliability of pressure sensors with operation temperatures up to 500 °C are analyzed with FE-simulations in this study. Platinum based, resistive pressure sensors, fabricated in thin film technology and bulk micro-machining are the subject of this study. The packaging technology combines ceramic substrates with low coefficients of thermal expansion (CTE) and a glass-solder process. The investigated sensor substrates were AlN, Si_3N_4 and a Low-Temperature-Cofired-Ceramic (LTCC). Two different assembly variants were chosen for the interconnection of the sensors: platinum thin wire bonding and gold micro bump interconnections. 3D FE-models of the sensor-assemblies, including temperature dependent materials properties were developed to analyze the distribution of mechanical stresses in the different assembly components. We measured the global chip-deformation at room temperature for verification of our FE-models. With combination of FE-simulations and metallographic device-cross-sections, cracks in the cavity sealing were identified as major failure mechanism of our sensors. According to the FE-simulations, devices assembled with our flip-chip method combined with LTCC-substrates showed an optimized performance regarding signal-shift and reliability. The sensor-signal drift after the assembly process was reduced from 27 % to 3 % for the optimized configuration.

1. Introduction

The application of micromechanical sensors in chemical reactors, aircraft or automobiles is an approach for condition monitoring of processes and machines in harsh environments. Electronics, integrated in Venus vehicles for NASA missions[1] and sensors in harsh, automotive environments [2] are attractive areas of application for MEMS-devices. MEMS based on temperature-stable functional layers like silicon-carbide (SiC) or platinum have been developed for temperatures above 400 °C [3], [4], [5]. The operation at high temperatures (HT) around 500 °C in the environment of motors or turbines generates thermal-mechanical stresses in the chip, the substrate and the die-attachment. A robust assembly and interconnection technology with HT-stable materials is necessary for reliable sensor operation. Several approaches for electronic

packaging in harsh environments have been published [6]. MEMS, based on silicon carbide or platinum showed large signal drifts and high offset voltages during and after operation [7]. Failure mode analysis of HT-sensors have been published in the past [8]. Assembly and packaging technologies for the existing HT-sensor-elements has to be further developed for reliability improvement. Finite-Element (FE)-simulation is an approved method for numerical calculations of stresses in sensors and assembly components for sensor optimization reasons has been shown in literature [9]. We defined the following goals for this study:

- Development of 3D FE-models of High-Temperature sensor-assemblies with variable geometric and material parameters
- Thermal-mechanical simulation of the assembly behavior under different thermal conditions
- Verification of the FE-results with optical measurements of the chip deformation for assembled assemblies
- Analysis of local stresses for identification of highly loaded regions in the different assemblies
- Mapping of the induced strains in the sensor membrane for estimations of the sensor-signal-behavior
- Experimental validation of the proposed failure modes with metallurgical investigations
- Selection of an optimized assembly configuration

2. Sensor and assembly technology for 500 °C

In earlier studies, we presented packaging technologies and reliability investigations of HT-pressure-sensors. We

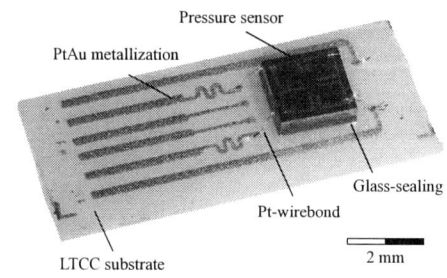

Figure 1. Robust platinum wire-bonded(WB)-pressure-sensor-assembly for operation up to 500 °C

developed a glass-solder-process with a boro-silicate-glass

978-1-4799-4789-8/14 $31.00 © 2014 IEEE

for die-attachment and thin wire bonding of platinum for interconnection [10]. Furthermore, we investigated techniques and materials for a robust flip-chip technology for operation temperatures up to 500 °C [11]. The sensor-element is based on a platinum thin film forming a Wheatstone-bridge on a 50 μm membrane, processed by drie-etching of silicon. Figure 1 depicts an assembled pressure sensor functional up to 500 °C, on a ceramic substrate interconnected with platinum wire bonds. We investigated AlN, Si_3N_4 and a LTCC as substrates. The LTCC, a special Si-matched ceramic based on Al_2O_3, boro-silicate-glass and cordierite, developed at the Herms-dorfer Institut für Technische Keramik (HITK), was provided by the University of Ilmenau and described in earlier publications [12], [13].

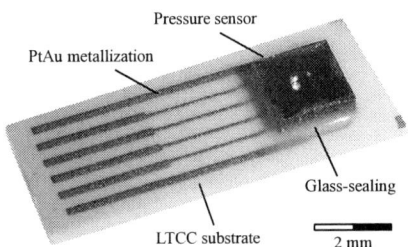

Figure 2. Robust flip-chip(FC)-pressure-sensor-assembly for operation up to 500 °C

The HT-platinum thin wire bond(WB)-package reveals the disadvantage of an exposed metallization layer and interconnection wires. Shorts through conductive particles like sood are possible. An approach that we made to overcome this issue, is a flip-chip technique which is robust up to 500 °C. Figure 2 shows a flip-chip (FC)-assembly on LTCC of a pressure sensitive device.

3. FE-model and utilized materials parameters

3D FE-models of the introduced HT-sensor-assemblies were built with ANSYS v.13. A quarter-model was realized by using symmetry to reduce the amount of FE-nodes. For the correct modelling of electronics with FEM, it is necessary to compare and match the developed models with real devices. For that reason we made metallographic cross-sections of different processed HT-assemblies. A micrograph of a FC-assembly cross-section in comparison to the FE-model is shown in Figure 3. Figure 4 depicts the FE-quartermodel of a WB-sensor-assembly. The sensor is glass-soldered on a ceramic substrate with a cavity under the circular membrane.

In Figure 5 the FC-assembly FE-model is depicted with a face-down mounted chip and glass-solder applied as a ring. For the stress-free temperature T_{ref}, we utilized for both assemblies the glass transition temperature of the

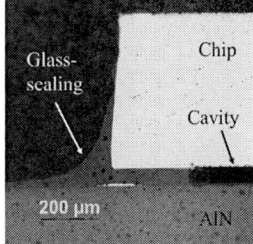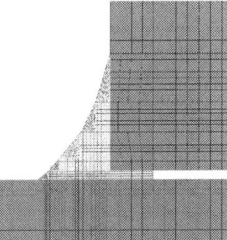

Figure 3. Cross section: FC-assembly mounted with glass-solder on AlN (left), FE-Model FC-assembly (right)

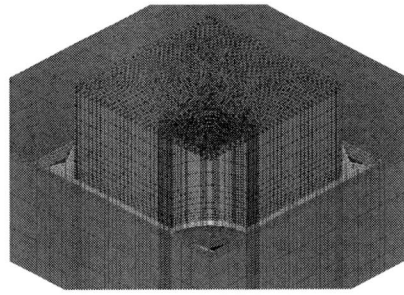

Figure 4. FE-quarter-model of the wire-bond assembly (WB)

glass-solder $T_g = 550 °C$, which we obtained from TMA-measurements of the glass.

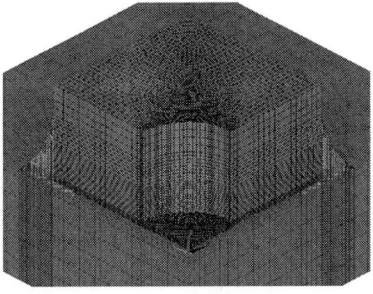

Figure 5. FE-quarter-model of the flip-chip (FC) assembly (FC)

In Table 1 the temperature-dependent material parameters, utilized for the FE-simulations in this study are given. The values for the ceramics are given by manufacturers, for Si taken from literature and the glass-solder was characterized by our group.

4. Results and discussion

A. Experimental verification of the FE-model

The developed FE-models can be approved by verification of numeric results with parameters obtained by measurements of real devices. The chip warpage after processing is a well detectable parameter that can be compared to simulation results. We performed optical chip deformation measurements with a White-Light-Interferometer (WLI) from Zygo. Due to CTE-mismatches the sensor chip deforms after cooling down from the

Table 1. Material properties for the FE-simulations of Chip, substrates and glass-sealing

Material property	Si	AlN	Si₃N₄	LTCC	Boro-glass
CTE in ppm/K					
-40 - 200 °C	2.6	5.0	2.9	2.8	3.6
200 - 400 °C	3.2	5.3	3.0	3.3	5.3
400 - 600 °C	3.8	5.6	3.1	3.9	6.7
E-modulus in GPa					
-40 - 200 °C	161	305	310	220	38
200 - 400 °C	157	303	300	200	37
400 - 600 °C	151	294	290	180	32

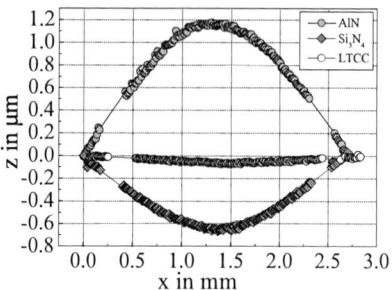

Figure 8. White-Light-Interferometry measurement results for the z-coordinate on a diagonal path x of the chip surface after WB-assembly

soldering temperature. The deflection and the shape of the chip surface is dependent on the CTE, Young's-modulus and geometric parameters [14]. Figure 6 shows results for the WLI measurements of the chip deformation in z direction at 20 °C for assemblies on AlN, Si₃N₄ and LTCC.

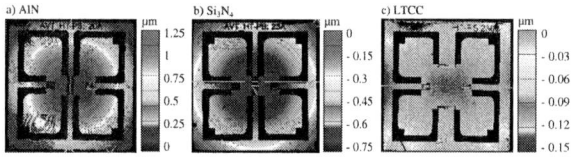

Figure 6. Z-coordinate of chip surface, measured with White-Light-Interferometry after processing: WB-assembly with a) AlN-substrate b) Si₃N₄-substrate and c) LTCC-substrate

In Figure 7 the FE-results for chip deformation of the 3 assemblies after cooling down from the process-temperature at 700 °C to 20 °C is shown.

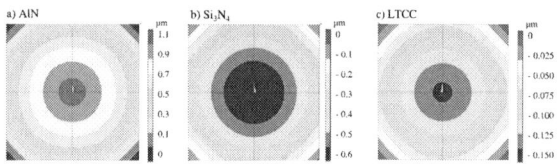

Figure 7. FE-Simulation, Z-coordinate of chip surface after processing: WB-assembly with a) AlN-substrate b) Si₃N₄-substrate and c) LTCC-substrate

The z-coordinate obtained on a diagonal path over the chip surface is presented in Figure 8. The path for AlN, Si₃N₄ and LTCC assemblies are compared.

Figure 9 depicts the results for the diagonal path across the chip of the FE-simulations.

The simulation results of the chip warpage after the assembly are in good agreement with the optical measurements. The deviation of FE and measurement data is for all assemblies below 10 %. The good correlation verifies the geometric and material parameters of our FE-models and the assumption we made for T_{ref}.

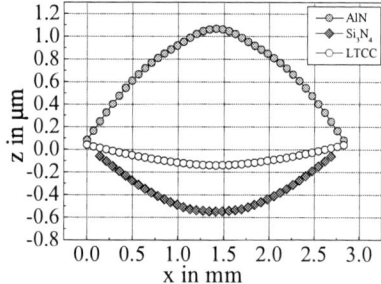

Figure 9. FE-Results for the z-coordinate on a diagonal path x of the chip surface after WB-assembly

B. Packaging stress in the sensor sealing and failure mode analysis

For the reliability of the pressure sensor the stress in the sensor and in the sealing fillet is of significant importance. A crack in the sealing of the sensor-cavity can lead to a failure of the whole device because the reference volume is not insulated any more. Cracks in the chip can also lead to a device failure. For brittle materials like glass and silicon the maximum first principle stress is crucial for crack ignition. The investigation of the local 1st principle stresses in the chip and the glass-solder revealed decreasing values for increasing temperature. Figure 10 depicts FE-results for the maximum stress values for WB-assemblies, dependent on the device temperature.

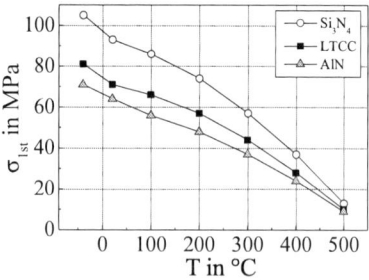

Figure 10. Results for the maximum, 1st principle stress in the glass solder fillet for the WB-assembly dependent on the device temperature

In Figure 11 FE-results for the maximum stress values

978-1-4799-4789-8/14 $31.00 © 2014 IEEE

in the glass-sealing of FC-assemblies is shown.

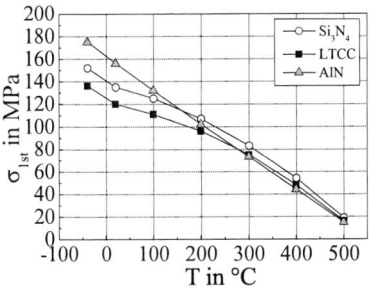

Figure 11. Results for the maximum, 1st principle stress in the glass solder fillet for the FC-assembly dependent on the device temperature

The simulation results for the stress in the sealing of WB-assemblies were significantly lower, for AlN in average by 60 %, for Si_3N_4 by 30 % and for LTCC by 42 %. The reason is the ring-shape of the sealing compared to the application as a die-attachment underneath the chip for WB-assemblies. AlN-substrate assemblies showed the highest results for stress in the sealing with values around 160 MPa at 20 °C. The values for the AlN-assemblies were at -40 °C below 140 MPa.

For AlN FC-sensor-assemblies for which we obtained the highest stress values with FEM, failed devices after processing were detected. Microscopic inspections revealed cracks in the glass-fillet, as shown in Figure 12.

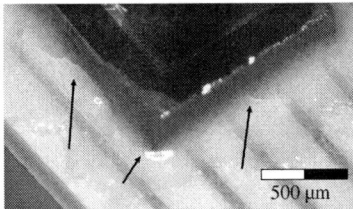

Figure 12. Cracks in the glass-solder fillet (arrows) after assembly FC on AlN

Metallographic cross-sections were made for a deeper investigation of the crack, especially to analyze the crack-depth inside the fillet. For several failed specimen cracks were observed that reach into the sensor cavity and therefore enable the sensor-sealing. Figure 13 depicts a cross-section of a FC-device on AlN.

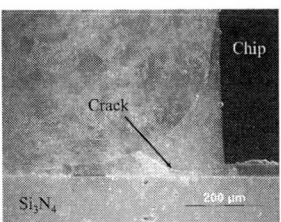

Figure 13. Crack in the glass solder fillet after assembly FC on Si_3N_4

In Figure 14 a crack in the glass-fillet of a failed sensor is compared to a contourplot of the FE-results for the 1st principle stress in the glass-solder ring of a FC-assembly on AlN. It is remarkable that the calculated position of

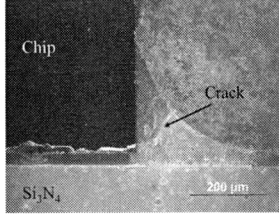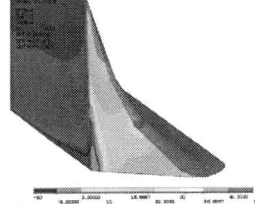

Figure 14. Crack in the glass solder fillet after assembly FC on AlN (left), FE-result for 1st principle stress in the glass-sealing σ_{1st} at 20 °C(right)

the maximum stress is equivalent to the crack position in the real glass-fillet. Therefore the assumed failure mode was an opening of the sensor cavity through cracking. For Si_3N_4 and LTCC-devices no cracks were observed and no failed devices after the assembly-process detected. This fact is in agreement with the simulation results, were for both assemblies lower stress values in the sealing than for WB-assemblies on AlN were observed.

C. Analysis of strain distributions in the sensor-membrane

In this section, the effect of the assembly process with regard to the induced strain un the sensor-membrane is investigated and correlated with experimental data. Figure 15 shows the meshed membrane of the chip with the schematic sketch of the sensor-meanders, compared to the result for strain in the membrane for an WB-assembly on AlN at 20 °C.

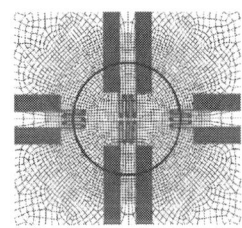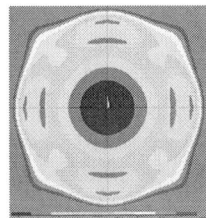

Figure 15. Meshed membrane with position of sensor meanders (left), FE-result for 1st principle strain ε_{1st} in the membrane surface after WB-assembly on AlN at 20 °C(right)

The meandrous resistors are sensitive for the strain ε_x at their position on the membrane. In Figure 16 ε_x is depicted on a path through the sensor-meanders across the center of the chip for 20 °C and 500 °C. The resitor position is indicated with a schematic cross-section through the membrane. Figure ?? depicts the strain in the membrane for for 20 °C and 500 °C.

In ?? the membrane-strain for sensor-assemblies on LTCC is presented. The difference of the induced strain ε_x for the first and the second meander is responsible for an virtual pressure on the membrane, thus unbalancing

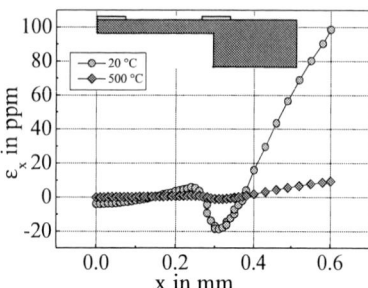

Figure 16. FE-results for WB-assemblies on AlN: ε_x on a path through the sensor-meanders across the center of the chip for 20 °C and 500 °C

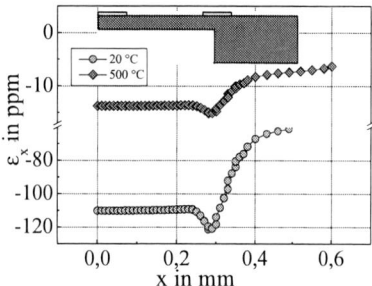

Figure 17. FE-results for WB-assemblies on Si_3N_4: ε_x on a path through the sensor-meanders across the center of the chip for 20 °C and 500 °C

of the Wheatstone-bridge and therefore causing a sensor-offset U_B.

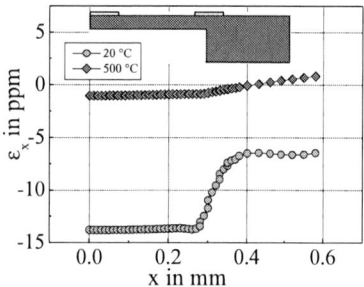

Figure 18. FE-results for WB-assemblies on LTCC: ε_x on a path through the sensor-meanders across the center of the chip for 20 °C and 500 °C

We calculated the difference of the strain at meander position 1 and position 2 on the membrane, named $\Delta\varepsilon_x$ and present the results in 19. The WB-assembly on AlN showed the highest values, FC-assembly on LTCC the lowest. The tendency for the obtained strain differences is correlating with the measured offset-voltage change ΔU_B depicted in figure 19. Bridge voltage results are not presented for FC-devices on AlN, due to the failing of the sensors. The differences of the measured offset-bridge-voltages match well with the FE-results for the strain differences in the sensor-resistors. Therefore, we were able to model the thermal-mechanical behavior of the sensor-membrane and prognosticate the offset-behavior of real devices from FE-simulations.

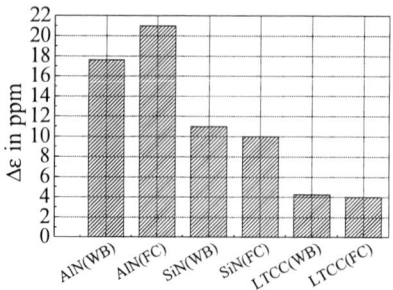

Figure 19. FE-results: Difference of strain in the sensor meanders $\Delta\varepsilon_x$ after assembly on different substrates with WB-wire bond and FC-flip-chip method

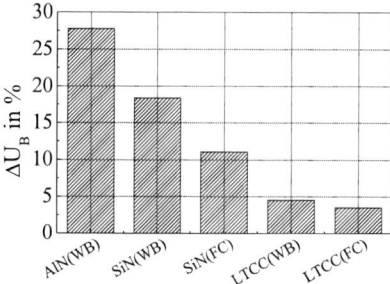

Figure 20. Change of the bridge voltage of HT-pressure sensors after assembly on different substrates with WB-wire bond and FC-flip-chip method

5. Conclusions and outlook

High temperature packaging has a strong effect on the pressure-sensor performance and reliability. Following, the main results of this paper are summarized:

- Development of 3D FE-models for wire-bond and flip-chip high-temperature pressure-sensor-assemblies
- Verification of FE-simulations with optical measurements deviation below 10 %
- High stress values up to 160 MPa at 20 °C in the glass-sealing for FC-AlN-substrates
- Cracking of the sensor-sealing identified as failure mode for FC-assemblies on AlN
- Stress reduction in glass sealing with LTCC substrate
- Strains in the membrane reduced with LTCC-substrate, approved by bridge voltage measurements

The FC-assemblies showed a lower influence on the sensor-signal, but seem to be more suceptible for failures than the WB-technique. The utilization of the Si-matched LTCC-substrate optimized the sensor performance in terms of reliability and reduction of cross-sensitivities. The FE-investigation of the mechanical strains in the chips membrane confirmed this statement. The induced stresses in the assembly components, were higher for the FC-technique.

In the next steps, we will investigate the long-term stability of both presented assembly technologies for 1000 h of operation at 500 °C and 2000 thermal-cycles. Furthermore, we will analyse the thermal-mechanical influence of the packaging technologies on the pressure sensitivity over the whole operation temperature range.

6. Acknowledgements

The investigations presented in this paper were funded by the German Federal Ministry of Education and Research, BMBF, within the Spitzencluster project SiC-Tech, FKZ: 16 SV5127. The authors are very grateful for this financial support. We also want to thank our project partner the Robert Bosch GmbH for the good cooperation and MicroTEC Worldwide for the financial support.

References

[1] T. George, K. A Son, R. A. Powers, L.Y. Del Castillo, and R. Okojie. Harsh environment microtechnologies for NASA and terrestrial applications. In *2005 IEEE Sensors*, pages 6 pp.–, 2005.

[2] R. Wayne Johnson, John L. Evans, Peter Jacobsen, James R. Thompson, and Mark Christopher. The changing automotive environment: high-temperature electronics. *Electronics Packaging Manufacturing, IEEE Transactions on*, 27(3):164–176, 2004.

[3] Robert S. Okojie, Steven M. Page, and Mitch Wolff. Performance of MEMS-DCA SiC pressure transducers under various dynamic conditions. In *2006 IMAPS International High Temperature Electronics Conference, Santa Fe, NM*, page 70, 2006.

[4] D.J. Young, Jiangang Du, C.A. Zorman, and W.H. Ko. High-temperature single-crystal 3C-SiC capacitive pressure sensor. *IEEE Sensors Journal*, 4(4):464–470, 2004.

[5] S. Fricke, A. Friedberger, H. Seidel, and U. Schmid. Micromachined pressure sensor based on sapphire for high temperature applications. *Procedia Engineering*, 5:1396–1400, 2010.

[6] P. McCluskey. Reliability of power electronics under thermal loading. In *2012 7th International Conference on Integrated Power Electronics Systems (CIPS)*, pages 1–8, 2012.

[7] R.S. Okojie, E. Savrun, Phong Nguyen, Vu Nguyen, and C. Blaha. Reliability evaluation of direct chip attached silicon carbide pressure transducers. In *Proceedings of IEEE Sensors, 2004*, pages 635–638 vol.2, 2004.

[8] R.S. Okojie, P. Nguyen, V. Nguyen, E. Savrun, D. Lukco, J. Buehler, and T. McCue. Failure mechanisms in MEMS based silicon carbide high temperature pressure sensors. In *Reliability physics symposium, 2007. proceedings. 45th annual. ieee international*, pages 429–432, 2007.

[9] E. Zukowski, E. Deier, and J. Wilde. Correct modelling of geometry and materials properties in the thermo-mechanical finite-elements-simulation of chip scale packages. In *Proceedings of the 6th International Conference on Thermal, Mechanical and Multi-Physics Simulation and Experiments in Micro-Electronics and Micro-Systems, 2005. EuroSimE 2005*, pages 545–552, 2005.

[10] R. Zeiser, P. Wagner, and J. Wilde. Assembly and packaging technologies for high-temperature SiC sensors. In *Electronic Components and Technology Conference (ECTC), 2012 IEEE 62nd*, pages 338–343, 2012.

[11] R. Zeiser, L. Lehmann, V. Fiedler, and J. Wilde. Reliability of flip-chip technologies for SiC-MEMS operating at 500°c. In *Electronic Components and Technology Conference (ECTC), 2013 IEEE 63rd*, pages 1538–1544, 2013.

[12] M. Fischer. SiCer - a substrate to combine ceramic and silicon based micro systems. In *IMAPS/ACerS 8th International Conference and Exhibition on Ceramic Interconnect and Ceramic Microsystems Technologies, CICMT 2012. Proceedings*, pages 158–161, 2012.

[13] Michael Fischer. Silicon on ceramics - a new integration concept for silicon devices to LTCC. *Journal of Microelectronics and Electronic Packaging*, (6):1 – 5, 2009.

[14] E. Suhir. Modeling of thermal stress in microelectronic and photonic structures: Role, attributes, challenges, and brief review. *Journal of Electronic Packaging*, 125(2):261–267, June 2003.

Reliability Study on Chip Capacitor Solder Joints
under Thermo-Mechanical and Vibration Loading

Karsten Meier, Mike Roellig[1], Andreas Schiessl[2], Klaus-Juergen Wolter

Technische Universität Dresden, Electronics Packaging Laboratory, D-01062 Dresden, Germany
[1]Fraunhofer IKTS-MD, Dresden, Germany; [2]Continental Automotive GmbH, Regensburg, Germany
karsten.meier@tu-dresden.de, phone +49 351 463 36594

Abstract

In this work we present the results on a reliability study on chip capacitor solder joints. The components were tested under three different loading conditions. First, temperature shock tests were conducted on a set of various chip capacitor components. Tested components were evaluated for the occurred damage and the causing damage mechanisms. Using finite element analysis (FEA) the accumulated solder joint creep strain per cycle was determined and used to establish a life time model based on the Coffin-Manson approach. Second, another set of components was exposed to vibration loading. These components were tested in the as cast and isothermally pre-aged condition. The vibration experiments were accomplished at room and elevated temperature. The evaluation focused on the occurred damage as well as the causing damage mechanisms again. FEA was utilised to determine the maximum von Mises stress of the solder joints. Life time and stress data were merged to define the parameters for a Basquin life time model for the vibration load cases. In a third step sequential experiments were accomplished. Temperature cycling with subsequent vibration loading and vice versa was done. Observed cycles to failure were compared to the results from the temperature shock and vibration experiments. A reduction in crack initiation as well as failure cycle count was observed. The damage mechanism was studied as for the single load experiments.

Temperature shock testing was proofed to cause dominant shear loads within the solder joints. Observed cracks appeared to be based on creep deformation. In contrast, vibration causes dominant tensile and compression within the solder joint. The cracks showed a refined grain zone at their boarder pointing to an at least partly plastic deformation cause. Combined loads revealed superposed damage mechanisms.

Both pre-ageing before as well as vibration experiments at elevated temperatures significantly enhance the solder joint damage. However, the combination of vibration and temperature cycling proposes the damage process even stronger. Solder joint life time reveals to be significantly shorter after vibration pre-ageing and subsequent temperature cycling tests than after temperature shock experiments.

1. Introduction

An electronic module is exposed to mechanical, thermal and thermo-mechanical loads during the use cycle in an automotive application. Vibrations from the engine and suspension cause mechanical loads beginning with the start of the engine until the end of the drive cycle. Heat from the engine, gear and exhaust system induce thermal loads after heat up of these systems which cause the mechanical loads named above to occur at high temperatures. After the end of the actual drive the heat from the mentioned car systems starts to heat up the electronic modules due to the missing passive cooling. This adds a thermo-mechanical load cycle to the electronic modules for each use cycle.

To cover the effects of these conditions on the reliability of typical electronic components one need to conduct single, superposed as well as sequential experiments introducing vibration, thermal and thermo-mechanical loads. Besides degradation effects as delamination, corrosion, migration etc. the damage of solder joints is still a major failure mode leading to system fails. Therefore, solder joints of ceramic capacitors were investigated under thermal, mechanical and thermo-mechanical loads within this study.

2. Experimental Procedure

Temperature Cycling

A set of three component sizes was used to vary the stress induced into the solder joints and therefore to force the solder joints to fail after different cycle counts (see Figure 1). Ceramic capacitors with a size of CC0603, CC0805 and CC1812 were mounted on a test board. The test board shown in Figure 1 had a simple layout to enable testing of a high number of components and easy cross sectioning. The applied test condition was a temperature shock profile as explained in Table 1. The used solder alloy was SnAg3.5Cu0.75. The test boards were stressed to a certain number of shock cycles and the solder joint damage was examined by cross sectioning afterwards (see Figure 2). Cracks were found within the solder joint stand off as wells as the meniscus. Hence, crack lengths were recorded for each damage site. To consider solder joint geometry variations the crack lengths l_c were related to the individual reference length l and h respectively.

Table 1: Temperature shock profile for testing of ceramic capacitors

T_{min}	T_{max}	t_{dwell}	cycles[A]
[°C]		[min]	[-]
-40	+125	15	0.1…1

[A]temperature shock cycles were normalised to the maximum cycle count due to confidence reasons

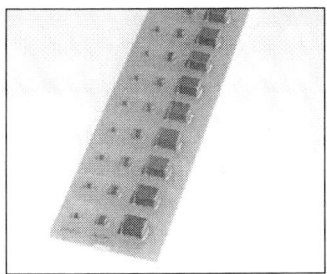

Figure 1: Test board for temperature shock testing of CC0603, CC0805 and CC1812 components

Figure 2: Cross section of a CC0603 component showing a typical stand-off crack after 2000 temperature shock cycles

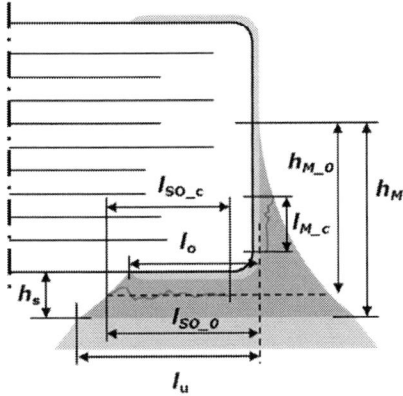

Figure 3: Crack length evaluation for a capacitor solder joint after TCT testing to consider geometry variations

Vibration Testing

Vibration tests were conducted using a specific test board design introduced in earlier work ([1], [2]). The used strap design shown in Figure 4 enables a defined and reproducible solder joint loading on CC0805 components. Solder joints were made using SnAg3.5Cu0.75 solder alloy again. Tests at elevated temperatures are possible due to an integrated local component heating option. The heater enables a homogeneous temperature distribution throughout the component mounting site. Vibration experiments were accomplished using a sine dwell with an excitation frequency f_{exc} close to the resonant frequency f_0 of the test board. f_0 was determined for each test board at the target test temperature. Strap deflection and cycles were varied to alter the applied solder joint loading.

The strap deflection was measured for each component multiple times during an experimental run. The detailed test conditions are given in Table 2. Crack evaluation was done considering individual solder joint geometries as shown in Figure 6.

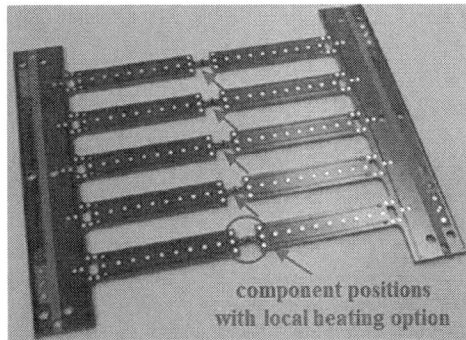

Figure 4: Test board for vibration testing of CC0805 components at room and elevated temperature

Table 2: Conditions of the vibration experiments

strap deflection[A] [-]	frequency [Hz]	T [°C]	cycles[A] [-]
0.05…1	$f_{exc} \approx f_0$ = 250…350	RT, 125	0.004…1

[A]strap deflections and cycle numbers were normalised to the maximum cycle count due to confidence reasons

Figure 5: Cross section of a CC0805 component showing characteristic crack growth from the outer meniscus after vibration loading

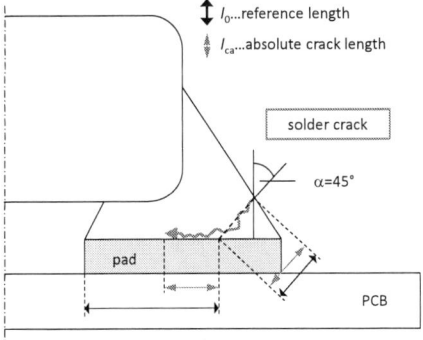

Figure 6: Crack length evaluation for a capacitor solder joint after vibration loading to consider geometry variations

978-1-4799-4789-8/14 $31.00 © 2014 IEEE

Combined Loading

For the combined loading experiments the test boards with CC0805 components from the vibration experiments were used. SnAg3.5Cu0.75 was used as solder alloy. The test boards were loaded with temperature cycles and vibration load sequentially. Two different sequences were considered. Either temperature cycling followed by vibration loading or vice versa. For the temperature cycling a profile as explained in Table 3 was used independent from the load sequence. The vibration loading was done in a different way. If the vibration loading was conducted first the experiments were done at elevated temperature. In case of vibration experiments as the secondary loading the experiments were accomplished at room temperature. Secondary vibration experiments at elevated temperatures are going to be conducted. An overview of the conditions of the vibration experiments is given in Table 4. In total 24 test boards were used to cover 16 load cases. Cross sectioning was applied to reveal damage locations and crack growth. Since three crack locations namely stand-off (SO), lower (LM) and upper meniscus (UM) were noticed to occur in the sequential experiments the relative crack length was determined as shown in Figure 7. The effective crack length was considered to be the average of the three relative crack lengths. As soon as two of the three cracks reach 100% length the effective crack length was considered to be 100% independent from the third crack length.

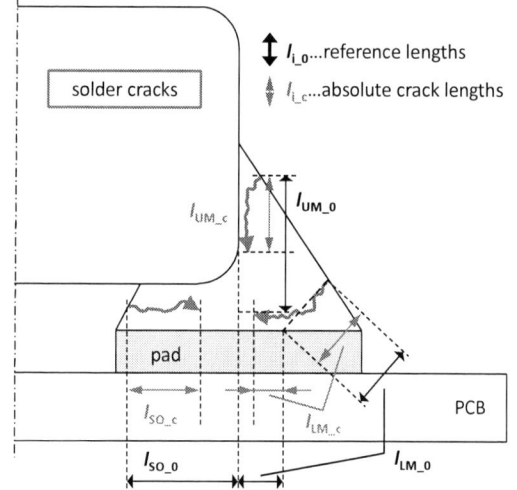

Figure 7: Crack length evaluation for a capacitor solder joint after sequential TCT and vibration loading to consider geometry variations

Table 3: Temperature cycling profile for testing of ceramic capacitors

T_{min}	T_{max}	t_{dwell}	cycles[A]
[°C]		[min]	[-]
-40	+125	20	0.5, 1

[A]temperature shock cycles were normalised to the maximum cycle count due to confidence reasons

Table 4: Conditions of initial and secondary vibration experiments

load case	strap deflection[A]	frequency	T	cycles[A]
	[-]	[Hz]	[°C]	[-]
Vib+ TCT	0.24…1	$f_{exc} \approx f_0$ = 250…350	125	0.5, 1
TCT+ Vib	0.1…1	$f_{exc} \approx f_0$ = 250…350	RT, 125	0.1…1

[A]strap deflections and cycle numbers were normalised to the maximum cycle count due to confidence reasons

3. Results

Temperature Cycling

The relative crack lengths at the solder joint stand-off l_{SO} and meniscus l_M of the different component sizes were recorded and used to determine the cycles to failure. Failure was proposed to be 100 %. It was seen that the stand-off part of the solder joint fails first (see Table 5 and Figure 10). This is due to the dominant shear load within that part of the solder joint caused by the mismatch of the coefficients of thermal expansion ($CTE_{BaTiO3} > CTE_{FR4}$). The meniscus fails second. Hence, a subsequent FEA was used to calculate the accumulated creep strain ε_{cr}^{acc} per temperature shock cycle within the solder joint stand-off first. In a second FEA-step the stand-off was assumed to be failed and the accumulated creep strain was determined for the meniscus accordingly.

Table 5: Temperature shock cycles to failure and corresponding accumulate creep strain of solder joint stand-off and meniscus of ceramic capacitors

component size	solder joint part	N_f[A]	ε_{cr}^{acc} [A]
		[-]	[-]
0603	stand-off	0.68	0.93
	meniscus	0.82	0.94
0805	stand-off	0.69	1.00
	meniscus	0.86	0.89
1812	stand-off	0.59	0.95
	meniscus	1.00	0.80

[A]cycles to failure and creep strain were normalised to its maximum values due to confidence reasons

It can be seen that the stand-off and meniscus life times N_f do not indicate a significant trend considering component size and failure location only. Therefore, the FEA-based calculation of the accumulated creep strains considered the actual average solder joint geometry of each component size instead of scaling the joint geometries with the component size (see Figure 8). The determined data was used to assess the parameter of a Coffin-Mason (1) model describing the lifetime of lead-free solder joints under low cycle conditions considering high temperature plasticity.

$$N_f = A \cdot \varepsilon_{cr}^{acc-B} \tag{1}$$

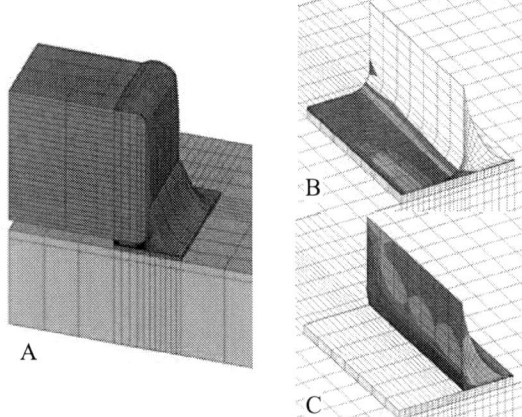

Figure 8: FEM model (A) and stand-off (B) and meniscus (C) creep strain results due to temperature shock loading

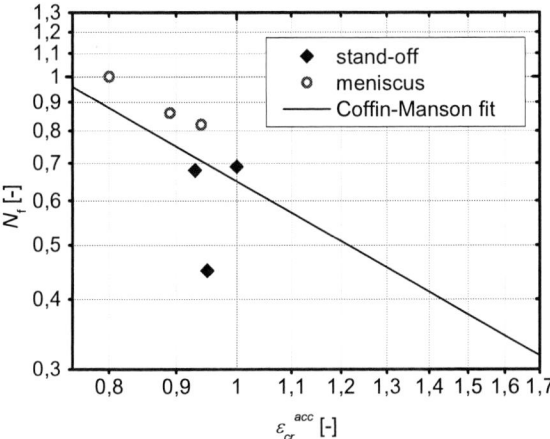

Figure 9: Life time modelling for temperature shock testing, experimentally determined normalised cycles to failure N_f are plotted against the calculated normalised accumulated creep strain per cycle within the solder joint stand-off or meniscus

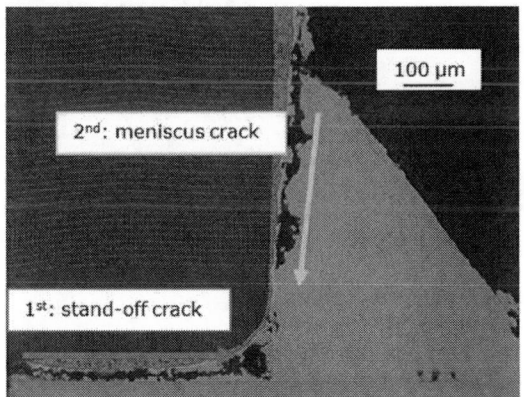

Figure 10: Damage occurring in a CC1812 component after temperature shock cycling: The solder joint stand-off fails first, the meniscus second. Rough crack boarders within the solder volume indicate creep based high temperature plastic deformation (low cycle fatigue)

Now a significant relation between accumulated creep strain and cycles to failure can be seen (see Figure 9). Only the CC1812 stand-off data seem to be an outlier. Furthermore, the results indicate to be independent from the solder joint geometry. Looking at the damage mechanism it is noticed that the occurring cracks propagate within the solder volume and show rough boarders. Both findings indicate creep deformation taking place which is characteristic for low cycle fatigue behaviour and proof the applicability of the Coffin-Manson model approach.

Vibration Testing

The observed relative crack lengths were plotted against the applied vibration cycles and strap deflection (see Figure 11). The vibration at room temperature result shows an increasing crack length for increased strap deflection and vibration cycles respectively. A crack growing during an extended number of cycles was seen instead of a spontaneous cracking – within a very few cycles – of the entire solder joint. This is in a good agreement with the damage seen in cross sections (see Figure 14). Along the crack path a refined grain structure was noticed which indicates plastic deformation in that zone. The solder joints start to crack at the outer meniscus due to tensile and compression stresses caused by the PCB bending. Subsequently, the crack grows towards the PCB pad and closely along the intermetallic interface onwards the stand-off and the edge of the joint below the component. In that state of damage the stand-off is stressed with superposed shear, tensile and compression loads.

Comparing the results from room temperature, pre-ageing and elevated temperature experiments (see Figure 12) the following can be stated: Both the isothermal pre-ageing as well as the elevated test temperature decrease the number of cycles until crack initiation and ultimate failure too. In detail, the isothermal pre-ageing caused a much earlier crack initiation but not much of a decrease of the failure cycle count. This can be explained by two facts. First, the isothermal pre-ageing does change the initial microstructure (grain coarsening and growth of intermetallic precipitations) and therefore eases plastic deformation to occur. Second, once the plastic deformation is started it locally refines the grain structure and the deformation resistance is increased again. Hence, crack growth is retarded and the failure cycle count is only little changed. Testing at elevated temperatures eases plastic deformation as well. However, even though grain refinement occurs the deformation resistance does not increase much. Both lead to even further reduced crack initiation as well as failure cycle counts.

As Figure 13 shows cracks with a length of 35 % to 99 % are very rarely noticed. A crack length of about 35 % is reached when the crack approaches the solder joint stand-off. These findings point out a change in the crack growth speed. During the period of cycles the crack propagates through the solder joint meniscus (crack length = 0...35 %) tensile and compression loads cause a mode I cracking (tensile opening). By the time the crack reaches the stand-off region its growth speed rises much due to superposed shear, tensile and compression loads. Hence, mode I and II (shear opening) are superposed as well.

To access the solder joint stress during the vibration testing a FEM model well matching the test board behaviour was used (see Figure 15). The maximum von Mises stress was calculated in correlation to the applied strap deflections. Both room temperature and elevated temperature conditions were considered. For the latter case the material modelling for the heated component mounting site was modified to account for the material temperature dependencies. Having the stress data at hand life time modelling based on a Basquin approach (2) was enabled (see Figure 16). The Basquin approach is valid for high cycle fatigue scenarios. The model parameters were determined for all three test conditions considering 100 % crack length as failure.

$$N_f = 10^C \cdot \sigma_{vM}^{-D} \qquad (2)$$

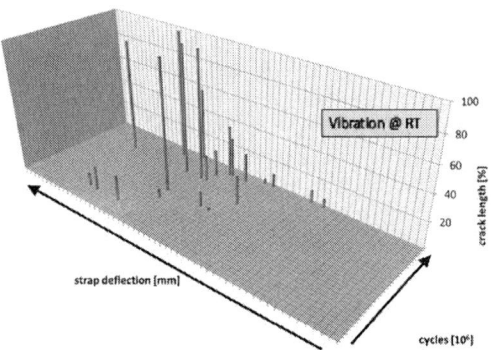

Figure 11: Vibration loading at room temperature causes increasing damage for higher strap deflection and vibration cycles

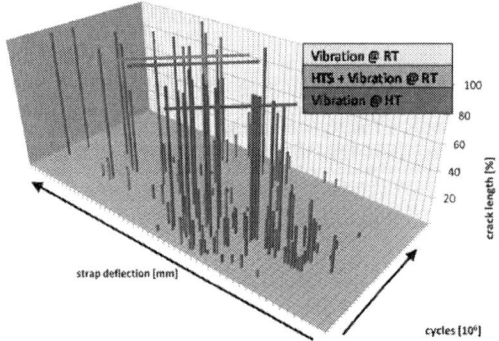

Figure 12: Vibration loading at room temperature without (green) or with HTS pre-ageing (blue) or at elevated temperature (red) causes increased damage, pre-ageing and elevated temperature cause reduced cycles to crack initiation and failure (coloured lines mark 100% crack length considered as failure)

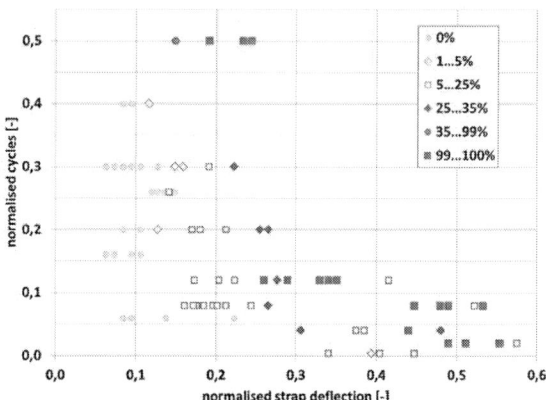

Figure 13: Classification of crack lengths after vibration at elevated temperature shows the minor presence of cracks with a length of 35 to 99 %

Figure 14: Partly plastic deformed solder joint of a CC0805 component after vibration loading

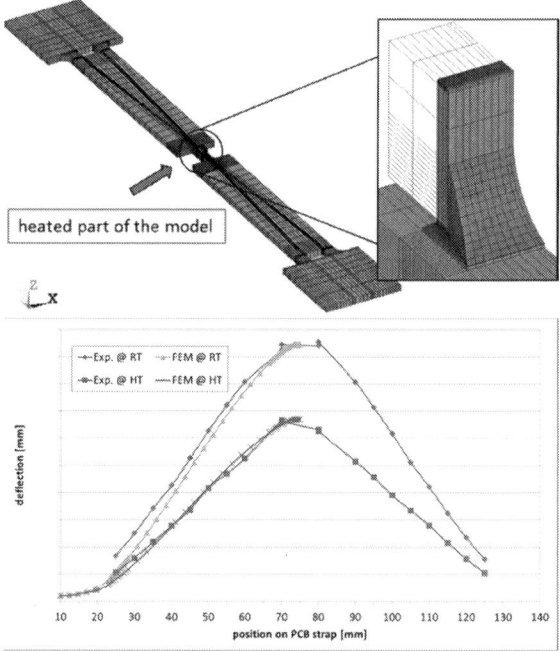

Figure 15: FEM model of a strap of the vibration test board used to determine the solder joint stress during vibration testing: closer view shows a quarter of the modelled component (top), comparison of model and test board deflection behaviour shows good agreement (bottom)

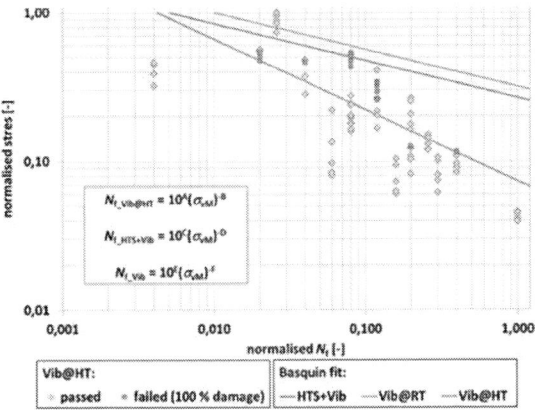

Figure 16: Life time modelling based on a Basquin model using the experimentally observed cycles to failure and the calculated von Mises stress

Combined Loading

Plotting the relative crack lengths of vibration experiments at room temperature with and without isothermal or temperature cycling pre-ageing against strap deflection and vibration cycles shows a significant influence of the pre-ageing steps (see Figure 17). Whereas isothermal pre-ageing reduces only the crack initiation cycle count temperature cycling reduces the cycles to failure (100 % crack length) too. Isothermal pre-ageing affects the microstructure as discussed above only. During temperature cycling the microstructure changes accordingly but due to the induced thermo-mechanical stresses damage (dislocation pile ups, grain boundaries, micro cracks) is accumulated as well. The latter fact eases the crack to propagate through the solder joint during vibration loading which leads to a shorter life time finally. At the same time the accumulated damage also shifts the damage sites. Temperature cycling pre-aged components show cracks initiated within the stand-off, the lower and upper meniscus after vibration loading (see Figure 18). This structurally weakens the solder joint and supports the damage progress.

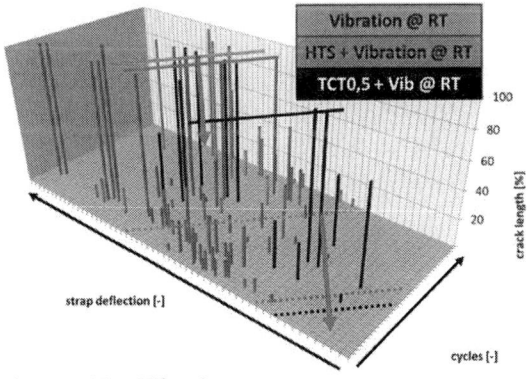

Figure 17: Vibration tests at room temperature without (green), with isothermal (blue) or temperature cycling pre-ageing (black): Pre-ageing causes decreased cycles counts for crack initiation (dashed lines) and failure (solid lines)

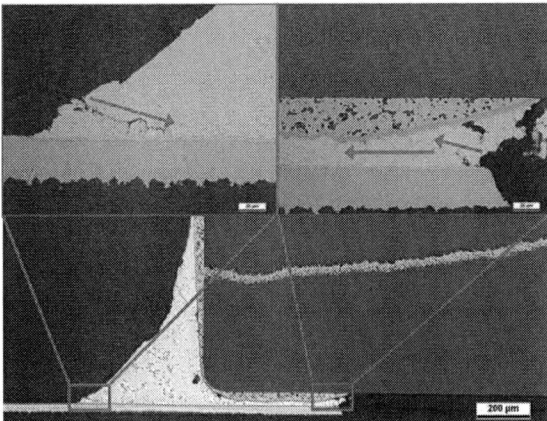

Figure 18: Cracks initiated at the stand-off and the lower meniscus after vibration at room temperature with temperature cycling pre-ageing

Comparing the results from vibration at elevated temperature and sequential vibration at elevated temperature and temperature cycling it can be seen that the failure cycle count is more or less unchanged if one looks for the vibration cycles (see Figure 19). However, the temperature cycling causes crack initiation and crack growth which is shown by a higher crack length at lower vibration cycle counts and strap deflections.

Assessing the cycles to failure after temperature cycling gives a significant lower life time then observed from temperature shock testing though the number of evaluated components is small yet (see Table 7). This is also remarkable cause of the less damage per cycle induced by temperature cycling than by temperature shock testing. The number of cycles to failure during temperature cycling with prior vibration is about less than 70 % of the number of cycles to failure during temperature shock testing. The vibration pre-ageing causes cracks initiated at the vibration characteristic lower meniscus site (see Figure 20). These cracks grow during the temperature cycling and therefore show the temperature cycling typical rough crack boarder. Due to the pre-damaged solder joint meniscus the stand-off faces an increased loading and fails earlier.

Table 6: Cycles to failure from temperature shock testing and vibration at elevated temperature with subsequent temperature cycling show a significant reduction due to the prior vibration

component size	load case	solder joint part	N_f [A] [-]
0805	TST	stand-off	0.69
		meniscus	0.86
	Vib@HT + TCT	stand-off	~0.5[B]
		meniscus	

[A]cycles to failure and creep strain were normalised to its maximum values due to confidence reasons; [B]assessed based on the currently existing results

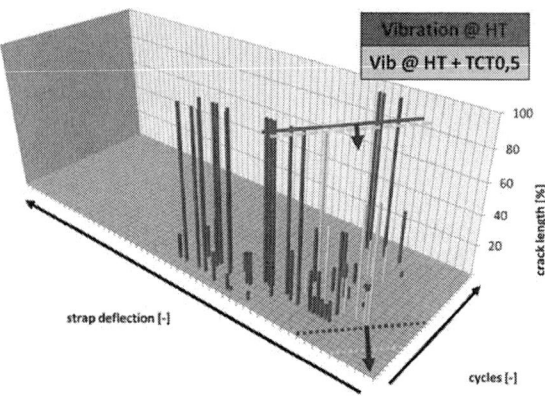

Figure 19: Vibration loading at elevated temperature without (red) or with temperature cycling post-stressing (beige): temperature cycle testing of prior vibrated PCBs shows increased crack lengths at lower cycle counts and strap deflections which is visualised by an earlier crack initiation (dashed lines)

Table 7: Cycles to failure from temperature shock testing and vibration at elevated temperature with subsequent temperature cycling show a significant reduction due to the prior vibration

component size	load case	solder joint part	N_f [A] [-]
0805	TST	stand-off	0.69
		meniscus	0.86
	Vib@HT + TCT	stand-off	~0.5 [B]
		meniscus	

[A]cycles to failure were normalised to its maximum value due to confidence reasons; [B]assessed based on the currently existing results

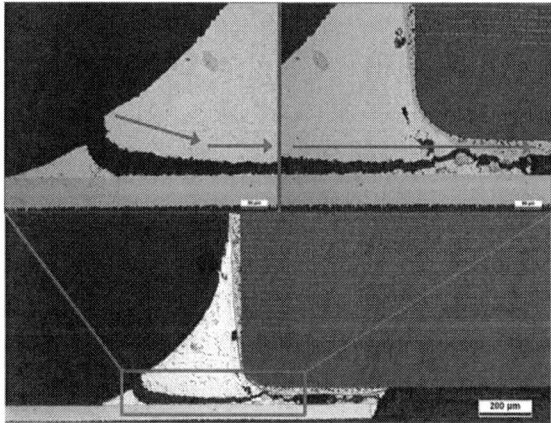

Figure 20: Crack initiated at the lower meniscus after vibration at elevated temperature and subsequent temperature cycling

4. Conclusions

Solder joints of ceramic capacitors were stressed with single temperature shock and vibration loads. Further experiments applied isothermal pre-ageing and increased temperature conditions. Finally combined loading was conducted using a sequence of temperature

cycling and vibration testing. Significant changes in the damage progress – damage initiation as well as occurrence of ultimate failure – and load cycles to failure were identified for each of these cases. Damage mechanisms were studied and found to agree with the observed differences in the failure behaviour. Superposed damage processes were seen for the combined load experiments leading to significantly shorter life times. Temperature cycling experiments with applied vibration pre-ageing showed lower life times than temperature shock experiments on unaged components. Life time models were determined for the single load cases without and with pre-ageing. To be able to establish life time models for the combined load cases further experimental and simulation work needs to be accomplished.

Since electronic components face combined loading conditions in the field use a combined life time testing for development and qualification purposes is recommended. To reproduce the field conditions best it is necessary to research which load is dominant during the field use. Based on this information a proper reliability testing can be done. In a further step life time models need to be established covering the different loads.

References

1. Meier, K.; Roellig, M.; Schiessl, A. Wolter, K.-J., "Life Time Prediction for Lead-free Solder Joints under Vibration Loads", *Proceedings of the 12th EuroSimE*, Linz, Austria, 2011.

2. Meier, K.; Roellig, M.; Schiessl, A. Wolter, K.-J., "Lifetime Assessment for Bipolar Components under Vibration and Temperature Loading", *Proceedings of the 14th EuroSimE*, Wroclaw, Poland, 2013.

Modeling of SiC Power Modules with Double Sided Cooling

Klas Brinkfeldt[1], Klaus Neumaier[2], Alexander Mann[1], Olaf Zschieschang[2], Alexander Otto[3], Eberhard Kaulfersch[4], Michael Edwards[1], Dag Andersson[1]

[1]Swerea IVF, Argongatan 30, 431 53 Mölndal, Sweden
[2]Fairchild Semiconductor GmbH, Einsteinring 28, D-85609 Aschheim, Germany
[3]Fraunhofer ENAS, Technologie-Campus 3, 09126 Chemnitz, Germany
[4]Berliner Nanotest und Design GmbH, Volmerstr. 9B, 12489 Berlin, Germany

Abstract

Silicon Carbide (SiC) based transistor devices have demonstrated higher efficiency switching operation compared to silicon-based, state-of-the-art solutions due to the superior electrical and thermal properties of the SiC material. The improved current density and thermal conductivity allows SiC-based power modules to be smaller than their silicon counterparts for comparable current densities. The active chip area can be reduced further by effectively cooling the devices. In this work, a new power module including SiC bipolar junction transistors (BJT) and diodes and integrated double sided cooling will be introduced. The target application of these modules is a new drive-train system for commercial electric vehicles.

The double sided cooling concept (named ^{2}Cool) is a feasibility study with the goal to further compact the inverter system. More efficient removal of heat from the junction leads to a higher power rating per die, which in turn leads to fewer die and reduced system volume. Since temperature is a main driver in expected failure modes an increase in cooling capability will also enhance margins of the SiC device reliability. In addition, the removal of wirebonds on the top side of the die will result in lower electrical inductance.

Several geometries of the heat exchanger cooling structures have been modeled in terms of thermal performance. The best geometry was a staggered pin-fin structure, which resulted in a junction temperature increase of 74 K at 400 W thermal loading. Also, thermo-mechanical modeling was used to make an estimation of stress in the power module materials.

1. Introduction

Electrification in the automotive field is continuing to gain momentum. Besides the energy storage system, the mechatronic drive train unit, consisting of a traction unit (motor and, if required, transmission modules), power module and control module is a critical technical part required for electrification of vehicles. Lately, the decentralization of the drive train into several, smart in- or near wheel units has been investigated [1].

For construction vehicles such as wheel loaders, the substitution of the central drive train by compact and smart electric drives attached to the individual wheels, coordinated and controlled by a central computer results in weight reduction and a more flexible vehicle due to the removal of wheel axles and transmission.

However, these types of vehicles are typically even more demanding with respect to power, performance, durability, and availability than other types of vehicles, which increase the demands on the drive train unit [2]. In the case of distributed drive systems, including in-wheel or near wheel architectures, the power inverter module needs to be highly integrated and compact.

SiC based power devices have many advantages compared to silicon devices. These include higher thermal conductivity, higher breakdown voltages, lower switching losses, and a capability to operate at higher switching frequencies and temperatures [3,4]. Due to improved current density and thermal conductivity, power modules based on SiC transistors and diodes can be smaller than their silicon counterparts for comparable current densities.

Further reduction in size can be achieved by effective cooling of the devices. Advantages of applying double sided cooling are expected to include an over-all lower on-state voltage and increases in the current carrying capability due to a larger heat exchange area and thereby reduction in thermal resistance [5-7]. In addition, removal of the wirebonds has been demonstrated to significantly reduce the switching cell inductance, which allows for high speed switching and even lower switching losses [8].

In this work, the fluid dynamics and thermal performance of a heat exchanger design based on novel 3D printing technology is presented. The direct manufacturing process gives a large freedom in available geometries and an opportunity to try different new cooling concepts. For example, sponge-like structures that are attached directly to the heat exchanger chassis or base plate without any added material interfaces. A sponge-like cooling structure is evaluated and compared to a more traditional pin-fin design. Thermal CFD simulations are used to estimate the flow parameters through the cooling structures and junction temperatures of the cooled switch devices. The results are then imported into a steady state thermo-mechanical model to make an estimation of the thermally induced stress.

2. Heat Exchanger Thermal CFD Analysis

This section describes the design and thermal performance evaluation of different cooling geometries for a 3D printed heat exchanger used for the double sided cooling of a SiC power module. Initially, two different cooling structures were evaluated. One was a simple inline pin-fin structure (shown in Figure 1a). The other was a sponge-like structure (shown in Figure 1b).

978-1-4799-4789-8/14 $31.00 © 2014 IEEE

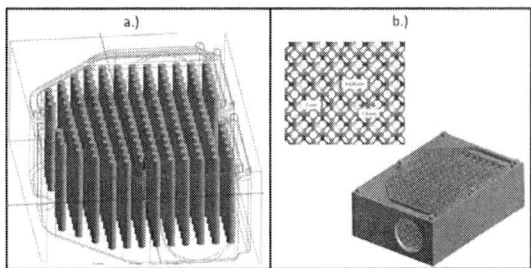

Figure 1: The two different cooling structures of the initial analysis.

For the thermal performance investigation, a reduced, single heat exchanger model with heat source regions at the bottom face was used. Only single-sided cooling was considered and the heat spreading layers in the DBC substrate of the power module to which the heat exchanger is attached were neglected.

Five individual regions (10 mm x 10 mm) of 2-D heat sources were used to mimic the heat load from the SiC switches and diodes. The sources are denoted as source 1 left/right (s1l, s1r), source 2 (s2), and source 3 left/right (s3l, s3r), as shown in Figure 2. The left and right denominations are from a top view perspective along the flow direction. In the initial analysis, each source was loaded by 30 W. The position of the sources and their power dissipation were not similar to the final design. However, they serve to compare the two cooling structures with each other.

All surfaces are assumed to be adiabatic and the fluid velocity at the inlet were assumed to be uniform (no profile was used). The fluid medium was water. The inlet fluid velocity was calculated for minimum (5 l/min) and maximum (15 l/min) fluid flow rates determined by the pump parameters. For pure water (1000 kg/m³) and an opening area of A = 75 mm² the equivalent inlet velocities become:

- 1.12 m/s (equivalent to dV/dt = 5 l/min)
- 2.24 m/s (equivalent to dV/dt = 10 l/min)
- 3.35 m/s (equivalent to dV/dt = 15 l/min)

A 2-equation model (standard k-ε model) was used to model turbulence.

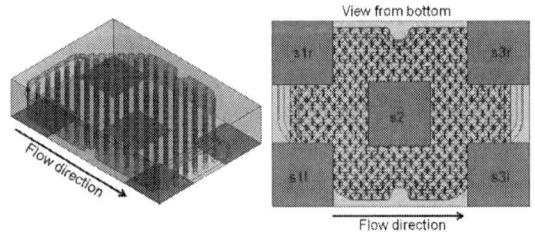

Figure 2: Assumed heat sources. Each hotspot was loaded by 30 W.

The results of initial thermal simulations are shown in Figure 3 and Figure 4. The maximum increase of the junction temperature for the pin-fin version was close to

11 K near the inlet and outlet, and 7-8 K at the center source s2.

For the sponge-like structure the temperature increases were slightly less at the coolant inlet and outlet, while slightly higher at the center hotspot. As shown in Figure 5, the flow for both the inline pin-fin and sponge structures were predominantly in straight lines from inlet to outlet with little mixing in the vertical direction.

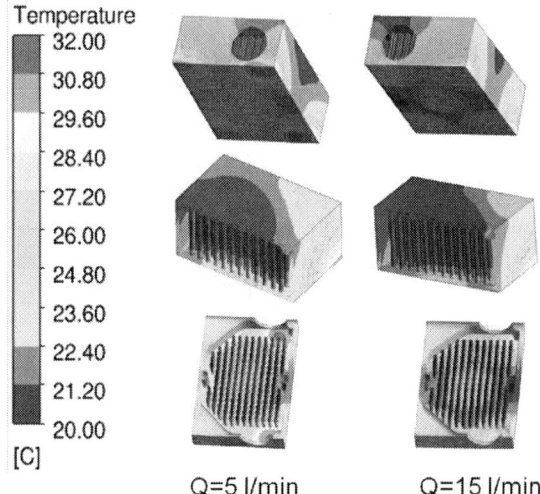

Q=5 l/min   Q=15 l/min

Figure 3: Temperature increase for the pin fin cooling structure.

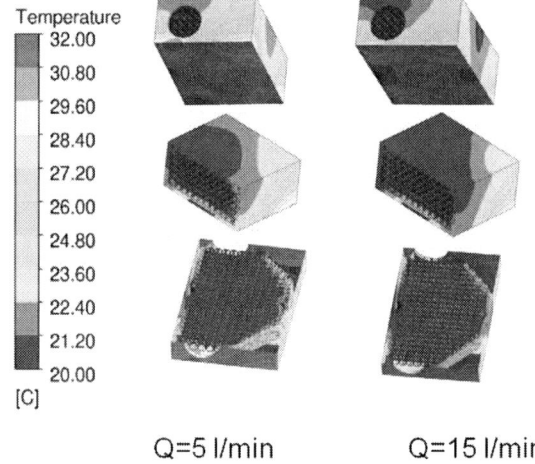

Q=5 l/min   Q=15 l/min

Figure 4: Temperature increase for the sponge cooling structure.

The conclusions of this initial analysis were that at least half of the structure height could be removed and that more vertical mixing of the coolant was needed. Therefore, a tilted sponge structure, which was designed to force the coolant liquid to flow more in the vertical directions was evaluated. Some improvements to the pin-fin was also made (staggering of the pins). In both of the new designs, the cooling structures were extended further, to the inlet and outlet openings. The openings themselves were shaped differently with flatter orifices in the new designs but with unchanged opening area. Figure 6 shows the new designs of the cooling structures.

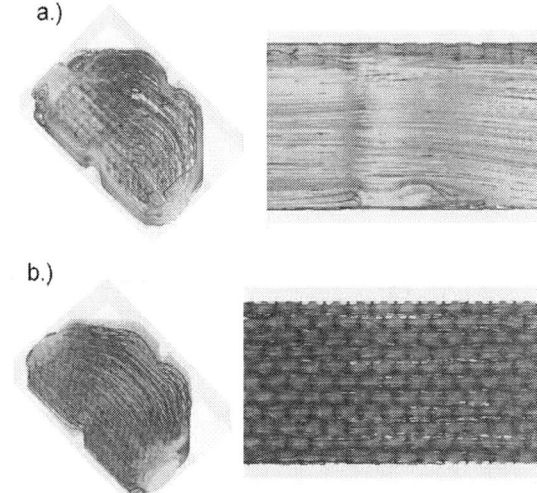

Figure 5: Flow lines for the a.) pin-fin and b.) sponge structures.

Thermal CFD analysis was performed on the new designs using identical boundary and load conditions as previously. The results are shown in Figure 7. The increase in junction temperature is lower for both of the new designs. It is clear that the heat spreads up into the pin structures in Figure 7a, but there is still very little thermal exchange with the cooling fluid, which remains very close to ambient temperatures. The situation is similar or even worse for the sponge structure. Here, the heat spreading from the bottom interface of the heat exchanger does not reach as far into the center of the structure as in the case of the pin-fin and the fluid temperature remains at or close to ambient.

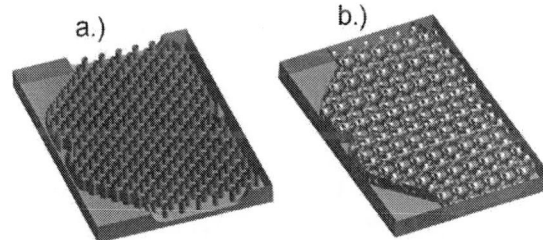

Figure 6: New heat exchanger pin-findesigns. Images are symmetric around the sectioned Z-plane.

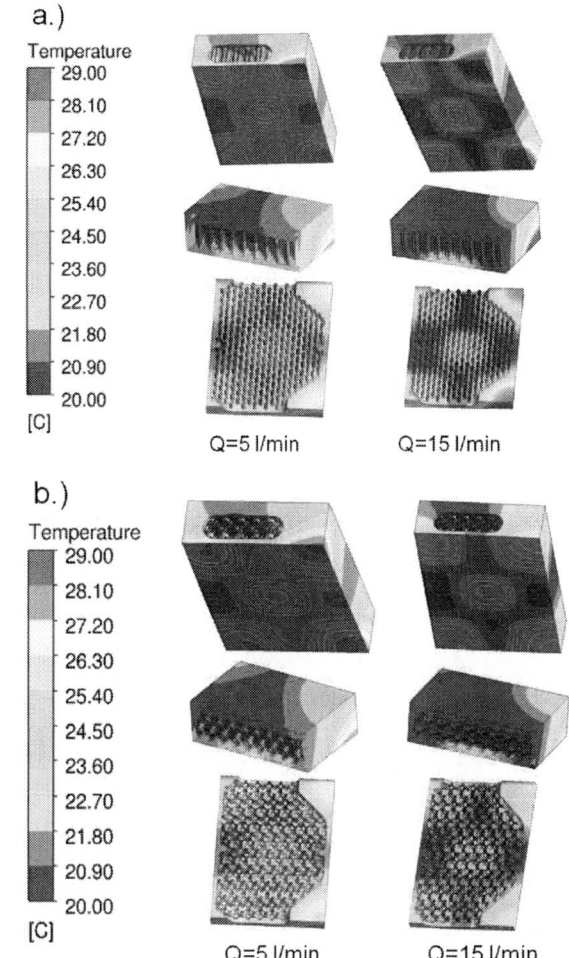

Figure 7: Temperature increase for the new pin-fin (a) and sponge (b) designs.

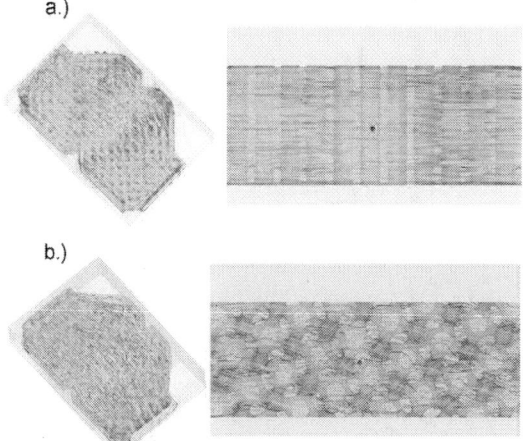

Figure 7: Flow velocity for the new pin-fin (a) and sponge (b) designs.

The pressure required to pump the fluid through the different structures was also simulated and compared. It was also found, for the new sponge-like structures in particular, that a significantly higher pressure drop occurs than for other structures (Figure 8). The limit set by the application is 100 - 200 mBar per heat exchanger.

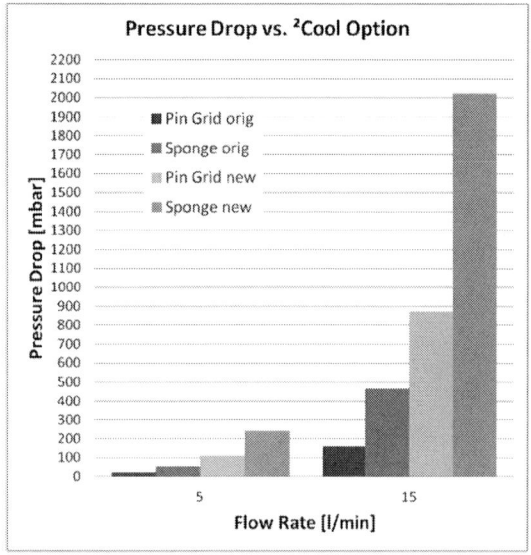

Figure 8: Pressure drop comparison between the different structures.

Figure 9 summarizes the thermal results for all of the designs. The junction temperature increases at the different heat source locations of the old inline pin-fin structure and the old version of the sponge are connected with dashed lines in the figure, and the new designs with solid lines. It can be seen that the new designs generate lower temperature increases than the old versions at all heat sources for minimum and maximum flow rates.

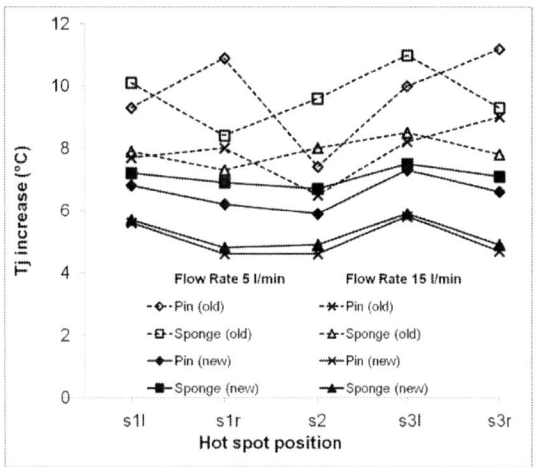

Figure 9: Modeled junction temperature increase with a load of 30 W and different flow rates.

To understand how the junction temperature increase is affected by higher thermal dissipation loads, the central heat source thermal loads were set to 100 W, 200 W, 300 W, and 400 W and the increase in temperature was modeled. The estimated increase in the junction temperature at the different heat dissipation levels for the new pin-fin and sponge designs is shown in Figure 10.

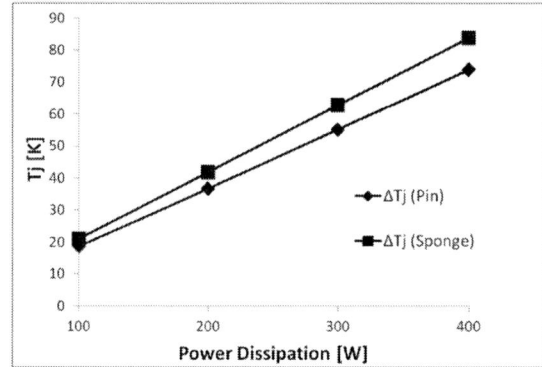

Figure 10: Modeled junction temperature increase at different loading of the central heat source.

3. Thermo-Mechanical Analysis

The results from the thermal CFD analysis of the new heat exchanger designs were imported into a thermo-mechanical model of the power module. The purpose was to get an initial estimation of the stresses in the module due to the expected thermal load.

The model was set up based on the module geometry shown in Figure 11. The modules consist of two AlN-based DBC (Directly Bonded Copper) substrates with SiC transistors and SiC diodes sandwiched in the middle. The mechanical boundary conditions consisted of fixed supports at one of the DBC as shown in Figure 11a. The thermal loading used in this analysis came from the CFD analysis and the junction temperatures from Figure 10 were set at the top surface of the SiC die. The ambient temperature was set to 20 °C. A third temperature, T, at the DBC outer surfaces where the boundary to the heat sink would be, was set to 28 °C, 50 °C, and 80 °C to get an estimation of how the stress levels in the system react to different temperature gradients.

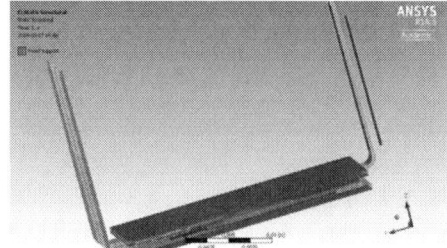

Figure 11: Module geometry and mechanical boundary condition.

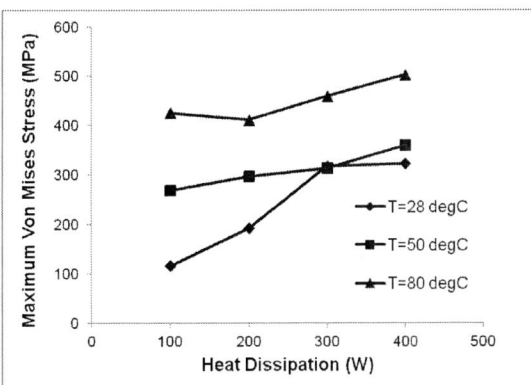

Figure 12: Results of the thermo-mechanical simulation for different thermal loads.

A summary of the results of the thermo-mechanical simulation is shown in Figure 12.

4. Discussion of Results

The results of the thermal analysis show that the new designs have slightly better heat exchanger performance compared to the old inline pin-grid and sponge designs. The old designs suffered from a lack of mixing of the fluid as shown in Figure 5. This was addressed in the new designs by staggering of the pins in the pin-fin version and a tilting of the sponge structure, which removes the possibility for the fluid to pass though the heat exchanger in straight lines. This may have contributed to a larger pressure drop, though the massive increase in pressure drop of the new designs is mostly a result of reducing the exchanger height to half of the old designs.

While the new sponge design forced the flow lines into a vertical wave motion, there was still not much mixing of the fluid. The unperturbed wave motion simply continues through the length of the exchanger to the outlet.

The staggering of the pins in the new pin-fin version showed a smaller increase in pressure drop and, as can be seen in Figure 9, a smaller increase of junction temperatures at all heat sources and for both minimum and maximum flow rates. This results in better over-all heat exchanger performance for the staggered pin-fin design than for the sponge design. The slight divergence of the lines in Figure 10 suggests that the pin-fin version improves even more compared to the sponge version when the heat dissipation increases.

Due to pressure drop limitations, the flow rate needs to be limited to the region around 5 l/min for the new designs and the tilted sponge structure exceeds this even at the lowest limit of the flow rate.

Figure 7 indicates that while the heat spreads well into the heat exchanger volume through the pin structures, the spreading through the sponge version is limited. Theoretical studies of porous heat exchangers conclude that a higher particle size increases the efficiency of the heat transfer [9,10], and analogous to this, the sponge structure could benefit from more material (smaller pores)

close to the heat source interface. This will be implemented in future designs.

The thermo-mechanical analysis shows that when the temperature at the heat exchanger boundary is locked at a lower value, the internal stress increases at a higher rate with dissipated heat due to a larger forced temperature gradient in the module. If the temperature gradient in the module is smaller, which is the case with T = 50 °C and T = 80 °C, the maximum induced stress changes less with changes in junction temperature. The absolute stress values are of less concern as this is an estimation. It should also be noted that only single sided cooling is assumed in estimating the junction temperatures in the thermal CFD analysis, on which the thermo-mechanical analysis is based.

5. Conclusions

Thermal CFD analysis has been performed on heat exchanger cooling structures. A pin-fin version was compared to a sponge-like structure. It was found that a staggered pin-fin version had slightly better thermal performance and significantly lower pressure drop than a sponge structure. An increase in junction temperature of 84 °C with the sponge structure and 74 °C with the pin-fin at a heat source power dissipation of 400 W were found. Thermo-mechanical simulations were also performed to make estimations of the stressed behavior of a SiC power module intended for double sided cooling. The target application of these modules is a new, distributed drive-train system for commercial electric vehicles which will benefit from smaller, efficient power modules.

Acknowledgements

The authors would like to acknowledge the European Commission for supporting these activities within the COSIVU project under grant agreement number 313980.

References

1. Rzepka, S., Otto, A., COSIVU – Compact, Smart and Reliable Drive Unit for Fully Electric Vehicles, Micromaterials and Nanomaterials, issue 15, 116-121, 2013.
2. Nord, S., Cortes, A., Electro-mobility: Key Technologies for Sustainable Transport Solutions, Micromaterials and Nanomaterials, issue 15, 16, 2013.
3. Weitzel, C. E., Palmour, J. W., Carter, C. H., Jr., Moore, K., Nordquist, K. J., et. al., IEEE Trans. on Electron Devices, Vol 43, No 10, 1996
4. Cooper, J.A., Jr., and Agarwal, A. Proc. of the IEEE, 07/2002; DOI:10.1109/JPROC.2002.1021561
5. Chang, H.-R., Bu, J., Kong, G., and Labayen, R., Proc. of the 23rd Int. Symp. on Power Semiconductior Devices and IC's, San Diego CA, May 23-26, 2011
6. Ning, P., Liang, Z., and Wang, F., Applied Power Electronics Conference and Exposition (APEC), Long Beach, CA, March 17-21, 2013

7. Zhang, H., Ang, S. S., Mantooth, H. A., and Krishnamurthy, S., Energy Conversion Congress and Exposition (ECCE), Denver, CO, Sept. 15-19, 2013

8. Hoene, E., Ostmann, A., Lai, B. T., Marczok, C., Müsing, A., and Kolar, J. W., PCIM Europe Conference for Power Electronics, Intelligent Motion, Renewable Energy and Energy Management, Nuremberg, Germany, 14-16 May 2013.

9. Hadim, H., North, M., Int. J. Thermal Science, 44 (33-42), 2005

10. Fu, W.-S., Huang, H.-C. and Liou, W.-Y., Int. J. Heat Mass Transfer, Vol. 39, No. 10, pp.2165-2175, 1996

Modeling and Simulation of Monolithic Integration of Rectifiers for Solid State Lighting Applications

M.R.Venkatesh, P. Liu, H.W. van Zeijl, G.Q. Zhang

Delft University of Technology, DIMES Research Center, Feldmannweg 17, 2628 CT, Delft, the Netherlands
Corresponding author :Manjunath.R.Venkatesh m.ramachandrappavenkatesh@student.tudelft.nl

Abstract

In this paper, a design methodology for modelling and simulation of monolithic integration of low power rectifier for solid state lighting application is presented. The power conversion is built based on schottky diode fabricated in a standard BiCMOS process. The process simulation is done by using TSUPREME4 to study the doping profiles and junction depth of the standard BiCMOS fabrication steps. The modelling and design of the schottky diode for rectifier design is then made with inputs from the process simulation tool. COMSOL Multiphysics environment is used to couple both semiconductor module and circuit analysis steps. The simulated IV characteristics of the schottky diode are used to build the spice model parameters of the fabricated device. A model for monolithically integrated rectifier application and analysis is presented and analysed.

1. Introduction

Solid State Lighting (SSL) system based on semiconductor light sources is one of the prominent light sources for future lighting applications. A complete SSL system consists of optical part, LED electrical driver and interconnects. The functionality of the electrical driver is to provide power to the optical part of the SSL system [1]. In many retrofit LED lamp designs like the MR16 and G4 lamps are usually powered by low voltage AC/DC input of 12V to 24V. The initial step down and power conversion is done by an external electrical or magnetic transformers. A simple block diagram of the SSL system of these designs is shown in Fig. 1. The next stage in design includes the rectifier block for AC to DC power conversion. DC supply is the main supply used for various driver architectures like buck or boost depending on the optical part present in the SSL system. The rectifier circuit block in this system mainly consists of schottky diodes and passive components which is about 50% of the PCB space. Since a SSL system is digital in nature, the lighting function can be combined with electrical functionality in a single wafer level for miniaturised and a multi functional system. The design proposal in this miniaturised system can have the complete SSL retrofit systems on a single wafer level with both the electrical power conversion block and optical blocks consisting of LEDs. Monolithic integration of power circuitry of rectifiers, passive components and sensors is a great challenge to reduce system space and create smaller products. In this work, the modelling and simulation for monolithic integration of the rectifier for the power conversion in the retrofit lamps mentioned is investigated. The diodes used for a bridge rectifier must have low

forward voltage drop, low leakage current and high breakdown voltages. Therefore, schottky diode is the most suitable diodes used for rectifier purpose due to their low forward voltage drop and high forward currents. The current is carried only by electrons (unipolar), is able to switch between the on and off states very fast [2].The disadvantages of the Schottky barrier diode is the high leakage current and low breakdown voltages. For low power rectifier architectures the modelling of schottky based rectifier with the required specification of conversion of 12V-24V AC input power supply to DC voltage is done by COMSOL Multiphysics 4.3b.

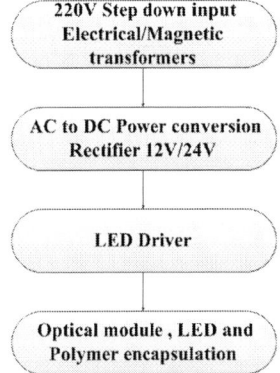

Figure 1: LED lighting system design.

2. Integrated rectifier design flow

In the analysis for modelling of the integration of rectifiers a systematic design flow is proposed in Fig. 2.

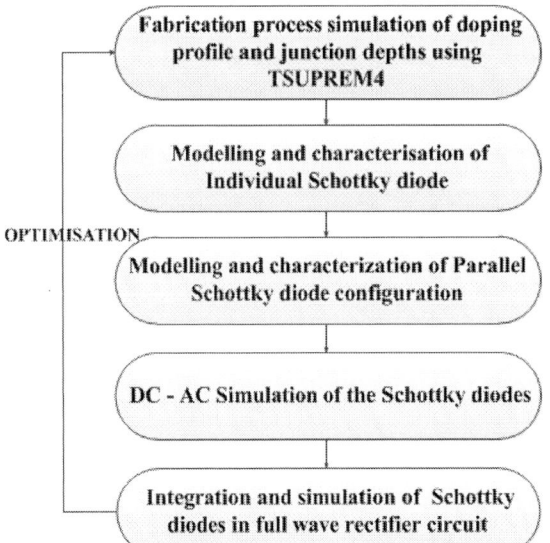

Figure 2: Design flow for Integration of rectifiers.

With thisdesign flow, it allows us to study and optimize the design steps at each stage. The simulated test data obtained at each design steps will be helpful in improving the design in the subsequent steps, as shown in Fig. 2..In this flow, the fabrication process is first studied to obtain information about the doping profile, doping concentrations and the junction depth of the fabrication process [4]. This data will serve as the input for the modelling of schottky diodes used in the rectifier designs. The simulation data of the doping profiles are obtained through TSUPREM4 process simulation tool. The doping concentrations data of the p-substrate, N-Well, shallow N+ used for ohmic contact and the P+ guard ring in the process simulation tool will yield useful information like the junction depths . Based on these data, COMSOL Multiphysics semiconductor module is used for modelling and analysis of schottky diodes. A standard BiCMOS process steps for bipolar transistor is used as the baseline to obtain the doping concentration of the layers used in the schottky diode fabrication process [5].

The substrate wafer used in this process is p-type <100> starting material doped with Boron at a concentration of $1 \times 10^{16} cm^{-3}$. The next stage is the implantation of N-type dopant for the N-Well formation in which the schottky diode would be fabricated. We have used a P-type substrate in this process in order to be able to integrate more devices that CMOS devices and BJT. With a P-substrate it is easier to build both NMOS and PMOS devices when compared to an n-substrate wafer. Ion Implantation process is used for the N-Well formation as shown in Fig. 3 and Fig. 4. The n-type dopant used in this process is phosphor beam of dopant ions. The phosphor implant dose was varied from $6 \times 10^{12} cm^{-3}$ to $9 \times 10^{12} cm^{-3}$ at implantation energy of 150keV. In this process step simulation, the wafer is also tilted by an angle of $7°$ to the incoming phosphor ion beam. A rotation of $22°$ from the wafer flat along the vertical axis is also performed, which is one of the standard process done to achieve better collision with the lattice of the incoming ion beam. This prevents channeling and achieves the desired dopant profile as per the calculated ion implantation. After this process, a drive in step is performed for the dopants to diffuse into the substrate, which gives the information for junction depths. The diffusion and drive in is then performed at high temperature and in oxygen and nitrogen environment. The variation in time in diffusion will result in different junction depths. The TSUPREM4 simulation for different NWell junction depth simulation is tabulated in Table 1.

Table 1: Junction depth for various drive-in diffusion time

N- Well Doping Concentration	Diffusion Time	Junction Depth [μm]
$3 \times 10^{16} cm^{-3}$	4 Hrs	2
$3 \times 10^{16} cm^{-3}$	8 Hrs	3
$5 \times 10^{16} cm^{-3}$	2 Hrs	1.6

The schottky metal contact is made directly on the N-Well region. The next step in the process is to have ohmic contact for the schottky diode.

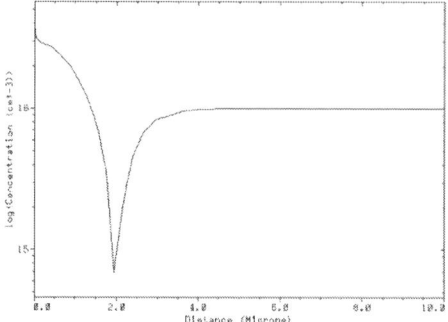

Figure 3. N- Well doping with 4Hrs diffusion

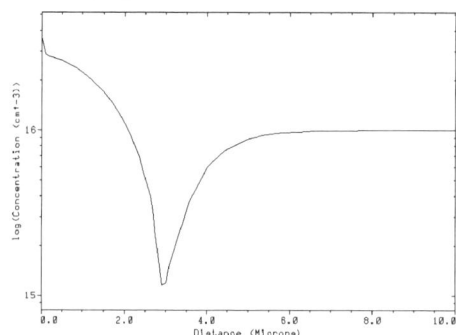

Figure 4. N- Well doping with 8Hrs diffusion

The schottky metal contact is made directly on the N-Well region. The next step in the process is to have ohmic contact for the schottky diode, which is obtained by interfacing the metal to a highly doped P or N region. For the N-type it is denoted as shallow N region and for the P-type (used for guard ring) it is denoted as shallow P region. Both the shallow N and P are highly doped regions in the order of $1 \times 10^{19} cm^{-3}$. This highly doped region results in increases the conductivity and a metal contact provided here will have behavior of small resistance. The shallow N and shallow P doping profile and the junction depth data are determined to be used in further simulation steps in modeling of the schottky diode.

3. Schottky diode modeling for rectifier application

In the next design step from Fig. 1, a model of schottky diode is done to study the behavior to be used for rectifier design using the COMSOL Multiphysics semiconductor module. The schottky diode is a unipolar device used mostly for power electronics conversion applications. It has low forward voltage drop and a high switching behavior. In comparison with P-N junction diode, schottky diode has approximately a turn on voltage which is 50% lesser in the order of 0.2V to 0.45V. The schottky diode is a majority carrier device, hence there is no diffusion capacitance associated in forward bias and during the switching from forward to reverse bias case there is no minority carrier storage time. This feature makes it suitable in fast switching applications. An analytical analysis of the schottky diode is first made and compared with modeled device.

For any semiconductor device, there should be mechanisms for current transfer to external components, which is achieved mainly by two ways:Ohmic contact and Schottky contact. For ohmic contact the metal is evaporated on the highly doped semiconductor region which can be p+ or n+ regions. The characteristics of this contact will be highly linear and the contact resistance is less, which allows current transfer without affecting the device performance. For schottky contact, the metal is evaporated directly on the lower doped semiconductor region. In this device the schottky contact is made on the N-Well region. The ohmic contact is linear and allows current in both ways but the schottky contact has rectifying behavior and it allows the flow of current only in one direction. The potential barrier seen by the electrons in the metal to move into the semiconductor in equilibrium is called as the schottky barrier φ_{bo} [6]. It is given by the below equation,

$$\varphi_{bo} = (\varphi_m - \chi) \qquad (1)$$

where φ_m is the metal work function and χ is the electron affinity of silicon. This is the barrier the electrons have to overcome to cross over to the semiconductor. In a similar way the electrons moving from semiconductor to the metal will also face a barrier called as built in voltage. The built in voltage V_{bi} for the schottky contact at equilibrium is given by,

$$V_{bi} = (\varphi_{bo} - \varphi_n) \qquad (2)$$

where φ_n is the work function of the n-type semiconductor. It is given as,

$$\varphi n = \frac{kT}{e} * \ln \frac{Nc}{Nd} \qquad (3)$$

In our fabrication process, we consider the schottky contact formed on the N-well. The schottky metal contact in this is made using aluminum having a work function of φ_m = 4.28eV. The electron affinity of silicon χ is 4.01eV. For different doping profiles as simulated in previous section of TSUPREM4 the built-in voltage is determined.

Table 2:Built in voltage for different doping concentration

N- Well Doping Concentration	N-type semiconductor work function φn (V)	Vbi – built in voltage (V)
1×10^{16} cm^{-3}	0.203	0.067
1×10^{17} cm^{-3}	0.144	0.126
1×10^{18} cm^{-3}	0.085	0.185
1×10^{18} cm^{-3}	0.026	0.243

From equation (1) it can be seen that schottky barrier height is dependent only on the metal work function and independent of the doping concentration of N-well [7]. However, the built in potential V_{bi} varies with the doping concentration of N-well. A N-well concentration of 10e17/cm^3 with a junction depth of 2um is modelled in

the design will have a barrier potential of 0.126V. This should be the applied forward voltage (Va) for obtaining current conduction in the schottky contact device.

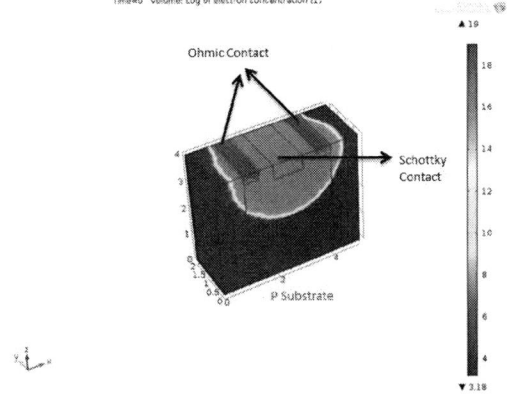

Figure 5. Volume Log of electron concentration

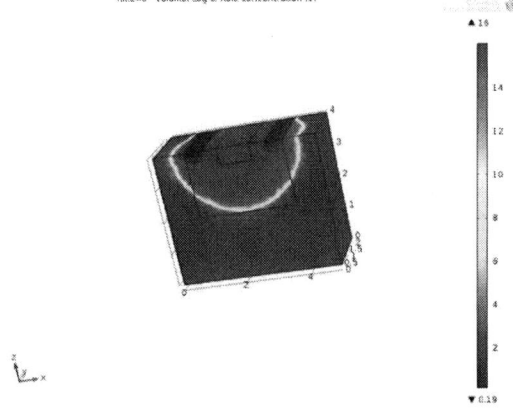

Figure 6. Volume Log of hole concentration

A comsol model for semiconductor with schottky metal contact is designed as shown in Fig. 5. The doping profile of the N-well region is kept at 1×10^{17}cm^{-3} with junction depth of 2μm. Ion implantation step with drive in diffusion is performed and a Gaussian profile of the doping concentration is obtained for the N-Well region. The schottky contact with aluminum as the metal has a width of 2μm. The ohmic contact is made at the interface of the highly doped N+ region which is the shallow N region having a junction depth of 0.3μm. The distribution of electron and hole concentration in the schottky diode in the entire semiconductor block is shown in Fig. 5 and 6.

The current transport in the schottky diode is mainly due to majority carriers (electrons in the n-type) described by the thermionic emission theory. The current density in the forward bias of the schottky contact is given as,

$$J = Js * \exp(\frac{eVa}{kT}) \qquad (4)$$

where Js is the reverse saturation current density given as

$$Js = A * T^2 * \exp(\frac{-e\varphi bn}{kT}) \qquad (5)$$

978-1-4799-4789-8/14 $31.00 © 2014 IEEE

The parameter A is called the Richardson constant for thermionic emission and $\varphi_{bn} = 0.65eV$ is the schottky barrier height of n-type silicon with aluminum contact. The current density J for different applied voltage is tabulated using the equations 4 and 5. The temperature is taken as 293.13 K. The Richardson's constant for Aluminum on n-type silicon is 110 A/cm^2K^2. The current density graphs are simulated with the comsol model as shown in Fig 7. In this model, the schottky contact is the cathode terminal and the Ohmic contact is the anode. The voltage at the anode of the diode is increased in steps of 0.1V.

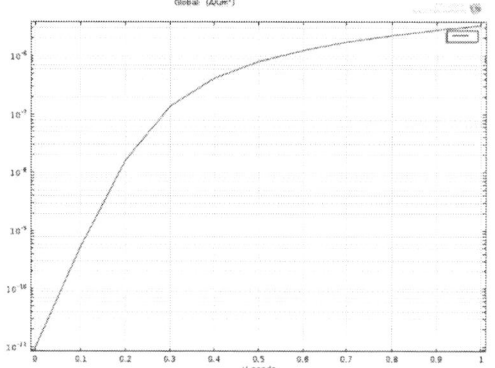

Figure 7: JV characteristics of schottky comsol model

The obtained results is tabulated and verified with the analytical solutions from the Table 3. The theoretical reverse saturation current density is 4.824×10^{-11} $A/\mu m^2$.

Table 3 : Current density vs applied voltage

Applied Voltage (Va) in V	Current density (J) A/um2 Theoretical	Current density (J) A/um2 Simulated
0.1	2.63×10^{-11}	5.36×10^{-10}
0.2	1.46×10^{-9}	1.6×10^{-8}
0.3	7.99×10^{-8}	1.37×10^{-7}
0.4	4.36×10^{-6}	4.16×10^{-7}
0.5	2.3×10^{-5}	1.2×10^{-6}

The simulated data of the schottky diode current is very close to the analytical solutions from Fig. 7. The current flow in the device is shown for the forward biased schottky diode Fig. 8. As shown in Fig. 8, it is clear that in the forward bias condition the total current flow is from the anode terminal to the cathode through N-Well.

In order to increase the current density of the schottky diode one method is to increase the area of the schottky diode. The other method is to design a parallel combination of anode and cathode contacts as multiple fingers. The current through the schottky contact to the Ohmic contact is limited by the series résistance in the n-well. The parallel combination of all the schottky diodes reduces the series resistance hence increasing the current density of the diodes. A parallel diode structure of the

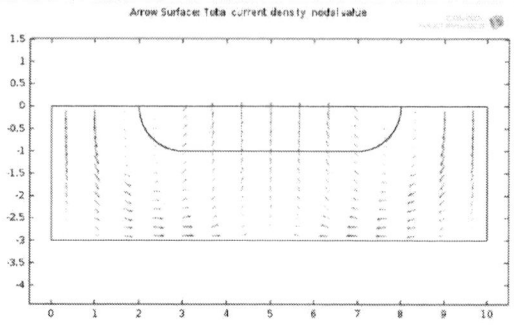

Figure 8. Current flow in Schottky diode

individual schottky diode was modeled and the behavior was analyzed. The structure of four parallel design of schottky is shown in Fig. 9 with the surface electron concentration at reverse bias of anode voltage. The blue region is the increase in the depletion region with decrease in the electron concentration when compared to the electron concentration at the Ohmic contacts.

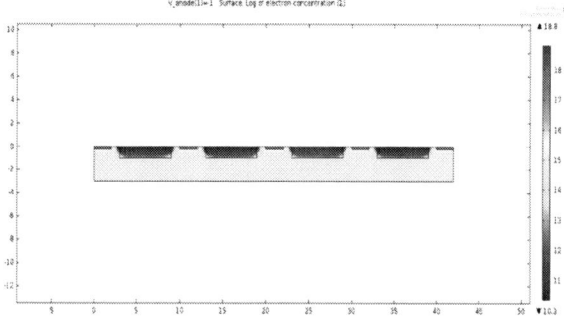

Figure 9. Surface electron concentration of parallel fingered Schottky diode in reverse bias condition at -1V

The IV curves of the parallel fingered schottky diode configuration are shown in Fig. 10. The current increased by a factor of 100 times but the drawback is that the reverse saturation current of this schottky diode also increased by a factor of 10.

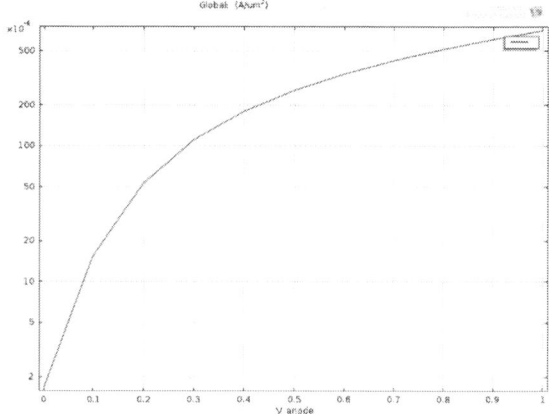

Figure 10: JV characteristics of parallel fingered Schottky diode model.

978-1-4799-4789-8/14 $31.00 © 2014 IEEE

For the case of low power rectifier diode application driving a single low power LED, the requirement for the breakdown voltage is in the range of 20-30V [8]. The above configuration of schottky diode incorporated with fingers was simulated for the range up to 30V and the diode performance was analyzed. The majority carrier current transport from the schottky contact to the anode terminal is less and determined by the reverse saturation current Js.

The main reason for choosing schottky diode in power converter applications than p-n diode is that the forward voltage drop of p-n diode will be in the range of 0.7-0.8V. For a AC voltage supply of 12V, a p-n diode with a full wave rectifier configuration there will be a voltage drop of about 1.4-1.6V in the output rectified voltage. This gives an efficiency loss of 20-30%. But a design of full wave rectifier with schottky diodes will allow a voltage drop of only 0.6-0.8V in the final output voltage. Hence, an integrated rectifier with schottky diodes would be a more efficient design than with p-n diodes.

An AC simulation for 12V, 50Hz power supply with a single schottky diode was carried out to check the behavior of the schottky diode in half–wave rectifier configuration. The AC/DC physics with electrical circuit feature is added to the semiconductor physics module in COMSOL [3]. It performs a complete analysis with the physics model of the parallel finger of schottky diode configuration. The half-wave rectified voltage of the schottky diode configuration is shown in Fig. 11.

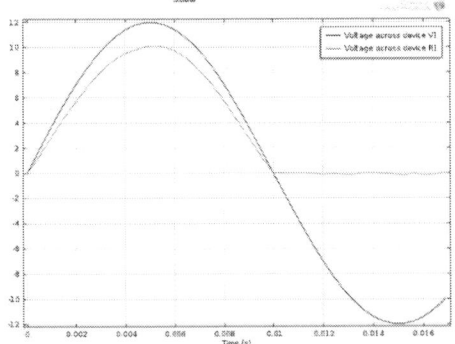

Figure 11. AC simulation of Half wave rectifier with Schottky diode.

The bridge rectifier configuration as shown in Fig. 12 is the most ideal full wave configuration, which consists of four diodes with 2 diodes always conducting during positive and negative half of the cycle.

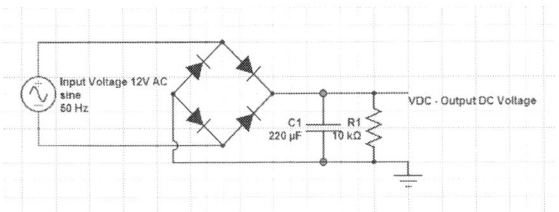

Figure 12. Bridge rectifier circuit

The filter part of the circuit converts the output ripple ac voltage into a smooth dc voltage. The schottky diodes modelled in the previous section was used and a layout for the integrated rectifier was designed.

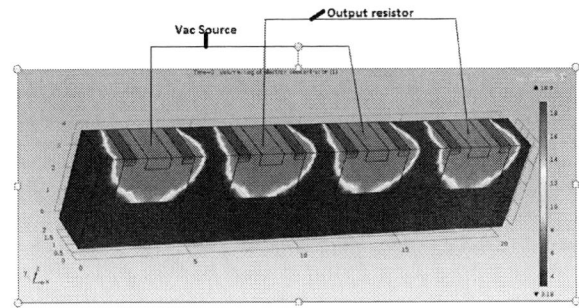

Figure 13 Model of Monolithically integrated Bridge rectifier circuit with initial electron concentration

The full wave rectifier configuration as per the circuit configuration was integrated as above design in Figure 12. This simulation result was tried in COMSOL but encountered many converging solution problems. In order to check if the modelled schottky diode with parallel finger configuration is suitable diode for the rectifier circuit, the diode model parameters was extracted from the IV characteristics. This was done by detailed IV characteristic simulation of the parallel finger schottky diode at different temperatures ranging from 20°C to 100°C. The parameter extraction model was used to build a model of the schottky device. The next step to simulate this was to obtain the model parameters from the schottky diode simulated. Important parameters were reverse saturation current (Is) and breakdown voltage. Which served as input parameters for the electric circuit simulation in the AC/DC module of the circuit simulation in COMSOL. The spice circuit was then simulated for the diode parameters obtained from the simulated schottky diode in full wave rectifier configuration and the output was obtained as shown in Fig. 14. The output DC voltage is used to power the driver circuit components and the LED module.

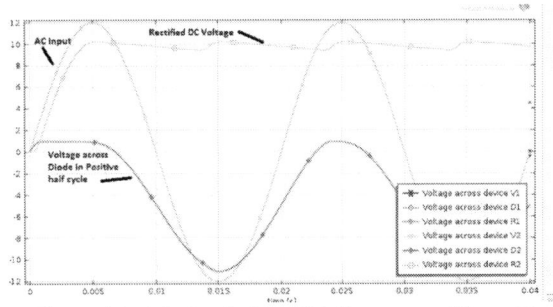

Figure 14. AC simulation of Full wave rectifier with Schottky diode.

In this design approach, the behaviour of the rectifier circuit from device fabrication simulation to circuit analysis is analysed . By using a highly well defined mesh size the multiphysics analysis can be improved and a more detailed study is required to understand the non-convergence issue of the 3D model of the diodes during circuit analysis step. Based on this design methodology, information for wafer level integration can be obtained for not only diodes and rectifiers, but also other active/passive components.

4. Conclusions

A design methodology for modelling and simulation of integrating rectifier circuit monolithically was presented. COMSOL Multiphysics environment for semiconductor and AC/DC modules. For low voltage solid state lighting applications, wafer level integration of power conversion modules can be integrated using schottky diode elements fabricated in standard BiCMOS process. This simulation environment provides a complete study of the fabrication process and circuit analysis for the modeled semiconductor devices. It links the processing parameters with device performance, which provides information for circuit design before processing. The input parameters such as the junction depths, doping profile, device structure can be changed at every step of the design to optimize the output circuit. The simulation results serves as a base for multifunctional LED module design with integrated power conversion circuitry, especially for wafer level integration.

Acknowledgments

The authors would like to thank Yi Wang and Willem van driel from Philips Lighting for the knowledge sharing and support in the rectifier and electrical module design for solid state lighting.

References

1. W.D. van Driel, X.J. Fan. Solid State Lighting Reliability. Fan Volume 1 2013.
2. H.-R. Chang, B.J. Baliga ,"High-current, low-forward-drop JBS power rectifiers",. *Solid-State Electronics,* Volume 29, Issue 3, March 1986, Pages 359–363
3. R. Millett, J. Wheeldon, T. Hall, and H. Schriemer. "Towards Modelling Semiconductor Heterojunctions", *Proceedings of the COMSOL Users Conference 2006*
4. Sami Franssila, Introduction to Microfabrication, 2nd Edition ,September 2010.
5. Alhan Farhanah Abdul Rahim, Ahmad Ismat Abdul Rahim, Md. Roslan Hashim*, Shahrul Aman Mohd. Saari , Mohd. Rais Ahmad, Mohd. Zahrin Abdul Wahab',Wan Sabeng Wan Adini' and Mohd. Ismahadi Syono "Fabrication and Electrical Characterization of Silicon Bipolar Transistors in a 0.5-pm based BiCMOS Technology", *ICSE2000 Proceeding.*
6. Donald. A. Neamen. Semiconductor Physics And Devices: Basic Principles 3rd Edition.
7. Tayel, M.B, El-Shawarby, A.M. "Characterization of barrier height due to metal for Schottky barrier diode", *ICCES '07 Proceeding.*
8. Braga, H.A.C. Dias, M.P. ; Almeida, P.S. On the use of a low frequency boost rectifier as a high power factor LED driver, *2012 10th IEEE/IAS international conference on industry applications*

978-1-4799-4789-8/14 $31.00 © 2014 IEEE

Microactuator modeling to develop a new template for the Braille

Sofiane Soulimane,[1,2] Med El Amine Brixi Nigassa,[1] Benyounes Bouazza,[1] Henri Camon.[3,4]

[1]Biomedical Engineering and Electronic Department, Tlemcen University, 13000, Algeria
[2]CDTA ; Centre de Développement des Technologies Avancées, Algies, 16000, Algeria
[3]CNRS ; LAAS, 7 avenue du colonel Roche, F-31077 Toulouse, France
[4]Université de Toulouse ; UPS, INSA, INP, ISAE ; LAAS ; F-31077 Toulouse, France

Abstract

The results of this work are contributions to the modeling and design a original microactuator system that allows braille reading. Here, we propose to use one matrix composed by six actuators to form the 63 Braille combinations. To define the geometry of the matrix we used the physiological limits of perception. . In this work, we use a finite element model with the Comsol software for modelling this type of miniature actuator for integration into a test device. In order to lower the operating voltage of such systems, we focus our study on microsystem based piezoelectric material. We demonstrated that micro-actuators can exhibit non uniform compression. This deformation depends on thin films thickness and design of membrane arms.

1. Introduction

Louis Braille developed the system of reading and writing used now by the blind. Developing new Braille technologies is key to improving the extremely low literacy rate of all blind people [1-2]. Micro Electro Mechanical System (MEMS) are needed in a wide range of applications, micro pump, micro mirror, artificial tactile skins smart textiles, artificial muscles...We feel that the field of MEMS) technology [3-4] has matured to the point where it can provide real solutions for Braille displays. Nowadays, the piezoelectric device give the better actuation [5-6] for low actuation voltage. In this study, we focus on actuator based on PZT sol gel material. The goal of this research was to simulate a Braille cells microactutor that could be fabricated using standard microfabrication methods.

2. Microactuator specifications

Braille system use the principle of the cell which corresponds to a matrix of six dots (figure 1) in relief on three rows and two columns. The number of combination thus obtained is 63 combinations.

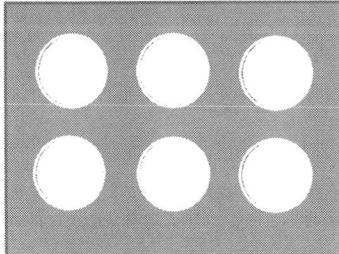

Figure 1: Braille matrix

Here we propose to use one matrix composed by six actuators to form the 63 combinations. To define the geometry of the matrix we used the physiological limits of perception. The standards dimensions used for construction of a Braille matrix are: The two columns constituting a Braille cell are separated by about 2.5 to 2.8 mm and their height is twice that distance. A write line in the right column of a character is distant from the left of the next character from about 3 to 3.2 millimeters. It is noted that the distance between characters is greater than that which separates the two columns of the same character. The threshold of sensitivity of the skin to a mechanical stimulus corresponds to a depression of 6μm and varies depending on the location of the stimulus. It is noted that the lowest thresholds are measured at the ends fingers. For the design of our actuator, a deflection greater than 6μm can be detected by the finger and a lower 6μm deflection cannot be detected by the finger (Figure 2).

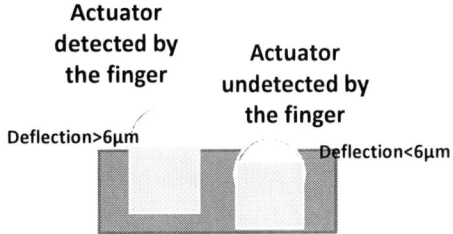

Figure 2: Detected and undetected actuator

The proposed actuators made of piezoelectric materials must have spherical form lake dots. The tops of Braille dots are generally domed, with a radius of curvature that is unspecified, but usually larger than the radius of the base of the dots.

3. Results and discussion

We use comsol multiphysics for three dimentionnal piezoelectric actuator. The detailed materials used in the microactuator stack are listed in Table 1.

Material	Thickness (μm)
Poly-Silicon	4
SiO2	2
Platine	1
PZT	360
Ruthenium	1
Au	0,5

Table 1: Thickness of the multilayer material of the actuator

Here, the polysilicon constitute the support membrane and the layer of oxide constitue orientation of the PZT piezoelectric material [3-5]. Platinum and Rhutenium are respectively bottom and top electrode. All simulations were performed with this materials thickness and standard diameter Braille dots about 1500µm. In first; we demonstrate that Braille micro-actuators can exhibit non uniform compression deformation with a maximal deflection at the center of the dots. The figure 3 shows a linear maximal deformation of the actuator composed of four arms with a length of 500µm and a width of 100 µm of each arms as function as the voltage.

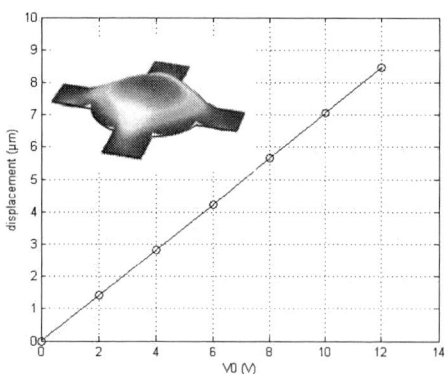

Figure 3 : Maximal deflexion of the actuator composed of four arms with a length of 500µm and a width of 100 µm of each arms.

We demonstrate that deformation for about 10 microns can be achieved for about 12-14 volts. Figure 4 shows the maximal deformation at the center of the actuator as function as arms numbers and width arms.

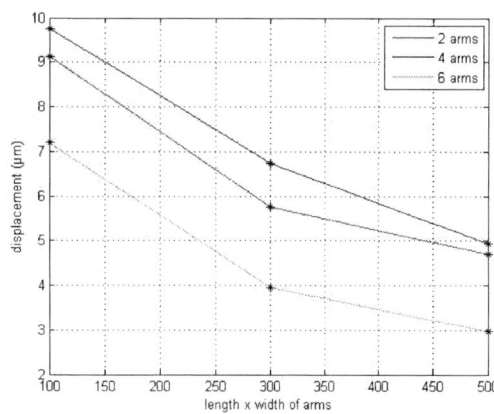

Figure 4 : Maximal deflexion of the actuator as function as arms number and width arms at 12 volts.

We note that the actuator composed of 2 and 4 arms give a large deformation in comparison with 6 arms actuators. The table 2 and 3 show respectively maximal deformation as function as width of the arm of the actuator composed by four and two arms at 12 volts. We note that in both cases of two and four arms, over the width arm is fine more the deformation is maximal at the center of the

actuator. In the case of two actuator arms, there is shown a large deflection at the two free edges of the actuator in proportion to the width of the arm

Arms length x width	Maximal deformation (µm)	3D result
500 x 500	4,93	
300 x 500	6,74	
100 x 500	9,76	

Table 2 : maximal deformation as function as width of the arm of the actuator composed by four arms at 12 volts.

We can say that the 4 arms actuator is the best design by giving the greater actuation and it's always obtained a domed deformation at the center of the device as in the case of the Braille system. In what follows, we will opt for this type of design to optimize other parameters.

Arms length x width	Maximal deformation (µm)	3D result
500 x 500	4,71	
300 x 500	5,76	
100 x 500	9,12	

Table 3 : maximal deformation as function as width of the arm of the actuator composed by two arms at 12 volts.

The figure 5 shows the deformation of the actuator according to the polysilicon thickness at 12 volts. We can observe that the deformation decreases when increasing the thickness of the polysilicon. For 4 μm thickness of polysilicon the deflection remains close to 10 microns necessary for detection of Braille.

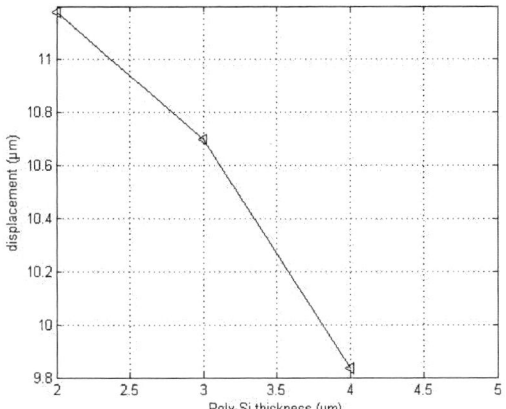

Figure 5 : Deformation of the actuator as function as polysilicon thickness at 12 volts.

To protect our actuator in use, we plan to use a polymer layer of polydimethylsiloxane (PDMS)[4]. This layer is the main component of our packaging system braille. Figure 6 shows the effect of the PDMS thickness layer.

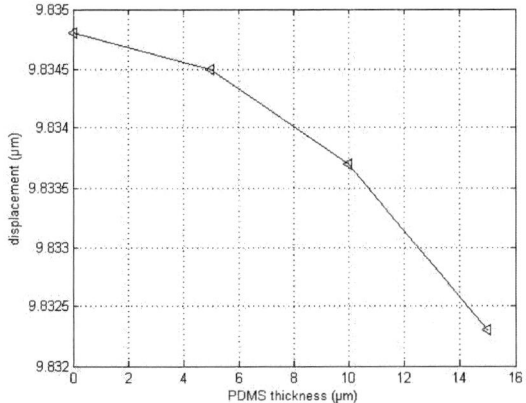

Figure 6 : Deformation of the actuator as function as polysilicon thickness at 12 volts.

We observe a slight decrease in deformation by increasing the thickness of the PDMS. This result confirms us in the choice of materials for protection.

4. Conclusions

In this paper, the design of a piezoelectric Braille microactuator is presented. Comsol multi-physics have been used to study the design of Braille cells. We demonstrate that the 4 arms actuator is the best design by giving the greater actuation and it's always give a domed deformation at the center of the device as in the case of the Braille system. Also, we demonstrate that polysilicon membrane and PDMS protection layers have little effect on the final deformation of the Braille microactuator.

The results of these simulations will provide quantitative characterizations concerning the actuation spatial resolution, and also allows optimizing the fabrication process in terms of materials and integrations steps.

Acknowledgments

Acknowledgments to the CDTA (Centre de Développement des Technologies Avancées) for allowing us to use their calculator for comsol multiphysics modeling.

References

1. Runyana N., Blazieb D.,"EAP actuatorsaid the quest for the "Holy Braille" of tactile displays," Proceedings of the SPIE, Volume 7642,2010, pp. 12-18.
2. Sergeant P.,Giraud F.,Lemaire-Semai B.,"Geometricaloptimization of an ultrasonic tactile plate," Sensors and Actuators A: Physical, Elsevier,June2010, volume 161 pp.91 - 100.
3. Casset, F., Danel, J.S., Chappaz, C., Civet, Y., "Low voltage actuated plate for haptic applications with PZT thin-film," Solid-State Sensors, Actuators and Microsystems (TRANSDUCERS & EUROSENSORS XXVII), 2013 Transducers & Eurosensors XXVII: The 17th International Conference on, 16-20 June2013, pp. 2733 – 2736.
4. Soulimane S., Al Ahmed M., Matmat M., Camon H., "Modeling of Smart Compliant Electro-Active Polymer Actuator," 9th. Int. Conf. on Thermal, Mechanical and Multiphysics Simulation and Experiments in Micro-Electronics and Micro-Systems, EuroSimE 2008, pp. 235 – 239.
5. Bozidar, M.,Tolga, K., and Hur K., "Characterization of ferroelectric material properties of multifunctional leadzirconate titanate for energy harvesting sensor nodes,"Journal of Applied Physics, Volume:109 , 2011, pp.014904 - 014904-5.
6. Smith, S.M., Talin A.A., Voight S., Hooper A., ConveyD., "Effect of annealing temperature on physical properties of thin epitaxial PZT films on STO/Si substrates" Proc. SPIE 4804, Sol-Gel Optics VI, 1 October 1, 2002.

Multiphysics Modelling of the Fabrication and Operation of a Micro-Pellistor Device

Ferenc Biró[1,2], Zoltán Hajnal[1,*], Andrea Edit Pap[1], István Bársony[1]

[1]MEMS Laboratory, Institute of Technical Physics and Materials Science,
Research Centre for Natural Sciences, Hungarian Academy of Sciences,
Konkoly Thege Miklós út 29-33, H-1121 Budapest, Hungary
[2]University of Pannonia, Egyetem u. 10, H-8200 Veszprém, Hungary
[*]corresponding author, e-mail: <hajnal.zoltan@ttk.mta.hu>

Abstract

Downsizing efforts in gas-sensing applications lead to ever smaller active elements. Integration with data processing circuitry requires the use of CMOS compatible fabrication technology, autonomous operation poses limits on energy consumption of the elements, whereas reliable catalytic detection often needs high temperatures that may otherwise be constrained by safety considerations. Under these conditions, development of active sensor elements proves to be a growing challenge for design and fabrication.

In this work we present a step-by-step study on a ≈ 500 µm diameter thermally isolated membrane element of a gas detecting microsensor device. Sensitivity is based on high temperature ($\approx 3\text{-}400$ °C) catalytic activity of a porous pellistor deposited on a multilayer SiO_2/SiN_x – filament heated – membrane that has to be durable enough for several thousand hours of operation, and as thin as possible to reduce heat conduction to the substrate. SiO_2 membranes tend to show high residual stress that can be significantly reduced by "sandwiching" with SiN_x.

We have used *COMSOL Multiphysics® 4.3a* [1] to assist the initial product design, and evaluation of operational constrains of the multi-layer thin film. The first part involved systematic thermo-mechanical iterations, while the latter consisted of a combination of gradual static thermo-electro-mechanical simulation steps.

As shown by simulating the steps of the deposition process in this work, the right combination of different techniques produces a stable 4-layer membrane with only a sub-micron deformation, and tolerable residual stresses after membrane forming (substrate removal) and during operation. Also, the pellistor filament heating power should be minimized and still reach the operating temperature of the catalyst hotspot. This design, supported by our model calculation was used to realize the device with targeted characteristics. The structure endures the distortion and thermal expansion and contraction during the heating cycles, whereas low power operation widens the range of possible applications.

1. Introduction

Multilayer membrane structures are becoming more and more ubiquitous as building elements in commercial MEMS based devices. Their composition and dimensions can be diverse, depending on the intended function and operating conditions of the device. Micro-heaters are also widely utilized MEMS elements in catalytic micropellistors and semiconductor gas sensors, or applied as micro sized infrared (IR) sources. These devices have to be able to operate between 150-600 °C depending on their functionality under both static and dynamic conditions, often with tight limits on power consumption and high reliability requirements.

Electronics integration assumes the use of CMOS compatible fabrication technologies, which – in practice – confine the choice of membrane materials to oxides and nitrides of the wafer base material, silicon. However, their thermal coefficients are quite different from that of silicon. Mechanical stability and several thousand hours of desired operating time require the reduction of residual stress of the membranes, either in perforated or full membrane type gas sensors. Composite membranes are prepared and post processed – usually at elevated temperatures – in multiple cycles by thermal oxidation, LPCVD, and/or other deposition techniques. These processes determine the final electrical and mechanical properties of the device, thus they must be tailored for the optimal operating temperature and lifetime.

Thin film structures can be regarded stress free if the residual mechanical stresses are below 100 MPa [2]. There are two possible ways to reduce the residual stress in these structures. In the application of non-stoichiometric SiN_x the residual stress could be controlled by changing the silicon-nitrogen ratio in the deposited film. Nevertheless, Si rich SiN_x films offer reduced thermal isolation.

The alternative is using a stacked composition of SiO_2–Si_3N_4 layers. Due to the differences in the magnitudes and the opposite sign of the residual stresses in SiO_2 and Si_3N_4, it is possible to tune a composite membrane to be almost stress-free.

Further, to increase power efficiency of the device, membranes need to be bad thermal conductors in order to minimize heat dissipation. Since heat conductivity in the freestanding part of the membrane is strongly depending on its thickness and composition, and the effect of the metallic filament must also be taken into account, accurate simulation of the thermal properties of the layer structures can be crucial during the design phase. As these devices operate either in ambient conditions, or in other gaseous medium, heat dissipation through diffusion and convection cannot be neglected either. Radiation loss may also reduce power efficiency.

In the following we present the basic design criteria of the active element of our micro-hotplate gas sensor device, describe the details of the multi-stage simulation of various steps of the membrane fabrication, concentrating on the thermo-mechanical performance. The simulation is completed then with a parametric study of the thermo-electro-mechanical behaviour of the whole device element. Validation of the models is provided by electrical, thermal, and optical interference measurements on several fabricated prototypes of the device.

2. Design and fabrication

The targeted sensor element, the micro-pellistor is to be realized through CMOS compatible technology. Thus the micro-heaters are formed on silicon wafers by multiple thin film growth and deposition techniques. First a 300 nm SiO_2 layer is grown thermally at 1200 °C, then 200 nm Si_3N_4 and 100 nm SiO_2 are deposited by Low Pressure Chemical Vapour Deposition (LPCVD) and atmospheric Chemical Vapour Deposition (CVD) at 800 °C and 430 °C, respectively. Between every deposition step, the sample is moved to a different chamber, thus cooled down to room temperature (RT). A heating filament (of 4 µm wire-width covering $\approx 100{\times}100$ µm^2 surface area) is then formed: using a masked deposition technique first a thin (< 100 nm) TiO_2 adhesion layer, and then Pt metal is sputtered to achieve a total of ≈ 270 nm thickness.

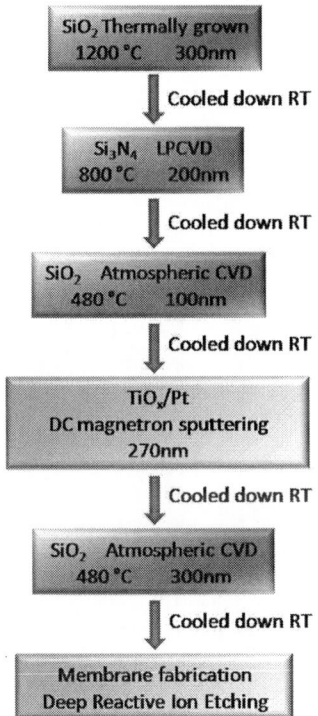

Figure 1: Process flow chart of micro-pellistor fabrication. Red boxes signalize high temperature treatment, blue ones close to room temperature (RT).

To provide chemical and electrical isolation, an additional 300 nm SiO_2 layer is deposited by CVD after removal of the sputtering mask. A freestanding membrane is produced by Deep Reactive Ion Etching (DRIE) from

the backside of the wafer that "drills" a 300-320 µm diameter cylindrical cavity centred under the filament. These basic processing steps are summarized in Figure 1. Further details about the device fabrication and the experimentally tested structures had been described elsewhere [3].

While wafer processing enables efficient mass production, the physical behaviour of the micro-heaters can be simulated for an individual element. Starting with a $400{\times}400$ µm^2 slab of a 370 µm thick pure silicon wafer, the oxide-nitride-oxide membrane "sandwich" was constructed using the model builder in COMSOL. Lacking specific empirical material parameters of the individual layer component materials, the default properties of the Materials Library of the MEMS module for Si, Pt, SiO_2 and Si_3N_4 have been used. The thin adhesion layer of TiO_2 was ignored in simulations to simplify the layer and mesh construction. Figure 2 shows the overview of the structure.

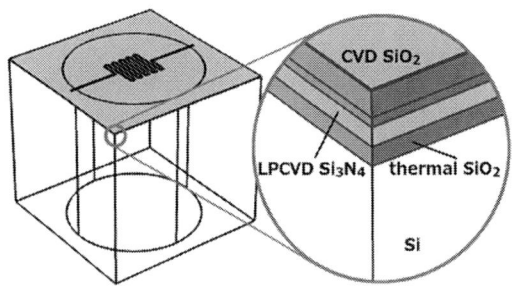

Figure 2: Overview of the model geometry with the layer composition of the membrane in the blow-up.

An appropriate mesh must be added to the geometry model, over which the coupled equations of the simulations are calculated. In our case, the thin film components require special attention, as the automated meshing would produce enormous amounts of mesh elements, adjusting sizes of mesh tetrahedra to the thickness of the films. We need to accurately simulate temperature dependent transfer processes of current generated heat and its dissipation where its gradient is expected to change most rapidly.

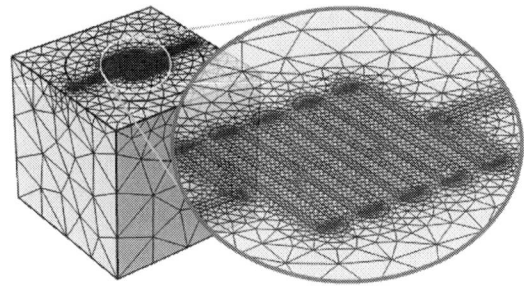

Figure 3: The structure of the adapted mesh (see text). Blow-up: refinement around the Pt filament.

Thus a reasonable way is to start from a triangular mesh on one lateral face of the filament, with the elements' lower (500 nm) and upper (2 μm) size limits commensurate with the layer thickness. A detail of the mesh is shown in Figure 3. Letting the triangular mesh run over the whole nitride surface, gradually expanding toward the edges, then sweeping this triangular distribution vertically through the membrane (allowing 2 divisions of each sub-layer), and finishing the bulk part with a tetrahedral mesh, results in ≈ 65 000 elements. This pushes the model to the computational limits affordable on a multicore workstation with 16GB memory. The final thermo-electro-mechanical simulation involves solutions for over 10^6 degrees of freedom.

To account for the fabrication recipe stages, each step is reproduced as a static "thermal stress" calculation in COMSOL. In the first 5 steps – as in Figure 1 – new layers are added on top of the structure one-by-one, while the layers below inherit the residual stresses from the previous step. The reference temperatures are set to the processing temperature of the given step, and the system is relaxed then to room temperature (RT). As it can be seen in Figure 4, local residual stress is strongly material and processing temperature dependent. In accordance with literature values [4], the first (bottom-most) thermal oxide layer sustains the most residual stress (≈ 0.3 GPa), which is relieved by the subsequent nitride and oxide layers.

plane of the sample to room temperature. Less pronounced – but certainly not negligible at higher temperatures – is the convective heat transfer to the surrounding medium. In our case normal ambient air was chosen, with the dissipation parameters as provided by COMSOL, using the whole surface of the sample. The least contribution at the typical operating temperatures is expected from radiation. For the simulations we chose a uniform 0.8 as radiative coefficient, and room temperature for the surroundings.

current [A]	max. T [K]	Δz [μm]
0.001	297	-5.0
0.002	311	-4.9
0.003	335	-4.7
0.004	375	-4.3
0.005	441	-3.7
0.006	553	-2.6
0.007	749	-0.6
0.008	1049	2.7
0.009	1382	6.8
0.010	1638	10.5

Table 1: Maximal temperature and vertical displacement values as function of the pellistor heating current.

Maximal values are listed in Table 1. Figure 5 shows the resulting temperature and stress distributions of the device at two different current values. Vertical displacement of the membrane is exaggerated to 5 times of the actual, and the original contour lines of the filament are left in place to emphasize membrane distortion (otherwise barely visible).

Figure 4: Von Mises stress after the five stages of simulated membrane formation.

3. Physics of device operation

Membrane formation and the simulation of operating conditions are finally unified in one parametric simulation stage. The necessary initial conditions are defined by the residual stresses from the last deposition step, and removal of the cylindrical middle part of the substrate. In this concluding stage a current is driven through the Pt filament. Besides other thermal properties, the propagation of the generated heat is calculated simultaneously with the thermal expansion of the membrane until a steady state is achieved for given current values, by taking the temperature dependence of the electrical conductivity also into account.

There are three ways of heat dissipation in this model system. The most important one is conduction through the substrate itself, which is modelled by setting the bottom

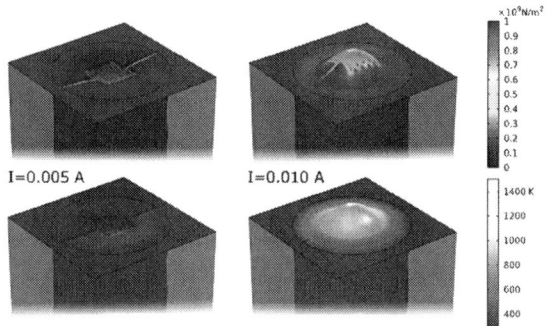

Figure 5: Von Mises stress (top) and temperature (bottom) distribution of the simulated micro-pellistor driven by 5 mA (left) and 10 mA (right).

4. Comparison with electrical measurements

To assess the electrical properties of the fabricated devices measurements have been carried out using a Keithley 2400 SourceMeter®, using the DC current generator for filament heating. Cold resistance of the micro-hotplate (measured at 100μA) was 181.2 Ω exceeding the 118 Ω of the same in the COMSOL model. As it can be clearly seen in Figure 6, under operating

978-1-4799-4789-8/14 $31.00 © 2014 IEEE   614

conditions the measured resistance rises with increasing temperature more slowly than the simulated one.

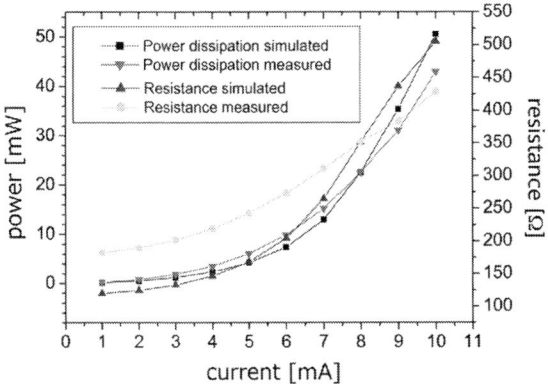

Figure 6: Resistance and power dissipation of the micro-hotplate in measurements and simulation.

The difference in the "cold" value can mostly be attributed to the larger cross-section of the filament in the simulation (as the TiO_2 sub-layer is ignored). As the Pt filament is deposited by magnetron sputtering it may contain more defects and voids, which may also alter its temperature coefficient of resistance (TCR). In our device, the measured TCR value is $\alpha_0 = 3.1 \cdot 10^{-3}$, whereas the ideal ($3.3 \cdot 10^{-3}$) is used in the COMSOL material properties definition.

5. Comparison with thermal measurements

One has to find a suitable indirect method to measure the temperature of the tiny micro-membranes with acceptable accuracy.

Salt	$T_{melting}$ [°C]
$AgNO_3$	212
KNO_3	334
$Pb(NO_3)_2$	470
V_2O_5	690
Na_2SO_4	884

Table 2: Transition temperatures of the inorganic compounds, used in micro-melting point measurements.

We have applied a *micro-melting point calibration technique* to determine the temperature / heating power relationship of our micro-hotplate. Interior areas of the filaments were coated by glycerol suspended inorganic compounds, known to exhibit sharp solid-liquid phase transition at their melting point (see Table 2). While slowly increasing the filament current, the phase transition was observed under optical microscope. Every measurement was repeated on five different samples and the measured power values were averaged to obtain the temperature of the hotplate. Figure 7 shows the average and maximum temperatures of the membrane surfaces from the COMSOL simulation, along with the measured micro-melting points of the 5 salts listed in Table 2. Simulated average temperatures show a fairly good

correspondence to the experimentally determined values, while maxima are well above them.

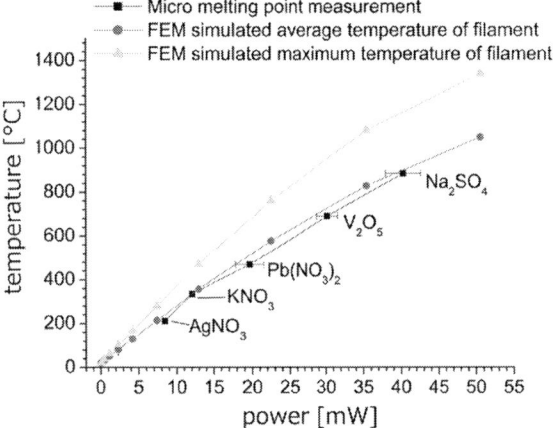

Figure 7: Simulated and measured temperature/heating power relationship of the micro-hotplate.

6. Thermo-mechanical distortion

For long-term stability and reliable operation of a device it can be crucial to assess thermal stress related mechanical load on the membrane in the duty cycle. Using a *Bruker ContourGT-I* Optical Microscope White Light Interference (WLI) based 3D topographic images were recorded at room temperature (Figure 8) as well as heated by currents of 6, 8, 10 mA (Figure 10).

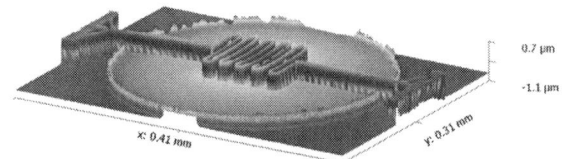

Figure 8: 3D optical microscopy image of a micro-hotplate at room temperature.

These topographic images are constructed from the interference patterns recorded digitally. In our case, we are dealing with a multilayer structure with several interfaces, all of which contribute somewhat to the interferences. In the 3D reconstruction, we can clearly identify the structural components of the micro-hotplate, with a certain offset caused by variation of the reflection signal at their boundaries. Observing a cross-section, the morphology of individual elements and even some layer thicknesses can be deduced from the regularities in their appearance. For better understanding, a line profile across the membrane area was plotted in the inset of Figure 9 indicating the path with a black line. At the outer edges of the membrane two peaks occur due to the sharp transition between silicon and cavity (etched out by DRIE to complete the membrane fabrication). Here a rapid change occurs in the intensity of interference lines that the instrument cannot process correctly, so these artefact peaks can be ignored. The level shift between the Si substrate and the membrane is approximately 300 nm that corresponds to the thickness of the first (thermal) SiO_2 layer. The height of the intense peaks in the filament area equals the sum of the layer thicknesses of Si_3N_4 (200 nm), SiO_2 (100 nm), TiO_x/Pt (15/270 nm), SiO_2 (300 nm).

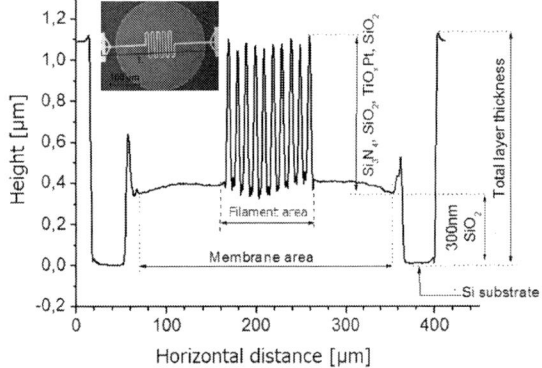

Figure 9: Line profile analysis of a room temperature device showing the components giving rise to the features in the height distribution. Inset: position of the line scan on the membrane.

At room temperature, the interior area of the filament is slightly bent downward, while the outer membrane regions show a slight upward buckling. At higher temperatures not only the filament area, but also the whole membrane increasingly buckles downward, as illustrated by Figure 10.

Figure 10: 3D reconstruction of the heated membrane driven by 10 mA.

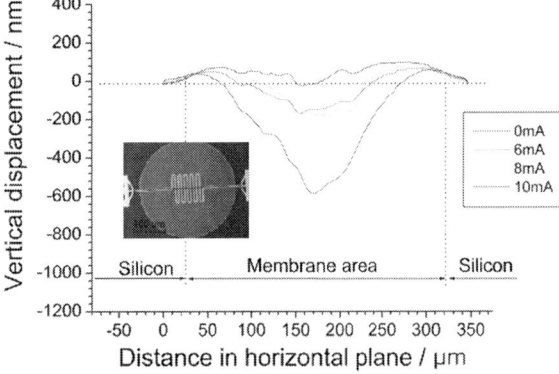

Figure 11: Line profiles at various heating currents corrected for the variation of interference peaks over the filament. Inset: scan line position on the membrane.

Using a combination of multiple line profiles (to subtract the intense level shifting around the filament) a filtering can be carried out to obtain a cross-sectional view of the distorted membrane surfaces at different driving currents. The results of this analysis are summarized in Figure 11.

This observed mechanical behaviour is in contrast to that predicted by the simulations (see Table 1 and Figure 5), both qualitatively and quantitatively. Several attempts were made to fine-tune the geometry parameters of the model, but keeping the given processing recipe of the layer deposition sequence does not allow for an easy remedy to this difference.

However, during the design of the device, many different options for the layer composition had been explored. There is a combination of the deposition sequence, which – besides resulting in comparable thermal behaviour – also exhibits the downward buckling behaviour upon heating described above.

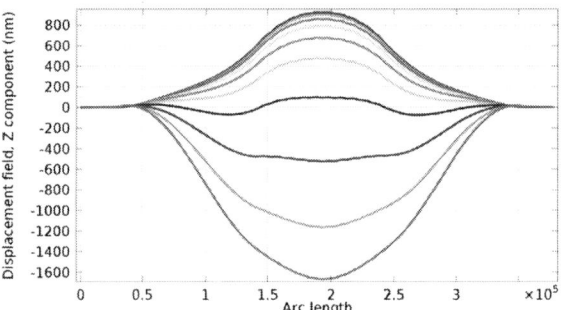

Figure 12: Vertical displacement in an alternative membrane composition (see text).

Such an example is shown in Figure 12. The different line colours correspond to heating currents from 0 to 10 mA, from the top (blue) to the bottom (red). The main difference with this latter structure is that after the thermal oxide layer, deposition of the filament follows immediately, then the nitride, and the other oxide layer.

The "flipping" of the buckling direction is actually representing a reliability hazard, since during operation the bending stress changes sign. It may lead to early fatigue failure, and has to be avoided.

In Figure 11 a saucer shaped line-scan profile is obtained for all loading conditions. It cannot be reproduced by any of the models. In the practical case the apparent buckling along the perimeter of the membrane might be a measurement artefact rather than a real phenomenon, as suggested by the simulations, where no such effect was visible close to the rim of the membrane. It is obvious that the deformation of the thick Si substrate around the membrane cannot reflect a line-scan maximum similar to Figure 11.

We are confident that the processing steps (also confirmed indirectly by the 3D topography) produce the intended structural composition of the membrane, thus the above-mentioned qualitative agreement can only be

978-1-4799-4789-8/14 $31.00 © 2014 IEEE   616

regarded as coincidental. Nevertheless, as the behaviour can be reproduced under different circumstances, a deeper analysis – and possibly optimization – of the temperature dependent material parameters is expected to show a way worth pursuing. The elastic parameters of the CVD oxide layers under different conditions certainly will have pronounced effects on the final thermo-mechanical simulation stage.

7. Conclusions

We constructed a parametrized model of a multilayer micro-pellistor device element, and carried out coupled electro-thermo-mechanical simulation of its operation. The electrical and thermal properties of the model provide a sufficiently close approximation of the device properties, enabling efficient analysis in the device design phase.

For the thermo-mechanical behaviour of the studied class of systems, extra care must be taken. The declaration of the parameters of the base materials – more specifically those of the CVD deposited SiO_2 – could, as in the presented case they certainly do, need project specific optimization and more thorough validation.

Acknowledgements

This work was partially supported by the Hungarian KMR 12-1-2012-0031 and KMR 12-1-2012-0107 projects.

References

1. http://www.comsol.com/
2. Laconte, J. *et al*, <u>Micromachined Thin-Film Sensors for SOI-CMOS Co-Integration</u>, Springer (Dordrecht, 2006), pp. 64-65.
3. Biró, F. et al, "Optimisation of low dissipation micro-hotplates - Thermo-mechanical design and characterization," *19th International Workshop on Thermal Investigations of ICs and Systems (THERMINIC)*, Berlin, Germany, September 2013, pp. 116-121, doi: 10.1109/THERMINIC.2013.6675223
4. MEMS Clearinghouse, *IEEE MEMS Workshop*, Florida, US, February 1993, p. 223, http://www.memsnet.org/material/silicondioxidesio2film/

Determination of Residual Stress with High Spatial Resolution at TSVs for 3D Integration: Comparison between HR-XRD, Raman Spectroscopy and fibDAC

D. Vogel[1], U. Zschenderlein[2], E. Auerswald[1], O. Hölck[2], P. Ramm[3], B. Wunderle[1,2], R. Pufall[4]

[1]Fraunhofer ENAS, Micro Materials Center (MMC), Chemnitz, Germany
[2]TU Chemnitz, Chair Materials and Reliability of Micro Systems, Chemnitz, Germany
[3]Fraunhofer EMFT, Heterogeneous System Integration, Munich, Germany
[4]Infineon Technologies AG, Munich, Germany
dietmar.vogel@enas.fraunhofer.de, +49-371-45001 412

Abstract

Three different experimental methods have been used to determine mechanical stresses in silicon nearby tungsten TSVs - HR-XRD performed at a synchrotron beamline, microRaman spectroscopy and stress relief techniques put into effect by FIB ion milling. All methods possess, to a different extend, high spatial resolution capabilities. However they differ in their sensitivity and response to the particular stress tensor components relevant for the residual stress state nearby TSV structures. Stress measurements were performed on test samples with TSVs in thinned dies, which were SLID bonded to a thicker Si substrate die. The measurements captured stresses introduced by the W-TSV as well as by the wafer bonding process. A stress range from several MPa to hundreds of MPa could have been covered with a spatial allocation ranging from 100 nm to tens of microns. Measurement results were compared to each other and to simulated stresses from finite element analysis.

1. Introduction

The heterogeneous integration of 3D system-in-package (SiP) is a challenge for reliability. At Through Silicon Vias (TSVs) the different thermo-mechanical behavior of the used materials may lead to residual stress with related failures. Therefore the determination of localized residual stresses is a key focus for industry.

Scanning X-ray micro diffraction (S-µXRD) and Raman spectroscopy are among the most often used methods to analyze residual stress around TSVs. Although S-µXRD is fast with high lateral resolution, its stress resolution is lower compared to the classical high resolution X-ray diffraction (HR-XRD) scans in reciprocal space. HR-XRD provides appropriate diffraction patterns for a profile analysis. Further, Raman spectroscopy is easy to perform, highly local and surface sensitive. But for non-uniaxial stress states it is very complicated to extract stress tensor components. Finite Element Analysis (FEA) is widely used. However, complex stress built-up processes have to be modeled adequately, which is cumbersome and sensitive to plenty of potential input errors.

In order to ensure reliable stress data, in this paper the residual stress in single-crystalline Si around W-filled TSVs was determined experimentally by three independent methods with high spatial resolution and compared to one another. Besides HR-XRD and microRaman, a recently developed new method on base of stress relief measurements (fibDAC) was extended to measure high stress gradient fields. Finally, all measured stress data was faced with respective FEA results. In contrast to Cu as TSV filler, W has the potential advantage of a lower CTE mismatch to Si resulting in lower thermal induced stress at the TSV-interface at room temperature. As test layout a cross-sectioned double-die stack was used consisting of a top die with TSVs which was bonded by Cu-Sn Solid Liquid Interdiffusion Bonding (SLID) to the bottom die.

2. Analyzed Samples and Finite Element Modeling

Analyzed samples. All experiments were carried out on cross sections of double die systems (Figure 1).

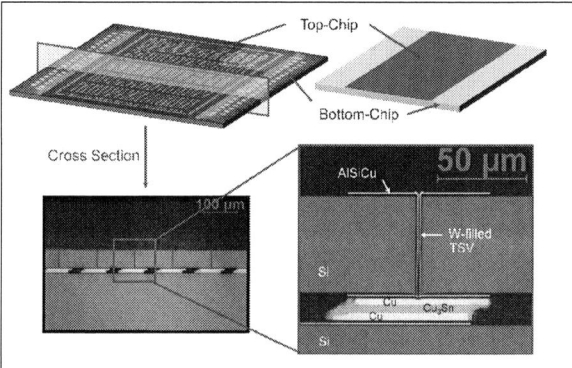

Figure 1: Cross section of double die (light microscope). Top die and bottom die are bonded by solid liquid interdiffusion (SLID) of Cu and Sn in chip to wafer process.

The layout is schematically pictured in Figure 2. It consists of a thin top die with W-TSV and contact pads and a thick bottom die. Both are bonded by a Cu-Sn-SLID as chip to wafer process. The cross sectioned W-TSV as well as the SLID bond are shown in Figure 1.

The cross section of a W-filled TSV is 10 x 3 μm^2 (Figure 2, top view). Its height is 50 µm, what equals the thickness of the top die.

Finite element analysis. An FE model of the double die system was built up in ANSYS to simulate the manufacturing steps of the double obtaining the stress state. The model and the processing parameters are shown in Figure 3. The applied materials, their parameters and the mechanical models are listed in Table 1. The volume shrinkage during the formation of Cu_3Sn is neglected.

The FE model is a quarter model with periodic boundary conditions. The cross section is considered by relaxation of nodes in z-direction after the last processing step. The orientation of the Si single crystal is indicated by the direction indices. The results of the FE analysis are discussed in detail during comparison with experimental results in the related sections. But in general we can draw two conclusions from that: At first the stress inside the Si within some microns distance to the TSV edge is strongly influenced by the interaction with the W metallisation. And secondly the influence of the SLID bond increases strongly by increasing the distance to the TSV edge.

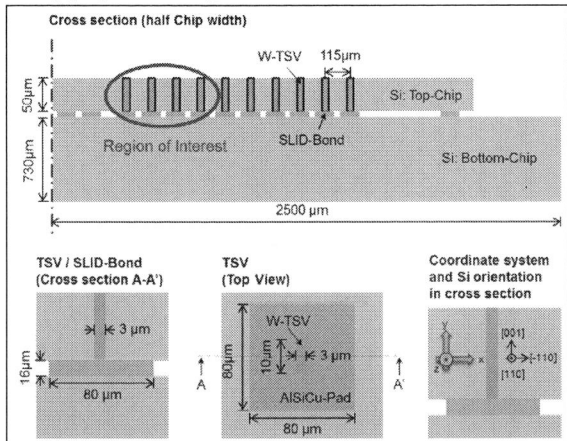

Figure 2: Double die sample layout with TSVs and SLID-Bonds (schematically).

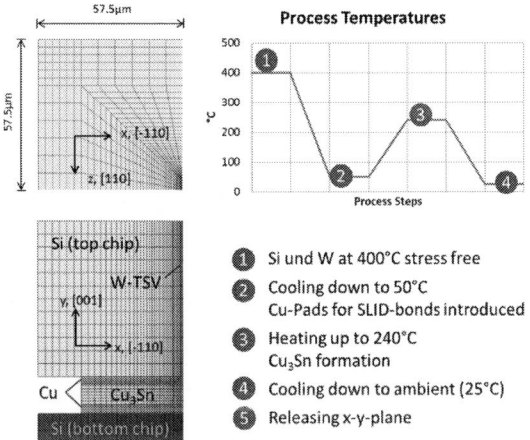

Figure 3: FE model und applied process parameters.

Table 1: Material properties and mechanical models used in FEA. [1][2][3].

Material	CTE (ppm/K)	E (GPa), sij (1/TPa)	Poissons ratio v	σ_y (MPa)	Mechanical Model
Si	2.8	s11: 7.74 s12: -2.14 s44: 12.56			anisotropic, elastic
CVD-W	4.4	E: 420	0.28		elastic
CVD/EPD-Cu	16.6	E: 90	0.36	200	elastic-plastic
Cu$_3$Sn	18	E: 100	0.30	480	elastic

3. Stress Measurements

HR-XRD. Experiments were performed for determination of residual stress inside Si of the top die nearby W-TSVs and near to SLID-bonds. All HR-XRD experiments were carried out at the beamline P08 at PETRA III (DESY). That beamline is described in detail elsewhere [4]. The beamline was run in *Micro Mode*. That mode provides a beam with fine line focus of a nominal cross section of 30 µm x 2 µm. Its divergence is in the order of 1/100°. Thus the width of a rocking curve obtained in a Si single crystal is in the same order. In Si wafers the typical misorientation of crystals are in the order of 1/10°. The die-to-wafer bonding process is usually even less accurate. At our samples we measured misorientations between top and bottom die of a few degrees. Since we do not perform diffraction on higher indexed lattice planes we can therefore assume that in our samples diffraction peaks of both dies can easily be separated.

Table 2: Properties of detectors used for HR-XRD

Detector	Material	Pixel Number	Detector Size	Pixel Depth	Read out Time
Mythen (SLS detector group)	Si	1 x 1280	50 µm x 8 mm	18 bit	250 µs
Cyberstar (FMB Oxford / Oxford Danfysik)	NaI	1 x 1	Ø 30 mm	-	-

We used a (6+2)-Circle-Diffractometer from Kohzu Precision, Kawasaki, Japan. The main axes provide a precision of von 2 x 10^{-5} deg. For registration of angle-dispersed X-rays a Mythen line detector was used. Some of its properties are shown in Table 2. The line was aligned to the tangential at the θ-circle on the detector side (Figure 4, side view), scanning the 2θ-space. Its resolution was determined to 1.63 x 10^{-3} deg per pixel. For easier peak search a Cyberstar Counter supported the adjustments and alignments (Table 2).

We determined an energy of 17.486±0.003 keV using a stress free Si analyser crystal. The diffraction geometry with relevant angles is shown in Figure 4. The poles of the lattice planes for the diffraction experiments are shown at Figure 5 in stereographic projection around the (1 1 0). These poles represent measuring directions and it is seen that all of them have a strong out-of-plane component. For every plane the reflexes from line scans were obtained, on a path over four TSVs along the x-direction with 10 µm distance to the SLID edge (Figure 6). The position of the TSVs were determined by intensity minimums of the detector signal. These minimums are due to shadowing of the W in TSVs. Analysing the diffraction peaks of different lattice planes an X-ray stress analysis (XSA) can be done to obtain strain or stress tensor ([5])..

Two of the measured reflexes are to be discussed in that paper. Figure 7 illustrates the influence of the W-TSV and the SLID bond on the spacing of the lattice planes (2 2 0)

978-1-4799-4789-8/14 $31.00 © 2014 IEEE

and Figure 8 the influence on (7 5 -3). The results of the XSA are obtained from a scan along the x-direction at 10 µm distance to the SLID-Bond (Figure 6). They are compared to two plots obtained from FEA at different depths with respect to the extinction depth of diffraction at specific lattice planes.

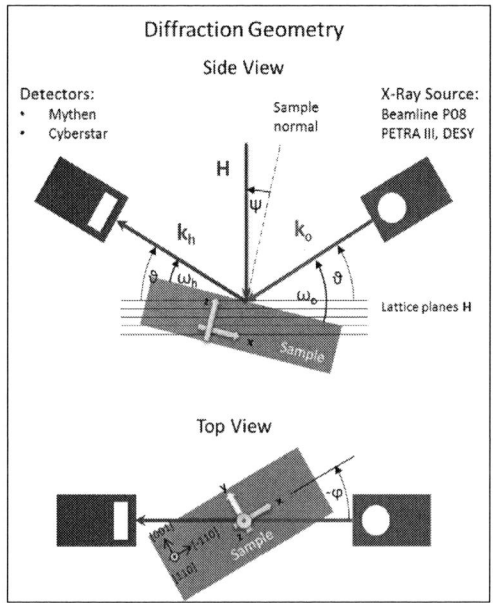

Figure 4: Diffraction Geometry for HR-XRD.

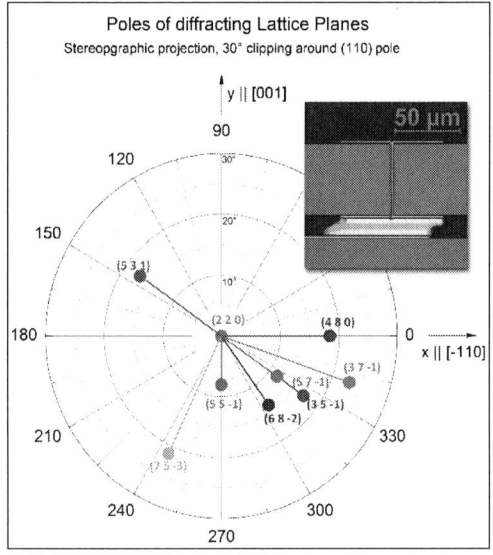

Figure 5: Reflex normals in stereographic projection around (1 1 0)-pole (30° clipping).

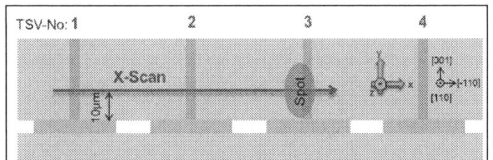

Figure 6: Path of line scan for HR-XRD.

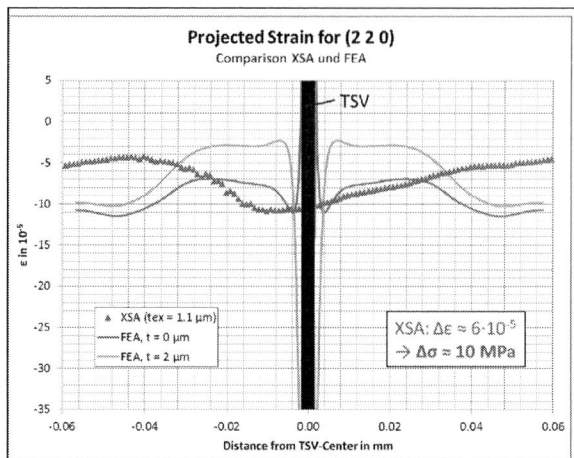

Figure 7: Projected strain for (2 2 0). Comparison between XSA and FEA for a scan along the x-direction at 10 µm distance to the SLID-Bond.

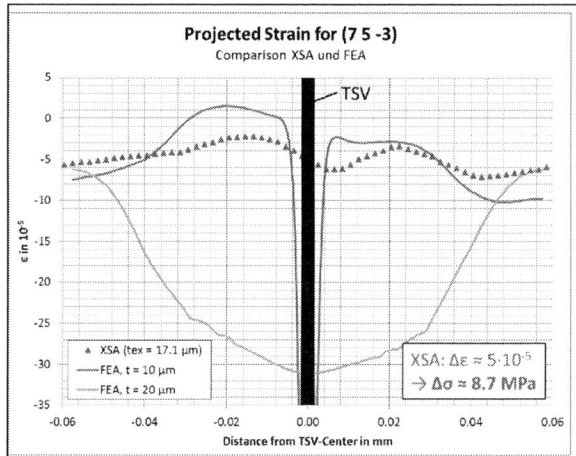

Figure 8: Projected strain onto (2 2 0)-plane (a) and (7 5 -3)-plane (b). Comparison between XSA and FEA for a scan along the x-direction at 10 µm distance to the SLID-Bond.

(2 2 0): The lattice plane for the (2 2 0) X-ray reflex lies in the surface of the (1 1 0) cross section. The measuring direction [1 1 0] lies in the TSV-centre (x=0), a structural mirror plane. The diffraction geometry and the strain distribution are therefore symmetrical to x=0. The projected spot size into the x-direction was 23.2 µm, which is relatively large due to the low incident angle and produces a strong convolution of the diffraction signal. The extinction depth (tex) is 1.1 µm. From FEA thus x-paths at surface and in depth of 2µm were chosen in Figure 7. They show a projection of strain onto the (2 2 0) normal. Both FEA graphs reveil small compression of some 10^{-5} and look similar at larger distance from TSV. The compression at the surface is slightly higher than just below the surface. The difference between both paths increases with decreasing distance to the TSV. Within 20 µm distance from the TSV centre the strains differ strongly and show a kind of reverse behaviour. This is due to rearrangement of stress during cross sectioning as well as the slight influence of stress around the 5 µm deep corners of the TSV. Due to the higher CTE of Cu and Cu_3Sn, the Si is under tension in x-direction and the surface is bent at material interfaces. Due to lateral contraction Si is under compression normal to (220). If the strong gradients of both FEA paths are neglected, one finds ε varying within 10^{-4}. The values of the XSA cover a slightly smaller range of 6×10^{-5}. This is plausible in consideration of convolution and corresponds to stress of 10 MPa, if shear is neglected. The global strain is satisfyingly matched. The strong gradients near the TSV couldn't be resolved with the XSA. The decreasing strain towards the TSV correspond with the FEA distribution below surface (t = 2 µm), but seems a little bit stronger than predicted. The relative compression can be explained by the higher Cu-CTE as well as the shrinkage of Cu_3Sn after the SLID formation. The XSA distribution doesn't show the maximum at distance of 30 µm and the decrease of strains at larger distance from TSV.

(7 5 -3): For the (7 5 -3) X-ray reflex the measuring direction is off any symmetry axis or plane, so that asymmetrical strain distributions were expected (Figure 8). The diffraction at (7 5 -3) yields an extinction depth of 17.1 µm. For comparison FEA-paths at depths of 10 and 20 µm were chosen. The W-TSV reaches only 5 µm deep so that its influence onto the strain distribution should be weaker with increasing depths. Both FEA-plots differ strongly and cross each other near the SLID edges at $x \approx \pm 40$ µm. The curve at 20 µm only shows influence of the SLID-bond. The projected spot size of the X-ray beam is 16.4 µm in x-direction, hence the convolution is less strong than for the (2 2 0) reflex resulting in less blurring of the strain distribution profile. The XSA results match the FEA results of the path in 10 µm depth well. Since the extinction is nonlinear with depth and the majority of X-rays are coming from depths below 16.4 µm, it is reasonable that the XSA results should match the FE-result in 10 µm depth better than in 20 µm. The XSA results cover a strain range of 5×10^{-5} which resembles about 9 MPa, if shear is neglected. If we keep in mind that the measuring signal is strongly convolved we can expect actually higher strain variations than predicted by the FEA. This is an indication that the FEA underestimates stresses in its model.

MicroRaman measurements. Stress measurements on electronics devices have been applied already in the 90th [6]. Although the method exhibits clear advantages like fast stress mapping, stress extraction from Raman line shift features challenges for many objects of interest. As far as Raman shifts commonly depend on more than one stress tensor component, straightforward stress extraction isn't possible from a measured single Raman shift [7]. Attempts have been made in recent years to overcome this handicap [8], basically incorporating the excitation of more than one phonon. In some cases combined data analysis from both Raman measurement and Finite

Element Simulation can be applied [9]. Utilizing this approach, Raman shift calculation from stress tensor values obtained by FEA has been used in this paper to compare FEA and measurement stress data.

Measuring stress in Si nearby TSVs, one has to mind the following circumstances:

- The essential normal stress tensor components in coordinate systems adapting the TSV symmetry often have opposite sign leading to mutual extinction in the Raman shifts. Consequently, captured absolute Raman shifts themselves do not necessarily represent the general stress level. In a bad case, separated normal stress components may be much higher than suggested by the Raman shift referred to an assumed uni-axial or equal-bi-axial stress state. Nevertheless, high Raman shifts indicate high normal stress levels.

- Measuring on cross sectioned samples, as in this paper, implies a severe strain/stress re-distribution at the new free surface. Stresses in the volume of the original untouched specimen must be adapted for cross sectioning by appropriate FEA.

Near surface stresses originating from strain mismatches at material interfaces often reveal a fast strain/stress re-distribution with increasing measurement depth. So, Raman shifts excited with visible laser light map a mixed stress field and do not represent the pure surface stress. Figure 9 illustrates this effect. It occurs most significant in the

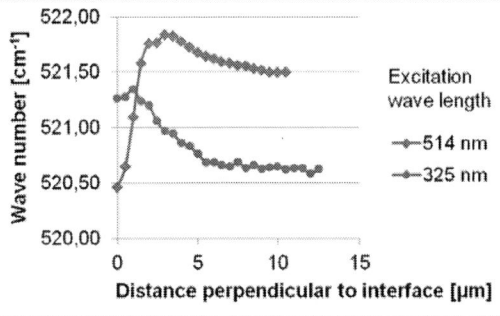

Figure 9: Raman shifts obtained with different excitation wave lengths in Si. Measurement along a line perpendicular to the material interface on a cross sectioned specimen (SiN layer on Si substrate, SiN bow stress: ~ 3.3 MPa tensile). Raman shift at 514 nm is caused by a depth mix of varying stresses. Nearby the interface at 0 nm it deviates significantly from that obtained by a 325 nm wave length, representing the pure surface stress state.

very vicinity of the material interface. With respect to stress measurements at TSVs, the region with highest stress levels is affected. For this reason all measurements on TSVs reported here have been excited with a He-Cd laser wavelength of 325 nm. The penetration depth in Si [10] is limited to approximately 10 nm avoiding the mentioned averaging effect.

- Approaching the material interface with the scanning laser spot, the size of the Raman scattering light source forming the spectrum peak is changing. Moreover, inner reflections of the incident laser beam at the material interface can alter the composition of phonon excitation, respectively the Raman peak shift. Both artificial peak shifts in the vicinity of the interface may distort the shift value over a final approach distance of ~ 1 μm. These errors cannot be avoided by plasma line drift corrections, which behave not the same way (see [9]).

Our Raman measurements in Si surrounding tungsten TSVs were carried out on cross sectioned specimen, as line scans perpendicular to the trench direction with different distances to the SLID bond. The cross section has a (110) orientation. Because we used a UV excitation by a He-Cd laser the light is absorbed within a surface-near layer of approx. 10 nm, i.e. measurements represent stresses from the very surface of the specimen.

Figure 10 shows the measured Raman shifts vs. the distance to the TSV center. In comparison, computed Raman shifts for different phonon excitation are given, which have been calculated from stress values of FEA.

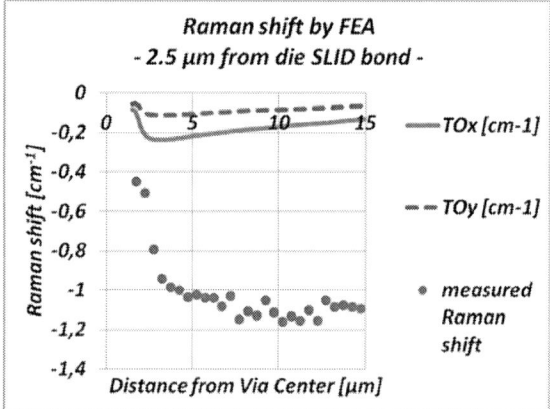

Figure 10: Raman shift along a line perpendicular to the TSV direction. Measurement on cross sectioned sample (110). Besides the measured data computed shifts are given, which would be expected for different phonon excitation, if simulated stresses from the described above FEA are used.

fibDAC stress relief measurements. Stress determination measuring deformation due to stress relaxation after local material removal is a classic experimental method. In the past years this method has been adopted on high spatial resolution treatment and imaging tools. So, focused ion beam (FIB) milling in combination with stress relief measurements by Digital Image Correlation (DIC) techniques were used, starting from a first publication in 2003 [11]. Several kinds of this approach have been reported later, e.g. [12] [13] [14]. The presented in this paper type of measurement uses the focused ion beam (FIB) to mill trenches of approximately 100 nm width into the surface of the specimen to trigger stress relief in its vicinity. Capturing the corresponding deformation on high resolution SEM micrographs by local digital image correlation (DIC), the original stress can be determined by simulating the elastic stress relief

process with the help of finite element analysis. This simulation starts from an arbitrarily chosen residual stress value. Stress relief displacements are scaled linearly until they fit the experimental values obtained by DIC. Because of the linear elastic material relaxation after ion milling, the same scaling factor can be used to determine the true stress at the place of ion milling. This trench technique has been established under the name "fibDAC stress relief technique" [15][12]. In the present work a new fibDAC technique is introduced allowing stress determination in gradient fields with a dedicated spatial resolution by using a single trench milling only.

Figure 11 shows a cross section image of the investigated tungsten TSVs. The stress level in the Si around the W-TSV is introduced by the W-TSV itself as well as by the bonding interconnect placed underneath the TSV. The diameter of the SLID bond (≈ 40 μm) is large in comparison to the thickness of the elongated TSV (≈ 3 μm). In the very vicinity of the TSV (a few μm), the SLID interconnects introduce a rather homogenous stress offset, whereas high stress gradients are caused by the TSV.

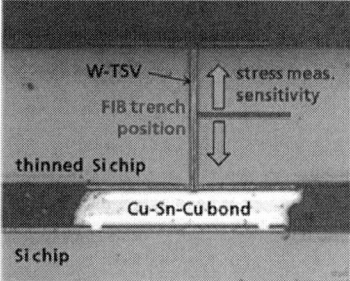

Figure 11: 3D stacked dies with marked fibDAC stress measurement trench. Bold arrows show the direction of determined normal stress component.

The fibDAC method has been used to measure the stress gradient nearby the TSV. The stress relief trench was placed perpendicular to the trench direction on cross the sectioned TSVs (see Figure 11), i.e. stress sensitivity occurs for the normal stress component in TSV direction. It has to be understood that the stress is measured on the cross sectioned specimen, which in general is different to the stress before cross sectioning. Careful cross sectioning as treatment process with subsequent polishing can be neglected to contribute to the stress state. However, the creation of a new free surface due to cross sectioning leads to severe re-distribution of stress and strain. If necessary, measurements have to be combined with appropriate FEA to access inner stresses around the TSV. Taking measurement on cross section alone, at least an idea on the magnitude of stresses can be obtained.

Figure 12 shows one of the real measurement trenches. It allows relief of vertical stress in the TSV and the adjacent silicon. After the images for DIC analysis have been picked up, a pit is milled into silicon to capture the trench profile, which is introduced in the finite element stress relief model.

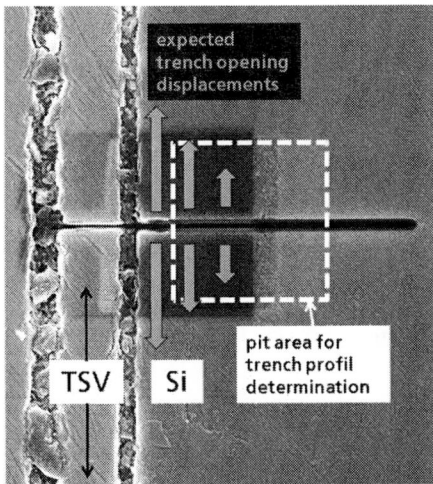

Figure 12: fibDAC trench cutting of the W-TSV and the adjacent Si. Bold arrows depict schematically trench opening displacements. After fibDAC measurement a small pit is ion milled in order to get imaging access to the real trench profile.

Figure 13 depicts the trench profile, Figure 14 the experimental DIC displacement field, respectively. The found displacement field indicates a trench opening after ion milling, which refers to tensile stress. The trench opening amount rapidly decreases with increasing distance from the W-Si interface, i.e. the measured stress is caused mostly by the strain mismatch from the TSV processing. The trench opening is caused by Si stress (and not by W stress), because similar opening takes place, if the trench is not extended into the W-TSV.

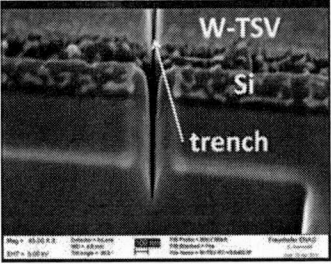

Figure 13: Stress relief trench of approx. 210 nm depth and 160 nm width, looked at under 30 degree to the milled pit front side.

Stress extraction from the DIC displacement field based on the described above fitting procedure between scaled FEA and measured DIC displacements. Standard fibDAC procedures commonly look for stress relaxation displacements in the middle of a longer trench [15], postulating relaxation displacements independent of the position along the trench. Because stresses change in our case within the evaluated area with increasing distance to the TSV, thin horizontal slices of the displacement field in Figure 14 have been analyzed separately under the assumption of approximately constant relaxation displacement in each of the slices.

Generally, this simplification introduces somewhat a smoothing effect on the obtained stress data, depicted as a function of distance from the W-Si interface. Nevertheless, this averaging is of minor importance if

compared to measurement noise and local stress fluctuation. Resultant stress values over distance to the TSV center is given in Figure 15. For comparison respective data from FEA for the same stress tensor component s_{yy} has been added.

Figure 14: Isoline plot of the trench opening displacement fields in pixel units. Decreasing displacement discontinuity between the trench edges with increasing distance from the TSV (at the top if the image). In comparison to Figure 12 this top view has been clockwise rotated over 90°.

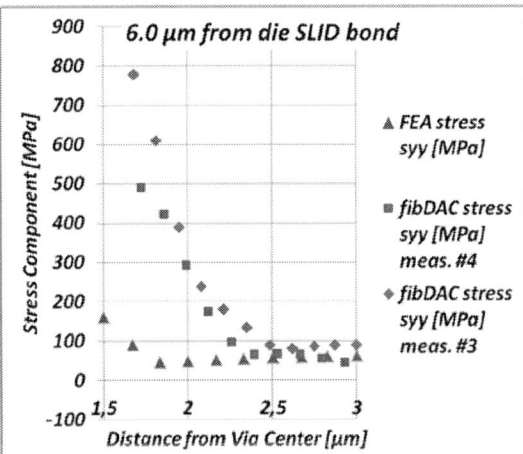

Figure 15: Normal stress along TSV direction (see Figure 11) in silicon approaching a W-TSV of 3 μm width. Most stress disappears over a 1 μm distance away from the TSV rim. The remaining far reaching stress of approx. 100 MPA (in experiment) probably is due to the Cu-Sn-Cu bond underneath the TSV. Measurement #3 is the resultant data referring to the displacement relaxation field in Figure 14 and the trench top view of Figure 12.

Measured stresses exceed significantly simulated. Namely the large increase of experimentally found stress approaching the TSV interface is not reflected by simulation data. Both measurements exhibit similar stress behavior and values. Distinct lower stress (point at 1.7 μm distance) of measurement #4 in comparison to measurement #3 probably is due to an insufficient stress release. The trench in measurement #4 stopped nearby the

TSV rim, which should lead to some additional constraint for stress relief and an underestimation of the stress value respectively.

4. Discussion

Stress values have been determined in Si around tungsten TSVs within an area of maximum 60 µm. The applied methods possess different spatial resolutions and realize dissimilar access to particular strain / stress components as well. As a matter of fact, comparison between results from different techniques has to take into account the distinguished capabilities of methods applied. For that reason Table 3 summarizes them, indicating Cartesian directions as shown in Figure 2. Listed available data refers to the particular experiments carried out.

Table 3: Overview on the conditions of stress determination

Measurement method	Lateral spatial resolution [µm]	Available data
Finite element analysis (FEA)	optional, depending on meshing	any of the strain/stress tensor components
HR-XRD	10 … 25 µm	strain projected on (220) and (75-3) directions in the unit cell system
microRaman	0.7 … 1 µm	Raman wavelength shift, being an arithmetic function of normal stress tensor components
fibDAC stress relief technique	100 nm	normal stress s_{yy}

A first look on Table 3 reveals that two of the measurement techniques allowed a rather coarse lateral spatial resolution in the micrometers region, whereas the fibDAC method provides a significant higher resolution of 100 nanometer. As a result the fibDAC tool only gives a realistic insight view to stress built-up due to the Si substrate/W-TSV mismatch, as far as this stress degrades over a distance ≤ 1 µm.

For HR-XRD measured changes of strain may cover the range of some 10^{-5} (Figure 7). This results in stress variation between 5 and 10 MPa if shear is neglected. This high stress resolution is achieved by HR-XRD only. In the opposite, microRaman and fibDAC measurements provide resolutions in the range > 10 MPa. XRD results show the high precision of the $\theta 2$ -Sans in combination with the XSA. The XSA enfolds its full potential at out-of-plane measuring direction. Strong gradients haven't been resolved by standard XSA due to the relatively large projected spot size. Here a profile analysis of the diffraction peaks gives better results.

One drawback of the HR-XRD by scanning single reflexes sequentially is its long adjusting time. Here the Scanning X-Ray Microdiffraction method is faster and provides therefore an alternative to map strains around TSVs [16]. But due to its smaller detector resolution the precision is somewhat worse. Another issue is that large in-plane components in XRD-measuring directions require low incident or exit angles. This results in very large projected spot sizes and therefore in low spatial resolution.

In case of small strain gradients the match between FEA and XSA is satisfying. The results give an indication that there is more stress in systems than predicted by FEA. Koseski showed that intrisic stress in W may be responsible for higher stress values in W-TSVs [17]. However, judging the graphs in Figure 7 and Figure 8 it has to be stated that the XSA measurements average over the penetration depth. I.e. they will not coincide with FEA data representing a particular depth under the surface. As far as both strain projections in XSA are equal or nearby to the surface normal, stress in the captured surface layer is small and changes rapidly starting from a plane stress state at the very surface to moderate stresses in projection direction, when the penetration depth is reached.

The comparison of the measured Raman shift with the simulated shifts (Figure 10) reveals much higher measured shift than simulated, independent on the selected phonon for the emulated peak shift. Namely, the steep shift jump for the 2 … 3 µm gap adjacent to the TSV is not reflected by the FEA data. The measured stress change over this distance is at least 5 times higher than the expected from FEA. It indicates normal stress levels not less than 500 MPa. For this estimation, a possible mutual compensation of stress tensor components in the resultant Raman shift is neglected and could lead to even higher values.

Analyzing the results of the fibDAC measurements (Figure 15) a similar conclusion is to be made as for Raman. The measured stress increase approaching the TSV interface exceeds the simulated at minimum by a factor of 5. This large stress gain appears over a 1 µm distance, which is somewhat faster than shown by Raman. Considering the lower lateral spatial resolution of Raman (see Table 3) it seems reasonable to assume, to a certain extend, a data smoothing of the steep increase by Raman scanning. fibDAC measurements are in well agreement with FEA for distances larger 2.5 µm from the center of the TSV.

Probably, stresses for larger distances from the TSV are induced rather by the SLID bonding underneath the TSV. In Raman shifts as well as in fibDAC stress they keep constant over a larger detected distance aside the narrow TSV and are of the magnitude of about 50 … 100 MPa for the s_{yy} stress component.. Concluding, the mutual comparison between Raman, fibDAC and FEA data indicates, that a severe increase of stresses in Si over the final 1 µm to the TSV exists. It reaches at least some hundreds of MPa and seems to originate from the TSV mismatch to the Si. It is not reflected by the FEA, which still is not well understood. I.e. the performed FEA modeling obviously does not describe the very near interface behavior sufficiently, but explains quit well the influence of the SLID bonding between the dies.

Comparing HR-XRD stress levels with that of Raman and fibDAC, it has to be taken into account that HR-XRD presents basically out-of-plane stresses with highest precision, which are rather low nearby the surface, whereas the other measurements correspond to the higher in-plane stresses.

5. Conclusions

Four different methods have been utilized to obtain a comprehensive view on stress development in silicon surrounding tungsten TSVs after via processing and SLID bonding of dies. Measurements were performed on cross sectioned specimens. While HR-XRD was used to study stresses in a nearby surface layer with emphasis on out-of-plane components, Raman and fibDAC measurements gave access to in-plane components. A severe stress increase in a 1 μm thick layer adjacent to the TSV was found, as well as a "bias" stress level above the SLID bonding width. The latter could have been well described by Finite Element Analysis, whereas the TSV induced stresses were not reflected accordingly by simulations. Future work will finalize the analysis by the FE-based computation of stresses in the non cross-sectioned sample, after a finished validating of FEA modeling by experimental results from the cross-sectioned specimen. Here the focus lies in possible formation of intrinsic stress in tungsten during its deposition as well as an analysis of the formation of the Cu_3Sn-phase.

Acknowledgments

The work was part of the project "Best-Reliable Ambient Intelligent Nano Sensor Systems" (eBRAINS). The project is co-funded under the Seventh Framework Programme for Research of the European Commission. We thank Oliver Seek and his team from the beamline P08 at PETRA III (DESY) for the support during our HR-XRD experiments in Hamburg. We thank Harishankaran Rajendran for his valuable work on the FEA. Raman measurements have been performed at Fraunhofer IZM. We acknowledge the assistance by A. Gollhardt.

References

[1] P. R. Raffo, „Yielding and fracture in tungsten and tungsten-rhenium alloys," *Journal of the Less Common Metals,* Bd. 17, Nr. 2, pp. 133-149, 1969.

[2] P. Ramm, M. J. Wolf und B. Wunderle, „Wafer-Level 3D System Integration," in s *Handbook of 3D Integration,* Wiley-VCH Verlag GmbH & Co. KGaA, 2008, p. 289–318.

[3] P. Garrou und C. Bower, „Overview of 3D Integration Process Technology," in s *Handbook of 3D Integration,* Wiley-VCH Verlag GmbH & Co. KGaA, 2008, p. 25–44.

[4] O. H. Seeck, C. Deiter, K. Pflaum, F. Bertam, A. Beerlink, H. Franz, J. Horbach, H. Schulte-Schrepping, B. M. Murphy, M. Greve und O. Magnussen, „The high-resolution diffraction beamline P08 at PETRA III," *Journal of Synchrotron Radiation,* Bd. 19, Nr. 1, pp. 30-38, 2012.

[5] L. Spieß, G. Teichert, R. Schwarzer, H. Behnken and C. Genzel, Moderne Röntgenbeugung : Röntgendiffraktometrie für Materialwissenschaftler,

Physiker und Chemiker, Wiesbaden: Teubner, 2008.

[6] I. d. Wolf, "Micro-Raman spectroscopy to study local mechanical stress in silicon integrated circuits," *Semicond. Sci. Technol.,* vol. 11, pp. 139-154, 1996.

[7] Qiu Zhao, J. Im, R. Huang and P. S. Ho, "Extension of Micro-Raman Spectroscopy for Full-Component Stress Characterization of TSV Structures," in *Proc. of ECTC,* Las Vegas, 2013.

[8] I. de Wolf, V. Simons, V. Cherman, R. Labie, B. Vandevelde and E. Beyne, „In-depth Raman Spectroscopy Analysis of Various Parameters Affecting the Mechanical Stress," in s *Proc. of IEEE,* San Diego, 2012.

[9] D. Vogel, E. Auerswald, J. Auersperg, S. Rzepka, B. Michel, "Measuring Techniques for Deformation and Stress Analysis in Micro-Dimensions," in *Proc. of EuroSimE,* Wroclaw, 2013.

[10] D.E. Aspnes, A.A. Studna, „Dielectric functions and optical parameters of Si, Ge, GaP, GaAs, GaSb,InP, InAs, and InSb from 1.5 to 6.0 eV," *Phys. Rev.,* Bd. B 27, p. 985, 1983.

[11] K.J. Kang, N. Yao, M.Y. He, A.G. Evans, "A method for in-situ measurement of the residual stress in thin films by using the focused ion beam," *Thin Solid Films,* vol. 443, pp. 71-77, 2003.

[12] N. Sabaté, D. Vogel, A. Gollhardt, J. Keller, B. Michel, "Measurement of residual stresses in micromachined structures in a micro region," *Appl. Physics Letters,* vol. 07910, 2006.

[13] A.M. Korsunsky, M. Sebastiani, E. Bemporad, "Focused ion beam ring drilling for residual stress evaluation," *Materials Letters,* vol. 63, pp. 1961-1963, 2009.

[14] D. Vogel, S. Rzepka, B. Michel, "Local stress measurement on metal lines and dielectrics of BEoL pattern by stress relief technique," in *Proc. of Semiconductor Conference,* Dresden, 2011.

[15] D. Vogel, A. Gollhardt, N. Sabate, J. Keller, B. Michel, H. Reichl, "Localized Stress Measurements – A New Approach Covering Needs for Advanced Micro and Nanoscale System Development," in *Proc. of 57th IEEE ECTC,* Reno, 2007.

[16] N. Tamura, A. A. MacDowell, R. Spolenak, B. C. Valek, J. C. Bravman, W. L. Brown, R. S. Celestre, H. A. Padmore, B. W. Batterman und J. R. Patel, „Scanning X-ray microdiffraction with submicrometer white beam for strain/stress and orientation mapping in thin films," *Journal of synchrotron radiation,* Bd. 10, Nr. Pt 2, pp. 137-143, 2003.

[17] R. P. Koseski, W. A. Osborn, S. J. Stranick, F. W. DelRio, M. D. Vaudin, T. Dao, V. H. Adams und R. F. Cock, „Micro-scale measurement and modeling of stress in silicon surrounding a tungsten-filled through-silicon via," *J. of Appl. Phys.,* 110 (7), 2011.

[18] J. R. Davis, ASM handbook. Vol. 2, Properties and selection : nonferrous alloys and special- purpose materials, Materials Park: ASM International, 1992.

Modelling, simulation and optimization for a SThm nanoprobe

Bin Yang[*], Michel Lenczner[†], Scott Cogan[*], Fabian Menges[‡], Heike Riel[‡], Bernd Gotsmann[‡]
Pawel Janus[§] and Guillaume Boetch[¶]

[*]FEMTO-ST, University of Franche-Comté, Belfort, France, bin.yang@femto-st.fr
[†]FEMTO-ST, University of Technology at Belfort-Montbeliard, France, michel.lenczner@utbm.fr
[‡]IBM Research-Zürich, Säumerstrasse 4, Ch8803 Rüschlikon, Switzerland, bgo@zurich.ibm.com
[§] Institute of Electron Technology al. Lotnikow 32/46 02-669 Warszawa, PL, janus@ite.waw.pl
[¶]Imina Technologie SA, EPFL Innovation Park Batiment E, 1015 Lausanne, Switzerland, guillaume.boetsch@imina.ch

Abstract

This paper presents modelling, simulation and optimization results for a novel SThm probe. The model takes into account thermo-electro-mechanical equations. Moreover, a tip-surface contact model derived by taking into account microscopic multi-asperity contact is proposed and discussed. Results of multi-objective optimization are reported, and finally a multi-scale model that should reduce the simulation time is stated.

KEY WORDS: *Scanning thermal microscopy, Nanoprobe, Tip-surface contact, Interface thermal conductance, Multi-objective optimization, Asymptotic model*

1. Introduction

Modern technology of micro/nanoelectronic components, sensors and MEMS/NEMS (Micro/Nano-Electro-Mechanical-Systems) requires increasingly the control of materials at the sub-micrometer down to the nanometer scale. Additionally, the heat transfer phenomena, including e.g. phonon heat conduction mechanism in micro- and nanostructures, may differ significantly from that on the macroscale. Therefore, micro- and nanometer resolution is required for most of the experiments.

Scanning Thermal Microscopy (SThM) is a versatile scanning probe technique allowing for high resolution mapping of the thermal properties and temperature of various substrates. SThM, as every AFM (Atomic Force Microscopy) related technique enables study at micro- and nanoscale which allows designers to better the understanding of heat transport in micro- and nanoelectronic devices.

The invention of the scanning tunneling microscope (STM) [2] and the atomic force microscope (AFM) [1] have allowed sub-micrometer and, at times, atomic scale spatially resolved imaging of surfaces. The spatial resolution of these near-field techniques is only limited by the active area of the sensor (which in the case of STM may only be a few atoms at the end of a metal wire). As described by Dinwiddie and Pylkki in 1994, first scanning thermal microscopy (SThM) probes employed resistance thermometry to measure thermal properties [3]. These probes were fashioned and made from Wollaston process wire consisting of a thin platinum core (ca. 5 μm in diameter) surrounded by a thick silver sheath (ca. 75 μm). Because of its high endurance, Wollaston probe is

attractive for microsystem diagnostics [16], however the active area in the range of a few micrometers does not allow quantitative thermal investigations at the nano-scale.

In this paper, we present and discuss a new thermal probe that is designed to achieve quantitative measurement in the range of few tenths nanometers. Two aspects of its modelling are discussed. Since the current models and theories are not sufficient to describe and predict the heat flow through the tip-sample interface and in the vicinity of the contact, a simple model of tip-sample heat transfer is stated. It is obtained by combining recent published results and experimental observations. In addition, a thin plate model is described that includes the three involved physical phenomena as well as the multilayered structure of the probe. It should be useful to reduce the simulation time and so to facilitate optimization. The latter has been carried out using an in-house developed software package SIMBAD for robust optimization and its connection with the FEM simulation software package COMSOL. Three objective optimization results are reported, namely to decrease the thermo-mechanical tip deflection, to increase the Joule heating effect in the tip and to increase the sensitivity of the piezoresistive sensor.

The paper is organized as follows. Section 2 present the probe design. In Section 3, the mathematical formulation of the physical problem is stated. Section 4 is dedicated to the tip-sample heat transfer model. Section 5 presents the SIMBAD-COMSOL package. In Section 6, a parametrization of the SThm probe is introduced allowing to perform the sensitivity analysis and then multi-objective optimizations. The thin plate model is presented in the last section.

2. The probe design

The novel type of nanosensor, described in this paper, is equipped with sharp, conductive tip, an integrated deflection sensor, and an actuation system. A modification of a double sided silicon micromachining process developed for manufacturing of piezoresistive AFM microprobes has been adapted to fabricate SThM sensors [7][8].

Proposed nanoprobes, as the only SThM/ECM probes on the market are integrated with deflection detection, which will significantly improve the system versatility and will enable new applications. As the NANOHEAT system

978-1-4799-4789-8/14 $31.00 © 2014 IEEE

as free of the bulk and complicated optical deflection sensors, it can be used in small SEM chambers. The

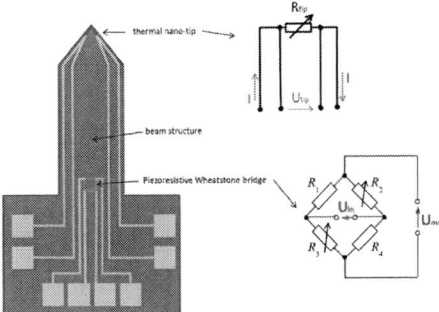

(a) Schematic view of NANOHEAT SThm probe

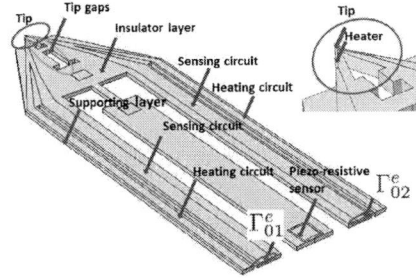

(b) First design of the cantilever

Figure 1.   NANOHEAT SThM probe

described SThM nanoprobes are designed to operate in two modes: a) as a passive thermosensing element or b) as an active heat flux meter. In the latter case, a larger current is passed through the resistive tip probe. The power that is required to maintain a constant temperature gradient between the tip and the sample corresponds to the local thermal conductivity of the sample. During active measurements temperature of the tip is increased by $20 - 30$ K above room temperature. In order to perform quantitative measurements of heat transport between the tip and the surface several crucial criteria have to be met:

-low thermal mass of the microtip allowing for AC thermal measurements (e.g. in the range of 10 kHz)

-high mechanical stiffness of the microtip. This ensures high endurance of the thermal sensor, which is brought into contact while surface scanning.

-low stiffness of the SThM cantilever, which is brought in contact with the investigated surface. The low stiffness of the SThM cantilever will enable surface measurements with relatively low load forces. As a consequence the tip wear is reduced and the sample is not modified.

-high thermal resistance of the SThM cantilever and tip's support. The high thermal resistance of the cantilever will reduce the heat transfer from the thermal tip to the cantilever supporting body. The effective thermal mass of the SThM sensor will be reduced, and its influence on the thermal behavior of the investigated structure will be

minimized.

Moreover, the heat transferred from the tip to the cantilever base causes parasitic deflection of the sensor and may influence signal from Wheatstone bridge. First results of modeling and simulations exhibit significant parasitic, 200 nm deflection of the cantilever due to tip's heating by 11 degrees above the room temperature.

According to the applications, developed SThM nanoprobe will enable surface measurements in contact scanning probe microscopy mode at load force ranging from 10 nN up to 1 microN. The load force will be detected with the resolution of 10 pN in the bandwidth of 100 Hz. The low load forces as well as sub-nanometer vertical spatial resolution in the range will be needed in investigations of graphene and molecular samples, whereas the high force will be applied in investigations of high-k insulators.

3. Mathematical model description

The SThm probe is designed as a three-layered structure. The silicon supporting layer has a thickness of 5 μm, it is covered by a 50 nm $SiO2$ layer where 100 nm thick platinum tracks are deposited. The latter consists of a heating circuit, a sensing circuit and a sharp resistive tip. The three corresponding domains are denoted by Ω_{Si}, Ω_{SiO2} and Ω_{Pt} and Ω denotes their union.

In Figure 1(b), the two inner platinum legs constitute the heating circuit and the two outer constitute the sensing circuit. A piezo-resistive sensor for stress measurement is located to the base of the probe, it is used to measure the tip displacement.

In all the paper, the Einstein summation convention is adopted. We use $\mathbf{C}^q$, $\mathbf{M}^q$, $\mathbf{k}^q$ where $q \in \{\mathrm{Pt}, \mathrm{SiO2}, \mathrm{Si}\}$ and $\mathbf{a}$, to denote the tensor of elasticity, the matrix of thermal expansion, the matrix of thermal conductivity in each layer and the matrix of electric conductivity in platinum. We denote by $\mathbf{u}$, T, and φ the vector of mechanical displacements, the difference of the temperature to the ambient temperature and the electric potential. The system is governed by the following equations,

$$\begin{cases} -\operatorname{div}(\sigma) = \mathbf{f} \text{ in } \Omega \\ -\operatorname{div}\mathbf{q} = (\nabla\varphi)^T \mathbf{a}\nabla\varphi \text{ in } \Omega_{\mathrm{Pt}} \\ -\operatorname{div}\mathbf{q} = 0 \text{ in } \Omega\backslash\Omega_{\mathrm{Pt}} \\ -\operatorname{div}(\mathbf{a}\nabla\varphi) = 0 \text{ in } \Omega_{\mathrm{Pt}} \end{cases} \quad (1)$$

where $\mathbf{a} = (1 + \alpha T)^{-1}\mathbf{a}^{ref}$, $\mathbf{f}$ is the body force load, $\sigma = \mathbf{C}^q\mathbf{s}(\mathbf{u}) + \mathbf{M}^q(T)$ is the tensor of mechanical stresses, $\mathbf{s}(\mathbf{u}) = \frac{1}{2}(\nabla u + \nabla u^T)$ is the tensor of strains, $\mathbf{q} = \mathbf{k}^q\nabla T$ is the heat flux, $\mathbf{a}^{ref}$ is the tensor of electric conductivity at ambient temperature, and α is the thermal coefficient.

Regarding the boundary conditions, the cantilever is clamped and with an imposed temperature on a part Γ_0 of the boundary, i.e. $\mathbf{u} = 0$ and $T = T_0$, and it is left free of load and thermally insulated on the other part, i.e. $\sigma\mathbf{n} = 0$ and $\nabla T.\mathbf{n} = 0$ where $\mathbf{n}$ denotes the outward unit normal vector to the boundary. Finally, a current source is applied to Γ^e_{01} and Γ^e_{02} is grounded, see Figure 1(b). The tip-sample interface condition is discussed in details in the next section, however it has not yet been taken into account in the other parts of the paper which are focused on the probe itself but not yet on its interaction with a sample.

4. Heat transfer through the tip surface contact

Scanning thermal microscopy probes the heat transfer between the integrated heater/sensor of a scanning probe cantilever and a sample surface. The heat flux across the tip-surface contact directly relates to the measured quantities, namely the thermal resistance into or out of the sample, which is to be related to thermal conductivity of the sample, and the sample temperature [14]. For a quantitative understanding and simulation of the tip-sample heat transfer, it is essential to quantify this thermal resistance and its dependence on the interface parameters. A schematic of the tip-surface contact and

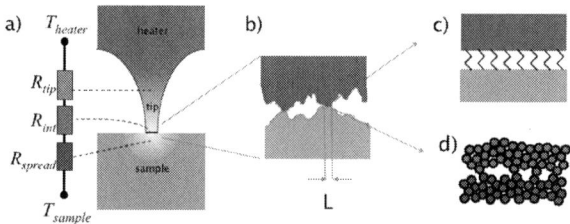

Figure 2. a) Schematic of the tip-sample heat transfer b) a microscopic multi-asperity contact c) an atomic scale interface d) atomic-scale roughness.

the associated thermal resistances are shown in Fig. 2 a). Scanning thermal microscopy tips are typically designed such that the thermal resistance of the tip R_{tip} is small compared the contribution of the sample R_{sample}. This is achieved by choosing good thermal conductors such as metals or silicon as tip materials. However, in quest of achieving high spacial resolution, the dimensions of the tip become smaller and as a consequence its thermal resistance becomes significant. In the case of silicon tips the phonon mean free path of about 100 nm at room temperature exceeds the diameter of the tip.

As a consequence, diffusive thermal transport as simulated in finite element modeling cannot be directly applied any longer. Instead, heat conduction through the tip is modeled as quasi-ballistic transport. For example, following the expression of thermal conductivity from kinetic theory: $k = Cv\Lambda$, with k being the thermal conductivity, C the heat capacity per unit volume, v the phonon velocity, and Λ the mean free path, one can apply Mathiessen's rule to calculate a mean free path modified from the bulk value through boundary scattering. For simple geometries such as conical, analytic expressions exist [5].

The thermal resistance of the sample can be calculated in a similar manner. If the characteristic length scale of the spreading resistance into the sample, the contact diameter, is much larger than the average mean free path of heat carriers in the sample, then diffusive transport leads to a scaling with the inverse of the contact diameter. In contrast, the resistance within the sample scales inversely proportional to the contact area in the ballistic transport regime. For spherical contact areas, both cases can be described with a single analytic interpolation formula.

The contributions from tip and sample to the resistance can be calculated with existing and established models, for simple geometries even using analytic equations. The contribution of the interface, however, is more complicated because both phonon mismatch and mechanical contact geometry have to be taken into account. Phonon mismatch, i.e. the difference in phonon dispersion between two solids, causes a thermal interface resistance (or thermal boundary resistance). Even for interfaces of perfect quality this leads to an appreciable resistance [15]. In reality, however, the interface quality is not perfect, due to oxide layers, non-continuous contact areas, or weak coupling bonds between the atoms of either solid. To account for the weak coupling, advanced contact models have been developed in which the transmission probability is related to the mechanical coupling spring between the two solids [12], [11].

In SThM there are further aspects to be considered. Due to the roughness of the tip and sample surfaces, the apparent contact area may in fact be divided into smaller contact spots. Appropriate modeling of this effect depends on the length scale L of the contact spots. If L is much larger than the mean free path of heat carriers than the solutions mentioned above can be applied. If L is smaller than the mean free path then ballistic solutions play a role. This has been pointed out for example by Prasher and Phelan [13]. As discussed recently [10], [4], the notion of finite contact spots may be extrapolated to the atomic scale. This can lead to a situation in which the contact spot diameter L is smaller even than the coherence length λ_{coh} of the phonons while the distance between individual contact spots may exceed λ_{coh}. In this regime, quantization of conductance may occur [4].

Including all these aspects into the modeling of the tip-surface contact requires taking all relevant length scales into account. Some of the effects mentioned above, namely the weak coupling between tip and sample as well as the effect of roughness on the contact area, are experimentally observed in the form of pressure dependence of the interface resistance. This effect will increase thermal conductance with applied load (together with an increased nominal contact area of a curved tip pressed into a surface). The magnitude of the load-dependence can therefore be an indication of the transport regime.

Following experimental data on the pressure dependence of conductance across weak interfaces [6] gives us

978-1-4799-4789-8/14 $31.00 © 2014 IEEE   628

an order of magnitude of the interface conductance

$$g_{int}(p) = g_0 + g_1 p$$

with g_{int} the interface conductance in units of W/(m²K), g_0 around 4×10^7 W/(m²K), p the pressure at the interface and g_1 on the order of 10^{-2} W/(NK). In contrast, for systems that may involve quantized transport pressure dependencies on the order of 1 W/(NK) have been reported. At sufficiently high pressures the pressure-dependence of the thermal conductance reduces again [6], [4]

5. Simbad a tool for optimization

The software SIMBAD provides a generic simulation-based design tool for investigating the behaviour of complex modeled systems. A MATLAB link has been set between COMSOL and SIMBAD so that COMSOL models may be used as an input for a design under SIMBAD. It includes the definition of the optimization problem: the initial value of parameters, the parameter relative ranges, the objective features and the constraints for geometry and objective features. It serves to transmit current parameters between the two software packages.

Three SIMBAD toolboxes have been used. The design sensitivity and effects analysis toolbox is used to quantify the impact of design variable modifications on the design objective of interest. It allows the design space to be reduced to the subset of influential variables. The multi-objective performance optimization toolbox is used to obtain an approximation of the Pareto front for the different design objectives. It provides the analyst with a useful indicator on the trade-offs between the objectives of interest. Finally, the model validation and uncertainty quantification is used to quantify the impact of both aleatory and epistemic (lack of knowledge) uncertainties in the design variables and system environment on the design objectives and constraints.

6. Optimization

The first step in optimization consists in building a full parametrization of the probe geometry, see Fig 3. The geometric parameters are updated during the sensitivity analysis as well as the optimization loops. According to the applications of this probe, three objectives are set, the maximization of the maximum temperature $T_{max} = \max_{x \in \Omega} T(x)$, the minimization of the maximum vertical mechanical displacement $u_{max} = \max_{x \in \Omega} u_3(x)$ for a fixed voltage source and no body force load and the maximization of the mean value σ_{mean} of the sensor stress i.e. the von-Mise stress in the piezoresistive sensor for a prescribed tip displacement. The multi-objective optimization is carried with two different simulation conditions. To optimize T_{max} and u_{max}, only the voltage source is imposed at 0.12 V. Then, the optimized value of σ_{mean} is found by fixing a 1 μm tip displacement.

In the following discussion, we distinguish between the original design shown in Figure 1(b) whose initial

parameters are provided by the probe designers, and the nominal design shown in 3 whose initial parameters come from a preliminary optimization. The parameters of the original design are splitted into three groups of layer thicknesses, leg dimensions and tip gaps. The latter includes the triangle-shaped and T-shaped gaps near the tip.

In the first group, both T_{max} and u_{max} are increasing with the platinum layer thickness and decreasing with the silicon layer thickness, and the sensor stress σ_{mean} is increasing with the silicon layer thickness. In fact, the tip displacement is more sensitive to the platinum layer thickness than the tip temperature because the thermal expansion coefficient of the platinum is three times larger than that of the silicon layer.

In the second group, only the width of the heating circuit and the gap around the middle leg are influential on the objectives. Since the fixed voltage is applied on the heating circuit, a wider heating circuit means a higher heating power and a larger thermal expansion, and consequently a higher tip temperature and a larger tip displacement. A larger gap means softer supporting layer and consequently a larger tip displacement and a lower sensor stress.

In the third group, T_{max} and u_{max} are increasing with the widths of the tip gaps and σ_{mean} is not sensitive to the parameters in this group. This is because wide tip gaps implies concentrated heat distribution in the tip and locally soft supporting layer. The tip temperature is also sensitive to the cross section of the heater.

By a preliminary optimization, the original design has been simplified as shown in Figure 3. According to a new sensitivity analysis, the influential parameters namely the width $Wg2$ of the middle gap, the widths Whb and Whu of the heating tracks, the widths Wcb and Wcu of the sensing tracks, the width WJn of the narrow part of the middle leg, the bottom end position $Hfin$ of the heater which also corresponds to the heater length, and the height $Hh2$ of the middle part of the tip gap are selected as active variables.

The results of optimization for T_{max} and u_{max} are shown in Figure 4. The coordinates of each plotted point are $\left(\frac{u_{max}}{u_{max}^0}, \frac{T_{max}}{T_{max}^0}\right)$ where u_{max}^0 and T_{max}^0 are related to the nominal design. In this approach, all the points along the pareto front correspond to optimal designs but for different compromises. Table 1 reports the chosen design in Figure 4. Compared with the original design, improvements by 300% for T_{max} (41.6 K vs 10 K) and 90% for u_{max} (20.67 nm vs 200 nm) were achieved. The piezoresistive sensor stress σ_{mean} is not presented in the two-dimensional graph, however, it has been improved by 3.59%. Those results are significant but simulation time was long, i.e. 12 hours for 1022 samples, each sample requiring a simulation. The average number of FEM elements in each simulation is about 3200 and the average time is 45 s. The simulation

Free Variable	Initial	Optimal	Relative value (%)
$Wg2$	22	4.744	21.55
Whb	18	9.979	55.44
Whu	10	9.642	96.42
Wcb	10	1.44	14.4
Wcu	10	2.589	25.89
WJn	5	1.923	38.45
$Hfin$	10	19.74	197.4
$Hh2$	10	3.624	36.24

Table 1. Optimization report

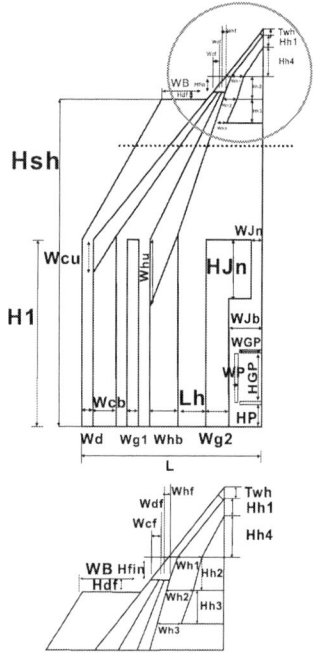

Figure 3. Parametrization of second design

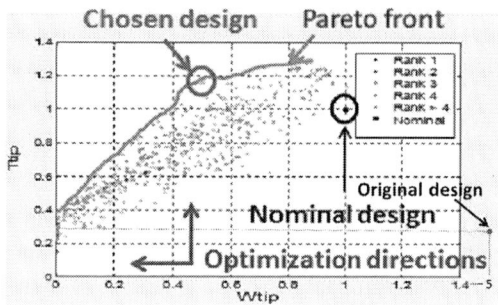

Figure 4. Two-objective optimization

time consists of about 3 s for the setting of COMSOL server environment, 2 s for building the geometry, 3 s for setting equations, 34 s for solving the equations and 3 s for extracting objective features. Considering the relative simplicity of the problem and the low number of active parameters, we consider the total optimization takes far too long so that to be routinely used. Since the number of finite elements cannot be reduced without degrading the accuracy, we have developed a thin plate model, presented in the last section, aimed at simplifying the model and thus to significantly reduce the computation time.

In addition, a robust analysis based on the Monte Carlo method has been applied to the chosen optimal design with 100 samples and a ±10% variation of active variables around the optimal value. Figure 5 shows the result where the optimal design is marked by a black square. The optimal design is considered as robust since it is located close to the center of the 95% confidence ellipse.

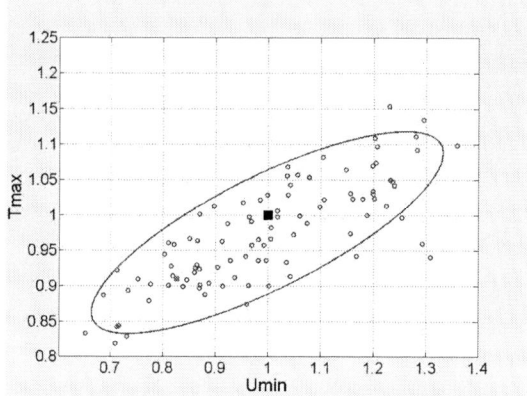

Figure 5. Scatter plot with 95% confidence ellipse

7. Thin plate model

The thin plate model introduced in this section is aimed at reducing the model 1 which is defined on the thin structure Ω. Its derivation is based on an asymptotic method related to the ratio ε between the order of the thickness and that of the width or of the length. Moreover, regarding to the small value of the thermal conductivity $k^{\mathrm{SiO2}} = 1.4$ compared to $k^{\mathrm{Pt}} = 71.6$ and to $k^{\mathrm{Si}} = 34$, we introduce the scaled coefficients $k^{\mathrm{Pt},0} = k^{\mathrm{Pt}}$, $k^{\mathrm{SiO2},0} = k^{\mathrm{SiO2}}/\varepsilon^2 = 140$ and $k^{\mathrm{Si},0} = k^{\mathrm{Si}}$, which are almost of the same order. The two-scale transform S used for this model is a linear transformation mapping a function v defined on the three-dimensional domain Ω into Sv a function defined on the product of the two-dimensional mid-plane ω of the plate with an interval J^1 which is a dilation of the thickness by the factor ε. The mid-plane ω is comprise of the projection ω_{Pt} of Ω_{Pt} on the $(x\text{-}y)$-plane and $\omega_S^0 = \omega\backslash\omega_{\mathrm{Pt}}$ the projection of the complementary part. The interval J^1 is also composed of three sub-intervals J^1_{Pt}, J^1_{SiO2} and J^1_{Si}, the scaled thickness of the three layers. We assume that the two-scale transform of each field v involved in the model admits an asymptotic expansion $Sv = \varepsilon^{m_0}v^0 + \varepsilon^{m_0+1}v^1$ where m_0 is determined by a preliminary mathematic analysis. For more detailed information on two-scale analysis, we refer to e.g. [9], but the detailed derivation of the present model differs by some points and will be published in a separate paper.

We introduce the temperature fields T_{Pt}, T_{SiO2} and T_{Si} to distinguish between the temperature fields in each layer.

The exponent m_0 is equal to -1 in the expansion of $S(u_3)$ and is equal to 0 in the expansion of the other terms.

In the two-scale model of the thermo-electro-mechanical thin probe, the temperatures T_{Pt}^0 and T_{Si}^0 in the platinum and the silicon layers, the voltage φ^0 and the mechanical displacement u_3^0 are constants in the thickness direction. Moreover, the temperature T_{SiO2}^0 of the SiO2 layer is equal to $\lambda(x_3^1 - b) + T_{\text{Si}}^0$ in ω_{Pt} with $\lambda = l^{-1}(T_{\text{Pt}}^0 - T_{\text{Si}}^0)$, l and b being the length of J_{SiO2}^1 and the bottom coordinate of J_{SiO2}^1, and $T_{\text{SiO2}}^0 = T_{\text{Si}}^0$ in the other part $\omega \backslash \omega_{\text{Pt}}$. Evidently, T_{Pt}^0 is defined in ω_{Pt} only and extended by zero in $\omega \backslash \omega_{\text{Pt}}$. The fields T_{Pt}^0, T_{Si}^0, φ^0 and u_3^0 are solution to the coupled two-dimensional partial differential equations,

$$
\begin{cases}
-k^{\text{Pt}}\Delta T_{\text{Pt}}^0 + r_{\text{SiO2}}k^{\text{SiO2}}\lambda = a\Delta\varphi^0 \text{ in } \omega_{\text{Pt}} \\
-\operatorname{div}(a\nabla\varphi^0) = 0 \text{ in } \omega_{\text{Pt}} \\
-k^{\text{Si}}\Delta T_{\text{Si}}^0 - r_{\text{Si}}^{-1}r_{\text{SiO2}}k^{\text{SiO2}}\lambda = 0 \text{ in } \omega_{\text{Pt}} \\
-k^{\text{Si}}\Delta T_{\text{Si}}^0 = 0 \text{ in } \omega \backslash \omega_{\text{Pt}} \\
-\partial_{\alpha\beta}^2(m_{\alpha\beta} + T_{\text{Pt}}^0 M_{\alpha\beta}^{\text{Pt}} + T_{\text{Si}}^0 M_{\alpha\beta}^{\text{Si}}) = \partial_\alpha q_\alpha + f_3^H \text{ in } \omega \\
m_{\alpha\beta} = -C_{\alpha\beta\theta\gamma}^H \partial_{x_\theta^0 x_\gamma^0}^2 u_3^0 \text{ in } \omega.
\end{cases}
$$

The temperature gradient λ in J_{SiO2}^1 plays the role of a negative heat source for the equation of T_{Pt}^0 and a positive heat source for this of T_{Si}^0. Denoting by $\oint_{J_X^1} f(x_3^1)\,dx_3^1$ the mean value over J_X^1, the parameters of the model are the volume ratio r_{SiO2} between Ω_{SiO2} and Ω, the volume ratio r_{Si} between Ω_{Si} and Ω, the electric conductivity $a = (1 + \alpha T_{\text{Pt}}^0)^{-1}a^{ref}$, the mean lateral body force $q_\alpha = \oint_{J^1} x_3^1 f_\alpha dx_3^1$ and the mean vertical body force $f_3^H = \oint_{J^1} f_3\,dx_3^1$. The piecewise constant elastic tensor is defined by $C_{\alpha\beta\theta\gamma}^H = \oint_{J^1}(C_{\alpha\beta\theta\gamma} + C_{\alpha\beta k3}(2 - \delta_{k3})L_{k3\theta\gamma}^M)(x_3^1)^2\,dx_3^1$ for $x^0 \in \omega_{\text{Pt}}$ and by $C_{\alpha\beta\theta\gamma}^H = \oint_{J_{\text{SiO2}}^1 \cup J_{\text{Si}}^1}(C_{\alpha\beta\theta\gamma} + C_{\alpha\beta k3}(2 - \delta_{k3})L_{k3\theta\gamma}^M)(x_3^1)^2\,dx_3^1$ for $x^0 \in \omega \backslash \omega_{\text{Pt}}$ where the coefficients $L_{k3\theta\gamma}^M$ are solution to $(2 - \delta_{k3})C_{i3k3}L_{k3\theta\gamma}^M = C_{i3\theta\gamma}$ for any θ, γ. The three matrices of thermal expansion coefficients are $M_{\alpha\beta}^{\text{Pt}} = \frac{|J_{\text{Pt}}^1|}{|J^1|}\oint_{J_{\text{Pt}}^1} x_3^1 M_{\alpha\beta}^h\,dx_3^1 + M^{\text{SiO2}}$, $M_{\alpha\beta}^{\text{Si}} = \frac{|J_{\text{Si}}^1|}{|J^1|}\oint_{J_{\text{Si}}^1} M_{\alpha\beta}^h x_3^1\,dx_3^1 - M^{\text{SiO2}}$ for $x^0 \in \omega_{\text{Pt}}$ and $M_{\alpha\beta}^{\text{Si}} = \oint_{J_{\text{SiO2}}^1 \cup J_{\text{Si}}^1} x_3^1 M_{\alpha\beta}^h\,dx_3^1$ for $x^0 \in \omega \backslash \omega_{\text{Pt}}$, where $M^{\text{SiO2}} = \frac{|J_{\text{SiO2}}^1|}{|J^1|}\oint_{J_{\text{SiO2}}^1} l^{-1}(x_3^1 - b)M_{\alpha\beta}^h x_3^1\,dx_3^1$ with $M_{\alpha\beta}^h = \delta_{\alpha\beta}M + 2C_{\alpha\beta\eta3}L_{\eta3}^H + C_{\alpha\beta33}L_{33}^H$ and the parameters $L_{\eta3}^H$ are solution to $(2 - \delta_{k3})C_{i3k3}L_{k3}^H = \delta_{i3}M_{i3}$.

The projection of the boundaries Γ^0, Γ_{01}^e and Γ_{02}^e on the $(x$-$y)$-plane are denoted by γ^0, γ_{01}^e and γ_{02}^e. The temperatures satisfy the boundary conditions $T_{\text{Pt}}^0 = T_{\text{Si}}^0 = 0$ on γ^0, the voltage source is imposed on γ_{01}^e and γ_{02}^e is grounded. The usual clamping conditions $u_3 = \partial_n u_3 = 0$ apply on γ^0 together with the free load boundary conditions $(m_{\alpha\beta} + T_{\text{Si}}^0 M^{\text{Si}})n_\beta n_\alpha = 0$ and $\partial_\tau((m_{\alpha\beta} + T_{\text{Si}}^0 M^{\text{Si}})n_\beta\tau_\alpha) + \partial_\beta(m_{\alpha\beta} + T_{\text{Si}}^0 M^{\text{Si}})n_\alpha = -q_\alpha n_\alpha$ on $\partial\omega \backslash \gamma^0$. Finally, the interfaces between ω_{Pt} and $\omega \backslash \omega_{\text{Pt}}$ are subjected to natural transmission conditions.

This model has not yet been implemented, but we observe that it is posed in a two-dimensional domain instead of a three-dimensional domain which implies a dramatic mesh reduction. Moreover, it involves one more temperature field but two less mechanical displacement fields. In total, we expect a significant simulation time reduction.

8. Conclusion

A novel SThm probe design has been presented with its functioning mode. Various aspects of its modeling, simulation and design have been reported: a tip-surface interaction model, a study of the design optimization, and a two-scale model of the thin probe which should shorten the simulation time.

Acknowledgements

This research effort is sponsored and funded by European FP7 NANOHEAT project and by the Labex ACTION program (contract ANR-11-LABX-01-01).

References

[1] Gerd Binnig, Calvin F Quate, and Ch Gerber. Atomic force microscope. *Physical review letters*, 56(9):930, 1986.

[2] Gerd Binnig and Heinrich Rohrer. Scanning tunneling microscope, August 10 1982. US Patent 4,343,993.

[3] RB Dinwiddie, RJ Pylkki, and PE West. Thermal conductivity contrast imaging with a scanning thermal microscope. *Thermal conductivity*, 22:668–668, 1993.

[4] B. Gotsmann and M. A. Lantz. Quantized thermal transport across contacts of rough surfaces. *Nat Mater*, 12(1):59–65, January 2013.

[5] B Gotsmann, MA Lanz, A Knoll, and U Dürig. Nanoscale thermal and mechanical interaction studies using heatable probes. *Nanotechnology, Volume 6: Nanoprobes*, Jan 2009.

[6] Wen-Pin Hsieh, Bin Chen, Jie Li, Pawel Keblinski, and David G Cahill. Pressure tuning of the thermal conductivity of the layered muscovite crystal. *Physical Review B*, 80(18):180302, 2009.

[7] P Janus, D Szmigiel, M Weisheit, G Wielgoszewski, Y Ritz, P Grabiec, M Hecker, T Gotszalk, P Sulecki, and E Zschech. Novel sthm nanoprobe for thermal properties investigation of micro-and nanoelectronic devices. *Microelectronic Engineering*, 87(5):1370–1374, 2010.

[8] G Jóźwiak, D Kopiec, P Zawieruch, T Gotszalk, P Janus, P Grabiec, and IW Rangelow. The spring constant calibration of the piezoresistive cantilever based biosensor. *Procedia Engineering*, 5:838–841, 2010.

[9] M. Lenczner and R. C. Smith. A two-scale model for an array of AFM's cantilever in the static case. *Mathematical and Computer Modelling*, 46(5-6):776–805, 2007.

[10] Yifei Mo, Kevin T. Turner, and Izabela Szlufarska. Friction laws at the nanoscale. *Nature*, 457(7233):1116–1119, February 2009.

[11] B N J Persson, A I Volokitin, and H Ueba. Phononic heat transfer across an interface: thermal boundary resistance. *Journal of Physics: Condensed Matter*, 23(4):045009, 2011.

[12] R. Prasher. Acoustic mismatch model for thermal contact resistance of van der Waals contacts. *Applied Physics Letters*, 94(4):041905, 2009.

[13] RS Prasher and PE Phelan. Microscopic and macroscopic thermal contact resistances of pressed mechanical contacts. *Journal of Applied Physics*, 100:063538, 2006.

[14] Li Shi and Arunava Majumdar. Thermal transport mechanisms at nanoscale point contacts. *Journal of Heat Transfer*, 124(2):329–337, July 2001.

[15] E. T. Swartz and R. O. Pohl. Thermal boundary resistance. *Rev. Mod. Phys.*, 61:605–668, Jul 1989.

[16] RF Szeloch, Teodor P Gotszalk, and P Janus. Scanning thermal microscopy in microsystem reliability analysis. *Microelectronics Reliability*, 42(9-11):1719–1722, 2002.

Combined Experimental- and FE-Studies on Sinter-Ag Behaviour and Effects on IGBT-Module Reliability

R. Dudek, R. Döring, P. Sommer, B. Seiler[1], K. Kreyssig[1], H. Walter[2], M. Becker[3], M. Günther[4]

Fraunhofer ENAS, Micro Materials Center Chemnitz
e-mail: rainer.dudek@enas.fraunhofer.de

[1] CWM GmbH, Chemnitz
[2] TU Berlin, Forschungsschwerpunkt „Technologien der Mikroperipherik"
[3] FH Kiel, Institut für Mechatronik
[4] Robert Bosch GmbH, CR/APJ3, Stuttgart

Abstract

For high temperature interconnection sintered silver can be used, however, it induces new demands on the thermo-mechanical design. That issue requires knowledge on the thermo-mechanical reliability of silver sintered devices, the subject of this paper. Material characteristics of the sinter layers are needed for simulation, which are addressed in the first part of the paper. Based on material properties of pure silver, for sintered silver with different porosities effective material characteristics have been derived by use of a micromechanical cell model. Shear loadings with in-situ deformation analyses have also been made to investigate sintered silver behavior. A complicated dependence on processing, temperature, and deformation rate is seen. Based on different effective constitutive models for the sintered interconnects, stress loadings are studied for a power module, an IGBT on DCB substrate, for passive and active thermal cycling. For the passive cycle complex interactions of the different layers of the stack are observed, which are not seen in a module with soft solder bonding. This result can be attributed to the missing decoupling by the soft soldering layer. Failure risks are evaluated by both conventional FEA and cohesive zone modeling. A quite different stress situation is depicted for active power cycling. The situation is even more complex and it is obvious from the simulations, that active power cycling can induce failure modes different from passive cycling.

1. Introduction

The increase in the allowable operating temperature represents an important prerequisite for increasing the efficiency of power modules and thus the energy efficiency of the electric drive and power control units. Silver sintering is one upcoming technology which meets several requirements in this field. Besides the development of the technology, questions arise concerning the thermo-mechanical reliability of the sinter-layers as well as the closely linked questions after their constitutive behavior in dependence on processing, time, and temperature. These issues are addressed in this paper.

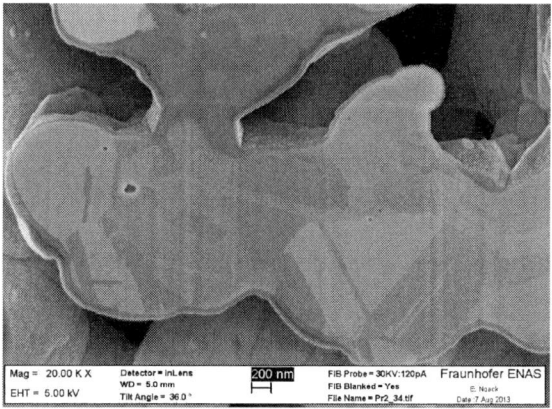

Fig. 1 FIB/SEM picture of a "sintering neck" in a pressure less sinter layer.

Various investigations have already shown that the behaviour of the Ag sinter material depend strongly on processing, in particular on the degree of porosity, e.g. [1], [2]. The latter depends on various influences, but in particular on the sintering pressure. A very heterogeneous microstructure occurs if no pressure is applied, as depicted in Fig. 1. However, as the microstructural spacings are low, effective material properties are applicable for simulation.

Mertens [1] studied the dependence of the effective elastic and plastic properties on processing conditions and temperature and found a strong dependence in both respects. Concerning material creep, significant differences from soft solders are expected for sintered Ag, as its homolgeous temperature is much higher than that of standard soft solders. However, creep effects at relatively low temperatures were already reported by Herboth et al. [3]. Although the status of the material characterization is far from beeing complete, additional results on sintered Ag have been achieved from different experiments as well as from a micromechanical cell model reported here.

Based on the effective Ag sinter layer characteristics, FE analyzes of an IGBT module aimed at findings to what extent the thermo-mechanical stress conditions change if a solder die attach is replaced by a sinter layer. As described above, the sintered Ag shows an elastic-plastic behavior without the strong creep of soft solder up to the high temperature range (150° C). Processing dependent change of the sinter-Ag material stiffness, in particular its decreasing ductility, the potential failure modes of an assembly made by sintering change compared to conventional soft solders. Besides low cycle fatigue of the layer itself, metallization fatigue but also brittle failure of die, DCB ceramics, or interfaces can occur. To account for the interface failure risk, cohesive zone modeling was included. By the latter, the effect of different material and geometric variants on the delamination risk of a metallization layer on the die can be characterized in a unique manner. It is finally shown that the kind of loading, that means passive thermal cycling versus active thermal cycling, greatly affects the module stress state and resulting failure risk.

2. Sintered silver as joining material

Meaningful simulation results for reliable failure prediction require material parameters, which were partly adopted from previous investigations, partly measured and partly derived from a micromechanical cell model.

2.1 Micromechanical cell model

Sintered layers have a degree of porosity, which depends on the sintering pressure. Analyses of the microstructure of sinter layers were made by FIB/SEM, as can be seen for a pressureless sinter layer in Fig. 2. Based on the analyses; micromechanical cell models were developed for theoretical modeling of effective properties of Ag sinter materials. Different void fractions were considered in those unit cell models by different arrangements of Ag-spheres similar to that used for transient liquid phase solder [4].

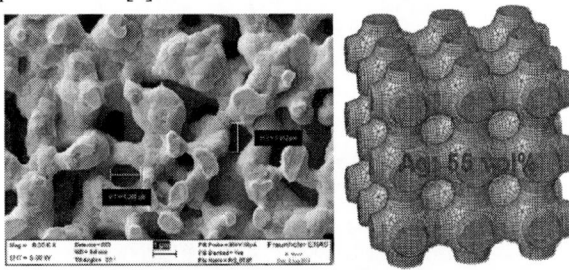

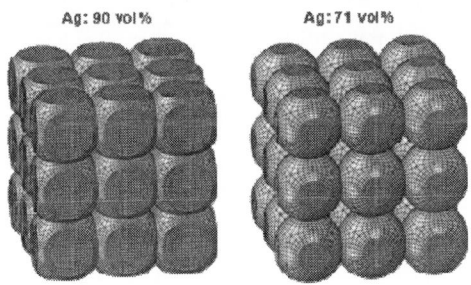

Fig. 2 FIB/SEM picture of a pressure less sinter layer and related FE- model (55vol% Ag) and other micromechanical cell models with sintered Ag particles packed differently dense.

Effective elastic-plastic properties were derived from the calculation results for different porosities as depicted in Fig. 3. For pure Ag an initial yield strenght of 160 MPa at RT was taken from the average of published data and own measurements. Conclusions concerning the effective material properties can be drawn with care, as various idealizations are used in the model. However, when compared to published measurements for sintered Ag, e.g. [1], [3], the calculated effective properties fit the measurements reasonably well. The T-dependence was also included in the elastic-plastic model in accordance with [1].

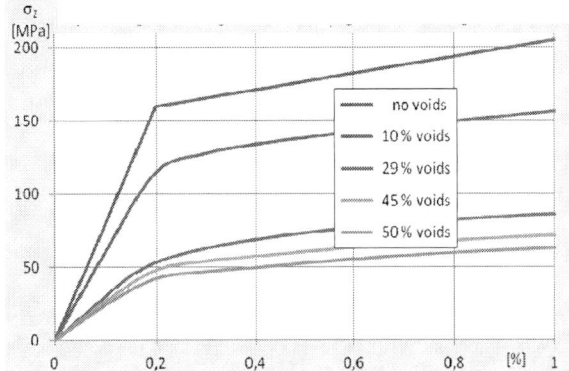

Fig. 3 Effective elastic-plastic models of the Ag-sinter layer with different void fractions.

Moreover, from simulations it was possible to understand the localized deformation mechanisms. The calculations show that the stress peaks are located at the contact areas between the particles, where the particles are sintered together, and the so-called sintering necks. Local plasticity at these locations is the major source of different effective plastic behavior dependent on the microstructure.

The stress distribution on the surface of the sinter particles are exemplarily depicted in Fig. 4. Shown is the stress component in the vertical direction, the direction of pull.

Fig. 4 Tensile stress distribution on the surface of the sintered particles.

It is clear that the stress peaks at the contact areas between the particles, regardless of their density. Plastic or creep strains show a concentration at the same location at the "sintering necks", as can be seen from Fig. 5.

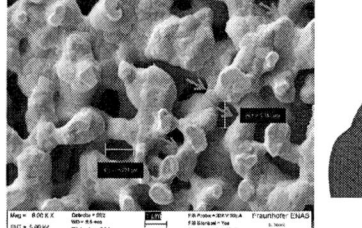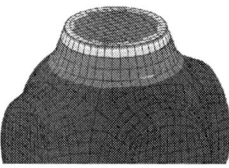

Fig. 5 "Sintering necks" as a location of inelastic strain concentrations.

Concerning material creep, differences from soft solders are expected for sintered Ag, as its homolgeous temperature is much higher than that of standard soft solders. For example, for SAC solders the homologeous temperatures are 0.47-0.95 in an operating temperature range from -40° C-200° C while they are 0.19-0.38 for sintered Ag. Hence, material creep which usually dominates the constitutive response for homologeous temperatures bigger than 0.5, is not expected for sintered Ag, what should make a sintered interconnect stiffer than a soft soldered one.

A trend of lower creep impact was also confirmed by recent material testing, however, Herboth et al. [3] have shown creep effects already at temperatures as low as 125°C. They propsed an additive strain rate decomposition in elastic, plastic and creep terms,

$$\dot{\varepsilon} = \dot{\varepsilon}_{el} + \dot{\varepsilon}_{pl} + \dot{\varepsilon}_{cr} \quad (1)$$

with a secondary creep rate:

$$\dot{\varepsilon}_{cr} = A \, [\sigma]^n \exp\left(\frac{-Q}{RT}\right) \quad (2)$$

The the measured creep rates were found to fit to a secondary creep law of Norton type (2) above a certain temperature and for low strain rates.

Additional measurements in the current project confirmed the occurrence of creep effects. Already for pure silver creep occurs at relatively low temperatures, as can be observed from Fig. 6. It is noted that for pure silver a certain threshold stress level seems to exist for the occurrence of creep, i.e. no creep strains were found for low stresses.

It is obvious from the figure that for sintered Ag creep tendencies become already visible at RT (measurements ongoing). This finding can be attributed to the high local stress at the sintereing necks, as seen from the cell model (Fig. 4), which leads to high effective creep rates of the material in connection with the power law dependence of the creep rate (2). However, as is also obvious from the figure, a secondary creep description is not capable of describing Ag creep for a broader loading scenario. In particular, for high strain rates as they occur in power

cycling, primary creep seems to play a dominant role. Further work is required in this area.

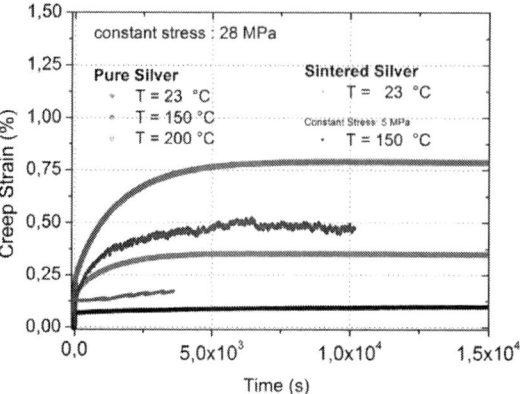

Fig. 6 Creep behavior of pure and sintered silver at elevated temperatures.

2.2 Sintered Ag deformation behavior in shear

To study the effective properties of the sinter layers, in situ deformation analyses were made for both thermal and mechanical shear deformations of the sintering layer in a chip on substrate configuration. In situ measurement of the shear strain distribution was made from the side view by use of optical correlation techniques, available in the microDAC software package [5], by help of a loading system which allows for in-situ deformation analysis.

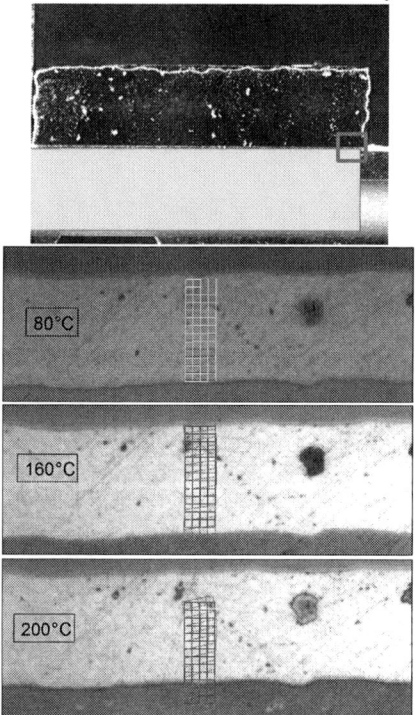

Fig. 7 In-situ Deformation Analysis for T- Loading up to 200 °C, Die on Cu-Leadframe with sintering layer sinter-pressure 10 MPa

978-1-4799-4789-8/14 $31.00 © 2014 IEEE

Thermally induced deformations were analysed at a 2.3x 2.3 mm² die on leadframe, sintered at 10 MPa with resultig layer thickness of 20 μm. The deformations in the layer close to the die outer edge are depicted in Fig. 7 for three temperatures as a deformed grid in relation of the undeformed grid at 30 °C. Obviously, the layer shear deformation is close to zero even at 200 °C, but only a tilting of the bi-material structure overlaid by a homogeneous thermal expansion occurs. This behaviour is in sharp contrast to the one seen with standard soldered dies, where the shear deformation dominates by far.

To study the properties of the sinter layers and their interfaces to metallizations, die shear tests were made, where the test specimens on DCB substrate moved against the shear chisel with different velocities down to 1 μm/s. Fig. 8 shows the different critical shear forces for a shear velocity 50 μm/s in dependence on sintering pressure. A nonlinear behaviour in the force-displacement curves are seen for the low pressure Ag, what points to plastic deformation.

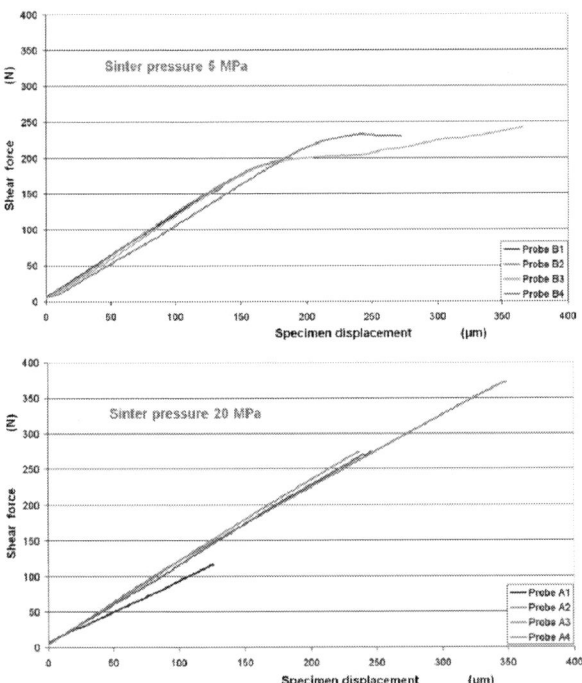

Fig. 8 Die shear tests force-displacement curves in dependence on the sinter pressure.

As shown in Fig. 9, the failure modes differ for the different shear tests: while the 5 MPa specimens failed in the interface to the substrate at lower forces, the 20 MPa specimens failed by die fracture. This type of brittle fracture explains also the huge scatter in critical shear forces to be observed from Fig. 8. It can be concluded that not only the stiffness of the sinteringe layer itself, but also the strenghts of the interfaces increase with sintering pressure.

Interesting results were also achieved concerning the shear-rate dependent strain distribution in the Ag layer and adjacent metallizations. For low shear rates 1 μm/s a relatively high and uniform shear strain was seen in the sinter layer, but for 10 μm/s the shear deformation moved to the DCB copper, as depicted in Fig. 10.

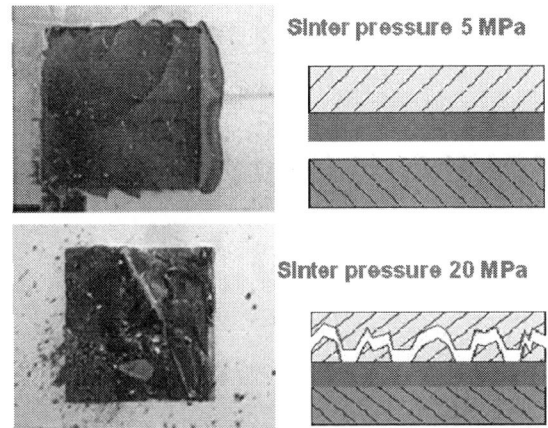

Fig. 9 Dominant failure modes from shear tests in dependence on the sinter pressure.

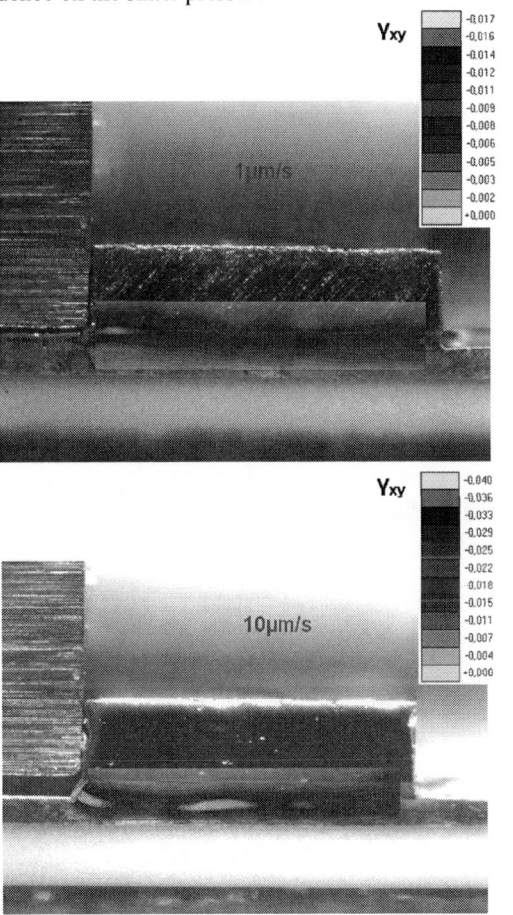

Fig. 10 Engineering shear strain in sinter layer and its vicinity, sinter pressure 10 MPa.

In both cases the high strains are mainly plastic deformation and the higher shear strain in DCB copper can be attributed to a lower or vanishing plastic deformation in the sintered Ag. That means, even sintered Ag with low porosity shows a rate dependent plastic behavior down to room temperature. Comparisons to numerical analyses of the tests are ongoing.

3. Stress analyses on a power module with sintered die bonds

Parametric FE-studies of a characteristic assembly, an IGBT on DCB substrate, were made to study the effect of Ag sinterered die bond on the stress in the assembly. Both measured constitutive data and the simulated homogenized response from the micromechanical model were considered in the analyses.

Plastic strain in the metallizations, stress in the die and an interface failure damage measure were used as failure indicators. To account for the interface failure risk, cohesive zone modeling was included. By the latter, the effect of different material and geometric variants on the delamination risk of a metallization layer on the die can be characterized in a unique manner, even if the actual adhesion properties may vary and are not yet fully known.

3.1 Modelling of the IGBT on DCB substrate

The characteristic assembly, an IGBT on DCB substrate on heatsink, was modelled in accordance with Fig. 11. Emphasis was laid on the inclusion of the different layers in the interconnection region. In particular, interface layers were modelled at the interfaces between the metalliazions, which could also be treated as cohesive layer applying the CZM. Care was taken to ensure a sufficient mesh refinement of these layers.

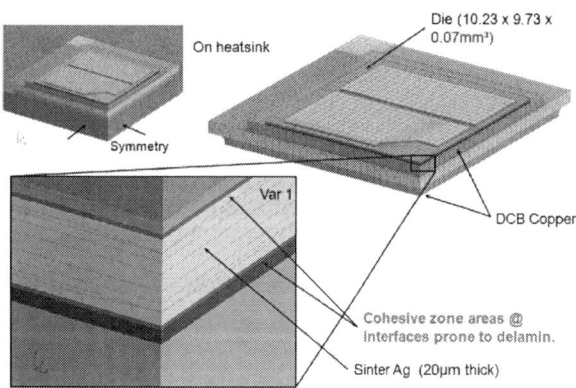

Fig. 11   Double symmetric FE model for IGBT soldered to DCB substrate, cutout with layer structure (bottom).

Different layer designs have been studied. Effects of different joining materials have also been investigated.

3.2 Passive thermal cycle

A temperature change or temperature cycle 30 min/30 min, 150° C/-65° C was simulated with initial stress free temperature of 200° C.

At first the passive cycle with IGBTs mounted on DCB substrate without heatsink is considered for different Ag sinter layer properties. Elastic-plastic models with kinematic hardening were used for high sinter pressure material (>90 vol% Ag, denoted as m1), pressureless sintered Ag (approx. 55 vol% Ag, m2) and the plastic/secondary creep model for the medium pressure material (approx. 85% Ag, m3) described above.

Parametric studies revealed that all metallic layers come in the plastic deformation range when cooled from stress free T to -65 °C and plastic property changes in one layer affecs frequently the stress in all the other parts of the assembly. This behaviour is generally different from the classical assembly with high lead solder die attach, which decouples the semiconductor from the substrate. Fig. 12 compares the plastic strain in the DCB copper with soft solder die attach versus sintered Ag (m1) die attach. Significant higher plastic straining occurs, i.e. DCB fatigue or delamination can become an issue in a sintered assembly. High plastic strains are also observed in the Al-metallization on top of the die, which concentrates at the outer pheripheral edges as depicted in Fig. 13.

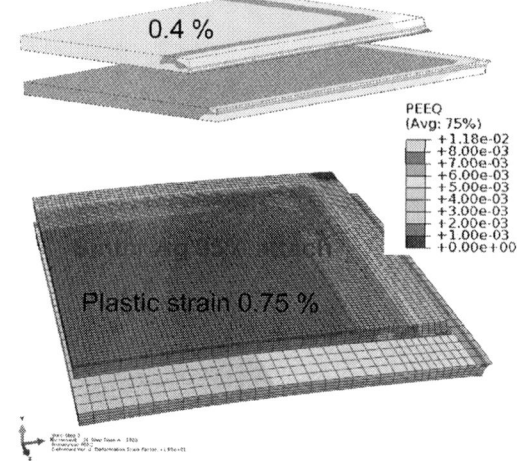

Fig. 12   Equivalent plastic strain in DCB copper after cooling.

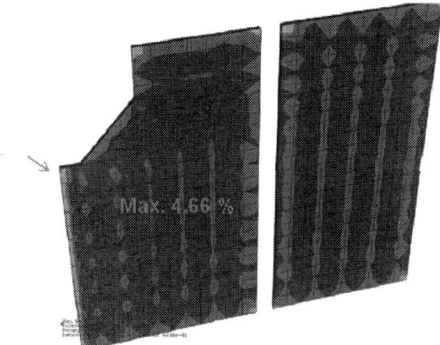

Fig. 13   Equivalent plastic strain in Al metallization on die top after cooling.

Multiple effects from the Ag porosity are also obvious because the effective layer stiffness decreases. If the inelastic strain in the joining layer itself is regarded, it depends on the effective plastic properties as well as on the constitutive model. With the effective plastic descriptions m1&m2, i.e. when the porosity increases to 45 %, an increase of 1.25 X in equivalent plastic straining is calculated in the Ag layer but a decreasing overall stress. Additionally, significant effects are seen from the die top metallization on the stresses in the die and the layers below. For example, maximum plastic straining in the sinter layer increases by 2X when a Cu top metallization is deposited, see Fig. 14.

The plastic strain always concentrates towards the outer peripheral edge of the layer, regardless of the constitutive model applied. Fig 15 shows both the equiv. plastic and creep strains, if secondary creep is included in the model (m3). The results for the two types of constitutive models (T-dependent elastic-plastic and the combined plastic-creep models) differ not significantly, neither in distribution nor in amplitude of the inelastic strains for the passive cycle considered.

It is noted that the greatest amount of the creep strain contribution to the total inelastic strain is accumulated in the high temperature part of the cooling steps, see Fig. 16. At the upper cyclic dwell of 150 °C the plastic contribution is zero, all inelastic strain is creep strain.

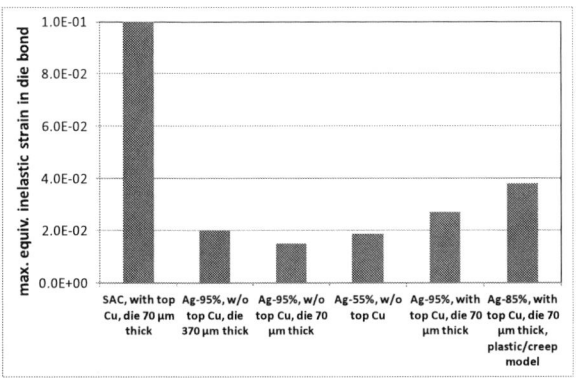

Fig. 14 Comparison of max. equivalent plastic (inelastic) strain in the sinter layer after cooling.

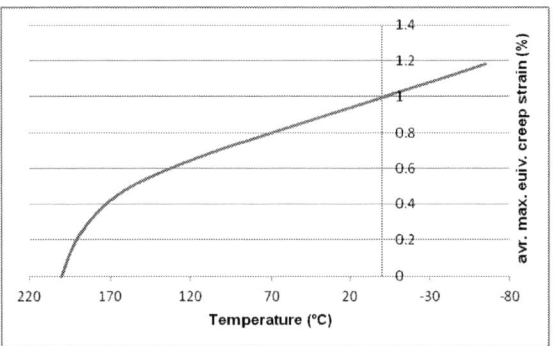

Fig. 16 Increase of averaged max. creep strain in the sinter layer during cooling to -65 ° C.

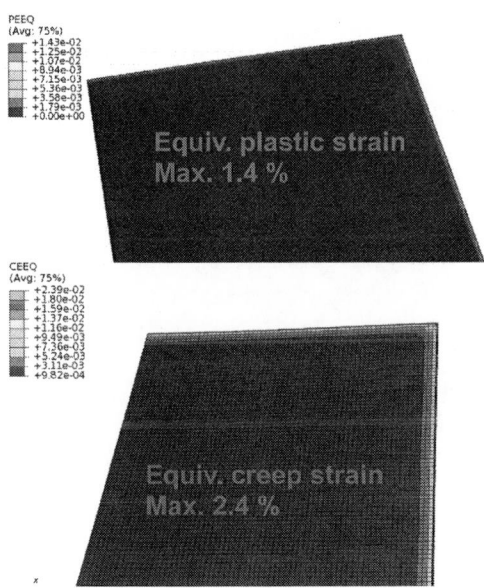

Fig. 15 Plastic and creep strain distribution pattern in the sinter layer after cooling, when material model m3 is applied.

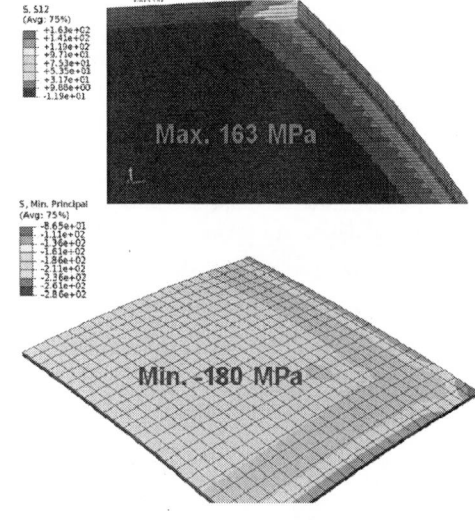

Fig. 17 Typical stress distributions (shear stress and minimum principal stress) in the Si (IGBT) without top metallization after cooling from the stress-free temperature to - 65 °C with dense sinter layer (m1).

The maximum principal stress in the die can be treated as failure indicator for die cracking or delamination. It can be observed from Fig. 17 that characteristic edge stress concentrations occur, which can cause brittle shear fracture. However, the stress level is not extremely high because of the plastic deformation of the metallization layers. It can be additionally observed from the figure that for thin dies the top surface comes under high compressive stresses, e.g. -180 MPa, which can affect the integrity of the structures situated there. If the die stress is treated as failure indication for either brittle die fracture or delamination at the interface, effects of different designs are depicted in Fig. 18. The evaluation of the failure mechanism die fracture/delamination is however limited when calculated by conventional means of finite element simulation based on continuum mechanics, because it is caused by the high stress concentrations at the free interface edges of the die and metallization layers, which depend on the mesh density.

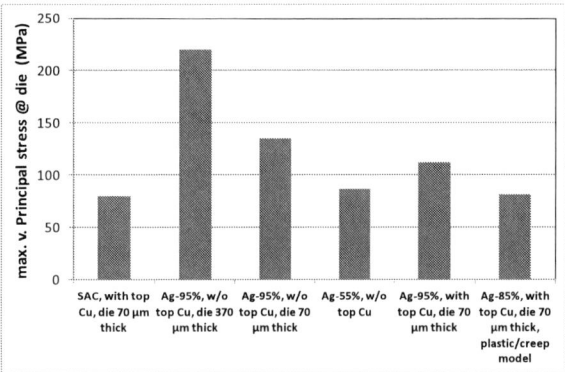

Fig. 18 Max. principal die stress after cooling to -65 ° C.

The cohesive zone method, a damage methodology better suited for damage prediction, was applied to the assembly. In particular it was proven previously that this methodology is very useful for the description of interface debonding [6]. An analogous linear traction-separation approach provided by ABAQUS 6.10 [7] is used again. In a first attempt, bad interface adhesion was assumed at the sinter layer top and bottom interfaces.

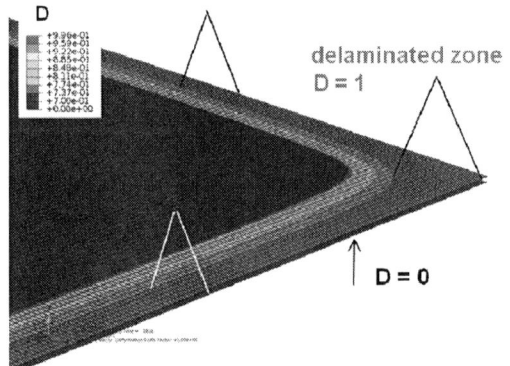

Fig. 19 Delamination results at interfaces Si-die/Ag-layer and Ag-layer/substrate using the damage parameter D after cooling to -65 ° (corner region, symmetric).

The resulting delamination after cooling to -65°C at the sinter layer top and bottom interfaces is depicted in Fig. 19 by the damage parameter D. Values of D close to 1 (red color) show the delaminated surface, while the rest of the colored surfaces can be interpreted as "potential" delamination area. Obviously, delamination occurs only at the sinter layer top surface to the die metallization, while for the same bad adhesion no delamination is seen on the bottom interface. As expected, interface delamination initiates from the corners. The delamination shows stable crack growth, i.e. the delamination leads to local stress relief and grows only gradually. With further stress, delamination from the outer periphery to the center of the IGBT is to be expected.

More studies are required with respect to different interface parameters and material constitutive data.

3.3 Active thermal cycle

Sequentially coupled transient temperature and mechanical analyses were performed to study the stress effects of active temperature loadings, i.e. the transient temperature fields were introduced in the mechanical analyses. The temperature load was set again at 200 °C stress free and intrinsic stress states at 30 °C were calculated and subsequently those during transient temperature cycles with $\Delta T_{junction}= 150$ K, 3 s power on/power off. Characteristic temperature fields are shown in Fig. 20 for different mounting conditions

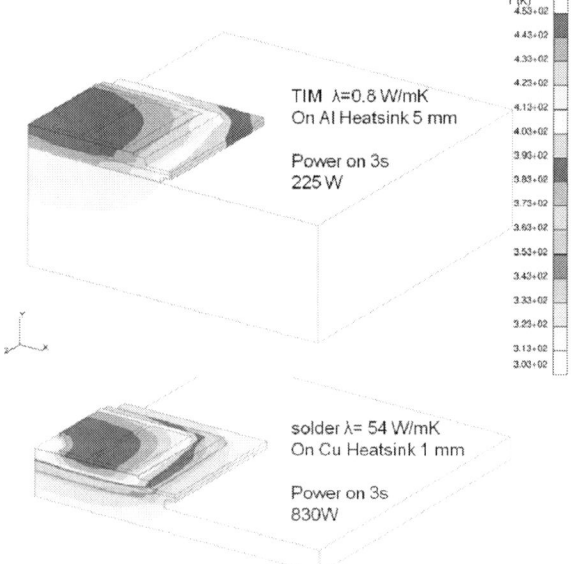

Fig. 20 Temperature fields for a $\Delta T_{junction}= 150$ K for different mountig conditions.

The deformation of the assembly soldered to a Cu heatsink overlaid by the v. Mises stress distribution pattern is depicted in Fig. 21. In the case shown assembly warpage is higher at elevated T, i.e. after cooling induced warpage increases from stress-free undeformed state, it increases further during reversed, but transient heating. The opposite occurs for the module on thick Al heatsink. Overall maximum stress tends to be higher at the power off state, because stress mainly results from the mismatch and increases with higher delta in reference to the stress-free state at 200°C.

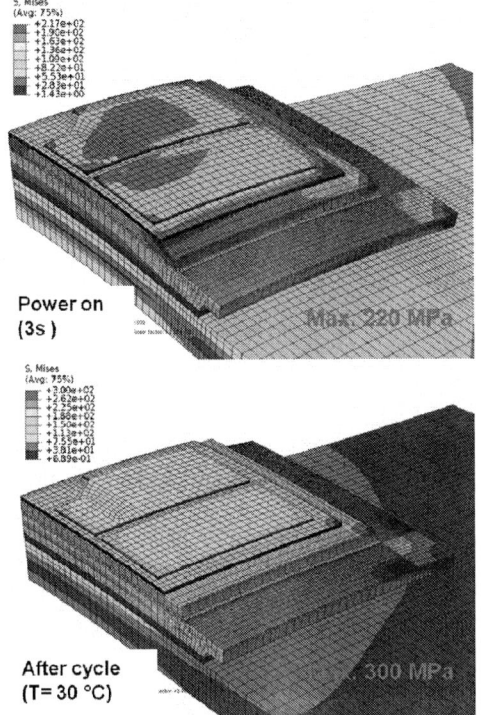

Fig. 21 Deformation (20x) and v. Mises stress (MPa) for mountig condition 2 (on Cu heatsink).

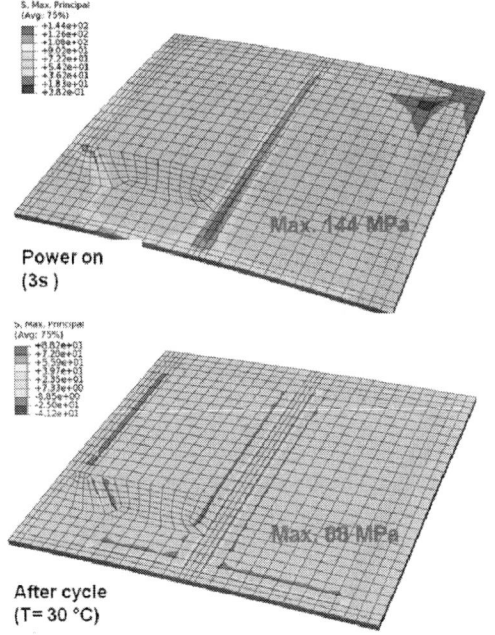

Fig. 22 Deformation (20x) and max. principal stress (MPa) @ IGBT for mountig condition 1 (on Al heatsink).

However, multiple plastic deformations and stress gradients due to temperature gradients as well as mounting affected warpage lead to a strong stress redistribution during cycling. For example, maximum principal stress in the die increases with active heating, as can be seen from Fig. 22 for the mounting condition 1 on Al heatsink. For mounting condition 2 on Cu heatsink the effect reverses.

If the cyclic inelastic strain in the Ag sinter layer is regarded it becomes very low or vanishes, dependent on the Ag material properties and other design parameters. A difference is also seen from different mounting conditions. Hence, the stress in the module from active cycling differs fundamentally from that due to passive cycling.

Even from these few results it is shown clearly that no simple rules apply considering thermo-mechanical stress of mounted popwer modules subjected to active loading. A strong interaction of various design- and material parameters occurs.

5. Summary

New challenges concerning the thermo-mechanical reliability of Ag sintered interconnects arise because of the change in material stiffness and decreasing ductility of the joining material. Furthermore, the constitutive behaviour of sintered Ag was shown to include complex dependencies from processing and temperature and includes creep and rate dependence.

Compared to a soldered assembly, stress decoupling in the power module stack is greatly reduced and much stronger interactions between the different layers of the assembly occur. The potential failure modes change compared to conventional soft solders.

Different failure modes have to be expected in dependence on the rigidity of the sintered Ag material, i.e. its processing dependent behavior, but also on dependece on loading being either passive ar active. Thus, no simple rule can be given for a save assembly design yet. FEA has shown a great potential for reliability prognostics of these assemblies.

Acknowledgements

The work was funded by the project ProPower. This funding by the German ministry BMBF was gratefully appreciated. The authors thank the entire project team for the good cooperation, but in particular the project leader, Prof. H.-J. Albrecht.

References

[1] Mertens, Chr. (2004): Niedertemperatur-Verbindungstechnik in der Leistungselktronik, Fortschrittsber. VDI , Reihe 21, Nr. 65, VDI Verlag, Düsseldorf, Germany

[2] R. Mroßko, H. Oppermann, B. Wunderle, B. Michel, "Reliability Analysis of Low Temperature Low Pressure Ag-Sinter Die Attach", NSTI-Nanotech 2011,Vol.2, Boston , p.149

[3] Thomas Herboth, Michael Guenther, Andreas Fix, Juergen Wilde, "Failure Mechanisms of Sintered Silver Interconnections for Power Electronic Applications", 2013 Electronic Components & Technology Conference, pp. 1621- 1627

[4] Dudek, R., et al. (2013), "Reliability Issues for High Temperature Interconnections Based on Transient Liquid Phase Soldering" Proceedings EuroSimE, Wroclaw, Poland, 2013

[5] *uniDAC Manual* (V. 5.1), CWM GmbH, Chemnitz, Feb. 2011

[6] Dudek, R.; Pufall, R.; Seiler, B.; Michel, B.: "Studies on the Reliability of Power Packages Based on Strength and Fracture Criteria" Proceedings ITherm 2012, San Diego

[7] SIMULIA (ABAQUS) Manuals (V. 6.10), Dassault Systemes Simula Corp., Providence, RI, USA, 2010

Fluid Damping in Compliant, Comb-Actuated Torsional Micromirrors

R. Mirzazadeh[1*], S. Mariani[1], A. Ghisi[1], M. De Fazio[2]

[1] Politecnico di Milano, Dipartimento di Ingegneria Civile e Ambientale
Piazza L. da Vinci 32, 20133 Milano, Italy
[2] STMicroelectronics, Advanced System Technology
Via C. Olivetti 2, 20041 Agrate Brianza, Italy
*Author to whom correspondence should be addressed:
E-Mail: ramin.mirzazadeh@polimi.it ; Tel.: +39-02-2399-4274; Fax: +39-02-2399-4300

Abstract

Fluid damping is studied for resonant torsional micromirrors, electrostatically actuated by comb fingers. A three-dimensional computational fluid dynamics (CFD) model of the air flow around the moving parts of the mirror is developed, coping with dynamic remeshing procedures to properly account for the large displacement setting required by the motion of the compliant structure. The time evolution of the damping torque contributions, due to shear at comb fingers and to drag over the surfaces of the micromirror plate, are computed. The relevant numerical estimation of the overall quality factor of the system is shown to compare well with available experimental results.

1. Introduction

Electrostatically, comb-actuated MEMS (micro-electro-mechanical system) mirrors are of particular interest for a broad range of light manipulation applications like flying spot projectors, imaging and telecommunication ones. They can meet high resolution, small size, low power consumption and high scanning speed requirements in such demanding context [1,2].

These resonating apparatuses are usually embedded in air at ambient pressure, mostly for two reasons. First, to prevent excessive dynamic amplification of the oscillations, which can lead to a structural failure of the compliant suspending springs. Second, to provide a smooth frequency response in the proximity of resonance, otherwise any small perturbation in the input signal frequency may lead to drastic system performance losses. The resulting interaction of the vibrating mechanical parts with the surrounding air obviously affects the dynamic energy dissipation of the system. In this work we specifically focus on this topic, providing a numerical frame tailored for large displacement amplitudes.

As the micromirrors are actuated through comb fingers, in addition to the drag damping linked to the moving masses inside the viscous fluid, two other dissipation sources have to be accounted for: squeeze film damping, and shear damping. The former dissipative mechanism takes place between two closely located plates, which have a relative motion toward each other, see e.g. [3]. The latter one takes instead place between two plates, still close to each other, having a relative sliding motion. At the microscale, Howe et al. [4] and

Zhang et al. [5] carried out experiments on MEMS devices and then suggested one dimensional Stokes and Couette models for the prediction of the relevant damping. Sudipto and Aluru [6] attacked the problem of computing the device dynamics and the damping in comb fingers by coupling two-dimensional compressible Navier-Stokes solutions with an electro-mechanical solver. Ye et al. [7] investigated the air damping by numerically solving the three-dimensional (3D) Stokes equations via a precorrected FFT boundary element method, while Frangi et al. [8] adopted a boundary element approach accounting for a corrected slip boundary condition induced by the surface traction term. Other studies, based on Navier-Stokes CFD models, have been conducted, e.g. in [9,10], but they are all limited to relatively small oscillations.

The other way around, in order to enhance the scanning and resolution characteristics of torsional micromirrors, large tilting angles, on the order of ±15° [11], have to be accounted for. In such cases, the solid boundaries confining the air flow change much their relative position during the oscillation, resulting in flow evolutions far more complex than those studied in the aforementioned research activities. It therefore emerges the necessity of approaching the problem through numerical methods.

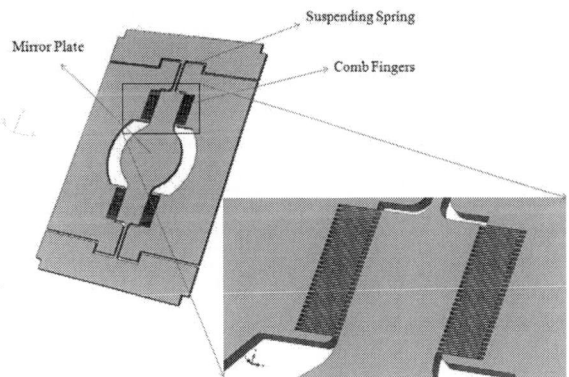

Figure 1: Schematic layout of the micromirror, and close-up of the comb fingers region

In this work, to capture the viscous dissipation mechanisms relevant to a compliant, resonant micromirror (see Figure 1), two different 3D models of the plate and of the gap between the comb fingers are built in the commercial finite element code ANSYS CFX [12]. As for the second model, the mentioned large tilting angle causes the air flow to continuously and smoothly switch between engaged and disengaged geometries, as the amplitude of the out-of-plane displacement at the tip of each finger exceeds the thickness of the actuation plate. Hence, dynamic mesh evolution, or remeshing methods have been adopted.

By deriving the damping torque evolution over a whole period, the quality factor corresponding to each dissipation mechanism and the overall device one have been obtained. The latter one has been compared to the experimental value, obtained through an optical test setup for image acquisition, which was specifically designed to get the micromirror angle response versus driving voltage. The said comparison testifies that the proposed numerical models well describe the dissipation mechanisms here accounted for.

2. Fluid flow modeling in the presence of a compliant resonant structure

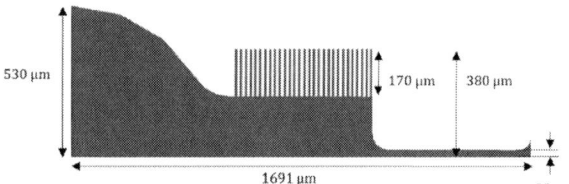

Figure 2: Micromirror dimensions

Parameter	Value
Mirror diameter	1060 μm
Spring length	579.5 μm
Spring width	44 μm
Finger length	170 μm
Finger width	6 μm
Finger span	760 μm
Finger gap	3 μm
Number of fingers	29
Thickness of layout	50 μm
Substrate depth	450 μm

Table 1: Micromirror design parameters

As discussed in the introduction, the compliant micromirror under study is embedded in air at atmospheric pressure and resonates torsionally at a frequency of around 20 kHz, with a maximum tilting angle of $\pm 12°$. The nominal dimensions of the moving parts of the device are gathered in Figure 2 and Table 1.

To understand the type of air flow to be modelled in (MEMS) applications like the current one, the Reynolds number $Re = \rho v L / \mu$ has to be first considered. Here, ρ is the air density at the considered pressure, v is the maximum value of fluid velocity, and L is a characteristic length or size of the domain wherein fluid flows. As $Re \ll 1$ at the micron length-scale and for the characteristic velocity v experienced during the torsional rotations at the desired frequency, the air flow turns out to be laminar, see [13]. Accounting for the working frequency of the system, even at the tip of the mirror plate, where the amplitude of velocity is maximum, the Mach number is in order of 0.01; so, compressibility effects can be neglected [14].

When moving toward micron length-scales, gas rarefaction also plays a role in the flow model. The nondimensional parameter catching this effect is the Knudsen number Kn, defined as the ratio between the mean free path λ of air molecules and a characteristic length L of the geometry. λ can be related to the physical state of the air through [15]:

$$\lambda = \frac{\Re T_K}{\sqrt{2}\pi p \mathfrak{D}^2} \tag{1}$$

where: $\Re = 1.380662 \cdot 10^{-23}$ J/K is the Boltzmann constant; T_K is the absolute temperature; $\mathfrak{D} = 3.66 \cdot 10^{-10}$ m is the averaged air molecule diameter; and p is the air pressure value. For the standard ambient temperature and pressure values, the mean free path of air molecules is obtained as $\lambda = 69$ nm [16]. Accounting for the gap between comb fingers as the narrowest segment characterizing the micromirror geometry, the Knudsen number can be evaluated as $Kn \approx 0.01$. Based on a classification given in [17], $Kn = 0.01$ is at the boundary between ordinary stick and slip flow regimes. This classification is based on empirical information, thus the mentioned threshold cannot be considered as an absolute one. Anyhow, we consider here the flow to be in the stick continuum regime, as slip along the boundaries is not expected to affect much the solution in the considered case.

To compute the effects of fluid damping over the micromirror in the aforementioned conditions, numerical models have been built in ANSYS CFX. The air is considered as an incompressible fluid, whose laminar flow is governed by the Navier-Stokes equations. To partially decouple the fluid-structure interaction problem, the micromirror has been considered as a moving rigid body, and only the fluid domain has been modeled. Due to the working frequency and large amplitude of the tilting angle under the foreseen working conditions, a stationary regime cannot be guaranteed during a single vibration cycle; so, transient analyses have been run coupled with a dynamic remeshing procedure. As for the remeshing needed to continuously update the space discretization and prevent excessive distortions, a workflow has been devised according to what shown in Figure 3.

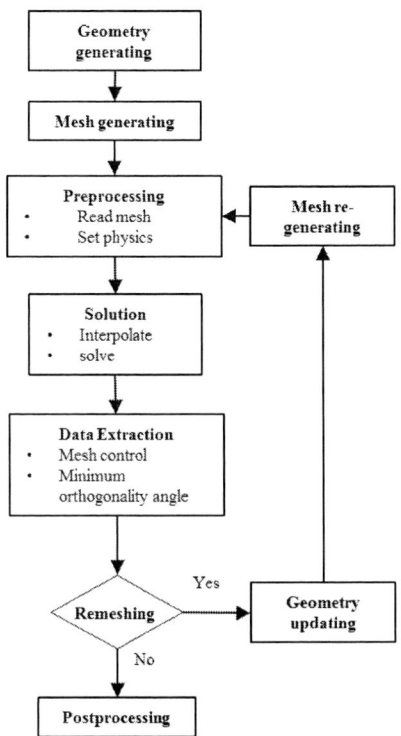

Figure 3: Simulation workflow with remeshing procedure

The mesh is automatically updated anytime its quality, at the beginning of a time step of the transient analysis, decreases below a critical threshold. The chosen index to control mesh distortion is the minimum orthogonality angle in the fluid domain. In fact, according to a rather standard rationale, angles between adjacent element faces or edges should be as close as possible to some optimal values, like e.g. 90° for quadrilateral face elements and 60° for triangular face elements. If the value of the chosen angle gets too far from the target one, ANSYS stops the solution process and triggers the remeshing step, wherein the geometry is updated, a new mesh is produced and all the previously generated field variables are transferred to the new discretization.

Because of the different length-scales characterizing the gap between the comb fingers and the overall geometry of the micromirror, CFD analyses are split into two sets: one considering the oscillating mirror plate, in the absence of comb arrays, to compute drag damping; the other one, accounting for the periodic geometry of the actuation plate and therefore including only a couple of comb fingers (see Figure 4), to compute shear damping. Obviously, the latter analysis is not appropriate to account for boundary effects, taking place at the ending comb fingers of the array. Due to the large number of comb fingers featured by the device, such end effects can be anyway neglected in the evaluation of the overall quality factor of the mirror.

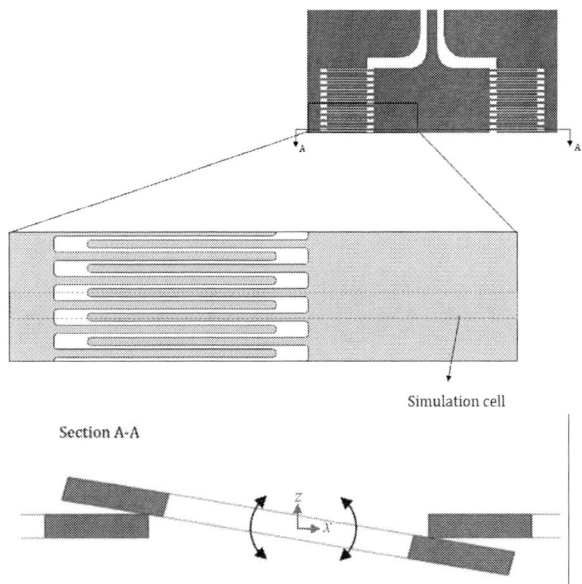

Figure 4: Modeled comb finger cell

3. Numerical results

Transient analyses have been carried out using ANSYS CFX, to compute the velocity and pressure fields during oscillations of the micromirror. Frequency and amplitude of harmonic oscillations have been set according to the working condition of the considered device. The (ambient) temperature and pressure of air have been respectively chosen as 25 °C and 1 atm. Since the system is assumed to behave under isothermal conditions, no thermal losses have been accounted for.

In all the analyses, each period of oscillation has been discretized with 1725 time steps, in order to guarantee a smooth mesh deformation and keep the required mesh quality, particularly at the tips of comb fingers.

Focusing first on the local analysis to compute the dissipation term due to comb fingers, it is worth mentioning that, during each oscillation, two distinct phases exist. A first one is when the comb finger is at rest and stator and rotor surfaces are perfectly facing each other: this phase is called the engaged one. A second one is when the torsion of the supporting springs and the relevant tilting angle are at their maximum: this phase is called the disengaged one. Through a continuous, smooth transition from engaged to disengaged conditions, the flow is not forming uniformly. As an example, the complex flow around the comb finger at the disengaged phase can be seen in Figure 5: here, the velocity field at the mid-plane of gap between comb fingers is shown.

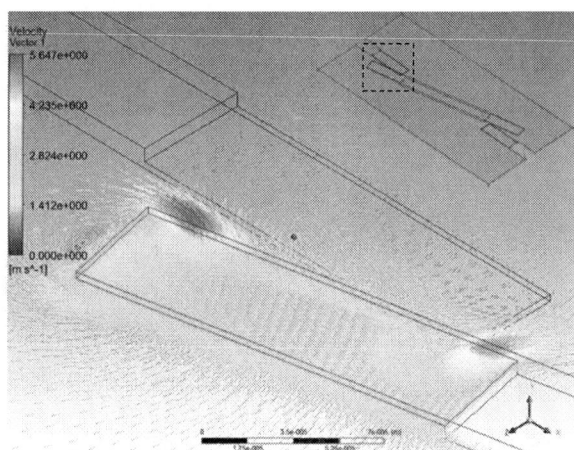

Figure 5: Velocity field between comb fingers at maximum tilting angle

Due to the evolution of the flow along a single oscillation period, characterized according to the preceding discussion by a non-stationary solution, a tracking of the dissipative force (or torque) over time is required. At any time, the damping torque linked to the air flow around the comb fingers is given by the integral of the moments of shear tractions about the mirror longitudinal axis, i.e. in vectorial form:

$$\bar{T}_\tau = \int_A \bar{r} \times \bar{\tau} \, dA \qquad (2)$$

where: A is the wetted surface; $\bar{r}$ is the position vector with respect to the rotation axis; $\bar{\tau} = \mu \frac{\partial \bar{u}}{\partial \bar{n}}$ is the shear traction vector acting over the comb finger because of the air; $\times$ stands for vector cross product; $\bar{u}$ is the velocity field close to the comb finger surface A; and $\bar{n}$ is the unit vector normal to the finger surface.

In Figure 6, the computed time evolution of the damping torque on a comb finger is reported. It can be noticed that the evolution of the damping torque over a single period is not harmonic. The two main features of this plot are: the maximum torque value corresponds to a completely engaged geometry; the asymmetry above and below the zero value baseline can be explained by the different flow formation in the engaged and disengaged phases. Referring to the latter feature, in the engaged phase, when the stator and rotor surfaces are closer and the air flow is highly confined between them, the viscous dissipation is the highest; the other way around, when the two surfaces are not perfectly aligned in the out-of-plane direction, like in the disengaged phase, the air flow is less confined and the relevant shear tractions is reduced in amplitude, thereby leading to smaller torque values. Close to peak values, when the maximum angular velocity occurs and mesh deformation takes place fast, the

remeshing process takes place several times; this is seen in the plot as small fluctuations of the torque value.

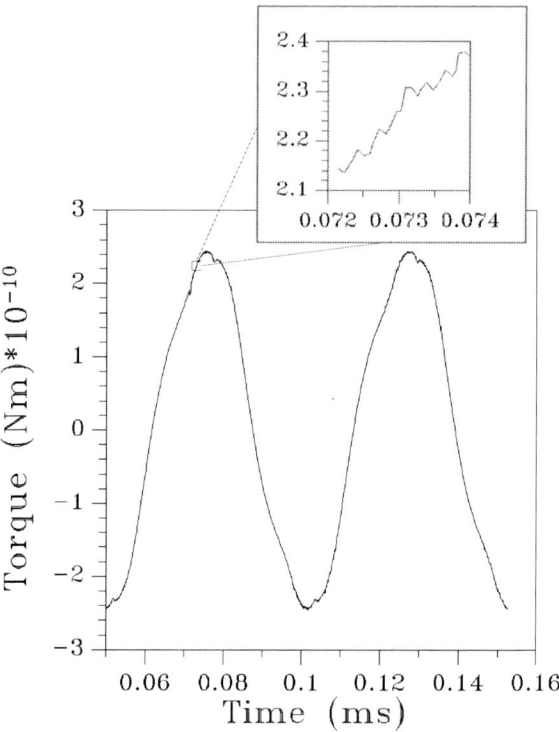

Figure 6: Time evolution of the imposed torque on a comb finger

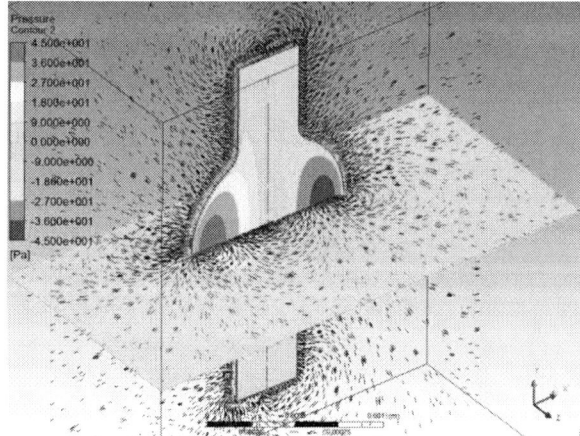

Figure 7: Pressure and velocity field around the micromirror plate

Moving now to the global analysis to compute the dissipation due to drag over the mirror surfaces, a similar procedure has been employed to simulate the relevant damping term. In this case, the air flow has been assumed less confined: all the surrounding surfaces have been set as open boundaries, except the bottom one, which represents the substrate and is located around 500 μm underneath the oscillating micromirror. As shown in

Figure 7, in this simulation the flow formation is simpler than the comb fingers one, the mesh quality indicator is never violated and remeshing is not needed.

As said, because of large surface of the mirror, the dominant dissipation mechanism is now drag, as caused by the fluid pressure over the plate surfaces. Accordingly, the damping torque is calculated as:

$$\bar{T}_p = \int_A \bar{r} \times p\,\bar{n}\,dA \tag{3}$$

where, p is the local pressure in the fluid. The time evolution of $\bar{T}_p$ is depicted in Figure 8 in terms of contributions provided by the top and bottom surfaces of the micromirror, along with the overall one. It can be seen that the difference between the torque values for the top and bottom surfaces, here related to squeeze film damping effects, is negligible owing to the relatively far substrate. The phase difference between the angular velocity and the damping torque, reported as well in Figure 8, implies that only the in-phase portion of the torque leads to energy dissipation.

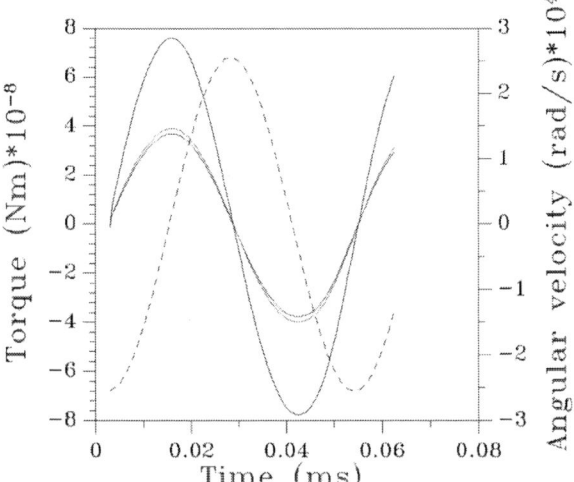

Figure 8: Time evolution of total torque acting on the mirror plate (continuous black line), torque acting on its top surface (blue line) and bottom surface (red line), and angular velocity of micromirror (dotted line)

The quality factor Q can now be computed on the basis of the ratio between the energy E_{st} stored in the system and the energy E_{loss} dissipated during one oscillation, namely:

$$Q = 2\pi \frac{E_{st}}{E_{loss}} \tag{4}$$

Since the natural frequencies and the corresponding vibration modes of the mechanical system are well separated, the dynamic behavior of the micromirror can be approximately considered as that of a single degree of freedom system. The relevant stored energy is thus given by:

$$E_{st} = \frac{1}{2}I_{yy}\dot{\theta}_{max}^2 = \frac{1}{2}I_{yy}\theta_{max}^2\omega^2 \tag{5}$$

where: I_{yy} is the mass moment of inertia of the micromirror about its axis; ω is the system angular frequency; $\dot{\theta}_{max}$ and θ_{max} are, respectively, the maximum amplitudes of the angular velocity and tilting angle.

The dissipated energy for each damping mechanism is given by the integral of the dissipated power $T\dot{\theta}$ over a period, i.e.:

$$E_{loss} = \int_{period} T\dot{\theta}\,dt \tag{6}$$

As far as the dissipation term provided by the comb fingers is concerned, the contribution obtained with the simulation cell has obviously to be multiplied by the number of fingers.

The two considered damping mechanisms linked to comb fingers and mirror plate surfaces correspond to the quality factors Q_{comb} and Q_{plate}, respectively. Accordingly, the overall quality factor Q is given by:

$$Q = \frac{1}{\dfrac{1}{Q_{comb}} + \dfrac{1}{Q_{plate}}} \tag{7}$$

Table 2 gathers the obtained energy loss and quality factor for each damping mechanism, and the total values. As expected, and according also to [18], the comb fingers provide the dominant contribution to the total energy dissipation.

	E_{loss} (J)	E_{st} (J)	Q
Comb fingers	1.78e-8	-	1074
Mirror plate	1.18e-8	-	1618
Total	2.96e-8	3.04e-6	645

Table 2: Energy values and corresponding quality factors

4. Experimental results

Due to the nonlinear terms characterizing the dynamics of the micromirror, its angle-vs-frequency response obtained in laboratory tests is different if the excitation frequency is swept up or down quasi-statically. The maximum oscillation amplitude is achieved when the drive frequency sweeps down from higher values as shown in Figure 9. When the frequency is decreased, the device shows a smooth amplification of the response up to its maximum value and then suffers a sudden drop. This behavior is the typical one of micromirrors adopting a nonlinear out-of-plane comb-drive actuation [19].

For such systems, the quality factor can be measured from the frequency response function through [20]:

$$Q \approx \frac{f_n}{\Delta f} \qquad (8)$$

where f_n is the resonance frequency and Δf is the frequency band at which the amplitude of oscillation is $1/\sqrt{2}$ times the resonant one. Using (8) and getting the relevant frequency values from the response function in Figure 9, the experimentally measured quality factor amounts to $Q = 623$.

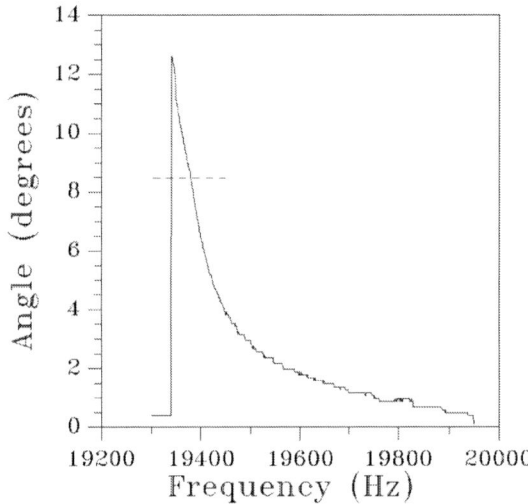

Figure 9: Experimentally measured frequency response of the micromirror

5. Conclusions

A numerical setup has been developed for predicting the quality factor of resonant micromirrors, featuring out-of-plane electrostatic actuation and undergoing large angles of tilting at high frequencies. The model accounts for the damping forces caused by shear at comb fingers and by drag over the mirror plate surfaces, both evolving nonlinearly during a cycle of oscillation. The simulation results have been validated through experimental data, in terms of overall system quality factor.

By means of the presented model, the fluid damping for arbitrary micromirror and comb finger geometries can be appropriately simulated, and relevant optimization procedures can be adopted for the future development of micromirrors characterized by enhanced dynamic properties.

Acknowledgments

The financial support provided by STMicroelectronics is gratefully acknowledged.

References

1. Van Kessel, P.F., Hornbeck, J., Meier, R.E., Douglass, M.R., "A MEMS-Based Projection Display," *Proceedings of the IEEE*, Vol. 86, Issue 8 (1998), pp. 1687-1704.
2. Ko, W., "Trends and Frontiers of MEMS," *Sensors and Actuators A: Physical*, Vol. 136 (2007), pp. 62-67.
3. Bao, M., Yang, H., "Squeeze Film Air Damping in MEMS," *Sensors and Actuators, A: Physical*, Vol. 136 (2007), pp. 3-27.
4. Cho, Y.H., Kwak, B.M., Pisano, A.P., Howe, R.T., "Viscous Energy Dissipation in Laterally Oscillating Planar Microstructures: a Theoretical and Experimental Study," *Proc IEEE Workshop on Microelectromechanical Systems,* Fort Lauderdale, USA, 1993, pp. 93–8.
5. Zhang, X., Tang, W., "Viscous Air Damping in Laterally Driven Microresonators," *Proc MEMS '94, IEEE Workshop on Micro Electro Mechanical Systems*, 1994, pp. 199-204.
6. De, S.K., Aluru, N.R., "Coupling of Hierarchical Fluid Models with Electrostatic and Mechanical Models for the Dynamic Analysis of MEMS," *Journal of Micromechanics and Microengineering*, Vol. 16 (2006), pp. 1705-1719.
7. Ding, J., Ye, W., "A Fast Integral Approach for Drag Force Calculation due to Oscillatory Slip Stokes Flows," *International Journal for Numerical Methods in Engineering*, Vol. 60 (2004), pp. 0235-1567.
8. Frangi, A, Spinola, G., Vigna, B., "On the Evaluation of Damping in MEMS in the Slip-Flow Regime," *International Journal for Numerical Methods in Engineering*, Vol. 68 (2006), pp.1031–1051.
9. Braghin, F., Leo, E., Resta, F., "The Damping in MEMS Inertial Sensors both at High and Low Pressure Levels" *Nonlinear Dynamics*, Vol. 54 (2008), pp. 79-92.
10. Sorger, A., Freitag, M., Shaporin, A., Mehner, J., "CFD Analysis of Viscous Losses in Complex Microsystems," *Proc Systems, Signals and Devices (SSD)*, 2012, pp. 1-4.
11. Grahman, J., Conrad, H., Sandner, T., Klose, T., Schenk, H., "Integrated Position Sensing for 2D Microscanning Mirrors Using the SOI-Device Layer as the Piezoresistive Mechanical-Elastic Transformer," *Proc SPIE 7208 MOEMS and Miniaturized Systems VIII*, San Jose, CA, January. 2009, pp. 720808-10.
12. ANSYS Reference guide, release 13.
13. Reynolds, O., "An Experimental Investigation of the Circumstances Which Determine Whether the Motion of Water Shall be Direct or Sinuous, and the Law of Resistance in Parallel Channels," *Philosophical Transactions of the Royal Society of London*, Vol. 174 (1883), pp. 935-982.

14. Graebel, W.P., _Engineering Fluid Mechanics_, Taylor & Francis (New York, 2011), p. 16.

15. Lide, D.L., _CRC Handbook of Chemistry and Physics_, CRC Press (Boca Raton, Florida, 2005), Internet Version, Sec. 6, p. 44.

16. Jenning, S., "The Mean Free Path in Air", _Journal of Aerosol Science_, Vol. 19 (1988), pp. 159-166.

17. Karniadakis, G.E., Beskok, A., Aluru, N., _Micro Flows and Nanoflows, Fundamentals and Simulation_, Springer (New York, 2002), p. 19.

18. Ye, W., Wang, X., Hemmert, W., Freeman, D., White, J., "Air Damping in Laterally Oscillating Microresonators: A Numerical and Experimental Study," _Journal of Microelectromechanical Systems_, Vol. 12 (2003), pp. 557–566.

19. Ataman, C., Urey, U., "Modelling and Characterization of Comb-Actuated Resonant Microscanners," _Journal of Micromechanics and Microengineering_, Vol. 16 (2006), pp. 9-16.

20. Rao, S.S., _Mechanical Vibrations_, Addison-Wesley (Reading, 1995), p. 204.

New equivalent stress describes the dicing caused anisotropic breaking strength of silicon dies

Matthias Steiert, Jürgen Wilde
Laboratory for Assembly and Packaging Technology,
Department of Microsystems Engineering (IMTEK), University of Freiburg,
Georges-Koehler-Allee 103, 79110 Freiburg, Germany
matthias.steiert@imtek.uni-freiburg.de

Abstract

Due to increased demands, reliability has become an important part of the research and development on electronics, in the last years. An important reliability factor is failure due to chip fracture. In several scientific papers three types of chip fracture are described: vertical cracks, horizontal cracks and mixed cracks. Till now, all approaches to explain the relations of these crack types have based on the deterministic fracture mechanics. But especially for statistical process control, it is better to use a probabilistic fracture mechanic based on the Weibull distribution.

In the present work damages caused by the dicing were analyzed by means of SEM-pictures. A special type of crack was found. Such cracks have a preferred orientation which is parallel to the dicing feet direction. The influence of those cracks on the breaking strength of silicon chips was investigated. The results showed a kind of dicing caused anisotropic breaking strength. This means the fracture relevant die strength depends on the local orientation of the stresses which acts on the chips. To detail this phenomenon a new equivalent stress was develop, which offers to directly compare stresses by means of theirs impact to fracture.

1. Introduction

New industrial guidelines e.g. from automotive industry pursue the 0 ppm strategy for electronic components, which means a failure probability below 0.1 ppm. Thus, the reliability of electronics has become an important part of research and development. Hence, also rare failure mechanisms like chip fracture, have become more and more of interest [1-3].

Chip fracture is essentially determined by two primary parameters: the first one is the stress, which acts on the chips during their life time and the second one is the breaking strength of the chips. Chip fracture occurs when the loaded stress exceeds the breaking strength of the chips.

Many investigations were performed in order to find out to which kind and level of stresses chips are exposed in electronic components [4-7]. An overview of previous investigations points out that typical die stresses are in the range of ±50 MPa up to ±200 MPa. It was also shown that stress values strongly scatter during the life cycle of electronics [4].

Investigations focused on the die strength showed characteristic breaking strengths in the range of 200 MPa up to 600 MPa depending on the used dicing technology [8, 9]. Moreover, a scattering of the chips breaking strength within a batch was shown. Both, the scattering of the stresses and the scattering of the breaking strength of the chips, leads in relevant risk of chip fracture as it is visible in Fig. 1.

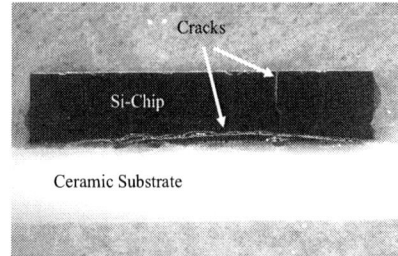

Fig. 1: Chip fractured after the mounting process with a vertical and a horizontal crack

Fig. 2: Fracture types due to the mounting process [1], vertical crack (left), horizontal crack (middle) and mixed crack (right)

Several publications [1, 10, 11] describe three types of chip fracture, a vertical crack and a horizontal crack as well as a mixed crack which peels off the chip corners (Fig. 2). It was shown that the horizontal and the mixed crack type are most common. Till now, all approaches to explain the relations behind the three crack types and there occurrence have based on the deterministic fracture mechanics. From this point of view and with this theory the propagation of an explicit flaw with a divined location and orientation in the material of one chip is analyzed. This offers to explain the conditions of chip fracture and it is possible to describe the basics of the three fracture types. But with the deterministic fracture mechanics it is only possible to describe exactly one case confined to one flaw and divined loading conditions.

This is problematic. Especially in the industrial process control it is necessary to characterize the process capability by means of statistical parameters which implicates all manufactured units. Thus, statistical parameters are needed which represents the breaking strength of a batch of chips and which also offers a prediction of the risk of chip fracture in an application. For this purpose the probabilistic fracture mechanics based on the Weibull distribution are used often. In this conception firstly the breaking strength of a few chips out of a production charge is measured by means of the different bending tests. Afterward a statistical evaluation with the Weibull distribution is performed and the characteristic breaking strength as well as the Weibull modulus can be received as statistical parameters which describe the breaking strength of the charge.

Previously, it was assumed that the breaking strength which has been measured with the bending tests gives a global value of a chip. In this case the area with the highest risk of fracture directly correlates with the area where the highest principal normal stress acts on the chips. But in the most cases this considerations leads in the assumption that the vertical crack should be the most common, because mostly the highest principal normal stress acts in the middle of the upper chip edges. This is not in accordance with the industrial reality where the horizontal and the mixed crack type are the most common.

We found that the assumption that the breaking strength measured with the bending tests gives a global value of a chip was not correct. The measurable breaking strength depends on the direction and the location of the measuring stresses. We call this phenomenon dicing caused anisotropic breaking strength. Due to an improvement of the probabilistic fracture mechanics by the introduction of a new equivalent stress it was possible to explain the causes of each fracture type.

2. Breaking Strength measurement

The three- and four-point bending tests are common methods to measure the fracture strength of brittle materials. In the present investigation they were used to measure the breaking strength of the chips.

As shown in Fig. 3, in both setups the sample is placed on two parallel supporter rollers. The mechanical load is deepened by other rollers at the up side of the sample. In case of three-point-bending, only one loading roller is used which is placed in the central position to the supporter rollers. In the four-point-bending, two loading rollers are used. Here, the distance between the supporter rollers and the upside loading rollers is a quarter of the distance which the supporter rollers have.

In bending the top surface is in a compressive stress state whereas the bottom surface is in a tensile stress state. As visible in Fig. 3, the tensile stress profile at the bottom surface in three- and four-point-bending is not the same. In the three-point-bending, there is a linear increase of stress from the first supporter roller to the loading roller and a linear decrease of stress from the loading roller to the second support roller. In four-point-bending, there are three different stress regions. The first region between the first supporter roller and the first loading roller is characterized by a linear stress ramp, followed from the second region with a constant stress state and the third region with a linear stress drop down.

In order to analyze the breaking strength of a sample, the maximum stress, that fractured the sample, has to be known. For three-point-bending, the maximum stress is located directly at the position of the loading roller and for the four-point-bending it occurs in the second stress region. By using the linear beam theory the maximum stress can be calculated with Eq. 8 in case of three-point bending and with Eq. 9 in case of four-point-bending [12]. Here, l_2 is the outer load span, l_1 the inner load span, b is the chip width, h is the thickness of the chip and F the applied force which is measure and stored during the measurement by the bending machine.

$$\sigma_{max} = \frac{3 \cdot F \cdot l_2}{2 \cdot b \cdot h^2} \tag{8}$$

$$\sigma_{max} = \frac{3 \cdot F}{2 \cdot b \cdot h^2}(l_2 - l_1) \tag{9}$$

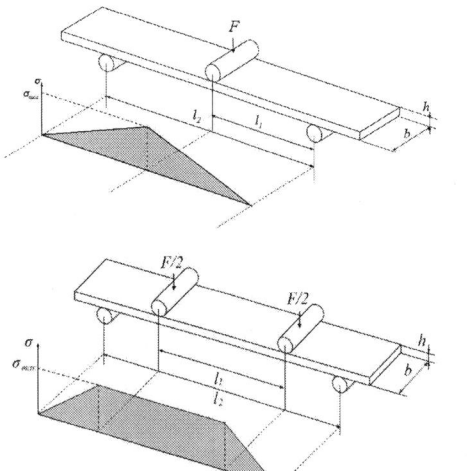

Fig. 3: Schematic of a three- and four- point-bending test and the induced bending stress along the sample axis

3. Statistical Data Evaluation

Basis of the data evaluation are the Weibull distribution Eq. 1 and the Weibull integral Eq. 2. The Weibull distribution characterises the overall correlation between the probability of failure and the stress that acts in a material. Herein σ_0 is the scaling parameter and m is the Weibull modulus.

$$F_A(\sigma) = 1 - \exp\left(-\frac{1}{A_0}\int_A \left(\frac{\sigma(A)}{\sigma_0}\right)^m dA\right) = 1 - \exp\left(-\frac{A_E}{A_0}\left(\frac{\sigma}{\sigma_0}\right)^m\right) \tag{1}$$

$$R = \frac{1}{A_0}\int_A \left(\frac{\sigma(A)}{\sigma_0}\right)^m dA = \frac{A_E}{A_0}\left(\frac{\sigma}{\sigma_0}\right)^m := \left(\frac{\sigma}{\sigma_\theta}\right)^m \tag{2}$$

The parameter A represents an area of the chips side wall. Mostly the volume is used instead of the side wall. But as it was shown in earlier publications [8] the location of the flaws, which influence the breaking strength of the chips, is decisive for the choice of this parameter. Dicing flaws are located at the chips side walls and thus, the Weibull integral Eq. 2 with the stress σ(A) of a certain loading case has to be solved for the side wall area. The solution of the Weibull integral gain accesses to the effective loaded side wall area A_E. A_0 is a normalisation parameter and a unit of measurement. Mostly it is given in 1 mm^2.

4. Location and orientation of Dicing Defects

First, to get an over view, six different dicing processes were compared by means of induced damages. The analyzed methods were diamond-scribing, blade-dicing, stealth-dicing, water-jet-laser-dicing, deep-reactive-ion-etching and thermal-laser-separation.

Depending on the working principle of a dicing method, the surfaces of the wafers and the edges of the chips are affected by mechanical, thermal or chemical loads during the dicing. This is why each dicing technology leads to specific damage types and characteristic strength distributions of the dies. In Fig. 4 the chip side walls are shown after dicing with the different techniques.

Fig. 4: SEM pictures of flaws and damages at the chips side walls induced due to the different dicing techniques [8]

By diamond-scribing and blade-dicing, mechanical effects are causative to the damages. They lead to small cracks at the chip edges, so-called chipping. By water-jet-laser-dicing and stealth-dicing, the silicon is heated up to temperatures above the melting point. After cooling down, polycrystalline silicon covers the side walls and internal stresses as well as micro-cracks are induced. A special case is the Thermal-Laser-Separation (TLS). It is a complex method which is based on thermo-mechanical stresses induced due to

heating up and rapidly cooling down the silicon. For the TLS first initializing-cracks are induced at the wafer surfaces using a diamond-tip. This is done at any starting point of the later dicing trenches. Due to the thermo-mechanical stresses an initializing crack starts to grow. By controlling the stresses the growing direction can be predetermined along the desired dicing trenches. Herein a problem is that the crack front jumps between crystallographic planes and this leads to sharp edges at the side walls which can be seen in Fig. 4.

The pictures in Fig. 4 give also an idea about the problem with the assumption of a die strength which is valid for the whole chip. It is visible that e.g. the chipping appears differently between the front side edges and the back side edges after the diamond scribing. The same is after the water-jet-laser-dicing and the DRIE with different levels of damages at front and back side. Additionally, layers of amorphous silicon are generated due to the stealth-dicing. These areas are strongly damaged, but the chip edges are quite good. These are the first indications of different strength depending on the area.

More detailed analyses of defects after the blade-dicing were done. The pictures in Fig. 5 show typical damages and cracks due to the blade-dicing. Additionally to the different sizes of the chipping at front and back side, micro-cracks located at the chip side-walls and side-edges were found. Such cracks are not directly observable, because after the dicing the side-walls are covered with layers of amorphous silicon. Hence, the chips were etched slightly to remove these cover layers.

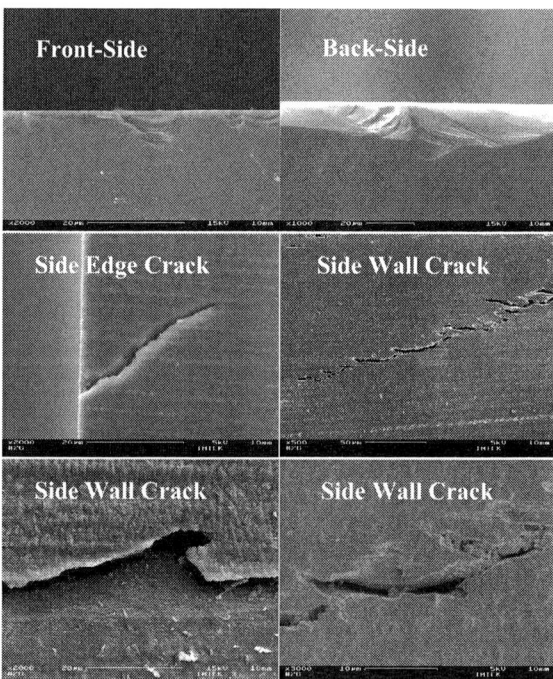

Fig. 5: SEM pictures of flaws and damages after the blade-dicing

Moreover, it was found that the shown micro-cracks are all orientated in parallel direction to the dicing feed direction and the chip edges (in Fig. 6 crack type 2, 3 and 4). We never observed cracks which are orientated vertical to the feed direction (type 1). In important fact is that crack type 3 and 4 only can be opened by stresses which act vertical to their orientation. But till now, breaking strength measurements have been done with stresses parallel to the chip edges which means also parallel to the crack orientation of type 3 and 4. Thus, those cracks never were implicated in the measurement results. In conclusion it can be said: it is very likely that the measurable breaking strength depends on the area which is loaded and additionally on the direction of the stresses.

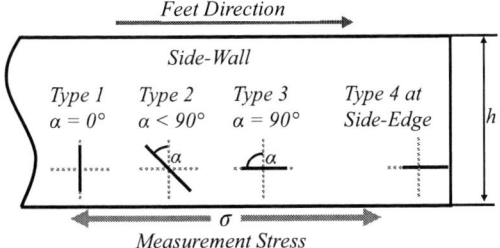

Fig. 6: Schematic of defect orientation and measurement stresses

5. Samples and Tools for Measurements with Different Stress Orientations

In order to minimize all other influences, which have nothing to do with the dicing like the wafer grinding, on the measured strength, all samples were based on the same starting material. We used standard prime double side polished (100)-silicon wafers with a size of 4 inch which all were from the same batch. In dependence on established test standards [12], a relatively large specimen size of 20 mm to 4 mm was used for measurements with stresses parallel to the dicing feet direction.

Fig. 7: Relation during measurements parallel to the feet direction using the example of stealth-dicing

For those parallel measurements a four-point bending test was used. A specimen insert in the bending test is shown in Fig. 7. As written due to the four-point bending test the bottom surface is in a tensile stress state. Hence, the measured breaking strength only applies to this chip side. Though this, it is possible to differentiate the breaking strength, which depending on defects close to the front or back side edge of a chip. For this purpose the desired chip side is inserted downwards in the bending test. The relations of the feet direction, the dicing damages and the measurement stress is shown in Fig. 8.

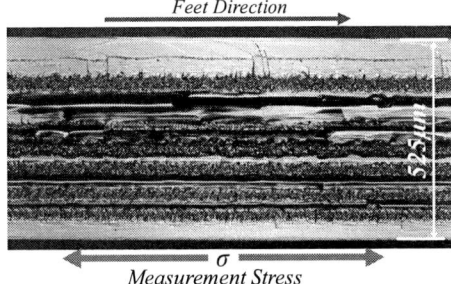

Fig. 8: Relation during measurements parallel to the feet direction using the example of stealth-dicing

For measurements with a stress orientation vertical to the dicing feet direction special samples were prepared. First the whole side wall was separated from the rest of the chips. The received strip was separated again. Specimens were received with a size of 525 μm (the thickness of the wafer) to 200 μm x 180 μm. The whole procedure is shown in Fig. 9 and some examples are shown in Fig. 10.

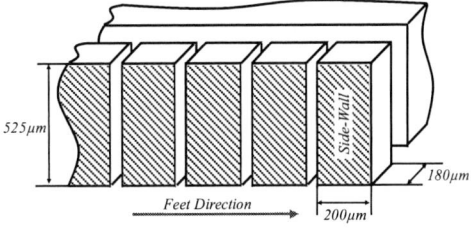

Fig. 9: Preparation of samples for measurements with a stress in vertical direction to the chip edges

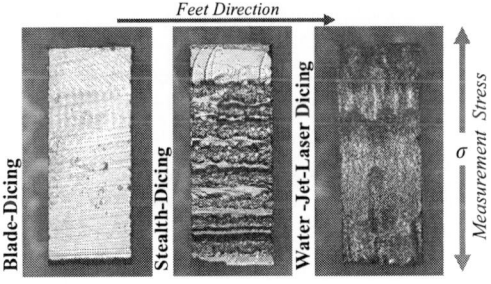

Fig. 10: Samples prepared for measurement

In order to measure such small sample sizes, a miniaturized three-point-bending test was developed which is shown in Fig.11. The rollers are approximated due to radii at the loading tip and the trench edges. The sample is inserted in the bending test with the damaged side wall downwards. Due to this the whole damaged area is in a tensile stress state.

978-1-4799-4789-8/14 $31.00 © 2014 IEEE   651

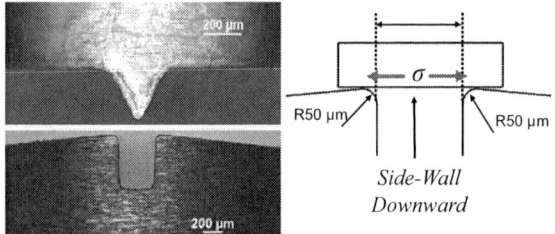

Fig. 11: Microscope pictures of the miniaturized three-point-bending (l) and measurement configuration (r)

6. Breaking Strength

For each dicing technology and for each measurement direction 31 samples were tested and statistically evaluated using the maximum-likelihood estimation which gives access to the characteristic breaking strength and the Weibull modulus. The char. breaking strength includes the size of the chips as well as the shape of the stresses the bending test induce in the specimen. To get a parameter which is independent to the chip size and measurement tools the effective loaded side wall of the test arrangement has to be calculated by solving the Weibull integral (Eq. 2).

For measurements in X-direction (see Fig. 12) with the four-point bending test the effective loaded side wall can by calculated with Eq. 3. For tests in Z-direction Eq. 4 can be used. Herein, L_2 is the outer load span, which is the distance between the supporter rollers, h is the thickness and b is the width of the chip and m is the Weibull modulus.

$$A_{E,4P} = 2 \cdot \frac{m+2}{2(m+1)^2} \cdot L_2 \cdot h \qquad (3)$$

$$A_{E,3P} = \frac{1}{(m+1)} \cdot L_2 \cdot b \qquad (4)$$

With Eq. 2 and the effective loaded side wall it is possible to calculate the scaling parameter which is independent to the test arrangement. The procedure is called scaling which is comparable with a kind of normalization with a divined chip size. Especially in our case it is important to scale the data, because of the big size differences by measurements in X and Z-direction.

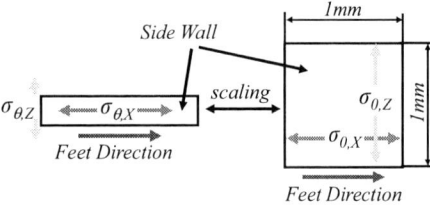

Fig. 12: Relation during measurements parallel to the feet direction using the example of stealth-dicing

In Fig. 13 and Fig. 14 all results are shown. First of all it is visible that the differences between the front and the back side also appear in the breaking strength

of the chips. E.g. due to the blade-dicing the scaling Parameter at the front side is 425 MPa and at the back side it is 335 MPa. Additionally a strong dependency of the breaking strength to the stress orientation was found. In case of the blade dicing the scaling parameter in Z-direction was 183 MPa. Also the scattering represented by the Weibull modulus strongly depends to the stress orientation.

In conclusion it can be said that the risk of chip fracture can't be predicted with only one global strength value. More than this, it is important to consider the differences of the die strength depending on the stresses which act on the chip in an application. This phenomenon we call dicing caused anisotropic breaking strength.

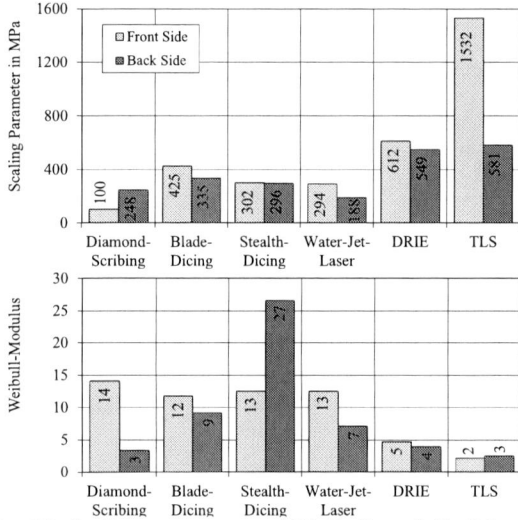

Fig. 13: Scaling parameter and Weibull moduli of the front and back sides measured in X-direction [9]

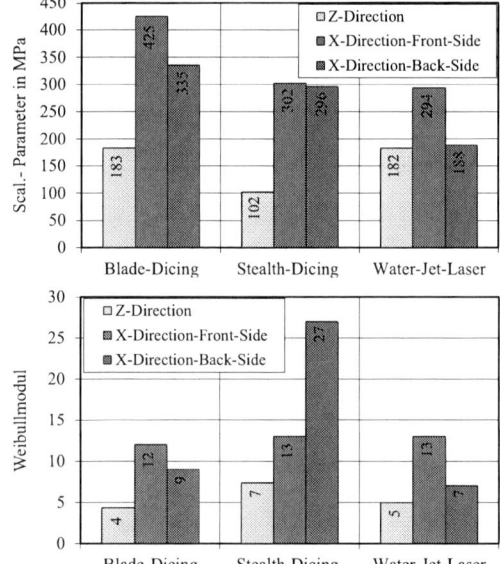

Fig. 14: Scaling parameter and Weibull moduli of the front and back sides and in different directions

7. FE-Simulation and Implementation of a New Equivalent Stress

For predictions of the risk of chip fracture adapted from FE-Simulation results, several software tools are available, e.g. CARES-Live or STAU. Those tools overs different failure criteria to calculate the risk of chip fracture like the ranking theory also called the Maximum Stress Theory. It assumes that failure will occur when the maximum principal stress at any point reaches the breaking strength of a chip. Also, other more detail failure criteria are implemented, but anyhow, the problem with the dicing caused anisotropic breaking strength is the same for all.

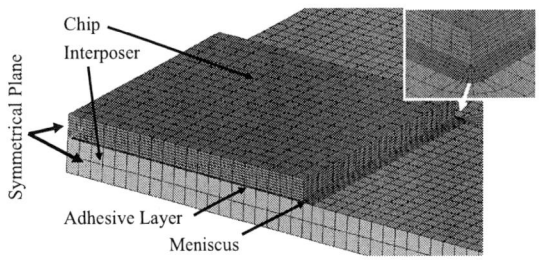

Fig. 15: Quarter view of the FE-model consisting of a Chip mounted on a FR4 interposer

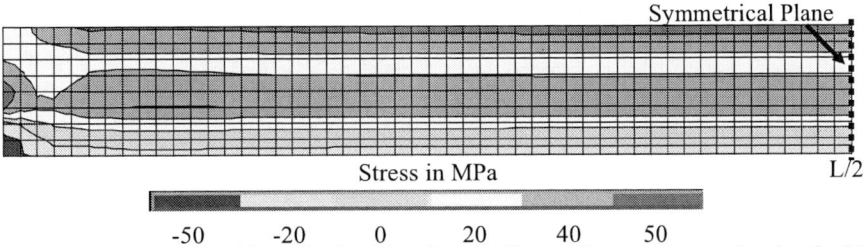

Fig. 16: First principal stresses at the side wall of a chip after the die attachment process simulated with the FE-model

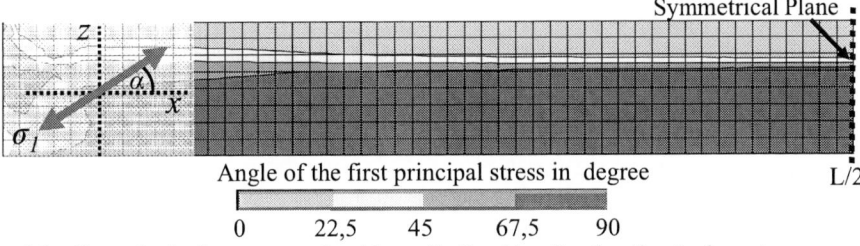

Fig. 17: Angle of the first principal stresses at the side wall of a chip after the die attachment process simulated with the FE-model

To figure out the problem and to describe our solution we develop a simplified FE-model (see Fig. 15) consisting of a chip mounted on a FR4 interposer. Linear elastic material models were used. Chip fracture often occurs after the chip mounting process. This process is divided in three steps: dispensing of the adhesive, placement of the chip and heating up for hardening of the adhesive. The hardening temperature mostly is close to 150°C. Because of different coefficient of thermal expansion of the materials, stresses will be induced in the chip by cooling down to room temperature after the hardening. With the FE-model the stresses were simulated at room temperature (25°C) after the hardening at 150°C.

Depending on the applied fracture criterion different equivalent stresses have to be calculated, e.g. the von Mises stress. Here the maximum principle stress theory was used and thus, the first principal stresses at the side walls of a chip after the mounting process were simulated. They are shown in Fig. 16.

The named software tools use the Weibull distribution and the simulated stresses to calculate the risk of chip fracture. For this purpose the risk of fracture of each fracture relevant element in the FE-model, these are all elements of the side wall, are calculated. Through this the Weibull integral (Eq. 1) can be written as a sum as shown in Eq. 5. Here $A_{eff,Element}$ is the effective side wall of the element and $\sigma_{1,i}$ is the maximum principle stress of the ith element (received due to the simulation). The scaling parameter σ_0 and the Weibull modulus m should be measured.

$$F_{Chip}(\sigma_1) = 1 - \exp\left(-\sum_i A_{eff,Element} \cdot \left(\frac{\sigma_{1,i}}{\sigma_0}\right)^m\right) \quad (5)$$

Up to now, the scaling parameter and the Weibull modulus have been assumed as valid for the whole chip. Then, the highest risk of chip is given directly at the upper chip edges. But as visible in Fig. 17, in a big area the maximum principle stresses act vertical to the dicing feet direction, which leads to smaller scaling parameters as well as smaller Weibull moduli.

Because of the dicing caused anisotropic breaking strength of silicon chips the probability of fracture $F_i(\sigma_1,\alpha,x,z)$ of the ith element depends not only on the stress value of the maximum principal stress of this element, it also depends on the orientation of the stress as well as the location of the element at the side wall. Then, the Weibull distribution of the ith element is given with Eq. 6. Here $\sigma_0(\alpha_i,x_i,z_i)$ and $m(\alpha_i,x_i,z_i)$ are the Weibull parameter of the element.

$$F_i(\sigma_1,\alpha,x,z)=1-\exp\left(-A_{eff,Element}\cdot\left(\frac{\sigma_{1,i}}{\sigma_0(\alpha_i,x_i,z_i)}\right)^{m(\alpha_i,x_i,z_i)}\right) \quad (6)$$

The new equivalent stress to describe the dicing caused anisotropic breaking strength of silicon dies is developed with the goal to adapt the common tools to calculate fracture probabilities. For the first this means that the Weibull parameter should be constant over the whole side walls. Thus, we first define a kind of norm chip with the parameters $\sigma_{0,norm}$ and m_{norm}. Then the probability of fracture of the ith element $F_i(\sigma_{equ})$ is calculable with Eq. 7.

$$F_i(\sigma_{equ})=1-\exp\left(-A_{eff,Element}\left(\frac{\sigma_{equ,i}}{\sigma_{0,norm}}\right)^{m_{norm}}\right) \quad (7)$$

Furthermore, we define that the fracture probability of the ith element of the real chip should be the same than of the ith element of the norm chip which means: $F_i(\sigma_1,\alpha,x,z) = F_i(\sigma_{equ})$. The definition of the equivalent stress can be written as Eq. 8.

$$\left(\frac{\sigma_{1,i}}{\sigma_0(\alpha_i,x_i,z_i)}\right)^{\frac{m(\alpha_i,x_i,z_i)}{m_{norm}}}\cdot\sigma_{0,norm}=\sigma_{equ,i} \quad (8)$$

In contrast to the simulated maximum principle stresses the equivalent stresses according to Eq. 8 are directly comparable in consideration of theirs impact to fracture the chip. Due to this Eq. 5 can be written as Eq. 9 and all common methods to predict the risk of chip fracture can be used as known.

$$F_{Chip}(\sigma_{equ})=1-\exp\left(-\sum_i A_{eff,Element}\cdot\left(\frac{\sigma_{equ,i}}{\sigma_{0,norm}}\right)^{m_{norm}}\right) \quad (9)$$

8. Modelling the dicing caused anisotropic breaking strength for the blade-dicing

The equivalent stress has to be solved element wise. Therefore mathematical expressions of $\sigma_0(\alpha_i,x_i,z_i)$ and $m(\alpha_i,x_i,z_i)$ are needed. Those expressions can be different according to the dicing technique. But, for the basic ideas of the equivalent stress it is not very important how such expressions are formulated at last.

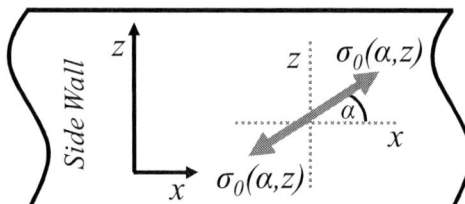

Fig. 18: Definitions of stress directions

In the present work an empiric expression in correlation to the found breaking strength values of the blade-dicing was developed. For this purpose some assumptions were made. The first assumption: The breaking strength is constant in X-direction (see Fig. 18). Furthermore, it is assumed that the scaling parameter as well as the Weibull modulus of stresses in X-direction increase linear from the back side to the front side of the chips. With these assumptions the Weibull parameters depending on the location can be described with Eq. 10 and Eq. 11. Here h_{chip} is the height of the chip. $\sigma_{0,front}$ and m_{front} are the Weibull parameters measured for the front side and $\sigma_{0,back}$ as well as m_{back} are the Weibull parameters measured for the back side.

$$\sigma_{0,x}(z)=\frac{\sigma_{0,front}-\sigma_{0,back}}{h_{Chip}}\cdot z+\sigma_{0,back} \quad (10)$$

$$m_x(z)=\frac{m_{front}-m_{back}}{h_{Chip}}\cdot z+m_{back} \quad (11)$$

Furthermore, the dependencies on the angel α, with which the first principal stress acts, have to be considered. It is assumed that the Weibull parameters are cyclically to the angel α.

$$\sigma_0(\alpha,z)=\sigma_{0,x}(z)\cdot\left(\left(1-\frac{\sigma_{0,\alpha=90°}}{\sigma_{0,x}(z)}\right)\cdot\cos(\alpha)+\frac{\sigma_{0,\alpha=90°}}{\sigma_{0,x}(z)}\right) \quad (12)$$

$$m(\alpha,z)=m_x(z)\cdot\left(\left(1-\frac{m_{\alpha=90°}}{m_x(z)}\right)\cdot\cos(\alpha)+\frac{m_{\alpha=90°}}{m_x(z)}\right) \quad (13)$$

Eq. 12 and Eq. 13 gives the empirically found expressions of $\sigma_0(\alpha_i,x_i,z_i)$ and $m(\alpha_i,x_i,z_i)$ for the blade-dicing. As visible in Fig. 19 the equations Eq. 12 and Eq 13 link all measured values in meaningful correlation.

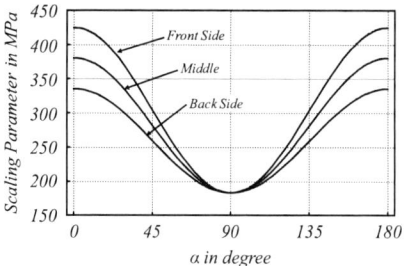

Fig. 19: Plot of Eq. 12 with the measured values of the blade-dicing

9. Calculation of the equivalent stress

With the simulation results, the first principal stress at each note of an element and the impact angle of the stress, it is possible to calculate the equivalent stress of any note of the chip side walls using Eq. 8, Eq. 12 and Eq. 13 as well as the measured breaking strength of blade diced chips.

978-1-4799-4789-8/14 $31.00 © 2014 IEEE

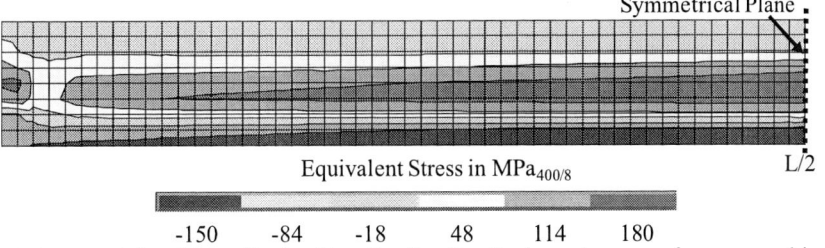

Equivalent Stress in MPa$_{400/8}$

-150	-84	-18	48	114	180

Fig. 20: Equivalent stresses at the side wall of a chip after the die attachment process for a norm chip with $\sigma_{0,norm} = 400$ MPa and $m_{norm} = 8$

In the present work the norm parameters $\sigma_{0,norm} = 400$ MPa and $m_{norm} = 8$ were used. This means, the equivalent stresses have the same impact to fracture this norm chip, than the real stresses on the real chips. The equivalent stresses for blade diced chips and with the shown simulation results are visible in Fig. 20.

Now the three fracture types (see Fig. 2) are explainable also with the probabilistic fracture mechanics. As can be seen (Fig. 20 and 21) the highest equivalent stresses are in the middle of the side wall height (see Fig. 21). This directly means that the highest risk of chip fracture is given in these areas. Here, the impact angle of the stresses with 90° is vertical to the crack shown in Fig. 5. Thus the crack propagation direction will be horizontal which leads in horizontal crack types. Also high stresses are given at the chip corners, which lead in mixed crack types.

It is also possible to estimate the risk of fracture for different areas of the chip side wall by means of Eq. 9. Such an extrapolation is visible in Fig. 21. Now it is visible that horizontal cracks are most common with a probability of fracture of approximately 60 ppm.

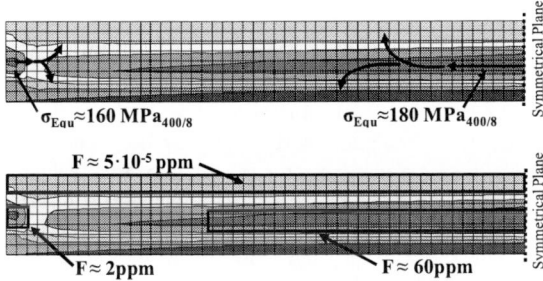

Fig. 21: Crack propagation and fracture types

10. Conclusions

In the present work the phenomena of an anisotropic breaking strength is shown, which depends on the dicing process. The description of this phenomenon is based on the probabilistic fracture mechanic using the Weibull distribution. For this purpose an equivalent stress was developed. It offers to compare stresses directly by means of theirs impact to fracture a chip. For the future the shown methods as well as the equivalent stress will be a very powerful tool of the research and development of the assembly and packaging technology.

Acknowledgments

The IGF-Project No. 16.672N of the research association "Forschungsvereinigung Schweißen und verwandte Verfahren e. V. des DVS, Aachener Straße 172, 40223 Düsseldorf, Germany" was, on the basis of a resolution of the German Bundestag, promoted by the German Ministry of Economic Affairs and Technology via AiF within the framework of the programme for the promotion of joint industrial research and development (IGF).

References

1. M.Ranjan, L. Gopalakrishnan, K. Srihari, C. Woychik: "Die cracking in flip chip assemblies", Proc. of the 48th IEEE ECTC, Seattle, WA, May 1998, pp. 729 - 733
2. D. Yang, J. Bielen, F. Theunis, W.D. van Driel, G.Q. Zhang: "Die fracture probability prediction and design guidelines for laminate-based over-molded packages", Proc. of the 9th IEEE EuroSimE, Freiburg i.B., Apr. 2008, pp.1-6
3. B. Vijayakumar, Y. Guo: "Failure rate prediction and prevention of die cracking in over molded plastic packages", Proc. of the IEEE ITherm, San Diego, CA, Jun. 2006, pp. 803-809
4. D. Pustan, E. Rastiagaev, J.Wilde: "In situ analysis of the stress development during fabrication processes of micro-assemblies", Proc. of the 59th IEEE ECTC, San Diego, CA, May 2009, pp. 117 - 124
5. Y. Zou, J.C. Suhling, R.W. Johnson, R.C. Jaeger: "Complete stress state measurements in chip on board packages", Proc. of the International Conferece on Multichip Modules and High Density Packaging, Denver, CO, Apr. 1998, pp. 405 - 415
6. P. Palaniappa, D.F. Baldwin: "In process stress analysis of flip-chip assemblies during underfill cure", Microelectronics Reliability. Vol. 40, Jul. 2000, pp. 1181–1190
7. S. Park, H.C. Lee, B. Sammakia, K. Raghunathan: "Predictive model for optimized design parameters in flip-chip packages and assemblies", Transactions on Components and Packaging Technologies, vol. 30 (2007), pp. 294-301
8. M. Steiert, J. Wilde: "New probabilistic reliability model describing the risk of chip fracture in the chip-on-board technology", Proc. of the 4th IEEE ESTC, Amsterdam, Sep. 2012
9. M. Steiert, J. Wilde: " Chip-Side-Healing as a Basis for Robust Bare-Chip Assemblies", Proc. of the 64th IEEE Electronic Components and Technology Conference, Mai 2013, Las Vegas
10. T.C. Taylor, F. L. Yuan: "Thermal stress and fracture in shear-constrained semiconductor device structures", IRE Transactions on Electron Devices, vol.9, no.3, pp.303-308, May 1962, doi: 10.1109/T-ED.1962.14987
11. J. Bolger, C. Mooney: "Die Attach in Hi-Rel P-Dips: Polyimides or low chloride Epoxies?", IEEE Transactions on Components, Hybrids, and Manu-facturing Technology, vol. 7, no.4, Dec 1984 pp.394-398, doi: 10.1109/TCHMT.1984.1136379
12. DIN EN 845: Advanced technical ceramics - Mechanical properties of monolithic ceramics at room temperature

Accurate Prediction of SnAgCu Solder Joint Fatigue of QFP Packages for Thermal Cycling

M. Niessner[1], G. Schuetz[2], C. Birzer[2], H. Preu[2] and L. Weiss[1]

[1]Infineon Technologies AG, Munich, Germany

[2]Infineon Technologies AG, Regensburg, Germany

Abstract

We present an approach for the accurate prediction of the solder joint fatigue of quad flat packages (QFPs) with gullwing-shaped leads exposed to thermal cycling on board (TCoB). The derived fatigue life model is experimentally validated against more than 25 legs that differ in package size, materials, thermal cycling temperature profile and exposed pad vs. non-exposed pad type. The fatigue life model shows an accuracy of 25% and is used to analyze the sensitivity of the solder joint lifetime of QFPs w.r.t. changes in material properties and geometry.

1. Motivation

Darveaux [1], Syed [2,3], Fan *et al.* [4] and several other authors presented modeling strategies and accurate fatigue life models of the solder joints, i.e. the solder balls, of ball grid array (BGA) type packages, but only few publications exist on the fatigue prediction of the solder joints of quad flat packages (QFPs) with gull-wing shaped leads (see Fig. 1).

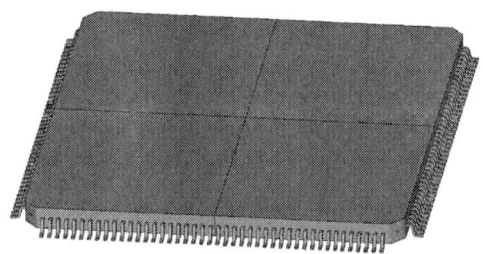

Figure 1: View of a low profile quad flat package (LQFP) with 176 pins (LQFP-176).

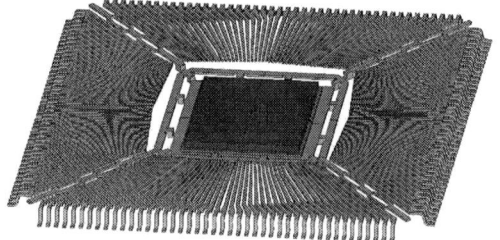

Figure 2: View of the leadframe with exposed pad (dark yellow), silicone die (blue) and glue (red) inside the LQFP-176 shown in Fig. 1.

On the one hand, researches might argue that working on the fatigue life modeling of QFP solder joints is of less interest compared to other packages as QFPs achieve a relatively high number of cycles in the thermal cycling on board (TCoB) reliability test – even with large body sizes of 24 x 24 x 1.4 mm³. The good performance of this type of package is due to the gullwing-shaped leads which serve as mechanical springs and, thus, buffer the thermal mismatch generated during thermal cycling between the QFP and the printed circuit board (PCB).

On the other hand, the requirements concerning the lifetime of solder joints of packaged products for the automotive market have been successively increased to such a high level that package developers are driven to perform end-of-life TCoB testing even of QFPs in order to demonstrate that the criteria are still met. Moreover, the solder joint fatigue life modeling of QFPs is also of high interest when the focus is moved away from a standalone package perspective to a system perspective. In the application, the QFP will be soldered on a PCB that will have, maybe on both sides, different other devices. Moreover, this PCB will be mounted in a module. Consequently, the forces experienced by the solder joints of the QFP during thermal cycling of the module will be different than during standard TCoB testing where the package is cycled on a PCB with approximately free edges. Schafet *et al.* [5] demonstrated by simulation that the solder joint life of a QFP is reduced by a factor of appx. 2 when a QFP is cycled in an automotive electronic control unit instead on a PCB with free edges.

Both the requirements of the automotive market and the finding by Schafet *et al.* [5] drives the development of QFPs with long solder joint lifetime. As a result, the experimental testing time is significant when a new QFP needs to be qualified for automotive applications. Consequently, simulation and a fatigue life model are needed to assess already at an early stage of the package development whether a new QFP has the potential to pass with its geometry and bill of material (BOM) a certain number of cycles and, thus, is worth the effort of qualification testing or not.

2. Fatigue Life Modeling Approach

For deriving a fatigue life prediction model of the solder joints of QFPs, the systematic workflow suggested by Darveaux in [1] was followed:

(1) Identify a material model suitable for describing the creep in your solder joint material.

(2) Generate a FEM model of the desired package type, solder joints and PCB.

(3) Identify a volume for the averaging of creep results and pay attention to the meshing.

(4) Fix a reference temperature for your calculations.

(5) Re-Simulate available experimental results and correlate the experimental lifetimes with the volume-averaged simulation results in order to obtain the coefficients of a Coffin-Manson type fatigue life model that can be later used for extrapolation of lifetimes of new package configurations.

Assuming that secondary creep is the phenomenon that correlates well with the overall solder joint fatigue mechanism in SnAgCu solder joints, the Garofalo model systematically extracted and verified by Schubert et al. [6] was chosen for this work.

In order to avoid running large and computationally expensive finite element (FE) models of QFPs with 144 and 176 pins, a two-step approach consisting of a global model and a sub-model similar to [4,7,8] is employed. Schafet et al. [5] showed that the solder joints at the corners are the most critical ones. This finding correlates with our experimental results and allows for focusing on these two critical solder joints. Eventually, a global model of the QFP package soldered on the PCB is used to calculate for different temperatures the displacements at the outermost pins and a sub-model encompassing the outermost pins, their solder joints and PCB pads is used to evaluate the creep in the solder throughout several cycles (see Fig. 2).

Figure 3 shows a characteristic fatigue crack that is fully propagated through the solder joint of a gullwing-shaped lead. Our experiments indicate that the crack starts near the pin on the side of the solder joint that is oriented towards the package. Simulation correlates with this experimental result as the highest accumulation of creep strain is calculated on the package side and near the lead (see Figs. 4 and 5). Hence, layers of solder elements below the pin are chosen as the control volume (see Fig. 6). Schafet et al. [5] and Che et al. [7] use similar control volumes.

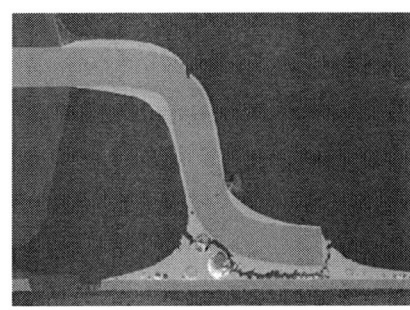

Figure 3: Cross-section of a gullwing-shaped lead and its solder joint after end-of-life thermal cycling. A crack has fully propagated through the solder joint.

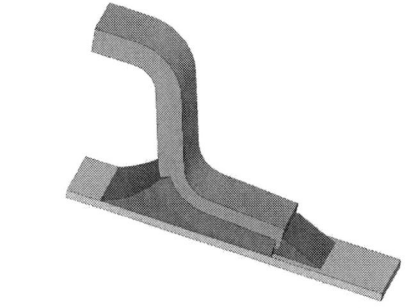

Figure 4: Geometry model of a gullwing-shaped lead, the solder joint and the copper PCB pad.

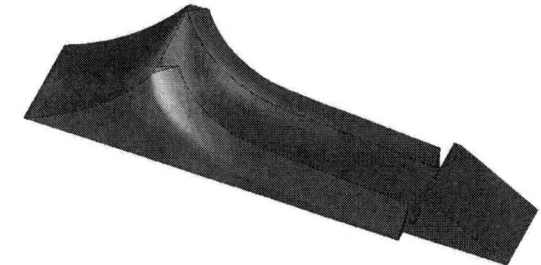

Figure 5: Accumulated creep strain in solder joint (characteristic result).

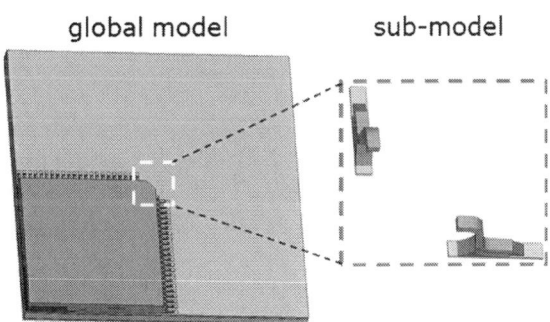

Figure 2: View of the global model (quarter model of a QFP on PCB, left) and of the sub-model (two outermost pins with solder joints and PCB pads). The quarter model uses symmetry boundary conditions at the symmetry axes and coupled boundary conditions at the free edges of the PCB.

Figure 6: Geometry model of a gullwing-shaped lead with the control volume for the averaging of creep results (left), control volume (right top), distribution of accumulated creep strain in the control volume (right bottom).

The choice of the reference or "initial stress-free" temperature is a topic that is not settled yet [4,8]. Experimental investigations of Mavoori *et al.* [9] indicate that residual stresses in lead-free solder relax within few hours by 50% at room temperature and within shorter time and by a higher percentage at higher temperatures. Che *et al.* [8] argues based on [9] that stresses should relax to their lowest value or almost completely during the first dwell at high cycling temperature. This is why Fan *et al.* [4] suggests the maximum cycling temperature as reasonable reference temperature. As the fatigue life model in this work is aimed at covering two different thermal cycling profiles, the maximum temperature was not an option here. In agreement with [4,5,7,8], room temperature is used as reference temperature for the correlation with experimental results.

3. Material Data

The PCB is modeled using linear-elastic material data as its glass transition temperature (Tg) of 175°C is outside the temperature range used for thermal cycling. The description of the PCB was obtained from averaging the measurements of several PCBs used for TCoB testing. The epoxy molding compounds (EMCs) were also modeled using linear-elastic properties. All molding compounds were measured in-house in order to have a comparable set of data. The Tg of the molding compounds was determined from the peak of the phase angle of the dynamic mechanical analysis (DMA). Table 1 summarizes the material data used in the simulation.

Material	CTE [ppm/K]	Tg [°C]	Young's Modulus [GPa]	Poisson's ratio
4-Layer PCB (below Tg) (Orthotropic)	X: 14 Y: 17 Z: 50	175	X,Y: 22…17 Z: 5	X,Y: 0.17 Z: 0.2
Copper	17.6	-	121	0.3
Adhesive	80 … 205	40	3.5 … 0.1	0.33
Silicon 100 (Orthotropic)	1.7…3.35	-	see [10]	see [10]
EMC A	8 … 34	120	28 … 1	0.3
EMC B	10 … 37	100	26 … 1	0.3
EMC C	6 … 30	115	34 … 1	0.3
SnAgCu Solder	21	-	48.5… 33 see [6] for creep mod.	0.36

Table 1: Material data. The Young's modulus data is continuously temperature dependent except for copper and silicon. For the adhesive, the molding compounds and the solder, only the Young's modulus for low (around -40°C) and high temperatures (around 260°C) are listed, but no values for the temperatures in between.

4. Correlation with Experimental Data

Experimental end-of-life data from 25 legs with low profile quad flat packages (LQFPs) were available for correlation. The end-of-life failure criterion was electrical failure of daisy-chain test structures. For all of these packages the characteristic lifetime N_f was calculated from the Weibull distribution to $N_f = MTTF-1.5\sigma$ (MTTF is the medium time to failure and σ is the standard deviation).

The most important differences in the set of 25 data points are the presence of three different molding compounds (see Table 1), two different thermal cycling temperature profiles (see Table 2) and three different LQFP geometries (see Table 3). Two of the tested LQFPs are of the same size, but one has an exposed pad (epad) which is soldered to the PCB whilst the other version has its pad completely inside the body (see Fig. 7).

All data points were modeled and simulated using the approach described in section 2. After three cycles, the change of the accumulated creep strain per cycle was less than 2% which is considered as stable. The resulting volume-averaged creep strain values per cycle were used to fit a Coffin-Manson type fatigue life model to the experimental lifetimes. The VAVG creep strain per cycle was used instead of the VAVG creep energy density per cycle as it allowed for a better correlation w.r.t. the experiments. Less than ten data points were sufficient for fitting a first fatigue life model that was able to extrapolate the lifetimes of most of the other data points within reasonable accuracy. This indicates that the simulation approach in combination with the fitted fatigue life model is able to cover the dominant physics of this class of experimentally tested packages.

Profile	Minimum Temperature [°C]	Maximum Temperature [°C]	Ramp Time [min]	Dwell Time [min]
TC1	-40	125	12	18
TC2	-40	150	15	15

Table 2: Thermal cycling profiles.

Package Type	LQFP-176 Non-exposed pad	LQFP-176 Exposed pad	LQFP-144 Exposed pad
Number of Pins	176	176	144
Body [mm³]	24x24x1.4	24x24x1.4	20x20x1.4
Exposed Pad / Diepad [mm²]	8.8x8.8	7.8x7.8	7.5 x 7.5
Die Size [mm³]	7.4x7.4x0.3	6.5x7.0x0.6	6.5x6.5x0.6
Pin width [mm]	0.2	0.22	0.2
Pin pitch [mm]	0.5	0.5	0.5

Table 3: Data of the three package geometries evaluated in the experiment.

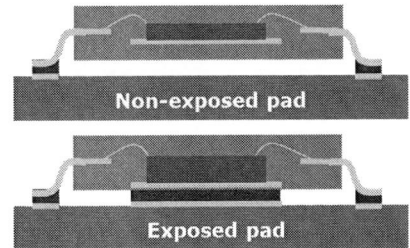

Figure 7: Schematic representations of the non-exposed pad and exposed pad LQFP geometries.

If all available experimental data points are used for fatigue life model fitting, the correlation plot shown in in Fig. 8 is obtained. The resulting fatigue life model is able to cover all except one data point within an accuracy of ±25% enabling the prediction of the lifetime of the analyzed class of LQFPs within certain boundary conditions.

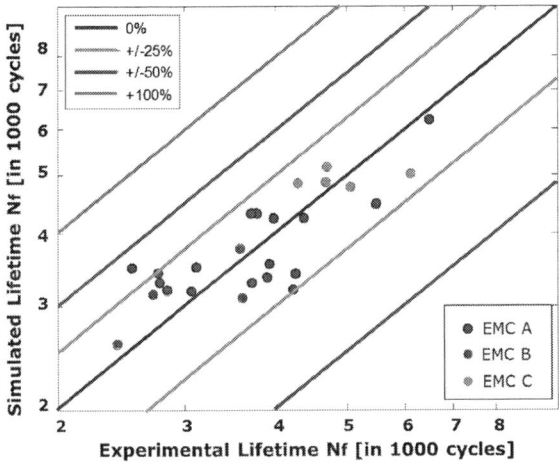

Figure 8: Comparison of the experimental lifetimes of the 25 analyzed data points and the lifetimes obtained for the data points from simulation and using the fatigue life model. The diagonal lines indicate the relative error of the predicted w.r.t. the experimental lifetime.

5. Results and Discussion

The validated fatigue life model allows for a systematic analysis of the solder joint fatigue of LQFPs by means of simulation.

First, the accumulation of creep strain during thermal cycling in a critical solder joint is analyzed. Figure 9 shows the volume-averaged (VAVG) creep strain accumulated in a control volume during the 3rd thermal cycle for an LQFP-176 epad with EMC type A and TC2 thermal cycling profile. The approach used in this work, i.e. the creep model from [6] in combination with the reference temperature of 25°C and the material data from section 3, shows that the predominant share (90%) of the creep strain is produced during the temperature ramps from low to high temperature and vice versa. During the dwell at low temperature, slightly more creep strain is produced than at high temperature. This behavior can be explained by analyzing the VAVG von Mises stress in the control volume which is used here as an indicator for the presence of shape changing stresses (see Fig. 10).

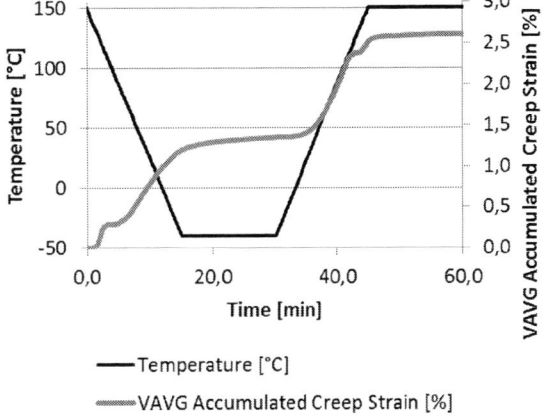

Figure 9: Temperature and volume-averaged (VAVG) creep strain accumulated in the control-volume of a corner solder joint of an LQFP-176 epad with EMC type A during a TC2 cycle.

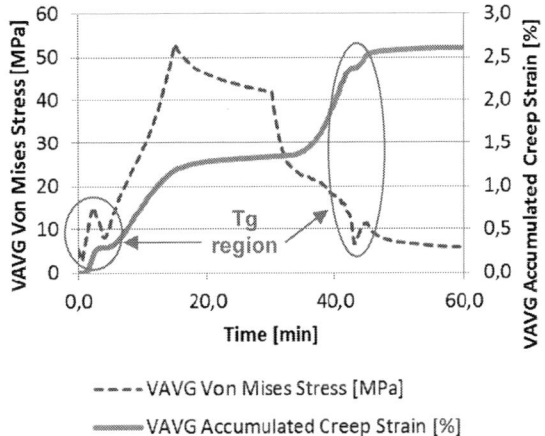

Figure 10: Volume-averaged (VAVG) von Mises stress and VAVG creep strain accumulated in the control-volume of a corner solder joint of an LQFP-176 epad with EMC type A during a TC2 cycle.

Recalling that the Garofalo model [6] includes a dependence on temperature that models one essential characteristic of creep mechanisms – the creep rate increases with higher temperature – the accumulation of creep strain can be explained as follows:

(1) During the dwell at high temperature, the creep mechanism relaxes the stresses in the solder joints

quickly. Within short time, the stresses are no longer high enough to activate the creep mechanism.

(2) When cooling from high to low temperature, new stresses are generated in the solder joint because of the thermal mismatch of package and PCB. These new stresses are high enough to activate the creep mechanism: creep strain is accumulated. The lower the temperature the more stress is needed to activate the creep mechanism, i.e. the accumulation is slowed down.

(3) During the dwell at low temperature, a comparably high stress level is present which can, however, only be slowly relaxed.

(4) When heating from low to high temperature, the stresses are reduced due to the decrease in the thermal mismatch of package and PCB, but at the same time the creep mechanism is activated and the stresses are relaxed faster with higher temperature.

Moreover, Figs. 9 and 10 show the influence of the glass transition region (Tg region) of the EMC. When heating from low to high temperature, the coefficient of thermal expansion (CTE) of the molding compound changes around the glass transition temperature (Tg) from a value below the CTE of the PCB to a value above the CTE of the PCB. This results into a situation where the thermal mismatch between package and PCB is temporarily minimized before it is increased again. This phenomenon results into temporary drop in the VAVG von Mises stress and in a temporary plateau in the VAVG creep strain.

Summarizing these first findings, the creep strain and, thus, the amount of damage produced in a solder joint mainly depends on the amount of stress that can be "accumulated" in the solder joint during cooling from high to low temperature. The main drivers of this accumulation are the delta between the highest and the lowest temperature as well was the thermal mismatch between PCB and QFP. Consequently, the dominant parameter for a package of the same size and type on the same PCB for a fixed thermal cycling profile should be the CTEs of the material with the largest volume in the package: the molding compound. Moreover, as the larger share of the cycling range is below Tg, the CTE of the molding compound below Tg (CTE1) should be the most influential CTE.

This hypothesis can be verified when calculating the VAVG creep strains produced for the three different molding compounds listed in Table 1 during a TC2 profile (see Fig. 11). Indeed, EMC type C with the lowest CTE below Tg produces the highest VAVG creep strain during a TC2 cycle whilst EMC type B with the highest CTE below Tg produces the least VAVG creep strain. This finding from simulation correlates with the experimental findings: packages of the same size exposed to the same cycling profile achieved the highest number of cycles in TCoB testing with EMC type B and the lowest number of cycles with EMC type C.

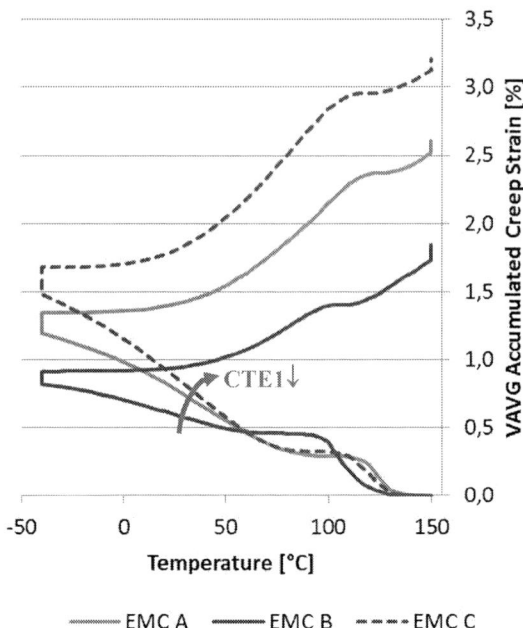

Figure 11: Volume-averaged (VAVG) creep strains accumulated in the control-volume of a corner solder joint of an LQFP-176 epad with EMC types A, B and C during a TC2 cycle.

The following findings were obtained from a systematic analysis of the influence of different factors on the TCoB performance of an LQFP with a fixed package body size of 24x24x1.4 mm³:

(1) **Material properties:** The material properties with the highest influences were the CTE of the molding compound below Tg and the CTEs of the PCB in lateral direction. A change of ±1 ppm/K in the CTE of the EMC or in both CTEs of the PCB leads to a change in the characteristic lifetime of ±15%. This trend correlates with experimental results. The second highest influences have the Tg of the molding compound and the Young's modulus of the PCB.

(2) **Exposed pad vs. non-exposed pad type LQFP-176:** Simulation predicts little difference between the two package types. This correlates with experimental findings. Even though the geometries of these two packages are different, the stresses experienced by the solder joints during thermal cycling are similar for the investigated body size of 24x24x1.4 mm³.

(3) **Size of exposed pad:** Simulation predicts increasing lifetime with increasing size of the exposed pad due to increased coupling of the QFP and the PCB. When going from a 7.8x7.8 mm² to a 9.2x9.2 mm² exposed pad, about 5% should be gained in lifetime. When going from a 7.8x7.8 mm² to an 11x11 mm² exposed pad, about 20% should be gained in lifetime. Unfortunately, no experimental end-of-life data was available to fully verify this finding.

(4) **Influence of die size and thickness:** Simulation predicts low influence of die size and thickness if the exposed pad is kept constant. No experimental data was available to support this finding.

The first finding is very helpful for explaining why there is quite some spread observed in the experimental lifetime testing of LQFPs even for very similar packages and BOM. Especially the CTEs of PCBs are usually not monitored by the PCB manufacturers and the purchase specifications usually allow for a variation of ±1 ppm/K or even more. Our characterization of PCBs performed for material data extraction (see Table 1) revealed that this tolerance range is well used. When also the possible variations of the CTEs of the molding compounds and the variations in the solder fillet formation are considered, a spread of more than 20% for "similar" legs is plausible.

On the other hand, finding 4 is a bit surprising. Of course, one could argue that the package center, where the silicon die is located, might be too far away from the solder joints in order to have an influence on solder joint fatigue life, but common sense would expect us to see at least some influence. The low influence of the silicon die size and thickness might also result from the assumption that the reference temperature of the whole simulation model is 25°C. Using this reference temperature means to ignore the history of the package and the solder joints, i.e. stresses from the package assembly process and from soldering the package on the PCB.

6. Conclusion and Outlook

Following the workflow proposed by Darveaux [1], both a simulation procedure and a fatigue life model for predicting the characteristic lifetime of LQFPs exposed to thermal cycling on board were successfully derived.

A systematic analysis of the TCoB performance of the LQFP showed that the CTE of the molding compound below Tg and the CTEs of the PCB are the material properties with the highest impact. Consequently, if the BOM of the package is to be changed, molding compounds with high CTE below Tg are preferable to molding compounds with low CTE below Tg. Of course, also other reliability requirements need to be considered when selecting the molding compound. If a BOM is fixed, the size of the exposed pad offers an option to increase the characteristic lifetime if needed.

Future work will focus on including visco-elastic material models of the molding compounds and on including the process history of the package and the solder joints.

Acknowledgments

The authors would like to thank their colleagues Heinz Pape, Ingolf Rau, Chan Lam Cha and Chet Hung Kong for their support during this work.

References

1. R. Darveaux, "Solder Joint Fatigue Life Model", Proc. TMS Annual Meeting 1997, Orlando, FL, 1997, pp. 213-218.
2. A. Syed, "ACES of finite element and life prediction models for solder joint reliability", Proc. TMS Annual Meeting 1997, Orlando, FL, 1997, pp. 347-358.
3. A. Syed, "Predicting Solder Joint Reliability for Thermal, Power, & Bend Cycle within 25% Accuracy", Proc. of ECTC 2001, Orlando, FL, 2001, pp. 255-263.
4. X. Fan *et al.*, "Effect of finite element modeling techniques on solder joint fatigue life prediction of flip-chip BGA packages", Proc. ECTC 2006, San Diego, CA, 2006.
5. N. Schafet *et al.*, "Development of a submodel technique for the simulation of solder joint fatigue of electronic devices mounted within an assembled ECU", Proc. EuroSimE 2009, Delft, The Netherlands, 2009, pp. 1 -8.
6. A. Schubert, R. Dudek *et al.* "Fatigue life models for SnAgCu and SnPb solder joints evaluated by experiments and simulation", Proc. ECTC 2003, New Orleans, LA, 2003, pp. 603-610.
7. F. Che *et al.*, "Lead free solder joint reliability characterization for PBGA, PQFP and TSSOP assemblies", Proc. ECTC 2005, Orlando, FL, 2005, pp. 916-921.
8. F. Che *et al.*, "Fatigue Reliability Analysis of Sn–Ag–Cu Solder Joints Subject to Thermal Cycling", IEEE Transactions on Device and Materials Reliability, Vol. 13, No. 1, 2013, pp. 36-49.
9. H. Mavoori *et al.*, "Creep, stress relaxation, and plastic deformation in Sn-Ag and Sn-Zn eutectic solders", J. Electron. Mater., vol. 26, no. 7, 1997, pp. 783-790.
10. M. Hopcroft *et al.*, "What is the Young's Modulus of Silicon?", J. MEMS, vol. 19, no. 2, 2010, pp. 229-238.

Thermo-Mechanical Properties of Underfills at Partial and Full Filler Percolation – Sub-Layering the Underfill.

Gerd Schlottig[*], Marie Haupt[†], Severin Zimmermann[*], Jonas Zürcher[*], Thomas Brunschwiler[*]

[*] IBM Research GmbH, Säumerstrasse 4, 8803 Rüschlikon, Switzerland
Email: erd@zurich.ibm.com, Phone: +41 44 724 8874
[†]AMIC Angewandte Micro-Messtechnik GmbH, Volmerstraße 9B, 12489 Berlin, Germany

Abstract

Underfill materials protect electronic package interconnects. They are designed to bridge the thermal expansion behavior of die, substrates and interconnects and couple their stiffness. However 3D integration requires higher thermal conductivities than conventional underfills can provide. Because percolating thermal underfills promise to offer substantially higher conductivities, this study investigates property gradients within percolating thermal underfills.

All sphere-filler composites in electronic packages have limited properties close to the interface: Fillers cannot penetrate the interface. Thus, structure and properties show a interface vicinity gradient. Although the composite behavior dependence on filler content and distribution has been well described for both disperse fillers and heterogeneous materials in general, especially of the CTE, moduli, and conductivity, the transition to communicating fillers both in bulk and contrained spaces is only covered insufficiently. The effective medium approach to thermo-mechanical underfill modelling faces a lack of geometry representation, especially when the ratio of filler size to gap size increases to 0.5 and beyond.

Therefore we sub-layer the underfill and apply unit cell model results to the ratios and fill fractions of the layers. A central layer of the linear buckling phase shows a high filler volume fraction and is approximated by a face centric cubic unit cell. It reveals a 30 % drop in Young's modulus where the share of the central layer is more than one third of the entire underfill thickness. To evaluate the model errors we compare a stacked-layer effective modulus to an effective modulus from a constrained space unit cell that is being published separately. We present bulk specimen preparation and relate first experimental results to the unconstrained space unit cell model. All values are shown for an example of monodisperse 45 μm diameter silica fillers in a 2 GPa matrix.

With this study we suggest that the ratio of filler size to gap height could be utilized to tailor the modulus along the height axis. Package simulations could potentially benefit from sublayering an underfill layer.

1. Introduction: Sub-Layering in Underfill Modelling

All composites in electronic packaging based on spherical fillers face a challenge in its heterogeneous properties close to the interface: By physical limitation the fillers cannot move their center of gravity close to the boundary wall, cannot penetrate. Thus, the vicinity layer (VL) is a matrix rich layer. The filler volume fraction increases monotonically travelling away from the interface normal, and towards the interface one observes a filler depletion. This means that the matrix dominates the material proper-

ties in the interface utmost vicinity, and looses influence with increasing distance from it.

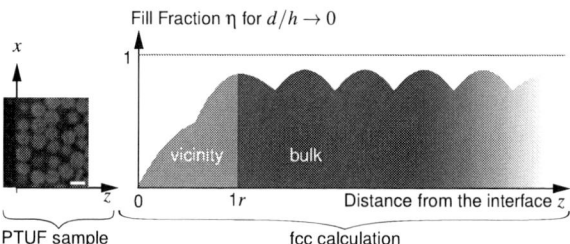

Figure 1. The filler volume fraction depends on the distance from the interface, especially in transition away from the interface into the composite bulk. The plot on the right shows η for fully percolating face centric cubic lattice packing (fcc_{perc}). The micrograph to the left shows a CT scan obtained from a PTUF sample (45 μm diameter fillers).

In thermoset-based composites, this involves an increased coefficient of thermal expansion (CTE) close to the interface, and a decreased Young's modulus close to the interface. To the contrary, in electronic packaging a low CTE mismatch is desired, and although a low Young's modulus could accommodate the stress consequences of such, interconnects are protected by a moderate Young's modulus and a close-to-interconnect CTE. The most sensitive z-position of the interconnect that needs protection has been subject of different studies [1], [2], and there are also partial-height underfills commercially available now [3]. Thus, any parameters to influence the Young's modulus and CTE of the underfill layer are of interest.

The recently published concept of sequentially created underfills - percolating thermal underfills (PTUF) opens a door to additional design space in this respect [4], [5]. A PTUF layer could have a ratio of filler diameter to cavity height d/h above 0.5 – less than two filler diameters to reach from substrate to chip, or from chip to chip. Depending on the ratio d/h the fill fraction varies significantly, and with it the local properties in the layer, especially in the interface-normal direction. Thus, we study in this work the thermomechanical behavior of different sublayers in differently dense packed PTUFs, with an emphasis on the linear buckling phase.

We apply the results of unit cell finite element (FE) simulations of unconstrained fcc lattices of varying filler volume fractions, and compare to both experimental readings and constrained space (cs) unit cell FE simulations

978-1-4799-4789-8/14 $31.00 © 2014 IEEE

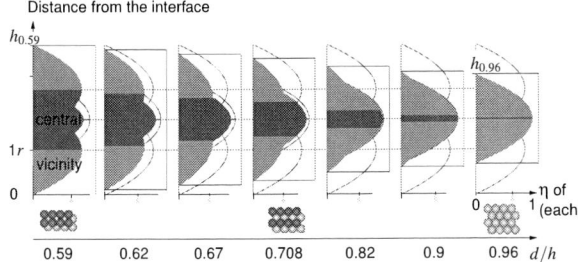

Figure 2. The filler volume fraction η plotted against the distance from the interfaces. The plot multiples correspond to different d/h ratios in the linear buckling phase. Note the decreasing height h with increasing d/h. Both extremal cases appear as thin outlines in each sub-plot.

[6], [7]. However, the effective medium approach becomes increasingly erroneous when the bodies' geometry scale is so small that it approaches the microstructure scale of the composite. In the linear buckling phase of the PTUF this is the case, as the filler diameter is more than half the layer thickness. Studies of two FE unit cell approaches take this risk to study the major influences on the effective properties [6], [7]. This study adds a sub-layering approach to extend the adjustment parameters.

The PTUF concept and the linear buckling phase
The PTUF concept has two aspects: the desired percolation between fillers in an underfill, and its sequential processing. True particle percolation improves the thermal conductivity significantly, but conventional underfills do not achieve true percolation, and even in conductive adhesives the contacts have a thin interlayer in the point contacts. When sequentially processing the composite, first microparticles, then the matrix penetration, a higher fill fraction of monodisperse particles and also true percolation can be achieved. It was shown before that percolation can be achieved with the sequential fabrication method, resulting in a 5fold improvement compared to the state-of-the-art. The point contacts can be further enhanced by reinforcing the point bonds by neck type collars between the fillers [4]. Figure 3 illustrates the concept.

To keep the porosity of the particle laden gap of height h large for a feasible backfilling, monodisperse fillers are an ideal system to study the influence of the particle diameter d. Thus, the ratio d/h is essential for the analysis. For small fillers compared to h the filler volume fraction η can potentially approach its maximum of $\eta \approx 0.74$ for the fcc crystalline packing of spheres, a pattern analyzed in detail for the PTUF Young's modulus in [6]. In real underfill cavities the volumes are strictly constrained, and this leads to lower filler volume fractions depending on the ratio d/h. Schmidt and Löwen[8] analyzed the potential packings in detail. Figure 4 shows the fill fraction against the ratio. Thermally, such an underfill will profit the most from a low d/h ratio in the *linear buckling* phase, even if real world filler packings will be affordable when exhibiting less regularly buckling patterns.

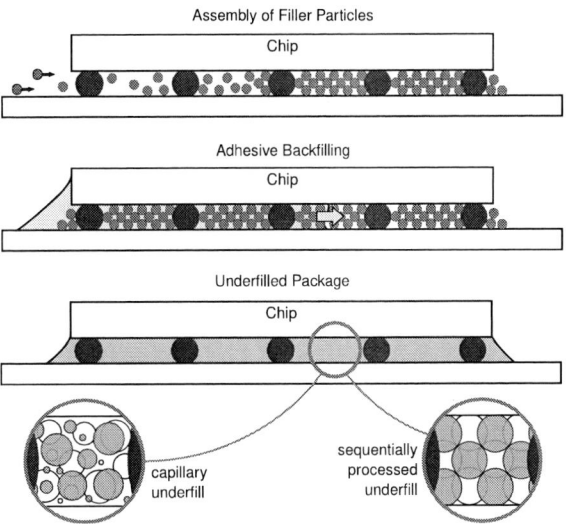

Figure 3. The PTUF concept: A sequential processing of fillers and matrix allows a different microstructure of the underfill layer, and thus properties beyond the pre-formulated capillary underfills.

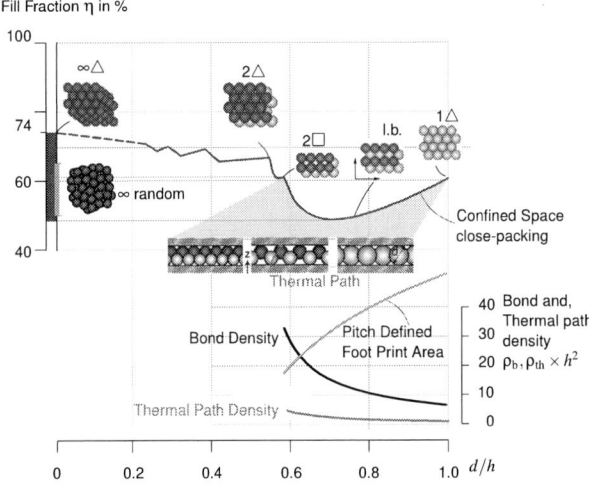

Figure 4. The filler volume fraction of the maximum possible packing in constrained space, and related to unconstrained space. The plot shows η against d/h, according to [8].

2. Methods: Three Stacked Layers

The fillers of the composite cannot penetrate the interface of the composite to adjacent materials, such as substrate or silicon. Thus, for monodisperse spherical fillers, the first composite layer of thickness $z_{\text{vicinity}} = d/2 = r$, parallel to the interface, cannot contain as many filler particles as the following bulk. (For polydisperse fillers z_{vicinity} needs to consider the diameter distribution, and could for instance take r_{average}.)

This different local content goes along with different local properties. When using an effective medium approach to model the underfill these local differences disappear. This necessary simplification might hide interesting behavior in the buckling phase, because the vicinity layer is not thin at moderate d/h ratios, compared to the bulk or a *central* layer (CL).

For the linear buckling phase we define therefore three layers parallel to the interfaces, and analyse them subsequently. The layers are the Vicinity and Central Layers, as shown in the following figure:

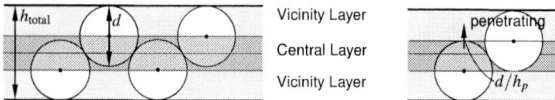

Figure 5. Sublayers of the PTUF: Vicinity Layers of constant height $r = d/2$, and the Central Layer of varying thickness. The right schematic shows the transition point, all filler particles start to take share in the vicinity layers.

Accordingly, Figure 6 shows the filler volume fractions η in the buckling phase for all three: the entire underfill layer η, the interface vicinity layer η_{vicinity}, and the central layer η_{central}. The blue shaded data marks all ratios below a transition point p: At smaller ratios d/h than at p the vicinity layer contains only filler hemispheres in interface contact. At larger ratios more fillers share the vicinity layers. At p the unit volume of the central layer becomes maximum, not its height. In terms of the transition point, the ratios d/h below yield more thermal paths, and the PTUF concept is thermally more attractive.

The fill fraction of the central layer η_{central} converges to impenetrable disk packing density of $\pi\sqrt{3}/6 \approx 0.907$ because with increasing d/h ratio the central layer thins down and approaches the 2dimensional limit. Towards lower d/h ratios, η_{central} approaches the maximal 3dimensional packing density of an fcc lattice $\eta_{\text{fcc,perc}} = \pi/(3\sqrt{2}) \approx 0.74$. Before, it encounters a significant minimum at the CL's maximum unit volume.

As a consequence of this minimum, the center layer reaches $\eta_{\text{fcc,perc}}$ twice along d/h, and it reaches a fill fraction very similar in the range between. Thus, a fcc$_{\text{perc}}$ model may approximate the center layer properties sufficiently well. The vicinity fill fraction η_{vicinity} approaches the monolayer densest packing density of $\eta_{\text{vicinity}} \approx 0.6$ because the central layer vanishes at $d/h = 1$.

For deriving an effective Young's modulus from the three-layer model in z-direction, normal to the interfaces, the stacking of the layers results in an in-series connection, where an applied stress field load remains uniform between the layers. Thus, the harmonic mean, such as of the lower Reuss bound, can be used. As a consequence of sublayering the underfill body, this does represent all point-contacts within the layers, although not fully between the layers. The effective modulus $E_{\text{eff,Reuss}}$ is then

$$E_{\text{eff,Reuss}} = \frac{E_{\text{vicinity}} E_{\text{central}}}{\eta_{\text{vicinity}} E_{\text{central}} + \eta_{\text{central}} E_{\text{vicinity}}}, \quad (1)$$

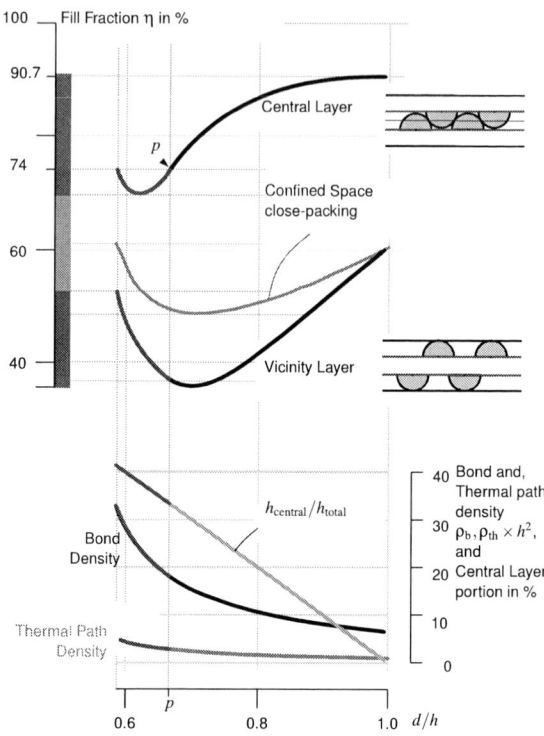

Figure 6. Filler Volume Fractions of both the Central and Vicinity Layers vs. d/h compared to the entire PTUF layer in the linear buckling phase. At the bottom of the plot the densities of bonds are thermal paths per unit area, as well as the central layer height portion. Blue indicates the portions for d/h below the transition point.

where E_{vicinity} and E_{central} are the respective layer moduli. A schematic of the three-layer model looks like this:

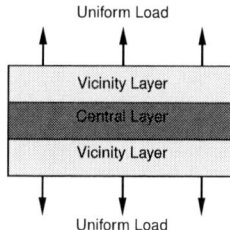

Figure 7. Schematic of the three-layer model: Two vicinity layers enclose a central layer, thermomechanically the connect in-series under a unifrom load normal to the interfaces.

The simulation parameters in the fcc unit cell When assuming isotropic behavior, both η_{vicinity} and η_{central} can be approximated using fcc unit cell derived effective Young's moduli $E_{\text{eff,fcc}}$. Above the transition point η_{central} increases steadily with η towards values that could be approximated by fcc inter-penetrating fillers. The inter-penetration leads to a steady decrease in matrix portion in the unit cell. The vicinity layer shows a strong η_{vicinity} decrease for lower d/h ratios, in a η range that has already been presented [6] for fcc lattice packings. A fcc unit cell of low η does not contain any point contacts (thermo-mechanically communicating bonds), whereas the vicinity layer does. Also, the vicinity layer offers parallel force paths through the different moduli of fillers and matrix. These paths are not present in the fcc model. Thus, the fcc might introduce errors into the vicinity layer behavior. To estimate such errors, we thus calculate an effective Young's modulus $E_{\text{eff,Reuss}}$ for the three-layer stack and compare it to an $E_{\text{eff,cs}}$ obtained from constrained space unit cells in [7].

Table 1 shows the Parameters varied in [6]. By varying the filler spacing in a range of $(+r\ldots+50\,nm\ldots-50\,nm\ldots-r/3)$ almost the entire range of η could be covered. Negative values thereby correspond to interpenetrating spheres. This study restricts the comparison to Silica fillers in a 2 GPa matrix.

Table 1. Properties varied to study their impact on the effective behavior of a fcc lattice packing over η, values from [6].

Material	Young's Modulus	CTE in ppm/K	Poisson's ratio
Polymer Matrix	10MPa to 5GPa	20 to 80	0.3
Silica Fillers	70GPa	0.6	0.20
Alumina Fillers	300GPa	5 to 8	0.21

The following schematic shows the unit cells used in the unconstrained fcc packing and the constrained space packing. For clarity the filler volumes are transparent in the plot. This reveals the size of the contact points. Figure 8 shows a penetration of 50 nm (-50 nm spacing) in the fcc model (a) and a neck contact (an enlarged contact point) at a 100 nm spacing of the fillers.

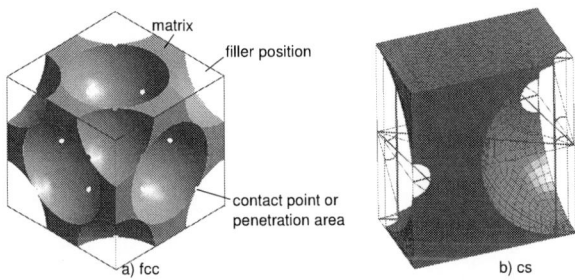

Figure 8. Unit cells of both the unconstrained fcc lattice (a), and the contrained space linear buckling phase (b) for the models in [6], [7].

Experimentals To proof the correctness of modulus tendencies of the earlier fcc results we sequentially fabricated bulk specimen using different fillers and matrix materials. Figure 9 shows the according tool made from Aluminum. The tool allows to mold 20 mm long specimen of a $2\times1\,mm^2$ cross section. We used a 45 µm filler diameter of silica spheres ($d/h = 0.045$) in two different epoxy resins with a Young's modulus of 2.1 and 2.3 GPa at room temperature. The tool was manually filled with the silica spheres and the particle laden tool then evacuated. Under vacuum we backfilled the cavities of the tool with

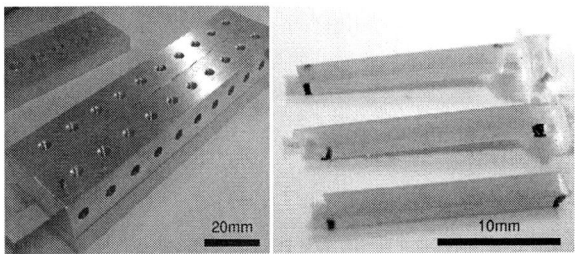

Figure 9. The mold tool shown on the left, typical specimen shown on the right.

the resin, and let the specimen cure at the resin specific temperatures and times. To obtain the effective Young's modulus we used a three-point bending experiment with a 16 mm wide support span in a strain controlled load frame (Thelkin SelMaxi) with a 50 N load cell to allow force readings. We recorded the load reactions twice for two samples of each material combination and fitted the results in the lower linear range. Two cross sections of the specimen allowed a fill fraction estimate. Specimen that reflect a constrained space buckling phase were not ready for measurements at the time of print.

3. Results: Moduli of Unit Cells, Three-Layer Stack and Experiments

Figure 10 shows the Young's moduli vs. fill fraction for the fcc and cs models. To achieve an η variation the fcc model changes the filler spacing, whereas the cs model keeps a filler spacing of 100 nm but changes the d/h ratio.

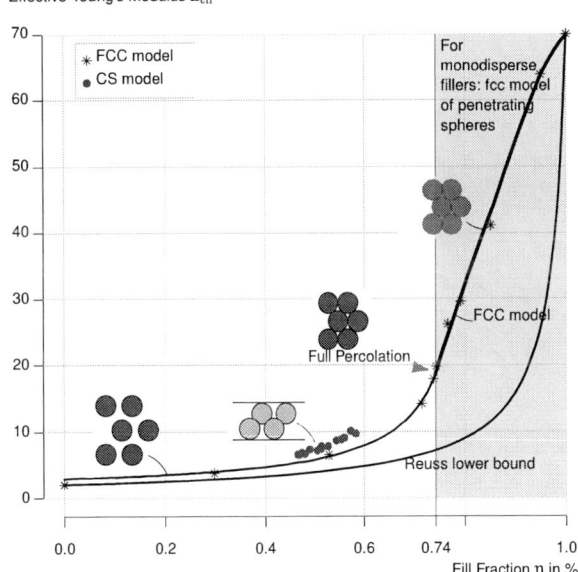

Figure 10. The effective Young's Modulus E_{eff} against Fill Fraction η for the fcc and cs unit cell models of [6], [7], both for Silica fillers in a 2 GPa matrix.

Using the fcc effective Young's modulus for the three-layer model dependent on the corresponding fill fractoin η, the resulting Young's moduli for the two different layers are plotted in Figure 11. Note that in the graph the fill fraction projects the buckling phase blue part only, that is all values below the transition point.

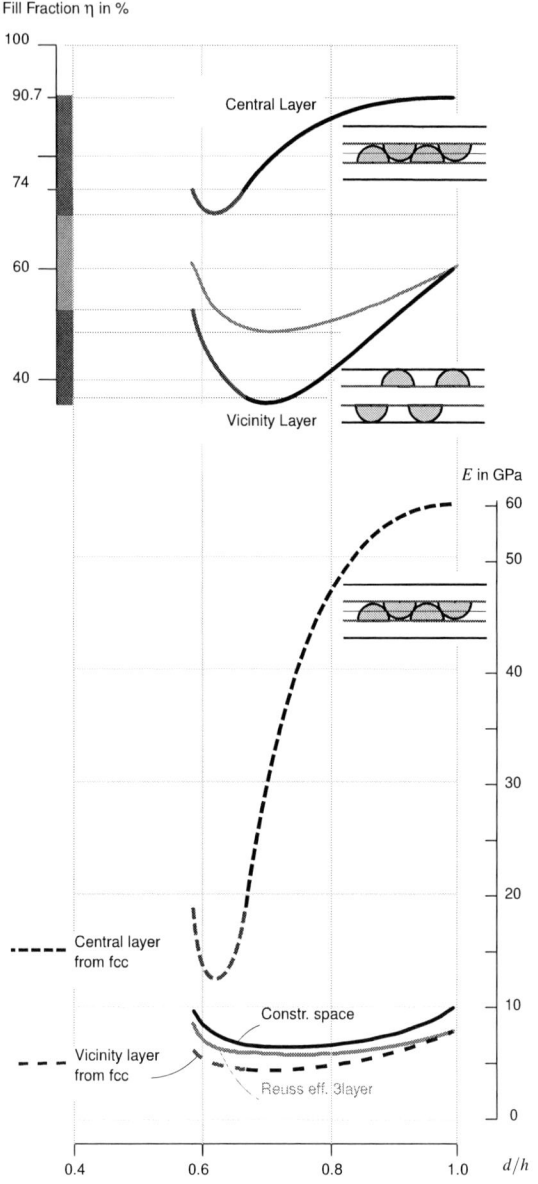

Figure 11. The Young's Moduli E against the d/h ratio for the different layers, compared to the effective Young's Moduli E_{eff} of the cs model (black solid), and according to the Reuss three-layer stacking (red).

The plot shows that both layers' moduli obviously follow their fill fraction, and as the fill fractions are below and above the effective value, also the moduli are below and above the effective value. The plot also shows that the central layer modulus undergoes the largest change throughout d/h ratios of the buckling phase. It shows

a local minimum in the range where the layer's height takes its largest share, a minimum with more about 30% modulus reduction compared to the lowest d/h ratio (silica filler in a 2 GPa epoxy matrix). This minimum occurs at the minimal filler volume fraction of the central layer, where the layer maximizes its unit volume, at a $d/h \approx 0.62$.

The vicinity layer shows much less variation along d/h, and converges into the effective modulus for the vanishing central layer towards $d/h = 1$. The effective Young's modulus is lower than, yet close to a unit cell modeling of constrained space packing, as presented in our parallel work at [7]. Its maximum deviation is ca. 21 % towards $d/h \approx 1$, the deviation at the minimum central layer modulus of $d/h \approx 0.62$ is ca.18 %. The systematically lower values indicate too much matrix dominance compared to filler modulus. This underlines the conditional choice of fcc for the vicinity layer, because the fcc cannot account for filler contacts at lower fill fractions.

Experimental readings In the Young's modulus plot in Figure 12 the experimental results are compared to the fcc model. The experimental spread along the E-axis reflects the spread of eight data points, the spread along the η-axis an estimated fill fraction from two specimen cross sections. The E_{exp} values settle at modulus values of 11.1

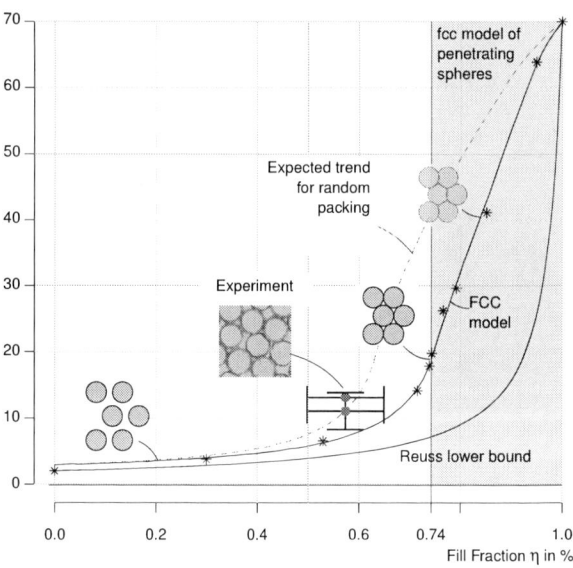

Figure 12. The experimentally determined Young's Modulus E_{exp} (three-point bending) compared to E_{fcc}, and plotted against the d/h ratio.

and 13.1 GPa, and thus higher than the fcc model data of 7.1 GPa at an estimated η = 0.57. In the bulk specimen fabrication the fillers packed in a random distribution and η thus must be lower than $\eta_{\mathrm{fcc,perc}}$. This has two consequences on the Young's modulus: The modulus is lower because of fewer thermo-mechanically communicating point contacts. And the onset of modulus increase towards higher filler fractions should be expected at lower values of η, because in a random close packing there is *partial* percolation before full percolation (there are more point contacts at low fill fractions than in a lattice of equidistant fillers).

4. Discussion

The three-layer model reveals a 33% modulus drop in the central of the three layers (E_c) at a low d/h ratio, towards a local minimum before the transition point. This minimum lies within the thickest central layer portions (more than a third of h_{total}). Also thermally, the lower d/h ratios are favorable. Although E_c increases 4fold towards $d/h = 1$ after this minimum, the impact of this increase is marginal due to the decrease in $h_{central}$. The 30% change in modulus remains hidden in the one-layer effective modulus. The fact that the minimum shows at the central layer's maximum volume indicates a potential usefulness. For a package, a lower modulus of the central layer could be thermo-mechanically attractive. Thus, a sub-layering of PTUF layers could have two advantages: a) Compared to the one-layer effective modulus, a three-layer approach adds a dependence on the microstructural geometry ($E = f(\eta(z))$) to the model. This dependence becomes important because the μm-sized features (fillers) reach the scale of the underfill layer thickness. b) When optimizing the underfill properties, a sub-layering could reveal a d/h ratio that offers a better compromise of CTE and Young's modulus for both the interconnect protection and thermal resistance.

The full-percolating fcc model (fcc$_{perc}$) seems appropriate for a thinned layer of similar filler fractions and similar sphere point-contacts, such as in the central layer. An fcc model at lower filler fractions cannot represent the filler-to-boundary contacts present in the vicinity layer. Thus, it creates a larger error for the vicinity layer. On the one hand, both fcc models together alterate a E_{eff} only by ca. 20 % compared to the contrained space model. On the other hand there is not much potential to tailor the modulus of the vicinity layer by the d/h ratio only, especially at lower d/h ratios that are thermally appealing. A dedicated constrained space unit cell model could be used to obtain the vicinity layer modulus.

The model does further not reflect anisotropy at this point, although it could be included as an effective property. Anisotropy needs to be expected in strictly crystalline filler networks. A strong anisotropy originates in dominant bond directedness, such as in the linear buckling phase, and as analyzed for the CTE in [7]. A weaker anisotropy originates for instance in the fcc lattice, as analyzed for the Young's modulus in [6] to cause 5 % difference in the lattice orientations 100 and 110. Strictly crystalline packings are very difficult to fabricate practically though. This three-layer study limits the analysis to a one-directional Young's modulus, calculated in z-tension normal to the interfaces.

Independently from the individual optimum, the d/h ratios of 0.58, 0.62, and 0.71 correspond to filler diameters of 35, 37 and 42 μm in a 60 μm cavity. Likewise, a 45 μm diameter filler corresponds to different cavity heights of 77, 73, and 64 μm. Thus, to tailor the d/h ratio in the discussed range the absolute measures require high monodispersity quality of the fillers.

The experimentally obtained Young's moduli support an expected slope-shift towards lower fill fractions, when departing from the fcc lattice model towards a randomly packed composite. However, for both the d/h ratio and η the bulk specimen are not suited to compare to the linear buckling phase composite. Such specimen are however crucial to verify the findings. By grinding the specimen thinner a buckling phase similar specimen could be generated. Preferably though, specimen should be created in their final height during the matrix cure already.

5. Conclusions and Outlook

Dividing the linear buckling phase of a constrained space packed Percolating Thermal Underfill into three layers shows significant changes of the central layer's Young's modulus over the ratio d/h. This z-position dependency remains hidden in models without $E = f(\eta(z))$ type resolution. Thus, package scale thermo-mechanical simulations may profit from sub-layering the underfill layer. Specifically before the transition point, where a central layer can take more than one third of the total layer thickness, the $E_{central}$ shows a minimum that is more than 30 % below the extremal value of the buckling phase of largest possible thickness ($d/h \approx 0.59$). For silica fillers in a 2 GPa matrix material $E_{central}$ drops from 18.8 GPa to 12.5 GPa, the latter elasticity at a layer thickness 38 % of the total underfill thickness. Future work can improve and extend the findings by chosing a better model for the vicinity layer, by modeling a neck based PTUF, by adding anisotropy, by investigating the CTEs, and finally by realizing modulus measurements of real buckling phase specimen.

6. Acknowledgements

This work was partly carried out at the Advanced Thermal Packaging Group of the IBM Zurich Research Laboratory, and we thank Bruno Michel and Walter Riess for their contineous support of this activity. Partly this work was also carried out within the European Hyperconnect Project under the Seventh Framework Program for Research and Technological Development, (FP7-NMP-310420). We thank Elke Noack and Remì Pantou of the FhG ENAS in Chemnitz for the computer tomography analysis, and Florian Schindler-Saefkow of AMIC in Berlin for the FE modelling support.

References

[1] P.N. An and P.A. Kohl. Thermal-mechanical stress modeling of copper chip-to-substrate pillar connections. *IEEE Transactions on Components and Packaging Technologies*, 33(3):621–628, 2010.

[2] G. Schlottig, T. Brunschwiler, J. Goicochea, W. Escher, and B. Michel. A multivariate parameter analysis of copper pillars eases the design of denser interconnects. In *13th International Conference on Thermal, Mechanical & Multi-Physics Simulation, and Experiments in Microelectronics and Microsystems (EuroSimE)*, 2012.

[3] G. Sears. LORD solderbrace™ for improved reliability and throughput in WLCSP. In *46th International Symposium on Microelectronics (IMAPS)*, 2013.

[4] J.V. Goicochea, Thomas Brunschwiler, J. Zürcher, H. Wolf, K. Matsumoto, and Bruno Michel. Enhanced centrifugal percolating thermal underfills based on neck formation by capillary bridging. In *2012 13th IEEE Intersociety Conference on Thermal and Thermomechanical Phenomena in Electronic Systems (ITherm)*, pages 1234–1241, 2012.

[5] T. Brunschwiler, G. Schlottig, S. Ni, L. Yu, J.V. Goicochea, J. Zürcher, and H. Wolf. Formulation of percolating thermal underfills using hierarchical self-assembly of microparticles and nanoparticles by centrifugal forces and capillary bridging. *Journal of Microelectronics and Electronic Packaging*, 9:149—159, 2012.

[6] G. Schlottig, F. Schindler-Saefkow, J. Zürcher, B. Michel, and T. Brunschwiler. Sequentially formed underfills: Thermomechanical properties of underfills at full filler percolation. In *Proceedings of the 15th Electronics Packaging Technology Conference (EPTC)*, pages 560—564, Singapore, 2013.

[7] T. Brunschwiler, F. Schindler-Saefkow, M. Haupt, R. Gordin, J. Zürcher, S. Zimmermann, and G. Schlottig. Compound properties study of percolating and neck-based thermal underfills. In *14th IEEE Intersociety Conference on Thermal and Thermomechanical Phenomena in Electronic Systems (ITherm)*, to be published, 2014.

[8] M. Schmidt and H. Löwen. Phase diagram of hard spheres confined between two parallel plates. *Physical Review E*, 55(6):7228, 1997.

Hidden Head-In-Pillow soldering failures

Bart Vandevelde[a], Geert Willems[a], Bart Allaert[b]

[a]imec, Kapeldreef 75, Leuven, B-3001, Belgium

[b]Connect Group, Frankrijklaan 22 - 8970 Poperinge, Belgium

e-mail: Bart.Vandevelde@imec.be

Abstract

One of the upcoming reliability issues which is related to the lead-free solder introduction, are the head-in-pillow solderability problems, mainly for BGA packages. These problems are due to excessive package warpage at reflow temperature. Both convex and concave warpage at reflow temperature can lead to the head-in-pillow problem where the solder paste and solder ball are in mechanical contact but not forming one uniform joint. With the thermo-Moiré profile measurements, this paper explains for two flex BGA packages the head-in-pillow. Both local and global height differences higher than 100 µm have been measured at solder reflow temperature. This can be sufficient to have no contact between the molten solder ball and solder paste. Finally, the impact of package drying is measured.

1. Introduction to Head-in-Pillow soldering issues

As an IC package consists of different materials (silicon, mold compound, underfill, BT substrate, copper leadrame, etc.) having different coefficients of thermal expansion (CTE), the materials expand or shrink differently under temperature changes resulting in mechanical stresses and global package warpage.

Excessive package warpage at temperatures above solder reflow temperature can lead to solder process failures. Both convex warpage (corners bends downwards) and concave warpage (corners bends upwards) can happen but they can lead to different kind of soldering failures.

When the package has an excessive **convex** deformation at reflow temperature (Figure 1), the outer solder joints can collapse such that two or more joints form one shorted joint. After cooling down below solder reflow temperature, this joint is frozen in and the component will indicate an electrical failure after assembly [6].

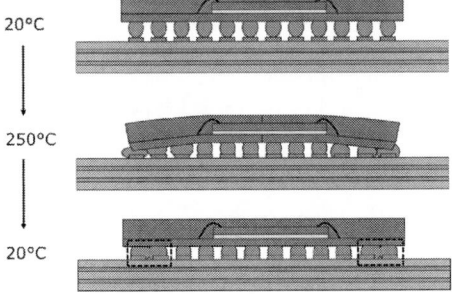

Figure 1: Schematic drawing indicating the impact of convex warpage at reflow temperature leading to shorts between balls

When the package has an excessive **concave** warpage, another soldering failure can be experienced. A too high concave warpage at solder solidification temperature can cause the so-called head-in-pillow where the solder paste deposit wets the pad, but does not fully wet the ball or the liquid solder connection gets disconnected without merging together again prior to solidification (Figure 2). This results in a solder joint with enough of a connection to provide temporary electrical interconnectivity but lacking sufficient mechanical strength (Figure 3). Due to the lack of joint strength, the joint will fail with little mechanical stress. This defect is usually not detected in electrical testing, and typically shows up as a field failure after the assembly has been exposed to some temperature increases where the BGA ball loses from the PCB solder pad due to package warpage.

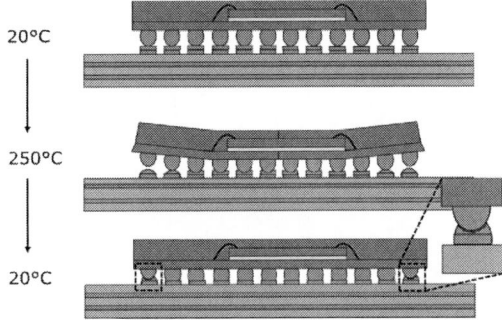

Figure 2: Schematic drawing depicting the consequences of concave warpage at reflow temperature leading to head-in-pillow failures.

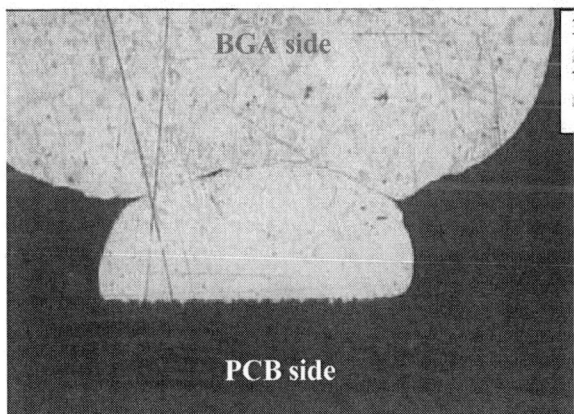

Figure 3: Cross-section visualising the head-in-pillow BGA joint seen after cooling down from soldering process. It actually looks like a head has pressed into a soft pillow.

978-1-4799-4789-8/14 $31.00 © 2014 IEEE

Convex warpage can also lead to head-in-pillow in the center area: this depends a lot on the shape of the deformation, determining which solder joints can keep the package attached to the board.

The head-in-pillow failure mode gained a lot of importance since the transition to lead-free assembly [1, 2, 3]. About 30°C higher reflow temperature causes more absolute deformation. And the use of low-CTE (7 to 10 ppm/°C) mould compounds increasing the CTE mismatch with the PCB (14-18ppm/°C) and the BT substrate (14 ppm/°C) [3]. In particular the latter one brings back the head-in-pillow as critical process failure issue (which was basically non-existing in the pre-leadfree/green mould era). Moisture uptake of the package and/or PCB can also have an impact on the warping behaviour of the package, as it will also be revealed later in this paper. Baking of the packages can be one of the solutions.

While warpage is the main source of the head-in-pillow problem and therefore should be well controlled, improving the flux chemistry in combination with an optimised thermal profile can also reduce the issue. When the BGA solder ball is not in contact with the solder paste (including the fluxing agent), the ball oxidises and when the ball comes back to the solder paste, the fluxing agent can be gone. The addition of flux dipping, and/or N_2 gas reflow both reduces the HiP defect rate.

Increasing the paste volume is also recommended. This can be done by using square aperture vs. round opening, or by enlarging overall deposition volume without jeopardizing bridging.

2. Sample description and measurement setup

This paper reports the head-in-pillow soldering failures experienced with two flex BGA packages. The basic geometrical properties of these two packages are summarised in Table 1.

Table 1: Description of two flex BGA packages measured in this work

17mm FBGA	1 mm pitch, 0.5 mm ball size
	256 pins (16x16 area array)
	17x17x1.4 mm³
27 mm FBGA	1 mm pitch, 0.64 mm ball size
	676 pins (26x26 area array)
	27x27x2.25 mm³

In order to be able to explain the head-in-pillow soldering issues, measurements of the warpage during a solder reflow profile are measured using the INSIDIX measurement system [1]. The warpage is measured experimentally by means of a topography and deformation measurement (TDM) and is based on the Projection Moiré principle. A light projector illuminates the sample with a striped light pattern consisting of equidistant parallel lines, under an angle of about 45°. The projected light pattern is recorded by a camera. If the sample surface is curved, there will be a variation in the recorded light pattern. By measuring the offset between the projected and recorded pattern, one can calculate the amount of curvature. In addition, the sample can be heated up and cooled down while performing TDM. This allows to simulate the temperature profile of manufacturing processes and to calculate the corresponding warpage evolution.

A typical reflow temperature profile has been applied to these packages, with a maximum of 245°C.

3. Hidden head-in-pillow solder failure for 17 mm flex BGA packages

As the 3D profile of the bottom side of the BGA is difficult to measure due to the balls, the warpage was measured at the top "mould" side of the package. The package was stored at room temperature and did not have any drying nor wetting process in advance. It is expected that the package took up some moisture.

Figure 4 shows the measured out-of-plane profiles of the 17 mm size packages at different temperature stages of the solder temperature profile. It can be seen that the package is rather flat at room temperature and a convex warpage is found in particular above 200°C, with a highest deformation at 245°C.

Figure 5 shows the difference between the topography profile at 245°C and at 23°C. The delta graph obviously indicates that this BGA package gets a convex deformation when it is at the highest temperature. The four corners are warping down. Figure 6 shows this out-of-plane deformation over the diagonal. The package warps almost **150 µm** and this explains why the inner balls are having a potential for having the head-in-pillow solder failure.

The JEITA ED7306 guideline sets that for package with 1 mm pitch and 0.5 mm ball height, the maximum permissable warpage is 220µm. Taking into account that the board itself also can warp and at room temperature, the flex side is already warped down several tens of microns, the measured warpage comes close this critical value.

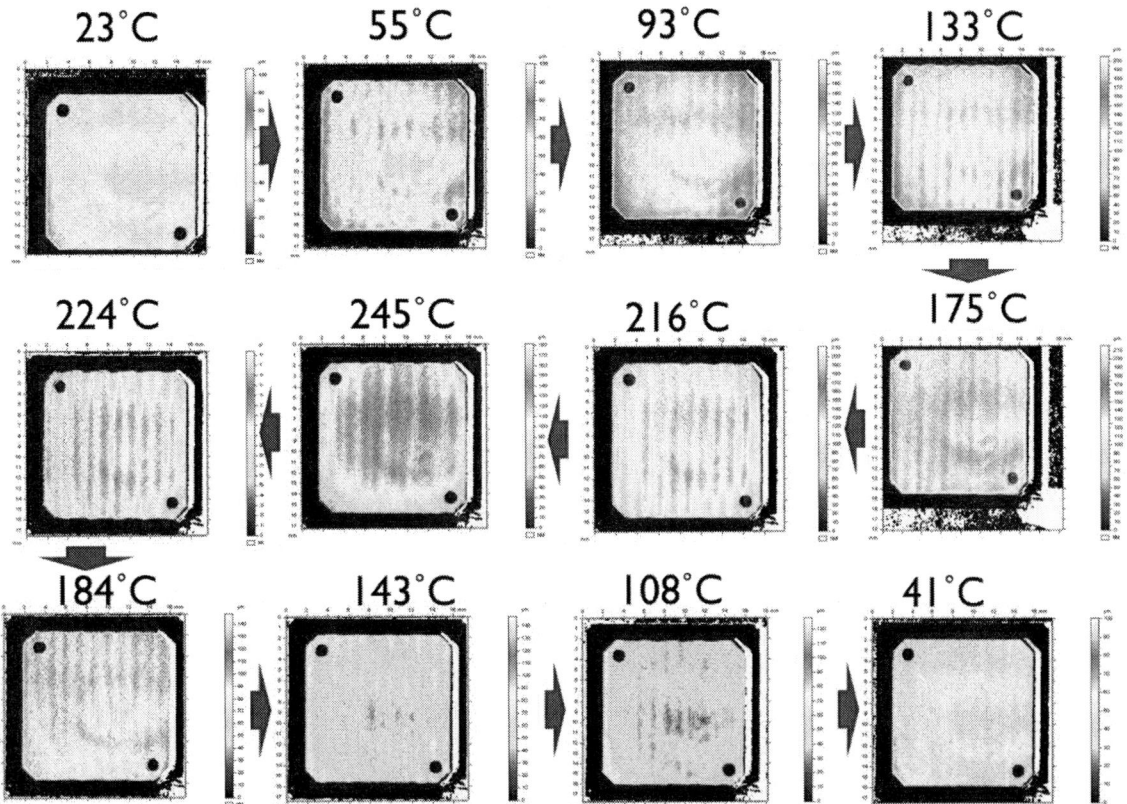

Figure 4: 3D profile of the 17 mm flexBGA package measured during a solder reflow profile. The package was stored at room temperature being able to take up some moisture

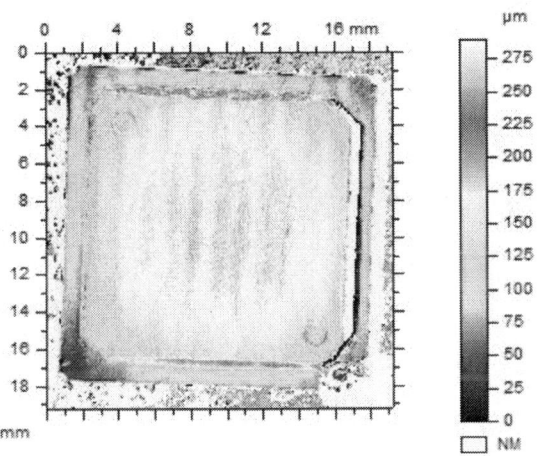

Figure 5: Difference in z-deformation between 245°C and 26°C. This plot indicates that the package has a convex deformation from room to reflow temperature (17 mm FBGA)

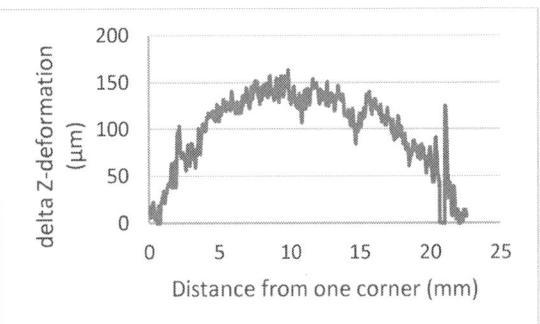

Figure 6: Delta warpage between 246°C and 26°C shown over one diagonal (17mm FBGA)

As shown in Figure 7, the convex warpage mainly occurs above 180°C. This is caused by the higher thermal expansion of the mould compound than the expansion of the laminate. The turning point around 110°C is probably related to the glass transition point Tg of the mould compound. Below this Tg, the CTE of the mould compound is lower than the one of the laminate.

978-1-4799-4789-8/14 $31.00 © 2014 IEEE

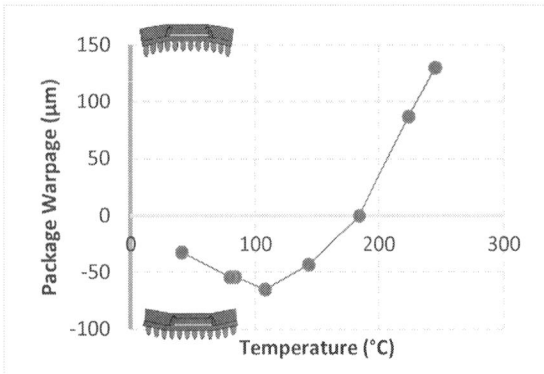

Figure 7: Warpage evolution of 17 mm FBGA as function of temperature

4. Hidden head-in-pillow solder failures for 27 mm flex BGA packages

For a 27x27 mm^2 flex BGA package with 1 mm ball pitch, another kind of "hidden" head-in-pillow was experienced. The non-wetted balls were not found in the inner area like with the 17 mm package, nor at the edge, however, they were found in the third and fourth outer row. Soldering issues for the same package type were also reported by reference [5].

In order to explain this result, the warpage versus temperature measurements were performed on this package with the same measurement approach. The topographies at different temperature stages are shown in Figure 8.

At room temperature, the package has some convex warpage. But at temperatures above solder reflow temperature, the package gets a concave shape. This is however only for the inner area. The edges and the four corners are still warping downward when the temperature rises, as it is indicated in the delta contour plot in Figure 9 and the z-deformation over the diagonal Figure 10.

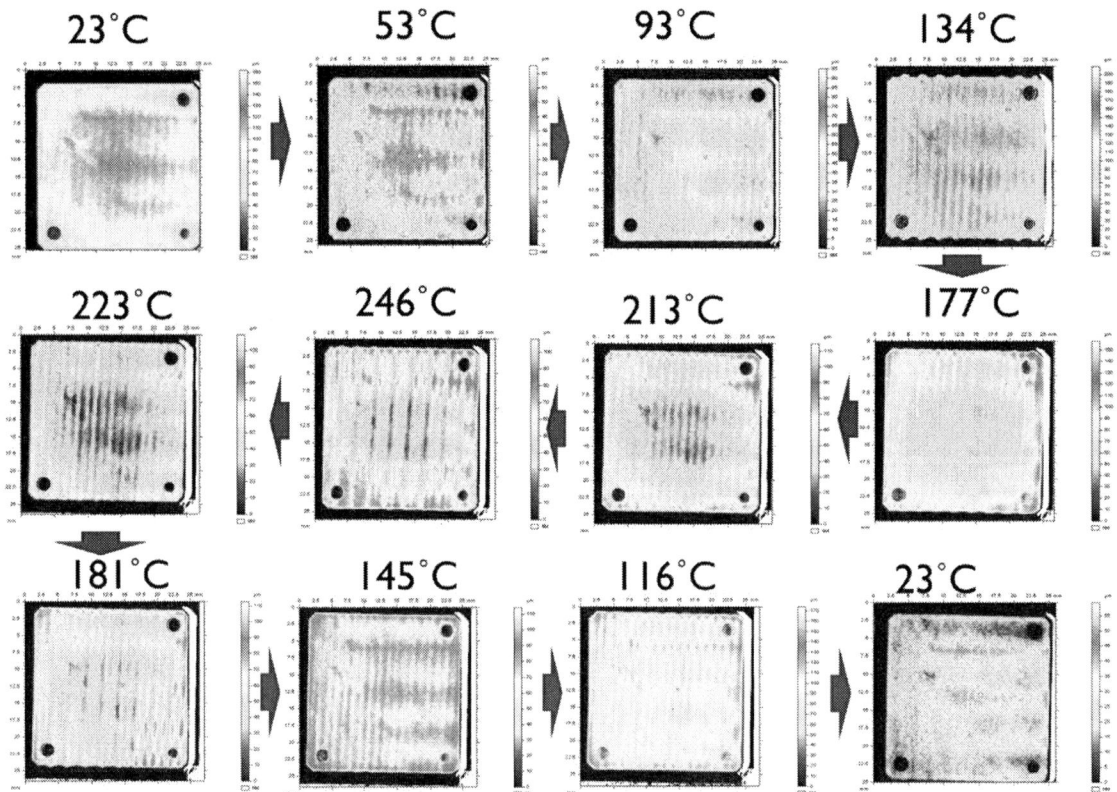

Figure 8: Warpage of the 27 mm FBGA package measured during a solder reflow profile.

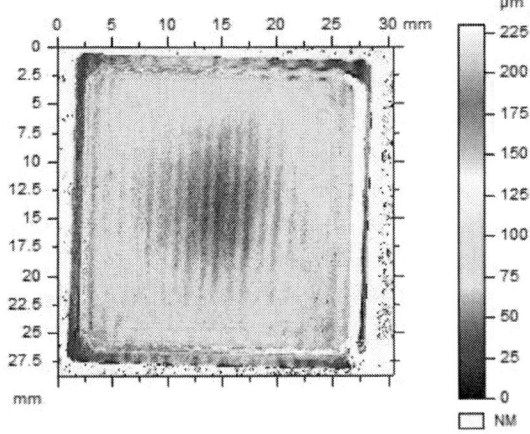

Figure 9: Difference in z-deformation between 245°C and 26°C for the 27 mm FBGA

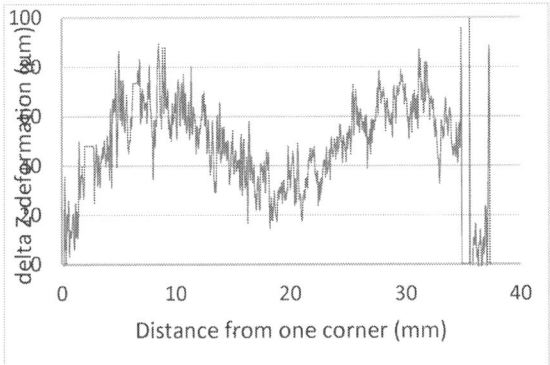

Figure 10: Delta warpage between 246°C and 26°C shown over one diagonal (for 27mm FBGA).

The consequence of such deformation profile at reflow temperature is that at about 3 to 4 mm from the edge, the package is at highest point. With a 1 mm pitch, this means that the third and fourth row of solder joints have the risk to have no contact with the printed circuit board leading to the head-in-pillow issue experienced with this package. This is schematically shown in Figure 11. Taking into account already some initial convex warpage of the package at room temperature and also the outer unmoulded flex substrate which is warping down, we can easily have a height difference of 100 µm between the third and the outer row of solder joints at reflow temperature. This can be sufficient to get the head-in-pillow issue.

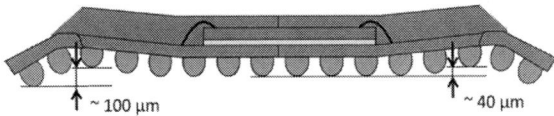

Figure 11: Schematic drawing showing the deformation of FBGA 27mm package at reflow temperature. Between the third and outer row of solder joints, about 100 µm height difference was measured.

5. Impact of moisture on risk for head-in-pillow

The warpage-temperature evolution has been measured for three 17mm FBGA packages with following pre-conditions:

- the package stored at room temperature: this is the one which has been presented in section 3
- the dried package which has been put for 5 days at 125°C
- the wetted sample which was put for 15 hours in the 85°C/85% oven

Figure 12 shows the warpage evolution for the three pre-conditions. As expected, the highest warpage was found for the wetted package, the lowest for the dried package. The difference is about a factor 4. The stored sample is lying in between the two curves. Drying these packages before the reflow process gives a reduction of about 2.5 in warpage and probably solves the head-in-pillow issue.

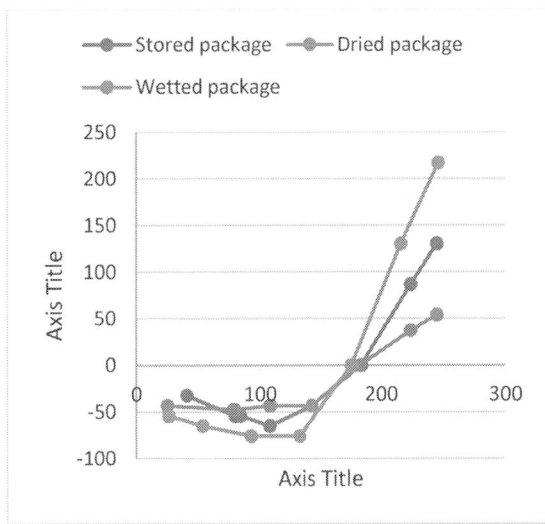

Figure 12: Warpage evolution of 17 mm FBGA as function of temperature.

6. Conclusions

Head-in-Pillow is a challenging and complex soldering problem with much risk for the OEM. A head-in-pillow defect is the incomplete coalescence of the solder joint between a BGA and the printed solder paste.

Following conclusions and recommendations are defined in this work:

- when head-in-pillow failures are experienced after soldering, Project Moiré measurements are very useful to visualise and quantify the warpage problem
- for two flex BGA packages, a high warpage at reflow temperature was measured which explains the head-in-pillow sensitivity of these package

- although both packages are flex BGA's, the HiP balls are found at different locations: the 17 mm has the highest problems in the center area, the 27 mm FBGA has the HiP balls in the third and fourth outer row
- besides improving solder paste and temperature profile conditions, drying the packages reduces the warpage and the risk for head-in-pillow. The reduction in warpage was a factor 2.5 for the 17 mm FBGA

Acknowledgments

The authors would like to thank Bjorn Debecker and the REMO group at imec for their support in this work. Also thanks to the Flemish funding agency IWT for their support to the I SEE project.

References

1. Scalzo, M., "Addressing the Challenge of Head-inPillow Defects in Electronics Assembly", Indium Corporation Technical Library, 2009.
2. Vandevelde, B. Excessive warpage of large packages during reflow soldering. In: The ELFNET Book on Failure Mechanisms, Testing Methods and Quality Issues of Lead Free Solder Interconnects. Springer; pp.283-296; 2011. (Chapter 13)
3. Vandevelde B., Deweerdt R., Duflos F., Gonzalez M., Vanderstraeten D., Blansaer E., Brizar G., Gillon R. (2009), Impact of Moisture Absorption on Warpage of Large BGA packages during a lead-free reflow process, pp. 162-165, Therminic Workshop, Leuven, Belgium.
4. M. Hertl, D. Weidmann, and J-C. Lecomte (2009), Process Optimization: Influence of Heating and Cooling Rate on the Thermo-Mechanical Stress Generated in Components, EMPC2009, Rimini, Italy.
5. A. Arazna, G. Koziol, W. Steplewski, K. Lipiec, Head on pillow defects in BGAs solder joints, ESTC conference, 13-16 Sept. 2010, Berlin, Germany.
6. B. Vandevelde, M. Lofrano and G. Willems, Green mold compounds: impact on second level interconnect reliability. In: Electronics Packaging Technology Conference - EPTC. ieee, 2011. (7-9 December 2011; Singapore, Singapore.)

Reliability and Accelerated Test Methods for Plastic Materials in LED-Based Products

M. Yazdan Mehr[1,2*], W.D. van Driel[2,3], G.Q. Zhang[2]

[1] Materials innovation institute (M2i), Delft, The Netherlands

[2] Delft University of Technology, EEMCS Faculty, Delft, The Netherlands

[3] Philips Lighting, Eindhoven, The Netherlands

Abstract:

In this study the effects of thermal ageing on the optical properties of both lens and the remote-phosphor samples, made from Bisphenol-A polycarbonate (BPA-PC) are investigated. The BPA-PC lens and remote phosphor plates are currently widely used in light conversion carriers and optical lenses in LED-based products. Lens and the remote phosphor BPA-PC samples of 3 mm thickness were thermally aged at temperature range 100 to 140 °C. The phosphor plates, combined with a blue LED light source, produce white light with a correlated colour temperature (CCT) of 4000 K. The colour shifting due to thermal ageing was studied by Integrated Sphere. Results show that thermal ageing leads to a significant decrease in the luminous flux and chromatic properties of plates. It is also shown that by increasing the temperature, the kinetics of degradation reaction becomes faster, inferring that lumen depreciation takes place at shorter time. Lumen depreciation up to 30% reduction is extrapolated to temperatures lower than 100 °C. It is shown that the lifetime, defined as 30% lumen depreciation at 40 °C, is around 35 khrs for remote phosphor and around 100 khrs for BPA-PC lens. A significant change both in the correlated colour temperature (CCT) and in the chromaticity coordinates (CIE x,y) is also observed in thermally aged specimens. Deterioration of the chromatic properties of the phosphor plates is correlated to the decrease in the luminous flux. Results also confirm the colour shifting of white light towards yellow region. Based on the observed decay of CCT and colour shifting, one could conclude that the thermal degradation of the remote phosphor plates affects both the efficiency and the colour of the LED products. The proposed thermal-ageing qualification method can be used by industries to efficiently select the proper phosphor materials and verify the product design, without many trial-error based interactions.

Introduction:

The BPA-PC lens and remote phosphor plates are currently widely used in light conversion carriers and optical lenses in LED-based products [1- 4]. In 1996, a new lighting device was invented by Nichia Chemical Co. which is a blue InGaN LED chip coated with yttrium aluminum garnet yellow phosphor ($Y_3Al_5O_{12}$:Ce, YAG:Ce). When the chip is biased under certain current, blue light is emitted by the InGaN chip through electron-hole recombination in the P-N junctions. Some of the blue light from the LED excites the YAG:Ce phosphor to emit yellow light, and then the rest of the blue light is mixed with the yellow light to generate white light [5].

To describe "colour", chromaticity coordinates (x,y) and colour temperature (CT) are more often used, with the former being defined by the Commission Internationale de l' Eclairage (CIE) System and the latter being the temperature of a blackbody whose chromaticity most nearly resembles that of a light source [5]. Low colour temperature implies warm while high colour temperature appears to be a colder (more blue) light. Daylight has a rather low colour temperature when the sun is dawn , and a higher one during the day.

The main target of this paper is to investigate the effect of ageing temperature on the optical properties and the reliability of remote phosphor. For this reason a set of accelerated thermal stress tests were applied with temperature level between 100 and 140 °C. The reliability studies and life time assessment at temperatures lower than 100 °C are done by extrapolation.

2. Experimental methods:

One type of 3 mm-thick remote phosphor plate with Correlated Colour Temperature (CCT) of 3000 K are used in this study. The specimens are kept in a furnace at 100, 120, and 140 ºC up to 3000 h. Testing temperatures for accelerated lumen depreciation test is determined in such a way that the temperature does not exceed the glass transition temperature of the plastics. Glass transition temperature (Tg) of BPA-PC is 150 ºC, so the maximum accelerated temperature is chosen 10 ºC below the Tg. Optical properties of thermally-aged plates, i.e. Luminous flux depreciation, were studied at room temperature, using an integrated sphere.

3. Reliability model:

The reliability model for the life time assessment is based on an exponential luminous decay equation, where the time-to-failure can be calculated as [10]

$$\phi(t) = \beta \exp(-at) \qquad (1),$$

where $\Phi(t)$ represents the lumen output, α is the rate of reaction or depreciation rate parameter, t is time and β is a pre-factor. According to alliance for solid state illumination system and technology (ASSIST), when lumen output, Φ, is equal to 70%, t is time-to-failure [6]. The rate of reaction, α, is related to the activation energy of the reaction and to the ageing temperature as follows [7]

$$a = A \exp(\frac{-E_a}{KT}) \qquad (2),$$

where A is a pre-exponential factor, E_a is the activation energy (ev) of the degradation reaction, K is the gas constant, and T is the absolute temperature (K).

4. Results:

It is noticeable that there is a significant decay both in the phosphor yellow emission and in the blue peak. As is shown in our previous work the yellowing of BPA-PC plates leads to the reduction in the light transmissivity of plates [8]. Reduction in yellow emission also illustrates the decay of phosphor conversion efficiency, Figure 1.

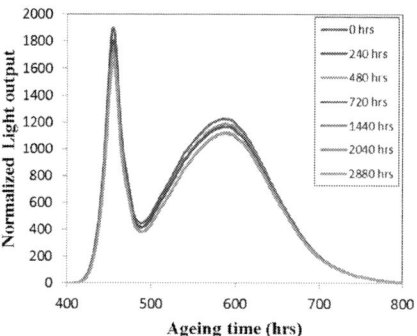

Figure 1: The evolution of spectral power distribution (SPD) of sample B at 140 ºC

A more explanation of the effects of thermal-ageing on the performance of remote is given in Figure 2. This Figure shows the progress of the normalized flux intensity and as a result the degradation kinetics of the phosphor plates. Obviously, the degradation rate shows a significant dependence on the stress temperature level; the higher the ageing temperature, the higher the lumen depreciation and the degradation kinetics.

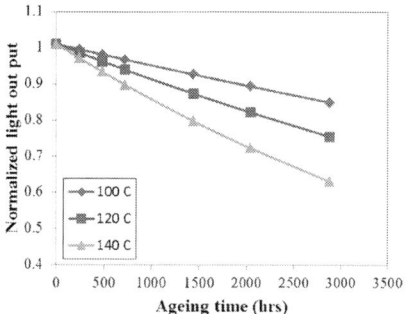

Figure 2: Normalized flux of remote phosphor plates at different thermal stress tests

The activation energy of the degradation reaction in LEDs depends on the materials and the working conditions. The activation energy can be calculated from Equation (2). In order to obtain the activation energy, the natural logarithm of the reaction rates is plotted against the inverse of the absolute temperature (see Figure 3). The slope is multiplied by the negative of the gas constant to obtain the activation energy, E_a, in the eV. Activation energy for remote phosphor is between 0.333 eV, which is in agreement with previous reported values [7].

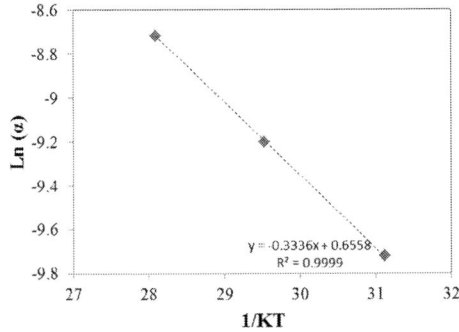

Figure 3: Plot of ln (α) vs E/KT for remote phosphor

Thermal stress test also has some significant effects on the CCT. In Table 1 the variation of CCT during high temperature stress test is shown for remote phosphor plate. It is obvious that by increasing the thermal ageing time and the temperature, the CCT decreases. One can also notice that the higher the ageing temperature, the higher the degradation kinetics.

Table 1: Correlated colour temperature (CCT) during high thermal-stress tests (temperature is in degree C and time is in hour)

T \ t	0	240	480	720	1440	2040	2880
100	3170	3167	3150	3128	3105	3090	3056
120	3170	3150	3120	3105	3065	3048	3018
140	3170	3130	3100	3080	3029	3010	2980

The reduction in Colour Temperature suggests that the degradation of the remote phosphor plates has consequences not only on the light extraction efficiency but also on the colour of the emitted light. Colour shifting of light is determined by variation of Chromaticity Coordinate (CIE x,y). The values of x and y of the aged remote phosphors at 140 °C is shown in Table 2. One can see that both x and y decreases by increasing ageing time.

Table 2: (x,y) values of aged samples at 140 ºC during high thermal-stress test

Ageing time (hrs)	x	y
0	0.4152	0.3547
240	0.4149	0.3538
480	0.4146	0.3528
720	0.4132	0.3518
1440	0.4068	0.3461
2040	0.4053	0.3442
2880	0.4032	0.3435

4.2. Prediction of time-to-failure at lower temperatures

The real working temperature of LEDs is much lower than the applied temperatures in the tests [13]. Therefore, the kinetics of lumen depreciation to 30% of its initial value by using exponential luminous decay model and Arrhenius equation should be extrapolated to temperatures lower than 100 °C. This can be done using Equation 1 by equating ϕ to 0.7, knowing that α can be obtained from Equation 2. The values of α, calculated for 40, 60 and 80 °C, are given in Table 3. As is seen the higher the temperature the faster the reaction rate.

Table 3: Reaction rate (α) for different temperatures

Temp (°C)	Remote Phosphor
40	7.95E-06
60	1.68E-05
80	3.28E-05
100	5.93E-05
120	1.01E-04
140	1.64E-04

Figure 4 illustrates time-to-failure (70% lumen decay) of remote phosphors, calculated at different temperatures. It is seen that at 40 °C the light output reduces to 70% of its initial value after 45 khrs.

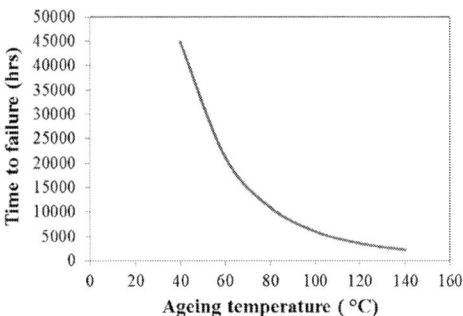

Figure 4: Time-to-failure (70% lumen decay) of remote phosphor at different temperatures

5. Discussions:

High temperature stresses can damage the optical properties of the package and of the material used for the encapsulation [1-5]. This can result in a significant reduction in the luminous flux, emitted by the devices. Spectral power distribution (SPD) method is used to study the effect of high temperature stress test on the optical degradation of remote phosphor. The goal was to study the effect of temperature on the lumen depreciation of LED-based products and on their CCTs. A significant change both in the correlated colour temperature (CCT) and in the chromaticity coordinates (CIE x,y) is also observed in thermally aged specimens Worsening of the chromatic properties of the phosphor plates is correlated to the decrease in the luminous flux. One can conclude that the thermal degradation of the remote phosphor plate affects both the efficiency and the colour of the LED products. It is shown that the degradation mechanisms is thermally activated and has activation energy of 0.333 eV (Figure 3). It is clearly seen that the lower the depreciation rate, the better the performance a remote phosphor could have. The results also show that there is a direct relation between the temperature and kinetics of degradation. It is also shown that decreasing the transsimity of PC plates together with the reduction in phosphor efficiency limits the reliability of remote phosphor light sources and there is a colour shift towards yellow. The life time of this kind of remote phosphor is calculated as 45 khrs which is slightly higher than that 2

other remote phosphors which were studied completely in our previous paper. The reason might be due to the lens material [9].

6. Conclusions:

The accelerated optical degradation of a remote phosphor, under elevated temperature stress, is studied. BPA-PC plates are exposed to temperature in the range of 100 to 140 °C. Exponential luminous decay model and Arrhenius equation are used to predict the lumen depreciation and the lifetime of plastic lens in LED lamps in real service conditions. The following conclusions can be drawn from this study:

- A significant decay both in the phosphor yellow emission and in the blue peak intensity, with blue emission being more influenced.

- During the stress thermal ageing tests a significant change both in the correlated colour temperature (CCT) and in the chromaticity coordinates (CIE x,y) take place.

- The lifetime of the remote phosphor, defined as 30% lumen depreciation at 40 °C, is around 45 khrs for the commercial grades plates.

Acknowledgments

This research was carried out under project number M71.9.10380 in the framework of the Research Program of the Materials innovation institute M2i (www.m2i.nl). The authors would like to thank M2i for funding this project. Authors would also like to acknowledge "TNO innovation for life" company for SPD measurements.

References:

[1] U. Zehnder, A . Weimar, U . Strauss, M . Fehrer, B. Hahn, H. J. Lugauer , V. Harle Industrial production of GaN and InGaN-light emitting diodes on SiC-substrates, Journal Crystal Growth 230 (2001) 497–502.

[2] M.H. Chang, D. Das, P.V. Varde, M. Pecht, Light emitting diodes reliability review, Microelectronics Reliability 52 (2012) 762-782.

[3] M. Meneghini, L. Trevisanello, S. Podda, S. Buso, G. Spiazzi, G. Meneghesso, E. Zanoni Stability and performance evaluation of high-

brightness light-emitting diodes under DC and pulsed bias conditions, Proc. SPIE. (2006) 633-70.

[4] L. Trevisanello, M. Meneghini, G. Mura, M. Vanzi, M. Pavesi, G. Meneghesso, E. Zanoni, Accelerated life test of high brightness light emitting diodes, IEEE Trans. Device Mater. Reliability 8 (2008) 304–311.

[5] S. Ye, F. Xiao, Y.X. Pan, Y.Y. Ma, Q.Y. Zhang, Phosphors in phosphor-converted white light-emitting diodes: Recent advances in materials, techniques and properties, Materials Science and Engineering R 71 (2010) 1-34

[6] Illuminating Engineering Society, TM-21-11 Projecting Long Term Lumen Maintenance of LED Light Sources, 2012

[7] S. Koh, C. Yuan, B. Sun, B. Li, X. Fan, G.Q. Zhang, Indoor SSL product level accelerated lifetime test, 13th EuroSimE Conference

[8] Yazdan Mehr M, van Driel W.D, Jansen K.M.B, Deeben P, Zhang G.Q. Lifetime Assessment of Plastics Lenses used in LED-based Products. Microelectronics Reliability 54 (2014) 138–142

[9] Yazdan Mehr M, van Driel, Zhang G.Q. Accelerated life time testing and optical degradation of remote phosphor plates. Journal Microelectronics Reliability 2013, submitted

Finite Element Multi-physics Modeling for Ohmic Contact of Microswitches

H.Liu[1,2], D. Leray[1,2], P. Pons[2], S. Colin[1]
[1]Institut Clément Ader, INSA, Univ de Toulouse
[2]CNRS, LAAS, Univ de Toulouse, INSA, LAAS
Toulouse, France
hliu317@gmail.com

Abstract

The purpose of this paper is to investigate the thermoelectrical behaviour of ohmic microcontacts under low force. The temperature in the contact zone is very important for the reliability of microswitches. As it is very difficult to measure the inner temperature, the numerical thermal modelling of electrical contacts offers interesting perspectives. A multi-physics modelling of electrical contact is accomplished with the finite element commercial package ANSYS[TM]. Two approaches for coupled-field analysis are investigated, namely *direct* and *load transfer*. The thermo-electro-mechanical modelling is firstly validated with a smooth sphere-plane contact, and then applied for a real rough contact computation, elastic-plastic material deformation is included in the modelling. The temperature distribution on the contact surface is plotted, and the maximum temperature is found around the asperities with the highest deformation. The multi-physics model offers a reliable method to investigate the steady-state thermal behaviour of electrical contact with rough surface included.

Keywords: multi-physics, finite element modelling, ohmic contact, microswitches

1. Introduction and Motivations

Reliability of ohmic contacts is a great challenge for microswitches [1]. Since Joule heating is extremely localized in the contact zone, the temperature there may reach the softening or melting temperature while the device remains at room temperature [2-3].

The numerical coupled-simulations have been proved to be an efficient method to study the thermoelectrical behaviour of contact, and have been carried out by many researchers (see [4-10]). Monnier *et al.* [5] investigated a macro smooth sphere-plane contact problem using a coupled-field finite element (FE) simulation. Based on this, Ghaednia *et al.* [7] studied the influence of thermal expansion and plastic deformation on an asperity contact with a 3D thermo-electro-mechanical model. However, for microcontacts, low force and weak current are required, which makes them different from the macro-scale contact behaviour. Also, the temperature reached more than 1000 K in [7] while the temperature-dependent material properties were not taken into account. The research of Leidner *et al.* ([9-10]) seems more pertinent, in which the current density distribution for contact with layered structure and 'real world' contact topographies was discussed, and the simulation results were confirmed with the thermal images by infrared thermography. However, the study focused on the connector system, in which the contacting spots were much larger than that of microswitches. Motivated by microswitches contact, a 2D elastic multiphysics contact model was developed using ANSYS[TM] in [8], and the study presented a distinct difference on contact radius between force control and displacement control with a voltage or current applied. However, the resulting temperature would reach to 662 K with the applied voltage V*=0.221, the temperature-dependent material properties were not considered in the modelling, and only elastic deformation was considered. Also for the electrical contact of microswitches, Shanthraj *et al.* [6] developed an iterative procedure to solve the electro-thermo-mechanical field equations with three-dimensional fractal surface included. Interesting results were obtained, such as the influence of the initial residual strain and ambient temperature. However, there were no experimental results to validate the model.

Summarizing the published works, a promising multi-physics model is still missing for microcontacts problem, especially with rough surface and elastic-plastic material deformation involved.

The organization of the paper is as follows: the introduction and motivations are presented in the first part, followed by the theoretical background of the electrical contact. The description of the FE model and the methodology of multiphysics calculation are presented in the third part, and then the model validation and the simulation results are discussed, finally comes the conclusion.

2. Theoretical Background

In the case of a unique contact spot, the electrical contact resistance (ECR) is calculated according to its radius a. Comparing the contact radius to the electron mean free path l, three electrical transport regimes are defined: diffusive, ballistic and intermediary, and the electrical resistance can be calculated with the following equations (1-3):

$$R_D = \rho / 2a \quad (1)$$

$$R_B = 4\rho l / 3\pi a^2 \quad (2)$$

$$R_{int} = f(l/a) \times R_D + R_B = \frac{1 + 0.83(l/a)}{1 + 1.33(l/a)} R_D + R_B \quad (3)$$

Where ρ is the electrical resistivity and l is the electron mean free path. R_D, R_B and R_{int} are the electrical resistance for diffusive, ballistic and intermediary transport regimes respectively. $f(l/a)$ is the interpolation

978-1-4799-4789-8/14 $31.00 © 2014 IEEE

function, which accounts for the transition from the ballistic to diffusive regime.

When multiple spots are in contact, the effective contact resistance depends on the radii of the spots and their distribution. Formulae used to calculate the ECR for multiple spot contact [11-12] do not deal with spots under different transport regimes, and this leads us to define a lower and an upper ECR limit, as [13] proposed.

For the lower limit, it is assumed that the contact spots are in parallel without interacting with each other. For the upper limit, the ECR is obtained by replacing all contact asperities with one single spot while keeping the contact area constant, and the effective radius a_{eff} is defined. The two ECR limits can be calculated with equations (4-5):

$$\frac{1}{R_l} = \sum_{i=1}^{N} \frac{1}{R_{ci}} \quad (4)$$

$$R_u = f \frac{\rho_{av}}{2a_{eff}} + \frac{4\rho_{av}\lambda}{3\pi a_{eff}^2} \quad (5)$$

Where N is the number of the asperities in contact and ρ_{av} is the average electrical resistivity, R_{ci} is the resistance of contact spot i, and the subscripts l and u refer to the lower and upper limits for the contact resistance, respectively.

In the thermal domain, the calculation of the maximum temperature of the contact, also called contact temperature T_C, was firstly proposed by Holm [14]:

$$T_C = \sqrt{\frac{V_C^2}{4L} + T_0^2} \quad (6)$$

Where $L=2.45\times10^{-8}$ (W×Ω/K^2) is the Lorentz constant and T_0 is the ambient temperature. However, the recent experiments [15-16] indicated the breakdown of the V-T relation. It was found that with small size contact spots whose radius is the order of or smaller than the mean free path, the contact "melting" was not observed although the contact exceeded the 'melting voltage' of the contact pairs. It is due to the ballistic transport of electrons in contact which is not responsible for the contact heating, Jensen *et al.* [2] then proposed a new expression:

$$T_C^2 - T_0^2 = \sqrt{\frac{f(l/a)R_B}{Rc} \times \frac{V_C^2}{4L}} \quad (7)$$

Considering the temperature distribution in the contact area, it was found that, most of the Joule heating is located close at the periphery of the contact area [17], this was also validated by the thermographic measurements [18-19]. While for the steady-state temperature, Greenwood and Williamson [17] predicted that it falls off with distance from the contact area, but the FEM by Hennessy *et al.* [8] predicted that the maximum temperature occurred slightly on the asperity side. Regarding this disagreement, the study presented focuses on the steady-state thermal behaviour, and zooms in the temperature distribution on the contact surface.

3. Finite Element Modelling

This work is based on the previous study [20], in which a determinist method was used to describe the rough surface of microcontacts. As the determinist method, an atomic force microscopy (AFM) was used to scan the topography of a microswitch bump, and this described well the topographies of micro-scale asperities and macro-scale bump shape. The AFM scan data was then imported to ANSYS™ (package 11.0 used), and a FE contact model was developed.

A. Topography of contact surface in microswitches

A real microswitch is used for AFM scanning, as shown in Figure 1(a). Contact pairs are as follows:
- Bridge is fabricated in gold, with thickness of 4 μm.
- Bump is also fabricated in gold, with a layer of 1 μm-thick coated on a silicon substrate, and it is with a hemi-sphere form of radius 4 μm.

Since microswitches usually work under low force, it is the highest asperities making contact, also called a-spots, which are in the dimension of few hundreds or even tens of nanometers [21], so high resolution is required to properly map the surface topography.

A 4 μm-wide square AFM scan is carried out on the microswitch bump. Scan X-Y grid is set to 256 lines, which gives a 15.6 nm horizontal resolution. This horizontal resolution was proved to be fine enough for modelling contact behaviour and evaluating ECR properly [22]. Figure 1(b) shows the surface topography, generated with AFM data and treated by Matlab.

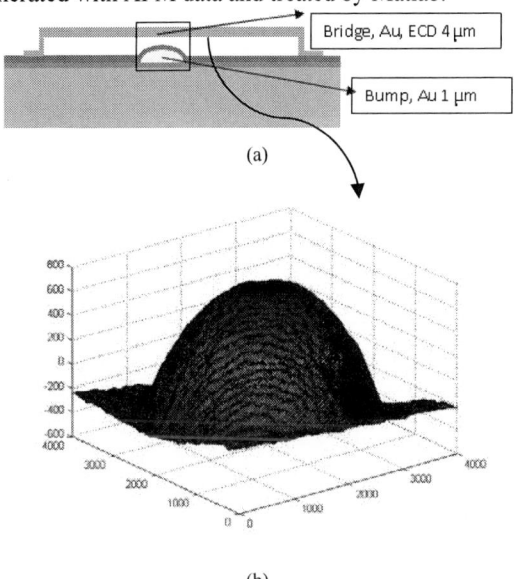

(a)

(b)

Figure 1: (a) Schematic of the microswitch, (b) Surface topography of the microswitch bump: obtained by AFM scanning

B. Geometry of the FE model

Both contact parts are modelled as deformable bodies. The top surface of the lower body is built using AFM data. The bottom surface of the upper body is defined as perfectly flat surface. Due to the coupling effect between the mechanical, thermal and electrical behaviors, modelling both volumes as deformable bodies with real material is more accurate and practical for multiphysics problems [7].

Instead of taking the whole bump in modelling, the FE model is developed with a reduced size as defined in the previous work [20] for the sake of calculating time. The dimension of the model is 1.2 μm ×1.2 μm×1 μm, and the AFM data is taken from the top of the bump (see Figure 2(a)) where the contact takes place.

For the validation of multiphysics analysis, a smooth-smooth (S-S) model is developed. It also consists of two deformable bodies, but the rough surface of the bump is replaced by a smooth sphere cap. Least squares fitting is used to define the sphere radius, and the resulting value is $Rs = 2.1 \times 10^3$ nm (see Figure 2(b)).

C. Material properties and boundary conditions

The mechanical property of the contact material, which is gold, is defined as elastic-plastic. Strain hardening is also considered with tangent module defined as 10 GPa, and MISO option is used in ANSYS™. Table 1 lists all material properties; temperature-dependent material properties and thermal expansion are not included in the modelling.

Table 1: Material properties in the FE model: Au

Properties (unit)	Value
Yong's modulus E (GPa)	80
Poisson's ratio v	0.42
Yield strength σ_y (GPa)	0.3
Ultimate strength σ_u (GPa)	0.36
Tangent modulus Et (GPa)	10
Electrical resistivity ρ (10^{-9} $\Omega{\times}$m)	22.14
Thermal conductivity λ (W/(m×K))	318
Mean free path of electrons l (nm)	38

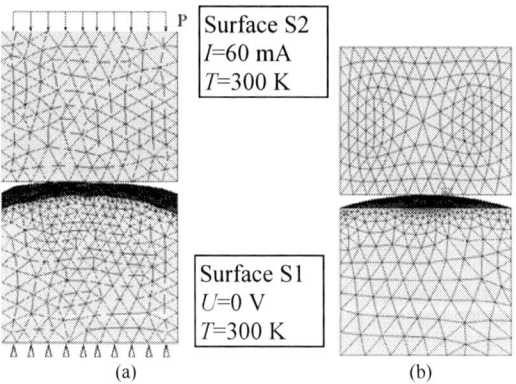

Figure 2: FE contact model with (a) rough surface based on the AFM data, and (b) smooth-smooth surfaces

Mechanical boundary conditions include (Figure 2):

- Bottom surface of the lower body (noted S1) is clamped.

- For all nodes of top surface of the upper body (noted S2), the degree of freedom UZ (DOF) is coupled using 'CP' command, so that they have the same displacement in the Z direction.

- Pressure is applied uniformly on surface S2, the maximum contact force is 145 μN, as the experiment did [3].

For thermoelectrical boundary conditions (Figure 2):

- Electrical current of 60 mA is applied vertically on surface S2, and voltage on surface S2 is coupled.

- Voltage on surface S1 is constraint to zero.

- Temperature is constrained as 300 K on surfaces S1 and S2, this is taken as the reference temperature.

D. Multiphysics computational methodology and finite element meshing

For multiphysics problem, ANSYS™ provides two possibilities:

- Direct analysis
- Sequential (load transfer) analysis

Both methods are used in the study.

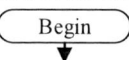

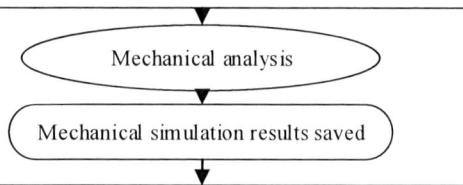

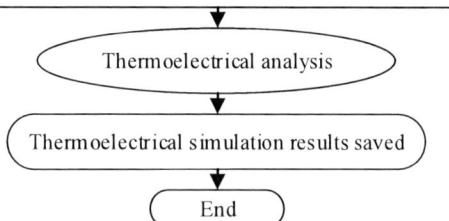

Figure 3: Algorithm of the sequential multi-physics modelling

The direct method usually involves only one analysis that uses a coupled-field element type containing all necessary degrees of freedom [23]. In the FE models presented, the 3D 10-node tetrahedral element SOLID227 is used to mesh the volume, and the DOF option is set as "111" to activate the fields of mechanical, thermal and electrical.

The principle of the sequential method is: the input of one physics analysis depends on the results from another analysis, the analyses are then coupled [23]. Figure 3 shows the algorithm of the method. For meshing the volume, instead of using one element SOLID227 in the direct method, SOLID92 is used for the structural analysis and SOLID98 for the thermoelectrical analysis. Both elements are 3D 10-node tetrahedral elements.

For the contact pair, 3D surface-to-surface contact element CONTA174 and target element TARGE170 are used to mesh the contact surfaces belonging to the bump and bridge respectively. These elements are selected to consider the large deflection and nonlinear behaviour of contact asperities. Table 2 and Table 3 summarize all of the elements used. Also, the augmented Lagrange method is used to seek contact, and large deformation is considered in the calculation.

Table 2: Element types for FE model: structural and direct multiphysics analysis

		Structure	Multiphysics
SOLID	Element	187	227
	DOF	UX, UY, UZ	UX, UY, UZ, Volt, Temp
TARGE	Element	170	170
CONTA	Element	174	174
	DOF	UX, UY, UZ	UX, UY, UZ, Volt, Temp
	Real constant		TCC=318×10^9 (W/(m^2×K) ECC=4.52×10^{16} (S/m^2)

Table 3: Solid element types for FE model: sequential multiphysics method

Analysis	Structure	Thermoelectric
Element	SOLID92	SOLID98
DOF	UX, UY, UZ	Volt, Temp, Mag

E. Electrical and thermal contact conditions

To take into account the surface interaction for electric contact, a parameter called ECC (Electrical Contact Conductance) is defined in ANSYS^{TM} by:

$$ECC = J/\Delta V \quad (8)$$

Since the electrical and heat transfer are much localized, conduction phenomena at the contact interface depend on the surface geometry, the roughness of the contact area, etc. ([23-25]), the theoretical determination of ECC is very difficult.

It is assumed in the study that the contact and target interfaces are perfect, that there is no potential jump at the interface, and ECC can be considered as the inverse of the electrical resistance of interface per unit area [24]. In the presented models, the value of ECC is defined as the interface resistance equaling to the resistance of a 1 nm-thick layer of the material, so that ECC = 1 / (ρ. e), where e = 1 nm.

Similarly, the conductive heat transfer between two contacting surfaces is defined by TCC (Thermal Conductance Contact):

$$TCC = Q/\Delta T \quad (9)$$

Its value is defined as the conductance equaling to that of a 1 nm-thick layer: TCC = λ / e with λ the thermal conductivity of the material.

4. Results with direct method

The multiphysics analysis was firstly developed with the smooth-smooth contact model. The meshing grid is about 40 nm-60 nm for the contact surfaces, and the whole FE model includes:

- 11500 solid elements
- 1210 contact elements
- 200 target elements.

A loading-unloading cycle is applied on the model with 10 steps for loading and 10 steps for unloading.

A. Validation of direct multiphysics modelling

The validation of multiphysics contact analysis is carried out in two terms:

- Mechanical behaviour
- Thermoelectrical behaviour

For the mechanical behavior, the multiphysics simulations are compared with the counterpart mechanical simulations in which the block is meshed with SOLID 187 (see Table 2), which is also a 3D 10-node tetrahedral structural solid element. The contact area as a function of load steps is plotted in Figure 4. It shows that the mechanical and multiphysics simulations predict the identical results, so the multiphysics simulations are validated in terms of mechanical behavior.

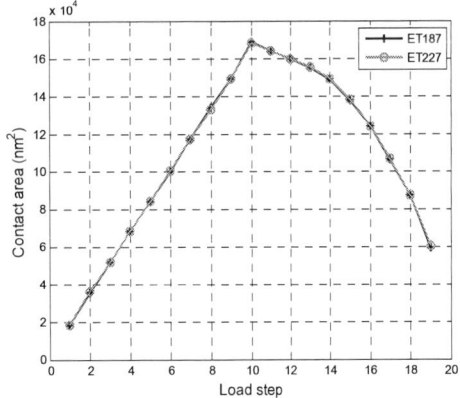

Figure 4: Contact area as a function of load steps: comparison between the direct multiphysics simulation and structural simulation

For the thermoelectrical behaviour, the criteria chosen are the contact resistance and the maximum temperature.

With FE simulations, contact occurs at a finite number of elements. After the calculation, contact results data are exported out as text files, then Matlab is used to calculate the number of spots and the area of each contact spot, and finally the constriction resistance is calculated with equations (1-3) for one spot or (4-5) for multiple spots. This method of contact resistance calculation is denoted as 'Method A'.

However, one issue should be addressed is that the contact radius is about 220 nm under the highest load (see Figure 4), compared to the dimension of the model which is of 1200 nm, the spot size is too large to satisfy the assumption of Holm resistance analytical calculation, which considers two semi-infinite spaces in contact on a disc [14]. So the 'cylindrical resistance' is then calculated instead as Timsit proposed [26]:

$$R_c = \left(\rho/2a\right)\begin{bmatrix} 1-1.41581\left(a/R\right)+0.06322\left(a/R\right)^2 \\ +0.15261\left(a/R\right)^3+0.19998\left(a/R\right)^4 \end{bmatrix} \quad (10)$$

where R is the radius of cylinder. In our models, the section geometry is a square, so an equivalent cross-sectional area is defined to adapt Timsit's formula with $R = \sqrt{L/\pi}$. Nevertheless, no available formulas are found for the ballistic and intermediary regimes in the 'cylindrical' condition, so equations (2-3) are still used. Considering that the spot is quite small in the ballistic and intermediary transport mode, the ratio a/R is small, so the effect of 'cylindrical resistance' should be ignorable.

With multiphysics simulation, on the other hand, an electrical resistance can be obtained directly with the resulting voltage divided by the current applied. This simulated resistance, also called total resistance R_t herein, is the sum of bulk resistance R_b and contact resistance R_c. Hence, contact resistance should be calculated by:

$$R_c = R_t - R_b \quad (11)$$

This method for obtaining the contact resistance is denoted as 'Method B'.

The validation of the thermoelectrical behaviours for multiphysics simulations is carried out firstly by comparing the contact resistance obtained by 'Method A' and 'Method B'. This is to validate whether the ECC value is proper, so that the electrical interface can be considered prefect in FE modelling. The contact resistance as a function of load steps is plotted in Figure 5(a), the resistances obtained by two methods are very close except for a sensible difference of 13% under very low force.

Similarly, for the TCC value, the simulated contact temperature as a function of load steps is compared with the ones calculated by equation (6) using the simulated voltage, labelled 'analytical' in Figure 5(b). A good agreement is found between the simulation and the analytical calculation, and this validates that the Joule heating calculation is well implanted in the simulation.

And thus the multiphysics analysis is validated in terms of thermoelectrical behaviour.

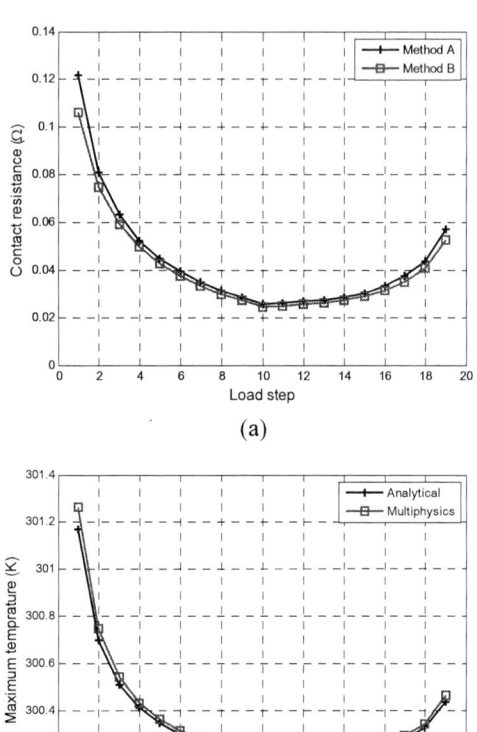

Figure 5: Validation of the direct multiphysics modelling: (a) contact resistance obtained with 'Method A' vs. 'Method B', (b) contact temperature calculated with the simulated voltage vs. obtained directly from the simulations

B. Problem with the direct method

The direct method could be the best choice since it can couple the mechanical, electrical and thermal behaviors together at one time, so that the temperature-dependent material properties and thermal expansion can be implanted in the model easily. However, the computation time is too long, it took 27 hours for the smooth contact simulation discussed above (Linux system with 4 CPU@2.33GHz, and 16 GB Memory), compared to only 38 minutes a mechanical simulation took. One mechanical simulation with rough surface and 32 nm meshing grid asked for 50 calculation hours, and it makes using the direct method for the rough contact modelling unreasonable.

5. Results with sequential method

A. Validation of sequential multiphysics modelling

With the sequential model, the parameters in the mechanical analysis is just the same as pure mechanical simulation, but for the thermoelectrical analysis, one parameter should be addressed is the PINB value (pinball region), which determines the contact status of the elements. It is either a 2-D circle or a 3-D sphere (see Figure 6), and the elements are considered to be near-field contact when they enter into the pinball region. In the thermoelectrical analysis, the contact surface behaviour is defined as bonded always, so a relatively small PINB value is required to prevent any false contact detection. Our simulations showed that the PINB value has much influence on the electrical and thermal interaction on contact surfaces. Indeed, the transfer of current and heating happens when the elements locate within the pinball region, so for the same mechanical analysis, a larger PINB value causes an easier electrical contact and a smaller contact resistance.

The influence of the PINB value on the total contact resistance is presented in Figure 7. The simulations were carried out with the S-S model and a coarse mesh grid of about 130 nm on the contact surface. As discussed above, with PINB decreases, the simulated resistance increases. Compared the electrical resistance and the contact temperature with the direct method, 0.01 was chosen for PINB value. Further simulation were then launched with finer meshing grid of 40 nm - 60 nm, and good results were found, so in the following simulations, PINB is set as 0.01.

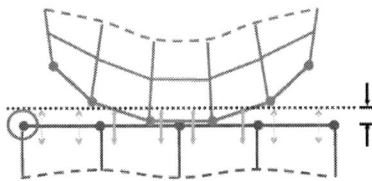

Figure 6: The definition of 'PINB region' in the contact model (ANSYS thermal analysis guide, 2009)

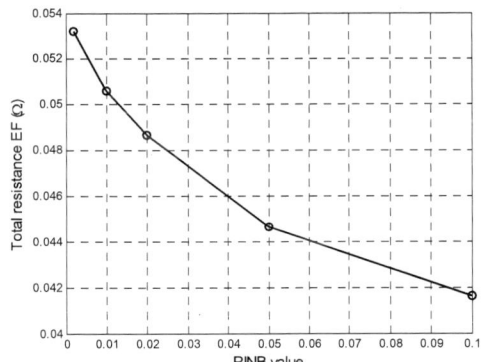

Figure 7: Influence of PINB value on the FE total resistance

The validation of the sequential method is also accomplished in two terms: mechanical behaviour and thermoelectrical behaviour. The presented model is with smooth surface, and the meshing is the same as described in the direct method part. For mechanical behaviour, the simulated contact deformation and contact area is compared with the pure structural model, and the results are listed in Table 4. For thermoelectrical behaviour, the simulated total electrical resistance and the contact temperature are compared with the ones from direct method, and the results are listed in Table 5. The results are pretty close in both terms, while a little more sensible difference is found for thermoelectrical behaviour. It should be taken into mind that the PINB value influences the resulting resistance and temperature. However, a great advantage with sequential method is that it takes much less time than the direct method, just about 1.6 times of a pure mechanical simulation.

Table 4: Mechanical results comparison: between the sequential model and structure model

Results	Structure model	Sequential model	Relative diff (%)
Deformation (nm)	17.89	18.41	2.9
Contact area (nm²)	168679	171422	1.6

Table 5: Thermoelectrical results comparison: between the sequential model and direct model

Results	Direct model	Sequential model	Relative diff (%)
R_t **- EF (Ω)**	0.0565	0.0545	3.6
ΔT **- EF (K)**	0.206	0.192	7.0

B. Results with the rough contact model

The sequential method was then applied for the rough contact modelling. Two simulations with different meshing grid size were launched, one with coarse meshing, which is 96 nm for the bump contact surface, and the other with finer meshing of 32 nm. The results for the contact force of 145 μN are presented in Table 6. Although the time cost is quite different for two simulations, the results are very close, so this confirms that the meshing grid of 32 nm is fine enough to predict the thermoelectrical behavior of electrical contact exactly.

The contact pressure and temperature distribution on the bump surface with fine meshing are shown in Figure 8. The highest temperature is found to be not located on the asperities with highest deformation, but around them. And it seems that the temperature distribution matches well with the contact pressure. Considering that the asperities tend to form a large spot, and it is easier for larger spots to dissipate the heating because of larger area, so it is reasonable that the highest temperature is found located on the outer rim of highest deformation zone, as the measurement presented [18]. Though there are some temperature differences between the asperities and the proximity, the temperature is almost uniform on the contact zone, and this explains why the meshing grid has little influence on the maximum temperature.

The results also present that, the temperature increases very little (about 0.2°C with current 60 mA) as the reason of low resistance, so the temperature dependence of the material properties can be ignored. As we discussed in the previous study [27], the measured high contact resistances were most likely due to the oxide or contamination film on the surface, so the next step is to take them into account in the FE modelling, to have the contact resistance and thermoelectrical behaviour close to the realistic case.

Table 6: Simulations results with rough contact model

Results	Meshing 96 nm	Meshing 32 nm	Relative diff (%)
R_t - EF (Ω)	0.0219	0.0225	2.7
ΔT - EF (K)	0.18	0.183	1.65
Time cost (mins)	73	2298	

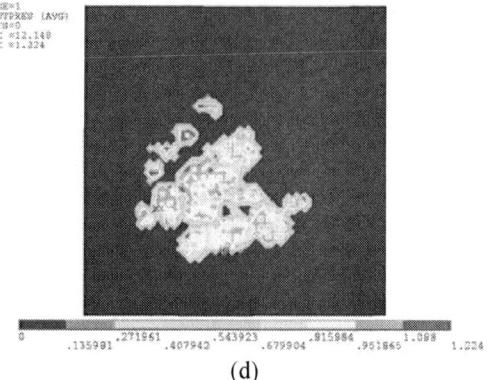

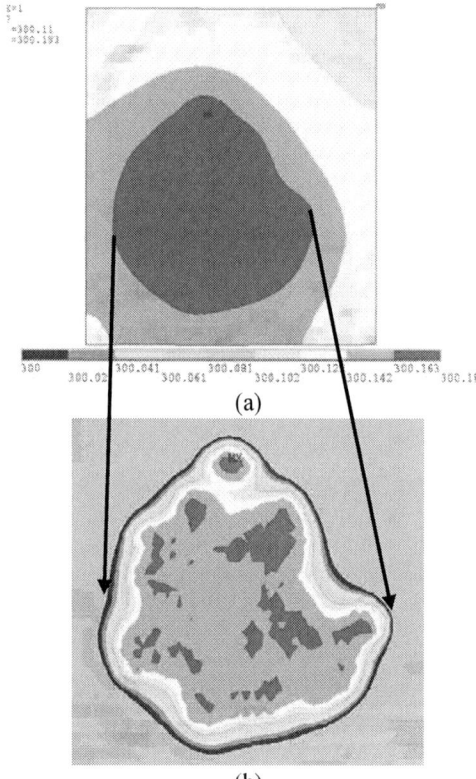

(a)

(b)

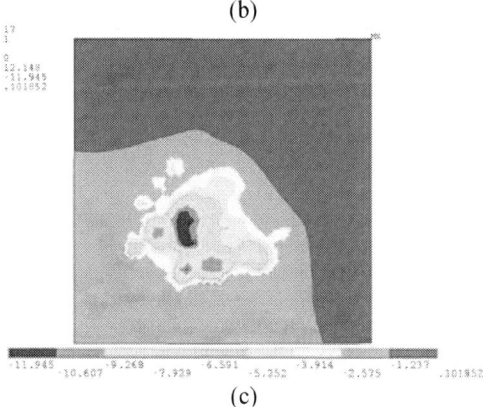

(c)

(d)

Figure 8: Simulation results with sequential method on rough contact surface: (a) temperature distribution: $T_{Min}=300$ K, $T_{Max}=300,183$ K (b) temperature distribution- zoom in the contact zone: $T_{Min}=300,168$ K, $T_{Max}=300,183$ K, (c) contact deformation, $UZ_{Max}=11.945$ nm, (d) contact pressure: 0-1.224 GPa

6. Conclusions

A multi-physics FE modelling was developed with commercial package ANSYS™ to investigate the thermo-electrical-mechanical behavior of electrical contact of microswitches. Two approaches namely, *direct* and *load transfer*, were studied. The multiphysics modelling was validated in terms of mechanical and thermoelectrical behavior of electrical micro-contact. Considering the computation cost, the load transfer method was applied to investigate the rough contact problem. The results indicate that the temperature distribution is almost uniform in the contact zone, while the maximum temperature is located around the asperities with the highest deformation.

In a general way, the finite element multi-physics modelling provides a reliable method to investigate the thermal-electrical-mechanical behavior of electrical contact, especially for the problem with rough surfaces and the elastic-plastic deformation included. Further study should be accomplished with the oxide film included, and with temperature-dependent material properties.

Acknowledgments

The authors would like to thank Dr. F. Pennec for the kindly support on ANSYS programming.

References

1. W. M. van Spengen, "MEMS reliability from a failure mechanisms perspective," *Microelectronics Reliability*, Vol. 43, No. 7 (2003), pp. 1049–1060.
2. B. D. Jensen, L. L.-W. Chow, K. Huang, K. Saitou, J. L. Volakis, and K. Kurabayashi, "Effect of nanoscale heating on electrical transport in RF MEMS switch contacts," *Journal of Microelectromechanical Systems*, Vol. 14, No. 5 (2005), pp. 935 – 946.
3. A. Broue, J. Dhennin, P. Charvet, P. Pons, N. B. Jemaa, P. Heeb, F. Coccetti, and R. Plana, "Multi-Physical Characterization of Micro-Contact Materials for MEMS Switches," *Proc IEEE 56th*

Holm Conference on Electrical Contacts, Charleston, SC, 2010, pp. 1–10.

4. C. L. Tsai, W. L. Dai, D. W. Dickinson, and J. C. Papritan, "Analysis and development of a real-time control methodology in resistance spot welding," *Welding Research Supplement* (1991), p. 339s–351s.

5. A. Monnier, B. Froidurot, C. Jarrige, P. Testé, and R. Meyer, "A Mechanical, Electrical, Thermal Coupled-Field Simulation of a Sphere-Plane Electrical Contact," *IEEE Transactions on Components and Packaging Technologies*, Vol. 30 (2007), pp. 787–795.

6. P. Shanthraj, O. Rezvanian, and M. A. Zikry, "Electrothermomechanical Finite-Element Modelling of Metal Microcontacts in MEMS," *Journal of Microelectromechanical Systems*, Vol. 20, No. 2 (2011), pp. 371–382.

7. H. Ghaednia, A. Rostami, and R. L. Jackson, "The Influence of Thermal Expansion and Plastic Deformation on a Thermo-Electro Mechanical Spherical Asperity Contact," *Proc IEEE 58th Holm Conference on Electrical Contacts*, Portland, OR, 2012, pp. 1–7.

8. R. P. Hennessy, N. E. McGruer, and G. G. Adams, "Modelling of a Thermal-Electrical-Mechanical Coupled Field Contact," *Journal of Tribology*, Vol. 134, No. 4 (2012), pp. 041402–1–8.

9. M. Leidner, H. Schmidt, M. Myers, and H. F. Schlaak, "A new simulation approach to characterizing the mechanical and electrical qualities of a connector contact," *EPJ. Applied physics*, Vol. 49, No. 2 (2010), pp. 22909-1-10.

10. M. Leidner, H. Schmidt, and M. Myers, "Simulation of the Current Density Distribution within Electrical Contacts," *Proc 56th IEEE Holm Conference on Electrical Contacts*, Charleston, SC, 2010, pp. 1–9.

11. J. A. Greenwood and J. B. P. Willamson, "Contact of nominally flat surfaces," *Proc. Roy. Soc (London), series A*, Vol. 295 (1966), pp. 300–319.

12. L. Boyer, S. Noel, and F. Houze, "Constriction resistance of a multispot contact: an improved analytical expression," *IEEE Transactions on Components, Hybrids, and Manufacturing Technology*, Vol. 14, No. 1 (1991), pp. 134–136.

13. S. Majumder, N. E. McGruer, G. G. Adams, P. M. Zavracky, R. H. Morrison, and J. Krim, "Study of contacts in an electrostatically actuated microswitch," *Sensors and Actuators A: Physical*, Vol. 93, No. 1 (2001), pp. 19–26.

14. R. Holm, <u>Electric Contacts-Theory and Applications</u>, *4th edn.* (Berlin: Springer, 1967), pp. 60-64, pp. 3-4.

15. R. S. Timsit, "On the Evaluation of Contact Temperature from Potential-Drop Measurements," *IEEE Transactions on Components, Hybrids, and Manufacturing Technology*, Vol. 6, No. 1 (1983), pp. 115–121.

16. C. Maul, J. W. McBride, and J. Swingler, "Intermittency phenomena in electrical connectors,"

IEEE Transactions on Components and Packaging Technologies, Vol. 24, No. 3 (2001), pp. 370–377.

17. J. A. Greenwood and J. B. P. Williamson, "Electrical Conduction in Solids. II. Theory of Temperature-Dependent Conductors," *Proc. R. Soc. Lond. A*, Vol. 246, No. 1244 (1958), pp. 13–31.

18. M. Myers, M. Leidner, and H. Schmidt, "Effect of Contact Parameters on Current Density Distribution in a Contact Interface," *Proc IEEE 57th Holm Conference on Electrical Contacts*, Minneapolis, MN, 2011, pp. 1–9.

19. M. Myers, M. Leidner, H. Schmidt, S. Sachs, and A. Baeumer, "Contact Resistance Reduction by Matching Current and Mechanical Load Carrying Asperity Junctions," *Proc IEEE 58th Holm Conference on Electrical Contacts*, Portland, OR, 2012, pp. 1–8.

20. H. Liu, D. Leray, P. Pons, S. Colin, and A. Broué, "Finite element based surface roughness study for ohmic contact of microswitches," *Proc IEEE 58th Holm Conference on Electrical Contacts*, Portland, OR, 2012, pp. 1-10.

21. R. S. Timsit, "Electrical Conduction Through Small Contact Spots," *Transactions on Components and Packaging Technologies*, Vol. 29 (2006), pp. 727–734.

22. F. Pennec, D. Peyrou, D. Leray, P. Pons, R. Plana, and F. Courtade, "Impact of the Surface Roughness Description on the Electrical Contact Resistance of Ohmic Switches Under Low Actuation Forces," *IEEE Transactions on Components, Packaging and Manufacturing Technology*, Vol. 2, No. 1 (2012), pp. 85–94.

23. ANSYS™ couple-field analysis guide, Release 11.0.

24. A. Beloufa, "Minimization by FEM of the transient electrical contact resistance and contact temperature of power automotive connector," *Proc European conference of chemical engineering, and European conference of civil engineering, and European conference of mechanical engineering, and European conference on Control*, Stevens Point, Wisconsin, USA, 2010, pp. 50–58.

25. P. Cavaliere, V. Dattoma, and F. W. Panella, "Numerical analysis of multipoint CDW welding process on stainless AISI304 steel bars," *Computational Materials Science*, Vol. 46, No. 4 (2009), pp. 1109–1118.

26. R. S. Timsit, "The potential distribution in a constricted cylinder," *J. Phys. D: Appl. Phys.*, Vol. 10, No. 15 (1977), p. 2011.

27. H. Liu, D. Leray, P. Pons, and S. Colin, "An Asperity-Based Finite Element Model for Electrical Contact of Microswitches," *Proc IEEE 59th Holm Conference on Electrical Contacts*, Newport, RI, 2013, pp. 1–10.

GaN-based LEDs: state of the art and reliability-limiting mechanisms

Enrico Zanoni, Matteo Meneghini, Nicola Trivellin, Matteo Dal Lago, and Gaudenzio Meneghesso

University of Padova, Department of Information Engineering

via Gradenigo 6/B 35131 Padova, Italy

email: zanoni@dei.unipd.it, matteo.meneghini@dei.unipd.it, gauss@dei.unipd.it

Abstract

This paper reviews the main characteristics of state-of-the art high-brightness light-emitting diodes (LEDs), and the mechanisms responsible for the degradation of these devices. After a description of the structure, advantages and limits of power LEDs, we describe the following, relevant, degradation mechanisms: (i) the degradation of the active region of the devices, due to the generation of non-radiative defects and/or of shunt leakage paths; (ii) the worsening of the optical properties of the package and phosphor, which may induce a gradual degradation of the optical signal emitted by the devices; (iii) the catastrophic failure due to electrical overstress (hot-plugging) and to ESD events.

The results described within this paper are critically compared to the data presented in the literature, quoted in the list of references.

1. State of the art LED systems

The first Light-Emitting Diodes (LEDs) were commercialized in the early sixties; for almost four decades, these devices were considered as low-signal optoelectronic devices, capable of optical powers in the range of milliwatt, which could be mostly used as indicators in electronic systems, for display and for low-power signaling applications. The early generations of LEDs had emission wavelengths in the green, yellow and red spectral range, due to the optical properties of the semiconductor material (AlInGaP). Thanks to the progress – in the early nineties – of gallium nitride technology, it become possible to fabricate also LEDs emitting in the blue and violet spectral range; these devices can have a very high wall-plug efficiency (in excess of 50 %), and are the basis for the realization of white solid-state light sources [1, 2].

Since LEDs have a monochromatic spectrum, the generation of white light is usually obtained through color mixing. A first solution is to use three LEDs (one red, one green and one blue, RGB approach), and to mix their light by means of a suitable package/lens system. This approach permits the user to modulate the spectrum of the white light, thus tuning its chromatic properties (correlated color temperature, CCT, and chromatic coordinates), by simply varying the current flowing through each of the LEDs. On the other hand, due to the strong difference between the spectrum of an RGB system and that of a black body, the color rendering index (CRI) of a RGB light source is significantly lower (60-70) than what is considered as acceptable for interior lighting applications (typically 80). Another approach for obtaining white light from LEDs is to use a combination of a violet or blue LEDs and a phosphorescent material

(phosphor) emitting in the visible spectral range. The most common phosphors are based on YAG (Yttrium Aluminium Garnet), and – when excited with a blue radiation (around 460 nm) – emit a broad spectrum in the yellow-green spectral range. Phosphors can be either dispersed in the silicon matrix used to cover the LEDs, or deposited on the LED chip via conformal coating (Fig. 1). Another interesting approach, that allows one to reduce the heating of the phosphors, is to place the phosphorescent layer far from the blue light source used for excitation: this approach, usually referred to as "remote phosphor" technology, can guarantee high efficiencies and longer lifetimes.

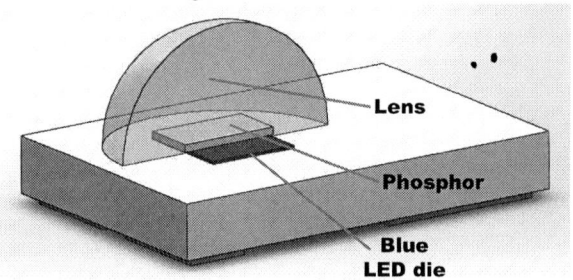

Fig. 1: typical structure of a packaged high power LED

By phosphor conversion it is possible to achieve very high efficacies (>200 lm/W); to obtain high color rendering indexes, multi-phosphor systems (involving two or three different phosphorescent elements) can be used; CRIs in excess of 95 have already been demonstrated based on this approach. More complex approaches involve the use of white (phosphor-converted) and monochromatic LEDs; commonly adopted solutions are lamps based on cool-white and red LEDs, or the red-green-blue-white (RGBW) configuration; also in this case it is possible to reach very high color rendering indexes, at the expense of a higher complexity of the circuit. In fact, each color channel must be driven individually, in order to guarantee the stability of the color coordinates. Fig. 2 shows the typical spectra of a RGB light source, a light source based on cool-white and red LEDs, and a RGBW LED lamp. The three light sources have the same correlated color temperature, CCT=3000 K; due to the different approaches used for the generation of white light, the values of CRI differ significantly for the different light sources. This is consistent with what presented by Narendran et al. in [3].

In a LED lamp (which usually comprises several LEDs), the LED chips can be individually mounted on a Printed Circuit Board (PCB), or included on the same package, via the so-called Chip on Board (COB) Technology. The latter approach allows one to reduce the

978-1-4799-4789-8/14 $31.00 © 2014 IEEE

size of a light source (Light Emitting Surface, LES), although it can limit the flexibility of the design (if only certain lumen packages are available).

Secondary optical elements can be placed on the LEDs, with the aim of shaping the output beam (lenses or reflectors) or to mix the spectra of the individual devices (mixing chambers). The design of the secondary optical elements must be optimized in order to reduce the size/cost of the lenses/reflectors, while maximizing the luminous output.

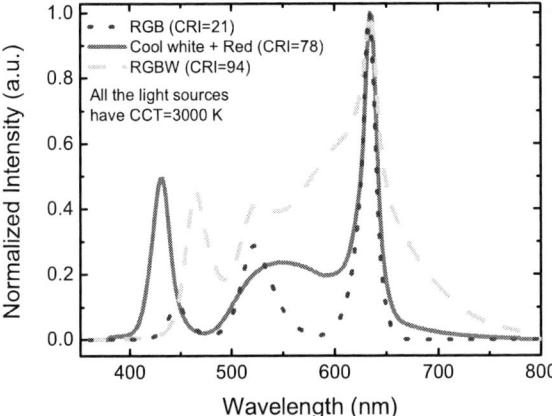

Fig. 2: spectral power distribution for three different light sources; (a) RGB light source; (b) LED lamp constituted by a cool white LED and a red LED; (c) RGBW system. All the analyzed lamps have a correlated color temperature of 3000 K. The CRI values of these lamps are listed in the legend

LED-based light engines may dissipate high power levels: when they reach high temperatures, LEDs show a significant decrease in their efficiency and a decrease in the operating voltage; moreover, high temperatures may significantly accelerate the degradation processes, thus limiting the lifetime of solid-state light sources.

Several strategies can be used to reduce the effects of self-heating: typically, high power devices are mounted on PCBs with metal core (usually referred to as insulated metal substrate, IMS); in this case, the thermal resistance between junction and ambient depends strongly on the conductivity of the insulating layer (whose typical thickness is 30-100 μm) placed between the copper and the thermally-conductive substrate. Other approaches include the use of a FR4 board with metallic vias (to ease heat spreading), or the use of chip-on-board technology.

Thanks to their very quick response, LEDs can be biased both under direct current (dc) and under pulsed conditions. In the first case, dimming is obtained by changing the input current; in the case of pulsed bias, the optical power of the LEDs can be regulated by pulse width modulation (PWM). The latter approach guarantees a better linearity of dimming, at the expense of a lower efficiency (at the LED side) and possible reduction of lifetime [4].

2. Degradation mechanisms of high power LEDs

The current generations of power LEDs – often based on semiconductor chips larger than 0.7 x 0.7 mm^2 or 1x1 mm^2 - can generate luminous fluxes in excess of 100-200 lm. Power dissipation is usually in the range between 1 and 3 W (for each LED chip), while efficacy ranges between 110 lm/W (for warm white LEDs) and 160 lm/W (for cool white LEDs). At least half of the power dissipated on an LED is converted into heat, due to the non-ideal quantum efficiency of the LED chips, to joule dissipation, and to the conversion losses of the phosphors. The latter are strongly dependent on the shift between the emission wavelength of the LED pump and of the phosphors (Stokes shift). Typical current densities of state-of-the-art power LEDs are in the range 50-100 A/cm^2.

Over the last decade, several research groups [4-28] demonstrated that LEDs can show a significant degradation when they are submitted to high temperature/high current operating conditions. An LED is a complex device, comprising several components (semiconductor chip, phosphor, lens, package, ...), many of which can degrade during ageing. The standard IES-LM-80 provides general guidelines for testing the lumen maintenance of LED light sources: the LEDs should be tested at a minimum of three case temperatures (55 °C, 85 °C, and a third temperature selected by the manufacturer), for several thousands of hours. Photometric measurements must be carried out (at 25 °C) repeatedly during ageing: the degradation tests should last at least 6000 hours, with data collection at a minimum of every 1000 hours. The lumen maintenance of an LED source can be projected based on the IES-TM-21 standard, based on the experimental data obtained by the procedures defined in IES-LM-80.

The two standards quoted above provide important information on the long-term degradation kinetics of LEDs; on the other hand, they do not give any information on the physical mechanisms responsible for degradation, and on the role of the various driving forces (current, temperature, optical power) in inducing the decrease in the efficiency of the devices. Moreover, IES-LM-80 and IES-TM-21 mostly analyze the lumen depreciation of the devices; other mechanisms – such as the shift of the chromatic properties [21], the increase in the operating voltage [9, 11, 13], and the increase in reverse current [11] should be also considered in order to better describe the degradation of LEDs.

In the following we give an overview of the main mechanisms responsible for the degradation of LEDs for lighting applications; we describe (i) the degradation of the blue semiconductor chip; (ii) the degradation of the phosphors and package; (iii) the catastrophic failure due to ESD and electrical overstress.

Fig. 3 reports typical degradation curves measured on white LEDs submitted to high temperature/high current stress conditions. The devices were stressed with a junction temperature of 160 °C (which is close to the maximum temperature recommended by several manufacturers, i.e. 150 °C), and several current levels. Results indicate that stress can induce a severe degradation of the optical characteristics of the devices (-30 % optical power after 1000 h of stress, for the hardest stress conditions). For these devices, the degradation rate increases with increasing stress current. Considering that all the devices are stressed with the same junction temperature, this result indicates that current is a major driving force for the degradation process. Current-induced degradation can be explained by considering that – at high current densities – part of the electrons may have enough energy to interact with the lattice thus inducing the generation of non-radiative defects within the active region of the devices [5-7, 14, 19, 22]. As a consequence, the non-radiative recombination rate (A) increases significantly during stress time: since the efficiency of LEDs depends on the ratio

$$\eta \propto \frac{Bn^2}{An + Bn^2 + Cn^3} \qquad (1)$$

an increase in coefficient A results in a significant decrease in the efficiency of LEDs, especially for low current densities. In (1), A, B, and C are the non-radiative, radiative, and Auger recombination coefficients, while n is the carrier density.

Changes in the concentration of defects within the active region of the devices may result in an increase in the leakage current components, especially in the reverse-bias region and in the sub turn-on part of the forward I-V (current-voltage) curves [11]. These modifications (see an example in Fig. 4) may be also due to the generation of defect-related shunt paths across the junction.

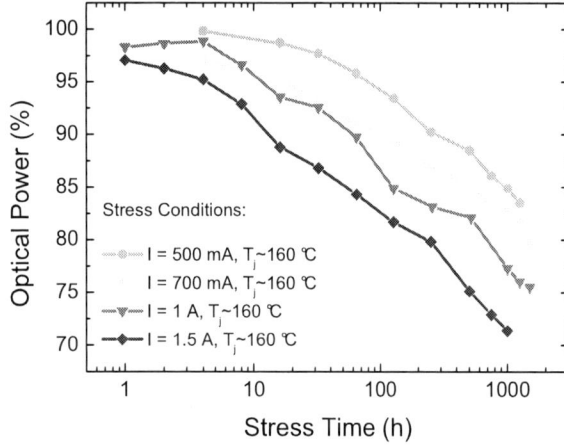

Fig. 3: optical degradation of high power white LEDs submitted to stress at high current/temperature levels

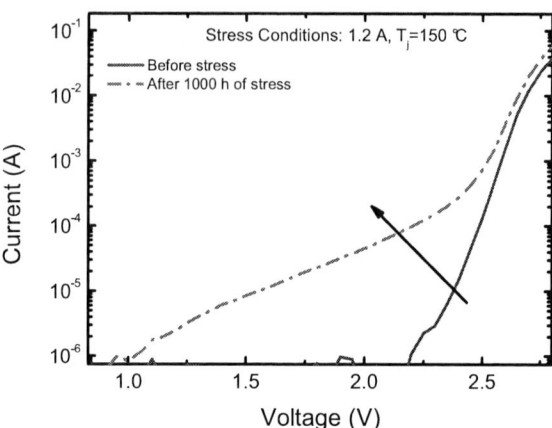

Fig. 4: degradation of the electrical characteristics of a LED submitted to stress at high temperature/current levels

Exposure to high temperature may significantly accelerate the degradation kinetics, either due to an increase in the rate of defect-generation within the semiconductor chip, or to the worsening of the optical properties of the package, phosphors and lenses [11]. The package is usually fabricated in plastic or ceramic material, and should maximize the light extraction efficiency. Plastic packages can show a significant darkening when submitted to high temperature stress (as described in [14, 23]); this can result in a decrease in the extraction efficiency, and in a shift of the chromatic properties. Package degradation may be also due to the interaction with volatile organic compounds (VOCs), which are chemical structures that are present in many components of an LED lamp (o-rings, glues, gaskets, …). When LEDs are operated in the presence of VOCs and with little air movement, VOCs can penetrate into the silicone lens (which is permeable), thus being trapped in the lens or encapsulant [26]. Further reactions may lead to the discoloration of the optical elements, and to the degradation of the efficiency and of the chromatic properties of the devices. These processes can be accelerated by the presence of high energy radiation (in the case of blue and white LEDs) and by the heat produced by the luminaire.

The exposure to high temperatures can also induce a severe degradation of the properties of the phosphor layer used for the generation of white light; the most important consequences of high temperature treatment are a decrease in the conversion efficiency of the phosphor layer, and a change in the chromatic properties [21]. Typical degradation kinetics of phosphor plates for lighting applications are reported in Fig. 5. The degradation of the phosphor layer can be ascribed to the darkening of the silicon or polycarbonate layer used as a carrier for the (very thin) phosphor layer. Phosphor and package degradation are usually thermally-activated, with activation energies in the range 0.5-1.5 eV [14]; an accurate selection of the materials, and the reduction of

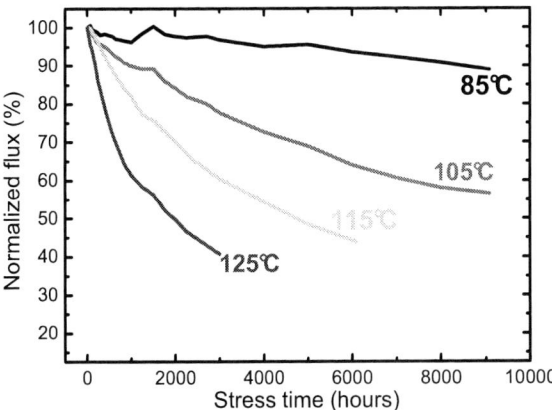

Fig. 5: results of purely thermal stress on phosphor plates for solid-state lighting. The phosphor plates were aged at four different temperatures

the operating temperatures, permits to significantly improve the lifetime.

The degradation mechanisms described above induce a gradual decrease in the efficiency of the devices; however, LEDs can show also a catastrophic (or sudden) failure, due to electrostatic discharges (ESD, [15, 16, 20]) or to electrical overstress (EOS). The first case is quite common both in the manufacturing environment (if proper precautions are not taken), during installation, and in the real-life applications (consider the case of lightning close to streetlamps). ESD events have a very short duration (few 100s nanoseconds), and high current levels; the effect of ESD can be studied by proper simulators, based on the HBM (Human Body Model) or TLP (Transmission Line Pulser) methods [16]. In most of the cases, ESD failure take place in proximity of pre-existing epitaxial or processing defects, where – at high voltage levels – current density can reach very high levels, thus leading to the failure of the devices. For InGaN-based LEDs (commonly used for lighting applications), structural defects –such as dislocations and V-shaped defects – are supposed to play an important role in the degradation [16]. Generally, LEDs are more robust towards forward-bias ESD, rather than in reverse-bias conditions [27]. To prevent ESD failure, a protection diode is usually placed in parallel to the LED chip. Moreover, the ESD robustness of LEDs is strongly dependent on the properties and quality of the semiconductor material: red LEDs (based on AlInGaP) have usually a higher robustness with respect to blue and green devices (based on InGaN), due to intrinsic differences between the two semiconductor materials (see and example in Fig. 6).

Catastrophic degradation may occur also due to EOS: in this case the devices are subjected to a current or voltage which is beyond the specification limits, for a certain time (from ms to seconds). A particularly interesting case occurs in the case of hot-plugging, i.e. when a LED chain is directly connected to an energized power supply [28]. In this case high current pulses can be generated, due to

the mismatch between the output voltage of the driver and the forward voltage of the LED chain. The amplitude of such pulses depends on the number of LEDs in the chain (Fig. 7), while their duration can be of several hundreds of microseconds. EOS and hot-plugging may lead to the failure of the LEDs, due to the extremely high power dissipation. EOS and hot-plugging can be prevented through the use of proper transient voltage suppression devices/systems, placed on (or in proximity of) the LED board [28].

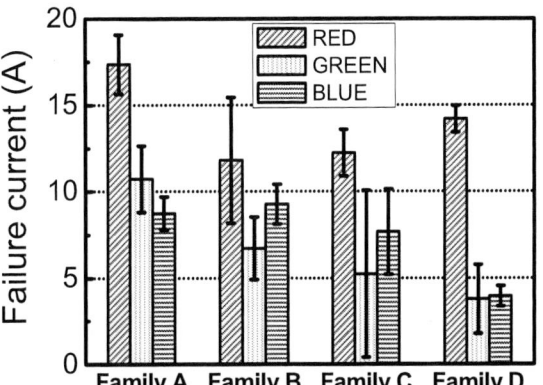

Fig. 6: failure rate for low power red, green, and blue LEDs fabricated by four different manufacturers; the devices were submitted to reverse-bias TLP test (pulse duration = 100 ns)

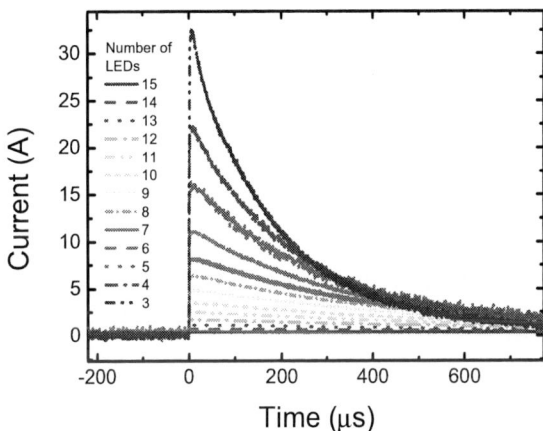

Fig. 7: typical pulseS measured after hot-plugging of LED chains with increasing number of LEDs. Hot-plugging occurs at t=0, spike current increases with decreasing number of LEDs in the chain

3. Conclusions

In summary with this paper we have summarized the relevant characteristics of state-of-the-art LEDs, and the most critical mechanisms responsible for LED failure.

The results summarized within this paper indicate that – although LEDs are excellent in terms of performance – the design of SSL systems must be carefully optimized in order to avoid failures and guarantee long lifetime. We have discussed how current, temperature, optical power may act as driving forces for the degradation processes; moreover, we have described the role of ESD and EOS in inducing the permanent failure of the devices.

References

1. 1. S. P. DenBaars, D. Feezell, K. Kelchner, S. Pimputkar, C.-C. Pan, C.-C. Yen, S. Tanaka, Y. Zhao, N. Pfaff, R.Farrell, M. Iza, S. Keller, U. Mishra, J. S. Speck, S. Nakamura, "Development of gallium-nitride-based light-emitting diodes (LEDs) and laser diodes for energy-efficient lighting and displays", Acta Materialia 61, 945–951, 2013

2. Available online: http://www.cree.com/news-and-events/cree-news/press-releases/2012/december/mkr-intro

3. N. Narendran, and L. Deng, "Color rendering properties of LED light sources", Proceedings of the SPIE, Volume 4776, pp. 61-67 (2002)

4. M. Meneghini, M. Dal Lago, L. Rodighiero, N. Trivellin, E. Zanoni, and G. Meneghesso, "Reliability issues in GaN-based light-emitting diodes: Effect of dc and PWM stress", Microelectronics Reliability 52, 1621-1626 (2012)

5. F. Manyakhin, A. Kovalev, and A. E. Yunovich, "Aging Mechanisms of InGaN/AlGaN/GaN Light-Emitting Diodes Operating at High Currents", MRS Internet J. Nitride Semicond. Res., 3, 53, 1998

6. O. Pursiainen, N. Linder, A. Jaeger, R. Oberschmid, and K. Streubel, "Identification of aging mechanisms in the optical and electrical characteristics of light-emitting diodes", Appl. Phys. Lett., Vol. 79, no. 18, pp. 2895-2897, 2001

7. X.A. Cao, P.M. Sandvik, S.F. LeBoeuf, and S.D. Arthur, "Defect generation in InGaN/GaN light-emitting diodes under forward and reverse electrical stresses", Microelectronics Reliability 43, 1987-1991, 2003

8. N. Narendran, Y. Gu, J. P. Freyssinier, H. Yu, and L. Deng, "Solid-state lighting: failure analysis of white LEDs", Journal of Crystal Growth 268, 449–456, 2004

9. M. Meneghini, L. Trevisanello, G. Meneghesso, E. Zanoni, F. Rossi, M. Pavesi, U. Zehnder, U. Strauss, "High-temperature failure of GaN LEDs related with passivation", Superlattices and Microstructures 40 (4-6 SPEC. ISS.) , pp. 405-411, 2006

10. S. Ishizaki, H. Kimura, and M. Sugimoto, "Lifetime estimation of high power LEDs," J. Light & Vis. Env., vol. 31, no. 1, pp. 11–18, 2007

11. M. Meneghini, L. Trevisanello, G. Meneghesso, and E. Zanoni, "A review on the reliability of GaN-based LEDs", IEEE Transactions on Material, Devices and Reliability, Vol. 8, No. 2, pp. 323-331, 2008

12. A. Uddin, A. C. Wei, and T. C. Andersson, "Study of degradation mechanism of blue light emitting diodes", Thin Solid Films 483, 378, 2005

13. M. Meneghini, L. Rigutti, L. Trevisanello, A. Cavallini, G. Meneghesso, E. Zanoni, "A model for the thermal degradation of metal/ (p-GaN) interface in GaN-based light emitting diodes", Journal of Applied Physics 103, 063703, 2008

14. G. Meneghesso, M. Meneghini, E. Zanoni, "Recent results on the degradation of white LEDs for lighting", Journal of Physics D: Applied Physics 43, 354007, 2010

15. S. K. Jeon, J. G. Lee, E. H. Park, J. H. Jang, J. G. Lim, S. K. Kim, and J. S. Park, "The effect of the internal capacitance of InGaN-light emitting diode on the electrostatic discharge properties", Appl. Phys. Lett. 94, 131106, 2009

16. M. Meneghini, A. Tazzoli, R. Butendeich, B. Hahn, G. Meneghesso, E. Zanoni, "Soft and hard failures of ingan-based LEDs submitted to electrostatic discharge testing", IEEE Electron Device Letters 31, 579, 2010

17. J. Hu, L. Yang, and M. W. Shin, "Electrical, optical and thermal degradation of high power GaN/InGaN light-emitting diodes", J. Phys. D: Appl. Phys. 41, 035107, 2008.

18. U. Zehnder, A. Weimar, U. Strauss, M. Fehrer, B. Hahn, H.-J. Lugauer, V. Härle, "Industrial production of GaNand InGaN-light emitting diodes on SiC-substrates", Journal of Crystal Growth 230, 497, 2001

19. M. Meneghini, C. de Santi, N. Trivellin, K. Orita, S. Takigawa, D. Ueda, G. Meneghesso, E. Zanoni, "Investigation of the deep level involved in InGaN laser degradation by deep level transient spectroscopy", Appl. Phys. Lett. 99, 093506, 2011

20. S.-C. Shei, J.-K. Sheu, and C.-F. Shen, "Improved reliability and ESD characteristics of flip-chip GaN-based LEDs with internal inverse-parallel protection diodes," IEEE Electron Device Lett., vol. 28, no. 5, pp. 346–349, 2007

21. M. dal Lago, M. Meneghini, N. Trivellin, G. Mura, M. Vanzi, G. Meneghesso, and E. Zanoni, "Phosphors for LED-based light sources: Thermal properties and reliability issues", Microelectronics Reliability 52, 2164, 2012

22. P. N. Grillot, M. R. Krames, H. Zhao, and S. H. Teoh, "Sixty Thousand Hour Light Output Reliability of AlGaInP Light Emitting Diodes", IEEE Trans. Device Mater. Rel., Vol. 6, No. 4, pp. 564-574, 2006.

23. M. Meneghini, L. Trevisanello, S. Podda, S. Buso, G. Spiazzi, G. Meneghesso, and E. Zanoni, "Stability and performance evaluation of high brightness light emitting diodes under DC and pulsed bias conditions ",Proceedings of SPIE Volume 6337, 2006, Article number 63370R

24. A. Y. Polyakov, et al, "Enhanced tunneling in GaN/InGaN multi-quantum-well heterojunction diodes after short-term injection annealing", J. Appl. Phys. 91 5203, 2002

25. M. Meneghini, U. Zehnder, B. Hahn, G. Meneghesso, and E. Zanoni, "Degradation of high-brightness green LEDs submitted to reverse electrical stress", IEEE Electron Device Letters 30, 1051-1053 (2009)

[26] available online: http://www.cree.com/~/media/Files/Cree/LED%20Compone nts%20and%20Modules/XLamp/XLamp%20Application%2 0Notes/XLamp_Chemical_Comp.pdf

[27] M. Meneghini, A. Tazzoli, G. Mura, G. Meneghesso, and E. Zanoni, "A review on the physical mechanisms that limit the reliability of GaN-based LEDs", IEEE Transactions on Electron Devices vol. 57, no. 1, pp. 108-118, 2010

[28] M. Dal Lago, M. Meneghini, N. Trivellin, G. Mura, M. Vanzi, G. Meneghesso, E. Zanoni, ""Hot-plugging" of LED modules: Electrical characterization and device degradation", Microelectronics Reliability 53 (2013) 1524-1528

System Reliability for LED-Based Products

J. Lynn Davis, Karmann Mills, Mike Lamvik, Robert Yaga, Sarah D. Shepherd, James Bittle, Nick Baldasaro, Eric Solano, Georgiy Bobashev, Cortina Johnson, and Amy Evans

RTI International

PO Box 12194; Research Triangle Park, NC 27709-2194

Contact: ldavis@rti.org

Abstract

Results from accelerated life tests (ALT) on mass-produced commercially available 6" downlights are reported along with results from commercial LEDs. The luminaires capture many of the design features found in modern luminaires. In general, a systems perspective is required to understand the reliability of these devices since LED failure is rare. In contrast, components such as drivers, lenses, and reflector are more likely to impact luminaire reliability than LEDs.

1. Background

Electric lighting technologies account for approximately 19% of worldwide electricity consumption, and the promotion of energy efficient lighting can have a positive impact on energy consumption and CO_2 emissions [1]. Although the lighting industry has become a ubiquitous part of modern culture, the disposable nature of conventional lighting products does not promote a close examination of system-level reliability. In a conventional luminaire, the device is operated until failure occurs, which is usually defined as the absence of light (i.e., catastrophic failure). When failure occurs, the lamp or ballast is often replaced and the device becomes operational again. In some instances, preventative maintenance schedules are established to replace lamps and ballasts at regular intervals often at significant labor and maintenance costs for lighting systems operator.

The advent of solid-state lighting (SSL) technologies provides the opportunity for both greater energy efficiency and longer lifetime from a lighting product. SSL luminaires are based on light emitting diodes (LED) that are integrated into the luminaire and can be difficult to replace. Electronic drivers, which power the LEDs, are often directly integrated into the luminaire and difficult to replace as well. A well designed SSL system can achieve luminous efficacies on the order of 160 lumen per Watt (lm/W) at the LED level and 100 lm/W at the luminaire level. Shifting from disposable lamps (e.g., fluorescent and incandescent lamps) to integrated LED sources and drivers in a luminaire that could be in use for 15 years or more requires a re-examination of the reliability paradigm of lighting devices.

The definition of failure for SSL devices is a subject of debate within the industry. Catastrophic failure in SSL luminaires is analogous to a blown filament in incandescent lamps and is clearly one indication of failure. However, the depreciation of lighting levels below some pre-determine threshold (e.g., 70% of initial luminous flux) is sometimes considered a lumen maintenance failure [2]. Likewise, in some applications such as museums or retail, excessive color shift can also be considered a failure, whereas in others it may not be [3].

This paper reports on a series of accelerated life tests (ALT) on SSL luminaires to identify potential failure modes that may arise during an expected 15+ year lifetime. These ALT protocols used various combinations of temperature, humidity, and electrical bias to accelerate aging in both the light management system (i.e., LEDs, lenses, reflectors) and power management system (i.e., driver) of the luminaire. The findings from these studies provide insights into the lifetime of SSL luminaires and help to establish an experimental basis for discussing the reliability of luminaire systems.

2. Previous Studies of SSL Luminaire Reliability

The reliability of high brightness LEDs (HBLEDs) used for illumination have been studied for the past ten years and major findings are reviewed elsewhere [4]. There has also been previous work on other SSL luminaire components such as drivers [4, 5], but in general there is a lack of publicly available data on ALT studies of SSL luminaires and lamps.

Pacific Northwest National Lab (PNNL) developed a stepped-stress methodology to compare the durability of the Philips L-Prize 60 Watt equivalent LED lamps with market-proven compact fluorescent lamps (CFL) [6]. In this study, simultaneous combinations of electrical, thermal, vibration, and humidity stresses of increasing magnitude were used. The stress tests were performed on specially built lamps, so the general characteristics of mass market lighting products were not determined. However, the findings from this study provide guidance into the potential behavior of well-designed SSL lamps. In this study, PNNL found that all of the LED lamps were still fully electrically functional at the end of testing, whereas none of the CFLs were. In addition, failure of the CFLs occurred well before the highest stressor conditions in the tests in all cases. In contrast, all the LED lamps were able to survive the most extreme test conditions and remain fully operable. In a separate report, PNNL published data on the lumen maintenance of the L-Prize lamps and found that the level of lumen depreciation was negligible, even after 25,000 hours at an ambient temperature of 45°C [7].

RTI International, in association with the LED Systems Reliability Consortium (LSRC), tested commercial luminaires using a highly accelerated life test (HALT) also known as the "Hammer Test." In this test, commercial 6" downlights were subjected to a series of sequential environmental stresses including temperature cycling (-50°C to 125°C), wet high temperature operational life test (WHTOL) at 85°C and 85% relative humidity (RH), and high temperature operational life test (HTOL) at 120°C. Environmental extreme conditions which were outside the typical use environment of these products were intentionally chosen to accelerate failure to less than 1000 hours of testing. In this experiment, commercially available, mass-produced SSL luminaires were found to be exceptionally robust even under the harsh conditions of the Hammer Test. When device failure did occur, it could generally be traced to electrical components such as printed circuit boards, solder joints, or other electrical connections [8]. Failure of the LEDs was rare, even under the highly accelerating conditions of the Hammer Test [8]. In a subsequent report, these authors focused on causes of lumen depreciation in 6" downlights and found that degradation of optical components such as lenses and reflectors can play a significant role in lumen depreciation in some designs [9].

3. Experimental Section

Four different models of mass produced, commercially available 6" downlights were chosen for testing and these luminaires were purchased from various distributors in order to ensure an unbiased sample selection. Three of the models were retrofit downlights that could be powered through standard Edison sockets. The new construction downlight is powered by directly connecting the electrical mains to the driver. Additional information on the luminaires in these tests is provided in **Table 1**.

Table 1. Details of the 6" downlights examined in this study.

	Luminaire Model			
	A	**B**	**C**	**D**
Lumi-naire Type	Retro-fit	New con-struct.	Retro-fit	Retro-fit
LED Type	HB-LED	HB-LED	Mid-Power LED	HB-LED
CCT (K)	2700	2700	2700	2800 and 4000
Number Tested	33	3	20	8
LE	61 lm/W	45 lm/W	61 lm/W	44 lm/W
Price	$20		$18	$70

LE = luminous efficacy

HBLED = high brightness LED

CCT = correlated color temperature

lm/W = lumen/Watt

Temperature and humidity (T-H) environments were used to accelerate aging of the luminaires and select components. Common protocols from the electronics industry were followed; for example, ALT tests under 85°C and 85% RH (hereafter termed 85/85) followed procedures outlined in JESD22-A101C. This procedure calls for a continuous soak at 85/85 with electrical power to the luminaires cycled on and off at one-hour intervals.

The T-H ALT procedure consisted of placing the luminaires under test in an environmental chamber (either Tenney T20RC-2.0 or Tenney TC10RS) set to a constant environment. After each 250-hour test, the luminaires were removed from the environmental chamber, allowed to equilibrate to room temperature, and their performance measured in a 65" integrating sphere using procedures outlined in LM-79 [10]. All measurements of the 6" downlights were performed in the 2π configuration with the luminaires mounted on the exterior of the sphere. The luminaires were operated for a minimum of 1 hour prior to testing to allow thermal equilibration. During the measurements, corrections were made for zero level and self-absorbance as outlined in LM-79. Calibrations were performed using a National Institute of Standards and Technology (NIST)-traceable forward flux standard from Labsphere (Model Number FFS-100-1000). Functional luminaires were returned to the environmental chamber after photometric testing for additional T-H exposure.

Testing was also performed on individual high-brightness LEDs using analogous procedures. For these tests, LEDs similar to those used in Luminaires B and D were chosen for testing. Luminaire A uses a hybrid LED light source with two different colors of LEDs combining to produce warm white light. The major LED used in Luminaire A is the same model as the LEDs used in Luminaires B and D, but the CCT value is different.

The individual LEDs were mounted separately on a metal-core board attached to a large heat sink using pre-cut thermal adhesive. The LEDs were connected in series, with five LEDs on the heat sink. The heat sink and LED assembly was placed in the environment chamber for 85/85 and the LEDs were powered by a driver that was kept external to the environmental chamber. The test population of LEDs consisted of 10 blue LEDs, 10 cool white LEDs, and 10 warm white LEDs. While all three test populations were from the sample product family, the die size of the blue LEDs (0.92 mm^2) was smaller than that of the white LEDs (1.92 mm^2). The white LEDs were operated at 500 mA during testing to match the conditions in the LM-80 report. After each round of testing, the individual LEDs were measured in a 10" integrating sphere using a 2π configuration with the LEDs and heat sink mounted external to the integrating sphere. This configuration allowed each LED to be measured separately. During the measurements, corrections were made for zero level and self-absorbance as outlined in LM-79. Calibrations were performed using a NIST-traceable forward flux standard from Labsphere (Model Number FFS-100-400).

4. LED Performance During 85/85

Individual LEDs mounted separately on metal core boards were attached to heat sinks via pre-cut thermal adhesives and secured with screws as shown in **Figure 1**. Although the luminaires studied during these tests contained warm white LEDs, the study of lumen maintenance of individual LEDs was expanded to include blue, warm white, and cool white LEDs in order to investigate the behavior of different phosphor formulations. The white LEDs (i.e., both cool white and warm white) were fabricated in the proximate phosphor configuration with the phosphor residing on top of the die. Silicone encapsulants are presumed to be used throughout the structure.

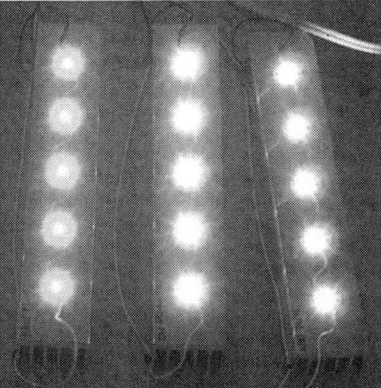

Figure 1: Image of the mounting configuration used for testing individual LEDs.

The flux maintenance of the LED populations as a function of time in 85/85 is shown in **Figure 2**. Luminous flux is reported for both cool white and warm white LEDs, whereas radiant flux is reported for the blue LED.

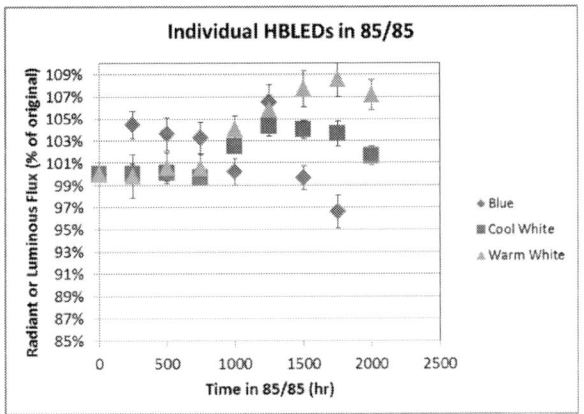

Figure 2: Change in light flux as a function of time in 85/85 for individual LEDs. All LEDs were equivalent products from the same manufacturer, although the color temperature of the populations was different. The data points represent the average value at each time for the population of 10 samples. The error bars represent one standard deviation.

Figure 2 demonstrates that the luminous flux of both the warm white and cool white LEDs actually increases during time in 85/85. This finding is in general agreement with the TM-21 extrapolation based on 10,080 hours of testing under the LM-80 test protocol [11, 12]. The TM-21 decay rate constant, which is reported for warm white LEDs only, was determined by the manufacturer to be negative at 85°C and 500 mA. A negative TM-21 decay rate constant indicates that the lumen maintenance is projected to increase.

In contrast, a general decline in radiant flux emission was observed for the direct emitter, blue LEDs. It is unclear whether this decrease in radiant flux is due to a reduction in emission intensity from the LED or whether the moisture in the 85/85 environment has affected the light extraction efficiency from the LED, perhaps due to the appearance of small voids at the encapsulant die interface, as suggested by Wu *et al.* [13].

The increase in luminous flux observed for the population of individual LEDs can be explained by the spectral changes that are occurring in the LEDs during the 85/85 experiment. As shown in **Figures 3** and **4**, the correlated color temperatures (CCTs) of both the warm white and cool white LEDs increase in a nearly linear fashion over time. As might be expected, a similar trend was observed with color shift as measured by Δu'v'. For these specific LEDs, the rate of change in CCT (or Δu'v') was higher for warm white LEDs than was observed for cooler colors. The difference in CCT and color point shift between these equivalent products may be due to different thermal stabilities of the phosphor mixtures used to achieve warm white and cool white colors. This finding indicates that significant spectral changes are occurring in the individual LEDs during 85/85 which may increase the lumen content of the emitted light.

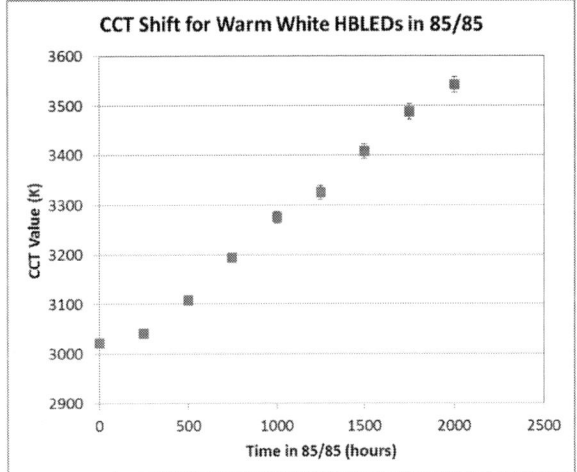

Figure 3: Change in CCT values for a population of 10 warm white LEDs as a function of time in 85/85. The error bars represent one standard deviation.

978-1-4799-4789-8/14 $31.00 © 2014 IEEE

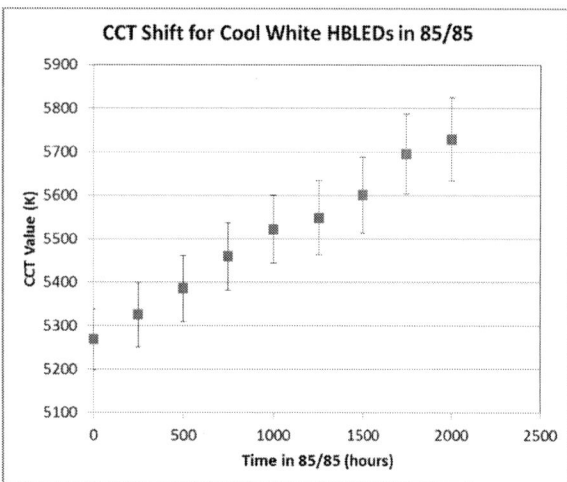

Figure 4: Change in CCT values for a popoulation of 10 cool white LEDs as a function of time in 85/85. The error bars represent one standard deviation from the mean.

Although CCT values were observed to increase for the white LED products that were examined in this test, we have also tested other LED products where the CCT values decreased in time. Hence, CCT shift (or alternatively shifts in Δu'v') in LED products is likely a function of LED design and the materials, including phosphors and encapsulants, used in their construction.

We have not conducted analogous 85/85 testing on mid-power LEDs similar to those used in Luminaire C. Therefore, the behavior of the individual mid-power LEDs in 85/85 is not currently known. However, as discussed below, findings from the lumen maintenance of the individual luminaire populations can provide some insights into this phenomenon.

5. Lumen Maintenance for SSL Luminaires

The commercially available, mass-produced luminaires described in Table 1 were placed in T-H environments (i.e., either 85/85 or 75/75) and their photometric properties measured for each 250 hours of exposure. Some results for the lumen maintenance of Luminaires A, B, and C in both 85/85 and 75/75 have been reported previously [9]. In this earlier study, significant lumen depreciation was observed for Luminaire A resulting in an average lumen depreciation of 30% at 1,500 hours of 85/85. The rate of light loss for Luminaire A in 75/75 was significantly lower and the average lumen depreciation value at 1,500 hours of testing was approximately 4%. The rapid lumen depreciation observed for Luminaire A in 85/85 has been traced to the oxidative degradation of the polymeric materials used to make the lenses and reflector [9]. In this previous study, Luminaires B and C were not observed to undergo more than 15% lumen depreciation in 2,000 hours of 85/85. This lower rate of lumen depreciation was shown to be due to the use of more stable polymers in the optical system [9].

The average lumen maintenance behavior of the four different 6" downlight products examined in this study are summarized in **Table 2**. Based on the findings from Davis *et al.*[9], the decision was made to test Luminaires B, C, and D without modification. However, test results are reported for Luminaire A with a fresh lens and reflector for each test. In this way, the lumen depreciation for Luminaire A reported in this paper is dominated by the characteristics of the LED light engine and is not influenced by the aging of the lens and reflector. The numbers in the chart represent the average lumen maintenance for all products whereas the number in parentheses represents the number of products still in test at each indicated time. When products exhibit catastrophic failure, they are removed from testing and no longer included in the data shown in Table 2.

Table 2: Average lumen maintenance of the downlights examined in this study after exposure to 85/85. (NOTE: The numbers in parentheses report the number of units still functional at that stage of the test).

Time (hours)	Lum. A	Lum. B	Lum. C	Lum. D
0	100 (11)	100 (3)	100 (10)	100 (4)
250		93 (3)	100 (10)	100 (3*)
500		92 (3)	98 (10)	90 (1*)
1000		89 (3)	92 (8)	- (0)
2000	86*	87 (3)	85 (1)	
3000	84*	85 (3)	- (0)	
4000	76*	80 (3)		

*Luminaire A was tested with a fresh lens and reflector for all times starting with 2,000 hours in 85/85.

Luminaire A is a hybrid luminaire containing two different color LEDs, and adaptive power control is used to adjust the power levels of the two LEDs to maintain color point and luminous flux. The change in CCT values that were observed for the population of this device is shown in **Figure 5**. The initial CCT values for this population of luminaires were 2734 K (with a standard deviation of 17). However, as these samples age, a significant variation in the CCT value is observed as indicated by the error bars in Figure 5. This variability may be due to the impact of aging on the adaptive power control circuit since the aging characteristics of the main LED used in this product can be expected to follow the behavior shown in Figures 3 and 4.

The change in CCT values for Luminaires B, C, and D are shown in **Figure 6**. Although Luminaires B and D use the small type of LEDs, the observed rate of CCT shift was quite different which is likely due to the difference in construction for the two devices. Luminaire B does not have an optical mixing chamber, but instead has individual secondary lenses affixed to each LED with a gasket retarding moisture ingress. The assembly is clamped with a bolted metal plate that keeps the gasket under compression. In contrast, Luminaire D did not contain a gasket to slow moisture ingress. Luminaire C does not have a gasket to slow moisture ingress, but the CCT stability of this population of luminaires was excellent. The average CCT value shifted by less than 100K during 2,000 hours of testing at 85/85. This finding indicates that the specific mid-power LED used in this product has good color stability under the test conditions.

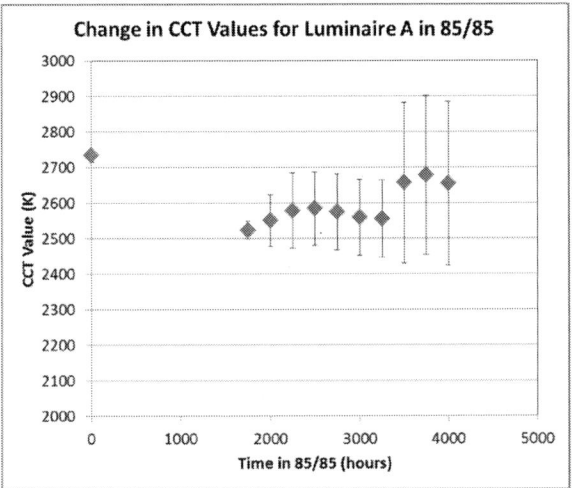

Figure 5: Change in CCT values for the population of Luminaire A as a function of time in 85/85. The data points indicated the population average, while the error bars represent one standard deviation.

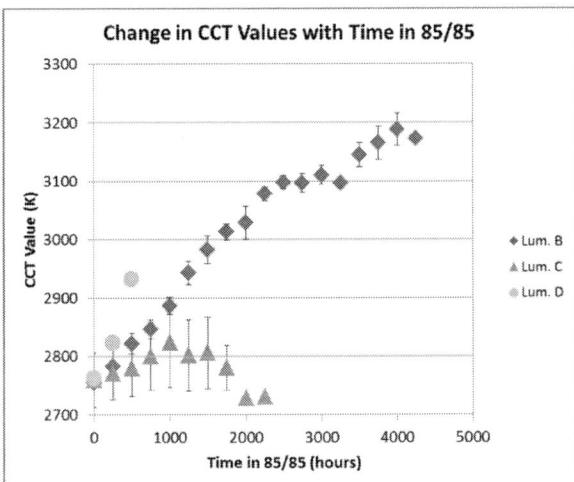

Figure 6: Change in CCT values for the population of Luminaire B, C, and D as a function of time in 85/85. The data points indicated the population average, while the error bars represent one standard deviation.

6. Catastrophic Failure of Luminaires During 85/85

A total of 44 samples of Luminaires A, B, C, and D were subjected to 85/85 and these samples were operated until failure occurred. For this study, failure is defined as the absence of light (i.e., catastrophic failure). Of the 44 samples in 85/85, 30 units (68%) have failed and 14 units (32%) are still functional. These 14 units will continue in

testing until failure occurs. Some of these units have been tested for more than 5,000 hours.

A total of 20 different samples were also inserted in 75/75 testing and this test group consisted of 10 samples from Luminaire A and 10 samples from Luminaire B. Of these 20 units, 7 units (35%) have failed through 3,250 hours of testing and the remaining 13 units (65%) are still functioning. The functional units will be tested until catastrophic failure occurs.

A variety of failure modes were observed for the units that have failed to date in 85/85 and 75/75. The most common point of failure is with the capacitors, with film caps more likely to fail than electrolytic capacitors. The higher failure rates observed in this testing for film capacitors compared to electrolytic capacitors may be due to the higher level of derating used by manufacturers for electrolytic capacitors. The performance of electrolytic capacitors in high temperature shelf life experiments has been examined by Lall *et al.* and it was determined that this test caused a general increase in equivalent series resistance (ESR) and a decrease in capacitance of electrolytic capacitors [14]. However, capacitor derating and circuit design allowed the driver to continue to operate even after significant degradation of some of the electrolytic capacitors in the device. Likewise, a literature review of failure modes of film capacitors in 85/85 revealed that increases in ESR and decreases in capacitance will occur in 85/85 environments [15].

Using the data from these 64 luminaires, Weibull probability plots can be constructed for the 85/85 and 75/75 data as shown in **Figure 7**. The analysis was conducted by accounting for the interval censored nature of the failures, since failure time is only known with the accuracy of the test interval (i.e., 250 hours). Maximum likelihood estimation methods were used for the right censored data (i.e., functional parts that are still being tested). In this analysis, the β value was calculated to be 1.31 indicating that the T-H environments used in this test are accelerating aging. This value is lower than that reported previous (1.935) in the Hammer Test [9].

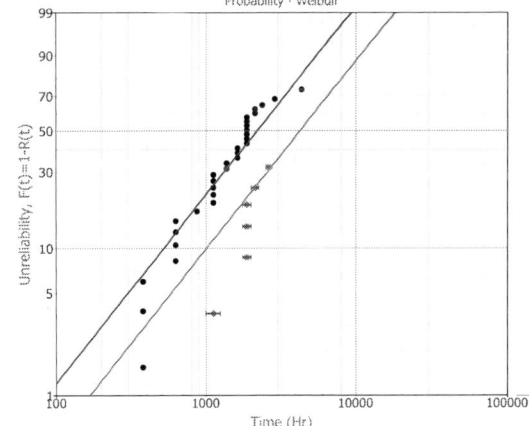

Figure 7: Weibull probability plot for the luminaires examined in 85/85 and 75/75.

7. Discussion

This work examined the performance of a broad cross-section of lighting technologies. Four different commercial luminaires were studied, each with different design attributes. Luminaires A, C, and D have similar structures with the LEDs residing at the base of an optical mixing cavity that is used to mix and diffuse the light. However, Luminaire A uses a hybrid LED light engine built around two different colors of HBLEDs, Luminaire C uses mid-power LEDs, and Luminaire D uses HBLEDs of the same color. In contrast, Luminaire B does not have an optical mixing cavity but instead utilizes a series of secondary lenses, which are sealed around each LED with a gasket, to diffuse and shape the light beam. One important commonality among these four products is that they were designed to serve similar markets (i.e., lighting applications needing 6" downlights). In this way the luminaires evaluated in this study provide a snapshot of the design options available to the lighting community.

LED products available on the market today are generally robust and able to withstand significant environmental stresses without failure. Due to this fact, testing protocols using LM-80 and similar approaches require long test times (6,000 hours minimum) to produce sufficient data to allow estimation of product lumen maintenance using TM-21. The long test times place a significant testing burden on LED manufacturers and slow the rate of new product introduction.

However, as shown above, T-H testing in 85/85 is able to provide information analogous to the most severe conditions tested in LM-80 in a significantly shorter period of time. For the HBLEDs examined in Figures 3-5, T-H testing was able to reproduce at least qualitatively the general behavior (i.e., lumen maintenance and color point shift) of this specific LED under the most extreme conditions examined in the product's LM-80 report. The extension of this approach to other LEDs is unknown at this time and caution is necessary before extending this comparison to other LED products. However, while general information on the behavior of other LEDs under T-H profiles has not been published, it is intriguing to consider whether a public repository of LED T-H profiles can be used to reduce LED testing burden.

Although LEDs have received a significant amount of attention in discussions on SSL product reliability, no LED failures were observed in this test nor have any LED failures been reported in testing of the L-Prize lamps [6, 7] or the Hammer Test on 6" luminaires [8]. This finding reinforces the general belief that the current generation of LED products is well characterized and has excellent performance.

The results discussed here and those presented previously [9] demonstrate that when proper accommodations are made for the oxidative photodegradation of materials such as polycarbonates and TiO_2-filled polymers, the lumen maintenance of SSL products in aggressive T-H testing can be expected to be excellent. The four different product architectures examined in this report all demonstrated excellent lumen maintenance performance in 85/85. Average lumen maintenance of the product populations was above 0.75 when catastrophic failure occurred. This finding suggests that lumen maintenance is less likely to be an issue in well-designed SSL devices compared to failure of the driver electronics.

In this study and in the Hammer Test report [8], when luminaire failure has been observed, it is generally traced to the driver circuit. Capacitor failures have occurred in many of the luminaires that we have tested in 85/85 and in the Hammer Test. We have observed a general trend for failure to occur more frequently in the thin film capacitors compared to electrolytic capacitors. The greater tendency for thin film capacitors to fail in our tests compared to electrolytic capacitors may be due in part to the higher derating generally applied to electrolytic capacitors and the greater impact of moisture on film capacitors compared to sealed electrolytic capacitors. In our tests, the slow degradation of capacitors gradually impacts the electrical characteristics of the driver including power consumption and power factor. Ultimately the electronic circuit can no longer compensate and the unit either fails outright or begins to exhibit signs of impending failure such as excessive flicker or reduced light output due to low drive voltages.

8. Conclusions

Well-designed SSL products utilizing LED-based light engines have the capability to deliver both energy efficiency and long lifetimes. While a significant amount of attention has been focused on the reliability of LEDs, this study combined with previous work by PNNL [6,7] and RTI International [8,9] have demonstrated that LED failure is rare even under extreme environmental conditions. Instead, other luminaire components such as drivers, lens, and reflectors are usually responsible for device failure. Consequently, a systems perspective that takes into account the reliability of all luminaire components is essential to understanding the lifetime of SSL devices.

References

1. International Energy Association, "Light's Labour's Lost: Policies for Energy-Efficient Lighting," (2006), Paris, France.
2. Next Generation Lighting Industry Alliance and the U.S. Department of Energy, "LED Luminaire Lifetime: Recommendations for Testing and Reporting." (2001).
3. S. Rosenfeld, "Lighting Art with LEDs at the Smithsonian American Art Museum," presentation at the DOE SSL Market Introduction Workshop (2011), Seattle, WA.
4. W.D. van Driel and X.J. Fan, ed., Solid State Lighting Reliability: Components to Systems, Springer (New York, 2013).

5. Han and Narendran, "An Accelerated Test Method for Predicting the Useful Life of an LED Driver," *IEEE Transactions on Power Electronics*, Vol. 26, No. 8 (2011), pp. 2249-2257.

6. PNNL, "L-Prize: Stress testing of the Philips 60 W replacement lamp entry," U.S. Department of Energy, April 2012.

7. PNNL, "L-Prize: Lumen maintenance testing of the Philips 60 W replacement lamp L Prize Entry," U.S. Department of Energy, July 2013.

8. RTI International, "Hammer test findings for solid-state lighting luminaires," U.S. Department of Energy, December 2013.

9. J.L. Davis *et al.*, "Insights into accelerated aging of SSL luminaires," *Proceedings of the SPIE. LED-based Illumination Systems.* Vol. 8835 (2013).

10. Illuminating Engineering Society, "Approved Method: Electrical and Photometric Measurements of Solid-State Lighting Products," LM-79-08 (2008).

11. Illumination Engineering Society, "LM-80-08 Approved Method: Measuring Lumen Maintenance of LED Light Sources," (2008).

12. Illumination Engineering Society, "TM-21-11 Projecting Long Term Maintenance of LED Light Sources," (2011).

13. B. Wu *et al.*, "Effect investigation of delamination on optical output of high power LEDs," 2011 *12th International Conference on Electronic Packaging Technology and High Density Packaging (ICEPD-HDT)*, (2011) pp. 1072.

14. P. Lall, P. Sakalaukus, and J.L. Davis, "Prognostics of damage accrual in SSL luminaires and drivers subjected to HTSL accelerated aging," *InterPACK 2013* (2013).

15. Q. Sun, Y. Tang, J. Feng, and T. Jin, "Reliability assessment of metallized film capacitors using reduced degradation test samples," *Quality and Reliability Engineering International* vol. 29 (2013) pp. 259 – 265.

Assessment of Microelectronics Interconnect Reliability – Current Practice and Trends

Peter Borgesen
Department of Systems Science & Industrial Engineering
Binghamton University
Binghamton, NY 13902
USA

When it comes to the long-term reliability of the most common microelectronics interconnects, too little effort is spent on investigating important aspects while a significant amount of day-to-day reliability testing may be more or less wasted. This would seem to be at least partially so because reliability managers often fail to ask themselves what it is that they *really* want to know. Notably, while rarely explicitly recognized even ongoing 'engineering tests' will usually be meaningless unless they reveal something about relative performance in service. This requires a model or at least a quantitative mechanistic understanding.

In spite of repeated predictions to the contrary wirebonding technology continues to evolve and remain competitive. However, the backbone of the highly automated microelectronics assembly industry remains soldering at levels ranging from through hole through SMT to flip chip. Solder joints are ubiquitous in microelectronics, and the ultimate life of a product is often limited by the wear-out and failure of one of them. Over decades massive efforts have therefore been dedicated to the optimization and assessment of solder joint life.

There is an ongoing urge, in the industry, to conclude that solder joint reliability is sufficiently understood and that all that remains is to evaluate materials, designs, and processes through standard accelerated testing. Depending on the specifics this may, however, be strongly misleading and surprises still abound. Notably, 'new' failure mechanisms such as cratering or IMC fracture may be worthy of more in-depth research.

To the extent that concerns remain with respect to defect levels, typically in high reliability military and aerospace applications, products may furthermore be subject to so-called Environmental Stress Screening (ESS) in which they are pre-stressed to eliminate major defects before shipment. This may, however, reduce the useful life of good parts much more than expected.

Alternatives to solder, such as sintered Ag or nanoparticle based Ag or Cu joints, are usually benchmarked against solder in terms of reliability. This is certainly easier than the direct prediction of long term reliability in service. However, accelerated testing can easily become useless or worse without interpretations accounting for the different mechanisms and factors involved.

Finally, the most common approach to 2.5D and 3D assembly is still solder based, lending the illusion that experience gathered from BGA, CSP and conventional flip chip assemblies can be somehow scaled to these much finer dimensions. This is clearly not the case, but so far the community often seems unclear as to what are the primary reliability concerns.

978-1-4799-4789-8/14 $31.00 © 2014 IEEE

This presentation will offer an overview and discussion of these issues.

IMC Failure and Cratering: Traditionally, intermetallic failure was only observed in shock testing or when something was wrong with the intermetallic. Rather than try to asses the long term life of intermetallic bonds the conventional approach to the latter has therefore been to try to prevent the problem. The switch to the less compliant lead free solder alloys and the more brittle PCB laminates compatible with this led to a strong increase in the occurrence of solder pad cratering in shock and vibration testing. Originally, indications were that solder fatigue would still dominate failure in thermal cycling, but more recently companies are starting to report an increased occurrence or even dominance of pad cratering failures in thermal cycling tests. Indications are that this may be a result of relatively mild, invisible damage induced in assembly or handling but it is questionable whether preventing this is going to remain a realistic solution for everyone.

Intermetallic bond properties often vary much more than those of solder, and laminate properties are very sensitive to materials and processing. The development of fatigue models of the type available for solder is therefore not a realistic approach. New thinking is required here.

Underfill: With the introduction of underfilling, first for flip chip and later for BGA and CSP assemblies as well, solder fatigue models which had been used to predict SnPb joints in accelerated testing and service without getting anyone fired suddenly did not apply. Considerable efforts have been applied to argue for or even prove their continued validity, but indications are that the preceding success of models was associated with the similarity between stress ***distributions***, as opposed to stress ***levels***, in the many different configurations tested without an underfill.

Thermal Cycling of Lead Free Solder: Major research efforts have focused on SnAgCu type solder joints with the result that we now seem to understand the materials science better than for SnPb. However, the behavior of these alloys is ***much*** more complicated and although several existing models have been carefully adapted to them, these models are certain to miss important trends.

One reason for complications is that each SnAgCu joint undergoes significant undercooling during cool-down from reflow and invariably ends up solidifying based on a single nucleation event. This is not necessarily true for larger (dog bone) test samples, so much research on these can be quite misleading. Realistic solder joints may have up to three Sn grain orientations which may in some configurations end up interlaced so that cross sectioning reveals a very large number of boundaries. However, the boundaries are all so-called twin boundaries which, unlike large angle grain boundaries, do not form preferred crack propagation paths. BGA and CSP joints, among other, have few or no boundaries and their properties therefore vary strongly because of the extreme anisotropy of individual Sn grains. This means that failure distributions cannot be Weibull (or log-normal, or so forth), and research would seem warranted to address consequences for the extrapolation of test results to the 'early failures' of actual concern to any of us.

A source of greater complication is that the properties of realistic SnAgCu are dominated by the distributions of the secondary precipitates, notably Ag_3Sn, formed at solidification. This has two major consequences. The one most commonly recognized is that the creep properties of the solder varies strongly with aging, even at room temperature, and even stronger in thermal cycling as the precipitates coarsen. The consequences for Finite Element Modeling are dramatic, and still commonly ignored. The interpretation of test results and, in particular, prediction of life in service, becomes complicated.

Another major consequence of the sensitivity to the precipitate distributions has so far been almost universally ignored. Solidification of the Sn during cool-down from reflow is a stochastic event and occurs faster for larger solder volumes and pad sizes. As a result the initial precipitate distributions tend to vary systematically, leading to different acceleration factors depending on solder volume, pad size, and pad finishes. If this is not properly accounted for, interpretations of accelerated test results may easily become misleading!

Solder properties are also sensitive to the reflow profile, notably the cooling rate, but this is primarily through effects on the solute concentration of Ag rather than on the precipitates. This, as well as the above effects of joint size and configuration, leads to different sensitivities to long term aging as well.

The life of realistic SnAgCu solder joints in thermal cycling has been shown to be controlled by the formation of a continuous network of high angle grain boundaries across the high strain region of the joint, except for very strain ranges and/or harsh cycling conditions, i.e. for life times of less than a couple of hundred cycles. The variation with recrystallization, as well as the factors outlined above, is not compatible with any of the current thermal fatigue models. Instead we propose a new picture and a first practical approximation with well defined limitations based on the prediction of continuous, but not completely dynamic, recrystallization.

Isothermal Cycling of Lead Free Solder: Much less effort has traditionally been directed towards the prediction of life in long term isothermal cycling. However, an increasing number of expensive microelectronics products are subjected to long term vibration, notably in automotive, military and aerospace applications.

The understanding of damage evolution and failure in isothermal cycling should in principle be easier, although interactions with aging are of course still a concern. Indeed, the fatigue life of a SnAgCu joint tends to scale quite well with the rate of work done on the joint. This should make the prediction of life in cycling with a fixed amplitude, and thus the calculation of acceleration factors, straightforward. The only concern in that respect is that the choice of constitutive relations is less trivial than often assumed.

Of much greater concern is, however, that realistic service conditions almost always involve significant variations in cycling amplitude. For those concerned with actual predictions of life this is commonly dealt with on the basis of the so-called Miner's rule of linear damage accumulation. Much more generally interpretations of accelerated test results, including those from random vibration testing, are based on the explicit or implicit assumption of such a rule. Systematic research has however shown that this can lead to orders of magnitude errors in absolute or relative predictions of life. Significant progress has been made towards a practical approach to dealing with this.

Realistic Service Conditions: Variations in thermal cycling amplitudes also lead to a break down of Miner's rule, but the consequences seem to be much less dramatic and our current understanding suggests that approaches to dealing with it will be simpler than for isothermal cycling.

More generally, realistic service does of course often involve combinations of, say, vibration and thermal cycling. A very limited number of reports have been published on the effects of somewhat randomly selected combinations. In agreement with these results our mechanistic understanding suggests that vibration will tend to have less effect on subsequent life in thermal cycling than predicted by Miner's rule. Depending on the specifics thermal cycling may, however, have a much stronger effect on the subsequent life in vibration than predicted.

ESS: Environmental Stress Screening protocols vary but may for example involve the subjection of the product to vibration and/or thermal cycling up to a level that is not supposed to reduce the life of non-defective parts by more than 10%. If the life in service is dominated by thermal fatigue, an accelerated test based assessment of the fraction of life remaining after ESS may be realistic. That will undoubtedly depend on the ESS conditions. We currently recommend emphasizing aggressive vibration for that case. However, if the life in service is dominated by vibration such preconditioning would reduce life in service by a much larger fraction than life in an accelerated vibration test. The more general case would seem to require more research before a proper protocol can be defined.

Solder Alternatives: There is a growing demand for higher operating temperature solder and/other an interconnect material with greater thermal, and perhaps electrical, conductivity. Current high temperature solder candidates would seem to be sufficiently different from the SnAgCu family alloys that we should expect different acceleration factors, and possibly different damage rate controlling mechanisms.

Solder alternatives include filled adhesives, which may not be superior and obviously involve different damage mechanisms and concerns, as well as silver and copper. Work is ongoing on sintered Ag, silver nano-particles, and mixtures of silver nano-particles and micron scale copper particles. Lockheed Martin has developed a Cu nano-particle paste which offers the potential for becoming a drop-in replacement for solder. The paste can be stencil printed like solder paste and the particles fuse abruptly at around $200^{\circ}C$ when sent through a conventional mass reflow oven in a nitrogen ambient. Importantly, a repair process appears to be feasible as well. Material and processes are still under development and the resulting properties continue to improve. The basic characteristics are, however, fundamentally different from those of solder. Copper is first of all much harder and less compliant than solder. The nano-Cu based joints are nanocrystalline and nanoporous. The small grain size alone may be responsible for enhanced ductility and creep at intermediate stresses, but the porosity appears to interact with this in a complex fashion as well as reducing the strength and enhancing fatigue crack growth. Sintered Ag obviously tends to lead to larger grains, but porosity is still certain to affect properties.

In view of the above it should be obvious that the mere fact that sintered Ag bonds have outperformed solder in accelerated thermal cycling of a particular configuration is far from sufficient to suggest that the actual reliability, under whatever service conditions, is superior as well.

2.5/3D Assemblies: Current and, in particular forthcoming, applications of chips assembled onto Si interposers or stacked onto each other based on soldering of microbumps vary substantially, and so do the resulting joint configurations. Many still use thick enough solder layers to leave unreacted Sn between the two intermetallic bond layers. In this case mass reflow is possible and reliability concerns include thermomigration and electromigration. Lateral thermal expansion mismatches are small, although power cycling is still a concern, but the chips are generally underfilled, making the thermal expansion mismatch between underfill and joint a potential concern in thermal cycling instead. As mentioned above a model would need to account for the very different stress distributions in joints confined within an underfill. More critically, perhaps, the solder microstructure may be very different from that formed in larger joints and the intermetallic layers constitute a larger fraction of the overall joint. Adding to the challenges is that the kinetics of intermetallic formation tends to be very different as well. Published results tend to be contradictory, some of which we have shown to be a result of interactions between reflow profile and subsequent thermal history. This all affects relative intermetallic layer thicknesses and properties in complex fashions. In general, joint properties vary significantly with dimensions and published results and observations are not easily generalized.

Finer pitches and dimensions may lead to final joints in which the Sn is completely reacted to form an intermetallic joint. This is also often done deliberately, either during assembly or in subsequent heat treatment, to stabilize the properties of the joints and allow them support a significant load in further compression bonding steps. This is among other believed to eliminate electromigration concerns. However, intermetallic reaction kinetics is as said different and this approach appears to raise the risk of significant void formation within or between the intermetallic layers. It has yet to be ascertained whether such voids may be affected by electromigration, thermomigration or stress migration as well.

So far, the relative importance of the different reliability concerns with respect to life in service does not often appear to be known. However, like for regular flip chips some manufacturers seem to be primarily focused on the survival of the assembly until completion of underfilling process.

Design for Thermo-Mechanical Reliability of a 3D Microelectronic Component Using 3D FEM

BELHENINI Soufyane[a], TOUGUI Abdellah[a] and DOSSEUL Franck[b]

a- Laboratoire de Mécanique et Rhéologie (LMR), Université de Tours. 7 Av Marcel Dassault, 37200, Tours. France

b- STMicroelectronics. 16 Rue Pierre et Marie Curie, 37100 Tours. France

Email: soufyane.belhenini@univ-tours.fr

Abstract

3D microelectronic components are exposed to electrical, thermal, mechanical and chemical stresses generated by storage, transport, manipulation, functioning and environment. The reliability of the components depends partially on the reliability of interconnections which insures the mechanical and the electrical junctions between components and printed circuits. Reliability has to be evaluated on mechanical demonstrators before the production stage. It is currently studied by employing standardized tests.

Modeling has been proven to be a very efficient tool in IC Packaging reliability, especially for designing and optimization, compared with experimental standardized tests, which are expensive and time-consuming.

In this work, the board level thermomechanical reliability of a 3D chip to wafer component was evaluated by using a coupled thermal-structural numerical analysis. 3D FEM results were employing in the design optimization of 3D components. The critical solder bump strain energy density is used as the principal reliability criteria. The influence of the internal architectures of 3D components and the TSVs locations on the thermomechanical reliability has been studied.

Keywords: microelectronic, 3D integration, reliability, thermal cycling test, FEM calculations.

1. Introduction

3D integration appears as a way ok keeping increasing density of microelectronic components to adapt them for portable and telecommunication products. It is inspired by the stacking technique of active components on passive layer components developed by R. Henry [1]. The 3D integration in microelectronics consists in the stacking of planar components one on top of the other. The electrical connection between dies forming the 3D component is done by using Through-Silicon-Vias "TSV" [2-4]. Like planar components, direct flip-chip attachment using solder bump has become the most efficient approach for 3D components at the second level assembly [5-6].

3D component reliability depends partially depended on interconnection reliability. It has to be evaluated on mechanical demonstrators with daisy chains before real production. Reliability is currently studied by employing standardized tests [7-8]. The accelerated thermal cycling test (TC) is used to evaluate thermomechanical reliability at the board level. The thermal cycling test results give information about component thermomechanical behavior under environmental temperature changes. The CTE difference between the chip and the PCB induces stresses in the interconnections, which, if excessive, can lead to failure. Many studies related to the thermal cycling reliability of microelectronic components show that thermal fatigue crack initiates and propagates through the bulk solder [9-11].

Modeling has proved to be a very efficient tool in IC packaging reliability, especially for designing and optimization, compared with experimental standardized tests, which are expensive and time-consuming. The finite element modeling approach can be used to understand the thermomechanical behavior of 3D components during thermal cyclic loading. It can give quantitative and qualitative information about critical areas of 3D components. Several modeling methods have been developed to satisfy the requirements in package design analysis. Two-dimensional thermomechanical models are used to reduce time calculations [2]. Three-dimensional thermomechanical models with geometrical simplifications, sub-modeling techniques and symmetry considerations are developed to predict the component fatigue life under TC loading [13-15].

In this study, design for the reliability of 3D chip to wafer component is studied. Numerical simulations of the thermal cycling test were carried out using a coupling thermomechanical model. Submodeling technique has been used to improve the precision of numerical calculations. Numerical results will be employed in the design optimization of 3D components and life prediction using a fatigue model. The effects of C2W internal architecture on mechanical behavior during temperature cycling analyzed. Face to face (F2F) and the back to face (B2F) bonding solutions have been studied. Two types of TSV repartition are modeled which are: bumps under TSV and no bumps under TSV. The critical areas in the C2W components studied in this paper are solder joints. A high risk of failure of external solder bumps is caused by the CTE mismatch between chip carrier and PCB. We choose the inelastic strain energy density of the corner solder as a principal comparison parameter. The maximum equivalent

978-1-4799-4789-8/14 $31.00 © 2014 IEEE

stress of silicon die is employed as a second comparison parameter.

2. Materials and methods

Used in this study, the 3D C2W components have a dimension of 3 x 3 x 0.2 mm³. The F2F and the B2F internal architectures are presented in Fig. 1. Die2 contain TSV of 100 µm external diameter, 4 µm thick and 100 µm of deep. Top die thickness is 20 µm for the B2F architecture and 35 µm for the F2F. Copper pillars based on Cu/Sn intermetallic soldering technology are used for the F2F stacking and die attach film is employed for the B2F assembly. The component is mounted on a PCB board by 49 SnAgCu-Ni (SAC125-5) solder bumps (7 x 7 full area arrays). The pitch between adjacent solder joints is 400 µm, and bump size is 250 µm.

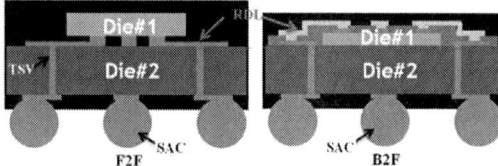

Fig. 1. Schematic Face-to-Face and Back-to-Face internal architectures for the 3D component.

A quarter of the assembly (component/PCB) is modeled taking symmetries into account. Solid-to-solid sub-modeling technique is employed to simulate the thermomechanical response of the component under TC loading. The numerical results of the global model are employed as boundary conditions in the local model. We have considered a three-dimensional coupled thermal-structural model for the global model and a structural model for local calculations. A finer mesh is assigned to the local model. The internal architectures are presented in Fig. 2. Table I gives the dimensions of the component's elements.

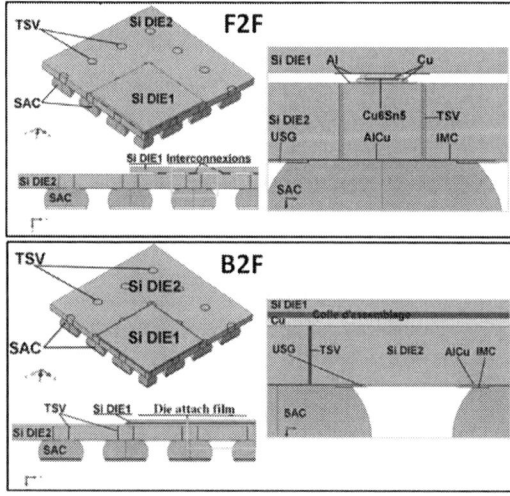

Fig. 2. 3D F2F and B2F internal architectures.

Table I. The package element sizes (D = diameter, e= thickness)

Element		Sizes
Die2		3 x 3 x 0.1 mm³
DIE1	F2F	1.6 x 1.6 x 0.035 mm³
	B2F	1.6 x 1.6 x 0.02 mm³
Die Attach Film		1.8 x 1.8 x 0.01 mm³
Copper pad		$D = 0.250$ mm, $e = 0.0015$ mm
USG		$e = 0.002$ mm
Aluminium		$D = 0.260$ mm, $e = 0.0015$ mm
IMC		$D = 0.220$ mm, $e = 0.001$ mm

All material properties excluding SAC125-5 and copper are assumed to be elastic. We consider that the mechanical properties excluding the solder material are insensitive to temperature variations. SAC125-5 Young's modulus varie between 46 and 39 GPa for temperature variations between -40 and +125 °C. A single hyperbolic sine creep model is used for solder material viscoplastic behavior.

$$\dot{\varepsilon} = C_1[sinh(C_2\sigma)]^{C_3}exp\left(-\frac{C_4}{T}\right)$$

Where C_1=277984 s⁻¹, C_2=0.02447 MPa⁻¹, C_3=6.41 and C_4=0.56 eV [3].

Material properties are shown in Table II. The SAC125-5 elastic modulus

Table II. Material properties of the 3D C2W components.

Element	E (GPa)	ν	CTE (ppm/K)
PCB	20	0.28	2.3
Si die	131	0.3	2.8
Cu PCB pad	110	0.34	18
SAC125-5	46	0.3	22
Passiv USG	90	0.3	2.6
Al	71	0.33	23
IMC	92	0.34	17
Die attach	1.24	0.4	80

Global and local models of the component/PCB assembly are shown in Fig. 3. The G condition of the JESD22-A104 standard [4] is used in this study. The temperature cycle profile shown in Fig. 4 (-40 °C to +125 °C) is applied to the 3D coupled thermal-structural global model. The PCB vertical displacement is blocked and then two planes of symmetry (x-z) and (y-z) are considered as adiabatic. Nodal displacements resulting from global model calculations are applied as a boundary condition in the local model. Processing, modeling, solving and post-processing operations are performed using the commercial code Abaqus.

A number of calculation assumptions are considered in order to reduce calculations cost:

- A stress free initial condition is applied for the initial temperature (125 °C). IMC growth during thermal cycling is ignored
- The IMC dendritic form is idealized to a simple layer form

- Microstructural changes of solder material are ignored
- Isothermal loading conditions are applied.

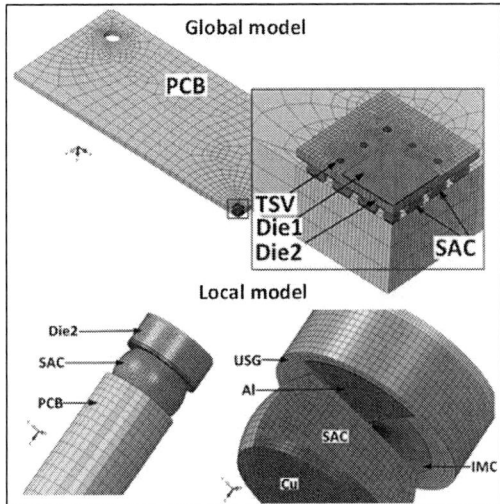

Fig. 3. Global and local models for the thermal cycling simulations.

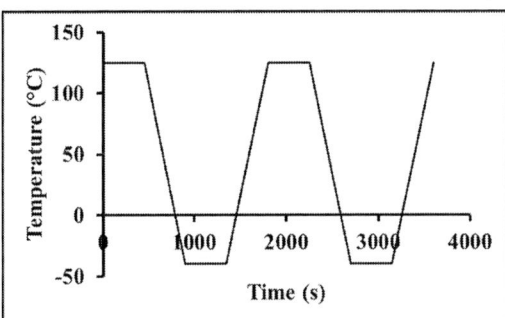

Fig. 4. Simulated TC profile (-40/125 °C).

Parametric study was first performed on reference models which have been validated by experimental results. The effects of the bonding solution and the TSV repartition in the die2 are studied. The four architectural solutions analyzed are:

- F2F architecture with tsv under bump
- F2F architecture with no tsv under bump
- B2F architecture with tsv under bump
- B2F architecture with no tsv under bump

3. Results and discussions

Global model results

Global thermomechanical models have been used to give qualitative information about critical areas in assembly. The maximum inelastic strain is obtained on solder bumps. The critical solder is located at the corner of the solder array as shown in fig. 5. This numerical result is validated by theoretical studies which indicate that the maximum strain is located on the solder furthest away from the neutral line. The four architectural solutions give the same qualitative results on the solder response under temperature cycling loading. The

critical area on the critical bump is located on the chip side. This location is due to the large CTE mismatch between the solder and the chip. Although the most common fractures are observed on the solder bumps, the silicon dies of 3D components may present a failure risk because of their fragility caused by tsv holes.

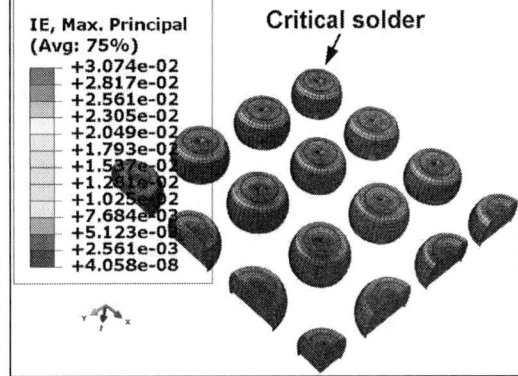

Fig. 5. Solder bumps maximum principal inelastic strain

Failure analysis has shown that some components are cracked in the silicon die around tsv (fig. 6). In the same components, solder cracks are observed but the order of cracking is not established. We assume that the first cracks appeared in the solder bump and in a second time, silicon die cracks appeared.

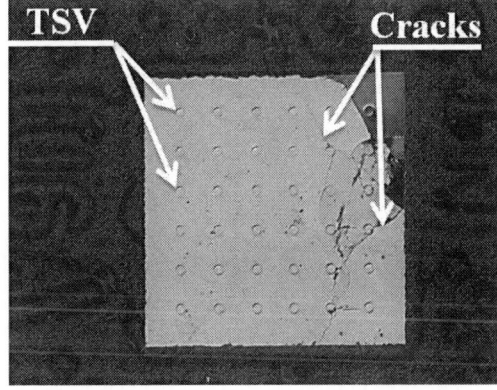

Fig. 6. Silicon die cracks (around tsv).

Silicon behavior under temperature cycling loading is obtained by numerical calculation to locate critical areas. Qualitative analysis of silicon dies maximum equivalent stress is presented in fig. 7. The maximum Coffin Manson equivalent stress is used to locate critical areas. For all architectural solutions studied, the silicon critical area is located on the top face of die2. For the F2F architecture, and no tsv under bump solution, critical areas are located around die1 corners. However, stress concentration is located around external tsv for the F2F architecture wiht tsv under bump repartition. TSV repartitions don't have an impact on the silicon die stress concentration area for B2F architecture. Critical areas obtained by numerical calculations correspond to the real failure zone.

978-1-4799-4789-8/14 $31.00 © 2014 IEEE   707

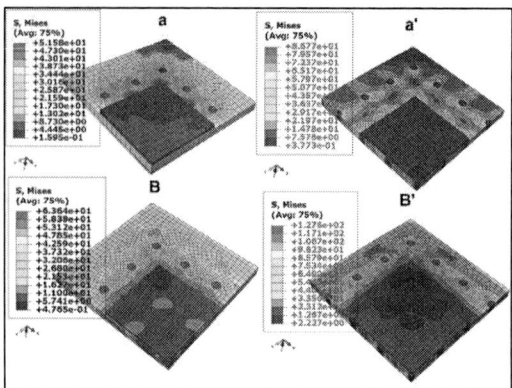

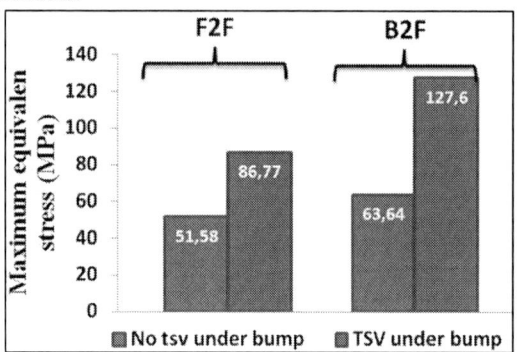

Fig. 7. Silicon dies maximum equivalent stress, a) F2F no tsv under bump, a') F2F tsv under bump, b) B2F no tsv under bump, b') B2F tsv under bump.

Quantitative analysis of silicon dies maximum equivalent stresse is presented in fig. 8. The B2F architectural solution is more stressed than the F2F solution. The no tsv under bump repartition is more resistant than the tsv under bump repartition. This result is due to the number of tsv which is 49 for the tsv under bump solution, and 36 for the no tsv under bump architectural solution. In addition, for the F2F solution, metallic studs used for mechanical and electrical connections absorb part of the energy caused by temperature variations. Based on the silicon die equivalent stress, the F2F component with no tsv under bump repartitions is the most reliable.

Fig. 8. Silicon dies maximum equivalent stress.

Local model results

Submodeling technique has been used to improve the accuracy of global calculations. The inelastic maximal principal strain calculated on the local model is presented by fig. 9. The strain concentration area of the critical solder is located at the SAC/IMC interface and it extends to the solder bump. This area obtained by numerical calculations corresponds to the crack growth area observed on failed flip-chip components. This location is the same for all architectural solutions studied.

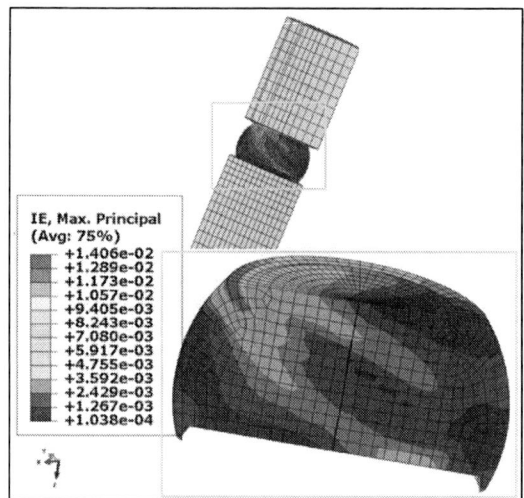

Fig. 9. Critical solder maximum inelastic strain.

The maximum inelastic strain energy density is used as reliability criteria. Based on this parameter, quantitative and qualitative comparisons are carried out to optimize the 3D component design. Total inelastic strain energy density is composed of creep strain energy density (ECDDEN) and plastic strain energy density (EPDDEN). Fig. 10 illustrates qualitative comparisons of the critical bump maximum inelastic strain energy density. The plastic strain and the creep strain concentration areas are the same for all solutions. The corner solder bump is the most distorted at the chip side. Plastic strain energy density is higher than the creep stain energy density and this in all architectural solutions.

Quantitative comparison presented in Fig. 11 show that the maximum strain energy density obtained on B2F components is higher than that obtained on the F2F configuration. For the B2F solution, Die Attach Film (DAF) which has 80 ppm CTE, causes high strain on the top side of silicon die (Fig. 7). Solder bump distortion is amplified by the die distortion. Whatever the bounding solution, the tsv under bump repartition is the most reliable. This result can be explained by the silicon behavior presented in fig. 8. Indeed, the bump under tsv repartition causes high strain of the silicon die. The silicon die dissipates part of the stain energy caused by thermomechanical loading. For the no tsv under bump repartition, the quantity of energy dissipated by silicon die is less than that dissipated when using the bump under tsv. Therefore, the part of energy dissipated by solder bumps is most important.

Based on critical solder bump maximum inelastic strain energy density, the F2F component with tsv under bump repartition is the most reliable.

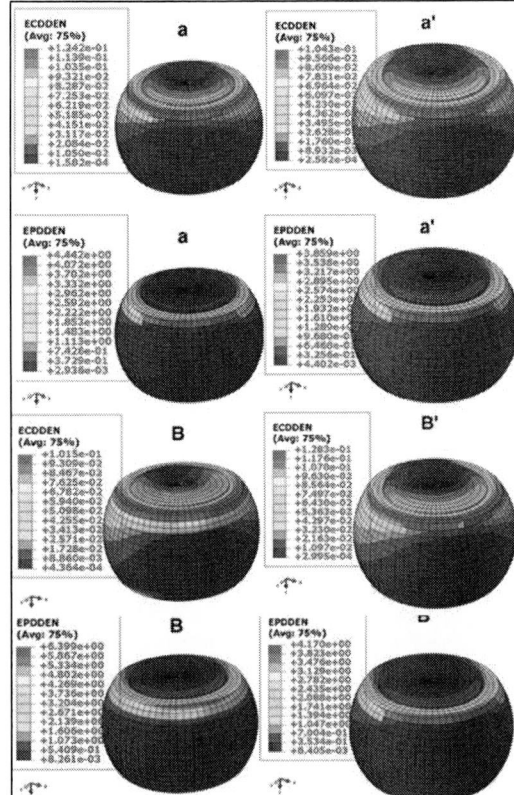

Fig. 10. Maximum inelastic strain energy density, a and a') F2F, no tsv under bumps and tsv under bumps, B and B') B2F, no tsv under bump and tsv under bump.

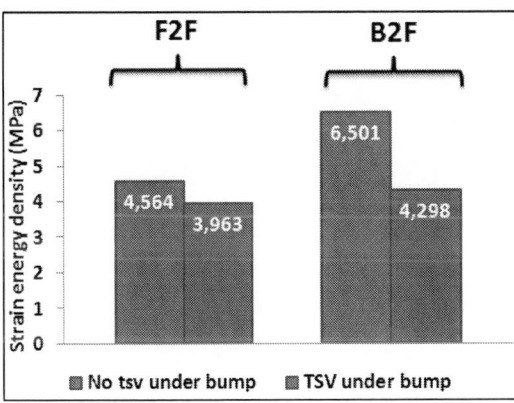

Fig. 11. Maximum inelastic strain energy density.

4. Conclusions

The design for the thermomechanical reliability of 3D C2W components is a complex process involving many elements. Modeling has been proven to be a very efficient tool for designing and optimization. In this study, 3D finite elements thermomechanical simulations are used to optimize the internal architectures of 3D components. Qualitative and quantitative comparisons are carried out in order to determine the thermomechanical resistance of different architectural solutions. The results of this numerical study lead to the following conclusions:

- The critical area with the greatest risk of failure is located on the corner bump in all architectural solutions.
- For all solutions, the chip side of the corner bump is the most distorted.
- For the solder bump, the tsv repartition in the silicon die doesn't have an impact on strain concentration
- For the silicon die, stress concentration depends on internal architecture and tsv repartition
- The F2F solution with tsv under bump repartition is the most reliable solution when using solder bump strain energy density as reliability criteria
- F2F with no tsv under bump solution is the most reliable solution when using the silicon die maximum equivalent stress as reliability criteria
- B2F architectural solutions are less resistant to thermomechanical loading

Acknowledgments

The authors would like to thank all the members of ''Laboratoire de Mecanique et Rhéologie-LMR'' for their support. Special thanks to Mrs. Freya COLIN and Luc LECROISEY for their help.

References

1. H. R, «Project Tinkertoy: A System of Mechanized Production of Electronics Based on Modular Design,» IRE Transactions on Production Techniques, vol. 1, N°1, p. 11, 1956.
2. T. H. Wang et Y.-S. Lai, «Optimization of Thermomechanical Reliability of Board-Level Flip-Chip Packages Implemented With Organic or Silicon Substrates,» IEEE Transaction on Electronics Packaging Manufacturing, vol. 31, N°2, pp. 174-179, 2008.
3. H. Shi, F. Che, C. Tian, R. Zhang, J. T. Park et T. Ueda, «Analysis of edge and corner bonded PSvfBGA reliability under thermal cycling conditions by experimental and finite element methods,» Microelectronics Reliability, vol. 52, N° 9-10, pp. 1870-1875, 2012.
4. JESD22-A104D, «Temperature Cycling,» JEDEC STANDARD, 2009.
5. L. Jian-Qiang et R. Ken, « 3D integration: why, what, who and when?,» Future Fab Int, N° 23, 2007.
6. S. Spiesshoefer, L. Schaper, S. Burkett, G. Vangara, Z. Rahman et P. Arunasalam, «Z Axis interconnects Using Fine Pitch, Nanoscale Through Silicon Vias : Process Development,» 54th Electronic Component and technology Conference, pp. 446-471, 2004.
7. R. Hon, S.-W. Lee, S. Zhang et C. Wong, «Multistack flip chip 3D packaging with copper

plated through-silicon vertical interconnection,» Proceedings of 7th Electronic Packaging Technology Conference (EPTC), vol. 2, 2005.

8. K. Takahashi, Y. Taguchi, M. Tomisaka, H. Yonemura, M. Hoshino, M. Ueno, Y. Egawa, Y. Nemoto, Y. Yamaji, H. Terao, M. Umemoto, K. Kameyama, A. Suzuki, Y. Okayama, T. Yonezawa et K. Kondo, «Process integration of 3D chip stack with vertical interconnection,» Proceeding 54th Electronic Components and Technology Conference ECTC, pp. 601-609, 2004.

9. J. Lau, R. Lee, M. Yuen et P. Chan, «3D LED and IC wafer level packaging,» Microelectronics International, vol. 27, N° 2, pp. 98-105, 2010.

10. B. Zhang, h. Ding et X. Sheng, «Reliability study of board-level lead-free interconnections under sequential thermal cycling and drop impact,» Microelectronics Reliability, vol. 49, N°5, pp. 530-536, 2009.

11. S. Ishikawa, H. Tohmyoh, S. Watanabe, T. Nishimura et Y. Nakano, «Extending the fatigue life of Pb-free SAC solder joints under thermal cycling,» Microelectronics Reliability, vol. 53, pp. 741-747, 2013.

12. X. Hui, L. Xiaoyan, y. Yongchang, L. Na et S. Yaowu, «Damage Behavior of SnAgCu Solder under Thermal Cycling,» Rare Metal Materials and Engineering, vol. 42, N°2, pp. 221-226, 2013.

13. JEDEC, «Global Standards for the Microelectronics Industry,» [En ligne]. Available: http://www.jedec.org.

14. MIL-HDBK-217F, Military handbook of Electronic Equipment, Department of Defense-USA éd., 1995.

15. A. Yeo, C. Lee et J. H. L. Pang, «Flip Chip Solder Joint Reliability Analysis Using Viscoplastic and Elastic-Plastic-Creep Constitutive Models,» IEEE Transactions on Components and Packaging technologies, vol. 29, N°2, pp. 355-363, 2006.

16. C. Noritake, P. Limaye, M. Gonzalez et B. Vandevelde, «Thermal Cycle Reliability of 3D Chip Stacked Package Using Pb-free Solder Bumps: Parameter Study by FEM Analysis,» 7th. Int. Conf: on Thermal, Mechanical and Multiphysics Simulation and Experiments in Micro-Electronics and Micro-Systems, EuroSimE, pp. 1-6, 2006.

17. Y. Ma, J. Luan, K. Goh, J. Whiddon, F. Che, G. Hu et X. Baraton, «Finite Element Analysis of Thermal Cycling Reliability of an Extra Large Thermally Enhanced Flip Chip BGA Package with Rotated Die,» 10th E lectron ics Packaging Techn o logy Con ference (EPTC), pp. 709-715, 2008.

NUMERICAL MODELING OF FLEXIBLE ACTUATOR FOR DYNAMIC LIGHTING

Teng Ma, Xueming Li, Jia Wei, G.Q. Zhang and P.M.Sarro
Electronic Components, Technology and Materials Laboratory (ECTM),
Delft University of Technology,
Delft, the Netherlands

Abstract

We presented the numerical modeling of flexible actuator in the application of dynamic lighting. In order to analyze the Lorentz force exerted on the actuator, we first modeled the surrounding magnetic flux of the magnet. With appropriate layout design, the Lorentz force density distribution was then analyzed. The modeling provides us the overview of the Lorentz force density distribution, and thus the movement of the actuator which is crucial for the dynamic lighting system.

1. Introduction

Dynamic lighting is widely used in both business and consumer applications, such as stage lighting in art performance and task ambient lighting for office lighting, which generally work with the idea of adapting illumination to where it is needed. However, current dynamic lightings are far from being compact and smart. Stage lighting uses big motors to drive the lamp in order to manipulate the light distribution, from which the noise generated by the motors will seriously affect the acoustic effect of the performance, while task ambient lighting uses lamp array and controls the light by simply controlling each lamp, which is inefficient and wasteful [2, 3].

A well-designed driver for the SSL can help to make the whole system more compact and smart. We developed tunable optics for the dynamic lighting system, in a way that an actuator driving the optics to the required position to meet the light distribution manipulation. Among all the actuation methods, electromagnetic actuation provides large force in a long range, giving more flexibility in designing the movable optics. The Lorentz force of magnetic actuation depends on the current through the coils and the density of the magnetic field, making the design and fabrication less complex. While other actuation methods need more complicated design processes, for example, electrostatic needs high voltage and piezoelectric actuation requires complicated material preparation, etc. Thus we employed electromagnetic actuation to drive the optics in our design [1].

In this work, we were going to design a 2D model to guide the design of the real dynamic lighting system based on electromagnetic theory. In this model, several parameters, which were essential in the practical system design, were going to be analyzed and discussed. Magnetic flux distribution was first studied. Furthermore, Lorentz force which can be treated as a key factor for the design of the driving system, was investigated by considering its relationship with other parameters (e.g. the amount of current flowing through coils and the position of the coils). Finally, in order to verify the feasibility of

the lighting system design, the calculations of Lorentz force exerted on the actuator were implemented.

2. System Description

As shown in Fig.1, a tunable system consists of several components, including coils as the driver, LED as the light source, ring shape magnet as the magnetic source and optics as the light controller. A constant current I_0 flows through the coils when a DC voltage V_{dc} drives on it. Additionally, the ring shape magnet provides magnetic field in the surrounding air space. According to Ampere's Law, Lorentz force will be generated in the direction which is perpendicular to both the magnetic flux and the flowing current. The horizontal component of Lorentz force will pull up the coils and optics. By setting up an LED right under the tunable optics, the light beam emitted from the LED can be controlled. Therefore, a dynamic lighting system is realized.

However, the distributions of magnetic flux and Lorentz force are invisible in a real system. An improper design without comprehensive considering of the distribution of magnetic flux, Lorentz force and other geometry parameters will result in the poor performance, or even a completely deactivated system. Therefore, a good simulation model for the magnetic derived Lorentz force is critical for the system design and the subsequent optimization. In the following sections, the details of how to establish and analyze this 2D model will be presented.

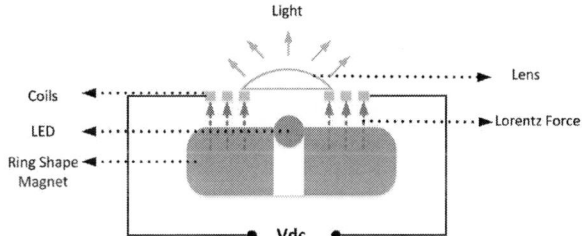

Fig. 1. Schematic view of the flexible electromagnetic actuator.

3. Model Definition

In modeling and simulation, COMSOL Multiphysics 4.3 program with the electromagnetic AC/DC module is used. The magnetic potential, current density and Lorentz force can be expressed as:

$$J^e = \nabla \times (\mu \nabla \times A), \qquad (a)$$

$$B = \mu_0 (H + M). \qquad (b)$$

And the relation between magnetic field and current density is given by:

$$F = J^e \times B. \qquad (c)$$

978-1-4799-4789-8/14 $31.00 © 2014 IEEE

Where μ and μ_0 are the permeability of the medium and air, respectively. $\mathbf{A}$ is the magnetic vector potential, and $\mathbf{J}^e$ is externally applied current. $\mathbf{H}$ is the electric field; $\mathbf{M}$ is the applied magnetic field, and $\mathbf{B}$ is the magnetic flux density. $\mathbf{F}$ is the Lorentz force density.

The 2D axisymmetric magnetic field (mf) application mode was used due to the symmetry of the system. A model of the cross-section of the system is shown in Fig.2. The model can be divided into three parts - ring shape magnet with a hole in the middle, copper windings and the surrounding air space which was large enough to avoid boundary effects. The geometry of the system is presented by Fig.2, with more detailed parameters collected in Table 1.

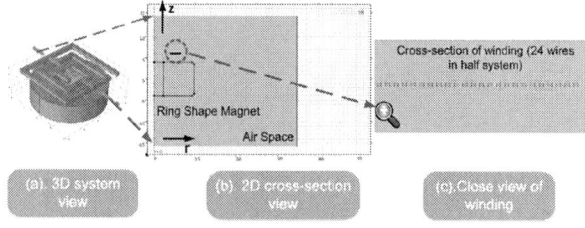

Fig.2. (a) Model of system geometry. The model consists of a ring shape magnet, a coil and optics mounted above; (b) A 2D axisymmetric model of cross section is made to study the magnetic actuation; (c) Close view of windings, of which z and r are indicated as the vertical and lateral-axis in the 2D model.

Table 1. Geometric and technical parameters description in the system.

Parameter description	value	units
Magnetization of magnet	900	KA/m
Wire conductivity	60	MS/m
Air permittivity	1	-
Applied current density(J_0)	0.112	GA/m^2
Magnetic radius(Mr)	7.5	mm
Magnetic height(Mh)	8	mm
Hole radius(Hr)	2.25	mm
Hole height(Hh)	8	mm
Wire width(Ww)	50	um
Wire height(Wh)	25	um
Number of wires on half system	24	-
Wire spacing(Ws)	50	um
Length of a segment wire(L_{seg})	8.2	mm
Total width of winding (width of 24 wires + spacing)	2.35	mm
Default position of windings located in r-axis	4	mm
Default position of windings located in z-axis	2	mm

4. Results of Simulation

Ring shape magnet was employed as the magnetic flux source due to its unique magnetic field distribution. The magnetic field around the ring shape magnet has a great horizontal component, which can be the source for the vertical Lorentz force exerted on the coil. As shown in Fig. 3 (a), most of the magnetic field distributed over the top of the magnet are obliquely upward, which can be further

decomposed into two components. One is the component in vertical direction $\mathbf{B}_{x,vertical}$ and the other component locates in the lateral direction $\mathbf{B}_{x,lateral}$ (because $\mathbf{M}$ and $\mathbf{B}$ are in the same direction when $\mathbf{H}$ equals to zero according to eq. (b)) as shown in Fig. 3 (b).

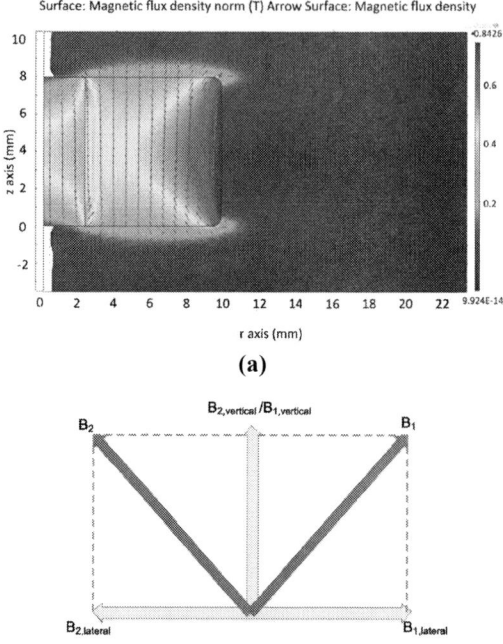

Fig.3. (a) Magnetic flux distribution of ring shape magnet; (b) Magnetic flux density decomposed into two components.

Due to the symmetric features of the windings (as shown in Fig.4.(b), the current flows through the left half and right half of windings has the opposite direction, but the magnetic flux-$\mathbf{B}_{x,vertical}$ is presented in the same direction in both sides), the lateral force generated by the vertical magnetic flux components will be canceled out by each other. On the other hand, the lateral components of magnetic flux on the left and right half of the windings will lead the Lorentz force in the upward direction at the same time due to the reason that the current and lateral magnetic flux of both half sides have the opposite sign, respectively. Therefore, in the following sections, we can focus on the vertical part of the Lorentz force $\mathbf{F}_z$ which is generated by $\mathbf{B}_{x,lateral}$.

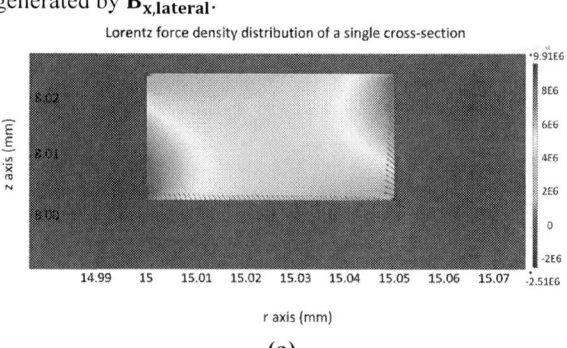

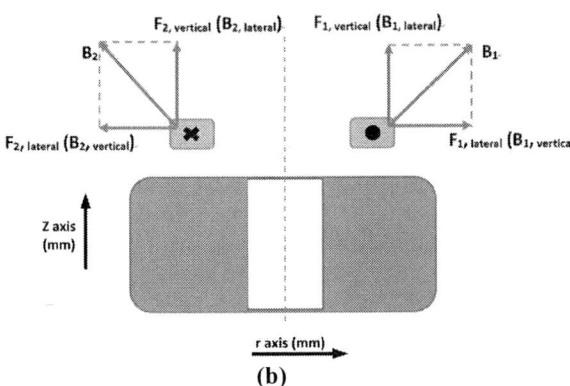

(b)

Fig. 4. (a) Lorentz force distributes over the cross-section of a wire mounted on the magnet; (b) The vertical component of the Lorentz force can be used to drive the optics, while the lateral components will be canceled due to the asymmetry of the magnetic field distribution.

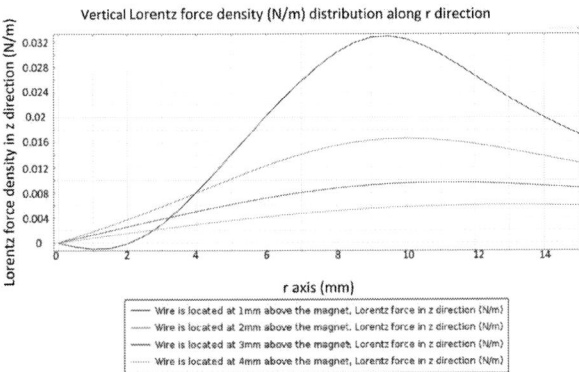

Fig.5. Single wire cross-section plot vertical Lorentz force density F_Z along r axis (lateral direction) for different heights above the magnet.

Fig.5, shows the vertical Lorentz force density distribution as a function of the position along r-axis for different heights above the magnet. In this case, we only simulate a single wire cross-section with constant current density J_0. It can be observed that the Lorentz force density is decreased as the height of the wires rises, and it reached the maximum value at the edge of the magnet. In the practical design, the windings should be located at the position where it has the relative large Lorentz force density. As shown in Table 1, the default positions of windings are not optimal, but they can be optimized during the design process.

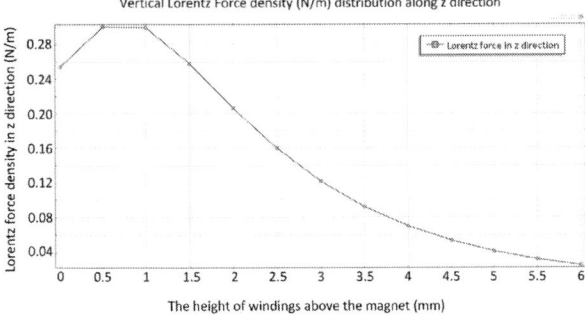

Fig.6. Windings with 24 cross-sections plot vertical Lorentz force density F_Z along z axis (position of windings located above the magnet) when the windings are located at default position r=4mm.

In Fig.6, it shows that the Lorentz force density has the maximum value at the range from 0.5mm to 1mm above the magnet and decays exponentially along the z direction. This is because the distribution of the magnetic field is attenuated as the coils rising. At the 0mm above the magnet, the lateral component of the magnetic field is very small, while most of the magnetic flux is composed by vertical components. As the windings move up, the lateral component of the magnetic field becomes larger and thus the Lorentz force density increases. After it reaches the peak, the intensity of the magnetic field drops, which in turn leads smaller Lorentz force density. This non-linear relationship can be used to roughly analyze the height of magnet and vertical Lorentz force density during design. The equations that describes the relationship between the vertical Lorentz force and the gravity force of tunable optics system are shown below:

$$F_Z = G_{total} \pm \boldsymbol{Spring\ Constant}. \qquad (d)$$

Where

$$G_{total} = [M_{optics} + M_{windings} + M_{sbustrate}] \times g, \quad (e)$$

and

$$F_z = 4 \times F_{Z,density} \times L_{seg}. \qquad (f)$$

M_{optics}, $M_{windings}$ and $M_{sbustrate}$ are the mass of optics, windings and flexible substrate (used to support optics and windings) material, respectively. g is the acceleration of gravity. $\boldsymbol{Spring\ Constant}$ is the force provided by the substrate. Its value and direction depend on the windings position and the other two forces together. For example, if F_Z (Lorentz force in vertical direction) is greater than the total gravity, the value of $\boldsymbol{Spring\ Constant}$ is equals to $|F_Z| - |G_{total}|$ and its direction is as same as the total gravity. L_{seg} is the length of a segment wire which is defined in our simulation process. It can be found in Table 1. $F_{Z,density}$ is obtained from Fig.6. In fact, the 3D system can be regarded as four repeated parts to the 2D structure as what we simulated above, the total Lorentz force of the 3D system can be calculated by the 2D model by just simply multiplied by the coefficient of 4. Additionally, we assume that the friction force is small in our system and it can be ignored in the calculations.

Once the value of F_Z is determined by choosing the magnet, windings and optics with the specific geometries, the dynamic range along the vertical direction can be found out. Thus, the performance of the tunable optics can be evaluated and optimized. Generally, when the height of windings beyond 5mm, the system will suffer from the insufficient Lorentz force, which makes the optics no longer move upward. The calculations will be presented in the next section with more details.

The dependencies between perpendicular Lorentz force density and current for the analyzed structure are shown in the Fig. 7. The increased current flowing through the winding causes the enhancement of vertical Lorentz force F_Z. In our case, the applying current can be

978-1-4799-4789-8/14 $31.00 © 2014 IEEE   713

tuned by the choices of the geometry of copper windings and applied voltage.

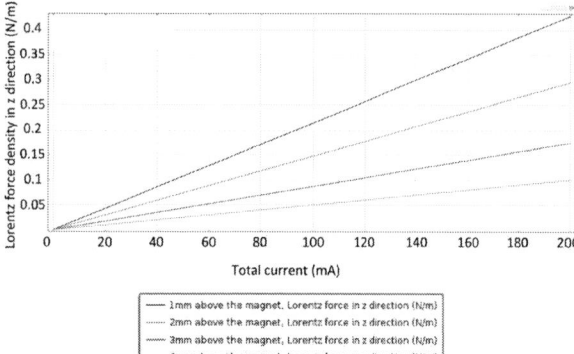

Fig.7. Windings with 24 cross-sections plot vertical Lorentz force density F_z as a function of winding current for different heights above the magnet when the windings are located at default position r=4mm.

5. Calculations of the Force Exerted on the Actuator

In this section, we calculated the Lorentz force during the actuation to evaluate the dynamic range of the moving based on the simulation results. According to some pre-determined parameters listed in Table 1 (which also used in COMSOL modeling), the line width and space are both 50um, and the winding is placed from r=4mm to r=6.35mm (because the total width of windings are 2.35mm). In the steady state, the optics and winding are located at z=2mm above the magnet. According to the simulation results shown in Fig. 4 and 5, and eq. (f), the Lorentz force exerted on the windings at default position is:

$$F_z(2mm) = 4 \times F_{z,density}(2mm) \times L_{seg} = 4 \times 0.2 \frac{N}{m} \times 8.6 \times 10^{-3} m \approx +7mN.$$

Similarly, we can get the results at z=5mm and z=5.5mm:

$$F_z(5mm) = 4 \times F_{z,density}(5mm) \times L_{seg} = 4 \times 0.04 \frac{N}{m} \times 8.6 \times 10^{-3} m \approx +1.38mN,$$

and

$$F_z(5.5mm) = 4 \times F_{z,density}(5.5mm) \times L_{seg} = 4 \times 0.03 \frac{N}{m} \times 8.6 \times 10^{-3} m \approx +1mN.$$

The mass of the optics is 0.1g and the mass of windings can be calculated as:

$$M_{windings} = V_{windings} \times \rho_{cu} = Ww \times Wh \times L_{seg} \times 24 \times 4 \times \rho_{cu} \approx 9 \times 10^{-3} g.$$

Therefore, we can assume that the mass of substrate equals to the mass of winding. Then, the total gravity can be written as:

$$G_{total} = -[M_{optics} + M_{windings} + M_{substrate}] \times g \approx -1.2mN.$$

By comparing the Lorentz force to the total gravity, we can determine the dynamic range of the lighting system.

a) When z=2mm, $|F_z(2mm)| \gg |G_{total}|$.

b) When z=5mm, $|F_z(5mm)| > |G_{total}|$. We can get:

$$Spring\ Cons. = F_z(5mm) + G_{total} \approx -0.18mN.$$

c) When z=5.5mm, $|F_z(5.5mm)| < |G_{total}|$.

Where "+" or "-" refer to the direction of the force. According to eq. (d), we can conclude that, in this case, the dynamic range of the moving optics is about 3mm (the default position of windings is 2mm in z-axis and it rises up to 5mm for the maximum height). Beyond this range, the Lorentz force is no longer sufficient to support the optics and windings.

6. Conclusion

In this work, a 2D numerical model was built up for guiding of the new dynamic lighting system design. Electromagnetic actuation mechanism was employed to provide the driving force of optics in our design due to its large force. In order to make the design smart and compact, we here studied the Lorentz force exerted on the actuator. A magnetic field distribution was first studied. Ring shape magnet is used as the magnetic flux source due to that the composition of the magnetic field above the ring shape magnet has great horizontal part, which promises large vertical Lorentz force. The distribution of the Lorentz force density was then studied. The Lorentz force density is strongly dependent on its position. The force is strong when close to the magnets and decreases fast as the height goes further than 5mm. Thus, the design can realize a dynamic range of 3mm, which is good enough for the dynamic lighting application.

References

1. G. Q. Zhang, "Simply enhancing life with light - More than illumination," 2010. [Online]. Available: http://ec.europa.eu/.
2. Yajiang Applied Technologies, "Dynamic stage effects in Shanghai international film festival," 2013. [Online]. Available: http://www.yajiang.cn/.
3. A. Steidle and L. Werth, "Freedom from constraints: Darkness and dim illumination promote creativity," Journal of Environmental Psychology, vol. 35, pp. 67–80, 2013.

Contact

*Xueming Li, Tel: 31-152787307; E-mail address: X.Li-5@tudelft.nl.

Thermal Performance of Embedded Heat Pipe in High Power Density LED Streetlight Module

Hongyu Tang*[1], Jia Zhao[1], Bo Li[2], Stanley Y Y Leung[1], Cadmus C A Yuan[2,3] and G Q Zhang[4]

[1]State Key Laboratory of Solid-State Lighting, Changzhou, Jiangsu, 213161, China

[2]State Key Laboratory of Solid-State Lighting, Haidian, Beijing, 100086, China

[3]Research and Development Center for Semiconductor Lighting, Institute of Semiconductors, Chinese Academy of Sciences, Haidian, Beijing, 100086, China

[4]Delft Institute of Microsystems and Nanoelectronics (DIMES), Delft University of Technology, Delft, the Netherlands

*hytang@sklssl.org

Abstract

The excellence of energy efficiency and reliability of LED attract the application of outdoor area lighting. The trends toward increase of power density while minimizing the structure, made the heat dissipation design challenging. The thermal performance of a novel streetlight module with embedded heat pipe is investigated in this study. The thermal performance of the new module is compared with a common module with only metal fins design. The thermal capabilities, including temperature uniformity and thermal resistance of heat pipe under different heat loads have been investigated experimentally and analyzed using finite element modeling. The comparison of the thermal properties is presented, and the implication on design capability and reliability is discussed.

1. Introduction

Light emitting diode (LED) as a solid state semiconductor device, which directly converts electrical energy into light. The energy efficiency of LED is much improved compared to high pressure sodium or metal halide technologies that was commonly used by conventional streetlight. The advantages of low energy consumption and long life span made LED light source as a mainstream for street lighting. Even though LED has a fairly satisfactory progress in the photoelectric conversion efficiency, typically more than 80% of the input power still wasted as heat [1, 2]. Thermal management design is important despite the luminous efficacies of LEDs are much improved recently. Like other electronic devices, a proper thermal management solution is critical to the operation and the reliability of a LED illumination application. A commercialized Chip-on-Board (COB) LED light source (Figure 1) used for streetlight application can be up to above 90W of nominal input power. If there is no proper thermal design, the life span of the light source could be reduced dramatically or even failure catastrophically.

Common LED streetlight is constructed by a modulated design. The LED light sources are mounted on a module structure. The modules are then mounted into the streetlight fixture. The module structure has dual function. It acted as the mechanical support, as well as a heat sink for thermal dissipation. The heat dissipation design of a LED streetlight module usually is a passive design using metal fin heat sinks (Figure 2). A more advanced design with heat pipe (HP) integrated into the metal fin heat sink were proposed [1–6]. Among the passive cooling technologies, metal fin heat sink structure is simple and having the highest reliability. The heat dissipation is relied on the temperature difference between the hot surface and the ambient. For the high power LEDs array, the metal fin heat sink with natural convection may not be capable to bear the heating load especially under high ambient temperature. Heat dissipation system using heat pipe assemblies (including heat pipe, vapor chamber, and thermal tower, etc.) are attracting more attention. Heat pipe comprises of a sealed tube with the inside wall having a porous wick structure, and partially filled working fluid. The heat dissipation mechanism is based on the release of latent heat of vaporization and condensation during the phase change of the working fluid. Therefore, allowing the high heat transfer at low temperature differences.

In this paper, the thermal performance of HP with parallel condensers used for LED streetlight module is investigated. The thermal performance of an aluminum based heat sink and a heat sink in the form of HP with parallel condensers were compared. The temperature profile under different heating load were measured experimentally. The thermal resistance of the different designs was evaluated using finite element analysis. The thermal performance is compared and the selection criteria is discussed.

Figure 1. A commercialized 90W COB LED light source (Copyright Cree, Inc).

978-1-4799-4789-8/14 $31.00 © 2014 IEEE

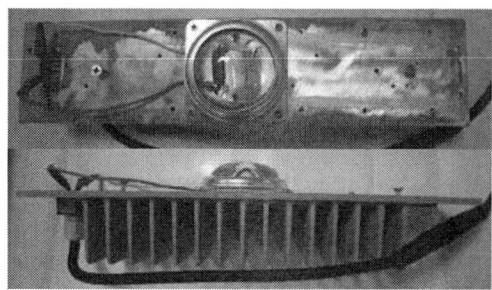

Figure 2. A common LED streetlight module with metal fin heat sink.

Figure 3. A LED streetlight module with 3 heat pipes embedded into the metal fin heat sink.

2. Design of LED streetlight module with heat pipe

The streetlight module used for this study is designed as a heat sink like structure embedded with a set of 3 heat pipes. The appearance of the module is shown in Figure 3. The light source was mounted on to a carrying platform that is positioned at the center of the module. The metal fins structure is directly connected to the opposite edges of the platform. The symmetric design is intended to comply with special required arrangement. The heat pipes are welded to the light source carrying platform and the metal fins of the heat sink. The design is aimed to dissipate the heat through the heat sink and the heat pipes simultaneously. Portion of heat generated from the light source is transferred by conduction to the metal fins and dissipated to the ambient through natural convection. And portion of heat is dissipated through providing latent heat energy to the working fluid inside the heat pipe for phase change. Since the metal fins can have a lower temperature than the light source carrying platform, it could also facilitate the re-condensation.

The light source carrying platform in the form of a rectangular plate with a dimension of 72 mm (L)×68 mm(W)×10 mm (H). The inner diameter of the heat pipe is 8 mm. The thickness of the wick layer is 0.6 mm. The metal fin structure is made of aluminum and having dimension of 89 mm×62 mm×68 mm. The total weight of the module is about 600 g. The heat pipe is targeted to operate at above 40 °C with water as working fluid. The casing and the wick of the heat pipe are made of copper. The working fluid will evaporate in the wick structure inside the heated location. The steam is traveled spontaneously to the cool side due to the pressure difference. The steam is condensed into water. The liquid water is driven by capillary force of the wick and flow back to heated location for compensation. The process is continued in order to sustain the circulation of working fluid under phase change.

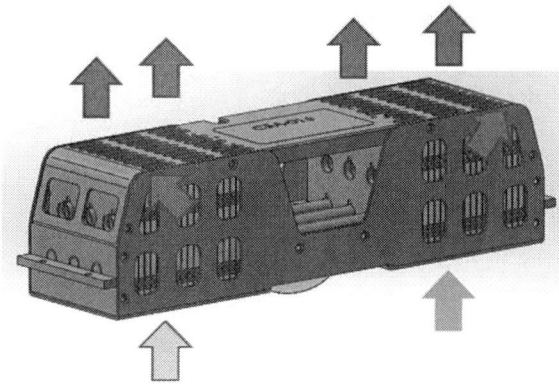

Figure 4. Air flow diagram of LED streetlight module with parallel condenser heat pipe.

The convective heat transfer from the metal fins to the ambient is through natural convection. The air flow path through the module from bottom to top is illustrated in Figure 4. The pitch of the fins has been analyzed for optimum performance [8]. In consideration of ease of manufacturing, fins in planner structure were adopted.

3. Experimental

The temperature distribution of the streetlight module with heat pipe was measured under different input power. A typical module without heat pipe was also examined for comparison. The measurement was performed under steady state condition. Constant current was applied to the light source during the experiments. The module was stabilized for more than 50 minutes to ensure the operation was in steady state before the data recording. The overall temperature distribution was recorded by taking the thermal image using an infra-red (IR) camera (Fluke Ti55FT). In addition, thermal couples were mounted on the position at the solder point, the metal fin near center, and the metal fin near the edge for key temperature measurement. The actual mounting positions were shown in Figure 5. The module was tested under room temperature of about 25 °C.

978-1-4799-4789-8/14 $31.00 © 2014 IEEE

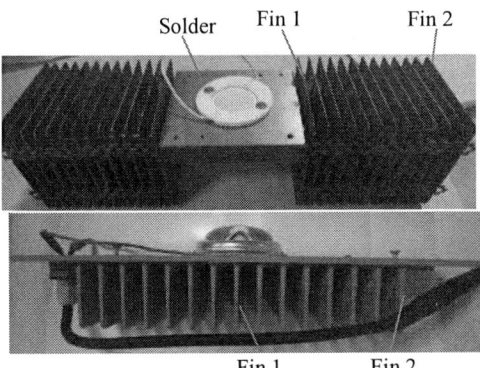

Figure 5. The thermal couple mounting positions on the LED streetlight module with heat sink *with HP* (upper photo) and heat sink *without HP* (lower photo).

4. Results

The measured temperatures using thermal couples at different locations are listed in Table 1. The highest temperature was located near the light source, with 70.9 °C for the module with heat pipe, and 92.2 °C for the module without heat pipe measured with 80W power input. The differences of the highest temperature point of the 2 modules is 21.3 °C, which is significant. The thermal resistance R_{th} of the modules are determined by

$$R_{th} = \frac{T_{solder} - T_{fin2}}{P_{th}} \quad (1)$$

with P_{th} is the thermal input power. The P_{th} is assumed to be 80% of the input power. The thermal resistance of the module with heat pipes is up to 0.28 °C/W, and the thermal resistance of the module without heat pipe is up to 0.61 °C/W. The heat pipes effectively reduced the thermal resistance by more than 50%. Furthermore, the variation of the thermal resistance of the module with heat pipes is small under different T_{solder}. It demonstrated that the heat pipe can be effectively operating at wide range of temperature.

Table 1. Tested temperatures (°C) in the LED streetlight module

	Heat sink with HP			Heat sink without HP		
Current(A)	2	1.5	1.1	2	1.5	1.1
Power(W)	80	60	40	80	60	40
T_{solder}(°C)	70.9	58.6	`49.3	92.2	73.0	58.4
T_{Fin1}(°C)	63.8	53.6	45.7	73.5	56.4	48.1
T_{Fin2}(°C)	56.0	47.0	40.4	57.2	43.8	40.0
T_{Air}(°C)	23.4	23.0	22.9	25.0	24.5	24.3
Thermal resistant(°C/W)	0.23	0.24	0.28	0.55	0.61	0.58

The thermal images of the modules operating under 40W input power is shown in Figure 6. It can be observed that the thermal gradient of the module without heat pipe is higher than the module with heat pipes. The temperature distribution is more uniform for the module with the heat pipe. It is implied that the heat transfer rate is enhanced by the heat pipes. There is no hotspot observed near the joint positions between the metal fins and the heat pipes indicated the welding does not created a substantial thermal resistance. It is also indicated that the embedding of the heat pipes does not have significant effect on the air flow path for natural convection.

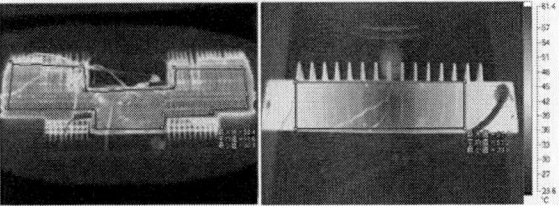

Figure 6. Temperature distribution of LED streetlight module by IR camera under 40W input power. Module with HP (left) and module without HP (right)

5. Thermal performance analysis
Modeling

The thermal performance of the modules with and without the heat pipes was analyzed by finite element model (ANSYS). The steady state temperature distribution of the modules can be numerically analysis by solving the differential function in equation [7],

$$\frac{\partial}{\partial x}\left(k_x \frac{\partial T}{\partial x}\right) + \frac{\partial}{\partial y}\left(k_y \frac{\partial T}{\partial y}\right) + \frac{\partial}{\partial z}\left(k_z \frac{\partial T}{\partial z}\right) = -\dot{q} \quad (2)$$

for k are thermal conductivities of materials in x, y, z direction, T is temperature, and q is the power. The built models are shown in Figure 7. Since the structure of the modules is in symmetry, the analysis of LED streetlight module with HP was based on a quarter model. The internal of the heat pipe was simplified to be a highly thermal conductive layer [7]. The thermal conductivity of the materials is listed in Table 2 [5, 6]. The meshing of the model is using Brick 8 node element. The meshed model is shown in Figure 8. The thermal loading was applied as power density from the LED light source. The boundary condition was assumed that the air flow is under natural convection with the convective heat transfer coefficients on different surface are listed in Table 3. The models were defined to be under ambient temperature of 25 °C.

Table 2. Thermal conductivity of different materials defined for the finite element analysis.

Parts	Material	k (W/m·°C))
LED Die	GaN	20
LED package	EMC	3
Heats spreader	Copper	387.6
Heatpipe wall	See [8]	100
Heatpipe	see [8]	6000
Fin	Aluminium	202.4

978-1-4799-4789-8/14 $31.00 © 2014 IEEE   717

Table 3. The convective heat transfer coefficient of each heat sink

Type	Region	h W/(m²·°C)
Module without heat pipe	Fin center	10
	Fin – two sides	15
Module with heat pipe	Fin - center	3
	Fin – two sides	8

Table 4. Simulated temperatures (°C) in the LED streetlight module

Position	Module with HP		Module without HP	
	FEM	Measured	FEM	Measured
Solder point	48.1	49.3	58.6	58.4
HP-middle	47.2	45.6	-	-
HP-end	41.0	40.5	-	-
Fin1	41.8	45.7	50.1	48.1
Fin2	39.2	40.4	41.0	40.0

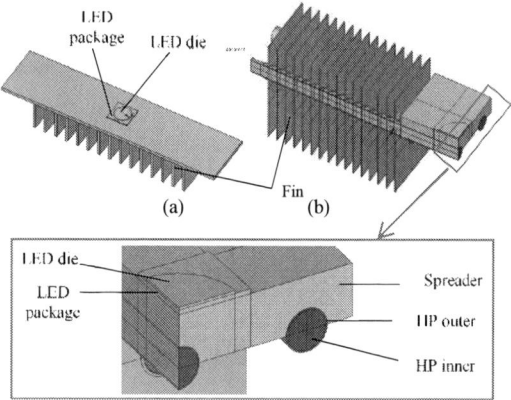

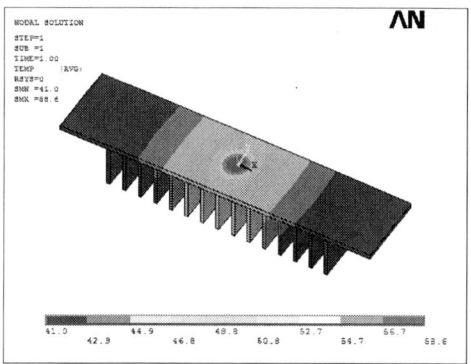

Figure 9. Simulated temperature distribution of LED streetlight module without heat pipe.

Figure 7. 3D model of LED streetlight module without heat pipe (left) and with heat pipes (right).

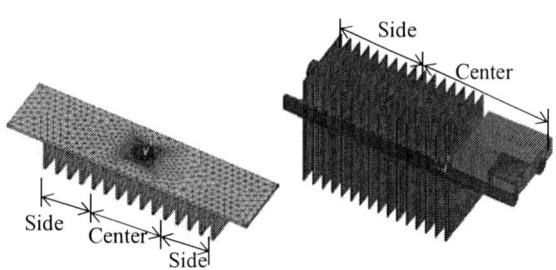

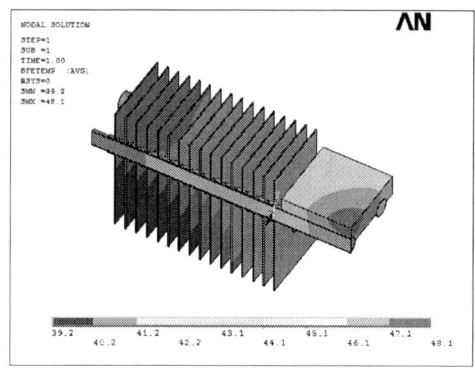

Figure 8. Meshing of the 3D model of LED streetlight module without heat pipe (left) and with heat pipes (right).

Figure 10. Simulated temperature distribution of LED streetlight module with heat pipe.

Benchmarking

Simulation of the temperature distribution of the module under 40W of input power was performed. The overall temperature distribution of the module without heat pipe is shown in Figure 9, and the module with heat pipes is shown in Figure 10. The temperatures at the key positions are listed in Table 4 and compared with the experimental result. The simulation results are in good match with the measurements. The small discrepancies are within the measurement tolerance.

Discussion

The thermal performance of the module with and without the heat pipe is compared using finite element modeling. The reliability of a streetlight modules is critically dependent on the junction temperature of the LED dies. Since the junction temperature cannot be directly measure, it is a common practice to take the solder point temperature (T_{solder}), which is close to the LED die, as estimation. The plot of T_{solder} as a function of power input of the modules design is shown in Figure 11. The result shows that the module installed with heat pipes having the lowest T$_{solder}$ than the module without heat pipe. The T$_{solder}$ difference is up to over 20 °C at the input power of 70W. In order to improve the thermal performance, heat sink made of high thermal conductivity material is usually considered. A module that with the structure identical to the one without heat pipe but made of copper was simulated as a comparison. Although the thermal performance can be improved by using copper module, the module with heat pipe still has lowest T$_{solder}$. With 80W of input power, the T$_{solder}$ difference between the two is about 12 °C. The module with heat pipes is capable to handle higher energy density light source compared to module without heat pipe.

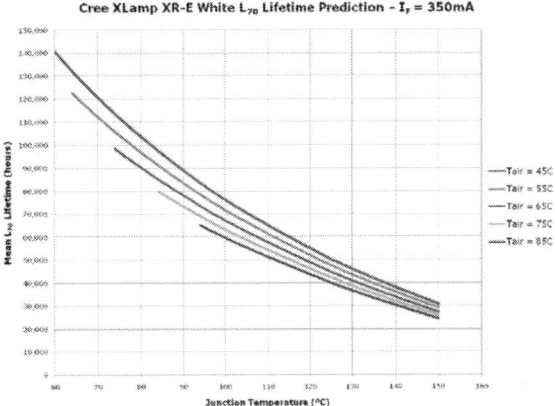

Figure 12. L70 lifetime of a commercial LED light source as function of junction temperature [9] (Copyright Cree, Inc).

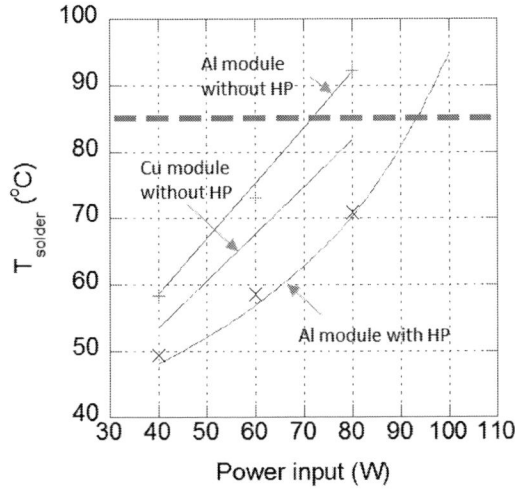

Figure 11. The solder point temperature of LED streetlight module under different input power. ✕ is the measurement result of module with heat pipe; + is the measurement result of module without heat pipe.

The lumen maintenance of the streetlight module is dependent on the junction temperature of the LED die. A L70 lifetime of a commercial LED light source was taken as an example (Figure 12). A decrease of 20 °C from 80 °C to 60 °C, the product life is increased from 105000 hours to 140000 hours (T$_{air}$=45 °C). It is corresponding to 8 years extension of useful life. While the cost of the module with heat pipes is expected to be increase, the added cost could be amortized through the extended life time and achieve a competitive total cost of ownership.

5. Conclusions

The thermal performance of streetlight modules with and without heat pipe was investigated. The improvement of the heat dissipation with the heat pipe is significant. The thermal resistance of the module with the heat pipes is reduced by over 50% compared with the module without heat pipe. Temperature distribution is more uniform with the heat pipe module. The heat dissipation capability of different module design was analyzed using finite element model. The module with heat pipe is capable for light source of about 90W. The heat dissipation capability of the heat pipe module is even better than a module made of copper. The solder joint temperature can be more than 20 °C lower under 70W of power input with the heat pipe module compared to the module without heat pipe. The lowering of the temperature is expected to improve the lumen maintenance and extend the useful life of the streetlight module by over 30%. The proposed streetlight module with heat pipe is a promising solution for the demanding of the high power LED light source.

Acknowledgments

This work is supported by the Sino-German cooperation project, LED beyond conventional lighting: Off grid application. Authors would like to thank Huaiyu Ye and Bo Sun from Delft University of Technology for the technical advises. The first author would also like to thanks to Lei Zhong for the assistance on thermal measurement.

References

[1] Ji Li, Feng Lin, Daming Wang, Wenkai Tian, "A loop-heat-pipe heat sink with parallel condensers for high-power integrated LED chips," *Applied Thermal Engineering*, 56 (2013) 18-26.

[2] Xiang-you Lu, Tse-Chao Hua, Mei-jing Liu, Yuan-xia Cheng, "Thermal analysis of loop HP used for high-power LED," *Thermochimica Acta*, 493 (2009) 25-29

[3] HP reliability documentation, November 10, 1999. Thermacore,Inc.

[4] Lijiang Bai. "The FEM Analysis of a Heat Pipe," Master thesis: Beijing University of Technology, 2004.

[5] Hui Huang Cheng, De-Shau Huang Ming-Tzer Lin, "Heat dissipation design and analysis of high power LED array using the finite element method," *Microelectronics Reliability*, 52 (2012) 905-911.

[6] Jifi Jakovenko, Robert Werkhoven, *et al*. "Thermal simulation and validation of 8W LED Luminaire," *Proc 12th EuroSimE*, Linz, April. 2011

[7] J. P. Holman (2009). <u>Heat Transfer</u>. Singapore: McGraw Hill.

[8] Huaiyu Ye, Bo Li, Hongyu Tang, Jia Zhao, Cadmus Yuan, Guoqi Zhang, "Design of vertical fin arrays with heat pipes used for high-power light-emitting diodes," *Microelectronics Reliability*, submitted.

[9] "XLamp X-RE Lighting-Class Lumen Maintenance," 2009, Cree Inc.

AUTHOR INDEX

AboRas, M. 11
Acconcia, D. 7
Adli, A. 425
Ahmar, J. 538
Allaert, B. 669
Allaf, K. 194
Altieri-Weimar, P. 77
Ancey, P. 27
Andersson, D. 263, 295, 597
Anees, S. 43
Angerer, P. 163
Antretter, T. 106, 121, 163, 394
Ardito, R. 348
Arrazat, B. 407
Auerswald, E. 618
Aydin, M. 418
Ayub, S. 584
Azzopardi, S. 283
Baccar, F. 283
Baets, J. 451
Bakowski, M. 302
Baldasaro, N. 693
Baldwin, T. 31
Bar, P. 372
Barsony, I. 612
Basrour, S. 27
Bechou, L. 138
Bechtold, T. 435
Becker, M. 632
Behlert, R. 331
Belcin, O. 225
Belhenini, S. 705
Belov, I. 302
Benhadjala, W. 138
Bermejo, R. 163
Berndt, M. 584
Bertarelli, E. 7
Beyer, G. 386
Beyne, E. 386, 494
Bieniek, T. 215
Birleanu, C. 225
Biro, F. 612
Birzer, C. 656
Bittle, J. 693
Biyik, S. 418
Blaudeck, T. 380
Blayac, S. 407
Bobashev, G. 693
Boetch, G. 626
Bord-Majek, I. 138, 231
Borgesen, P. 700
Bossuyt, F. 451
Bouazza, B. 609
Bouchard, P. 245
Brinkfeldt, K. 263, 597

Brizoux, M. 163
Broll, M. 577
Brunner, R. 106
Brunschwiler, T. 662
Buhl, P. 32
Bulashevich, K. 314
Cadiou, S. 295
Camon, H. 609
Casset, F. 27
Chappaz, C. 27
Chausse, P. 372
Chen, N. 473
Chenniki, W. 231
Cherman, V. 386
Chernyakov, A. 314
Chiang, K. 253, 469
Chung, S. 111
Cocchetti, G. 7
Cogan, S. 626
Colin, S. 680
Corigliano, A. 7, 348
Coutellier, D. 38
Croes, K. 494
Danel, J. 27
Dareys, S. 319
Davis, J. 693
Defay, E. 27
Dellaert, D. 199
Devos, A. 27
Dieppedale, C. 27
Dinh, T. 215
Diot, J. 231
Doring, R. 632
Dorwarth, M. 73
Dosseul, F. 705
Doutreloigne, J. 199
Driel, W. 47, 210, 675
Dudek, R. 128, 632
Dudescu, C. 225
Durand, C. 38
Eckhaut, D. 7
Eder, H. 440
Edwards, M. 263, 295, 597
Ernst, L. 18
Escoubas, S. 372
Evans, A. 693
Ewuame, K. 245, 372
Eyert, V. 502
Fan, X. 96, 363
Fanget, S. 27
Farghaly, M. 429
Fazio, M. 641
Fellner, K. 121
Filipovic, L. 23
Fink, M. 400

AUTHOR INDEX

Fiori, V. 245, 372
Fix, A. 531
Fornara, P. 407
France-Lanord, A. 502
Freeman, C. 502
Fremont, H. 55
Fuchs, P. 116, 121
Gafforelli, G. 348
Galasso, G. 440
Gallois-Garreignot, S. 245, 372
Ganser, H. 106
Geckeler, C. 73
Gessner, T. 380, 481
Ghisi, A. 641
Giacomo, A. 407
Goldbeck, B. 55
Gonzalez, M. 326, 343, 386, 451, 494
Gorisse, M. 27
Goroll, M. 100
Gotsmann, B. 626
Grams, A. 366
Gromala, P. 18, 170, 413
Gunther, M. 632
Gunther, S. 73
Guyenot, M. 32, 531
Guziewicz, M. 310
Hajnal, Z. 612
Han, B. 413
Han, C. 68
Hartmann, S. 380, 459
Haupt, M. 662
Heinkel, U. 481
Hempel, J. 43
Henaff, F. 283
Hermann, S. 380
Hiller, K. 459
Holck, O. 380, 618
Hossfeld, M. 32
Hotellier, N. 372
Huber, S. 577
Hung, T. 469
Hutzler, A. 278
Iannacci, J. 272
Inal, K. 245, 407
Ivankovic, A. 386
Iwamoto, N. 31
Janczyk, G. 215
Janicki, M. 498
Janke, W. 310
Jankowski, K. 82
Jansen, K. 18, 240, 400, 425
Janus, P. 626
Jaouen, H. 245
Johnson, C. 693
Kabakchiev, A. 32, 569

Kaltenbacher, M. 440
Kaltwasser, A. 577
Karpov, S. 314
Karunamurthy, B. 394, 440
Kaulfersch, E. 128, 597
Kehrberg, S. 73
Keller, J. 240
Keymeulen, B. 451
Kiener, D. 106
Kim, D. 413
Kisiel, R. 310
Klingler, M. 38
Kljucar, L. 343
Kludt, J. 55
Koh, S. 96, 210
Kolchuzhin, V. 481
Kooten, W. 47
Korvink, J. 435
Kozic, D. 106
Kozlov, A. 174
Kraemer, F. 510, 518, 524
Krasniewski, J. 310
Kreyssig, K. 632
Krivec, T. 116, 163
Krol, D. 194
Kudryavtsev, M. 435
Kuenzig, T. 272
Kunzig, T. 331
Kuo, T. 253
Kwok, S. 290
Lago, M. 688
Lamvik, M. 693
Lan, J. 446
Lang, K. 366, 577
Le Rhun, G. 27
Leal, G. 235
Lee, T. 111
Leisner, P. 302
Lenczner, M. 626
Leray, D. 680
Leung, S. 338, 715
Levrier, B. 138, 231
Li, B. 715
Li, X. 711
Liao, C. 1
Liao, K. 1
Liao, L. 469
Lim, J. 302
Lin, C. 253
Lionti, S. 245
Liu, C. 469
Liu, H. 680
Liu, P. 603
Liu, S. 1
Liu, Y. 181, 181, 187

AUTHOR INDEX

Lobur, M. 357
Lofrano, M. 326
Lu, G. 363
Ma, T. 711
Ma, X. 87, 92
Macurova, K. 163
Maia, W. 163
Maj, C. 357
Manier, C. 547
Mann, A. 597
Manz, J. 558
Mao, S. 487
Mariani, S. 641
Markert, E. 481
Markisch, S. 73
Maus, I. 400
Mavromaras, A. 502
May, D. 11, 459, 547
Mayer, D. 170
Mehner, J. 73, 459, 481
Mehr, M. 210, 675
Meier, K. 569, 590
Meinshausen, L. 55
Melz, T. 170
Meneghesso, G. 688
Meneghini, M. 688
Menges, F. 626
Messemaeker, J. 494
Meszmer, P. 459
Metais, B. 32, 569
Metasch, R. 32
Metaseh, R. 569
Michel, B. 11, 128, 156, 240, 400
Michel, L. 319
Middendorf, A. 366
Mills, K. 693
Mirzazadeh, R. 641
Mitova, R. 547
Moelans, N. 494
Mohammed, U. 429
Montmitonnet, P. 407
Morianz, M. 163
Moujbani, A. 55
Mrossko, R. 547
Muller, J. 584
Muschol, T. 272
Mysliwiec, M. 310
Nabiollahi, N. 494
Naceur, H. 38
Napieralski, A. 357, 498
Nauwelaers, B. 487
Nee, H. 302
Neumaier, K. 597
Niessner, M. 400, 656
Nigassa, M. 609

Nowottnick, M. 128
Oh, S. 111
Onsia, B. 257
Oppermann, H. 547
Orellana, S. 407
Orio, R. 23
Osmolovskyi, S. 61
Otremba, R. 394
Otto, A. 597
Ousten, Y. 138, 231
Ozturk, B. 18
Palczynska, A. 170, 215
Pantou, R. 400
Pap, A. 612
Parain, J. 319
Park, M. 111
Park, S. 68
Pasquet, D. 215
Pelisset, T. 394
Pesth, F. 170
Peter, E. 524
Pichler, P. 278
Pin, S. 319
Pinter, G. 116, 121
Pletz, M. 163
Polster, T. 440
Pons, P. 680
Poshtan, E. 156
Preu, H. 400, 656
Prewitz, T. 366
Procopio, F. 348
Pufall, R. 100, 618
Pustan, M. 225
Qian, Q. 181
Qu, J. 144
Quendo, C. 295
Quere, Y. 295
Raghavan, S. 235
Ramm, P. 618
Ras, M. 547
Ratchev, R. 32, 569
Reindl, L. 43
Renaux, P. 27
Reuther, G. 100
Rezaie-Adli, A. 240
Riel, H. 626
Rigby, D. 502
Rivero, C. 407
Rochus, V. 257, 429, 487
Roellig, M. 32, 61, 569, 590
Rola, K. 194
Rost, F. 240, 425
Rottenberg, X. 257, 429, 487
Rudnyi, E. 435
Rusu, F. 225

AUTHOR INDEX

Rzepka, S. 11, 128, 156, 240
Samara, V. 257
Sanchez-Soriano, M. 295
Sarro, P. 711
Sartor, M. 319
Saxe, P. 502
Schacht, R. 11
Schiessl, A. 590
Schindler-Saefkow, F. 240, 425
Schletz, A. 278
Schlottig, G. 662
Schmadlak, I. 235
Schmitz, S. 366
Schongrundner, R. 106, 121, 163
Schrag, G. 272, 331, 558
Schuetz, G. 656
Schuld, M. 47
Schuler, X. 32
Schulz, S. 380
Schuster, J. 481
Schwerz, R. 61
Seib, J. 524
Seiler, B. 632
Selberherr, S. 23
Sergeev, V. 314
Shaporin, A. 459
Shepherd, S. 693
Shirangi, H. 128
Silber, C. 18, 156
Silva, M. 263
Simo, G. 128
Sitaraman, S. 235
Smirnov, V. 314
Solano, E. 693
Sommer, P. 632
Soulimane, S. 609
Soussan, P. 257
Springborn, M. 547
Steiert, M. 648
Su, Y. 253, 469
Suhir, E. 138
Sun, B. 96
Swierczynski, R. 152
Szermer, M. 357, 498
Tang, H. 715
Tavernier, C. 245
Theolier, L. 283
Thomas, O. 372
Tilmans, H. 429, 487
Todri-Sanial, A. 563
Tokei, Z. 343
Tougui, A. 705
Treml, R. 106
Trivellin, N. 688
Tunaboylu, B. 543

Urbanski, K. 152
Vallauri, R. 7
Valzasina, C. 348
Vandepitte, D. 386
Vandevelde, B. 343, 386, 669
Vanfleteren, J. 451
Veknatesh, M. 603
Vereecke, B. 257
Vianne, B. 372
Vogel, D. 618
Wachutka, G. 272, 331, 558
Wagner, C. 481
Walter, H. 577, 632
Wang, M. 205
Watzke, S. 77
Wei, J. 210, 711
Weide-Zaage, K. 55
Weiss, L. 656
Wiese, S. 510, 518, 524, 538
Wijgers, R. 47
Wilde, J. 43, 584, 648
Willems, G. 669
Wilson, C. 494
Wimmer, E. 502
Winkler, T. 11
Wittler, O. 366, 577
Woirgard, E. 283
Wolf, I. 343, 386, 494
Wolter, K. 61, 569, 590
Wong, C. 338
Wongtimnoi, K. 231
Wright, A. 278
Wu, M. 205, 446
Wunderle, B. 11, 156, 240, 380, 400, 459, 547, 618
Wymyslowski, A. 82, 152, 194, 215
Xiong, Y. 338
Xu, Y. 187
Yadur, A. 413
Yaga, R. 693
Yang, B. 626
Yang, S. 144
Ye, H. 210
Youssef, A. 18
Yuan, C. 96, 338, 363, 715
Yuen, M. 290
Zafer, B. 543
Zajac, P. 357, 498
Zakgeim, A. 314
Zanoni, E. 688
Zeijl, H. 603
Zeiser, R. 584
Zhang, G. ..87, 92, 96, 210, 338, 363, 603, 675, 711, 715
Zhang, J. 187
Zhang, Y. 302
Zhao, J. 715

AUTHOR INDEX

Zhao, L..96, 210
Zimmermann, S. 662
Zschenderlein, U. 618
Zschieschang, O. 597
Zubel, I...194
Zurcher, J..662

IEEE
445 Hoes Lane
Piscataway, NJ 08854-4141

ISBN 978-1-4799-4789-8